Encyclopedia of
Plant Physiology

New Series Volume 9

Editors

A. Pirson, Göttingen
M. H. Zimmermann, Harvard

Hormonal Regulation of Development I

Molecular Aspects of Plant Hormones

Edited by
J. MacMillan

Contributors

J.R. Bearder A. Crozier D. Gross M.A. Hall
N.P. Kefford H.W. Liebisch J. MacMillan N. Murofushi
D.R. Reeve G. Schneider G. Sembdner J.L. Stoddart
N. Takahashi M.A. Venis T. Yokota M. Zeroni

With 126 Figures

Springer-Verlag Berlin Heidelberg New York 1980

Editor:
Professor Jake MacMillan
School of Chemistry
University of Bristol
Cantock's Close
Bristol BS8 1TS/United Kingdom

Planning Volumes 9–11:
Professor N.P. Kefford
Institute of Tropical Agriculture
and Human Resources
University of Hawaii
Honolulu, Hawaii 96822/USA

ISBN-13: 978-3-642-67706-9 e-ISBN-13: 978-3-642-67704-5
DOI: 10.1007/978-3-642-67704-5

Typesetting, printing and bookbinding: Universitätsdruckerei H. Stürtz AG, Würzburg.
2131/3130-543210

Foreword

This is the first of the set of three volumes in the Encyclopedia of Plant Physiology, New Series, that will cover the area of the hormonal regulation of plant growth and development. The overall plan for the set assumes that this area of plant physiology is sufficiently mature for a review of current knowledge to be organized in terms of unifying principles and processes. Reviews in the past have generally treated each class of hormone individually, but this set of volumes is subdivided according to the properties common to all classes. Such an organization permits the examination of the hypothesis that differing classes of hormones, acting according to common principles, are determinants of processes and phases in plant development. Also in keeping with this theme, a plant hormone is defined as a compound with the properties held in common by the native members of the recognized classes of hormone.

Current knowledge of the hormonal regulation of plant development is grouped so that the three volumes consider advancing levels of organizational complexity, viz: molecular and subcellular; cells, tissues, organs, and the plant as an organized whole; and the plant in relation to its environment.

The present volume treats the molecular and subcellular aspects of hormones and the processes they regulate. Although it deals with chemically distinct classes of hormone, this volume stresses properties and modes of studying them, that are common to all classes.

In a second volume of the set, the roles of hormones at levels of organization from the cell up to the whole plant are traced. The cellular processes of increase and change and the interrelations of cells in tissues, of tissues in organs, and of organs in the whole plant, are considered in turn. During this progressive treatment of levels of organization, the relevant basic properties of hormones are introduced and illustrated.

A third volume addresses the interrelationships of hormones with factors in the environments of the tissues, the organs and the whole plants, within which the hormones are functioning. When this volume touches upon wide-reaching topics such as photomorphogenesis or plant movements, only those aspects that relate to principles of hormonal regulation are treated. Separate volumes of the Encyclopedia of Plant Physiology, New Series, provide comprehensive treatments of topics such as a forthcoming volume on photomorphogenesis and one on plant movements (Vol. 7).

My role in the preparation of these volumes has been to propose a theme and prepare a plan to cover the current status of the field of hormonal regulation, then to circumscribe the portions of the plan that form logical volumes. Thereafter, the editors of the individual volumes have determined the manner in which the domain, for which they accepted responsibility, was treated. The editor of the present volume is Professor J. MacMillan and, in an Introduction to his volume, he outlines his approach and that of his authors.

The base from which these volumes have developed is the old series of the Encyclopedia of Plant Physiology. The volumes in the New Series of the Encyclopedia may, therefore, concentrate on principles that may be derived from the mass of older information and on the findings of the past twenty years. The length of each volume has been deliberately restricted, but effective organization of topics and their succinct treatment assures the reader of a concise but comprehensive statement of current knowledge and thought in the field.

I thank Professor Kenneth V. Thimann for reviewing the theme and initial plan for these volumes with me.

October, 1980 N.P. KEFFORD

Contents

Introduction
J. MacMillan 1

1 Plant Hormones and Other Growth Substances – Their Background, Structures and Occurrence
J.R. Bearder (With 10 Figures)

1.1 Introduction 9
1.2 Ethylene 9
1.3 Auxins 12
1.4 Gibberellins 22
1.5 Cytokinins 38
1.6 Abscisic Acid and Related Compounds 48
1.7 Other Plant Constituents Which Affect Plant Growth 56
a) Aromatic Compounds 58
b) Nitrogen-Containing Compounds 74
c) Terpenoid Compounds 76
d) Aliphatic Compounds 79
e) Other Compounds 80
References 80

2 Extraction, Purification, and Identification
T. Yokota, N. Murofushi, and N. Takahashi (With 38 Figures)

2.1 Methods of Extraction, Purification, and Isolation 113
2.1.1 Extraction of Active Principles from Plant Materials 113
a) General Remarks 113
b) Auxins 114
c) Gibberellins 116
d) Cytokinins 117
e) Abscisic Acid and Related Compounds 117
2.1.2 Fractionation Based on Solvent Partitioning and Ion Exchange Resin . . 118
a) General Remarks 118
b) Auxins 120
c) Gibberellins 124
d) Cytokinins 125
e) Abscisic Acid and Related Compounds 128
2.1.3 Column Chromatography and Other Purification Techniques 129
a) General Remarks 129
b) Adsorption Column Chromatography 130
c) Partition Column Chromatography 132
d) Sephadex Column and Gel Permeation Column Chromatography . . 138
e) Insoluble Polyvinylpyrrolidone Column Chromatography 139
f) Countercurrent Distribution 142
g) Other Techniques 143

2.1.4 Examples of Purification of Plant Hormones 144
a) Isolation of Auxins from Young Citrus Fruits 144
b) Isolation of Indole-3-Ethanol from Cucumber Seedlings 145
c) Isolation of Gibberellins A_1, A_5, A_6, and A_8 from Immature *Phaseolus* Seeds 146
d) Isolation of Gibberellins and Gibberellin Conjugates from Mature *Phaseolus* Seeds 147
e) Isolation of Cytokinins from Immature Sweet Corn 150
f) Isolation of Glucosylzeatin and Glucosyl Ribosylzeatin from *Vinca rosea* Crown Gall 154
g) Isolation of Abscisic Acid from Young Cotton Fruits 154
2.2 Identification Without Isolation 154
2.2.1 Criteria of Identification and Reliability 154
2.2.2 Paper and Thin-Layer Chromatography 156
a) General Remarks 156
b) Identification of Plant Hormones by Paper and Thin-Layer Chromatography 157
2.2.3 Gas-Liquid Chromatography 168
a) General Remarks 168
b) Identification of Plant Hormones by Gas-Liquid Chromatography . . 169
2.2.4 High Performance (Pressure) Liquid Chromatography 175
a) General Remarks 175
b) Identification of Plant Hormones by High Performance Liquid Chromatography 176
2.2.5 Combined Gas-Liquid Chromatography-Mass Spectrometry 181
a) General Remarks 181
b) Identification of Plant Hormones by GC-MS 182
c) Selected Ion Monitoring 184
2.2.6 Optical Rotatory Dispersion and Circular Dichroism 190
References 190

3 Quantitative Analysis of Plant Hormones

D.R. REEVE and A. CROZIER (With 13 Figures)

3.1 Introduction 203
3.2 Theoretical Considerations 203
3.2.1 Basic Analytical Errors – Accuracy and Precision 203
3.2.2 Analysis of Samples from Open-Ended Systems 209
3.2.2.1 Open Versus Closed Systems 209
3.2.2.2 Accuracy as a Probability Term 209
3.2.2.3 Verification of Accuracy 214
a) Successive Approximation 215
b) Information Content 226
c) Some General Considerations 236
3.3 Assays 240
3.3.1 Bioassays 240
3.3.2 Immunological Assays 245
3.3.3 Physicochemical Detectors 247
3.4 Chromatographic Procedures 254
3.5 Internal Standards 262
3.6 Analytical Procedures 264
3.6.1 Indole-3-Acetic Acid 264
3.6.2 Gibberellins 267
3.6.3 Cytokinins 269
3.6.4 Abscisic Acid 270
3.7 Summary 272
3.8 List of Symbols 274
References 276

4 Biosynthesis and Metabolism of Plant Hormones

G. SEMBDNER, D. GROSS, H.-W. LIEBISCH and G. SCHNEIDER (With 34 Figures)

4.1 Biosynthetic Pathways . . . 281
4.1.1 Auxins . . . 281
a) Tryptophan as Primary Precursor . . . 281
b) Pathways of IAA Formation from Tryptophan . . . 283
c) Indole-3-Pyruvic Acid Pathway . . . 283
d) Tryptamine Pathway . . . 287
e) Other Possible Pathways . . . 288
f) Chlorinated Auxins . . . 289
4.1.2 Cytokinins . . . 289
a) Biosynthesis of tRNA-Cytokinins . . . 289
b) Biosynthesis of Free Cytokinins . . . 292
4.1.3 Plant Hormones of Terpenoid Origin: General Terpenoid Pathway . . . 293
4.1.4 Abscisic Acid . . . 295
a) Mevalonic Acid as Precursor of ABA . . . 295
b) Compounds with a Pre-Formed Carbon Skeleton of ABA . . . 298
c) Formation of ABA by Degradation of Carotenoids . . . 300
4.1.5 Gibberellins . . . 301
a) General View . . . 301
b) Enzymic Cyclization of Geranylgeranyl Pyrophosphate to ent-Kaurene 304
c) Oxidation of ent-Kaurene . . . 307
d) Contraction of Ring B in ent-Kaurene . . . 309
e) Conversion of C_{20}-Gibberellins into C_{19}-Gibberellins . . . 311
f) Predominant Steps in Regulation of Biosynthesis . . . 314
4.1.6 Ethylene . . . 317
a) Ethylene Formation in Model Systems . . . 318
b) Synthesis in Vivo – Physiological Pathways . . . 319
c) Enzymes – Reaction Mechanisms . . . 324
d) Regulation of Ethylene Biosynthesis . . . 326
4.2 Metabolism (Interconversion and Catabolism) . . . 336
4.2.1 Auxins . . . 336
a) Routes and Products of IAA Catabolism . . . 336
b) Enzymes of IAA Catabolism . . . 340
c) Physiological Significance of IAA Catabolism . . . 345
d) Control of IAA Catabolism in Vivo . . . 346
4.2.2 Cytokinins . . . 347
4.2.3 Abscisic Acid . . . 350
a) 6′-Hydroxymethyl ABA . . . 350
b) Phaseic Acid . . . 352
c) 4′-Dihydrophaseic Acid and Its 4′-Epimer . . . 353
4.2.4 Gibberellins . . . 354
a) Interconversion of C_{19}-Gibberellins in the Fungus . . . 354
b) Metabolism of Gibberellins in Higher Plants . . . 355
c) Catabolism of Gibberellins . . . 359
d) Control of Gibberellin Metabolism by Other Plant Hormones and by Environmental Factors . . . 360
4.2.5 Ethylene . . . 362
4.3 Conjugation . . . 363
4.3.1 Auxins . . . 363
a) Peptidyl IAA Conjugates . . . 364
b) Glycosyl IAA Conjugates . . . 366
c) myo-Inositol IAA Conjugates . . . 367
4.3.2 Cytokinins . . . 368
a) Metabolic Conjugation Pathways . . . 368
b) Enzymic Investigations . . . 371
4.3.3 Abscisic Acid . . . 373

4.3.4 Gibberellins . . . 376
a) Endogenous Gibberellin Conjugates . . . 376
b) Chemical Synthesis of Gibberellin Conjugates . . . 376
c) Metabolism of Gibberellin Conjugates . . . 377
d) Enzymatic Hydrolysis and Biological Activity of Gibberellin Conjugates 381
e) Biological Function of Gibberellin Conjugates . . . 382
4.4 Localization of Biosynthesis and Metabolism . . . 382
4.4.1 Auxins . . . 383
4.4.2 Cytokinins . . . 384
4.4.3 Abscisic Acid . . . 387
4.4.4 Gibberellins . . . 387
4.4.5 Ethylene . . . 389
References . . . 390

5 Molecular and Subcellular Aspects of Hormone Action

J.L. Stoddart and M.A. Venis (With 21 Figures)

5.1 Introduction . . . 445
5.1.1 Receptor Characteristics . . . 445
5.1.2 General Methodology of Receptor Studies . . . 446
a) Analysis of Binding Data . . . 446
b) Acquisition of Binding Data . . . 446
5.2 Auxins . . . 448
5.2.1 Structure-Activity Relationships . . . 448
5.2.2 Potential Location of Auxin Receptors . . . 450
5.2.3 Nuclear and Cytoplasmic Receptors . . . 453
a) Auxin Mediator Proteins . . . 453
b) Auxin-Binding Proteins . . . 455
5.2.4 Membrane-Bound Receptors . . . 457
a) The Corn Coleoptile System . . . 457
b) Membrane-Bound Sites in Other Tissues . . . 469
5.3 Cytokinins . . . 470
5.3.1 Structure-Activity Relationships . . . 470
5.3.2 Cytokinins in Transfer RNA (tRNA) . . . 472
5.3.3 Cytokinin Binding to Sub-Cellular Components . . . 474
5.4 Gibberellins . . . 478
5.4.1 Structure-Activity Relationships . . . 478
5.4.2 Transcriptional Control: Analogies with Animal Steroid Hormones . . . 483
a) Steroid Hormones . . . 484
b) Gibberellins . . . 486
5.4.3 Effects on Membrane Organization . . . 490
5.4.4 Action in Artificial Membrane Systems . . . 492
5.4.5 Changes in Cell Extensibility . . . 497
5.5 Abscisic Acid . . . 498
5.6 Ethylene . . . 498
5.6.1 Structure-Activity Relationships . . . 499
5.6.2 Aspects of Molecular Action . . . 499
5.7 Concluding Remarks . . . 501
References . . . 501

6 Molecular Effects of Hormone Treatment on Tissue

M. Zeroni and M.A. Hall (With 10 Figures)

6.1 Introduction . . . 511
6.2 Effects of Hormones on Processes Involved in Growth . . . 513

6.2.1 Effects of Auxins on Cell Extensibility 514
a) Kinetic Studies on Cell Extension 516
b) Effects on Nucleic Acids and Proteins 519
c) The Acid Growth Effect 521
d) Effect on Metabolism of the Cell Wall 524
6.2.2 Other Growth Regulators 530
a) Gibberellins 530
b) Cytokinins 530
c) Ethylene 532
d) Abscisic Acid 532
e) 3-Methyleneoxindole 532
f) Fusicoccin 533
6.2.3 Conclusions 534
6.3 Effects of Growth Regulators on Ion Transport and Regulation of Membrane Properties 537
6.3.1 Hormones and Stomata 541
6.3.2 Hormone-Directed Transport 544
6.4 Effects of Hormones on Non-Growing Systems 544
6.4.1 The Aleurone Layer System 544
6.5 Effects of Hormones on Differentiating Systems 553
6.6 Conclusions 567
References 568

Author Index 587

Subject Index 651

List of Contributors

J.R. Bearder
School of Chemistry
University of Bristol
Cantock's Close
Bristol BS8 1TS/United Kingdom
Present address:
Shell Research Ltd.
Shell Biosciences Laboratory
Sittingbourne Research Centre
Sittingbourne, Kent ME9 8AG/
United Kingdom

A. Crozier
Botany Department
The University of Glasgow
Glasgow G12 8QQ/United Kingdom

D. Gross
Akademie der Wissenschaften der DDR
Institut für Biochemie der Pflanzen
Postfach 250
4010 Halle/GDR

M.A. Hall
Dept. of Botany and Microbiology
The University College of Wales
Aberystwyth, Dyfed,
Wales SY23 IRX/United Kingdom

N.P. Kefford
Institute of Tropical Agriculture
and Human Resources
University of Hawaii
Honolulu, Hawaii 96822/USA

H.W. Liebisch
Akademie der Wissenschaften der DDR
Institut für Biochemie der Pflanzen
Postfach 250
4010 Halle/GDR

J. MacMillan
School of Chemistry
University of Bristol
Cantock's Close
Bristol BS8 1TS/United Kingdom

N. Murofushi
Dept. of Agricultural Chemistry
The University of Tokyo
Bunkyo-ku
Tokyo/Japan

D.R. Reeve
Botany Department
The University of Glasgow
Glasgow G12 8QQ/United Kingdom

G. Schneider
Akademie der Wissenschaften der DDR
Institut für Biochemie der Pflanzen
Postfach 250
4010 Halle/GDR

G. Sembdner
Akademie der Wissenschaften der DDR
Institut für Biochemie der Pflanzen
Postfach 250
4010 Halle/GDR

J.L. Stoddart
Plant Biochemistry Department
University College of Wales
Welsh Plant Breeding Station
Aberystwyth, Dyfed SY23 3EB/
United Kingdom

N. TAKAHASHI
Dept. of Agricultural Chemistry
The University of Tokyo
Bunkyo-ku
Tokyo/Japan

M.A. VENIS
Shell Research Ltd.
Shell Biosciences Laboratory
Sittingbourne Research Centre
Sittingbourne, Kent ME9 8AG/
United Kingdom

T. YOKOTA
Dept. of Agricultural Chemistry
The University of Tokyo
Bunkyo-ku
Tokyo/Japan

M. ZERONI
Closed System Agriculture
The Jacob Blaustein Institute
for Desert Research
Sede Boger/Israel

List of Abbreviations

Cytokinin abbreviations are given in Fig. 4.2 (p. 290).
Abbreviations, used in Figures and Tables only, are given therein.
Symbols used in Chapter 3 are listed at end of chapter.

ABA	abscisic acid
ACC	1-aminocyclopropane-1-carboxylic acid
AFID	alkali flame ionisation detector
AD-Co-Thr	N-[9-(β-D-ribofuranosyl)-purine-6-carbamoyl]threonine (Fig. 1.7, p. 39)
Alar	B-995, Succinic acid 2,2-dimethylhydrazide
AMO-1618	2′-isopropyl-4′-(trimethylammonium chloride)-5′-methylphenyl piperidine-1-carboxylate (Fig. 4.18, p. 315)
AMP	adenosine-5′-monophosphate
cAMP	cyclic adenosine-3′,5′-monophosphate
ATP	adenosine-5′-triphosphate
AVG	L-2-amino-4-(2′-aminoethoxy)-trans-3-butenoic acid (aminoethoxyvinylglycine)
BI	buffer-insoluble (cellulase)
B-995	Alar, succinic acid 2,2-dimethylhydrazide
BA	N^6-benzyladenine
BAR	6-(o-hydroxybenzylamino)-9β-D-ribofuranosylpurine
BOA	2-benzoxazoline
BS	buffer-soluble (cellulase)
BSA	bis-trimethylsilylacetamide
BSTFA	bis-trimethylsilyltrifluoroacetamide
c	cis
C-value	relative amount of DNA per nucleus; normal diploid cell after DNA synthesis and before mitosis has a 4C-value
C_n	number of carbon atoms
C-n	position of carbon atom in molecule
CAPA	2-chloro-4-aminophenoxyacetic acid
CCC	2-chloroethyltrimethyl ammonium chloride
CCD	counter current distribution
CD	circular dichroism
CCCP	carbonylcyanide m-chlorophenylhydrazone
Chloramben	2,5-dichloro-3-aminobenzoic acid
CI	chemical ionization
Ci	Curie
CoASH	coenzyme A
cv	cultivar
2,4-D	2,4-dichlorophenoxyacetic acid
DEAE	diethylaminoethyl
DMBOA	6,7-dimethoxy-2-benzoxazoline
2,4-DNP	2,4-dinitrophenol
DP	degree of polymerization
2,4-DP	2-(dichlorophenoxy)propionic acid
DTE	dithioerythritol
EEi	elastic extensibility
EI	electron impact
ent	enantiomer
EtIAA	ethyl indole-3-acetate
ER	endoplasmic reticulum
FID	flame ionization detector
FMN	flavin mononucleotide
FUDR	5-fluorodeoxyuridine
G1	period of cell cycle following mitosis and prior to DNA synthesis
G2	period of cell cycle following DNA synthesis and prior to mitosis

xg	X acceleration due to gravity
GA	gibberellin
GC-MS	combined gas chromatography-mass spectrometry
GC-CIMS	combined gas chromatography-chemical ionization mass spectrometry
GLC	gas liquid chromatography
GLP	growth-limiting protein
GMP	guanosine-5′-monophosphate
cGMP	cyclicguanosine-3′,5′-monophosphate
GPC	gel permeation chromatography
HFB	heptafluorobutyl
HPLC	high performance liquid chromatography
IAA	indole-3-acetic acid
IAAld	indole-3-acetaldehyde
IAcry	indole-3-acrylic acid
IAld	indole-3-carboxaldehyde
IAM	indole-3-acetamide
IAN	indole-3-acetonitrile
IBA	indole-3-butyric acid
ICA	indole-3-carboxylic acid
ICI	Imperial Chemical Industries Ltd.
IEt	indole-3-ethanol
ILA	indole-3-lactic acid
IPA	indole-3-propionic acid
IPyA	indole-3-pyruvic acid
IR	infrared
c-IRP	chromosomal IAA receptor protein
n-IRP	nuclear IAA receptor protein
KBM	α-keto-γ-(methylthio)butyric acid
KP	pellet obtained by centrifugation at n × 1000 g.
LC	liquid chromatography
M	molar
MBOA	6-methoxy-2-benzoxazolinone
MCPA	2-methyl-4-chlorophenoxyacetic acid
MDMP	(D)-2-(4-methyl-2,6-dinitroanilino)-N-methylpropionamide
Me	methyl
MeneOx	methyleneoxindole
MeOx	methyloxindole
MF	mass fragmentography
MID	multiple ion detection
MTA	5′-methylthioadenosine
MTR	5′-methylthioribose
NAA	naphthyl-1(or 2)-acetic acid
NAD	nicotinamide-adenine dinucleotide
NADH	reduced nicotinamide-adenine dinucleotide
NADP	nicotinamide-adenine dinucleotide phosphate
NADPH	reduced nicotinamide dinucleotide phosphate
NMR	nuclear magnetic resonance
NPA	N-(naphth-1-yl)phthalamic acid
o	ortho
ORD	optical rotatory dispersion
p	para
PAA	phenylacetic acid
PAL	phenylalanine ammonia lyase
PC-C (PCT)	phosphorylcholine-cytidyltransferase
PC-G (PGT)	phosphorylcholine-glyceride transferase
PEi	plastic extensibility
PEP	phosphoenol pyruvate
Phosphon D	tributyl-2,4-dichlorobenzylphosphonium chloride (Fig. 4.18, p.315)
Phosphon S	tributyl-2,4-dichlorobenzylammonium chloride (Fig. 4.18, p. 315)
PM	plasma membrane
PMB	p-mercuribenzoate
PMBS	p-mercuribenzenesulphonate
PMSF	phenylmethylsulphonylfluoride
PP	pyrophosphate
ppb	parts per billion

PPC	paper chromatography
ppm	parts per million
PVP	polyvinylpyrrolidone
RER	rough endoplasmic reticulum
RF	radio-frequency
R_F	ratio of travel of compound to travel of solvent front
RHS	right hand side
SAM	S-adenosylmethionine
SDS	sodium dodecylsulphate
SF	supernatant factor
SIM	selected ion monitoring
SKF-525A	2,2-diphenylpentanoic acid 2-(dimethylamino)ethyl ester
SKF-7997	tris-(N,N-diethylaminoethyl)-phosphate trichloride
SMM	S-methylmethionine
S	period of cell cycle during DNA synthesis
SID	single ion detection
SPC	silica gel chromatography
t	trans
TBA	trichlorobenzoic acid
TIBA	2,3,6-triidobenzoic acid
TLC	thin layer chromatography
TMCS	trimethylchlorosilane
TMS	trimethylsilane
TMSi	trimethylsilyl
Trimarol	2,4-dichlorophenylphenyl-5-pyrimidinylmethanol
UDP	uridinediphosphate
UV	ultra-violet light
WCOT	wall-coated open tubular (column for GLC)
WEC	continuous extensibility (of cell-wall)
Y	yield threshold

Nomenclature and Numbering

Abscisic Acid and Derivatives. The numbering of carbon atoms is shown in Table 1.16 (p. 49).

Auxins. Auxins are named thus, indole-3-acetic acid (IAA), naphthalene-1-acetic acid (NAA) etc. Numbering of the indole ring system is shown in Table 1.1 (p. 13).

Gibberellins and Related Diterpenes. The ent-gibberellane and ent-kaurane system of nomenclature is used and is explained in Chapter 1.4 (pp. 34–35). The numbering of the carbon skeleton is given in Fig. 1.4 (p. 38) and Fig. 1.5 (p. 38). Since the use of the ent-operator is confusing, in relation to drawn structures, when specifying α- and β-stereochemistry for substituents, the authors of Chapter 4 have used the device of specifying the substituent and its stereochemistry outside the ent-operator. For example, ent-7α-hydroxykaurenoic acid becomes 7β-hydroxy-ent-kaurenoic acid. Although strictly incorrect, this device has the advantage that the specified stereochemistry corresponds to the drawn structures.

Cytokinins. The nomenclature, numbering, and abbreviations are given in Fig. 4.2 (pp. 290–291). The abbreviations are based upon those established for nucleic acid derivatives (IUPAC-IUB Commission on Biochemical Nomenclature, 1970, J. Biol. Chem., 245, 5171–6).

Introduction

J. MacMillan

The scope of this volume has been deliberately limited and selective for the following reasons. Firstly, in accord with the underlying philosophy of the New Series of the Encyclopedia of Plant Physiology, a prime consideration has been to provide a volume of reasonable size enabling it to be found on the bookshelves of individuals as well as on library shelves. Secondly, since the publication in 1961 of the original Encyclopedia Volume XIV – Growth and Growth Substances – almost 10000 research papers on plant hormones and plant growth regulation have been published and literature citations alone could have filled a reasonably sized volume. Length has been kept within reasonable proportions in two other ways. In the early stages of planning it was intended to include chapters on bio-assay methods and on structure-activity relationships. However it became evident after discussing these plans with colleagues and prospective authors that there was little that could be added on these topics, per se, that had not been adequately covered in existing reviews. Thus the planned chapter on bio-assays has been replaced by one on quantitative analysis, including bio-assay methods, and the discussion on structure-activity relationships has been restricted to those aspects which are of direct relevance to the discussion of hormone receptors.

In the original Encyclopedia, Volume XIV on Growth and Growth Substances (1961) contained 1357 pages, over 500 of which were devoted to auxins and only 35 pages to the gibberellins; ethylene received passing mention and the cytokinins and abscisic acid were, of course, not included. In contrast, the present volume is concerned, in roughly equal measure, with the five presently recognized groups of plant hormones and Chapter 1 includes a brief review of other plant constituents which affect plant development. Since the five main groups of hormones may be presumed to act on basic processes of plant growth and development, either concertedly, consequentially, or separately, it seems logical to discuss the main groups of hormones together in relation to the properties common to each group. The six chapters in this volume have been organized in this way, rather than in the traditional method of discussing each group of hormones in separate chapters.

In Chapter 1 the five groups of hormones are introduced with a brief history. The individual members of each group, and their occurrence, are then listed. Only those hormones which have been unequivocally identified, either by isolation or by mass spectrometry, are given. As stated earlier the chapter also includes plant constituents which have been found to affect plant development.

As pointed out by the authors of Chapter 2, the recognition of the existence of a particular hormone ultimately depends upon its isolation and the determination of the biological and chemical properties of the pure hormones. The methods by which plant extracts are fractionated and by which the pure hormones can be obtained are therefore the subject of Chapter 2. Examples of the isolation of each group are included. Once the hormones have been isolated and their properties have been determined, they may then be identified without isolation by methods which are discussed in this chapter. These methods, particularly combined gas chromatography-mass spectrometry, are being increasingly used to quantify the hormones in plant extracts and many examples of the use of deuterium-labelled hormones, as internal standards, in quantitative analysis by mass spectrometry have been described in the last two years.

The topic of quantitative analysis is critically examined in Chapter 3, in which the criteria required for accurate analysis are assessed. This subject is cogent, not only in the context of the present volume, but is equally important for the general problem of quantitative analyses of single organic compounds in bulk samples. Two approaches to the problem are discussed. One is that of successive approximation towards an accurate value. The other is the analysis of an open-ended system in terms of information theory to determine the number of bits of information, required for a given accuracy, and to assess the number of bits of information provided by the various analytical procedures. The chapter is deliberately provocative but it is intended to encourage critical appraisal of the accuracy of the methods used. Conversely it may stimulate workers to examine critically the accuracy which is required to answer the questions which they are asking.

Biosynthesis of plant hormones and their further metabolism have been areas of very rapid progress in the last two decades. These topics are reviewed in depth in Chapter 4. In Chapter 5, the important topic of receptor sites of the hormones is discussed and, in the final Chapter 6, the molecular effects of the hormones on plant tissue are reviewed.

There are many inherent short-comings in a multi-author volume. Some overlap of material between chapters is unavoidable. It is hoped that unnecessary overlap has been eliminated and that any overlap which remains is helpful to the reader. Also, the greater the number of authors the longer is the gestation period. Some papers came to our attention too late to be included in the main body of the text; others were published too late to be included; and some are known to be in press and not yet published. To mitigate these omissions, the opportunity has been taken to include stop-press items in this introduction.

General

The Symposia papers, delivered at the Tenth International Conference on Plant Growth Substances, Madison, July 1979, will appear in Proceedings in Life Sciences (ed. by SKOOG, Springer 1980) and summaries of the other papers and poster demonstrations are contained in the Abstracts of that meeting. Papers presented at a symposium on the gibberellins, organized jointly by the British Plant Growth Regulator Group and the Society of Experimental Biology, are to be published (LENTON, 1980). The chemistry (HEDDEN, 1979) and the metabolism (PHINNEY, 1979) of the gibberellins have been reviewed. Methods for the chemical preparation of isotopically labelled GAs, a subject not covered in this volume, have been briefly reviewed (MACMILLAN, 1980).

The biochemistry and physiology of abscisic acid has been reviewed recently (WALTON, 1980).

Chapter 1

The number of GAs has increased by three. Gibberellin A_{58} (12α-hydroxyGA_4) has been characterized (BEARDER et al., 1980); it was isolated from seed of *Cucurbita maxima* in which it occurs mainly as a conjugate. Gibberellin A_{59} (2,3-didehydroGA_{21}) has been isolated from immature seed of *Canavalia gladiata* (TAKAHASHI et al., unpublished). Another of the five putative 1β-hydroxy GA's, present in wheat germ (GASKIN et al., 1980) has been identified as 1β-hydroxyGA_{20} by the preparation of the methyl ester from GA_3 (KIRKWOOD et al., unpublished).

Gibberellin A_1 has been identified in the suspensors of seed of *Phaseolus coccineus* (ALPI et al., 1979). Gibberellins A_1, A_4 and A_7, together with the Δ-(1,10)-open lactone from GA_1, have been found in somatic cell embryo cultures of carrot and anise (NOMA et al., 1979).

The long-suspected role of GAs in controlling differentiation in ferns has received some direct support by YAMANE et al. (1979). The occurrence of GA_9 methyl ester in secretions from prothallia of *Lycopodium japonicum* was indicated by GC-CIMS; exogeneous $^3HGA_9$ was rapidly converted into $^3HGA_9$ methyl ester by the prothallia; and authentic GA_9 methyl ester induced antheridia at 10^{-10} M and inhibited archegonial formation at 10^{-9}. However the native GA_9 methyl ester was present in much lower concentrations and may not be the principle endogenous regulator.

A decisive advance in the brassin story (MITCHELL et al., 1970) has been made by the isolation of brassinolide from the pollen of *Brassica napus*, and the determination of its structure as (22R, 23R, 24S)-2α,3α,22,23-tetrahydroxy-24-methyl-6,7-seco-5α-cholestano-6,7-lactone (GROVE et al., 1979). The biological activities of this natural steroidal lactone and of two synthetic isomers are briefly described in the Abstracts of the Tenth International Conference on Plant Growth Substances by MEUDT et al., MANDAVA et al., and YOPP et al. The synthesis of the biologically active (22R, 23R, 24R) and (22S, 23S, 24R) stereo-isomers from ergosterol (THOMPSON et al., 1979) provides useful quantities of these steroidal lactones for detailed biological study.

Chapter 2

The application of droplet counter current chromatography (DCCC) to the gibberellins has been described (BEARDER and MACMILLAN, 1980). This technique (HOSTETTMANN, 1980) which is convenient, economic in solvents and can be used with a wide range of sample weights, should provide a useful addition to the existing separatory methods for plant hormones.

Methods for the isolation of IAA, ABA, GA's and cytokinins from one sample of plant material have been described (RADEMACHER, 1978; RADEMACHER and GRAEBE, 1979).

The use of GC-SICM using deuterated internal standards for the quantitation of plant hormones increases. Two recent examples are the quantitation of the cytokinins in *Zea mays* kernels (SUMMONS et al., 1979a) and of cytokinin glycosides in lupin pods (SUMMONS et al., 1979b). SUMMONS et al. (1979c) have demonstrated the excellence of CI-MS, in conjunction with HPLC, for the identification of zeatin and its metabolites. The technique of computerized mass spectrometer linked scan system for recording metastable ions has been discussed (HADDON, 1979). In this method, which is complementary to mass analyzed kinetic energy spectroscopy (MIKES), daughter ions from a common parent ion (B/E scans) or the parent ions of daughter ions (B^2/E scans) can be recorded. The technique is especially suited to the analysis of mixtures and its application to the identification of ABA in a crude sugar cane extract is illustrated.

The use of permethylated derivatives for the identification of cytokinins (MCLEOD et al., 1976; MORRIS, 1977) by GC-MS has been extended to the gibberellins and their conjugates (ZAERR and MORRIS, 1979; RIVIER et al., 1980).

A recent study of the stability of IAA (YAMAKAWA et al., 1979) has confirmed that IAA is rapidly decomposed by red and blue light, especially the latter at high intensity (3200 lux), and that it is stable to autoclaving and to aeration.

Chapter 4

Further details of the GA-biosynthetic pathway in cell-free preparations from immature seed of *Cucurbita maxima* have been described by GRAEBE et al. (1980). These authors have shown that GA_{36}, and not GA_{13}, is a precursor of GA_4, thus defining the point in the pathway at which the C_{19}-gibberellins (e.g., GA_4) and the C_{20}-gibberellins (e.g., GA_{43}) diverge. They also provide evidence that 7β-hydroxykaurenolide is formed, in this system, from ent-kaura-6,16-dien-19-oic acid and not from ent-7α-hydroxykaur-16-en-19-oic acid (see also GRAEBE et al., 1980).

Recently it has been shown that GA_{53}, GA_{44}, GA_{19}, GA_{17}, and GA_{20} are common to *Vicia faba* (SPONSEL et al., 1979), *Spinacea oleracea* (METZGER and ZEEVAART, 1980), *Agrostemma githago* (JONES and ZEEVAART, 1980), and *Zea mays* (HEDDEN et al., 1980). Gibberellins A_1, A_5, A_8, and A_{29} have either been detected or tentatively identified, either singly or in combination, in these plants. The results suggest a common biosynthetic pathway, possibly from GA_{53} via GA_{19} to the C_{19}-GAs which is consistent with the recent results of GRAEBE et al. (1980).

Two further examples of the use of cell-free systems in studying the control and sites of GA-biosynthesis have been published. Firstly, the potential of the suspensors in the seed of *Phaseolus coccineus* to biosynthesize GAs has been explored (ALPI et al., 1979; CECCARELLI et al., 1979, 1980). Secondly, HEDDEN and PHINNEY (1979) have compared the kaurene-synthetase activity in cell-free preparations from etiolated shoots of normal and d-5 maize seedlings and provided evidence that the B-activity of this cyclase is controlled by the d-5 gene.

Several metabolic studies have been reported recently. SPONSEL et al. (1979) have discussed the differences in GA-metabolism between the *Phaseoleae* and *Viciae*. The latter tribe are not hydroxylated at C-3 nor are they readily conjugated. The Phaseoleae differ from the Viciae in both these respects. RAILTON (1980) has reported kinetic studies exploring a connection between the conversion of $^3HGA_1$ to $^3HGA_8$ and the leaf sheath elongation in Tan-ginbozu rice seedlings. The data indicate a negative correlation which, it is suggested, may be caused by the rapid turn-over of $^3HGA_8$. In a different approach to the investigation of the possible role of 2β-hydroxylation in the regulation of native plant GA's, HOAD et al. (1980) have bio-assayed several 2α- and 2β-mono substituted derivatives of GA_9 and GA_4 in seven bio-assay systems and shown that the prevention of 2β-hydroxylation by a substituent does not affect biological activity. These authors did, however, find that 2,2-dimethylGA_4 was more active by two orders of magnitude than GA_4 in bio-assays using monocotyledons.

PARKER et al. (1979) have shown that zeatin is metabolized by excised leaves of *Populus alba*, principally to adenosine, O-β-D-glucosylzeatin, O-β-D-glucosyl dihydrozeatin, and O-β-D-glucosyl-9-β-D-ribofuranosyldihydrozeatin.

HARTUNG et al. (1980) have concluded that, in *Spinacia oleracea*, ABA is metabolized in the cytoplasm of mesophyll cells and not in the chloroplasts.

Chapter 5

Two recent papers report on GA_1 binding. KEITH and SRIVASTAVA (1980) have incubated $^3HGA_1$ with slices of dwarf pea epicotyls at 0° for 3 days, conditions under which GA_1 is not metabolized. They observed binding of the GA_1 to two soluble proteins with MW 6×10^5 daltons and kd 6×10^{-8} and MW $4\text{–}7\times10^4$ daltons and kd 1.4×10^{-6} M. However the total binding is less than 0.5% of the tissue content and the bound GA_1 is non-exchangeable. The Satchard plot does not take into account the covalent attachment of GA_1 to protein which occurs in pea (STODDARD, unpublished) and which may be their so-called high affinity site.

STODDART and WILLIAMS (1980) have extended their studies on the relationship between the elongation response of lettuce hypocotyl sections to GA_1 and the incorporation of radio-activity in $^3HGA_7$ into a 2000 g pelletable fraction (2KP) from homogenates of the same tissue. From data obtained at 0° C (arrested growth of hypocotyl sections) and 30° C (permitted growth) they conclude that 2KP labelling is not a consequence of growth but must either precede growth or be an unconnected concurrent process.

Aducci et al. (1980) have found that binding of the fungal toxin, fusicoccin, to membranes, prepared from maize roots, increases several-fold if roots are washed in water prior to homogenization. Binding is inhibited if the wash solution is added to the assay mixture in vitro. The inference is that washing removes the natural ligand that normally binds to the fusiccoccin receptor sites. Characterization of this putative ligand should help clarify the physiological role of the binding sites.

Finally, I warmly thank the contributors to this volume and appreciate their gentlemanly responses to my proddings. I also thank Professor A. Pirson and Mrs. L. Teppert for their boundless patience and help – and my wife, with some initial help from Dr. V.M. Sponsel, for preparing the subject index.

References

Aducci, P., Crosetti, G., Federico, R., Ballio, A.: Fusicoccin receptors. Evidence for endogenous ligand. Planta *148*, 208–10 (1980)

Alpi, A., Lorenzi, R., Cionini, P.G., Bennicia, A., D'Amato, F.: Identification of GA_1 in the embryo suspensor of *Phaseolus coccineus*. Planta *147*, 225–227 (1979)

Bearder, J.R., MacMillan, J.: Separation of gibberellins and related compounds by droplet counter-current chromatography. In: British Plant Growth Regulator Group, Monograph 5 Lenton, J.R. (Ed.). 1980

Bearder, J.R., Bleckschmidt, S., Gaskin, P., Graebe, J.E., MacMillan, J.: unpublished (1980) but see Bearder and MacMillan (1980) and Graebe et al. (1980)

Ceccarelli, N., Lorenzi, R., Alpi, A.: Kaurene and kaurenol biosynthesis in cell-free systems of *Phaseolus coccineus* suspensor. Phytochemistry *18*, 1657–8 (1979)

Ceccarelli, N., Lorenzi, R., Alpi, A.: Kaurene metabolism in cell-free extracts of *Phaseolus coccineus* suspensors. Submitted to Planta (1980)

Gaskin, P., Kirkwood, P.S., Lenton, J.R., MacMillan, J., Radey, M.E.: Identification of gibberellins in developing wheat grain. Agric. Biol. Chem. in press (1980)

Graebe, J.E., Hedden, P., Rademacher, W.: Gibberellin biosynthesis. In: British Plant Growth Regulator Group, Monograph 5 Lenton, J.R. (Ed.). 1980

Grove, M.D., Spenser, G.F., Rohwedder, W.K., Mandava, N., Worley, J.F., Warthen, J.D. Jr., Steffens, G.L., Flippen-Anderson, J.L., Cook, J.C. Jr.: Brassinolide, a plant growth promoting steroid isolated from *Brassica napus* pollen. Nature (London) *281*, 216–7 (1979)

Haddon, W.F.: Computerised mass spectrometer linked scan system for recording metastable ions. Anal. Chem. *51*, 983–8 (1979)

Hartung, W., Gimmler, H., Heilmann, B., Kaiser, G.: The site of abscisic acid metabolism in mesophyll cells of *Spinacea oleracea*. Plant. Sci. Lett. *18*, 359–64 (1980)

Hedden, P.: Aspects of Gibberellin Chemistry. In: Plant Growth Substances Mandava, N.B. (Ed.). A.C.S. Symp. Ser. 111. Am. Chem. Soc. Washington DC, 1979, pp. 19–56

Hedden, P., Graebe, J.E.: Kaurenolide biosynthesis in a cell-free system from *Cucurbita maxima* seeds. Phytochemistry submitted (1980)

Hedden, P., Phinney, B.O.: Comparison of *ent*-kaurene and *ent*-isokaurene synthesis in cell-free systems from etiolated shoots of normal and dwarf-5 maize seedlings. Phytochemistry *18*, 1475–9 (1979)

Hedden, P., Phinney, B.O., Heupel, R., Fujii, D., Cohen, H., Gaskin, P., MacMillan, J., Graebe, J.E.: Identification of plant hormones from young tassels of Zea mays L. Phytochemistry submitted (1980)

Hoad, G.V., Phinney, B.O., Sponsel, V.M., MacMillan, J.: The biological activity of sixteen gibberellin A_4 and gibberellin A_7 derivatives using seven bio-assays. Phytochemistry in press (1980)

Hostettmann, K.: Droplet counter-current chromatography and its application to the preparative scale separation of natural products. Planta Med. J Med Plant Res. *39*, 1–18 (1980)

Jones, M.G., Zeevaart, J.A.D.: Gibberellins and the photoperiodic control of stem elongation in *Agrostemma githago* L. Abstr. 10th Int. Conf. Plant Growth Substances, Madison, U.S.A. (1979) Jones, M.G., Zeevaart, J.A.D.: Plant Physiol. submitted (1980)

Keith, B., Srivastava, L.: In vivo binding of gibberellin A_1 in dwarf pea epicotyls. Plant Physiol. in press (1980)

Lenton, J.R.: British Plant Growth Regulator Group Monograph 5, 1980. British Plant Growth Regulator Group, Dr. M.B. Jackson, A.R.C. Letcombe Laboratory, Wantage, England, 1980

MacMillan, J.: Partial synthesis of isotopically labelled gibberellins. In: Proceedings of Life Sciences Skoog, F. (Ed.). Springer, Berlin-Heidelberg-New York, 1980, pp. 161–168

McLeod, J.K., Gummons, R.E., Letham, D.S.: Mass spectrometry of cytokinin metabolites. Per(trimethylsilyl) and permethyl derivatives of glucosides of zeatin and β-benzylaminopurine. J. Org. Chem. *41*, 3959–68 (1976)

Metzger, J.D., Zeevaart, J.A.D.: Identification of six endogenous gibberellins in spinach shoots. Plant Physiol. *65*, 623–626 (1980)

Mitchell, J.W., Mandava, N., Worley, J.F., Plummer, J.R., Smith, M.V.: Brassins – a new family of plant hormones from Rape pollen. Nature (London) *225*, 1065–6 (1970)

Morris, R.O.: Mass spectrometric identification of cytokinins. Glucosyl zeatin and glucosyl ribozeatin from *Vinca rosea* crown gall. Plant Physiol. *59*, 1029–33 (1977)

Noma, M., Huber, J., Pharis, R.P.: Occurrence of $\Delta^{1(10)}$ gibberellin A_1 counterpart, GA_1, GA_4 and GA_7 in somatic cell embryo cultures of carrot and anise. Agric. Biol. Chem. *43*, 1793–4 (1979)

Parker, C.W., McLeod, J.K., Summons, R.E.: The complex of O-glucosylzeatin derivatives formed in *Populus* species. Phytochemistry *18*, 819–24 (1979)

Phinney, B.O.: Gibberellin biosynthesis in the fungus, *Gibberella fujikuroi* and in higher plants. In: Plant Growth Substances (ed. Mandava, N.B.), ACS Symp. Ser. 111. Am. Chem. Soc. Washington DC, 1979, pp 57–72

Rademacher, W.: Gaschromatographische Analyse der Veränderungen im Hormongehalt des wachsenden Weizenkorns. Dissertation, Georg-August-Universität, Göttingen (1978)

Rademacher, W., Graebe, J.E.: Isolation of IAA, ABA, GAs and cytokinins from one sample of plant material for analysis by GC and GC-MS. Abstr. 10th Int. Conf. Plant Growth Substances. Madison, U.S.A., 1979

Railton, I.D.: A study of the role of 2β-hydroxylation in the control of C-19 gibberellin levels during seedling growth. Z. Pflanzenphysiol. *96*, 103–114 (1980)

Rivier, L., Gaskin, P., Albone, K.S., MacMillan, J.: GC-MS identification of endogenous gibberellins and gibberellin conjugates as their permethylated derivatives. Phytochemistry in press (1980)

Skoog, F.: Plant Growth Substances 1979. Proceedings in Life Sciences. Springer, Berlin-Heidelberg-New York, 1980

Sponsel, V.M., Gaskin, P., MacMillan, J.: The identification of gibberellins in immature seeds of *Vicia faba* and some chemotaxonomic considerations. Planta *146*, 101–105 (1979)

Stoddart, J.L., Williams, P.D.: Interaction of ^3H gibberellin A_1 with sub-cellular fractions from lettuce (*Lactuca sativum* L.) hypocotyls. Planta *148*, 485–490 (1980)

Summons, R.E., Duke, C.C., Eichholzer, J.V., Entsch, B., Letham, D.S., McLeod, J.K., Parker, C.W.: Mass spectrometric analysis of cytokinins in plant tissue II. Quantitation of cytokinins in *Zea mays* kernels using deuterium labelled standards. Biomed. Mass Spectr. *6*, 407–412 (1979a)

Summons, R.E., Entsch, B., Parker, C.W., Letham, D.S.: Mass spectrometric analysis of cytokinins in plant tissue III. Quantitation of the cytokinin glycoside complex of lupin pods by stable isotope dilution. FEBS Lett. *107*, 21–25 (1979b)

Summons, R.E., Entsch, B., Letham, D.S., Gollnow, B.I., McLeod, J.K.: Regulators of cell division in plant tissue XXVIII. Metabolites of zeatin in sweet corn kernels. Purification and identification using high performance liquid chromatography and chemical ionisation mass spectrometry. Planta *147*, 422–434 (1979c)

Thompson, M.J., Mandava, N., Flippen-Anderson, J.L., Worley, J.F., Dutky, S.R., Robbins, W.E., Lusby, W.: Synthesis of brassino steroids. New plant growth promoting steroids. J. Org. Chem. *44*, 5002–4 (1979)

Walton, D.C.: Biochemistry and physiology of abscisic acid. Annu. Rev. Plant Physiol. *31*, 453–489 (1980)

Yamakawa, T., Kurahashi, O., Ishida, K., Kato, S., Komoda, T., Minoda, Y.: Stability of indole-3-acetic acid to autoclaving, aeration and light illumination. Agric. Biol. Chem. *43*, 879–880 (1979)

Yamane, H., Takahashi, N., Taneko, K., Furuya, M.: Identification of gibberellin A_9 methyl ester as a natural substance regulating formation of reproductive organs in *Lycopodium japonicum*. Planta *147*, 251–256 (1979)

Zaerr, J.B., Morris, R.O.: Proc. 5th N. American Forest and Biol Workshop (University of Florida) p. 401 (1979)

1 Plant Hormones and Other Growth Substances – Their Background, Structures and Occurrence

J.R. BEARDER

1.1 Introduction

The purpose of this chapter is to acquaint the reader with the chemical structures of various plant growth regulators and reports of their detection from plant sources. In general only plant growth substances from higher plants are considered. Growth substances from algae have been recently reviewed by AUGIER (see AUGIER, 1978). KOCHERT (1978) has reviewed sexual pheromones in algae and fungi. This topic will be dealt with in detail also in a forthcoming volume of this Encyclopedia on *Interactions Between Plant Cells*. Secondary metabolites from bacteria and fungi that affect the growth of higher plants have been reviewed by STROBEL (1974, 1977), PATIL (1974), RUDOLPH (1976), HARBORNE (1977) and TAMURA (1979). LETHAM (1978b) has reviewed growth substances (other than the principal hormones) from both higher and lower plants.

The first five sections of this chapter contain a brief history of the discovery of the plant hormones – ethylene, auxins, gibberellins, cytokinins, and abscisic acid and related compounds. In each case – apart from ethylene – tables on their identification in plants are provided. Preliminary communications are not included in the tables. Citations in parentheses concern the structural elucidation of the relevant compounds. In order to keep these lists to a reasonable size rigorous criteria of identification have been employed. Reports of the identification of plant hormones have been tabulated when they have been either isolated and characterized by traditional methods (melting point and spectroscopic techniques) or identified by mass spectrometry or combined gas chromatography-mass spectrometry. Mass spectrometry is a very reliable and sensitive method of detection; its main limitation is that it cannot establish the absolute stereochemistry of a compound. Furthermore the possibility of detecting hormones at very low concentration also increases the risk of erroneous identifications due to extraneous contamination where quantities of the more abundant hormones are handled in the same laboratory.

The sixth section is a catalogue of additional plant constituents which affect plant growth.

1.2 Ethylene

The discovery of ethylene as a plant growth regulator differs from that of the other plant hormones in that it was known as an exogenous effector of plant growth long before it was detected as a natural product and accepted as an endogenous hormone. Details of the history of the physiological action of ethylene have been reviewed by BURG (1962), PRATT and GOESCHL (1969) and ABELES (1973). Other aspects of ethylene in plants have also been discussed by these authors and by MAPSON and HULME (1970), YANG (1974) and OSBORNE (1978).

The early work concerning the effect of ethylene on plant growth is inextricably entangled with the toxic effects of smoke and lighting gas on plants. The first report of this kind was by GIRARDIN (1864) who observed the defoliation of trees in the vicinity of leaking gas-mains. Similar observations are reviewed by CROCKER and KNIGHT (1908) and KNIGHT and CROCKER (1913) who also attributed the premature closure of greenhouse-grown blossoms to leaks of lighting gas. It was not until 1901 that ethylene was implicated as the active factor in lighting gas. NELJUBOV (1901, 1913) noticed that pea seedlings which were germinated in the laboratory air grew in a horizontal direction. In a series of experiments he showed that the abnormal growth was due to a substance in the laboratory air and not other cultural conditions. If the laboratory air was first passed over hot copper oxide the seedlings grew normally. NELJUBOV suspected that the phenomenon was due to lighting gas in the laboratory atmosphere, so he systematically examined several constituents of coal gas for their effect on seedling growth. While many compounds caused injury and inhibited growth, only ethylene, and to a lesser extent acetylene, caused the horizontal growth syndrome. As little as 0.06 ppm of ethylene caused horizontal growth and removal of the gas caused a return to normal vertical growth.

The beneficial effects of smoke on some crops has been known for a long time. RODRIGUEZ (1932) has described how Puerto Rican pineapple farmers and Phillipino mango growers used to light fires close to their fields believing that smoke helped to initiate and synchronize the flowering of their crops. One of the most important observations was the effect of combustion products on the ripening of the lemon crop in California. The lemon growers had assumed that heat and humidity were the agents which hastened de-greening of lemons during storage. However when traditional kerosene-burning stoves were replaced by new forms of heating the lemons failed to ripen. SIEVERS and TRUE (1912) showed that the fumes of incompletely combusted kerosene were essential for the ripening of citrus fruits. The active component was later identified as ethylene by DENNY (1924a, b). Ethylene was soon shown to cause the ripening of a wide range of fruits (HARVEY, 1928; see BURG, 1962).

Many other physiological effects were discovered particularly by investigators at the Boyce Thompson Institute. Amongst other effects discovered, ethylene caused the breaking of potato dormancy (ROSA, 1925), the induction of flowering in pineapple plants (RODRIGUEZ, 1932), and the abscission of leaves (ZIMMERMAN and CROCKER, 1931).

The significance of these novel effects was increased when it became apparent that ethylene was a common plant product. The earliest evidence of this was provided by COUSINS (1910) who found that gases from oranges caused the ripening of bananas in mixed commercial shipments. (It is now thought that the ethylene was evolved from fungi growing on the oranges, as oranges are now known to produce very little ethylene.) The significance of COUSINS' observation was not appreciated and it was 22 years before this phenomenon was rediscovered. ELMER (1932, 1936) noticed that germinating potatoes were inhibited in their sprouting development when ripe apples or pears were stored close by, and he attributed the effect to fruit volatiles. Both BOTJES (1933) and HUELIN (1933) demonstrated that ethylene could simulate the apple vapours, and chemical proof that ethylene was the active gas was first provided by GANE (1934). He passed air over apples for four weeks and trapped the olefinic volatiles by brominating them at −65° C. The crude ethylene dibromide was characterized by reaction with aniline to provide N,N′-diphenylethylenediamine. Subsequently other methods were developed for the qualitative analysis of ethylene from plants (PRATT et al., 1948; NIEDERL and BRENNER, 1938; NIEDERL et al., 1938; GIBSON and CRANE, 1963; see BURG, 1962; WARD et al., 1978). Within a few years ethylene was detected by bioassay in numerous plants both from fruits and vegetative tissue. It is now known that ethylene is a ubiquitous plant product produced in essentially all parts of higher plants such as leaves, stems, roots, flowers, fruits, and tubers (BURG, 1962; OSBORNE, 1978). BURG (1962) has tabulated plants in which ethylene has been identified. Since then the routine use of gas chromatographic analysis has demonstrated the almost universal presence of ethylene among plants, making an up-to-date list neither practicable nor worthwhile. The quantities of ethylene produced by fruits range from about 50 µl kg^{-1} h^{-1} for oranges to 350 µl kg^{-1} h^{-1} for passion fruit. Possibly the highest ethylene production measured from a plant is in the fading blossoms of *Vanda* orchids, which produce 3400 µl kg^{-1} h^{-1} (AKAMINE, 1963). The rates of production of ethylene are highly susceptible to environmental factors. The suppression of ethylene evolution at extremes of temperature and at low oxygen tension (BURG and THIMANN, 1959) are usefully exploited for the storage of fruit. Bruising or injury is another influence on ethylene production. Such damage can cause bursts of ethylene production which can last for up to 24 h (ABELES and RUBINSTEIN, 1964; MCGLASSON and PRATT, 1964). This "stress ethylene" production is another significant factor in effective fruit storage. Higher plants are not the only biological sources of ethylene. The classic discovery of the production of ethylene by a fungus (*Penicillium digitatum*) was made independently by MILLER et al. (1940) and BIALE (1940). Since then many fungi and some bacteria have been shown to produce ethylene (see BURG, 1962; YANG, 1974; OSBORNE, 1978).

The first suggestion (CROCKER et al., 1935) that ethylene was a plant hormone (endogenous plant growth regulator), initiating fruit ripening and regulating vegetative growth was strongly criticized (WENT and THIMANN, 1937). At that time auxin was the central plant hormone and it seemed reasonable to attribute the effects of ethylene to changes in auxin content (e.g., MICHENER, 1938). This attitude and the exceptional experimental difficulty of identifying and measuring the concentrations of ethylene seem to have contributed to the decline in status of ethylene to the role of an exogenous compound which caused interesting physiological effects. Interest in ethylene was only maintained by fruit physiologists for whom ethylene remained an important endogenous ripener (BIALE, 1950). In 1959 this state of affairs changed abruptly with the introduction of gas chromatography as an analytical technique (BURG and STOLWIJK, 1959; HUELIN and KENNETT, 1959; MEIGH, 1959). The adoption of this simple technique initiated a new wave of research into ethylene physiology which has still not diminished.

The metabolism of ethylene is discussed in Chapter 4. Although ethylene oxide has yet to be identified as a natural product in plants it does seem likely to occur as it has been recently detected as a metabolite in *Vicia faba* when ethylene was applied at physiological concentrations (JERI and HALL, 1978).

PEGG (1976b) has recently reviewed the role of ethylene in plant pathogenesis.

1.3 Auxins

Research into the auxins can be said to have originated with the experiments of CHARLES DARWIN (1880) published in his last book, *Power of Movement in Plants*. DARWIN showed that the phototropic response of a grass coleoptile (*Phalaris canariensis*) did not occur if the tip was removed. From this work he concluded "that when seedlings are freely exposed to a lateral light some influence is transmitted from the upper to the lower point, causing the latter to bend". This hypothesis was expanded by studies of the phototropic response in *Avena* coleoptiles by ROTHERT, FITTING, BOYSEN JENSEN, PAÁL, STARK and SEUBERT (see BOYSEN JENSEN, 1936, SÖDING, 1961).[1] These studies led to the idea that the transmitted influence was a substance. The experiments culminated with the work of WENT (1928) who demonstrated that a growth substance diffused from severed coleoptile tips into agar. When a small block of this agar was placed unilaterally on a decapitated *Avena* coleoptile, the resulting curvature was found to be roughly proportional to the concentration of growth substance present. Thus he was able to explain both the correlative nature of the tropistic response and the endogenous control of growth rates in terms of a specific diffusible substance.

Since plant material contained only very small quantities of this substance, other sources were investigated. In a series of publications by KÖGL and coworkers, the isolation of three compounds was reported which were exceptionally active in the *Avena* coleoptile test. These compounds were named auxins (from the Greek, auxein, to increase). Auxin a and auxin a lactone were isolated from human urine (KÖGL et al., 1933) and auxin b from malt (KÖGL et al., 1934a). These compounds were characterized and assigned structural formulae (KÖGL and ERXLEBEN, 1934). Many attempts have been made to repeat these isolations but without success. The existence of these auxins can now be discounted on the basis of mass spectrometric analysis of the original samples (VLIEGENTHART and VLIEGENTHART, 1966), and of a synthesis of "auxin b" lactone (HWANG and MATSUI, 1968) which possessed little biological activity.

[1] The history of research on auxins is presented in more detail in Vol. XIV (1961) of the old Encyclopedia

Table 1.1. Indole-3-acetic acid

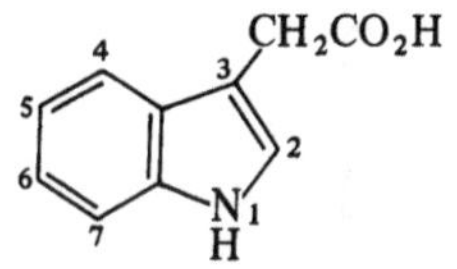

Plant material	Reference
Zea mays	
Kernels	HAAGEN-SMIT et al. (1946)
Coleoptile tips	GREENWOOD et al. (1972)
Roots	BRIDGES et al. (1973)
Roots	ELLIOTT and GREENWOOD (1974)
Root cap and apex	RIVIER and PILET (1974)
Seedlings	BANDURSKI and SCHULZE (1974)
Phaseolus mungo Etiolated seedlings	OKAMOTO et al. (1967)
Phaseolus vulgaris Shoot apices	WHITE et al. (1975)
Phaseolus vulgaris Shoot apices	MCDOUGALL and HILLMAN (1978)
Undaria pinnatifida Thalli	ABE et al. (1972)
Avena sativa	
Seedlings	BANDURSKI and SCHULZE (1974)
Grains	PERCIVAL and BANDURSKI (1976)
Fagales castanea Insect gall	YOKOTA et al. (1974)
Ricinus communis	
Phloem and root pressure sap	HALL and MEDLOW (1974)
Xylem sap	ALLEN et al. (1979)
Pinus radiata Buds	ZABKIEWICZ and STEELE (1974)
Citrus unshiu Fruit	TAKAHASHI et al. (1975)
Prunus cerasus Seed	HOPPING and BUKOVAC (1975)
Gossypium hirsutum Ovules	SHINDY and SMITH (1975)
Panax ginseng Callus	NISHIO et al. (1976)
Nicotiana tabacum Callus	NISHIO et al. (1976)
Pseudotuga menziesii	
Shoots	DEYOE and ZAERR (1976)
Shoot tips and seedlings	CARUSO et al. (1978)
Tropaeolum majus Embryo and suspensor	PRZYBYLLOK and NAGL (1977)
Oryza sativa Bran	SUZUKI et al. (1977)
Picea sitchensis Cambial region	LITTLE et al. (1978)
Triticum aestivum (Developing grain)	GASKIN et al. (1980)

Indole-3-acetic acid (IAA) (Table 1.1) is now regarded as the most significant auxin in plants. The important function of IAA as a plant growth hormone was revealed by KÖGL et al. (1934b), who isolated it from human urine, and by KÖGL and KOSTERMANNS (1934) who found the compound in yeast. A little later THIMANN (1935) repeated the isolation of the biologically active metabolite (rhizopin), first isolated by NIELSEN (1930), from the fungus *Rhizopus suinus* and showed that it was IAA. The first isolation of IAA from a plant was achieved by HAAGEN-SMIT et al. (1942) who isolated it from an alkali hydrolysate of corn meal (*Zea mays*). This result was corroborated by BERGER and AVERY (1944). Subsequently free IAA was shown to occur in immature kernels of *Zea mays* by HAAGEN-SMIT et al. (1946).

Table 1.2. Indole-3-acetyl derivatives

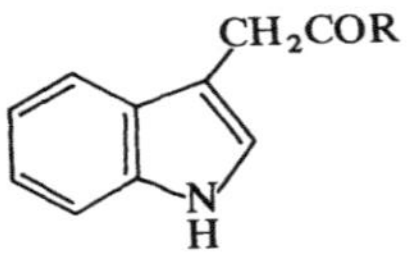

Compound	Name	Plant material	Reference
R = OCH_3	Methyl indole-3-acetate	*Citrus unshiu* Fruit	TAKAHASHI et al. (1975)
		Brassica pekinensis Club root	TAMURA et al. (1972)
			NOMOTO and TAMURA (1970)
		Fagales castanea Insect gall	YOKOTA et al. (1974)
R = OC_2H_5	Ethyl indole-3-acetate	*Gossypium hirsutum* Ovules	SHINDY and SMITH (1975)
R = NH_2	Indole-3-acetamide	*Citrus unshiu* Fruit	TAKAHASHI et al. (1975)
		Brassica pekinensis Club root	TAMURA et al. (1972)
			NOMOTO and TAMURA (1970)
		Phaseolus mungo Etiolated seedlings	ISOGAI et al. (1967a)
		Fagales castanea Insect gall	YOKOTA et al. (1974)
R = 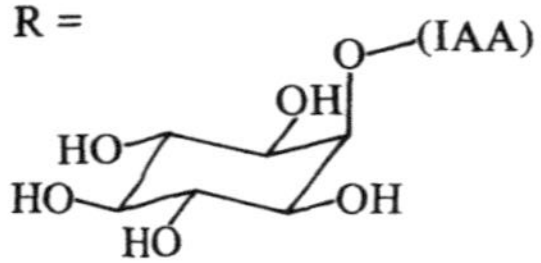 (+2 isomers)	2-O-(Indole-3-acetyl)-myo-inositol	*Zea mays* Kernels	UEDA et al. (1970)
			NICHOLLS (1967)
			(UEDA and BANDURSKI, 1974)
			(NICHOLLS et al., 1971)
R = 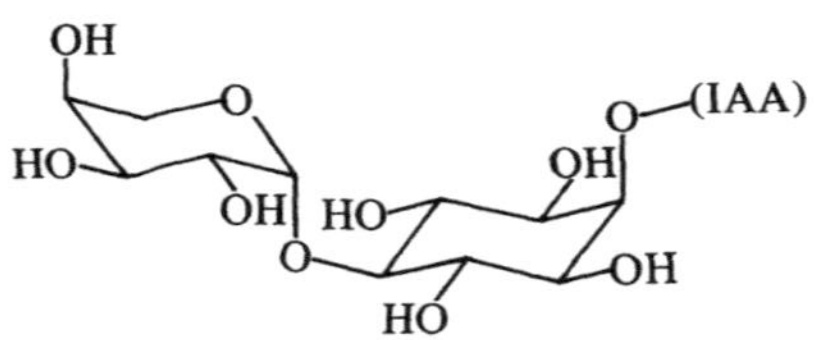 (+2 isomers)	5-O-β-L-Arabinopyranosyl-2-O-indole-3-acetyl-myo-inositol	*Zea mays* Kernels	UEDA et al. (1970)
			(UEDA and BANDURSKI, 1974)

R = 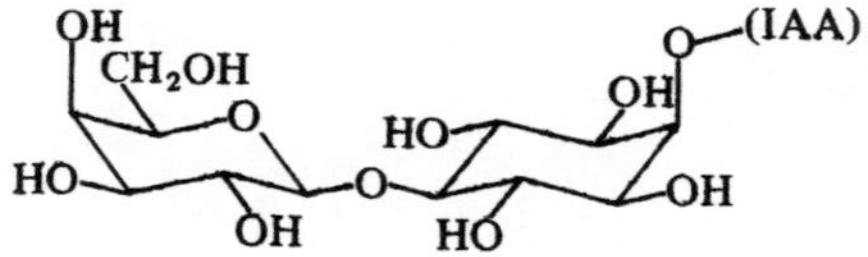 (+2 isomers)	5-O-β-D-Galactopyranosyl-2-O-indole-3-acetyl-myo-inositol	*Zea mays* Kernels	UEDA et al. (1970) (UEDA and BANDURSKI, 1974)
R = 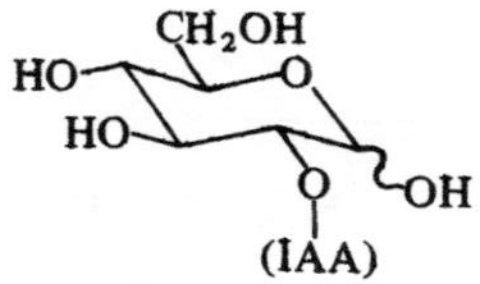	2-O-(Indole-3-acetyl)-D-glucopyranose	*Zea mays* Kernels	EHMANN (1974)
R = 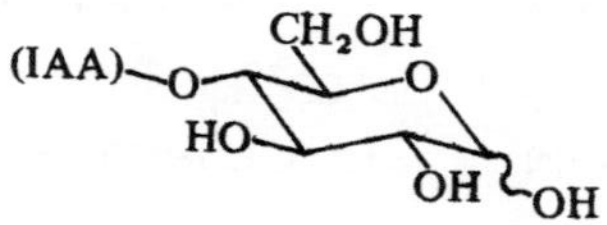	4-O-(Indole-3-acetyl)-D-glucopyranose	*Zea mays* Kernels	EHMANN (1974)
R = 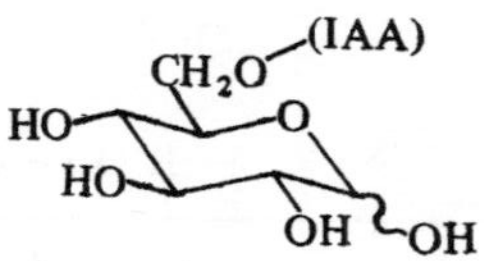	6-O-(Indole-3-acetyl)-D-glucopyranose	*Zea mays* Kernels	EHMANN (1974)
	di-O-(Indole-3-acetyl)-myo-inositol	*Zea mays* Kernels	EHMANN and BANDURSKI (1974)
	tri-O-(Indole-3-acetyl)-myo-inositol	*Zea mays* Kernels	EHMANN and BANDURSKI (1974)

Table 1.3. Chloroindoles

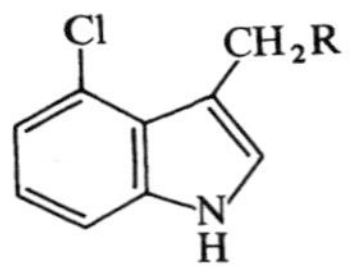

Compound	Name	Plant material	Reference
R = CO_2H	4-Chloroindole-3-acetic acid	*Pisum sativum* Immature seeds *Vicia faba* Immature seeds	MARUMO et al. (1968a) HOFINGER and BÖTTGER (1979)
R = CO_2CH_3	Methyl 4-chloroindole-3-acetate	*Pisum sativum* Immature seeds	MARUMO et al. (1968b), GANDAR and NITSCH (1968), ENGVILD et al. (1978)
		Vicia faba Immature seeds	HOFINGER and BÖTTGER (1979)
R = $CONHCH(CO_2H)CH_2CO_2CH_3$	Monomethyl 4-chlorindole-3-acetyl-L-aspartate	*Pisum sativum* Immature seeds	HATTORI and MARUMO (1972)
R = $CH(CO_2H)NHCH_2CO_2R'$	α-N-Carbomethoxyacetyl-D-4-chlorotryptophan	*Pisum sativum* Immature seeds	MARUMO and HATTORI (1971)
R′=CH_3 and C_2H_5	α-N-Carboethoxyacetyl-D-4-chlorotryptophan	*Pisum sativum* Immature seeds	MARUMO and HATTORI (1971)

Table 1.4. Indole-3-acetonitriles

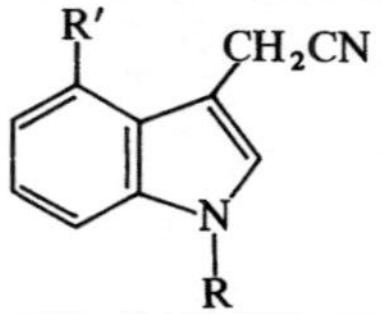

Compound	Name	Plant material	Reference
R = R′ = H	Indole-3-acetonitrile	*Brassica oleracea*	HENBEST et al. (1953), PROCHAZHA and SANDA (1960)
		Brassica pekinensis Club root	TAMURA et al. (1972), NOMOTO and TAMURA (1970)
R = H,R′ = OCH_3	4-Methoxyindole-3-acetonitrile	*Brassica pekinensis* Club root	TAMURA et al. (1972), NOMOTO and TAMURA (1970)
R = OCH_3,R′= H	1-Methoxyindole-3-acetonitrile	*Brassica pekinensis* Club root	TAMURA et al. (1972), NOMOTO and TAMURA (1970)

Table 1.5. Indole derivatives

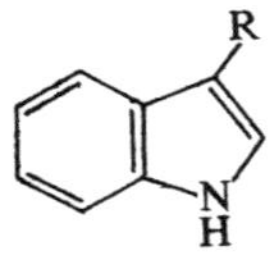

Compound	Name	Plant material	Reference
R = CH_2CH_2OH	Indole-3-ethanol	*Cucumis sativus* Seedlings	Rayle and Purves (1967)
R = CH_2CHO	Indole-3-acetaldehyde	*Fagales castanea* Insect gall	Yokota et al. (1974a) (as dimethyl acetal)
R = $CH_2CH{:}NOH$	Indole-3-acetoxime	*Brassica oleracea*	Kindl (1968)
R = $CH_2CH_2NH_2$	Tryptamine	*Lycopersicon esculentum*	Schneider et al. (1972), see also Smith (1977)
R = $CH_2CH(NHCO.CH_2CO_2H).CO_2H$	α-N-Malonyl-D-tryptophan	*Triticum aestivum* Roots	Elliott (1971)
R = CHO	Indole-3-carboxaldehyde	*Brassica oleracea*	Jones and Taylor (1957), Prochazha and Sanda (1960)
		Phaseolus mungo Etiolated seedlings	Isogai et al. (1967a)
		Equisetum telmateia Aerial stems	Bourdoux et al. (1971)
		Murraya exotica Stem bark	Chowdhury and Chakraborty (1971)
		Gossypium hirsutum Ovules	Shindy and Smith (1975)
R = CO_2H	Indole-3-carboxylic acid	*Brassica oleracea*	Jones and Taylor (1957), Prochazha and Sanda (1960)

Table 1.6. Other indole complexes

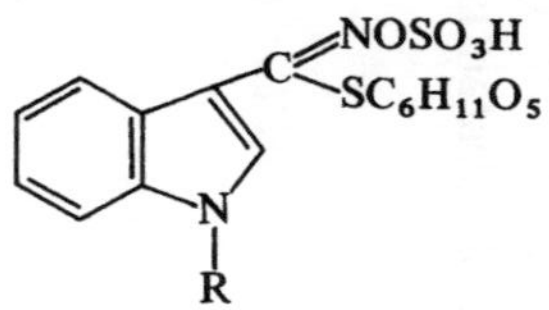

	Name	Plant	Reference
R = H (glucobrassicin)	Indole-3-methylglucosinolate	*Brassica oleracea*	GMELIN and VIRTANEN (1961)
		Raphanus sativus	GMELIN and VIRTANEN (1961)
		Isatis tinctoria	ELLIOTT and STOWE (1971)
R = OCH_3 (neoglucobrassicin)	1-Methoxyindole-3-methylglucosinolate	*Brassica napus*	GMELIN and VIRTANEN (1962)
		Isatis tinctoria	ELLIOTT and STOWE (1971)
R = SO_3^- (1–sulphoglucobrassicin)	1-Sulphoindole-3-methylglucosinolate	*Isatis tinctoria*	ELLIOTT and STOWE (1971) (ELLIOTT and STOWE, 1970)
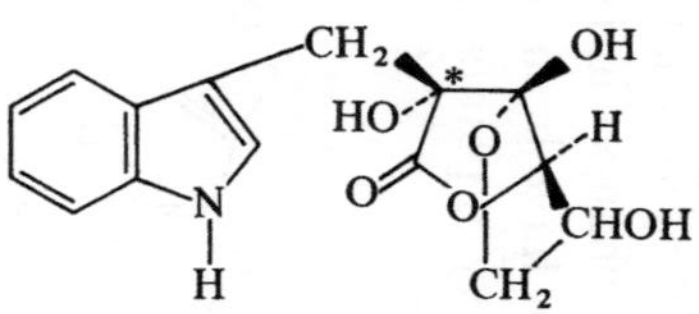 Ascorbigen A (B is epimer at *)		*Brassica oleracea*	PROCHAZHA and SANDA (1960) (KISS and NEUKOM, 1966)

In the next two decades numerous reports appeared on the detection of IAA in various plants using bioassay and paper chromatography with semi-specific chromogenic sprays (see LARSEN, 1951; GORDON, 1965; BENTLEY, 1958; BENTLEY, 1961; SCHNEIDER et al., 1972; SCHNEIDER and WIGHTMAN, 1974). However, it was not until 1967 that IAA was isolated and characterized from a second plant source, *Phaseolus mungo* (OKAMOTO et al., 1967). Since then IAA has been isolated from only a few plant sources and conclusively shown to be present in others by mass spectrometry (see Table 1.1). SHELDRAKE (1973) has described the occurrence of IAA in many other groups of living organism. Many derivatives of IAA and related compounds have now been identified in plant extracts (see Tables 1.2–1.6). While IAA and its methyl ester were not detected in *Pisum sativum* (HATTORI and MARUMO, 1972) 4-chloroindole-3-acetic acid (MARUMO et al., 1968a) and its methyl ester (see Table 1.3) (MARUMO et al., 1968b; GANDAR and NITSCH, 1968; ENGVILD et al., 1978) were identified.

Members of the Brassicaceae contain a variety of indole compounds (see Table 1.6) many of which may be implicated in auxin metabolism (KUTÁCEK and KEFELI, 1968; SCHNEIDER and WIGHTMAN, 1974). The goitrogenic glucosinolates glucobrassicin and neoglucobrassicin have been found in the overwhelming majority of Cruciferae examined (SCHRAUDOLF, 1968) and they have also been detected in species of Capparidaceae, Resedaceae, and Tovariaceae (SCHRAUDOLF, 1965). Since indole-3-acetonitrile is readily formed at low pH by myrosinase-catalyzed breakdown of glucobrassicin during tissue extraction, it is not certain that indole-3-acetonitrile is a normal constituent of *Brassica* plants (VIRTANEN, 1965), but it is probably present in small amounts (KUTÁCEK and KEFELI, 1968). Likewise although the major portions of indole-3-carboxaldehyde and indole-3-carboxylic acid (see Table 1.5), found in plant extracts, seem to be artefacts in *Brassica* extracts, they are also thought to be native in small quantities (KUTÁCEK and KEFELI, 1968). Ascorbigen [a mixture of the epimers A and B (Table 1.6)] is also thought to arise predominantly as an artefact during the work-up of *Brassica* plants. Myrosinase-catalyzed cleavage of glucobrassicin at pH 7 yields glucose, sulphate, and presumably indole-3-methyl-isothiocyanate (see Fig. 1.1) as an unstable intermediate which breaks down into thiocyanate and 3-hydroxymethylindole (see Fig. 1.1) (GMELIN and VIRTANEN, 1961). The 3-hydroxymethylindole may then condense with endogenous ascorbic acid to form ascorbigen (PIIRONEN and VIRTANEN, 1962).

The concentration of free IAA in plants is usually between one and 100 μg kg^{-1} fresh weight tissue (SCHNEIDER and WIGHTMAN, 1974; BANDURSKI and SCHULZE, 1977). Quantitative changes in IAA concentration during the development of the *Citrus unshiu* fruit have been monitored by TAKAHASHI et al. (1975). Recent examination of a variety of plants has revealed the presence of substantial

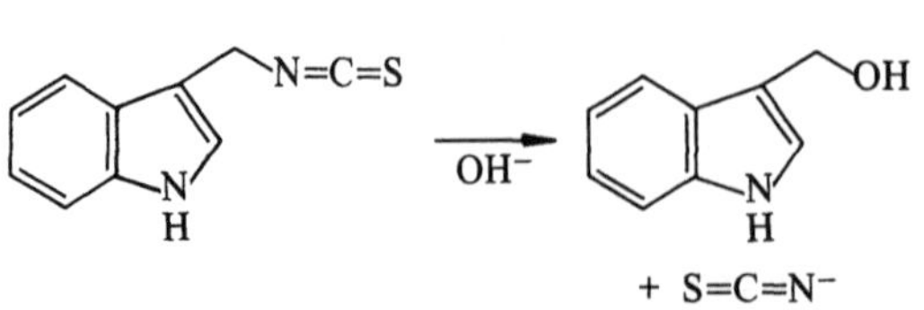

Fig. 1.1

quantities of IAA bound as ester and/or peptidyl complexes (BANDURSKI and SCHULZE, 1977). These complexes occur in all the species analyzed. Cereal grains (e.g., *Avena*) contain mainly esterified IAA, whereas peptidyl IAA is the principle derivative in legume seeds. In contrast to the seeds, *Avena* tissue contains mainly peptidyl IAA. In more detailed earlier work, BANDURSKI and co-workers characterized a number of IAA conjugates (SEMBDNER, 1974) from *Zea mays* (see Table 1.2). These include three IAA-myo-inositol esters, six IAA-myo-inositol glycosides and three IAA glucosides. Trace quantities of di-O-(indole-3-acetyl)-myo-inositol and tri-O-(indole-3-acetyl)-myo-inositol were also detected. A 1,4β-linked cellulosic glucan containing IAA was partially characterized from the same source (PISKORNIK and BANDURSKI, 1972). PERCIVAL and BANDURSKI (1976) detected an IAA ester glucoprotein in *Avena* grains. BANDURSKI et al. (1977) have suggested that light may control the ratio of ester to free IAA. A rhamnose-bound IAA has been described by GANGULY et al. (1974). Two important conjugates of exogenously supplied IAA are indole-3-acetylaspartate (see Fig. 1.2) (TILLBERG, 1974; see MOLLEN et al., 1972) and indole-3-acetyl-1-β-glucose (ZENK, 1961) (see Fig. 1.2). However they have yet to be unambiguously identified as endogenous substances.

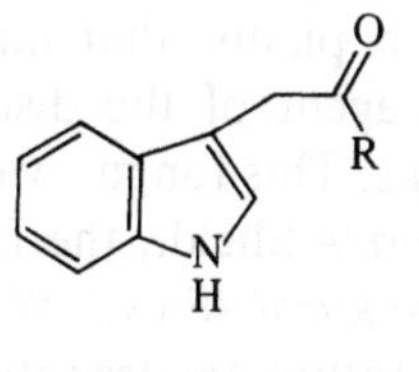

R= NHCH(CO_2H)CH_2CO_2H Indole–3–acetylaspartate

Fig. 1.2 R= 1–β–glucosyl Indole–3–acetyl–1–β–glucose

There are several other indoles implicated in the biosynthesis and metabolism of IAA (see Chap. 4) which have only been partially characterized as natural constituents of higher plants (BENTLEY, 1958, 1961; STOWE, 1959; SCHNEIDER and WIGHTMAN, 1974).

There are some reports of the existence of non-indole auxins (see SCHNEIDER and WIGHTMAN, 1974) but most of these remain to be confirmed. Investigation of the growth promoting factor in the bulb of *Lycoris radiata* yielded p-hydroxyphenylacetic acid (see Fig. 1.3), a compound with considerable auxin activity (ISOGAI et al., 1974). This compound has also been isolated from the alga *Undaria*

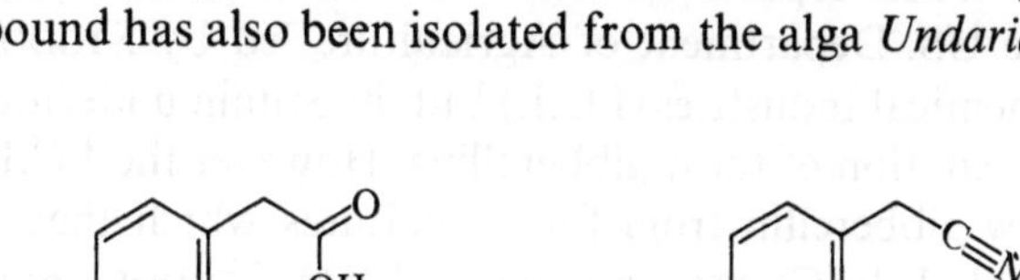

R= H Phenylacetic acid Phenylacetonitrile

Fig. 1.3 R= OH p–Hydroxyphenylacetic acid

pinnatifida (ABE et al., 1974) together with phenylacetic acid (see Fig. 1.3) which also has high auxin activity (MILBORROW et al., 1975). Phenylacetic acid is also present in *Phaseolus mungo* (OKAMOTO et al., 1967). On the basis of identification of phenylacetic acid by its chromatographic parameters in many plant tissues, and from metabolic studies, WIGHTMAN has suggested that phenylacetic acid may be a commonly occurring natural auxin (WIGHTMAN, 1973; WIGHTMAN and RAUTHAN, 1974; WIGHTMAN, 1977). The related compound phenylacetonitrile (see Fig. 1.3) may also be involved (WHEELER, 1977).

PEGG (1976a) has reviewed the role of auxins in plant disease.

1.4 Gibberellins

The "bakanae" (foolish seedling) disease of rice was first described in Japan in 1809 (KONISHI, 1828). This disease is endemic in the Orient, having caused serious crop damage in the past (STODOLA, 1958). Up to 40% decreases in yields have been reported. The principal symptom of the disease is the appearance of tall, pale-green, thin plants, that markedly overgrow their uninfected neighbours. The causative agent of the disease was first identified by HORI (1898) as an imperfect fungus. This fungus was later classified by WÖLLENWEBER (1931) as *Fusarium moniliforme* Sheld., the imperfect form of the rarely found perfect stage, *Gibberella fujikuroi* (Saw.) Wr. However both names are now commonly used in the literature to describe the fungus. A detailed account of the history of the disease and fungus has been compiled by STODOLA (1958), and a more concise account by STOWE and YAMAKI (1957).

All research on the gibberellins stems from the work of KUROSAWA (1926), who succeeded in producing the overgrowth effect in rice and maize seedlings solely by treating them with a cell-free filtrate of the culture medium in which *G. fujikuroi* had been grown. The active substance was shown to be non-enzymic as it survived autoclaving. Several years of preliminary work on this material culminated in the isolation of two biologically active compounds by YABUTA and SUMIKI (1938). These compounds were named gibberellins A and B. The nature of the gibberellin carbon skeleton was established by the identification of the selenium dehydrogenation product of both gibberellins as a fluorene derivative (YABUTA et al., 1941).

When reports of this work reached the West ten years later a group at the US. Department of Agriculture led by STODOLA, and workers at Imperial Chemical Industries (I.C.I.) Ltd. in Britain undertook the isolation and structural elucidation of these gibberellins. However the I.C.I. workers isolated an entirely new gibberellin from fungal cultures which they called gibberellic acid (GA_3, Table 1.7) (CURTIS and CROSS, 1954). STODOLA et al. (1955) also isolated a new metabolite gibberellin X, which was shown to be identical with gibberellic acid. The work of the British group was simplified by the selection by BRIAN and co-workers (BORROW et al., 1955), of a strain of *G. fujikuroi* and conditions of culture which produced mainly one compound. The strain used by the Ameri-

Table 1.7. Gibberellins from higher plants and fungi

	Plant material	Reference
GA₁ 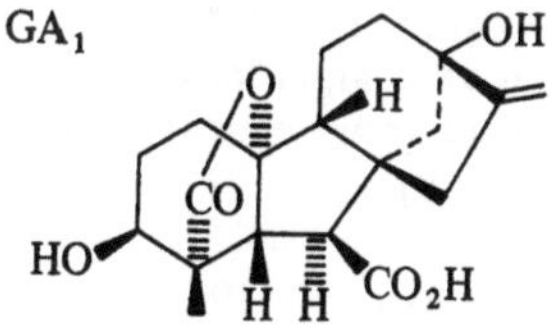	*G. fujikuroi*	
	Mycelial filtrate	TAKAHASHI et al. (1955)
		STODOLA et al. (1957)
		GROVE et al. (1958)
	Citrus unshiu Water sprouts	KAWARADA and SUMIKI (1959)
	Phaseolus vulgaris	
	Immature seed	WEST and PHINNEY (1959)
		(WEST, 1961)
		YAMANE et al. (1977)
		HIRAGA et al. (1974b)
	Mature seed	HIRAGA et al. (1974a)
	Phaseolus coccineus[a]	
	Immature seed	MACMILLAN et al. (1960)
		DURLEY et al. (1971)
	Light grown seedlings	BOWEN et al. (1973)
	Althaea rosea Shoot apices	HARADA and NITSCH (1967)
	Cucumis sativus Seed	HEMPHILL et al. (1972)
	Cucumis melo Seed	HEMPHILL et al. (1972)
	Corylus avellana Seed	WILLIAMS et al. (1974)
	Sonneratia apelata Leaves	GANGULY and SIRCAR (1974)
	Gossypium hirsutum Ovules	SHINDY and SMITH (1975)
	Vigna unguiculata Seed	ADESOMOJU (1977)
	Triticum aestivum Grains	ECKERT et al. (1978)
	Germinating grains	LENTON (unpublished work)
	Secale cereale Grains	ECKERT et al. (1978)
GA₂ 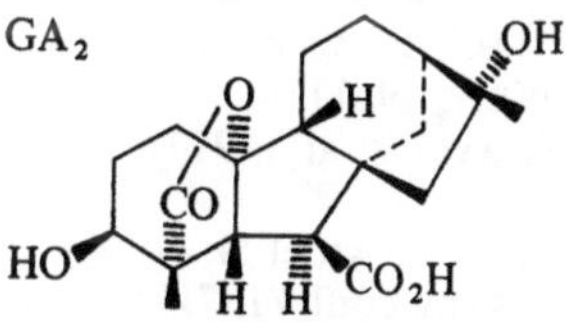	*G. fujikuroi*	TAKAHASHI et al. (1955)
		GROVE (1961b)
GA₃ 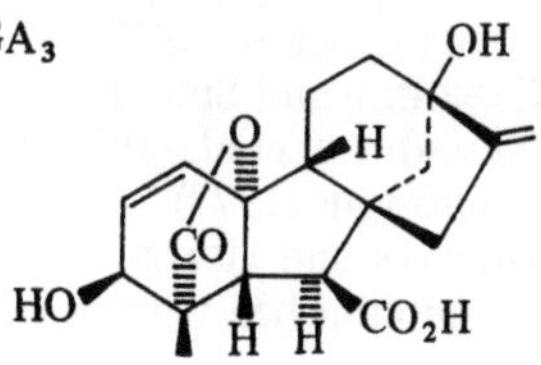	*G. fujikuroi*	CROSS (1954)
		STODOLA et al. (1955)
		TAKAHASHI et al. (1955)
		(HARTSUCK and LIPSCOMB, 1963)
		(MCCAPRA et al., 1966)
	Althaea rosea Shoot apices	HARADA and NITSCH (1967)
	Cassia fistula Flowers	SIRCAR et al. (1970)
	Phaseolus coccineus[a] Seed	DURLEY et al. (1971)
	Cucumis sativus Seed	HEMPHILL et al. (1972)
	Cucumis melo Seed	HEMPHILL et al. (1972)
	Sonneratia apelata Leaves	GANGULY and SIRCAR (1974)
	Rhizophora mucranata Leaves	GANGULY and SIRCAR (1974)
	Gossypium hirsutum Ovules	SHINDY and SMITH (1975)
	Avena sativa Inflorescence	KAUFMAN et al. (1976)
	Pinus attenuata Pollen	KAMIENSKA et al. (1976)
	Triticum aestivum Grains	ECKERT et al. (1978)
	Germinating grains	LENTON (unpublished work)
	Secale cereale Grains	ECKERT et al. (1978)

Table 1.7 (continued)

	Plant material	Reference
GA_4 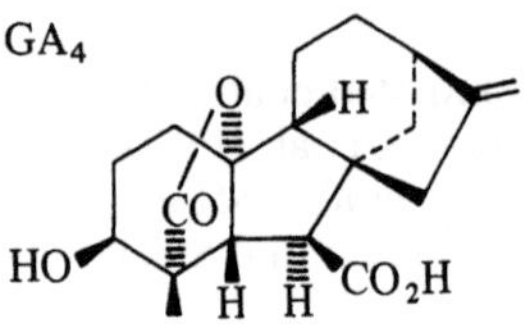	*G. fujikuroi*	TAKAHASHI et al. (1957)
		GROVE et al. (1960)
	Sphaceloma manihoticola	RADEMACHER and GRAEBE (1979)
	Phaseolus coccineus[a]	
	Dark-grown seedlings	CROZIER et al. (1971)
	Light-grown seedlings	BOWEN et al. (1973)
	Seed	SPONSEL (unpublished work)
	Phaseolus vulgaris	
	Immature seed	HIRAGA et al. (1974b)
		YAMANE et al. (1977)
	Cucumis sativus Seed	HEMPHILL et al. (1972)
	Pyrus malus Seed	SINSKA et al. (1973)
		HOAD (1978)
	Gossypium hirsutum Ovules	SHINDY and SMITH (1975)
	Marah macrocarpus[b]	
	Immature seed	BEELEY et al. (1975)
	Pinus attenuata Pollen	KAMIENSKA et al. (1976)
	Triticum aestivum	
	Chloroplasts	BROWNING and SAUNDERS (1977)
	Vigna unguiculata Seed	ADESOMOJU (1977)
	Cucurbita maxima Cotyledons	GRAEBE et al. (unpublished work)
GA_5 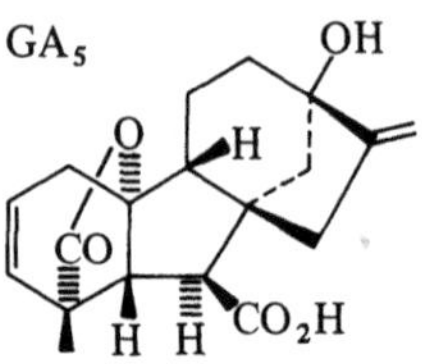	*Phaseolus vulgaris*	
	Immature seed	WEST and PHINNEY (1959)
		(MACMILLAN et al., 1960)
		(WEST, 1961)
		YAMANE et al. (1977)
	Phaseolus coccineus[a]	
	Immature seed	MACMILLAN et al. (1960)
		DURLEY et al. (1971)
	Light-grown seedlings	BOWEN et al. (1973)
	Cucumis melo Seed	HEMPHILL et al. (1972)
	Rhizophora mucranata Leaves	GANGULY and SIRCAR (1974)
	Prunus persica Immature seed	YAMAGUCHI et al. (1975a)
	Vigna unguiculata Seed	ADESOMOJU (1977)
	Vicia faba Seed	HORGAN and HEALD (unpublished work)
GA_6 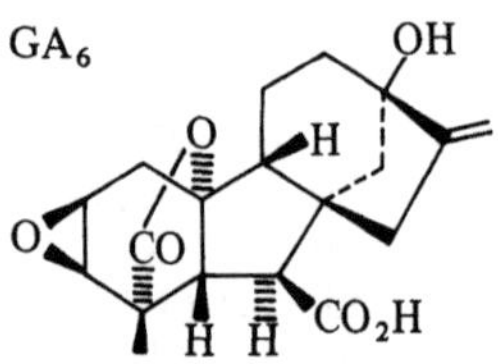	*Phaeolus coccineus*[a]	
	Immature seed	MACMILLAN et al. (1962)
		DURLEY et al. (1971)
	Phaseolus vulgaris	
	Immature seed	DURLEY et al. (1971)
		YAMANE et al. (1977)
	Vigna unguiculata Seed	ADESOMOJU (1977)

Table 1.7 (continued)

	Plant material	Reference
GA_7 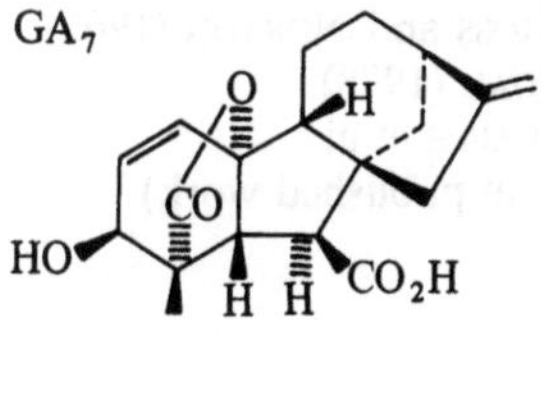	*G. fujikuroi*	Cross et al. (1962)
	Cucumis sativus Seed	Hemphill et al. (1972)
	Pyrus malus Seed	Sinska et al. (1973)
		Hoad (1978)
	Gossypium hirsutum Ovules	Shindy and Smith (1975)
	Marah macrocarpus[b]	
	Immature seed	Beeley et al. (1975)
	Pinus attenuata Pollen	Kamienska et al. (1976)
GA_8 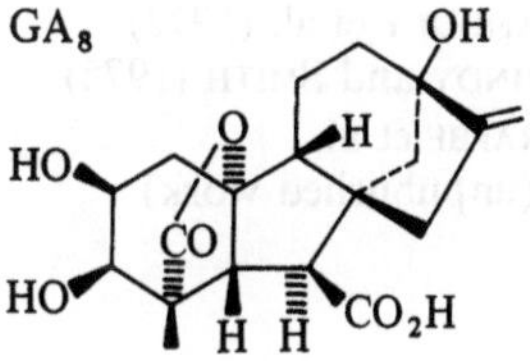	*Phaseolus coccineus*[a]	
	Immature seed	MacMillan et al. (1962)
		Durley et al. (1971)
	Phaseolus vulgaris	
	Immature seed	Durley et al. (1971)
		Hiraga et al. (1974b)
		Yamane et al. (1977)
	Mature seed	Hiraga et al. (1974a)
	Pharbitis nil Immature seed	Yokota et al. (1971a)
	Calonyction aculeatum	
	Immature seed	Murofushi et al. (1973)
	Vigna unguiculata Seed	Adesomoju (1977)
	Secale cereale Immature fruit	Dathe et al. (1978a)
GA_9 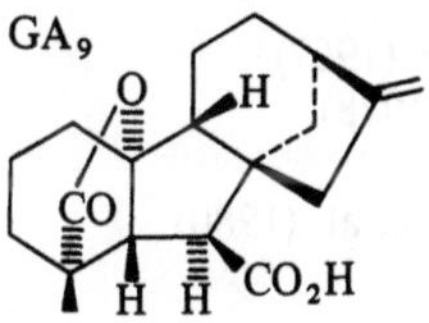	*G. fujikuroi*	Cross et al. (1962)
	Althaea rosea Shoot apices	Harada and Nitsch (1967)
	Enhydra fluctuans	Ganguly et al. (1972)
	Pyrus malus Seed	Sinska et al. (1973)
		Hoad (1978)
	Pisum sativum Seed	Frydman et al. (1974)
	Rhizophora mucranata Leaves	Ganguly and Sircar (1974)
	Corylus avellana Seed	Williams et al. (1974)
	Gossypium hirsutum Ovules	Shindy and Smith (1975)
	Triticum aestivum	
	Chloroplasts	Browning and Saunders (1977)
(Δ-15,16 isomer)	*Picea sitchensis* Needles	Lorenzi et al. (1977)
GA_{10} 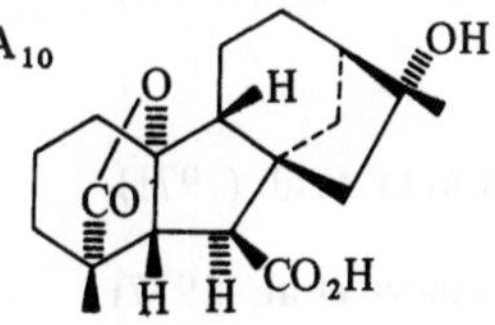	*G. fujikuroi*	Hanson (1966)
GA_{11} 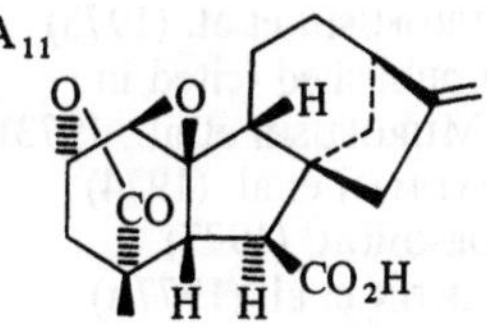	*G. fujikuroi*	Brown et al. (1967)

Table 1.7 (continued)

	Plant material	Reference
GA_{12} 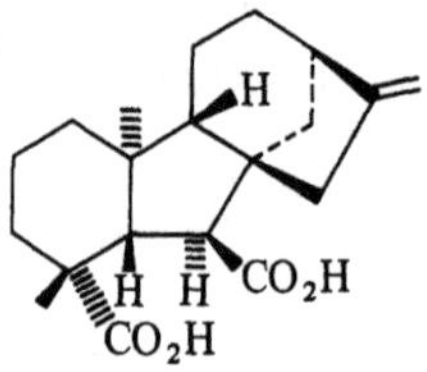	*G. fujikuroi*	Cross and Norton (1965)
	Pyrus malus Seed	Hoad (1978)
	Cucurbita maxima Seed	Graebe et al. (unpublished work)
GA_{13} 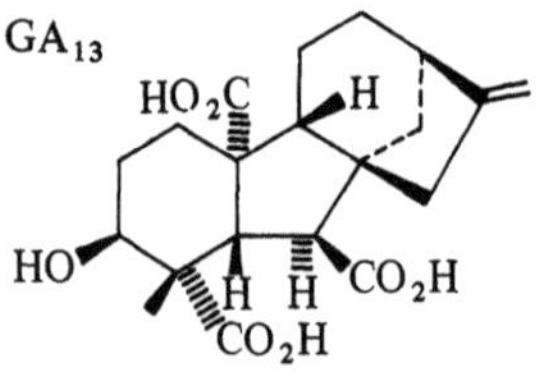	*G. fujikuroi*	Galt (1965)
	Enhydra fluctuans	Ganguly et al. (1972)
	Gossypium hirsutum Ovules	Shindy and Smith (1975)
	Cucurbita maxima Seed	Graebe et al. (unpublished work)
GA_{14} 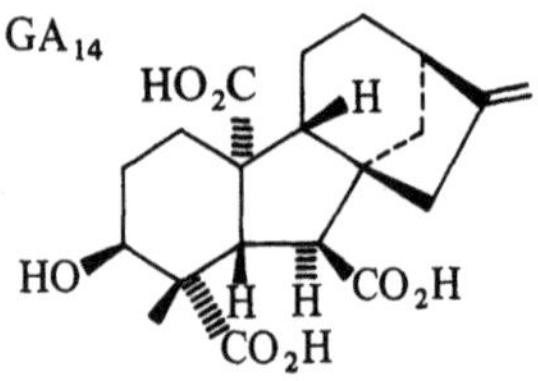	*G. fujikuroi*	Cross (1966)
		Jones et al. (1968)
GA_{15} 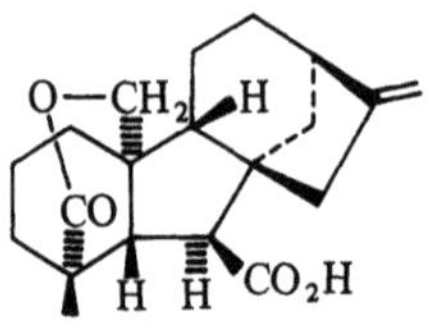	*G. fujikuroi*	Hanson (1967)
	Pyrus malus Seed	Hoad (1978)
	Triticum aestivum Developing grain	Gaskin et al. (1980)
GA_{16} 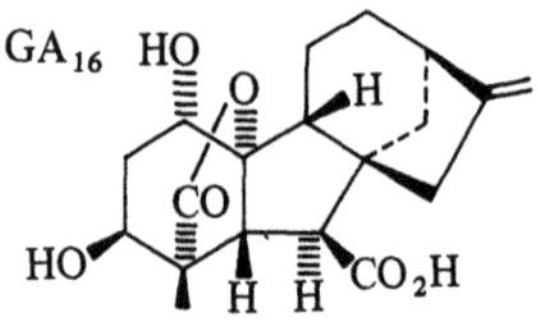	*G. fujikuroi*	Galt (1968)
		Bearder and MacMillan (1973)
		McInnes et al. (1973)
	Secale cereale Immature fruit	Dathe et al. (1978a)
GA_{17} 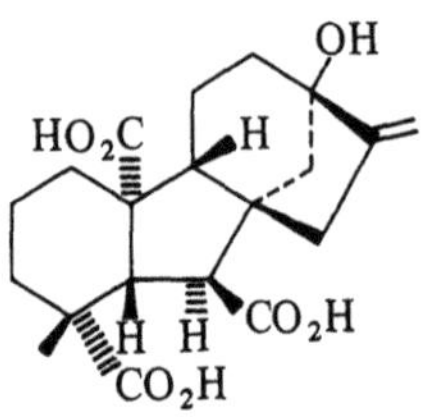	*Phaseolus coccineus*[a] Immature seed	Durley et al. (1971)
	Phaseolus vulgaris Immature seed	Yamane et al. (1977)
	Calonyction aculeatum Immature seed	Murofushi et al. (1973)
	Pharbitis nil	Unpublished (cited in Murofushi et al., 1973)
	Pisum sativum Seed	Frydman et al. (1974)
	Vigna unguiculata Seed	Adesomoju (1977)
	Pyrus communis Seed	Martin et al. (1977a)
	Pyrus malus Seed	Hoad (1978)
	Triticum aestivum Developing grain	Gaskin et al. (1980)
	Vicia faba Seed	Sponsel et al. (1979)
	Spinacia oleracea Shoots	Metzger and Zeevaart (1980)

Table 1.7 (continued)

	Plant material	Reference
GA_{18} 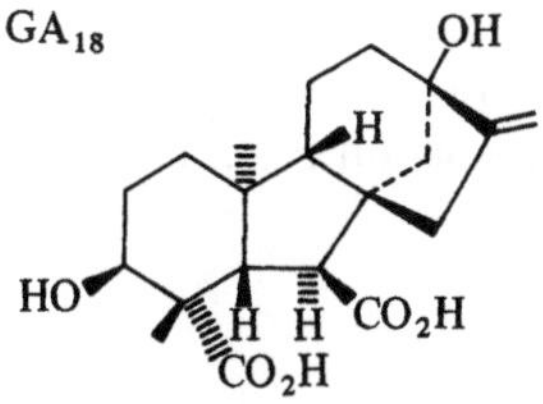	*Lupinus luteus*	
	Immature seed	Koshimizu et al. (1968)
	Wisteria floribunda	
	Immature seed	Koshimizu et al. (1972)
GA_{19} 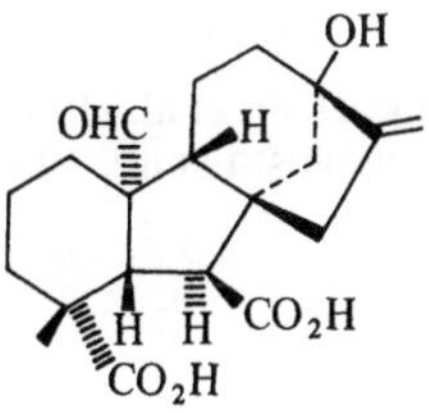	*Phyllostachys edulis* Shoots	Murofushi et al. (1966)
	Phaseolus coccineus[a]	
	Immature seed	Durley et al. (1971)
	Lupinus luteus Immature seed	Fukui et al. (1972)
	Calonyction aculeatum	
	Immature seed	Murofushi et al. (1973)
	Vigna unguiculata Seed	Adesomoju (1977)
	Oryza sativa Shoots	Kurogochi et al. (1978)
	Humulus lupulus Cones	Watanabe et al. (1978b)
	Triticum aestivum	
	Developing grain	Gaskin et al. (1980)
	Germinating grains	Lenton (unpublished work)
	Vicia faba Seed	Sponsel et al. (1979)
	Beta vulgaris Leaf tips	Lenton (unpublished work)
	Spinacia oleracea Shoots	Metzger and Zeevaart (1980)
GA_{20} 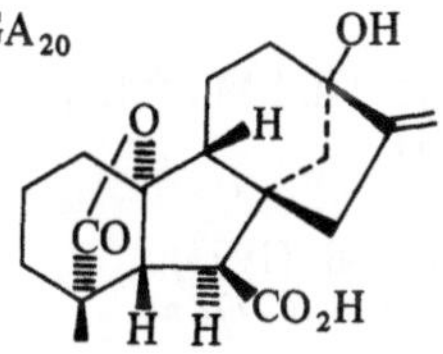	*Pharbitis nil* Immature seed	Murofushi et al. (1968)
		Yokota et al. (1971a)
	Phaseolus coccineus[a]	
	Immature seed	Durley et al. (1971)
	Light-grown seedlings	Bowen et al. (1973)
	Phaseolus vulgaris	
	Immature seed	Yamane et al. (1977)
	Bryophyllum daigremontianum	
	Shoot tips and upper leaves	Gaskin et al. (1973)
	Pisum sativum	
	Pods	Komoda et al. (1968)[c]
	Seed	Frydman and MacMillan (1973)
		Eeuwens et al. (1973)
		Frydman et al. (1974)
	Vigna unguiculata Seed	Adesomoju (1977)
	Pyrus malus Seed	Hoad (1978)
	Stevia rebaudiana Shoots	Alves and Ruddat (1979)
	Spinacia oleracea Shoots	Metzger and Zeevaart (1980)
	Vicia faba Seed	Horgan and Heald (unpublished work)
		Sponsel et al. (1979)
	Beta vulgaris Leaf tips	Lenton (unpublished work)

Table 1.7 (continued)

	Plant material	Reference
GA_{21} 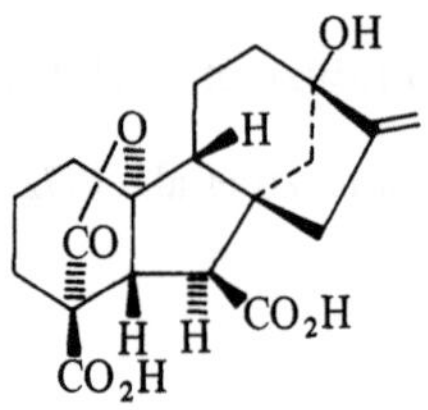	*Canavalia gladiata* Immature seed	MUROFUSHI et al. (1969a) (MUROFUSHI et al., 1969b)
GA_{22} 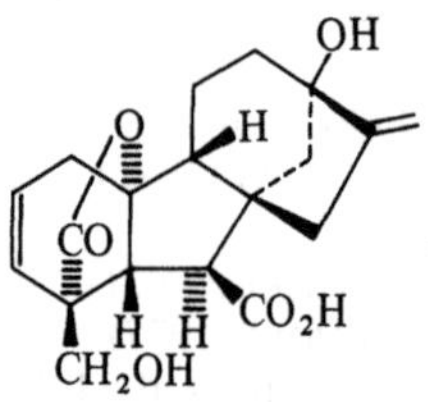	*Canavalia gladiata* Immature seed	MUROFUSHI et al. (1969a) (MUROFUSHI et al., 1969b)
GA_{23} 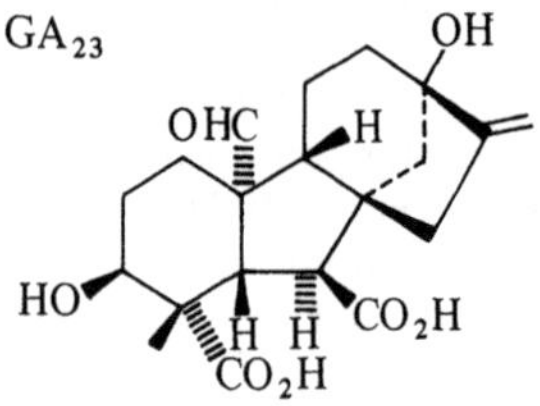	*Lupinus luteus* Immature seed *Wisteria floribunda* Immature seed	FUKUI et al. (1972) KOSHIMIZU et al. (1972)
GA_{24} 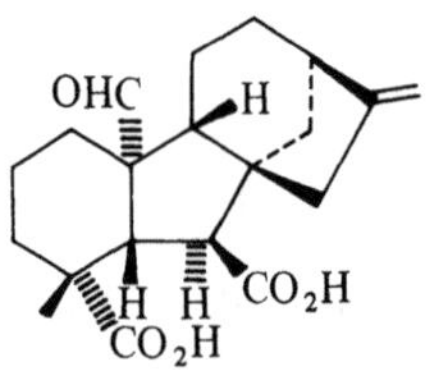	*G. fujikuroi* *Marah macrocarpus*[b] Immature seed *Secale cereale* Immature fruit *Triticum aestivum* Developing grain	HARRISON and MACMILLAN (1971) BEELEY et al. (1975) DATHE et al. (1978a) GASKIN et al. (1980)
GA_{25} 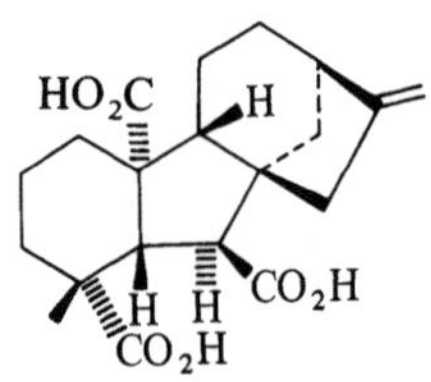	*G. fujikuroi* *Marah macrocarpus*[b] Immature seed *Pyrus communis* Seed *Cucurbita maxima* Cotyledons	HARRISON and MACMILLAN (1971) BEELEY et al. (1975) MARTIN et al. (1977a) GRAEBE et al. (unpublished work)
GA_{26} 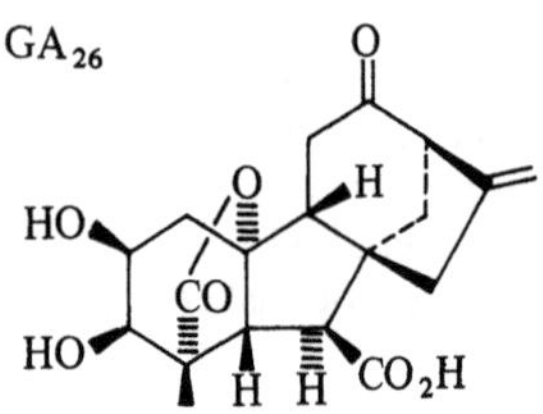	*Pharbitis nil* Immature seed	YOKOTA et al. (1971a)

Table 1.7 (continued)

	Plant material	Reference
GA_{27} 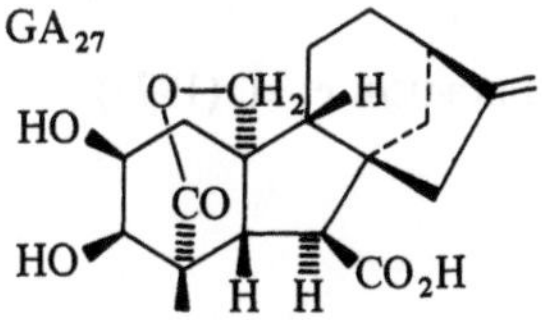	*Pharbitis nil* Immature seed	YOKOTA et al. (1971a)
	Calonyction aculeatum Immature seed	MUROFUSHI et al. (1973)
GA_{28} 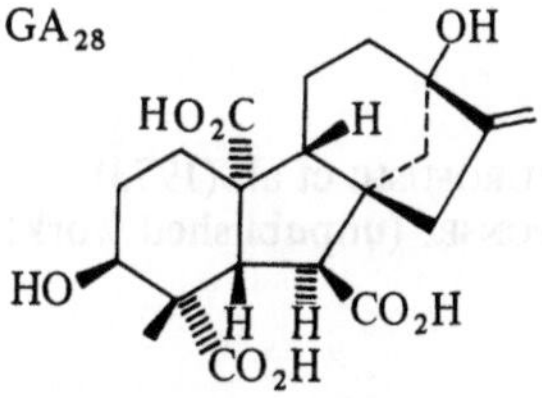	*Lupinus luteus* Immature seed	FUKUI et al. (1971)
	Phaseolus coccineus Seed	SPONSEL (unpublished work)
GA_{29} 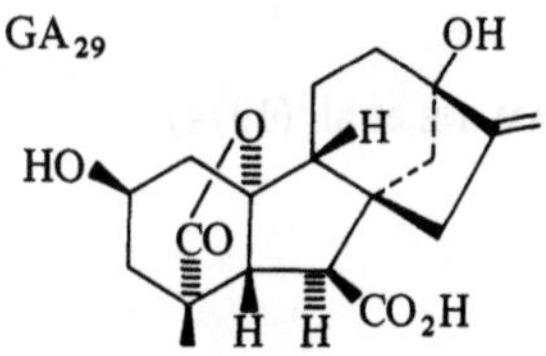	*Calonyction aculeatum* Immature seed	MUROFUSHI et al. (1973) (YOKOTA et al., 1971b)
	Pisum sativum Seed	FRYDMAN and MACMILLAN (1973)
		FRYDMAN et al. (1974)
	Prunus domestica Fruits	REED and MARTIN (1976)
	Phaseolus vulgaris Immature seed	YAMANE et al. (1977)
	Vigna unguiculata Seed	ADESOMOJU (1977)
	Vicia faba Seed	SPONSEL et al. (1979)
	Bryophyllum daigremontianum Shoot tips and upper leaves	ZEEVAART (unpublished work)
GA_{30} 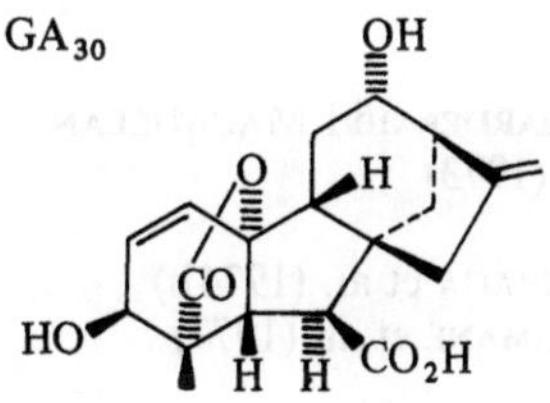	*Calonyction aculeatum* Immature seed	MUROFUSHI et al. (1973)
GA_{31} 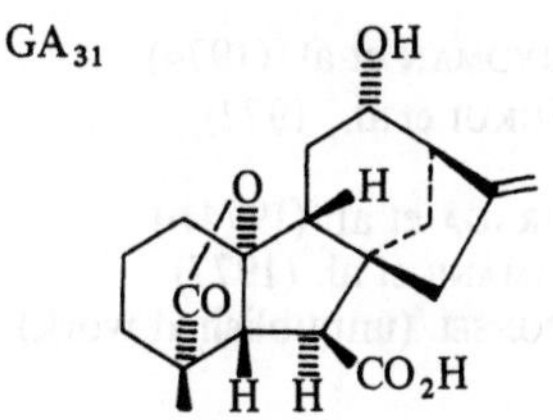	*Calonyction aculeatum* Immature seed	MUROFUSHI et al. (1973)
GA_{32} 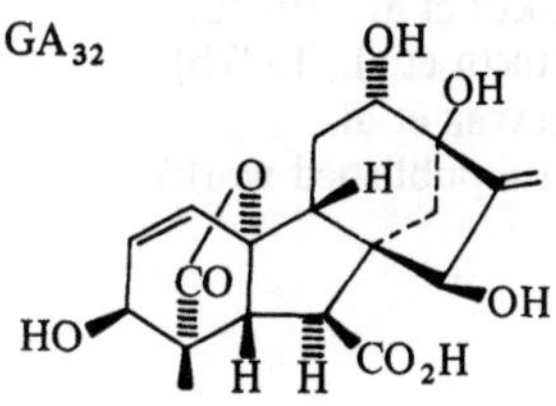	*Prunus armenica* Immature seed	COOMBE (1971)
	Prunus persica Immature seed	YAMAGUCHI et al. (1975a) (YAMAGUCHI et al., 1975b)
	Prunus cerasus Developing fruit	BUKOVAC et al. (1979)

Table 1.7 (continued)

	Plant material	Reference
GA_{33} 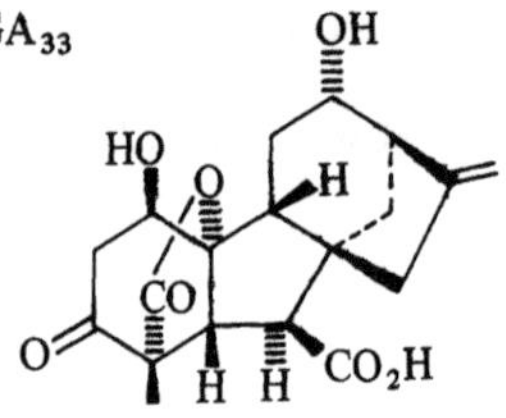	*Calonyction aculeatum* Immature seed	MUROFUSHI et al. (1973)
GA_{34} 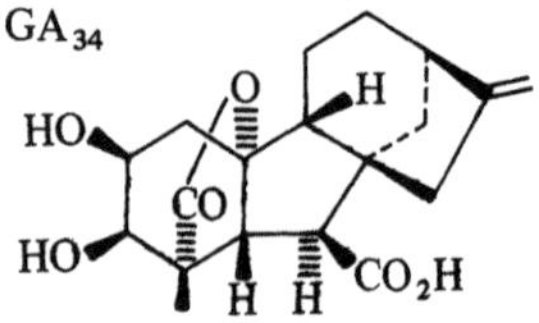	*Calonyction aculeatum* Immature seed *Phaseolus coccineus* Seed	 MUROFUSHI et al. (1973) SPONSEL (unpublished work)
GA_{35} 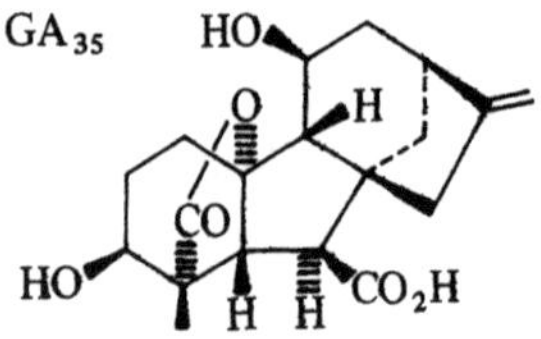	*Cytisus scoparius* Immature seed	YAMANE et al. (1974)
GA_{36} 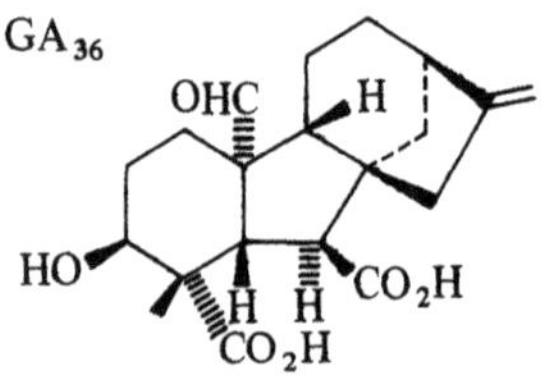	*G. fujikuroi*	BEARDER and MACMILLAN (1973)
GA_{37} 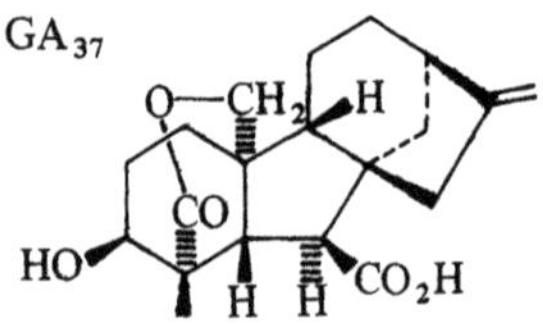	*G. fujikuroi* *Phaseolus vulgaris* Immature seed	BEARDER and MACMILLAN (1973) HIRAGA et al. (1974b) YAMANE et al. (1977)
GA_{38} 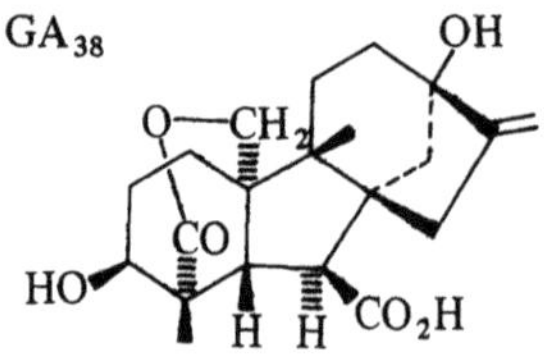	*Pisum sativum* Seed *Phaseolus vulgaris* Immature seed *Phaseolus coccineus* Seed	FRYDMAN et al. (1974) (FUKUI et al., 1972) HIRAGA et al. (1974b) YAMANE et al. (1977) SPONSEL (unpublished work)
GA_{39} 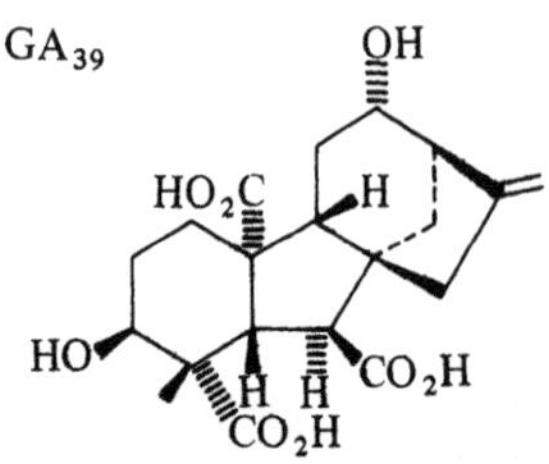	*Cucurbita pepo* Immature seed *Cucurbita maxima* Seed	FUKUI et al. (1977a) (FUKUI et al., 1977b) GRAEBE et al. (unpublished work)

Table 1.7 (continued)

	Plant material	Reference
GA_{40} 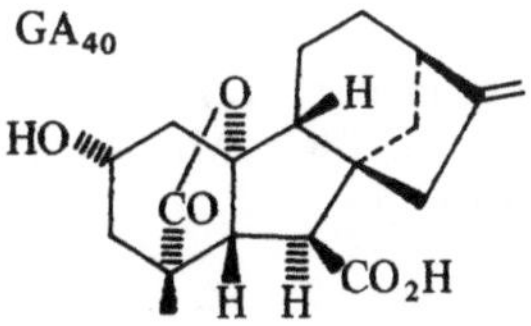	*G. fujikuroi*	YAMAGUCHI et al. (1975c)
GA_{41} 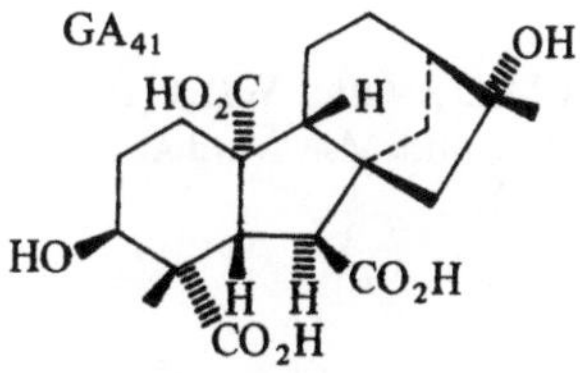	*G. fujikuroi*	BEARDER and MACMILLAN (1973)
GA_{42} 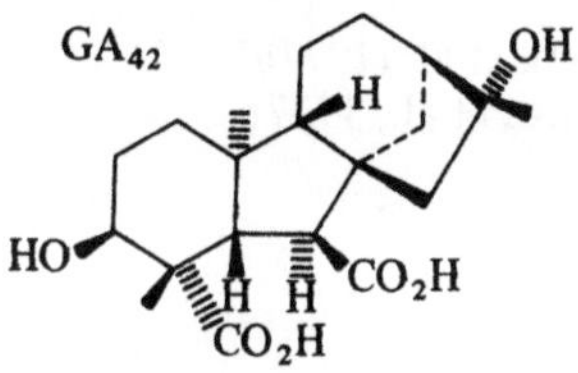	*G. fujikuroi*	BEARDER and MACMILLAN (1973)
GA_{43} 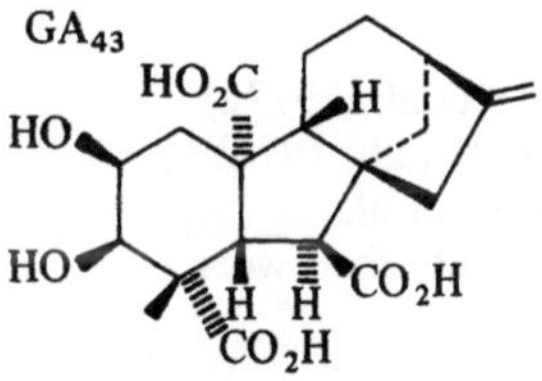	*Cucurbita maxima* Immature seed	GRAEBE et al. (1974)
	Marah macrocarpus[b] Immature seed	BEELEY et al. (1975)
GA_{44} 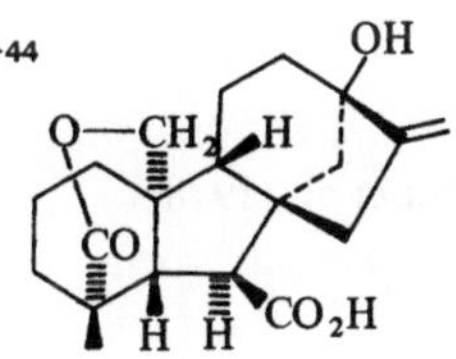	*Pisum sativum* Seed	FRYDMAN et al. (1974) (FUKUI et al., 1972)
	Phaseolus vulgaris Immature seed	YAMANE et al. (1977)
	Phaseolus coccineus Seed	SPONSEL (unpublished work)
	Pyrus malus Seed	HOAD (1978)
	Triticum aestivum Wheat germ	LENTON (unpublished work)
	Vicia faba Seed	HORGAN and HEALD (unpublished work)
		SPONSEL et al. (1979)
	Beta vulgaris Leaf tips	LENTON (unpublished work)
	Spinacia oleracea Shoots	METZGER and ZEEVAART (1980)
	Triticum aestivum Developing grain	GASKIN et al. (1980)
GA_{45} 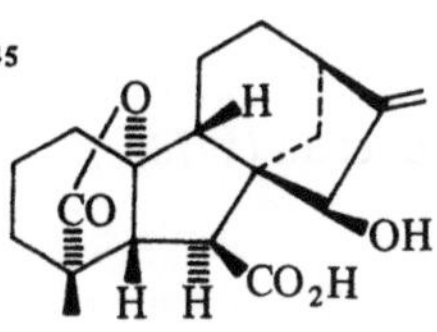	*Pyrus communis* Immature seed	MARTIN et al. (1977a)

Table 1.7 (continued)

	Plant material	Reference
GA_{46} 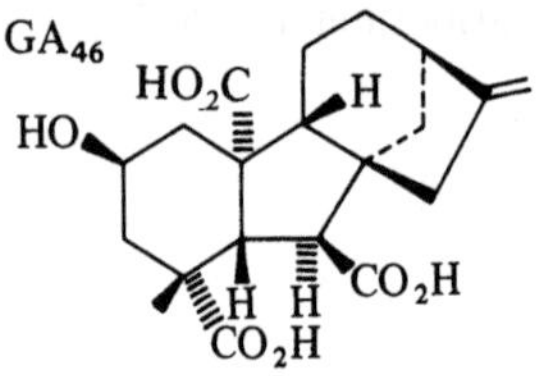	*Marah macrocarpus*[b] Immature seed	BEELEY et al. (1975) (BEELEY and MACMILLAN, 1976)
GA_{47} 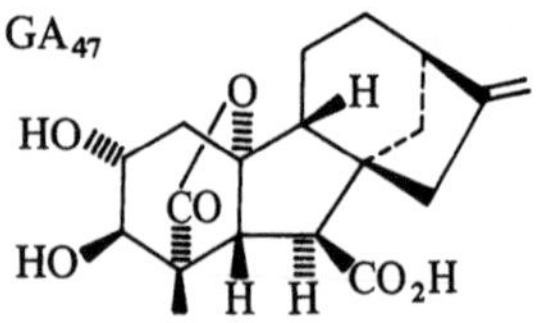	*G. fujikuroi*	MACMILLAN and WELS (1974) (BEELEY and MACMILLAN, 1976)
GA_{48} 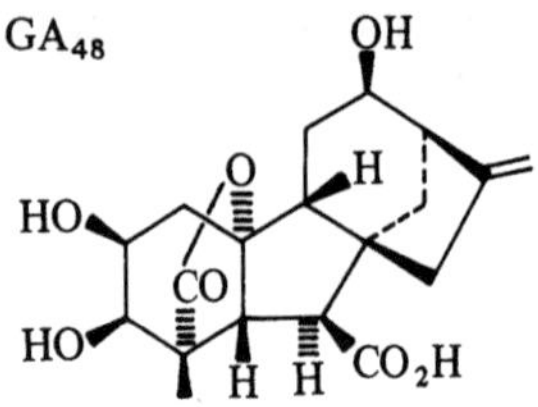	*Cucurbita pepo* Immature seed	FUKUI et al. (1977a) (FUKUI et al., 1977b)
GA_{49} 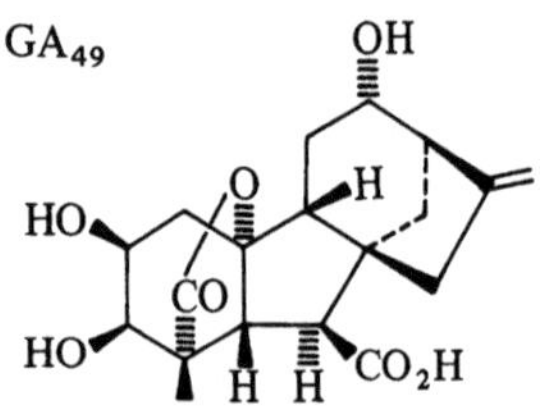	*Cucurbita pepo* Immature seed *Cucurbita maxima* Seed	FUKUI et al. (1977a) (FUKUI et al., 1977b) GRAEBE et al. (unpublished work)
GA_{50} 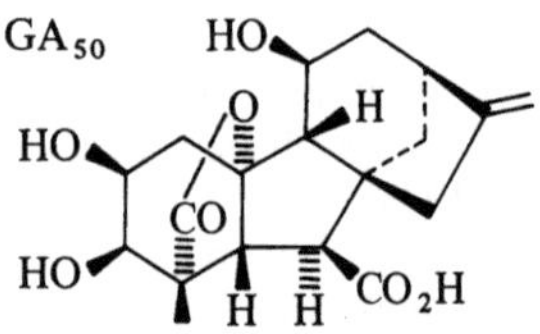	*Lagenaria leucantha* Immature seed	FUKUI et al. (1978)
GA_{51} 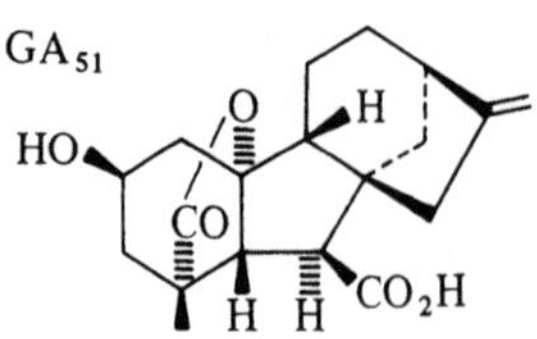	*Pisum sativum* Immature seed	SPONSEL and MACMILLAN (1977)
GA_{52} 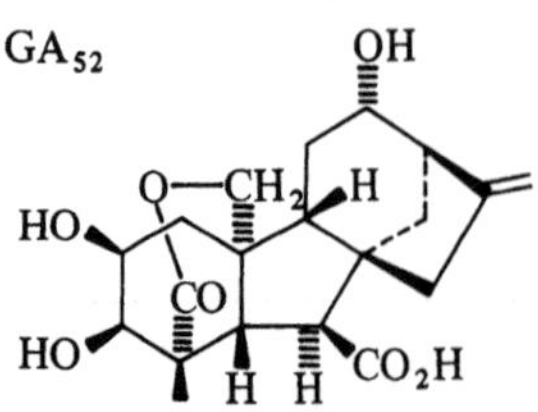	*Lagenaria leucantha* Immature seed	FUKUI et al. (1978)

Table 1.7 (continued)

	Plant material	Reference
GA_{53} 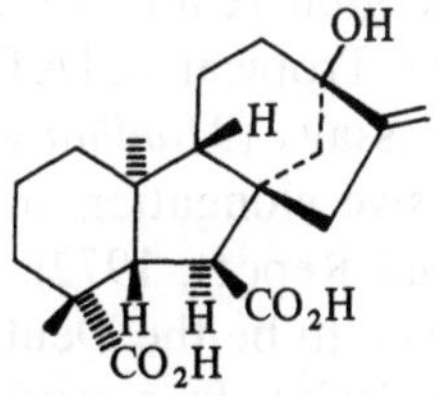	*Vicia faba* Seed	SPONSEL et al. (1979) (BEARDER et al., 1975)
	Spinacia oleracea Shoots	METZGER and ZEEVAART (1980)
GA_{54} 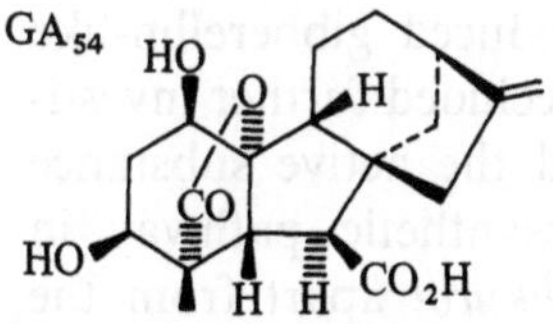	*G. fujikuroi*	MUROFUSHI et al. (1979)
	Triticum aestivum Developing grain	GASKIN et al. (1980)
GA_{55} 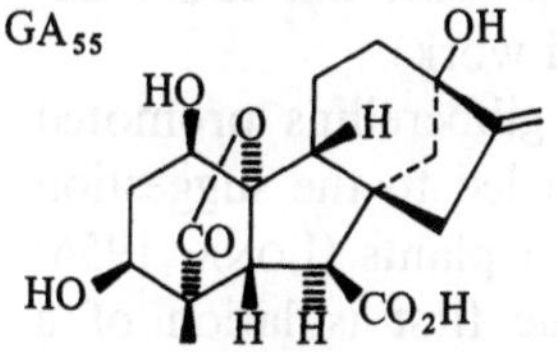	*G. fujikuroi*	MUROFUSHI et al. (1979)
	Triticum aestivum Developing grain	GASKIN et al. (1980)
GA_{56} 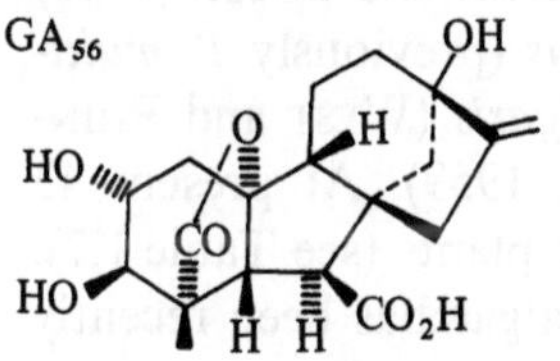	*G. fujikuroi*	MUROFUSHI et al. (1979)
GA_{57} 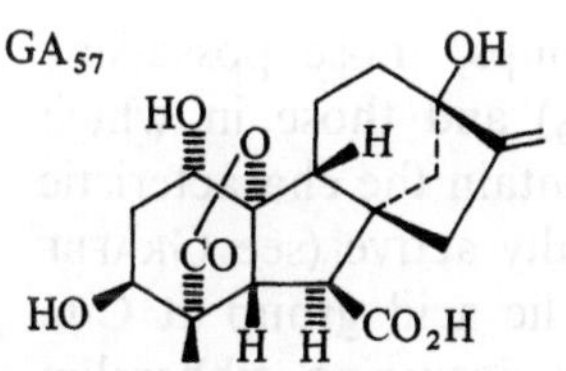	*G. fujikuroi*	MUROFUSHI et al. (1980)

[a] Formally *Phaseolus multiflorus*
[b] Formally *Echinocystis macrocarpa*
[c] Identified by its IR spectrum

can workers gave mixtures of gibberellic acid and gibberellin A_1. The final structure of gibberellic acid (GA_3) was established by chemical methods in a long series of publications by the I.C.I. group published in the Journal of the Chemical Society (see BRIAN et al., 1960; GROVE, 1961a; HANSON, 1968). This structure was confirmed by X-ray analyses (HARTSUCK and LIPSCOMB, 1963; MCCAPRA et al., 1966). A reinvestigation of the gibberellins originally isolated by the Japanese workers led to the characterization of gibberellic acid (GA_3) and three other compounds GA_1, GA_2, and GA_4 (TAKAHASHI et al.,

1955, 1957). Further examinations of minor metabolites of *G. fujikuroi* by British and Japanese chemists have now established the presence of 24 gibberellins (see Table 1.7).

Recently one other fungus has been shown to produce gibberellins. In 1972 J.C. Lozano of the Centro Internacional de Agricultura Tropical (CIAT) in Columbia described a previously unreported disease of cassava (*Manihot esculenta*). Among other symptoms the disease caused extensive elongation of the internodes of young infected cassava plants (CIAT Annual Report, 1972). The causative agent of this "superelongation disease" was shown to be the Deuteromycete *Sphaceloma manihoticola* (CIAT Annual Report, 1975). In a study of the disease KRAUSZ (1976) showed that the fungus produced gibberellin-like substances in liquid culture. But unreproducible results precluded further investigation. RADEMACHER and GRAEBE (1979) first identified the active substance as gibberellin A_4. Subsequent studies indicate the biosynthetic pathway in *Sphaceloma manihoticola* to be similar to that in *G. fujikuroi* apart from the absence of 1,2-dehydro- and 13-hydroxy-gibberellins (RADEMACHER et al., unpublished work; BEARDER and MACMILLAN, unpublished work).

The observation that the exogenous application of gibberellins promoted many normal processes of plant growth and development led to the suggestion that gibberellins might occur as native hormones in higher plants (LONA, 1956; RADLEY, 1956; PHINNEY et al., 1957; BRIAN, 1957). The first isolation of a gibberellin from a higher plant was achieved by MACMILLAN and SUTER (1958) who found GA_1 in immature seeds of *Phaseolus coccineus* (previously *P. multiflorus*). Soon after, GA_1 was isolated from *Phaseolus vulgaris* (WEST and PHINNEY, 1959) and *Citrus unshiu* (KAWARADA and SUMIKI, 1959). At present 43 of the 57 known gibberellins have been detected in higher plants (see Table 1.7). The current status of naturally occurring gibberellins in algae has been recently reviewed by TAYLOR and WILKINSON (1977).

The structures of the gibberellins fall into two groups, those possessing the full complement of diterpenoid carbon atoms (C_{20}) and those in which the angular C-20 has been lost (C_{19}). The latter group contain the characteristic γ-lactone of ring A and are generally the most biologically active (see GRAEBE and ROPERS, 1978). All the gibberellins have a carboxylic acid group at C-6. The trivial nomenclature by which each new naturally occurring gibberellin is given an A number (GA_1 .... GA_∞) as it is chemically characterized was established by MACMILLAN and TAKAHASHI (1968). A systematic nomenclature was originally based on the gibbane skeleton (Fig. 1.4) but this has now been superceded by the gibberellane skeleton (Fig. 1.4) (ROWE, 1968).

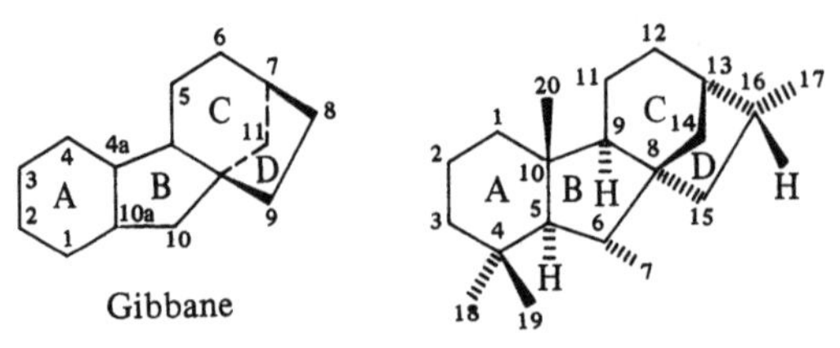

Fig. 1.4

Unfortunately all known gibberellins possess the absolute stereochemistry enantiomeric to the gibberellane skeleton. This requires that the systematic name of a natural gibberellin be prefixed with the operator ent (enantiomeric) which reverses the stereochemical designation at each chiral centre. This may give rise to some confusion as substituents which are α or β with reference to traditionally drawn structural formulae (Table 1.7) become ent-β and ent-α respectively in the systematic nomenclature. Thus GA_3 (see Table 1.7) is ent-3α,10,13-trihydroxy-20-norgibberella-1,16-diene-7,19-dioic acid 19,10-lactone.

As might be anticipated, taxonomically related species often contain structurally similar gibberellins. The pattern of hydroxylation at positions 3 and 13 for species within 4 families is illustrated in Table 1.9. The enzymatic basis for this pattern is emerging from biosynthetic studies (see Chap. 4).

Conjugated gibberellins (SEMBDNER, 1974) were first found in higher plants in 1967 when GA_8 glucoside was isolated from mature seeds of *Phaseolus coccineus* (SCHREIBER et al., 1967). Glucosyl ethers and esters continue to be found in higher plants particularly from mature tissue (Table 1.8). At present glucose is the only sugar known to be complexed with gibberellins. The function of these conjugates is unknown.

It is no coincidence that most gibberellins have been isolated from immature seed as this tissue can contain up to 16 mg kg^{-1} fresh weight of gibberellin (MACMILLAN et al., 1960). In other tissues gibberellin content is often 1000 times less. Qualitative and quantitative changes in gibberellins have been monitored in developing seed of *Phaseolus coccineus* (DURLEY et al., 1971), *Pisum sativum* (FRYDMAN et al., 1974) and *Prunus domestica* (MARTIN et al., 1977b). HIRAGA et al. (1974a, b) have examined the gibberellin content of immature, mature, and germinated seeds of *Phaseolus vulgaris* and found that as the seeds mature the concentration of free gibberellins is reduced and the proportion of conjugated gibberellins increases.

A number of naturally occurring compounds with some gibberellin-like activity are known which possess the related ent-kaurane carbon skeleton. These compounds (illustrated in Fig. 1.5) are ent-kaurene (JONES, 1969; CROSS et al., 1970; MURAKAMI, 1972), ent-kaurenol, ent-kaurenoic acid (KATSUMI et al., 1964; CROSS et al., 1970; MURAKAMI, 1972), steviol (RUDDAT et al., 1963; KATSUMI et al., 1964; BRIAN et al., 1967; MURAKAMI, 1972; VALIO and ROCHA, 1976), ent-6α,7α-dihydroxykaurenoic acid (CROSS et al., 1970), and grandiflorenic acid (BECKER and KEMPF, 1976). The first three compounds listed are known precursors to gibberellins (see Chap. 4), so they may owe their activity to conversion to gibberellins within the bioassay organism. The antheridium-inducing factor (antheridiogen A_{An}) from the fern *Anemia phyllitidis* (NAKANISHI et al., 1971, 1972) has a structure closely related to the gibberellins (see Fig. 1.6). It also has gibberellin-like activity in several bioassays (SHARP et al., 1975; BÜRCKY, 1977). The fungal metabolite helminthosporal (see Fig. 1.6) (TAMURA et al., 1965) exhibits gibberellin-like activity in a range of bioassays (see KATO et al., 1968). This has been attributed to its structural similarity to the gibberellin C/D rings (BRIGGS, 1966, COOMBE et al., 1974).

The chemical synthesis of gibberellins has been recently reviewed by FUJITA and NODE (1977). The most notable achievement in this field is the recent total

Table 1.8. Gibberellin derivatives from higher plants and the fungus *G. fujikuroi*

		Plant material	Reference
GA_1	β–D–Glucopyranosyl ester	*Phaseolus vulgaris* Mature seed	HIRAGA et al. (1974a)
GA_1	n–Propyl ester	*Cucumis sativus* Seed	HEMPHILL et al. (1973)
GA_1	3–O–β–D–Glucopyranoside	*Dolichos lablab* Seed	YOKOTA et al. (1978)
GA_3	3–O–β–D–Glucopyranoside	*Pharbitis nil* Immature seed	YOKOTA et al. (1971b)
	3–O–β–Acetate	*G. fujikuroi*	SCHREIBER et al. (1966)
GA_3	n–Propyl ester	*Cucumis sativus* Seed	HEMPHILL et al. (1973)
GA_4	β–D–Glucopyranosyl ester	*Phaseolus vulgaris* Mature seed	HIRAGA et al. (1974a)
GA_8	3–O–β–D–Glucopyranoside	*Phaseolus coccineus* Seed	SCHREIBER et al. (1970)
		Althaea rosea Shoot apices	HARADA and YOKOTA (1970)
		Pharbitis nil Immature seed	YOKOTA et al. (1971b)
		Phaseolus vulgaris Immature and mature seed	HIRAGA et al. (1974a, b)
GA_8	"Conjugate"	*Phaseolus coccineus* Immature seed	GASKIN and MACMILLAN (1975)
GA_9	β–D–Glucopyranosyl ester	*Picea sitchensis* Needles	LORENZI et al. (1976)
GA_{17}	Methyl ester	*Calonyction aculeatum* Immature seed	MUROFUSHI et al. (1973)
GA_{17}	"Conjugate"	*Phaseolus coccineus* Immature seed	GASKIN and MACMILLAN (1975)
GA_{20}	"Conjugate"	*Phaseolus coccineus* Immature seed	GASKIN and MACMILLAN (1975)
GA_{26}	2–O–β–D–Glucopyranoside	*Pharbitis nil* Immature seed	YOKOTA et al. (1971b)
GA_{27}	2–O–β–D–Glucopyranoside	*Pharbitis nil* Immature seed	YOKOTA et al. (1971b)
GA_{28}	"Conjugate"	*Phaseolus coccineus* Immature seed	GASKIN and MACMILLAN (1975)
GA_{29}	2–O–β–D–Glucopyranoside	*Pharbitis nil* Immature seed	YOKOTA et al. (1971b)
GA_{35}	11–O–β–D–Glucopyranoside	*Cytisus scoparius* Immature seed	YAMANE et al. (1974)
GA_{37}	β–D–Glucopyranosyl ester	*Phaseolus vulgaris* Mature seed	HIRAGA et al. (1974a)
GA_{38}	β–D–Glucopyranosyl ester	*Phaseolus vulgaris* Mature seed	HIRAGA et al. (1974a)
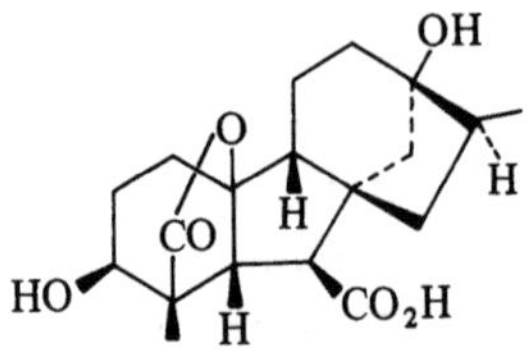 "Tetrahydro GA_3"[a]		*Sonneratia apelata* Leaves	GASKIN et al. (1972)
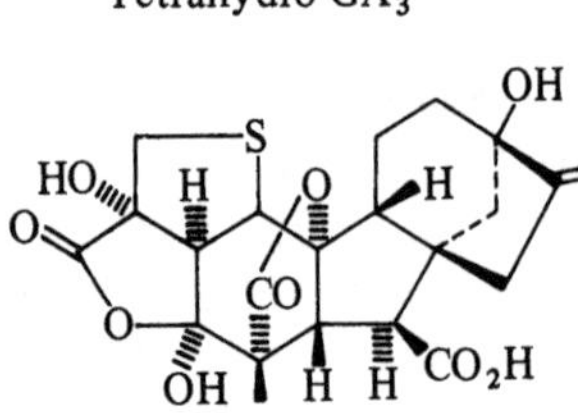 Gibberethione[b]		*Pharbitis nil* Immature seed	YOKOTA et al. (1974b)

[a] This compound was at first erroneously given the name GA_{25}
[b] Formally called pharbitic acid

Table 1.9. Comparison of gibberellins (GAs) from Leguminosae, Convolvulaceae, Cucurbitaceae and Rosaceae with respect to 3- and 13-hydroxylation[a]

	3-OH[b]		3-H		13-OH		13-H	
	C_{19}-GAs	C_{20}-GAs	C_{19}-GAs	C_{20}-GAs	C_{19}-GAs	C_{20}-GAs	C_{19}-GAs	C_{20}-GAs
Leguminosae								
Phaseolus coccineus	1, 3, 4, 5, 6, 8, 34	28, 38	20	17, 19	1, 3, 5, 6, 8, 20	17, 19, 28, 38	4, 34	—
Phaseolus vulgaris	1, 4, 5, 6, 8	37, 38	20, 29	17, 44	1, 5, 6, 8, 20, 29	17, 38, 44	4	37
Vigna unguiculata	1, 4, 5, 6, 8	—	20, 29	17, 19	1, 5, 6, 8, 20, 29	17, 19	4	—
Pisum sativum	—	38	9, 20, 29, 51	17, 44	20, 29	17, 38, 44	9, 51	—
Vicia faba	5	—	20	44, 53	5, 20	44, 53	—	—
Lupinus luteus	—	18, 23, 28	—	19	—	18, 19, 23, 28	—	—
Wisteria floribunda	—	18, 23	—	—	—	18, 23	—	—
Cytisus scoparius	35	—	—	—	—	—	35	—
Cassia fistula	3	—	—	—	3	—	—	—
Canavalia gladiata	22	—	21	—	21, 22	—	—	—
Convolvulaceae								
Pharbitis nil	3, 8, 26	27	20, 29[c]	17	3, 8, 20, 29[c]	17	26	27
Calonyction aculeatum	8, 30, 31, 33, 34	27	29	17, 19	8, 29	17, 19	30, 31, 33, 34	27
Cucurbitaceae								
Marah macrocarpus	4, 7	43	—	24, 25, 46	—	—	4, 7	24, 25, 43, 46
Cucurbita maxima	49	13, 39, 43	—	12	—	—	49	12, 13, 39, 43
Cucurbita pepo	48, 49, 50	39	—	—	—	—	48, 49, 50	39
Lagenaria leucantha	50	52	—	—	—	—	50	52
Cucumis sativus	1, 3, 4, 7	—	—	—	1, 3	—	4, 7	—
Cucumis melo	1, 3, 5	—	—	—	1, 3, 5	—	—	—
Rosaceae								
Pyrus malus	4, 7	—	9, 20	12, 15, 17, 44	20	17, 44	4, 7, 9,	12, 15
Pyrus communis	—	—	45	17, 25	—	17	45	25
Prunus persica	5, 32	—	—	—	5, 32	—	—	—
Prunus armenica	32	—	—	—	32	—	—	—
Prunus domestica	—	—	29	—	29	—	—	—

[a] Published data from Table 1.7 [b] Including 2,3-olefins [c] As a glucoside

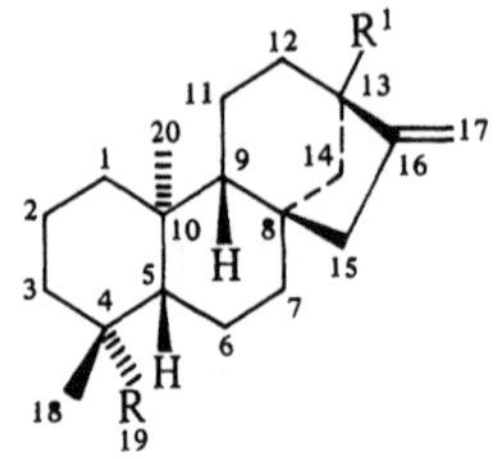

$R = CH_3$ $R^1 = H$ ent–Kaurene
$R = CH_2OH$ $R^1 = H$ ent–Kaurenol
$R = CO_2H$ $R^1 = H$ ent–Kaurenoic acid
$R = CO_2H$ $R^1 = OH$ Steviol

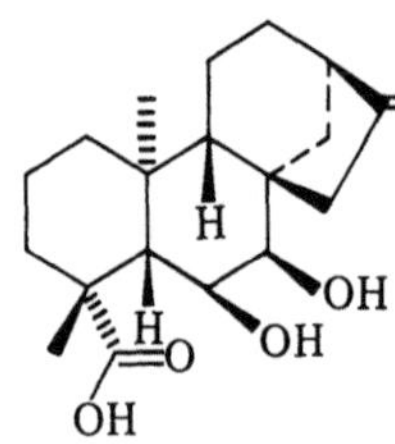

ent–6α,7α–Dihydroxykaurenoic acid

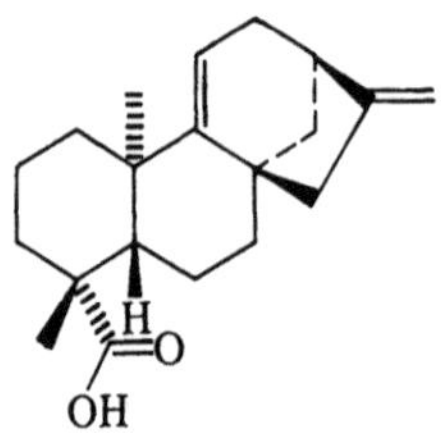

Grandiflorenic acid

Fig. 1.5

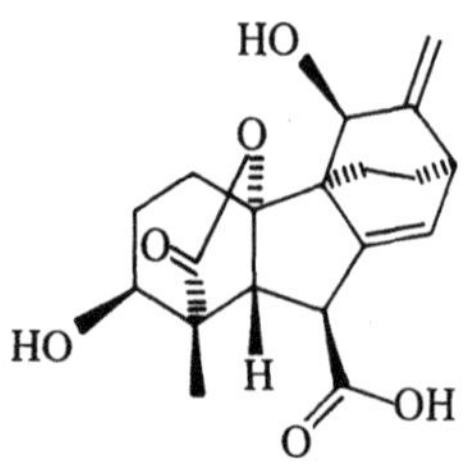

Antheridiogen A_{An}

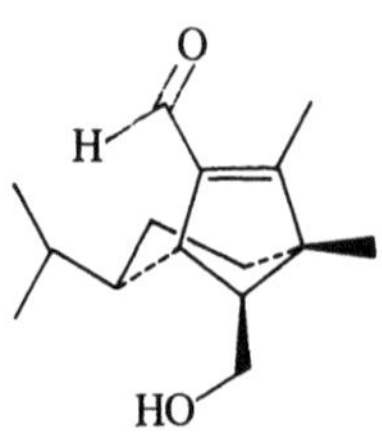

Helminthosporol

Fig. 1.6

synthesis of GA_3 by COREY et al. (1978) the culmination of one of the longest quests in synthetic organic chemistry. As many of the gibberellins have been chemically interconverted the total synthesis of a few gibberellins has provided formal relay total syntheses of about fifteen gibberellins.

PEGG (1976c) has reviewed the role of gibberellins in plant disease.

1.5 Cytokinins

The idea that cell division in plants might be controlled by specific chemicals dates back at least to WIESNER (1892). Evidence for this hypothesis was provided by the work of HABERLANDT (1913) who showed that phloem diffusates could

Fig. 1.7

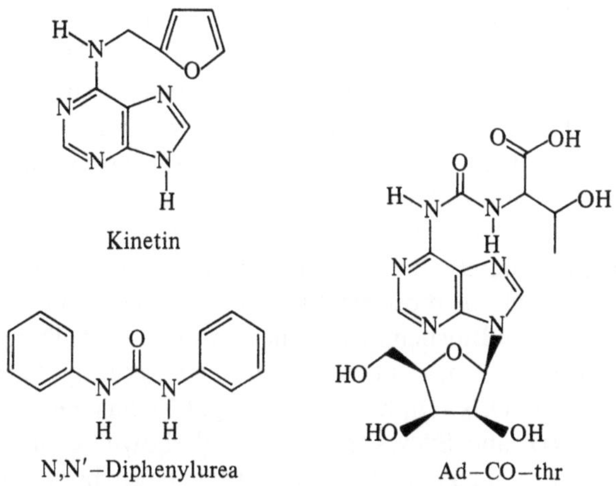

induce cell division in potato parenchyma. Subsequently (1921) he found that cell division induced by wounding in several plant tissues could be prevented by rinsing the wounded tissue and restored by the application of macerated tissue to the wound surface. HABERLANDT coined the word Wundhormone to describe the substance responsible for the promotion of cell division.

The modern era of cytokinin research originates directly from studies by SKOOG into the chemical control of differentiation. He began work in 1941 using the newly developed technique of plant tissue culture. The stem internode tissue culture of tobacco (*Nicotiana tabacum* cv. Wisconsin No. 38) did not grow in the absence of auxin. When indole-3-acetic acid was supplied cell enlargement and proliferation ensued, particularly in the pith (SKOOG and TSUI, 1948). JABLONSKI and SKOOG (1954) incubated pith alone in the auxin-supplemented medium and observed "... an enormous cell enlargement entirely unaccompanied by cell division". Cell division was restored when vascular tissue was put in contact with the pith sections. This result initiated the search for chemicals which would induce cell division in a way similar to that of the unknown substance from vascular tissue.

Attempts to isolate the active factor from tobacco stems gave inconsistent results so other natural sources were explored. Coconut milk, malt extract (JABLONSKI and SKOOG, 1954), yeast extract, and autoclaved DNA (MILLER et al., 1955a) all furthered cell division, by promoting activity in the tobacco pith assay. The first active compound to be isolated and characterized from a natural source was kinetin (Fig. 1.7), extracted by MILLER et al. (1955b, 1956) from autoclaved herring sperm DNA. The structure was confirmed by chemical synthesis (MILLER et al., 1955b, 1956). The synthetic material stimulated cell division in the tobacco pith system at 0.01 ppm. Kinetin itself is not a natural product; it is derived from the breakdown of DNA during autoclaving. It is either formed from reaction between adenine and deoxyribose (HALL and DE ROPP, 1955) or by a complex process involving dehydration and rearrangement of deoxyadenosine residues at the polynucleotide level (SCOPES et al., 1976).

As similar molecules were shown to have the same type of activity the generic name kinin was proposed (MILLER et al., 1956). However this name was subsequently withdrawn since it had previously been used in another sense in mammalian biochemistry. Instead the term cytokinin was suggested (SKOOG et al., 1965) and is now generally accepted to describe compounds with cell division (cytokinesis) promoting activity in plants.

All the known naturally occurring cytokinins possess the N^6-substituted adenine moiety. It is appropriate therefore that the abbreviations used for cytokinins should be those established for nucleic acid derivatives (IUPAC-IUB Commission on Biochemical Nomenclature, 1970). This system is used in this chapter. However, a new nomenclature has recently been put forward by LETHAM (1978a). N,N′-Diphenylurea (Fig. 1.7) has been isolated from coconut milk (SHANTZ and STEWARD, 1955) and shown to have cytokinin-like activity (see BRUCE et al., 1965). However, it has never been isolated since, so there is some doubt as to its natural occurrence.

The discovery of kinetin stimulated work on the synthesis of structural analogues (see SKOOG and ARMSTRONG, 1970; SKOOG, 1973) and accelerated the search for cytokinins in plant tissue. It was not until 1963 that a cytokinin was isolated and identified from plant sources. LETHAM (1963) isolated zeatin (t-io^{6}Ade) (Table 1.10) from immature kernels of *Zea mays* and later determined its structure (LETHAM et al., 1964, 1967). Zeatin was found to be even more active than kinetin. The same compound was subsequently also isolated from *Zea mays* by LETHAM and MILLER (1965). MILLER (1965) also obtained evidence for other cytokinins in *Zea mays* kernels. Detailed studies by LETHAM (1973) have revealed the presence of zeatin riboside (t-io^{6}A) (Table 1.10) zeatin-ribotide (the 5-phosphate) and small proportions of other cytokinins (see Table 1.14). Zeatin and zeatin riboside have now been conclusively identified in a range of plants (see Table 1.5) and assumed to be present in a great many others from bioassay and chromatographic methods (see KENDE, 1971). Zeatin and/or zeatin riboside have also been identified in fungi (MILLER, 1967; CRAFTS and MILLER, 1974) and in the phytopathogenic bacteria *Corynebacterium fascians* (SCARBROUGH et al., 1973; EINSET and SKOOG, 1977) and *Agrobacterium tumefaciens* (CHAPMAN et al., 1976; KAISS-CHAPMAN and MORRIS, 1977). Both cis and trans isomers (about the side chain double bond) occur naturally, however the cis isomer is almost exclusively obtained from RNA hydrolysates, whereas the free cytokinin is normally the more active trans isomer. The 2-methylthio-substituted zeatin riboside (t-ms^{2}io^{6}A) has also been detected in plants (see Table 1.11) and the bacteria *C. fascians* (ARMSTRONG et al., 1976; see CHERAYIL and LIPSETT, 1977) and *A. tumefaciens* (KAISS-CHAPMAN and MORRIS, 1977; see CHERAYIL and LIPSETT, 1977).

The second cytokinin to be isolated from natural sources was N^6-(Δ^2-isopentenyl)adenine (i^{6}Ade) (see Table 1.12) which was found as a constituent base of two serine tRNA's of yeast (BIEMANN et al., 1966). It and its riboside (i^{6}A) (see Table 1.12) had been previously synthesized and shown to be active by LEONARD and FUJI (1964) and BEAUCHESNE and GOUTAREL (1963) as part of a project to make zeatin analogues. i^{6}A(de) is probably the most widely distributed cytokinin in nature as it has been shown to occur in RNA from

Table 1.10. Zeatin and zeatin riboside in plants

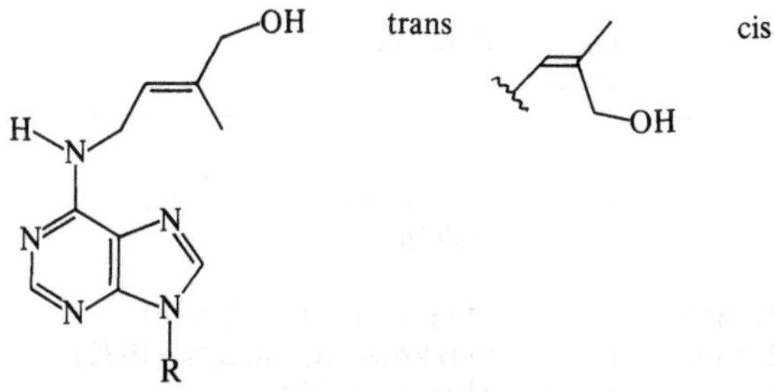

Zeatin (R=H); Zeatin riboside (R=ribosyl)

Plant material	From sRNA	R	cis/ trans	Reference
Zea mays				
Kernels	×	H	t	LETHAM and MILLER (1965) (LETHAM et al., 1967) (SHAW et al., 1966)
Kernels	×	H	t	LETHAM (1973)
Kernels	×	Ribosyl	t	LETHAM (1973)
Kernels	√	Ribosyl	c	HALL et al. (1967) (LEONARD et al., 1971)
Helianthus annuus				
Leaves	×	H	t	KLÄMBT (1968)
Cichorium intybus				
Roots	×	Ribosyl	t	BUI-DANG-HA and NITSCH (1970)
Pisum sativum				
Roots	√	Ribosyl	c	BABCOCK and MORRIS (1970)
Epicotyl		Ribosyl	c/t	PLAYTIS and LEONARD (1971)
Shoots	√	Ribosyl	c/t	VREMAN et al. (1972)
Seed	√	Ribosyl	c	EINSET et al. (1976)
Triticum aestivum				
Wheat germ	√	Ribosyl	c	BURROWS et al. (1970) (PLAYTIS and LEONARD, 1971)
Nicotiana tabacum				
Callus	√	Ribosyl	c	BURROWS et al. (1971)
Callus	√	Ribosyl	c	BURROWS (1976)
Callus	√	Ribosyl	c	PLAYTIS and LEONARD (1971)
Callus	√	Ribosyl	c	DYSON and HALL (1972)
Shoot apices	×	Ribosyl	c	HASHIZUME et al. (1978)
Phaseolus vulgaris				
Fruit	×	H	t	KRASNUK et al. (1971)
Acer pseudoplatanus				
Sap	×	Ribosyl	t	HORGAN et al. (1973)
Spring sap	×	Ribosyl	t	PURSE et al. (1976)
Spring sap	×	H	t	PURSE et al. (1976)
Cocos nucifera				
Milk	×	Ribosyl	t	LETHAM (1974)
Milk	×	Ribosyl	t	VAN STADEN and DREWES (1975b)
Milk	×	H	t	VAN STADEN and DREWES (1975b)

Table 1.10 (continued)

Plant material	From sRNA	R	cis/trans	Reference
Hordeum vulgare				
Malt extract	×	H	t	VAN STADEN and DREWES (1975a)
Vinca rosea				
Crown gall tissue	×	Ribosyl	t	MILLER (1974, 1975a, b)
Crown gall tissue	×	Ribosyl	t	PETERSON and MILLER (1976)
Crown gall tissue	×	H	t	MILLER (1975a)
Crown gall tissue	×	H	t	PETERSON and MILLER (1976)
Gossypium hirsutum				
Ovules	×	H	t	SHINDY and SMITH (1975)
Ovules	×	Ribosyl	t	SHINDY and SMITH (1975)
Lycopersicon esculentum				
Xylem sap	×	Ribosyl	t	VAN STADEN and MENARY (1976)
Xylem sap	×	H	t	VAN STADEN and MENARY (1976)
Euglena gracilis	√	Ribosyl	c	SWAMINATHAN and BOCK (1977)
Actinidia chinensis				
Fruit	×	Ribosyl	t	YOUNG (1977)
Fruit	×	H	t	YOUNG (1977)
Prunus cerasus				
Fruit	×	Ribosyl	t	YOUNG (1977)
Fruit	×	H	t	YOUNG (1977)
Fruit	×	H	t	HOPPING et al. (1979)
Mercurialis annua				
Buds	×	Ribosyl	c/t	DAUPHIN et al. (1977)
Buds	×	H	t	DAUPHIN et al. (1977)
Humulus lupulus				
Cone	×	Ribosyl	c/t	WATANABE et al. (1978a, c)
Cone	×	H	t	WATANABE et al. (1978a)
Shoots	×	Ribosyl	t	WATANABE et al. (1978b)
Spinacia oleracea				
Leaves	√	Ribosyl	c/t	VREMAN et al. (1978)
Daucus carota				
Roots	×	Ribosyl	t	MIZUNO and KOMAMINE (1978)
Raphanus sativus				
Roots	×	H	t	KOYAMA and KAWAI (1978)
Solanum tuberosum				
Aerial and subterranean parts	×	Ribosyl	c	MAUK and LANGILLE (1978)
Brassica oleracea				
Hearts	×	Ribosyl	t	HASHIZUME et al. (1979)
Mercurialis ambigua				
Shoot apices	×	H	t	DAUPHIN et al. (1979)
Shoot apices	×	Ribosyl	c/t	DAUPHIN et al. (1979)

Table 1.11. 2-Methylthiozeatin riboside in plants

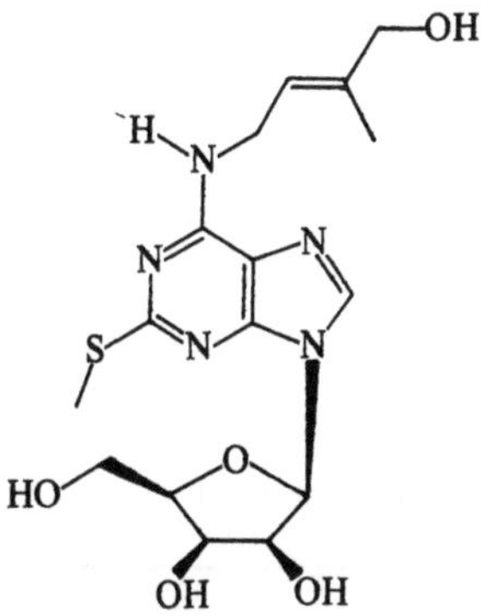

Plant material	From sRNA	Reference
Triticum aestivum Wheat germ	√	BURROWS et al. (1970)
Nicotiana tabacum Callus	√	BURROWS et al. (1971)
Pisum sativum Shoots	√	VREMAN et al. (1972)
Pisum sativum Shoots	√	VREMAN et al. (1974)[a]
Euglena gracilis	√	SWAMINATHAN and BOCK (1977)
Spinacia oleracea Leaves	√	VREMAN et al. (1978)[a]
Spinacia oleracea Chloroplasts	√	VREMAN et al. (1978)[b]

[a] cis isomer also present
[b] cis isomer only

mammals (ROBINS et al., 1967; STAEHELIN et al., 1968), bacteria (BURROWS et al., 1969; ARMSTRONG et al., 1970; CHAPMAN et al., 1976), fungi (HALL et al., 1966) and plants (see Table 1.12). Its occurrence as a free cytokinin has been reported in *C. fascians* (KLÄMBT et al., 1966; HELGESON and LEONARD, 1966; ARMSTRONG et al., 1976), a fungus (TANAKA et al., 1978), sea water (PEDERSEN, 1973), and several species of higher plants (see Table 1.12). The 2-methylthiosubstituted derivative (ms^2i^6A) (see Table 1.13) has been found in wheat germ RNA (BURROWS et al., 1970) as well as a number of bacterial RNA's (HARADA et al., 1968; NISHIMURA et al., 1969; BURROWS et al., 1969; ARMSTRONG et al., 1970; CHAPMAN et al., 1976).

The detailed pioneering work of ZACHAU and co-workers (BIEMANN et al., 1966; ZACHAU et al., 1966) resulted in the identification of i^6A as a constituent of two serine tRNA species from yeast. They demonstrated that i^6A occurred only once in the polynucleotide chain and was placed specifically adjacent to the 3′ end of the anticodon in both species. Later studies showed that cytokinin activity was only associated with hydrolysates from tRNA species corresponding to codons with the initial letter U (uridine). Furthermore, in all such species which have been sequenced the cytokinin base has been found adjacent to the 3′ end of the anticodon (see SKOOG and ARMSTRONG, 1970; HALL, 1970, 1973; BURROWS, 1975; NISHIMURA, 1972; MCCLOSKEY and NISHIMURA, 1977). It is still not known whether the existence of cytokinin bases in tRNA's is

Table 1.12. N^6-(Δ^2-Isopentenyl)adenine and N^6-(Δ^2-isopentenyl)adenosine in plants

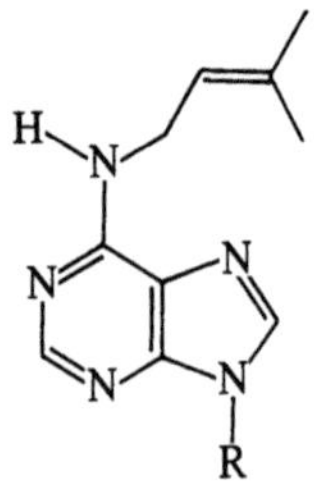

N^6–(Δ^2–isopentenyl) adenine (R=H); N^6–(Δ^2–isopentenyl) adenosine (R=ribosyl).

Plant material	From sRNA	R	Reference
Triticum aestivum			
Wheat germ	√	Ribosyl	BURROWS et al. (1970)
Nicotiana tabacum			
Callus	√	Ribosyl	BURROWS et al. (1971)
Callus	√	Ribosyl	BURROWS (1976)
Callus	√	Ribosyl	DYSON and HALL (1972)
Callus	×	Ribosyl	DYSON and HALL (1972)
Pisum sativum			
Shoots	√	Ribosyl	VREMAN et al. (1972)
Gossypium hirsutum			
Ovules	×	Ribosyl	SHINDY and SMITH (1975)
Ovules	×	H	SHINDY and SMITH (1975)
Funaria × Physcomitrium			
Callus	×	H	BEUTELMANN and BAUER (1977)
Euglena gracilis	√	Ribosyl	SWAMINATHAN and BOCK (1977)
Spinacia oleracea			
Leaves	√	Ribosyl	VREMAN et al. (1978)
Chlorplasts	√	Ribosyl	VREMAN et al. (1978)
Mercurialis annua			
Buds	×	Ribosyl	DAUPHIN et al. (1977)
Mercurialis ambigua			
Shoot apices	×	Ribosyl	DAUPHIN et al. (1979)
Humulus lupulus			
Shoots	×	Ribosyl	WATANABE et al. (1978b)

of importance in relation to the function or metabolism of cytokinins in plants (see SKOOG, 1973; BURROWS, 1975, 1976, 1978a, b).

N-[9-(β-D-Ribofuranosyl)purine-6-carbonoyl]threonine (Ad-CO-thr) (Fig. 1.7) has an analogous status to i^6A in tRNA. This base only occurs adjacent to the 3′ end of the tRNA species which correspond to codons with the initial letter A (adenine) (see SKOOG and ARMSTRONG, 1970; HALL, 1973; NISHIMURA, 1972; MCCLOSKEY and NISHIMURA, 1977). Ad-CO-thr has no cytokinin activity itself but lipophilic derivatives do exhibit activity (DYSON et al., 1970). For this reason Ad-CO-thr has been considered as a cytokinin (DYSON

Table 1.13. 2-Methylthio-N^6-(Δ^2-isopentenyl)adenosine in plants

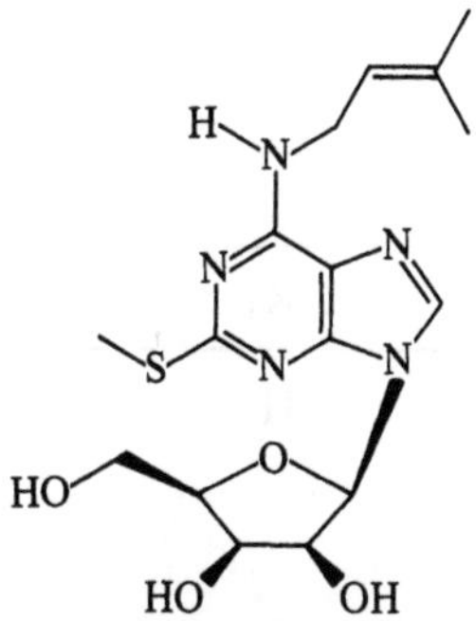

2-Methylthio-N^6-(Δ^2-isopentenyl) adenosine

Plant material	From sRNA	Reference
Triticum aestivum Wheat germ	√	Burrows et al. (1970)
Brassica oleracea Hearts	×	Hashizume et al. (1979)

et al., 1970). It has been suggested that N,N′-diphenylurea owes its activity to its structural similarity to Ad-CO-thr.

The qualitative distribution of tRNA cytokinins among living organisms it not clear-cut (see Chapman et al., 1976; Swaminathan and Bock, 1977). However one can generalize by saying that i^6A occurs in animals, ms^2i^6A is predominant in bacteria, and in plants io^6A is most abundant.

There are a number of cytokinins which have been isolated from non-RNA plant sources only (Table 1.15). These include dihydrozeatin, dihydrozeatin riboside, glucosyldihydrozeatin, glucosylzeatin and glucosylribosylzeatin. o-Hydroxybenzyladenosine is the only natural cytokinin known with a non-isoprenoid side chain (Horgan et al., 1975). In vivo the level of this novel cytokinin appears to be under phytochrome control (Thompson et al., 1975). Raphanatin (see Table 1.15) which had been known as a metabolite of zeatin in radish (Parker et al., 1972) has recently been shown to be endogenous in that plant (Summons et al., 1977). There are a number of commonly occurring purines which have some cytokinin-like activity but are not normally considered as cytokinins. Taylor et al. (1974) detected adenine and adenosine in fruitlets and tracheal fluid of cotton and ascribed most of the biological activity of the extracts to these compounds. N^6-Methyladenine is widely occurring in tRNA (Dunn et al., 1960; Nishimura, 1972; McCloskey and Nishimura, 1977) and has some cytokinin-like activity (Kuraishi, 1959; Miller, 1962; Fox, 1969).

The occurrence of cytokinins in plants and in specific tissue in the plant have been discussed by Kende (1971), Sheldrake (1973) and Letham (1978a). Wareing et al. (1977) have reviewed cytokinin relations in the whole plant. The role of cytokinins in plant diseases has been summarized by Dekhuijzen (1976). The structures and occurrence of cytokinins has recently been reviewed by Horgan (1978).

Table 1.14. Other cytokinins in *Zea mays* kernels (LETHAM, 1973)

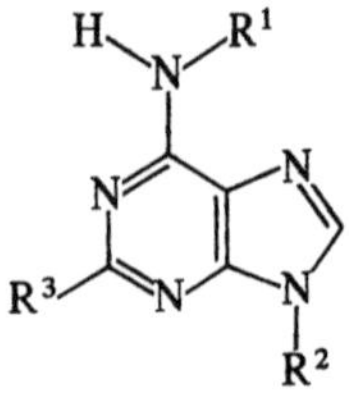

R_1	R_2	R_3
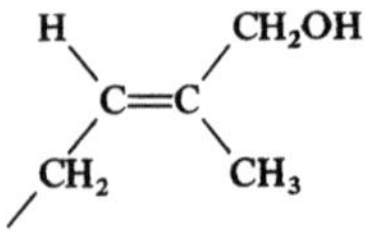 (Zeatin ribotide)	Ribosyl-5′-phosphate	H
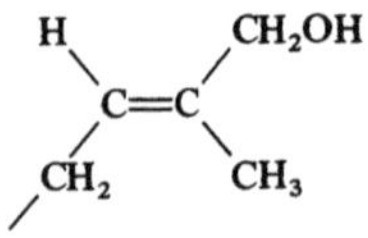 (2-Hydroxyzeatin)	H	OH
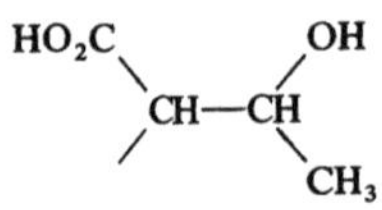	Ribosyl	H
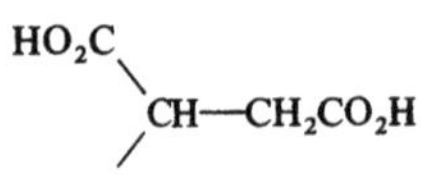	Ribosyl	H
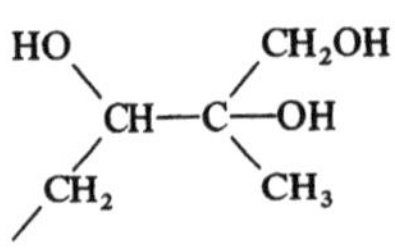	H	H
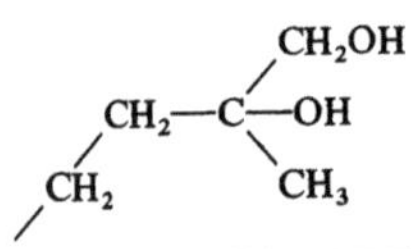	H	H

Table 1.15. Other cytokinins from plants[a]

Compound	Plant material	Reference
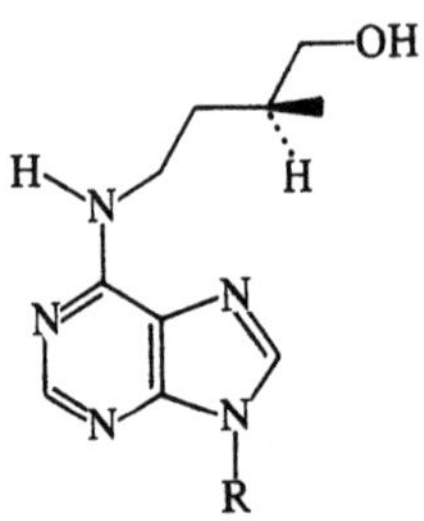 Dihydrozeatin (R=H)	*Lupinus luteus* Immature seed	KOSHIMIZU et al. (1967) (FUJI and OGAWA, 1972)
	Phaseolus vulgaris Fruit	KRASNUK et al. (1971)
	Gossypium hirsutum Ovules	SHINDY and SMITH (1975)
	Acer pseudoplatanus Spring sap	PURSE et al. (1976)

Table 1.15 (continued)

Compound	Plant Material	Reference
Dihydrozeatin riboside (R = ribosyl)	*Phaseolus vulgaris* Leaves	WANG and HORGAN (1978)
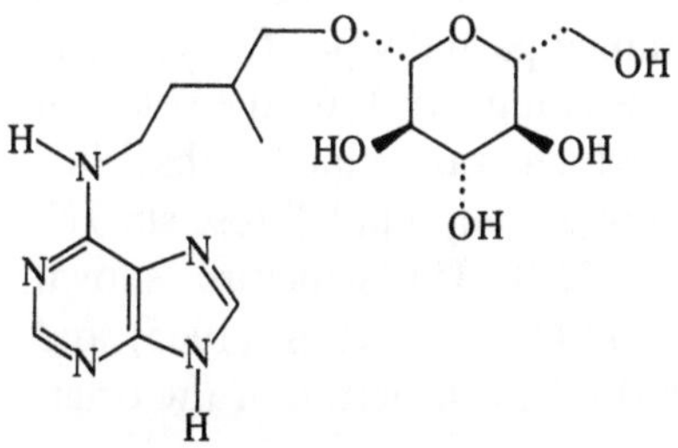 O-β-D-Glucosyldihydrozeatin	*Phaseolus vulgaris* Leaves	WANG et al. (1977)
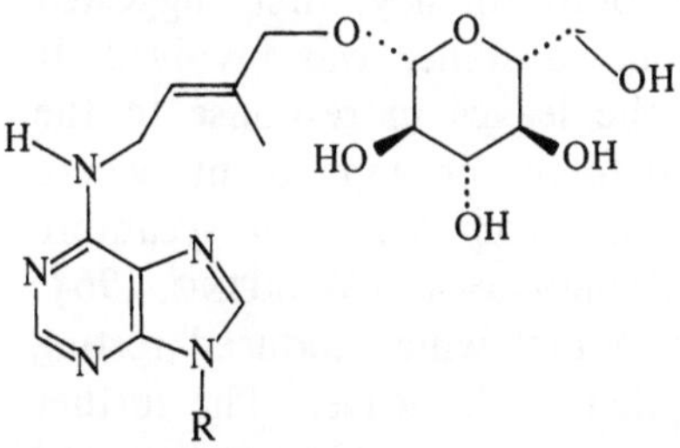 O-β-D-Glucosylzeatin (R=H) O-β-D-Glucosylzeatin riboside (R=ribosyl)	*Vinca rosea* Crown gall	PETERSON and MILLER (1977) (MORRIS, 1977)
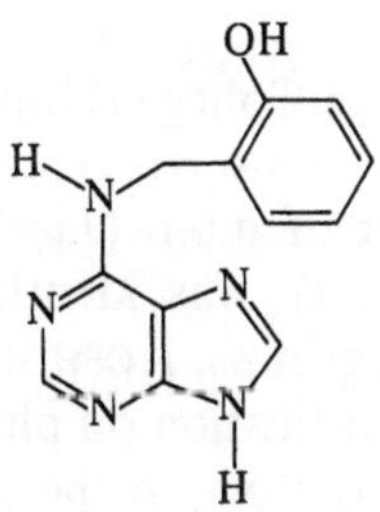 N[6]-(o-Hydroxybenzyl) adenine	*Populus × robusta* Leaves	HORGAN et al. (1975)
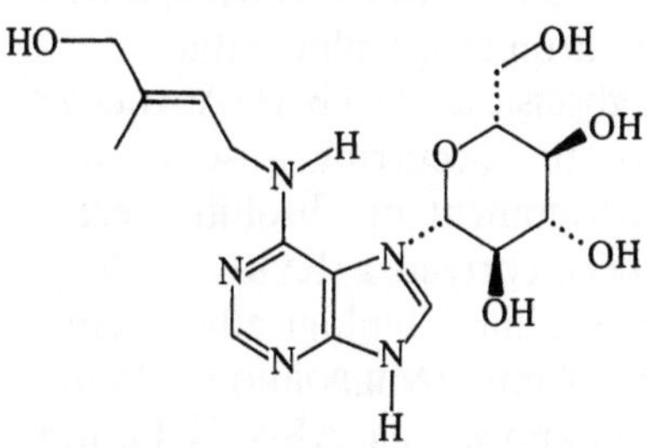 Raphanatin (7β-D-Glucopyranosylzeatin)	*Raphanus sativus* Seed	SUMMONS et al. (1977) (DUKE et al., 1975)

[a] All from non-RNA sources

1.6 Abscisic Acid and Related Compounds

The discovery of abscisic acid stemmed from concurrent work in several independent laboratories concerned with quite different research projects.

A group at the University of California led by CARNS and ADDICOTT was investigating the possible existence of a substance in plants which accelerates leaf abscission. From young cotton bolls they obtained partially purified extracts which accelerated leaf abscission in young cotton seedlings. In 1961 the research resulted in the isolation of an active compound which was named "abscisin" (LUI and CARNS, 1961). Later a second more active compound "abscisin II" was isolated and characterized (OHKUMA et al., 1963). The structure, shown in Table 1.16, was determined for "abscisin II" by OHKUMA et al. (1965) and confirmed by synthesis by CORNFORTH et al. (1965b). The structure of the original "abscisin" (abscisin I) remains unknown.

Concurrently WAREING's group at Aberystwyth were investigating the cause of dormancy in trees. The simplest hypothesis of dormancy, first suggested by HEMBERG (1949), was that a growth-inhibitory substance was involved. It was proposed that an inhibitor is formed in the leaves in response to the short-day conditions of autumn and transported to the growing points where it would halt growth and induce bud formation. From leaves of sycamore (*Acer pseudoplatanus*) WAREING and co-workers (ROBINSON and WAREING, 1964; WAREING et al., 1964) obtained a highly active extract which induced resting bud formation in sycamore seedlings when applied to the leaves. The further analysis of this material named "dormin", was pursued by CORNFORTH and his colleagues at the Milstead Laboratory of Shell Research Ltd. CORNFORTH et al. (1965a) isolated the active substance in crystalline form and showed that it was identical to "abscisin II".

At about the same time ROTHWELL and WAIN (1964) at Wye College (University of London) were following up preliminary studies by VAN STEVENINCK (1959) and attempting to isolate a substance from the fruit of lupin (*Lupinus luteus*) which was responsible for flower drop. This substance was identified as "abscisin II" almost simultaneously by three independent groups. KOSHIMIZU et al. (1966) obtained crystalline material and based their identification on physical properties, CORNFORTH et al. (1966a) used optical rotatory dispersion (O.R.D.) to identify "abscisin II" in the highly active semi-crystalline concentrate obtained by ROTHWELL and WAIN (1964), and PORTER and VAN STEVENINCK (1966) based their identification on bioassay and chromatographic data.

By general consent "abscisin II" was renamed abscisic acid (ABA) (ADDICOTT et al., 1968). The absolute configuration of naturally occurring abscisic acid is (+)S-abscisic acid (Table 1.16), the original assignment of absolute stereochemistry (CORNFORTH et al., 1967) having now been corrected (RYBACK, 1972; see MILBORROW, 1974a). The 2-trans-isomer which is almost biologically inactive (MILBORROW, 1966) has been also detected in plants (MILBORROW, 1970). Whether or not it is derived enzymatically is unknown, as ABA is known to isomerize to a 1:1 mixture with 2-trans-ABA in light (see MILBORROW, 1974a). This property has been used to increase the reliability of gas chromatographic identifications of ABA (LENTON et al., 1971).

Table 1.16. Abscisic acid and its metabolites in plants

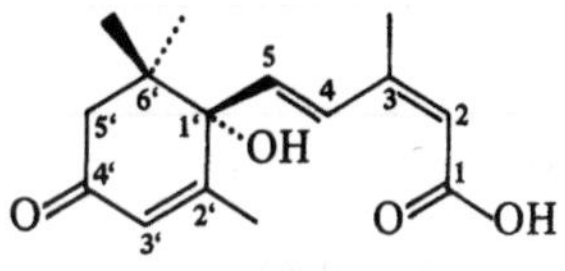

(+)−Abscisic acid

Plant material	Reference
Gossypium hirsutum	
Young fruits	OHKUMA et al. (1963) (OHKUMA et al., 1965) (CORNFORTH et al., 1965b) (KIENZLE et al. 1978)
Ovules	SHINDY and SMITH (1975)
Acer pseudoplatanus Leaves	CORNFORTH et al. (1966b)
Lupinus luteus Immature seed	KOSHIMIZU et al. (1966)
Lupinus albus Phloem exudate	HOAD (1978)
Pisum sativum	
Seed	ISOGAI et al. (1967b)
Seedlings	KOMOTO et al. (1972)
Chloroplasts	RAILTON et al. (1974)
Seed	FRYDMAN et al. (1974)
Dioscorea batatas Aerial tubers	HASHIMOTO et al. (1968)
Pyrus malus Juice	GASKIN and MACMILLAN (1968)
Pyrus domestica Seed	BALBOA-ZAVALA and DENNIS (1977)
Pyrus communis Immature seed	MARTIN et al. (1977a)
Taxus baccata	
Seed	LE PAGE-DEGIVRY et al. (1969)
Fruit	LENTON et al. (1971)
Ceratonia siliqua Immature fruit	MOST et al. (1970)
Coffea arabica Flower buds	BROWNING et al. (1970)
Vernonia anthelmintica Leaves	SANYAL et al. (1970)
Vitis rotundifolia Leaves	RAPP and ZIEGLER (1971)
Vitis vinifera	LOVEYS and KRIEDEMANN (1974)
Pinus radiata Young stems	JENKINS and SHEPHERD (1972)
Pinus densiflora Pollen	SHIBUYA et al. (1978)
Pinus sylvestris Shoots	ANDERSSON et al. (1978)
Abies balsamea Dormant buds	LITTLE et al. (1972)
Betula pubescens	
Apical organs	LENTON et al. (1972)
Wood bark and bleeding sap	DATHE et al. (1978b)
Solanum laciniatum Tubers	BIALEK et al. (1973)
Citrus sinensis	
Peel	GOLDSCHMIDT et al. (1973)
Dormant buds	JONES et al. (1976)
Bryophyllum daigremontianum	
Shoot tips and upper leaves	GASKIN et al. (1973)
Corylus avellana Seed and fruit	WILLIAMS et al. (1973)
Hedera helix Leaves	HILLMAN et al. (1974)
Phaseolus vulgaris Immature seed	HIRAGA et al. (1974b)
Chrysanthemum morifolium Leaves	SENGUPTA et al. (1974)
Persea gratissima Fruit	MILBORROW (1974b)
Hibiscus rosa-sinensis	
Abscission zone explants	SWANSON et al. (1975)

Table 1.16 (continued)

Plant material	Reference
Marah macrocarpus[a] Seed	BEELEY et al. (1975)
Prunus cerasus	
Buds	MIELKE and DENNIS (1975, 1978)
Seed and pericarp	DAVISON et al. (1976)
Prunus persica	
Immature seed	YAMAGUCHI et al. (1975a)
Seed	BONAMY and DENNIS (1977)
Prunus domestica Fruit	SHAYBANY et al. (1977)
Lactuca sativa	
Fruits	McWHA and HILLMAN (1975)
Fruits	BERRIE and ROBERTSON (1976)
Glycine max	
Seed and pod walls	QUEBEDEAUX et al. (1976)
Seed	CIHA et al. (1977)
Triticum aestivum Developing grains	KING (1976)
Cucurbita pepo Seed	FUKI et al. (1977a)
Cucurbita maxima Seed	GRAEBE et al. (unpublished work)
Lolium temulentum Leaves and apices	KING et al. (1977)
Zea mays Root cap and apices	RIVIER et al. (1977)
Juglans nigra Leaves	SHAYBANY and MARTIN (1977)
Juglans hindsii Leaves	SHAYBANY and MARTIN (1977)
Juglans regia Leaves	SHAYBANY and MARTIN (1977)
Juglans hindsii × *J. regia* Leaves	SHAYBANY and MARTIN (1977)
Ricinus communis	
Leaves, shoot tips and phloem sap	ZEEVAART (1977)
Carum carvi Fruit	MENDEZ (1978)
Foeniculum vulgare Fruit	MENDEZ (1978)
Pimpinella anisum Fruit	MENDEZ (1978)
Thapsia villosa Fruit	MENDEZ (1978)
Angelica pachycarpa Fruit	MENDEZ (1978)
Secale cereale Immature fruit	DATHE et al. (178a)
Picea sitchensis Bark peelings	LITTLE et al. (1978)
Picea abies Shoots	ANDERSSON et al. (1978)
Euphorbia lathyrus Leaves	SIVAKUMARAN and HALL (1978)
Oryza sativa Seedlings	KUROGUCHI et al. (1978)
Humulus lupulus Shoots	WATANABE et al. (1978b)
Robinia pseudacacia Seed	HIRAI et al. (1978)
Pseudotuga menziesii Dormant shoots	WEBBER et al. (1979)
Vigna unguiculata Seed	ADESOMOJU et al. (1980)
Vicia faba Seed	SPONSEL et al. (1979)
Spinacia oleracea Leaves	ZEEVAART (unpublished work)

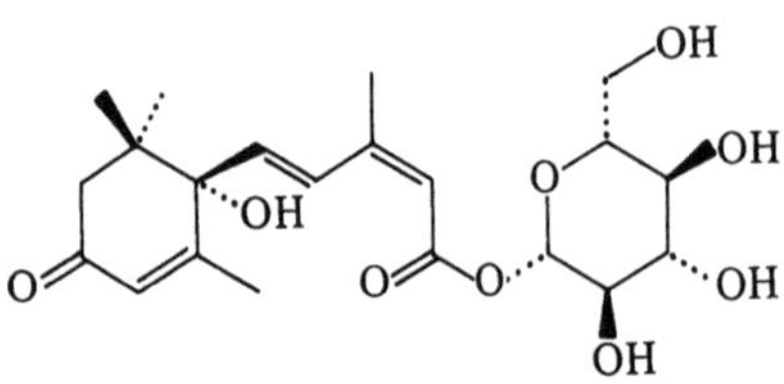

(+)-Abscisyl-β-D-Glucopyranoside

Lupinus luteus Immature fruit	KOSHIMIZU et al. (1968a)
Rosa arvensis Pseudocarp	MILBORROW (1970)

Table 1.16 (continued)

Plant material	Reference
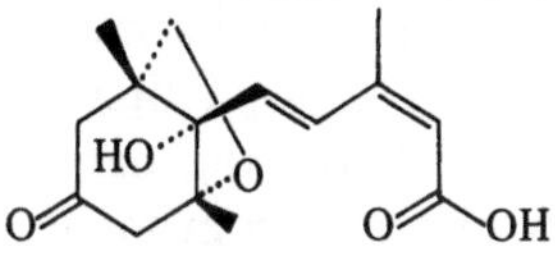	
Phaseic acid	
Phaseolus coccineus[b]	
Seed	MacMillan and Pryce (1969a, b) (Milborrow, 1969, 1975)
Seed	Durley et al. (1971)
Bryophyllum daigremontianum	
Shoot tips and upper leaves	Gaskin et al. (1973)
Vitis vinifera	
Leaves	Loveys and Kriedemann (1974)
Leaves	Kriedemann et al. (1975)
Pyrus communis Seed	Martin et al. (1977a)
Ricinus communis	
Leaves, shoot tips and phloem sap	Zeevaart (1977)
Vigna unguiculata Seed	Adesomoju et al. (1980)
Prunus domestica Fruit	Martin (unpublished work)
Vicia faba Seed	Sponsel et al. (1979)
Beta vulgaris Leaf tips	Lenton (unpublished work)
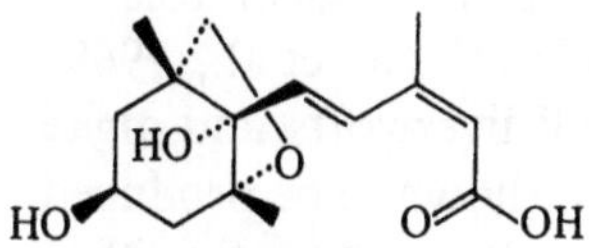	
Dihydrophaseic acid	
Phaseolus vulgaris	
Seed	Tinelli et al. (1973) (Milborrow, 1975)
Seed	Walton et al. (1973)
Seed and wilted plants	Zeevaart and Milborrow (1976)
Roots	Takasugu et al. (1973)
Phaseolus coccineus[b] Seed	Sponsel (unpublished work)
Pisum sativum Seed	Frydman et al. (1974)
Marah macrocarpus[a] Seed	Beeley et al. (1975)
Ricinus communis	
Leaves, shoot tips and phloem sap	Zeevaart (1977)
Pyrus communis Seed	Martin et al. (1977a)
Secale cereale Mature fruit	Dathe et al. (1978a)
Vigna unguiculata Seed	Adesomoju et al. (1980)
Prunus domestica Fruit	Martin (unpublished work)
Cucurbita maxima Seed	Graebe (unpublished work)
Vicia faba Seed	Sponsel et al. (1979)
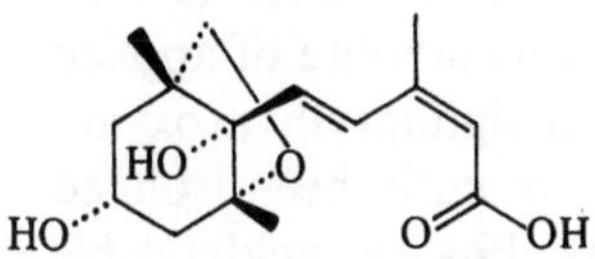	
epi–Dihydrophaseic acid	
Phaseolus vulgaris	
Seed and wilted plants	Zeevaart and Milborrow (1976)

[a] Formally *Echinocystis macrocarpa*

[b] Formally *Phaseolus multiflorus*

ABA has been detected (see Table 1.16) in angiosperms, gymnosperms, a fern (MILBORROW, 1967), a horse-tail (D.R. ROBINSON, unpublished) and a moss (HIRON, cited by PRYCE, 1972b). In 1967 MILBORROW (1967) showed that a major constituent of "inhibitor β" was ABA. "Inhibitor β" was the name given by BENNET-CLARK and KEFFORD (1953) to inhibitory material from many plant extracts contained in an R_F zone higher than indoleacetic acid on paper chromatography. The concentration of ABA in plant tissue normally lies in the range 0.01–1 ppm (see MILBORROW, 1978). In rose hips up to 4 ppm has been detected (MILBORROW, 1967). ABA occurs in all parts of plants but is most concentrated in seed and young fruits. The great variety of research into the correlative changes in ABA concentration with various physiological events has been reviewed by ADDICOTT and LYON (1969), MILBORROW (1974a, 1978) and RASCHKE (1975). The huge increase in ABA levels in wilted plants was first described by WRIGHT (WRIGHT, 1969; WRIGHT and HIRON, 1969). Excised wheat seedlings which were wilted, losing about 10% of their fresh weight, were shown to contain forty-fold more ABA than controls. Similar results to these and the fact that exogenously applied ABA causes rapid stomatal closure (MITTELHEUSER and VAN STEVENINCK, 1969) imply that ABA is intimately involved in the regulation of transpiration in plants (see RASCHKE, 1975; ITAI et al., 1978). Below the taxonomic level of mosses the physiological role of ABA appears to be taken by lunularic acid (Table 1.17) (VALIO et al., 1969; VALIO and SCHWABE, 1970) which has been detected in all liverworts and algae investigated (PRYCE, 1971, 1972b). Recently ABA has been shown to be produced by the fungus *Cercospora rosicola* (ASSANTE et al., 1977). It is interesting that this fungus is frequently found on rose trees – one of the richest sources of ABA.

The status of ABA as a plant hormone has recently been reviewed by WAREING (1978). PEGG (1976d) has reviewed the involvement of ABA in plant diseases.

The metabolism of ABA is discussed in Chapter 4. However there are a number of metabolites of ABA which have been detected as endogenous substances in plant material. The glucosyl ester (Table 1.16) appears to be a rapid-storage product for excess ABA, as it is a major metabolite of externally applied ABA in plants (see MILBORROW, 1970). Although the glucosyl ester has only been identified in two species (Table 1.16) it may well be widespread in plants as free ABA is released from many plant extracts on alkaline hydrolysis. ABA released in this way can amount to between one tenth and ten times the quantity of free ABA (see MILBORROW, 1974a). Phaseic acid which was first isolated from *Phaseolus vulgaris* by MACMILLAN and PRYCE (1968) was shown by MILBORROW (1969, 1975) to possess the structure shown in Table 1.16. It may well be an artefact of "metabolite C" (Fig. 1.8) which is a metabolite of applied ABA in tomato (MILBORROW, 1969), β-Hydroxy-β-methylglutarylhydroxyabscisic acid (Fig. 1.8), a conjugate of "metabolite C" has recently been isolated from seeds of *Robinia pseudacacia* (HIRAI et al., 1978). Phaseic acid (Table 1.16 has a much lower biological activity than ABA (MACMILLAN and PRYCE, 1969b; MILBORROW, 1969; DAVIS et al., 1972) so it may be a deactivation product. Evidence has been presented that phaseic acid might regulate photosynthesis (KRIEDEMANN et al., 1975) but this has been challenged (SHARKEY and

Table 1.17. Naturally occurring compounds related to abscisic acid

Plant material	Reference
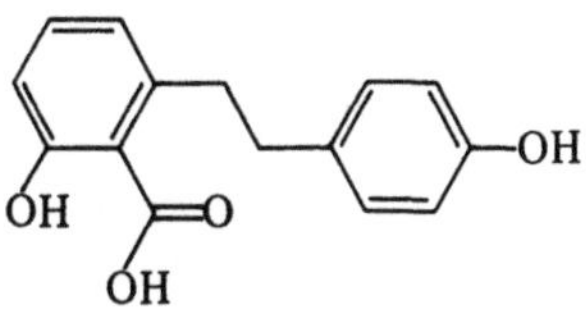	
Lunularic acid	
Lunularia cruciata Thalli	VALIO et al. (1969) VALIO and SCHWABE (1970)
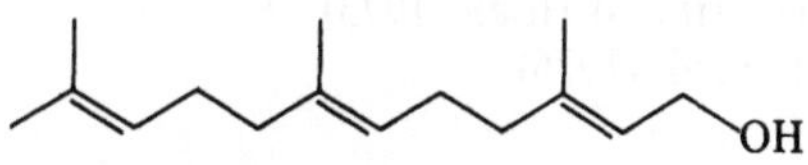	
All–trans–farnesol[a]	
Sorghum sudanese Leaves	WELLBURN et al. (1974)
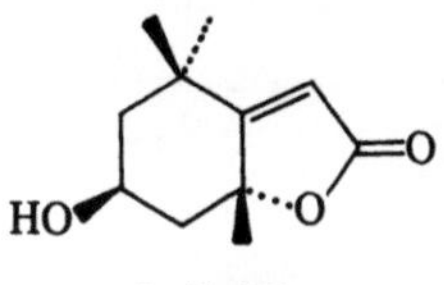	
Loliolide	
Fumaria officinalis	MANSKE (1938) (HODGES and PORTE, 1964)
Lolium perenne Leaves	HODGES and PORTE (1964) (ISOE et al., 1972) (KIENZLE et al., 1978)
Digitalis purpurea Leaves	WADA (1965)
Digitalis lanata	TSCHESCHE and BUSCHAUER (unpublished work) (HODGES and PORTE, 1964)
Menyanthes trifoliata	SAKAN et al. (1969)
Arnica montana Leaves	HOLUB et al. (1975)
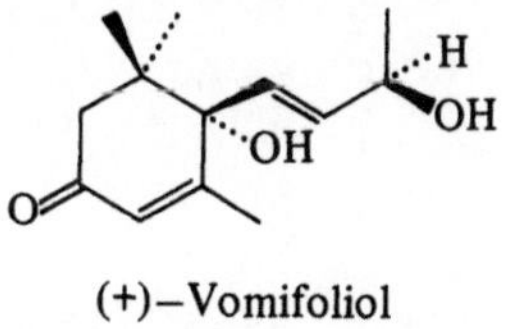	
(+)–Vomifoliol	
Rauwolfia vomitoria Flowers	POUSSET and POISSON (1969) (WEISS et al., 1973) (GALBRAITH and HORN, 1973)
Podocarpus blumei Leaves	GALBRAITH and HORN (1972)
Croton sparsifolius Leaves and stems	SATISH and BHAKUNI (1972)
Croton lobatus Leaves and stems	STUART and WOO-MING (1975)
Croton trinitatis Leaves and stems	STUART and WOO-MING (1975)
Phaseolus vulgaris Roots	TAKASUGU et al. (1973)
Vinca rosea Leaves and stems	BHAKUNI et al. (1974)
Palicourea alpina Leaves and twigs	STUART and WOO-MING (1975)
Cannabis sativa Leaves and stems	BERCHT et al. (1976)
Cucurbita pepo Seed	FUKUI et al. (1977a)

Table 1.17 (continued)

Plant material	Reference
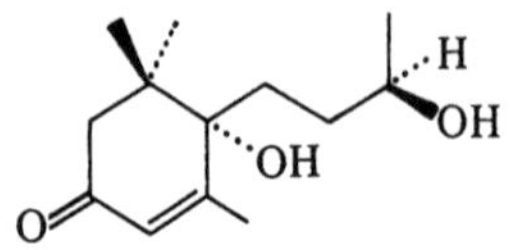	
(+)–3,4–Dihydrovomifoliol	
Podocarpus blumei Leaves	GALBRAITH and HORN (1972) (WEISS et al., 1973) (GALBRAITH and HORN, 1973)
Cannabis sativa Leaves and stems	BERCHT et al. (1976)
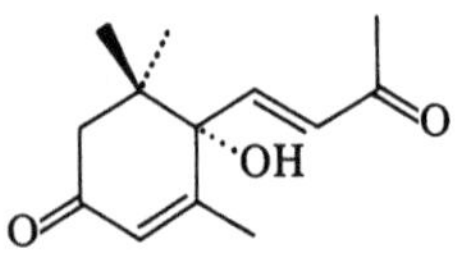	
(+)-Dehydrovomifoliol	
Phaseolus vulgaris Roots	TAKASUGU et al. (1973)
Cucurbita pepo Seed	FUKUI et al. (1977a)
Oryza sativa Husks	KATO et al. (1977)
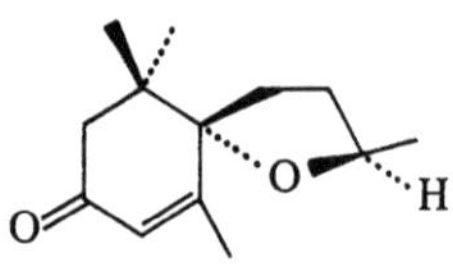	
(–)–Theaspirone	
Camellia sinensis "Tea leaves"	INA et al. (1968) (WEISS et al., 1973)

[a] Widely distributed in nature, see SIMONSEN and BARTON (1952); WINDHOLZ (1976)

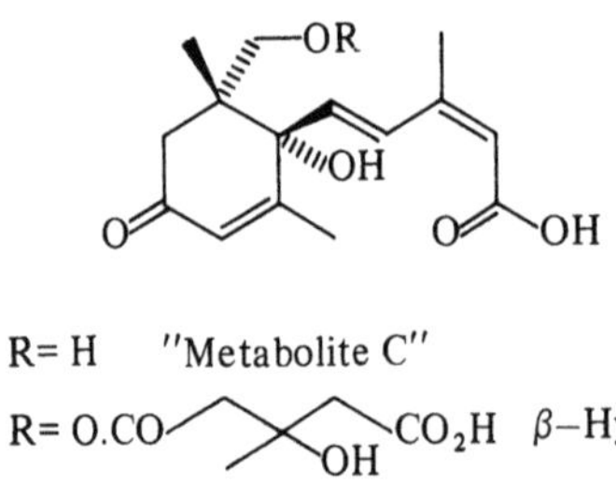

R= H "Metabolite C"

R= O.CO... CO_2H β–Hydroxy–β–methylglutarylhydroxyabscisic acid

Fig. 1.8

RASCHKE, 1980). A list of plants in which phaseic acid has been detected is given in Table 1.16. Dihydrophaseic acid (MILBORROW, 1975) is another metabolite of ABA which has been shown to occur in many plants (see Table 1.16). The epimer at the 4′-position, epi-dihydrophaseic acid (Table 1.16) is also a natural product (ZEEVAART and MILBORROW, 1976).

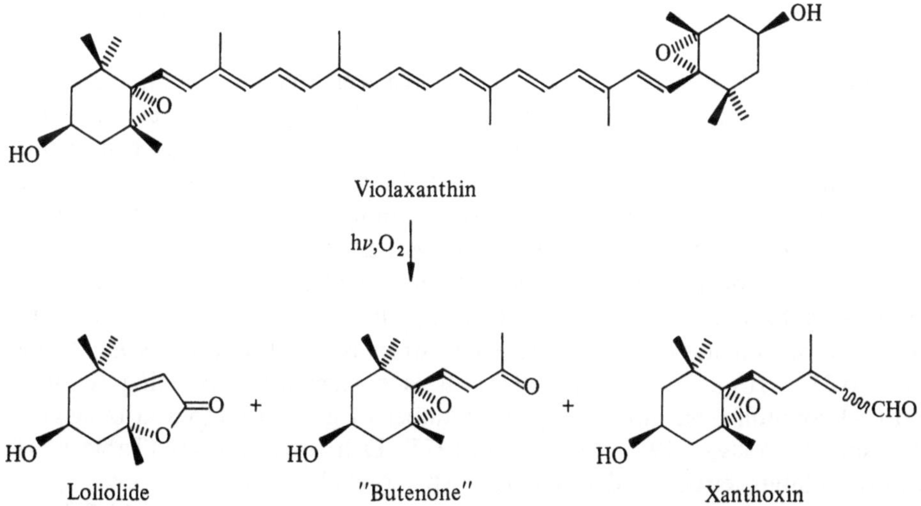

Fig. 1.9

When the structure of ABA was published its obvious structural resemblance to the terminal portions of many carotenoids was apparent. Other evidence also gave rise to the hypothesis that carotenoids might be the natural precursors of ABA (TAYLOR and SMITH, 1967). For instance, phototropic organs of fungi were known to contain carotenoids and their action spectrum was similar to the absorption spectrum of carotenoids. Light-grown plants were known to contain more "inhibitor β" than etiolated ones (WRIGHT, 1954; MASUDA, 1962). SIMPSON and WAIN (1961) obtained results suggesting that an inhibitor is formed in plants in light. Stimulated by these observations TAYLOR and SMITH showed that exposure of the carotenoid, violaxanthin, to a bright light on a damp filter paper gave rise to a neutral substance which inhibited the germination of cress seeds. This inhibitor, named xanthoxin (see Fig. 1.9) was purified and identified by TAYLOR and BURDEN (1970b, 1972). The 2-cis-compound which has similar biological activity to ABA was admixed with approximately an equal quantity of the less active 2-trans-isomer. In addition to xanthoxin the "butenone" and loliolide were also formed (see Fig. 1.9). The same workers also detected 2-cis- and 2-trans-xanthoxin in dwarf bean (*Phaseolus vulgaris*) and wheat (*Triticum aestivum*) by gas chromatography and bioassay. The "butenone" and loliolide were apparently also present (TAYLOR and BURDEN, 1970a). More recently FIRN and FRIEND (1972) have shown that soybean lipoxygenase is capable of cleaving violaxanthin to form a similar range of products. Xanthoxin has been shown to be converted efficiently into ABA in tomato and dwarf bean shoots (TAYLOR and BURDEN, 1973) but it is not thought to be on the main biosynthetic pathway to ABA in plants (see Chap. 4). Although xanthoxin has since been detected chromatographically in a wide range of plant species (FIRN et al., 1972; ZEEVAART, 1974; ANSTIS et al., 1975; KING et al., 1977; THOMPSON and BRUINSMA, 1977; BÖTTGER, 1978; DÖRFFLING et al., 1978), its existence in plants has yet to be rigorously established. ORITANI and YAMASHITA (1973) and KIENZLE et al. (1978) have reported syntheses of (±)-

xanthoxin. Loliolide has been isolated from a number of plants (see Table 1.17).

Sorghum plants subjected to water stress have been found to contain increased quantities of all-trans-farnesol (OGUNKANMI et al., 1974; WELLBURN et al., 1974). Furthermore all-trans-farnesol (Table 1.17) has been shown to induce stomatal closure slightly faster and more completely than ABA. The possible role of all-trans-farnesol as an endogenous antitranspirant in *Sorghum* plants has been recently discussed (FENTON et al., 1977; MANSFIELD et al., 1978).

(+)-Vomifoliol, a naturally occurring relative of ABA (see Table 1.17) is also a powerful antitranspirant. It was as equally efficient as ABA in the rate and extent of stomatal closure in epidermal strips of *Eichhornia crassipes* (COKE et al., 1975; STUART and COKE, 1975). It is particularly interesting to note that (+)-vomifoliol has no activity in a number of growth inhibition bioassays (POUSSET and POISSON, 1969; COKE et al., 1975). Dehydrovomifoliol (Table 1.17) however inhibits lettuce seed germination (KATO et al., 1977).

A number of other naturally occurring compounds are known which have some structural resemblance to ABA (see Table 1.17) but their biological activity is unknown.

1.7 Other Plant Constituents Which Affect Plant Growth[2]

The substances described in this section are plant products which have been shown to modify plant growth in various bioassay systems. The majority of these compounds are inhibitors. In very few cases has a natural role been established for these compounds in vivo. In some cases interaction with hormones has been postulated. Thus it has been suggested that phenolic compounds mediate the fine control of IAA levels by interaction with IAA oxidase (WOLF et al., 1976). In other instances an allelopathic[3] function is indicated, e.g., the *Podocarpus* lactones (BROWN and SANCHEZ, 1974). The mechanisms of action of inhibitors have been discussed by RICE (1974).

In the space available a comprehensive list of these compounds and of their occurrence in plants is not possible. Furthermore many of the compounds such as phenolics and fatty acids are ubiquitous. The wide distribution of growth inhibitors in plants has recently been demonstrated by SCHREIBER (1975), who showed that of a selection of 524 secondary plant constituents of known structure representing a range of structural types, 197 had significant inhibitory activity. For much of the original literature on this topic the reader is referred to the following reviews: BONNER (1950), HEMBERG (1961), MORELAND et al. (1966), KEFELI and KADYROV (1971), STEWARD and KRIKORIAN (1971), THIMANN (1972), GROSS (1975) and LETHAM (1978b). There are also a number of useful reviews

2 This section complements and supplements Encyclopedia of plant physiology, vol. 8 (BELL, E.A., CHARLWOOD, B.V., eds.), Secondary plant products. Berlin, Heidelberg, New York: Springer 1980

3 Allelopathy is the biochemical interaction between plants (MOLISCH, 1937)

Fig. 1.10. Some common phenolic constituents of plants

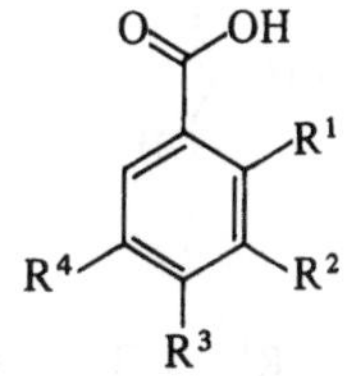

	R^1	R^2	R^3	R^4
Salicylic acid	OH	H	H	H
Gallic acid	H	OH	OH	OH
p–Hydroxybenzoic acid	H	H	OH	H
Protocatechuic acid	H	OH	OH	H
Gentisic acid	OH	H	H	OH
Vanillic acid	H	OCH_3	OH	H
Syringic acid	H	OCH_3	OH	OCH_3

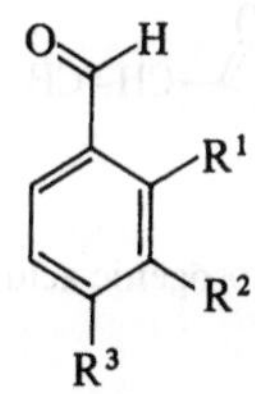

	R^1	R^2	R^3
Salicylaldehyde	OH	H	H
Vanillin	H	OCH_3	OH

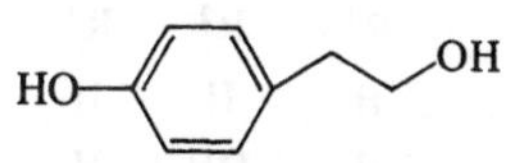

Dihydroconiferyl alcohol

concerned primarily with germination inhibitors (EVENARI, 1949; TOOLE et al., 1956; VAN SUMERE, 1960; WAREING, 1965; VAN SUMERE et al., 1972; MAYER and POLJAKOFF-MAYBER, 1975). The ecological role of growth-affecting plant products (allelopathy) has been discussed by BÖRNER (1960), GRÜMMER (1961), EVENARI (1961), MOJÉ (1966), WHITTAKER (1970), MULLER (1970), WENT (1970), National Academy of Sciences (1971), WHITTAKER and FEENY (1971), MULLER and CHOU (1972), RICE (1974, 1979), HARBORNE (1977), SWAIN (1977) and PUTNAM and DUKE (1978). For reference to the occurrence of many of these compounds the reader is directed to books by KARRER (1958), ROBINSON (1975), DEVON and SCOTT (1972, 1975), and WINDHOLZ (1976).

In the present treatment of this topic the relevant compounds are either included in Fig. 1.10 or listed (structures 1–140) at the end of this section. The compounds are discussed according to structural type under the following head-

Fig. 1.10 (continued)

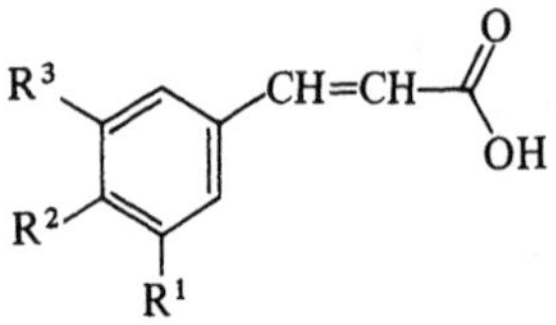

	R^1	R^2	R^3
Cinnamic acid	H	H	H
p-Coumaric acid	H	OH	H
Caffeic acid	OH	OH	H
Ferulic acid	OCH_3	OH	H
Sinapic acid	OCH_3	OH	OCH_3

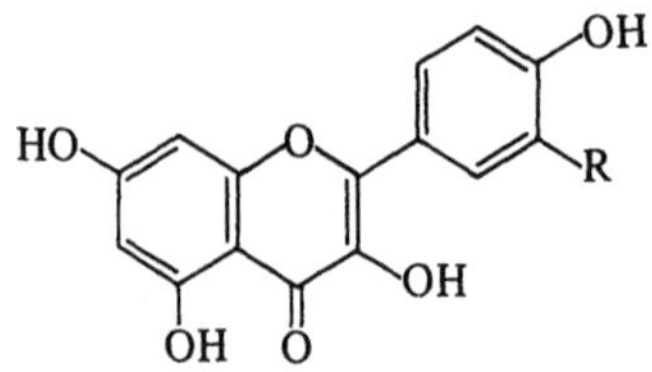

Kaempferol (R=H)
Quercetin (R=OH)

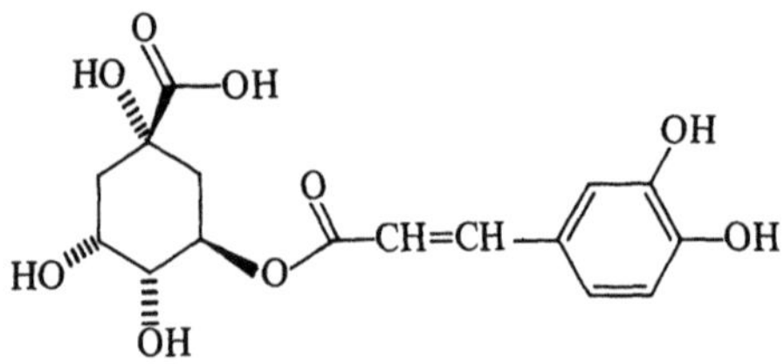

Chlorogenic acid

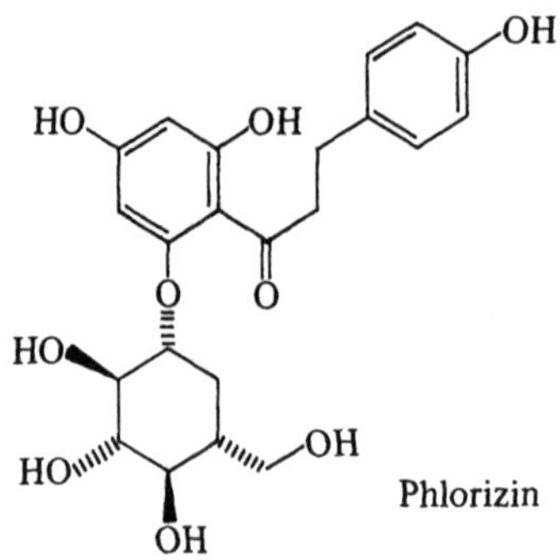

Phlorizin

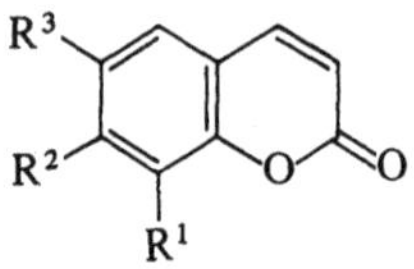

	R^1	R^2	R^3
Coumarin	H	H	H
Umbelliferone	H	OH	H
Aesculetin	H	OH	OH
Scopoletin	H	OH	OCH_3
Daphnetin	OH	OH	H

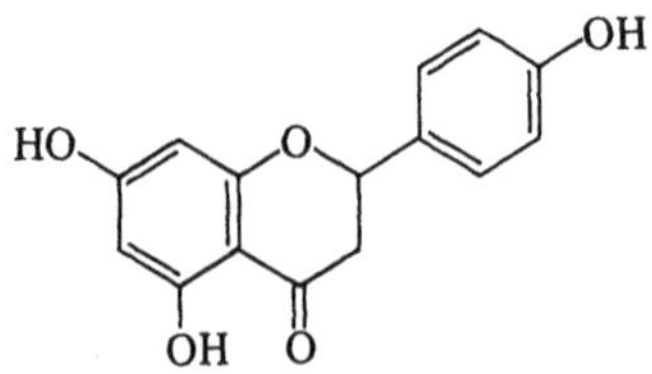

Naringenin

ings: (a) aromatic compounds; (b) nitrogen-containing compounds; (c) terpenoid compounds; (d) aliphatic compounds; and (e) other compounds.

a) Aromatic Compounds

Aromatic phenols form the main group of naturally occurring inhibitors among plant products. These range from simple phenols and benzoic acids (REYNOLDS, 1978) to the polymeric tannins (CORCORAN, 1971; CORCORAN et al., 1972; HARADA and NAKAYAMA, 1974; JACOBSON and CORCORAN, 1977) and humic acids (VAUGHAN and LINEHAN, 1976). This group also includes the cinnamic acids, the coumarins (VAN SUMERE, 1960; VAN SUMERE et al., 1972) and the flavonoids (McCLURE, 1975). It has been suggested that these compounds are second in

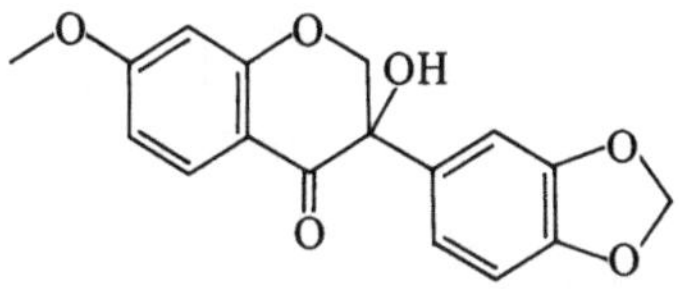

(1) Pisatin

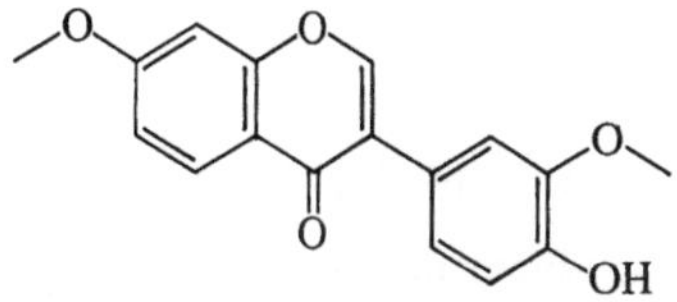

(2) Sayanedine

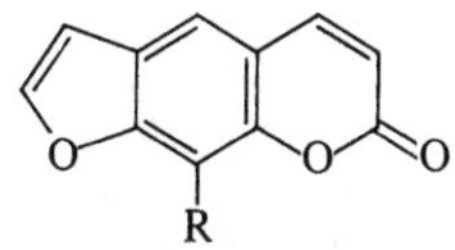

(3) Psoralen (R=H)
(4) Xanthotoxin ($R=OCH_3$)

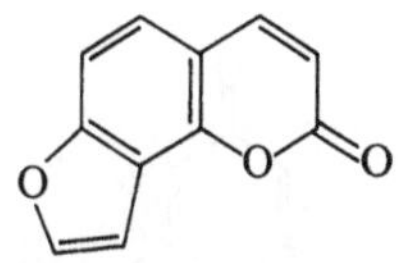

(5) Angelicin

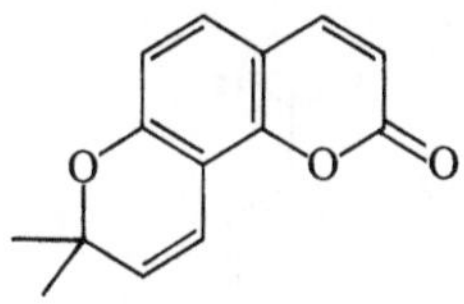

(6) Seselin

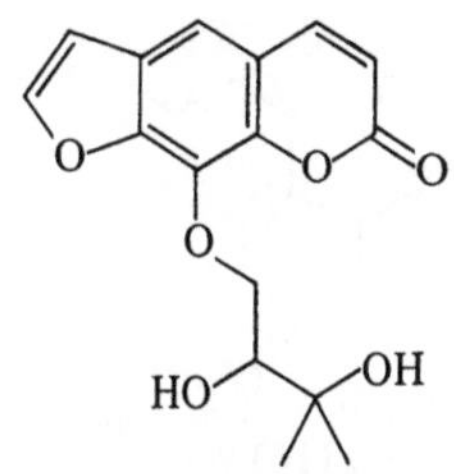

(7) Heraclenol

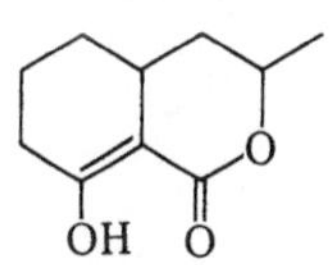

(8) Ramulosin

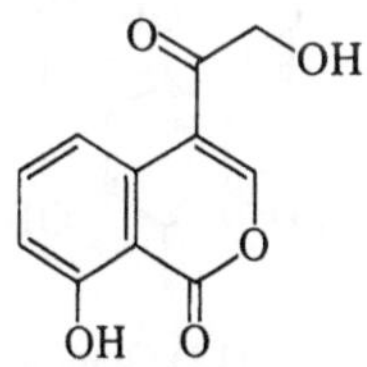

(9) Oosponol

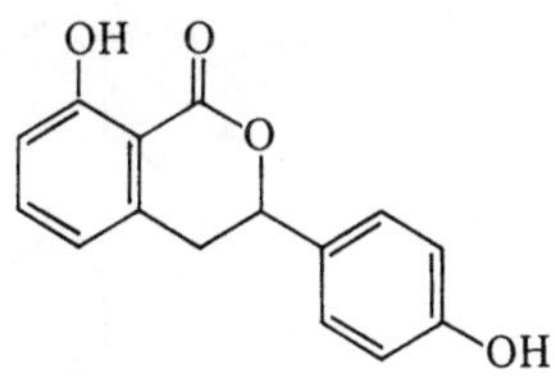

(10) Hydrangenol

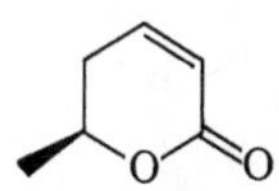

(11) Parasorbic acid

abundance in plants only to carbohydrates (PRIDHAM, 1965). The structures of some commonly encountered phenolic inhibitors are given in Fig. 1.10. These compounds often occur in plants as glycosides. The possible role of such compounds in plant growth regulation has recently been discussed by KEFELI and KUTACEK (1977) and KEFELI (1978). There are numerous reports of the isolation of hydroxybenzoic and hydroxycinnamic acids as endogenous inhibitors in plants, e.g., BONNER and GALSTON (1944), VARGA and KÖVES (1959), WANG

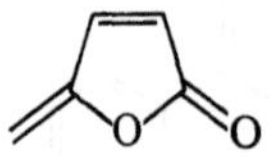

(12) Protoanemonin

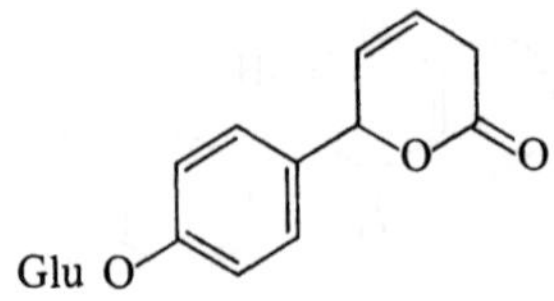

(13) Psilotin

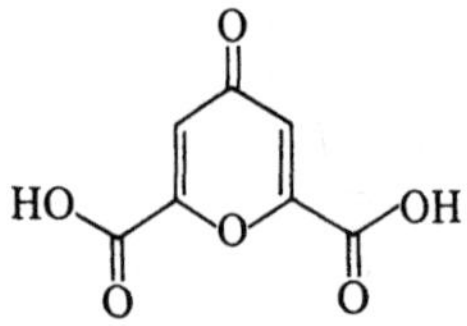

(14) Chelidonic acid

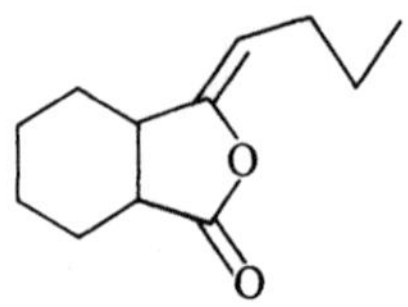

(15)

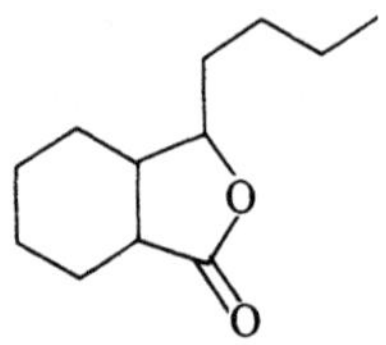

(16)

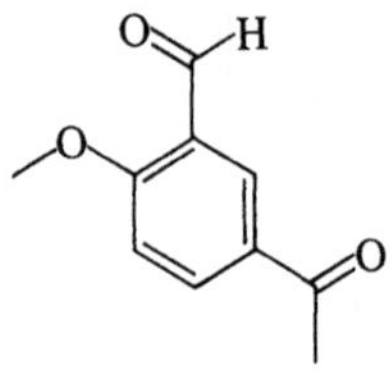

(17)

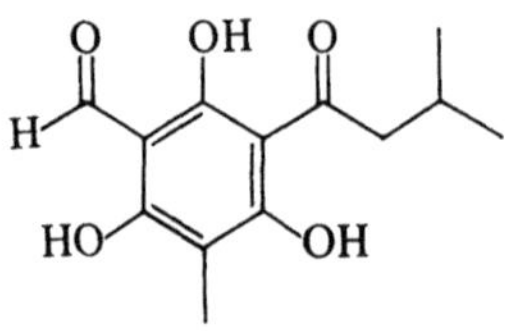

(18) Grandinol

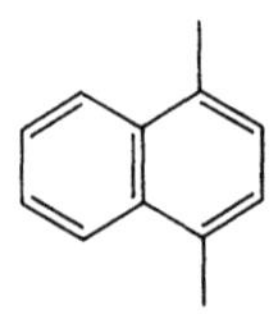

(19)

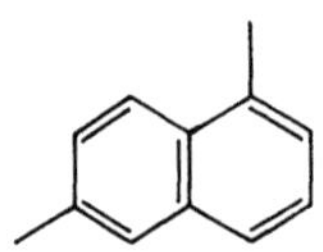

(20)

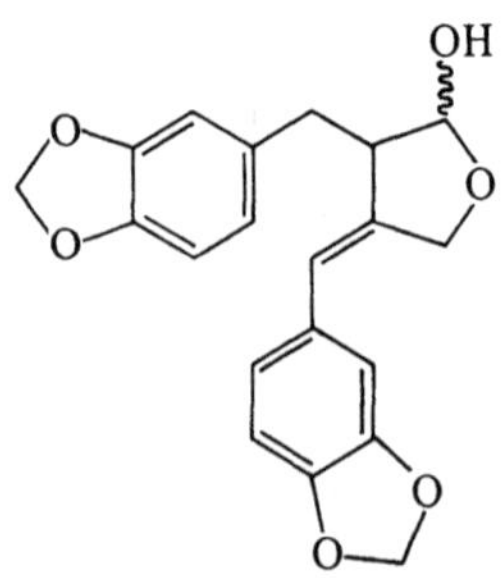

(21) Acanthotoxin

et al. (1967), MENDEZ et al. (1968), RUDNICKI et al. (1973), SANCHEZA TAMES et al. (1973), MATSUSHIMA et al. (1973), ALTREE-WILLIAMS et al. (1975), KIMURA et al. (1976), and THAKUR (1977). Compounds of this type were originally thought to account for much of the biological activity of "inhibitor β" before the discovery of abscisic acid (see MORELAND et al., 1966). In addition to growth

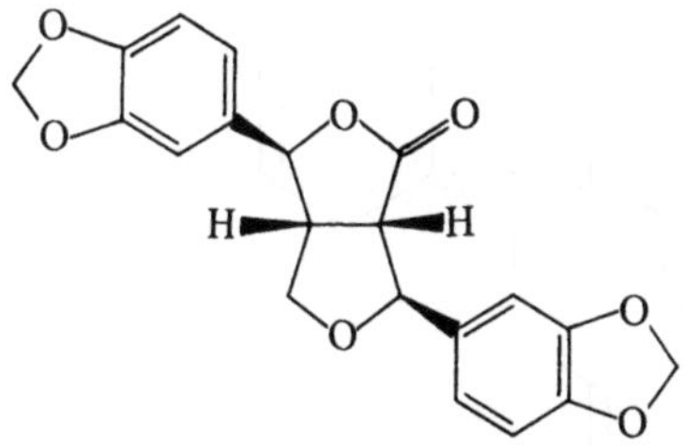

(22)

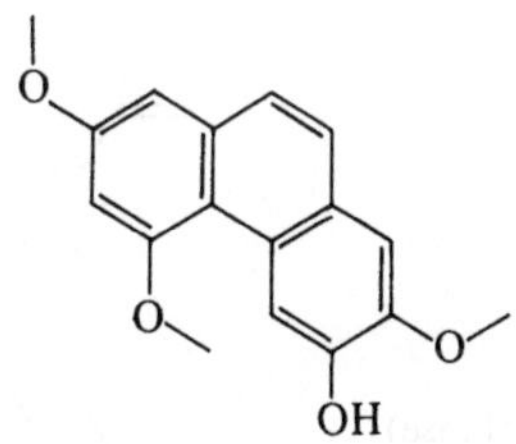

(23) Batatasin I

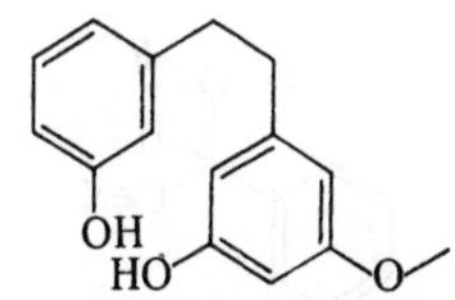

(24) Batatasin III

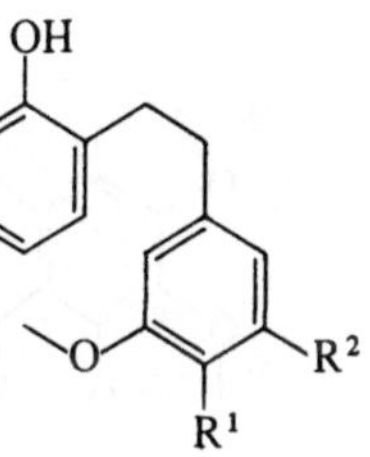

(25) Batatasin IV (R^1=H,R^2=OH)
(26) Batatasin V (R^1=R^2=OCH_3)

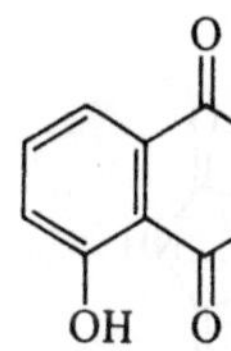

(27) Juglone

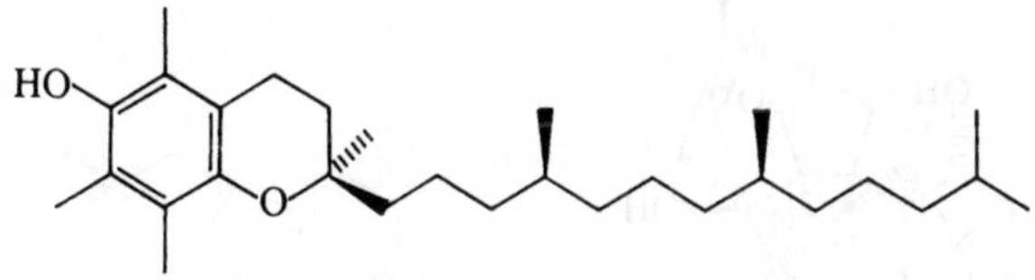

(28) α–Tocopherol

and germination inhibition many of these compounds display other regulatory properties such as suppression of mitochondrial metabolism (DESMOS et al., 1975) and stimulation of rooting (ROY et al., 1972). Phenolic acids have also been shown to affect flowering. Salicyclic acid (Fig. 1.10) induced flowering in *Lemna* species (CLELAND and AJAMI, 1974; KHURANA and MAKESHWARI, 1978) and gallic acid (Fig. 1.10) was identified as the specific flowering-inhibitory substance isolated from the leaves of *Kalanchoe blossfeldiana* (PRYCE, 1972a). The "lettuce cotyledon factor", a synergist of gibberellin action in the induction of lettuce hypocotyl elongation, has been identified (SHIBATA et al., 1974) as

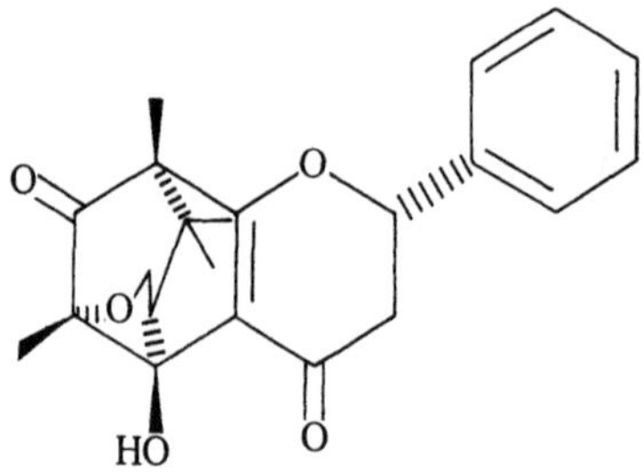

(29) Louisfieserone

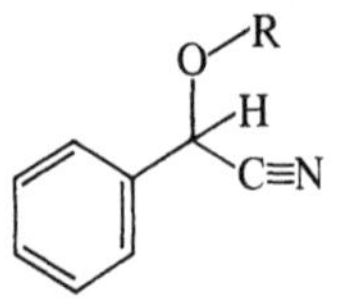

(30) Amygdalin (R=β–Gentiobiose)
(31) Prunasin (R=β–Glucose)
(32) Mandelonitrile (R=H)

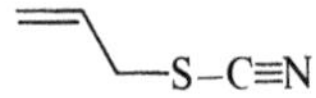

(33)

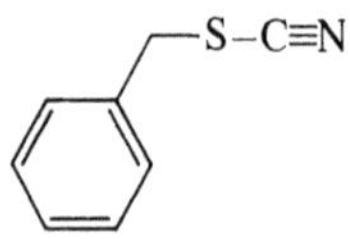

(34)

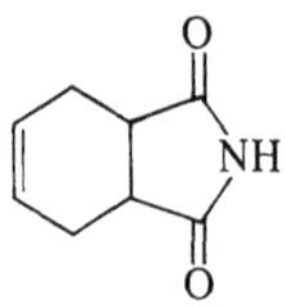

(35)

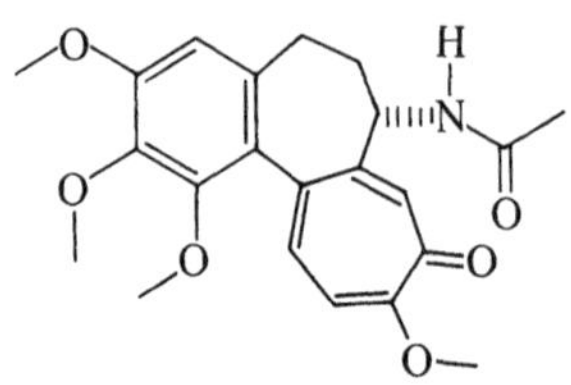

(36) Colchicine

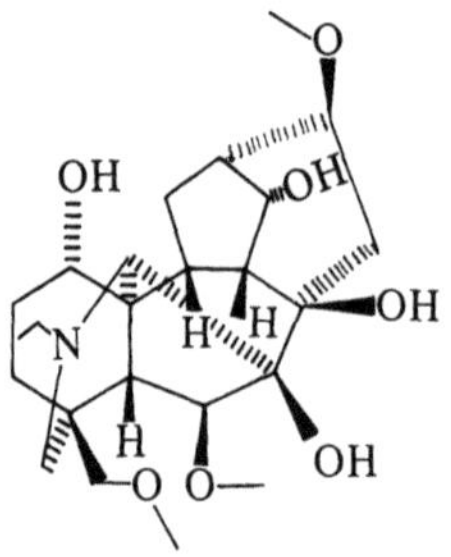

(37) Delcosine

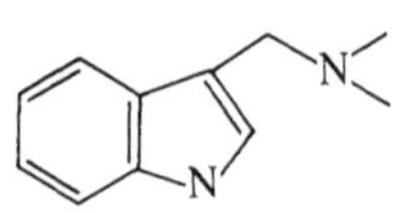

(38) Gramine

dihydroconiferyl alcohol (Fig. 1.10) and has been shown to stimulate cucumber hypocotyl elongation and to be a constituent of pea, cucumber, and artichoke (SAKURAI et al., 1974., see also NAKAMURA et al., 1977).

The action of phenolic compounds on plant growth is frequently attributed to their interaction with IAA oxidase, thus regulating IAA levels in vivo (see SCHNEIDER and WIGHTMAN, 1974; THIMANN, 1972; VAN SUMERE et al., 1975; WOLF et al., 1976; LETHAM, 1978b). In general monophenolic compounds stimu-

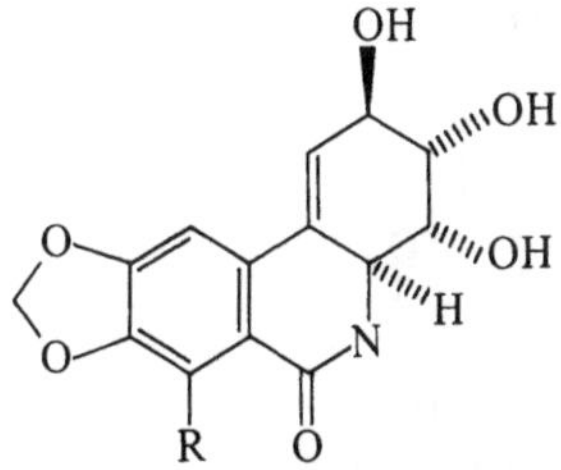

(39) Lycoricidine (R=H)
(40) Lycoricidinol (R=OH)

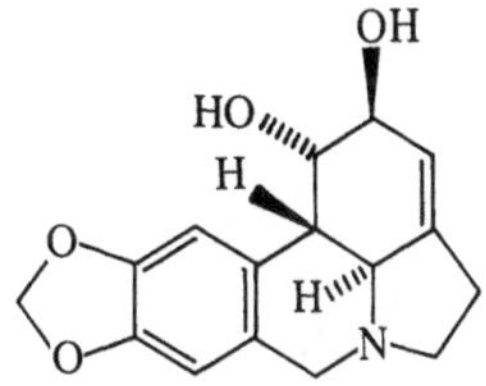

(41) Lycorine

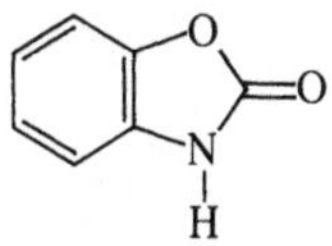

(42)

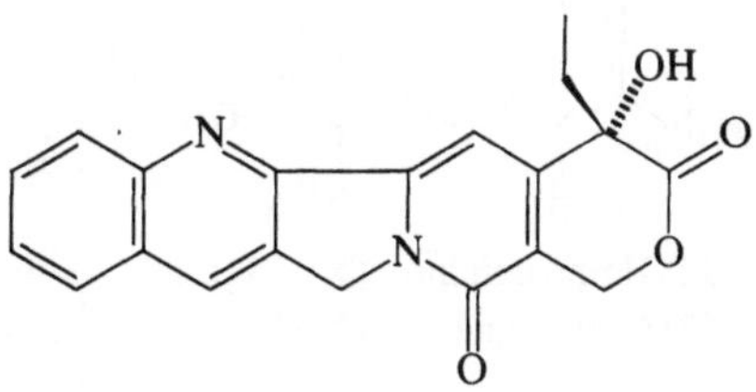

(43) Camptothecin

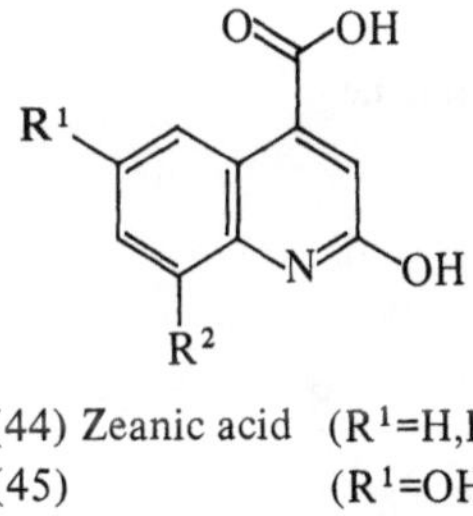

(44) Zeanic acid (R^1=H,R^2=OH)
(45) (R^1=OH,R^2=H)

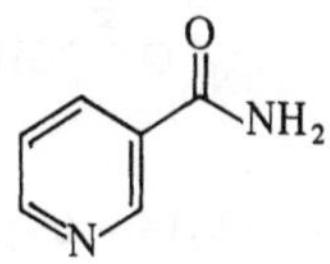

(46) Nicotinamide

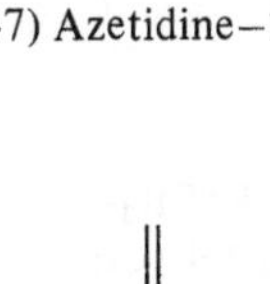

(47) Azetidine–2–carboxylic acid

(48) Pipecolic acid

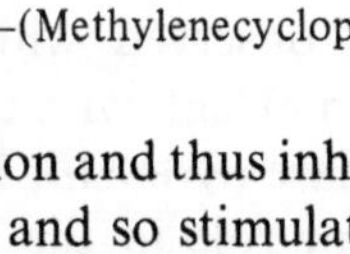

(49) a–(Methylenecyclopropyl) glycine

late IAA decarboxylation and thus inhibit growth whereas o-dihydroxy phenolics inhibit IAA oxidation and so stimulate growth. However at high concentration most phenolics inhibit growth irrespective of hydroxylation pattern. Thus caffeic acid (Fig. 1.10) (an inhibitor of IAA oxidase) stimulated mesocotyl growth and p-coumaric acid (Fig. 1.10) (a co-factor for IAA oxidase) inhibited growth.

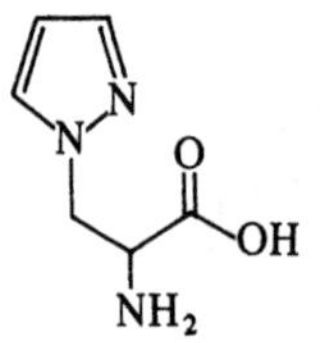

(50) β–Pyrazol–1–ylalanine

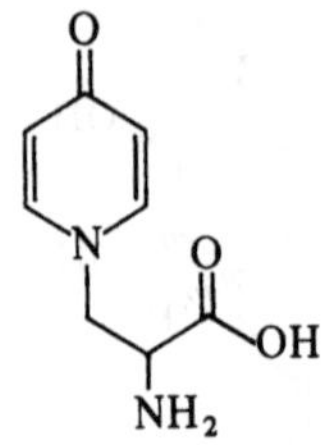

(51) Mimosine

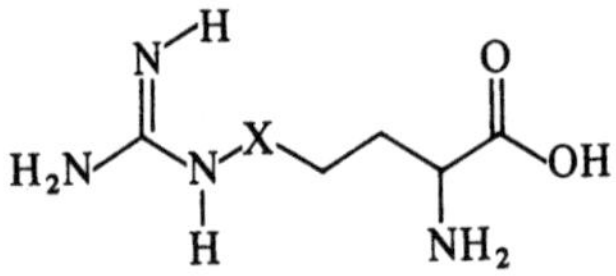

(52) Homoarginine $(X=CH_2)$
(53) Canavanine $(X=O)$

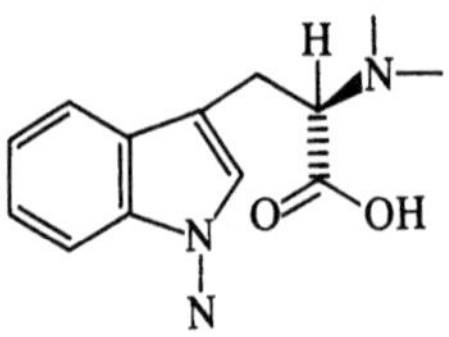

(54) N,N–Dimethyl–L– tryptophan

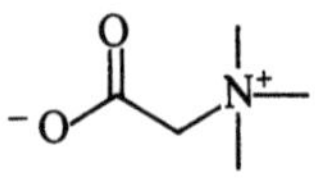

(55) Betaine

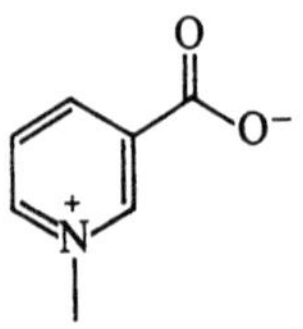

(56) Trigonelline

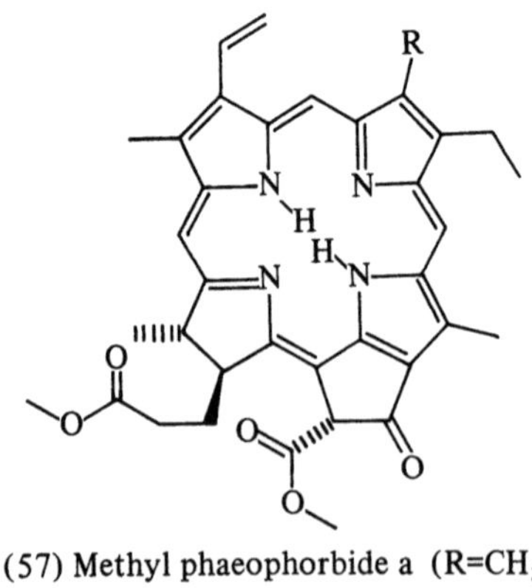

(57) Methyl phaeophorbide a $(R=CH_3)$
(58) Methyl phaeophorbide b $(R=CHO)$

But at $>10^{-4}$ M caffeic acid was also inhibitory (NITSCH and NITSCH, 1962). This structure/activity relationship is also observed with hydroxyflavonoids (STENLID, 1963, 1968, 1976b) and, to a certain extent, with coumarins (THIMANN, 1972). Betalains also interact with IAA oxidase (STENLID, 1976a).

Flavonoids apparently play an important role in the photomorphogenesis in *Pisum sativum* (FURUYA et al., 1962; GALSTON, 1969). Light alters the balance of glucosides of kaempferol and quercitin (Fig. 1.10), thus mediating the activity of IAA oxidase. As with the phenolic acids, the action of flavonoids is not

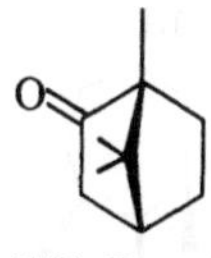

(59) Camphor

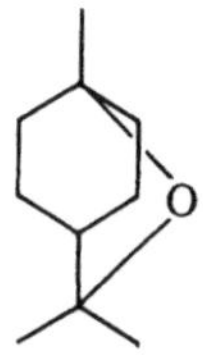

(60) Cineole

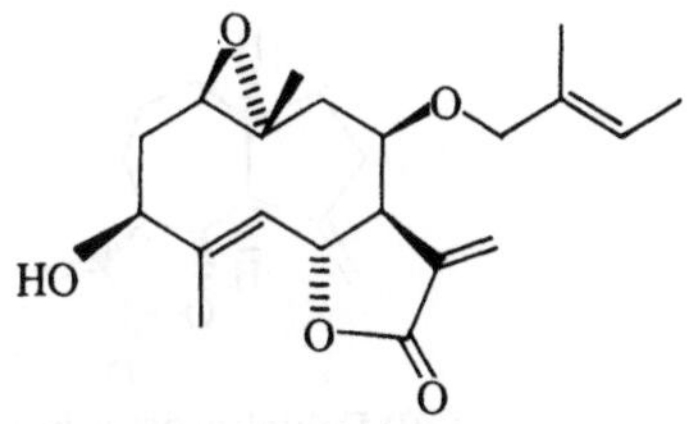

(61) Heliangine

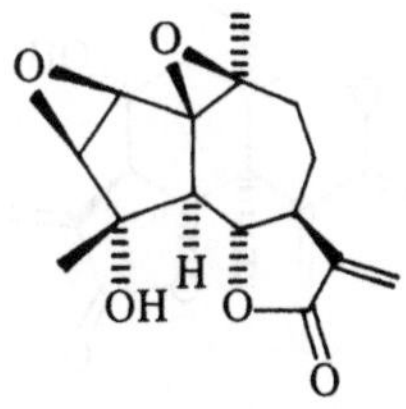

(63) Chrysartemin A

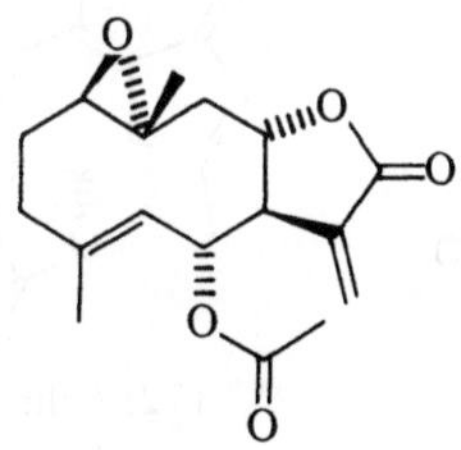

(62) Pyrethrosin

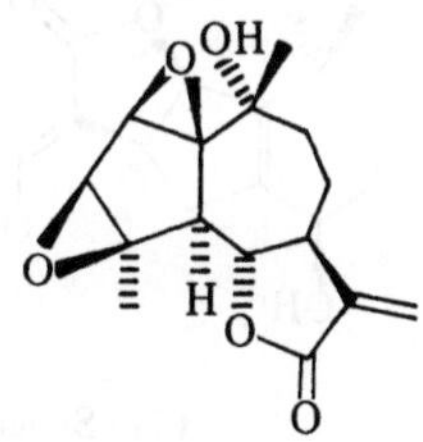

(64) Chrysartemin B

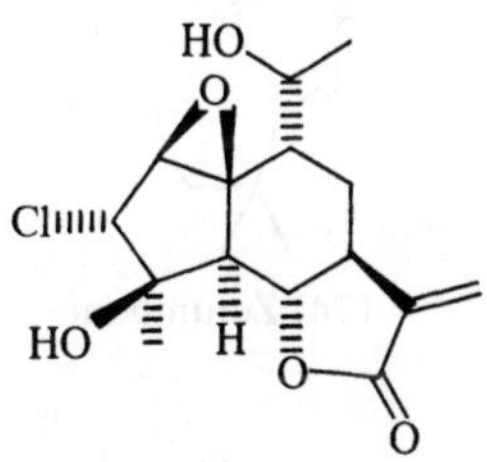

(65) Chlorochrymonin

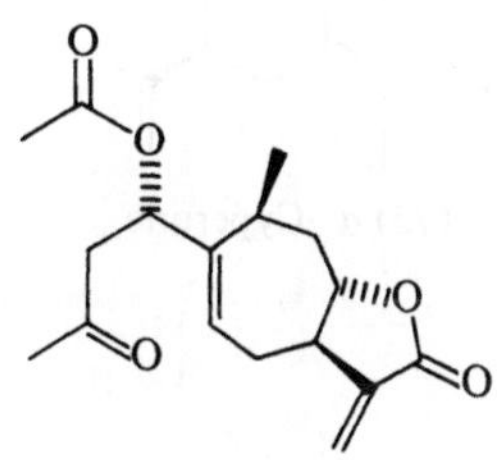

(66) Xanthinin

restricted to interaction with IAA oxidase. For instance the flavanone naringenin (Fig. 1.10), a constituent of dormant peach buds, antagonized the action of gibberellins (PHILLIPS, 1962). Flavonoids have been shown to inhibit mitochondrial phosphorylation (STENLID, 1970; KOEPPE and MILLER, 1974), and the flavonoid sayanedine (2) from *Pisum sativum* exhibited cytokinin-like activity (ISOGAI et al., 1970). The dihydrochalcone phlorizin (Fig. 1.10) from apple roots, and its decomposition products have been implicated as the major toxins in the "apple re-plant problem" (see MORELAND et al., 1966; STENLID, 1968; RICE, 1974).

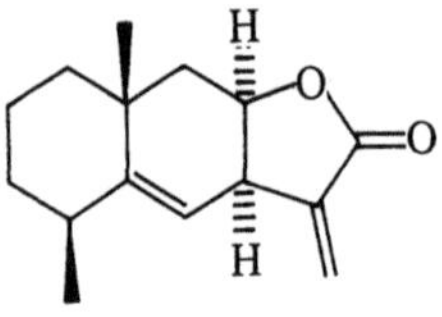

(67) Alantolactone

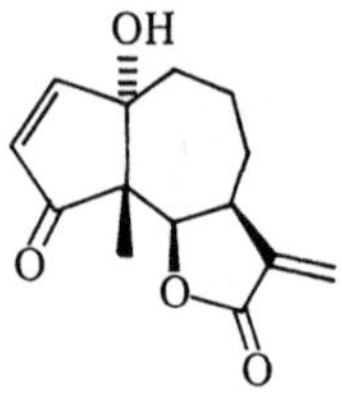

(68) Parthenin

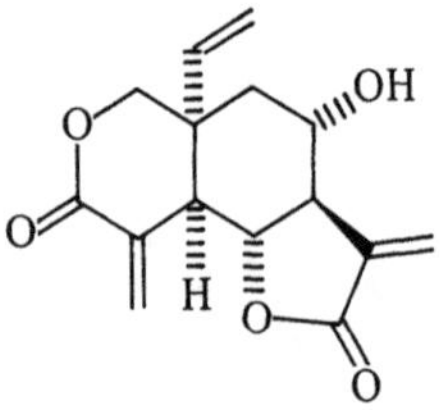

(69) Vernolepin

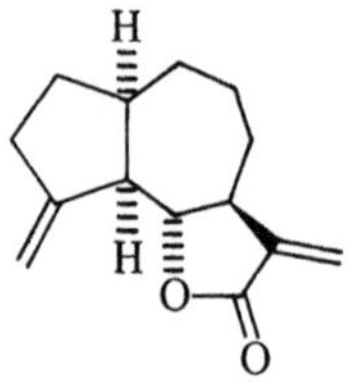

(70) Dehydrocostus lactone

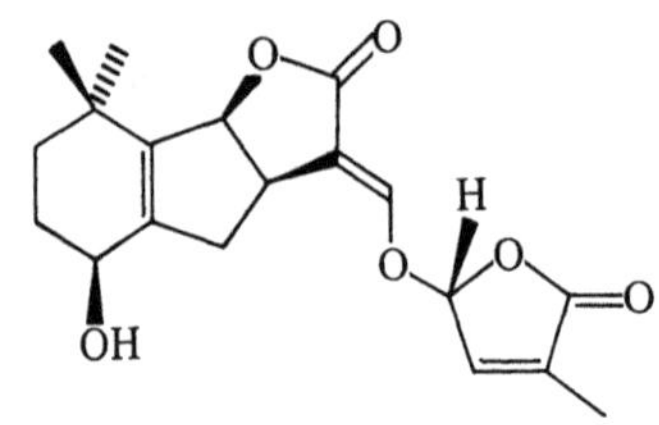

(71) Strigol

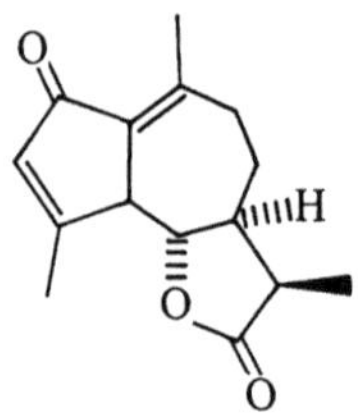

(72) Achillin

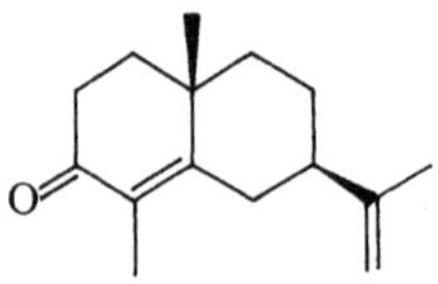

(73) α–Cyperone

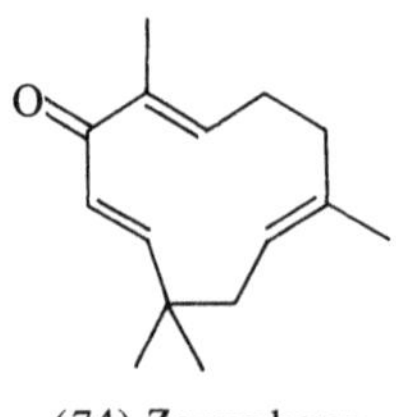

(74) Zerumbone

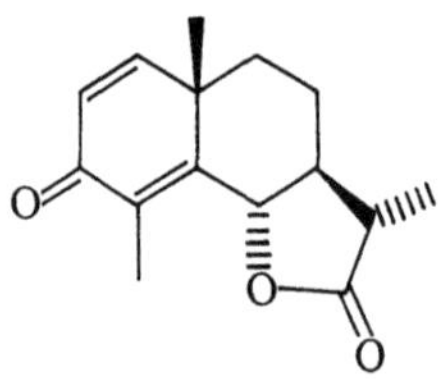

(75) Santonin

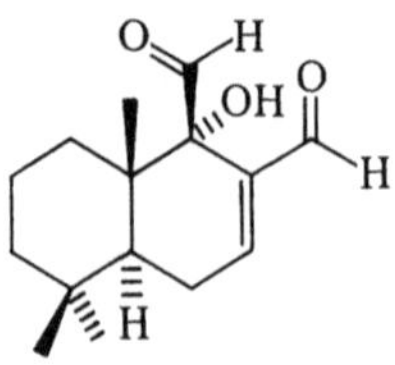

(76) Warburganal

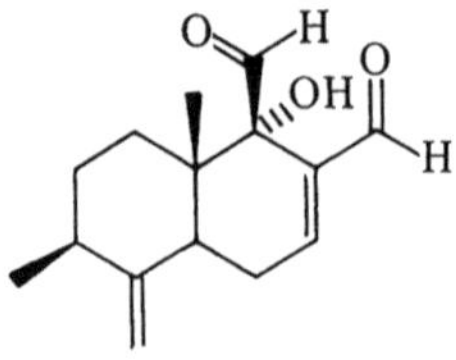

(77) Muzigadial

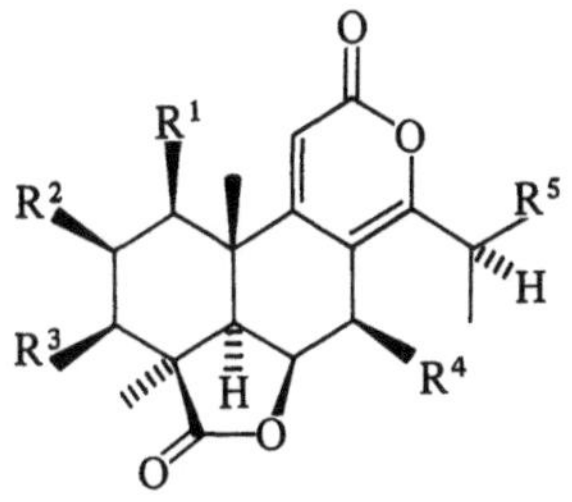

	R^1	R^2	R^3	R^4	R^5
(78) Nagilactone A	OH	H	H	OH	CH_3
(79) Nagilactone B	OH	OH	H	OH	CH_3
(80) Nagilactone C	–O–		OH	OH	CH_3
(81) Nagilactone D	–O–		OH	H	H

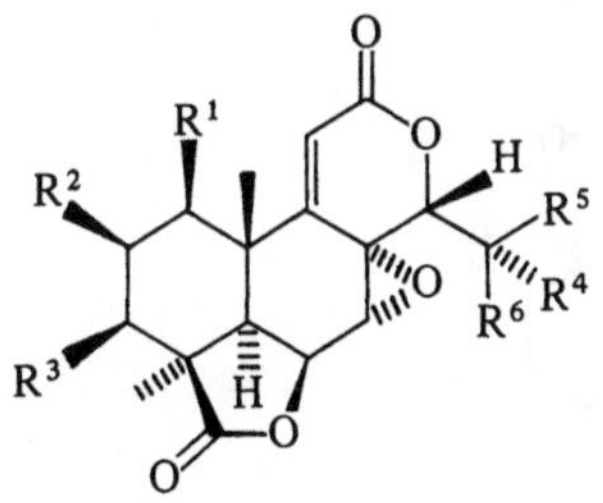

	R^1	R^2	R^3	R^4	R^5	R^6
(82) Nagilactone E	H	H	OH	H	CH_3	CH_3
(83) Inumakilactone A	–O–		OH	H	CH_3	OH
(84) Inumakilactone B	–O–		OH	$-CH_2-$		H
(85) Sellowin A	H	–O–		H	CH_3	CH_2OH
(86) Podolactone A	H	–O–		OH	CH_3	CH_2OH
(87) Podolactone B	–O–		OH	OH	CH_3	CH_2OH
(88) Podolactone C	H	–O–		OH	CH_3	CH_2SOCH_3
(89) Podolactone D	H	Δ		OH	CH_3	CH_2SOCH_3

Simple coumarins such as umbelliferone, aesculetin, daphnetin and scopoletin (Fig. 1.10) have long been known to be strong inhibitors of germination (MAYER and EVENARI, 1952; MAYER and POLJAKOFF-MAYBER, 1975; VAN SUMERE et al., 1972) and root growth (GOODWIN and TAVES, 1950; POLLOCK et al., 1954; AVERS and GOODWIN, 1956). More complex naturally occurring and inhibitory coumarins have been found, such as psoralen (3) from *Psoralea subauculis* (BASKIN et al., RODIGHIERO, 1954), angelicin (5) and xanthotoxin (4) from *Archangelica officinalis* (KOMISSARENKO et al., 1971; RODIGHIERO, 1954), seselin (6) from *Citrus* roots (GOREN and TOMER, 1971), and heraclenol (7) from parsley (KATO et al., 1978). Inhibitory effects of coumarins on organogenesis and callus growth have been reported (BAGNI and SERAFINI FRACASSINI, 1971). Under specific

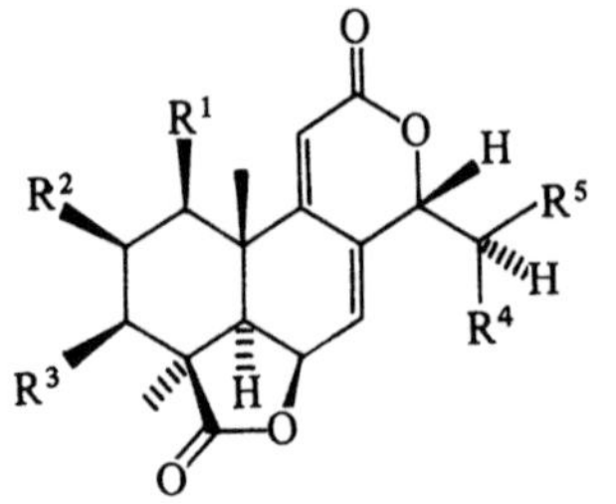

	R^1	R^2	R^3	R^4	R^5
(90) Nagilactone F	H	H	H	CH_3	CH_3
(91) Ponalactone A	–O–		OH	CH_3	CH_3
(92) Podolactone E	–O–		OH	$-CH_2-$	

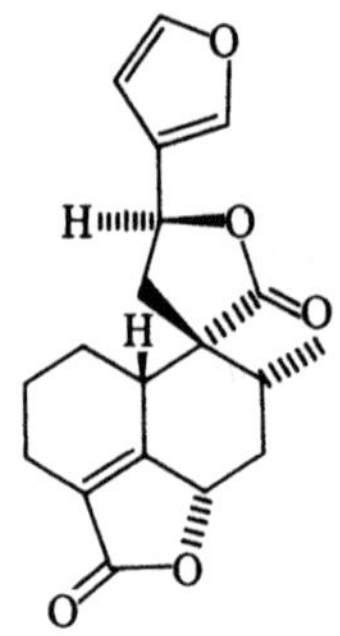

(93) Eugarzasadone (teucvin)

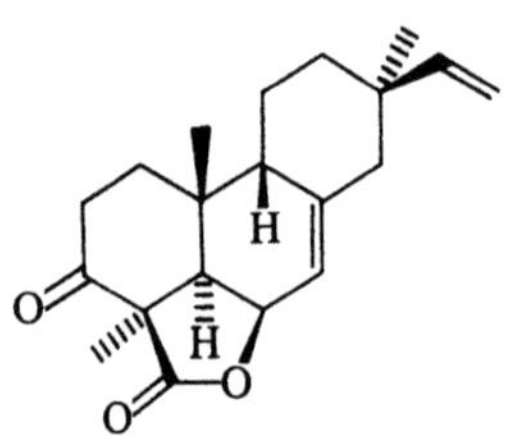

(94) Momilactone A

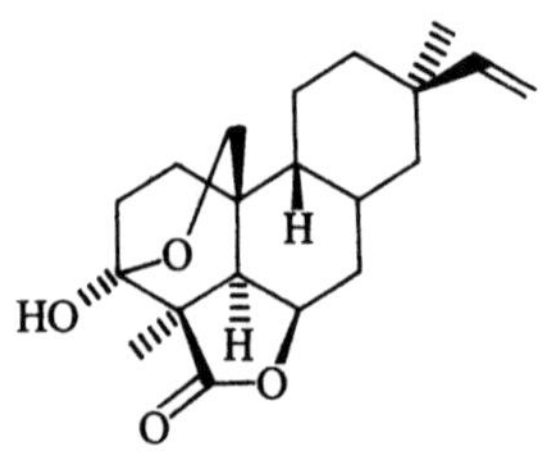

(95) Momilactone B

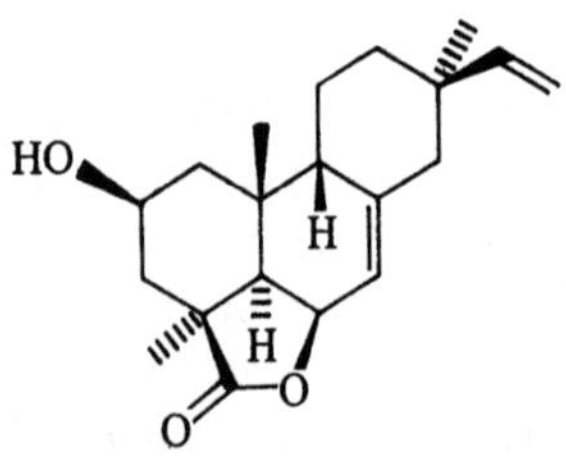

(96) Momilactone C

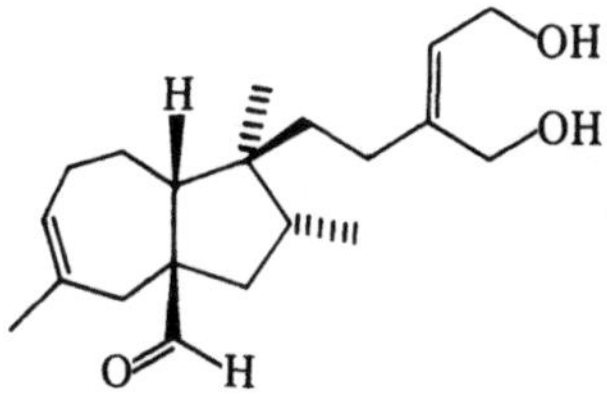

(97) Portulal

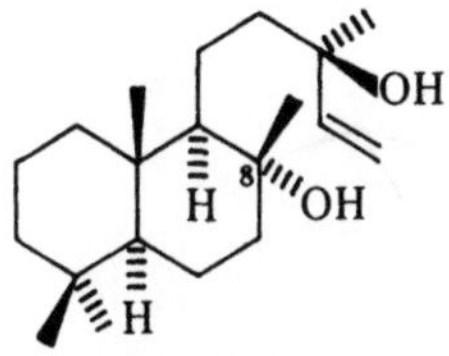

(98) Sclareol
(99) epi–Sclareol

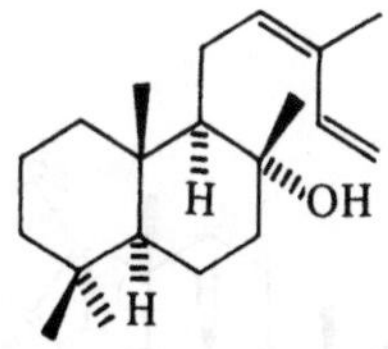

(100) cis–Abienol

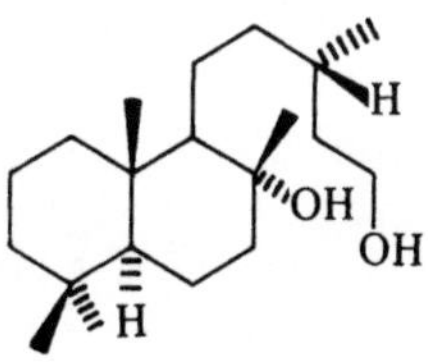

(101) Labdanediol

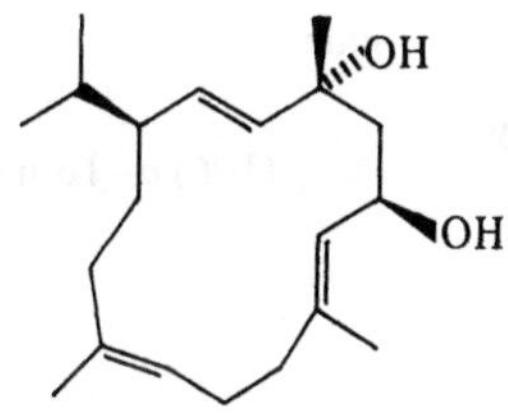

(102) 4,8,13–Duvatriene–1,3–diol

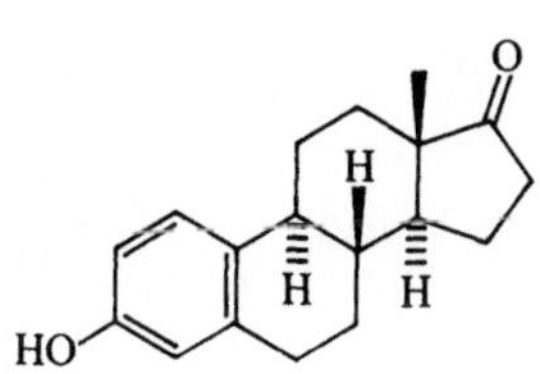

(103) Oestrone

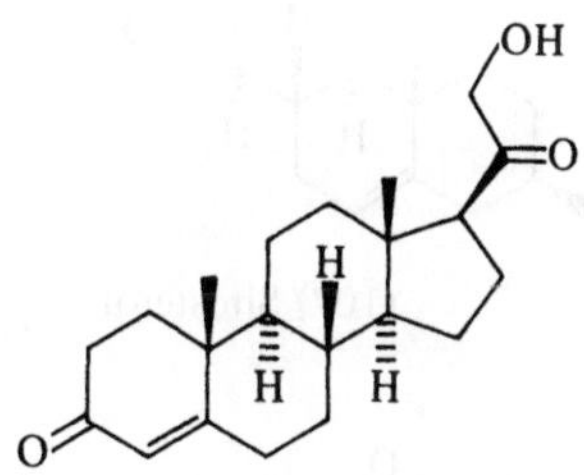

(104) 11–Desoxycorticosterone

conditions the *promotion* of root growth by coumarins has been observed (VASQUEZ, 1973). The growth-inhibitory effects of coumarins have been attributed to reaction with enzyme sulphydryl groups (THIMANN and BONNER, 1949) although this theory has been challenged (MAYER and EVENARI, 1952). The induction of light-sensitive dormancy in lettuce seeds by coumarins has been suggested as being due to anti-gibberellin properties (BERRIE et al., 1968). In this study the naturally occurring iso-coumarins ramulosin (8) and oosponol (9) were also found to be active. In contrast the isocoumarin hydrangeol (10) (isolated from the leaves of *Hydrangea macrophylla* and *H. hortensia*) enhanced the activity

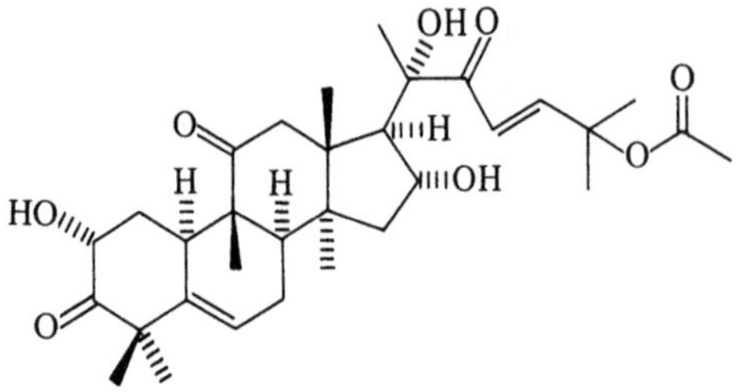

(105) Cucurbitacin B

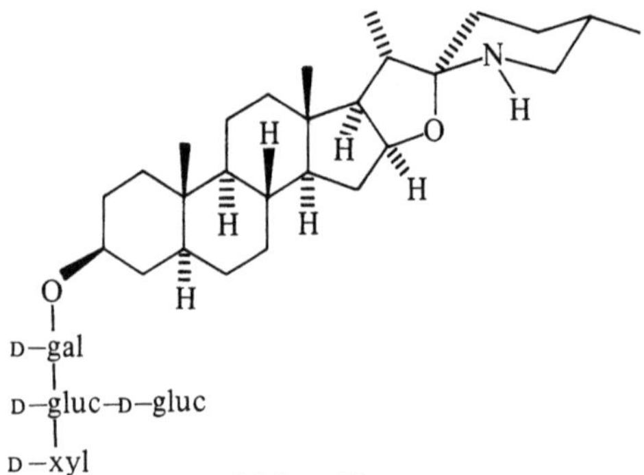

(106) α-Tomatine

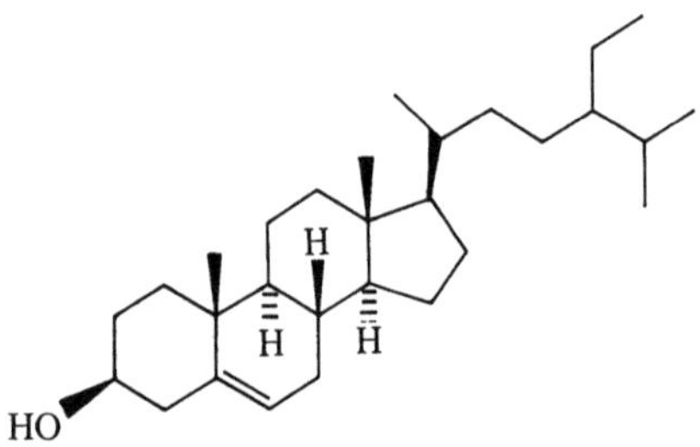

(107) Sitosterol

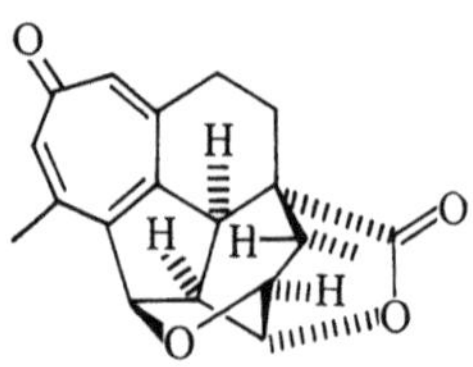

(108) Harringtonolide

of low concentrations of applied gibberellin (ASEN et al., 1960) in a growth assay.

Besides the coumarins there are other growth-inhibiting unsaturated lactones. Parasorbic acid (11) from *Sorbus aucuparia* and protoanemonin (12) from members of the Ranunculaceae have been shown (like coumarin) to inhibit

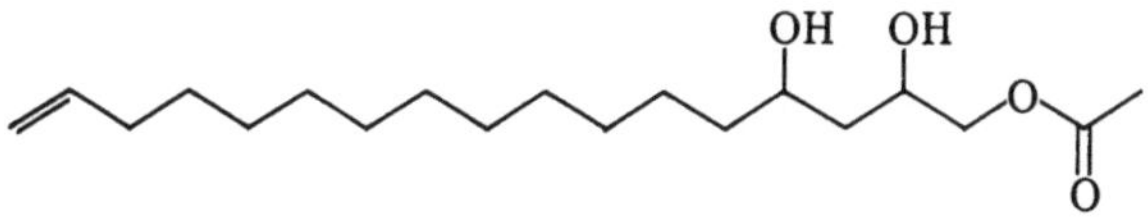

(109)

(110) Docosanol

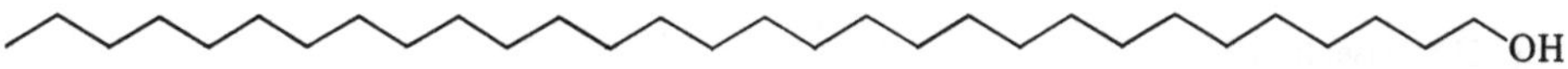

(111) Triacontanol

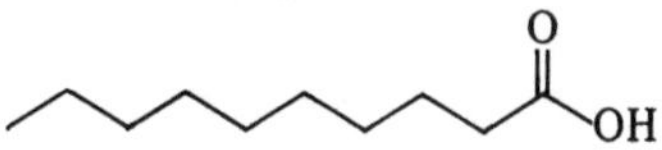

(112) Capric acid

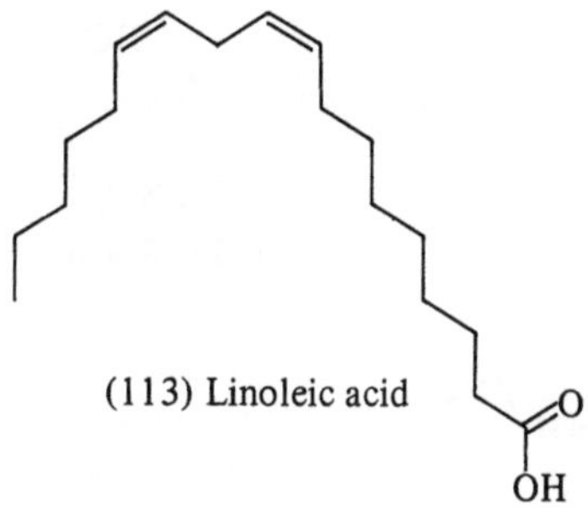

(113) Linoleic acid

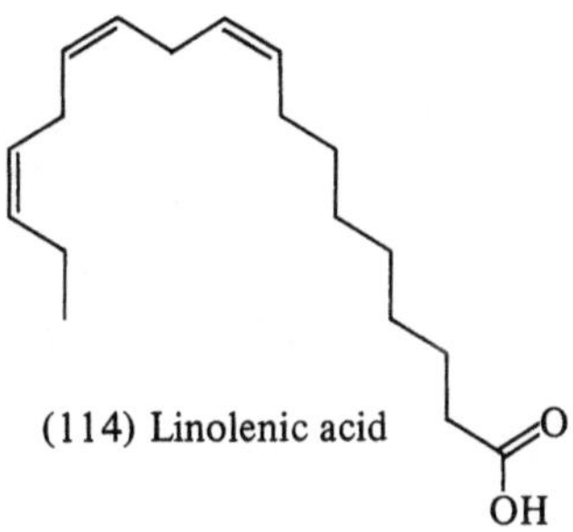

(114) Linolenic acid

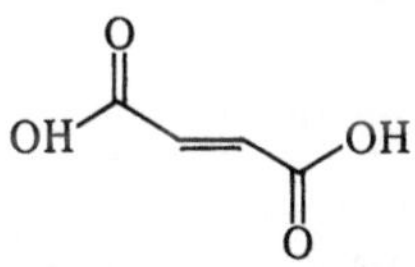

(115) Fumaric acid

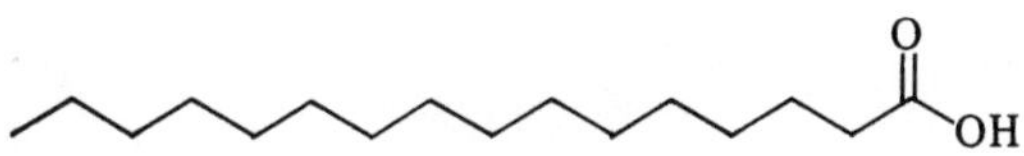

(116) Palmitic acid

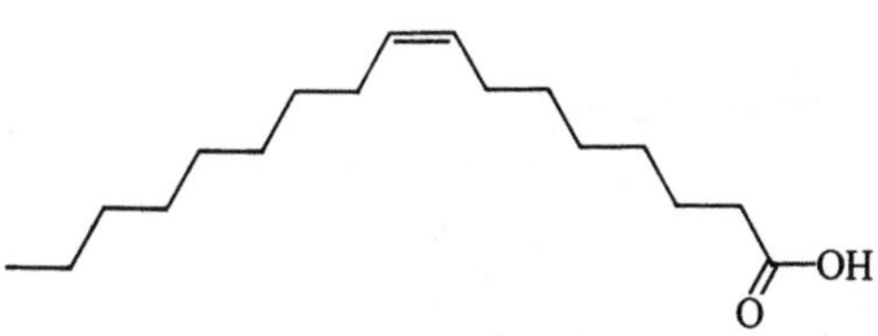

(117) Oleic acid

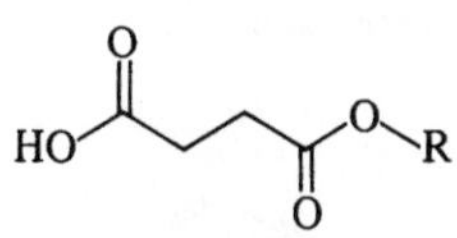

(118) $R{=}CH_3$

(119) $R{=}CH_2CH_3$

root growth and block mitosis (BUSTON and ROY, 1949; CORNMAN, 1946, 1947; ERICKSON and ROSEN, 1949). Psilotin (13), a recently discovered lactone from *Psilotum nudum* and *Tmesipteris tannensis*, has an activity similar to protoanemonin (SIEGEL, 1976). The γ-pyrone chelidonic acid (14) is a constituent of many plants (RAMSTAD, 1953; ATKINSON and ECKERMANN, 1965; BOUGH and GANDER,

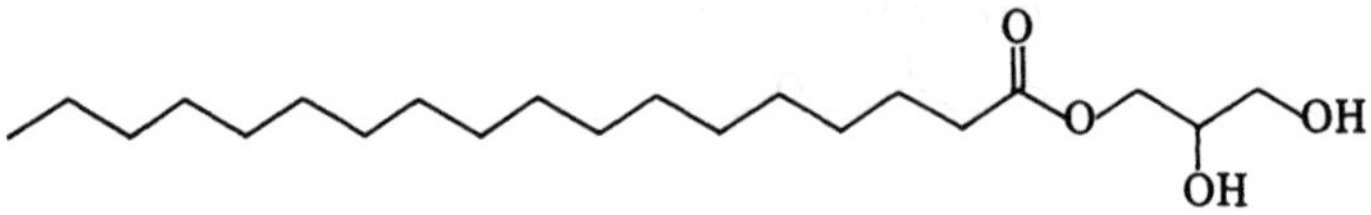

(120) α–Stearoyl glycerol

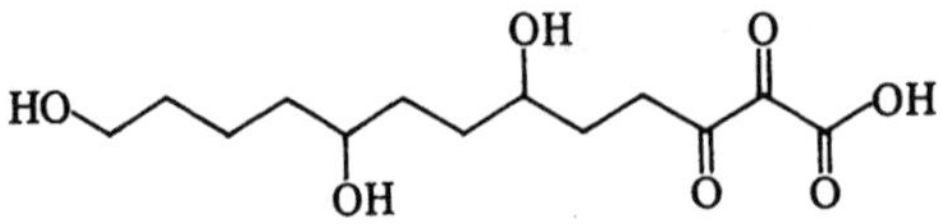

(121) Phaseolic acid

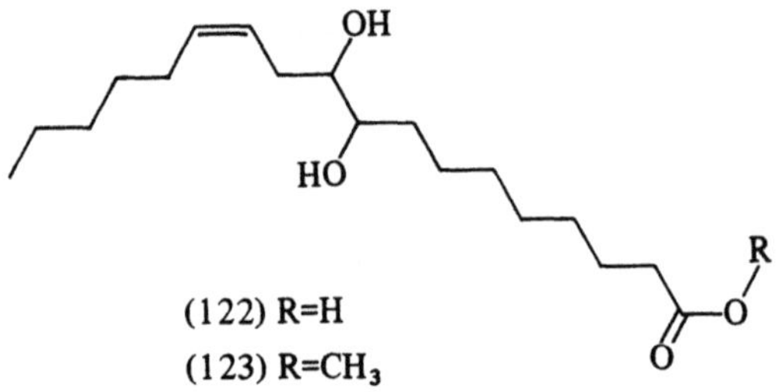

(122) R=H
(123) R=CH_3

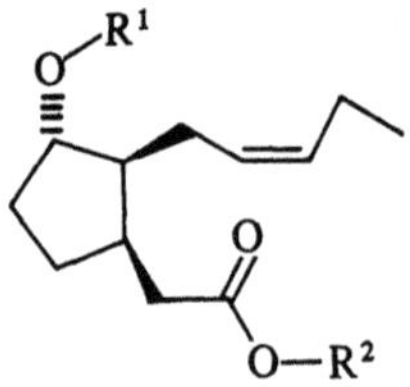

(124) Cucurbic acid (R^1=R^2=H)
(125) R^1=Glucosyl,R^2=H
(126) R^1=Glucosyl,R^2=CH^3

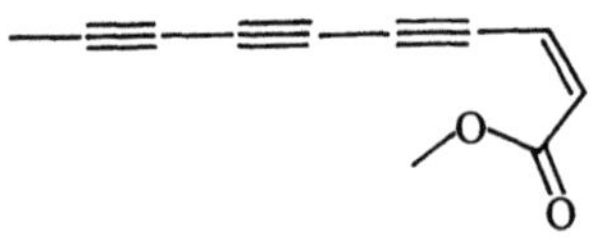

(127) Dehydromatricaria ester

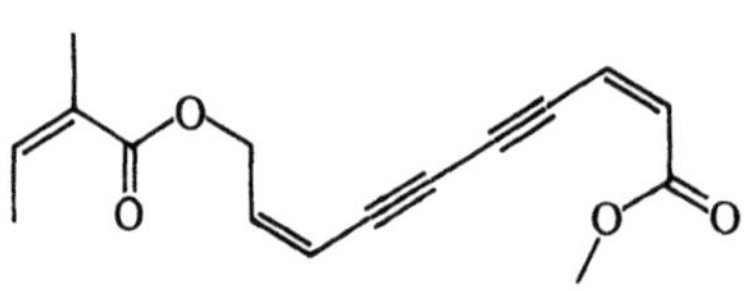

(128)

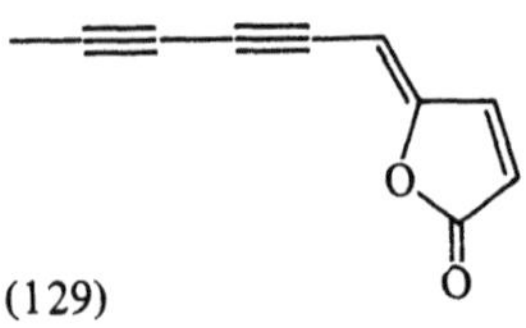

(129)
Dehydromatricaria lactone

1972). Leopold et al. (1952) have reported that it has a spectrum of activity very similar to coumarin and other unsaturated lactones. Notably the inhibitory effect could be alleviated by treatment with a sulphydryl reagent. The large number of terpenoid unsaturated lactones, which will be discussed later, share many of the biological properties of the lactones mentioned above. Two of

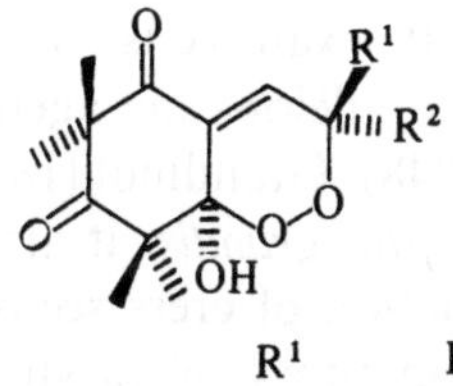

	R^1	R^2
(130) G_1	CH_3	CH_2CH_3
(131) G_2	CH_2CH_3	CH_3
(132) G_3	CH_3	CH_3

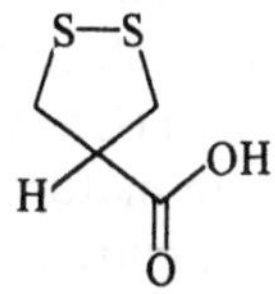

(133) Asparagusic acid

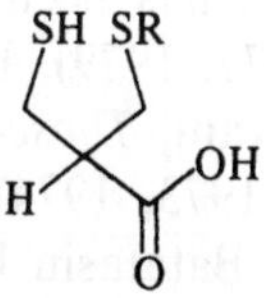

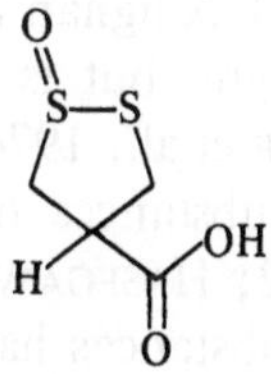

(134) Dihydroasparagusic acid (R=H) (136) Asparagusic acid S–oxide
(135) S–Acetyldihydroasparagusic acid ($R=CH_3CO$)

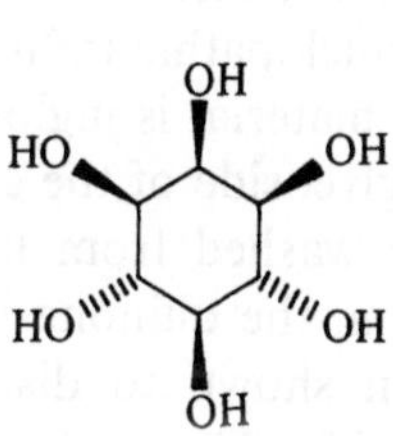

(137) myo–Inositol

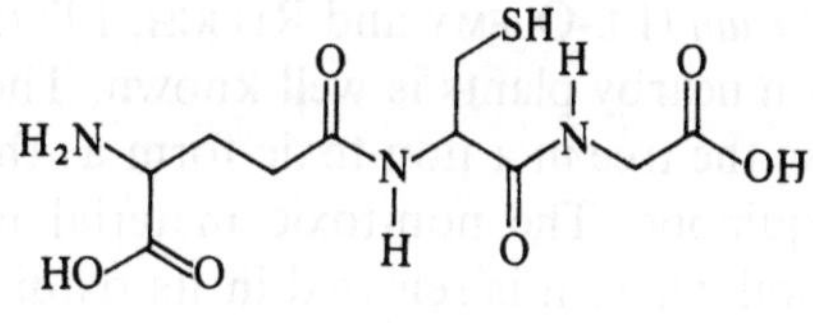

(138) Glutathione

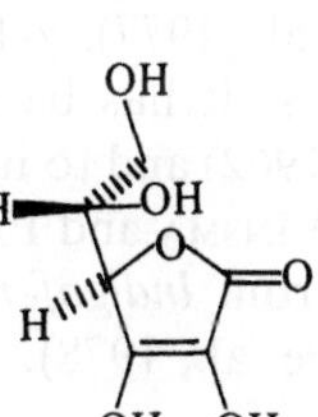

(139) L–Ascorbic acid

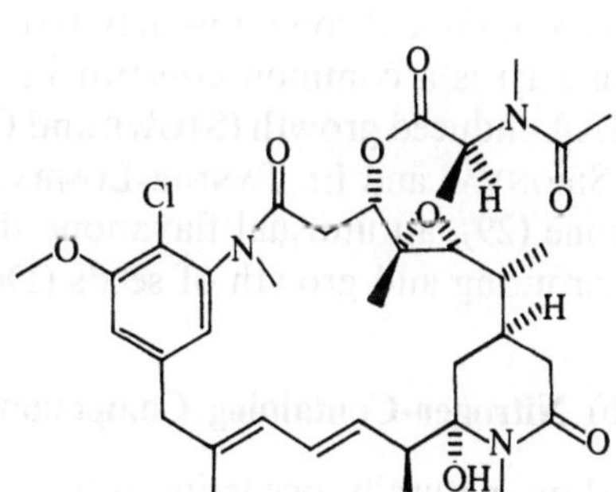

(140) Maytansine

the most active germination inhibitors present in seeds of many of the Umbelliferae (e.g., *Levisticum officinale*) have been identified as the lactones (15) and (16) (MOEWUS and SCHADER, 1951).

Many naturally occurring aromatic aldehydes such as benzaldehyde, salicylaldehyde, and vanillin (Fig. 1.10) have growth-inhibitory properties (MOJÉ, 1966;

VAN SUMERE et al., 1972). A more specific example is 3-acetyl-6-methoxybenzaldehyde (17) which is believed to be the allelopathic agent in the desert shrub *Encelia farinosa* (GRAY and BONNER, 1948). Grandinol (18) is a rooting inhibitor extracted from mature leaves of *Eucalyptus grandis*. It also inhibits the rooting of mung bean cuttings and the germination of cress seeds (CROW et al., 1977).

There are many other aromatic inhibitors of varying complexity. Tubers of *Solanum tuberosum* produce potent inhibitors of sprouting which have been characterized as 1,4- and 1,6-dimethylnaphthalene (19) and (20) (MEIGH et al., 1973). There are two lignans with germination-inhibiting properties. Acanthotoxin (21) from *Zanthoxylum acanthopodium* inhibits lettuce seed germination (ROY et al., 1977). A lignan from *Aegilops ovata* which inhibits germination of lettuce seeds in light, but is inactive in darkness, has been reassigned the structure (22) (LAVIE et al., 1974; COOPER et al., 1977, 1979). Batatasins are dormancy-inducing substances from bulbils of the yam, *Dioscorea batatas* (HASHIMOTO et al., 1972; HASEGAWA and HASHIMOTO, 1973, 1974). The structures of four of these substances have been established. Batatasin I has the phenanthrenoid structure (23) (LETCHER, 1973) and batatasins III, IV and V were shown to be the dihydrostilbenes (24), (25) and (26) (HASHIMOTO et al., 1974; HASHIMOTO and TAJIMA, 1978). Batatasin I (23) has also been isolated from *Dioscorea dumetorum* (EL-OLEMY and REISCH, 1979). The allelopathic influence of *Juglans nigra* on nearby plants is well known. The toxic material is juglone (27) which occurs in the tree in a non-toxic form as the 4-β-glycoside of the corresponding hydroquinone. The non-toxic material may be washed from living leaves into the soil where it is released in its oxidized form, the quinone juglone (WHITTAKER, 1970; THOMSON, 1971). Juglone has been shown to disrupt the metabolism in isolated corn mitochondria (KOEPPE, 1972). The seed germination stimulant of the parasite *Orobanche crenata* exuded from the roots of *Vicia faba* is believed to contain a chroman moiety (see DAVIES et al., 1977). α-Tocopherol (28) (Vitamin E) is a common constituent of seed oils. It has been shown to stimulate IAA-induced growth (STOWE and OBREITER, 1962) and to induce flower initiation (SIRONVAL and EL TANNIR-LOMBA, 1960; BRUINSMA and PATIL, 1963). Lousfieserone (29), an unusual flavanone derivative from *Indigofera suffruticosa* inhibits sprouting and growth of seeds (DOMINGUEZ et al., 1978).

b) Nitrogen-Containing Compounds

Many naturally occurring nitrogen-containing compounds affect plant growth. Hydrogen cyanide, a strong inhibitor of growth processes, can be liberated enzymatically from cyanogenic glucosides (CONN and BUTLER, 1969; JONES, 1972, SEIGLER, 1977). Thus amygdalin (30), of common occurrence in seeds of the Rosaceae, may be hydrolyzed via prunasin (31) and mandelonitrile (32) (which is also inhibitory) to hydrogen cyanide and benzaldehyde (JONES and ENZIE, 1961). Amygdalin is thought to be indirectly responsible for the "peach re-plant problem", as it also occurs in woody tissue of peach root (RICE, 1974). In a similar manner the mustard oil glycosides (glucosinolates) of the Cruciferae (KJAER, 1960) are hydrolyzed to the volatile mustard oils which may yield isothiocyanate. Allyl- and β-phenethylisothiocyanate, (33) and (34), are germination and growth inhibitors of this type (ETTLINGER and KJAER, 1968). Ammonia, which is a strong inhibitor of germination and growth, may be formed by bacterial degradation of nitrogenous compounds. This effect may account for the inhibition of germination of sugar beet seeds (REHM, 1953) although the tetrahydrophthalamide (35) has been isolated as a germination inhibitor from sugar beet fruit (MITCHELL and TOLBERT, 1968).

Many alkaloids have been shown to affect plant growth processes (ROBINSON, 1974; WALLER and NOWACKI, 1978). Perhaps the best known is colchicine (36) (DEWAR, 1945) which disrupts microtubule formation and halts mitosis (see FRAGATA, 1972)[4]. The diterpenoid alkaloid delcosine (37) from *Delphinium ajacis* (WALLER and BURSTROM, 1969) inhibited the growth of plant tissue and apparently interacts with gibberellins. Consideration of the similarity in structure of delcosine with the gibberellins led the authors to suggest that it may act as a competitive inhibitor of gibberellin action. The indole, gramine (38), isolated as an active secretion from the roots of *Hordeum sativum*, inhibited the germination of the weed *Stellaria media* (OVERLAND, 1966). It was suggested that such an allelopathic influence might account for the success of barley as a "smother crop". Growth-inhibitory extracts from the bulbs of *Lycoris radiata* were shown by OKAMOTO et al. (1968) to contain the active compounds lycoricidine (39) and lycoricidinol (40). The latter compound is probably identical with narciclasine from *Narcissus* species (PIOZZI et al., 1968). As well as exhibiting growth-inhibitory activity in *Avena* and rice seedling assays, these compounds also inhibited cell division in tobacco tissue culture. The related alkaloid lycorine (41) from various Amaryllidaceae (including *L. radiata*) also possesses inhibitory effects on plant growth (DELEO et al., 1973a, b). EVENARI (1949) has listed the following alkaloids as strong germination inhibitors: cocaine, berberine, codeine, physostigmine, caffeine, quinone, strychnine, cinchonin, cinchonidin and tropa acid. The benzoxazolinone (42) first isolated as an antifungal factor from maize and wheat plants (VIRTANEN et al., 1956) was later shown to be an inhibitor of wheat coleoptile elongation (WILKINS et al., 1974). More recently related compounds have been isolated from *Zea mays* and some of them have been shown to inhibit auxin-induced growth of oat coleoptile sections and inhibit auxin-receptor binding (VENIS and WATSON, 1978; WOODWARD et al., 1979). Camptothecin (43) was first isolated from the tree *Camptotheca acuminata* as a leukemia and tumour inhibitor (WALL et al., 1966). BUTA and WORLEY (1976) have shown that this alkaloid acts as a selective plant growth regulator. A 10^{-4} M solution caused 100% inhibition of tobacco growth, whereas beans and sorghum were unaffected at this concentration. 2-Hydroxylated cinchoninic acids show plant growth stimulatory activity. Zeanic acid (44), isolated from corn-steep liquor (MATSUSHIMA et al., 1973) was shown to promote the growth of dwarf rice and radish cotyledons (MATSUSHIMA and ARIMA, 1973). An isomer (45), isolated from rice bran, also had growth-promoting properties (SAHASHI, 1925, 1926, 1927). A plant growth regulator recently isolated from rice-hulls has been showed to be nicotinamide (46) (TAKEUCHI et al., 1975). It stimulated

4 Effects of this alkaloid on cell structures are treated in several chapters of Vol. 7 of this Encyclopedia: *Plant Movements* (HAUPT, W. and FEINLEIB, M.E., eds.) 1979, and also in a forthcoming volume on carbohydrates (TANNER, W. and LOEWUS, F.A., eds.)

the growth of dwarf rice in the presence of NH_4^+. The effects of nicotinamide on several crop plants have recently been reported (SANA et al., 1977; SANA and OTA, 1977).

There are over 200 uncommon non-protein amino acids which occur in plants (FOWDEN, 1970, 1974; BELL, 1976, 1980). Many of these compounds, which have a restricted distribution, are toxic and growth-inhibitory to microorganisms (FOWDEN et al., 1967). A number of these amino acids also inhibit the growth of plants. Azetidine-2-carboxylic acid (47), the lower homologue of proline, is found only in members of the Liliaceae. Pipecolic acid (48), the higher homologue of proline, has been isolated from various legumes. α-Methylene-cyclopropylglycine (49) has been characterized as a component of *Litchi chinensis*. β-Pyrazol-l-ylalanine (50) has been found in seed extracts of several of the Cucurbitaceae. Mimosine (51) is restricted to the genera *Mimosa* and *Leucaena*. These five amino acids all inhibit the growth of *Phaseolus aureus* seedlings (FOWDEN, 1963; FOWDEN and RICHMOND, 1963; SMITH and FOWDEN, 1966). Homoarginine (52) from *Lathyrus* species inhibits the growth of *Chlorella*. The arginine isostere canavanine (53), limited to the legume subfamily Papilionoideae, also has inhibitory properties (BONNER, 1949; STEWARD et al., 1958). Some of the natural protein amino acids are phytotoxic (MORELAND et al., 1966; WILSON and BELL, 1978). The inhibitory effect of non-protein amino acids is thought to be due to their inhibition of protein synthesis or to their incorporation into protein with consequent disruption of protein function. The amino acid N,N-dimethyl-L-tryptophan (54) has been isolated as an endogenous inhibitor from seeds of *Abrus precatorius* (MANDAVA et al., 1974). Betaine (55), a quaternary ammonium compound from mature sugar beet leaves inhibits the growth of leaf discs (WHEELER, 1963). The betaine trigonelline (56) which is widely distributed in nature (see WINDHOLZ, 1976) has been identified as the factor in cotyledons of *Pisum sativum* which blocks mitosis at the G2 stage (LYNN et al., 1978; EVANS et al., 1979).

In a study of the role of the calyx in the development of the fruits of persimmon (*Diospyros kaki*), two compounds were isolated which inhibited the growth of cytokinin-dependent tobacco callus. These compounds were identified as the porphyrin derivatives methyl phaeophorbide a (57) and b (58) (HAYASHI et al., 1977).

Several aliphatic diamines and polyamines which are of widespread occurrence in plants have been shown to stimulate cellular proliferation in explants of dormant tubers of *Helianthus tuberosus* (BERTOSSI et al., 1965; BAGNI, 1966).

c) Terpenoid Compounds

There are a large number of naturally occurring terpenoids from plants which affect plant growth. Many monoterpene-essential oils have growth-inhibitory properties (RICE, 1974). The allelopathic influence of Southern Californian chaparral on adjacent grassland has been investigated by MULLER (MULLER et al., 1964; MULLER, 1965, 1970). For instance the inhibition of a variety of annual species in the vicinity of *Salvia leucophylla* shrubs was attributed to the volatile terpene camphor (59) and cineole (60) contained in the leaves.

A number of α-methylene lactones isolated from various Compositae have plant growth-inhibiting properties, sometimes accompanied by stimulation of adventitious rooting. As in the case of the previously discussed unsaturated lactones, biological activity can sometimes be reduced by treatment with HS-containing compounds (RODRIGUEZ et al., 1976).

Examples of such lactones include:

1. Heliangine (61) from *Helianthus tuberosus* (SHIBAOKA, 1961; MORIMOTO et al., 1966; NISHIKAWA et al., 1966; SHIBAOKA et al., 1967a, b, c).
2. Pyrethrosin (62) from *Chrysanthemum cinerariaefolium* (SHIBAOKA et al., 1967b; IRIUCHIJIMA and TAMURA, 1970; GABE et al., 1971).
3. Chrysartemin A (63) from *Chrysanthemum morifolium* (OSAWA et al., 1971), *C. parthenium, Artemisia mexicana,* and *A. klotzchiana* (ROMO et al., 1970).
4. Chrysartemin B (64) from *Chrysanthemum morifolium* (ROMO et al., 1970; OSAWA et al., 1971, 1977) and *Handelia trichophylla* (TARASOV et al., 1976).
5. Chlorochrymonin (65) from *C. morifolium* (OSAWA et al., 1973).
6. Xanthinin (66) from *Xanthium pennsylvanicum (X. strumarium)* (GEISSMAN et al., 1954; DEUEL and GEISSMAN, 1957; GEISSMAN, 1962; WINTERS et al., 1969; KHAN, 1975) and *X. orientale* (DOMINGUEZ et al., 1971).
7. Alantolactone (67) from *Inula helenium* (MARSHALL and COHEN, 1964; DALVI et al., 1971).
8. Parthenin (68) from *Parthenium hysterophorus* (HERZ et al., 1962; KANCHAN, 1975; ROJAS GARCIDUENAS and DOMINGUEZ, 1976).
9. Vernolepin (69) from *Vernonia hymenolepis* (KUPCHAN et al., 1968; SEQUEIRA et al., 1968).
10. Dehydrocostus lactone (70) from *Saussurea lappa* (PAUL et al., 1960; MATHUR et al., 1965; MAÇAIRA et al., 1977; KALSI et al., 1977).

The biological activity of these compounds has stimulated investigations of the activities of model α-methylene-γ-butyrolactones (IINO et al., 1972) and α-methylene-δ-valerolactones (TANAKA et al., 1973). The α,β-unsaturated γ-lactone functionality is also present in the potent germination stimulator strigol (71) (COOK et al., 1966, 1972; see JOHNSON, 1978). This compound is exuded from the roots of *Gossypium hirsutum* and stimulates the germination of seeds of the parasitic weed *Striga lutea*. In vitro 10^{-11} M solutions of strigol cause 50% stimulation of germination. Several sesquiterpene lactones without α,β-unsaturation have been shown to possess inhibitory activity. Absinthin, a constituent of the shrub *Artemisia absinthium* is such a compound (VOKÁČ et al., 1968). This substance has growth- and germination-inhibitory properties (SCHWAER, 1962) and is thought to contribute to the allelopathic influence of the plant (FUNKE, 1943; BODE, 1964). Achilline (72) from *Achillea* and *Artemisia* species (MARX and WHITE, 1969; DOMINGUEZ and CARDENAS, 1975) inhibits radicle development in germinating seeds (ROJAS GARCIDUENAS and DOMINGUEZ, 1976). A number of sesquiterpenoid ketones exhibit plant growth-regulatory behaviour. α-Cyperone (73) from *Cyperus rotundus* (MCQUILLIN, 1955; KOMAI et al., 1977) is inhibitory in various bioassays. Some cross-conjugated ketons also show biological activity. Zerumbone (74) from *Zingiber zerumbet* (DEV,

1960) and santonin (75) from *Artemisia* species (see SIMONSEN and BARTON, 1952; NAKAZAKI and ARAKAWA, 1964) promote rooting in mung bean cuttings (KALSI et al., 1978). Other terpenoid α,β-unsaturated ketones also promote rooting (KALSI et al., 1979). The East African tree *Warburgia ugandensis* contains two dialdehydes, warburganal (76) and muzigadial (77), which exhibit various biological activities including plant growth-regulating activity (KUBO et al., 1977).

As with the sesquiterpenoids, the majority of diterpenoid plant growth-effecting compounds are unsaturated lactones. Nor- and bisnor-diterpene dilactones are found in a variety of *Podocarpus* species (BROWN and SANCHEZ, 1974; ITO and KODAMA, 1976). About 30 of these compounds are known, of which at least 15 strongly inhibit the expansion and mitosis of plant cells. Structurally the lactones fall into three groups, those with an α-pyrone ring (78)–(81), those with a dihydro-α-pyrone ring (82)–(89), and those with a dihydro-α-pyrone ring with heteroannularly extended unsaturation (90)–(92). It should be noted that the structures of many of these compounds have been revised (as illustrated) in the light of recent X-ray analyses (POPPLETON, 1975; GODFREY and WATERS, 1975; ARORA et al., 1976). These changes affect the stereochemistry and/or position of epoxide groups in ring A. The biological properties of these compounds have been systematically evaluated (GALBRAITH et al., 1972; HAYASHI et al., 1972; HAYASHI and SAKAN, 1974). Generally the less polar compounds have the least bioactivity. Thus nagilactones D and F, (81) and (90), inumakilactone B (84) and podolactone E (92) are the most active compounds. Podolactone E (92) causes 50% growth inhibition in *Avena* coleoptiles at 10^{-7} M. At very low concentration (10^{-7}–10^{-8} M) however, the α-pyrones (78)–(81) show growth-*promoting* activity (HAYASHI and SAKAN, 1974). The norditerpene dilactone eugarzasadine (93) from *Teucrium cubense* is also inhibitory to plant growth (DOMINGUEZ et al., 1974a, b; ROJAS GARCIDUENAS and DOMINGUEZ, 1976). Eurgarzasadine is identical to teucvin (FUJITA et al., 1974). The diterpene lactones momilactones A (94), B (95) and C (96) have been isolated from rice husks (KATO et al., 1973; TSUNAKAWA et al., 1976). They inhibit the growth of rice roots at <100 ppm (KATO et al., 1977).

A few non-lactone diterpenoid inhibitors are known. Portulal (97) from *Portulaca grandiflora* has a similar activity to heliangine (61) (YAMAZAKI et al., 1969; MITSUHASHI and SHIBAOKA, 1965). Immature tobacco leaves have yielded a number of compounds which inhibit the growth of wheat coleoptiles including sclareol (98), episclareol (99), cis-abienol (100), labdanediol (101) and 4,8,13-duvatriene-1,3-diol (102) (SPRINGER et al., 1975; CUTLER et al., 1977).

Steroid and triterpenoid plant products which affect plant growth will not be considered here as they are the subject of a recent review by GEUNS (1978) (but see also VAN ROMPUY and ZEEVAART, 1979). Examples of this type include oestrone (103), 11-desoxycorticosterone (104), cucurbitacin B (105), α-tomatine (106), and sitosterol (107).

The lactone harringtonolide (108) may be of terpenoid origin. This compound, isolated from *Cephalotaxus harringtonia*, inhibits bud growth of decapitated tobacco and bean plants (BUTA et al., 1978).

d) Aliphatic Compounds

Many naturally occurring aliphatic alcohols, aldehydes, ketones, acids, and esters have been shown to affect plant growth. A variety of lipids ("oleanimins") have been shown by STOWE and co-workers to promote the elongation of pea sections in the presence of auxins (see LETHAM, 1978b). The inhibitory effect of long-chain alcohols, aldehydes, and ketones on lettuce seed germination has been shown to increase with increasing lipophilicity (REYNOLDS, 1977). Inhibition of seed germination by low molecular weight monohydric alcohols has also been observed by THIESS and LICHTENTHALER (1973). Fatty alcohols of chain lengths C_9, C_{10}, and C_{11} show strong activity in the inhibition of axillary and terminal bud growth (STEFFENS et al., 1967). The diol (109) isolated from avocado mesocarp inhibited the elongation of wheat coleoptiles and the growth of soybean callus in the presence of cytokinins (BITTNER et al., 1971). Growth-promoting effects have been reported for two aliphatic alcohols. An "auxin" from Maryland Mammoth tobacco was identified as docosanol (110) (CROSBY and VLITOS, 1961). Triacontanol (111) caused an increase in the growth and yields of corn, barley and rice (RIES et al., 1977, 1978; RIES and WERT, 1977). Fatty acids of various chain lengths inhibit seed germination (LE POIDEVIN, 1965; SCHUMAN and MCCALLA, 1976; BERRIE et al., 1976; WANNER et al., 1977). Short-chain fatty acids also inhibit gibberellin-induced amylolysis of barely endosperm (BULLER et al., 1976). They may also have an antitranspirant function in some plants (WILLMER et al., 1978). Capric acid (112), an inhibitor in the *Avena* coleoptile bioassay, has been isolated from dormant *Iris hollandica* bulbs (ANDO and TSUKAMOTO, 1974). Linoleic acid and linolenic acid (113) and (114), from etiolated leaves of *Hordeum vulgare* have been shown to inhibit the unrolling of these leaves, but they are not thought to have this function in vivo (MEINHENETT and CARR, 1973). A number of acidic growth inhibitors have been obtained from dwarf peas (KOMOTO et al., 1972a). These compounds include fumaric (115), palmitic (116), oleic (117) and abscisic acids and methyl and ethyl hydrogen succinates (118) and (119). Neutral inhibitors from the same source included sitosterol (107), pisatin (1), and α-stearoyl glycerol (120) (KOMOTO et al., 1972b). All these inhibitors interfered with the response of pea stems to GA_3. Phaseolic acid from *Phaseolus vulgaris* has been assigned the structure (121). It has been reported to have gibberellin-like activity in dwarf maize and barley endosperm bioassays (REDEMAN et al., 1968). Several inhibitors of dwarf rice growth have been isolated from *Cucurbita pepo* (FUKUI et al., 1977a, c). Among these were 9,10-dihydroxy-12-octadecenoic acid (122) and its methyl ester (123) cucurbic acid (124), its glucosyl ether (125), and its methyl ester glucosyl ether (126).

Three polyacetylenic esters have been isolated from subterranean stems of *Solidago altissima*. These compounds strongly inhibit the growth of seedlings of barnyard millet (*Panicum crus-galli* L. var *frumentaceum* TRIN) and may have an allelopathic function. The three compounds have been characterized as dehydromatricaria ester (127), the angelate (128), and dehydromatricaria lactone (129) (ICHIHARA et al., 1976, 1978; KOBAYASHI et al., 1976).

e) Other Compounds

Extracts of mature leaves of *Eucalyptus grandis* have yielded three rooting inhibitors besides the previously mentioned grandinol (18). These compounds have been identified as G_1 (130), G_2 (131) and G_3 (132) (CROW et al., 1971, 1976; STERNS, 1971). A series of inhibitors have been extracted from etiolated shoots of *Asparagus officinale*. These compounds have been identified as asparagusic acid (133), dihydroasparagusic acid (134), S-acetyldihydroasparagusic acid (135) (YANAGAWA et al., 1972; KITAHARA et al., 1972) and the syn and anti isomers of asparagusic acid-S-oxide (136) (YANAGAWA et al., 1973). These compounds show inhibitory effects on lettuce germination and the growth of lettuce and other seedlings. Dihydroasparagusic acid (134) promotes the rooting of mung bean cuttings (KUHNLE et al., 1975). The widely distributed hexitols, and in particular myo-inositol (137), have been known as growth factors for some time (POLLARD et al., 1961). myo-Inositol is an important component of cell walls but its synergistic growth-promoting activity may be associated with its previously mentioned complexes with IAA (STEWARD and KRIKORIAN, 1971). The peptide glutathione (138) which is present in plant cells inhibits the growth of soybean callus tissue (BERGMANN and RENNENBERG, 1975). Ascorbic acid (139) has an inhibitory action on plant growth and it has even been suggested as a hormone (TONZIG and MARRÈ, 1961). The antileukaemic macrolide maytansine (140) from *Maytenus ovatus* (KUPCHAN et al., 1972) was shown to inhibit the growth of tobacco callus and rice seedlings yet to promote growth in an *Avena* assay (KOMADA and ISOGAI, 1978).

Acknowledgements. I wish to thank Professor J. MACMILLAN F.R.S. for helpful discussions throughout the preparation of the manuscript and Professor B.O. PHINNEY for his critical comments on the final draft. Some useful material for this chapter was kindly provided by Professor N. TAKAHASHI. Finally, I am particularly grateful to Miss M. PANES who typed the whole manuscript.

References

Abe, H., Uchiyama, M., Sato, R.: Isolation and identification of native auxins in marine algae. Agric. Biol. Chem. *36*, 2259–2260 (1972)

Abe, H., Uchiyama, M., Sato, R.: Isolation of phenylacetic acid and its *p*-hydroxy derivative as auxin substances from *Undaria pinnatifida*. Agric. Biol. Chem. *38*, 897–898 (1974)

Abeles, F.B.: Ethylene in Plant Biology. New York, London: Academic Press 1973

Abeles, F.B., Rubinstein, B.: Regulation of ethylene evolution and leaf abscission by auxin. Plant Physiol. *39*, 963–969 (1964)

Addicott, F.T., Lyon, J.L.: Physiology of abscisic acid and related substances. Annu. Rev. Plant Physiol. *20*, 139–164 (1969)

Addicott, F.T., Carns, H.R., Cornforth, J.W., Lyon, J.L., Milborrow, B.V., Ohkuma, K., Ryback, G., Smith, O.E., Thiessen, W.E., Wareing, P.F.: Abscisic acid: a proposal for the designation of abscisin II (dormin). In: Biochemistry and physiology of plant growth substances. Wightman, F., Setterfield, G. (eds.), pp. 1627–1529. Ottawa: The Runge Press 1968

Adesmoju, A.A.: An investigation of some hormonal bases for abscission in cowpea (*Vigna unguiculata* L. Walp.). Ph.D. thesis. University of Ibadan, Nigeria (1977)

Adesomoju, A.A., Okugun, J.I., Ekong, D.E.U., Gaskin, P.: GC-MS identification of abscisic acid and abscisic acid metabolites in seed of *Vigna unguiculata*. Phytochemistry *19*, 223–225 (1980)

Akamine, E.K.: Ethylene production in fading *Vanda* orchid blossoms. Science *140*, 1217–1218 (1963)

Allen, J.R.F., Greenway, A.M., Baker, D.A.: Determination of indole-3-acetic acid in xylem sap of *Ricinus communis* L. using mass fragmentography. Planta *144*, 299–303 (1979)

Altree-Williams, S., Howden, M.E.H., Keegan, J.T., Malcolm, H.D.R., Wyllie, S.G.: Acidic growth inhibitors from peach buds. Aust. J. Plant Physiol. *2*, 105–109 (1975)

Alves, L.M., Ruddat, M.: The presence of gibberellin A_{20} in *Stevia rebaudiana* and its significance for the biological activity of steviol. Plant Cell Physiol. *20*, 123–130 (1979)

Andersson, B., Haggstrom, N., Andersson, K.: Identification of abscisic acid in shoots of *Picea abies* and *Pinus sylvestris* by combined gas chromatography-mass spectrometry. A versatile method for clean-up and quantification. J. Chromatogr. *157*, 303–310 (1978)

Ando, T., Tsukamoto, Y.: Capric acid: a growth inhibiting substance from dormant *Iris hollandica* bulbs. Phytochemistry *13*, 1031–1032 (1974)

Anstis, P.J.P., Friend, J., Gardner, D.C.J.: The role of xanthoxin in the inhibition of pea seedling growth by red light. Phytochemistry *14*, 31–35 (1975)

Armstrong, D.J., Evans, P.K, Burrows, W.J., Skoog, F., Petit, J.-F., Dahl, J.L., Steward, T., Stominger, J.L., Leonard, N.J., Hecht, S.M., Occolowitz, J.: Cytokinins. Activity and identification in *Staphylococcus epidermidis* transfer ribonucleic acid. J. Biol. Chem. *245*, 2922–2926 (1970)

Armstrong, D.J., Scarbrough, E., Skoog, F., Cole, D.L., Leonard, N.J.: Cytokinins in *Corynebacterium fascians* cultures. Isolation and identification of 6-(-4-hydroxy-3-methyl-*cis*-2-butenylamino)-2-methylthiopurine. Plant Physiol. *58*, 749–752 (1976)

Arora, S.K., Bates, R.B., Chou, P.C.C., Sanchez, W.E., Brown, K.S., Galbraith, M.N.: Structures of norditerpene lactones from *Podocarpus* species. J. Org. Chem. *41*, 2458–2461 (1976)

Asen, S., Cathey, H.M., Stuart, N.W.: Enhancement of gibberellin growth-promoting activity by hydrangeol isolated from leaves of *Hydrangea macrophylla*. Plant Physiol. *35*, 816–819 (1960)

Assante, G., Merlini, L., Nasini, G.: (+)-Abscisic acid, a metabolite of the fungus *Cercospora rosicola*. Experientia *33*, 1556–1557 (1977)

Atkinson, M.R., Eckermann, G.: Identification of Cherry and Hageman's "compound X1" from maize as chelidonic acid. Aust. J. Biol. Sci. *18*, 437–439 (1965)

Augier, H.: Les hormones des algues. État actuel des connaissances. VII – Applications, conclusion, bibliographie. Bot. Mar. *21*, 175–197 (1978)

Avers, C.J., Goodwin, R.H.: Studies on roots IV. Effects of coumarin and scopoletin on the standard root growth pattern of *Phleum pratense*. Am. J. Bot. *43*, 612–620 (1956)

Babcock, D.F., Morris, R.O.: Quantitative measurements of isoprenoid nucleosides in *t*-RNA. Biochemistry *9*, 3701–3705 (1970)

Bagni, N.: Aliphatic amines and a growth-factor of coconut milk as stimulating cellular proliferation of *Helianthus tuberosus* (Jerusalem artichoke) in vitro. Experientia *22*, 737–733 (1966)

Bagni, N., Serafini Fracassini, D.: Effect of coumarin derivatives on organo-genesis and callus growth in *Cichorium intybus* roots and *Helianthus tuberosus* tubers in vitro. Experientia *27*, 1239–1241 (1971)

Balboa-Zavala, O., Dennis, F.G.: Abscisic acid and apple seed dormancy. J. Am. Soc. Hortic. Sci. *102*, 633–637 (1977)

Bandurski, R.S., Schulze, A.: Concentration of indole-3-acetic acid and its esters in *Avena* and *Zea*. Plant Physiol. *54*, 257–262 (1974)

Bandurski, R.S., Schulze, A.: Concentration of indole-3-acetic acid and its derivatives in plants. Plant Physiol. *60*, 211–213 (1977)

Bandurski, R.S., Schulze, A., Cohen, J.D.: Photo-regulation of the ratio of ester to free indole-3-acetic acid. Biochem. Biophys. Res. Commun. *79*, 1219–1223 (1977)

Baskin, J.M., Ludlow, C.J., Harris, T.M., Wolf, F.T.: Psoralen, an inhibitor in the seeds of *Psoralea subacaulis*. Phytochemistry *6*, 1209–1213 (1967)

Bearder, J.R., MacMillan, J.: Gibberellins A_{16}, A_{36}, A_{37}, A_{41} and A_{42} from *Gibberella fujikuroi*. J. Chem. Soc. Perkin *1*, 2824–2830 (1973)
Bearder, J.R., MacMillan, J., Wels, C.M., Phinney, B.O.: The metabolism of steviol to 13-hydroxylated *ent*-gibberellanes and *ent*-kauranes. Phytochemistry *14*, 1741–1748 (1975)
Beauchesne, G., Goutarel, R.: Activité de certaines purines substituée sur le développement des cultures de tissus de moelle de Tabac en présence d'acide indoléacétique. Physiol. Plant. *16*, 630–631 (1963)
Becker, H., Kempf, T.: Untersuchung der Grandiflorensäure (Kaura Δ9 (11), 16 dien-18-Carbonsäure) auf Gibberellin-Aktivität. Z. Pflanzenphysiol. *80*, 87–91 (1976)
Beeley, L.J., MacMillan, J.: Partial synthesis of 2-hydroxy gibberellins; characterisation of two new gibberellins, A_{46} and A_{47}. J. Chem. Soc. Perkin *1*, 1022–1028 (1976)
Beeley, L.J., Gaskin, P., MacMillan, J.: Gibberellin A_{43}, and other terpenes in endosperm of *Echinocystis macrocarpa*. Phytochemistry *14*, 779–783 (1975)
Bell, E.A.: 'Uncommon' amino acids in plants. FEBS Lett. *64*, 29–35 (1976)
Bell, E.A.: Non-protein amino acids. In: Encyclopedia of plant physiology, New Series. Bell, E.A., Charlwood, B.V. (eds.) Vol. 8, pp. 403–432 (1980)
Bennet-Clark, T.A., Kefford, N.P.: Chromatography of the growth substances in plant extracts. Nature (London) *171*, 645–647 (1953)
Bentley, J.A.: The naturally-occurring auxins and inhibitors. Annu. Rev. Plant Physiol. *9*, 47–80 (1958)
Bentley, J.A.: Chemistry of the native auxins. In: Encyclopedia of Plant Physiology. Ruhland, W. (ed.), Vol. XIV, pp. 485–500. Berlin-Göttingen-Heidelberg: Springer 1961
Bercht, C.A., Samrak, H.M., Lousberg, R.J.J.Ch., Theuns, H., Salemink, C.A.: Isolation of vomifoliol and dihydrovomifoliol from *Cannabis*. Phytochemistry *15*, 830–831 (1976)
Berger, J., Avery, G.S.: Isolation of an auxin precursor and an auxin (indoleacetic acid) from maize. Am. J. Bot. *31*, 199–203 (1944)
Bergmann, L., Rennenberg, H.: Growth inhibition of soybean callus tissues by glutathione. Naturwissenschaften *62*, 349–350 (1975)
Berrie, A.M.M., Robertson, J.: ABA as an endogenous component in lettuce fruits *Lactuca sativa* L. cv. Grand Rapids. Does it control thermodormancy? Planta *131*, 211–215 (1976)
Berrie, A.M.M., Parker, W., Knights, B.A., Hendrie, M.R.: Studies on lettuce germination. I. Coumarin induced dormancy. Phytochemistry *7*, 567–573 (1968)
Berrie, A.M.M., Don, R., Buller, D., Alam, M., Parker, W.: The occurrence and function of short chain fatty acids in plants. Plant Sci. Lett. *6*, 163–173 (1976)
Bertossi, F., Bagni, N., Moruzzi, G., Caldarera, C.M.: Spermine as a new growth-promoting substance for *Helianthus tuberosus* (Jerusalem artichoke) in vitro. Experientia *21*, 80–81 (1965)
Beutelmann, P., Bauer, L.: Purification and identification of a cytokinin from moss callus cells. Planta *133*, 215–217 (1977)
Bhakuni, D.S., Joshi, P.P., Uprety, H., Kapil, R.S.: Roseoside a C_{13} glucoside from *Vinca rosea*. Phytochemistry *13*, 2541–2543 (1974)
Biale, J.B.: Effect of emanations from several species of fungi on respiration and color development of citrus fruits. Science *91*, 458–459 (1940)
Biale, J.B.: Postharvest physiology and biochemistry of fruits. Annu. Rev. Plant Physiol. *1*, 183–206 (1950)
Bialek, K., Bielinska-Czarnecka, M., Gaskin, P., MacMillan, J.: The levels of ABA in inhibitor-β complex from potato tubers. Bull. Acad. Pol. Sci. Biol. *21*, 781–784 (1973)
Biemann, K., Tsunakawa, S., Sonnenbichler, J., Feldman, H., Dütting, D., Zachau, H.G.: The structure of an odd nucleoside from serine-specific transfer RNA. Angew. Chem. Int. Ed. Engl. *5*, 590–591 (1966)
Bittner, S., Gazit, S., Blumenfeld, A.: Isolation and identification of a growth inhibitor from Avocado. Phytochemistry *10*, 1417–1421 (1971)
Bode, H.R.: Allelopathy of wormwood *(Artemisia absinthium)* gas secretions on adjacent plants. Naturwissenschaften *51*, 117 (1964)

Bonamy, P.A., Dennis, F.G.: ABA levels in seeds of peach. I. Changes during maturation and storage. J. Am. Soc. Hortic. Sci. *102*, 23–26 (1977)
Bonner, J.: Limiting factors and growth inhibitors in the growth of *Avena* coleoptile. Am. J. Bot. *36*, 323–332 (1949)
Bonner, J.: The role of toxic substances in the interactions of higher plants. Bot. Rev. *16*, 51–65 (1950)
Bonner, J., Galston, A.W.: Toxic substances from the culture media of guayule which may inhibit growth. Bot. Gaz. *106*, 185–198 (1944)
Börner, H.: Liberation of organic substances from higher plants and their role in the soil sickness problem. Bot. Rev. *26*, 393–424 (1960)
Borrow, A., Brian, P.W., Chester, V.E., Curtis, P.J., Hemming, H.G., Henehan, C., Jeffreys, E.G., Lloyd, P.B., Nixon, I.S., Norris, G.L.F., Radley, M.: Gibberellic acid, a metabolic product of the fungus *Gibberella fujikuroi:* some observations on its production and isolation. J. Sci. Food Agric. *6*, 340–348 (1955)
Botjes, J.O.: Aethyleen aus vermoedelijke oorzaak von de groeiremmende werking van rijpe appels. Plantenziek. *39*, 207–211 (1933)
Böttger, M.: The occurrence of cis, trans- and trans,trans-xanthoxin in pea roots. Z. Pflanzenphysiol. *86*, 265–268 (1978)
Bough, W.A., Gander, J.E.: Isolation and characterisation of chelidonic acid from *Sorghum vulgare*. Phytochemistry *11*, 209–213 (1972)
Bourdoux, P., Vandervorst, D., Hootelé, C.: Isolation of indole-3-carboxaldehyde from *Equisetum telmateia*. Phytochemistry *10*, 1934–1935 (1971)
Bowen, D.H., Crozier, A., MacMillan, J., Reid, D.M.: Characterization of gibberellins from light-grown *Phaseolus coccineus* seedlings by combined G.C.-M.S. Phytochemistry *12*, 2935–2941 (1973)
Boysen Jensen, P.: Growth hormones in plants (transl. by Avery, G.S., Burkholder, P.R.). New York: McGraw-Hill 1936
Brian, P.W.: The effects of some microbial metabolic products on plant growth. Symp. Soc. Exp. Biol. *11*, 166–182 (1957)
Brian, P.W., Grove, J.F., MacMillan, J.: The gibberellins. Fortschr. Chem. Org. Naturst. *18*, 350–433 (1960)
Brian, P.W., Grove, J.F., Mulholland, T.P.C.: Relationship between structure and growth promoting activity of the gibberellins and some allied compounds in four test systems. Phytochemistry *6*, 1475–1499 (1967)
Bridges, I.C., Hillman, J.R., Wilkins, M.B.: Identification and localisation of auxin in primary roots of *Zea mays* by mass spectrometry. Planta *115*, 189–192 (1973)
Briggs, D.E.: Gibberellin-like activity of helminthosporol and helminthosporic acid. Nature (London) *210*, 418–419 (1966)
Brown, J.C., Cross, B.E., Hanson, J.R.: Two gibbane 1→3-lactones. Tetrahedron *23*, 4095–4103 (1967)
Brown, K.S., Sanchez, W.E.: Distribution and functions of norditerpene dilactones in *Podocarpus* species. Biochem. Syst. Ecol. *2*, 11–14 (1974)
Browning, G., Saunders, P.F.: Membrane localised gibberellins GA_9 and GA_4 in wheat chloroplasts. Nature (London) *265*, 375–377 (1977)
Browning, G., Hoad, G.V., Gaskin, P.: Identification of abscisic acid in flower buds of *Coffea arabica* L. Planta *94*, 213–219 (1970)
Bruce, M.I., Zwar, J.A., Kefford, N.P.: Chemical structure and plant kinin activity: the activity of urea and thiourea derivatives. Life Sci. *4*, 461–466 (1965)
Bruinsma, J., Patil, S.S.: The effects of indole-3-acetic acid, gibberellic acid and vitamin E on flower initiation in unvernalized Petkus winter rye plants. Naturwissenschaften *50*, 505 (1963)
Bürcky, K.: Gibberellinaktivität verschiedener Schizaeaceen-Antheridiogene im Zwergerbrenntest. Z. Naturforsch. *32c*, 652–653 (1977)
Bui-Dang-Ha, D., Nitsch, J.P.: Isolation of zeatin riboside from chicory root. Planta *95*, 119–126 (1970)
Bukovac, M.J., Yuda, E., Murofushi, N., Takahashi, N.: Endogenous plant growth sub-

stances in developing fruit of *Prunus cerasus* L. VII. Isolation of gibberellin A_{32}. Plant Physiol. *63*, 129–132 (1979)
Buller, D.C., Parker, W., Reid, J.S.G.: Short chain fatty acids as inhibitors of gibberellin induced amylolysis in barley endosperm. Nature (London) *260*, 169–170 (1976)
Burg, S.P.: The physiology of ethylene formation. Annu. Rev. Plant Physiol. *13*, 265–302 (1962)
Burg, S.P., Stolwijk, J.A.J.: A highly sensitive katharometer and its application to the measurement of ethylene and other gases of biological importance. J. Biochem. Microbiol. Technol. Eng. *1*, 245–259 (1959)
Burg, S.P., Thimann, K.V.: The physiology of ethylene formation in apples. Proc. Natl. Acad. Sci. *45*, 335–344 (1959)
Burrows, W.J.: Mechanism of action of cytokinins. Curr. Adv. Plant. Sci. *7*, 837–847 (1975)
Burrows, W.J.: Mode of action of N,N′-diphenylurea: the isolation and identification of the cytokinins in the *t*-RNA from tobacco callus grown in the presence of N,N′-diphenylurea. Planta *130*, 313–316 (1976)
Burrows, W.J.: Cytokinins. Biochem. Soc. Trans. *6*, 1395–1400 (1978a)
Burrows, W.J.: Evidence in support of biosynthesis *de novo* of free cytokinins. Planta *138*, 53–57 (1978b)
Burrows, W.J., Armstrong, D.J., Skoog, F., Hecht, S.M., Boyle, J.T.A., Leonard, N.J., Occolowitz, J.: The isolation and identification of two cytokinins from *E. coli t*-RNA. Biochemistry *8*, 3071–3076 (1969)
Burrows, W.J., Armstrong, D.J., Kaminek, M., Skoog, F., Bock, R.M., Hecht, S.M., Dammann, L.G., Leonard, N.J., Occolowitz, J.L.: Isolation and identification of four cytokinins from wheat germ transfer RNA. Biochemistry *9*, 1867–1872 (1970)
Burrows, W.J., Skoog, F., Leonard, N.J.: Isolation and identification of cytokinins located in the transfer RNA of tobacco callus grown in the presence of 6-benzylaminopurine. Biochemistry *10*, 2189–2194 (1971)
Buston, H.W., Roy, S.K.: The physiological activity of some simple unsaturated lactones. II. Effect on certain tissues of higher plants. Arch. Biochem. *22*, 269–274 (1949)
Buta, J.G., Warley, J.F.: Camptothecin, a selective plant growth regulator. J. Agric. Food Chem. *24*, 1085–1086 (1976)
Buta, J.G., Flippen, J.L., Lusby, W.R.: Harringtonolide, a plant growth inhibitory tropone from *Cephalotaxus harringtonia* (Forbes) K. Koch. J. Org. Chem. *43*, 1002–1003 (1978)
Caruso, J.L., Smith, R.G., Smith, L.M., Cheng, T.-Y., Daves, G.D.: Determination of indole-3-acetic acid in Douglas fir using a deuterated analog and selected ion monitoring. Comparison of microquantities in seedling and adult tree. Plant Physiol. *62*, 841–845 (1978)
Chapman, R.W., Morris, R.O., Zaerr, J.B.: Occurrence of trans-ribosylzeatin in *Agrobacterium tumefaciens t*-RNA. Nature (London) *262*, 153–154 (1976)
Cherayil, J.D., Lipsett, M.N.: Zeatin ribonucleosides in the transfer ribonucleic acid of *Rhizobium leguminosarum, Agrobacterium tumefaciens, Corynebacterium fascians* and *Erwinia amylovora*. J. Bacteriol. *131*, 741–744 (1977)
Chowdhury, B.K., Chakraborty, D.P.: 3-Formylindole from *Murraya exotica*. Phytochemistry *10*, 481–483 (1971)
CIAT Annual Report: Centro Internacional de Agricultura Tropical, Cali, Columbia (1972)
CIAT Annual Report: Centro Internacional de Agricultura Tropical, Cali, Columbia (1975)
Ciha, A.J., Brenner, M.L., Brun, W.A.: Rapid separation and quantitation of ABA from plant tissues using H.P.L.C. Plant Physiol. *59*, 821–826 (1977)
Cleland, C., Ajami, A.: Identification of the flower-inducing factor isolated from aphid honeydew as being salicyclic acid. Plant Physiol. *54*, 904–906 (1974)
Coke, L.B., Stuart, K.L., Whittle, Y.G.: Further effects of vomifoliol on stomatal aperture and on the germination of lettuce and growth of cucumber seedlings. Planta *127*, 21–25 (1975)
Comman, I.: The responses of onion and lily mitosis to coumarin and parasorbic acid. J. Exp. Bot. *23*, 292–297 (1947)
Conn, E.E., Butler, G.W.: The biosynthesis of cyanogenic glycosides and other simple

nitrogen compounds. In: Perspectives in phytochemistry. Harborne, J.B., Swain, T. (eds.), pp. 47–74. New York, London: Academic Press 1969
Cook, C.E., Whichard, L.P., Turner, B., Wall, M.E., Egley, G.H.: Germination of witchweed (*Striga lutea* Lour.): isolation and properties of a potent stimulant. Science *154*, 1189–1190 (1966)
Cook, C.E., Whichard, L.P., Wall, M.E., Egley, G.H., Coggon, P., Luhan, P.A., McPhail, A.T.: Germination stimulants. II. The structure of strigol – a potent seed germination stimulant for Witchweed (*Striga lutea* Lour.). J. Am. Chem. Soc. *94*, 6198–6199 (1972)
Coombe, B.G.: GA_{32}. A polar gibberellin with high biological potency. Science *172*, 856–857 (1971)
Coombe, B.G., Mander, L.N., Paleg, L.G., Turner, J.V.: Gibberellin-like activity of helminthosporic acid analogs. Aust. J. Plant Physiol. *1*, 473–481 (1974)
Cooper, R., Levy, E.C., Lavie, D.: Novel germination inhibitors from *Aegilops ovata* L. J.Chem. Soc. *D*, 794–795 (1977)
Cooper, R., Gottlieb, H.E., Lavie, D., Levy, E.C.: Lignans from *Aegilops ovata* L. Synthesis of a 2,4- and a 2,6-diaryl monoepoxylignanolide. Tetrahedron *35*, 861–868 (1979)
Corcoran, M.R.: Tannins, a newly discovered group of gibberellin antagonists. Plant Physiol. Suppl. *47*, 34 (1971)
Corcoran, M.R., Geissman, T.A., Phinney, B.O.: Tannins as gibberellin antagonists. Plant Physiol. *49*, 323–330 (1972)
Corey, E.J., Danheiser, R.L., Chandrasekaran, S., Keck, G.E., Gopolan, B., Larsen, S.D., Siret, P., Gras, J.-L.: Stereospecific total synthesis of gibberellic acid. J. Am. Chem. Soc. *100*, 8034–8036 (1978)
Cornforth, J.W., Milborrow, B.V., Ryback, G., Wareing, P.F.: Identity of sycamore 'dormin' with abscisin II. Nature (London) *205*, 1269–1270 (1965a)
Cornforth, J.W., Milborrow, B.V., Ryback, G.: Synthesis of abscisin II. Nature (London) *206*, 715 (1965b)
Cornforth, J.W., Milborrow, B.V., Ryback, G., Rothwell, K., Wain, R.L.: Identification of the yellow lupin growth inhibitor as (+)-abscisin II ((+)-dormin). Nature (London) *211*, 742–743 (1966a)
Cornforth, J.W., Milborrow, B.V., Ryback, G., Wareing, P.F.: Isolation of sycamore dormin and its identity with abscisin II. Tetrahedron *8*, Suppl. Part II, 603–610 (1966b)
Cornforth, J.W., Draber, W., Milborrow, B.V., Ryback, G.: Absolute stereochemistry of (+)-abscisin II. Chem. Commun. 114–116 (1967)
Cornman, I.: Alteration of mitosis by coumarin and parasorbic acid. Am. J. Bot. *33*, 217 (1946)
Cousins, H.H.: Agricultural experiment. Jam. Dep. Agric. Annu. Rep. *7*, 6–9 (1910)
Crafts, C.B., Miller, C.O.: Detection and identification of cytokinins produced by mycorrhizal fungi. Plant Physiol. *54*, 586–588 (1974)
Crocker, W., Knight, L.I.: Effect of illuminating gas and ethylene upon flowering carnations. Bot. Gaz. *46*, 259–276 (1908)
Crocker, W., Hitchcock, A.E., Zimmerman, P.W.: Similarities in the effects of ethylene and the plant auxins. Contrib. Boyce Thompson Inst. *7*, 231–248 (1935)
Crosby, D.G., Vlitos, A.J.: New auxins from Maryland Mammoth tobacco. In: Plant growth regulation. 4th Int. Conf. pp. 57–59. New York: Yonkers 1961
Cross, B.E.: Gibberellic acid, Part I. J. Chem. Soc. 4670–4676 (1954)
Cross, B.E.: Gibberellin A_{14}. J. Chem. Soc. 501–504 (1966)
Cross, B.E., Norton, K.: Gibberellin A_{12}. J. Chem. Soc. 1570–1572 (1965)
Cross, B.E., Galt, R.H.B., Hanson, J.R.: Gibberellin A_7 and gibberellin A_9. Tetrahedron *18*, 451–459 (1962)
Cross, B.E., Stewart, J.C., Stoddart, J.L.: 6β,7β-Dihydroxykaurenoic acid: its biological activity and possible role in the biosynthesis of gibberellic acid. Phytochemistry *9*, 1065–1071 (1970)
Crow, W.D., Nicholls, W., Sterns, M.: Root inhibitors in *Eucalyptus grandis*. Naturally occurring derivatives of the 2,3-dioxobicyclo [4.4.0] decane system. Tetrahedron Lett. 1353–1356 (1971)
Crow, W.D., Osawa, T., Platz, K.M., Southerland, D.S.: Root inhibitors in *Eucalyptus*

grandis II. (Synthesis of the inhibitors and origin of the peroxide linkage). Aust. J. Chem. *29*, 2525–2531 (1976)

Crow, W.D., Osawa, T., Paton, D.M., Willing, R.R.: Structure of grandinol: a novel root inhibitor from *Eucalyptus grandis*. Tetrahedron Lett. 1073–1074 (1977)

Crozier, A., Bowen, D.H., MacMillan, J., Reid, D.M., Most, B.H.: Characterization of gibberellins from dark-grown *Phaseolus coccineus* seedlings by gas-liquid chromatography and combined G.C.-M.S. Planta *97*, 142–154 (1971)

Curtis, P.J., Cross, B.E.: Gibberellic acid. A new metabolite from the culture filtrates of *Gibberella fujikuroi*. Chem. Ind. 1066 (1954)

Cutler, H.G., Reid, W.W., Délétang, J.: Plant growth inhibiting properties of diterpenes from tobacco. Plant Cell Physiol. *18*, 711–714 (1977)

Dalvi, P.R., Singh, B., Salunkhe, D.K.: A study on phytotoxicity of alantolactone. Chem. Biol. Interact. *3*, 13–18 (1971)

Darwin, C., Darwin, F.: The power of movement in plants. London: John Murray 1880

Dathe, W., Schneider, G., Sembdner, G.: Endogenous gibberellins and inhibitors in caryopses of rye. Phytochemistry *17*, 963–966 (1978a)

Dathe, W., Sembdner, G., Kefeli, V.I., Vlasov, P.V.: Gibberellins, abscisic acid and related inhibitors in branches and bleeding sap of birch (*Betula pubescens* Ehrh.). Biochem. Physiol. Pflanz. *173*, 238–248 (1978b)

Dauphin, B., Teller, G., Durand, B.: Mise au point d'une nouvelle méthode de purification, caracterisation et dosage de cytokinines endogènes, extraites de bourgeons de Mercuriales annuelles. Physiol. Vég. *15*, 747–762 (1977)

Dauphin, B., Teller, G., Dward, B.: Identification and quantitative analysis of cytokinins from shoot apices of *Mercuralis ambigua* by gas chromatography-mass spectrometry computer system. Planta *144*, 113–119 (1979)

Davis, L.A., Lyon, J.L., Addicott, F.T.: Phaseic acid: occurrence in cotton fruit acceleration of abscission. Planta *102*, 294–301 (1972)

Davis, M., Pettet, M., Scanlon, D.B., Ferrito, V.: Synthesis of some benzopyran derivatives related to the seed germination stimulant of *Orobanche crenata*. 1. 3,3a, 4, 9b-tetrahydro-2*H*-furo[3,2-*C*] [1] benzopyrans. Aust. J. Chem. *30*, 2289–2292 (1977)

Davison, R.M., Rudnicki, R.M., Bukovac, M.J.: Endogenous plant growth substances in developing fruit of *Prunus cerasus* L. V. Changes in inhibitor (ABA) levels in the seed and pericarp. J. Am. Soc. Hortic. Sci. *101*, 519–523 (1976)

Dekhuijzen, H.M.: Endogenous cytokinins in healthy and diseased plants. In: Encyclopedia of plant physiology. Heitefuss, R., Williams, P.H. (eds.) Vol. IV, pp. 526–559. Berlin-Heidelberg-New York: Springer 1976

De Leo, P., Dalessandro, G., De Santis, A., Arrigoni, O.: Inhibitory effect of lycorine on cell division and cell elongation. Plant Cell Physiol. *14*, 481–486 (1973a)

De Leo, P., Dalessandro, G., De Santis, A., Arrigoni, O.: Metabolic responses to lycorine in plants. Plant Cell Physiol. *14*, 487–496 (1973b)

Denny, F.E.: Hastening the coloration of lemons. J. Agric. Res. *27*, 757–768 (1924a)

Denny, F.E.: The effect of ethylene upon respiration of lemons. Bot. Gaz. *77*, 322–329 (1924b)

Desmos, E.K., Woolwine, M., Wilson, R.H., McMillan, C.: The effect of ten phenolic compounds on hypocotyl growth and mitochondrial metabolism of mung bean. Am. J. Bot. *62*, 97–102 (1975)

Deuel, P.G., Geissman, T.A.: The structure of xanthinin and xanthatin. J. Am. Chem. Soc. *79*, 3778–3783 (1957)

Dev, S.: Studies in sesquiterpenes – XVI. Zerumbone, a mono-cyclic sesquiterpene ketone. Tetrahedron *8*, 171–180 (1960)

Devon, T.K., Scott, A.I.: Handbook of naturally occurring compounds. Vol. II. Terpenes 596 pp. New York, London: Academic Press 1972

Devon, T.K., Scott, A.I.: Handbook of naturally occurring compounds. Vol. I. Acetogenins, shikimates and carbohydrates. 644 pp. New York, London: Academic Press 1975

Dewar, M.J.S.: The structure of colchicine. Nature (London) *155*, 141–142 (1945)

Deyoe, D.R., Zaerr, J.R.: Indole-3-acetic acid in Douglas fir. Analysis by gas-liquid chromatography and mass spectrometry. Plant Physiol. *58*, 299–303 (1976)

Dominguez, X.A., Cardenas, E.: Achillin and deacetylmatricarin from two *Artemesia* species. Phytochemistry *14*, 2511–2512 (1975)
Dominguez, X.A., Perez, F.M., Leyter, L.: Xanthinin and β-sitosterol from *Xanthium orientale*. Phytochemistry *10*, 2828 (1971)
Dominguez, X.A., Merijanian, A., Gonzalez, B.L.: Medicinal plants of Mexico XXIV. Terpenoids of *Teucrium cubense*. Phytochemistry *13*, 754–755 (1974a)
Dominguez, X.A., Merijanian, A., Gonzalez, B.I., Zamudio, A., Salazar, A.L.: Structure of eugarzasadine, a new norditerpene lactone isolated from *Teucrium cubense*. Rev. Latinoam. Quim. *5*, 225–229 (1974b)
Dominguez, X.A., Martinez, C., Calero, A., Dominguez Jr., X.A., Hinojosa, M., Zamudio, A., Zabel, V., Smith, W.B., Watson, W.H.: Louisfieserone, an unusual flavanone derivative from *Indigofera suffruticosa* Mill. Tetrahedron Lett. 429–432 (1978)
Dörffling, K., Böttger, M., Martin, D., Schmidt, V., Borowski, D.: Physiology and chemistry of substances accelerating abscission in senescent petioles and fruit stalks. Physiol. Plant. *43*, 292–296 (1978)
Duke, C.C., Liepa, A.J., MacLeod, J.K., Letham, D.S., Parker, C.W.: Synthesis of raphanatin and its 6-benzylaminopurine analogue. J. Chem. Soc. *D*, 964–965 (1975)
Dunn, D.B., Smith, J.D., Spahr, P.F.: Nucleotide composition of soluble ribonucleic acid from *Escherichia coli*. J. Mol. Biol. *2*, 113–117 (1960)
Durley, R.C., MacMillan, J., Pryce, R.J.: Investigation of gibberellins and other growth substances in the seed of *Phaseolus multiflorus* and of *Phaseolus vulgaris* by gas chromatography and by gas chromatography-mass spectrometry. Phytochemistry *10*, 1891–1908 (1971)
Dyson, W.H., Hall, R.H.: N^6-(Δ^2-isopentenyl)adenosine: its occurrence as a free nucleoside in an autonomous strain of tobacco tissue. Plant Physiol. *50*, 616–621 (1972)
Dyson, W.H., Chen, C.M., Alam, S.N., Hall, R.H., Hong, C.I., Chheda, G.B.: Cytokinin activity of ureidopurine derivatives related to a modified nucleoside found in transfer RNA. Science *170*, 328–330 (1970)
Eckert, H., Schilling, G., Podesak, W., Franke, P.: Extraction and identification of gibberellins (GA_1 and GA_3) from *Triticum aestivum* L. and *Secale cereale* L. and changes in contents during autogenesis. Biochem. Physiol. Pflanz. *172*, 475–486 (1978)
Eeuwens, C.J., Gaskin, P., MacMillan, J.: Gibberellin A_{20} in seed of *Pisum sativum* cv. Alaska. Planta *115*, 73–76 (1973)
Ehmann, A.: Isolation of 2-O-(indole-3-acetyl)-D-glucopyranose, 4-O-(indole-3-acetyl)-D-glucopyranose and 6-O-(indole-3-acetyl)-D-glucopyranose from kernels of *Zea mays* by gas-liquid chromatography-mass spectrometry. Carbohydr. Res. *34*, 99–114 (1974)
Ehmann, A., Bandurski, R.S.: The isolation of di-O-(indole-3-acetyl)-*myo*-inositol and tri-O-(indole-3-acetyl)-*myo*-inositol from mature kernels of *Zea mays*. Carbohydr. Res. *36*, 1–12 (1974)
Einset, J.W., Skoog, F.K.: Isolation and identification of ribosyl-*cis*-zeatin from transfer RNA of *Corynebacterium fascians*. Biochem. Biophys. Res. Commun. *79*, 1117–1121 (1977)
Einset, J.W., Swaminathan, S., Skoog, F.: Ribosyl-*cis*-zeatin in a leucyl transfer RNA species from peas. Plant Physiol. *58*, 140–142 (1976)
Elliott, M.C.: α-N-Malonyl-D-tryptophan in seedling wheat roots. New Phytol. *70*, 1005–1015 (1971)
Elliott, M.C., Greenwood, M.S.: Indol-3-yl-acetic acid in roots of *Zea mays*. Phytochemistry *13*, 239–241 (1974)
Elliott, M.C., Stowe, B.B.: A novel sulphonated natural indole. Phytochemistry *9*, 1629–1632 (1970)
Elliott, M.C., Stowe, B.B.: Indole compounds related to auxins and goitrogens of wood (*Isatis tinctoria* L.). Plant Physiol. *47*, 366–372 (1971)
Elmer, O.H.: The growth inhibition of potato sprouts by the volatile products of apples. Science *75*, 193 (1932)
Elmer, O.H.: Growth inhibition in the potato caused by a gas emanating from apples. J. Agric. Res. *52*, 609–626 (1936)

El-Olemy, M.M., Reisch, J.: Isolation of batatasin I from *Dioscorea dumetorum* rhizomes. Z. Naturforsch. *34 C*, 288–289 (1979)

Engvild, K.C., Egsgaard, H., Larsen, E.: Gas chromatographic-mass spectrometric identification of 4-chloroindolyl-3-acetic acid methyl ester in immature green peas. Physiol. Plant. *42*, 365–368 (1978)

Erickson, R.O., Rosen, G.V.: Cytological effects of protoanemonin on the root tip of *Zea mays*. Am. J. Bot. *86*, 317–322 (1949)

Ettlinger, M.G., Kjaer, A.: Sulphur compounds in plants. Rec. Adv. Phytochem. *1*, 59–144 (1968)

Evans, L.S., Almeida, M.S., Lynn, D.G., Nakanishi, K.: Chemical characterisation of a hormone that promotes cell arrest in G_2 in complex tissues. Science *203*, 1122–1123 (1979)

Evenari, M.: Germination inhibitors. Bot. Rev. *15*, 153–194 (1949)

Evenari, M.: Chemical influences of other plants (allelopathy). In: Encyclopedia of plant physiology. Ruhland, W. (ed.), Vol. XVI, pp. 691–736. Berlin-Göttingen-Heidelberg: Springer 1961

Fenton, R., Davies, W.J., Mansfield, T.A.: The role of farnesol as a regulator of stomatal opening in *Sorghum*. J. Exp. Bot. *28*, 1043–1053 (1977)

Firn, R.D., Friend, J.: Enzymatic production of the plant growth inhibitor xanthoxin. Planta *103*, 263–266 (1972)

Firn, R.D., Burden, R.S., Taylor, H.F.: Detection and estimation of the growth inhibitor xanthoxin in plants. Planta *102*, 115–126 (1972)

Fowden, L.: Amino-acid analogues and the growth of seedlings. J. Exp. Bot. *14*, 387–398 (1963)

Fowden, L.: The non-protein amino-acids of plants. Prog. Phytochem. *2*, 203–266 (1970)

Fowden, L.: Non-protein amino acids from plants: Distribution, biosynthesis and analog function. Rec. Adv. Phytochem. *8*, 95–122 (1974)

Fowden, L., Richmond, M.H.: Replacement of proline by azetidine-2-carboxylic acid during biosynthesis of protein. Biochim. Biophys. Acta *71*, 459–461 (1963)

Fowden, L., Lewis, D., Tristram, H.: Toxic amino acids: their action as anti-metabolites. Adv. Enzymol. *29*, 89–163 (1967)

Fox, J.E.: The cytokinins. In: Physiology of plant growth and development. Wilkins, M.B. (ed.), pp. 85–123. London: McGraw-Hill 1969

Fragata, M.: Endogenous recovery from colchicine induced inhibition of growth. Experientia *28*, 506–507 (1972)

Frydman, V.M., MacMillan, J.: Identification of gibberellins A_{20} and A_{29} in seed of *Pisum sativum* cv. Progress No. 9 by combined gas chromatography mass spectrometry. Planta *115*, 11–15 (1973)

Frydman, V.M., Gaskin, P., MacMillan, J.: Qualitative and quantitative analysis of gibberellins throughout seed maturation in *Pisum sativum* cv. Progress No. 9. Planta *118*, 123–132 (1974)

Fujii, T., Ogawa, N.: The absolute configuration of (−)-dihydrozeatin. Tetrahedron Lett. 3075–3078 (1972)

Fujita, E., Node, M.: Synthesis of gibberellins. Heterocycles *7*, 709–752 (1977)

Fujita, E., Uchida, I., Fujita, T.: Terpenoids. Part XXXII. Structure and stereochemistry of teucvin, a novel norclerodane-type diterpene from *Teucrium viscidum var. Miquelianum*. J. Chem. Soc. Perkin *I*, 1547–1555 (1974)

Fukui, H., Koshimizu, K., Mitsui, T.: Gibberellin A_{28} in the fruits of *Lupinus luteus*. Phytochemistry *10*, 671–673 (1971)

Fukui, H., Ishii, H., Koshimizu, K., Katsumi, M., Ogawa, Y., Mitsui, T.: The structure of gibberellin A_{23} and the biological properties of 3,13-dihydroxy C_{20}-gibberellins. Agric. Biol. Chem. *36*, 1003–1012 (1972)

Fukui, H., Koshimizu, K., Usuda, S., Yamazuki, Y.: Isolation of plant growth regulators from seeds of *Curcurbita pepo* L. Agric. Biol. Chem. *41*, 175–180 (1977a)

Fukui, H., Nemori, R., Koshimizu, K., Yamazuki, Y.: Structures of gibberellins A_{39}, A_{48}, A_{49} and a new kaurenolide in *Cucurbita pepo* L. Agric. Biol. Chem. *41*, 181–187 (1977b)

Fukui, H., Koshimizu, K., Yamazaki, Y., Usuda, S.: Structures of plant growth inhibitors in seeds of *Cucurbita pepo* L. Agric. Biol. Chem. *41*, 189–194 (1977c)
Fukui, H., Koshimizu, K., Nemori, R.: Two new gibberellins A_{50} and A_{52} in seeds of *Lagenaria leucantha*. Agric. Biol. Chem. *42*, 1571–1576 (1978)
Funke, G.L.: The influence of *A. absinthium* on neighbouring plants. An essay of experimental sociology. III. Blumea *5*, 281–293 (1943)
Furuya, M., Galston, A.W., Stowe, B.B.: Isolation from peas of cofactors and inhibitors of indolyl-3-acetic acid oxidase. Nature (London) *193*, 456–457 (1962)
Gabe, E.J., Neidle, S., Rogers, D., Nordman, C.E.: X-ray determination of the molecular structure of a 3-*o*-chlorophenyl-isoxazoline derivative of pyrethrosin. J. Chem. Soc. D. 559–560 (1971)
Galbraith, M.N., Horn, D.H.S.: Structures of the natural products blumlenols A, B and C. J. Chem. Commun. 113–114 (1972)
Galbraith, M.N., Horn, D.H.S.: Stereochemistry of the blumenols: conversion of blumenol A into *(S)*-(+)-abscisic acid. J. Chem. Commun. 566–567 (1973)
Galbraith, M.N., Horn, D.H.S., Ito, S., Kodama, M., Sasse, J.M.: Structure-activity relationships of podolactones and related compounds. Agric. Biol. Chem. *36*, 2393–2396 (1972)
Galston, A.W.: Flavanoids and photomorphogenesis in peas. In: Perspectives in phytochemistry. pp. 193–204. New York, London: Academic Press 1969
Galt, R.H.B.: Gibberellin A_{13}. J. Chem. Soc. 3143–3151 (1965)
Galt, R.H.B.: Gibberellin A_{16} methyl ester. Tetrahedron *24*, 1337–1339 (1968)
Gandar, J.C., Nitsch, C.: Isolement de l'ester methylique d'un acide chloro-3-indolylacétique a partir graines immature de pois *Pisum sativum*. C.R. Acad. Sci. Ser. D *265*, 1795–1798 (1968)
Gane, R.: Production of ethylene by some ripening fruit. Nature (London) *134*, 1008 (1934)
Ganguly, S.N., Sircar, S.M.: Gibberellins from mangrove plants. Phytochemistry *13*, 1911–1913 (1974)
Ganguly, S.N., Ganguly, T., Sircar, S.M.: Gibberellins of *Enhydra fluctuans*. Phytochemistry *11*, 3433–3434 (1972)
Ganguly, T., Ganguly, S.N., Sircar, P.K., Sircar, S.M.: Rhamnose bound indole-3-acetic acid in the floral parts of *Peltophorum fernigineum*. Physiol. Plant. *31*, 330–332 (1974)
Gaskin, P., MacMillan, J.: Identification and estimation of abscisic acid in a crude plant extract by combined gas chromatography-mass spectrometry. Phytochemistry *7*, 1699–1701 (1968)
Gaskin, P., MacMillan, J.: Polyoxygenated *ent*-kauranes and water soluble conjugates in seed of *Phaseolus coccineus*. Phytochemistry *14*, 1575–1578 (1975)
Gaskin, P., MacMillan, J., Ganguly, S.N., Sanyal, T., Sircar, P.K., Sircar, S.M.: Identification of the gibberellin from *Sonneratia apelata* Ham as tetrahydrogibberellin A_3. Chem. Ind. 424–425 (1972)
Gaskin, P., MacMillan, J., Zeevaart, J.A.D.: Identification of gibberellin A_{20}, abscisic acid and phaseic acid from flowering *Bryophyllum daigremontianum* by combined gas chromatography-mass spectrometry. Planta *111*, 347–352 (1973)
Gaskin, P., Kirkwood, P.S., Lenton, J.R., MacMillan, J., Radley, M.E.: Identification of gibberellins in developing wheat grain. Agric. Biol. Chem. in press (1980)
Geissman, T.A.: The structure of xanthinin. J. Org. Chem. *27*, 2692–2693 (1962)
Geissman, T.A., Deuel, P., Bonde, E.K., Addicott, F.T.: Xanthinin. A plant growth regulating compound from *Xanthium pennsylvanicum*. J. Am. Chem. Soc. *76*, 685–687 (1954)
Geuns, J.M.C.: Steroid hormones and plant growth and development. Phytochemistry *17*, 1–14 (1978)
Gibson, M.S., Crane, F.L.: Paper chromatography method for identification of ethylene. Plant Physiol. *38*, 729–730 (1963)
Girardin, J.P.: Einfluß des Leuchtgases auf die Promenaden und Straßenbäume. Jahresber. Agric. Chem. *7*, 199–200 (1864)
Gmelin, R., Virtanen, A.I.: Glucobrassicin, the precursor of the thiocyanate ion, 3-indol-

ylacetonitrile, and ascorbigen in *Brassica oleracea* (and related species). Ann. Acad. Sci. Fenn. Ser. A2 *107*, 1–25 (1961)

Gmelin, R., Virtanen, A.I.: Neoglucobrassicin, ein zweiter SCN^--Precursor von Indoltyp in *Brassica*-Arten. Acta Chem. Scand. *16*, 1378–1384 (1962)

Godfrey, J.E., Waters, J.M.: The crystal and molecular structure of the bisnorditerpenoid, inumakilactone. Aust. J. Chem. *28*, 745–753 (1975)

Goldschmidt, E.E., Goren, R., Evenchen, Z., Bittner, S.: Increase in free and bound ABA during natural and ethylene-induced senescence of citrus fruit peel. Plant Physiol. *51*, 879–882 (1973)

Goodwin, R.H., Taves, C.: The effect of coumarin derivatives on the growth of *Avena* roots. Am. J. Bot. *37*, 224–231 (1950)

Gordon, S.A.: Occurrence, formation and inactivation of auxins. Annu. Rev. Plant Physiol. *5*, 341–378 (1954)

Goren, R., Tomer, E.: Effects of seselin and coumarin on growth, indoleacetic acid oxidase and peroxidase with special reference to cucumber (*Cucumis sativa* L.) radicles. Plant. Physiol. *47*, 312–316 (1971)

Graebe, J.E., Ropers, H.J.: Gibberellins. In: Phytohormones and related compounds: a comprehensive treatise. Letham, D.S., Goodwin, P.B., Higgins, T.J.V. (eds.) Vol. 1, pp. 107–204. Amsterdam, Oxford, New York: Elsevier/North Holland Biomedical Press 1978

Graebe, J.E., Hedden, P., Gaskin, P., MacMillan, J.: The biosynthesis of a C_{19}-gibberellin from mevalonic acid in a cell-free system from a higher plant. Planta *120*, 307–309 (1974)

Gray, R., Bonner, J.: Structure determination and synthesis of a plant growth inhibitor, 3-acetyl-6-methoxybenzaldehyde found in leaves of *Encelia farinosa*. J. Am. Chem. Soc. *70*, 1249–1250 (1948)

Greenwood, M.S., Shaw, S., Hillman, J.R., Ritchie, A., Wilkins, M.B.: Identification of auxin from *Zea* coleoptile tips by mass spectrometry. Planta *108*, 179–183 (1972)

Gross, D.: Growth regulating substances of plant origin. Phytochemistry *14*, 2105–2112 (1975)

Grove, J.F.: The gibberellins. Quart. Rev. *15*, 56–70 (1961 a)

Grove, J.F.: Gibberellin A_2. J. Chem. Soc. 3545–3547 (1961 b)

Grove, J.F., Jeffs, P.W., Mulholland, T.P.C.: The relation between gibberellin A_1 and gibberellic acid. J. Chem. Soc. 1236–1240 (1958)

Grove, J.F., MacMillan, J., Mulholland, T.P.C., Turner, R.B.: The stereochemistry of gibberic and epigibberic acid. J. Chem. Soc. 3049–3057 (1960)

Grümmer, G.: The role of toxic substances in the relationship between higher plants. Symp. Soc. Exp. Biol. *15*, 209–228 (1961)

Haagen-Smit, A.J., Leach, W.D., Bergren, W.R.: The estimation, isolation and identification of auxins in plant materials. Am. J. Bot. *29*, 500–506 (1942)

Haagen-Smit, A.J., Dandliker, W.B., Wittwer, S.H., Murneek, A.E.: Isolation of 3-indoleacetic acid from immature corn kernels. Am. J. Bot. *33*, 118–120 (1946)

Haberlandt, G.: Zur Physiologie der Zellteilung. Sitzungsber. K. Preuss. Akad. Wiss. 318–345 (1913)

Haberlandt, G.: Wundhormone als Erreger von Zellteilungen. Beitr. Allg. Bot. *2*, 1–53 (1921)

Hall, R.H.: N^6-(Δ^2-Isopentenyl)adenosine: chemical reactions, biosynthesis, metabolism and significance to the structure and function of *t*-RNA. Prog. Nucl. Acid. Res. Mol. Biol. *10*, 57–86 (1970)

Hall, R.H.: Cytokinins as a probe of developmental processes. Annu. Rev. Plant. Physiol. *24*, 415–444 (1973)

Hall, R.H., de Ropp, R.S.: Formation of 6-furfurylaminopurine from DNA breakdown products. J. Am. Chem. Soc. *77*, 6400 (1955)

Hall, R.H., Robins, M.J., Stasiuk, L., Thedford, R.: Isolation of N^6-(γ,γ-dimethylallyl)adenosine from soluble ribonucleic acid. J. Am. Chem. Soc. *88*, 2614–2615 (1966)

Hall, R.H., Csonka, L., David, H., McLennan, B.: Cytokinins in the soluble RNA of plant tissues. Science *156*, 69–71 (1967)

Hall, S.M., Medlow, G.C.: Identification of IAA in phloem and root pressure saps of *Ricinus communis* L. by mass spectrometry. Planta *119*, 257–261 (1974)

Hanson, J.R.: Gibberellin A_{10}. Tetrahedron *22*, 701–703 (1966)

Hanson, J.R.: Gibberellin A_{15}. Tetrahedron *23*, 733–735 (1967)

Hanson, J.R.: The gibberellins. In: The tetracyclic diterpenes. pp. 41–59. New York: Pergamon Press 1968

Harada, F., Gross, H.J., Kimura, F., Chang, S.H., Nishimura, S., RajBhandary, U.L.: 2-Methylthio N^6-(Δ^2-isopentenyl)adenosine. A component of *E. coli* tyrosine *t*-RNA. Biochem. Biophys. Res. Commun. *33*, 299 (1968)

Harada, H., Nitsch, J.P.: Isolation of gibberellins A_1, A_3, A_9 and a fourth growth substance from *Althea rosea* CAV. Phytochemistry *6*, 1695–1703 (1967)

Harada, H., Yokota, T.: Isolation of gibberellin A_8-glucoside from shoot apices of *Althea rosea*. Planta *92*, 100–104 (1970)

Harada, J., Nakayama, H.: The inhibitory effect of tannic acid on the gibberellic acid-induced growth of rice. Proc. Crop. Sci. Soc. Jpn. *43*, 493–497 (1974)

Harborne, J.B.: Introduction to ecological biochemistry. New York, London: Academic Press 1977

Harrison, D.M., MacMillan, J.: Two new gibberellins, A_{24} and A_{25}, from *Gibberella fujikuroi*; their isolation structure and correlation with gibberellins A_{13} and A_{15}. J. Chem. Soc. *C*, 631–636 (1971)

Hartsuck, J.A., Lipscomb, W.N.: Molecular and crystal structure of the di-*p*-bromobenzoate of the methyl ester of gibberellic acid. J. Am. Chem. Soc. *85*, 3414–3419 (1963)

Harvey, R.B.: Ethylene is a ripener of fruit and vegetables. Science *67*, 421–422 (1928)

Hasegawa, K., Hashimoto, T.: Quantitative changes of batatasins and abscisic acid in relation to the development of dormancy in yam bulbils. Plant Cell Physiol. *14*, 369–377 (1973)

Hasegawa, K., Hashimoto, T.: Gibberellin-induced dormancy and batatasin content in yam bulbils. Plant Cell. Physiol. *15*, 1–6 (1974)

Hashimoto, T., Tajima, M.: Structures and synthesis of the growth inhibitors batatasin IV and V, and their physiological activities. Phytochemistry *17*, 1179–1184 (1978)

Hashimoto, T., Ikai, T., Tamura, S.: Isolation of (+)-abscisin II from dormant aerial tubers of *Dioscorea batatas*. Planta *78*, 89–92 (1968)

Hashimoto, T., Hasegawa, K., Kawarada, A.: Batatasins: new dormancy-inducing substances of yam bulbils. Planta *108*, 369–374 (1972)

Hashimoto, T., Hasegawa, K., Yamaguchi, H., Saito, M., Ishimoto, S.: Structure and synthesis of batatasins, dormancy-inducing substances of yam bulbils. Phytochemistry *13*, 2849–2852 (1974)

Hashizume, T., Kimura, K., Sugiyama, T.: Identification of *cis*-zeatin-D-riboside from the top of tobacco plants. Heterocycles *10*, 139–146 (1978)

Hashizume, T., Sugiyama, T., Imura, M., Cory, H.T., Scott, M.F., McCloskey, J.A.: Determination of cytokinins by mass spectrometry based on stable isotope dilution. Anal. Biochem. *92*, 111–122 (1979)

Hattori, H., Marumo, S.: Monomethyl-4-chloroindolyl-3-acetyl-L-aspartate and absence of indolyl-3-acetic acid in immature seeds of *Pisum sativum*. Planta *102*, 85–90 (1972)

Hayashi, H., Koshimizu, K., Asahira, T., Matsubara, S.: Callus growth inhibitors from *Persimmon calyx*. Agric. Biol. Chem. *41*, 2495–2496 (1977)

Hayashi, Y., Sakan, T.: Nagilactones, plant growth regulators with an antiauxin-like activity. In: Plant growth substances 1973, pp. 525–532. Tokyo: Hirokawa Publishing Company 1974

Hayashi, Y., Yokai, J., Watanabe, Y., Sakan, T.: Structures of nagilactones E and F, and biological activity of nagilactones as plant growth regulators. Chem. Lett. 759–762 (1972)

Helgeson, J.P., Leonard, N.J.: Cytokinins: identification of compounds isolated from *Corynebacterium fascians*. Proc. Natl. Acad. Sci. USA *56*, 60–63 (1966)

Hemberg, T.: Significance of the growth inhibitory substances and auxins for the rest period of the potato tuber. Physiol. Plant *2*, 24–36 (1949)

Hemberg, T.: Biogenous inhibitors. In: Encyclopedia of plant physiology. Ruhland, W. (ed.) Vol. XIV, pp. 1162–1184. Berlin-Göttingen-Heidelberg: Springer 1961
Hemphill, D.D., Baker, L.R., Sell, H.M.: Isolation and identification of the gibberellins of *Cucumis sativus* and *Cucumis melo*. Planta *103*, 241–248 (1972)
Hemphill, D.D., Baker, L.R., Sell, H.M.: Isolation of novel conjugated gibberellins from *Cucumis sativus* seed. Can. J. Biochem. *51*, 1647–1653 (1973)
Henbest, H.B., Jones, E.R.H., Smith, G.F.: Isolation of a new plant growth hormone, 3-indoleacetonitrile. J. Chem. Soc. 3796–3801 (1953)
Herz, W., Watanabe, H., Miyazaki, M., Kishida, Y.: The structures of parthenin and ambrosin. J. Am. Chem. Soc. *84*, 2601–2610 (1962)
Hillman, J.R., Young, I., Knights, B.A.: ABA in leaves of *Hedera helix* L. Planta *119*, 263–266 (1974)
Hiraga, K., Kawake, S., Yokota, T., Murofushi, N., Takahashi, N.: Isolation and characterisation of gibberellins in mature seeds of *Phaseolus vulgaris*. Agric. Biol. Chem. *38*, 2511–2520 (1974a)
Hiraga, K., Kawabe, S., Yokota, T., Murofushi, N., Takahashi, N.: Isolation and characterisation of plant growth substances in immature and etiolated seedlings of *Phaseolus vulgaris*. Agric. Biol. Chem. *38*, 2521–2527 (1974b)
Hirai, N., Fukui, H., Koshimizu, K.: A novel abscisic acid metabolite from seeds of *Robinia pseudacacia*. Phytochemistry *17*, 1625–1627 (1978)
Hoad, G.V.: Effect of water stress on abscisic acid levels in white lupin (*Lupinus albus* L.) fruit, leaves and phloem exudate. Planta *142*, 287–290 (1978)
Hodges, R., Porte, A.L.: The structure of loliolide: a terpene from *Lolium perenne*. Tetrahedron *20*, 1463–1467 (1964)
Hofinger, M., Böttger, M.: Identification by GC-MS of 4-chloroindolylacetic acid and its methyl ester in immature *Vicia faba* seeds. Phytochemistry *18*, 653–654 (1979)
Holub, M., Samek, Z., Poplawski, J.: Loliolide from *Arnica montana*. Phytochemistry *14*, 1659 (1975)
Hopping, M.E., Bukovac, M.J.: Endogenous plant growth substances in developing fruit of *Prunus cerasus* L. III. Isolation of indole-3-acetic acid from the seed. J. Am. Soc. Hortic. Sci. *100*, 384–386 (1975)
Hopping, M.E., Young, H., Bukovac, M.J.: Endogenous plant growth substances in developing fruit of *Prunus cerasus* L. VI. Cytokinins in relation to initial fruit development. J. Am. Soc. Hortic. Sci. *104*, 47–52 (1979)
Horgan, R.: Nature and distribution of cytokinins. Philos. Trans. R. Soc. London Ser. B *284*, 439–447 (1978)
Horgan, R., Hewett, E.W., Purse, J.G., Horgan, J.M., Wareing, P.F.: Identification of a cytokinin in sycamore sap by gas chromatography-mass spectrometry. Plant Sci. Lett. *1*, 321–324 (1973)
Horgan, R., Hewett, E.W., Horgan, J.M., Purse, J., Wareing, P.F.: A new cytokinin from *Populus robusta*. Phytochemistry *14*, 1005–1008 (1975)
Hori, S.: Some observations on "bakanae" disease of the rice plant. Mem. Agric. Res. Stn. *12*, (1), 110–119 (1898)
Huelin, F.E.: Effects of ethylene and of apple vapours on the sprouting of potatoes. Rep. Food Invest. Bd. Gr. Br. *1932*, 51–53 (1933)
Huelin, F.E., Kennett, B.H.: Nature of the olefines produced by apples. Nature (London) *184*, 996 (1959)
Hwang, Y.-S., Matsui, M.: Synthesis of the stereoisomeric mixture of the compound having the proposed structure for "auxin b lactone". Agric. Biol. Chem. *32*, 81–87 (1968)
Ichihara, K., Kawai, T., Kaji, M., Noda, M.: A new polyacetylene from *Solidago altissima* L. Agric. Biol. Chem. *40*, 353–358 (1976)
Ichihara, K., Kawai, T., Noda, M.: Polyacetylenes of *Solidago altissima* L. Agric. Biol. Chem. *42*, 427–431 (1978)
Iino, Y., Tanaka, A., Yamashita, K.: Plant growth inhibitors I. Plant growth inhibitory activities of synthetic α-methylene-γ-butyrolactones. Agric. Biol. Chem. *36*, 2505–2509 (1972)
Ina, K., Sakato, Y., Fukami, H.: Isolation and structural elucidation of theaspirone, a component of tea essential oil. Tetrahedron Lett. 2777–2780 (1968)

Iriuchijima, S., Tamura, S.: Stereochemistry of pyrethrosin, cyclopyrethrosin acetate and isocyclopyrethrosin acetate. Agric. Biol. Chem. *34*, 204–209 (1970)

Isoe, S., Hyeon, S.B., Katsumura, S., Sakan, T.: Photooxygenation of carotenoids. II. The absolute configuration of loliolide and dihydroactinidiolide. Tetrahedron Lett. 2517–2520 (1972)

Isogai, Y., Okamoto, T., Koizumi, T.: Isolation of indole-3-acetamide, 2-phenylacetamide and indole-3-carboxaldehyde from etiolated seedlings of *Phaseolus*. Chem. Pharm. Bull. *15*, 151–158 (1967a)

Isogai, Y., Okamoto, T., Komoda, Y.: Isolation of a plant growth inhibitory substance from garden peas (*Pisum sativum* L.) and its identification wth (+)-Abscisin II. Chem. Pharm. Bull. *15*, 1256–1257 (1967b)

Isogai, Y., Komoda, Y., Okamota, T.: Plant growth regulators in the pea plant (*Pisum sativum* L.) Chem. Pharm. Bull. *18*, 1872–1879 (1970)

Isogai, Y., Nomoto, S., Noma, T., Okamoto, T.: Isolation of parahydroxyphenylacetic acid from scape-sprouting bulbs of *Lycoris radiata* Herb. In: Plant growth substances 1973. pp. 9–14. Tokyo: Hirokawa Publishing Co. Inc. 1974

Itai, C., Weyers, J.D.B., Hillman, J.R., Meidner, H., Willmer, C.: Abscisic acid and guard cells of *Commelina communis* L. Nature (London) *271*, 652–653 (1978)

Ito, S., Kodama, M.: Norditerpene dilactones from *Podocarpus* species. Heterocycles *4*, 595–624 (1976)

IUPAC-IUB Commission on Biochemical Nomenclature: Abbreviations and symbols for nucleic acids, polynucleotides and their constituents. J. Biol. Chem. *245*, 5171–5176 (1970)

Jablonski, J., Skoog, F.: Cell enlargement and cell division in excised tobacco pith tissue. Physiol. Plant *7*, 16–24 (1954)

Jacobson, A., Corcoran, M.R.: Tannins as gibberellin antagonists in the synthesis of α-amylase and acid phosphatase by barley seeds. Plant Physiol. *59*, 129–133 (1977)

Jenkins, P.A., Shepherd, K.R.: Identification of abscisic acid in young stems of *Pinus radiata*. New Phytol. *71*, 501–511 (1972)

Jerie, P.H., Hall, M.A.: Identification of ethylene oxide as a major metabolite of ethylene in *Vicia faba*. L. Proc. R. Soc. London Ser. B *200*, 87–94 (1978)

Johnson, A.W.: The interface of academic and industrial chemistry. Chem. Brit. *14*, 332–337 (1978)

Jones, D.A.: Cyanogenic glycosides and their function. In: Phytochemical ecology. Harborne, J.B. (ed.), pp. 103–124. New York, London: Academic Press 1972

Jones, E.R.H., Taylor, W.C.: Some indole constituents of cabbage. Nature (London) *79*, 1138 (1957)

Jones, K.C.: Similarities between gibberellins and related compounds in inducing acid phosphatase and reducing sugar release from barley endosperm. Plant Physiol. *44*, 1695–1700 (1969)

Jones, K.C., West, C.A., Phinney, B.O.: Isolation, identification and biological properties of gibberellin A_{14} from *Gibberella fujikuroi*. Phytochemistry *8*, 283–291 (1968)

Jones, M.B., Enzie, J.V.: Identification of a cyanogenetic growth-inhibiting substance in extracts from peach flower buds. Science *134*, 284 (1961)

Jones, W.W., Coggins, C.W., Embleton, T.W.: Endogenous ABA in relation to bud growth in alternate bearing 'Valencia' orange. Plant Physiol. *58*, 681–682 (1976)

Kaiss-Chapman, R.W., Morris, R.O.: Trans-zeatin in culture filtrates of *Agrobacterium tumefaciens*. Biochem. Biophys. Res. Commun. *76*, 453–459 (1977)

Kalsi, P.S., Vij, V.K., Singh, O.S., Wadia, M.S.: Terpenoid lactones as plant growth regulators. Phytochemistry *16*, 784–786 (1977)

Kalsi, P.S., Singh, O.S., Chhabra, B.R.: Cross conjugated terpenoid ketones: a new group of plant growth regulators. Phytochemistry *17*, 576–577 (1978)

Kalsi, P.S., Chhabra, B.R., Singh, O.S.: Conjugated terpenoid ketones: a new group of plant growth regulators. Experientia *35*, 481–482 (1979)

Kamienska, A., Durley, R.C., Pharis, R.P.: Isolation of gibberellins A_3, A_4 and A_7 from *Pinus attenuata* pollen. Phytochemistry *15*, 421–424 (1976)

Kanchan, S.D.: Growth inhibitor from *Parthenium hysterophorus* Linn. Curr. Sci. *44*, 358–359 (1975)

Karrer, W.: Konstitution und Vorkommen der organischen Pflanzenstoffe. Basel: Birkhäuser Verlag 1958

Kato, J., Katsumi, M., Tamura, S., Sakurai, A.: Plant growth-regulating activities of helminthosporol and its derivatives. In: Biochemistry and physiology of plant growth substances. Wightman, F., Setterfield, G. (eds.), pp. 347–359. Ottawa: The Runge Press 1968

Kato, T., Kabuto, C., Sasaki, N., Tsunagawa, M., Aizawa, H., Fujita, K., Kato, Y., Kitahara, Y.: Momilactones, growth inhibitors from rice, *Oryza sativa* L. Tetrahedron Lett. 3861–3864 (1973)

Kato, T., Tsunakawa, M., Sasaki, N., Aizawa, H., Fujita, K., Kitahara, Y., Takahashi, N.: Growth and germination inhibitors in rice husks. Phytochemistry *16*, 45–48 (1977)

Kato, T., Kobayashi, M., Sasaki, N., Kitahara, Y., Takahashi, N.: The coumarin heraclenol as a growth inhibitor in parsley seeds. Phytochemistry *17*, 158–159 (1978)

Katsumi, M., Phinney, B.O., Jefferies, P.R., Henrick, C.A.: Growth response of the d-5 and an-1 mutants of maize to some kaurene derivatives. Science *144*, 849–850 (1964)

Kaufman, P.B., Ghosheh, N.S., Nakosteen, L., Pharis, R.P., Durley, R.C., Morf, W.: Analysis of native gibberellins in the internode, nodes, leaves and inflorescence of developing *Avena* plants. Plant Physiol. *58*, 131–134 (1976)

Kawarada, A., Sumiki, Y.: The occurrence of gibberellin A_1 in water sprouts of citrus. Bull. Agric. Chem. Soc. Jpn. *23*, 343–344 (1959)

Kefeli, V.I.: Natural plant growth inhibitors and phytohormones. The Hague: Junk 1978

Kefeli, V.I., Kadyrov, C.Sh.: Natural growth inhibitors, their chemical and physiological properties. Annu. Rev. Plant Physiol. *22*, 185–196 (1971)

Kefeli, V.I., Kutacek, M.: Phenolic substances and their possible role in plant growth regulation. In: Plant growth regulation. Pilet, P.E. (ed.), pp. 181–188. Berlin-Heidelberg-New York: Springer 1977

Kende, H.: The cytokinins. Int. Rev. Cytol. *31*, 301–338 (1971)

Khan, A.A.: Primary, preventative and permissive roles of hormones in plant systems. Bot. Rev. *41*, 391–420 (1975)

Khurana, J.P., Makeshwari, S.C.: Induction of flowering in *Lemna paucicostata* by salicyclic acid. Plant Sci. Lett. *12*, 127–131 (1978)

Kienzle, F., Mayer, H., Minder, R.E., Thommer, H.: Synthese von optisch aktiven, natürlichen carotinoiden und strukturell verwandten Verbindungen. III. Synthese von (+)-Abscisinsäure, (−)-xanthoxin, (−)-loliolid, (−)-actinidiolid and (−)-dihydroactinidiolid. Helv. Chim. Acta *61*, 2616–2627 (1978)

Kimura, Y., Takesako, K., Takahashi, Y., Tamura, S.: Isolation, identification and biological activities of growth inhibitors in peanut. Agric. Biol. Chem. *40*, 1183–1188 (1976)

Kindl, H.: Occurrence of indole-3-acetaldehyde oxime. Hoppe Seylers Z. Physiol. Chem. *349*, 519–520 (1968)

King, R.W.: ABA in developing wheat grains and its relationship to grain growth and maturation. Planta *132*, 43–51 (1976)

King, R.W., Evans, L.T., Firn, R.D.: Abscisic acid and xanthoxin contents in the long-day plant *Lolium temulentum* L. in relation to daylength. Aust. J. Plant Physiol. *4*, 217–223 (1977)

Kiss, G., Neukom, H.: Über die Struktur des Ascorbigens. Helv. Chim. Acta *49*, 989–992 (1966)

Kitahara, Y., Yanagawa, H., Kato, T., Takahashi, N.: Asparagusic acid, a new plant growth inhibitor in etiolated young asparagus shoots. Plant Cell Physiol. *13*, 923–925 (1972)

Kjaer, A.: Naturally derived *iso*-thiocyanates (mustard oils) and their parent glucosides. Fortschr. Chem. Org. Naturst. *18*, 122–176 (1960)

Klämbt, D.: Cytokinine aus *Helianthus annuus*. Planta *82*, 170–178 (1968)

Klämbt, D., Thies, G., Skoog, F.: Isolation of cytokinins from *Corynebacterium fascians*. Proc. Natl. Acad. Sci. USA *56*, 52–59 (1966)

Knight, L.I. Crocker, W.: Toxicity of smoke. Bot. Gaz. *55*, 337–371 (1913)

Kobayashi, A., Kaiya, S., Yamashita, K.: Synthesis of a new acetylenic compound isolated from *Solidago altissima*. Agric. Biol. Chem. *40*, 2257–2260 (1976)

Kochert, G.: Sexual pheromones in algae and fungi. Annu. Rev. Plant Physiol. *29*, 461–486 (1978)

Kögl, F., Erxleben, H.: Über die Konstitution der Auxine a und b. Hoppe-Seylers Z. Physiol. Chem. *227*, 51–73 (1934)

Kögl, F., Kostermanns, D.G.F.R.: Hetero-auxin aus Stoffwechselprodukt niederer pflanzlicher Organismen. Isolierung aus Hefe. Hoppe-Seylers Z. Physiol. Chem. *228*, 113–121 (1934)

Kögl, F., Haagen-Smit, A.J., Erxleben, H.: Über ein Phytohormon der Zellstrechung. Reindarstellung des Auxins aus menschlichem Harn. Hoppe Seylers Z. Physiol. Chem. *214*, 241–261 (1933)

Kögl, F., Erxleben, H., Haagen-Smit, A.J.: Über die Isolierung der Auxine a und b aus pflanzlichen Materialien. Hoppe Seylers Z. Physiol. Chem. *225*, 215–229 (1934a)

Kögl, F., Haagen-Smit, A.J., Erxleben, H.: Über ein neues Auxin ("Hetero-auxin") aus Harn. Hoppe Seylers Z. Physiol. Chem. *228*, 90–103 (1934b)

Koeppe, D.E.: Reactions of isolated corn mitochondria influenced by juglone. Physiol. Plant. *27*, 89–94 (1972)

Koeppe, D.E., Miller, R.J.: Kaempferol inhibitions of corn mitochondrial phosphorylation. Plant Physiol. *54*, 374–378 (1974)

Komai, K., Iwamura, J., Ueki, K.: Isolation, identification and physiological activities of sesquiterpenes in purple nutsedge tubers. Zasso Kenkyo *22*, 14–18 (1977)

Komissarenko, N.F., Zhamba, G.E., Garshtya, L.Ya., Bukolova, T.P.: Phytotoxic properties of some coumarins and furocoumarins. Fiziol. Biokhim. Osn. Vzaimodeistviya Rast. Fitotsenozakh. *2*, 69–74 (1971)

Komoda, Y., Isogai, Y.: Biological activities of maytansine on plants. Sci. Pap. Coll. Gen. Educ. Univ. Tokyo *28*, 129–134 (1978)

Komoda, Y., Isogai, Y., Okamoto, T.: Isolation of gibberellin A_{20} from pea pods. Sci. Pap. Coll. Gen. Educ. Univ. Tokyo *18*, 221–230 (1968)

Komoto, N., Isogai, S., Tamura, S.: Isolation of acidic growth inhibitors in dwarf peas. Agric. Biol. Chem. *36*, 2547–2553 (1972a)

Komoto, N., Noma, M., Ikegami, S., Tamura, S.: Isolation of neutral growth inhibititors in dwarf peas. Agric. Biol. Chem. *36*, 2555–2561 (1972b)

Konishi, T.A.: Empty-fruited seedling (Shunanae). In: Agricultural reminiscences (Nogyo-Yowa). Edo (Tokyo) (1828)

Koshimizu, K., Fukui, H., Mitsui, T., Ogawa, Y.: Identity of lupine inhibitor with abscisin II and its biological activity on growth of rice seedlings. Agric. Biol. Chem. *30*, 941–943 (1966)

Koshimizu, K., Matsubara, S., Kasaki, T., Mitsui, T.: Isolation of a new cytokinin from immature yellow lupin seeds. Agric. Biol. Chem. *31*, 795–801 (1967)

Koshimizu, K., Inui, M., Fukui, H., Mitsui, T.: Isolation of (±)-abscisyl-β-D-glucopyranoside from immature fruits of *Lupinus luteus*. Agric. Biol. Chem. *32*, 788–791 (1968a)

Koshimizu, K., Fukui, H., Kusaki, T., Ogawa, Y., Mitsui, T.: Isolation and structure of gibberellin A_{18} from immature seeds of *Lupinus luteus*. Agric. Biol. Chem. *32*, 1135–1140 (1968b)

Koshimizu, K., Ishii, H., Fukui, H., Mitsui, T.: Gibberellin A_{18} and A_{23} from immature seeds of *Wisteria floribunda*. Phytochemistry *11*, 2355 (1972)

Koyama, S., Kawai, H.: Isolation and identification of *trans*-zeatin from the roots of *Raphanus sativus* L. cv. Sakurajima. Agric. Biol. Chem. *42*, 1997–2001 (1978)

Krasnuk, M., Witham, F.H., Tegley, J.R.: Cytokinins extracted from Pinto bean fruit. Plant Physiol. *48*, 320–324 (1971)

Krausz, J.P.: Ph.D. Dissertation, Cornell University, USA (1976)

Kriedemann, P.E., Loveys, B.R., Downton, W.J.S.: Internal control of stomatal physiology and photosynthesis II. Photosynthetic responses to phaseic acid. Aust. J. Plant Physiol. *2*, 553–567 (1975)

Kubo, L., Muira, L., Pettei, M.J., Lee, Y.W., Pilkiewicz, F., Nakanishi, K.: Muzigadial and warburganal, potent antifungal, antiyeast and African army worm antifeedant agents. Tetrahedron Lett. 4553–4556 (1977)

Kuhnle, J.A., Corse, J., Chan, B.G.: Promotion of rooting of mung bean cuttings by

dihydroasparagusic acid synergistic interaction with IAA. Biochem. Physiol. Pflanz. *167*, 541–552 (1975)
Kupchan, M.S., Hemingway, R.J., Wasner, D., Karim, A.: Vernolepin, a novel elamolide dilactone tumor inhibitor from *Vernonia hymenolepis*. J. Am. Chem. Soc. *90*, 3596–3597 (1968)
Kupchan, M.S., Komoda, Y., Court, W.A., Thomas, G.J., Smith, R.M., Karim, A., Gilmore, C.J., Haltiwanger, R.C., Bryan, R.F.: Maytansine, a novel antileukemic ansa macrolide from *Maytenus ovatus*. J. Am. Chem. Soc. *94*, 1354–1356 (1972)
Kuraishi, S.: Effect of kinetin analogs on leaf growth. Sci. Pap. Coll. Gen. Educ. Univ. Tokyo *9*, 67–104 (1959)
Kurogochi, S., Murofushi, N., Ota, Y., Takahashi, N.: Gibberellins and inhibitors in the rice plant. Agric. Biol. Chem. *42*, 207–208 (1978)
Kurosawa, E.: Experimental studies on the secretion of *Fusarium heterosporum* on rice plants. Trans. Natl. Hist. Soc. Formosa *16*, 213–227 (1926)
Kutáček, M., Kefeli, V.I.: The present knowledge of indole compounds in plants of the Brassicaceae family. In: Biochemistry and physiology of plant growth substances. Wightman, F., Setterfield, G. (eds.), pp. 127–152. Ottawa: The Runge Press 1968
Larsen, P.: Formation, occurrence and inactivation of growth substances. Annu. Rev. Plant Physiol. *2*, 169–198 (1951)
Lavie, O., Levy, D.C., Cohen, A., Evenari, M., Gutterman, Y.: New germination inhibitor from *Aegilops ovata*. Nature (London) *249*, 388 (1974)
Lenton, J.R., Perry, V.M., Saunders, P.F.: The identification and quantitative analysis of abscisic acid in plant extracts by gas liquid chromatography. Planta *96*, 271–280 (1971)
Lenton, J.R., Perry, V.M., Saunders, P.F.: Endogenous abscisic acid in relation to photoperiodically induced bud dormancy. Planta *106*, 13–22 (1972)
Leonard, N.J., Fuji, T.: The synthesis of compounds possessing kinetin activity. The use of a blocking group at the 9-position of adenine for the biosynthesis of 1-substituted adenines. Proc. Natl. Acad. Sci. USA *51*, 73–75 (1964)
Leonard, N.J., Playtis, A.J., Skoog, F., Schmitz, R.Y.: A stereoselective synthesis of *cis*-zeatin. J. Am. Chem. Soc. *93*, 3056–3058 (1971)
Leopold, A.C., Scott, F.I., Klein, W.H., Ranstad, E.: Chelidonic acid and its effect on plant growth. Physiol. Plant. *5*, 85–90 (1952)
Le Page-Degivry, M.T., Bulard, C., Milborrow, B.V.: Mise en évidence de l'acide (+)-abscissique chez une gymnosperme. C. R. Acad. Sci. D *269*, 2534–2536 (1969)
Le Poidevin, N.: Inhibition of the germination of mustard seeds by saturated fatty acids. Phytochemistry *4*, 525–526 (1965)
Letcher, R.M.: Structure and synthesis of growth inhibitor batatasin I from *Dioscorea batatas*. Phytochemistry *12*, 2789–2790 (1973)
Letham, D.S.: Zeatin, a factor inducing cell division from *Zea mays*. Life Sci. *8*, 569–573 (1963)
Letham, D.S.: Cytokinins from *Zea mays*. Phytochemistry *12*, 2445–2455 (1973)
Letham, D.S.: The cytokinins of coconut milk. Physiol. Plant. *32*, 66–70 (1974)
Letham, D.S.: Cytokinins. In: Phytohormones and related compounds: a comprehensive treatise. Letham, D.S., Goodwin, P.B., Higgins, T.J.V. (eds.), Vol. I, pp. 205–263. Amsterdam, Oxford, New York: Elsevier/North Holland Biomedical Press 1978a
Letham, D.S.: Naturally occurring plant growth regulators other than the principal hormones of higher plants. In: Phytohormones and related compounds: a comprehensive treatise. Letham, D.S., Goodwin, P.B., Higgins, T.J.V. (eds.), Vol. I. pp. 349–417. Amsterdam, Oxford, New York: Elsevier/North Holland Biomedical Press 1978b
Letham, D.S., Miller, C.O.: Identity of kinetin-like factors from *Zea mays*. Plant Cell. Physiol. *6*, 355–359 (1965)
Letham, D.S., Shannon, J.C., MacDonald, I.R.C.: The structure of zeatin, a (kinetin-like) factor inducing cell division. Proc. Chem. Soc. London 230–231 (1964)
Letham, D.S., Shannon, J.S., McDonald, I.R.C.: The identity of zeatin. Tetrahedron *23*, 479–486 (1967)
Little, C.M.A., Strunz, G.M., La France, R., Bonga, J.M.: Identification of abscisic acid in *Abies balsamea*. Phytochemistry *11*, 3535 (1972)

Little, C.H.A., Heald, J.K., Browning, G.: Identification and measurement of indoleacetic acid and abscisic acid in the cambial region of *Picea sitchensis* (Bong.) Carr. by combined gas chromatography-mass spectrometry. Planta *139*, 133–138 (1978)

Liu, W.-C., Carns, H.R.: Isolation of abscisin and abscission accelerating substance. Science *134*, 384–385 (1961)

Lona, F.: L'azione dell'acido gibberellico sull' accrescimento caulinare di talune piante erbacea in condizioni estene controllate. Nuovo G. Bot. Ital. *63*, 61–76 (1956)

Lorenzi, R., Horgan, R., Wareing, P.F.: Cytokinins in *Picea sitchensis* Carriere – identification and relation to growth. Biochem. Physiol. Pflanz. *168*, 333–340 (1975)

Lorenzi, R., Horgan, R., Heald, J.K.: Gibberellin A_9 glucosyl ester in needles of *Picea sitchensis*. Phytochemistry *15*, 789–790 (1976)

Lorenzi, R., Saunders, P.F., Heald, J.K., Horgan, R.: A novel gibberellin from needles of *Picea sitchensis*. Plant Sci. Lett. *8*, 179–182 (1977)

Loveys, B.R., Kriedemann, P.E.: Internal control of stomatal physiology I. Stomatal regulation and associated changes in endogenous levels of ABA and phaseic acids. Aust. J. Plant Physiol. *1*, 407–415 (1974)

Lynn, D.G., Nakanishi, K., Patt, S.L., Occolowitz, J.L., Almeida, S., Evans, L.S.: Isolation and characterisation of the first mitotic cycle hormone that regulates cell proliferation. J. Am. Chem. Soc. *100*, 7759–7760 (1978)

Maçaira, L.A., Garcia, M., Rabi, J.A.: Chemical transformations of abundant natural products. 3. Modifications of Eremanthin leading to other naturally occurring guaianolides. J. Org. Chem. *42*, 4207–4209 (1977)

MacMillan, J., Pryce, R.J.: Phaseic acid, a putative relative of abscisic acid, from seed of *Phaseolus multiflorus*. Chem. Commun. 124–126 (1968)

MacMillan, J., Pryce, R.J.: Phaseic acid, a relative of abscisic acid from seed of *Phaseolus multiflorus*. Possible structures. Tetrahedron *25*, 5893–5901 (1969a)

MacMillan, J., Pryce, R.J.: The constitution of phaseic acid; a relative of abscisic acid from *Phaseolus multiflorus*. An interpretation of the mass spectrum of phaseic and a probable structure. Tetrahedron *25*, 5903–5914 (1969b)

MacMillan, J., Suter, P.J.: The occurrence of gibberellin A_1 in higher plants: isolation from the seed of runner bean (*Phaseolus multiflorus*). Naturwissenschaften *45*, 46 (1958)

MacMillan, J., Takahashi, N.: Proposed procedure for the allocation of trivial names to the gibberellins. Nature (London) *217*, 170–171 (1968)

MacMillan, J., Wels, C.M.: Detailed analysis of metabolites from mevalonic lactone in *Gibberella fujikuroi*. Phytochemistry *13*, 1394–1417 (1974)

MacMillan, J., Seaton, J.C., Suter, P.J.: Isolation of gibberellin A_1 and gibberellin A_5 from *Phaseolus multiflorus*. Tetrahedron *11*, 60–66 (1960)

MacMillan, J., Seaton, J.C., Suter, P.J.: Isolation and structures of gibberellin A_6 and gibberellin A_8. Tetrahedron *18*, 349–355 (1962)

Mandava, N., Anderson, J.D., Dutky, S.R.: Indole plant-growth inhibitor from *Abrus precatorius* seeds. Phytochemistry *13*, 2853–2856 (1974)

Mansfield, T.A., Wellburn, A.R., Moreira, T.J.S.: The role of abscisic acid and farnesol in the alleviation of water stress. Philos. Trans. R. Soc. London Ser. B *284*, 471–482 (1978)

Manske, R.H.F.: The alkaloids of fumariaceous plants. XVIII. *Fumaria officinalis* L. Can. J. Res. *16*, B, 438–444 (1938)

Mapson, L.W., Hulme, A.C.: The biosynthesis, physiological effects and mode of action of ethylene. Prog. Phytochem. *2*, 343–384 (1970)

Marshall, J.A., Cohen, N.: The structure of alantolactone. J. Org. Chem. *29*, 3727–3729 (1964)

Martin, G.C., Dennis, F.G., Gaskin, P., MacMillan, J.: Identification of gibberellins A_{17}, A_{25}, A_{45}, abscisic acid, phaseic acid, and dihydrophaseic acid in seeds of *Pyrus communis*. Phytochemistry *16*, 605–607 (1977a)

Martin, G.C., Dennis, F.G., MacMillan, J., Gaskin, P.: Hormones in Pear Seeds. I. Levels of gibberellins, abscisic acid, phaseic acid, dihydrophaseic acid, and two metabolites of dihydrophaseic acid in immature seeds of *Pyrus communis* L. J. Am. Soc. Hortic. Sci. *102*, 16–19 (1977b)

Marumo, S., Hattori, H.: Isolation of D-4-chlorotryptophan derivatives as auxin-related metabolites from immature seeds of *Pisum sativum*. Planta *90*, 208–211 (1971)

Marumo, S., Hattori, H., Abe, H., Munakata, K.: Isolation of 4-chloroindolyl-3-acetic acid from immature seeds of *Pisum sativum*. Nature (London) *219*, 959–960 (1968a)

Marumo, S., Abe, H., Hattori, H., Munakata, K.: Isolation of a novel auxin, methyl 4-chloroindoleacetate from immature seeds of *Pisum sativum*. Agric. Biol. Chem. *32*, 117–118 (1968b)

Marx, J.N., White, E.H.: The stereochemistry and synthesis of achillin. Tetrahedron *25*, 2117–2120 (1969)

Masuda, Y.: Effect of light on a growth inhibitor in wheat roots. Plant Physiol. *31*, 780–790 (1962)

Mathur, S.B., Hiremath, S.V., Kulkarni, G.H., Kelkar, G.R., Bhattacharyga, S.C.: Terpenoids – LXX. Structure of dehydrocostus lactone. Tetrahedron *21*, 3575–3590 (1965)

Matsuhima, H., Arima, K.: Physiological activities of zeanic acid, a new plant-growth promoter from corn steep liquor. Agric. Biol. Chem. *37*, 1873–1880 (1973)

Matsushima, H., Fukumi, H., Arima, K.: Isolation of zeanic acid, a natural plant growth-regulator from corn-steep liquor and its chemical structure. Agric. Biol. Chem. *37*, 1865–1872 (1973)

Mauk, C.S., Langille, A.R.: Physiology of tuberisation in *Solanum tuberosum* L. *cis*-Zeatin riboside in the potato plant: its identification and changes in endogenous levels as influenced by temperature and photoperiod. Plant Physiol. *62*, 438–442 (1978)

Mayer, A.M., Evenari, M.: The relation between the structure of coumarin and its derivatives, and their activity as germination inhibitors. J. Exp. Bot. *3*, 246–252 (1952)

Mayer, A.M., Poljakoff-Mayber, A.: The germination of seeds, 2nd Ed. Elmsford, New York: Pergamon Press 1975

McCapra, F., McPhail, A.T., Scott, A.I., Sim, G.A., Young, D.W.: Stereochemistry of gibberellic acid: X-ray analysis of methyl bromogibberellate. J. Chem. Soc. *C*, 1577–1585 (1966)

McCloskey, J.A., Nishimura, S.: Modified nucleosides in transfer RNA. Acc. Chem. Res. *10*, 403–410 (1977)

McClure, J.W.: Physiology and function of flavonoids. In: The flavonoids. Harborne, J.B.., Mabry, T.J., Mabry, H. (eds.), pp. 970–1055. London: Chapman and Hall 1975

McDougall, J., Hillman, J.R.: Purification of IAA from shoot tissues of *Phaseolus vulgaris* and its analysis by GC-MS. J. Exp. Bot. *29*, 375–386 (1978)

McGlasson, W.G., Pratt, H.K.: Effects of wounding on respiration and ethylene production by canteloupe fruit tissue. Plant Physiol. *39*, 128–132 (1964)

McInnes, A.G., Smith, D.G., Arsenault, G.P., Vining, L.C.: Biosynthesis of gibberellins in *Gibberella fujikuroi*. Gibberellin A_{16}. Can. J. Biochem. *51*, 1470–1474 (1973)

McQuillin, F.J.: The structure of Cyperone. Part III. Natural and synthetic cyperones. J. Chem. Soc. 528–534 (1955)

McWha, J.A., Hillman, J.R.: Endogenous ABA in lettuce fruits. Z. Pflanzenphysiol. *74*, 292–297 (1975)

Meigh, D.F.: Nature of the olefines produced by apples. Nature (London) *184*, 1072–1073 (1959)

Meigh, D.F., Filmer, A.A.E., Self, R.: The production of growth suppressing volatile substances by stored potato tubers. II Growth-inhibitory volatile aromatic compounds produced by *Solanum tuberosum* tubers. Phytochemistry *12*, 987–993 (1973)

Meinhenett, R., Carr, D.J.: Growth inhibitors from etiolated leaves of barley (*Hordeum vulgare* L.). Aust. J. Biol. Sci. *26*, 527–538 (1973)

Mendez, J.: Endogenous abscisic acid in umbelliferous fruits. Z. Pflanzenphysiol. *86*, 61–64 (1978)

Mendez, J., Gesto, M.D.V., Vasquez, A., Vieitez, E., Secare, E.: Growth substances isolated from woody cuttings of *Alnus glutinosa* Medic. and *Fraxinus excelsior* L. Phytochemistry *I*, 575–579 (1968)

Metzger, J.D., Zeevaart, J.A.D.: Identification of six endogenous gibberellins in spinach shoots. Plant Physiol. *65*, 623–626 (1980)

Michener, H.D.: The action of ethylene on plant growth. Am. J. Bot. *25*, 711–720 (1938)

Mielke, E.A., Dennis, F.G.: Hormonal control of flower bud dormancy in sour cherry (*Prunus cerasus* L.). I. Identification of abscisic acid. J. Am. Soc. Hortic. Sci. *100*, 285–287 (1975)

Mielke, E.A., Dennis, F.G.: Hormonal control of flower bud dormancy in sour cherry (*Prunus cerasus* L.). III. Effects of leaves, defoliation and temperature on levels of abscisic acid in flower primordia. J. Am. Soc. Hortic. Sci. *103*, 446–449 (1978)

Milborrow, B.V.: The effects of synthetic *dl*-dormin (abscisin II) on the growth of the oat mesocotyl. Planta *20*, 155–171 (1966)

Milborrow, B.V.: The identification of (+)-abscisin II ((+)-dormin) in plants and measurement of concentration. Planta *76*, 93–113 (1967)

Milborrow, B.V.: Identification of "metabolite C" from abscisic acid and a new structure for phaseic acid. J. Chem. Soc. *D*, 966–967 (1969)

Milborrow, B.V.: The metabolism of abscisic acid. J. Exp. Bot. *21*, 17–29 (1970)

Milborrow, B.V.: The chemistry and physiology of abscisic acid. Annu. Rev. Plant Physiol. *25*, 259–308 (1974a)

Milborrow, B.V.: Chemistry and biochemistry of abscisic acid. Rev. Adv. Phytochem. *7*, 57–91 (1974b)

Milborrow, B.V.: The absolute configuration of phaseic and dihydrophaseic acids. Phytochemistry *14*, 1045–1054 (1975)

Milborrow, B.V.: Abscisic acid. In: Phytohormones and related compounds: a comprehensive treatise. Letham, D.S., Goodwin, P.B., Higgins, T.J.V. (eds.), Vol. I, pp. 295–347. Amsterdam, Oxford, New York: Elsevier/North Holland Biomedical Press 1978

Milborrow, B.V., Purse, J.G., Wightman, F.: On the auxin activity of phenylacetic acid. Ann. Bot. (London) *39*, 1143–1146 (1975)

Miller, C.O.: Interaction of 6-methylaminopurine and adenine in division of soy bean callus cells. Nature (London) *194*, 787–788 (1962)

Miller, C.O.: Evidence for the natural occurrence of zeatin and derivatives: compounds from maize which promote cell division. Proc. Natl. Acad. Sci. USA *54*, 1052–1058 (1965)

Miller, C.O.: Zeatin and zeatin riboside from a mycorrhizal fungus. Science *157*, 1055–1057 (1967)

Miller, C.O.: Ribosyl-trans-zeatin, a major cytokinin produced by crown gall tumor tissue. Proc. Natl. Acad. Sci. USA *71*, 334–338 (1974)

Miller, C.O.: Cell division factors from *Vinca rosea* L. crown gall tumor tissue. Proc. Natl. Acad. Sci. USA *72*, 1883–1886 (1975a)

Miller, C.O.: Revised methods for purification of ribosyl *trans*-zeatin from *Vinca rosea* L. crown gall tumor tissue. Plant Physiol. *55*, 448–449 (1975b)

Miller, C.O., Skoog, F., von Saltza, M.H., Strong, F.M.: Kinetin a cell division factor from deoxyribonucleic acid. J. Am. Chem. Soc. *77*, 1329 (1955a)

Miller, C.O., Skoog, F., Okumura, M.H., von Saltza, M.H., Strong, F.M.: Structure and synthesis of kinetin. J. Am. Chem. Soc. *77*, 2662 (1955b)

Miller, C.O., Skoog, F., Okomura, F.S., von Saltza, M.H., Strong, F.M.: Isolation, structure and synthesis of kinetin, a substance promoting cell division. J. Am. Chem. Soc. *78*, 1345–1350 (1956)

Miller, E.V., Winston, J.R., Fisher, D.F.: Production of epinasty by emanations from normal and decaying citrus fruits and from *Penicillium digitatum*. J. Agric. Res. *60*, 269–277 (1940)

Mitchell, E.D., Tolbert, N.E.: Isolation from sugar beet fruit and characterisation of *cis*-4-cyclohexene-1,2-dicarboximide as a germination inhibitor. Biochemistry *7*, 1019–1025 (1968)

Mitsuhashi, M., Shibaoka, H.: Isolation of an inhibitor of growth and root formation from *Portulaca grandiflora* leaves. Plant Cell Physiol. *6*, 87–99 (1965)

Mittelheuser, C.J., van Steveninck, R.F.M.: Stomatal closure and inhibition of transpiration induced by (RS)-abscisic acid. Nature (London) *221*, 281–282 (1969)

Mizuno, K., Komamine, A.: Isolation and identification of substances inducing formation of tracheary elements in cultured carrot-root slices. Planta *138*, 59–62 (1978)

Moewus, F., Schader, E.: Über die keimungs- und wachstumshemmende Wirkung einiger Phthalide. Ber. Dtsch. Bot. Ges. *64*, 127–132 (1951)

Mojé, W.: Organic soil toxins. In: Diagnostic criteria for plants and soils. Chapman, H.D. (ed.), pp. 533–569. Abilene: Quality Printing Company Inc. 1966
Molisch, H.: Der Einfluss einer Pflanze auf die andere Allelopathie. Fischer: Jena 1937
Mollan, R.C., Donnelly, D.M.X., Harmey, M.A.: Synthesis of indole-3-acetylaspartic acid. Phytochemistry *11*, 1485–1488 (1972)
Moreland, D.E., Egley, G.H., Worsham, A.D., Monaco, T.J.: Regulation of plant growth by constituents from higher plants. Adv. Chem. Ser. *53*, 112–141 (1966)
Morimoto, H., Sanno, Y., Oshio, H.: Chemical studies on heliangine, a new sesquiterpenoid lactone isolated from the leaves of *Helianthus tuberosus* L. Tetrahedron *22*, 3173–3179 (1966)
Morris, R.O.: Mass spectroscopic identification of cytokinins. Glucosyl zeatin and glucosyl ribosylzeatin from *Vinca rosea* crown gall. Plant Physiol. *59*, 1029–1033 (1977)
Most, B.H., Gaskin, P., MacMillan, J.: The occurrence of abscisic acid in inhibitors B and C from immature fruit of *Ceretonia siliqua* L. (carob) and in commercial carob syrup. Planta *92*, 41–49 (1970)
Muller, C.H.: Inhibitory terpenes volatilized from *Salvia* shrubs. Bull. Torrey Bot. Club *92*, 38–45 (1965)
Muller, C.H.: Phytotoxins as plant habitat variables. Rec. Adv. Phytochem. *3*, 105–121 (1970)
Muller, C.H., Chou, C.-H.: Phytotoxins: an ecological phase of phytochemistry. In: Phytochemical ecology, Harborne, J.B. (ed.), pp. 201–216. New York, London: Academic Press 1972
Muller, C.H., Muller, W.H., Haines, B.L.: Volatile growth inhibitors produced by aromatic shrubs. Science *143*, 471 (1964)
Murakami, Y.: Dwarfing genes in rice and their relation to gibberellin biosynthesis. In: Plant growth substances 1970. Carr, D.J. (ed.), pp. 166–174. Berlin, Heidelberg, New York: Springer 1972
Murofushi, N., Iriuchijima, S., Takahashi, N., Tamura, S., Kato, J., Wada, Y., Watanabe, E., Aoyama, E.: Isolation and structure of a novel C_{20}-gibberellin in bamboo shoots. Agric. Biol. Chem. *30*, 917–924 (1966)
Murofushi, N., Sugimoto, M., Itoh, K., Takahashi, N.: A novel gibberellin, GA_{57}, produced by *Gibberella fujikwoi*. Agric. Biol. Chem. in press (1980)
Murofushi, N., Takahashi, N., Yokota, T., Tamura, S.: Gibberellins in immature seeds of *Pharbitis nil*. Part I. Isolation and structure of a novel gibberellin, gibberellin A_{20}. Agric. Biol. Chem. *32*, 1239–1245 (1968)
Murofushi, N., Takahashi, N., Yokota, T., Kato, J., Shiotani, Y., Tamura, S.: Gibberellins in immature seeds of *Canavalia*. Part I. Isolation and biological activity of gibberellins A_{21} and A_{22}. Agric. Biol. Chem. *33*, 592–597 (1969a)
Murofushi, N., Takahashi, N., Yokota, T., Tamura, S.: Gibberellins in immature seeds of *Canavalia*. Part II. Structures of gibberellins A_{21} and A_{22}. Agric. Biol. Chem. *33*, 598–609 (1969b)
Murofushi, N., Yokota, T., Watanabe, A., Takahashi, N.: Isolation and characterisation of gibberellins in *Calonyction aculeatum* and structures of gibberellins A_{30}, A_{31}, A_{33} and A_{34}. Agric. Biol. Chem. *37*, 1101–1113 (1973)
Murofushi, N., Sugimoto, M., Itoh, K., Takahashi, N.: Three novel gibberellins produced by *Gibberella fujikuroi*. Agric. Biol. Chem. (Tokyo) *43*, 2179–2185 (1979)
Nakamura, T., Kikuta, M., Watanabe, M., Takahashi, N.: Effect of dihydroconiferyl alcohol on gibberellic acid-induced hook elongation in decotylized pea seedlings. Plant Cell Physiol. *18*, 227–282 (1977)
Nakanishi, K., Endo, M., Näf, U.: Structure of the antheridium-inducing factor of the fern *Anemia phylliditis*. J. Am. Chem. Soc. *93*, 5579–5581 (1971)
Nakanishi, K., Endo, M., Näf, U., McKeon, W., Walker, R.: Isolation of the antheridiogen of *Anemia phyllitidis*. Physiol. Plant. *26*, 183–185 (1972)
Nakazaki, M., Arakawa, H.: The absolute configuration at C_{11} of Santonin. Bull. Chem. Soc. Jpn. *37*, 464–467 (1964)
National Academy of Sciences: Biochemical interactions among plants. Washington D.C. 1971

Neljubov, D.: Über die horizontale Nutation der Stengel von *Pisum sativum* und einigen anderen Pflanzen. Beih. Bot. Zentralber. *10*, 128–138 (1901)

Neljubov, D.: J. Imp. Acad. Sci. (St. Petersburg) Ser VIII *31*, 1–163 (1913)

Nicholls, P.B.: The isolation of indole-3-acetyl-2-0-*myo*-inositol from *Zea mays*. Planta *72*, 258–264 (1967)

Nicholls, P.B., Ong, B.L., Tate, M.E.: Assignment of the ester linkage of 2-0-indoleacetyl-*myo*-inositol isolated from *Zea mays*. Phytochemistry *10*, 2207–2208 (1971)

Niederl, J.B., Brenner, M.W.: Micromethod for the identification and estimation of ethylene in ripening fruits. Mikrochemie *24*, 134–145 (1938)

Niederl, J.B., Brenner, M.W., Kelly, J.N.: The identification and estimation of ethylene in the volatile products of ripening bananas. Am. J. Bot. *25*, 357–361 (1938)

Nielsen, N.: Untersuchungen über einen neuen wachstumsregulierenden Stoff: Rhizopin. Jahrb. Wiss. Bot. *73*, 125–191 (1930)

Nishikawa, N., Kamiya, K., Takabatake, A., Oshio, H., Tomiie, Y., Nitta, I.: The X-ray analysis of dihydroheliangine monochloroacetate. Tetrahedron *22*, 3601–3606 (1966)

Nishimura, S.: Minor components in transfer RNA: their characterization, location and function. Prog. Nucleic Acid Res. Mol. Biol. *12*, 49–85 (1972)

Nishimura, S., Yamada, Y., Ishikura, H.: The presence of 2-methylthio-N^6-(Δ^2-isopentenyl)adenosine in serine and phenylalanine *t*-RNAs from *Escherichia coli*. Biochem. Biophys. Acta *179*, 517–520 (1969)

Nishio, M., Zushi, S., Ishii, T., Furuga, T., Syono, K.: Mass fragmentographic determination of indole-3-acetic acid in callus tissue of *Panax ginseng* and *Nicotiana tabacum*. Chem. Pharm. Bull. *24*, 2038–2042 (1976)

Nitsch, J.P., Nitsch, C.: Composés phenoliques et croissance végétale. Ann. Physiol. Veg. *4*, 211–225 (1962)

Nomoto, M., Tamura, S.: Isolation and identification of indole derivatives in club roots of Chinese cabbage. Agric. Biol. Chem. *34*, 1590–1592 (1970)

Ogunkanmi, A.B., Wellburn, A.R., Mansfield, T.A.: Detection and preliminary identification of endogenous antitranspirants in water-stressed *Sorghum* plants. Planta *117*, 293–302 (1974)

Ohkuma, K., Lyon, J.L., Addicott, F.T., Smith, O.E.: Abscisin II, an abscision-accelerating substance from young cotton fruit. Science *142*, 1592–1593 (1963)

Ohkuma, K., Addicott, F.T., Smith, O.E., Thiessen, W.E.: Structure of abscisin II. Tetrahedron Lett. 2529–2535 (1965)

Okamoto, T., Isogai, Y., Koizumi, T.: Isolation of indole-3-acetic acid, phenylacetic acid and several plant growth inhibitors from etiolated seedlings of *Phaseolus*. Chem. Pharm. Bull. *15*, 159–163 (1967)

Okamoto, T., Torii, Y., Isogai, Y.: Lycoricidinol and lycoricidine, new plant growth regulators in the bulbs of *Lycoris radiata* Herb. Chem. Pharm. Bull. *16*, 1860–1864 (1968)

Oritani, T., Yamashita, K.: Synthesis of (±)-xanthoxin. Agric. Biol. Chem. *37*, 1215–1217 (1973)

Osawa, T., Suzuki, A., Tamura, S.: Isolation of chrysartemins A and B as rooting cofactors in *Chrysanthemum morifolium*. Agric. Biol. Chem. *35*, 1966–1972 (1971)

Osawa, T., Suzuki, A., Tamura, S., Ohashi, Y., Sasada, Y.: Structure of chlorochrymonin, a novel sesquiterpene lactone from *Chrysanthemum morifolium*. Tetrahedron Lett. 5135–5138 (1973)

Osawa, T., Taylor, D., Suzuki, A., Tamura, S.: Revised structure and stereochemistry of chrysartemin B. Tetrahedron Lett. 1169–1172 (1977)

Osborne, D.S.: Ethylene. In: Phytohormones and related compounds: a comprehensive treatise. Letham, D.S., Goodwin, P.B., Higgins, T.J.V. (eds.), Vol. I, pp. 265–294. Amsterdam, Oxford, New York: Elsevier/North Holland Biomedical Press 1978

Overland, L.: The role of allelopathic substances in the 'smother crop' barley. Am. J. Bot. *53*, 423–432 (1966)

Parker, C.W., Letham, D.S., Cowley, D.E., MacLeod, J.K.: Raphanatin, an unusual purine derivative and a metabolite of zeatin. Biochem. Biophys. Res. Commun. *49*, 460–466 (1972)

Patil, S.S.: Toxins produced by phytopathogenic bacteria. Annu. Rev. Phytopathol. *12*, 259–277 (1974)
Paul, A., Bawdekar, A.S., Joshi, R.S., Kalkarni, G.H., Rao, A.S., Kellar, G.R., Bhattacharyya, S.C.: Terpenoids XX. Examination of *Costus* root oil. Perfum. Essent. Oil Rec. *57*, 115–120 (1960)
Pedersen, M.: Identification of cytokinin, 6-(3-methyl-2-butenylamino) purine in sea water and the effect of cytokinins on brown algae. Physiol. Plant. *28*, 101–105 (1973)
Pegg, G.F.: Endogenous auxins in healthy and diseased plants. In: Encyclopedia of plant physiology. Heitefuss, R., Williams, P.H. (eds.), Vol. IV, pp. 560–581. Berlin-Heidelberg-New York: Springer 1976a
Pegg, G.F.: The involvement of ethylene in plant pathogenesis. In: Encyclopedia of plant physiology. Heitefuss, R., Williams, P.H. (eds.), Vol. IV, pp. 582–591. Berlin-Heidelberg-New York: Springer 1976b
Pegg, G.F.: Endogenous gibberellins in healthy and diseased plants. In: Encyclopedia of plant physiology. Heitefuss, R., Williams, P.H. (eds.), Vol. IV, pp. 592–606. Berlin-Heidelberg-New York: Springer 1976c
Pegg, G.F.: Endogenous inhibitors in healthy and diseased plants. In: Encyclopedia of plant physiology. Heitefuss, R., Williams, P.H. (eds.), Vol. IV, pp. 607–616. Berlin-Heidelberg-New York: Springer 1976d
Percival, F.W., Bandurski, R.S.: Esters of indole-3-acetic acid from *Avena* seeds. Plant Physiol. *58*, 60–67 (1976)
Peterson, J.B., Miller, C.O.: Cytokinins in *Vinca rosea* L. crown gall tumor tissue as influenced by compounds containing reduced nitrogen. Plant Physiol. *57*, 393–399 (1976)
Peterson, J.B., Miller, C.O.: Glucosylzeatin and glucosylribosylzeatin from *Vinca rosea* L. crown gall tumor tissue. Plant Physiol. *59*, 1026–1028 (1977)
Phillips, I.D.J.: Some interactions of gibberellic acid with naringenin (5,7,4′-trihydroxy flavanone) in the control of dormancy and growth in plants. J. Exp. Bot. *13*, 213–226 (1962)
Phinney, B.O., West, C.A., Ritzel, M., Neely, P.M.: Evidence for gibberellin-like substances from flowering plants. Proc. Natl. Acad. Sci. USA *43*, 398–404 (1957)
Piironen, E., Virtanen, A.I.: The synthesis of ascorbigen from ascorbic acid and 3-hydroxymethylindole. Acta Chem. Scand. *16*, 1286–1287 (1962)
Piozzi, F., Fuganti, C., Mondelli, R., Geriotti, G.: Narciclasine and narciprimine. Tetrahedron *24*, 1119–1131 (1968)
Piskornik, Z., Bandurski, R.S.: Purification and partial characterization of a glucan containing indole-3-acetic acid. Plant Physiol. *50*, 176–182 (1972)
Playtis, A.J., Leonard, N.J.: The synthesis of ribosyl-*cis*-zeatin and t.l.c. separation of the *cis* and *trans* isomers of ribosyl zeatin. Biochem. Biophys. Res. Commun. *45*, 1–5 (1971)
Pollard, J.K., Shantz, E.M., Steward, F.C.: Hexitols in coconut milk: their role in nurture of dividing cells. Plant Physiol. *36*, 492–501 (1961)
Pollock, B.M., Goodwin, R.H., Greene, S.: Studies on roots. II. Effects of coumarin, scopoletin and other substances on growth. Am. J. Bot. *41*, 521–529 (1954)
Poppleton, B.J.: Podolactone A *p*-bromobenzoate, stereochemistry and absolute configuration. Cryst. Struct. Commun. *4*, 101–106 (1975)
Porter, N.G., van Steveninck, P.F.M.: An abscission-promoting factor in *Lupinus luteus* L. Life Sci. *5*, 2301–2308 (1966)
Pousset, J.-L., Poisson, J.: Vomifoliol: alcool terpenique isolé des feuilles du *Rauwolfia vomitoria* Afz. Tetrahedron Lett. 1173–1174 (1969)
Pratt, H.K., Goeschl, J.D.: Physiological roles of ethylene in plants. Annu. Rev. Plant Physiol. *20*, 541–584 (1969)
Pratt, H.K., Young, R.E., Biale, J.B.: The identification of ethylene as a volatile product of ripening avocados. Plant Physiol. *23*, 526–531 (1948)
Pridham, J.B.: Low molecular weight phenols in higher plants. Annu. Rev. Plant Physiol. *16*, 13–36 (1965)
Prochazha, Z., Sanda, V.: Isolation of pure ascorbigen and some other indole derivatives from Savoy cabbage. Collect. Czech. Chem. Commun. *25*, 270–280 (1960)

Pryce, R.J.: Lunularic acid, a common endogenous growth inhibitor of liverworts. Planta *97*, 354–357 (1971)
Pryce, R.J.: Gallic acid as a natural inhibitor of flowering in *Kalanchoe blossfeldiana*. Phytochemistry *11*, 1911–1918 (1972a)
Pryce, R.J.: The occurrence of lunularic acid and abscisic acids in plants. Phytochemistry *11*, 1759–1761 (1972b)
Przbyllok, T., Nagl, W.: Auxin concentration in the embryo and suspensors of *Tropaeolum majus* as determined by mass fragmentation (single ion detection). Z. Pflanzenphysiol. *84*, 463–465 (1977)
Purse, J.G., Horgan, R., Horgan, J.M., Wareing, P.F.: Cytokinins of sycamore spring sap. Planta *132*, 1–8 (1976)
Putnam, A.R., Duke, W.B.: Allelopathy in agroecosystems. Annu. Rev. Phytopathol. *16*, 431–451 (1978)
Quebedeaux, B., Sweetser, P.B., Powell, J.C.: ABA levels in soybean reproductive structures during development. Plant Physiol. *58*, 363–366 (1976)
Rademacher, W., Graebe, J.E.: Gibberellin A_4 produced by *Sphaceloma manihoticola*, the cause of the superelongation disease of cassava (*Manihot esculenta*). Biochem. Biophys. Res. Commun. *91*, 35–40 (1979)
Radley, M.: Occurrence of substances similar to gibberellic acid in higher plants. Nature (London) *178*, 1070–1071 (1956)
Railton, I.D., Reid, D.M., Gaskin, P., MacMillan, J.: Characterization of ABA in chloroplasts of *Pisum sativum* L. cv. Alaska by combined gas chromatography-mass spectrometry. Planta *117*, 179–182 (1974)
Ramstad, E.: The presence of distribution of chelidonic acid in some plant families. Helv. Physiol. Pharmacol. Acta *28*, 45–57 (1953)
Rapp, A., Ziegler, A.: Nachweis von Abscisinsäure in Weinreben. Vitis *10*, 111–119 (1971)
Raschke, K.: Stomatal action. Annu. Rev. Plant Physiol. *26*, 309–340 (1975)
Rayle, D.L., Purves, W.K.: Isolation and identification of indole-3-ethanol (tryptophol) from cucumber seedlings. Plant Physiol. *42*, 520–524 (1967)
Redeman, C.T., Rappaport, L., Thompson, R.H.: Phaseolic acid: a new plant growth regulator from bean seeds. In: Biochemistry and physiology of plant growth substances. Wightman, F., Setterfield, G. (eds.). Ottawa: The Runge Press 1968
Reed, W., Martin, G.C.: Identification of gibberellin A_{29} and evidence for abscisic acid in extracts from fruits of immature French prune (*Prunus domestica* L.). J. Am. Soc. Hortic. Sci. *101*, 527–531 (1976)
Rehm, S.: The origin of toxic ammonia in germinating garden beet seed. J. Hortic. Sci. *28*, 1–13 (1953)
Reynolds, T.: Comparative effects of aliphatic compounds on inhibition of lettuce fruit germination. Ann. Bot. (London) *41*, 637–648 (1977)
Reynolds, T.: Comparative effects of aromatic compounds on inhibition of lettuce fruit germination. Ann. Bot. (London) *42*, 419–427 (1978)
Rice, E.L.: Allelopathy. New York, London: Academic Press 1974
Rice, E.L.: Allelopathy – an update. Bot. Rev. *45*, 15–109 (1979)
Ries, S.K., Wert, V.: Growth responses of rice seedlings to triacontanol in light and dark. Planta *135*, 77–82 (1977)
Ries, S.K., Wert, V., Sweeley, C.C., Leavitt, R.A.: Triacontanol: a new naturally occurring plant growth regulator. Science *195*, 1339–1340 (1977)
Ries, S.K., Richman, T.L., Wert, V.F.: Growth and yield of crops treated with triacontanol. J. Am. Soc. Hortic. Sci. *103*, 361–364 (1978)
Rivier, L., Pilet, P.-E.: Indolyl-3-acetic acid in cap and apex of maize roots: identification and quantitation by mass fragmentography. Planta *120*, 107–112 (1974)
Rivier, L., Milon, H., Pilet, P.-E.: Gas chromatography-mass spectrometric determination of ABA levels in the cap and apex of maize roots. Planta *134*, 23–27 (1977)
Robins, M.J., Hall, R.H., Thedford, R.: N^6-(Δ^2-isopentenyl)-adenosine. A component of *t*-RNA of yeast and of mammalian tissue, methods of isolation and characterization. Biochemistry *6*, 1837–1848 (1967)

Robinson, P.M., Wareing, P.F.: Chemical nature and biological properties of the inhibitor varying with photoperiod in sycamore (*Acer pseudoplatanus*). Physiol. Plant *17*, 314–323 (1964)
Robinson, T.: Metabolism and function of alkaloids in plants. Science *184*, 430–435 (1974)
Robinson, T.: The organic constituents of higher plants. 3rd Ed. North Amherst: Cordus Press 1975
Rodighiero, G.: Influence of natural furanocoumarins on the germination of seeds and on the growth of lettuce sprouts and roots. G. Biochim. *3*, 138–146 (1954)
Rodriguez, A.G.: Influence of smoke and ethylene on the fruiting of the pineapple (*Ananas sativus* Shult). J. Dep. Agric. Puerto Rico *16*, 5–18 (1932)
Rodriguez, E., Towers, G.H.N., Mitchell, J.C.: Biological activities of sesquiterpene lactones. Phytochemistry *15*, 1573–1580 (1976)
Rojas Garciduenas, M., Dominguez, X.A.: Partenina, achilina y eugarzasadina tres nuevos inhibidores lactónicos del desarrollo vegetal. Turrialba *26*, 10–13 (1976)
Romo, J., Romo de Vivar, A., Trevino, R., Joseph-Nathan, P., Diaz, E.: Constituents of *Artemesia* and *Chrysanthemum* species – the structures of chrysartemins A and B. Phytochemistry *9*, 1615–1621 (1970)
Rosa, J.T.: Shortening the rest period of potatoes with ethylene gas. Potato News Bull. *2*, 363–365 (1925)
Rothwell, K., Wain, R.L.: Growth inhibitor in yellow lupine. Colloq. Int. C.N.R. S. *123*, 363–375 (1964)
Rowe, J.W.: The common and systematic nomenclature of cyclic diterpenes, 3rd revision. U.S. Department of Agriculture 1968
Roy, B.N., Roychondhury, N., Bose, T.K., Baeu, R.N.: Endogenous phenolic compounds as regulators of rooting in cuttings. Phyton (Buenos Aires) *30*, 147–151 (1972)
Roy, S., Guha, R., Chakraborty, D.F.: Acanthotoxin: a germination inhibiting lignan from *Zanthoxylum acanthopodium* D.C. Chem. Ind. 231–232 (1977)
Ruddat, M., Lang, A., Mosettig, E.: Gibberellin activity of steviol, a plant terpenoid. Naturwissenschaften *50*, 23–24 (1963)
Rudnicki, R., Hammond, R.K., Bukovac, M.J.: Endogenous plant growth substances in developing fruits of *Prunus cerasus* L. II. Level of extractable *p*-coumaric acid in the pericarp. J. Am. Soc. Hortic. Sci. *98*, 225–229 (1973)
Rudolph, K.: Non-specific toxins. In: Encyclopedia of plant physiology. Heitefuss, R., Williams, P.H. (eds.), Vol. IV, pp. 270–315. Berlin-Heidelberg-New York: Springer 1976
Ryback, G.: Revision of the absolute stereochemistry of (+)-abscisic acid. J. Chem. Soc. *D*, 1190–1191 (1972)
Sahashi, Y.: Über das Vorkommen von Di-hydroxychinolin-carbonsäure (β-Säure von U. Suzuki) in der Reiskleie. Biochem. Z. *159*, 221–234 (1925)
Sahashi, Y.: Über die Konstitution der durch Hydrolyse von Roh-Oryzanin entstehenden β-Säure (Dioxychinolincarbonsäure). II. Biochem. Z. *168*, 69–72 (1926)
Sahashi, Y.: Synthese der durch Hydrolyse des Roh-Oryzanins entstehenden β-Säure (2,6-Dioxychinolin-4-carbonsäure). Biochem. Z. *189*, 208–215 (1927)
Sakan, T., Murai, F., Isoe, S., Hyeon, S.B., Hayashi, Y.: Biologically active C_9-, C_{10}-, and C_{11}-terpenes from *Actinidia polygama, Boschniakia rossica,* and *Menyanthes trifoliata.* Nippon Kagaku Zasshi *90*, 507–528 (1969)
Sakurai, N., Shibata, K., Kaminska, S.: Stimulation of cucumber hypocotyl elongation by dihydroconiferyl alcohol. Interactions between dihydroconiferyl alcohol and auxin or gibberellin. Plant Cell Physiol. *15*, 709–716 (1974)
Sana, I.S., Ota, Y.: Plant growth-regulating activities of nicotinamide. II. Effect of nicotinamide on growth of several crops. Jpn. J. Crop Sci. *46*, 8–13 (1977)
Sana, I.S., Nakayama, M., Ota, Y.: Plant growth-regulating activities of nicotinamide. I. Effect of nicotinamide on growth of rice seedlings. Jpn. J. Crop Sci. *46*, 1–7 (1977)
Sancheza Tames, R., Gesto, M.D.V., Vieitez, E.: Growth substances isolated from tubers of *Cyperus esculentus* var. aureus. Physiol. Plant. *28*, 195–200 (1973)
Sanyal, T., Ganguly, S.N., Sircar, P.K., Sircar, S.M.: Abscisic acid in the leaf of *Vernonia anthelmintica.* Planta *92*, 282–284 (1970)

Satish, S., Bhakuni, D.S.: Constituents of Indian and other plants. Phytochemistry *11*, 2888–2890 (1972)

Scarbrough, E., Armstrong, D.J., Skoog, F., Frihart, C.R., Leonard, N.J.: Isolation of *cis*-zeatin from *Corynebacterium fascians* cultures. Proc. Natl. Acad. Sci. USA *70*, 3825–3829 (1973)

Schneider, E.A., Wightman, F.: Metabolism of auxin in higher plants. Annu. Rev. Plant Physiol. *25*, 487–513 (1974)

Schneider, E.A., Gibson, R.A., Wightman, F.: The native indoles of barley and tomato shoots. J. Exp. Bot. *23*, 152–170 (1972)

Schraudolf, H.: Zur Verbreitung von Glucobrassicin und Neoglucobrassicin in höheren Pflanzen. Experientia *21*, 520–522 (1965)

Schraudolf, H.: Untersuchungen zur Verbreitung von Indolgluosinolaten in Cruciferen. Experientia *24*, 434–435 (1968)

Schreiber, K.: Plant growth inhibitors of plant origin. Environ. Qual. Saf. Suppl. *3* 483–485 (1975)

Schreiber, K., Schneider, K., Sembdner, G., Focke I.: Isolierung von O(2)-Acetyl-Gibberellinsäure als Stoffwechselproduckt von *Fusarium moniliforme* Sheld. Phytochemistry *5*, 1221–1225 (1966)

Schreiber, K., Weiland, J., Sembdner, G.: Isolierung und Struktur eines Gibberellinglucosids. Tetrahedron Lett. 4285–4288 (1967)

Schreiber, K., Weiland, J., Sembdner, G.: Isolierung von Gibberellin-A_8-O(3)-β-D-Glucopyranosid aus Früchten von *Phaseolus coccineus*. Phytochemistry *9*, 189–198 (1970)

Schuman, G.E., McCalla, T.M.: Effect of short-chain fatty acids extracted from beef cattle manure on germination and seedling development. Appl. Environ. Microbiol. *31*, 655–660 (1976)

Schwaer, C.: The effects of isolated components of wormwood on *Foeniculum vulgare*, *Lepidium sativum*, and *Lactuca sativa* var. longifolia. Flora (Jena) *152*, 509–515 (1962)

Scopes, D.I.C., Zarnack, U., Leonard, N.J., Schmitz, R.Y., Skoog, F.: Alternative routes for the genesis of kinetin: a synthetic intramolecular route for 2′-deoxyadenosine to kinetin. Phytochemistry *15*, 1523–1526 (1976)

Seigler, D.S.: The naturally occurring cyanogenic glycosides. Prog. Phytochem. *4*, 83–120 (1977)

Sembdner, G.: Conjugates of plant hormones. In: Biochemistry and chemistry of plant growth regulators. Schreiber, K., Schmitte, H.R., Sembdner, G. (eds.), pp. 283–302. Halle (Saale) G.D.R.: Academy of Sciences of the German Democratic Republic 1974

Sengupta, S.K., Rogers, M.N., Lovak, E.J.: Effects of photoperiod and ethephon treatments on abscisic acid levels in *Chrysanthemum morifolium* Ramat. J. Am. Soc. Hortic. Sci. *99*, 416–420 (1974)

Sequeira, L., Hemingway, R.J., Kupchan, S.M.: Vernolepin: a new, reversible plant growth inhibitor. Science *161*, 789–790 (1968)

Shantz, E.M., Steward, F.C.: The identification of compound A from coconut milk as 1,3-diphenylurea. J. Am. Chem. Soc. *77*, 6351–6353 (1955)

Sharkey, T.D., Raschke, K.: Effects of phaseic acid and dihydrophaseic acid on stomata and the photosynthetic apparatus. Plant Physiol. *65*, 291–297 (1980)

Sharp, P.B., Keitt, G.W., Clum, H.H., Näf, U.: Activity of antheridiogen from the fern *Anemia phyllitidis* in three flowering plant bioassays. Physiol. Plant. *34*, 101–105 (1975)

Shaw, S., Smallwood, B.M., Wilson, D.V.: Synthesis of zeatin, a naturally occurring adenine derivative with plant cell-division promoting activity and its 9β-D-ribofuranoside. J. Chem. Soc. *C*, 921–924 (1966)

Shaybany, B., Martin, G.C.: Abscisic acid identification and its quantitation in leaves of *Juglans* seedlings during waterlogging. J. Am. Soc. Hortic. Sci. *102*, 300–302 (1977)

Shaybany, B., Weinbaum, S.A., Martin, G.C.: Identification of ABA stereoisomers in French prune seeds and association of ABA with ethylene-enhanced prune abscission. J. Am. Soc. Hortic. Sci. *102*, 501–503 (1977)

Sheldrake, A.R.: The production of hormones in higher plants. Biol. Rev. *48*, 509–559 (1973)

Shibaoka, H.: The mechanism of the growth inhibiting effect of light. Plant Cell Physiol. *2*, 175–196 (1961)
Shibaoka, H., Mitsuhashi, M., Shimokoriyama, M.: Promotion of adventious root formation by heliangine and its removal by cysteine. Plant Cell Physiol. *8*, 161–170 (1967a)
Shibaoka, H., Shimokoriyama, M., Iriuchijima, S., Tamura, S.: Promoting activity of terpenic lactones in *Phaseolus* rooting and their reactivity toward cysteine. Plant Cell Physiol. *8*, 297–305 (1967b)
Shibaoka, H., Anzai, T., Mitsuhashi, M., Shimokoriyama, M.: Interaction between heliangine and pyrimidines in adventitious root formation of *Phaseolus* cuttings. Plant Cell Physiol. *8*, 647–656 (1967c)
Shibata, K., Kubota, T., Kamisaka, S.: Isolation and chemical identification of a lettuce cotyledon factor, a synergist of the gibberellin action in inducing lettuce hypocotyl elongation. Plant Cell Physiol. *15*, 191–194 (1974)
Shibuya, T., Funamizu, M., Kitahara, Y.: Abscisic acid from *Pinus densiflora* pollen. Phytochemistry *17*, 322–323 (1978)
Shindy, W.W., Smith, O.E.: Identification of plant hormones from cotton ovules. Plant Physiol. *55*, 550–554 (1975)
Siegel, S.M.: Inhibitory activity of the phenolic glucoside psilotin and its reversal by gibberellic acid and thiols. Phytochemistry *15*, 566–567 (1976)
Sievers, A.F., True, R.H.: A preliminary study of the forced curing of lemons as practiced in California. U.S. Dep. Agric. Bur. Plant Ind. Bull. *232*, 1–38 (1912)
Simonsen, J., Barton, D.H.R.: The terpenes. Vol. III. Cambridge: Cambridge University Press 1952
Simpson, G.M., Wain, R.L.: A relationship between gibberellic acid and light in the control of internode extension of dwarf peas (*Pisum sativum*). J. Exp. Bot. *12*, 207–216 (1961)
Sinska, I., Lewak, S., Gaskin, P., MacMillan, J.: Reinvestigation of apple-seed gibberellins. Planta *114*, 359–364 (1973)
Sircar, P.K., Dey, B., Sanyal, T., Ganguly, S.N., Sircar, S.M.: Gibberellic acid in the floral parts of *Cassia fistula*. Phytochemistry *9*, 735–736 (1970)
Sironval, G., El Tannir-Lomba, J.: Vitamin E and flowering of *Fragaria vesca* L. var. *Semperflorens* Duch. Nature (London) *185*, 855–856 (1960)
Sivakumaran, S., Hall, M.A.: Effects of age and water stress on endogenous levels of plant growth regulators in *Euphorbia lathyrus* L. J. Exp. Bot. *29*, 195–205 (1978)
Skoog, F.: A survey of cytokinins and cytokinin antagonists with reference to nucleic acid and protein metabolism. Biochem. Soc. Symp. *38*, 195–215 (1973)
Skoog, F., Armstrong, D.J.: Cytokinins. Annu. Rev. Plant Physiol. *21*, 359–384 (1970)
Skoog, F., Tsui, C.: Chemical control of growth and bud formation in tobacco stem segments and callus cultured in vitro. Am. J. Bot. *35*, 782–787 (1948)
Skoog, F., Strong, F.M., Miller, C.O.: Cytokinins. Science *148*, 532–533 (1965)
Smith, I.K., Fowden, L.: A study of mimosine toxicity in plants. J. Exp. Bot. *17*, 750–761 (1966)
Smith, T.A.: Tryptamine and related compounds in plants. Phytochemistry *16*, 171–175 (1977)
Söding, H.: Die Auxine. Historische Übersicht. In: Encyclopedia of plant physiology. Ruhland, W. (ed.), Vol. XIV, pp. 450–484. Berlin-Göttingen-Heidelberg: Springer 1961
Sponsel, V.M., MacMillan, J.: Further studies on the metabolism of gibberellins (GAs) A_9, A_{20} and A_{29} in immature seeds of *Pisum sativum* cv. Progress No. 9. Planta *135*, 129–136 (1977)
Sponsel, V.M., Gaskin, P., MacMillan, J.: The identification of gibberellins in immature seeds of *Vicia faba*, and some chemotaxonomic considerations. Planta *146*, 101–105 (1979)
Springer, J.P., Clardy, J., Cox, R.H., Cutler, H.G., Cole, R.J.: The structure of a new type of plant growth inhibitor extracted from immature tobacco leaves. Tetrahedron Lett. 2737–2740 (1975)
Staehelin, M., Rogg, H., Baguley, B.C., Ginsburg, T., Wehrli, W.: Structure of mammalian serine *t*-RNA. Nature (London) *219*, 1363–1365 (1968)

Steffens, G.L., Tso, L.C., Spaulding, D.W.: Fatty alcohol inhibition of tobacco axillary and terminal bud growth. J. Agric. Food Chem. *15*, 972–975 (1967)
Stenlid, G.: The effects of flavonoid compounds on oxidative phosphorylation and on the enzymatic destruction of indoleacetic acid. Physiol. Plant. *16*, 110–120 (1963)
Stenlid, G.: On the physiological effects of phloridzin, phloretin and some related substances upon higher plants. Physiol. Plant. *21*, 882–894 (1968)
Stenlid, G.: Flavonoids as inhibitors of the formation of adenosine triphosphate in plant mitochondria. Phytochemistry *9*, 2251–2256 (1970)
Stenlid, G.: Physiological effects of betalains upon higher plants. Phytochemistry *15*, 661–663 (1976a)
Stenlid, G.: Effects of substituents in the A-ring on the physiological activity of flavones. Phytochemistry *15*, 911–912 (1976b)
Sterns, M.: Crystal and molecular structure of a root inhibitor from *Eucalyptus grandis,* 4-ethyl-1-hydroxy-4,8,8,10,10-pentamethyl-7,9-dioxo-2,3-dioxobicyclo[4.4.0]decene-5. J. Cryst. Mol. Struct. *1*, 373–381 (1971)
Steward, F.C., Krikorian, A.D.: Plants, chemicals and growth. New York, London: Academic Press. 232 pp. 1971
Steward, F.C., Pollard, J.K., Patchett, A.A., Witkop, B.: The effects of selected nitrogen compounds on the growth of plant tissue cultures. Biochim. Biophys. Acta *28*, 308–317 (1958)
Stodola, F.H.: Source book on gibberellin 1828–1957. Agricultural Research Service. U.S. Department of Agriculture 1958
Stodola, F.H., Roper, K.B., Fernel, D.I., Conway, H.F., Sohns, V.E., Langford, C.E., Jackson, R.W.: The microbiological production of gibberellins A and X. Arch. Biochem. Biophys. *54*, 240–245 (1955)
Stodola, F.H., Nelson, G.E.N., Spence, D.J.: The separation of gibberellin A and gibberellic acid on buffered partition columns. Arch. Biochem. Biophys. *66*, 438–443 (1957)
Stowe, B.B.: Occurrence and metabolism of simple indoles in plants. Fortschr. Chem. Org. Naturst. *17*, 248–297 (1959)
Stowe, B.B., Obreiter, J.B.: Growth promotion in pea stem sections. II. By natural oils and vitamins. Plant Physiol. *37*, 158–164 (1962)
Stowe, B.B., Yamaki, T.: The history and physiological action of the gibberellins. Annu. Rev. Plant Physiol. *8*, 181–216 (1957)
Strobel, G.A.: Phytotoxins produced by plant parasites. Annu. Rev. Plant Physiol. *25*, 541–566 (1974)
Strobel, G.A.: Bacterial phytotoxins. Annu. Rev. Microbiol. *31*, 205–224 (1977)
Stuart, K.L., Coke. L.B.: The effect of vomifoliol on stomatal aperture. Planta *122*, 307–310 (1975)
Stuart, K.L., Woo-Ming, R.B.: Vomifoliol in *Croton* and *Palicourea* species. Phytochemistry *14*, 594–595 (1975)
Summons, R.E., MacLeod, J.K., Parker, C.W., Letham, D.S.: The occurrence of raphanatin as an endogenous cytokinin in radish seed. Identification and quantitation by gas chromatographic-mass spectrometric analysis using deuterium labelled standards. FEBS Lett. *82*, 211–214 (1977)
Suzuki, Y., Kinashi, H., Takeuchi, S., Kawarada, A.: (+)-5-Hydroxy-dioxindole-3-acetic acid, a synergist from rice bran of auxin induced ethylene production in plant tissue. Phytochemistry *16*, 635–637 (1977)
Swain, T.: Secondary compounds as protective agents. Annu. Rev. Plant Physiol. *28*, 479–501 (1977)
Swaminathan, S., Bock, R.M.: Isolation and identification of cytokinins from *Euglena gracilis t*-RNA. Biochemistry *16*, 1355–1363 (1977)
Swanson, B.T., Wilkins, H.F., Weiser, C.F., Klein, I.: Endogenous ethylene and abscisic acid relative to phytogerontology. Plant Physiol. *55*, 370–376 (1975)
Takahashi, N., Kitamura, H., Kawarada, A., Seta, Y., Takai, M., Tamura, S., Sumiki, Y.: Isolation of gibberellins and their properties. Bull. Agric. Chem. Soc. Jpn. *19*, 267–277 (1955)

Takahashi, N., Seta, Y., Kitamura, H., Sumiki, Y.: A new gibberellin, gibberellin A_4. Bull. Agric. Chem. Soc. Jpn. *21*, 396 (1957)
Takahashi, N., Yamaguchi, I., Kono, T., Igoshi, M., Hirose, K., Suzuki, K.: Characterisation of plant growth substances in *Citrus unshiu* and their change in fruit development. Plant Cell Physiol. *16*, 1101–1111 (1975)
Takasugu, M., Anetai, M., Katsui, N., Masamune, T.: The occurrence of vomifoliol, dehydrovomifoliol and dihydrophaseic acid in the roots of "kidney bean" (*Phaseolus vulgaris* L.). Chem. Lett. 245–248 (1973)
Takeuchi, S., Kono, Y., Kawarada, A., Ota, Y., Nakayama, M.: Nicotinamide as a plant growth regulator isolated from rice-bulbs. Agric. Biol. Chem. *39*, 859–878 (1975)
Tamura, S.: Plant growth regulators produced by microorganisms. Adv. Pestic. Sci. Plenary Lect. Symp. Pap. Int. Congr. Pestic. Chem. 4th. Geissbuehler, H. ed. *2*, 356–365 (1979)
Tamura, S., Sakurai, A., Kainuma, K., Takai, M.: Isolation of helminthosporol, a natural plant-growth regulator and its chemical structure. Agric. Biol. Chem. *29*, 216–221 (1965)
Tamura, S., Nomoto, M., Nagao, M.: Isolation and characterization of indole derivatives in club roots of Chinese cabbage. In: Plant growth substances 1970. Carr, D.J. (ed.), pp. 127–132. Berlin-Heidelberg-New York: Springer 1972
Tanaka, A., Nakata, T., Yamashita, K.: A facile synthesis of conjugated α-methylene-δ-valerolactones. Agric. Biol. Chem. *37*, 2365–2371 (1973)
Tanaka, Y., Abe, H., Uchiyama, M., Taya, Y., Nishimura, S.: Isopentenyl adenine from *Dictyostelium discoideum*. Phytochemistry *17*, 543–544 (1978)
Tarasov, V.A., Abdullaev, N.D., Kasymov, Sh.Z., Sidyakin, G.P.: Chrysartemin B, a sesquiterpene lactone from *Handelia trichophylla*. Khim. Prir. Soedin. 667–668 (1976)
Taylor, H.F., Burden, R.S.: Xanthoxin, a new naturally occurring plant growth inhibitor. Nature (London) *227*, 302–304 (1970a)
Taylor, H.F., Burden, R.S.: Identification of plant growth inhibitors produced by photolysis of violaxanthin. Phytochemistry *9*, 2217–2223 (1970b)
Taylor, H.F., Burden, R.S.: Xanthoxin a recently discovered plant growth inhibitor. Proc. R. Soc. London Ser. B *180*, 317–346 (1972)
Taylor, H.F., Burden, R.S.: Preparation and metabolism of 2-[^{14}C]-*cis, trans*-xanthoxin. J. Exp. Bot. *24*, 873–880 (1973)
Taylor, H.F., Smith, T.A.: Production of plant growth inhibitors from xanthophylls. Nature (London) *215*, 1513–1514 (1967)
Taylor, I.E.P., Wilkinson, A.J.: The occurrence of gibberellins and gibberellin-like substances in algae. Phycologia *16*, 37–42 (1977)
Taylor, P.A., Kosuge, T., Devay, J.E.: Compounds associated with cytokinin activity in fruitlets and tracheal fluid of *Gossypium hirsutum*. Physiol. Plant. *30*, 119–124 (1974)
Thakur, M.L.: Phenolic growth inhibitors isolated from dormant buds of sugar maple (*Acer saccharum* Marsh). J Exp. Bot. *28*, 795–803 (1977)
Thiess, D.E., Lichtenthaler, H.K.: Inhibition of seed germination by low-molecular-weight monohydric alcohols. Naturwissenschaften *60*, 302 (1973)
Thimann, K.V.: On the plant growth hormone produced by *Rhizopus suinus*. J. Biol. Chem. *109*, 279–291 (1935)
Thimann, K.V.: Physiology of development: The hormones. In: Plant physiology. Steward, F.C. (ed.), Vol. VIB. New York, London: Academic Press 1972
Thimann, K.V., Bonner, W.D.: Inhibition of plant growth by protoanemonin and coumarin and its prevention by BAL. Proc. Natl. Acad. Sci. USA *35*, 272–276 (1949)
Thompson, A.G., Bruinsma, J.: Xanthoxin: a growth inhibitor in light-grown sunflower seedlings, *Helianthus annuus* L. J. Exp. Bot. *28*, 804–810 (1977)
Thompson, A.G., Horgan, R., Heald, J.K.: A quantitative analysis of cytokinin using single-ion-current-monitoring. Planta *124*, 207–210 (1975)
Thomson, R.H.: Naturally occurring quinones. New York, London: Academic Press 1971
Tillberg, E.: Occurrence of endogenous indol-3-yl-aspartic acid in light and dark grown bean seedlings (*Phaseolus vulgaris*). Physiol. Plant. *31*, 271–274 (1974)
Tinelli, E.T., Sondheimer, E., Walton, D.C., Gaskin, P., MacMillan, J.: Metabolites of 2-^{14}C-abscisic acid. Tetrahedron Lett. 139–140 (1973)

Tonzig, S., Marrè, E.: Ascorbic acid as a growth hormone. In: Plant growth regulation. 4th Int. Conf. pp. 725–734. Yonkers, New York 1961
Toole, E.H., Hendricks, S.B., Borthwick, H.A., Toole, V.K.: Physiology of seed germination. Annu. Rev. Plant Physiol. *7*, 299–324 (1956)
Tsunakawa, M., Ohba, A., Sasaki, N., Kabuto, C.: Momilactone-C, a minor constituent of growth inhibitors in rice husk. Chem. Lett. 1157–1158 (1976)
Ueda, M., Bandurski, R.S.: Structure of indole-3-acetic acid *myo*-inositol esters and pentamethyl *myo*-inositols. Phytochemistry *13*, 243–248 (1974)
Ueda, M., Ehmann, A., Bandurski, R.S.: Gas-liquid chromatographic analysis of indole-3-acetic acid *myo*-inositol esters in maize kernels. Plant Physiol. *46*, 715–719 (1970)
Valio, I.M.F., Rocha, R.F.: Physiological effects of steviol. Z. Pflanzenphysiol. *78*, 90–94 (1976)
Valio, I.M.F., Schwabe, W.W.: Growth and dormancy in *Lunularia cruciata* (L.) Dum. J. Exp. Bot. *21*, 138–150 (1970)
Valio, I.M.F., Burden, R.S., Schwabe, W.W.: New natural growth inhibitor in the liverwort *Lunularia cruciata* (L.) Dum. Nature (London) *223*, 1176–1178 (1969)
Van Rompuy, L.L.L., Zeevaart, J.A.D.: Are steroidal estrogens natural plant constituents? Phytochemistry *18*, 863–865 (1979)
Van Staden, J., Drewes, S.E.: Isolation and identification of zeatin from malt extract. Plant Sci. Lett. *4*, 391–394 (1975a)
Van Staden, J., Drewes, S.E.: Identification of zeatin and zeatinriboside in coconut milk. Physiol. Plant. *34*, 106–109 (1975b)
Van Staden, J., Menary, R.C.: Identification of cytokinins in the xylem sap of tomato. Z. Pflanzenphysiol. *78*, 262–265 (1976)
Van Steveninck, R.F.M.: Abscission-accelerators in lupins (*Lupinus luteus* L.). Nature (London) *183*, 1246–1248 (1959)
Van Sumere, C.F.: Germination inhibitors. In: Phenolics in plants in health and disease. Pridham, J.B. (ed.), pp. 25–34. Oxford: Pergamon Press 1960
Van Sumere, C.F., Cottenie, J., De Greef, J., Kint, J.: Biochemical studies in relation to the possible germination regulatory role of naturally occurring coumarin and phenolics. Rec. Adv. Phytochem. *4*, 165–221 (1972)
Van Sumere, C.F., Albrecht, J., Dedonder, A., De Pooter, H., Pé, I.: Plant proteins and phenolics. In: The chemistry and biochemistry of plant proteins. Harborne, J.B., van Sumere, C.F. (eds.), pp. 211–264. New York, London: Academic Press 1975
Varga, M., Köves, E.: Phenolic acids as growth and germination inhibitors in dry fruits. Nature (London) *183*, 401 (1959)
Vaughan, D., Linehan, D.J.: The growth of wheat plants in humic acid solutions under axenic conditions. Plant Soil *44*, 445–449 (1976)
Vazquez, A.: Effect of umbelliferone on rooting in bean cuttings. Plant Sci. Lett. *1*, 433–438 (1973)
Venis, M.A., Watson, P.J.: Naturally occurring modifiers of auxin-receptor interaction in corn: identification as benzoxazinolinones. Planta *142*, 103–107 (1978)
Virtanen, A.E., Hietala, P.K., Wahlroos, O.: An antifungal factor in maize and wheat plants. Suom. Kermistil. B *29* 143 (1956)
Virtanen, A.I.: Studies on organic sulphur compounds and other labile substances in plants. Phytochemistry *4*, 207–228 (1965)
Vliegenhart, J.A., Vliegenhart, J.F.G.: Reinvestigation of authentic samples of auxins A and B, and related products by mass spectrometry. Recl. Trav. Chim. Pays Bas *85*, 1266–1272 (1966)
Vokáč, K., Samek, Z., Herout, V., Šorm, F.: The structure of artabsin and absinthin. Tetrahedron Lett. 3855–3857 (1968)
Vreman, H.J., Skoog, F., Frihart, C.R., Leonard, N.J.: Cytokinins in *Pisum* transfer ribonucleic acid. Plant Physiol. *49*, 848–851 (1972)
Vreman, H.J., Schmitz, R.Y., Skoog, F., Playtis, A.J., Frihart, C.R., Leonard, N.J.: Synthesis of 2-methylthio-*cis*- and *trans*-ribosylzeatin and their isolation from *Pisum t*-RNA. Phytochemistry *13*, 31–37 (1974)
Vreman, H.J., Thomas, R., Corse, J., Swaminathan, S., Murai, N.: Cytokinins in tRNA

obtained from *Spinacia oleracia* L. leaves and isolated chloroplasts. Plant Physiol. *61*, 296–306 (1978)
Wada, T.: Structure of digiprolactone. Chem. Pharm. Bull. *13*, 43–49 (1965)
Wall, M.E., Wani, M.C., Cook, C.E., Palmer, K.H., McPhail, A.T., Sim, G.A.: Plant antitumor agents 1. Isolation and structure of camptothecin, a novel alkaloidal leukemia and tumor inhibitor from *Camptotheca acuminata*. J. Am. Chem. Soc. *88*, 3888–3890 (1966)
Waller, G.R., Burstrom, H.: Diterpenoid alkaloids as plant growth inhibitors. Nature (London) *222*, 576–578 (1969)
Waller, G.R., Nowacki, E.K.: Alkaloid biology and metabolism in plants. New York: Plenum Press 1978
Walton, D.C., Dorn, B., Fey, J.: The isolation of an ABA metabolite 4′-dihydrophaseic acid from non-imbibed *Phaseolus vulgaris* seed. Planta *112*, 87–90 (1973)
Wang, T.L., Horgan, R.: Dihydrozeatin riboside, a minor cytokinin from the leaves of *Phaseolus vulgaris* L. Planta *140*, 151–153 (1978)
Wang, T.L., Thompson, A.G., Horgan, R.: A cytokinin glucoside from the leaves of *Phaseolus vulgaris*. Planta *135*, 285–288 (1977)
Wang, T.S.C., Yang, T.-K., Chuang, T.-T.: Soil phenolic acids as plant growth inhibitors. Soil Sci. *103*, 239–246 (1967)
Wanner, H., Santos, A., Stocker, H.: Different effects of long chain fatty acids on seed germination. Biochem. Physiol. Pflanz. *171*, 391–399 (1977)
Ward, T.M., Wright, M., Roberts, J.A., Self, R., Osborne, D.J.: Analytical procedures for the assay and identification of ethylene. In: Isolation of plant growth substances. Hillman, J.R. (ed.), pp. 135–151. Cambridge: Cambridge University Press 1978
Wareing, P.F.: Endogenous inhibitors in seed germination and dormancy. In: Encyclopedia of plant physiology. Ruhland, W. (ed.), Vol. XV/2, pp. 909–924. Berlin-Göttingen-Heidelberg: Springer 1965
Wareing, P.F.: Abscisic acid as a natural growth regulator. Philos. Trans. R. Soc. London Ser. B *284*, 483–498 (1978)
Wareing, P.F., Eagles, C.F., Robinson, P.M.: Natural inhibitors as dormancy agents. Colloq. Int. C.N.R.S. *123*, 376–386 (1964)
Wareing, P.F., Horgan, R., Henson, I.E., Davis, W.: Cytokinin relations in the whole plant. In: Plant growth regulation. Pilet, P.E. (ed.), pp. 147–153. Berlin-Heidelberg-New York: Springer 1977
Watanabe, N., Yokota, T., Takahashi, N.: Identification of zeatin and zeatin riboside in cones of the hop plant and their possible role in cone growth. Plant Cell Physiol. *19*, 617–625 (1978a)
Watanabe, N., Yokota, T., Takahashi, N.: Identification of N^6-(3-methyl-but-2-enyl) adenosine, zeatin, zeatin riboside, gibberellin A_{19} and abscisic acid in shoots of the hop plant. Plant Cell Physiol. *19*, 1263–1270 (1978b)
Watanabe, N., Yokota, T., Takahashi, N.: *cis*-Zeatin riboside: its occurrence as a free nucleoside in cones of the hop plant. Agric. Biol. Chem. *42*, 2415–2416 (1978c)
Webber, J.E., Laver, M.L., Zaerr, J.B., Lavender, D.P.: Seasonal variation of abscisic acid in the dormant shoots of Douglas-fir. Can. J. Bot. *57*, 534–538 (1979)
Weiss, G., Koreeda, M., Nakanishi, K.: Stereochemistry of theaspirone and the blumenols. J. Chem. Soc. *D*, 565–566 (1973)
Wellburn, A.R., Ogunkanmi, A.B., Fenton, R., Mansfield, T.A.: All-*trans*-farnesol: a naturally occurring antitranspirant? Planta *120*, 255–263 (1974)
Went, F.W.: Wuchsstoff und Wachstum. Recl. Trav. Bot. Néerl. *25*, 1–116 (1928)
Went, F.W.: Plants and the chemical environment. In: Chemical ecology. Sondheimer, E., Simeone, J.B. (eds.), pp. 71–82. New York, London: Academic Press 1970
Went, F.W., Thimann, K.V.: Phytohormones. 294 pp. New York: MacMillan 1937
West, C.A.: The chemistry of gibberellins from flowering plants. In: Plant growth regulation. Klein, R.M. (ed.), pp. 473–482. Iowa: Ames 1961
West, C.A., Phinney, B.O.: Isolation and properties of a gibberellin from *Phaseolus vulgaris* L. J. Am. Chem. Soc. *81*, 2424–2427 (1959)
Wheeler, A.W.: Betaine: a plant-growth substance from sugar beet (*Beta vulgaris*). J. Exp. Bot. *14*, 265–271 (1963)

Wheeler, A.W.: Auxin-like growth activity of phenylacetonitrile. Ann. Bot. (London) *41*, 867–872 (1977)
White, J.C., Medlow, G.C., Hillman, J.R., Wilkins, M.B.: Correlative inhibition of lateral bud growth in *Phaseolus vulgaris* L. Isolation of indoleacetic acid from the inhibitory region. J. Exp. Bot. *26*, 419–424 (1975)
Whittaker, R.H.: The biochemical ecology of higher plants. In: Chemical ecology. Sondheimer, E., Simeone, J.B. (eds.), pp. 43–70. New York, London: Academic Press 1970
Whittaker, R.H., Feeny, P.P.: Allelochemics: Chemical interactions between species. Science *171*, 757–770 (1971)
Wiesner, J.: Die Elementarstruktur und das Wachstum der lebenden Substanz. Vienna: A. Holder 1892
Wightman, F.: Biosynthesis of auxins in tomato shoots. Biochem. Soc. Symp. *38*, 247–275 (1973)
Wightman, F.: Gas chromatographic identification and quantitative estimation of natural auxins in developing plant organs. In: Plant growth regulation. Pilet, P.E. (ed.), pp. 75–90. Berlin-Heidelberg-New York: Springer 1977
Wightman, F., Rauthan, B.S.: Evidence for the biosynthesis and natural occurrence of the auxin, phenylacetic acid in shoots of higher plants. In: Plant growth substances 1973. pp. 15–27. Tokyo: Hirokawa Publishing Co. Inc. 1974
Wilkins, H., Burden, R.S., Wain, R.L.: Growth inhibitors in roots of light- and dark-grown seedlings of *Zea mays*. Ann. Appl. Biol. *78*, 337–338 (1974)
Williams, P.M., Ross, J.D., Bradbeer, J.W.: The abscisic acid content of the seeds and fruits of *Corylus avellana* L. Planta *110*, 303–310 (1973)
Williams, P.M., Bradbeer, J.W., Gaskin, P., MacMillan, J.: The identification and determination of gibberellins A_1 and A_9 in seeds of *Corylus avellana* L. Planta *117*, 101–103 (1974)
Willmer, C.M., Don, R., Parker, W.: Levels of short-chain fatty acids and of abscisic acid in water-stressed and non-stressed leaves and their effects on stomata in epidermal strips and excised leaves. Planta *139*, 281–287 (1978)
Wilson, M.F., Bell, E.A.: Amino acids and β-aminopropionitrile as inhibitors of seed germination and growth. Phytochemistry *17*, 403–406 (1978)
Windholz, M. (ed.): The Merck Index. 9th Ed. Rahway, New Jersey: Merck and Co. Inc. 1976
Winters, T.E., Geissman, T.A., Safir, D.: Sesquiterpene lactones of *Xanthium* species. Xanthanol and isoxanthanol, and correlation with xanthinin and ivalbin. J. Org. Chem. *34*, 153–155 (1969)
Wolf, F.T., Tilford, R.H., Martinez, M.L.: Effects of phenolic acids and their derivatives upon the growth of Avena coleoptiles. Z. Pflanzenphysiol. *80*, 243–250 (1976)
Wöllenweber, H.W.: Fusarium monograph. Parasitic and saprophytic fungi. Z. Parsitenkd. *3*, 269–516 (1931)
Woodward, M.D., Corcuera, L.J., Schnoes, H.K., Helgeson, J.P., Upper, C.D.: Identification of 1,4-benzoxazin-3-ones in maize extracts by gas liquid chromatography-mass spectrometry. Plant Physiol. *63*, 9–13 (1979)
Wright, S.T.C.: A chromatographic study of auxins in relation to fruit morphogenesis and fruit drop in blackcurrant (*Ribes nigrum*). PhD. thesis. Univ. Bristol (1954)
Wright, S.T.C.: An increase in the 'inhibitor β' content of detached wheat leaves following a period of wilting. Planta *86*, 10–31 (1969)
Wright, S.T.C., Hiron, R.W.P.: (+)-Abscisic acid, the growth inhibitor induced in detached wheat leaves by a period of wilting. Nature (London) *224*, 719–720 (1969)
Yabuta, T., Sumiki, Y.: Communication to the editor. J. Agric. Chem. Soc. Jpn. *14*, 1526 (1938)
Yabuta, T., Sumiki, Y., Aso, K., Tamura, T., Igarashi, H., Tamari, K.: The chemical constitution of gibberellin (3). J. Agric. Chem. Soc. Jpn. *17*, 975–984 (1941)
Yamaguchi, I., Yokota, T., Murofushi, N., Takahashi, N., Ogawa, Y.: Isolation of gibberellins A_5, A_{32}, A_{32} acetonide and (±)-abscisic acid from *Prunus persica*. Agric. Biol. Chem. *39*, 2399–2403 (1975a)
Yamaguchi, I., Yokota, T., Murofushi, N., Takahashi, N.: Structure elucidation of gibberellin A_{32} and its acetonide. Agric. Biol. Chem. *39*, 2405–2410 (1975b)

Yamaguchi, I., Miyamoto, M., Yamane, H., Murofushi, N., Takahashi, N., Fujita, K.: Elucidation of the structure of gibberellin A_{40} from *Gibberella fujikuroi*. J. Chem. Soc. Perkin *I*, 996–999 (1975c)

Yamane, H., Yamaguchi, I., Murofushi, N., Takahashi, N.: Isolation and structures of gibberellin A_{35} and its glucoside from *Cytisus scoparius*. Agric. Biol. Chem. *38*, 649–655 (1974)

Yamane, H., Murofushi, N., Osada, H., Takahashi, N.: Metabolism of gibberellins in early immature bean seeds. Phytochemistry *16*, 831–835 (1977)

Yamazaki, S., Tamura, S., Muramo, F., Saito, Y.: Structure of portulal. Tetrahedron Lett. 359–362 (1969)

Yanagawa, H., Kato, T., Kitahara, Y., Takahashi, N., Sato, Y.: Asparagusic acid, dihydroasparagusic acid and S-acetyldihydroasparagusic acid. New plant growth inhibitors in etiolated young *Asparagus officinalis*. Tetrahedron Lett. 2549–2552 (1972)

Yanagawa, H., Kato, T., Kitahara, Y.: Asparagusic acid-S-oxides. New plant growth regulators in etiolated young asparagus shoots. Tetrahedron Lett. 1073–1075 (1973)

Yang, S.F.: The biochemistry of ethylene: biogenesis and metabolism. Rec. Adv. Phytochem. *7*, 131–164 (1974)

Yokota, T., Murofushi, N., Takahashi, N., Tamura, S.: Gibberellins in immature seeds of *Pharbitis nil*. Part II. Isolation and structures of novel gibberellins, gibberellins A_{26} and A_{27}. Agric. Biol. Chem. *35*, 573–582 (1971a)

Yokota, T., Murofushi, N., Takahashi, N., Tamura, S.: Gibberellins in immature seeds of *Pharbitis nil*. Part III. Isolation and structures of gibberellin glycosides. Agric. Biol. Chem. *35*, 583–595 (1971b)

Yokota, T., Okabayashi, M., Takahashi, N., Shimura, I., Umeya, K.: Plant growth regulators in chestnut gall tissue and wasps. In: Plant growth substances 1973. pp. 28–38. Tokyo: Hirokawa Publishing Co. Inc. 1974a

Yokota, T., Yamazaka, S., Takahashi, N., Iitaka, Y.: Structure of pharbitic acid, a gibberellin-related diterpenoid. Tetrahedron Lett. 2957–2960 (1974b)

Yokota, T., Kobayashi, S., Yamane, H., Takahashi, N.: Isolation of a novel gibberellin glucoside 3-O-β-D-glucopyranosylgibberellin A_1 from *Dolichos lablab* seed. Agric. Biol. Chem. *42*, 1811–1812 (1978)

Young, H.: Identification of cytokinins from natural sources by G.C.-M.S. Anal. Biochem. *79*, 226–233 (1977)

Zabkiewicz, J.A., Steele, K.D.: Root-promoting activity of *P. radiata* bud extracts. In: Mechanism of regulation of plant growth. Bieleski, R.L., Ferguson, A.R., Cresswell, M.M. (eds.), pp. 687–692, Bull. 12. Wellington: The Royal Society of New Zealand 1974

Zachau, H.G., Dütting, D., Feldman, H.: Nucleotide sequences of two serine-specific transfer ribonucleic acids. Angew. Chem. Int. Ed. *5*, 422 (1966)

Zeevaart, J.A.D.: Levels of (+)-abscisic acid and xanthoxin in spinach under different environmental conditions. Plant Physiol. *53*, 644–648 (1974)

Zeevaart, J.A.D.: Sites of ABA synthesis and metabolism in *Ricinus communis* L. Plant Physiol. *59*, 788–791 (1977)

Zeevaart, J.A.D., Milborrow, B.V.: Metabolism of ABA and the occurrence of *epi*-dihydrophaseic acid in *Phaseolus vulgaris*. Phytochemistry *15*, 493–500 (1976)

Zenk, M.H.: 1-(indole-3-acetyl)-β-D-glucose, a new compound in the metabolism of indole-3-acetic acid in plants. Nature (London) *191*, 493–494 (1961)

Zimmerman, P.W., Crocker, W.: The effect of ethylene and illuminating gas on roses. Contrib. Boyce Thompson Inst. *3*, 459–481 (1931)

2 Extraction, Purification, and Identification

T. Yokota, N. Murofushi, and N. Takahashi

As shown in Chapter 1, plant hormones and other compounds which show physiological effects on plants are of widespread occurrence in the plant kingdom. The discovery and characterization of these compounds has depended upon isolating them in a pure state. Recent advances in purification techniques, especially the development of chromatography, enable the purification of even small amounts of components contained in plants. Furthermore the chromatographic techniques can now be applied as analytical tools for the detection or identification of plant components without isolation. In this chapter these various techniques including extraction, purification, and identification are discussed, centering on plant hormones, i.e., auxins, gibberellins, cytokinins, abscisic acid and related compounds, and ethylene. Some of these techniques have been reviewed recently (Hillman, 1978).

2.1 Methods of Extraction, Purification, and Isolation

2.1.1 Extraction of Active Principles from Plant Materials

a) General Remarks

The types of plant hormones and their concentrations vary with plant species and with tissues or organs such as fruit, seed, leaf, and stem (Sheldrake, 1973). Fluctuation in the levels of hormones is also observed during growth and development. Therefore, if one wants to isolate an active principle in a certain plant and subsequently to determine the chemical structure, it is recommended to harvest an organ rich in the active principle at an appropriate stage. (This obviously does not apply when plant homones, contained in a specified tissue, are being analyzed in physiological studies). In fact, most of the plant hormones have been isolated originally from rich sources. Generally, seeds of various plants are rich in plant hormones and most of the plant gibberellins have been initially isolated from seeds at an immature stage. GA_{19} is one of the exceptions which was originally isolated from young shoot of bamboo (*Phyllostachys edulis*) (Murofushi et al., 1966). The discovery of auxins and cytokinins as plant hormones was also demonstrated by the isolation of indole-3-acetic acid (Haagen-Smit et al., 1942, 1946) and zeatin (Letham, 1963) from seeds of *Zea mays,* which were subsequently found to contain a wide range of compounds, e.g., IAA glycosides (Labarca et al., 1965; Nicholls, 1967; Piskornik and Bandurski, 1972; Ueda and Bandurski, 1974; Ehmann, 1974; Ehmann and Bandurski, 1974), cytokinins (Letham, 1973) and gibberellins (Jones, 1964). It should be noted that extraordinarily high levels of glucobrassicin have been found in leaf of *Brassica oleracea* var. *sabauda* (GMELIN, 1964) and seed of *Isatis tinctoria* L. (ELLIOTT and STOWE, 1971): the amounts were 0.5% and 0.23% respectively.

Thus the first step of purification prior to solvent extraction should be carried out by separating a specified plant tissue in which active principles are localized. Otherwise further purification procedure will be hampered by impurities contained in unnecessary tissues. For example immature seeds of *Prunus persica* were excised from the fruitlets and used for the extraction and purification of GA_{32} since the pericarp contains a large amount of impurity and inhibitors (YAMAGUCHI et al., 1975). However, when active principles are equally distributed over all parts of the plant extraction of the whole plant material cannot be avoided.

Solvent extraction of plant tissue should be carried out immediately after harvesting. If not, the plant material should be immediately frozen and stored in a freezer to prevent any enzymatic and chemical changes of the compounds in the tissue. Plant materials are usually extensively extracted with water-miscible solvents such as methanol, ethanol, and acetone and, at the same time, homogenization of plant materials should be carried out to increase extraction efficiency. After two or three extractions, the combined extract is concentrated to an aqueous solution which is then subjected to solvent partitioning. It is often an advantage to use water-miscible solvents which can extensively extract a variety of compounds. However, in some cases this may cause difficulty in purification of the active principle because of unwanted contaminants. Such a problem, experienced in the unsuccessful purification of GA_{19} from methanolic extracts of bamboo shoots, was overcome by extraction with boiling water which allowed selective extraction of GA_{19} relative to impurities (MUROFUSHI et al., 1966). Portulal (structure 97, p. 69) a factor inducing adventitious root formation, is also effectively extracted with hot water from leaves of *Portulaca grandiflora* (MITSUHASHI and SHIBAOKA, 1965).

The aqueous solution, obtained either by direct extraction with water or by evaporation of an extract made with an aqueous organic solvent, should be immediately subjected to further solvent partitioning and, if not, it should be frozen or stored at low temperature with toluene covering the surface in order to prevent proliferation of microorganisms.

Sometimes there is the problem of deciding whether an isolated compound is a plant constituent or an artefact formed during the isolation procedure. Precautions must therefore be taken to minimize the possibility of artefact formation. Specific examples of artefact formation are discussed in the following section under the individual plant hormones.

b) Auxins

Extraction of auxins from plant tissue is carried out with methanol, ethanol, acetone, or ether (peroxide-free). Alcohols and acetone provide more thorough extraction than ether, since direct ether extraction of solid plant material may lose part of the total auxins (SRIVASTAVA, 1963). This was substantiated by the findings that the ether extract of citrus fruits contained only half the total

IAA, whilst subsequent extraction of the residue with methanol could recover the remaining IAA along with indoleacetamide (IGOSHI et al., 1971; TAKAHASHI et al., 1975).

Precautions must be taken to exclude chemical or enzymatic changes of auxins during extraction. The following enzymes are known: oxidase (SCOTT, 1972; SEQUEIRIA, 1973; SCHNEIDER and WIGHTMAN, 1974), esterase (SRIVASTAVA, 1963; KOPCEWICZ et al., 1974) and myrosinase (LANGER and MICHAJLOVSKIJ, 1958; GMELIN, 1964; SCHLÜTER and GMELIN, 1972).

Corn kernels (UEDA and BANDURSKI, 1969) and seedlings (HAMILTON et al., 1961) contain a large amount of bound auxins relative to free auxins, and ether extraction of such tissues may result in hydrolysis of the bound auxins to give IAA, presumably as a result of esterase activity that persists during the ether extraction (SRIVASTAVA, 1963). In this sense alcohol, which is believed to terminate or inactivate most of the enzymes, is an excellent solvent for auxin extraction. SRIVASTAVA (1963) reported that the ethanol extraction of corn kernel at $-10°$ minimized the production of artefacts due to chemical or enzymatic reaction. However alcohol extraction sometimes produces alcoholysis products such as ethyl indoleacetate (EtIAA) (REDEMANN et al., 1951; FUKUI et al., 1957). In such cases the alcohol can be replaced by acetone (MARUMO et al., 1968b; TAKAHASHI et al., 1975). It is ironical that indole-3-acetaldehyde (IAAld), which could not be isolated because of its labile nature, was isolated as its dimethyl acetal, presumably an artefact formed during methanol extraction from chestnut gall (YOKOTA et al., 1974).

IAA glucosyl ester which was isolated as an IAA metabolite from *Colchicum* leaves (ZENK, 1961) and wheat coleoptile (KLÄMBT, 1961) has been found to be present as a natural constituent of *Avena* coleoptile (KEGLEVIĆ, 1969) and corn kernel (EHMANN, 1974), suggesting that this glucosyl ester may be widely distributed in the plant kingdom. ZENK (1961) found that IAA glucosyl ester is decomposed in ammoniacal solvent to yield IAM and IAA, suggesting that some of the isolations of IAM and IAA were possibly derived by degradation of IAA glycosyl esters. Chestnut gall also contains a polar auxin, extractable into n-butanol fraction. However, IAA was also isolated from this fraction, indicating that hydrolysis must have taken place during the purification (YOKOTA et al., 1974).

Glucosinolates[1] found in the Cruciferae and other families are susceptible to myrosinase and chemicals (acids and bases), forming a variety of auxins as artefacts such as IAA, indole-3-acetonitrile (IAN), indole-3-carboxylic acid (ICA), indole-3-carboxaldehyde (IAld) and ascorbigen (GMELIN, 1964). GMELIN (1964) claimed that most of the earlier work on this plant material (cabbage) must have involved such artefact formation (HENBEST et al., 1953; WELLER et al., 1954; JONES and TAYLOR, 1957; PROCHÁZKA and SǍNDA, 1960). The instability of glucosinolates can be overcome by extracting the plant tissue with boiling methanol which deactivates myrosinase (GMELIN, 1964; ELLIOTT and STOWE, 1970, 1971).

IAA itself is a chemically fragile compound which, when its solution is exposed to air and light, produces coloured materials and finally completely

1 This group is extensively treated by E.W. UNDERHILL in Vol. 8 of this Encyclopedia (1980).

decomposes. In order to prevent such oxidation, reagents such as sodium diethyldithiocarbamate, santoquin (MANN and JAWORSKI, 1970) ascorbic acid (NIEDERWIESER and GILIBERTI, 1971) and carbon dioxide (RAJ and HUTZINGER, 1970a) are used for stabilization of IAA during extraction and further purification steps. It is obvious that careful experimentation in dim light is preferable.

MANN and JAWORSKI (1970) found that considerable loss of IAA occurs through sublimation when an ether solution is evaporated to dryness by using a rotary evaporator. This loss of IAA can be prevented by evaporating the ether at atmospheric pressure.

c) Gibberellins

Extraction of gibberellins from plant tissue is usually carried out with methanol or acetone. BROWNING and SAUNDERS (1977) reported that the extraction of chloroplast membranes from wheat seedlings using non-ionic detergent Triton X100 gave about 1000 times more gibberellin activity than methanol extraction did. However, this work could not be reproduced with gibberellins in pea seed (MACMILLAN, 1977) and rice plants (our unpublished results). Recently SAUNDERS has indicated (personal communication) that the original results with wheat cannot readily be reproduced.

Gibberellins seem to be relatively stable as compared with auxins. However, artefact formation should also be taken into account during extraction and further purification steps as described below. Generally gibberellins are susceptible to acidic conditions, which can cause C/D ring rearrangement of 13-hydroxygibberellins into ketoacids (MACMILLAN et al., 1960) and hydration of exocyclic methylene in gibberellins lacking a 13-hydroxy group (GROVE, 1961; HANSON, 1966). Methanol extracts of certain plant tissues show considerable acidity ranging to pH 3. The C/D ring rearrangement due to such acidity has been experienced in our laboratory by prolonged storage of the methanol extract of *Pharbitis nil* seed at room temperature (YOKATA, unpublished). Gibberellin A_3 forms gibberellenic acid and other compounds under mild acidic conditions and isoGA_3 under mild basic conditions and these transformations have been experienced during the purification of GA_3 glucoside (YOKOTA et al., 1971b). Gibberellins having α-glycol functions react with acetone to form acetonides under mild acidic condition. For example GA_8 (JONES, 1964; YOKOTA, unpublished) and GA_{32} (YAMAGUCHI et al., 1975) have been converted into acetonides during charcoal chromatography using aqueous acetone as eluant. GA_6 has been shown to form a chlorohydrin derivative by action of hydrochloric acid during purification (JONES, 1964; DURLEY et al., 1971). Hydrolysis of GA_1 glucosyl ester has been found during extraction and further purification (HIRAGA et al., 1974b). It should be stated that aqueous solution of pure GA_3 glucoside, when kept at room temperature for a few days, has been found to release free gibberellin, presumably because of contaminating microorganisms (unpublished). Formation of artefacts, as discussed earlier, can be minimized by careful handling of the sample although some artefacts will inevitably be formed to some extent during purification.

d) Cytokinins

Cytokinins are extracted with aqueous ethanol or aqueous methanol. Less polar solvents such as ether cannot be used for cytokinin extraction because of low solubility of cytokinins. The extraction is performed very frequently at low temperature to minimize enzymic or chemical degradation because of possibilities that ribotide cytokinins release riboside cytokinins by the action of phosphatase and that sometimes the free bases are formed from riboside cytokinins by hydrolysis (DEKHUIJEN and GEVERS, 1975; MILLER, 1965). However, it is known that enzymes such as phosphatase and ribonuclease can survive to some extent in an alcohol solution at low temperature (BIELSKI, 1964).

The extraction of cytokinins from plant material has been critically discussed by HORGAN (1978).

Degradation of cytokinins during ion exchange chromatography is discussed in Section 2.1.2d.

e) Abscisic Acid and Related Compounds

Extraction of abscisic acid (ABA) and its related compounds has been reviewed recently by SAUNDERS (1978). It is usually carried out with aqueous acetone or methanol. This procedure should be performed in dim light since ABA, in solutions exposed to light, is readily converted to the trans, trans-isomer although trans, trans-ABA is confirmed to be a natural constituent in some cases (MILBORROW, 1970; GASKIN and MACMILLAN, 1968). MILBORROW and MALLABY (1975) reported that methyl abscisate undergoes nearly 50% conversion to the trans, trans-isomer when methyl abscisate, in acetone or methanol, is placed for 12 days inside a window during cloudy weather. When exposed to ultraviolet light only 4 h are enough to obtain 50% conversion (LENTON et al., 1971). Methyl phaseate has also been known to be converted to the trans, trans-isomer in an aged solution (GASKIN and MACMILLAN, 1968).

The ABA conjugate, (+)-1-abscisyl-β-D-glucopyranoside, which was isolated by KOSHIMIZU et al. (1968a) from *Lupinus* seeds is believed to represent a considerable proportion of the total amount of abscisic acid in plants (RUDNICKI and PIENIAZEK, 1971; GOLDSCHMIDT et al., 1973). MILBORROW and MALLABY (1975) reported that this compound releases methyl abscisate as an artefact during methanol extraction. Therefore most of the reports claiming the presence of neutral inhibitors in methanolic extracts of plant material may have been due to such artefact formation as well as to other types of inhibitors (KEFELI and KADYROV, 1971; TAKAHASHI et al., 1973; FIRN et al., 1972; MILBORROW, 1974). Methanol which is made slightly basic by adding ammonia or sodium bicarbonate, or even methanol itself, causes loss of the ABA conjugate followed by the formation of an equivalent amount of methyl abscisate. ZEEVAART and MILBORROW (1976) found that methyl esters of ABA, phaseic acid, dihydrophaseic acid, and epidihydrophaseic acid are also derived from their conjugates during neutral, and particularly basic, methanol extraction. ABA conjugate undergoes rapid methanolysis relative to other conjugates. Such susceptibility of the conjugates to methanolysis is ascribable to the dienoic ester grouping.

This methanolysis can be overcome by use of methanol containing 1% acetic acid or acetone alone, where no methyl abscisate is found in the extract (MILBORROW and MALLABY, 1975). In contrast there have been no reports of the methanolysis of gibberellin conjugates.

An antioxidant 2,6-di-t-butyl-4-methylphenol can be used to prevent oxidation during extraction and further fractionation (MILBORROW, 1972; MILBORROW and MALLABY, 1975; ZEEVAART and MILBORROW, 1976).

During extraction and further fractionation extreme pH and high temperature should be avoided since the conjugates are easily hydrolyzed under mild conditions such as pH 11 at 60° for 0.5–1 h (MILBORROW, 1970; SWEESTER and VATVAUS, 1976).

2.1.2 Fractionation Based on Solvent Partitioning and Ion Exchange Resin

a) General Remarks

Organic compounds are classified into neutral, strongly acidic (carboxylic acid), weakly acidic (phenol), basic, and amphoteric compounds. Solvent partitioning of neutral compounds is directly concerned with distribution (partition) coefficient $Kd = C_0/C_a$ between organic phase and aqueous phase. Therefore neutral compounds can be extracted with an appropriate solvent irrespective of the pH of the aqueous phase. However, distributions of acidic and basic compounds are affected by the pH of the aqueous phase and pKa values. For example, the Kd of an undissociated carboxylic acid is expressed as:

$$Kd = \frac{[RCOOH]_0}{[RCOOH]_a}$$

This acid is dissociated as following:

$$RCOOH \rightleftarrows RCOO^- + H^+$$

Since $RCOO^-$ is soluble in aqueous phase but not in organic phase, the distribution ratio (Kd.r.) is given by (in practice this value is usually used as Kd).

$$Kd.r. = \frac{[RCOOH]_0}{[RCOOH]_a + [RCOO^-]_a}$$

At pH 3, a carboxylic acid of pKa=4–5 is mostly undissociated because the equilibrium shifts to the undissociated form at a pH lower than the pKa value. This means that the Kd.r. at pH 3 is near to Kd. Thus acidic compounds can be extracted into an organic phase at pH 3 since its undissociated form (RCOOH) has a high solubility in the organic solvent. In the case of a polar acidic compound which has a low Kd value for non-polar solvent, polar solvent must be used to favour the partitioning into the organic phase. At a pH higher than the pKa value the dissociated form ($RCOO^-$) is predominant. Thus, carboxylic acids can be transferred into an aqueous phase at pH 7–8 from an organic phase.

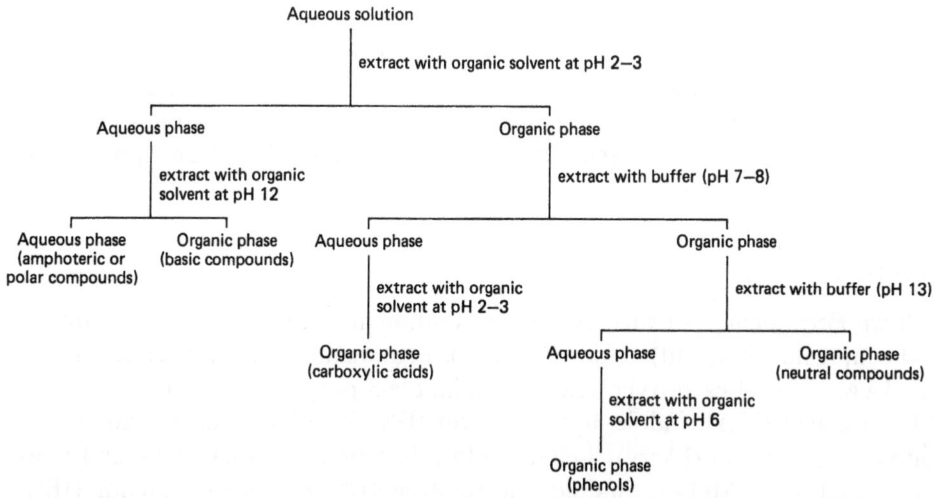

Fig. 2.1. Fundamental fractionation procedure based on solvent partitioning

Similarly this theory can be extended to phenolic compounds (pKa= 9.5–11.0) and basic compounds (pKb=3–10). Amphoteric compounds remain in an aqueous phase because they always take dissociated forms irrespective of the pH value.

The fundamental procedure for solvent partitioning is illustrated in Fig. 2.1. However, for practical purpose, modified procedures are usually used, and they are discussed in later sections under the individual plant hormones.

In practice, effective extraction can be achieved by multiple extractions with minimum amounts of solvent. When V ml of aqueous solution containing W mg of solute is equilibrated with V′ ml of organic solvent and W′ mg of the solute remains in the aqueous phase, the distribution coefficient (Kd) is expressed as:

$$Kd=\frac{\frac{W-W'}{V'}}{\frac{W'}{V}}, \quad \text{thus} \quad W'=W\cdot\frac{V}{Kd\cdot V'+V}$$

The amount of solute which remains in the aqueous phase after extracting n-times with equal volumes of solvent is expressed as:

$$Wn=W\left(\frac{V}{Kd\cdot V'+V}\right)^{n}$$

From this equation it is found that two extractions with 5 ml of solvent are more effective than a single extraction with 10 ml of solvent. When 10 ml of aqueous solution containing 1 mg of the solute with Kd=10 is subjected to the above procedures, the amount of the solute remaining in the aqueous phase is 1/11 mg after 1×10 ml extraction and only 1/36 mg after 2×5 ml extraction.

Ion exchange resins can effect the separation of acidic, basic, and neutral compounds and examples are described later. This separation may be simply explained by the fact that a cation (acidic) exchange resin sorbs basic compounds whilst an anion (basic) exchange resin sorbs acidic compounds. Chromatographic usage of ion exchange resins is also possible by the choice of an appropriate resin and eluant as described later.

b) Auxins

Solvent Partitioning. Auxins can be conventionally grouped into free auxins and conjugate (or bound) auxins (Chap. 1). Free auxins comprise acidic auxins, i.e., IAA, indole-3-carboxylic acid (ICA), indole-3-propionic acid (IPA), indole-3-butyric acid (IBA), indole-3-pyruvic acid (IPyA), indole-3-lactic acid (ILA), indole-3-acrylic acid (IAcry), 4-chloroindole-3-acetic acid (4-Cl-IAA), and neutral auxins, i.e., MeIAA, indole-3-acetonitrile (IAN), indole-3-ethanol (IEt), indole-3-acetamide (IAM), indole-3-acetaldehyde (IAAld), indole-3-carboxaldehyde (IAld), and Me4-Cl-IAA. These acidic and neutral auxins can be partitioned into acidic and neutral fractions respectively based on the procedures shown in Fig. 2.2 or 2.3. Ether (peroxide-free) and ethyl acetate are the most commonly used solvents. Benzene can also extract IAA and MeIAA from the aqueous phase (YOKOTA et al., 1974) but is not recommended because of its toxicity.

POWELL (1964) reported a fractionation procedure for simple indoles including neutral, acidic, basic, and water-soluble compounds. These indoles were separated into four fractions by using methylene dichloride, and neutral indoles were transferred into acetonitrile by partitioning between acetonitrile and hexane (Fig. 2.4). However, this procedure must be carefully employed since 1 N ammonium hydroxide, used in the initial extraction, has been found to cause undesirable effects on auxins as described later (see also Sect. 2.1.1.b).

ATSUMI et al. (1976) found that the auxin content in tobacco crown gall is estimated to be extraordinarily high after the usual extraction procedure. This increase of IAA content was ascribed to the occurrence of IPyA which has been found to produce IAA and other indoles during extraction and purification steps, especially under basic conditions (BENTLEY et al., 1956; SHELDRAKE, 1973; ATSUMI et al., 1976). ATSUMI et al. (1976) devised an extraction procedure which can eliminate IPyA from the IAA fraction as shown in Fig. 2.5. Selective partitioning of IAA into the organic phase was successfully carried out with methylene dichloride instead of ether because the extractability of IPyA with ether is nearly four times as much as that with methylene dichloride. This method gives 91% recovery of IAA and excludes 80% of IPyA.

Sweet corn kernel contains a variety of IAA glycosides which cannot be fractionated by the usual solvent partitioning because of their low solubility. EHMANN and BANDURSKI (1972) devised an effective procedure shown in Fig. 2.6. The water-insoluble residue, termed crude A fraction, contained IAA esters of cellulosic glucan (PISKORNIK and BANDURSKI, 1972). The aqueous filtrate was partitioned with n-butanol to give aqueous and n-butanol fractions. The former, termed crude B fraction, contained IAA esters of myoinositol, arabino-

Fig. 2.2. Fractionation procedure for neutral and acidic compounds – I

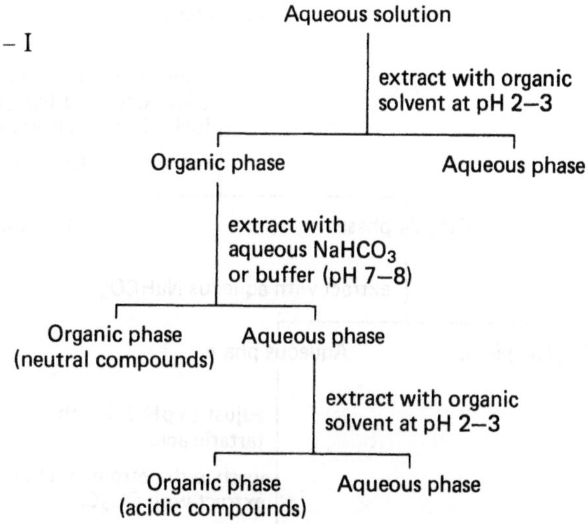

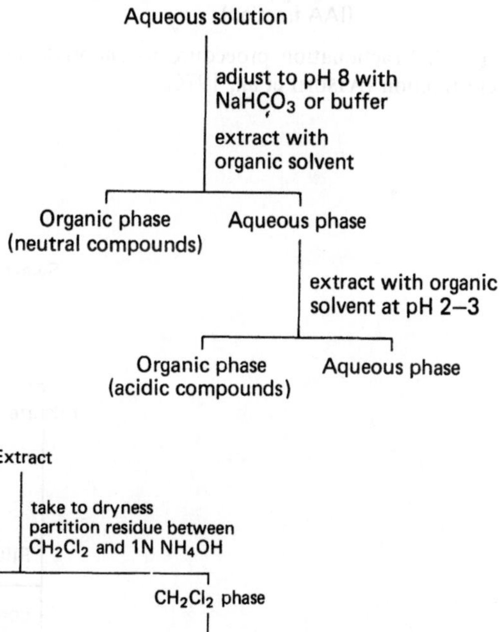

Fig. 2.3. Fractionation procedure for neutral and acidic compounds – II

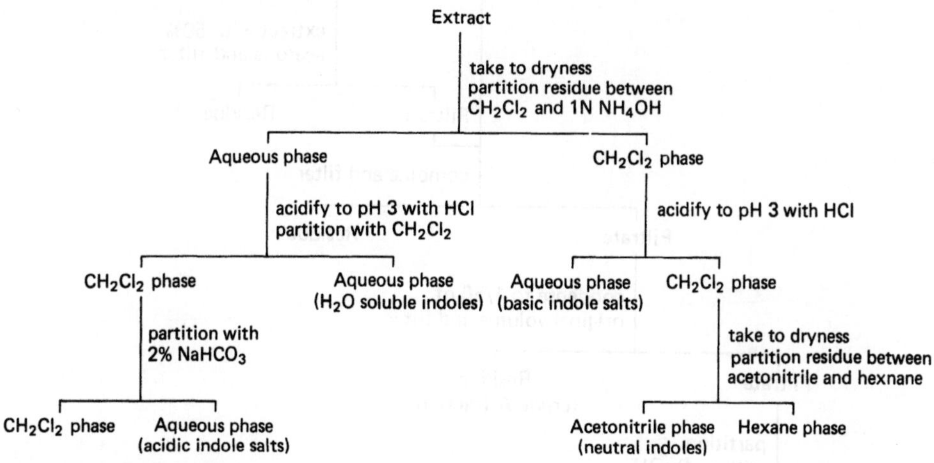

Fig. 2.4. Fractionation procedure for neutral, acidic, basic and water-soluble indoles. Water-soluble indoles include 5-hydroxyindole derivatives, tryptophan, abrine, and hypaphorine. (Powell, 1964)

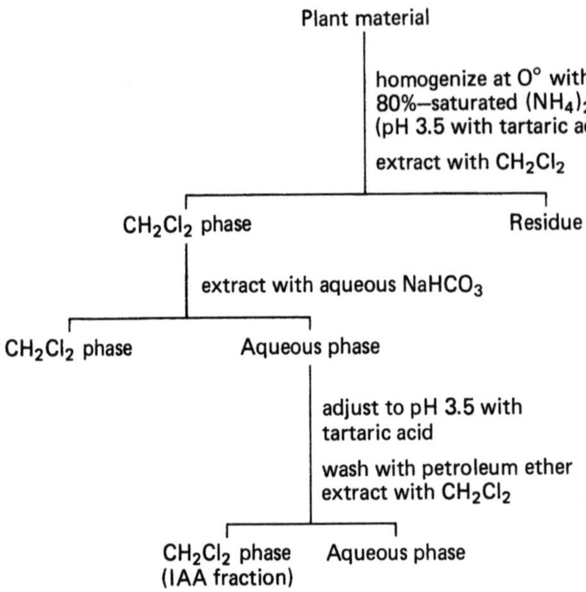

Fig. 2.5. Fractionation procedure to climinate indole-3-pyruvic acid from indole-3-acetic acid fraction (ATSUMI et al., 1976)

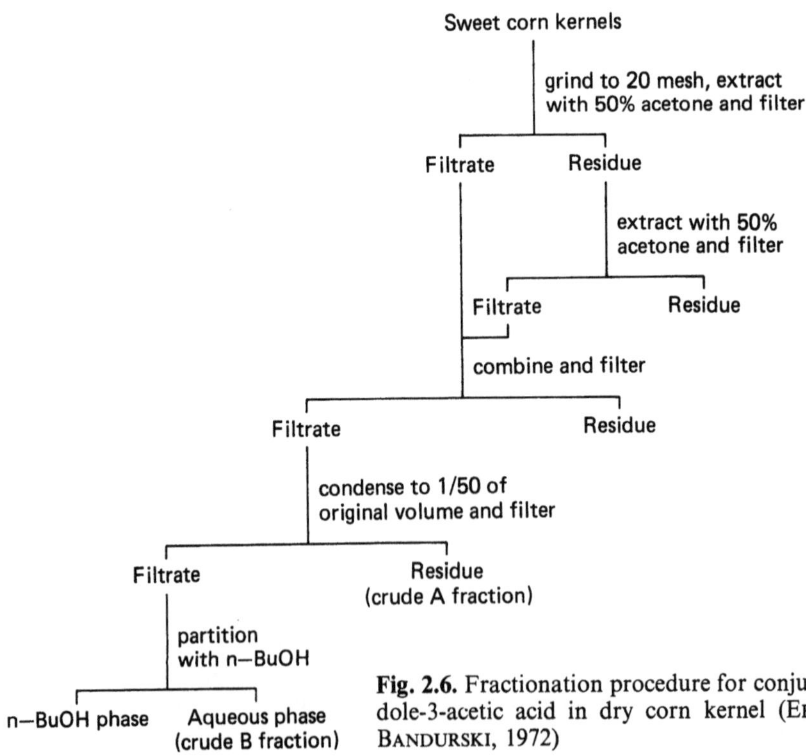

Fig. 2.6. Fractionation procedure for conjugates of indole-3-acetic acid in dry corn kernel (EHMANN and BANDURSKI, 1972)

sylmyoinositol and galactosylmyoinositol (UEDA and BANDURSKI, 1974). The n-butanol fraction contained three isomers of IAA glucosyl esters, i.e., 2-O-, 4-O- and 6-O-(indole-3-acetyl)-D-glucopyranose (EHMANN, 1974) in addition to two myoinositol esters of IAA, i.e., di- and tri-O-(indole-3-acetyl)-myoinositol (EHMANN and BANDURSKI, 1974).

ZENK (1961) showed that continuous extraction using a Soxhlet type apparatus with ethyl acetate but not with ether can extract 1-O-(indole-3-acetyl)-β-D-glucose, a metabolite of IAA in *Colchicum* leaves.

Glucosinolates, acidic water-soluble conjugate auxins, are not subjected to solvent partitioning, but directly purified by anionotropic (acidic) alumina column after boiling methanol extraction (GMELIN, 1964; ELLIOTT and STOWE, 1970).

Ion Exchange Resin. RAJ and HUTZINGER (1970a) indicated that separation of 22 indole compounds into neutral, acidic, basic, and amphoteric fractions can be effected by a combination of Dowex-50WX2 (triethylammonium$^+$), Dowex-1X2 (formate$^-$) and DEAE-Sephadex A-25 (acetate$^-$) as shown in Fig. 2.7. These ion exchange resins are used in the salt and not in the free acid or base form, and the use of ammonia, alkali, and mineral acids as eluants is avoided because these chemicals may cause undesirable chemical modifications of the indoles. A labile indole, IPyA, was eluted from DEAE Sephadex column with 5% ammonium acetate in 50% ethanol instead of the solvent used in Fig. 2.7. All indoles are stable under the conditions described.

Dowex-50WX2 has been successfully used by EHMANN and BANDURSKI (1972) to purify and concentrate IAA myoinositol esters as a group. They could be eluted with 1 mM citrate buffer (pH 3.3. or 6.2) resulting in 54-fold

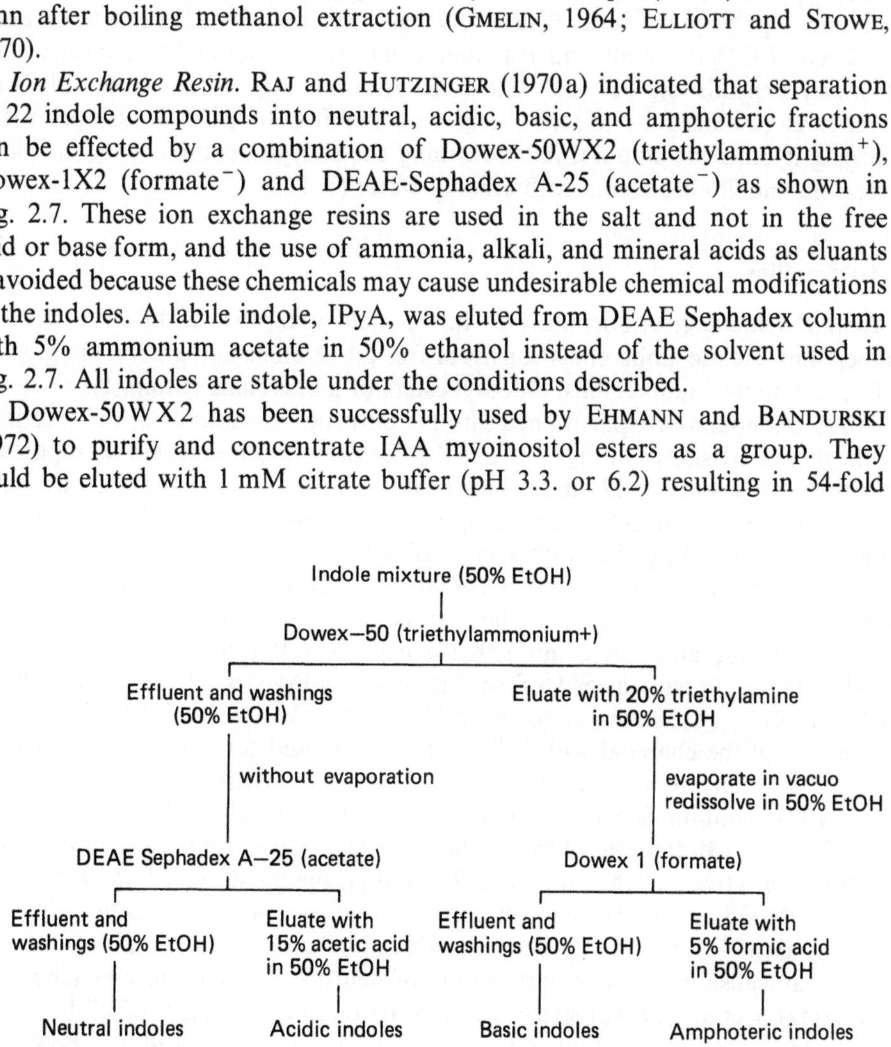

Fig. 2.7. Fractionation of indoles by ion-exchange chromatography. The following compounds are used in this procedure. Neutral indoles: EtIAA, IAN, IEt, IAM, IAld, melatonin. Acidic indoles: IAA, ICA, IPA, IPyA, IAcry, ILA, 5-hydroxy-IAA. Basic indoles: gramine, tryptamine, Nω-methyltryptamine, Nω, Nω-dimethyltryptamine, 5-methoxytryptamine, serotonin. Amphoteric indoles: Tryptophan, Nε- (indole-3-acetyl)-L-lysine, Nδ-(indole-3-acetyl)-L-ornithine. (RAJ and HUTZINGER, 1970a)

purification in a single column step. On the other hand a 20% sulphonated polystyrene divinyl benzene copolymer resin was shown to be more effective than Dowex-50, esters being eluted with aqueous acetone with 125-fold purification.

Dowex-50 (triethylammonium$^+$ or brucinium$^+$) and Amberlite IR 120 (trimethylammonium$^+$) have been used to make crystalline glucosinolate salts in the final isolation steps of glucosinolates (GMELIN, 1964; ELLIOTT and STOWE, 1970).

DEAE cellulose has also been used to obtain a neutral fraction in the isolation procedure of IEt (RAYLE and PURVES, 1967).

Neutral polystyrene resin Polapak Q has been used by NIEDERWIESER and GILIBERTI (1971) to desalt and fractionate indoles including IAA, desorption of which depends on pH and salt concentration. Aqueous sample solution which was stabilized by ascorbic acid was acidified to pH 1–2 just prior to charging it onto the column. After washing the charged column with distilled water, the indoles are eluted with ethanol-water (1:1).

c) Gibberellins

Solvent Partitioning. Gibberellins so far isolated comprise acidic and neutral gibberellins. Acidic gibberellins consist of free gibberellins and gibberellin glucosides, and neutral gibberellins, glucosyl esters of gibberellins (Chap. 1).

Free gibberellins are partitioned into the ethyl acetate-soluble acidic fraction by the procedure shown in Figs. 2.1 and 2.2. DURLEY and PHARIS (1972) reported partition coefficients of 27 gibberellins (Table 2.1). Less polar gibberellins, i.e., GA_4, GA_7, GA_9, and GA_{12} (especially the latter two) were found to be significantly partitioned into the ethyl acetate phase from a 1.5 M phosphate buffer of pH 8. Such gibberellins may therefore behave partially as if they were neutral compounds. On the other hand the extraction of polar-free gibberellins such as GA_{21}, GA_{23} and GA_{28} into ethyl acetate is relatively low. FUKUI et al. (1971) used charcoal, instead of solvent partitioning, to adsorb gibberellins, including GA_{28}, from the aqueous residue from an extract of *Lupinus* seeds. The eluate of the charcoal with 70% acetone was used for further purification.

Seeds of *Pharbitis nil* (YOKOTA et al., 1971 a, b) and *Cytisus scoparius* (YAMANE et al., 1974) contain free gibberellins and acidic gibberellin glucosides, which were fractionated into the ethyl acetate acidic fraction and n-butanol acidic fraction respectively as shown in Fig. 2.8. GA_{32}, the most polar free gibberellin found in *Prunus persica*, is also partitioned into acidic n-butanol fraction (YAMAGUCHI et al., 1975). Partition procedures which include back extraction of organic layers may cause incomplete partitioning of neutral glucosyl esters of gibberellins. HIRAGA et al. (1974a) succeeded in fractionating a variety of gibberellins contained in mature seed of *Phaseolus vulgaris* by the procedure shown in Fig. 2.9. The exclusion of the back extraction effected clear partitioning of glucosyl esters of GA_4 and GA_{37} into the neutral ethyl acetate fraction, GA_1 and GA_8 into the acidic ethyl acetate fraction, glucosyl esters of GA_1 and GA_{38} into the neutral n-butanol fraction and GA_8 glucoside into the acidic n-butanol fraction.

Table 2.1. Partition coefficients (Kd=C aq./C org.)[a] of the gibberellins and ent-kaurenoic acid between ethyl acetate and 1.5 M phosphate buffer solution at five pH values (DURLEY and PHARIS, 1972)

Gibberellin	pH				
	8.0	6.5	5.0	3.5	2.5
A_1	∞	∞	1.2	0.17	0.11
A_2	∞	7.9	0.97	0.19	0.15
A_3	∞	∞	1.2	0.21	0.17
A_4	2.2	0.29	0.05	0	0
A_5	∞	4.8	0.19	0	0
A_6	∞	5.4	0.49	0.05	0
A_7	3.2	0.56	0.10	0	0
A_8	∞	∞	4.9	0.64	0.45
A_9	0.34	0.06	0	0	0
A_{10}	11.3	1.6	0.33	0	0
A_{12}	0.56	0.04	0	0	0
A_{13}	∞	7.1	0.06	0	0
A_{14}	∞	0.41	0	0	0
A_{16}	∞	3.2	0.16	0	0
A_{17}	∞	∞	0.50	0.04	0
A_{18}	∞	∞	0.42	0	0
A_{19}	∞	4.6	0.81	0.10	0
A_{20}	∞	2.1	0.09	0	0
A_{21}	∞	∞	9.1	0.89	0.08
A_{22}	∞	15.1	1.4	0.51	0.19
A_{23}	∞	∞	19.4	1.0	0.17
A_{24}	∞	0.83	0	0	0
A_{25}	13.1	0.66	0	0	0
A_{26}	∞	∞	3.2	0.44	0.21
A_{27}	∞	1.6	0.18	0.05	0
A_{28}	∞	∞	12.7	0.81	0.07
A_{29}	∞	∞	1.9	0.20	0.15
ent-kaurenoic acid	0.24	0.04	0	0	0

[a] Kd's < 0.02 are taken as 0; values > 20 are taken as ∞

Ion Exchange Resin. Ion exchange resins are usually not used in fractionating gibberellins although a few examples are published. ASAKAWA et al. (1974) indicated that GA_3 metabolites, including GA_3 glucoside, could be separated into three fractions when Dowex 1 (formate$^-$) column chromatography was conducted by using a linear gradient of water and 1.3 M formic acid. Cation exchange resin has been used in the purification of gibberellin A_9 glucosyl ester (LORENZI et al., 1976).

d) Cytokinins

Solvent Partitioning. Representative cytokinins are classified into free-base cytokinins, riboside cytokinins and ribotide cytokinins (Chap. 1). 6-(3-Methylbut-2-enylamino)purine (i^6Ade, see Fig. 4.2 for cytokinin abbreviations), one of the free base cytokinins, has two pKa values, 3.4 (basic) and 10.4 (acidic) (LEONARD

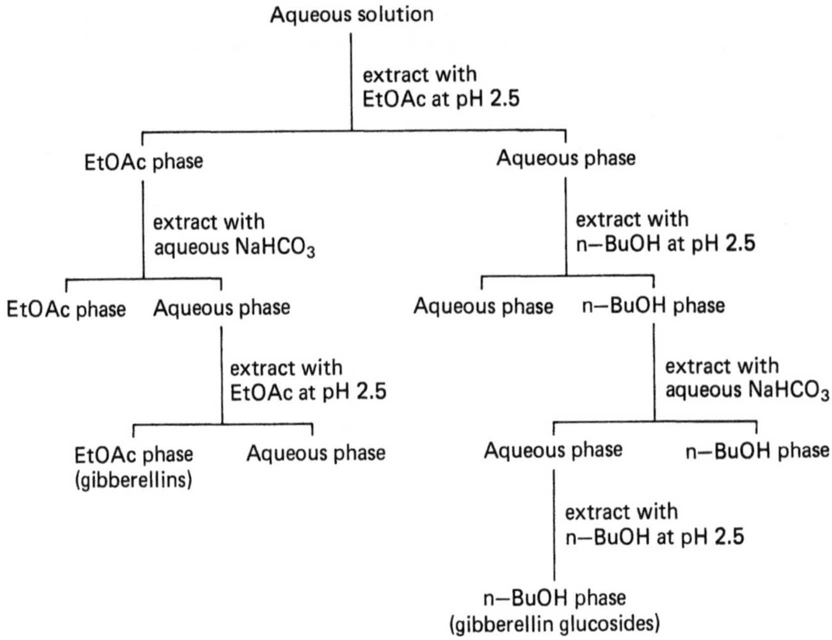

Fig. 2.8. Fractionation procedure for gibberellins and gibberellin glucosides (YOKOTA et al., 1971a, b)

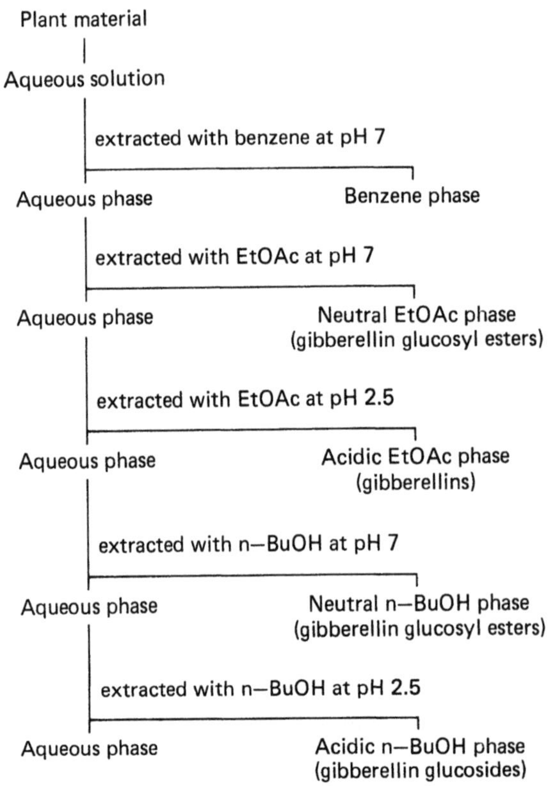

Fig. 2.9. Fractionation procedure for gibberellins, gibberellin glucosides and gibberellin glucosyl esters. (HIRAGA et al., 1974a)

Table 2.2. Partition coefficients (Kd=C org./C aq.) of cytokinin bases. (LETHAM, 1974)

Solvent and pH of aqueous phase	Partition coefficients		
	Zeatin	Kinetin	i^6Ade
Petroleum (b.p. 60–80°), pH 7.0	0.0004	0.0006	0.003
Petroleum (b.p. 60–80°), pH 3.0	0.0003	0.0004	0.001
Diethyl ether, pH 7.0	0.032	0.810	2.33
Diethyl ether, pH 3.0	0.011	0.237	0.322
Ethyl acetate, pH 7.0	0.240	3.29	6.88
Ethyl acetate, pH 3.0	0.049	1.78	1.49
n-Butanol, pH 7.0	6.26	20.6	40.4
n-Butanol, pH 3.0	1.59	8.51	10.7

and FUJII, 1964). The pKa 3.4 is attributable to the protonated exocyclic nitrogen, whilst the pKa 10.4 is attributable to dissociation of the NH group in the imidazole ring (LETHAM et al., 1967). Thus free-base cytokinins are amphoteric compounds. On the other hand, riboside cytokinins show weak basicity since the imidazole-NH-groups are blocked by the ribosyl groups. For example, 6-(3-methylbut-2-enylamino)-9-β-D-ribofuranosylpurine (i^6A) shows the pKa 3.8 (basic) value (MARTIN and REESE, 1968). Ribotide cytokinins again are amphoteric because of the phosphate group.

Because of their amphoteric nature and their low solubility in organic solvents, the cytokinins cannot be fractionated using the procedures shown in Figs. 2.1 and 2.2. LETHAM (1974) has reported the distribution coefficients of cytokinins (Table 2.2). Kinetin and i^6Ade can be easily extracted into ether at pH 7, and into ethyl acetate both at pH 3 and 7. i^6A, whose partition coefficient is about 30% of the value for i^6Ade, is also transferred mainly into the organic phase. Thus extraction with ether or ethyl acetate at pH 3 which is frequently used to remove impurities should be carefully employed (HEMBERG and WESTLIN, 1973). On the other hand n-butanol extraction is of great value in cytokinin purification because of the very high n-butanol partition coefficient at pH 7. Such extraction also has the advantage of separating cytokinin bases and nucleosides from nucleotide cytokinins and nucleoside cytokinins carrying carboxyl groups (MILLER, 1965; LETHAM, 1973) which remain in aqueous solution at pH 7.

On the other hand, polar cytokinins such as zeatin riboside can be extracted into ethyl acetate by using salting out procedures. MILLER (1975a) found that zeatin riboside could be transferred into ethyl acetate by extensive extraction of frozen crown gall tissue to which high concentration of phosphate buffer was added. This procedure, however, may be unsuitable for handling a large-scale extract.

Ion Exchange Resin. Ion exchange resins are widely used in purification of cytokinins because this is one of the most useful procedures to remove inhibitors and impurities from crude plant extract. Cation exchange resins (strong acids) such as Dowex-50W (MILLER, 1974), Zerolit 225 (WANG et al., 1977) and Zeokarb 225 (HORGAN, 1973a; LETHAM, 1973) retain most cytokinins except cytokinins having an acidic side chain. These resins are used in the proton form or in the ammonium form and both forms show similar binding and elution patterns for cytokinins (VREMAN and CORSE, 1975). Cytokinins which are retained in a strong acid resin cannot be eluted with water or aqueous alcohol such as 70% or 95% ethanol, but eluted with aqueous ammonia ranging from 1 N to 6 N. VREMAN and CORSE (1975) studied the recovery of 2-methylthiozeatin (ms^2-t-io^6Ade) and zeatin (t-io^6Ade) in the model experiment using Dowex 50 W. The elution with 1 N ammonia resulted in recovering 20% of ms^2-t-io^6Ade and 70% of t-io^6Ade. However, better recoveries were obtained by eluting with 1 N ammonia in 70% of ethanol to give about 55% and 80% recoveries respectively. Elution with ammoniacal ethanol had been applied by other workers in the isolation of several cytokinins from *Zea mays* (LETHAM, 1973) and of zeatin riboside from crown gall tissue (MILLER, 1974).

There is a possibility that nucleosides and nucleotides liberate the corresponding cytokinin free bases in strong acid resins because the elution with ammoniacal solution causes localized heating. This problem has been experienced with zeatin riboside (DEKHUIJEN and GEVERS, 1975; TEGLEY et al., 1971). This heating seems to be prevented by careful elution with ammoniacal solution at low temperature (MILLER, 1965, 1974; DYSON and HALL, 1972).

Cellulose phosphate (ammonium$^+$, equilibrated to pH 3) is useful in later stages of purification. Cytokinins are eluted from this low capacity resin with 0.1–0.3 N ammonia with good recovery (PARKER et al., 1972, 1973; LETHAM, 1973; DEKHUIJEN and GEVERS, 1975). Duolite CS-101, a weak acid resin (COOH), also gives an excellent recovery (VREMAN and CORSE, 1975).

Anion exchange resins (strong base), e.g., De-Acidite FF (acetate$^-$), were used by LETHAM (1973). Most cytokinins can be eluted with 0.04 N ammonia and 1.5 N acetic acid. Nucleotide cytokinins and nucleoside cytokinins carrying carboxyl groups, which are most strongly retained in the resin, can be eluted with 2 N formic acid. Dowex 1 (formate$^-$), which has a similar nature to De-Acidite FF, also retained nucleotide cytokinins and nucleoside cytokinins carrying carboxyl groups (LETHAM, 1973) as well as zeatin glycoside (YOSHIDA and ORITANI, 1972). These cytokinins are eluted with aqueous formic acid ranging from 0.5 N to 1.5 N. Zeatin and zeatin riboside are not retained in this resin.

DEAE cellulose (formate$^-$) has been used in the purification of the metabolites of zeatin, i.e., 7-β-D-glucosyl zeatin and 9-β-D-glucopyranosyl zeatin (PARKER et al., 1972, 1973). These compounds are not retained by this resin and are found in the water effluent.

e) Abscisic Acid and Related Compounds

Solvent Partitioning. The extraction procedures shown in Figs. 2.1 and 2.2 are entirely applicable for abscisic acid and related compounds. Ethyl acetate and ether have been used in most cases. The acidic fraction contains acidic inhibitors, i.e., ABA, phaseic acid, dihydrophaseic acid, and epidihydrophaseic acid, whilst the neutral fraction contains xanthoxin. Conjugates remain in the aqueous fraction and are, without further solvent partitioning using polar solvents, subjected to charcoal treatment (KOSHIMIZU et al., 1968a). GOLDSCHMIDT and MONSELISE

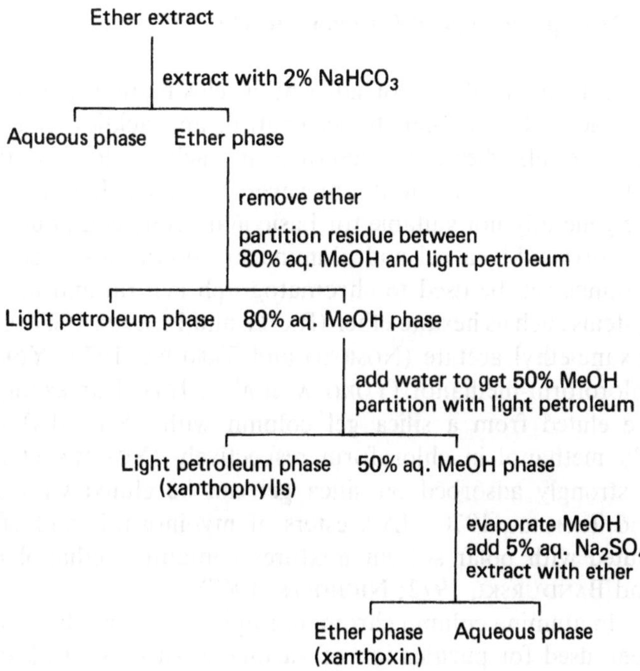

Fig. 2.10. Fractionation procedure for xanthoxin (FIRN et al., 1972)

(1968) found that ABA and IAA can be largely partitioned into ether from aqueous solution adjusted to pH 7.0. Therefore such a procedure should not be employed in the solvent partitioning procedures except for special purposes.

TAYLOR and BURDEN (1970) and FIRN et al. (1972) reported that xanthoxin could be extracted with ether directly from crushed plant tissues. This ether extract was subjected to solvent partitioning shown in Fig. 2.10 which removed acids, lipophilic compounds, and xanthophylls. The partitioning between light petroleum and 50% aqueous methanol removed xanthophylls which were partitioned into the petroleum ether phase. Xanthoxin has been known to be derived in vitro by chemical or photochemical oxidation of plant xanthophylls (TAYLOR and BURDEN, 1970; BURDEN and TAYLOR, 1970).

2.1.3 Column Chromatography and Other Purification Techniques

a) General Remarks

Chromatography has proved to be a powerful purification technique for the isolation of plant hormones. In the field of plant physiology chromatographic purification is also used to obtain test samples for biological assay and instrumental analysis which are hampered by contaminating inhibitors and other impurities.

In this section purification techniques based on column chromatography and some other techniques are discussed. Thin-layer and paper chromatography, which are extensively used for both purification and identification, are discussed in Section 2.2.1.

b) Adsorption Column Chromatography

Adsorbents available for adsorption chromatography include silica gel (acidic and neutral), alumina (basic, neutral, and acidic), charcoal, and others. Uses of these adsorbents are described in each section for the different groups of plant hormones. It should, however, be stated that acidic and basic adsorbents are generally not suitable for basic and acidic compounds respectively.

Auxins. Silica gel and alumina are useful adsorbents for auxins. Silica gel columns can be used to chromatograph neutral and acidic auxins with solvent systems such as hexane-ether (RAYLE and PURVES, 1967; OKAMOTO et al., 1967a), hexane-ethyl acetate (NOMOTO and TAMURA, 1970; YOKOTA et al., 1974), and chloroform-methanol (YOKOTA et al., 1974). For example, MeIAA and IAA are eluted from a silica gel column with 15% ethyl acetate in hexane and 4% methanol in chloroform respectively (YOKOTA et al., 1974). IAM which is strongly adsorbed on silica gel can be eluted with ethyl acetate (NOMOTO and TAMURA, 1970). IAA esters of myoinositols and of cellulosic glucans are eluted with polar solvent mixtures containing ethanol or acetone (PISKORNIK and BANDURSKI, 1972; NICHOLLS, 1967).

In alumina column chromatography, basic, neutral, and acidic alumina have been used for purification of various auxins. Neutral and basic alumina have been used to purify neutral auxins which can be eluted with various solvent systems (REDEMANN et al., 1951; HENBEST et al., 1953; ISOGAI et al., 1967a; OKAMOTO et al., 1967a, b; MARUMO et al., 1968a; YOKOTA et al., 1974). For example, MeIAA and IAM can be eluted from a neutral alumina column with 5% ethyl acetate in hexane (YOKOTA et al., 1974) and ca. 5% methanol in ethyl acetate (IGOSHI et al., 1971) respectively. Me4-Cl-IAA can be eluted from a basic alumina column with 15% ethyl acetate in hexane (MARUMO et al., 1968a). Acid alumina which has anion exchange properties is a useful adsorbent to purify glucosinolates including glucobrassicin, neoglucobrassicin, and sulphoglucobrassicin (GMELIN, 1964; ELLIOTT and STOWE, 1970). These are eluted with 1% potassium sulphate in water.

Other adsorbents, i.e., magnesium silicate (RAYLE and PURVES, 1967), calcium sulphate (HENBEST et al., 1953), and charcoal (YOKOTA et al., 1974) have been used in some isolation experiments.

Gibberellins. Charcoal is one of the most suitable materials to deal with a large amount of crude material because of its large sample capacity. Column chromatography using granular charcoal or a mixture of Celite-charcoal has been used frequently for the isolation of gibberellins. Elution is usually carried out with increasing acetone content in water. A variety of gibberellins are eluted with 30–90% acetone. The acetone concentrations necessary for elution are 35–40% for GA_8, 45–60% for GA_1 and GA_3, and 65–70% for GA_5 and GA_{20}. This elution pattern, however, varies with the column size, eluant volume, and amount of sample. Gibberellins are eluted in the following order: GA_8, GA_{26}, GA_3, $GA_5 = GA_{20}$, GA_{27} (YOKOTA et al., 1971a); GA_8, GA_1, $ABA \approx GA_{38}$, $GA_4 \approx GA_5 \approx GA_6 \approx GA_{37}$ (HIRAGA et al., 1974b); GA_8, $GA_1 \approx$ phaseic acid, GA_6, $GA_5 = GA_{20}$, GA_{19}, GA_{17} (DURLEY et al., 1971). Gibberellin glucosides can also be purified by charcoal chromatography using aqueous acetone (YO-

KOTA et al., 1971b; HIRAGA et al., 1974a, b; YAMANE et al., 1974) and aqueous methanol (YOKOTA et al., 1971b; HARADA and YOKOTA, 1970). Charcoal has low affinity for inorganic salt and sugar, and therefore can be used for removing such materials from gibberellin glucoside fractions.

Silicic acid column chromatography has been usually applied to partially purified gibberellin fraction. Silica gel is sometimes mixed with Celite to obtain a smooth solvent flow. General eluting solvents are ethyl acetate mixed with less polar solvents such as light petroleum (CROSS et al., 1962), benzene (YOKOTA et al., 1971a), or chloroform (MACMILLAN et al., 1960). A wide range of gibberellins is eluted in order of polarity by increasing ethyl acetate concentration.

Silica gel impregnated with silver nitrate can effect the separation of double-bond isomers because of the affinity between silver-ion and double-bond π-electrons. Separation varies with the nature of the double bonds. MUROFUSHI et al. (1968) used silica gel containing 25% silver nitrate to separate GA_{20} methyl ester from a mixture of four compounds obtained by partial hydrogenation of GA_5 methyl ester. This system can also be used in thin-layer chromatography.

Alumina (basic) column chromatography can separate a mixture of methyl esters of GA_1, GA_2 and GA_3 by using an ethyl acetate-benzene mixture. Increasing ethyl acetate concentration allows the separation of methyl esters of GA_1, GA_3, and GA_2 in that order of elution (TAKAHASHI et al., 1955).

Cytokinins. Charcoal has been used, although not chromatographically, for purifying dihydrozeatin from immature seeds of *Lupinus luteus* (KOSHIMIZU et al., 1967). The charcoal was successively eluted with 70% acetone, ethanolic ammonia, and a mixture of pyridine and ethanolic ammonia. Dihydrozeatin was eluted in the last fraction. According to LETHAM (1973), cytokinins which are not extractable by n-butanol at pH 7, such as ribotide cytokinins, are eluted from charcoal with pyridine-water (1:9) in good recovery, while poor recoveries of n-butanol-extractable compounds including zeatin are obtained. Charcoal is not so useful for cytokinins as for gibberellins.

Column chromatography using silica gel or alumina is usually not used to purify cytokinins although silica gel thin-layer chromatography has been frequently used (MILLER, 1974; LETHAM, 1973).

Abscisic Acid and Related Compounds. Charcoal-Celite (1:2) column chromatography has been frequently used by many workers. In the study of ABA metabolism in *Phaseolus vulgaris* seed WALTON et al. (1973) reported that phaseic acid and dihydrophaseic acid could be eluted from the column with 40% acetone in water and ABA with 60% acetone. However, this elution profile is variable, depending on the chromatographic conditions. For example ABA has been reported to be eluted with 20–30% acetone (ISOGAI et al., 1967b), 30–40% acetone (KOMOTO et al., 1972) and 50–60% acetone (OHKUMA et al., 1963; KOSHIMIZU et al., 1966; DAVIS et al., 1972).

ZEEVAART (1974) used charcoal chromatography to prepare ABA and xanthoxin fractions from acidic and neutral ether fractions of spinach respectively. The ABA fraction eluted with 60% acetone could be analyzed by gas chromatography after thin-layer chromatography and methylation. The xanthoxin fraction eluted with 50% acetone was also subjected to gas chromatographic analysis after thin-layer chromatography and acetylation.

Silica gel or silica gel-Celite column chromatography has been frequently used to isolate ABA. ABA can be eluted from the column with various solvents such as 10–30% ethyl acetate in chloroform (OHKUMA et al., 1963; KOSHIMIZU et al., 1966), 25–30% ethyl acetate in benzene (HASHIMOTO et al., 1968) and 1–5% methanol in methylene dichloride (ISOGAI et al., 1967b; KOMOTO et al., 1972). Phaseic acid has been eluted with 20–30% ethyl acetate in chloroform (MACMILLAN et al., 1960) and xanthoxin with ethyl acetate-benzene (1:2) mixture (TAYLOR and BURDEN, 1970).

c) Partition Column Chromatography

Partition column chromatography is frequently used for a variety of plant hormones because of both the excellent sample recovery relative to adsorption chromatography and the good resolution power.

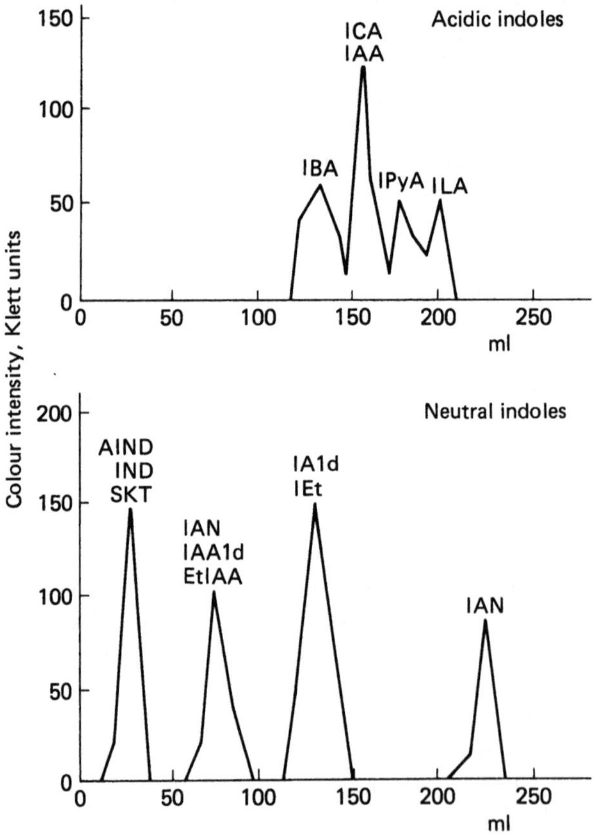

Fig. 2.11. Elution patterns of acidic and neutral indoles from silica gel partition column. Acidic indoles: column, 8.0 g of silica gel hydrated with 5.0 ml of 0.5 M formic acid; fraction size 5.4 ml; solvents, Varigrad chamber No. 1–100 ml n-hexane saturated with 0.5 M formic acid, Varigrad chamber No. 2–100 ml n-hexane saturated with 0.5 M formic acid, Varigrad chamber No. 3–75 ml ethyl acetate saturated with 0.5 M formic acid. Neutral indoles; same as for acidic indoles, except fraction size is 9.4 ml. *AIND* N-acetyl indole; *IND* indole; *SKT* skatole. (POWELL, 1964)

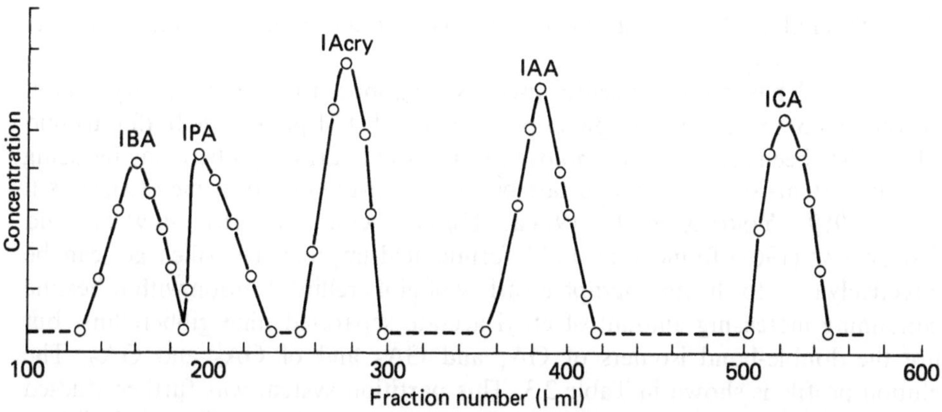

Fig. 2.12. Separation of acidic indoles by Sephadex G-25 partition chromatography. Column, Sephadex G-25 M swollen with aqueous phase of benzene-dioxane-water (1:1:1), 150×2.5 cm; eluent, organic phase of the solvent system, 6–8 ml/h, 22–25° C; fraction size, 1 ml. Sephadex G-25 M (1 g) impregnated with a solution of five indole acids (1 mg each) in aqueous phase (2.5 ml) was placed on top of the column. (RAY and HUTZINGER, 1970b)

Auxins. POWELL (1960) reported partition column chromatography for indole compounds. This partition system comprised 0.5M formic acid-impregnated silica gel and stepwise solvent system using n-butanol-petroleum ether mixtures. Afterwards POWELL (1964) devised a more elaborate technique which comprised 0.5M formic acid-impregnated silica gel and gradient elution using ethyl acetate and n-hexane. This partition column allowed reasonable separation of acidic and neutral indoles (Fig. 2.11). For basic and water-soluble auxins concentrated ammonia was used in place of 5% formic acid. These techniques have been successfully used to prepare auxin extracts from plants for gas chromatography and spectrofluorometry (POWELL, 1964; DEYOE and ZAERR, 1976).

Other solvent systems for silica gel partition chromatography have been successfully used in the isolation of IAA, 4-Cl-IAA (pH 6.2–6.9 phosphate buffer; ethyl acetate-n-hexane) (MARUMO et al., 1968b; IGOSHI et al., 1971; ABE et al., 1972).

Partition column chromatography using Sephadex G-25 as a support for the stationary phase can effectively separate indole acids in two-phase systems such as benzene-dioxane-water (1:1:1) (RAJ and HUTZINGER, 1970b). As shown in Fig. 2.12 the separation of each compound seems much better than the silica gel partition system. This technique seems promising because of the high sample recoveries and availability of neutral partition system for labile compounds.

Gibberellins. STODOLA et al. (1957) indicated that partition chromatography can separate the double-bond isomers GA_1 and GA_3 which had not been chromatographically separable. The column, comprising Hyflo Supercel (diatomonous earth) impregnated with ca. 2M phosphate buffer of pH 6.2, could separate GA_1 and GA_3 in that order of elution. However, the method is inconvenient since, in this system, a large amount of solvent (ether) is required to elute gibberellins. According to Pitel et al. (1971) this column does not separate isoGA_3 and GA_3.

Silica gel has been frequently used as a support for the stationary phase. A column composed of silica gel impregnated with 1 M phosphate buffer around pH 5.5 has been successfully employed in isolating various gibberellins by using solvent systems such as ethyl acetate-benzene or n-butanol-benzene (Takahashi et al., 1959; Yokota et al., 1971a; Hiraga et al., 1974b). Powell and Tautvydas (1967) found that 0.5M formic acid-impregnated silica gel can be effectively used to chromatograph a variety of gibberellins. Elution with n-hexane containing increasing amount of ethyl acetate separated nine gibberellins, but not the double-bond isomers of GA_1 and GA_3 and of GA_4 and GA_7. The elution profile is shown in Table 2.3. This partition system was further studied by Durley et al. (1972) who reported the chromatographic profiles of 33 gibberellins and ABA on a gradient-eluted silica gel partition column as shown in Table 2.4. They claimed that the silica gel-formic acid column gives excellent resolution in contrast with other methods such as silica gel adsorption. This silica gel-formic acid system has been extended to high-performance liquid chromatography by which Reeve et al. (1976) and Crozier and Reeve (1977) analyzed radioactive gibberellins, acidic and neutral indoles, and cis/trans-ABA (see Sect. 2.2.4).

Sephadex can retain a larger amount of stationary phase than any other support, enabling a large amount of sample to be applied to the Sephadex partition column. This is the reason why Sephadex has been used for isolation of a number of gibberellins and gibberellin glucosides. Murofushi et al. (1969) used Sephadex G-50 impregnated with 1 M phosphate buffer of pH 5.4 and n-butanol as the eluant in the isolation of GA_{21} and GA_{22} from *Canavalia* seeds. Koshimizu et al. (1968b) and Fukui et al. (1971, 1972) also used Sephadex LH 20-Celite mixture impregnated with the same buffer and n-butanol-benzene as the eluant in the isolation of GA_{18}, GA_{19}, GA_{23}, and GA_{28} from *Lupinus* seeds. Yokota et al. (1971b) applied Sephadex G-50 partition chromatography for purifying gibberellin glucosides from immature seeds of *Pharbitis nil*. Increasing n-butanol content in ethyl acetate gave a clear separation of three fractions containing GA_{27} glucoside, GA_3 glucoside/GA_{26} glucoside, and GA_8 glucoside/GA_{29} glucoside in that order of elution. It should be noted that this reasonable resolution was obtainable for 13 g of crude material charged onto 50 g of Sephadex G-50 impregnated with 230 ml of phosphate buffer.

Solvent systems, known to separate double bond isomers on thin-layer chromatography (Kagawa et al., 1963; MacMillan and Suter, 1963), were introduced into Sephadex partition chromatography by Pitel et al. (1971). The column, prepared with Sephadex G-25 impregnated with the aqueous phase of a two-phase solvent system, is developed with the organic phase. A pair of solvent systems, carbon tetrachloride-acetic acid-water (8:3:5) and benzene-petroleum ether-acetic acid-water (6:2:5:3), can clearly separate GA_4, GA_7 and isoGA_7 in that order of elution (Fig. 2.13). On the other hand, the solvent systems, benzene-ethyl acetate-acetic acid-water (55:25:30:50 or 14:7:10:10), can separate GA_1, GA_3, and isoGA_3 in that order of elution (Fig. 2.14). GA_1 and dihydroGA_1, which are not resolved by this technique, can be separated

Table 2.3. Separation of GA_{1-9} on silica gel partition column with solvent system 0.5 M formic acid-ethyl acetate in n-hexane. (POWELL and TAUTVYDAS, 1967)

Ethyl acetate %	Gibberellin	Ethyl acetate %	Gibberellin	Ethyl acetate %	Gibberellin
0		22.5		45.0	A_2
1.0		24.0	A_6	46.0	A_2 (trace)
3.0		25.5		48.0	
6.0	A_9	27.0		51.0	
9.0		30.0		54.0	
12.0		33.0		57.0	
15.0	A_4	36.0		60.0	
16.5	A_4, A_7	39.0		63.0	A_8
18.0		40.5		66.0	
19.5	A_5	42.0	A_1, A_3		
21.0		43.5	A_1, A_3		

Column: 8 g silica gel hydrated with 5 ml of 0.5 M formic acid; eluents, 0.5 M formic acid-saturated solution of ethyl acetate-hexane; elution volume for each fraction, 25 ml.

Table 2.4. Separation of 33 gibberellins and abscisic acid on Woelm silica gel partition column with solvent system, 0.5M formic acid-ethyl acetate in n-hexane (gradient elution). (DURLEY et al., 1972)

Fraction no.	Giberellin	Fraction no.	Gibberellin
2	A_9 A_{12}	14	A_1 A_3 A_{19}
3	A_9 A_{11} A_{14} A_{24} A_{31}	15	A_2 A_{13} A_{19}
4	A_4 A_5 A_6 A_7 A_{14} A_{15} A_{20} A_{25} A_{31} ABA	16	A_2 A_{13} A_{22}
		17	A_{18} A_{22} A_{26} A_{29}
5	A_6 A_{10} A_{15}	18	A_{18} A_{26} A_{29}
6	A_{10}	19	A_{17}
8	A_{27} A_{34}	20	A_{17} A_{23}
9	A_{16} A_{27} A_{34}	21	A_{21} A_{23}
10	A_{16} A_{27} A_{34} A_{33}	22	A_{21}
11	A_{33}	23	A_8 A_{28}
12	A_1 A_3 A_{30}	24	A_8 A_{28}
13	A_1 A_3 A_{30}		

Column: Woelm Silica Gel for Partition Chromatography (20 g) equilibrated with 0.5 M formic acid-saturated solution of ethyl acetate-hexane (10:90), 20×1.3 cm. Varigrad gradient system (0.5 M formic acid-saturated solvents are used); chamber 1, ethyl acetate-hexane 65:35 (129 ml); chamber 2, ethyl acetate-hexane 20:80 (147 ml); chambers 3 and 4, 100% ethyl acetate (114 ml); Fraction size, 20 ml.

by argentation partition chromatography on Sephadex G-25 (VINING, 1971). These excellent techniques have a clear advantage in obtaining pure specimens of some gibberellin double-bond isomers which are difficult to prepare by other techniques. However, these specified solvent systems are not suitable for the separation of a wide range of gibberellin homologues.

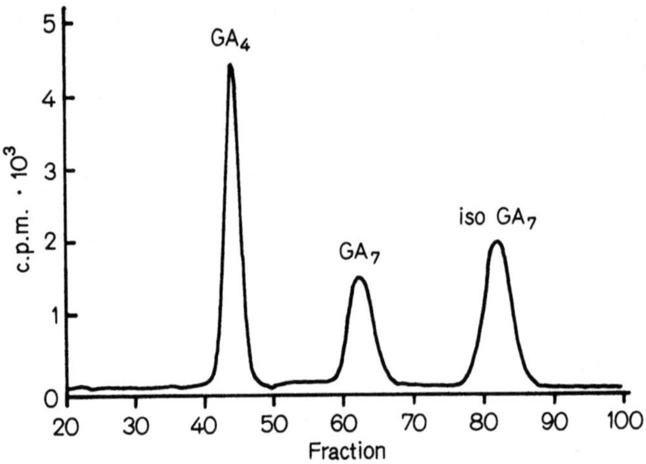

Fig. 2.13. Separation of ^{14}C-labelled gibberellins GA_4, GA_7 and iso-GA_7 by Sephadex G-25 partition chromatography. Column, Sephadex G-25 swollen with aqueous phase of benzene-petroleum ether (b.p. 60–80°)-acetic acid-water (6:2:5:3), 95 × 2.6 cm; eluent, organic phase of the solvent system, 1 ml/min; fraction size, 20 ml. A 200 mg sample of ^{14}C-labelled mixture was applied in 2.5 ml aqueous phase absorbed in lg dry Sephadex. (PITEL et al., 1971)

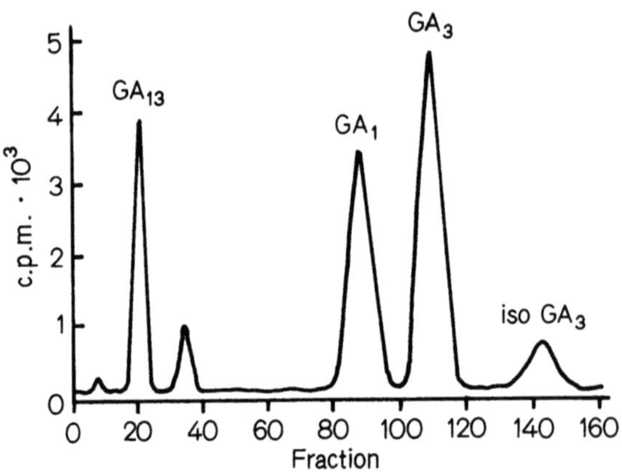

Fig. 2.14. Separation of ^{14}C-labelled gibberellins GA_1, GA_3, *iso*-GA_3 and GA_{13}. Column, Sephadex G-25 swollen with aqueous phase of benzene-ethyl acetate-acetic acid-water (55:25:30:50), 95 × 2.5 cm; eluent, organic phase of the solvent system, 1 ml/min; fraction size, 20 ml. A 500 mg sample was applied in the minimum volume (5 ml) of aqueous phase absorbed in 2 g dry Sephadex. (PITEL et al., 1971)

MACMILLAN and WELS (1973) devised a versatile Sephadex LH-20 partition chromatography using three two-phase solvent systems. The partition column using a solvent system of petroleum ether-ethyl acetate-acetic acid-methanol-water (100:80:5:40:7) can separate a wide range of gibberellins and ABA (Fig. 2.15). This carefully packed column, termed a wide-range column, reached a resolution of 5500 theoretical plates for GA_3, although some gibberellins still overlapped. The condensed region between GA_{12}-aldehyde and GA_{14}-aldehyde can be expanded by a narrow-range column using a solvent system of light petroleum-ethyl acetate-acetic acid-methanol-water (50:15:10:10:2). Non-polar compounds such as ent-kaurene, ent-kaurenol and ent-kaurenoic acid can be completely separated by using a solvent system of light petroleum-acetic acid-methanol (100:1:40). These columns can accept up to 200 mg loadings with

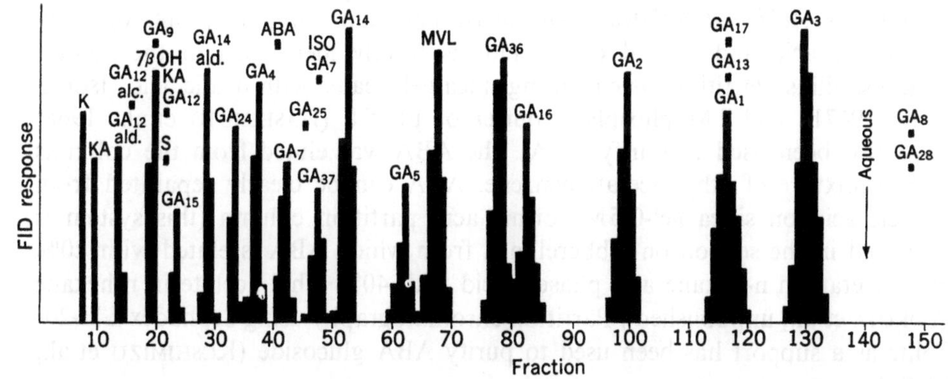

Fig. 2.15. Separation of gibberellins, abscisic acid and other compounds by Sephadex LH-20 partition chromatography. Column, Sephadex LH-20 swollen with aqueous phase of light petroleum-ethyl acetate-acetic acid-methanol-water (100:80:5:40:7), 147×1.5 cm; eluent, organic phase of the solvent system, 50 ml/h; fraction size, 10 ml. The 29 compounds (160–330 µg of each) were applied in the minimum volume of aqueous phase. *K*, *ent*-kaurene; *KA*, ent-kaurenoic acid; *GA_{12} ald.*, GA_{12}-aldehyde; *GA_{12} alc.*, GA_{12}-alcohol; *7βOHKA*, *ent*-7-α-hydroxykaurenoic acid; *7αOHKA*, *ent*-7β-hydroxykaurenoic acid; *S*, steviol; *GA_{14} ald.*, GA_{14}-aldehyde; *MVL*, (2-^{14}C)-mevalonic acid lactone (detected by liquid scintillation counting). (MacMillan and Wels, 1973)

little loss in resolution. Terpenoid metabolites of *Gibberella fujikuroi* could be successfully fractionated by the wide-range column and narrow-range column and a number of fractions thus obtained were subjected to combined GC-MS analysis which enabled the detection of 72 compounds including 15 gibberellins (MacMillan and Wels, 1974). Fukui et al. (1977) used the wide-range column which could separate isomeric 12-hydroxygibberellins, i.e., GA_{48} and GA_{49}.

Cytokinins. Partition chromatography using Celite 545 was devised by Hall (1962, 1965) to separate minor nucleosides from nucleic acid hydrolyzate and has been widely applied to the purification of cytokinins from tRNA hydrolyzates. The column composed of purified Celite 545 mixed with the lower phase of a two-phase solvent system is developed with the upper phase. The following solvent systems have been used: ethyl acetate-water, ethyl acetate-n-propanol-water (4:1:2), ethyl acetate-methyl cellosolve-water (4:1:2). Cytokinins such as i^6A (Robins et al., 1967; Burrows et al., 1969), c-io^{6}A (Hall et al., 1967) and ms^2-i^{6}A (Burrows et al., 1969) move fast through the column and elute before the ribonucleosides appear.

Sephadex partition chromatography which has been successfully applied to auxins and gibberellins is also effective for cytokinins. Watanabe et al. (1978) found that the Sephadex LH-20 partition column was effective in purifying cytokinins from plant extract by using 0.2M phosphate buffer of pH 7.2 as a stationary phase and 10% n-butanol in ethyl acetate as a mobile phase. Purification of the extract of *Humulus lupulus* by this partition system resulted in up to 70-fold reduction of the dry weight to give zeatin- and zeatin riboside-rich fractions. Zeatin and zeatin riboside were eluted with 2.5 and 4.8 column volumes respectively.

Abscisic Acid and Related Compounds. Partition column chromatography of abscisic acid and related compounds is briefly mentioned in the section on gibberellins. Partition systems using silica gel treated with oxalic acid (ISOGAI et al., 1967b) and 1M phosphate buffer of pH 5.8 (HASHIMOTO et al., 1968) have also been used to purify ABA. The ABA was eluted from the columns with a mixture of ethyl acetate-benzene. ABA can be clearly separated from phaseic acid on silica gel-0.5M formic acid partition column (this system is discussed in the section on gibberellins), from which ABA is eluted with 20% ethyl acetate in n-hexane and phaseic acid with 40% ethyl acetate in n-hexane (YOKOTA et al., unpublished). Partition chromatography using Sephadex LH-20-Celite as a support has been used to purify ABA glucoside (KOSHIMIZU et al., 1968a).

d) Sephadex Column and Gel Permeation Column Chromatography

Sephadex G-10, G-15, G-25 and LH-20 (alkylated form of G-25), classified as molecular sieves, have been frequently used for a variety of plant hormones but the mechanism of separation involves, in addition to molecular sieving, reversed phase partition in the case of Sephadex LH-20 and ion exchange-adsorption effect in the case of Sephadex G-10, G-15 and G-25 (REEVE and CROZIER, 1976). Recently, gel permeation technique based on the molecular size separation has been reported (see section on gibberellins).

Auxins. Sephadex LH-20 has been used to purify various auxins such as IEt (RAYLE and PURVES, 1967), IAA (DEYOE and ZAERR, 1976), IAAld dimethylacetal (YOKOTA et al., 1974), 4-Cl-IAA (MARUMO et al., 1968b), and 4-Cl-IAA aspartate (HATTORI and MARUMO, 1972). Developing is usually performed with polar solvents such as methanol and ethanol. STEEN and ELIASSON (1969) used 96% or 70% ethanol to which was added hydrochloric acid to a concentration of 0.001 M, resulting in a clear separation of IAA from ABA which interferes with the *Avena* straight growth test. A 1% solution of 1.0 mM hydrochloric acid in 95% ethanol has been also used to purify IAA from Douglas fir (DEYOE and ZAERR, 1976). Generally the addition of acid or a buffer solution is helpful in obtaining reproducible elution patterns. Sephadex LH-20-100 has been used to purify cellulosic glucan esters of IAA with the solvent system of ethanol-water (1:1) (PISKORNIK and BANDURSKI, 1972).

Sephadex G-10 and G-25 have been used to purify IAA myoinositol esters (LABARCA et al., 1965) and sulpho-glucobrassicin (ELLIOTT and STOWE, 1970) with water as the eluant.

Gibberellins. Sephadex chromatography seems to be used very rarely for purification of gibberellins, although Sephadex G-10 has been once used for gibberellin purification by CROZIER et al. (1969). In spite of this, Sephadex column chromatography seems to be very promising for gibberellin purification, since recently Sephadex LH-20 column was found to separate a mixture of GA_4, GA_7 and GA_3 with the solvent system of water-n-propanol-n-butanol (20:4:1) (YAMAGUCHI, unpublished).

Gel permeation chromatography based on molecular size separation has been examined by REEVE and CROZIER (1976) by using porous polystyrene beads

(Bio-Beads SX-12, SX-8 and SX-4) which have the molecular exclusion limits of 400, 1000, and 1500 molecular units respectively. Analogous compounds within a group, for example, the gibberellins, showed a linear relationship between elution volumes and log of molecular weights. Such a relationship seems, however, not to hold for compounds with totally different structures such as gibberellins, auxins, cytokinins, and ABA. This separation technique was demonstrated to be effective for purification of crude plant extracts.

Cytokinins. Sephadex LH-20 column chromatography, which was developed during the study on ribonucleic acid constituents (ARMSTRONG et al., 1969), is now widely used in the purification of cytokinins. Cytokinins are almost quantitatively eluted with water or aqueous ethanol from this column. By increasing the ethanol content analysis time can be shortened (Fig. 2.16). However, excellent separation of individual cytokinins is attained by use of water or 35% ethanol. The elution volume relative to the column volume for each cytokinin eluted with 35% ethanol is as follows: zeatin riboside (1.1), t-zeatin (1.4), i^6A (1.7), kinetin riboside (1.8), benzyladenine riboside (2.1), i^6Ade (2.1), kinetin (2.2), benzyladenine (2.7), ms^2i^6A (2.9), phenylaminopurine (3.4) and ms^2-i^6Ade (5.3). This elution pattern in which polar cytokinins move faster than less polar ones indicates that a reversed phase partition mechanism is operating in Sephadex LH-20 chromatography. Although the polar zeatin riboside in tRNA hydrolyzate is not separated from the early large nucleoside peak, Sephadex G-10 instead of Sephadex LH-20 can be used for separation of this compound (BURROWS et al., 1971).

Sephadex LH-20 chromatography is used not only in analyzing tRNA constituents (EINSET et al., 1976; VREMAN et al., 1972; BURROWS et al., 1971) but also in purifying free cytokinins contained in plants (DYSON and HALL, 1972; HORGAN et al., 1973a; PETERSON and MILLER, 1976, 1977; WANG et al., 1977). HEWETT and WAREING (1973) detected seven cytokinins in mature leaves of *Populus robusta* Schneid after chromatography on Sephadex LH-20 using 35% ethanol elution. The most slowly moving compound was a new cytokinin, whose structure was later determined to be 6-(o-hydroxybenzylamino)-9-β-D-ribofuranosyl purine (HORGAN et al., 1973b).

Abscisic Acid and Related Compounds. SWEETSTER and VATVARS (1976) prepared ABA fractions from acidic ether fractions of plant tissues by using Sephadex G-25 column chromatography. Elution with 20% methanol adjusted to pH 3.0 with sulphuric acid not only removed many contaminants from the ABA fraction, but also separated IAA which is eluted in later fractions. The ABA fraction thus obtained was directly subjected to high performance liquid chromatographic analysis in order to estimate ABA levels. The separation of ABA and IAA can also be effected by Sephadex LH-20 column chromatography, which is discussed in the section on auxins.

e) Insoluble Polyvinylpyrrolidone Column Chromatography

Insoluble polyvinylpyrrolidone (PVP) has been used for the chromatography of various compounds including nucleic acid base components (LERNER et al., 1968; LAMMI and LERNER, 1969; DOUGHERTY and SCHEPARTZ, 1969a, c), amino

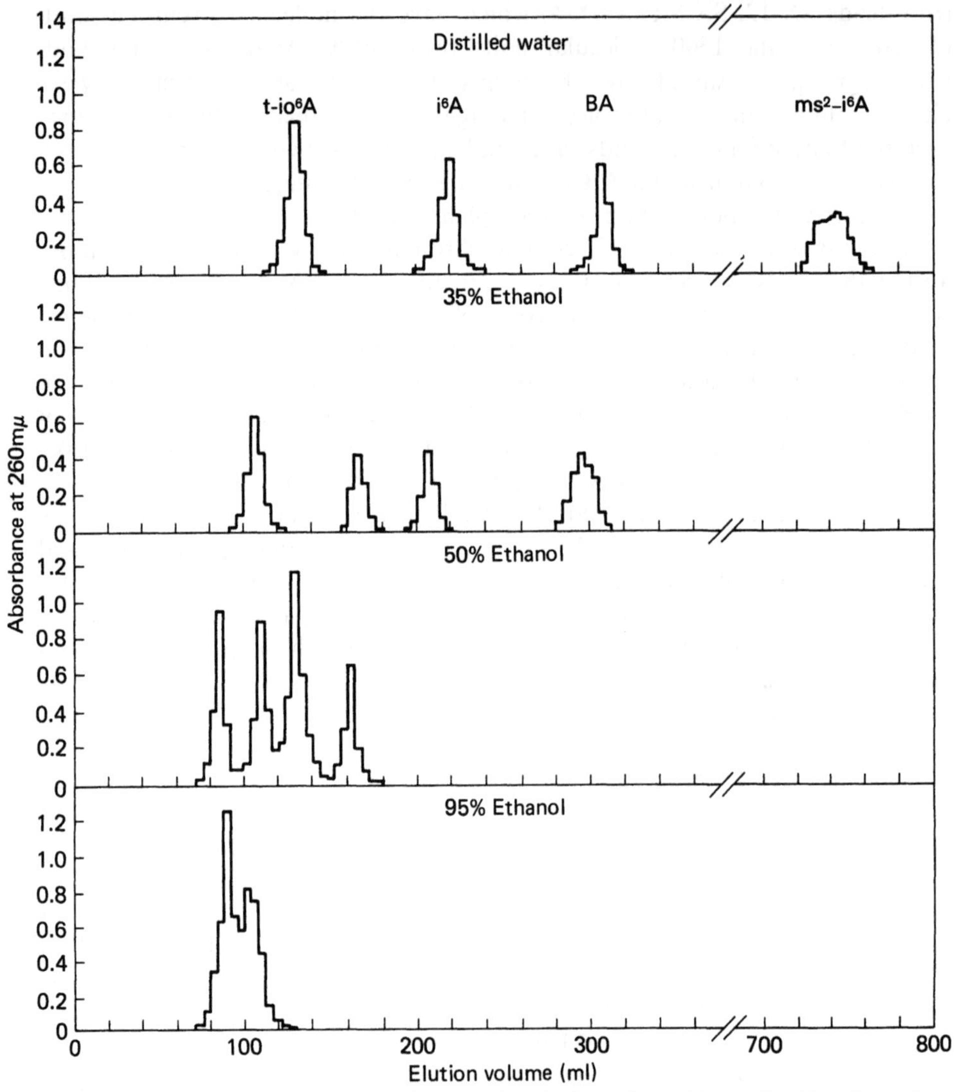

Fig. 2.16. Effect of ethanol concentration on the elution of cytokinin ribosides from Sephadex LH-20 columns. One ml samples containing a mixture of benzylaldenine (BA) i^6A, ms^2-i^6A and t-io^6A were fractionated on Sephadex LH-20 columns (*upper row*, 2.4 × 15 cm, 20 g; others, 2.4 × 20 cm, 25 g). Fractions of 4 ml were collected. (ARMSTRONG et al., 1969)

acids (DOUGHERTY and SCHEPARTZ, 1969b) and phenolic compounds (CLIFFORD, 1974). It has also been found to be an excellent material for the purification of plant hormones including gibberellins, cytokinins, IAA, and ABA as described below.

Gibberellins. GLENN et al. (1972) reported the elution profiles of eight gibberellins, ABA, IAA and zeatin on PVP columns using 0.1 M phosphate buffer. At both pH 8.0 and 5.0 some degree of selectivity was found among the eight gibberellins and ABA, whilst apparent selectivity between these compounds and IAA or zeatin was observed. The advantages of using PVP rest not only

Table 2.5. Elution volumes of cytokinins on PVP columns (25 × 1.7 cm, except where otherwise stated) using a 0.013 M phosphate buffer at different pH values (BIDDINGTON and THOMAS, 1976)

Cytokinin	Peak elution volume (ml)			Elution range (ml)		
	pH3.5	pH6.4	pH9.5	pH3.5	pH6.4	pH9.5
Dihydrozeatin	65	150 590[a]	195	50–75	120–190 540–640[a]	145–240
Zeatin	68	170 700[a]	225	55–85	140–215 630–770	175–280
Zeatin riboside	87	110 440[a]	120	70–100	80–150 380–510[a]	90–150
6-(3-Methylbut-2-enylamino) purine	75	225		60–90	210–300	
6-(3-Methylbut-2-enylamino)-9-β-D-furanosylpurine	105	155 590[a]	160	80–130	120–190 530–660[a]	120–195
Kinetin	125	360		100–150	285–475	
Kinetin riboside	160	185		125–200	145–230	
N⁶-Benzyladenine (BA)	150	620		120–190	500–710	
BA riboside	210	310		150–280	230–400	
N^6-(O-hydroxybenzyl) adenine (hyd-BA)	735[b] 105[c]	2800[b] 400[c]		420–1260[b] 60–180[c]	1680–3990[b] 240–570[c]	
Hyd-BA riboside	875[b] 125[c]	1225[b] 185[c]		420–1540[b] 60–220[c]	630–1890[b] 90–270[c]	

[a] Elution from a 60 × 2.2 cm column
[b] Estimated for a 25 × 1.7 cm column based on the results obtained with a 10 × 1.0 cm column
[c] Elution from a 10 × 1.0 cm column

on the essentially quantitative recovery (90–99%), but also on the fact that PVP chromatography can greatly reduce the dry weight of the plant extract because PVP forms insoluble complexes with phenols and presumably other compounds under appropriate conditions. In fact a 50- to 70-fold reduction of the dry weight of the plant extracts was attained with essentially no loss of gibberellins.

Cytokinins. PVP column chromatography was found to be very suitable for purifying and separating individual cytokinins. BIDDINGTON and THOMAS (1973, 1976) reported the elution profiles of several cytokinins which are shown in Table 2.5. Elution patterns of cytokinins using 0.013 M phosphate buffer are highly affected by the acidity of the buffer. At pH 6.4 nucleoside cytokinins are eluted faster than the corresponding free bases. At pH 3.5 this relationship is reversed and the elution is more rapid, especially of cytokinin free bases. o-Hydroxybenzyladenine and its riboside, both of which move very slowly at pH 6.4 presumably because of their phenolic nature, can be eluted at reasonable positions at pH 3.5. THOMAS et al. (1975a) demonstrated the occurrence of nine cytokinin-active compounds in cabbage head by using this technique. MILLER (1975a) and PETERSON and MILLER (1977) used PVP chromatography to purify ribosylzeatin (elution with 0.1 M potassium biphosphate), glucosylzeatin and its riboside (elution with water) from *Vinca rosea* L. crown gall.

Abscisic Acid. PVP column chromatography is also effective in concentrating ABA from crude acidic fractions. LENTON et al. (1971) purified acidic ether fractions of some plant tissues through PVP column using water as the eluant, and found that 95% of the dry weight was retained by the column whilst ABA was eluted with the void volume. The ABA fraction was purified by thin-layer chromatography and, after methylation, was subjected to quantitative gas chromatographic analysis. The overall recovery of ABA was determined to be of the order of 47% and negligible interconversion of the isomers was found.

f) Countercurrent Distribution

Countercurrent distribution depends on the partitioning of mixed compounds between two solvents and separations are based on differences in the partition coefficients. The merits of countercurrent distribution are that the distribution patterns can be predicted if the partition coefficients in a certain solvent system are known, and also that mild solvent systems can be selected for unstable compounds. The procedure can be operated by separating funnels or commercially available instruments.

Countercurrent distribution is usually used at an early stage in the purification of plant hormones because it can deal with a large amount of sample.

Auxins. Countercurrent distribution has been employed for various auxins. Solvent systems used for auxin purification are: water-2:3 mixture of ethyl acetate and ether for ascorbigen (PROCHÁZKA and ŠANDA, 1960); 75% ethanol-benzene, 80% methanol-1:1 mixture of benzene and light petroleum for IAN (PROCHÁZCA and SǍNDA, 1960); 50% methanol-1:1 mixture of ether and light petroleum for ICA and IAld (PROCHÁZKA and SǍNDA, 1960); pH 6.5 phosphate buffer-ethyl acetate for IAA (IGOSHI et al., 1971; ABE et al., 1972); pH 7.1 phosphate buffer-ethyl acetate for 4-Cl-IAA (MARUMO et al., 1968b); pH 4.1 tartarate buffer-ethyl acetate for 4-Cl-IAA aspartate (HATTORI and MARUMO, 1972). MARUMO and HATTORI (1970) indicated that chlorinated analogues of IAA in *Pisum sativum* seeds can be effectively purified by countercurrent distribution followed by Sephadex LH-20 column and silica gel partition chromatography.

Gibberellins. A number of examples are found in the reports of gibberellin isolations from various plant materials, e.g., *Phyllostachys edulis* (MUROFUSHI et al., 1966), *Canavalia gladiata* (MUROFUSHI et al., 1969), *Pharbitis nil* (YOKOTA et al., 1971a), *Cytisus scoparius* (YAMANE et al., 1974), *Calonyction aculeatum* (MUROFUSHI et al., 1973), and *Phaseolus coccineus* (BOWEN et al., 1973). Partition systems between ethyl acetate and 1 M or 1.5 M phosphate buffer around pH 5.5 have been used in most cases for gibberellin purifications. With these systems most gibberellins are located in the middle tubes and a high proportion of the dry weight of a plant extract is located in and around the first and last tubes. It is possible to change the distribution pattern by altering the pH and molarity of the phosphate buffer in order to suit particular gibberellins.

CROZIER et al. (1969) reported the effectiveness of countercurrent distribution, Sephadex G-10 and silicic acid partition chromatography in purification of gibberellin-like substances from plant tissues without apparent loss of gibberellin activity. The crude acidic extract of *Phaseolus coccineus* seedlings (48 kg), which showed no biological activity because of impurity, revealed an activity equivalent to 87 μg of GA_3 after two successive countercurrent distribution procedures, the activity reaching 230 μg of GA_3 equivalent after Sephadex G-10 and silicic acid partition chromatography. Extracts of the same seedlings when purified with ion exchange resins, basic lead acetate treatment, and phosphate-buffered Celite column, were shown to lose most of their biological activity. This means that such drastic procedures as ion exchange and lead acetate treatment are not suitable for gibberellins although they were used in the early history of the research on plant hormones.

g) Other Techniques

Precipitation Reagents. Cytokinins can be precipitated or crystallized as the complexes with some reagents. This property has been utilized for purification techniques. Zeatin and dihydrozeatin precipitate as silver complexes from acidic silver nitrate solution. Free bases can be recovered by extracting their silver complexes with 0.2 N hydrochloric acid (LETHAM, 1963; KOSHIMIZU et al., 1967). Ribosyl zeatin has been found to be precipitated as a mercury complex from which ribosyl zeatin is recovered by hydrogen sulphide treatment (MILLER, 1975b).

Picric acid forms crystalline complexes with cytokinins including zeatin, dihydrozeatin, and 6-(3,4-dihydroxy-3-methylbutylamino)purine, whilst 3-iodo-picric acid has been used to make crystalline complex with 6-(2,3,4-trihydroxy-3-methylbutylamino)purine (LETHAM, 1963, 1973; KOSHIMIZU et al., 1967). Free bases can be regenerated by passing the picrate through a Dowex 1 ($formate^-$) column. IAN has been isolated as a crystalline picrate (PROCHÁZKA and ŠANDA, 1960).

Zeatin ribotide has been isolated as crystalline barium salt (LETHAM, 1973), whilst glucosinolate auxins have been crystallized as brucinium or triethylammonium salts as discussed in Section 2.1.2.b.

Other precipitation reagents which are now rarely used in plant hormone research are not referred to here.

Sublimation and Distillation. ABA sublimes at 120° C as noted by OHKUMA et al. (1963) in the original isolation of ABA from young cotton fruits. This property was found by LITTLE et al. (1972) to be effective in the purification of ABA from dormant buds of balsam fir.

IAA has been purified by distillation in the early investigation of auxins (e.g., THIMANN, 1935). This relatively volatile nature of IAA, and especially of neutral auxins, may cause partial loss during evaporation and drying procedures under vacuum (MANN and JAWORSKI, 1970; POWELL, 1960).

2.1.4 Examples of Purification of Plant Hormones

a) Isolation of Auxins from Young Citrus Fruits

The procedure, adopted by IGOSHI et al. (1971) for the isolation of IAA and IAM from *Citrus unshiu* fruits, is shown in Fig. 2.17. The fruits were extracted five times with ether with homogenizing. However, the ether did not extract the total auxins present in the fruits. Auxins could be completely recovered from the residue by subsequent extraction with methanol as was discussed in the Section 2.1.1.a. The isolated IAA and IAM were characterized by melting points and GC-MS. Although it seems unusual that IAM was isolated from the acidic ethyl acetate fraction, this may be due to its polar nature which causes incomplete solvent partitioning. This polar nature was substantiated by the fact that IAM showed a lower Rf value on TLC than IAA and was more slowly eluted than IAA from the partition column. For the identification of neutral auxin, the fruits were extracted with acetone and subjected to the usual

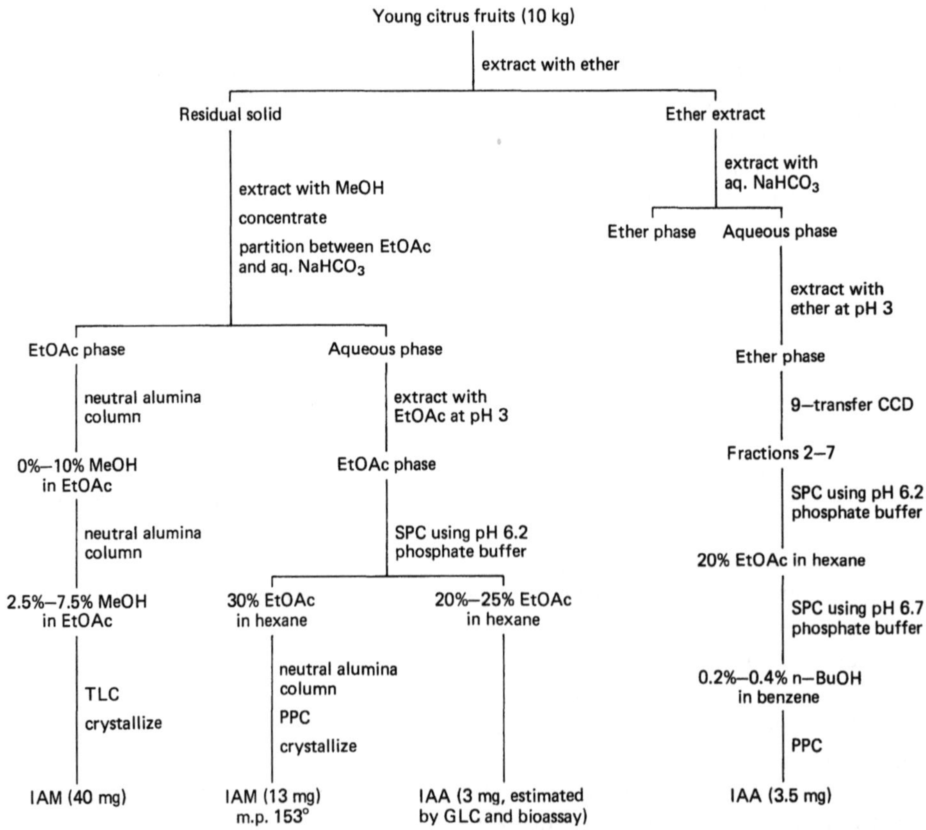

Fig. 2.17. Isolation of auxins from young citrus fruits. *CCD*, countercurrent distribution between pH 6.5 phosphate buffer and ethyl acetate. *SPC*, silica gel partition chromatography. *PPC*, paper chromatography with solvent system, isopropanol-conc. ammonia-water (10:1:1). *TLC*, thin-layer chromatography with solvent system, isopropyl ether-acetic acid (95:5). (IGOSHI et al., 1971)

solvent partitioning (TAKAHASHI et al., 1975). The neutral ether fraction, after several purification procedures was analyzed by mass chromatography to confirm the presence of MeIAA. The use of acetone as extraction solvent excludes the possibility that the MeIAA identified is an artefact derived from methanolysis.

b) Isolation of Indole-3-Ethanol from Cucumber Seedlings

The ether extract of cucumber seedling stems was first passed through a DEAE-cellulose pad. This procedure removed, from the extract, pigments and IAA which were retained by DEAE-cellulose. Pure indole-3-ethanol, isolated after the purification procedures shown in Fig. 2.18, was identified by direct comparison with an authentic specimen by Rf values on TLC and by several physicochemical analyses including NMR, IR, mass and UV spectra (RAYLE and PURVES, 1967).

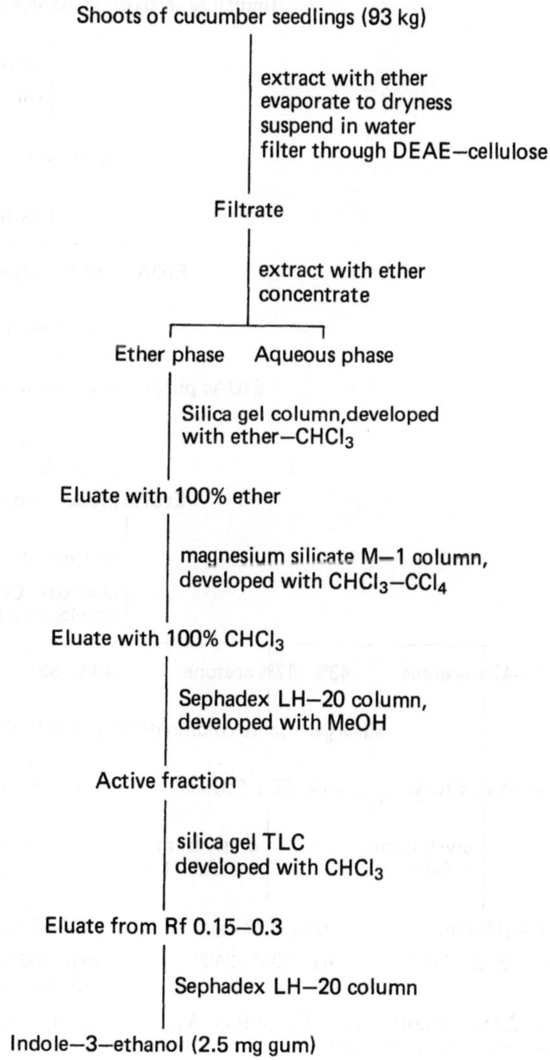

Fig. 2.18. Isolation of indole-3-ethanol from cucumber seedlings. (RAYLE and PURVES, 1967)

c) Isolation of Gibberellins A_1, A_5, A_6, and A_8 from Immature *Phaseolus* Seeds

Immature seed of *Phaseolus coccineus (multiflorus)* is a rich source of gibberellins. The acidic ethyl acetate fraction was purified by charcoal-Celite column chromatography and silica gel-Celite column chromatography and finally crystallization as shown in Fig. 2.19 (MacMillan et al., 1962). Four gibberellins, GA_1, GA_5, GA_6, and GA_8, were obtained in good yields. It should be stated that such simple purification procedure may be insufficient to deal with other plant tissues which have lower gibberellin levels. In such cases several further techniques such as countercurrent distribution, partition chromatography, and thin-layer chromatography must be included in the purification procedure.

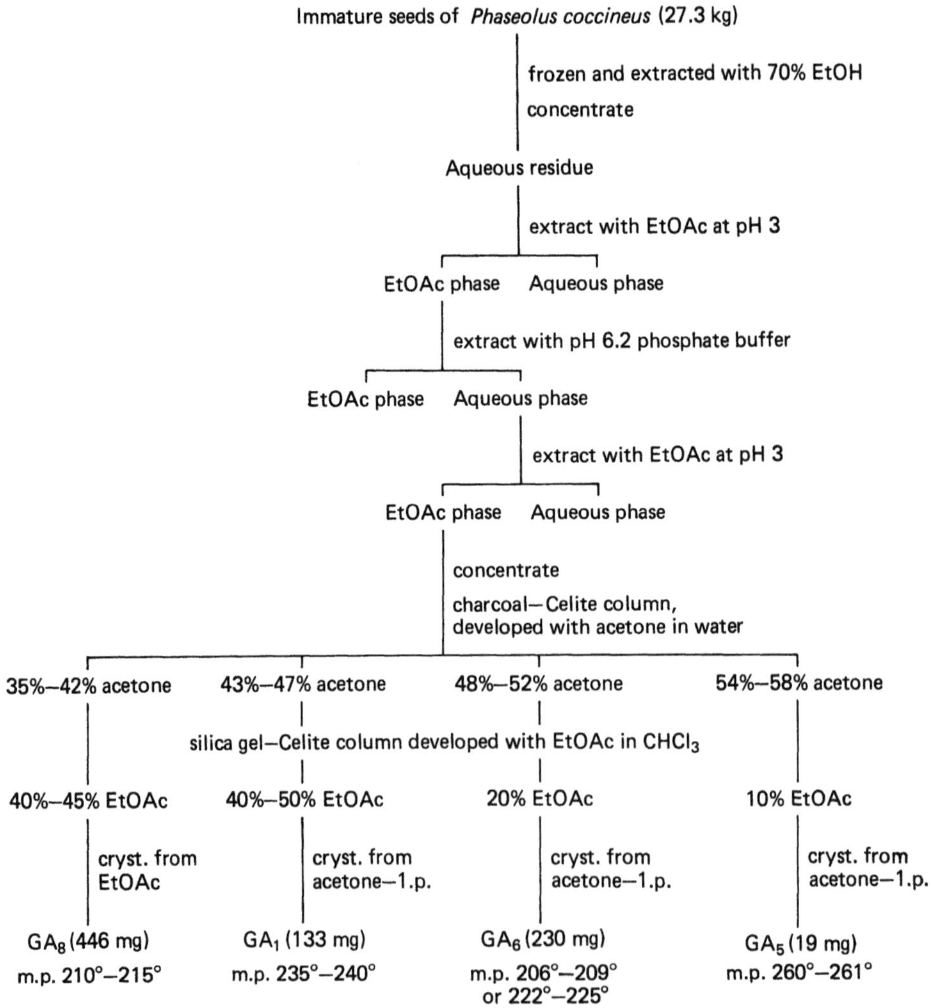

Fig. 2.19. Isolation of gibberellins A_1, A_5, A_6 and A_8 from immature *Phaseolus* seeds. (MacMillan et al., 1962)

d) Isolation of Gibberellins and Gibberellin Conjugates from Mature *Phaseolus* Seeds

The mature seed of *Phaseolus vulgaris* contains a variety of gibberellins including free gibberellins, gibberellin glucosides, and gibberellin glucosyl esters (HIRAGA et al., 1974a). However, mature seed generally contains lower levels of gibberellins than immature seed. Therefore, purification procedures are not so simple as those used in the case of immature seed of *P. coccineus*. The mature seeds (100 kg) were extracted with methanol and the extract was then subjected to solvent partitioning which has been discussed in Section 2.1.2.c. The neutral ethyl acetate fraction was first subjected to silica gel chromatography in order to eliminate a large amount of non-polar impurity. Elution with benzene containing increasing amounts of ethyl acetate (up to 100%) did not give any bio-active fraction. An active fraction was obtained by elution with ethyl acetate containing 5–20% methanol. Successive chromatography on charcoal column, silica gel column and silica gel thin layer gave a pair of mixtures: one was a mixture of glucosyl esters of GA_4 and GA_{37}, and the other a mixture of glucosyl esters of GA_1 and GA_{38} (Fig. 2.20). GA_1 and GA_{38} glucosyl esters although present in a minor amount in this fraction, were found to be mostly distributed into neutral n-butanol fraction. The neutral n-butanol fraction was also purified by similar techniques as shown in Fig. 2.21. The separation of glucosyl esters of GA_1 and GA_{38} was accomplished by thin-layer chromatography using the solvent system, acetone-benzene (4:1).

The acidic ethyl acetate and n-butanol fractions were purified by analogous chromatographic techniques using silica gel column, charcoal column, buffer-

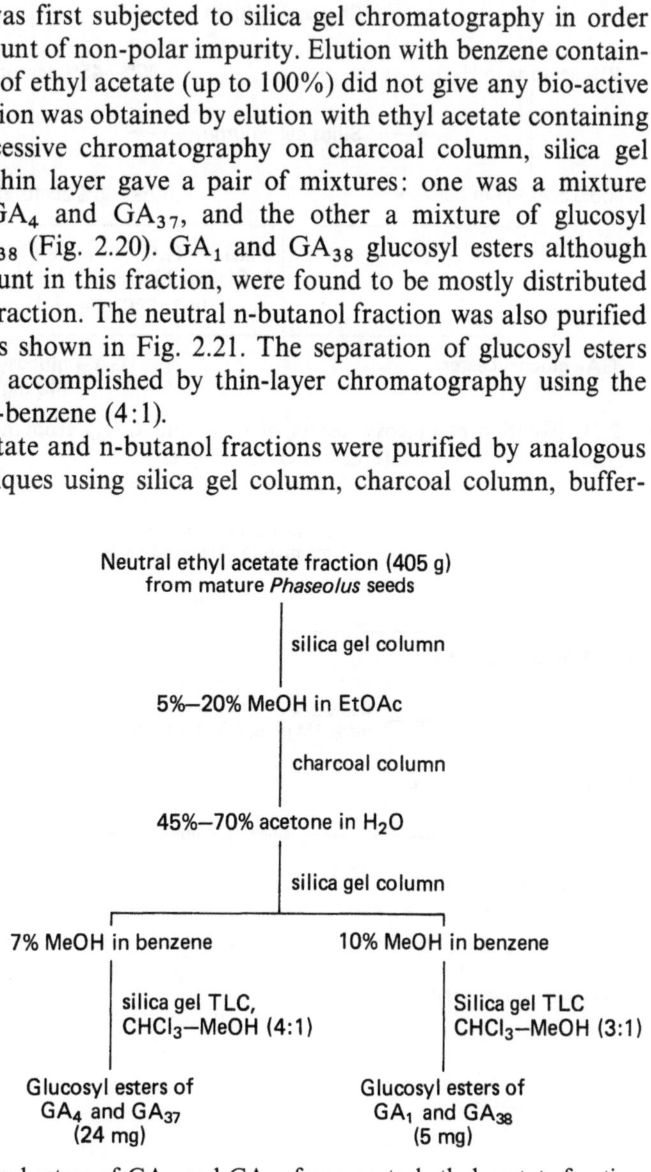

Fig. 2.20. Isolation of glucosyl esters of GA_4 and GA_{37} from neutral ethyl acetate fraction of mature *Phaseolus* seeds. (HIRAGA et al., 1974a)

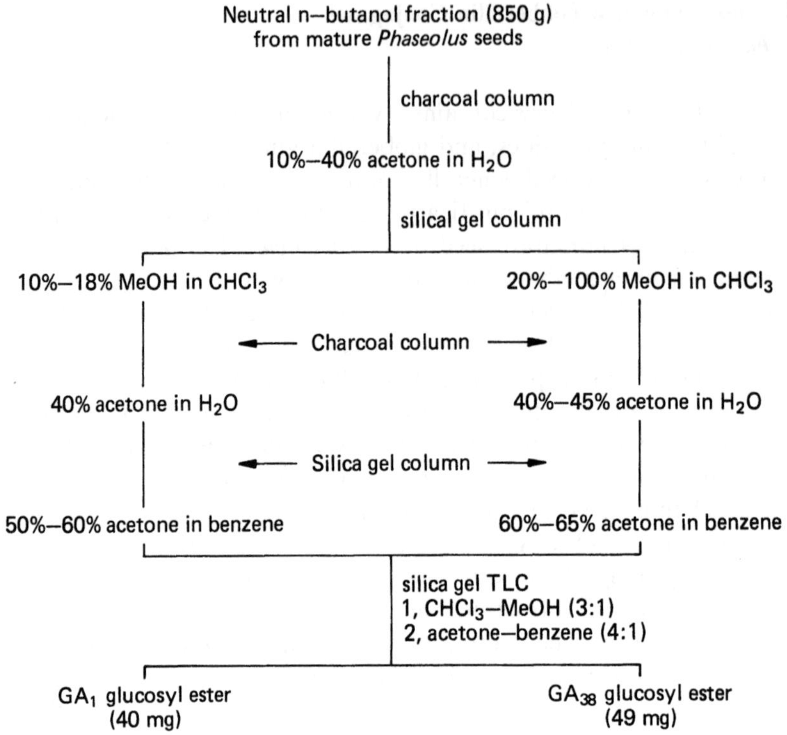

Fig. 2.21. Isolation of glucosyl esters of GA_1 and GA_{38} from neutral n-butanol fraction of mature *Phaseolus* seeds. (HIRAGA et al., 1974a)

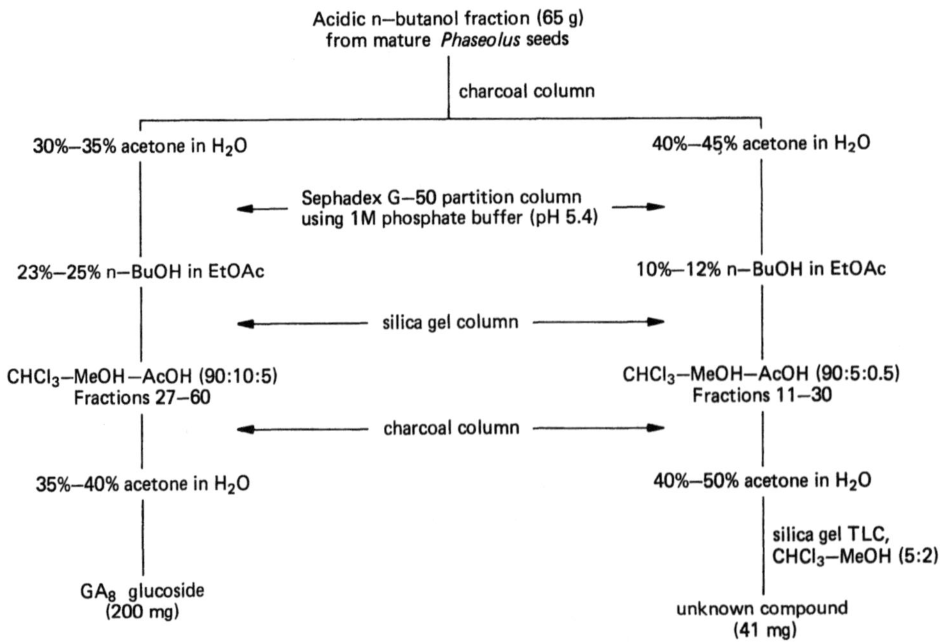

Fig. 2.22. Isolation of GA_8 glucoside from acidic n-butanol fraction of mature *Phaseolus* seeds. (HIRAGA et al., 1974a)

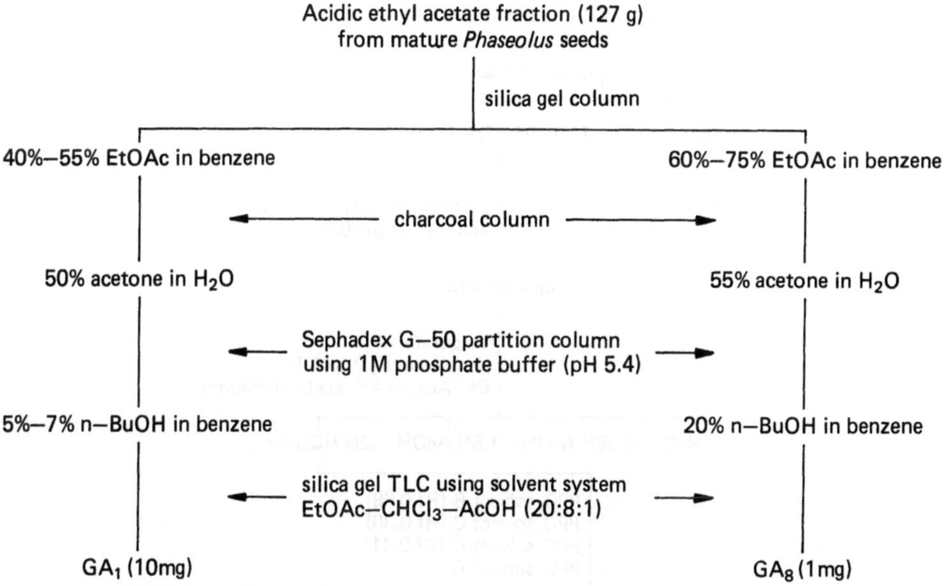

Fig. 2.23. Isolation of GA_1 and GA_8 from acidic ethyl acetate fraction of mature *Phaseolus* seeds. (HIRAGA et al., 1974a)

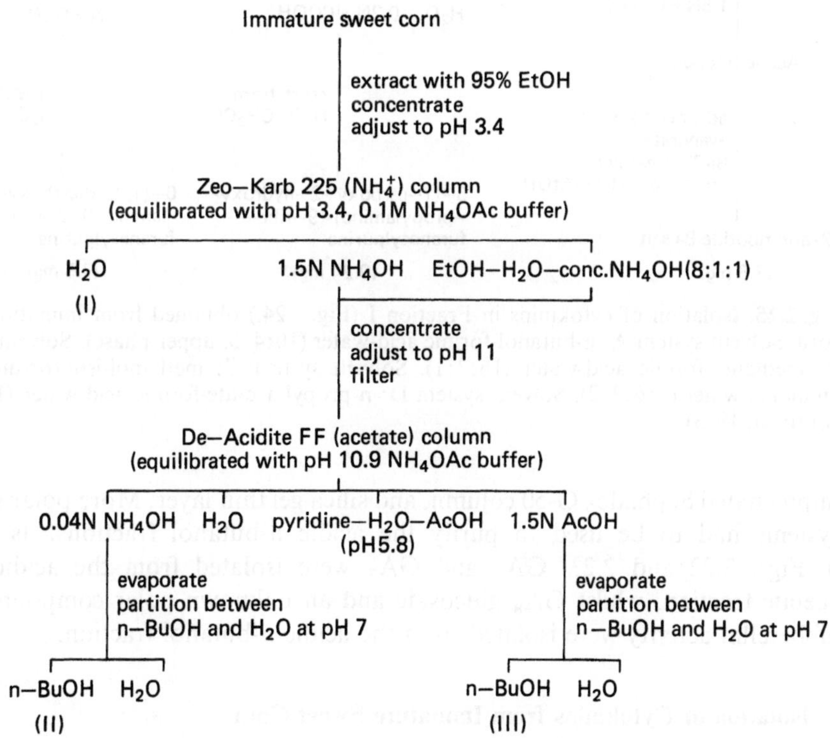

Fig. 2.24. Fractionation procedure of cytokinins in immature sweet corn using ion exchange resins and solvent partitioning. (LETHAM, 1973)

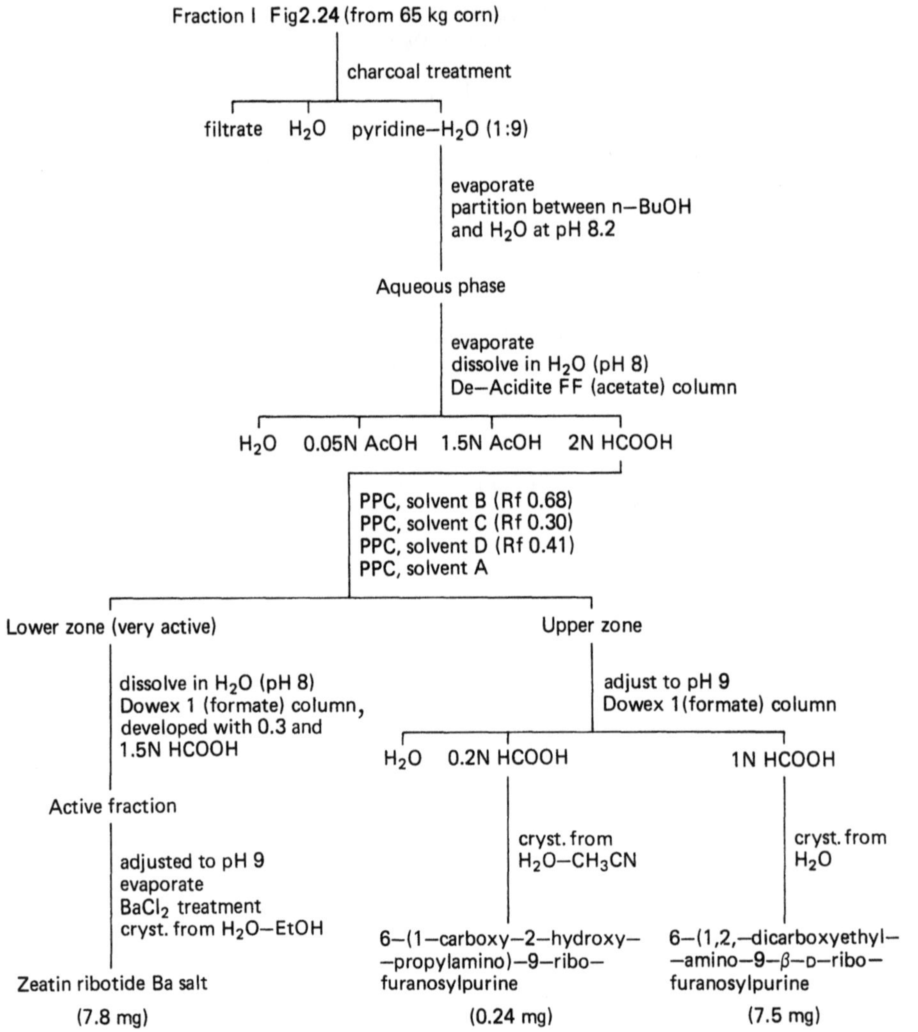

Fig. 2.25. Isolation of cytokinins in Fraction I (Fig. 2.24.) obtained from immature sweet corn. Solvent system A: n-butanol-formic acid-water (10:4:5, upper phase). Solvent system B: methanol-formic acid-water (16:3:1). Solvent system C: methanol-isopropanol-conc. ammonia-water (9:6:3:2). Solvent system D: n-propyl acetate-formic acid-water (11:5:3). (LETHAM, 1973)

impregnated Sephadex G-50 column, and silica gel thin layer. More polar solvent systems had to be used to purify the acidic n-butanol fractions, as shown in Figs. 2.22 and 2.23. GA_1 and GA_8 were isolated from the acidic ethyl acetate fraction, whilst GA_8 glucoside and an unknown polar compound with gibberellin activity were isolated from the acidic n-butanol fraction.

e) Isolation of Cytokinins from Immature Sweet Corn

Eight cytokinins were isolated from immature sweet corn by LETHAM (1973). The purification procedures are outlined in Figs. 2.24–2.27. The aqueous extract

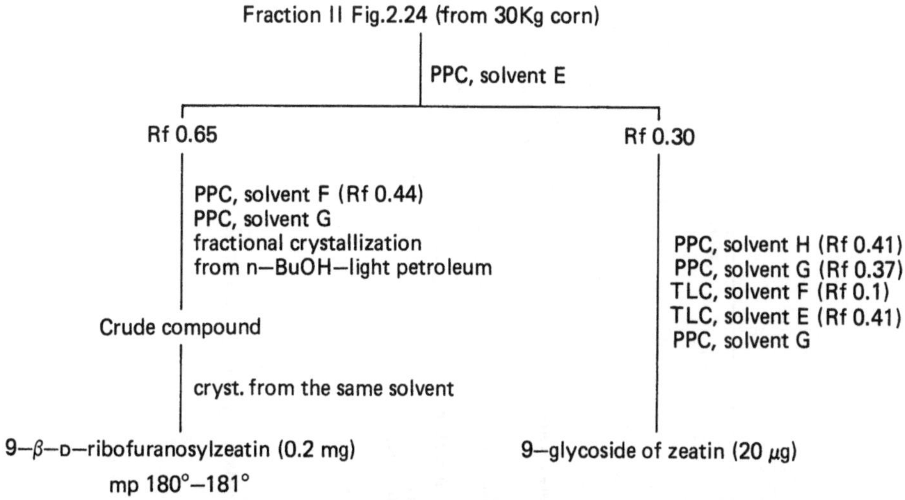

Fig. 2.26. Isolation of cytokinins in Fraction II (Fig. 2.24) obtained from immature sweet corn. Solvent system E: n-butanol-acetic acid-water (12:3:5). Solvent system F: methyl ethyl ketone saturated with water. Solvent system G: n-butanol saturated with water. Solvent system H: n-butanol-conc. ammonia-water (6:1:2, upper phase). TLC: silica gel thin layer chromatography. (LETHAM, 1973)

was successively treated with cation and then anion exchange resin to afford three cytokinin-active fractions (Fig. 2.24). Fraction I, which was not retained by a Zeo-Karb 225 column, contained cytokinins with an acidic side chain. The fraction eluted with ammoniacal solvents from this column was further fractionated by a De-Acidite FF column and n-butanol extraction to afford fractions II and III.

Fraction I was purified by the procedure shown in Fig. 2.25 to isolate three cytokinins, i.e., zeatin ribotide, 6-(1-carboxy-2-hydroxypropylamino)-9-ribofuranosylpurine and 6-(1,2-dicarboxyethylamino)-9-β-D-ribofuranosylpurine. It should be noted that charcoal treatment and partition between n-butanol and water are effective procedures to purify these cytokinins with an acidic side chain. They were isolated by further procedures including anion exchange resin and paper chromatography. Zeatin ribotide was crystallized as the barium salt which was converted to the ammonium salt by passing through a cellulose phosphate column (ammonium$^+$).

Fraction II was purified by repeated paper and silica gel thin-layer chromatography to give the 9-glycoside of zeatin (identity of sugar moiety not established) and 9-β-D-ribofuranosylzeatin, as shown in Fig. 2.26. The latter was obtained as crystals after fractional crystallization from n-butanol-light petroleum ether.

Fraction III was subjected to chromatography on paper, silica gel thin layer, and cellulose phosphate, which allowed the isolation of the 9-glycoside of zeatin and 2-hydroxy-6-(4-hydroxy-3-methyl-trans-but-2-enylamino) purine. Purification of 6-(2,3,4-trihydroxy-3-methylbutylamino)purine was completed by the following further steps: precipitation as the 3-iodo-2,4,6-trinitrophenolate, conversion to the base and, finally, crystallization from water. 6-(3,4-Dihydroxy-3-

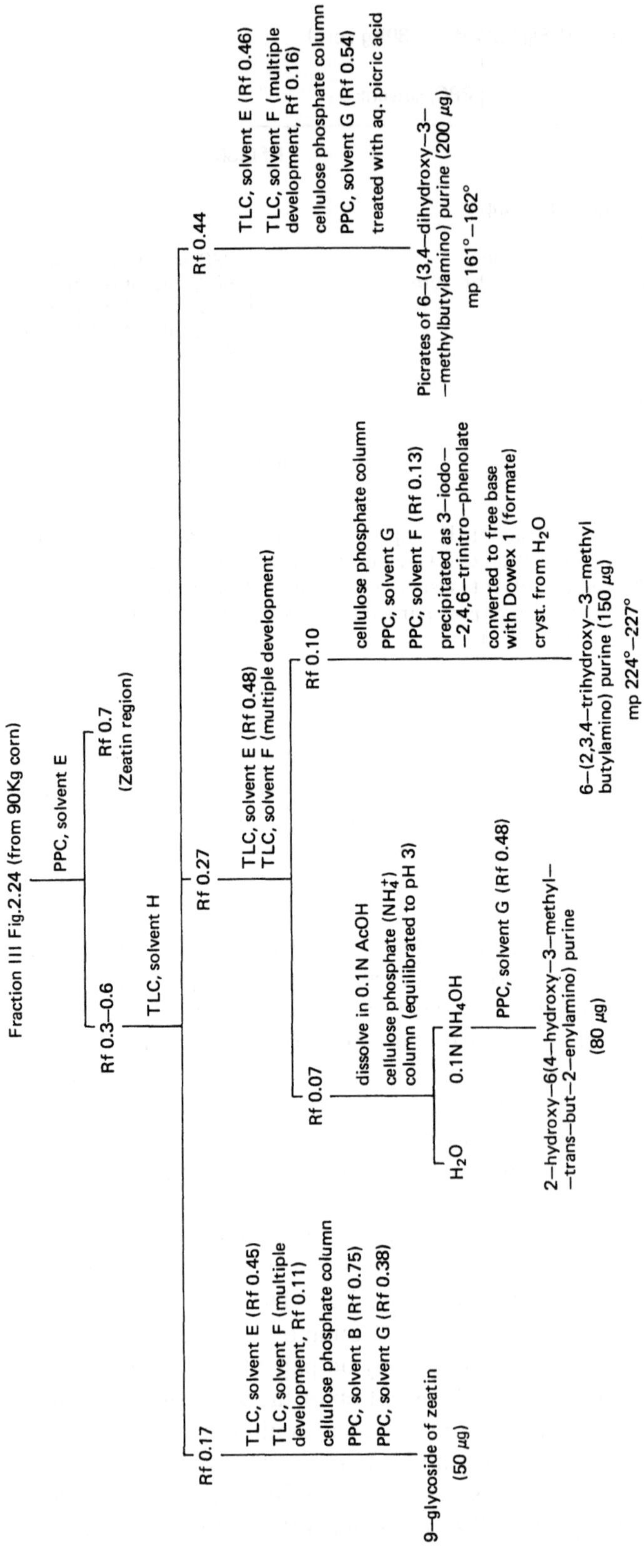

Fig. 2.27. Isolation of cytokinins in Fraction III (Fig. 2.24) obtained from immature sweet corn. Solvent systems, see Fig. 2.25 and 2.26. (LETHAM, 1973)

Fig. 2.28. Isolation of glucosyl zeatin and glucosyl ribosylzeatin from *Vinca rosea* crown gall. (PETERSON and MILLER, 1977)

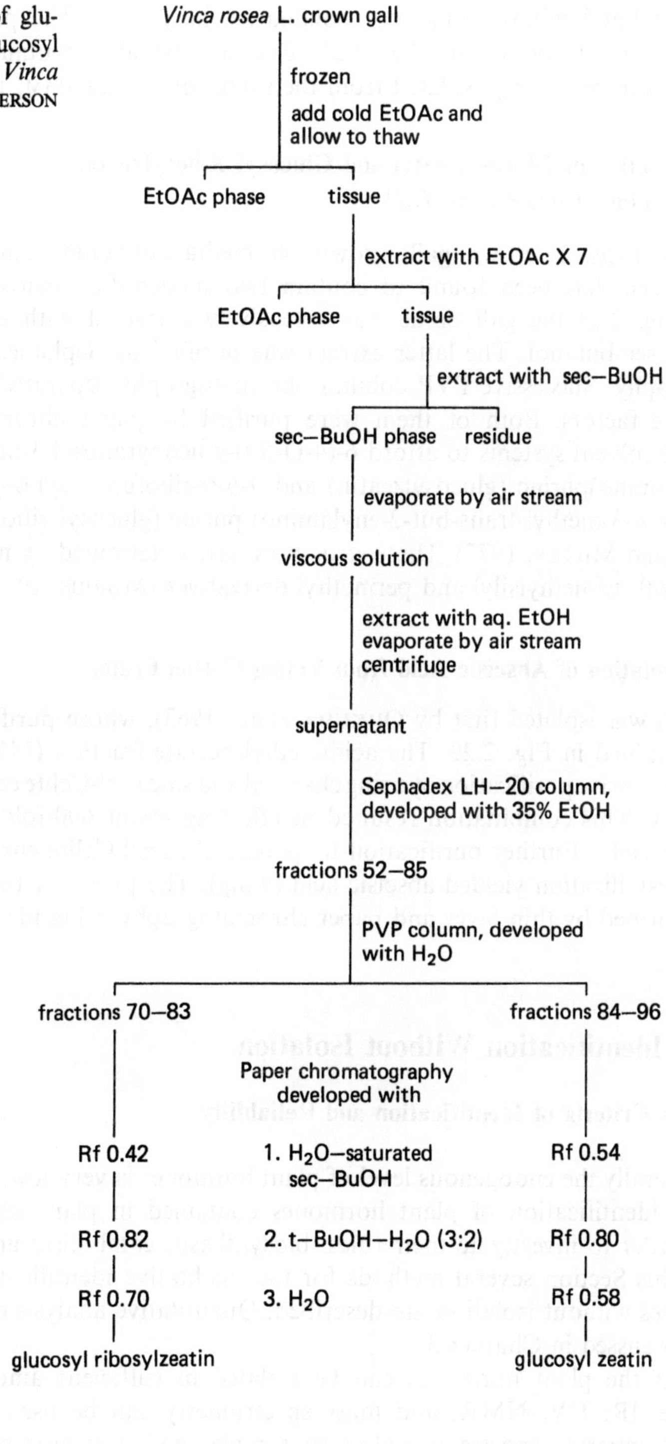

methylbutylamino) purine was isolated as the picrate. The purification procedures are illustrated in Fig. 2.27. Fraction III also contained zeatin which had been previously isolated from the same source (LETHAM, 1963).

f) Isolation of Glucosylzeatin and Glucosyl Ribosylzeatin from *Vinca rosea* Crown Gall

Vinca rosea L. crown gall, grown on media containing sources of reduced nitrogen, has been found to contain two unidentified cytokinins. As shown in Fig. 2.28 the gall tissue was extensively extracted with ethyl acetate and then sec-butanol. The latter extract was purified by Sephadex LH-20 chromatography. Successive PVP column chromatography separated two biologically active factors. Both of them were purified by paper chromatography using three solvent systems to afford 6-(4-O-β-D-glucopyranosyl-3-methyl-trans-but-2-enylamino)purine (glucosylzeatin) and 9-β-D-ribofuranosyl-6-(4-O-β-D-glucopyranosyl-3-methyl-trans-but-2-enylamino) purine (glucosyl ribosylzeatin) (PETERSON and MILLER, 1977). These structures were determined by mass spectrometry of both trimethylsilyl and permethyl derivatives (MORRIS, 1977).

g) Isolation of Abscisic Acid from Young Cotton Fruits

ABA was isolated first by OHKUMA et al. (1963), whose purification procedure is outlined in Fig. 2.29. The acidic ethyl acetate fraction (147 g) was subjected to successive purification by using charcoal and silica gel-Celite column chromatography. This combination resulted in effecting about 600-fold reduction of the dry weight. Further purification by paper, silica gel-Celite chromatography and recrystallization yielded abscisic acid (9 mg). The purity of this compound was confirmed by thin-layer and paper chromatography using nine solvent systems.

2.2 Identification Without Isolation

2.2.1 Criteria of Identification and Reliability

Generally the endogenous level of plant hormones is very low, but the detection and identification of plant hormones contained in plant tissues is necessary in order to investigate their roles, biosynthesis, catabolism and transportation. In this Section several methods for the qualitative identification of plant hormones without isolation are described. Quantitative analysis of plant hormones is discussed in Chapter 3.

If the plant hormones can be isolated in sufficient amount and in pure state, IR, UV, NMR, and mass spectrometry can be used for identification by comparing spectra of unknown samples and authentic specimens, as well as the classical mixed melting point test. However, it happens quite often that samples in such quantity and purity cannot be easily obtained in plant physiological studies. So other analytical methods, such as paper chromatography, thin-

Fig. 2.29. Isolation of abscisic acid from young cotton fruits. (OHKUMA et al., 1963)

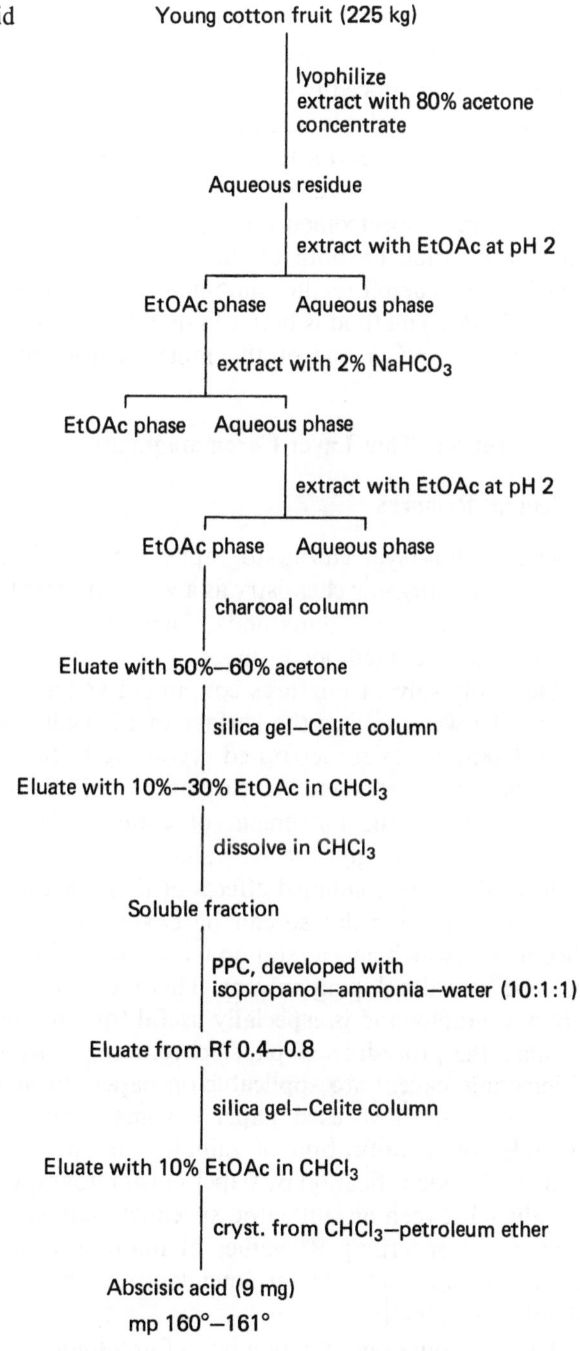

layer chromatography, gas-liquid chromatography and combined gas-liquid chromatography-mass spectrometry are often used for the identification of plant hormones in minute amounts of samples without isolation.

Purification procedures are always necessary prior to identification. Usually fractionation of a crude extract into several fractions, namely, acidic, basic, neutral, and amphoteric, by solvent extraction and purification by various kinds of chromatographies are carried out. Samples thus obtained can be subjected to identification procedures. In many cases a combination of several methods is preferable for reliable identification. If the number of the methods applied is increased, the identification becomes more reliable, and the combination of different types of method is preferable. For instance, gas-liquid chromatography and mass spectrometry are quite different types of identification methods and identification by both of them is usually more reliable than either alone (see Chap. 3). Therefore, the combined gas-liquid chromatography-mass spectrometry (GC-MS) method is better than either gas-liquid chromotography or mass spectrometry and is one of the most convenient methods for identification.

2.2.2 Paper and Thin-Layer Chromatography

a) General Remarks

Paper and thin-layer chromatography are very frequently used in the field of biological and organic chemistry as a very important technique for the separation and identification of compounds. There is little difference between these two chromatographic methods in theory and in practice.

Normally solvent mixtures containing water are used in paper chromatography. The water covers the surface of the cellulose and acts as the stationary phase. Compounds are separated according to the difference of partition coefficients between water and developing solvent. However, there are some other factors affecting the movement of compounds, namely, adsorption and ion exchange. Accordingly, the Rf value in paper chromatography is believed to be decided by the combined effects of these factors.

In some cases cellulose can be coated with non-polar substances such as silicone wax which acts as stationary phase, and polar solvents such as methanol are used as a developing solvent. This method is called "reverse-phase" paper chromatography and is especially useful for compounds with low polarity.

Since the procedure of paper chromatography is simple and various kinds of detecting reagent are applicable on paper chromatograms after development, many scientists have used paper chromatography as one of the most handy methods for identification of minute amounts of samples from early in this century. The identification by paper chromatography is conducted by comparing Rf values between an authentic specimen and an unknown sample. There are many data concerning Rf values of important compounds but co-chromatography with an authentic specimen and an unknown sample is preferable for reliable identification.

Paper chromatography requires a fairly long time for development and corrosive reagents cannot be used to detect spots on chromatograms, e.g., sulphuric acid which is frequently used for the detection of many organic compounds is not applicable. In recent years, crystalline cellulose powder with good separating ability came to be used in thin-layer chromatography.

Thin-layer chromatography relies mainly on adsorption, but partition chromatography can also play a part in some cases. As adsorbent or carrier of stationary phase silica gel is the most commonly used, and alumina, kieselguhr, and cellulose powder are used in some cases. There are many types of adsorbents and thin layers can be prepared in laboratories by use of an applicator. Also many types of thin layers on glass or plastic plates, which show good separation and reproducibility, are commercially available. Adsorbent and developing solvent should be selected according to the chemical and physical properties of samples. Thin-layer chromatography is superior to paper chromatography in the following points: (1) rapid development (2) good separation (3) applicability of many kinds of reagent for detection. Thin-layer chromatography, however, shows poor reproducibility, and in identification it is always necessary to carry out co-chromatography as in the case of paper chromatography.

Combination of bioassay and paper or thin-layer chromatography is frequently used at an early stage of purification or identification of biologically active compounds. Chromatograms are divided into 10–20 fractions according to Rf value and each fraction is extracted with solvent to allow the corresponding samples to be bioassayed. The Rf value of a bio-active zone in the histograms gives important information on the identification of the compound. By this method bio-activity can sometimes be detected which cannot be found without such treatment because of the removal of some inhibitors contained in the crude sample. Generally bioassay combined with paper or thin-layer chromatography is carried out for the presumption of compounds and then instrumental analyses are applied for further confirmation of the identification of biologically active compounds.

Paper and thin-layer chromatography are used not only for identification but also for purification and isolation of compounds, where thicker papers and layers are sometimes used for larger amount of samples.

b) Identification of Plant Hormones by Paper and Thin-Layer Chromatography

Auxins. Paper chromatography has been used for identification of auxins. Stowe and Thimann (1954) reported paper chromatography of 12 indole derivatives and Weller et al. (1954) reported the detection of indole-3-acetic acid (IAA) in plant extract by paper chromatography as well as paper chromatographic data of 22 IAA-related compounds. The Rf values and reaction colours of these indole derivatives are shown in Tables 2.6 and 2.7. The developing solvents usually used for auxins consist of isopropanol or n-butanol and aqueous ammonium hydroxide. Although few solvent systems have been described, the combination of several solvent systems is preferable for more definite identification. Thus, Goldschmidt et al. (1971) identified the main auxin in *Citrus* tissue as IAA by co-paper chromatography using eight solvent systems. Since some auxins are not stable under acidic or basic condition, paper chromatography using such developing solvents should be carried out carefully.

Thin-layer chromatography of indole derivatives including auxins has been investigated very well not only in the field of plant physiology but also in the field of medical science where the analysis of urine is very important. Silica gel is the most generally used adsorbent. Byrd et al. (1974) surveyed the thin-

Table 2.6. Rf values on paper chromatography and colour reactions of indole derivatives. (STOWE and THIMANN, 1954)

Compound	Rf	Salkowski	Ehrlich	Diazotized p-nitroaniline		Diazotized Sulphanilic acid
				Acidic	Basic	Acidic
Allylindole	0.9	Pink	Purple	Yellow-brown	Yellow-brown	Brown
Skatole	0.9	Brown	Purple	Yellow	Yellow	Yellow
Methyl indole-3-acetate	0.84	Yellow-orange	Purple	Yellow-orange	Yellow	Purple
Ethyl indole-3-acetate	0.81	Yellow-orange	Purple	Yellow-orange	Yellow	
N(β-3-Indolylethyl)pyrrolidine	0.81	Yellow	Purple	Yellow	Yellow	Yellow
Gramine	0.80	Blue-brown	None	None	None	None
Gramine methosulphate	0.80	Blue-brown	None	None	None	None
Indole	0.79	Pink	Red-purple	Brown	Brown	Brown
N^{12}-Ethyltryptamine	0.79	Yellow	Purple	Yellow-orange		Yellow
2-Phenylindole	0.77	Blue-green	Purple	Red	Red	Orange
N^{12}-Acetyl-N^{12}-ethyltryptamine	0.77	Yellow	Purple	Yellow	Orange	Yellow
Indole-3-acetonitrile	0.75	Blue	Purple	Orange		
Methyl indole-3-butyrate	0.73	Yellow-brown	Purple	Yellow		
N^{12}-Acetyltryptamine	0.73	Yellow	Purple	Orange	Yellow	Orange-brown
Indole-3-aldehyde	0.72	Pink	Purple	Orange		Orange
Tryptophol	0.70	Yellow-brown	Purple	Orange	Yellow	Yellow
N-Acetylindoxyl	0.67	Blue	Orange-brown	Orange	Orange	Orange
Tryptamine	0.65	Yellow-brown	Purple	Orange		Yellow
Oxyindole	0.64	None	None	Yellow		Yellow
Indole-3-acetamide	0.59	Crimson	Blue	Orange	Yellow	
5-Hydroxytryptamine	0.52	Blue	Purple	Red-orange	Purple	Red
Indole-3-butyric acid	0.44	Yellow-brown	Purple	Orange	Orange	
Indole-3-propionic acid	0.35	Yellow	Purple	Orange	Orange	Yellow
N^{12}-Succinyltryptamine	0.33	Yellow-brown	Purple	Orange		Orange
Indole-3-lactic acid	0.30	Orange	Purple	Yellow	Yellow	
Abrine	0.29	Yellow	Purple	Yellow-orange	None	
Hypaphorine	0.26	Yellow	Purple	None	None	
Indole-3-acetic acid	0.25	Crimson	Purple	Brown	Yellow	Yellow
Indole-3-carboxylic acid	0.20	Crimson	Red-purple	Orange-brown	Brown	Orange
Tryptophan	0.18	Yellow	Purple	Yellow-orange	None	
5-Hydroxyindole-3-acetic acid	0.17	Green-blue	Green-blue	Red	Orange	Orange
Indole-3-pyruvic acid	0.12	Crimson	Purple	Orange	Yellow	Orange
5-Hydroxytryptophan	0.11	Green-blue	Slate	Red	Purple	Red
7-Hydroxyindole-3-acetic acid	0.10	Green-blue	Green-blue	Brown	Purple	Red
Indole-3-thioacrylic acid	0.09	Brown	Blue	Orange	Yellow	Yellow

Solvent: Iso-propanol-ammonia-water (10:10:1).

Table 2.7. Rf values of indole compounds on paper chromatography (Rf values × 10^2), temp. 30° C. (WELLER et al., 1954)

Compound	Solvent (a)	Solvent (b)	Colour of spot with *p*-dimethylamino-benzaldehyde
Indole-3-acetic acid	25	75	Purple
Indole-3-acetonitrile	85	89	Purple
Indole-3-acetaldehyde	88	88	Yellow brown (streak)
Indole-3-acetohydrazide	91	69	Purple
Indole-3-acetohydroxamic acid	58	71	Yellow
Indole-3-acetamide	84	87	Blue-purple
Indole-3-propionic acid	30	77	Purple
Indole-3-butyric acid	37	84	Purple
Indole-3-carboxaldehyde	87	87	Purple
Indole-3-carboxylic acid	15	68	Pink
Ethyl indole-3-acetate	84	89	Purple
Ethyl indole-3-carboxylate	88	96	Yellow
2-Phenylindole-3-acetic acid	90	88	Yellow-purple
2-Methylindole-3-acetic acid	26	78	Purple
2-Methylindole	86	96	Red
Tryptophol	88	90	Purple
L-Tryptophan	23	74	Purple
Tryptamine	79	90	Purple
1-Hydroxyindole-3-acetic acid	22	97	Brown
N,N′-Diindolyl-3,3′-diacetic acid	24	80	Brown
Indole	95	95	Pink
Isatin	74	80	Yellow

Solvents: (a) 1-Butanol saturated with 5% NH_4OH. (b) 1-Propanol-conc. NH_4OH-H_2O (60:30:10, v/v).

layer chromatography of 27 compounds by using silica gel and nine solvent systems, as shown in Table 2.8. Sometimes cellulose powder is used as reported by RAJ and HUTZINGER (1970b), who examined 14 indole derivatives by using cellulose thin layer and nine solvent systems (Table 2.9). Polyamide is used also in order to improve the separation of auxins from phenolic compounds which are contained in the extracts from plant tissue (RAILTON, 1970).

In order to detect auxins on thin-layer chromatograms Salkowski reagent and Ehrlich reagent have been used (STOWE and THIMANN, 1954; WELLER et al., 1954). Salkowski reagent consists of strong acid, ferric chloride, and water, revealing various colour reactions with most indole derivatives. The lower limit for detection is 100 ng in the case of IAA. The Salkowski reaction is very sensitive but is disturbed by the presence of ferrous ion. The Ehrlich and van Urk reagents, which consist of p-dimethylaminobenzaldehyde and hydrochloric acid in ethanol, give a red-purple colour with indole derivatives. Although they are less sensitive than the Salkowski reagent, they produce more stable colours than the Salkowski reagent. There are some other reagents newly devised or modified for the improvement of sensitivity, development time, and stability of colour with indole derivatives. For instance. HUTZINGER and HEACOCK (1972)

Table 2.8. Rf values of indole derivatives on silica gel F_{254} plates. (BYRD et al., 1974).

Compound	Rf (× 100) in solvent system								
	1	2	3	4	5	6	7	8	9
Indole-3-acetic acid	74	96	38	93	35	76	0	22	70
Indole-3-carboxylic acid	84	96	56	85	35	78	0	19	39
Indole-3-propionic acid	80	90	47	95	40	83	0	30	68
Indole-3-lactic acid	6	33	2	40	0	83	0	23	7
5-Hydroxyindole-3-acetic acid	45	50	17	70	13	76	0	12	10
N-Acetyltryptophan	7	44	2	58	0	83	0	23	8
Indole-3-acrylic acid	71	85	45	93	30	81	0	29	57
Indole-3-pyruvic acid	0	24[a]	0	19[a]	0	–[a]	0	19	0
Indole-3-acetamide	78	73	28	90	7	26	0	85	43
Indole-3-acetaldehyde	88	78	48	96	50	100	6	86	40
Oxindole	88	95	60	10	37	100	8	86	65
N-Acetyltryptamine	84	84	30	96	–	100	9	27	89[a]
N-Acetyl-5-hydroxytryptamine	83	72	23	91	–	100	0	–	18
Tryptophol	92	89	50	96	40	100	0	93	58
Indole-3-acetonitrile	95	97	58	95	73	100	30	95	84
5-Hydroxyindole	95	88	55	96	–	100	10	93	42
Indole	100	100	81	95	95	100	27	100	95
Skatole	100	100	87	95	95	100	35	100	95
Tryptophanol	2	14	0	28	43[b]	88	0	56	2
Tryptamine	0	22	0	30	55[b]	79	0	61	3
5-Hydroxytryptamine	0	9	0	20	7[b]	40–65	0	40	0
5-Methoxytryptamine	0	22	0	33	47[b]	78	0	58	1
Bufotenine	0	7	0	13	7[b]	60–80	0	82	0
N-Methyltryptamine	0	33	0	36	–	79	0	55	2
Tryptophan	0	17	0	25	0	50	0	15	0
5-Hydroxytryptophan	0	2	0	6	0	46	0	2	0
Indican	0	25	0	33	0	90	0	36	0

[a] Decomposition
[b] Ammonia atmosphere
Developing solvent systems: (1) Dichloromethane-ethanol-ethyl acetate (80:10:10). (2) Chloroform-methanol-glacial acetic acid (80:15:5). (3) 2-Propanol-n-heptane (25:75). (4) Chloroform-methanol-glacial acetic acid (75:20:5). (5) 2-Butanone-n-hexane (35:65), silica gel impregnated with 0.05 M ammonium formate (pH 4.5). (6) 2-Propanol-water-25% ammonia solution (75:20:5). (7) Chloroform-n-heptane (65:35). (8) 1-Butanol-ethanol-25% ammonia solution (80:10:10). (9) Chloroform-glacial acetic acid (95:5).

reported the colour reaction of 47 indole derivatives with eight-electron acceptor reagents such as tetracyanoethylene, and EHMANN (1977) reported the colour reaction of 79 indole derivatives with a modified reagent, the van Urk-Salkowski reagent. The characteristic colour on chromatograms gives useful information for identification together with Rf value in paper and thin-layer chromatography.

Gibberellins. Paper chromatography is not often used for the identification of gibberellins because the usual reagents for detection contain sulphuric acid and cannot be used. Gibberellins on paper chromatograms are detected by spraying 1% potassium permanganate solution or by bioassay. MACMILLAN et al. (1961) reported the paper chromatography of seven gibberellins by using

Table 2.9. Rf values (×100) of neutral and acidic indoles on cellulose TLC plates (RAJ and HUTZINGER, 1970b)

Compound	Solvent system								
	1	2	3	4	5	6	7	8	9
Indole-3-carboxylic acid	89	64	95	98	67	–	–	–	–
Indole-3-acetic acid	93	75	94	81	87	–	–	–	–
Indole-3-butyric acid	96	92	98	92	98	–	–	–	–
5-Hydroxyindole-3-acetic acid	47	00	31	53	08	–	–	–	–
DL-Indole-3-lactic acid	64T	07	66	53T	22T	–	–	–	–
Indole-3-pyruvic acid	M	20	77	M	21T	–	–	–	–
Indole-3-acrylic acid	89	58	93	88	67	–	–	–	–
Indole-3-propionic acid	96	80	95	89	93	–	–	–	–
Ethyl indole-3-acetate	99	97	–	–	–	99	97	96	97
Indole-3-acetaldehyde	99	M	–	–	–	93T	95T	98	97
Indole-3-carboxaldehyde	89	63	–	–	–	75	88	95	90
Indole-3-ethanol	95	75	–	–	–	87	95	98	95
Melatonin	76	51	–	–	–	40	32	87	73
Indole-3-acetamide	51	47	–	–	–	13	11	58	47

Solvent systems: (1) Benzene-dioxane-water (1:1:1), (2) Benzene-acetic acid-water (8:3:5), (3) Benzene-ethyl acetate-acetic acid-water (11:5:6:10), (4) Benzene-pyridine-water (1:1:1), (5) Chloroform-methanol-water (6:4:5), (6) Benzene-dioxane-petroleum ether (boiling range 60–80°)-water (2:2:1:2), (7) Benzene-dioxane-petroleum ether (boiling range 60–80°)-water (1:2:2:2), (8) Benzene-dioxane-water-34% ammonia (10:10:10:0.5), (9) Benzene-dioxane-water-triethylamine (10:10:10:0.5).
T: tailing, M: multiple spots or streaking indicating decomposition of the sample.
Merck cellulose F precoated thin-layer plates are exposed to the vapours of the aqueous phases of the two-phase systems (1 h) and then developed with the organic phases.

four solvent systems, as shown in Table 2.10. It should be noted that di- and tricarboxylic gibberellins show much lower Rf value in comparison with monocarboxylic ones on paper chromatography when developed with the solvents containing ammonium hydroxide, while there is no large difference in Rf values due to the number of carboxylic groups in thin-layer chromatography using solvent systems containing an organic acid, as shown in the case of GA_{19} (MUROFUSHI et al., 1966).

Thin-layer chromatography was carried out as a convenient method for identification in earlier studies of gibberellin chemistry. Although identification of gibberellins is becoming more difficult because of the increasing number of gibberellins (59 gibberellins in 1980), thin-layer chromatography is still an important technique as a preliminary test for identification. Gibberellins are separated by adsorption thin-layer chromatography in many cases but a mixture of GA_1 and GA_3, and a mixture of GA_4 and GA_7 are only separated clearly by partition thin-layer chromatography (KAGAWA et al., 1963; MACMILLAN and SUTER, 1963; CAVELL et al., 1967; PITEL et al., 1971). It is necessary for chromatography plates to be equilibrated in the vapour of lower or upper phase of the solvent systems for a long time prior to development in partition chromatography. To reduce the time JONES (1970) devised a method of rapid chromatography using a glass fibre impregnated with silica gel.

Table 2.10. Rf values of gibberellins on paper chromatography. (MACMILLAN et al., 1961)

Gibberellin	Solvent system			
	1	2	3	4
A_8	0.15	0.40	0.25	0.35
A_3	0.29	0.50	0.40	0.48
A_1	0.31	0.52	0.40	0.49
A_6	0.33	0.51	0.42	0.51
A_5	0.45	0.59	0.49	0.56
A_4	0.60	0.72	0.57	0.72
A_7	0.61	0.71	0.57	0.71
A_9	0.725	0.77	0.65	0.78

Solvent systems: (1) n-Butanol-1.5 N ammonium hydroxide (3:1), descending. (2) Isopropanol-water (4:1), ascending. (3) n-Butanol-tert-amyl alcohol-acetone-ammonia-water, (5:5:5:2:3), descending. (4) Isopropanol-7N-ammonium hydroxide (5:1), ascending.
Paper: Whatman No. 1, Temperature: 20° C.

Table 2.11. Rf values of gibberellins and their methyl esters on thin layer chromatography. (KAGAWA et al., 1963)

Gibberellin	Free acid						Methyl ester	
	Silica gel G			Kieselguhr G			Silica gel G	
	Solvent system			Solvent system			Solvent system	
	1	2	4	3	4	5	6	7
A_1	0.20	0.49	0.0	0.0	0.28	0.49	0.31	0.29
A_2	0.17	0.40	0.0	0.0	0.23	0.37	0.23	0.13
A_3	0.19	0.54	0.0	0.0	0.18	0.40	0.35	0.32
A_4	0.63	0.95	0.67	0.67	0.90	1.00	0.73	0.75
A_5	0.53	0.87	0.27	0.45	0.85	0.90	0.60	0.69
A_6	0.59	0.87	0.11	0.33	0.86	0.84	0.66	0.67
A_7	0.60	0.90	0.57	0.45	0.85	0.91	0.71	0.72
A_8	0.04	0.30	0.0	0.0	0.06	0.10	0.17	0.12
A_9	0.87	0.95	1.00	1.00	1.00	1.00	0.98	0.96

Solvent systems: (1) Benzene-n-butanol-acetic acid (80:15:6). (2) Benzene-n-butanol-acetic acid (70:25:5). (3) Carbon tetrachloride-acetic acid-water (8:3:5), lower phase. (4) Carbon tetrachloride-acetic acid-water (8:3:5), lower phase plus 10% ethyl acetate. (5) Carbon tetrachloride-acetic acid-water (8:3:5), lower phase plus 20% ethyl acetate. (6) Ethyl ether-benzene (4:1). (7) Ethyl ether-petroleum ether (4:1), developed twice.
With solvent systems 3, 4 and 5, plates are equilibrated overnight with upper phase then developed with lower phase or lower phase plus ethyl acetate.

Rf values of free gibberellins are listed in Tables 2.11–2.13. The value of movement relative to that of the known gibberellins, such as R_{GA_3}, is sometimes used (SEMBDNER et al., 1962; ELSON et al., 1964) because absolute Rf values are not reproducible.

Table 2.12. Thin layer chromatographic properties of the gibberellins. (ELSON et al., 1964)

Gibberellin	R_{GA_3}(reproducibility in brackets)				Colour of induced fluorescence
	Solvent system				
	1(±0.02)	2(±0.03)	3(±0.05)	4(±0.03)	
A_1	1	1	1	1	Blue
A_2	0.92	0.91	1.1	0.90	Purple
A_3	1	1	1	1	Green-blue
A_4	1.17	1.13	1.48	0	Purple
A_5	1.0	1.12	1.35	0.80	Blue
A_6	1	1.07	0.95	0.95	Blue
A_7	1.17	1.13	1.48	0	Yellow
A_8	0.89	0.84	0.52	1.05	Blue
A_9	1.19	1.25	1.55	0	Purple
Gibberellenic acid	0.50	0.42	0.10	1.03	Green-blue

Solvent systems: (1) Isopropanol-water (4:1). Rf of GA_3:0.55. (2) Isopropanol-4.5 N ammonium hydroxide (3:1). Rf of GA_3:0.55. (3) n-Butanol-4.5 N ammonium hydroxide (3:1). Rf of GA_3:0.55. (4) Phosphate buffer (0.1 M, pH 6.3) on silica impregnated with capryl alcohol. Rf of GA_3:0.70.
Equilibration prior to partition chromatography was carried out for 16 h. Impregnation of silica gel G with capryl alcohol was accomplished by developing the chromoplate in a 7% solution of capryl alcohol in light petroleum (40–60° C) and drying at room temperature.

Table 2.13. TLC Rf values of the gibberellins. (CAVELL et al., 1967)

Gibberellin	Kieselgel				Kieselguhr	
	Solvent system				Solvent system	
	1	2	4	5	2	3
A_1	0.06	0.00	0.37	0.95	0.20	0.55
A_2	0.01	0.00	0.19	0.85	0.24	0.67
A_3	0.06	0.00	0.37	0.95	0.13	0.45
A_4	0.20	0.69	0.61	0.80	1.00	1.00
A_5	0.16	0.29	0.59	0.80	0.87	1.00
A_6	0.13	0.16	0.57	0.90	0.82	1.00
A_7	0.19	0.60	0.62	0.80	1.00	1.00
A_8	0.01	0.00	0.24	1.00	0.04	0.20
A_9	0.59	1.00	0.78	0.65	1.00	1.00
A_{10}	0.06	0.36	0.33	0.60	0.91	1.00
A_{11}	0.49	1.00	0.74	0.80	1.00	0.88
A_{12}	0.67	1.00	0.78	0.70	1.00	1.00
A_{13}	0.14	0.11	0.46	0.90	0.39	1.00
A_{14}	0.26	0.75	0.63	0.80	0.86	0.72
A_{15}	0.44	1.00	0.78	0.57	1.00	1.00
A_{18}	0.04	0.02	0.35	0.95	0.34	1.00
A_{19}	0.08	0.00	0.42	0.95	0.12	0.82

Solvent systems: (1) Di-isopropyl ether-acetic acid (95:5). (2) Benzene-acetic acid-water (8:3:5). (3) Benzene-propionic acid-water (8:3:5). (4) Ethyl acetate-chloroform-acetic acid (15:5:1). (5) Water.
With solvent systems 2 and 3 plates were equilibrated overnight with lower phase then developed with upper phase.

Table 2.14. Rf values of paper and thin layer chromatography of gibberellin glucosides (YOKOTA et al., 1971b)

Gibberellin glucoside	PPC Solvent system a	TLC Solvent system b	c
3-O-β-Glucosyl-gibberellin A_3	0.59	0.28	0.37
3-O-β-Glucosyl-gibberellenic acid	0.22	0.20	0.43
3-O-β-Glucosyl-iso-gibberellin A_3	0.57	0.24	0.47
2-O-β-Glucosyl-gibberellin A_{26}	0.55	0.33	0.51
2-O-β-Glucosyl-gibberellin A_8	0.51	0.20	0.46
2-O-β-Glucosyl-gibberellin A_{27}	0.63	0.36	0.55
2-O-β-Glucosyl-gibberellin A_{29}	0.54	0.24	0.43

PPC: Toyo filter paper No. 51. Solvent system: (a) Isopropanol-7N ammonium hydroxide-water (8:1:1).
TLC: Merck Silica gel G. Solvent systems: (b) Chloroform-methanol-acetic acid-water (40:15:3:2), (c) Acetone-acetic acid (97:3).

Table 2.15. Rf values of gibberellin glucosyl esters on TLC. (HIRAGA et al., 1974c)

Compound	Solvent system A[a]	B
GA_1 glucosyl ester	0.47	0.19
GA_3 glucosyl ester	0.47	0.17
GA_4 glucosyl ester	0.62	0.33
GA_{37} glucosyl ester	0.62	0.35
GA_{38} glucosyl ester	0.47	0.12

Adsorbent: Silica gel G.
Solvent systems: (A) Chloroform-methanol (3:1). (B) Benzene-acetone (1:5).
[a] Data in the original paper are revised (HIRAGA et al., unpublished)

Conjugate gibberellins, glucosides and glucosyl esters of gibberellins, cannot be chromatographed with developing solvents used for free gibberellins due to their high polarity. Usually the resolution between some gibberellin glucosides or glucosyl esters is difficult because the glucosyl group has a dominant influence on the polarity of conjugate gibberellins. However, glucosyl esters of GA_1 and GA_{38} can be separated by using silica gel and acetone-benzene mixture (5:1) (HIRAGA et al., 1974c). Rf values of conjugate gibberellins in paper and thin-layer chromatography are shown in Tables 2.14 and 2.15.

Detection of gibberellins on thin-layer plates is conducted by spraying with sulphuric acid-ethanol (5:95) or sulphuric acid-water (70:30). After heating at 100–110° C for 10 min gibberellins on chromatograms can be detected as fluorescent spots under UV light (e.g., 254 nm, 320 nm). This detection method

Table 2.16. Rf values of cytokinins on paper chromatography. (ROBINS et al., 1967)

Compound	Solvent system				
	A	B	C	D	E
6-(3-Methylbut-2-enylanino)-9-β-D-ribofuranosylpurine	0.80	0.87	0.80	0.87	0.80
6-(3-Methylbut-2-enylamino) purine	0.86	0.90	0.83	0.88	0.86
N^1-(Δ^2-Isopentenyl)adenosine	0.69	0.67	0.64	0.78	0.24
N^1-(Δ^2-Isopentenyl)adenine	0.77	0.80	0.53	0.77	
N^6, N^6-Dimethyladenosine	0.63	0.75	0.52	0.73	0.54
N^6, N^6-Dimethyladenine	0.71	0.78	0.43	0.73	0.67
Zeatin	0.75	0.87	0.61	0.77	0.61
Zeatin riboside	0.63	0.79	0.56	0.77	0.41

Solvent systems: (A) 1-Butanol-water-conc. ammonium hydroxide (86:14:5). (B) 1-Butanol-glacial acetic acid-water (5:3:2). (C) 2-Propanol-conc. hydrochloric acid-water (680:170:144). (D) 2-Propanol-water-conc. ammonium hydroxide (7:2:1). (E) Ethyl acetate-1-propanol-water (4:1:2).

has excellent sensitivity, the lower limit of detection being below 0.1 μg for most gibberellins. Gibberellins A_3, A_7, A_{30} and A_{32} having ring A structure (1) can be detected as fluorescent spots under UV light after treatment with sulphuric acid *without heating*. After heating even 0.5 ng of GA_3 is detectable. The colour of fluorescent spots varies depending on different structural features. Most gibberellins reveal blue-purple fluorescence, but a few exceptional gibberellins give other fluorescence colours, i.e., GA_3 gives green fluorescence without heating, GA_{30} white, GA_{34} and GA_{40} bluish brown. GA_{17} cannot be detected in high sensitivity by this method, its lower limit of detection being over 100 times more than those of other gibberellins. GA_{32} gives blue colour (not fluorescence) under visible light by spraying with sulphuric acid and heating. These colours and tints, as well as Rf values, are very useful for identification of gibberellins.

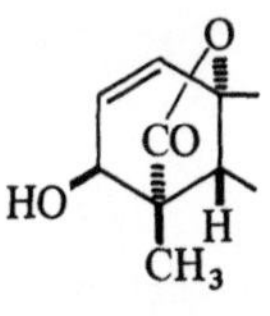

structure (1)

Cytokinins. Paper and thin-layer chromatography are indispensable for identification and purification of cytokinins. There are many reports concerning paper and cellulose thin-layer chromatography, some of which are shown in Tables 2.16–2.20. Rf value of zeatin riboside is greater than that of zeatin when developed with water. The mobility of zeatin riboside is enhanced in 0.03 M borate buffer (pH 8.4), but that of zeatin is not. This behaviour in paper chromatography has been used to distinguish between zeatin and zeatin riboside. Rf values of dihydrozeatin are very close to that of zeatin, but a clear difference between them is observed when developed on paper with n-buta-

Table 2.17. Rf values of adenine-related compounds on paper chromatography. (CHEN and HALL, 1969)

Compound	Solvent system					
	1	2	3	4	5	6
6-(3-Methylbut-2-enylamino)-9-β-D-ribofuranosylpurine	0.82	0.80	0.77	0.86	0.75	0.91
6-(3-Methylbut-2-enylamino)-9-β-D-ribofuranosylpurine 2′(3′)-phosphate	0.04	0.08	0.88	0.54	0.77	0.80
Adenosine 2′ (3′)-phosphate	0.01	0.02	0.83	0.16	0.43	0.46
Zeatin riboside	0.64	0.43	0.80	0.77	0.62	0.82
Zeatin	0.74	0.62	0.63	0.77	0.73	0.80
N^6-(3-Hydroxy-3-methylbutyl)-adenine	0.73	0.56	0.59	0.79	–	–
3H-7,7-Dimethyl-7, 8, 9-trihydropyrimido-[2,1-i]purine	0.49	0.07	0.59	0.61	–	–

Solvent systems: (1) n-Butanol-water-conc. ammonium hydroxide (86:14:5). (2) Ethyl acetate-1-propanol-water (4:1:2). (3) Ethanol-0.1 M ammonium borate (pH 9) (1:9). (4) 2-Propanol-water-conc. ammonium hydroxide (7:2:1). (5) t-Butanol-formate-water (20:5:8). (6) 1-Propanol-conc. ammonium hydroxide-water (55:10:35).

Table 2.18. Rf and Ra[a] values of zeatin and dihydrozeatin on paper chromatography (KOSHIMIZU et al., 1967)

Solvent system	Zeatin		Dihydrozeatin	
	Rf	Ra[a]	Rf	Ra[a]
1	0.55	3.23	0.63	3.67
2	0.83	1.43	0.82	1.43
3	0.68	1.16	0.69	1.18
4	0.73	1.39	0.76	1.43
5	0.81	1.86	0.83	1.92

[a] Ra: movement relative to adenine
Paper: Whatman No. 1
Solvents: (1) n-Butanol-water-conc. ammonium hydroxide (172:18:10, v/v). (2) n-Butanol-acetic acid-water (12:3:5, v/v). (3) 2N Ammonium hydroxide. (4) n-Butanol saturated with water. (5) Isopropanol-water (4:1, 0.1 ml of conc. ammonium hydroxide for each litre of tank volume was added to a beaker on the tank bottom).

nol-water-conc. ammonium hydroxide (KOSHIMIZU et al., 1967) or developed on cellulose-coated film (MN-Polygram, Cel 300) with water (NITSCH, 1968). Ribotide cytokinins can be distinguished from riboside cytokinins since the former remains nearly at the origin when appropriate solvent systems are used. The Rf values of trans-ribosylzeatin are quite similar to those of cis isomer. The same relation is observed for the cis and trans isomers of zeatin. These isomers can be separated on silica gel thin-layer chromatography using a solvent system of chloroform-methanol (9:1), the Rf values of these compounds being

Table 2.19. Rf values of natural cytokinins on thin layer chromatography. (NITSCH, 1968)

	Solvent 1	Solvent 2	Solvent 3
6-(3-Methylbut-2-enylamino)purine	0.28	0.90	0.32
Zeatin	0.37	0.80	0.43
Dihydrozeatin	0.48	0.83	0.50
6-(3-Methylbut-2-enylamino)-9-β-D-ribofuranosylpurine	0.50	–	0.76
Zeatin riboside	0.65	–	0.87

Thin layer: cellulose-coated film (MN-Polygram, Cel 300) pre-washed with the solvent.
Solvents: (1) Water. (2) Butanol-28% ammonia-water (86:5:9). (3) Boric acid 0.03 M, pH adjusted to 8.4 with NaOH.

Table 2.20. Rf values of cytokinins and nitrogen bases separated by chromatography on $PVP/CaSO_4$ thin layers. (THOMAS et al., 1975b)

	Solvent system	
	A	B
Zeatin	0.52	0.23
6-(3-Methylbut-2-enylamino)purine	0.40	0.16
Kinetin	0.32	0.12
Benzyladenine	0.43	0.17
Dihydrozeatin	0.55	0.24
Zeatin riboside	0.60	0.78
6-(3-Methylbut-2-enylamino)-9-β-D-ribofuranosylpurine	0.51	0.64
Kinetin riboside	0.45	0.60
Benzyladenine riboside	0.34	0.46
Adenine	0.52	0.34
Guanine	0.20	0.36
N^6,N^6-Dimethylaminopurine	0.56	0.32
Adenosine	0.63	0.98
N^6,N^6-Dimethylaminopurine riboside	0.60	0.98
Cytosine	0.81	0.80
Uracil	0.77	0.80
Thymidine	0.76	0.80

Solvent systems: (A) 0.013 M Phosphate, pH 6.4+25% acetone. (B): 0.03 M Borate, pH 8.8+10% acetone.

0.25 (trans-zeatin), 0.32 (cis-zeatin) (LEONARD et al., 1971) and 0.14 (trans-ribosylzeatin), 0.20 (cis-ribosylzeatin) (PLAYTIS and LEONARD, 1971). Thin-layer plates prepared from insoluble polyvinylpyrrolidone and calcium sulphate are effective for the separation of cytokinins and phenolic compounds, and further, riboside cytokinins can be separated from free-base cytokinins by using the solvent system of 0.03 M borate buffer (pH 8.8)-10% acetone (THOMAS et al., 1975b).

Cytokinins on paper or thin-layer plates can be detected as blue spots by spraying with bromophenol blue/silver nitrate reagent. Since cytokinins show a strong UV absorption around 250 nm, cytokinins on paper or thin layer containing a fluorescent substance can be detected as dark spots against a fluorescent background under UV light (e.g., 254 nm). BURROWS et al. (1969) used thin layers of cellulose containing a fluorescent substance (Cellulose MN 300 F_{254}) to identify cytokinins [6-(3-methylbut-2-enylamino)-9-β-D-ribofuranolsylpurine and 2-methylthio-6-(3-methylbut-2-enylamino)-9-β-D-ribofuranosylpurine] in the hydrolysate of tRNA obtained from *Escherichia coli*.

Abscisic Acid. Abscisic acid is not often analyzed by paper chromatography. It is usually chromatographed on thin layers using the solvent systems, benzene-acetone-acetic acid (70:30:1) (ANTOSZEWSKI and RUDNICKI, 1969), benzene-ethyl acetate-acetic acid (15:3:1), toluene-ethyl acetate-acetic acid (25:15:2) (MILBORROW and NODDLE, 1970) and most of the solvent systems available for the thin-layer chromatography of free gibberellins. Since abscisic acid reveals a very strong UV absorption maximum around 255 nm, silica gel F_{254} is used very effectively for detection and abscisic acid can be detected as a dark spot against a green fluorescent background. On silica gel G ABA is also detected by spraying with aqueous 10% sulphuric acid and heating at 130 °C for 8 min, giving a yellowish fluorescent spot under UV light. This method can be applied to the determination of abscisic acid fluorometrically at 525 nm (ANTOSZEWSKI and RUDNICKI, 1969).

2.2.3 Gas-Liquid Chromatography

a) General Remarks

Gas-liquid chromatography (GLC) depends upon the partition between carrier gas and a stationary phase coating the surface of a support. Since GLC is much superior to paper and thin-layer chromatography in resolution and reproducibility, it is an excellent method for the identification of plant hormones in partially purified samples. The following points must be noted in the practical identification procedure. Since there remains a possibility that different compounds show the same retention time, analyses with different types of column packing are necessary. On the contrary it happens sometimes that the same compound shows different retention times, depending on the amount of impurities or the difference in the amount of samples injected. Accordingly, in order to compare samples the retention times of an authentic specimen and an unknown sample should be determined several times at the same detector response.

It is always necessary to vaporize samples in GLC. The region of sample injection is heated for this purpose above the temperature of the rest of the column, but there are many compounds which do not vaporize even at high temperature or are apt to decompose. In such cases the compounds have to be converted to more stable, volatile derivatives. Carboxylic acids and phenols are usually derivatized by conversion to the corresponding methyl esters and methyl ethers, respectively, by treatment with ethereal diazomethane, by which carboxylic acids and phenols are quantitatively methylated in a short time. Alcohols are converted to acetates, methyl ethers or trimethylsilyl (TMSi) ethers. Trimethylsilylation is frequently used in derivatization of compounds containing hydroxyl, amino, and carboxyl groups. As the trimethylsilylating reagent N,O-bis(trimethylsilyl)acetamide (BSA), N,O-bis(trimethylsilyl)trifluoroacetamide (BSTFA), hexamethyldisilazane and trimethylsilylimidazole are frequently used solely or with trimethylchlorosilane in acetonitrile or pyridine solution. There are many other trimethylsilylating reagents and techniques (PIERCE, 1968). The trimethylsilylation reaction should be carried out in anhydrous conditions and TMSi derivatives must be kept free from moisture to avoid hydrolysis. Hydroxyl and carboxyl groups are converted to TMSi derivatives within 1 min at room temperature but amino groups need higher temperature (~60 °C) and longer (~10 min).

GLC can be used as an isolation procedure. Each component separated with a packed column is collected into traps by cooling. Sometimes larger columns for preparative purposes are used for the isolation of a fairly large amount of compound. Preparative GLC is monitored with a thermal conductivity detector, or with a flame ionization detector connected with a splitting device. Gas chromatographs, specially designed for preparative use, are available commercially.

b) Identification of Plant Hormones by Gas-Liquid Chromatography

GLC is used very frequently for the identification of plant hormones. Especially it is a very important method for the analysis of ethylene which cannot be analyzed by other methods currently used for plant hormones.

Usually a flame ionization detector is used for the detection of plant hormones in GLC. This detector is highly sensitive, the lower limit of detection for ordinary organic compounds being below 10 ng. However, when more sensitive detection is required, an electron capture detector is known to be very effective, especially for the detection of the compounds which contain atoms or partial structures having a strong electron affinity, e.g., halogens. BROOK et al. (1967) devised the application of this detector for the analysis of IAA, for the first time, by converting it into the trifluoroacetate. SEELEY and POWELL (1970a) developed this method to the detection of IAA and gibberellins in plant extract successfully, the lower limit of detection being about 100 pg. Abscisic acid can be detected with an electron capture detector without introduction of halogen atoms, the lower limit of detection being only 1 pg (SEELEY and POWELL, 1970b). SWARTZ and POWELL (1979) have used an alkali flame ionization detector (AFID) which is selective for nitrogen-containing compounds to detect nanogram quantities of indole-3-acetic acid in extracts of *Fragaria* and *Malus* species. This detector was more than 50 times more sensitive for ethyl indole-3-acetate than the ordinary FID detector.

The introduction of radioactive nuclei into molecules is sometimes used to investigate their metabolism by GLC. The radioactive metabolites can be analyzed with a gas chromatograph equipped with a gas-flow radio-monitor (DURLEY et al., 1973). However, the sensitivity of this type of detector is not

Table 2.21. Relative retention times for various indole derivatives on three different columns. (DEDIO and ZALIK, 1966)

Compound	Column		
	A	B	C
Indole	0.16	0.22	0.11
Skatole	0.21	0.22	0.11
Indole-3-acetaldehyde	0.63	–	–
Tryptamine	0.71	0.69	–
Tryptophol	0.78	0.69	0.88
Indole-3-acetonitrile	0.86	0.78	1.57
Indole-3-acetamide	0.88	–	–
Indole-3-aldehyde	0.87	1.02	–
Ethyl indole-3-acetate	1.00	1.00	1.00

A: 10% SE-52 on Chromosorb W, 60/80 mesh, HMDS; $6^1/_2' \times {}^1/_4''$ stainless-steel column; oven temperature, 195° C; injector block temperature, 280° C; thermal conductivity cell detector; carrier gas, helium at 45 ml/min.
B: 5% SE-30 on Chromosorb W, 60/80 mesh; $5' \times {}^1/_8''$ stainless steel column; oven temperature, 200° C; injector block temperature, 270° C; hydrogen flame ionization detector; carrier gas, nitrogen at 10 ml/min.
C: 5% Neopentyl glycol succinate on Chromosorb W, 60/80 mesh, HMDS; $2^1/_2' \times {}^1/_4''$ stainless steel column; oven temperature, 205° C; injector block temperature, 280° C; thermal conductivity cell detector; carrier gas, helium at 90 ml/min.

very high and a more sensitive method is to trap the ^3H_2O and $^{14}CO_2$ from the FID detector for analysis with a liquid scintillation counter (MACMILLAN and WELS, 1974; FRYDMAN and MACMILLAN, 1975; YAMANE et al., 1977).

Auxins. GLC of auxins and other indole derivatives have been extensively investigated and there are some papers reporting GLC data for many auxin-related compounds, as shown in Tables 2.21 and 2.22 (DEDIO and ZALIK, 1966; GRUNWALD and LOCKARD, 1970). Acidic auxins such as IAA are usually analyzed after methylation. GRUNWALD and LOCKARD (1970) successfully performed GLC of methyl esters of ten acidic indole derivatives after trimethylsilylation for the improvement of resolution. Conjugate IAA's, such as IAA-inositol, cannot be subjected to GLC without derivatization because they are highly polar due to the presence of a polyalcoholic moiety. UEDA et al. (1970) conducted GLC of myoinositol glycosides of IAA by converting them into TMSi derivatives.

Gibberellins. Gibberellins can be identified by GLC with fairly high reliability. IKEKAWA et al. (1963) carried out GLC by using gibberellin A_1–A_9 methyl esters and suggested the effectiveness of GLC for the identification of gibberellins. However, identification of gibberellins by this method is becoming more difficult because of the increasing number of naturally occurring gibberellins with similar structural features and physical properties. Accordingly, determination should

Table 2.22. Comparison of relative retention (r), effective plate value (N), and resolution (R) of various indole TMSi derivatives and indole methyl esters on column substrate OV-101. (GRUNWALD and LOCKARD, 1970)

Indole, TMSi derivative	r	N	R	Indole, methyl ester	r	N	R
Indole-2-carboxylate	0.72	935		Indole-2-carboxylate	0.42	318	
			1.91				3.92
Indole-1-propionate	1.09	848		Indole-3-acetate	0.80	1063	
			2.30				0.18
Indole-3-acetate	1.49	905		Indole-5-carboxylate	0.82	653	
			1.69				0.52
Indole-3-carboxylate	1.85	1095		Indole-3-carboxylate	0.88	1290	
			1.98				0.30
Indole-3-propionate	2.32	1272		Indole-1-propionate	0.91	1373	
			0.20				1.87
Indole-5-carboxylate	2.37	1278		Indole-3-propionate	1.12	1110	
			3.92				3.28
Indole-3-butyrate	3.52	1765		Indole-3-butyrate	1.59	1745	
			0.61				0.38
Indole-3-lactate	3.73	1214		Indole-3-lactate	1.65	1628	
			2.19				4.73
Indole-3-glyoxylate	4.59	2490		Indole-3-glyoxylate	2.61	1834	
			3.79				1.42
Indole-3-acrylate	6.31	2190		Indole-3-acrylate	2.97	1695	

Column characteristics: column temperature, 200° C; detector temperature, 250° C; carrier gas, helium; column, 5% OV-101 on Anakrom ABS 80/90 mesh; retention time of ethylindole-3-acetate, 4.8 min.

be carried out by use of several column packings. Some gibberellins can be determined by GLC as methyl esters but gibberellins carrying two or more hydroxyl groups do not always show good resolution. Some of them, GA_3 methyl ester for example, tend to decompose, showing only a collapsed peak. Such breakdown can be overcome by trimethylsilylation of methyl esters. The column packings most frequently used for gibberellin derivatives are SE-30, SE-33, QF-1, OV-1 and OV-17 with low liquid phase (2%–3%). Besides, XE-60 is also successfully used for separating some pairs of gibberellin derivatives with similar structures, such as methyl esters of GA_5 and GA_{20}. CAVELL et al. (1967) reported retention times on GLC of methyl esters and methyl ester TMSi ethers of 17 gibberellins, as shown in Table 2.23. SCHNEIDER et al. (1975) reported GLC data on methyl ester TMSi ethers of GA_3 and GA_8 glucosides as well as TMSi ester TMSi ethers of ten gibberellins (Table 2.24), and Hiraga et al. (1974c) also reported GLC data of five gibberellin glucosyl esters (Table 2.25).

Cytokinins. Cytokinins are usually analyzed as TMSi derivatives by GLC. Hydroxyl groups in the N^6-substituent and in the sugar moiety attached to the 9-position in the purine skeleton are readily trimethylsilylated. A TMSi group is also introduced to the 9-position if no substituent is attached there. The trimethylsilylating conditions for cytokinins are summarized in Table 2.26.

Table 2.23. GLC retention times of gibberellins (CAVELL et al., 1967)

Gibberellin	Retention time (min)			
	Methyl ester		TMSi ether of methyl ester	
	(1) 2% QF-1	(2) 2% SE-33	(3) 2% QF-1	(4) 2% SE-33
A_1	19.7	14.7	16.3	16.8
A_2	20.7	15.1	23.1	8.2
A_3	20.6	17.6	19.1	18.3
A_4	8.4	7.4	11.2	8.9
A_5	10.4	6.7	11.3	8.1
A_6	17.1	9.4	19.1	11.4
A_7	9.1	7.9	12.8	9.5
A_8	38.6	30.7	20.7	29.3
A_9	4.3	3.9	(7.4)[a]	(4.5)[a]
A_{10}	10.4	6.9	16.0	10.8
A_{11}	6.9	5.1	(12.3)[a]	(6.0)[a]
A_{12}	2.0	4.0	(3.5)[a]	(4.9)[a]
A_{13}	6.2	11.9	6.2	12.2
A_{14}	4.8	8.4	4.5	8.7
A_{15}	14.6	9.9	(24.9)[a]	(12.4)[a]
A_{18}	10.7	16.5	6.9	13.3
A_{19}	8.7	9.9	9.8	12.9

[a] Methyl ester only

Silanized glass columns, 5 ft × 1/4 in. i.d., were packed with QF-1 or SE-33 on silanized Gaschrom A. The 2% QF-1 and 2% SE-33 columns had efficiencies of 450 and 1060 theoretical plates respectively, as calculated with 5-cholestane. Column temperature and carrier gas (N_2) flow rate: (1) 201° C, 60 ml min^{-1}; (2) 190° C, 80 ml min^{-1}; (3) 179° C, 84 ml min^{-1}; (4) 187° C, 75 ml min^{-1}.

Table 2.24. Retention times (t) and relative retention (r) of TMSi derivatives of gibberellin-O-glucosides and their methyl esters on 3% QF-1 (SCHNEIDER et al., 1975)

TMSi derivative of	Free acid (TMSi-gibberellin glucoside-TMSi)		Methyl ester (TMSi-gibberellin glucoside-Me)	
	t (min)	r	t (min)	r
GA_1-O(3)-glucoside	15.9	2.49	15.6	2.45
GA_1-O(13)-glucoside	15.4	2.41	15.1	2.37
GA_3-O(3)-glucoside	12.8	2.01	14.0	2.20
GA_3-O(13)-glucoside	10.0	1.57	11.3	1.76
GA_8-O(2)-glucoside	12.1	1.90	12.8	2.00
Allogibberic acid-O(13)-glucoside	3.25	0.51	3.20	0.50
Progesterone	6.4	1.00	6.4	1.00
5α-Cholestane	0.9	0.14		

Column characterstics: column, 1.5 m × 4 mm, silanized glass column, packed with 3% QF-1 on Gas-Chrom Q (125–160 μm); temperature, 245° C; carrier gas, 175 ml N_2/min.

Table 2.25. GLC retention times of gibberellin glucosyl ester TMSi ethers. (HIRAGA et al., 1974c)

Compound	Retention time (min)	
	(a)	(b)
Gibberellin A_1 glucosyl ester	14.8	18.8
Gibberellin A_3 glucosyl ester	16.5	20.7
Gibberellin A_4 glucosyl ester	13.1	14.7
Gibberellin A_{37} glucosyl ester	22.0	23.0
Gibberellin A_{38} glucosyl ester	25.7	28.0

(a) Column, 2% QF-1 (3 mm × 1 m); column temp., 224° C; carrier gas, N_2 (34 ml/min)
(b) Column, 2% OV-1 (3 mm × 1 m), column temp., 243° C; carrier gas, N_2 (33 ml/min).

Table 2.26. Trimethylsilylating conditions for cytokinins

Reagent	Temp.	Time	References
(1) BSA-acetonitrile (1:2)	60° C	5 min	MOST et al. (1968)
(2) BSTFA-pyridine (1:2)	65° C	5 min	DYSON and HALL (1972)
(3) BSA	90° C	1 h	WANG et al. (1977)
(4) BSTFA-acetonitrile (1:4)	90° C	30 min	HORGAN et al. (1973)
(5) BSTFA-1% TMCS	60° C	2 h	MORRIS (1977)
(6) BSA-TMCS-acetonitrile (10:1:20)	70° C	5 min	WATANABE et al. (1978)

Usually the N^6-position is not trimethylsilylated in most cases but it can be trimethylsilylated by using BSA-trimethylchlorosilane-acetonitrile (10:1:20) (WATANABE et al., 1978a), yielding fully trimethylsilylated cytokinins.

After completion of trimethylsilylation, the reaction mixture is directly injected into a gas chromatograph. When a flame ionization detector is used, a minimum detectable amount is around 5 ng, the relative detector response varying for each cytokinin (e.g., 100 for i^6A, 75 for ms^2-i^6A, 60 for zeatin riboside) (BABCOCK and MORRIS, 1970).

UPPER et al. (1970) performed GLC of TMSi derivatives of 22 cytokinins and related compounds including nucleosides, using a column of QF-1 on Gas-Chrom Q, and successfully separated them with the exception of ms^2-i^6A and zeatin riboside (Table 2.27). ALAM and HALL (1971) succeeded in resolving the Δ^2- and Δ^3-isomers of the side chain in N^6-isopentyladenosine by use of 10% DC-11 on Gas-Chrom Q.

Permethyl derivatives of cytokinins can also be effectively analyzed by GLC (see Section 2.2.5.b).

Abscisic Acid and Related Compounds. Abscisic acid and related compounds are easily analyzed by GLC. LENTON et al. (1968) identified cis, trans-abscisic acid by GLC in a plant extract after purification and methylation. These authors also discussed a standard procedure for purification of samples prior to GLC

Table 2.27. Retention times of TMSi derivatives of cytokinins and related compounds. (UPPER et al., 1970)

Compound	Retention time (min)	Relative retention time[a]
Methyladenine[b]	1.1–1.3	0.082
Adenine[b]	1.3–1.4	0.121
Kinetin (first peak)	3.8	0.335
6-(3-Methylbut-2-enylamino)purine	4.5–4.7	0.398
Guanine	5.0–5.7	0.404
Kinetin (second peak)	6.5	0.57
Dihydrozeatin	7.0	0.62
Zeatin	8.5–8.9	0.75
6-Benzylaminopurine	8.8	0.78
Adenosine (first major peak)[c]	9.2–9.4	0.81
2-Methylthio-6-(3-methylbut-2-enylamino)purine[b]	9.4–9.6	0.83
Adenosine (second peak)[c]	9.9–10.2	0.87
6-(3-Methyl-3-hydroxybutylamino)purine	9.95	0.88
Guanosine[b]	11.3–11.6	1.00
6-(3-Methylbut-3-enylamino)-9-β-D-ribofuranosylpurine	11.55	1.00
6-(3-Methylbutylamino)-9-β, D-ribofuranosylpurine[b]	11.7	1.01
6-(3-Methylbut-2-enylamino)-9-β-D-ribofuranosylpurine	11.8–12.25	1.04–1.08
Kinetin riboside	13.0	1.14
6-(3-Methyl-3-hydroxybutylamino)-9-β,D-ribofuranosylpurine	13.9	1.23
Zeatin riboside[b]	14.3	1.27
2-Methlythio-6-(3-methylbut-2-enylamino)-9-β-D-ribofuranosylpurine	14.8	1.31
6-Benzylamino-9-β,D-ribofuranosylpurine	15.25	1.34

[a] Retention time of guanosine = 1.00
[b] Only a single peak was obtained when nearly equal quantities of these groups of compounds were injected simultaneously
[c] Three distinct peaks were obtained when nearly equal quantities of these compounds were injected simultaneously

Chromatographic conditions: a 1.5 m × 1.5 mm (internal diameter) glass column of 2% QF-1 on Gas-Chrom Q, N_2 flow rate = 28 ml/min. Temperature program: isothermal at 150° C, 0 to 6 min; linear increase at 8° C/min, 6 to 9 min; linear increase at 4° C/min, 9 to 18 min.

analysis. Usually abscisic acid is analyzed by GLC after methylation or trimethylsilylation (DAVIS et al., 1968; GASKIN and MACMILLAN, 1968). GLC of abscisic acid is performed with packings used for gibberellin derivatives, such as SE-30 and QF-1, and with a flame ionization detector. However, the most sensitive GLC for minute amounts of abscisic acid is performed by using an electron capture detector as already mentioned (SEELEY and POWELL, 1970b).

Both cis, trans- and trans, trans-xanthoxin are usually analyzed by GLC as their acetylated derivatives (TAYLOR and BURDEN, 1970; FIRN et al., 1972;

ZEEVAART, 1974). Many liquid phases such as OV-17, SE-30, XE-60, and OV-225 can completely separate both isomers.

Ethylene. Since the concentration of ethylene in the air, which shows bioactivity on plants, is as low as 0.1–1 ppm, very high sensitivity is required to detect and identify ethylene present in, or given off by plants. GLC is the most powerful method for this purpose and has been extensively investigated (e.g., GINZBURG, 1959; LYONS et al., 1962; WARD et al., 1978).

Usually alumina, which acts by an adsorption mechanism, is used as a column packing for GLC of ethylene. To reduce the tailing of peaks, which occurs inevitably in adsorption GLC, alumina is sometimes treated with silicone oil or liquid paraffin (SHIMOKAWA and KASAI, 1966). It is usually necessary to remove the interfering volatile compounds, particularly water, when analyzing ethylene in low concentration evolved from plants or other materials. GALLIARD and GREY (1969) described a method using a pre-column by which interfering volatile compounds can be removed. The use of a pre-column was developed for the determination of ethylene in very low concentration by DEGREEF and DEPROFT (1976). In this method ethylene existing in very low level in the air can be trapped with 100% efficiency on a pre-column packed with Porapak-S at −90 °C, and then ethylene is released by heating the pre-column and introduced into the analytical GLC column. By this method ethylene at 0.1–0.01 ppb can be detected and determined.

2.2.4 High Performance (Pressure) Liquid Chromatography

a) General Remarks

In liquid column chromatography the components of a mixture are separated and eluted successively according to their properties. This technique is used not only for isolation but also for identification of compounds by comparing the elution volume of an unknown compound with that of an authentic specimen. Co-chromatography is also effective for identification. However, ordinary liquid column chromatography needs a long time and a rather large number of samples, while the separation is not good enough. High performance (pressure) liquid chromatography (HPLC) was devised to solve these problems. The columns in HPLC are prepared with packings, which should be homogenous and very fine. High pressure must be applied mechanically to force the developing solvent through the tightly packed column and to allow the elution of samples in a short time. In GLC the column temperature is programmed and, in HPLC, the composition of the eluting solvent is programmed either stepwise or by a gradient generator.

The most serious problem in HPLC is the method of detection. There is no versatile and sensitive detector available for most organic compounds, such as a flame ionization detector in GLC. The most commonly used detectors are a UV spectrometer and a refractometer. Measurement of the refractive index of the effluent is a general but insensitive method and the refractive index varies with temperature. UV detection is more sensitive and, when a particular chromophore can be selected, it is more specific. However, there are quite a few compounds, e.g., the gibberellins, showing no significant UV absorption except in the region of very short wave length (end absorption). For these compounds UV detection can only be performed by monitoring at 210–230 nm, which seriously limits the use of many kinds of solvent showing an end absorption. HPLC is suitable for the analysis of radioactive compounds, which are detected by a radio-monitor or by more sensitive means such as liquid scintillation counting after fractionation of exudates.

HPLC is superior to GLC in several respects, namely, broad applicability to a large variety of compounds including non-volatile or compounds unstable to heat, and applicability of many types of chromatographic techniques using commercially available packings for adsorption, partition (including reverse-phase), ion exchange and gel filtration. It is also an advantage that samples can be recovered without any breakdown and subjected to other methods of analysis.

b) Identification of Plant Hormones by High Performance Liquid Chromatography

The history in application of HPLC in the field of plant hormone researches is rather short. However, intensive investigations are being carried out in this field because many scientists are looking for new powerful identification methods for the analysis of plant hormones.

Auxins. Up to now auxins have been separated by HPLC using ion exchange packing (CARNES et al., 1974) and partition columns (CROZIER and REEVE, 1977), as illustrated in Fig. 2.30. Auxins can be easily detected by a UV detector due to their indole chromophore.

Gibberellins. Gibberellins are separated effectively by ordinary partition column chromatography using silica gel (DURLEY et al., 1972) and Sephadex LH-20 (MACMILLAN and WELS, 1973), being identified according to their retention volumes. However, these results cannot be developed directly to HPLC because gibberellins show no significant UV absorption except an end absorption which restricts the use of such solvents as acetic acid and ethyl acetate. Three methods have been devised to solve this problem.

One is the introduction of a chromophore into the GA molecule. CROZIER and REEVE (1977) prepared benzyl esters and successfully analyzed them on Partisil 10 (silica gel) by monitoring the eluate at 254 nm with a UV detector. They also published the mass spectra of the benzyl esters of GA_1, GA_3–GA_5, GA_7, GA_9, GA_{13}–GA_{15}, and GA_{20} (REEVE and CROZIER, 1978). HEFTMANN et al. (1978) used the p-nitrophenyl esters, detected at 245 nm, from a Partisil column impregnated with 5% (w/w) silver nitrate. However, long columns were used and retention volumes were very large; for example, it required 8 h to elute the separated pair, GA_1 and GA_3, a separation which can be achieved in a few minutes by the methods described below.

A second method of detection is by end absorption of the carboxyl group in the region 200–210 nm. This method has the advantage that the free GA's can be analyzed, but it has the disadvantage that the solvent choice is restricted

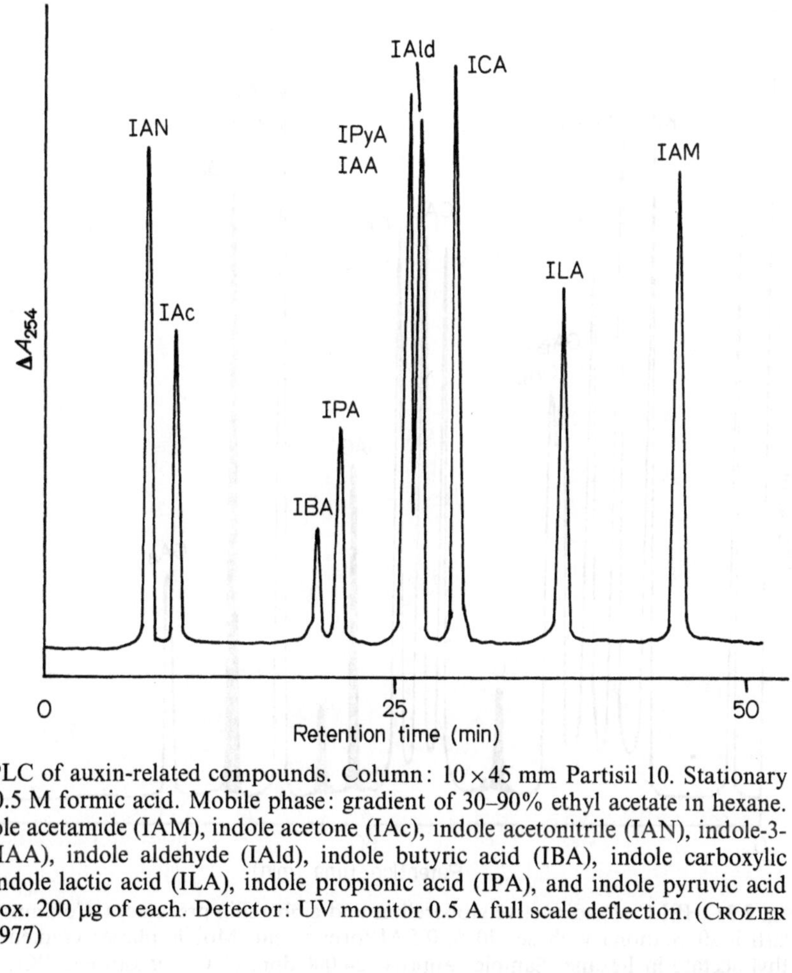

Fig. 2.30. HPLC of auxin-related compounds. Column: 10×45 mm Partisil 10. Stationary phase: 40% 0.5 M formic acid. Mobile phase: gradient of 30–90% ethyl acetate in hexane. Sample: Indole acetamide (IAM), indole acetone (IAc), indole acetonitrile (IAN), indole-3-acetic acid (IAA), indole aldehyde (IAld), indole butyric acid (IBA), indole carboxylic acid (ICA), indole lactic acid (ILA), indole propionic acid (IPA), and indole pyruvic acid (IPyA). Approx. 200 µg of each. Detector: UV monitor 0.5 A full scale deflection. (CROZIER and REEVE, 1977)

to those which are transparent at 200–210 nm. However, reverse-phase partition chromatography with silylated silica columns and aqueous methanol as the elution solvent has been used successfully. YAMAGUCHI et al. (1979) used dimethylsilylated and octadecylsilylated silica, eluted with aqueous ammonium chloride buffers and methanol, to separate free GA's and GA glucosides. They applied these methods to detect the 3-glucosyl ethers of GA_3 and gibberellenic acid in extracts of immature seed of *Quamoclit pennata*. Free GA's can also be analyzed on C_{18}-silica reverse-phase columns using methanol −0.8% aqueous phosphoric acid and UV detection at 205 nm (MACMILLAN et al., unpublished).

A third and indirect method of detection, was used by JONES et al. (1980). Aliquots of fractions from a C_{18}-silica column eluted with a linear gradient of methanol −1% aqueous acetic acid, were bio-assayed for bio-active GA's and were analyzed by GLC for bio-inactive GA's. These authors applied this method to confirm the presence of GA_3, GA_5, GA_{17}, GA_{20}, and GA_{29} in extracts of immature seed of *Pharbitis nil*, and to detect GA_{19} and GA_{44} which were tentatively identified by mass spectrometry.

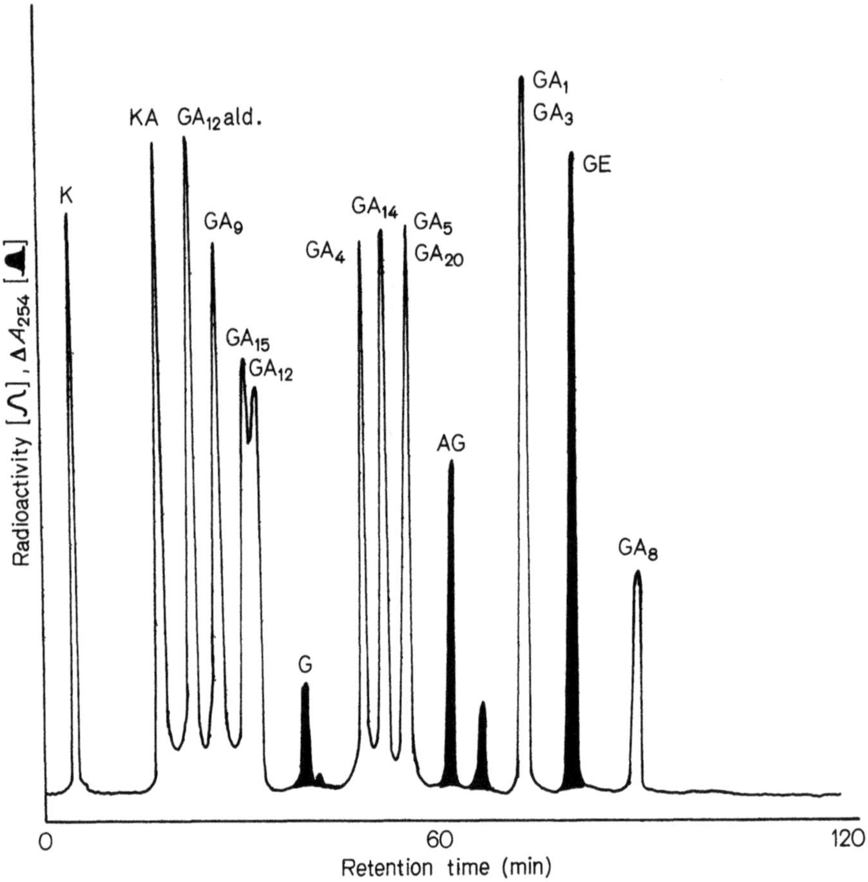

Fig. 2.31. HPLC of radioactive gibberellins and gibberellin precursors. Column: 10×45 mm Partisil 20. Stationary phase: 40%, 0.5 M formic acid. Mobile phase: gradient of 0–100% ethyl acetate in hexane. Sample: approx. 24,000 dpm (^{14}C)-*ent*-kaurene (K); 50,000 dpm (^{14}C)-GA_3; (^{3}H)-GA_5; (^{14}C)-GA_{12}; (^{14}C)-GA_{15}; and (^{3}H)-GA_{20}; 100,000 dpm (^{3}H)-*ent*-kaurenoic acid (KA); (^{3}H)-GA_1; (^{3}H)-GA_4; (^{3}H)-GA_8; (^{3}H)-GA_9; (^{3}H)-GA_{12} aldehyde and (^{3}H)-GA_{14}; and uncalibrated amounts of gibberic acid (G), allogibberic acid (AG), and gibberellenic acid (GE). Detectors: radioactivity monitor 1800 cpm full scale deflection (FSD), UV monitor 0.5 A FSD. (Reeve et al., 1976; Crozier and Reeve, 1977)

Radioactive gibberellins can be analyzed with an HPLC instrument equipped with a radioactivity detector (Reeve and Crozier, 1977). A number of radioactive gibberellins could be separated with excellent resolution with Partisil-20 – 5% formic acid and a gradient of a mobile phase (ethyl acetate-hexane), as shown in Fig. 2.31, by using a sensitive on-stream radioactivity monitor (Reeve et al., 1976; Crozier and Reeve, 1977).

Cytokinins. HPLC has been investigated more intensively for cytokinins than for the other plant hormones. Carnes et al. (1974) used an ion exchange column and separated several cytokinins and their ribosides at the 10 ng level. Pool and Powell (1974) also used an ion exchange resin for the analysis of cytokinins. Challice (1975) used both ion exchange resin and polyamide (Pellidon). The polyamide column showed excellent results and most cytokinins

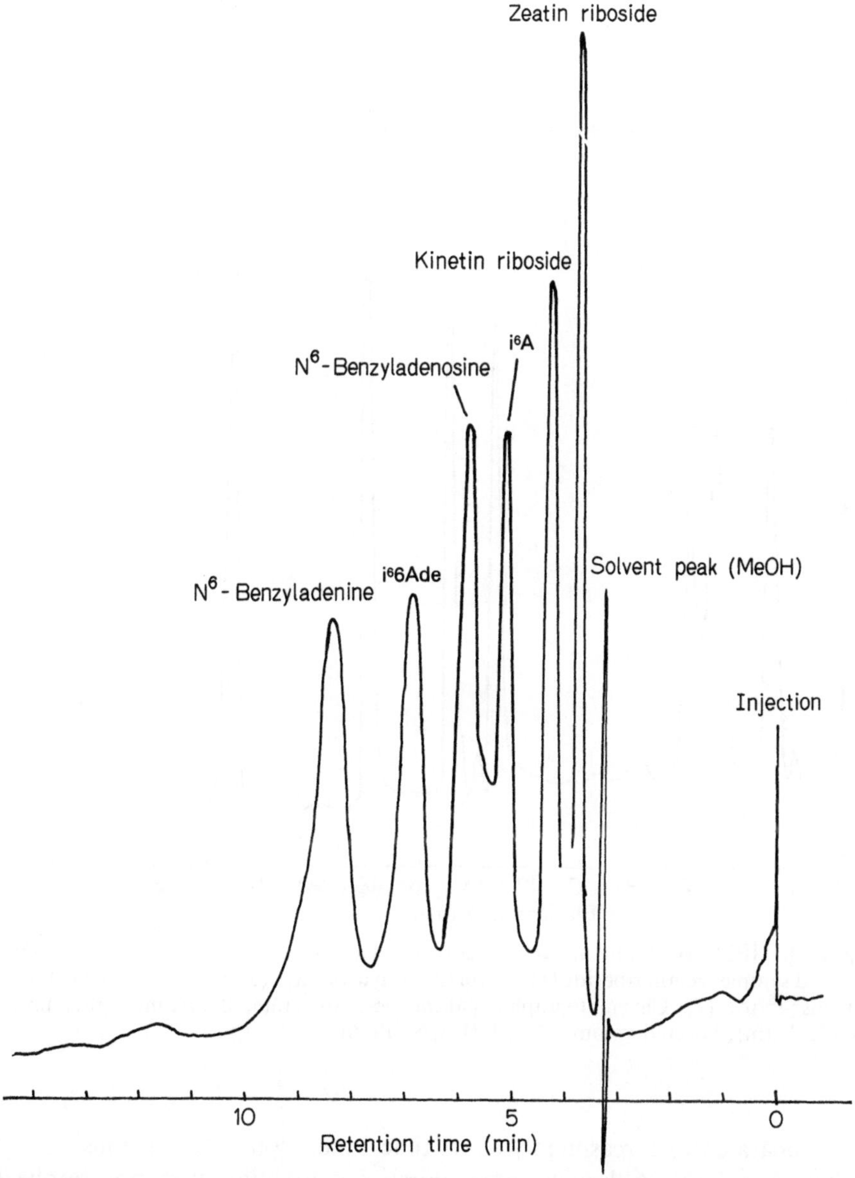

Fig. 2.32. HPLC separation of purines on polyamide with an aqueous buffer eluent. In this system N^6-(β-hydroxyethyl) adenine, N^6-bis(hydroxyethyl) adenine, adenosine, adenine and N^6-methyladenine ran between the solvent peak and zeatin riboside; dihydrozeatin and zeatin ran between zeatin riboside and kinetin riboside; kinetin ran between N^6-(Δ^2-isopentenyl)adenosine(i^{6}A) and N^6-benzyladenosine. Operating parameters-column: Pellidon polyamide (pellicular) 1 m × 4 mm diam. (glass), eluent: 0.05 M KH_2PO_4 (pH 4.5), temperature: 40° C, flow rate: 1.5 ml per min, pressure: ~500 psi, detector: UV 254 nm, absorbance range: 0–0.01, chart: 1 cm per min, sample loadings: 100 ng, injection vols: 5 µl. (CHALLICE, 1975)

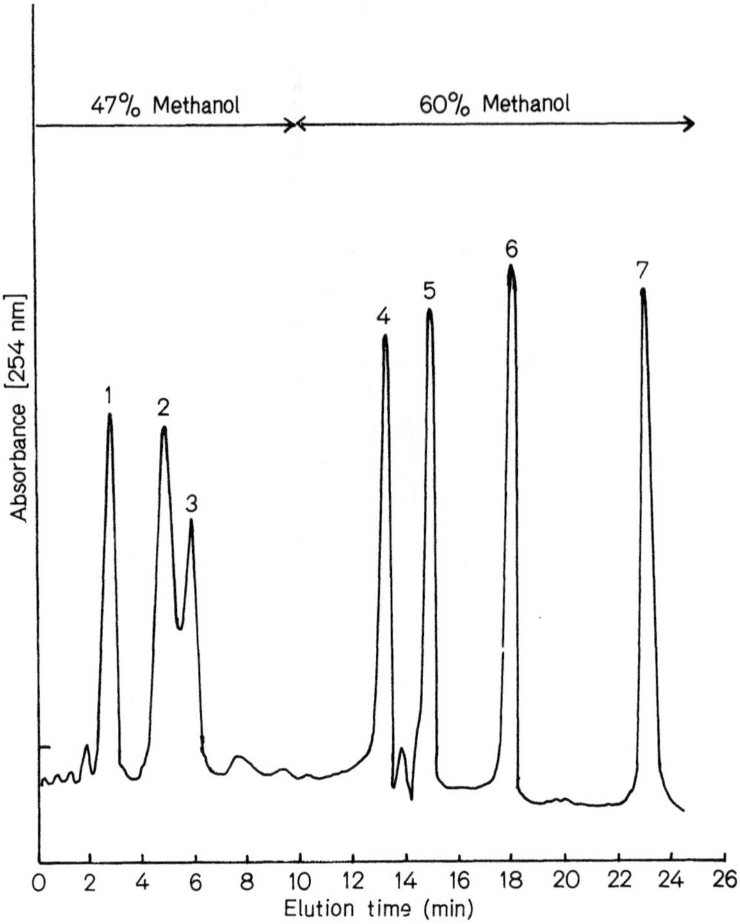

Fig. 2.33. HPLC of seven natively occurring cytokinins. Column: silica gel ODS/C18. Injected sample; zeatin riboside (1), zeatin (2), dihydrozeatin (3), i^6A, (4), i^6Ade (5), ms^2-i^6A (6), ms^2-i^6Ade (7). Chromatographic parameters: flow rate, 0.8 ml/min; pressure, 1075 psi (75.7 atm); injected volume, 3.5 µl. (HAHN, 1976)

were separated in a reasonably short time by elution with aqueous phosphate buffer (Fig. 2.32). Although some pairs of cytokinins were not resolved on the polyamide column, they were resolved clearly by the cation exchange system using Zipax SCX. CARNES et al. (1975) used a chemically bonded reverse-phase packing (Bondapak D_{18}/Porasil B) for the HPLC of cytokinins and achieved the same resolution as in the case of ordinary liquid column chromatography using LH-20. HAHN (1976) also used a reverse-phase packing (ODS/C18-010S) and resolved seven cytokinins eluting with the water-methanol system, as shown in Fig. 2.33, where the minimum detectable amount of cytokinins was 2.2 ng by monitoring at 254 nm with a UV detector.

Columns of insoluble polyvinylpyrrolidone can be inexpensively packed in laboratories. They give good resolution of cytokinins by using borate buffer/n-butanol as eluting solvent under isocratic conditions (THOMAS et al., 1975c).

Abscisic Acid. High performance chromatographic studies on abscisic acid have been frequently reported in papers concerning general identification methods for plant hormones by HPLC. CROZIER and REEVE (1977) separated cis, trans- and trans-trans-abscisic acid with Partisil 10-formic acid and ethyl acetate-hexane. SWEETSER and VATVARS (1976) reported the identification of abscisic acid in the extract from plant tissues by HPLC using SCX-Zipax, which also resolved the cis and trans isomer. CIHA et al. (1977) have described the purification of abscisic acid from soybeans with 98.9% recovery using a Bondapak C_{18}-Porasil B column and linear gradient elution with methanol in 0.2 N acetic acid. Abscisic acid can be analyzed well by methods other than HPLC. However, HPLC can be more readily applied to abscisic acid than to other plant hormones, because abscisic acid has high solubility in many organic solvents, a very strong UV absorption around 254 nm, and an acidic function suitable for partition chromatography.

2.2.5 Combined Gas-Liquid Chromatography-Mass Spectrometry

a) General Remarks

Combined gas chromatography-mass spectrometry (GC-MS) is an analytical method whereby a mixture is separated by gas chromatography and the separated components are led directly into the source of a mass spectrometer. Identification of the components is based upon both their GLC retention times and their mass spectra. Both criteria can be obtained on sub-microgram quantities. Thus GC-MS is a convenient and reliable method of identification and it is used extensively in the field of biological chemistry. The methodology has been described in relation to the GA's by GASKIN and MACMILLAN (1978).

GLC is performed at pressure slightly higher than atmospheric, while mass spectrometry is performed at a pressure below 10^{-5} mm Hg. Accordingly, a device, called a molecular separator, is necessary to remove carrier gas between the gas chromatogram and the mass spectrometer. The effluent from the column is transferred into the mass spectrometer through the molecular separator where most of the carrier gas (helium is usually used in GC-MS) is removed and enriched samples are introduced into the ionization chamber of the mass spectrometer. Three types of molecular separator are currently used. One is the Biemann-Watson separator which consists of a porous glass tube in which the light molecules of carrier gas are readily dispersed and are removed under vacuum. Another is the Ryhage separator (jet-type separator) composed of two fine tubes set closely to each other. When discharged from one tube, the lighter helium particles are dispersed by a high vacuum, while heavier molecules go straight into the mouth of another tube and then to a mass spectrometer. These two types of separator are commonly used, and it is said that the Biemann-Watson separator is more effective for the separation of compounds of lower molecular weight, while the Ryhage separator is better for compounds of higher molecular weight. The third separator, a silicone membrane, is a quite different type. Carrier gas cannot pass through the silicone membrane, while organic compounds in the carrier gas filter through the membrane and are introduced into the source of a mass spectrometer.

Compounds introduced into a mass spectrometer afford molecular ions and fragment ions formed from molecular ions by fragmentation. In most mass spectrometers currently used, ionization is performed by a high energy electron beam (electron impact) (EI). Recently, chemical ionization (CI) has been developed and put into practical use. In this method ionization is performed by the aid of a reagent gas, such as methane, isobutane or ammonia, and has proved a good method of obtaining more stable molecular ions than the electron impact method. Chemical ionization is conducted under higher pressure (0.1–a few mm Hg), which is very suitable for GC-MS because carrier gas need not be removed so exhaustively with a molecular separator as is the case with ordinary GC-MS instruments. GC-MS using chemical ionization (GC-CIMS) is a very promising method and instruments for GC-CIMS are now commericially available.

Two types of mass spectrometer, a magnetic sector type and a quadrupole type, are used in GC-MS. The latter is small and easy to operate but is inferior to the former in resolution.

A mass spectrum corresponding to each GLC peak is obtained by rapid scanning. Sometimes the peaks of the compounds to be determined are not observed in gas-liquid chromatograms, obtained by ion monitoring, being hidden in peaks of a comparatively large amount of impurities. In such cases, the spectral information might be obtained by scanning at the retention time corresponding to those of the compounds, if they reveal very intense and characteristic ion peaks in mass spectra.

Spectral data can be acquired and processed by an on-line computer where they can be stored. The stored data can be recalled from the computer and be projected on an oscilloscope. Then, the selected data are printed by a photocopier or drawn on paper with a pen recorder.

b) Identification of Plant Hormones by GC-MS

Gibberellins. At present GC-MS is probably the most reliable technique for the identification of minute amounts of plant hormones, especially of gibberellins which consist of many homologues. Indeed, all gibberellins, even those having very similar physical or structural properties, can be distinctly identified by GC-MS which utilizes the merits of both GLC and mass spectrometry. For instance, the methyl esters of GA_1 and its C-3 epimer can not be definitely distinguished by mass spectrometry due to the very similar mass spectral features, but they show peaks with clearly different retention times on GLC. On the other hand, methyl esters of GA_5 and GA_{20} show GLC peaks with almost the same retention times on some GLC columns, while they give completely different spectra.

Successful application of GC-MS in the field of plant hormones was demonstrated by MACMILLAN et al. (1967) for the first time. Fig. 2.34 shows the gas-liquid chromatogram of the methylated and trimethylsilylated crude sample obtained from immature seed of *Phaseolus coccineus*. The chromatogram was

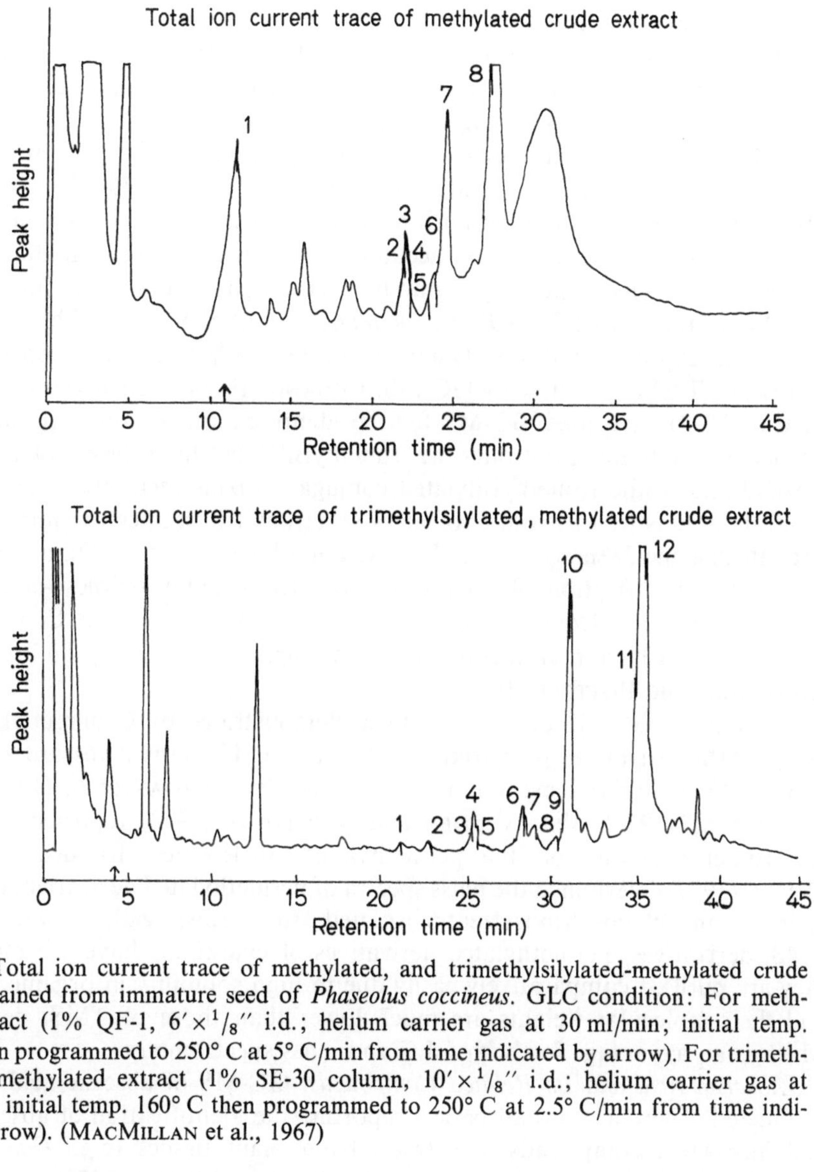

Fig. 2.34. Total ion current trace of methylated, and trimethylsilylated-methylated crude extract obtained from immature seed of *Phaseolus coccineus*. GLC condition: For methylated extract (1% QF-1, $6' \times {}^1/_8''$ i.d.; helium carrier gas at 30 ml/min; initial temp. 140° C, then programmed to 250° C at 5° C/min from time indicated by arrow). For trimethylsilylated-methylated extract (1% SE-30 column, $10' \times {}^1/_8''$ i.d.; helium carrier gas at 30 ml/min; initial temp. 160° C then programmed to 250° C at 2.5° C/min from time indicated by arrow). (MacMillan et al., 1967)

obtained by total ion monitoring and the rapid scanning with a mass spectrometer at GLC peaks afforded the mass spectra of several gibberellin methyl ester TMSi ethers. Since then there have been many papers reporting the identification of gibberellins in the extract from plant tissue by GC-MS (e.g., Bowen et al., 1973).

When performing GC-MS, the success of GLC is the most important factor. However, even if GLC goes well, there may be some problems in utilizing GC-MS. Conjugate gibberellins (gibberellin glucosides and glucosyl esters) are a case in point. The glucosides of GA_3 and GA_8 can be analyzed by GLC after methylation and trimethylsilylation (Schneider et al., 1975), but molecular weights of these derivatives are too big (GA_3 glucoside methyl ester TMSi ether: 882, GA_8 glucoside methyl ester TMSi ether: 972) to detect their molecular ions with good resolution. In such cases, identification must be done based on the analysis of fragment ions arising from each moiety.

Detailed mass spectrometric studies on gibberellins have been reported (ZARETKII et al., 1968; BINKS et al., 1969; TAKAHASHI et al., 1969; YOKOTA et al., 1975). Since data on the relationship between structures and mass spectra of gibberellin derivatives have been accumulated, structural determination of unknown gibberellins is possible without isolation in pure state as in the case of GA_{45} in the extract from *Pyrus communis* (MARTIN et al., 1977b).

Auxins. BRIDGES et al. (1973) have identified IAA in primary roots of *Zea mays* as its TMSi derivative by GC-MS. EHMANN (1974) and EHMANN and BANDURSKI (1974) have used GC-MS for the identification of conjugate auxins in *Zea mays*, determining not only the trimethylsilylated hydrolysate of conjugate auxins but also the trimethylsilylated conjugate auxins themselves. Their mass spectral data, as well as results of mass spectrometric studies by other authors (HUTZINGER and JAMIESON, 1970; JAMIESON and HUTZINGER, 1970), are expected to aid the identification of auxins by GC-MS. 4-Chloroindole-3-acetic acid, previously isolated from immature seed of *Pisum sativum* (see Chap. 1) and its methyl ester have been identified in immature seed of *Vicia faba* by GC-MS (HOFFINGER and BÖTTGER, 1979).

Cytokinins. GC-MS of cytokinins was demonstrated by UPPER et al. (1971), using TMSi derivatives of cytokinins. Usually TMSi derivatives are subjected to GC-MS analysis of cytokinins, as reported by HORGAN et al. (1973b) and WANG et al. (1977). VON MINDEN and MCCLOSKY (1973) reported the mass spectrometric studies on the permethylated nucleosides. By using GC-MS MORRIS (1977) determined the mass spectra of permethylated derivatives of glucosylzeatin and glucosylribosylzeatin isolated from crown gall, as well as their TMSi derivatives. Permethylated derivatives of cytokinins have the merit that they are purified comparatively easily due to high solubility in organic solvents and their molecular weights are much lower than those of TMSi derivatives which is an advantage for both GLC and mass spectrometry.

Abscisic Acid and Its Related Compounds. GC-MS of abscisic acid is easily conducted. There are several papers reporting the identification of abscisic acid and its related compounds in extracts from plant tissues (e.g., GASKIN and MACMILLAN, 1968; HILLMAN et al., 1974; SINDY and SMITH, 1975).

c) Selected Ion Monitoring

When samples for GLC contain a comparatively large amount of impurities, the peaks of minor but interesting compounds are sometimes concealed by those of impurities. This is inevitable in GLC, although some special detectors, such as an electron capture and an alkali flame ionisator detector, can be effectively used to detect some specified compounds carrying particular atoms. However, detectors, such as a thermal conductivity detector, a flame ionization detector, and a total ion monitoring in GC-MS, detect all organic compounds without selectivity. If GLC is conducted by monitoring the ion current from

Fig. 2.35. Gas-liquid chromatograms by total ion monitoring and single ion monitoring for the sample containing 6-(p-hydroxy-benzylamino)-9-β-D-ribofuranosyl-purine (P) and 6-(o-hydroxy-benzylamino)-9-β-D-ribofuranosyl-purine (O) (THOMPSON et al., 1975)

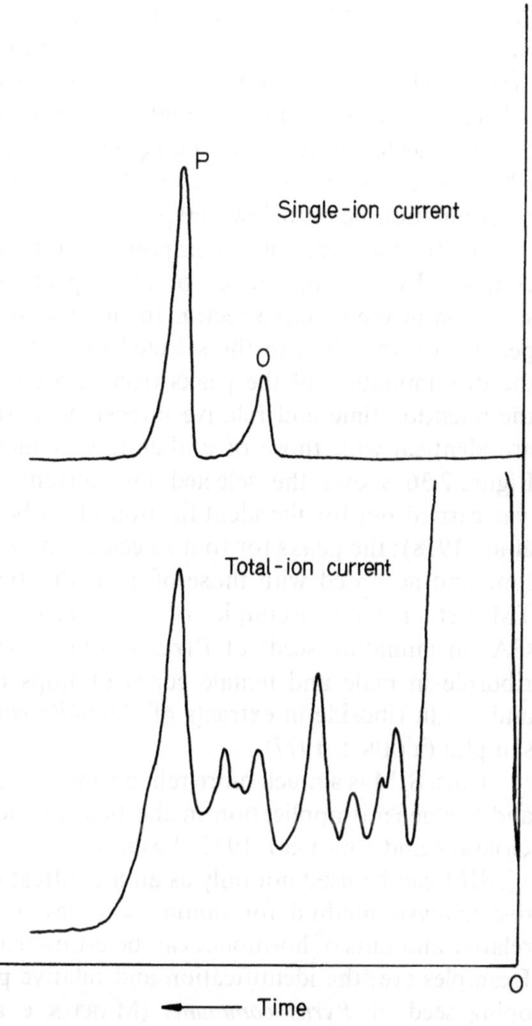

a specific and characteristic ion in the mass spectrum of a compound, the gas-liquid chromatograms are simplified and significant GLC peaks of compounds containing the selected ion appear clearly, as shown in Fig. 2.35 (THOMPSON et al., 1975). This technique has been developed as a modification in GC-MS. Although there has been some confusion in terminology, such an analytical method is usually called *Selected Ion Monitoring* (SIM) or *Mass Fragmentography* (MF) and the record is called *selected ion current profile* or *mass fragmentogram*. In SIM the monitoring of a single ion is called *Single Ion Mass Detection* (SID) and the monitoring of multiple ions is called *Multiple Ion Mass Detection* (MID). The selection of ions can be made by setting the correct ion-accelerating voltage or the correct strength of magnetic field. It is mechanically easier to change the ion-accelerating voltage at a fixed magnetic field than to change the power of the magnetic field at fixed ion accelerating voltage. Thus, SIM is usually carried out by the former method by operating an accelerating voltage alternator, which quickly switches the accelerating voltage corresponding to the several selected ions for MID. SIM can attain very high sensitivity by opening the slit before the ion collector as wide as possible and by amplifying the electric current generated by the selected ions to the maximum if required. The sensitivity of SIM is 100–1000 times that of usual GC-MS, the lower limit of sample amount for determination being below 100 pg.

In SID the peak of a compound under investigation sometimes becomes confused by the appearance of other peaks due to the impurities giving the same ion in their mass spectra. In the case of MID, the appearance of several peaks corresponding to the selected ions at the same retention time facilitates the discrimination of the peaks from those of impurities. Accordingly, if both the retention time and relative intensities of the peaks due to the selected ions are identical with those of authentic specimen, identification is accomplished. Figure 2.36 shows the selected ion current profile obtained by MID which was carried out for the identification of gibberellins in rice plants (KUROGOCHI et al., 1978); the peaks for four selected ions were detected at the same retention time and accorded with those of an authentic specimen of GA_{19} methyl ester TMSi ether. Other examples of the detection of plant hormones by SIM are: GA_4 in immature seeds of *Pyrus serotina* (NAKAGAWA et al., 1979); cis-zeatin-riboside in male and female cones of hops (WATANABE et al., 1978b); zeatin and zeatin riboside in extracts of *Actinidia chinensis* by MID of permethylated samples (YOUNG, 1977).

Thus, SIM is a much more reliable identification method than ordinary GLC, and has general application in the field of biological research (see the reviews, GORDON and FRIGERIO, 1972; FALKNER et al., 1975).

SIM can be used not only as an identification method but also as a quantitative analysis method for minute samples (see Chap. 3). By this method the relative amounts of hormones can be estimated without using internal standards. Examples are: the identification and relative proportions of hormones in developing seed of *Pyrus communis* (MARTIN et al., 1977a); the relative changes in the levels of GA_{19} throughout the life-cycle of *Oryza sativa* L. *japonica* (KUROGOCHI et al., 1979); and the qualitative and quantitative analysis of cytokinins from short apices of *Mercurialis ambigua* (DAUPHIN et al., 1979).

Using external standards, FRYDMAN et al. (1974) estimated the levels of GA_{17}, GA_9, GA_{20}, and GA_{29} in developing seed of *Pisum sativum*. However, the use of internal standards is preferable for the determination of absolute amounts of plant hormones. For example, indole-3-acetic acid has been quantified in root tips of *Zea mays* using 5-methylindole-3-acetic acid as an internal standard (RIVIER and PILET, 1974) and in shoot tips of Douglas fir (CARUSO et al., 1978), and in xylem sap of *Ricinus communis* (ALLEN et al., 1979), using $[^2H_2]$indole-3-acetic acid as internal standard. The quantitative analysis of cytokinins by mass spectrometry and isotope dilution has been described by SUMMONS et al. (1977) and by HASHIZUME et al. (1979). The former workers used $[^2H_2]$raphanatin to identify and quantify raphanatin (Chap. 1) in radish seed. The latter group prepared $[^2H]$-labelled i^6A, ms^2-i^6A, t-io^6A, and ms^2-t-io^6A and used them to test the sensitivity and linearity of response and applied the method

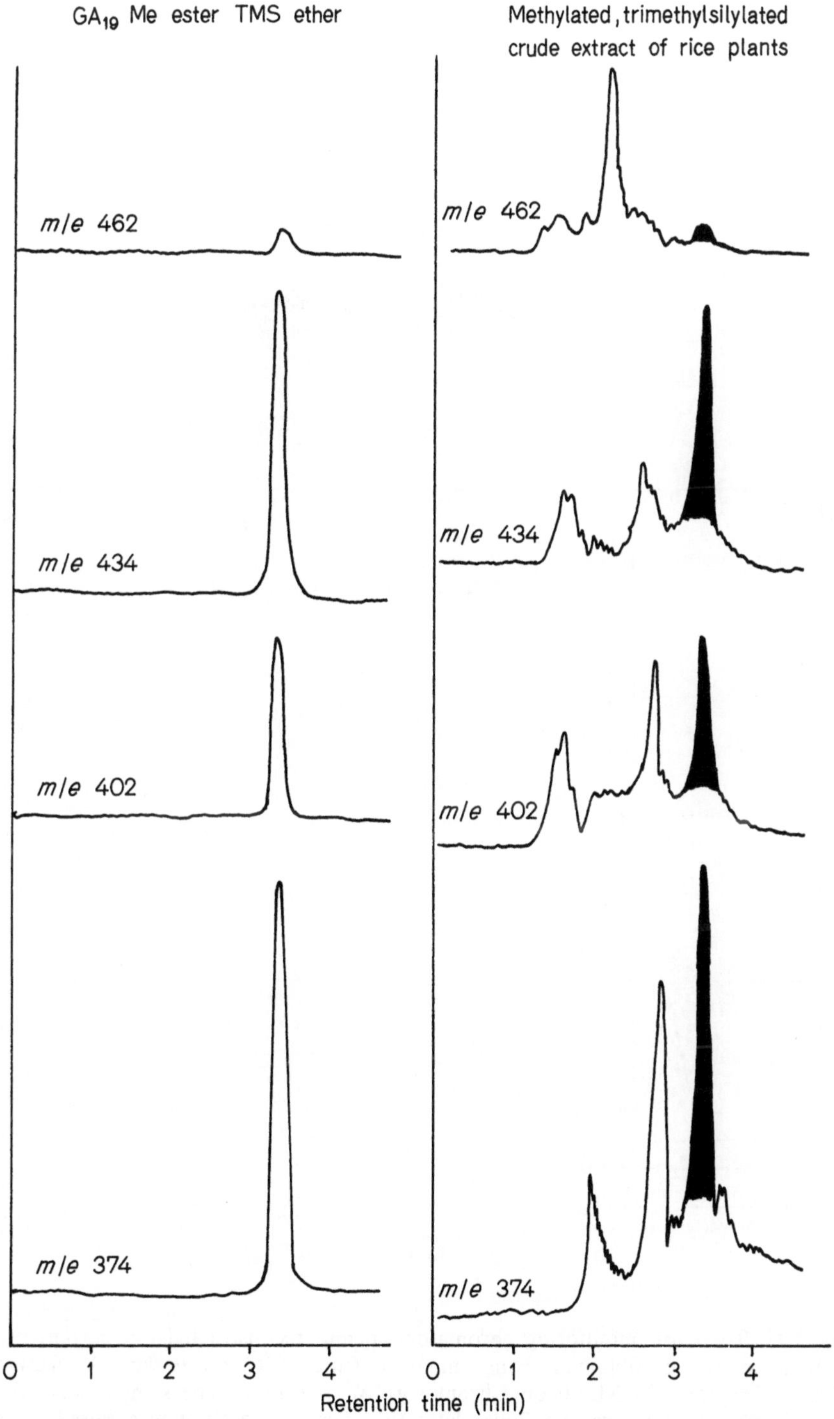

Fig. 2.36. Identification of GA_{19} in the extract of rice plants by MID. Peaks shaded accord with those of authentic GA_{19} methyl ester TMSi ether. (KUROGOCHI et al., 1978)

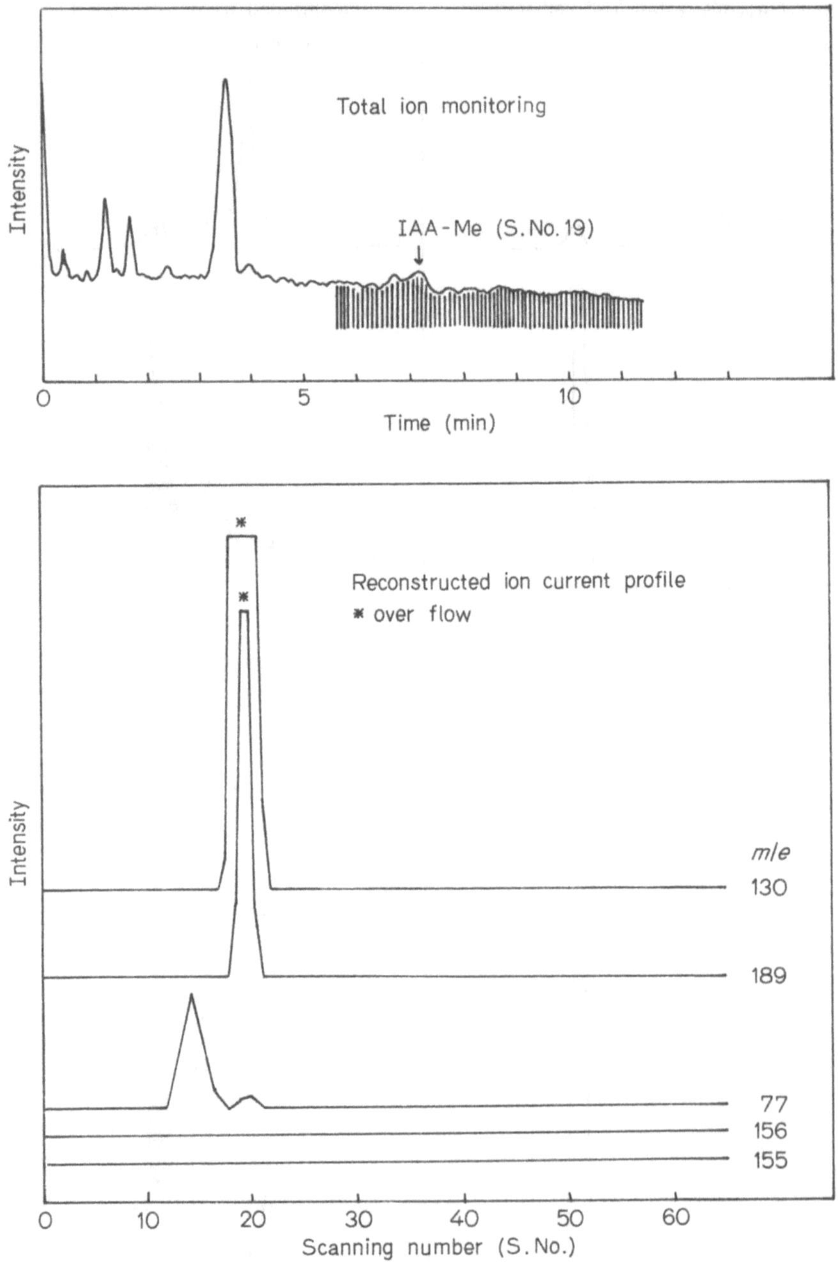

Fig. 2.37. Total ion monitoring chromatogram and reconstructed ion current profile of methylated extract obtained from immature fruit of *Citrus unshiu*. A glass column (3 mm × 1 m) on 2% XE-60 on Chromosorb W was used. The sample was injected at 160° C oven temperature and 2 min later the oven temperature was raised at the rate of 5° C/min. Repetitive scanning started at retention time = 5.5 min with 5 s intervals. The retention time of IAA-Me was 7.2 min and correlated with scan No. 19. (TAKAHASHI et al., 1975)

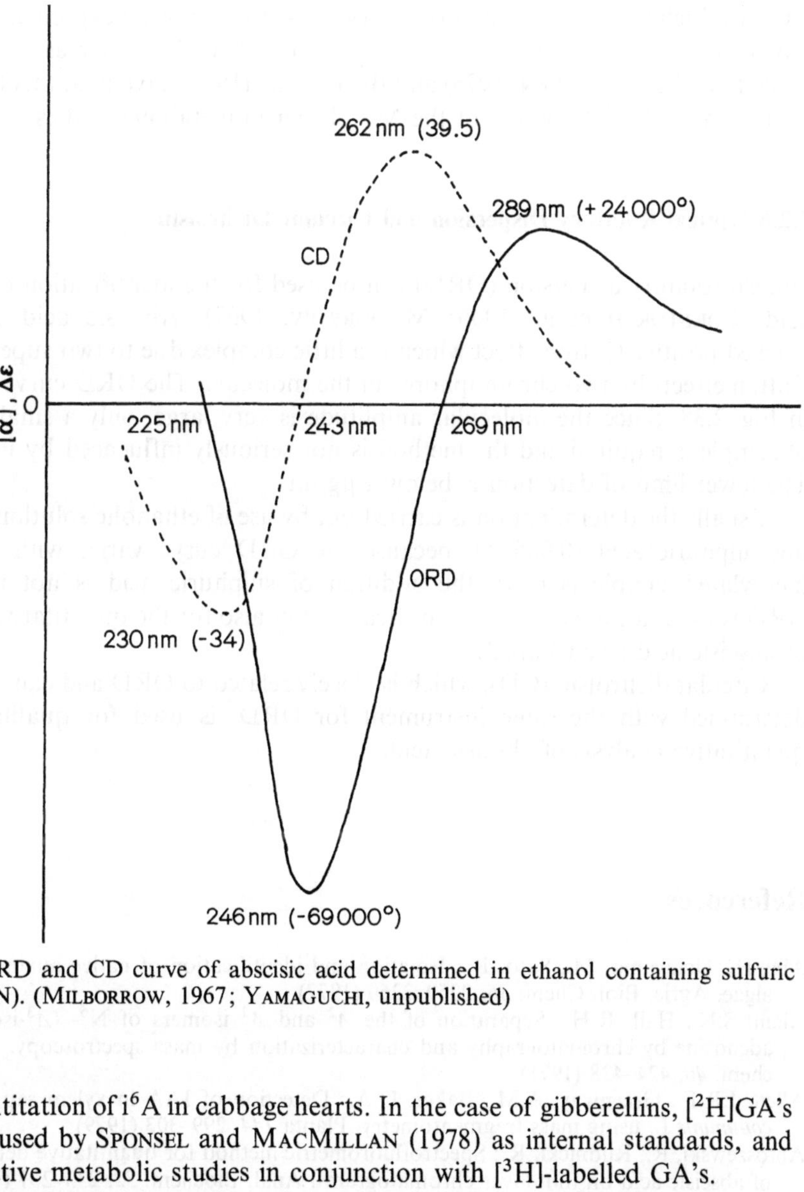

Fig. 2.38. ORD and CD curve of abscisic acid determined in ethanol containing sulfuric acid (0.005 N). (MILBORROW, 1967; YAMAGUCHI, unpublished)

to the quantitation of i^6A in cabbage hearts. In the case of gibberellins, [^{2}H]GA's have been used by SPONSEL and MACMILLAN (1978) as internal standards, and in quantitative metabolic studies in conjunction with [^{3}H]-labelled GA's.

Modern GC-MS instruments are equipped with a computer and all data, acquired from successive mass spectral scanning every few seconds during GLC can be stored in a data-system. By plotting the intensity of particular ion peaks contained in the stored mass spectra, a kind of GLC-trace can be reconstituted. This technique is called *repetitive scanning* or *mass chromatography* and the out-put is termed *reconstructed ion current profile* or *mass chromatogram*. This is a very convenient technique because any chromatogram can be obtained by selecting ion peaks after GLC has been performed. An example of the application of the method in the identification of a plant hormone is described by TAKAHASHI et al. (1975) who identified indole-3-acetic acid in extracts from immature fruits of *Citrus unshiu* (Fig. 2.37). Since repetitive scanning is basically

GC-MS, higher sensitivity than ordinary GC-MS cannot be expected. However repetitive scanning over a portion of the spectrum does increase sensitivity. SPONSEL and MACMILLAN (1978) and BEALE et al. (1979) have used this technique to determine the ^{2}H-content of the M^+ cluster in metabolic studies.

2.2.6 Optical Rotatory Dispersion and Circular Dichroism

Optical rotatory dispersion (ORD) can be used for the identification of abscisic acid (CORNFORTH et al., 1966; MILBORROW, 1967). Abscisic acid reveals a marked positive Cotton effect which is a little complex due to two superimposed Cotton effects by two chromophores in the molecule. The ORD curve is shown in Fig. 2.38. Since the molecular amplitude is very large, only a small amount of sample is required and the method is not seriously influenced by impurities. The lower limit of detection is below 1 µg/ml.

Usually the determination is carried out by use of ethanolic solution containing sulphuric acid (0.005 N), because the ORD curve varies with pH. If a methylated sample is used, the addition of sulphuric acid is not necessary. ORD is used not only for the identification but also for the quantitative analysis of abscisic acid (see Chap. 3).

Circular dichroism (CD), which is closely related to ORD and can be usually determined with the same instrument for ORD, is used for qualitative and quantitative analysis of abscisic acid.

References

Abe, H., Uchiyama, M., Sato, R.: Isolation and identification of native auxins in marine algae. Agric. Biol. Chem. *36*, 2259–2260 (1972)

Alam, S.N., Hall, R.H.: Separation of the Δ^2 and Δ^3 isomers of N^6- (Δ^3-isopentenyl)-adenosine by chromatography and characterization by mass spectroscopy. Anal. Biochem. *40*, 424–428 (1971)

Allen, J.R.F., Greenway, A.M., Baker, D.A.: Detection of IAA in xylem sap of *Ricinus communis* L. using mass fragmentometry. Planta *144*, 299–303 (1979)

Antoszewski, R., Rudnicki, R.: Spectrofluorometric method for quantitative determination of abscisic acid on thin layer chromatograms. Anal. Biochem. *32*, 233–237 (1969)

Armstrong, D.J., Burrows, W.J., Evans, P.K., Skoog, F.: Isolation of cytokinins from t-RNA. Biochem. Biophys. Res. Commun. *37*, 451–456 (1969)

Asakawa, Y., Tamari, K., Shoji, A., Kaji, J.: Metabolic products of gibberellin A_3 and their interconversion in dwarf kidney bean plants. Agric. Biol. Chem. *38*, 719–725 (1974)

Atsumi, S., Kuraishi, S., Hayashi, T.: An improvement of auxin extraction procedure and its application to cultured plant cells. Planta *129*, 245–247 (1976)

Babcock, D.F., Morris, R.O.: Quantitative measurement of isoprenoid nucleosides in transfer ribonucleic acid. Biochemistry *9*, 3701–3705 (1970)

Beale, M.H., Gaskin, P., Kirkwood, P.S., MacMillan, J.: Partial synthesis of gibberellin A_9 and [3α- and 3β-^{2}H] gibberellin A_9, gibberellin A_5 and [1β, 3-$^2H_2$- and $^3H_2$] gibberellin A_5: and gibberellin A_{20} and [1β, 3α-$^2H_2$- and $^3H_2$] gibberellin A_{20}. J.C.S. Perkin I, 885–891 (1980)

Bentley, J.A., Farrar, K.R., Housley, S., Smith, G.F., Taylor, W.C.: Some chemical and physiological properties of 3-indolylpyruvic acid. Biochem. J. *64*, 44–49 (1956)
Biddington, N.L., Thomas, T.H.: Chromatography of five cytokinins on an insoluble polyvinylpyrrolidone column. J. Chromatogr. *75*, 122–123 (1973)
Biddington, N.L., Thomas, T.H.: Effect of pH on the elution of cytokinins from polyvinylpyrrolidone columns. J. Chromatogr. *121*, 107–109 (1976)
Bielski, R.L.: The problem of halting enzyme action when extracting plant tissues. Anal. Biochem. *9*, 431–442 (1964)
Binks, R., MacMillan, J., Pryce, R.J.: Plant hormones VIII. Combined gas chromatography-mass spectrometry of the methyl esters of gibberellins A_1 to A_{24} and their trimethylsilyl ethers. Phytochemistry *8*, 271–284 (1969)
Bowen, D.H., Crozier, A., MacMillan, J., Reid, D.M.: Characterization of gibberellins from light-grown *Phaseolus coccineus* seedlings by combined GC-MS. Phytochemistry *12*, 2935–2941 (1973)
Bridges, I.G., Hillman, J.R., Wilkins, M.B.: Identification and localisation of auxin in primary roots of *Zea mays* by mass spectrometry. Planta *115*, 189–192 (1973)
Brook, J.L., Biggs, R.H., StJohn, P.A., Anthony, D.S.: Gas chromatography of several indole derivatives. Anal. Biochem. *18*, 453–458 (1967)
Browning, G., Saunders, P.F.: Membrane localized gibberellins A_9 and A_4 in wheat chloroplasts. Nature (London) *265*, 375–377 (1977)
Burden, R.S., Taylor, H.F.: The structure and chemical transformation of xanthoxin. Tetrahedron Lett. 4071–4074 (1970)
Burrows, W.J., Armstrong, D.J., Skoog, F., Hecht, S.M., Boyle, J.T.A., Leonard, N.J., Occolowitz, J.: The isolation and identification of two cytokinins from *Escherichia coli* transfer ribonucleic acids. Biochemistry *8*, 3071–3076 (1969)
Burrows, W.J., Skoog, F., Leonard, N.J.: Isolation and identification of cytokinins located in the transfer ribonucleic acid of tobacco callus grown in the presence of 6-benzylaminopurine. Biochemistry *10*, 2189–2194 (1971)
Byrd, D.J., Kochen, W., Idzko, D., Knorr, E.: The analysis of indolic tryptophan metabolites in human urine. J. Chromatogr. *94*, 85–106 (1974)
Carnes, M.G., Brenner, M.L., Andersen, C.R.: Rapid separation and identification of plant growth substances using high pressure liquid chromatography. In: Plant growth substances 1973. Proc. 8th. Int. Conf. of Plant Growth Subst. pp. 99–110. Tokyo: Hirokawa Publishing Co. Inc. 1974
Carnes, M.G., Brenner, M.L., Andersen, C.R.: Comparison of reversed-phase high-pressure liquid chromatography with Sephadex LH-20 for cytokinin analyses of tomato root pressure exudate. J. Chromatogr. *108*, 95–106 (1975)
Caruso, I.L., Smith, R.G., Smith, L.M., Cheng, T.-Y.: Determination of indole-3-acetic acid in Douglas fir using a deuterated analog and selected ion monitoring. Plant Physiol. *62*, 841–845 (1979)
Cavell, B.D., MacMillan, J., Pryce, R.J., Sheppard, A.C.: Plant hormone. V. Thin-layer and gas-liquid chromatography of the gibberellins; direct identification of the gibberellins in a crude plant extract by gas-liquid chromatography. Phytochemistry *6*, 867–874 (1967)
Challice, J.S.: Separation of cytokinins by high pressure liquid chromatography. Planta *122*, 203–207 (1975)
Chen, C.M., Hall, R.H.: Biosynthesis of N^6- (Δ^2-isopentenyl)adenosine in the transfer ribonucleic acid of cultured tobacco pith tissue. Phytochemistry *8*, 1687–1695 (1969)
Ciha, A.J., Brenner, M.L., Brun, W.A.: Rapid separation and quantification of abscisic acid from plant tissues using high performance liquid chromatography. Plant Physiol. *59*, 821–826 (1977)
Clifford, M.N.: The use of poly-N-vinylpyrrolidone as the adsorbent for the chromatographic separation of chlorogenic acids and other phenolic compounds. J. Chromatogr. *94*, 261–266 (1974)
Cornforth, J.W., Milborrow, B.V., Ryback, G.: Identification and estimation of (+)-abscisin II ('dormin') in plant extracts by spectropolarimetry. Nature (London) *210*, 627–628 (1966)

Cross, B.E., Galt, R.H.B., Hanson, J.R.: New metabolites of *Gibberella fujikuroi*. I. Gibberellin A_7 and gibberellin A_9. Tetrahedron *18*, 451–459 (1962)
Crozier, A., Reeve, D.R.: The application of high performance liquid chromatography to the analysis of plant hormones. In: Plant growth regulation. Proc. 9th. Int. Conf. Plant Growth Substances. Pilet, P.-E. (ed.), pp. 67–76. Berlin-Heidelberg-New York: Springer (1977)
Crozier, A., Aoki, H., Pharis, R.P.: Efficiency of countercurrent distribution, Sephadex G-10, and silicic acid partition chromatography in the purification and separation of gibberellin-like substances from plant tissue. J. Exp. Bot. *20*, 786–795 (1969)
Dauphin, B., Teller, G., Durrand, B.: Identification and quantitative analysis of cytokinins from shoot apices of *Mercurialis ambigua* by gas chromatography-mass spectrometry computer system. Planta *144*, 113–119 (1979)
Davis, L.A., Heinz, D.E., Addicott, F.T.: Gas-liquid chromatography of trimethylsilyl derivatives of abscisic acid and other plant hormones. Plant Physiol. *43*, 1389–1394 (1968)
Davis, L.A., Lyon, J.L., Addicott, F.T.: Phaseic acid: occurrence in cotton fruit; acceleration of abscission. Planta *102*, 294–301 (1972)
Dedio, W., Zalik, S.: Gas chromatography of indole auxins. Anal. Biochem. *16*, 36–52 (1966)
DeGreef, J., DeProft, M.: Gas chromatographic determination of ethylene in large air volumes at the fractional parts-per-billion level. Anal. Chem. *48*, 38–41 (1976)
Dekhuijen, H.M., Gevers, E.Ch.Th.: The recovery of cytokinins during extraction and purification of club root tissue. Physiol. Plant. *35*, 297–302 (1975)
DeYoe, D.R., Zaerr, J.B.: Indole-3-acetic acid in Douglas-fir. Analysis by gas-liquid chromatography and mass spectrometry. Plant Physiol. *58*, 299–303 (1976)
Dougherty, T.M., Schepartz, A.I.: Desalting of nucleic acid hydrolysates, nucleosides and bases by chromatography on poly-N-vinylpyrrolidone. J. Chromatogr. *40*, 299–302 (1969a)
Dougherty, T.M., Schepartz, A.I.: Separation of phenylalanine, tyrosine and tryptophan by chromatography on poly-N-vinylpyrrolidone. J. Chromatogr. *42*, 415–416 (1969b)
Dougherty, T.M., Schepartz, A.I.: Poly-N-vinylpyrrolidone chromatography. Effect of pH on elution of bases and a proposed mechanism for the adsorption of purines. J. Chromatogr. *43*, 397–399 (1969c)
Durley, R.C., Pharis, R.P.: Partition coefficients of 27 gibberellins. Phytochemistry *11*, 317–326 (1972)
Durley, R.C., MacMillan, J., Pryce, R.J.: Investigation of gibberellins and other growth substances in the seed of *Phaseolus multiflorus* and *Phaseolus vulgaris* by gas chromatography and by gas chromatography-mass spectrometry. Phytochemistry *10*, 1891–1908 (1971)
Durley, R.C., Crozier, A., Pharis, R.P. McLaughlin, G.E.: Chromatography of 33 gibberellins on a gradient eluted silica gel partition column. Phytochemistry *11*, 3029–3033 (1972)
Durley, R.C., Railton, I.D., Pharis, R.P.: Interconversion of gibberellin A_5 to gibberellin A_3 in seedlings of dwarf *Pisum sativum*. Phytochemistry *12*, 1609–1612 (1973)
Dyson, W.H., Hall, R.H.: N^6-(Δ^2-isopentenyl)adenosine: its occurrence as a free nucleoside in an autonomous strain of tobacco tissue. Plant Physiol. *50*, 616–621 (1972)
Ehmann, A.: Identification of 2-O-(indole-3-acetyl)-D-glucopyranose, 4-O-(indole-3-acetyl)-D-glucopyranose and 6-O-(indole-3-acetyl)-D-glucopyranose from kernels of *Zea mays* by gas-liquid chromatography-mass spectrometry. Carbohyd. Res. *34*, 99–114 (1974)
Ehmann, A.: The van Urk-Salkowski reagent – a sensitive and specific chromogenic reagent for silca gel thin-layer chromatographic detection and identification of indole derivatives. J. Chromatogr. *132*, 267–276 (1977)
Ehmann, A., Bandurski, R.S.: Purification of indole-3-acetic acid myoinositol esters on polystyrene-divinylbenzene resins. J. Chromatogr. *72*, 61–70 (1972)
Ehmann, A., Bandurski, R.S.: The isolation of di-O-(indole-3-acetyl)-*myo*-inositol and tri-O-(indole-3-acetyl)-*myo*-inositol from mature kernels of *Zea mays*. Carbohydr. Res. *36*, 1–12 (1974)

Einset, T.W., Swaminathan, S., Skoog, F.: Ribosyl-cis-zeatin in a leucyl transfer RNA species from peas. Plant. Physiol. *58*, 140–142 (1976)
Elliott, M.C., Stowe, B.B.: A novel sulphonated natural indole. Phytochemistry *9*, 1629–1632 (1970)
Elliott, M.C., Stowe, B.B.: Distribution and variation of indole glucosinolates in woad (*Isatis tinctoria* L.). Plant Physiol. *48*, 498–503 (1971)
Elson, G.W., Jones, D.F., MacMillan, J., Suter, P.J.: Plant hormones IV. Identification of the gibberellins of *Echinocystis macrocarpa* Greene by thin layer chromatography. Phytochemistry *3*, 93–101 (1964)
Falkner, F.C., Sweetman, B.J., Watson, J.T.: Biomedical applications of selected ion monitoring. Appl. Spectrosc. Rev. *10(1)*, 51–116 (1975)
Firn, R.D., Burden, R.S., Taylor, H.F.: The detection and estimation of the growth inhibitor xanthoxin in plants. Planta *102*, 115–126 (1972)
Frydman, V.M., MacMillan, J.: The metabolism of gibberellins A_9, A_{20} and A_{29} in immature seeds of *Pisum sativum* cv. Progress No. 9. Planta *125*, 181–195 (1975)
Frydman, V.M., Gaskin, P., MacMillan, J.: Qualitative and quantitative analyses of gibberellins throughout seed maturation in *Pisum sativum* cv. Progress No. 9. Planta *118*, 123–132 (1974)
Fukui, H.N., DeVries, J.E., Wittwer, S.H., Sell, H.M.: Ethyl-3-indole-acetate: an artifact in extracts of immature corn kernels. Nature (London) *180*, 1205 (1957)
Fukui, H., Koshimizu, K., Mitsui, T.: Gibberellin A_{28} in the fruits of *Lupinus luteus*. Phytochemistry *10*, 671–673 (1971)
Fukui, H., Ishii, H., Koshimizu, K., Katsumi, M., Ogawa, Y., Mitsui, T.: The structure of gibberellin A_{23} and the biological properties of 3,13-dihydroxy C_{20}-gibberellins. Agric. Biol. Chem. *36*, 1003–1012 (1972)
Fukui, H., Koshimizu, K., Usuda, S., Yamazaki, Y.: Isolation of plant growth regulators from seeds of *Cucurbita pepo* L. Agric. Biol. Chem. *41*, 175–180 (1977)
Galliard, T., Grey, T.C.: A rapid method for the determination of ethylene in the presence of other volatile natural products. J. Chromatogr. *41*, 442–445 (1969)
Gaskin, P., MacMillan, J.: Identification and estimation of abscisic acid in a crude plant extract by combined gas chromatography-mass spectrometry. Phytochemistry *7*, 1699–1701 (1968)
Gaskin, P., MacMillan, J.: GC and GC-MS techniques for gibberellins. In: Isolation of plant growth substances. Hillman, J.R. (ed.), pp. 79–95. London, New York, Melbourne: Cambridge University Press, 1978
Ginzburg, B.: Nature of the olefins produced by apples. Nature (London) *184*, 1072–1073 (1959)
Glenn, J.L., Kuo, C.C., Durley, R.C., Pharis, R.P.: Use of insoluble polyvinylpyrrolidone for purification of plant extracts and chromatography of plant hormones. Phytochemistry *11*, 345–351 (1972)
Gmelin, R.: Occurrence, isolation and properties of glucobrassicin and neoglucobrassicin. Colloq. Int. C.N.R.S. *123*, 159–168 (1964)
Goldschmidt, E.E., Monselise, S.P.: Native growth inhibitors from citrus shoot. Partition, bioassay, and characterization. Plant Physiol. *43*, 113–116 (1968)
Goldschmidt, E.E., Monselise, S.P., Goren, R.: On the identification of native auxins in citrus tissue. Can. J. Bot. *49*, 241–245 (1971)
Goldschmidt, E.E., Goren, R., Even-Chen, Z., Bittner, S.: Increase in free and bound abscisic acid during natural and ethylene-induced senescence of citrus fruit peel. Plant. Physiol. *51*, 879–882 (1973)
Gordon, A.E., Frigerio, A.: Mass fragmentography as an application of gas-liquid chromatography-mass spectrometry in biological research. J. Chromatogr. *73*, 401–417 (1972)
Grove, J.F.: Gibberellin A_2. J. Chem. Soc. 3545 (1961)
Grunwald, C., Lockard, R.G.: Analysis of indole acid derivatives by gas chromatography using liquid phase OV-101. J. Chromatogr. *52*, 491–493 (1970)
Haagen-Smit, A.J., Leech, W.D., Bergren, W.R.: The estimation, isolation and identification of auxins in plant materials. Am. J. Bot. *29*, 500–506 (1942)

Haagen-Smit, A.J., Dandliker, W.B., Wittwer, S.H., Murneek, A.E.: Isolation of 3-indoleacetic acid from immature corn kernels. Am. J. Bot. *33*, 118–120 (1946)

Hahn, H.: High-performance liquid chromatography and its use in cytokinin determination in *Agrobacterium tumefaciens* B6. Plant Cell Physiol. *17*, 1053–1058 (1976)

Hall, R.H.: Separation of nucleic acid degradation products on partition columns. J. Biol. Chem. *237*, 2283–2288 (1962)

Hall, R.H.: A general procedure for the isolation of minor nucleosides from ribonucleic acid hydrolysates. Biochemistry *4*, 661–670 (1965)

Hall, R.H., Csonka, L., David, H., Mclenman, B.: Cytokinins in the soluble RNA of plant tissues. Science *156*, 69–71 (1967)

Hamilton, R.H., Bandurksi, R.S., Grigsby, B.H.: Isolation of indole-3-acetic acid from corn kernels and etiolated corn seedlings. Plant Physiol. *36*, 354–359 (1961)

Hanson, J.R.: New metabolites of *Gibberella fujikuroi*. Tetrahedron *22*, 701–703 (1966)

Harada, H., Yokota, T.: Isolation of gibberellin A_8-glucoside from shoot apices of *Althaea rosea*. Planta *92*, 100–104 (1970)

Hashimoto, T., Ikai, T., Tamura, S.: Isolation of (+)-abscisin II from dormant aerial tubers of *Dioscorea batatas*. Planta *78*, 89–92 (1968)

Hashizume, T., Sugiyama, T., Imura, M., Cory, H.J., Scott, M.F., McCloskey, J.A.: Determination of cytokinins by mass spectrometry based on stable isotope dilution. Anal. Biochem. *92*, 111–112 (1979)

Hattori, H., Marumo, S.: Monomethyl-4-chloroindolyl-3-acetyl-L-aspartate and absence of indolyl-3-acetic acid in immature seeds of *Pisum sativum*. Planta *102*, 85–90 (1972)

Heftmann, E., Saunders, G.A., Haddon, W.F.: Argentation high pressure liquid chromatography and mass spectrometry of gibberellin esters. J. Chromatogr. *156*, 71–77 (1978)

Hemberg, T., Westlin, P.E.: The quantitative yield in purification of cytokinins. Model experiments with kinetin, 6-furfurylaminopurine. Physiol. Plant. *28*, 228–231 (1973)

Henbest, H.B., Jones, E.R.H., Smith, G.F.: Isolation of a new plant growth hormone, 3-indolylacetonitrile. J. Chem. Soc. 3796–3801 (1953)

Hewett, E.W., Wareing, P.F.: Cytokinins in *Populus robusta* Schneid: a complex in leaves. Planta *112*, 225–233 (1973)

Hillman, J.R. (ed.): Isolation of plant growth substances. London-New York-Melbourne: Cambridge University Press 1978

Hillman, J.R., Young, I., Knights, B.A.: Abscisic acid in leaves of *Hedera helix* L. Planta *119*, 263–266 (1974)

Hiraga, K., Yokota, T., Murofushi, N., Takahashi, N.: Isolation and characterization of gibberellins in mature seeds of *Phaseolus vulgaris*. Agric. Biol. Chem. *38*, 2511–2520 (1974a)

Hiraga, K., Kawabe, S., Yokota, T., Murofushi, N., Takahashi, N.: Isolation and characterization of plant growth substances in immature seeds and etiolated seedlings of *Phaseolus vulgaris*. Agric. Biol. Chem. *38*, 2521–2527 (1974b)

Hiraga, K., Yamane, H., Takahashi, N.: Biological activity of some synthetic gibberellin glucosyl esters. Phytochemistry *13*, 2371–2376 (1974c)

Hoffinger, M., Böttger, M.: Identification by GC-MS of 4-chloroindole-3-acetic acid and its methyl ester in immature seed of *Vicia faba* seeds. Phytochemistry *18*, 653–654 (1979)

Horgan, R.: Analytical procedures for cytokinins. In: Isolation of plant growth substances. Hillman, J.R. (ed.), pp. 97–114. London-New York-Melbourne: Cambridge University Press 1978

Horgan, R., Hewett, E.W., Purse, J.G., Wareing, P.F.: A new cytokinin from *Populus robusta*. Tetrahedron Lett. *30*, 2827–2828 (1973a)

Horgan, R., Hewett, E.W., Purse, J.G., Horgan, J.M., Wareing, P.F.: Identification of a cytokinin in sycamore sap by gas chromatography-mass spectrometry. Plant Sci. Lett. *1*, 321–324 (1973b)

Hutzinger, O., Heacock, R.A., MacNeil, J.D., Frei, R.W.: Indoles and auxins XIII. Identification and analysis of naturally occurring indoles via electron donor-acceptor complexes. J. Chromatogr. *68*, 173–182 (1972)

Hutzinger, O., Jamieson, W.D.: Indoles and auxins IX. Mass spectrometric identification and isolation of indoles as polynitrofluorenone complexes. Anal. Biochem. *35*, 351–358 (1970)

Igoshi, M., Yamaguchi, I., Takahashi, N., Hirose, K.: Plant growth substances in the young fruit of *Citrus unshiu*. Agric. Biol. Chem. *35*, 629–631 (1971)

Ikekawa, N., Kagawa, T., Sumiki, Y.: Biochemistry of bakanae fungus LXVII. Determination of nine gibberellins by gas and thin layer chromatographies. Proc. Jpn. Acad. *39*, 507–512 (1963)

Isogai, Y., Okamoto, T., Koizumi, T.: Studies on plant growth regulators. I. Isolation of indole-3-acetamide, 2-phenylacetamide, and indole-3-carboxaldehyde from etiolated seedlings of *Phaseolus*. Chem. Pharm. Bull. *15*, 151–158 (1967a)

Isogai, Y., Okamoto, T., Komoda, Y.: Isolation of a plant growth inhibitory substances from garden pea (*Pisum sativum* L.) and its identification with (+)-abscisin-II. Chem. Pharm. Bull. *15*, 1256–1257 (1967b)

Jamieson, W.D., Hutzinger, O.: Identification of simple naturally occurring indoles by mass spectrometry. Phytochemistry *9*, 2029–2036 (1970)

Jones, D.F.: Examination of the gibberellins of *Zea mays* and *Phaseolus multiflorus* using thin-layer chromatography. Nature (London) *202*, 1309–1310 (1964)

Jones, E.R.H., Taylor, W.C.: Some indole constituents of cabbage. Nature (London) *179*, 1138 (1957)

Jones, K.C.: A method for the rapid chromatography of gibberellins. J. Chromatogr. *52*, 512–516 (1970)

Jones, M.G., Metzger, J.D., Zeevaart, J.A.D.: Fractionation of gibberellins in plant extracts by reverse phase high performance liquid chromatography. Plant Physiol. *65*, 218–221 (1980)

Kagawa, T., Fukinbara, T., Sumiki, Y.: Thin layer chromatography of gibberellins. Agric. Biol. Chem. *27*, 598–599 (1963)

Kefeli, V.I., Kadyrov, C.Sh.: Natural growth inhibitors, their chemical and physiological properties. Annu. Rev. Plant Physiol. *22*, 185–196 (1971)

Keglević, D., Pokorny, M.: The chemical synthesis of 1-O-(indol-3′-ylacetyl)-β-D-glucopyranose. Biochem. J. *114*, 827–832 (1969)

Klämbt, H.D.: Wachstumsinduktion und Wuchsstoffmetabolismus im Weizenkoleoptilzylinder II. Mitteilung Stoffwechselprodukte der Indol-3-essigsäure und der Benzoesäure. Planta *55*, 618–631 (1961)

Komoto, N., Ikegami, S., Tamura, S.: Isolation of acidic growth inhibitors in dwarf peas. Agric. Biol. Chem. *36*, 2547–2553 (1972)

Kopcewicz, J., Ehmann, A., Bandurksi, R.S.: Enzymatic esterification of indole-3-acetic acid to *myo*-inositol and glucose. Plant Physiol. *54*, 846–851 (1974)

Koshimizu, K., Fukui, H., Kusaki, T., Mitsui, T., Ogawa, Y.: Identity of lupin inhibitor with abscisin II and its biological activity on growth of rice seedlings. Agric. Biol. Chem. *30*, 941–943 (1966)

Koshimizu, K., Matsubara, S., Kusaki, T., Mitsui, T.: Isolation of a new cytokinin from immature yellow lupin seeds. Agric. Biol. Chem *31*, 795–801 (1967)

Koshimizu, K., Inui, M., Fukui, H., Mitsui, T.: Isolation of (+)-abscisyl-β-D-glucopyranoside from immature fruits of *Lupinus luteus*. Agric. Biol. Chem. *32*, 789–791 (1968a)

Koshimizu, K., Fukui, H., Kusaki, T., Ogawa, Y., Mitsui, T.: Isolation and structure of gibberellin A_{18} from immature seeds of *Lupinus luteus*. Agric. Biol. Chem. *32*, 1135–1140 (1968b)

Kurogochi, S., Murofushi, N., Ota, Y., Takahashi, N.: Gibberellins and inhibitors in the rice plant. Agric. Biol. Chem. *42*, 207–208 (1978)

Kurogochi, S., Murofushi, N., Ota, Y., Takahashi, N.: Identification of gibberellins in the rice plant and quantiative changes of gibberellin A_{19} throughout its life cycle. Planta *146*, 185–191 (1979)

Labarca, C., Nicholls, P.B., Bandurski, R.S.: A partial characterization of indole acetylinositols from *Zea mays*. Biochem. Biophys. Res. Commun. *20*, 641–646 (1965)

Lammi, C.J., Lerner, J.: The influence of eluent salt composition on the elutions of riboflavin and adenine in poly-N-vinylpyrrolidone column chromatography. J. Chromatogr. *43*, 395–396 (1969)

Langer, P., Michajlowskij, N.: Präformiertes Rhodanid in Nahrungsmitteln als Hauptursache der Rhodanidausscheidung im Harn bei Tier und Mensch. Hoppe-Seylers Z. Physiol. Chem. *312*, 31–36 (1958)
Lenton, J.R., Bowen, M.R., Saunders, P.F.: Detection of abscisic acid in the xylem sap of willow (*Salix viminalis*) by gas-liquid chromatography. Nature (London) *220*, 86–87 (1968)
Lenton, J.R., Perry, V.M., Saunders, P.F.: The identification and quantitative analysis of abscisic acid in plant extracts by gas-liquid chromatography. Planta *96*, 271–280 (1971)
Leonard, N.J., Fujii, T.: The synthesis of compounds possessing kinetin activity. The use of a blocking group at the 9-position of adenine for the synthesis of 1-substituted adenines. Proc. Natl. Acad. Sci. USA *51*, 73–75 (1964)
Leonard, N.J., Playtis, A.J., Skoog, F., Schmitz, R.: A stereoselective synthesis of cis-zeatin. J. Am. Chem. Soc. *93*, 3056–3058 (1971)
Lerner, J., Dougherty, T.M., Schepartz, A.I.: Selectivity properties of poly-N-vinylpyrrolidone in column chromatography of nucleotides, their derivatives, and related compounds: A preliminary report. J. Chromatogr. *37*, 453–457 (1968)
Letham, D.S.: Zeatin, a factor inducing cell division isolated from *Zea mays*. Life Sci. *2*, 569–573 (1963)
Letham, D.S.: Cytokinins from *Zea mays*. Phytochemistry *12*, 2445–2455 (1973)
Letham, D.S.: Regulators of cell division in plant tissues XXI. Distribution coefficients for cytokinins. Planta *118*, 361–364 (1974)
Letham, D.S., Shannon, J.S., McDonald, I.R.C.: Regulators of cell division in plant tissues. III. The identity of zeatin. Tetrahedron *23*, 479–486 (1967)
Little, C.H.A., Strunz, G.M., France, R.La, Bonga, J.M.: Identification of abscisic acid in *Abies balsamea*. Phytochemistry *11*, 3535–3536 (1972)
Lorenzi, R., Horgan, R., Heald, J.R.: Gibberellin A_9 glucosyl ester in needles of *Picea sitchensis*. Phytochemistry *15*, 789–790 (1976)
Lyons, J.M., McGlasson, W.B., Pratt, H.K.: Ethylene production, respiration, and internal gas concentrations in cantaloupe fruits at various stages of maturity. Plant Physiol. *37*, 31–36 (1962)
MacMillan, J.: Oral presentation in 26th. Int. Congr. Pure Appl. Chem. (Session I). Tokyo 1977
MacMillan, J., Suter, P.J.: Thin layer chromatography of gibberellins. Nature (London) *197*, 790 (1963)
MacMillan, J., Wels, C.M.: Partition chromatography of gibberellins and related diterpenes on columns of Sephadex LH-20. J. Chromatogr. *87*, 271–276 (1973)
MacMillan, J., Wels, C.M.: Detailed analysis of metabolites from mevalonic lactone in *Gibberella fujikuroi*. Phytochemistry *13*, 1413–1417 (1974)
MacMillan, J., Seaton, J.C., Suter, P.J.: Plant hormone. I. Isolation of gibberellin A_1 and gibberellin A_5 from *Phaseolus multiflorus*. Tetrahedron *11*, 60–66 (1960)
MacMillan, J., Seaton, J.C., Suter, P.J.: Isolation and structures of gibberellins from higher plants. In: "Gibberellins" advances in chemistry series, No. 28. Gould, R.G. (ed.), pp. 18–25. Washington: American chemical society 1961
MacMillan, J., Seaton, J.C., Suter, P.J.: Plant hormone. II. Isolation and structures of gibberellin A_6 and gibberellin A_8. Tetrahedron *18*, 349–355 (1962)
MacMillan, J., Pryce, R.J., Eglinton, G., McCormick, A.: Identification of gibberellins in crude plant extracts by combined gas chromatography-mass spectrometry. Tetrahedron Lett. 2241–2243 (1967)
Mann, J.D., Jaworski, E.G.: Minimizing loss of indole acetic acid during purification of plant extracts. Planta *92*, 285–291 (1970)
Martin, D.M.G., Reese, C.B.: Some aspects of the chemistry of *N*(1)- and *N*(6)-dimethylallyl derivatives of adenosine and adenine. J. Chem. Soc. (C) 1731–1738 (1968)
Martin, G.C., Dennis, F.G.Jr., MacMillan, J., Gaskin, P.: Hormones in pear seeds. I. Levels of gibberellins, abscisic acid, phaseic acid, dihydrophaseic acid and two metabolites of dihydrophaseic acid in immature seeds of *Pyrus communis* L. J. Am. Soc. Hortic. Sci. *102*, 16–19 (1977a)

Martin, G.C., Dennis, F.G.Jr., Gaskin, P., MacMillan, J.: Identification of gibberellins A_{17}, A_{25}, A_{45}, abscisic acid, phaseic acid, and dihydrophaseic acid in seeds of *Pyrus communis*. Phytochemistry *16*, 605–607 (1977b)

Marumo, S., Hattori, H.: Isolation of D-4-chlorotryptophan derivatives as auxin-related metabolites from immature seeds of *Pisum sativum*. Planta *90*, 208–211 (1970)

Marumo. S., Abe, H., Hattori, H., Munakata, K.: Isolation of a novel auxin, methyl 4-chloroindoleacetate from immature seeds of *Pisum sativum*. Agric. Biol. Chem. *32*, 117–118 (1968a)

Marumo, S., Hattori, H., Abe, H., Munakata, K.: Isolation of 4-chloroindolyl-3-acetic acid from immature seeds of *Pisum sativum*. Nature (London) *219*, 959–960 (1968b)

Milborrow, B.V.: The identification of (+)-abscisin II [(+)-dormin] in plants and measurement of its concentrations. Planta *76*, 93–113 (1967)

Milborrow, B.V.: The metabolism of abscisic acid. J. Exp. Bot. *21*, 17–29 (1970)

Milborrow, B.V.: Stereochemical aspects of the formation of double bonds in abscisic acid. Biochem. J. *128*, 1135–1136 (1972)

Milborrow, B.V.: The chemistry and physiology of abscisic acid. Annu. Rev. Plant. Physiol. *25*, 259–307 (1974)

Milborrow, B.V., Mallaby, R.: Occurrence of methyl (+)-abscisate as an artefact of extraction. J. Exp. Bot. *26*, 741–748 (1975)

Milborrow, B.V., Noddle, R.C.: Conversion of 5-(1,2-epoxy-2,6,6,-trimethylcyclohexyl)-3-methylpenta-cis-2-trans-4-dienoic acid in plants. Biochem. J. *119*, 727–734 (1970)

Miller, C.O.: Evidence for the natural occurrence of zeatin and derivatives: compounds from maize which promote cell division. Proc. Natl. Acad. Sci. USA *54*, 1054–1058 (1965)

Miller, C.O.: Ribosyl-*trans*-zeatin, a major cytokinin produced by crown gall tumor tissue. Proc. Natl. Acad. Sci. USA *71*, 334–338 (1974)

Miller, C.O.: Revised methods for purification of ribosyl-*trans*-zeatin from *Vinca rosea* L. crown gall tumor tissue. Plant Physiol. *55*, 448–449 (1975a)

Miller, C.O.: Cell-division factors from *Vinca rosea* L. crown gall tumor tissue. Proc. Natl. Acad. Sci. USA *72*, 1883–1886 (1975b)

Mitsuhashi, M., Shibaoka, H.: Isolation of an inhibitor of growth and root formation from *Portulaca grandiflora* leaves. Plant Cell Physiol. *6*, 87–99 (1965)

Morris, R.O.: Mass spectroscopic identification of cytokinins. Glucosylzeatin and glucosylribosylzeatin from *Vinca rosea* crown gall. Plant Physiol. *59*, 1029–1033 (1977)

Most, B.H., Williams, J.C., Parker, K.J.: Gas chromatography of cytokinins. J. Chromatogr. *38*, 136–138 (1968)

Murofushi, N., Iriuchijima, S., Takahashi, N., Tamura, S., Kato, J., Wada, Y., Watanabe, E., Aoyama, T.: Isolation and structure of a novel C_{20} gibberellin in bamboo shoots. Agric. Biol. Chem. *30*, 917–924 (1966)

Murofushi, N., Takahashi, N., Yokota, T., Tamura, S.: Gibberellins in immature seeds of *Pharbitis nil*. Part I. Isolation and structure of a novel gibberellin, gibberellin A_{20}. Agric. Biol. Chem. *32*, 1239–1245 (1968)

Murofushi, N., Takahashi, N., Yokota, T., Kato, J., Shiotani, Y., Tamura, S.: Gibberellins in immature seeds of *Canavalia*. Part I. Isolation and biological activity of gibberellins A_{21} and A_{22}. Agric. Biol. Chem. *33*, 592–597 (1969)

Murofushi, N., Yokota, T., Watanabe, A., Takahashi, N.: Isolation and characterization of gibberellins in *Calonyction aculeatum* and structures of gibberellins A_{30}, A_{31}, A_{33} and A_{34}. Agric. Biol. Chem. *37*, 1101–1113 (1973)

Nakagawa, S., Matsui, H., Yuda, E., Murofushi, N., Takahashi, N., Akimori, N., Hishida, S.: Biologically active gibberellins in immature seed of *Pyrus serotina*. Phytochemistry *18*, 1695–1697 (1979)

Nicholls, P.B.: The isolation of indole-3-acetyl-2-O-*myo*-inositol from *Zea mays*. Planta *72*, 258–264 (1967)

Niederwieser, A., Giliberti, P.: Simple extraction of indole derivatives from aqueous solution by adsorption on neutral polystyrene resin. J. Chromatogr. *61*, 95–99 (1971)

Nitsch, J.P.: Natural cytokinins. In: S.C.I. Monograph No. 31, pp. 111–123. Society of Chemical Industry 1968

Nomoto, M., Tamura, S.: Isolation and identification of indole derivatives in clubroots of chinese cabbage. Agric. Biol. Chem. *33*, 1590–1592 (1970)

Ohkuma, K., Lyon, J.L., Addicott, F.T., Smith, O.E.: Abscisin II, an abscission-accelerating substance from young cotton fruit. Science *142*, 1592–1593 (1963)

Okamoto, T., Isogai, Y., Koizumi, T.: Studies on plant growth regulators. II. Isolation of indole-3-acetic acid, phenylacetic acid, and several plant growth inhibitors from etiolated seedlings of *Phaseolus*. Chem. Pharm. Bull. *15*, 159–163 (1967a)

Okamoto, T., Isogai, Y., Koizumi, T., Fujishiro, H., Sato, Y.: Studies on plant growth regulators. III. Isolation of indole-3-acetonitrile and methyl indole-3-acetate from the neutral fraction of the Moyashi extract. Chem. Pharm. Bull. *15*, 163–168 (1967b)

Parker, C.W., Letham, D.S., Cowley, D.E., MacLeod, J.K.: Raphanatin, an unusual purine derivative and a metabolite of zeatin. Biochem. Biophys. Res. Commun. *49*, 460–466 (1972)

Parker, C.W., Wilson, M.M., Letham, D.S., Cowley, D.E., MacLeod, J.K.: The glucosylation of cytokinins. Biochem. Biophys. Res. Commun. *55*, 1370–1376 (1973)

Peterson, J.B., Miller, C.O.: Cytokinins in *Vinca rosea* L. crown gall tumor tissue as influenced by compounds containing reduced nitrogen. Plant Physiol. *57*, 393–399 (1976)

Peterson, J.B., Miller, C.O.: Glucosyl zeatin and glucosyl ribosylzeatin from *Vinca rosea* L. crown gall tumor tissue. Plant Physiol. *59*, 1026–1028 (1977)

Pierce, A.E.: "Silylation of organic compounds". Rockford, Illinois: Pierce Chemical Co. 1968

Piskornik, Z., Bandurksi, R.S.: Purification and partial characterization of a glucan containing indole-3-acetic acid. Plant Physiol. *50*, 176–182 (1972)

Pitel, D.W., Vining, L.C., Arsenault, G.P.: Improved methods for preparing pure gibberellins from cultures of *Gibberella fujikuroi*. Isolation by adsorption or partition chromatography on silicic acid and by partition chromatography on Sephadex columns. Can. J. Biochem. *49*, 185–193 (1971)

Playtis, A.J., Leonard, N.J.: The synthesis of ribosyl-*cis*-zeatin and thin layer chromatographic separation of the *cis* and *trans* isomers of ribosylzeatin. Biochem. Biophys. Res. Commun. *45*, 1–5 (1971)

Pool, R.M., Powell, L.E.: Cytokinin analysis by high pressure liquid chromatography. In: Plant growth substances 1973. Proc. 8th. Int. Conf. Plant Growth Subst., pp. 93–98. Tokyo: Hirokawa Publishing Co. Inc. 1974

Powell, L.E.: Separation of plant growth regulating substances on silica gel columns. Plant Physiol. *35*, 256–261 (1960)

Powell, L.E.: Preparation of indole extracts from plants for gas chromatography and spectrophotofluorometry. Plant Physiol. *39*, 836–842 (1964)

Powell, L.E., Tautvydas, K.J.: Chromatography of gibberellins on silica gel partition columns. Nature (London) *213*, 292–293 (1967)

Procházka, Ž., Šanda, V.: On the bound form of ascorbic acid. XII. Isolation of pure ascorbigen and some other indole derivatives from savoy cabbage. Collect. Czech. Chem. Commun. *25*, 270–280 (1960)

Pryce, R.J., MacMillan, J., McCormick, A.: The identification of bamboo gibberellin in *Phaseolus multiflorus* by combined gas chromatography-mass spectrometry. Tetrahedron Lett. 5009–5011 (1967)

Railton, I.D.: Purification of plant auxin by polyamide thin-layer chromatography. J. Chromatogr. *70*, 202–205 (1970)

Raj, R.K., Hutzinger, O.: Indoles and auxins VI. Separation of naturally occurring indoles into acidic, basic, amphoteric, and neutral fractions by ion-exchange chromatography. Anal. Biochem. *33*, 43–46 (1970a)

Raj, R.K., Hutzinger, O.: Indoles and auxins VIII. Partition chromatography of naturally occurring indoles on cellulose thin layers and Sephadex columns. Anal. Biochem. *33*, 471–474 (1970b)

Rayle, D.L., Purves, W.K.: Isolation and identification of indole-3-ethanol (tryptophol) from cucumber seedlings. Plant Physiol. *42*, 520–524 (1967)

Redemann, C.T., Wittwer, S.H., Sell, H.M.: The fruit-setting factor from the ethanol extracts of immature corn kernels. Arch. Biochem. Biophys. *32*, 80–84 (1951)

Reeve, D.R., Crozier, A.: Purification of plant hormone extracts by gel permeation chromatography. Phytochemistry *15*, 791–793 (1976)
Reeve, D.R., Crozier, A.: Radioactivity monitor for high-performance liquid chromatography. J. Chromatogr. *137*, 271–282 (1977)
Reeve, D.R., Crozier, A.: The analysis of gibberellins by high performance liquid chromatography. In: Isolation of plant growth substances. Hillmann, J.R. (ed.), pp. 41–77. London-New York-Melbourne: Cambridge University Press (1978)
Reeve, D.R., Yokota, T., Nash, L.J., Crozier, A.: The development of a high performance liquid chromatograph with a sensitive on-stream radioactivity monitor for the analysis of ^{3}H- and ^{14}C-labelled gibberellins. J. Exp. Bot. *27*, 1243–1258 (1976)
Rivier, L., Pilet, P.-E.: Indolyl-3-acetic acid in cap and apex of maize roots: identification and quantification by mass fragmentography. Planta *120*, 107–112 (1974)
Robins, M.J., Hall, H.S., Thedford, R.: N^6- (Δ^2-Isopentenyl)adenosine. A component of the transfer ribonucleic acid of yeast and of mammalian tissue, methods of isolation, and characterization. Biochemistry *6*, 1837–1848 (1967)
Rudnicki, R., Pieniazek, J.: Free and bound abscisic acid in developing and ripe strawberries. Bull. Acad. Pol. Sci. *19*, 421–423 (1971)
Saunders, P.F.: The identification and quantitative analysis of abscisic acid in plant extracts. In: Isolation of plant growth substances. Hillman, J.R. (ed.), pp. 115–134. London-New York-Melbourne: Cambridge University Press 1978
Schlüter, M., Gmelin, R.: Abnormale enzymatische Spaltung von 4-Methylthiobutylglucosinolat in Frischpflanzen von *Eruca sativa*. Phytochemistry *11*, 3427–3431 (1972)
Schneider, E.A., Wightman, F.: Metabolism of auxin in higher plants. Annu. Rev. Plant Physiol. *25*, 487–513 (1974)
Schneider, G., Jänicke, S., Sembdner, G.: Gibberelline XXXIV. Mitt. Beitrag zur Gaschromatographie von Gibberellinen und Gibberellin-O-Glucosiden-N,O-bis(trimethylsilyl)-acetamid als Silylierungsreagens. J. Chromatogr. *109*, 409–412 (1975)
Scott, T.K.: Auxins and roots. Annu. Rev. Plant Physiol. *23*, 235–258 (1972)
Seeley, S.D., Powell, L.E.: The employment of electron capture detectors in the gas chromatography of plant hormones – an ultra sensitive assay. HortScience *5*, 340 (1970a)
Seeley, S.D., Powell, L.E.: Electron capture-gas chromatography for sensitive assay of abscisic acid. Anal. Biochem. *35*, 530–533 (1970b)
Sequeira, L.: Hormone metabolism in diseased plants. Annu. Rev. Plant Physiol. *24*, 353–380 (1973)
Sembdner, G., Gross, G., Schrieber, K.: Die Dünnschichtchromatographie von Gibberellinen. Experientia *18*, 584 (1962)
Sheldrake, A.R.: The production of hormones in higher plants. Biol. Rev. *48*, 509–559 (1973)
Shimokawa, K., Kasai, Z.: Biogenesis of ethylene in apple tissue. I. Formation of ethylene from glucose, acetate, pyruvate, and acetaldehyde in apple tissue. Plant Cell Physiol. *7*, 1–9 (1966)
Sindy, W.W., Smith, O.E.: Identification of plant hormones from cotton ovules. Plant Physiol. *55*, 550–554 (1975)
Sponsel, V.M., MacMillan, J.: Metabolism of gibberellin A_{29} in seeds of *Pisum sativum* cv. Progress No. 9: use of [^{2}H] and [^{3}H] GAs and the identification of a new GA catabolite. Planta *144*, 69–78 (1978)
Srivastava, B.I.S.: Ether-soluble and ether-insoluble auxins from immature corn kernels. Plant Physiol. *38*, 473–478 (1963)
Steen, I., Eliasson, L.: Separation of growth regulators from *Picea abies* Karst. on Sephadex LH-20. J. Chromatogr. *43*, 558–560 (1969)
Stodola, F.H., Nelson, G.E.N., Spence, D.J.: The separation of gibberellin A and gibberellic acid on buffered partition columns. Arch. Biochem. Biophys. *66*, 438–443 (1957)
Stowe, B.B., Thimann, K.V.: The paper chromatography of indole compounds and some indole-containing auxins of plant tissue. Arch. Biochem. Biophys. *51*, 499–516 (1954)
Summons, R.E., MacLeod, J.K., Parker, C.W., Letham, D.S.: The occurrence of raphanatin as an endogenous cytokinin in radish seed. Identification and quantitation by gas chro-

matographic-mass spectrometric analysis using deuterium labelled standards. FEBS Lett. *82*, 211–214 (1977)
Swartz, H.J., Powell, L.E.: Determination of indole acetic acid from plant samples by an alkali flame ionization detector. Physiol. Plant. *47*, 25–28 (1979)
Sweester, P.B., Vatvars, A.: High-performance liquid chromatographic analysis of abscisic acid in plant extracts. Anal. Biochem. *71*, 68–78 (1976)
Takahashi, N., Kitamura, H., Kawarada, A., Seta, Y., Takai, M., Tamura, S., Sumiki, Y.: Biochemical studies on "Bakanae" fungus Part XXXIV Isolation of gibberellins and their properties. Bull. Agric. Chem. Soc. Jpn. *19*, 267–277 (1955)
Takahashi, N., Seta, Y., Kitamura, H., Sumiki, Y.: Biochemical studies on "Bakanae" fungus. Part 48. A new gibberellin, gibberellin A_4. Bull. Agric. Chem. Soc. Jpn. *23*, 405–407 (1959)
Takahashi, N., Murofushi, N., Tamura, S., Wasada, N., Hoshino, H., Tsuchiya, T., Sasaki, S., Aoyama, T., Watanabe, E.: Mass spectrometric studies on gibberellins. Org. Mass Spectrom. *2*, 711–722 (1969)
Takahashi, N., Marumo, S., Otake, N.: The chemistry of biologically active natural products, pp. 63–71. Tokyo University Press 1973 (in Japanese)
Takahashi, N., Yamaguchi, I., Kōno, T., Igoshi, M., Hirose, K., Suzuki, K.: Characterization of plant growth substances in *Citrus unshiu* and their change in fruit development. Plant Cell Physiol. *16*, 1101–1111 (1975)
Taylor, H.F., Burden, R.S.: Xanthoxin, a new naturally occurring plant growth inhibitor. Nature (London) *227*, 302–303 (1970)
Tegley, J.R., Witham, F.H., Krasnuk, M.: Chromatographic analysis of a cytokinin from tissue cultures of crown-gall. Plant Physiol. *47*, 581–585 (1971)
Thimann, K.V.: On the plant growth hormone produced by *Rhizopus suinus*. J. Biol. Chem. *109*, 279–291 (1935)
Thomas, T.H., Carroll, J.E., Isenberg, F.M.R., Pendergrass, A., Howell, L.: Separation of cytokinins from Danish cabbage by column chromatography on insoluble polyvinylpyrrolidone. Physiol. Plant. *33*, 83–86 (1975a)
Thomas, T.H., Carrol, J.E., Isenberg, F.M.R., Pendergrass, A., Howell, L.: Thin-layer chromatography of cytokinins on a mixed layer of polyvinylpyrrolidone and calcium sulphate. J. Chromatogr. *103*, 211–215 (1975b)
Thomas, T.H., Carrol, J.E., Isenberg, F.M.R., Pendergrass, A., Howell, L.: A simple inexpensive high pressure chromatographic method for separating cytokinins in plant extracts. Plant Physiol. *56*, 410–414 (1975c)
Thompson, A.G., Horgan, R., Herald, J.K.: A quantitative analysis of cytokinin using single-ion current-monitoring. Planta *124*, 207–210 (1975)
Ueda, M., Bandurski, R.S.: A quantitative estimation of alkali-labile indole-3-acetic acid compounds in dormant and germinating maize kernels. Plant Physiol. *14*, 1175–1181 (1969)
Ueda, M., Bandurski, R.S.: Structure of indole-3-acetic acid myoinositol esters and pentamethylmyoinositols. Phytochemistry *13*, 243–253 (1974)
Ueda, M., Ehmann, A., Bandurski, R.S.: Gas-liquid chromatographic analysis of indole-3-acetic acid myoinositol esters in maize kernels. Plant Physiol. *46*, 715–719 (1970)
Upper, C.D., Helgeson, J.P., Kemp, J.D., Schmidt, C.J.: Gas-liquid chromatographic isolation of cytokinins from natural sources. 6-(3-Methyl-2-butenylamino)purine from *Agrobacterium tumefaciens*. Plant Physiol. *45*, 543–547 (1970)
Upper, C.D., Helgeson, J.P., Schmidt, C.J.: Identification of cytokinins by gas-liquid chromatography and gas-liquid chromatography-mass spectrometry. In: Plant growth substances 1970. Proc. 7th. Int. Conf. Plant Growth Subst. Carr. D.J. (ed.), pp. 798–807. Berlin-Heidelberg-New York: Springer 1972
Vining. L.C.: Separation of gibberellin A_1 and dihydrogibberellin A_1 by argentation chromatography on a Sephadex column. J. Chromatogr. *60*, 141–143 (1971)
von Minden, D.L., McCloskey, J.A.: Mass spectrometry of nucleic acid components. N,O-permethyl derivatives of nucleosides. J. Am. Chem. Soc. *95*, 7480–7490 (1973)
Vreman, H.J., Corse, J.: Recovery of cytokinins from cation exchange resins. Physiol. Plant. *35*, 333–336 (1975)

Vreman, H.J., Skoog, F., Frihart, C.R., Leonard, N.J.: Cytokinins in *Pisum* transfer ribonucleic acid. Plant Physiol. *49*, 848–851 (1972)

Walton, D.C., Dorn, B., Fey, J.: The isolation of an abscisic acid metabolite, 4′-dihydrophaseic acid, from non-imbided *Phaseolus vulgaris* seed. Planta *112*, 87–90 (1973)

Wang, T.L., Thompson, A.G., Horgan, R.: A cytokinin glucoside from the leaves of *Phaseolus vulgaris* L. Planta *135*, 285–288 (1977)

Ward, T.M., Wright, M., Roberts, J.A., Self, R., Osborne, D.J.: Analytical procedures for the assay and identification of ethylene. In: Isolation of plant growth substances. Hillman, J.R. (ed.), pp. 135–151. London-New York-Melbourne: Cambridge University Press 1978

Watanabe, N., Yokota, T., Takahashi, N.: Identification of zeatin and zeatin riboside in cones of the hop plant and their possible role in cone growth. Plant Cell Physiol. *19*, 617–625 (1978a)

Watanabe, N., Yokota, T., Takahashi, N.: Cis-zeatin riboside as free nucleoside in cones of the hop plant. Agric. Biol. Chem. *42*, 2415–2416 (1978b)

Weller, L.E., Wittwer, S.H., Sell, H.M.: The detection of 3-indolylacetic acid in cauliflower heads. Chromatographic behavior of some indole compounds. J. Am. Chem. Soc. *76*, 629–630 (1954)

Yamaguchi, I., Yokota, T., Murofushi, N., Takahashi, N., Ogawa, Y.: Isolation of gibberellins A_5, A_{32}, A_{32} acetonide and (+)-abscisic acid from *Prunus persica*. Agric. Biol. Chem. *39*, 2399–2403 (1975)

Yamaguchi, I., Yokota, T., Yoshida, S., Takahashi, N.: High pressure liquid chromatography of conjugated gibberellins. Phytochemistry *18*, 1699–1702 (1979)

Yamane, H., Yamaguchi, I., Murofushi, N., Takahashi, N.: Isolation and structures of gibberellin A_{35} and its glucoside from immature seed of *Cytisus scoparius*. Agric. Biol. Chem. *38*, 649–655 (1974)

Yamane, H., Murofushi, N., Osada, H., Takahashi, N.: Metabolism of gibberellins in early immature bean seeds. Phytochemistry *16*, 831–835 (1977)

Yokota, T., Murofushi, N., Takahashi, N., Tamura, S.: Gibberellins in immature seeds of *Pharbitis nil*. Part II. Isolation and structures of novel gibberellins, gibberellins A_{26} and A_{27}. Agric. Biol. Chem. *35*, 573–582 (1971a)

Yokota, T., Murofushi, N., Takahashi, N., Tamura, S.: Gibberellins in immature seeds of *Pharbitis nil*. Part III. Isolation and structures of gibberellin glucosides. Agric. Biol. Chem. *35*, 583–595 (1971b)

Yokota, T., Okabayashi, M., Takahashi, N., Shimura, I., Umeya, K.: Plant growth regulators in chestnut gall tissue and wasps. In: Plant growth substances 1973. Proc. 8th. Int. Conf. Plant Growth Subst., pp. 28–38. Tokyo: Hirokawa Publishing Co. Ltd. 1974

Yokota, T., Hiraga, K., Yamane, H., Takahashi, N.: Mass spectrometry of trimethylsilyl derivatives of gibberellin glucosides and glucosyl esters. Phytochemistry *14*, 1569–1574 (1975)

Yoshida, R., Oritani, T.: Cytokinin glucoside in roots of the rice plant. Plant Cell Physiol. *13*, 337–343 (1972)

Young, H.: Identification of cytokinins from natural sources by gas-liquid chromatography/mass spectrometry. Anal. Biochem. *79*, 226–233 (1977)

Zaretkii, V.I., Wulfson, N.S., Papernaĵa, I.B., Gurvich, I.A., Kucherov, V.F., Milstein, I.M., Serebryakov, E.P., Simolin, A.V.: Mass spectrometry of gibberellins-II. The location of the double bond in the gibbane system. Tetrahedron *24*, 2327–2337 (1968)

Zeevaart, J.A.D.: Levels of (+)-abscisic acid and xanthoxin in spinach under different environmental conditions. Plant Physiol. *53*, 644–648 (1974)

Zeevaart, J.A.D., Milborrow, B.V.: Metabolism of abscisic acid and the occurrence of *epi*-dihydrophaseic acid in *Phaseolus vulgaris*. Phytochemistry *15*, 493–500 (1976)

Zenk, M.H.: 1-(Indole-3-acetyl)-β-D-glucose, a new compound in the metabolism of indole-3-acetic acid in plants. Nature (London) *191*, 493–494 (1961)

3 Quantitative Analysis of Plant Hormones

D.R. REEVE and A. CROZIER

3.1 Introduction

Estimates of endogenous plant hormone levels are usually obtained from analyses of solvent extracts of plant tissues. The relevance of such estimates to the true hormonal status of the plant at the time of extraction is open to question and will undoubtedly be an area of much future investigation. However, in the absence of pertinent data the scope of this article will have to be confined to questions relating to the analysis of plant extracts. The typical extract is an exceedingly complex mixture containing trace amounts of hormones at concentrations rarely greater than one part in 10^6. Meaningful quantitative analysis is therefore a formidable technical problem, not only in terms of the requirement for methodology possessing low limits of detection, but also because of the absolute necessity of having to distinguish the hormone of interest from the overwhelming number of other compounds present. In the following article these considerations will be discussed from a theoretical viewpoint which is then used as a basis to assess the effectiveness of procedures currently employed in the quantitative analysis of plant hormones.

3.2 Theoretical Considerations

3.2.1 Basic Analytical Errors – Accuracy and Precision

The nature of all forms of quantitative analysis is statistical rather than absolute. Regardless of the exactness of the method, the instrumentation or the operator, all analyses involve a number of uncertainties. As a consequence the results of quantitative analysis tend to approach, but never equal, the true value of the parameter under investigation, and accordingly are best termed *estimates*. The magnitude and sign of the deviation of an estimate from the true value is the *absolute error* which can be expressed as a percentage of the true value to give the *percentage relative error*. *Accuracy* is defined as the concordance between estimated and true value (VOGEL, 1939), and it follows that both the absolute error and the percentage relative error provide a measure of accuracy. Hence it has become common practice to use the terms absolute error and absolute accuracy interchangeably, although it would be more logical to talk of inaccuracy rather than accuracy. Likewise, percentage relative error and *percentage accuracy*, or more simply accuracy, tend to be synonymous. However, due to the nature of the errors which predominate in plant hormone analysis, it is convenient to ascribe a more

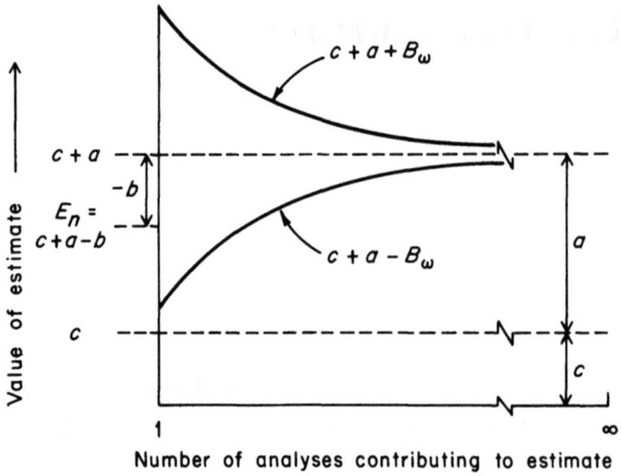

Fig. 3.1. Influence of the number of individual analyses contributing to an estimate of the random and non-random error of the estimate. True concentration of compound of interest is *c*. Non-random error is $+a$. Random error associated with a particular estimate, E_n, is $-b$. Random error associated with any estimate, confidence level ω, is $\pm B_\omega$

limited meaning to the word accuracy, and in order to do this, analytical errors will be discussed in greater detail.

Two fundamental types of error can be readily recognized, random and non-random. *Random error*, sometimes referred to as *indeterminate error*, can be looked upon as the background "noise" of the analytical system. It is not only generated by instrumental noise, but also arises from the uncertainties associated with such manipulations as reading the scales on volumetric glassware and microlitre syringes. With *precision* defined as the concordance between a series of measurements of the same quantity (VOGEL, 1939), it is apparent that precision will provide a measure of the random or indeterminate error of the analytical system. Since such errors are normally distributed, they can be quantified in terms of the width of the Gaussian error curve, i.e., $\pm B_\omega$ for a two-tailed confidence level ω[1]. Furthermore, it is evident that an averaging process can be applied to improve the precision of the method in proportion to the square root of the number of estimates averaged. However, regardless of the number of estimates contributing to the final average, there is no guarantee that accurate results will be obtained, since *non-random error* (also known as *constant* or *determinate error*) will apply the same bias to each estimate and hence will be present, undiminished, in the averaged result. This distinction between random and non-random error is vital, and is illustrated in Fig. 3.1, where an estimate E_n, based on a single analysis, deviates from the true value c, due to the net effect of a non-random error $+a$, and a contribution $-b$ arising from random error processes. By increasing the number of analyses contributing to each estimate, the probability of incurring a random error of magnitude $-b$ will diminish, but the error term $+a$ will remain unaffected. Thus non-random error ultimately limits the accuracy that can be achieved by an analytical technique. In practice non-random errors are minimized by employing a calibration procedure, which, if correctly designed, will offset non-random effects to such an extent that they become insignificant compared with the random error. Under such favourable circumstances, one of two statements of accuracy is

[1] The confidence limits $\pm B_\omega$ are defined such that the integral of individual probabilities from $B = -B_\omega$ to $B = +B_\omega$ is ω

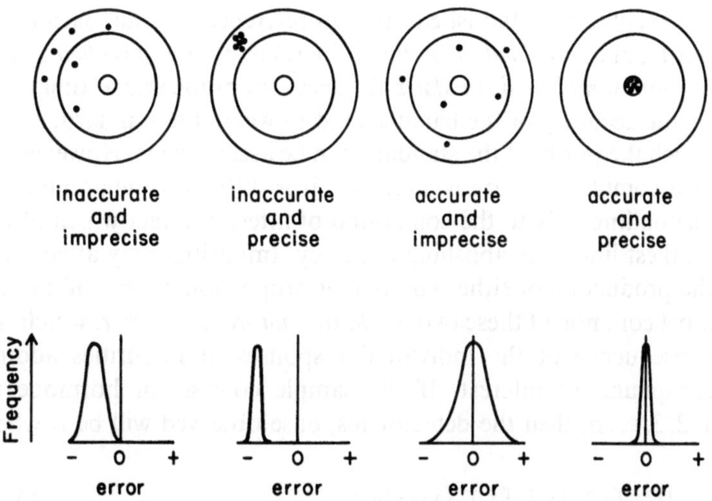

Fig. 3.2. Target analogy demonstrating the independence of the error terms, accuracy and precision. Accuracy refers to the non-random error and precision to the random error

commonly employed. Either, (i) "the method is accurate, its precision is $\pm B_\omega$", or, more simply, (ii) "the accuracy of the method is $\pm B_\omega$". It must be carefully noted that statement (ii), which expresses accuracy in terms of precision, will be invalid and, what is worse, highly misleading, if the effectiveness of the calibration is other than perfect. Indeed, plant hormone analysis usually is so far from ideal that the random error is small compared to the non-random error, and consequently precision measures only a minor proportion of the total error. This being so, it is less misleading to quote accuracy in terms of the total non-random error, and to use the precision term only as an indicator of the scatter in the values of individual estimates. Thus, for the purposes of this article, *accuracy* will be defined as *the sign and magnitude of the total non-random error associated with an estimate or estimates, expressed as a percentage of the true value of the parameter under investigation.* A "target" analogy, Fig. 3.2, is a useful means of illustrating the total independence of the terms accuracy and precision as defined above. From this it is evident that a rifle must not only be aimed accurately, but also must be designed so as to group its shots closely, i.e., it must be precise, if there is to be a high probability of hitting the bulls eye. In the case of plant hormone analysis this point can be made more formally. If the concentration of the substance of interest is c parts per million (ppm), the accuracy of the method is $+A\%$ and its precision $\pm B_\omega$ ppm, then the probability that an estimate of c will lie within the range $c + \frac{A \cdot c}{100} + B_\omega$ and $c + \frac{A \cdot c}{100} - B_\omega$ is ω. That is, both the accuracy and the precision terms are required to approach zero if an estimate is to be a reliable measure of the concentration c. The ensuing discussion will concentrate almost exclusively on accuracy rather than precision, as this single error term is the most crucial, yet least understood, factor limiting the performance of contemporary plant hormone analysis.

Accuracy is established through reference to a standard of known weight and purity, and, *provided the nature of the detector response being quantified is identical for both sample and standard*, the accuracy of the final estimate will be limited only by the accuracy of the balance used to weigh the standard, and the accuracy with which the purity of the standard can be ascertained[2]. However, when a constituent of a complex mixture is analyzed, it is difficult to guarantee that the detector is responding only to the compound of interest; a fact which often proves to be the weakest link in establishing accuracy. Impurities may affect the analysis through the production of either additive or proportional types of non-random error. The most common of these two is *additive non-random error*, which is a straightforward consequence of the individual responses of impurities adding to that of the compound of interest. If the sample consists of hormone u, and impurities 1, 2, 3, ... n, then the detector response observed will be r_t where

$$r_t = r_u + r_1 + r_2 + r_3 \ldots r_n$$

and $r_u, r_1, r_2, r_3, \ldots r_n$ are the absolute, discrete responses evoked by u, 1, 2, 3, ... n. Thus the non-random error incurred, either positive or negative, will be the algebraic sum of $r_1, r_2, r_3, \ldots r_n$. However, where extract components interact with u a *proportional non-random error* will arise. A good example of a detector system in which such errors are rife is the bioassay. Plant extracts contain substances which react synergistically or antagonistically with the hormone of interest, and consequently either positive or negative non-random errors, proportional to the hormone concentration, can result. This type of error is by no means restricted to bioassays, and can be present in analytical methods which are entirely physicochemical in nature. For example, a frequently employed fluorescence assay for indole-3-acetic acid (IAA) involves the formation of the indolo-α-pyrone (STOESSL and VENIS, 1970). Should an extract component interfere with the efficiency of the derivatization procedure, it is apparent that a negative non-random error proportional to the IAA concentration will be incurred. Thus, the observed detector response, r_t, will be

$$r_t = r_u + f_1(c) + f_2(c) + f_3(c) \ldots f_n(c)$$

where $f_1, f_2, f_3, \ldots f_n$ are discrete functions describing the proportional errors due to impurities 1, 2, 3, ... n and u is present at a concentration c.

The appa.ent simplicity of non-random error belies the subtlety of its influence on the analytical process, and, at the risk of labouring the obvious, this point will be further considered by means of a specific example. Consider an extract which has been split into a number of aliquots of a size suitable for the analysis of IAA by the indolo-α-pyrone fluorescence assay. If each aliquot contains M_u g of IAA, and the sensitivity of the method to IAA is R_u relative fluorescence units per gram, then the fluorescence due to IAA alone will be $M_u \cdot R_u$ relative units. However, fluorescence may arise from sources additional to IAA. Even if the extract is highly pure,

[2] For the moment we are assuming purity can be ascertained in absolute terms. Strictly speaking this is not the case, and this point will be dealt with more fully under Sections 3.2.2.2 and 3.2.2.3 (b)

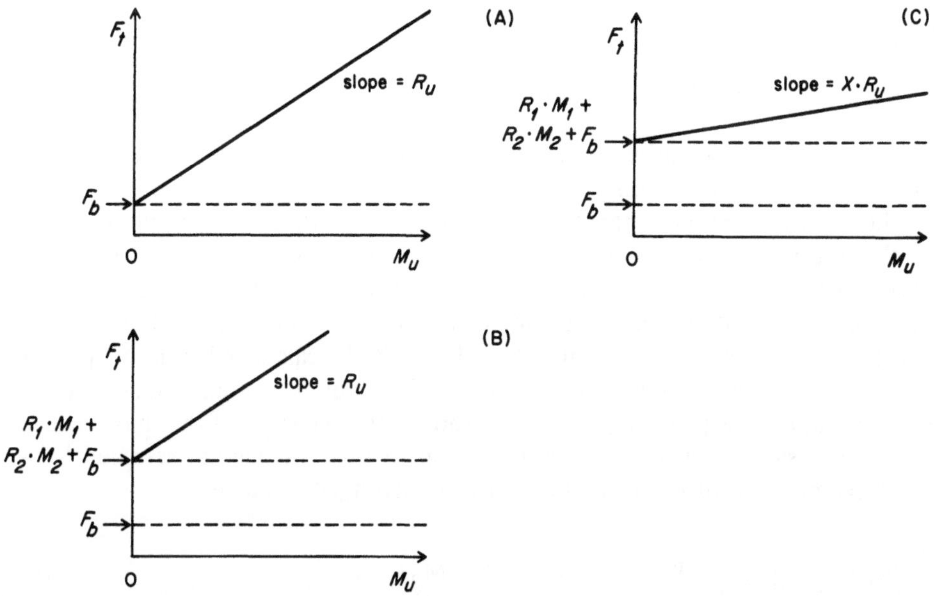

Fig. 3.3A–C. Hypothetical response curves for the fluorometric analysis of IAA, where F_t is the observed relative fluorescence, M_u is the amount of IAA in each aliquot of the sample, and R_u is the sensitivity of the method to IAA. (**A**) Sample in which the only source of error is that due to the background fluorescence F_b. (**B**) Sample which, in addition to the background fluorescence, incurs an additive non-random error $R_1M_1 + R_2M_2$, due to interference by impurities 1 and 2. (**C**) Sample similar to B except that it contains a third impurity which invokes a proportional non-random error of magnitude X

background fluorescence due to the reagents will contribute F_b relative units to the observed response. Thus, in the absence of extract impurities, the response obtained, F_t, can be described as

$$F_t = R_u \cdot M_u + F_b \tag{1}$$

that is a straight line of slope R_u intercepting the fluorescence axis at F_b (Fig. 3.3A). However, for the moment assume the extract contains two other components besides IAA, and for convenience label them 1 and 2. If, for each of these impurities, the method displays a sensitivity of R_1 and R_2 relative units per gram, and there are M_1 and M_2 grams of each present, then the fluorescence observed will be

$$F_t = R_u \cdot M_u + F_b + R_1 \cdot M_1 + R_2 \cdot M_2 \tag{2}$$

Under these circumstances a plot of F_t against M_u will be parallel to the standard curve of Eq. (1) but will intercept the fluorescence axis at $F_b + R_1 \cdot M_1 + R_2 \cdot M_2$ rather than F_b (Fig. 3.3B). Next, consider the effect of a third impurity for which the method is relatively insensitive. It is assumed that this compound has no significant influence on the additive non-random error of the analysis, but nevertheless interferes through reducing the indolo-α-pyrone yield by a factor X. This is an example of the simplest type of proportional non-random error. When its effects are taken into account, Eq. (2) becomes

$$F_t = X \cdot R_u \cdot M_u + F_b + R_1 \cdot M_1 + R_2 \cdot M_2 \tag{3}$$

and the new response curve has a slope of $X \cdot R_u$, although the intercept on the fluorescence axis will remain as $F_b + R_1 \cdot M_1 + R_2 \cdot M_2$ (Fig. 3.3C).

Having described the behaviour of our hypothetical four-component extract in the fluorescence assay, it is informative to consider ways in which the total non-random error of the analysis may be determined. Proportional errors can be accommodated relatively easily. Prior to analysis known amounts of IAA may be added to aliquots of the extract, and the value of $X \cdot R_u$ calculated from the plot of the response against amount of IAA added. This technique of standard addition is popularly upheld as providing rigorous proof of the accuracy of the fluorescence and other assays. However, the shortcomings of this practice are revealed by a closer examination of Eq. (3), which when re-arranged produces

$$M_u = \frac{1}{X \cdot R_u}(F_t - F_b - R_1 \cdot M_1 - R_2 \cdot M_2) \tag{4}$$

It follows that all terms on the right-hand side must be ascertained if the amount of IAA in the extract is to be estimated with definable accuracy. Thus, although the accuracy of any estimate will be improved as a result of subtracting the reagent background F_b from the total fluorescence F_t, and at the same time compensating for proportional non-random error by using $X \cdot R_u$ in place of R_u, the additive non-random error terms $R_1 \cdot M_1$ and $R_2 \cdot M_2$ will remain as undefined sources of error. In other words two conditions have to be fulfilled before a categorical statement of accuracy can be made. First, the actual values of M_1, M_2, R_1, R_2, must be determined to allow complete solution of Eq. (4). Secondly, proof must be provided that, of all the substances present in the extract, only 1 and 2 contribute significantly to the additive non-random error, i.e., proof that Eq. (4) is indeed a complete description of the analytical situation. *Thus it is inescapable that the amount and relevant properties of every impurity in an extract must be determined before accuracy can be defined.* This requirement arises because verification of accuracy and identity are mutually related processes. If a quantified response provides an accurate reflection of, for instance, the IAA level of a certain mixture, then this same response must furnish a unique characteristic by which IAA can be identified within that mixture. However, in common with all empirical sciences, the raw data of analytical chemistry are derived from the observation of differences, and thus establish non-identity rather than identity. Only when the property being quantified is known to distinguish the substance of interest from every other component of the system is identity and thus accuracy implied.

Some readers may feel the above criterion to be so stringent that, should it be applied, it will surely result in the demise of analytical chemistry as a practical tool. However, plant hormone analysis apart, there exist many situations where distinct limits can be placed upon the number of compounds likely to affect the response of the chosen detector. In such circumstances, freedom from interference is often self-evident and thus formal proof of accuracy superfluous. For example, where trace metals are analyzed by atomic absorption spectrophotometry, interfering impurities are restricted to the naturally occurring metals, many of which can be discounted on the basis either of low abundance, or of spectral characteristics markedly different from those of the metal under study. Should an interfering element be located, it may be independently assayed and its effects compensated for by an appropriate standard curve. Since all potential sources of interference can be dealt with in this fashion, a simple calibration procedure is sufficient to guarantee accurate results. Unfortunately the analysis of plant hormones is not so straightforward, because it is extremely difficult to place absolute limits upon the number of compounds present in an extract. In such a situation, freedom from interference is far from self-evident, and formal proof of accuracy assumes crucial importance.

3.2.2 Analysis of Samples from Open-Ended Systems

3.2.2.1 Open Versus Closed Systems

Prior to this point the discussion of analytical errors has been based upon the assumption that the sample has a finite number of components, and thus is a closed system which can be characterized in terms of absolute quantities. In the strictest sense this is true of all samples, as organic matter is not infinitely subdivisible and the mass of an extract, therefore, must constrain the number of different types of substances that can be present at any one time. However, bearing in mind that the mass of a molecule having a molecular weight of 100 is approximately 10^{-21} g, and that extracts often weigh as much as 1 g, it is apparent that for all practical purposes the number of compounds actually present might as well be infinite. Therefore absolute characterization is impossible and the extract will appear open-ended. This is true of even quite simple artificial mixtures. For example, a sample containing 99% compound A approximates a closed system of one component, but what of the 1% that is not A? Further analysis may reveal the presence of, say, 0.9% compound B, thus giving a description which now accounts for 99.9% of the mass of the sample, and consequently very nearly closes the system. However, the remaining 0.1% could be subdivided still further, and it is evident that this process would have to be continued virtually ad infinitum in order to describe all of the sample. It follows then, that the analysis of even highly pure compounds is still open-ended, and the possibility of interference to the detector response cannot be totally eliminated. Even so, it is equally apparent that in such cases the high concentration of the component of interest would reduce the likelihood of an estimate containing a significant non-random error. From the foregoing, it has to be concluded that, in most practical circumstances, accuracy does not have a discrete value, but assumes the form of a continuous distribution. This being so, the only sensible limits we can hope to place upon the accuracy of analysis will be in terms of probability.

3.2.2.2 Accuracy as a Probability Term

The chances of incurring a non-random error are governed by the relative ability of the analytical procedure to "see" the plant hormone against the background of

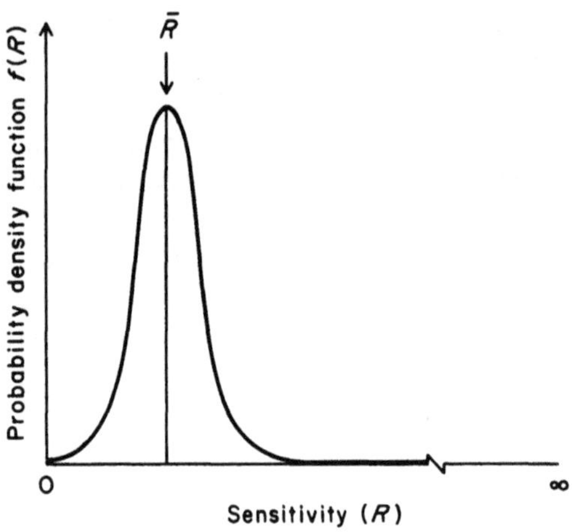

Fig. 3.4. Sensitivity distribution of a hypothetical detector. Most compounds elicit a response of about $\bar{R}$ per unit mass. However, the detector displays exceptional sensitivity and thus selectivity for the small number of substances whose characteristics place them in the right-hand limits of the curve

extract impurities. The ability to respond to some compounds and not others is often loosely termed selectivity, and is a crucial factor in determining accuracy.

If a very large random selection of substances is analyzed by a particular method, and the response per unit mass (R), i.e., sensitivity, recorded for each compound, then the compounds will be distributed with respect to R in a manner similar to that illustrated in Fig. 3.4, where f (R) is the probability density function[3]. Certain assumptions have been made with regard to the form of the sensitivity distribution. Firstly, since we are dealing with an open-ended system, the distribution must be continuous and no limit can be placed upon the magnitude of R. Secondly, all compounds must interact with the system in which they are placed, and consequently it is assumed all substances respond to some degree, i.e., $f(R) \rightarrow 0$ as $R \rightarrow 0$. Thirdly, to simplify the discussion which follows it is assumed that the occurrence of extract components which induce negative responses is insignificant compared with those invoking a positive response. In practice this assumption would appear valid for all but a few analytical techniques such as the bioassay, and, should it become necessary, the argument can be relatively easily modified to cater for those cases where negative responses are encountered. In effect, we are postulating that, given the infinite variety of substances possibly present in the extract, the response per unit mass of individual compounds as measured by any one detector will range from zero to plus infinity[4]. The analytical method, therefore, displays high selectivity for the small proportion of compounds whose characteristics place them in the upper tail of the distribution where R is relatively large. The parameter of interest, however, is the mean response per unit mass, $\bar{R}$.

[3] The probability density function can be looked upon as representing the number of compounds distributed within any one infinitesimally narrow response class centered on R

[4] Some readers may charge that this postulate is absurd in that it suggests compounds with, say, a UV absorbance of 10^{200} per g are possible. However, this is a problem in language rather than logic that often arises when continuous distributions are being considered. Perhaps, in view of what is known about this particular distribution, a more appropriate statement would be, "the occurrence of compounds with UV absorbances in excess of 10^{200} is *extremely unlikely* "

Obviously this is a measure of the average sensitivity of the detector to a mixture comprised of *equal amounts* of a large number of randomly selected substances, and should not differ from that obtained with a natural mixture, where substances are present in *differing amounts*, since the concentrations of extract components and their responses per unit mass are quite independent, random phenomena. Thus every analytical procedure will have a unique value of $\bar{R}$ which describes its overall sensitivity to any natural mixture comprised of a very large number of components. If the sensitivity of the chosen analytical method to u, the hormone of interest, is R_u, then the selectivity S, can be defined as the ratio of R_u to $\bar{R}$, i.e.

$$S = \frac{R_u}{\bar{R}} \qquad (5)$$

For example, assume mass fragmentography (MF) has a selectivity of 10^6 for IAA, then, *on a per gram basis*, the IAA in an extract will result in a response one million times greater than the combined response of all the other extract components.

The main reason for defining selectivity in the above manner is that its relationship to accuracy can be greatly simplified. Assume for the moment that the detector of choice displays identical selectivity to all extract impurities and u, the compound under analysis, in which case $R_u = \bar{R}$ and $S = 1$. If the concentration (w/w) of u in the extract is c ppm, then it follows that an impurity concentration of a ppm will result in a positive non-random error of a ppm. Since only positive, additive, non-random error is being considered as contributing to accuracy, it is evident the percentage accuracy will be $\frac{a}{c} \times 100\,\%$. Alternatively, if an accuracy of A% or better is to be achieved, the impurity concentration must not exceed A percent of c, i.e., $\frac{c \cdot A}{100}$ ppm. However, the selectivity of analytical procedures is nearly always greater than unity, and thus the overall sensitivity to impurities is only $\frac{1}{S}$ of that to the component u. Thus the actual impurity concentration which can be tolerated at an accuracy level of A% is $\frac{c \cdot A \cdot S}{100}$ ppm. Impurities represent all of the extract apart from u, and consequently always have the concentration 10^6–c ppm, which usually can be approximated by 10^6 ppm as the concentration of plant hormones in most extracts is negligible. It follows that for hormone concentrations less than, say, 0.1%

$$\frac{c \cdot A \cdot S}{100} = 10^6 \text{ ppm} \qquad (6)$$

and

$$A = \frac{10^8}{c \cdot S} \text{ percent} \qquad (7)$$

Bearing in mind that good accuracy figures are numerically small, it will be seen that accuracy can be directly improved by an increase in either the concentration of the hormone of interest or the selectivity of the method. Returning to the example of analysis of IAA by MF, where we arbitrarily assumed a selectivity of 10^6, substitution of this value in Eq. (7) dictates that analysis of an extract containing 100 ppm will achieve an accuracy of 1%, i.e., an estimate of 101 ppm. By most standards, this degree of accuracy is quite acceptable. However, consider the case of an extract containing only 0.1 ppm of IAA. Here the accuracy obtained will be $1000^0/_{00}$ which is far from adequate, and results in an IAA estimate of 1.1 ppm. Thus an important feature of Eq. (7) is that it indicates that all analytical methods are associated with a minimum permissible concentration of the substance of interest, below which the contribution of non-random errors to the estimate becomes excessive.

At the end of Section 3.2.2.1 it was concluded that the accuracy of analysis in open-ended systems cannot be absolute and must have the form of a continuous distribution, yet Eq. (7) would appear to provide a figure for accuracy which is free of probability. The reason for this lies in the nature of $\bar{R}$. It will be remembered that $\bar{R}$ was defined as the mean sensitivity of the chosen detector to a sample comprised of an infinite number of components. Such a sample necessarily would include all the organic matter in the universe, and obviously, as there can be only one such sample, $\bar{R}$ is invariant, and can be viewed as a universal constant describing one particular characteristic of the chosen detector. Thus Eqs. (5) to (7) will be correct only for the hypothetical case of an infinitely large sample. Where practical samples are involved, the mean sensitivity of the detector must deviate from $\bar{R}$, since no finite sample will be truly representative of all organic matter. For example, suppose the sensitivity of the chosen detector to each of the components of a bean extract is measured, and the sensitivity distribution plotted (Fig. 3.5A). In this particular case the mean sensitivity is R'_1. Repeating the process with the components of, say, a pea extract might give a new distribution with a mean sensitivity of R'_2 (Fig. 3.5B), while further analysis of extracts from other sources could yield R'_3 and R'_4 (Fig. 3.5C and 3.5D). Thus, in practice, we are not dealing with a single value of mean sensitivity $\bar{R}$, but a range of R' values continuously distributed about $\bar{R}$. The universal sensitivity distribution of Fig. 3.4 is reproduced in Fig. 3.6A along with the type of distribution one would anticipate for $\bar{R}'$ (Fig. 3.6B). In common with the sensitivity distribution, R' values will range from zero to plus infinity, with a mean of $\bar{R}$. However, as we are dealing with a distribution of means, the standard deviation of R' values will be $\frac{1}{\sqrt{N}}$ of that for R values, where N is the number of compounds in each sample. Consequently the width of the distribution of mean sensitivities will depend upon the complexity of the mixture under analysis. A further consideration is that the components of extracts are present in a random array of concentrations, and this fact, while not influencing the mean $\bar{R}$, will nevertheless add variance to R'. Thus the distribution of R' about $\bar{R}$ can be represented by a family of continuous distributions, the members of which differ in terms of (a) the number of components in the mixture, and (b) the distribution of these components with respect to their concentrations.

When presented with a particular extract there must be a unique value of R' which, if substituted in Eqs. (5) and (7) in place of $\bar{R}$, will give the actual selectivity

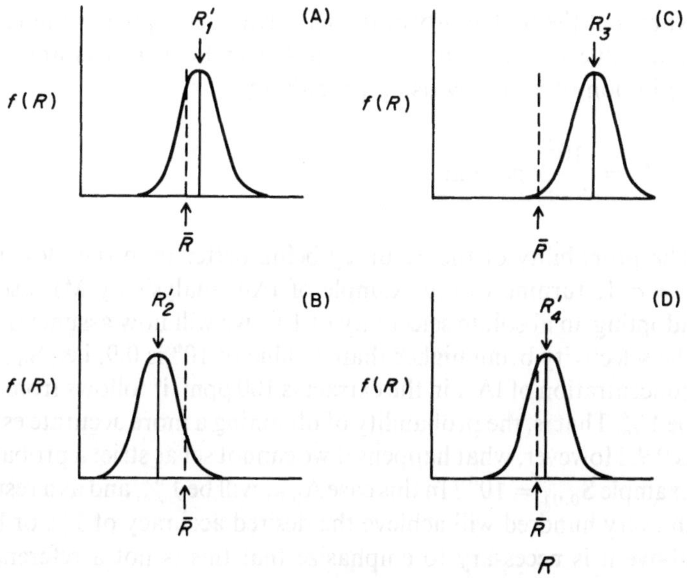

Fig. 3.5. A–D. Differing response distribution obtained from a range of sample types. For samples of equivalent complexity the main difference will be in terms of the estimate R′ of the actual mean sensitivity $\bar{R}$

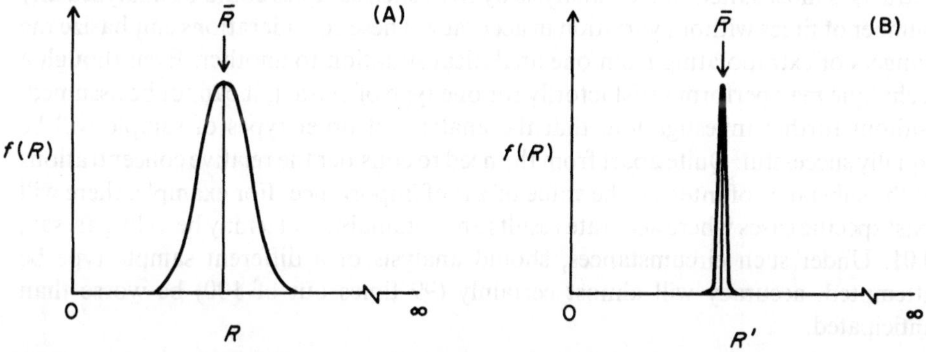

Fig. 3.6. A Sensitivity distribution of hypothetical detector. **B** Distribution of actual mean sensitivities R′ for a range of samples of equivalent complexity

and hence the accuracy obtainable. In theory, the value of R′ could be calculated after characterizing every component of the extract, but, as has already been pointed out, the complexity of the typical plant extract is likely to obstruct such a closed-system approach. However, the problem is amenable to open-ended analysis through the definition of an upper confidence limit R'_σ, where σ is the probability of encountering an extract giving rise to a R′ value less than R'_σ. This limiting value of R′ now can be substituted in Eq. (5) to give

$$S_\sigma = \frac{R_u}{R'_\sigma} \tag{8}$$

where σ is the probability of the selectivity being greater than a value S_σ. When this new value of selectivity is substituted in Eq. (7), it is apparent that we have a definition of accuracy using probability.

$$A_\alpha = \frac{10^8}{c \cdot S_\sigma} \text{ percent} \tag{9}$$

The probability of the accuracy being better than (i.e., less than) A_α is α, where $\alpha = \sigma$. Returning to the example of IAA analysis by MF used earlier, instead of adopting an absolute selectivity of 10^6, we will now assume that the probability of the selectivity being higher than a value of 10^6 is 0.9, i.e., $S_{0.9} = 10^6$. If the actual concentration of IAA in the extract is 100 ppm, it follows from Eq. (9) that $A_{0.9}$ will be 1%. That is, the probability of obtaining a more accurate estimate than 101 ppm is 0.9. However, what happens if we cannot set as strict a probability level as 0.9, for example $S_{0.01} = 10^6$? In this case $A_{0.01}$ will be 1%, and as a result, only one analysis in every hundred will achieve the desired accuracy of 1% or better. In stating the above it is necessary to emphasize that this is not a reference to precision. The definition of accuracy on p. 205 dispenses with such a possibility. In fact what is meant is that, should one hundred *different* types of extract chosen for analysis, the chances are that only one of these will be amenable to assay within the desired limit of accuracy. Obviously, given some means of discerning exactly which of the extracts is most suited to IAA analysis by MF, such samples could be analyzed any number of times without variation in accuracy. These considerations emphasize the dangers of extrapolating from one analytical situation to another. Even though a technique may perform satisfactorily for one type of extract, it cannot be assumed, without further investigation, that the analysis of other types of sample will be equally successful. Quite apart from the need to consider the relative concentrations of the substance of interest, the value of α is of importance. For example, there will exist specific cases where accurate results are obtainable, yet α may be as low as, say, 0.01. Under such circumstances, should analysis of a different sample type be attempted, accuracy will almost certainly (99 times out of 100) be worse than anticipated.

3.2.2.3 Verification of Accuracy

Establishing accuracy is an unavoidable facet of the development of any quantitative analytical technique. In theory this could be done by demonstrating that the required accuracy, available selectivity, and the lowest hormone concentration to be analyzed jointly fulfill Eq. (9). However, in practice, S_σ is difficult to obtain. It follows from the discussion of the distribution of R' in Section 3.2.2.2, that even if one did experimentally determine the sensitivity distribution of the chosen detector, it would still be necessary to know the number of components and the concentration of each in the extract, thereby defeating the original purpose of the open-ended approach. However, two viable alternatives exist. The first of these we have called successive approximation, and is based upon the fact that extracts can be simplified in a controlled fashion by the use of chromatographic procedures. With increasing purity the properties of the sample approach those of a much simpler system whose characteristics can be safely assumed. The second method requires that a probability limit be placed upon the number of compounds potentially present in the extract. The analysis is then conducted in such a way that sufficient information is obtained from the quantified response to distinguish the hormone of interest from all other possibilities within the chosen confidence limits. Verification of accuracy through the accumulation of information is quite distinct from, and should not be confused with, successive approximation. Although both approaches are open-ended, the former reveals, whilst the latter eliminates, sources of interference within definable limits of probability.

a) Successive Approximation

If we continue to assume that interfering compounds result in predominantly positive additions to the detector response, the actual value of an estimate E of the concentration c of a substance in a plant extract is related to the accuracy of the method by

$$E = c + \frac{c \cdot A}{100} \qquad (10)$$

Substitution of Eq. (7) in Eq. (10) produces

$$E = c + \frac{10^6}{S} \text{ ppm} \qquad (11)$$

Thus the absolute deviation, $\frac{10^6}{S}$, of a concentration estimate from its true value depends solely upon the reciprocal of the selectivity[5], and with increasing selectivity E asymptotically approaches c (Fig. 3.7A). In practice, however, it is necessary to allow for the random error associated with the analysis as indicated in Eq. (12)

$$E_\omega = c + \frac{10^6}{S} \pm B_\omega \text{ ppm} \qquad (12)$$

where the precision of the method is such that a proportion ω of all estimates will lie within $\pm B_\omega$ ppm of the true mean. With increasing selectivity, therefore, $\frac{10^6}{S}$ will become small and $E_\omega \to c \pm B_\omega$, (Fig. 3.7B). In other words, given sufficient selectivity the analysis will be dominated by random error and Eq. (12) simplifies to

$$E_\omega = c \pm B_\omega \qquad (13)$$

The above relationship suggests a basis for the verification of accuracy. If aliquots of an extract are analyzed by a number of methods each possessing progressively

[5] For the moment we are treating selectivity as being absolute. The need to consider S as a probability term will arise later in the discussion

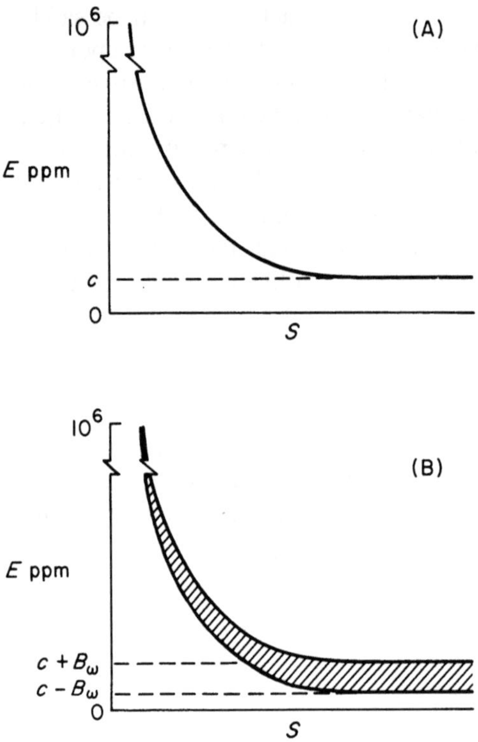

Fig. 3.7. A Relationship between the value of an estimate, *E*, and the selectivity, *S*, of the method used to obtain that estimate. The true concentration of the compound of interest is c ppm. **B** Relationship between *E* and *S* in the presence of a random error term of magnitude $\pm B_\omega$ ppm. The lower asymptote of *E* is $c \pm B_\omega$ ppm

greater selectivity, the estimates of hormone concentration so obtained must show an overall downward trend, until a point is reached where further improvement in the selectivity of analysis has no perceptible effect on the value of E. At this point it can be assumed that non-random error is likely to be small compared with random error, and, in accordance with the principles discussed in Section 3.2.1, the final estimate considered to be accurate.

The overall selectivity of an analytical procedure can be improved by including a number of chromatographic purification steps prior to analysis. Provided the concentration of the compound of interest is low, for example less than 0.1%, the effectiveness of the purification procedure can be readily defined in terms of the *purification factor*, Q, as follows,

$$Q = \frac{M_i}{M_f} \cdot \frac{Z}{100} \tag{14}$$

where M_i is the initial weight of the extract, M_f the weight subsequent to purification and, Z the percentage recovery of the compound of interest. Consider an extract containing c ppm of hormone. Eq. (11) indicates that analysis of this extract by a method of selectivity S will result in an estimate of $c + \frac{10^6}{S}$ ppm. However, when the same extract is chromatographed using a method of purification factor Q, the concentration of the hormone of interest will increase to $c \cdot Q$ ppm, while the impurity level will remain unchanged, i.e., 10^6 ppm, unless either c and/or Q is exceptionally large. Thus analysis of the purified extract will result in an estimate

$$E' = c \cdot Q + \frac{10^6}{S} \text{ ppm}$$

which then must be divided by the factor Q to obtain the concentration of hormone in the original extract, i.e.,

$$E = \frac{E'}{Q} = c + \frac{10^6}{S \cdot Q} \text{ ppm} \tag{15}$$

Eq. (15) indicates that sample prepurification will increase the overall selectivity of analysis by a factor Q, and it is this relationship which is exploited in the successive approximation.

Assume that analysis of an extract by a particular method results in an estimate of hormone concentration of E_1 ppm. In order that the accuracy of E_1 be determined, the extract can be chromatograpically purified by a factor Q_1, and re-analyzed to yield a second estimate E_2 ppm. Subtraction of Eq. (15) from Eq. (11) suggests that we can expect E_2 to differ from E_1 by a factor of $\left(1 - \frac{1}{Q_1}\right)$ of the total non-random error associated with E_1, i.e.,

$$E_1 - E_2 = \frac{10^6}{S} \left(1 - \frac{1}{Q_1}\right) \text{ppm} \tag{16}$$

Thus, provided Q_1 is greater than, say, 3, a significant non-random error in E_1 will be revealed as a discrepancy between E_1 and E_2. Should this happen E_1 must be rejected and E_2 tested for accuracy by subjecting the already purified extract to an additional chromatographic procedure of purification factor Q_2. Analysis subsequent to the two purification steps allows a third estimate E_3 to be made, and the difference $E_3 - E_2$ tested. If once again a difference is found, E_2 must also be rejected and the process repeated until an estimate E_n is obtained which remains constant with further purification. At this point in the successive approximation non-random error can be assumed to be insignificant compared with random error, and the estimate E_n taken as accurate.

The foregoing procedure makes it possible to determine the purification stage at which the overall selectivity of the method is sufficient for accurate analysis of the extract in question. However, just how accurate is the estimate E_n so obtained, and what level of confidence is associated with it? The answer lies in (i) the smallest difference in the term $E_n - E_{n+1}$ that can be conclusively attributed to non-random error processes, and (ii) the degree of purification Q_n carried out between E_n and E_{n+1}. In practice, E_n and E_{n+1} will be mean estimates, each derived from m individual analyses, and therefore associated with variances V_n^2 and V_{n+1}^2, respectively. However, the precision of analysis can be safely assumed to be constant throughout the approximation due to the necessity of employing a radioactive internal standard for the measurement of Q values. Therefore the common variance V_p^2 will provide a better indicator of the dispersion of individual estimates about both E_n and E_{n+1}, where

$$V_p^2 = \frac{V_n^2 + V_{n+1}^2}{2} \tag{17}$$[6]

The combined sample will have $2(m-1)$ degrees of freedom, and can be used to estimate the smallest difference likely to be detected between E_n and E_{n+1}. Eq. (16) indicates that the difference between E_n and E_{n+1} arising from the non-random error will be $\frac{10^6}{S}\left(1 - \frac{1}{Q_n}\right)$ ppm. This is the expected value, or the *expectation* of the difference between any pair of mean estimates, and is designated ΔE. Since the variance of the method has been estimated as V_p^2, the dispersion of pairs differences $E_n - E_{n+1}$ about ΔE will be $\frac{2V_p^2}{m}$. Consequently it is reasonable to assume that the individual pairs differences will be distributed about ΔE in accordance with a Student distribution of standard deviation $\sqrt{\frac{2V_p^2}{m}}$ and $2(m-1)$ degrees of freedom, as illustrated in Fig. 3.8A. A lower confidence limit $(E_n - E_{n+1})_\tau$, can be defined as

$$(E_n - E_{n+1})_\tau = \Delta E - t_\tau \cdot \sqrt{\frac{2V_p^2}{m}} \text{ ppm} \tag{18}$$

where t_τ is the t value corresponding to the probability level τ of a standardized Student distribution of $2(m-1)$ degrees of freedom. That is, a proportion $1-\tau$ of all pairs differences will lie above $(E_n - E_{n+1})_\tau$. Rearrangement yields

$$\Delta E = (E_n - E_{n+1})_\tau + t_\tau \cdot \sqrt{\frac{2V_p^2}{m}} \text{ ppm} \tag{19}$$

where ΔE now can be looked upon as the smallest expectation in $E_n - E_{n+1}$ which can be detected through setting a confidence limit of $(E_n - E_{n+1})_\tau$. Therefore, if

$$\Delta E \geqq (E_n - E_{n+1})_\tau + t_\tau \cdot \sqrt{\frac{2V_p^2}{m}} \text{ ppm} \tag{20}$$

then in a proportion $1-\tau$ of all cases

$$E_n - E_{n+1} \geqq (E_n - E_{n+1})_\tau \text{ ppm}$$

[6] Eq. (17) is valid only when V_n^2 and V_{n+1}^2 are calculated from the same number of individual estimates

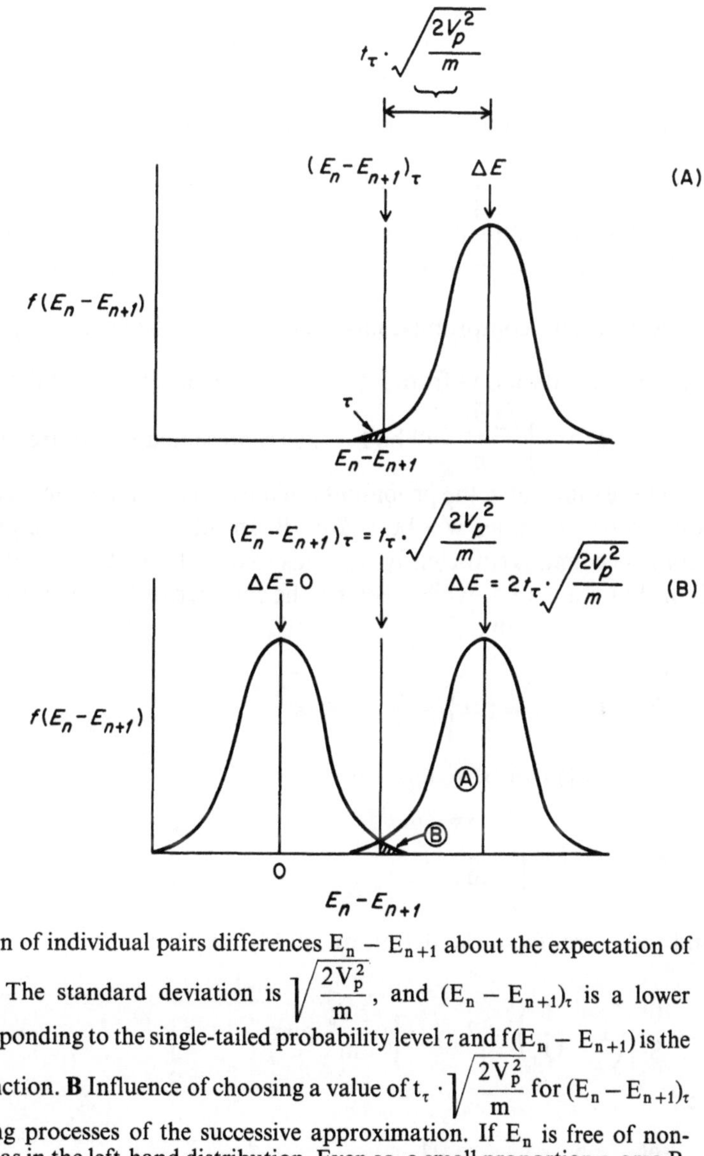

Fig. 3.8. A Distribution of individual pairs differences $E_n - E_{n+1}$ about the expectation of pairs differences ΔE. The standard deviation is $\sqrt{\frac{2V_p^2}{m}}$, and $(E_n - E_{n+1})_\tau$ is a lower confidence limit corresponding to the single-tailed probability level τ and $f(E_n - E_{n+1})$ is the probability density function. **B** Influence of choosing a value of $t_\tau \cdot \sqrt{\frac{2V_p^2}{m}}$ for $(E_n - E_{n+1})_\tau$ on the decision-making processes of the successive approximation. If E_n is free of non-random error, $\Delta E = 0$ as in the left-hand distribution. Even so, a small proportion τ, area B, of $E_n - E_{n+1}$ values will exceed the confidence limit $(E_n - E_{n+1})_\tau$. However, should a significant non-random error be present and ΔE exceed $2t_\tau \cdot \sqrt{\frac{2V_p^2}{m}}$, then a proportion of at least 1-τ, area A + B, of all $E_n - E_{n+1}$ values will be detected as being discrepant

Conversely, if no significant difference between E_n and E_{n+1} can be found, i.e., where

$$E_n - E_{n+1} < (E_n - E_{n+1})_\tau \text{ ppm}$$

then the probability that

$$\Delta E < (E_n - E_{n+1})_\tau + t_\tau \cdot \sqrt{\frac{2\,V_p^2}{m}}\ \text{ppm} \tag{21}$$

will be $1-\tau$. It follows from Eq. (20) that by setting a confidence limit of zero, ΔE values as low as $t_\tau \cdot \sqrt{\frac{2\,V_p^2}{m}}$ will be detected. However, if the limit chosen is too small, the probability of uncovering a difference, where in fact none exists, becomes sufficiently large to render the test impractical. Figure 3.8B illustrates a situation where a suitable compromise has been found. If the value $t_\tau \cdot \sqrt{\frac{2\,V_p^2}{m}}$ is chosen for $(E_n - E_{n+1})_\tau$, it follows from Eq. (20) that the probability of detecting ΔE values in excess of $2 \cdot t_\tau \cdot \sqrt{\frac{2\,V_p^2}{m}}$ will be $1-\tau$, as represented by the area A + B in the figure, and at the same time, the probability of *incorrectly* uncovering a difference between consecutive estimates will be τ, area B. In other words, this particular choice of confidence limit is sufficiently rigorous to detect, say, 95% of all ΔE values greater than that stipulated, while keeping the number of "false alarms" down to 5%.

On condition that $E_n - E_{n+1} < (E_n - E_{n+1})_\tau$ where

$$(E_n - E_{n+1})_\tau = t_\tau \cdot \sqrt{\frac{2\,V_p^2}{m}} \tag{22}$$

Equation (21) can be reduced to

$$\Delta E < 2 \cdot t_\tau \cdot \sqrt{\frac{2\,V_p^2}{m}}$$

and therefore, from the definition of ΔE we have

$$\frac{10^6}{S}\left(1 - \frac{1}{Q_n}\right) < 2 \cdot t_\tau \cdot \sqrt{\frac{2\,V_p^2}{m}}\ \text{ppm}$$

i.e.,

$$\frac{10^6}{S} < \frac{2 \cdot t_\tau}{1 - \frac{1}{Q_n}} \cdot \sqrt{\frac{2\,V_p^2}{m}}\ \text{ppm} \tag{23}$$

Thus, when $E_n - E_{n+1}$ falls below the value stipulated in Eq. (22), it can be assumed, probability $1-\tau$, that the non-random error associated with E_n is less than that indicated by the RHS of Eq. (23). Furthermore, as this expression is an upper limit, the concentration of hormone, c, must lie within the range

$$E_n > c > E_n - \frac{2 \cdot t_\tau}{1 - \frac{1}{Q_n}} \cdot \sqrt{\frac{2\,V_p^2}{m}}$$

and consequently the accuracy of E_n can be no worse than

$$\frac{\dfrac{2 \cdot t_\tau}{1-\dfrac{1}{Q_n}} \cdot \sqrt{\dfrac{2 V_p^2}{m}} \cdot 10^2}{E_n - \dfrac{2 \cdot t_\tau}{1-\dfrac{1}{Q_n}} \cdot \sqrt{\dfrac{2 \cdot V_p^2}{m}}} = \frac{2 \cdot 8 t_\tau \cdot \sqrt{V_p^2} \cdot 10^2}{E_n\left(1-\dfrac{1}{Q_n}\right) \cdot \sqrt{m} - 2.8 t_\tau \cdot \sqrt{V_p^2}} \text{ percent}$$

Therefore, from the definition of A_α in Section 3.2.2.2, we now have

$$A_\alpha = \frac{2.8 t_\tau \cdot \sqrt{V_p^2} \cdot 10^2}{E_n \cdot \left(1-\dfrac{1}{Q_n}\right) \cdot \sqrt{m} - 2.8 t_\tau \cdot \sqrt{V_p^2}} \text{ percent} \tag{24}$$

where $\alpha = 1 - \tau$.

An example of a hypothetical successive approximation is presented in Table 3.1. In this particular example, analysis by the indolo-α-pyrone fluorescence assay indicated that the concentration of IAA in a particular methanolic extract was $E_1 = 36.854$ ppm. The extract was then purified by solvent partitioning, and subsequent analysis provided a second estimate $E_2 = 15.310$ ppm. The possibility that the discrepancy between E_1 and E_2 might result from non-random error in E_1 was checked by calculating a confidence limit on $E_1 - E_2$ as follows. The fluorescence assay was performed five times per estimate, and as a result yielded two measures of the inherent variance of the method, $V_1^2 = 0.133$ and $V_2^2 = 0.200$. These can be combined according to Eq. (17), and the resulting V_p^2 value substituted in Eq. (22). The value of t_τ which excludes a single-tailed probability level of 0.05 for a standardized Student distribution of $2(5-1) = 8$ degrees of freedom is 1.859 [7], and hence $(E_1 - E_2)_{0.05}$ is 0.480 ppm. The actual value of $E_1 - E_2$ far exceeds $(E_1 - E_2)_{0.05}$, and consequently there can be little doubt that E_1 is inaccurate. Accordingly, the extract from stage 2 was chromatographed on a column of PVP and reanalyzed, $E_3 = 8.671$ ppm. Again it was found that the difference $E_2 - E_3$ far exceeded the corresponding confidence limit of 0.469 ppm, Table 3.1, and E_2 was rejected as being inaccurate. Reanalysis following DEAE cellulose column chromatography indicated that E_3 also was not reliable, and therefore E_4 was checked by subjecting the stage 4 extract to thin layer chromatography (TLC). This time the difference $E_4 - E_5$ was found to be 0.062 ppm while $(E_4 - E_5)_{0.05}$ was calculated as 0.441 ppm. Consequently there was no reason to believe that E_4 was any different from E_5, and thus E_4 can be considered accurate. The purification factor of the TLC procedure was calculated according to Eq. (14) as 4.5, using the respective weights of the extract at stage 4 and 5, and the percentage recovery of the radioactive internal standard. Substitution of this Q value into Eq. (24) gives an

[7] The integral of the discrete probabilities between $t = t_\tau$ and $t = \infty$ is 0.05

Table 3.1. Hypothetical successive approximation

	Purification factor Q_n	Mean estimate E_n ppm	Variance v_n^2	$E_n - E_{n+1}$	$(E_n - E_{n+1})_{0.05}$
Stage 1 (MeOH extract)		36.854	0.133		
↓ Partitioning	3.2			21.544	0.480
Stage 2		15.310	0.200		
↓ PVP	3.4			6.639	0.469
Stage 3		8.671	0.118		
↓ DEAE	4.8			2.867	0.368
Stage 4		5.804	0.078		
↓ TLC	4.5			0.062	0.441
Stage 5		5.742	0.204		

estimate of $A_{0.95}$ of 24%. The precision of E_4 can be found from the following formula

$$B_\omega = t_\omega \cdot \sqrt{\frac{V_P^2}{m}} \tag{25}$$

where t_ω is the t value of a standardized Student distribution of $2(m-1)$ degrees of freedom which excludes the two-tailed probability level of $1-\omega$ [8]. From Eq. (25) we find that $B_{0.95} = 0.772$ ppm, i.e., 13%. It will be noted that $A_{0.95}$ is approximately twice the percentage precision, as would be expected for Q values greater than, say, three. Indeed there is little point in attempting to generate very large Q values in the hope of improving A_α, since it can be seen, even intuitively, that the percentage accuracy cannot be assigned a value much less than twice the percentage precision. Furthermore there is no practical advantage in knowing that the non-random error of an estimate is greatly less than twice B_ω, as in any event differences between estimates that are less than this value will not be significant. What then are the limits we can place upon the IAA concentration of this particular extract? A proportion (0.95) of all estimates will lie between $E_4 + B_{0.95}$ and $E_4 - B_{0.95}$, i.e., 5.0 to 6.6 ppm. However, E_4 may contain a non-random error which from the available evidence cannot be worse than $\frac{A_{0.95} \cdot E_4}{100} = 1.4$ ppm. Thus the absolute upper and lower limits of E_4, confidence level $\omega \cdot \alpha = 0.9$, will be 3.6 to 6.6 ppm.

[8] From the definition of precision in Section 3.2.1, the integral of the discrete probabilities between $t = -t_\omega$ and $t = +t_\omega$ is ω. Thus the t-table value of interest is t_ω, where the integral from $t = t_\omega$ to $t = \infty$ is $1/2(1-\omega)$

Up till this point, the discussion of successive approximation has been simplified by the assumption that the selectivity of analysis S in Eq. (15) is constant throughout the approximation procedure. However, it is inevitable that the composition of an extract will differ from one stage of purification to another, and thus according to the concepts discussed in Section 3.2.2.2, S need not be constant. Consequently, unless we can demonstrate either by argument or direct measurement, that the variation in S is small, the validity of Eq. (16) and by inference Eqs. (22) and (24) must be questioned.

If S_n and S_{n+1} are the actual selectivities obtained during analysis of the extract at stage n and n+1, respectively, Eq. (16) can be represented more correctly as

$$E_n - E_{n+1} = \frac{10^6}{S_n} - \frac{10^6}{S_{n+1} \cdot Q_n} \text{ ppm} \tag{26}$$

In accordance with the definition of selectivity in Section 3.2.2.2, S_n and S_{n+1} can be replaced by $\frac{R_u}{R'_n}$ and $\frac{R_u}{R'_{n+1}}$, where R'_n and R'_{n+1} are the actual mean sensitivities of the chosen analytical method to the impurities present in the extract at stages n and n+1, while R_u is the sensitivity to the compound of interest. Thus

$$E_n - E_{n+1} = \frac{10^6}{R_u}\left(R'_n - \frac{R'_{n+1}}{Q_n}\right) \text{ ppm} \tag{27}$$

Limits can be placed on the R′ values in terms of the true mean detector sensitivity $\bar{R}$ and a two-tailed confidence limit R_σ[9]. Consequently the expectation of $E_n - E_{n+1}$, ΔE, may be defined using the confidence level σ, as

$$\Delta E_{\sigma^2} = \frac{10^6}{R_u}\left(\bar{R} \pm R'_\sigma - \frac{\bar{R} \pm R'_\sigma}{Q_n}\right) \text{ ppm}$$

which if rearranged produces

$$\Delta E_{\sigma^2} = 10^6 \cdot \frac{\bar{R}}{R_u}\left(1 - \frac{1}{Q_n}\right) \pm 10^6 \cdot \frac{R'_\sigma}{R_u}\left(1 - \frac{1}{Q_n}\right) \text{ ppm} \tag{28}$$

Equation 28 differs from Eq. (16) in that it has an extra term, $\pm 10^6 \cdot \frac{R'_\sigma}{R_u}\left(1 - \frac{1}{Q_n}\right)$, which provides a measure of the variance in ΔE arising through extract to extract differences in selectivity. It can be thought of as allowing for the possibility that a group of exceptionally interfering impurities could co-chromatograph with the compound of interest during the purification Q_n and thus appear undiminished in the estimate E_{n+1}.

If allowance is made for the precision of the method, $\pm B_\omega$, confidence limits for any pairs difference can be obtained from Eq. (28) as

[9] The integral of the discrete probabilities from $R = \bar{R} - R_\sigma$ to $R = \bar{R} + R_\sigma$ is σ

$$(E_n - E_{n+1})_{\sigma^2 \cdot \omega} = 10^6 \cdot \frac{\bar{R}}{R_u} \cdot \left(1 - \frac{1}{Q_n}\right) \pm \left[10^6 \cdot \frac{R'_\sigma}{R_u} \cdot \left(1 - \frac{1}{Q_n}\right) + \sqrt{2} \cdot B_\omega\right] \text{ppm} \tag{29}$$

Consequently, valid successive approximation would require the expectation of pairs differences, $10^6 \cdot \frac{\bar{R}}{R_u} \cdot \left(1 - \frac{1}{Q_n}\right)$, to be detected against a background variation of $\pm 10^6 \cdot \frac{R'_\sigma}{R_u} \cdot \left(1 - \frac{1}{Q_n}\right) + \sqrt{2} \cdot B_\omega$, rather than simply $\sqrt{2} \cdot B_\omega$, as was assumed to be the case for Eq. (22) and (24). Hence if we are to use the successive approximation procedure with impunity, there must be good reasons for assuming $10^6 \cdot \frac{R'_\sigma}{R_u} \left(1 - \frac{1}{Q_n}\right)$ to be small compared with $\sqrt{2} \cdot B_\omega$, and the analysis of relatively impure samples using a detector of high selectivity may be one case where this is possible. From the discussion in Section 3.2.2.2, it follows that R'_σ will be an inverse function of the square root of the number of components in the extract, and thus, when dealing with impure extracts containing a great number of components, R'_σ could well be negligible. Furthermore, as methods of high selectivity tend to display very large R_u values, R'_σ, whatever its value, could be reduced by a considerable factor. Consequently it may be possible to ignore the variation in selectivity in many practical situations.

However, it is not necessary to make too many assumptions about the magnitude of R'_σ as the successive approximation procedure can be designed to test not only for the condition

(a) $\Delta E \approx 0$

as in Eqs. (22) and (24), but also for

(b) $10^6 \cdot \frac{R'_\sigma}{R_u} + B_\omega \approx B_\omega$

In the example of IAA analysis by the indolo-α-pyrone fluorescence method, Table 3.1, it was stated that E_4 was accurate since it did not differ significantly from E_5, thus fulfilling condition (a). However, in order to test for condition (b) it is necessary to obtain further data. Aliquots of the stage 3 extract were purified by three non-correlated[10] high performance liquid chromatography (HPLC) systems, and then analyzed for IAA to give the estimates $E_{4(a)}$, $E_{4(b)}$ and $E_{4(c)}$, Tables 3.2 and 3.3. By combining the individual variances V_4^2, $V_{4(a)}^2$, $V_{4(b)}^2$, $V_{4(c)}^2$, and V_5^2, we can obtain a good estimate of the precision of the method as follows

$$V_p^2 = \frac{V_4^2 + V_{4(a)}^2 + V_{4(b)}^2 + V_{4(c)}^2 + V_5^2}{5}$$

[10] The problem of correlation between chromatographic procedures will be discussed in Section 3.4. For the moment, non-correlated can be taken to mean "as different, mechanistically, as possible"

Table 3.2. Successive approximation. Additional test for stage to stage variation in the selectivity of the fluorometric assay

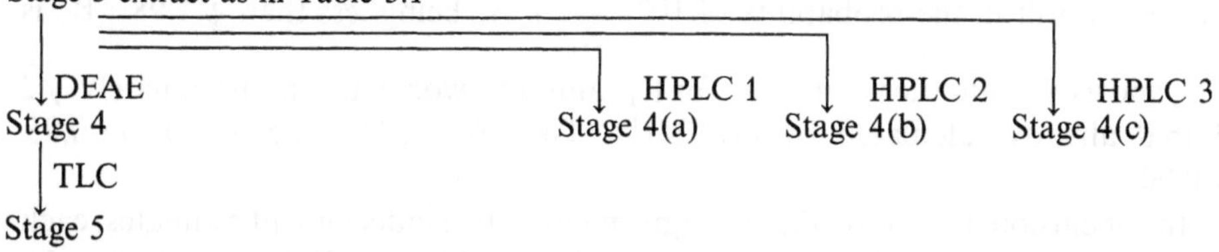

Table 3.3. Set of data illustrating the type of result that might be obtained by the procedure in Table 3.2

Stage	Individual estimates E′ ppm 1	2	3	4	5	Mean estimate E_n	Variance V_n^2
4	5.93	5.99	5.34	6.01	5.75	5.804	0.078
4(a)	5.25	6.12	5.87	6.19	5.28	5.742	0.204
4(b)	5.69	6.01	5.39	6.00	5.62	5.742	0.070
4(c)	5.38	5.74	5.21	6.14	6.04	5.702	0.163
5	5.51	5.65	5.24	5.26	5.81	5.494	0.061

$$V_p^2 = \frac{0.078 + 0.204 + 0.070 + 0.163 + 0.061}{5} = 0.115$$

$$V_t^2 = \frac{\sum E'^2 - \frac{1}{25} \cdot (\sum E')^2}{24} = 0.109$$

On the other hand, if we calculate the variation amongst all the individual analyses which contribute to E_4, $E_{4(a)}$, $E_{4(b)}$, $E_{4(c)}$, and E_5, it will be apparent that this total variance V_t^2 will contain any variation in R′ in addition to the random error V_p^2. Consequently, if V_t^2 is found to be significantly greater than V_p^2, condition (b) cannot be fulfilled and there are strong grounds for rejecting the successive approximation.

In the particular example under discussion, V_t^2 was calculated from the data in Table 3.3 as 0.109, which is just less than the value 0.115 obtained for V_p^2. Thus, on the available evidence, we have no reason to suspect that condition (b) has been violated. However, it must be noted that V_t^2 and V_p^2 are mere estimates of the true total variance s_t^2, and the true precision s_p^2, based on 24 and 20 degrees of freedom, respectively. Consequently the finding $V_t^2 \approx V_p^2$ does not necessarily imply that condition (b) has been fulfilled. What then are the chances of s_t^2 being greatly in excess of s_p^2 when $V_t^2 \approx V_p^2$? The answer lies in the properties of an F distribution of type $F_{(24,20)} = \frac{V_t^2}{V_p^2}$. The lower confidence limit for $F_{(24,20)}$ is 0.48 at a confidence level of 0.95[11], and thus, if $F = \frac{0.109}{0.115} = 0.948$ as was the case for the example in

[11] ie. the integral of the individual probabilities from $F = 0.48$ to $F = \infty$ is 0.95

question, then $\frac{s_t^2}{s_p^2}$ cannot be greater than $\frac{0.948}{0.48} = 1.98$. This means that if condition (b) is not fulfilled, the probability of $10^6 \cdot \frac{R'_\sigma}{R_u} + B_\omega$ being less than $\sqrt{1.98} \cdot B_\omega$ is 0.95, and as a result, the value of A_α for E_4 cannot be worse than approximately $\sqrt{2}$ of that already calculated by Eq. (24)[12]. Thus $A_\alpha \approx 34\%$ and $\alpha = 0.95 \cdot 0.95 = 0.90$.

In conclusion, it appears that the agreement of five independent estimates, each composed of five individual analyses, could provide a sufficient basis for the assumption that the non-random error of an average estimate is unlikely to be worse than two to three times the precision of the method. However, to be pedantic, it must be pointed out that we have assumed the variation in R′ to be normal, and in order to cover the possibility of its being otherwise the number of purification steps should be of the order of 50 to 100, rather than 5. In fact, as the name successive approximation implies, it is possible to keep on purifying and reanalyzing the extract ad infinitum, all the while obtaining a better and better estimate of hormone concentration. This poses the practical question: at what stage do we obtain a sufficient estimate? The answer is complex as it must involve not only the nature of the experiment but also that of the experimenter, and discussion of this point will be left to Section 3.2.2.3. c. For the moment it can be noted that all estimates are guesses, and that a guess based upon a single analysis without any corroborative evidence whatsoever could be a very poor one. However, by the simple expedient of performing one purification step and further analysis, the tentative estimate can be either rejected or transformed into a far more substantial piece of evidence. Situations where such a course of action can not be pursued because of practical constraints are rare.

b) Information Content

A quite different approach to the verification of accuracy involves the use of the information contained within the actual detector response. In the event of interference from impurities the change in the characteristics of the response serves to indicate that accuracy can no longer be guaranteed. Such an approach requires that the information be quantified in objective terms, and this can be achieved through the use of some of the basic concepts of information theory. These concepts are now briefly described and their application to tests of sample purity and analytical accuracy discussed.[13]

All observations yield information which is normally recorded in a coded form appropriate to the technique employed. For example, the height of a plant may be measured in millimeters whereas the response of a flame ionization detector is more conveniently expressed as Coulombs per gram. Regardless of the form of coding

[12] This is a simplification. In practice, the value of A_α would be obtained by substituting the average of E_4, $E_{4(a)}$, ... E_5, for E_n in Eq. (24), using the V_p^2 value calculated from Table 3.3B, and taking t_τ as applying to a Student distribution of 20 degrees of freedom.

[13] For a detailed discourse on information theory consult references 5, 6 and 7 in section 3.10.

Table 3.4A–C. Descriptions of simply hypothetical systems by observation of the properties blackness and roundness

A		B		C		
	Only bit (blackness)		Only bit (blackness)		First bit (blackness)	Second bit (roundness)
●	1	●	1	●	1	1
○	0	○	0	○	0	1
		■	1	■	1	0
				□	0	0

used all information can be defined in terms of a fundamental unit the Binary digit, or the bit. One bit of information is just sufficient to give a yes/no answer on whether a specified phenomenon is observable, and thus allows a distinction to be drawn between two alternatives. For example, consider a system containing one black ball and one white ball. By testing for blackness/non-blackness it is possible to distinguish completely between the individuals of the system, designating the black ball 1, and the white ball 0 (i.e., non-black), see Table 3.4A. However, if a black cube is added to the system it is indistinguishable from the black ball, on the basis of the one bit of information available (Table 3.4B). In order to guard against this possibility a second bit of information must be obtained by employing a test for roundness/non-roundness. From Table 3.4C it is apparent that by obtaining two bits of information it becomes feasible to differentiate between four alternatives. That is, the black cube can be designated the unique code 10, to distinguish it from the black ball (11), the white ball (01), and a white cube (00). Thus, for a system of n components it follows that the amount of information, I, required to furnish a full description will be given by

$$n = 2^I$$

or

$$I = \frac{\log n}{0.3} \text{ bits} \tag{30}$$

If it is decided to record the extent to which a specified phenomenon can be observed, rather than give a simple yes/no answer, the amount of information accrued will have to be greater than 1 bit. For example, when the height of a plant is measured with a one-meter rule and found to be 124 mm, this does not imply that the measurement is in fact exactly 124.0000000 ... mm, but rather, that a decision has been made to equate the height of the plant with one of the thousand discrete size classes available, namely 123.5 to 124.5 mm. Thus the information yielded by measurement with a one-meter rule is just sufficient to describe a system of 1000 components, that is I = 10 bits. By analogy, the number of perceptibly different response classes available from a physicochemical detector should be

$$n = \frac{\text{linear range}}{2 \times \text{precision}}$$

However, the precision is not necessarily constant over the entire analytical range. For example, by employing a suitable standard curve, amounts of IAA from 1 to 100 ng may be analyzed with a precision of ± 1 ng. However, on recalibration for operation over the range 1 to 100 µg, the precision may well increase to around ± 1 µg. Thus a better practical measure of the number of response classes available from an analytical procedure is represented by

$$n = \frac{100}{2 \cdot q} \tag{31}$$

where q is the percentage precision of the greatest estimate catered for by the calibration. It follows that the information content of an analytical procedure of percentage precision q is

$$I = \frac{1.7 - \log q}{0.3} \text{ bits} \tag{32}$$

From Eq. (32) it can be seen that by working to a precision of ± 3%, the information obtained per analysis will be approximately 4 bits.

Having obtained I_1 bits of information about a sample by analyzing it with detector 1, it is possible to gain a further I_2 bits by analysis with detector 2. The total information acquired will be $I_t = I_1 + I_2$, but if the detectors are identical, I_t cannot be greater than I_1 or I_2. Consequently, when summing information arising from a number of sources, it is imperative that the individual measurements be independent, and where this is not the case, allowance must be made for the degree of *correlation*. If each of the components of a sample is first analyzed by detector 1 to give a response of X units per gram, and then by detector 2 giving Y units per gram, the correlation coefficient $z = \frac{\sigma_{x \cdot y}}{\sigma_x \cdot \sigma_y}$ provides a measure of the fraction of two-dimensional variance, $\sigma_x \cdot \sigma_y$ that can be accounted for by the covariance $\sigma_{x \cdot y}$. Thus, the total information arising from the two analyses can be estimated as

$$I_t = I_1 + I_2 (1 - z) \tag{33}$$

and for more complex situations

$$\begin{aligned} I_t = {} & I_1 + I_2 (1 - z_{2,1}) + I_3 (1 - z_{3,1}) (1 - z_{3,2}) + \cdots \\ & I_n (1 - z_{n,1}) (1 - z_{n,2}) \cdots (1 - z_{n,n-1}) \end{aligned} \tag{34}$$

where $z_{n,1}$, $z_{n,2} \cdots z_{n,n-1}$ are the coefficients expressing the correlation between detector n and detector 1, detector n and detector 2 etc. It will be apparent from the form of Eq. (34) that for combinations composed of a large number of detectors, even a small degree of intercorrelation in their responses will result in the actual information obtained being a fraction of that potentially available.

Correlation is not the only phenomenon that prevents the potential information content of an assay from being fulfilled. Another major reason is that much of the information need not necessarily be "informative" in the context in which it is employed. For example, a flame ionization detector will respond to almost any organic molecule so the information contained in its response can be used as a basis for distinguishing between, say, organic and inorganic gases. However, the same information will be of little value when dealing with a sample entirely composed of either organic or inorganic molecules. These considerations make it necessary to account for the information distribution of the source. If a proportion ϱ_1 of the components of the system evokes a response of level 1, i.e., zero to two times the precision [see Eq. (31)], and a proportion ϱ_2 a response of level 2, i.e., two times to four times the precision and so on, then an information source of n levels will provide an amount of information

$$I = -\sum_{l=1}^{l=n} \varrho_l \cdot \log_2 \varrho_l \text{ bits} \tag{35}$$

It follows from the form of Eq. (35) that the responses of the components of a sample must be uniformly distributed across all of the available levels, if the potential information content of $\frac{\log n}{0.3}$ is to be achieved, i.e., $\varrho_1 = \varrho_2 = \varrho_3 \cdots = \varrho_n = \frac{1}{n}$. As a practical example of the need to allow for distribution of information, consider the case of gibberellin (GA) analysis by the lettuce hypocotyl bioassay. Hypocotyl lengths will most likely lie within the range 5 to 25 mm, and can be readily measured with a ruler 32 mm in length, which according to Eq. (30) will yield five bits of information. By the same token, a one-meter (1000 mm) rule could be used to obtain ten bits of information from the assay. This would seem to imply that the longer the rule used to measure hypocotyl length the more "informative" the bioassay. Obviously this is not the case, as much of the ruler's one-meter length would provide information that does not distinguish between individual members of the population of hypocotyls. Common sense dictates that there is no need to employ a ruler of such length and, indeed, if allowance is made for the distribution of information and Eq. (35) applied to both the situation with the one-meter rule and that with the 32 mm rule, the actual information content of the assay will be found to be approximately four bits in both cases (Table 3.5).

In Section 3.2.2.2 it was argued that the more selective the quantification technique the more accurate the results. However, even if a totally unselective method of analysis were to be employed, the non-random error cannot exceed a certain maximum value which is dictated by the purity of the sample, and in this regard an important consequence of information theory is that it allows purity to be defined in objective terms. If a sample is purified beyond a value P_π, where π is the probability of the purity being better than P, and analyzed by a system with an identical selectivity to the technique used to establish purity, the probability, α, of the percentage accuracy, A, being better than $100 - P$ will be equal to π. However, should the selectivity of the technique used for quantification be greater than that of the method employed to establish purity, α will be greater than π. In such a situation, the criterion of purity, P_π, can be used to determine a maximum value of

Table 3.5. Effect of ruler length on information content of a lettuce hypocotyl bioassay. It has been assumed that hypocotyl lengths are normally distributed over the range 5 to 25 mm

Response class mm	Proportion of sample occupying class ϱ_l	$-\varrho_l \cdot \log_2 \varrho_l$
0.5– 1.5	0	0
1.5– 2.5	0	0
2.5– 3.5	0.006	0.044
3.5– 4.5	0.009	0.061
4.5– 5.5	0.013	0.082
5.5– 6.5	0.019	0.109
6.5– 7.5	0.026	0.137
7.5– 8.5	0.034	0.167
8.5– 9.5	0.043	0.197
9.5–10.5	0.053	0.226
10.5–11.5	0.062	0.250
11.5–12.5	0.070	0.271
12.5–13.5	0.076	0.283
13.5–14.5	0.079	0.292
14.5–15.5	0.079	0.292
15.5–16.5	0.076	0.283
16.5–17.5	0.070	0.271
17.5–18.5	0.062	0.250
18.5–19.5	0.053	0.226
19.5–20.5	0.043	0.197
20.5–21.5	0.034	0.167
21.5–22.5	0.026	0.137
22.5–23.5	0.019	0.109
23.5–24.5	0.013	0.082
24.5–25.5	0.009	0.061
25.5–26.5	0.006	0.044
26.5–27.5	0	0
27.5–28.5	0	0
.	.	.
.	.	.
.	.	.
.	.	.
31.5–32.5	0	0 $-\sum_{l=1}^{l=32} \varrho_l \cdot \log_2 \varrho_l \approx 4.2$ bits
.	.	.
.	.	.
.	.	.
.	.	.
999.5–1000.5	0	0 $-\sum_{l=1}^{l=1000} \varrho_l \cdot \log_2 \varrho_l \approx 4.2$ bit

A_α. The magnitude of π will be governed by the amount of information acquired during the analysis, since the certainty with which the compound of interest can be distinguished from the virtually infinite number of potential contaminants will depend upon how comprehensively it can be described.

The test for purity involves taking a sample U which is purported to be compound u and analyzing it by the method of choice. Assuming that this process results in a total response r_t from which I_t bits of information can be extracted, then that part of the information I_u, which correlates with the characteristics of a standard sample of u, can in turn be related to a portion r_u of the total response. In this instance

$$P = \frac{r_u}{r_t} \times 100 \text{ percent} \tag{36}$$

where P is the purity of the response. Although response purity, P, need not reflect the true mass purity it is, nevertheless, the only criterion of purity required to establish accuracy as being less than 100 − P, provided the selectivity of the method used for quantification is equivalent to, or exceeds, that used in the test of purity. The information, I_u, forms the basic statistic with which to test the null hypothesis

$$H_0\colon U = u$$

For example, when analyzing for IAA by mass spectrometry, it might be found that only I_u out of the total I_t bits of the sample spectrum can be correlated with IAA [14]. Thus, on the basis of the information I_t we should reject H_0, and *deduce* that the sample is not IAA. However, since an amount r_u of the total ion charge r_t can be attributed to that part of the spectrum characteristic of IAA, an alternative outcome would be to accept H_0 on the basis of I_u bits of information and *induce* that the sample contains $\frac{r_u}{r_t} \times 100$ percent IAA. As the aim of analysis almost invariably is to demonstrate the presence of a certain substance rather than its absence, the inductive interpretation is the one most often chosen. It is inductive, not deductive, because it asserts that $\frac{r_u}{r_t} \times 100$ percent of the sample is identical to IAA, not only in terms of the information I_u, but also in all other respects, including properties of IAA yet to be elucidated. Thus the verification of purity reduces to the classic philosophical problem of scientific inference, a detailed discussion of which is beyond the scope of this article, and the interested reader is referred to references [1] and [2] of the selected bibliography in Section 3.10. However, it can be noted that, since we are dealing with an open-ended system, the amount of information required to justify the assumption U = u is infinite. One way of looking at this situation is that the information I_u can be used to *deduce* that U is different from every other component of the system with the exception of u. Thus, if the system has n components, and I_u is greater than $\frac{\log n}{0.3}$ bits [Eq. (30)], then it is possible to *induce* with impunity that U = u. However, as no absolute limits can be placed upon the variety of organic compounds, n is infinite, and likewise the amount of information required to define purity absolutely will be infinite. [15]

If purity cannot be defined in absolute terms, even in the case of standard reference compounds, then what exactly is meant by the null hypothesis? In

[14] The elucidation of the information content of mass spectra is discussed in Section 3.3.3

[15] This result can be obtained intuitively. An absolute value has an infinite number of significant figures, e.g., 3.612895768 ... and thus can be described only by an infinite amount of information

practice H_0 can be taken to assert that the sample, U, is identical to the standard, u, thereby possessing all the *known* properties of u. Under these circumstances the amount of information required to validate H_0 is not infinite, but is related to the magnitude of the body of knowledge describing u. In other words, if we assert that a substance extracted from a plant source is IAA, it is necessary to demonstrate that the putative IAA possesses every *known* property of IAA. A question of fundamental importance therefore is, how much information is contained in a description such as IAA? At first sight it might appear that the information content of so short a word cannot be large, but, here we are dealing with a semantic coding of the original chemical data, and need to interpret the information content of u by translation from the semantics of plant physiology to those of chemistry as follows, IAA ≡ indole-3-acetic acid. Likewise, the term indole-3-acetic acid will not mean very much to the majority of the English-speaking population, yet to a person familiar with the language of chemistry, this very simple phrase can be expanded until ultimately it embraces the entire state of knowledge of the physical sciences. Thus the information content of the concept coded for by the word IAA is not infinite, but none the less very, very large.

If we exclusively address ourselves to the problem of identification within a system completely composed of organic compounds, it is possible to estimate the information content of a semantically coded phrase such as IAA. The basic description of a compound lies in its physical characteristics such as mass, infra-red (IR), ultra-violet (UV), nuclear magnetic resonance (NMR), and optical rotatory dispersion (ORD) spectra, melting point, and crystal structure, and provides at least 1000 bits of information (see Section 3.3.3). However, when an unknown substance is characterized it is not sufficient simply to list its physical properties. The process of semantic reduction requires that the properties of the unknown be correlated with those of effectively every known organic compound through partial or total synthesis. As the number of characterized organic compounds is approximately 4×10^6 [16], the information content of a semantic phrase that describes a substance characterized to a level corresponding to the present state of knowledge is $4 \times 10^6 \times 10^3$, i.e., 4×10^9 bits. When armed with such a vast amount of information it is evident that today's organic chemist can distinguish one compound from $10^{1,000,000,000}$ other possible substances. Thus, although the inductive process of chemical characterization cannot be applied to the open-ended situation with total safety, it is apparent that there are means with which to analyze systems of virtually incomprehensible complexity.

Because chemical characteristics require the acquisition of large amounts of information, a considerable quantity of sample must be available for analysis. For example, the discovery of many of the GA's required the extraction of tens and even hundreds of kilograms of plant material. In the case of routine plant hormone analysis it is quite impractical to use samples of such size, and, consequently, the test of H_0 has to be conducted with a reduced amount of information. As a result, I_u is almost invariably very much less than 10^9 bits, and it is possible to fail to reject H_0 when it is untrue. It is, therefore, essential to have some measure of the probability

[16] The Chemical Abstracts service recorded the four millionth unique compound in December 1977 (Chemical and Engineering News, 1977, Dec. 5th, p. 37)

of wrongly inducing U = u. Indeed, provided purity is defined as a response purity according to Eq. (36) the probability, π, of the purity being greater than P will be the probability of rejecting H_0 when it is untrue. The term π is therefore of great importance as it objectively quantifies the subjective probability assessment contained in such statements as,

"On the basis of its retention behaviour in three TLC systems we can be *almost certain* that compound X is IAA".

or,

"The appearance in the sample mass spectrum of an ion peak corresponding to the molecular ion of the trimethylsilyl ester of IAA (TMSi-IAA) provides *conclusive* evidence that the sample contains IAA".

In these examples relatively small amounts of information have failed to reject the hypothesis U = IAA, and as a result H_0 has been taken as being essentially true. However, do the values of π applicable to these two cases justify the descriptions "almost certain" and "conclusive"? There is a considerable margin for error considering that the abbreviation IAA can distinguish one compound from $10^{1\,000\,000\,000}$ others, while the amount of information acquired from three TLC procedures is unlikely to embrace a system of more than 10^3 components.[17] The answer to the question lies not so much in a particular dogma as in the relationship between I_u and π.

The thermodynamics of the biosphere dictate that not all aggregates of matter will be equally stable, and while an infinite number of organic compounds may be possible, only a limited number of the total possibilities will in fact be present in significant amounts. Thus, if I_u can account for all the major components which together represent a proportion π of the total terrestrial biomass, then the probability of *not* making an error in inducing U = u will be π. Several approaches could be taken to elucidate a confidence limit on the number of compounds present in the biosphere or, more specifically from our point of view, in a plant extract. A semi-empirical method will be discussed, which, although far from exact, is useful in that it requires a minimum of assumptions and involves concepts of considerable importance.

A proportion, δ, of the mass of an extract can be represented by the molecular formula $C_vH_wO_xN_y$, where v is unlimited, x and y are both < 6 and $2v + 2 > w > \frac{6v - 5y - 16x + 18}{9}$. The latter expression is an approximation which limits the degree of saturation to between full saturation and that typical of condensed aromatics. Since both very small and very large combinations of C, H, O, and N are not as stable as compounds of intermediate size, a proportion ε of the mass of the extract will lie between the molecular weight limits a and b. For any nominal molecular weight, n, there will be K_n molecular formulae that comply with the rules of combination of C, H, O, and N, and thus the total number of formulae possible in the range a to b amu is $K_a + K_{a+1} + K_{a+2} \ldots K_{b-1} + K_b$. Because any one formula of molecular weight n could be satisfied by as many as H_n isomers, the total

[17] The information content of chromatographic procedures is discussed in Section 3.4

number of compounds that could occur in the molecular weight range a to b is J, where

$$J = K_a \cdot H_a + K_{a+1} \cdot H_{a+1} + K_{a+2} \cdot H_{a+2} \ldots K_{b-1} \cdot H_{b-1} + K_b \cdot H_b \qquad (37)$$

It follows that a proportion $\delta \cdot \varepsilon$ of the mass of a plant extract cannot contain more than J possible compounds, in which case, the probability of U being one or more of these when $U \neq u$ will be $\delta \cdot \varepsilon$. Therefore provided $I_u < \frac{\log J}{0.3}$ bits [Eq. (30)], the probability of rejecting H_0 when it is not true is $\delta \cdot \varepsilon$. Thus

$$\pi = \delta \cdot \varepsilon \qquad (38)$$

However, experience of the composition of a great number of natural products would indicate that the limits placed on w, x, and y are sufficiently unrestrictive to permit the assumption that δ has a value very close to unity.[18] Consequently, π can be taken as being equivalent to ε, and the amount of information required to fix π at an acceptable level can be calculated from the molecular weight distribution of the extract.

The foregoing argument can be applied to the practical situation in the following manner. Figure 3.9 illustrates a gel permeation chromatogram (GPC) which has been used to assess the molecular weight distribution of an acidic, ethyl acetate-soluble extract from *Phaseolus coccineus* seedlings. The data provide an accurate reflection of the molecular weight distribution of the extract since it has been experimentally verified that (1) mechanisms other than gel permeation do not contribute significantly to the separation (2) the differential refractive index response is proportional to the mass of extract eluted and (3) the molecular weight calibration is an average for a variety of molecular shapes. Figure 3.9 indicates that 90% of the extract is composed of compounds whose molecular weights range from 100 to 1400 amu. Thus ε can be taken as 0.9, and Eq. (37) solved for J between the limits a = 100 and b = 1400 amu.

By inspection of the molecular formulae presented in the Mass and Abundance Tables of Beynon and Williams (1963), the approximate relationship between K_n and n can be derived as

$$K_n = 0.46\,n - 18 \qquad (39)$$

However, the relationship between H_n and n is not so easily obtained. Despite the attention this problem has received over the last 60 years (see review by Rouvray, 1974), no one simple method exists for the exact calculation of H_n. However, most authors are in agreement that for a given homologous series, log H is quite closely proportional to the carbon number v. Furthermore, the homologous series of different classes of compounds produce parallel log H versus v curves, which differ

[18] In the case of very large carbon numbers, the limits on x and y may well be too restrictive to maintain an acceptable value of δ. This need not be a problem since we have the means to measure and manipulate the molecular weight distribution of extracts. The practical importance of this point will be discussed in Section 3.2.2.3 c

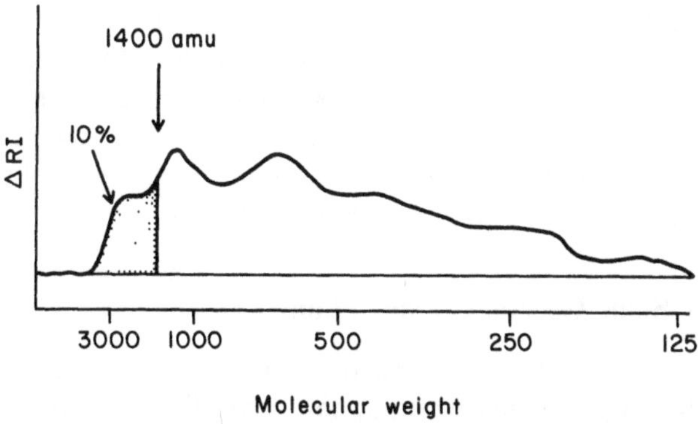

Fig. 3.9. Gel permeation chromatogram of an acidic, ethyl acetate extract from seedlings of *Phaseolus coccineus* cv. Prizewinner. *Column:* 2 m × 25 mm Biobeads SX-4. Solvent: tetrahydrofuran. Detector: differential refractometer. *Shaded area* represents 10% of the total sample and corresponds to all compounds with molecular weights greater than 1400 amu

from one another by no more than $\log H = \pm 2$ for a given carbon number (see FRANCIS, 1947; HENZE and BLAIR, 1934; and more recently LEDERBERG et al., 1969). Because carbon number and thus molecular weight are related to log H, rather than H, the RHS of Eq. (37) is dominated by the highest molecular weight term in H, and can be reduced to

$$J \approx K_b \cdot H_b \tag{40}$$

Thus, in the case of the *Phaseolus* extract only the values of K and H pertinent to compounds of molecular weight 1400 need be considered. Using the data of HENZE and BLAIR (1931) for isomeric hydrocarbons of the methane series between C_{10} and C_{40}, the relationship between H and v can be calculated as

$$\log H = 0.41\,v - 2.6 \tag{41}$$

From this equation it can be predicted that the compound $C_{100}H_{202}$, or for that matter, almost any compound of molecular weight 1400, will have approximately 10^{39} potential isomers. However, there is some evidence that, for reasons of symmetry, H will asymptotically approach a limiting value of about 10^{40}. For example, ROUVRAY (1974) quotes GONCHOROV (1973) as calculating the number of isomers of $C_{400}H_{802}$ as only 10^{40}, and therefore the estimate of 10^{39} for $C_{100}H_{202}$ is most likely an overestimate rather than underestimate. Even so it is of use in that it provides an upper limit for H_b.

Since the number of possible molecular formulae having a nominal molecular weight of 1400 can be calculated from Eq. (39) as $10^{2.8}$, an approximate upper limit of 10^{42} can be placed on J via Eq. (40). Thus, in the case of the *Phaseolus* extract, any test of H_o capable of distinguishing between 10^{42} compounds will account for 90% of the mass of the extract, or, put another way, if the substance under analysis is not, say, IAA, the probability of it being one of *only* 10^{42} other possibilities cannot be less than 0.9. The amount of information required to describe a system of 10^{42} components completely can be calculated to be 140 bits from Eq. (30), and thus I_u must exceed some 140 bits if the identity and hence purity of a substance derived from the *Phaseolus* extract is to be defined with a probability of $\pi = 0.9$. It could be argued that in view of the crudity of the approximations, the figure obtained for the minimum information requirement is unlikely to be very meaningful. However, even if J were underestimated by one million-fold, the minimum acceptable value of I_u would increase by only 20 bits. When this point is considered along with the fact that some analytical techniques can generate many hundreds of bits of information, it is apparent that practical means are available to establish purity and quantitative accuracy with a high degree of certainty.

c) Some General Considerations

Two methods have been described, successive approximation and information content, by which accuracy can be verified. Although absolute accuracy cannot be established, regardless of the approach taken, by expending greater and greater amounts of labour and sample, it is possible to obtain estimates of ever-increasing reliability. It is, therefore, of practical importance to be able to ascertain when the successive approximation comprises an adequate number of stages or at what point sufficient information has been accrued. Decision-making processes involving the setting of confidence levels are almost entirely subjective and relate not only to the nature of the experiment but also to that of the experimenter. In this regard, Science is no more free from manifestations of Man's propensity for gambling than any other sphere of human activity. Indeed, one of the most wide-ranging variables in the total experimental system is the degree of personal involvement of the analyst in the outcome of the analysis, and while for a given situation one group of people may reject estimates where α lies below 0.7, for others, incentives may be such that even absurdly low odds appear to be "almost certain". Consequently, if personal bias is to be reduced, there is no choice but to estimate and quote A_α for every determination of hormone concentration.

Regardless of the uncertainties associated with the interpretation of α, one fact remains abundantly clear. In plant physiology, measurements of hormone level are rarely made for their own sake; instead the estimates are deployed in support of some physiological model, where α will represent but one of a number of imponderables pendent to a central hypothesis. For example, one may hypothesize that the geotropic control of root growth is exerted by an upward diffusion of abscisic acid (ABA) from the root cap, and might test this by analyzing the ABA levels in tissue sectioned from ten different regions of the root. The probability of not making an error in accepting the proposed hypothesis will be the combined probability of the ten α values plus a number of other probabilities associated with parasitic hypotheses, such as that relating the ability of ABA to modify the rate of root extension. It is apparent that should any one of these uncertainties have an exceptionally low confidence level, then the combined probability also must be low. However, equally important is the fact that while an α value of 0.8 may be acceptable for a single estimate, the combined probability of ten α values each of 0.8 is 0.11 (Fig. 3.10) and sufficiently low to discourage all but the most committed gambler. In practice, therefore, it is difficult to justify the use of an analytical method when α values fall below 0.9 and in many situations there is a good case for working towards even higher levels of confidence.

In Sections 3.2.2.3. a and 3.2.2.3. b it was argued that the complexity of a typical plant extract is such that a five-stage successive approximation or 140 bits of

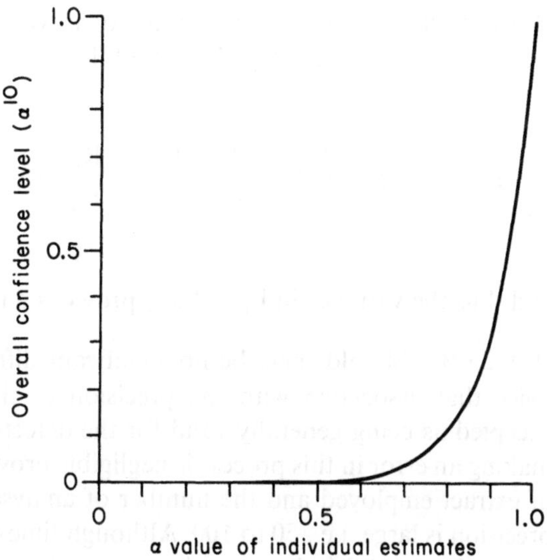

Fig. 3.10. Plot of the equation $y = x^{10}$ which illustrates the influence of the α value of individual estimates upon the confidence level of a physiological model relying on ten such estimates

information are necessary to guarantee an α value greater than 0.9. In favourable cases it is feasible to achieve such criteria, but even the most powerful analytical techniques would require the sample to contain about a microgram of the substance of interest. Bearing in mind that the level of plant hormone in vegetative tissue is typically of the order of a 1 ng per gram fresh weight, it is apparent that for routine analyses less stringent criteria must be applied. Therefore, given the present state of technology, is it a practical proposition to analyse small samples of vegatative tissue with acceptable accuracy? To answer this question it is necessary to re-examine the ways in which the criteria of Section 3.2.2.3. a and 3.2.2.3. b were derived. The quoted α values are "worst case" estimates and, in practice, the actual values could be considerably better. For example, α will account for all uncertainties in the successive approximation provided the number of stages is sufficiently large to yield a good measure of the variance in R′. However, under certain circumstances the properties of the chosen detector may be such that the term $10^6 \cdot \frac{R'_\sigma}{R_u}$ is always small compared with B_ω [condition (b), p. 224] and consequently a two-stage approximation is all that is necessary to guarantee accuracy. In Section 3.2.2.3. a it was argued that this type of behaviour is likely to occur when a highly selective detector is used for the analysis of a relatively crude extract, and it is exactly this situation which typifies the most effective plant hormone assays. In practice, therefore, it may be permissible to reduce the number of stages in the approximation to only two, thus saving considerable sample and labour.

In view of the paucity of knowledge concerning sensitivity distributions, it would seem premature to assume condition (b) to be a general property of an analytical method, without first providing evidence. If 50 to 100 different extracts, typical of the range of samples to be analyzed, are taken and a two-stage successive approximation performed on each, according to Eq. (28) the individual pairs differences, $E_1 - E_2$, will be inversely proportional to the purification factor Q_1.

However, if Q_1 is greater than 10 to 20, as will be the case when a suitable HPLC technique is employed, Eq. (28) simplifies to

$$(E_n - E_{n+1})_{\sigma^2\omega} \approx 10^6 \cdot \frac{\bar{R}}{R_u} \pm \left[10^6 \cdot \frac{R'_\sigma}{R_u} + \sqrt{2} \cdot B_\omega\right] \text{ppm} \tag{42}$$

and thus the variance in $E_n - E_{n+1}$ provides a measure of the deviation, $10^6 \cdot \frac{R'_\sigma}{R_u} + \sqrt{2} \cdot B_\omega$. Should there be no significant difference between this variance and twice that associated with the precision of the method, condition (b) can be accepted as being generally valid for the detector in question. The probability of making an error in this process is negligible provided the number of different types of extract employed and the number of analyses contributing to the estimate of precision is large, i.e., 50 to 100. Although time-consuming to carry out, the effort required to establish the generality of condition (b) is compensated for by the ease with which all subsequent samples can be processed.

Just as the five-stage criterion for successive approximation might prove to be excessively stringent in many cases, verification of accuracy by the accumulation of 140 bits of information from the quantified response applies to "worst case" rather than typical analyses. In Section 3.2.2.3. b it was argued that where molecular weight constrains the *potential* number of compounds in a proportion ε of an extract to J [Eq. (40)], then provided $I_u > \frac{\log J}{0.3}$ bits, the probability of U being one of the J compounds when $U \neq u$ is ε. However, this relationship will be valid only when the sample contains equal amounts of each of the J possible compounds. In practice, the extract will be dominated by a relatively small number of major constituents, and consequently the *actual* number of compounds making up a proportion ε of the extract will be only a fraction, $\frac{1}{G}$, of that given by Eq. (40). Hence the amount of information required to maintain a chosen value of ε and thus π [see Eq. (38)], is not $\frac{\log J}{0.3}$ but rather $\frac{\log J - \log G}{0.3}$ bits. Conversely, if 140 bits of information can be obtained, it is apparent that greater than 90% of the mass of the extract can be adequately described and π will exceed 0.9. The empirical deduction that the variety of organic matter is distinctly limited by factors apart from those of a purely thermodynamic nature is of fundamental importance to any open-ended analysis of biological phenomena. For example, what reasons lie behind the dearth of D-amino acids compared with the abundance of their L-analogues, or L-sugars compared with D-sugars, etc? A detailed discussion of this point is beyond the scope of this article,[19] but it suffices to say that, at present, there is no way in which this phenomenon can be quantified and G calculated, although experience of a range of natural products would suggest that G is very large. Thus there is no option but to err on the side of caution and employ the J of Eq. (40) rather than any lesser number.

---

[19] The interested reader should consult references [3] and [4] in the Selected Bibliography

In the analytical context, natural processes are not the only ones which restrict the variety of organic matter in the sample. Any purification procedure, if it is at all effective, must reduce the number of compounds that U might be when $U \neq u$; therefore the amount of information required to guarantee the accuracy of an estimate is reduced subsequent to purification. For example, of the J possible compounds given by Eq. (40), a proportion D will be acids, and thus by dealing exclusively with an acid fraction the amount of information required for an acceptable value of π can be reduced by $\frac{\log D}{0.3}$ bits. However, apart from inspecting every one of the 10^{42} possible structures, how is it possible to ascertain the value of D? It seems reasonable that the average purification factor, Q, of a technique capable of separating acids from neutral and basic species will be a function of D, but the exact nature of this relationship is far from clear. One complicating factor is that Q is not absolute and contains sampling variance. Furthermore, it is not possible to separate acids completely from the neutral and basic components of the extract, and thus the value of D as "seen" by the purification procedure will be uncertain. There would, therefore, seem to be no exact method for deriving the amount of information required to guarantee the accuracy of an estimate made subsequent to a chromatographic step of known average purification factor. Consequently it is necessary to err on the side of caution once again and cater for the J possible compounds of Eq. (40) regardless of the degree of purification to which the sample has been subjected.

Considering that J is of the order of 10^{42} while typical purification factors range from only 2 to 50, it can be intuitively seen that, even if an exact value of D could be determined, for the great majority of chromatographic procedures it would not ameliorate the severity of the information requirement. However, this is not the case with GPC where, by selecting for a narrow range of molecular sizes, it is possible to greatly restrict the number of compounds *potentially* present in the sample. For example, IAA has a molecular weight of 175 so that by selecting only that portion of the eluent from the GPC column which contains IAA it is possible to obtain a sample where a large proportion, ε, of the mass is accounted for by compounds with a molecular weight of less than, say, 200. This represents a great simplification of the analytical situation, since by calculation from Eqs. (39), (40) and (41), J cannot be larger than 7.5×10^4 and thus *only 16 bits of information are required to guarantee a value of* π *greater than* ε. The amounts of information pertinent to the analysis of other groups of plant hormones following GPC are listed in Table 3.6. In all cases molecular weight limits have been chosen to just include the substance of interest and thus ε usually will be greater than 0.9, though the exact value will depend upon the molecular weight resolution of the column and has to be determined by experiment.

There are other advantages to working with a sample of restricted molecular size. Apart from the dramatic reduction in the amount of information needed to test H_o, the limits placed upon the molecular formula $C_vH_wO_xN_y$ are less likely to be restrictive at low carbon numbers, and thus, under such special circumstances, the assumption $\delta \approx 1.0$ (p. 234) is readily justified. Furthermore, since the number of possible compounds is comparatively small, it should be feasible to employ structure-generating algorithms, such as the Heuristic Dendral of LEDERBERG et al.

Table 3.6. Amount of information necessary to guarantee accuracy when gel permeation chromatography is used to restrict the molecular weight range of the extract

	MW	MW cut-off	I_u (bits)
IAA	175	200	16
Zeatin	219	250	22
ABA	264	300	27
GA_1	348	400	36
Zeatin riboside	349	400	36

(1969), to furnish more exact J values. This particular programme has the advantage that entire groups of unlikely compounds such as the D-amino acids and their derivatives can be "bad listed" and thus excluded from the computation. It is ironic therefore that GPC should be the chromatographic technique which is least frequently employed in plant hormone analysis. This state of affairs probably represents little more than a fashionable pattern of use, although it could be argued that the technique has not found widespread application because of the sheer size and instability of classical gel columns, as well as the fact that purification factors are quite low, e.g., 2 to 3. However, with the advent of GPC supports specifically designed for HPLC, the first reason need no longer be a problem. The new generation of HPLC supports renders this technique as rapid and convenient as any facet of HPLC. The low purification factors are to be expected in view of the statement made earlier that, of the infinite number of compounds potentially present in plant extracts, very few, especially higher molecular weight structures, are abundant. The importance of GPC, therefore, lies not so much in its ability to limit the number of compounds *actually* present in an extract through purification, but rather in restricting the number of substances that *potentially* might be U when $U \neq u$.

3.3 Assays

3.3.1 Bioassays

Historically, the main impact of bioassays has been to establish the existence of endogenous plant growth regulators and to aid in the isolation of these substances in a pure form. Many cytokinins and GA's, as well as IAA and ABA, were originally detected in plant extracts by means of bioassays. Without biological activity to serve as an indicator it is unlikely that the extensive purification, which must precede rigorous chemical analysis, could have been carried out. Bioassays also have been extensively used as a means of analyzing the hormone content of plant extracts. Despite a growing awareness of their distinct limitations in this role and the potential of physicochemical methodology, the sheer simplicity of the bioassay is likely to ensure its popularity as an analytical tool for some time to come.

Table 3.7 Auxin bioassays

Method	Reference	Minimum detectable level of IAA	Range of linear response to IAA
Avena curvature bioassay	WENT (1928) KALDEWEY et al. (1969)	10^{-7} M	10^{-7}–5×10^{-6} M
Split pea-stem curvature bioassay	WENT (1934) VAN OVERBEEK and WENT (1937)	10^{-6} M	10^{-6}–4×10^{-5} M
Pea-stem section bioassay			
a) Dark-grown	THIMANN and SCHNEIDER (1939)	10^{-7} M	10^{-7}–10^{-4} M
b) Light-grown	OCKERSE and GALSTON (1967)	10^{-7} M	10^{-7}–10^{-4} M
Cucumber hypocotyl section bioassay	SAKURAI et al. (1974)	10^{-7} M	10^{-7}–5×10^{-5} M
Pea root section bioassay	AUDUS and THRESH (1953)	10^{-11} M	3×10^{-8}–10^{-5} M
Cress root growth bioassay	MOEWUS (1949)	10^{-11} M	5×10^{-8}–5×10^{-6} M
Avena coleoptile straight-growth bioassay	BONNER (1949) NITSCH and NITSCH (1956)	3×10^{-8} M	3×10^{-8}–5×10^{-6} M
Triticum coleoptile straight-growth bioassay	HANCOCK et al. (1964) NITSCH and NITSCH (1956)	5×10^{-8} M	5×10^{-8}–5×10^{-5} M
Avena first internode bioassay	NITSCH and NITSCH (1956)	5×10^{-9} M	5×10^{-9}–5×10^{-6} M

Over the years many different types of plant hormone bioassay have been devised and the essential features of a selection of the available test systems are presented in Tables 3.7–3.10. In most cases the detection limits and linear range are quite adequate but, unfortunately, the same cannot be said for the selectivity. Even though bioassays are moderately selective for classes of plant hormones when compared with many physicochemical assays, no known system is entirely free from interaction with extract impurities. Thus, regardless of the repeatability of bioassay data, the accuracy is always open to question until such time as verification is achieved by reference to a more definitive technique.

The following example involves the estimation of GA levels in *Phaseolus coccineus* seedlings and illustrates the problems that can severely restrict the interpretation of bioassay data. The points raised do, however, apply equally well to the analysis of other hormones, as well as to extracts from other plant sources. The acidic, ethyl acetate-soluble fraction obtained from a methanolic extract of light-grown *Phaseolus* seedlings was partially purified by Sephadex G-10 and charcoal-celite column chromatography, and then divided into two. One portion was subjected to TLC and the other to liquid chromatography (LC) on a silica gel partition column eluted with a hexane-ethyl acetate gradient (POWELL and TAUTVYDAS, 1967). When a 1/60 aliquot of each chromatographic fraction was

Table 3.8 Cytokinin bioassays

Method	Reference	Minimum detectable level of kinetin	Range of linear response to kinetin
Tobacco callus bioassay	MURASHIGE and SKOOG (1962)	5×10^{-9} M	5×10^{-9}–10^{-7} M
Soybean callus bioassay	MILLER (1963) MANOS and GOLDTHWAITE (1976)	2×10^{-8} M	2×10^{-8}–5×10^{-5} M
Carrot explant bioassay	CAPLIN and STEWARD (1949) LETHAM (1967)	5×10^{-9} M	5×10^{-9}–5×10^{-7} M
Radish cotyledon bioassay	LETHAM (1968)	5×10^{-8} M	5×10^{-8}–10^{-6} M
Lemna minor dark-growth bioassay	HILLMAN (1957)	3×10^{-7} M	3×10^{-7}–3×10^{-6} M
Amaranthus betacyanin bioassay	BIGOT (1968) BIDDINGTON and THOMAS (1973)	10^{-7} M	10^{-7}–10^{-5} M
Xanthium senescence bioassay	OSBORNE and McCALLA (1961)	5×10^{-7} M	5×10^{-7}–5×10^{-5} M
Barley senescence bioassay	KENDE (1965)	10^{-8} M	10^{-8}–10^{-5} M
Cucumber chlorophyll formation bioassay	FLETCHER and McCULLAGH (1971)	5×10^{-9} M	5×10^{-9}–5×10^{-5} M
Barley root growth inhibition	VAN ONCKELEN and VERBEEK(1972)	10^{-8} M	10^{-8}–10^{-5} M
Etiolated bean leaf disc	MILLER (1962)	10^{-7} M	10^{-7}–10^{-4} M
Funaria bud induction	BRANDES and KENDE (1968)	10^{-8} M	10^{-8}–10^{-5} M

tested in the Tanginbozu dwarf rice bioassay the LC fractions gave rise to a greater overall response and revealed more zones of biological activity than did the TLC fractions (Fig. 3.11). Thus, it would appear that the higher peak capacity of LC has resulted in a better separation of the GA's from one another as well as from impurities. Support for this view was obtained when a 1/120 aliquot of each chromatographic fraction was assayed. The LC fractions showed the anticipated reduction in biological activity, whereas the TLC fractions actually displayed enhanced activity at the lower dose (Fig. 3.11). Such anomalous dose-response behaviour clearly indicates that the selectivity of the bioassay is insufficient to cope with the level of interfering substances.

Even if it were possible to establish that the growth promotion in Fig. 3.11 was exclusively due to the action of GA's, it would still be very difficult to obtain a meaningful measure of the actual amount of GA present. To express the data as micrograms of GA_3 equivalents is misleading because the dose-response

Table 3.9 Gibberellin bioassays

Method	Reference	Minimum detectable level of GA_3	Range of linear response to GA_3
Tanginbozu dwarf rice leaf-sheath bioassay	MURAKAMI (1968)	3×10^{-13} mol	3×10^{-13}–3×10^{-9} mol
Progress No. 9 dwarf pea epicotyl bioassay	KÖHLER and LANG (1963)	3×10^{-12} mol	3×10^{-12}–3×10^{-9} mol
Dwarf maize leaf sheath bioassay	PHINNEY (1956)	10^{-11} mol	10^{-11}–10^{-8} mol
Cucumber hypocotyl bioassay	BRIAN et al. (1964)	3×10^{-12} mol[a]	3×10^{-12}–3×10^{-9} mol
Lettuce hypocotyl bioassay	FRANKLAND and WAREING (1960)	3×10^{-9} M[b]	3×10^{-9} –3×10^{-6} M
Barley α-amylase half seed bioassay	NICHOLS and PALEG (1963) JONES and VARNER (1967)	3×10^{-10} M	3×10^{-10}–3×10^{-6} M
Rumex leaf senescence bioassay	WHYTE and LUCKWELL (1966)	3×10^{-10} M	3×10^{-10}–3×10^{-7} M

[a] Test compound GA_4
[b] Test compound GA_7

characteristics vary markedly from one GA to another. Not only can the threshold dose differ by several orders of magnitude but there is often no parallelism between the slopes of the response curves, and, as a further complication, the size of the dose required to saturate the response can vary substantially (REEVE and CROZIER, 1974). These factors must exclude the use of a biological response as the basis for the quantification of an unknown GA. However, estimates of biological activity are

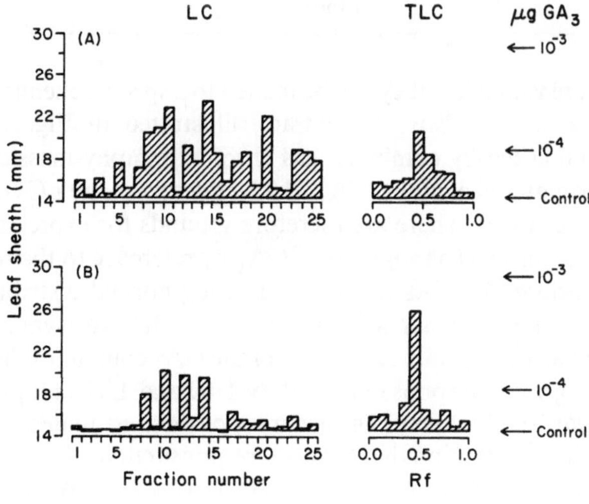

Fig. 3.11 A, B. Tanginbozu dwarf rice bioassays of eluates from a silica gel partition column (*LC*) and a silica gel G thin layer chromatogram (*TLC*) developed with ethyl acetate-chloroform-formic acid (50:50:1) from 60 red light-grown seedlings of *Phaseolus coccineus* cv. Prizewinner. Eluates were tested at (**A**) 60- and (**B**) 120-fold dilutions

Table 3.10 Abscisic acid bioassays

Method	Reference	Minimum detectable level of ABA	Range of linear response to ABA
Commelina stomatal closure bioassay	TUCKER and MANSFIELD (1971)	10^{-10} M	10^{-10}–10^{-4} M
Hordeum stomatal closure bioassay	CUMMINS et al. (1971)	$<10^{-7}$ M	–
Lemna growth bioassay	VAN STADEN and BORNMAN (1970) TILLBERG (1975)	10^{-11} M	10^{-11}–10^{-6} M
Avena first internode bioassay	NITSCH and NITSCH (1956) TAYLOR and BURDEN (1972)	$<4\times10^{-8}$ M	$<4\times10^{-8}$–4×10^{5} M
Avena coleoptile bioassay	NITSCH and NITSCH (1956) TILLBERG (1975)	10^{-7} M	10^{-7}–10^{-5} M
Triticum coleoptile bioassay	NITSCH and NITSCH (1956) TAYLOR and BURDEN (1972)	$<4\times10^{-8}$ M	$<4\times10^{-8}$–4×10^{-6} M
Lettuce germination bioassay	McWHA et al. (1973)	10^{-6} M	10^{-6}–10^{-5} M
Lettuce hypocotyl bioassay	TAYLOR and BURDEN (1972)	2×10^{-6} M	2×10^{-6}–$>2\times10^{-5}$
Barley half seed bioassay	CHRISPEELS and VARNER (1966) SIVORI et al. (1971)	5×10^{-8} M	5×10^{-8}–5×10^{-6} M
Cotton explant abscission bioassay	ADDICOTT et al. (1964) DAVIS et al. (1972)	4×10^{-12} mol	4×10^{-12}–4×10^{-10} mol
Rice seedling bioassay	OGAWA (1963) KOSHIMIZU et al. (1966)	3×10^{-7} M	3×10^{-7}–10^{-5} M

more valuable if they can be related to a specific chemical entity. For instance, in the case of the *Phaseolus* extract, illustrated in Fig. 3.11, it has previously been established by combined gas chromatography-mass spectrometry (GC-MS) that the material contains GA_1 which elutes from the LC column in fraction 14 (BOWEN et al., 1973). There are therefore grounds for expressing the biological activity in fraction 14 as nanograms of GA_1 by reference to the regression of the response on log-dose GA_1. As this involves a log-normal distribution the estimate will be a median rather than a mean value and will have asymmetric 95% confidence limits. Table 3.11 contains estimates of the GA_1 content of the *Phaseolus* extract based on the growth response induced by 1/60 and 1/120 aliquots of LC fraction 14 in the dwarf rice bioassay. The accuracy of these estimates is dependent upon the validity of the assumption that the dose-response curve of fraction 14 exactly mirrors that of the GA_1 dose-response curve. As halving the dose size had no significant effect on

Table 3.11 Estimated GA_1 content of an extract from 60 light-grown *Phaseolus coccineus* seedlings

Bioassay	Estimated GA_1 levels (ng)	
	Median value	Upper and lower 95% confidence limits
Dwarf Rice	18 [a]	5–60
	20 [b]	4–48
Lettuce hypocotyl	600 [c]	130–1400
Barley α-amylase half-seed	700 [c]	300–1200

[a] 1/60 aliquot assayed
[b] 1/120 aliquot assayed
[c] 1/6 aliquot assayed

the GA_1 estimates it would seem that the assumption is at least partially justified and that LC fraction 14 is acceptably pure. However, interference from extraneous material need not be revealed by testing at more than one dose. It was therefore of interest to analyze the extract in other bioassays which offered different selectivities. When LC fraction 14 was tested in the lettuce hypocotyl and barley α-amylase half-seed bioassays, the GA_1 estimates obtained were much higher than those based on the dwarf rice bioassay (Table 3.11). Without further investigation it is impossible to establish which figure is the most accurate, so under the circumstances the best estimate of the GA_1 content of the *Phaseolus* extract is 4–1400 ng. It should be noted that at no stage has rigorous proof of accuracy been obtained and thus there is no guarantee that the actual GA_1 content of the extract lies within even this broad range.

The problems discussed above are typical of those encountered in the analysis of trace components in complex mixtures when the selectivity of the detector system is inadequate, and they arise whenever a bioassay procedure is applied to the analysis of hormones in plant extracts. While an estimate of hormone content derived from a single bioassay may be precise, it is of little value, as the accuracy will always be open to question. By combining the results of several different types of bioassay, the effects of extract impurities can, to some degree, be randomized, resulting in a more accurate but less precise estimate. However, if the extract is at all impure, the improvement in accuracy will be offset by the loss in precision and in practice this usually means that the combined estimate is of little more use than that based on a single bioassay. Since bioassay specificity clearly cannot be altered, the only way in which accuracy can be enhanced is to purify the extract further to such a stage that the criteria for establishing A_α can be met (see Section 3.2.2.3. a).

3.3.2 Immunological Assays

Although immunological assays are extensively employed in the field of mammalian endocrinology, reports of their application to plant hormone analysis are limited in number (Fuchs, S. et al., 1971; Fuchs, Y. et al., 1972; Fuchs, Y. and

GERTMAN, 1974; PENGELLY and MEINS, 1977). This should not be taken to indicate their general unsuitability in this role, as the selectivity, limits of detection, and simplicity of a well-designed radioimmunoassay at least rival, and often exceed, those of many of the techniques currently used to analyze plant hormones.

Radioimmunoassay is based on the fact that when a saturable amount of antibody is reacted with a fixed quantity of radioactively labelled hormone, any unlabelled hormone present will compete with the labelled species for binding sites on the protein. After equilibration, bound and free hormone are separated and the radioactivity in either fraction is related to the hormone content of the extract via an appropriate standard curve. Typically, such curves are sigmoidal and this restricts the linear range of the assay to less than two orders of magnitude. The limit of detection depends upon a number of experimental variables, including the concentration and antiactivity of the serum and the concentration and specific radioactivity of the hormone. However, it is the magnitude of the association constant of the hormone-protein complex which ultimately limits detection and it is possible to obtain high affinity antisera that permit analysis in the picogram range. PENGELLY and MEINS (1977) have described a rabbit anti-IAA serum which has an association constant of $1.9 \times 10^7 \, l \, mol^{-1}$ facilitating quantification of IAA in the range 0.2–12 ng. However, unless all stages of the assay are carefully controlled, especially those manipulations involving the separation of bound and free hormone, inadequate precision may rule out the use of the lower end of the standard curve. For example, PENGELLY and MEINS (1977) obtained a precision of $B_\omega \pm 0.9$ ng ($\omega = 0.95$), which limited quantification to between one and two nanograms.

The accuracy of a radioimmunoassay depends upon its selectivity and the nature of the analytical situation in which it is used. High selectivity cannot always be assumed as this is very much a feature of the way in which the antibody is prepared. Small molecules such as plant hormones tend not to be antigenic in their own right and so must be covalently bound to a protein before injection into the host animal. There is considerable choice in the type of protein that can be used for this purpose, as well as in the manner of attachment of the hormone to the protein. For instance, FUCHS, S. et al. (1971) prepared a rabbit antiserum to indole-acetylhaemocyanin, the IAA being coupled at the side chain carboxyl group by a dicyclohexyl carbodiimide reaction. This particular anti-IAA serum exhibited low selectivity, cross-reacting extensively with a number of other indoles, and it was concluded that the antigenic determinant must have been considerably larger than IAA. PENGELLY and MEINS (1977) also prepared a rabbit anti-IAA serum but coupled the IAA at the indole nitrogen to bovine serum albumin by means of a formaldehyde condensation reaction. The antibody elicited by this particular antigen would appear to have been far more specific for IAA than that prepared by FUCHS, S. et al. (1971), as very little cross-reaction was observed with a range of related indoles, modification at either C-3 or C-5 dramatically reduced binding and α- and β-naphthaleneacetic acid were the only synthetic auxins to display any reactivity. Even though a good radioimmunoassay is highly selective, it is by no means free of non-random error. PENGELLY and MEINS (1977) noted that a number of compounds which adversely affected IAA analysis did so through the induction of a proportional non-random error. In subsequent work with plant extracts a change in assay sensitivity was used as an indicator of interference. Known amounts of IAA were added to aliquots of the extract prior to radioimmunoassay, the IAA measured was plotted against the IAA added and the slope of the regression line used as a criterion of accuracy. If the slope was within 15% of unity, the data were accepted and the quantity of endogenous IAA was calculated from the y intercept-slope product. However, in order to eliminate additive non-random errors it is essential that the successive approximation procedure of Section 3.2.3.3. a be carried out. Provided the assay is used in well-defined circumstances with regard to sample purity and type it should be possible to establish the generality of condition (b) according to the method described in Section 3.2.3.3. c and thereafter employ only a two-stage approximation for each estimate.

Although radioimmunoassay requires very little sophisticated equipment apart from a liquid scintillation counter, relatively large amounts of standards are required for preliminary immunization work, as well as standards of high specific radioactivity for the actual assay. While this is not a great problem in the analysis of IAA and ABA, supplies of many GA's and cytokinins are distinctly limited. However, there is an ever-increasing body of information on new and efficient methods of synthesis of these compounds and there is, therefore, no reason why radioimmunoassay should not begin to supersede the bioassay in plant hormone analysis as it has already done in medicine and pharmacology.

3.3.3 Physicochemical Detectors

The general suitability of a particular technique for quantitative analysis depends not only upon its limit of detection but also on the information content and selectivity of the response. The potential information content of a number of physicochemical detectors is presented in Table 3.12. It will be noted that only the spectrometric techniques can generate more than 4 bits, and in turn only two of these, mass and IR spectrometry, have the potential to provide large amounts of information. The mass spectrum can be thought of as being made up of n individual measurements or information channels, where n is the number of discrete m/e values contained within the spectrum. If the intensity of each m/e value is estimated with a precision of ± 3% it can be described by 4 bits [Eq. (32)] and hence the information content of the entire spectrum amounts to 4 n bits. However, ion intensities are commonly recorded as a fraction of the base peak and, to allow for this, information equivalent to one m/e value must be subtracted from the total, i.e., $I_t = 4(n-1)$ bits.

The underlying assumption in these calculations is that all sections of the mass range provide equivalent amounts of information regardless of the actual form of the mass spectrum. This assumption is sometimes unjustifiably criticized due to confusion over the way in which mass spectral data relate to other bodies of chemical information. The basic data of the mass spectrum is descriptive in that it can be used to test the null hypothesis, $H_0: U = u$. It has already been mentioned, in Section 3.2.2.3. b, that there are only two possible outcomes from this test, either (i) substance U is different from substance u or (ii) within the scope of the data supplied there is no reason to believe U is different from u. Should it be necessary to

Table 3.12 Physicochemical detectors

Technique	Property quantified	Mode	Limits of detection[a]				Linear range	Selectivity	Potential information content (bits)
			IAA	ABA	GAs	Cytokinins			
UV spectrometry	Absorption of UV light	Scan 230 – 400 nm	100 ng	10 ng	–	10 ng	10^3	medium	24
		Fixed wavelength	1 ng	100 pg	–	100 pg	10^3	medium	4
Spectro-fluorometry	Emission of UV/visible light in response to UV irradiation	Excitation scan 230–400 nm Emission scan 300–500 nm	100 pg	–	–	–	10^3	medium	28
		Fixed excitation and emission wavelengths	1 pg	–	–	–	10^3	high	4
Spectro-phosphorimetry	Emission of UV/visible light subsequent to UV irradiation	Excitation scan 230–400 nm Emission scan 300–500 nm	1 ng	–	–	–	10^3	medium	32
		Fixed excitation and emission wavelengths	10 pg	–	–	–	10^3	high	4
IR spectrometry (Fourier transform)	Absorption of IR radiation	Scan 300–3500 cm^{-1}	100 ng[b]	100 ng[b]	100 ng[b]	100 ng[b]	10^3	low	320
		fixed wavelength	1 ng[b]	1 ng[b]	1 ng[b]	1 ng[b]	10^3	medium	4
^1H NMR spectrometry (pulsed Fourier transform)	Absorption of RF radiation by ^1H nuclei held within a uniform magnetic field	Scan $\frac{\Delta\nu_{TMS}}{F}$ 1–10 ppm	10 µg[b]	10 µg[b]	10 µg[b]	10 µg[b]	?	low	124
		fixed $\frac{\Delta\nu_{TMS}}{F}$	100 ng[b]	100 ng[b]	100 ng[b]	100 ng[b]	?	medium	4

Mass spectrometry	Charge arising from fragment ions of specified m/e ratio	Scan 30–530 amu	100 ng	100 ng	100 ng	100 ng	10^2	low	2000
		Fixed m/e value	100 pg	1 ng	1 ng	1 ng	10^2	high	4
Optical Rotary Dispersion	Rotation of plane polarized light as a function of wavelength	Scan 210–300 nm	–	100 µg	1 mg	–	10^3	medium	8
		Limited wavelength	–	1 µg	10 µg	–	10^3	medium	4
Refractometry	Change in refractive index of solvent as function of solute concentration	–	1 µg	1 µg	1 µg	1 µg	10^3	low	4
Electrochemical	Current due to electrochemical oxidation or reduction at a polarizable electrode	–	10 ng[b]	10 ng[b]	10 ng[b]	10 ng[b]	10^5	medium	4
Ionization devices	Charge arising from total number of ions collected	a) flame	10 ng	10 ng	10 ng	10 ng	10^6	low	4
		b) alkali flame	100 pg	100 ng[b]	100 ng[b]	10 pg[b]	10^3	medium	4
		c) photo	1 ng[b]	1 ng[b]	1 ng[b]	1 ng[b]	10^7	low	4
Electron capture	Reduction in standing current associated with β source due to electron capture by sample	–	–	100 pg	–	–	10^2	high	4

[a] Estimated to the nearest order of magnitude. Data acquisition period assumed < 30 seconds. Concentration dependent responses based on use of smallest practical cell volume

[b] Estimated from the response to compounds other than the listed hormones

prove two substances are identical, the certainty with which this can be done will depend upon the amount of information available. The form of the information is irrelevant, as each independent information source can be looked upon as providing a unique test of H_0. It should be noted that at no stage during this process is it necessary to consider u as being composed of C, H, and O arranged in a particular form. The true description of u lies in its basic physical and chemical data, i.e., the mass, IR, UV, NMR, and ORD spectra, melting point, and crystal structure of u and its derivatives. However, this unwieldy list can be reduced by a series of rules to a conceptual form, the chemical structure. This is, in effect, a semantic interpretation of the basic data and like all semantic processes has the advantage of considerably reducing the amount of effort involved in transferring information. Due to the complexity of the mass spectrum, only a small part of the total information is ever useful in the process of structure elucidation. Particular importance is given to the molecular ion and some of its more abundant decomposition products, as these are more readily interpreted in terms of a structural concept. However, if the proposed structure is to be considered valid, it should be ultimately possible to rationalize all features of the spectrum. Thus, the difference between the "characteristic" ions and the other ions of a spectrum lies not in the amount of information they carry, but in the ease with which this information can be manipulated by the semantic process. For example, bis TMSi IAA has characteristic fragment ions at m/e 319 and 202 which can be correlated with the 189 and 130 ions of IAA Me, and, by this process, lend strength to the validity of the structural concept of the compound known as IAA. However, when the hypothesis H_0:U = IAA is being tested, the description of IAA used is not the structural concept but the basic data from which it is drawn. Thus, the 319 and 202 ions are no better a description, in non-semantic terms, than, say, 201 and 129 which quite coincidentally have relative intensities of zero.

Having argued that the particular form of an individual mass spectrum does not influence the amount of information carried per m/e value, it is now necessary to consider the general case of a sample population. According to the nature of the population, ions of certain masses will be more commonly encountered than others and thus only a proportion of the potential information content (4 bits) is available at each m/e value. WANGEN et al. (1971) have looked at the ion distribution of a sample population of 6800 compounds drawn from the *Atlas of Mass Spectral Data* (STENHAGEN et al., 1969). Using a relative intensity resolution of 1 bit they demonstrated that the information distribution of a scan 0–352 amu was such that only one quarter of the information potentially available was, in fact, accrued. They also investigated correlation between the various m/e values and found that mass positions differing by 1, 2, 13, 14, and 15 amu tend to be dependent. Such a finding seems reasonable since the commonest differences between ions will be 1H, 2H, CH, CH_2, and CH_3. The effect of such correlation would appear sufficient to halve the potential information content. Therefore, because of the combined influences of information distribution and correlation, it can be anticipated that the actual information content of a mass spectrum will be somewhere between one quarter and one eighth of that potentially available. However, it should be noted that the population of compounds contained within the *Atlas of Mass Spectral Data* is hardly typical of a plant extract. The molecular weight distribution is far lower and narrower than that common to plant material and, although this may have little effect on the above findings, it does indicate a general need for caution in extrapolation from "known" sample populations to those occurring in nature.

A sample as small as 100 ng will permit a scan from 30–530 amu with the acquisition of 2000 bits of information. Since, typically, some 500 bits of this are available after correction for distribution/correlation, it is evident that even the strictest requirements for I_u can easily be met. However, when insufficient sample is available and an incomplete spectrum is obtained, it is clearly of importance to know exactly what constitutes an acceptable mass spectrum. As was discussed in Sections 3.2.2.3.b and 3.2.2.3.c, this will depend upon the molecular weight distribution of the sample, and whether or not steps such as GPC were utilized to reduce its potential complexity. If the *Phaseolus* extract illustrated in Fig. 3.9 can be taken as typical, the analysis of plant extracts would appear to demand approximately 140 bits of information, and this means that after allowances for distribution/correlation within the spectrum, a potential of greater than 560 bits should be provided. This could be achieved if the intensities of 141 ions agreed with those of the standard to within $\pm 3\%$.

In practice, the mass spectrum of the sample rarely exactly matches that of the standard, and, of the total information, I_t, contained in the sample spectrum, only a portion, I_u, will correspond with the spectral characteristics of the standard. By inspection of Eq. (32) it is apparent that I_u can be calculated according to

$$I_u = \sum_{n=a}^{n=b} \frac{1.7 - \log |L_U - L_u|_n}{0.3} \text{ bits} \tag{43}$$

where $|L_U - L_u|_n$ is the difference, regardless of sign, between the percentage relative intensities of ion peaks having a nominal mass n in the sample and standard spectra and a and b are the m/e limits of the scans. It should be noted that $|L_U - L_u|_n$ cannot be smaller than the precision with which the percentage relative intensities of the ion peaks are measured and account should be taken of this fact when employing Eq. (43). For example, a partial mass spectrum of the reference compound GA_{20} benzyl ester is illustrated in Fig. 3.12 along with the mass spectrum of a substance isolated from an esterified *Phaseolus* extract. If it is assumed that the precision of the ion intensity measurements is $\pm 3\%$, it follows from Eq. (32) that 4.1 bits of information are provided for every nominal mass recorded. Thus the total potential information content of the spectra is $I_t = 4.1 \times (220 - 1) = 898$ bits. Of this, the amount of information which correlates the spectrum of the putative GA_{20} benzyl ester with that of the standard can be calculated via Eq. (43) as $I_u = 813$ bits (Table 3.13). Thus, it would appear that the putative GA_{20} benzyl ester spectrum closely matches that of the reference compound. However, it must be remembered that even quite different substances may have many features of their mass spectra in common because of distribution/correlation, and the data of WANGEN et al. (1971) suggest that as much as $75\%\ I_t$ bits of the information I_u is likely to be of little value in the test of H_0. In the case under discussion this means that approximately 673 bits of the 813 bits of information which correlate the putative GA_{20} benzyl ester spectrum with that of the standard are "non-informative", leaving only 140 bits for the test of $U = GA_{20}$

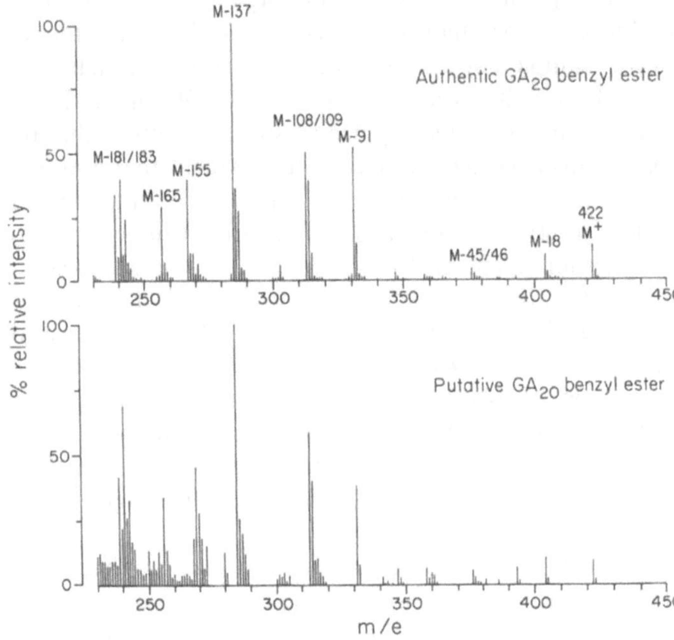

Fig. 3.12. Mass spectrum of a putative GA_{20} benzyl ester sample compared with the spectrum of the authentic reference compound

benzyl ester. Obviously the weakest link in this calculation is the allowance made tor information distribution and correlation and this will be so until such time as larger and more comprehensive libraries of spectra are available for analysis. However, in spite of the drawbacks, there would seem to be a good case for quoting I_t and I_u in all those situations where restrictions on space prevent the publication of full line spectra. For example, instead of quoting:

"GC-MS analysis of fraction X revealed a number of total ion current peaks, one of which had a mass spectrum similar to that of TMSi-IAA"

it would be more informative to state:

"GC-MS analysis of fraction X revealed a number of total ion current peaks. One of these provided a mass spectrum of 1500 bits, 1400 of which could be correlated with the spectrum of authentic TMSi-IAA".

The 140 bits criterion that has been applied to the identification and quantification of a particular substance should not be confused with the criterion of structural elucidation whereby selected portions of the spectrum are interpreted in the light of a structural concept. Structure fitting programmes invariably require that spectra be coded in a condensed form so that practical memory sizes and rapid access times can be realized. Because a number of investigators have shown that such abbreviations of spectral information need not affect the success of the programme (see GROTCH, 1970; KNOCK et al., 1970; WANGEN et al., 1971), there is a tendency to look upon mass spectral codings, such as the eight-peak system used by

Table 3.13 Calculation of the amount of information in the mass spectrum of a putative GA_{20} benzyl ester sample which matches the spectrum of an authentic sample of GA_{20} benzyl ester. L_U and L_u are the percentage relative ion intensities of the sample and reference spectra, respectively

m/e	L_U (percent)	L_u (percent)	$\lvert L_U - L_u \rvert_n$ (percent)	Set 3% minimum	$\frac{1.7 - \log\lvert L_U - L_u\rvert_n}{0.3}$ (bits)
230	11.0	3.7	7.3	7.3	2.8
231	11.9	2.5	9.4	9.4	2.4
232	9.3	1.2	8.1	8.1	2.6
233	9.4	1.2	8.2	8.2	2.6
234	6.9	0	6.9	6.9	2.9
235	8.1	0	8.1	8.1	2.6
236	9.4	0	9.4	9.4	2.4
237	9.4	0	9.4	9.4	2.4
238	7.5	0	7.5	7.5	2.7
239	41.8	33.7	8.1	8.1	2.6
240	21.9	10.0	11.9	11.9	2.1
.	.	.	.	.	.
.	.	.	.	.	.
.	.	.	.	.	.
440	0	0	0	3.0	4.1
441	0	0	0	3.0	4.1
442	0	0	0	3.0	4.1
443	0	0	0	3.0	4.1
444	0	0	0	3.0	4.1
445	0	0	0	3.0	4.1
446	0	0	0	3.0	4.1
447	0	0	0	3.0	4.1
448	0	0	0	3.0	4.1
449	0	0	0	3.0	4.1
450	0	0	0	3.0	4.1
				Total	813

the Aldermaston Mass Spectrometry Data Centre (RIDLEY, 1971), as a minimum requirement for the identification of an unknown compound. This implies that considerably less than 140 bits of information are sufficient for positive identification. However, when examined closely it is evident that the assumption is without foundation, as the success rate of the programme will be equivalent to π only when the compound of interest is drawn from a population identical to that contained in the spectral library used for the search process. Since natural sample populations make it necessary to contend with approximately 10^{42} possible compounds, it is evident that success rates, based on even large libraries of 10^4, will grossly overestimate π.

A further reason for confusion in this issue is due to the fact that newer structural elucidation programmes are "intelligent" and can make inferences based upon "experience". Although the success rate of such programmes still depends upon the amount of information supplied and the size of the reference population from which machine experience is derived, the better search algorithms can suggest a number of likely structures for the test compound when it is beyond the immediate experience of the machine. Such information is of great value when attempting to characterize a new substance. However, the knowledge that U has some structural similarities to u cannot be considered to be a strict test of the null hypothesis H_0: U = u, since such semantic interpretations are deduced from, and cannot add to, the information content of the basic data.

At present, too little is known of sensitivity distributions to permit anything but the broadest of interpretation to be placed upon the relative selectivities of the detectors listed in Table 3.12. As was discussed in Section 3.2.2.3. a, there are problems in defining confidence limits for S, since the variance in R' depends upon the number of compounds in the extract. However, with every successive approximation performed, some measure of $\bar{R}$ and R' is obtained [see Eq. (42)], and thus, with experience extending to a wider and wider range of sample types, eventually it should be possible to acquire practical parameters by which the selectivity of different detectors can be judged. The little knowledge of $\bar{R}$ that is available, however, indicates that the average selectivity can range from unity to around 10^6.

Detection limit, selectivity, and information content are to some extent interrelated. Systems of high selectivity tend to offer low limits of detection and a low information content, while those of low selectivity usually exhibit a higher detection limit and can, although do not always, yield a greater amount of information. This principle can be seen to be operating in the case of spectrometric detectors. For instance, in order to obtain a full scan mass spectrum of 2000 bits, approximately 100 ng of sample is a typical minimum requirement. When used in this mode the selectivity of the mass spectrometer is very low, as virtually all organic compounds will evoke a response. However, if only one m/e value is monitored, the selectivity is greatly increased and the limit of detection falls by about 500-fold. The price paid for this improvement is a reduction in the information content from 2000 to 4 bits. Clearly, the actual values of these parameters depend to a great extent upon the way in which the device is operated, and, in practice, the amount of sample available dictates the extent to which information content has to be compromised for a higher selectivity and a lower detection limit.

3.4 Chromatographic Procedures

Chromatographic techniques are firmly established as key processes in the accurate analysis of endogenous plant hormones, and in this role can add either information or selectivity. GIDDINGS (1967) has defined peak capacity, ϕ, as the maximum number of components that can be simultaneously separated from each other with unit resolution. Although not currently in vogue, this is a particularly useful chromatographic parameter as it is a direct measure of discriminating power. This is because $I = \phi$ when the standard deviation of the variation in retention time is small compared with one quarter of the related peak width. However, in addition to retention behaviour it is normal practice to quantify the areas of the various peaks

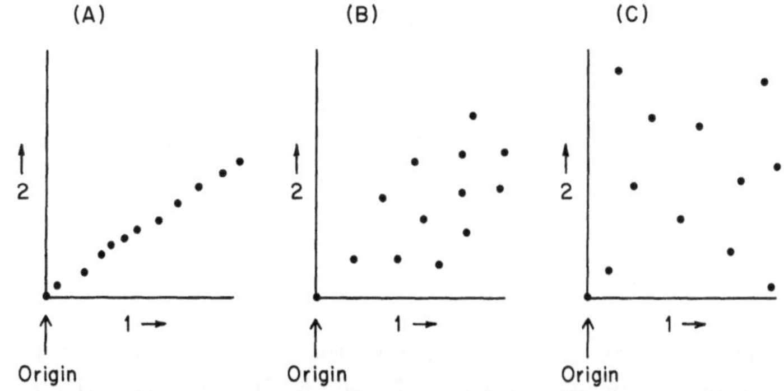

Fig. 3.13. Hypothetical two-dimensional chromatograms in which the degree of correlation between solvent systems 1 and 2 varies from unity (**A**) to zero (**C**)

on the chromatogram. If I_q bits of information are required to describe peak area adequately, then the information acquired from the entire chromatogram must be

$$I = I_q \cdot \phi \text{ bits} \tag{44}$$

Thus, when analysis is carried out using a number of different chromatographic procedures in *parallel*, the information contributed by each step is a direct function of both its quantitative precision and peak capacity. This information is additive after due allowance has been made for correlation between the retention characteristics of the different procedures, i.e., by analogy with Eq. (34),

$$I_t = I_q [\phi_1 + \phi_2 (1 - j_{2,1}) + \phi_3 (1 - j_{3,1})(1 - j_{3,2}) + \ldots \phi_n (1 - j_{n,1}) \ldots (1 - j_{n,n-1})] \tag{45}$$

where $\phi_1, \phi_2, \phi_3 \ldots \phi_n$ are the peak capacities of procedures 1, 2, 3 ... n, and $j_{n,1}$, $j_{n,2}$, $j_{n,n-1}$ are coefficients expressing the correlation between systems n and 1, between systems n and 2 etc. The concept of chromatographic correlation is best visualized in terms of two-dimensional TLC or paper chromatography and Fig. 3.13 illustrates three hypothetical chromatograms in which the degree of correlation between the first and second solvent system ranges from unity (A) to zero (C). It is evident that the discriminating power of the two-dimensional technique is heavily reliant on the ability of the second solvent system to generate a retention pattern which is completely different to that of the first.

On the basis of Fig. 3.13 it is tempting to suggest that j is identical to the classical correlation coefficient z, used in Eq. (34). However, it must be remembered that the distribution of retention indices is ideally rectangular and certainly never normal. Thus, if j is calculated as $\frac{\sigma_{x \cdot y}}{\sigma_x \cdot \sigma_y}$, its meaning will be obscure. Whatever the form of j, its solution will require access to the retention behaviour of a very large population of reference compounds, and, considering the number of subtly different chromatographic systems which could be employed in any one analytical situation, it is apparent that exact calculation of j values is hardly a practical proposition.[20] It would seem better to employ two or three chromatographic procedures that are mechanistically quite different and which offer the highest possible peak capacities. Under these circumstances the effects of correlation will be minimal. High peak capacities are ensured by using systems of high efficiency, N, and by employing relatively long analysis times, since, for the isocratic/isothermal case:

$$\phi = 1 + 0.6\,N^{\frac{1}{2}} \log(1 + k') \tag{46}$$

where k′ is the capacity factor of the last solute to be eluted. The information acquired can be used as a basis for establishing purity and verifying the accuracy of quantitative estimates in accordance with the criteria discussed in Section 3.2.2.3. b.

The situation is quite different when chromatographic procedures are used in series for the purification of an extract. Under these circumstances a specific fraction of each chromatogram is passed on to the next procedure, and, as a consequence, the potential information content falls by a factor of ϕ as it is directly exchanged for selectivity. For all but the most poorly optimized chromatographic systems, it can be assumed that the retention indices of the individual components of an extract are randomly and rectangularly distributed. This being so, any section of the chromatogram having a peak capacity $\Delta\phi$ [21] will contain, on average, a fraction, $\bar{F}$, of the total number of compounds in the extract so that

$$\bar{F} = \frac{\Delta\phi}{\phi} \tag{47}$$

where ϕ is the total peak capacity of the technique. The relationship can be further extended to describe the dry weight reduction resulting from such a selection process, since it can be safely assumed that the variation in the relative abundances of extract components is in no way dependent upon their chromatographic distribution. Where M_i is the initial weight of the extract and $\bar{M}_f$ the average weight of the selected fraction, the average dry weight reduction is

$$\frac{M_i}{\bar{M}_f} = \frac{1}{\bar{F}} = \frac{\phi}{\Delta\phi} \tag{48}$$

However, when chromatography is being employed for purification, the fraction of interest in the chromatogram ideally occupies no more than one peak width, consequently $\Delta\phi$ can be taken as unity and thus

$$\frac{M_i}{\bar{M}_f} = \phi \tag{49}$$

---

[20] MOFFAT et al. (1974), while recognizing the meaning of the classical correlation coefficient would be obscure when applied to anything but a normal, bivariant distribution, nevertheless demonstrated its usefulness in optimizing small combinations of chromatographic procedures

[21] For the isothermal/isocratic case, $\Delta\phi$ for the interval k'_1 to k'_2 is $0.6\,N^{\frac{1}{2}} \log \frac{(1+k'_2)}{(1+k'_1)}$

Substitution of Eq. (49) in Eq. (14) yields the average purification factor as

$$\bar{Q} = \frac{\phi \cdot Z}{100} \tag{50}$$

Thus, the average purification factor of a chromatographic step is directly related to the peak capacity and recovery figure, Z, of the procedure.

In practice, actual purification factors, Q, will vary from extract to extract about the mean, $\bar{Q}$. Therefore, in order to generalize the relationship between the chromatographic efficiency and the practical effectiveness of a certain purification procedure, it is necessary to consider a lower confidence level, Q_θ, where θ is the probability of Q being greater than Q_θ. When a chromatographic technique of purification factor, Q_θ, is used prior to an analytical method of selectivity, S_σ, it follows from the argument of Eq. (15) that

$$S^t_{\sigma_t} = S_\sigma \cdot Q_\theta \tag{51}$$

where $S^t_{\sigma_t}$ is the combined selectivity of the two techniques and

$$\sigma_t = \sigma \cdot \theta \tag{52}$$

Since $S^t_{\sigma_t}$ can be related to A_α via Eq. (9) it is evident that there is a mathematical description which facilitates the calculation of the minimum chromatographic performance necessary to achieve an acceptable level of non-random error when given the selectivity of the final analytical procedure.

However, θ must account for variance from two sources: (a) the natural variation in the number of compounds selected by the interval $\Delta\phi = 1$ and (b) the natural variation in the relative abundances of the various extract components. As very little is known about these two distributions it is difficult to establish Q_θ other than by an empirical means, although it should be noted that if both distributions are symmetrical, $\bar{Q} = Q_{0.5}$. As has already been discussed at various points in Section 3.2.2.3. a, the calculation of actual S_σ values is also difficult. Consequently, it is not currently feasible to place a lower limit on the chromatographic performance necessary to carry out a specified analysis. It is possible, however, and of practical value, to consider an average situation. By analogy with Eq. (9),

$$\bar{A} = \frac{10^8}{c \cdot \bar{S}} \text{ percent} \tag{53}$$

where $\bar{A}$ is the average percentage accuracy obtained when analyzing samples containing c ppm of hormone by a method of average selectivity $\bar{S}$. Accuracy can be improved by using a chromatographic method of average purification factor $\bar{Q}$ prior to analysis. Thus Eq. (53), in accordance with the argument of Eq. (15), becomes

$$\bar{A} = \frac{10^8}{c \cdot \bar{S} \cdot \bar{Q}} \text{ percent} \tag{54}$$

which on rearrangement and substitution for $\bar{Q}$ from Eq. (50) then yields

$$\phi \cdot Z = \frac{10^8}{c \cdot \bar{S} \cdot \bar{A}} \tag{55}$$

The importance of Eq. (55) is that it provides a measure of the chromatographic performance necessary for accurate analysis in a given situation.

Substitution of Eq. (46) in Eq. (55) and further rearrangement allows accuracy to be equated with efficiency, the most crucial chromatographic performance parameter.

$$N = \left\{\frac{1}{0.6 \log(1+k')} \cdot \left[\frac{10^8}{c \cdot \bar{S} \cdot \bar{A} \cdot Z} - 1\right]\right\}^2 \text{ theoretical plates} \tag{56}$$

It can be predicted from chromatographic theory that when k′ values exceed 4, any improvement in peak capacity, obtained through the use of longer retentions, is increasingly acquired at the expense of speed of analysis. Thus, regardless of the actual form of the chromatographic technique employed, $k' = 4$ provides a reasonable measure of the largest capacity factor likely to be employed[22]. Substitution of this value in Eq. (56) permits some simplification as follows:

$$N = \left(\frac{2.4 \times 10^8}{c \cdot \bar{S} \cdot \bar{A} \cdot Z} - 2.4\right)^2 \text{ theoretical plates} \tag{57}$$

The value of this equation can be seen in the following example of IAA analysis by the indolo-α-pyrone fluorescence assay. $\bar{S}$ for the fluorescence procedure is derived from Eq. (42), since the average of the pairs differences, $E_n - E_{n+1}$, must reflect the non-random error term, $10^6 \cdot \frac{R'}{R_u}$ ppm, i.e., $\frac{10^6}{S}$ ppm. For example, suppose a value of 10^5 is obtained for $\bar{S}$ and the required average accuracy of analysis must be at least 10%. Provided there is no interest in samples containing less than 0.1 ppm IAA, the chromatographic efficiency required, assuming 100% recovery of the sample, is

$$N = \left(\frac{2.4 \times 10^8}{10^{-1} \times 10^5 \times 10 \times 10^2} - 2.4\right)^2$$
$$= 470 \text{ theoretical plates}$$

This degree of efficiency can be provided by any number of chromatographic procedures. However, when samples containing as little as 0.01 ppm IAA have to be investigated, the efficiency required increases to 56,000 theoretical plates. Such efficiency is difficult to achieve by LC using even the most sophisticated procedures. Consequently, if the desired level of accuracy were to be achieved, more than one chromatographic technique would have to be used.

[22] Classical gas chromatography (GLC) is a notable exception to this rule, as capacity factors commonly range from 10 to 30

When a number of chromatographic processes are used in series, the total purification factor is the product of the actual Q values obtained at each stage, i.e.,

$$Q^t = Q_1 \cdot Q_2 \cdot Q_3 \dots Q_n \tag{58}$$

It must be noted, though, that the average purification factor of a chromatographic series, $\bar{Q}^t$, cannot be measured as the product of the respective peak capacities multiplied by the product of the recoveries, because $\bar{Q}^t \neq \bar{Q}_1 \cdot \bar{Q}_2 \cdot \bar{Q}_3 \dots \bar{Q}_n$. The issue is further confused since the degree of correlation which exists between the stages must also be taken into account and, as a consequence, it becomes difficult to generalize the relationship between chromatographic performance and purification factor. In practice, the guiding principal must be the same as when parallel combinations are used to provide information, i.e., maximum selectivity will be obtained by using a small number of highly efficient, mechanistically different, chromatographic procedures.

Regardless of whether chromatography is used to provide selectivity or information it can be seen that efficiency, through its relationship with peak capacity, is the key performance parameter. Obtaining high efficiencies is not without its problems, as it involves a dual compromise with the other chromatographic performance parameters, sample capacity, and speed of analysis. On the assumption that the speed of analysis is fixed, the compromise is such that systems of high efficiency have small sample capacities, while, in contrast, those with a high sample capacity usually generate only low efficiencies. The weight of a typical crude plant extract necessitates that the first purification step used in plant hormone analysis be a procedure of high sample capacity. However, with increasing purification the weight of the extract falls, making it possible to reduce the sample capacity, thereby increasing the efficiency of each ensuing chromatographic step. Ultimately a state of purity is reached whereby use can be made of high performance methodology.

Chromatographic separations can be carried out by a number of techniques, the most important of which are summarized in Table 3.14. Each represents a different

Table 3.14 Typical performance parameters for gas and liquid chromatography techniques

Technique	Support	Efficiency (eff. plates)	Plate Height (μm)	Speed of Analysis (eff. plates s^{-1})	Sample capacity
GLC	100–120 μm diatomite	2000 (k′ = 10)	600	10	10 μg
GLC	WCOT column	200000 (k′ = 5)	500	5000	0.1 μg
HPLC	5 μm silica	3000 (k′ = 5)	40	10	100 μg
Prep. HPLC	10 μm silica	2000 (k′ = 5)	120	1	10 mg
LC	60–200 μm silica	100 (k′ = 2)	1000	0.1	30 mg
LC	LH-20 Sephadex	2000 (k′ = 4)	250	0.05	30 mg
TLC	0.1–5 μm silica	900 (k′ = 0.5)	20	0.2	1 mg
Paper chromatography	Paper	100 (k′ = 0.5)	400	0.001	10 mg

point in the compromise between efficiency, speed of analysis, and sample capacity, and, as mentioned above, this is an important consideration when deciding which technique to use at a particular point in a multi-stage purification procedure. For instance, in the case of wall-coated open tubular (WCOT) capillary GLC columns, sample capacity is traded to a large degree for speed of analysis, and this enables high efficiencies to be achieved through the use of long columns while still maintaining practical retention times. The restricted sample capacity of this technique means, however, that it is of value only as the ultimate step in an analytical procedure. It is not surprising therefore that capillary columns are being increasingly used to advantage in combined GC-MS systems where their effect is to greatly enhance either the information content or the selectivity of the analysis. Classical LC procedures represent the other extreme, being well suited for use in the early stages of purification because of their high sample capacity. However, this is achieved by either sacrificing efficiency, as is the case for silica gel-based systems, or by reducing the speed of analysis, as is typical for dextran gel columns, such as LH-20 Sephadex. The silica gel systems are best suited to group separatory procedures where low efficiencies are not a drawback. When high resolution is required, for instance, in the separation of individual members of a group of hormones, dextran gel columns can be used provided that the long analysis times can be tolerated. The recent development of high efficiency LC chromatographic supports represents an important advance, as they allow the construction of preparative HPLC systems. These procedures have a sample capacity equivalent to that of classical LC, yet can generate high efficiencies at moderate speeds of analysis. Consequently, preparative HPLC may soon supersede many of the LC procedures currently in use.

After samples have been adequately purified it becomes feasible to use GLC and HPLC to advantage. As commonly practised, these techniques are broadly equivalent in that they display similar efficiencies and speeds of analysis, with the sample capacity of HPLC being about ten times greater than that of GLC. However, HPLC permits operation at ambient temperatures and simplifies sample recovery, both of which can be advantageous when samples are difficult to volatilize. The major difference between HPLC and GLC lies in the thermodynamics of the partitioning process. In liquid-solid and liquid-liquid processes the differences in the free energies of distribution of solutes $\Delta(\Delta G^\circ)$ are usually far greater than for gas-solid or gas-liquid systems. Thus, all other factors being equal, HPLC will almost always give a superior separation to GLC. In addition, $\Delta(\Delta G^\circ)$ is much more dependent upon the properties of the mobile and stationary phases in HPLC than it is in GLC, and thus HPLC is able to offer a much wider variety of column selectivities. This is of great importance when dealing with combinations of chromatographic procedures, as the low degree of correlation between LC systems makes it feasible to employ anything from six to ten different steps without sacrificing much of the potential selectivity or information content. On the other hand, there is little point in combining more than two or three GLC systems, because any potential gain in selectivity or information is offset by the high degree of intersystem correlation.

Table 3.15 gives an indication of the basic types of HPLC systems currently available. The range is, in fact, far greater than is suggested, as the retention characteristics of any one system can be dramatically altered by a relatively simple

Table 3.15 Polarity range of typical HPLC systems[a]

Polarity	Hormones	HPLC systems: Straight phase partition chromatography	Reverse phase partition chromatography	Ion exchange/ ion pair chromatography	Soap chromatography	Silica or alumina adsorption chromatography	Polyamide adsorption chromatography	Gel permeation chromatography
Low		—	—			—		—
	IAA and nonhydroxylated GA's	—	—		- - - / —	—	- - -	—
Medium	Monohydroxy GA's and ABA	—	- - -	- - - / —	—	- - -	- - -	—
	Di- and trihydroxy GA's	- - -	- - -	—	—		- - -	—
High	Zeatin, dihydrozeatin and tetrahydroxy GA's	- - -	- - -	—	—		- - - / —	—
	GA glucosyl conjugates ABA glucosyl conjugates Cytokinin ribosides and glucosides		- - -	—	—		—	—

[a] Usual applications fall within the area indicated by the *solid line*; however, in certain instances this can be extended as shown by the *dotted line*

modification of solvent conditions. Table 3.15 also shows that HPLC systems suitable for the analysis of GA's, ABA, IAA, and cytokinins must operate in the medium-to-high polarity range. This requirement is not as restrictive as it might seem because the polarity of most of these compounds can be readily altered by the formation of suitable derivatives.

A range of detectors is available for HPLC and GLC. The use of certain detectors, such as electrochemical and spectrofluorometric devices, is confined to HPLC because they exploit solute properties peculiar to liquid systems. All GLC detectors are at least theoretically compatible with HPLC, although in practice sensitivity and detection limits are adversely affected and the problems accompanying phase transformation have discouraged their widespread application. It is possible to link a mass spectrometer directly to HPLC systems via a wire transport interface (McFadden et al., 1977; Scott, 1977; McFadden, 1979) but at present this technique and other procedures require further refinement before widespread application becomes feasible (see Arpino and Guiochon, 1979). It is a much simpler practical proposition to couple a mass spectrometer to a gas chromatograph. It is in this mode, as the ultimate purification step prior to mass spectrometry, that GLC has been used most effectively in plant hormone analysis. Packed GLC columns are normally employed for this purpose, but with the availability of HPLC to facilitate rapid sample purification, opportunities now exist for the use of high efficiency capillary columns in combined GC-MS instrumentation.

3.5 Internal Standards

Because of the involvement of chromatographic procedures in the analysis of plant hormones, sample losses will inevitably occur which must adversely affect the accuracy of quantitative estimates. This error can be corrected by applying a conversion factor derived from the percentage recovery of a standard. However, in cases where run-to-run variation in recovery is unacceptably high, an internal marker must be added to every sample prior to analysis. The most suitable markers are isotopically labelled analogues of the compound of interest. Although these can be differentiated by techniques such as mass spectrometry and radioassay, in other respects they tend to behave in the same manner as their endogenous counterparts. When labelled compounds are not available, the best alternative is a closely related isomer of the hormone under study from which it can be distinguished only at the final stage of analysis. The more removed the structure of the internal marker from that of the endogenous hormone, the greater the risk of some degree of separation occuring before the ultimate chromatographic step. However, as it is known for even isotopes to be chromatographically separable, preliminary experiments are always necessary to determine the relative recoveries of the hormone and its marker.

A further problem, often overlooked, is that the final estimate results from two independent measurements, one of the internal marker and the other of the endogenous hormone. Thus, the accuracy of such estimates, A_α, will contain an error term, $A'\beta$, due to the uncertainty in the quantification of the response of the detector to the endogenous hormone, as well as a second term, $A''\gamma$, associated with the internal marker, i.e.,

$$A_\alpha = A'\beta + A''\gamma \tag{59}$$

where

$$\alpha = \beta \cdot \gamma \tag{60}$$

It follows that, no matter how accurately the detector response to the endogenous hormone is quantified, the accuracy of the final estimate is ultimately limited by the accuracy with which the internal standard can be measured. Consequently, for those procedures whose accuracy depends upon the provision of a minimum level of selectivity, it is imperative to demonstrate that adequate selectivity is available for both the hormone of interest and the internal standard. This can be easily done if estimates based on the use of a marker are subjected to successive approximation according to the criteria of Section 3 2.2.3a. Failing this, in those cases employing a deuterated analogue or closely related isomer, it is always possible to analyse each sample with and without the internal standard and so eliminate the possibility of interference by extract components. The use of radioactive markers is a special case. Ideally the labelled hormone should be the only radioactive component in the extract and thus the selectivity of a scintillation counter is more than sufficient to guarantee accuracy. However, should the compound of interest be unstable, safeguards must be taken to ensure the radioactive standard is distinguished from its decomposition products. In most circumstances, this can be satisfactorily achieved through the use of either a gas or high performance liquid chromatograph fitted with an on-stream radioactivity monitor.

When accuracy is being established by the acquisition of information, it is clear that the purity of the recovered internal standard must be verified to the same degree as the endogenous hormone. This should present few problems when the marker is a structurally related isomer of the compound of interest. However, for deuterated compounds, it is an unavoidable fact that the mass spectra of the hormone and its marker must be recorded simultaneously. The disadvantage of this is that it is difficult to ascertain unequivocally what proportion of the total information content is associated with the hormone and what proportion with the deuterium labelled marker, as any subtraction technique must make an initial assumption as to the form, and thus the information content, of one of the spectra. When radioactively labelled internal standards are used in conjunction with information content-based analyses, any decomposition of label, excluding self-radiolysis, will be accompanied by decomposition of the compound of interest. Thus, provided the radioassay is carried out on an aliquot of the same fraction used for quantification and the acquisition of information, the marker should function satisfactorily.

3.6 Analytical Procedures

3.6.1 Indole-3-Acetic Acid

Mass spectrometry has been used for a number of years to characterise naturally occurring indoles (see JAMIESON and HUTZINGER, 1970). IAA can be readily subjected to GLC after conversion to a volatile derivative such as IAA methyl ester (IAA Me), bis trimethylsilyl IAA (TMSi IAA), or heptafluorobutyl IAA methyl ester (HBF IAA Me). It is therefore an ideal candidate for analysis by GC-MS. However, the number of reports of the procedure being used for quantitative analysis of IAA is very limited.

BRIDGES et al. (1973) estimated IAA levels in the cortical, stelar, and apical regions of *Zea mays* roots by GC-MS. Tissue was extracted with methanol and the acidic, diethyl ether fraction purified by DEAE cellulose, paper chromatography, and TLC. The bisTMSi derivative was formed prior to GC-MS, and quantification was achieved by repeated scanning over the mass range 190–220 for the period during which the presumptive TMSi IAA was eluting from the column. By this means the amount of IAA present was related to the ion charge associated with the m/e 202 base peak. A larger extract of intact roots was subjected to an identical purification procedure and examined by GC-MS. A 140–330 amu spectrum was obtained which indicated that the TMSi IAA fragmentation products were the only contributors to the ion current of the presumptive TMSi IAA peak. The spectrum matched that of authentic TMSi IAA to within $\pm 3\%$ at each m/e value, allowing the information content to be calculated as $4 \times (330\text{-}140\text{-}1) = 756$ bits and thus providing evidence that the quantified response was associated with a pure compound. Furthermore, α would have exceeded π since quantification was based on the more selective m/e 202 ion charge rather than the total ion current. It should be noted, however, that quantification was carried out on extracts of specific regions of the root, whereas the mass spectrum was obtained from an extract of intact roots. If the mass spectrum is to be used to support accuracy it must be assumed that the extracts were identical in all respects apart from IAA content. This type of assumption is difficult to uphold and ideally both the information used to establish sample purity and the data upon which quantification is based should be derived from the same sample.

The sample losses that are an unavoidable feature of extensive purification procedures are a source of error which can adversely affect both the precision and accuracy of IAA estimates. In order to account for these losses BANDURSKI and SCHULZE (1974) used an isotopic dilution process when investigating IAA levels in etiolated shoots of *Zea mays* and *Avena sativa*. A known weight, M µg, of ^{14}C-IAA with a specific activity C_0 was added to an acetone extract of the plant tissue. This was processed by partitioning and back extraction to yield an acidic, diethyl ether-soluble fraction which was purified by DEAE cellulose, LH-20 Sephadex, and TLC. The TLC zone corresponding to IAA was eluted, silylated, and aliquots subjected to GC-MS analysis, while the remainder was run on a preparative GLC column and compounds emerging in the presumptive TMSi IAA zone were trapped. The amount of radioactivity in this highly purified fraction was determined by liquid scintillation counting, while the IAA level was estimated by UV spectrometry at 225 and 282 nm and colorimetrically using Salkowski reagent. If the quantitative estimates so obtained were in agreement, it was assumed that the sample was pure and the specific radioactivity of the purified IAA (C) was calculated. The quantity of endogenous IAA (d) in the sample was then derived from the expression:

$$d = \left(\frac{C_0}{C} - 1\right) M$$

The accuracy of the isotopic dilution procedure is limited by the accuracy with which the level of IAA in the purified extract can be determined. BANDURSKI and SCHULZE (1974) employed a number of techniques to quantify IAA, and their data conveniently illustrate the way in which the selectivity and information content contribute to accuracy. These authors state that they "... established the purity of the IAA isolated by agreement between Salkowski assay, UV spectrophotometry at two wavelenghts, and GC peak area". The use of four essentially independent assays provides a certain amount of information and thus evidence of sample purity. However, π is unlikely to be high because the information content calculated on the basis of a $\pm 3\%$ variation among the estimates results in a figure of only $4 \times 4 = 16$ bits. If purity is to be established with any certainty, more information is required, and this is in fact provided by mass spectra of 1570 bits. In the cases of the presumptive IAA from *Avena* and *Zea*, ca. 1540 bits of the supplied information can be correlated with the spectrum of authentic IAA and this process accounts for 40% of the total ion current. Given the value of I_u, the probability of the purity being better than 40% must be very high. Consequently, it is unlikely that there would have been a sufficient number of impurities present in large enough quantities to affect the accuracy of the more selective UV and colorimetric assays seriously. However, as a flame ionization detector exhibits very low specificity, it would be anticipated that the GLC data should have exceeded rather than agreed with the other estimates. This anomaly may be the result of baseline corrections carried out when the GLC peak areas of the presumptive IAA were calculated. It must be emphasized that, although the IAA estimates from the four assays were in close agreement, the 1540 bits of information provided by the mass spectra are the only strong proof of purity and hence acuracy. This is important, as in a subsequent investigation of IAA levels in a wide variety of plant tissues. BANDURSKI and SCHULZE (1977) used similar procedures but did not verify their estimates by mass spectrometry. If their data are to be taken as accurate, it is implied that the extracts were identical to those of *Zea* and *Avena* in all essential respects bar IAA content. Such an assumption can be upheld only when the validity of the analytical procedures has been demonstrated for a statistically meaningful number of plant tissues of different types, i.e., α must be estimated according to the criteria in Section 3.2.2.3. c.

There are many physiological systems whose very nature limits the availability of tissue for analysis, and with such material it is extremely difficult to use mass spectrometry to confirm the accuracy of quantitative estimates of hormone content. The only recourse is to employ methods with lower limits of detection, and this almost always involves compromising information content for selectivity. Highly selective procedures which have been used for the analysis of IAA include

MF (RIVIER and PILET, 1974) and the indolo-α-pyrone fluorescence assay (STOESSL and VENIS, 1970).

RIVIER and PILET (1974) used MF to measure nanogram amounts of IAA extracted from root tips of *Zea mays*. An internal marker of 5-methyl IAA was added to ca. 2.0 gm of tissue before maceration in water, acidification, and extraction into ethyl acetate. The ethyl acetate extract was methylated, treated with heptafluorobutylimidozole, the products extracted from acid solution into hexane, and aliquots analyzed by MF. The HFB IAA Me was monitored at m/e 385 and m/e 326 and the corresponding 5-methyl IAA derivative at m/e 340. A calibration curve was established by treating a series of solutions containing varying amounts of IAA together with 1 μg aliquots of 5-methyl IAA in an identical manner to the root material. The ratios of the heights of m/e 326 and m/e 340 peaks were plotted against the amounts of IAA originally present. By this means it was possible to analyze the IAA content of very small samples with a precision of $\pm 40\,\mu g\,kg^{-1}$ of tissue ($\omega = 0.95$) and, at the same time, to exclude any errors arising through discrimination between IAA and 5-methyl IAA in the purification procedure.

RIVIER and PILET (1974) claim the "unambiguous identification of the IAA derivative" on the basis of the GLC retention time of the putative HFB IAA Me and the relative intensities of the m/e 326 and 385 peaks. The validity of this assertion rests on the amount of information acquired which, assuming a GLC peak capacity of 5 and a correlation between the two peak heights to within $\pm 3\%$, can be calculated as $4\phi(n-1) + \phi = 4 \times 5 \times 1 + 5 = 25$ bits, where n is the number of channels monitored. This figure is the absolute maximum value and, in practice, must be reduced by 4 bits for every impurity peak present in the chromatogram. Due allowance must also be made for the distribution of information amongst the channels monitored so that the actual information content is well below the 140 bit minimum calculated in Section 3.2.2.3. b, and on these grounds must be considered insufficient evidence to establish the homogeneity of the quantified response. This being so, verification of accuracy must be sought in the overall selectivity of the method which could have been increased by including a number of purification procedures, improving the efficiency of the GLC column and by using longer retention times.

The acid-catalyzed condensation reaction of IAA with acetic anhydride to yield the highly fluorescent indolo-α-pyrone provides the basis of a selective assay for nanogram quantities of IAA (STOESSL and VENIS, 1970). LIEBERMAN and KNEGT (1977) used this procedure to determine IAA levels in etiolated epicotyls of *Pisum sativum*. An internal marker of ^{14}C-IAA was added to the ethanolic extract of ca. 20 g of tissue, and the acidic diethyl ether fraction purified by alumina and paper chromatography and TLC before estimation of radioactivity and IAA. In order to compensate for proportional non-random error due to the effects of extract impurities on pyrone formation and fluorescence yield proper, known amounts of IAA were added to aliquots of the purified extract before treatment with the trifluoracetic acid/acetic anhydride reagent. The relative fluorescence of the aliquots was measured at 490 nm using an excitation wavelength of 440 nm. After correction for background fluorescence, the level of endogenous IAA could be calculated from the y-intercept of the regression of relative fluorescence on the amount of IAA added. The quantity of IAA originally present in the ethanolic extract was readily derived from the amount of ^{14}C-IAA recovered and its specific activity.

The accuracy of the α-pyrone assay is a function of its specificity. This has not been defined but apparently depends to some extent on the experimental conditions. While the formation of indolo-α-pyrone may be a feature unique to indole-3-acetic acids, there are many ways in which an increase in relative fluorescence could be recorded at 490 nm without the involvement of indolo-α-pyrone. Fluorescence activation and emission bands tend to be broad so that exclusion of fluorescence arising from extraneous material is difficult. A blank consisting of unreacted extract plus reagent alleviates some of the problems, but, as the data of ELIASSON, et al. (1976) demonstrate, the reagent may react with components of the extract other than IAA to produce a considerable amount of fluorescent material and/or an increase in turbidity. Both these processes contribute to the apparent fluorescence of the reaction mixture and cannot be corrected by the usual blank procedures. However, ELIASSON et al. (1976) have suggested that the rapid first-order decomposition of indolo-α-pyrone under UV light could be used as a basis for blank correction. One way this might be done, if the half-life ($t_{\frac{1}{2}}$) of the decomposition can be shown to be independent of the nature of the extract, is as follows: A figure for the corrected fluorescence (F_c) can be obtained by doubling the difference between the fluorescence initially present (F_i) and that remaining ($F_{t\frac{1}{2}}$) after a time, $t_{\frac{1}{2}}$, of UV light treatment, i.e., $F_c = 2(F_i - F_{t\frac{1}{2}})$. Although this procedure would not account for the fluorescence of unstable reaction products other than indolo-α-pyrone, it should nevertheless eliminate many sources of interference and increase specificity, thereby ensuring greater accuracy. However, in common with other selective assays of low information content, the accuracy actually achieved in any one practical situation is best gauged through successive approximation (Sect. 2.2.3. a and c.)

Even though steps can be taken to eliminate the deleterious effects of the interfering components of plant extracts, there is strong evidence to suggest that considerable overestimation of IAA levels still might occur, due to the inadvertant hydrolysis of IAA esters (BANDURSKI and SCHULZE, 1974). Although attempts can be made to reduce the degree of hydrolysis, it is difficult to eliminate satisfactorily the possibility of such an occurrence and, considering the fact that the ratio of ester to free IAA can be as high as 1000:1, it would appear essential to have some knowledge of the IAA ester content of any plant material, prior to analysis for free acid. Ideally such knowledge should extend to chemical characterization of the esters involved and synthesis of labelled standards of these compounds to act as internal markers in quantitative analyses of the endogenous ester content. Given such data, investigation of the free IAA level of the extract can be undertaken using the labelled esters to monitor hydrolysis occurring during purification.

3.6.2 Gibberellins

A few reports of estimates of endogenous GA levels have appeared in the literature, but none of these can be considered strictly quantitative because of the absence of even basic criteria of accuracy and precision. The reason for this dearth of information may well be the difficulties associated with GA analysis. The GA's lack any peculiar physical property to facilitate selective detection and, in addition, the range of naturally occurring compounds of this type is very large indeed, making the complete measurement of the GA status of a plant a daunting prospect.

Ross and Bradbeer (1971) used GLC to estimate the levels of GA_1, GA_4, GA_5, GA_6, GA_7, GA_8, GA_{17}, and GA_{20} in acidic, diethyl ether-soluble extracts from *Corylus avellana* seeds. The selectivity of the method used by these workers can be calculated relatively easily, since no purification procedures were employed and quantification was based solely upon the non-selective response of a flame ionization detector. From Eq. (50) (Sect. 3.4) it can be seen that $\bar{Q}$ and thus $\bar{S}$ can be approximated by the peak capacity of the gas chromatogram, which from the data supplied would appear to be about ten. Thus, according to Section 3.2.2.2, the maximum permissible impurity concentration for an average accuracy of $\bar{A} = 5\%$ will be

$$\frac{c \cdot A \cdot \bar{S}}{100} = \frac{c \times 5 \times 10}{100} = \frac{c}{2} \text{ ppm.}$$

Since, by definition, the concentration of the compound of interest is c ppm, it follows that $c + \frac{c}{2} = 10^6$ ppm, i.e., $c = 6.6 \times 10^5$ ppm or 66%. In other words, unless the concentration of the GA of interest in the extract exceeded 66%, the probability of the accuracy being 5% or better would be less than 0.5. However, since the bioassay data presented indicate that the GA content of *Corylus* seed was less than 1 part in 10^3, the estimates of GA levels afforded by GC are unlikely to be even remotely accurate. This contention is borne out by a later publication (Williams et al., 1974) in which it is stated with reference to the earlier GLC analysis of Ross and Bradbeer (1971) that "...application of GLC-MS techniques to similar extracts subsequently failed to confirm any of the GA identifications as the relevant peaks contained components which masked any GA derivatives present".

A major requirement of quantitative GA analysis is that the detector be highly selective yet have the ability to respond to all the GA's. Such a paradoxical requirement for both specificity and versatility is best met by a mass spectrometer operating in the MF mode. This procedure was used when Frydman et al. (1974) determined the levels of GA_9, GA_{17}, GA_{20}, and GA_{23} during maturation of *Pisum sativum* seed; Martin et al. (1977) measured the GA_{17} and GA_{25} content of immature seeds of *Pyrus communis*, and Browning and Saunders (1977) investigated yields of GA_4 and GA_9 from *Triticum aestivum* chloroplast membrane preparations. It is difficult to comment in detail on the effectiveness of the analytical procedures described in these reports, as insufficient data are provided. However, some general remarks are in order. The accuracy of the estimates is open to question as in no instance was the quantified response demonstrated to be free from interference. Although full scan mass spectra were used to identify the GA's, they were obtained either from different extracts or under analytical conditions different from those used for MF and thus do not provide evidence of the purity of the MF peaks. The GA content of the *Pisum*, *Pyrus*, and *Triticum* extracts appeared to be quite high, in most instances, so the selectivity of the MF may well have been sufficient to allow accurate estimates to be made. However, as there is no actual evidence, such as the effects or further sample purification, to support this contention, there are no grounds for assuming even the crudest levels of accuracy. A further limitation is that internal standards were not used, thus sample handling losses, variations in derivatizing efficiency, and short-term drifts in mass spectrometer sensitivity would all have contributed to the error of the method. However, in contrast to IAA and ABA, labelled GA's suitable for use as internal markers are not available from commercial sources, so that investigators are faced with the choice of either accepting such errors or investing time in the synthesis of appropriate ^{2}H-, ^{3}H- or ^{14}C-labelled GA's. The problem is further compounded as unlabelled reference standards can be equally difficult to obtain and this is a major restraint in the quantitative analysis of a great many GA's.

3.6.3 Cytokinins

The number of publications reporting the use of physicochemical methodology for the analysis of endogenous cytokinins from higher plant sources is also limited. THOMPSON et al. (1975) employed MF to measure the levels of 6 (o-hydroxybenzylamino)-9-β-D-ribofuranosylpurine (o-OH BAR) in leaves of *Populus robusta*. Sample losses that occurred during the extensive purification procedures were monitored by including a known amount of the synthetic para-isomer, p-OH BAR in each extract. This served as a relatively effective internal marker because it did not separate from the o-OH BAR until the MF stage of the analysis. Although the precision of the method was quoted as 10%, the accuracy was not defined, but some indications can be obtained from a comparison of the mass fragmentogram and the corresponding total ion current trace. After correction for background, the area ratios of the presumptive o- and p-OH BAR peaks are identical for both the total ion current and m/e 661 traces. It can, therefore, be argued that, as quantification using the more selective MF response had little apparent effect on the accuracy of measurement, the purification factor of the clean-up procedures must have been sufficient to guarantee a reasonable value of A_{α}. Although this is a good example of the type of simple check that nearly always can be made, it nevertheless still falls far short of the verification of accuracy that is obtained by successive approximation procedures (Sect. 3.2.2.3.a).

SUMMONS et al. (1977) also used MF to determine the raphanatin (7-β-D-glucopyranosylzeatin) content of *Raphanus* seeds. $^{2}H_{2}$-raphanatin was synthesized for use as an internal marker. The main advantage of this deuterium-labelled standard is that it has essentially the same GC retention as its endogenous counterpart. Consequently, errors attributable to extract components altering mass spectrometer sensitivity are eliminated, since both the standard and the sample are present in the mass spectrometer at the same time. However, as pointed out in Section 3.5, there are distinct disadvantages in the use of deuterated labels when accuracy is validated via the acquisition of information. SUMMONS et al. (1977) state, "Mass spectra recorded during elution of the TMSi [$^{2}H_{2}$] raphanatin peak, showed significant contribution (approximately 20%) from unlabelled TMSi-raphanatin to all diagnostic ions... This result confirmed earlier indications that raphanatin was present in radish seed". The strength of this contention lies in the amount of spectral information that can be unambiguously assigned to the putative endogenous raphanatin. Because the label is not carried by all of the fragmentation products of [$^2H_2$]-raphanatin, it is difficult to partition the total information content of the spectrum between the labelled carrier, the endogenous hormone, and the inevitable background contaminants. For this reason, it is normal practice only to consider the molecular ion cluster and perhaps one or two other ion groups, and, as a consequence, the information with which the endogenous compound may be described is severely restricted. On the basis of the M^+ and $[M\text{-}CH_2OTMSi]^+$ ion groups presented, the amount of information attributable to the putative endogenous TMSi raphanatin is less than 16 bits. This description is grossly inadequate for establishing the identy and purity of an unknown extract component (see Sect. 3.2.2.3.b). Although it follows that π cannot be large, the accuracy probability term, α, would have exceeded π, as quantification was based on the selective m/e 188 and m/e 190 responses. Thus the quantitative estimates obtained by SUMMONS et al. (1977) may be accurate, but, without a reasonable value of π to limit α, there is inadequate evidence to support such a view.

3.6.4 Abscisic Acid

In contrast to the situation that exists with other hormones, the number of reports of physicochemical quantitative analysis of ABA is very large, presumably because of the apparent ease with which the compound can be detected in extracts from small amounts of plant tissue. However, out of the plethora of so-called "new/rapid/sensitive" assays for ABA very few, if any, provide rigorous proof of accuracy. This is not to suggest that most of the data available on ABA levels are in error, rather that there is insufficient evidence to warrant carte blanche assumptions of accuracy.

One of the earliest methods employed exploited the ORD properties of ABA (MILBORROW, 1967). Extensive sample purification was necessary to reduce the optical density at 250–310 nm to less than 1.5 before the level of ABA present could be obtained from the rotation at the first (+) extremum. One of the novel features of this method is that it facilitated the use of synthetic, racemic ABA as an internal standard, as both the racemate and the endogenous (+) ABA could be measured by UV spectrometry, yet were readily distinguishable in terms of optical rotation. MILBORROW (1967) set out four criteria for establishing the presence of (+) ABA in plant extracts. Firstly, the putative ABA must be derived from an acidic, ether fraction, a description equivalent to 1 bit. The second and third requirements are that the inhibitory activity must resemble ABA in terms of both biological symptoms and chromatographic properties. Normally only one TLC was used, therefore, assuming a typical peak capacity of six and the location of a characteristic, inhibitory bioassay response only at the specified Rf, 6 bits of information would be generated. The fourth criterion is that the ORD curve of the material eluted from the inhibitory zone of the TLC resembles authentic (+) ABA in all respects. Since most features of the ORD response can be characterized by the

(+) and (−) extrema, the spectrum can be considered to consist of two independent measurements. From the typical curves provided by MILBORROW (1967) it is apparent that 4 bits constitute an adequate description of the wavelength, and thus the total information content acquired by this process is $2 \times 4 = 8$ bits. Therefore, the four criteria provide no more than a 15 bit description of the presumptive (+) ABA, and, in view of the arguments presented in Section 3.2.2.3. b, this would seem much too small to ensure an acceptable value of the accuracy probability term α.

One of the more common ABA assays involves methylation of the sample and analysis of the ABA methyl ester (Me ABA) by GLC. Initial studies made use of a flame ionization detector (LENTON et al., 1971, 1972), but, with the demonstration of the marked electron affinity of ABA by SEELEY and POWELL (1970), more recent investigations have employed an electron capture detector (HARRISON and SAUNDERS, 1975). The high selectivity and low limits of detection of this device make it ideally suited to the accurate analysis of trace quantities of ABA. It is, however, still essential to verify accuracy, and successive approximation is the most convenient way of obtaining a demonstratably error-free estimate (see Sect. 3.2.2.3. a).

A number of investigators have attempted to establish the accuracy of their estimates by a UV isomerization technique which makes use of the fact that endogenous cis-ABA forms a 1:1 equilibrium mixture of the cis- and trans-isomers when irradiated with UV light for an adequate period of time. LENTON et al. (1971), who devised the procedure, provide the following description: "To obtain positive identification of a presumptive Me ABA peak on a chromatogram of a plant extract, a sample was irradiated as above (methanolic solution exposed to UV light for 4 h) and rechromatographed. The appearance of a new peak with the same retention time as trans Me ABA was taken to indicate the presence of ABA in the extract. When the two peaks present after irradiation had areas in the same ratio as the two peaks produced by irradiation of authentic Me ABA it was concluded that the presumptive Me ABA consisted entirely of Me ABA".

It is evident that the above process is descriptive, and the probability of such a description being unique and thus providing evidence of identity, purity, and hence accuracy depends upon its information content. The UV treatment can be thought of as supplying an additional, independent chromatogram in much the same manner as if a different chromatographic system, derivative of ABA or detector were to be used. The information acquired prior to UV irradiation is given by $1(\phi - n)$ bits where ϕ is the peak capacity and n is the number of impurity peaks present in the chromatogram. Following UV treatment a further $4(\phi - 1 - n)$ bits are obtained on the assumption that the two presumptive isomer peaks have an area ratio of within $\pm 3\%$ of the required value. However, as most of the compounds detected do not change retention on UV irradiation, much of the information of the second chromatogram is highly correlated with that of the first. Traces presented by HARRISON and SAUNDERS (1975) for a birch bud extract clearly demonstrate this point. Thus the amount of information contributed by rechromatography after UV treatment is probably as low as $4(2-1) = 4$ bits and the total usable information obtained by HARRISON and SAUNDERS (1975) can be calculated as $1(10-5) + 4 = 9$ bits. Obviously this is insufficient to guarantee an acceptable value of α according to the criteria outlined in Section 3.2.3. b. There is, however, no reason why the UV isomerization technique cannot be used to increase the selectivity of the analysis. If E_i^t and E_f^t are estimates of the quantities of trans-ABA present before and after UV light treatment, respectively, and E_i^c and E_f^c the corresponding estimates for the cis-isomer, then $E = E_f^t - E_i^t + E_i^c - E_f^c$ will, in the main, measure only those electron-capturing compounds which co-chromatograph with Me cis-ABA and are converted by UV light to a product chromatographically similar to Me trans-ABA.

Quantitative analysis of ABA has been carried out by MF (HILLMAN et al., 1974; RAILTON et al., 1974; BERRIE and ROBERTSON, 1976; RIVIER et al., 1977) and the same comments apply to these analyses as those expressed with regard to MF of IAA in Section 3.6.1. Basically, far too little information is supplied by MF when only one or two ions are monitored, and, in the absence of data to verify the purity of the quantified response, it is necessary to provide evidence that the procedure possesses adequate selectivity. Unfortunately this was not done so the accuracy of the method remains undefined.

Recently HPLC has begun to be used for quantitative analysis of ABA (QUEBEDEAUX et al., 1976; SWEETSER and VATVARS, 1976; CIHA et al., 1977; DURLEY et al., 1978). ABA absorbs strongly at 254 nm hence when the technique is used in conjunction with a UV monitor the limits of detection are only slightly inferior to those achieved by GLC-electron capture or MF, although the selectivity is much lower than offered by either of these GLC procedures. Both QUEBEDEAUX et al. (1976) and CIHA et al. (1977) used GLC-MS to demonstrate that their extracts contained ABA and employed HPLC to assess the amounts present. It must be stressed, however, that mass spectrometric data per se does not provide a guarantee of accuracy unless sufficient information can be unambiguously assigned to the actual peak being quantified. This is not too difficult a task with HPLC, as the UV absorbing peak of interest can be easily collected and analyzed by mass spectrometry to obtain an estimate P_π and thus A_α.

3.7 Summary

It has been our intention to provide a theoretical framework with which to rationalize the uncertainties associated with the analysis of samples from open-ended systems. In doing so, we have mainly concentrated on the way in which the accuracy probability term, α, is influenced by the presence of interfering substances in the extract, as this appears to be one of the most crucial, yet neglected, aspects of plant hormone analysis. It has been argued that there are only two ways of assessing the probability and magnitude of such interference. One requires that a percentage, P, of the total quantified response contain sufficient information to give a probability, π, of detecting interference should it occur. Under the circumstances accuracy can be taken as 100-P% and π will give a lower limit for α. The other approach involves demonstrating that the selectivity of the method is adequate for the particular sample concentration in question. This is achieved by means of a procedure referred to as successive approximation, and results in an estimate whose accuracy is small compared with the precision of the method. The value of α is determined by the significance level, τ, chosen for the approximation.

It must be remembered that estimates of α obtained by either of the above methods need not be correct, as interference is not the only mechanism by which non-random errors can be generated. Sample losses encountered during purification can result in marked inaccuracies which have to be accounted for by internal standardization. However, a more serious problem, briefly touched upon in Section 3.6.1, is posed by facile hydrolysis of hormone conjugates during extraction and purification. Such an event is a distinct possibility, not only for IAA, but also for the other hormones under discussion, and it follows that, if free hormone from this source is to be discounted, attempts must be made to characterize the relevant hormone conjugates that are present in an extract.

One of the inescapable features of biological systems is that their attributes can be described only in terms of probability density functions. Thus, in order to obtain a reasonable measure of a variable, such as an extractable quantity of plant hormone, it is evident that a large number of individual estimates must be made and such a requirement places severe demands upon the speed of the analytical process. However, if analysis is to be restricted to a definite sample population, it is unnecessary to validate the accuracy of each estimate, since the selectivity of the method can be adjusted to give an acceptable value of α, applicable not just to an individual extract but to the entire sample population. By this means the amount of plant material required for each analysis is greatly reduced, and, as a consequence, there is a marked improvement in the efficiency and speed of analysis of the procedures that can be employed. A further advantage is that it becomes possible to analyze plant tissues, such as root caps and apices, which are difficult to obtain in large amounts. The current needs of plant hormone analysis are, therefore, most likely to be fulfilled by high-speed systems, each designed to have sufficient selectivity to allow the accurate analysis of a specific sample type.

Modern LC is ideally suited to the early stages of extract purification. By judicious selection of the column type and the sequence in which a number of columns are interconnected, it is possible to improve the purity of crude samples to a high degree, and, in favourable instances, even substitute for such processes as solvent partitioning. The best way of achieving the actual quantification depends upon the individual properties of the hormone. In the case of IAA, the most obvious choice would be to exploit either the native fluorescence of the molecule or that of the indolo-α-pyrone derivative or, alternatively, the colour reaction with Salkowski reagent, as there is no good reason why any of these properties could not be detected on-line from an HPLC system. ABA is probably best monitored by means of an electron capture detector as no other device of equivalent simplicity affords such high selectivity. This detector could be very effective indeed when used with a packed GC column coupled via a cold trap to a WCOT capillary system. GA's and cytokinins pose more of a problem because it would be impractical to set up a specific analytical procedure for each member of these two groups. In view of this, MF is the most attractive method as it permits quite high levels of selectivity to be obtained without a concomitant loss in versatility.

Trends towards purpose-built automatic and semi-automatic analysers are now firmly established in the fields of clinical chemistry and pharmacology. However, because of the large investment in time, money, and expertise that is necessary, the introduction of similar methodology to plant hormone analysis is unlikely to be a rapid process. Nevertheless it is quite evident that the technology required for the development of such instrumentation is already available and, until such times as this potential is exploited, attempts to relate physiological phenomena to biochemical and chemical events will be severely restricted.

3.8 List of Symbols

A	Percentage accuracy.
$\bar{A}$	Mean percentage accuracy.
A_α	Maximum value of percentage accuracy likely at a confidence level of α.
B_ω	Absolute analytical precision in ppm for a confidence level of ω.
D	Fraction of the potential number of compounds in a proportion ε of an extract which is selected by a particular chromatographic procedure.
E	Estimate of hormone concentration in ppm.
E_n	nth Estimate of hormone concentration in ppm.
E_ω	Range of values for an individual estimate due to random error at a confidence level ω.
ΔE	Expectation of the pairs difference $E_n - E_{n+1}$.
$\bar{F}$	Average fraction of extract components selected by taking a zone $\Delta\phi = 1$ from a chromatogram.
F_b	Relative fluorescence due to background.
F_t	Total relative fluorescence observed.
G	Actual number of compounds representing a proportion ε of an extract as a fraction of the number potentially present.
H_n	Average number of isomers associated with a particular molecular formula of molecular weight n.
I	Information in bits.
I_q	Information arising from the quantitative aspects of a chromatogram.
I_t	Total of the information accrued from a number of sources.
I_u	That part of I_t which can be related to the characteristics of the reference compound u.
J	Potential number of compounds having the molecular formula $C_vH_wO_xN_y$ and lying within specified limits of molecular weight and v, w, x and y.
K_n	Number of molecular formulae of the type $C_vH_wO_xN_y$ that have a molecular weight of n and lie within specified limits of v, w, x and y.
L_u	Percentage relative intensity of ions in the mass spectrum of the reference compound u.
L_U	Percentage relative intensity of ions in the mass spectrum of the sample under test.
M_i	Weight of extract prior to purification.
M_f	Weight of extract subsequent to purification.
M_u	Weight of substance u in the extract.
N	Efficiency of a chromatographic process in theoretical plates.
P	Percentage purity.
P_π	Minimum value of percentage purity at a confidence level of π.
Q	Purification factor of a chromatographic procedure.
$\bar{Q}$	Average purification factor of a chromatographic procedure.
Q_n	Purification factor of the nth chromatographic procedure.
Q_θ	Minimum value of the purification factor likely at a confidence level of θ.
Q^t	Total purification factor of a combination of chromatographic procedures.
R	Sensitivity of a certain detector to a certain sample.
R_n	Sensitivity of a certain detector to the nth component of a sample.
R_u	Sensitivity of a certain detector to the reference compound u.
$\bar{R}$	Absolute mean sensitivity of a certain detector.

R'_n	Actual mean sensitivity of a certain detector to extract n.
R'_σ	Highest value of R′ likely at a confidence level σ.
S	Selectivity of certain analytical procedure.
S_n	Actual selectivity obtained when analysing the nth extract.
S^t	Actual total selectivity obtained through the use of a number of procedures.
$\bar{S}$	Mean selectivity of a certain analytical procedure.
S_σ	Lowest value of S likely at a confidence level σ.
U	Substance isolated from extract and purported to be the same as reference compound u.
V_n^2	Variance associated with the nth estimate.
V_p^2	Variance associated with the precision of the method.
X	Error factor associated with proportional non-random error.
Z	Percentage recovery of the compound of interest following purification.
c	Actual value of the concentration (w/w) of the hormone of interest in ppm.
$f_n(c)$	Function describing the proportional non-random error due to the nth component of the sample.
f(R)	Probability density function of the distribution of sensitivities for a given detector.
j	Correlation coefficient of chromatographic procedures.
k′	Chromatographic capacity factor – a dimensionless index of retention.
m	Number of discrete analyses contributing to an estimate.
q	Percentage analytical precision of the largest estimate catered for by the calibration procedure.
r_n	Absolute response of a certain detector to the nth component of an extract.
r_t	Total absolute response of a certain detector to a multicomponent sample.
r_u	Absolute response of a certain detector to the reference compound u.
s_p^2	True variance associated with the precision of an analytical method.
s_t^2	True total variance associated with a group of estimates.
t_τ	t value associated with the probability level τ of a standardized Student distribution of 2(m-1) degrees of freedom.
t_ω	t value of standardized student distribution of 2(m-1) degrees of freedom which excludes the two-tailed probability level 1-ω.
v	Number of carbon atoms per molecule for a particular substance.
w	Number of hydrogen atoms per molecule for a particular substance.
x	Number of oxygen atoms per molecule for a particular substance.
y	Number of nitrogen atoms per molecule for a particular substance.
z	Classical correlation coefficient.
α	Probability of accuracy being better than A_α percent.
δ	Fraction of the mass of an extract which can be accounted for by compounds of molecular formula $C_vH_wO_xN_y$.
ε	Fraction of the mass of an extract lying within specified limits of molecular weight.
θ	Probability of the purification factor being better than Q_θ.
π	Probability of the purity being better than P_π percent.
ϱ	Proportion of compounds in the sample population which occupy a given information channel.
σ	i) Probability of the selectivity being better than S_σ, i. e., probability of R′ being less than R'_σ. ii) Probability of R′ lying within the range $\bar{R} \pm R'_\sigma$.
τ	Confidence level used for the test, $\Delta E = 0$, in the successive approximation.
ϕ	Chromatographic peak capacity.
ω	Probability that for a given extract, estimates will be within $\pm$ B_ω of the mean.

Acknowledgements. D.R.R. was supported by a Science Research Council (UK) Grant to A.C. The authors wish to thank Professor J. MacMillan, F.R.S., Professor R.P. Pharis, Dr. J.L. Stoddart and Dr. R. Horgan for the detailed and helpful comments they made when reviewing a preliminary draft of this paper. We are especially grateful to our colleagues Margaret McLaughlin, Huw Powell, Linda Nash, David Wilcox (Department of Botany, University of Glasgow), Karen Loferski and Dean Regier (Department of Agricultural

Chemistry, Oregon State University). Their participation in numerous discussions and the time they devoted to reading and assessing assorted draft manuscripts was invaluable in the evolution and refinement of many of the concepts put forward in this chapter.

References

Addicott, F.T., Carns, H.R., Lyon, J.L., Smith, O.E., McMeans, J.L.: On the physiology of abscisins. In: Regulateurs naturels de la croissance vegetale. Nitsch, J.P. (ed.), pp 687–703. Paris: Centre National de la Recherche Scientifique 1964

Arpino, P.J., Guiochon, G.: LC/MS coupling. Anal. Chem. *51*, 682A–701A (1979)

Audus, L.J., Thresh, R.: A method of plant growth substance assay for use in paper partition chromatography. Physiol. Plant. *6*, 451–465 (1953)

Bandurski, R.S., Schulze, A.: Concentrations of indole-3-acetic acid and its esters in *Avena* and *Zea*. Plant Physiol. *54*, 257–262 (1974)

Bandurski, R.S., Schulze, A.: Concentration of indole-3-acetic acid and its derivatives in plants. Plant Physiol. *60*, 211–213 (1977)

Berrie, A.M.M., Robertson, J.: Abscisic acid as an endogenous component in lettuce fruits, *Lactuca sativa* L. cv Grand Rapids. Does it control thermodormancy? Planta *131*, 211–215 (1976)

Beynon, J.E., Williams, A.E.: Mass and abundance tables for use in mass spectrometry. Amsterdam: Elsevier Publishing Company 1963

Biddington, N.L., Thomas, T.H.: A modified *Amaranthus* betacyanin bioassay for the rapid determination of cytokinins in plant tissues. Planta *111*, 183–186 (1973)

Bigot, C.: Action d'adénines substituées sur la synthèse des bétacyanines dans la plantule d'*Amaranthus caudathus* L. Possibilité d'un test biologique de dosage des cytokinines. C.R. Acad. Sci. *266*, 349–352 (1968)

Bonner, J.: Limiting factors and growth inhibitors in the growth of the *Avena* coleoptile. Am. J. Bot. *36*, 323–332 (1949)

Bowen, D.H., Crozier, A., MacMillan, J., Reid, D.M.: Characterization of gibberellins from light-grown *Phaseolus coccineus* seedlings by combined GC-MS. Phytochemistry *12*, 2935–2941 (1973)

Brandes, H., Kende, H.: Studies on cytokinin-controlled bud formation in moss protenemata. Plant Physiol. *43*, 827–837 (1968)

Brian, P.W., Hemming, H.G., Lowe, D.: Comparative potency of nine gibberellins. Ann. Bot. (London) *28*, 369–389 (1964)

Bridges, I.G., Hillman, J.R., Wilkins, M.B.: Identification and localisation of auxin in primary roots of *Zea mays* by mass spectrometry. Planta *115*, 189–192 (1973)

Browning, G., Saunders, P.F.: Membrane localised gibberellins A_9 and A_4 in wheat chloroplasts. Nature (London) *265*, 375–377 (1977)

Caplin, S.M., Steward, F.C.: A technique for the controlled growth of excised plant tissue in liquid media under aseptic conditions. Nature (London) *163*, 920–921 (1949)

Chrispeels, M.J., Varner, J.E.: Inhibition of gibberellic acid-induced formation of α-amylase by abscisin II. Nature (London) *212*, 1066–1067 (1966)

Ciha, A.J., Brenner, M., Brun, W.A.: Rapid separation and quantification of abscisic acid from plant tissues using high performance liquid chromatography. Plant Physiol. *59*, 821–826 (1977)

Cummins, W.R., Kende, H., Raschke, K.: Specificity and reversibility of the rapid stomatal response to abscisic acid. Planta *99*, 347–351 (1971)

Davis, L.A., Lyon, J.L., Addicott, F.T.: Phaseic acid; occurrence in cotton fruit; acceleration of abscission. Planta *102*, 294–301 (1972)

Durley, R.C., Kannangara, T., Simpson, G.: Analysis of abscisins and 3-indolylacetic acid in leaves of *Sorghum bicolor* by high performance liquid chromatography. Can. J. Bot. *56*, 157–161 (1978)

Eliasson, L., Strömquist, L.-H., Tillberg, E.: Reliability of the indolo-α-pyrone fluorescence method for indole-3-acetic acid determination in crude plant extracts. Physiol. Plant. *36*, 16–19 (1976)

Fletcher, R.A., McCullagh, D.: Cytokinin-induced chlorophyll formation in cucumber cotyledons. Planta *101*, 88–90 (1971)

Francis, A.W.: Numbers of isomeric alkylbenzenes. J. Am. Chem. Soc. *69*, 1536–1537 (1947)

Frankland, B., Wareing, P.F.: Effect of gibberellic acid on hypocotyl growth of lettuce seedlings. Nature (London) *185*, 255–256 (1960)

Frydman, V.M., Gaskin, P., MacMillan, J.: Qualitative and quantitative analyses of gibberellins throughout seed maturation in *Pisum sativum* cv. Progress No. 9. Planta *118*, 123–132 (1974)

Fuchs, S., Haimovich, J., Fuchs, Y.: Immunological studies of plant hormones. Detection and estimation by immunological assays. Eur. J. Biochem. *18*, 384–390 (1971)

Fuchs, Y., Gertman, E.: Insoluble antibody column for isolation and quantitative determination of gibberellins. Plant Cell Physiol. *15*, 629–633 (1974)

Fuchs, Y., Mayak, S., Fuchs, S.: Detection and quantitative determination of abscisic acid by immunological assay. Planta *103*, 117–125 (1972)

Giddings, J.C.: Maximum number of components resolvable by gel filtration and other elution chromatographic methods. Anal. Chem. *39*, 1027–1028 (1967)

Grotch, S.L.: Matching of mass spectra when peak height is encoded to one bit. Anal. Chem. *42*, 1214–1222 (1970)

Hancock, R.C., Barlow, H.W.B., Lacey, H.J.: The East Malling coleoptile straight-growth test method. J. Exp. Bot. *15*, 166–176 (1964)

Harrison, M.A., Saunders, P.F.: The abscisic acid content of dormant birch buds. Planta *123*, 291–298 (1975)

Henze, H., Blair, C.: Number of isomeric hydrocarbons of the methane series. J. Am. Chem. Soc. *53*, 3077–3085 (1931)

Henze, H., Blair, C.: The number of structural isomers of the more important types of aliphatic compound. J. Am. Chem. Soc. *56*, 157 (1934)

Hillman, J.R., Young, I., Knights, B.A.: Abscisic acid in leaves of *Hedera helix* L. Planta *119*, 263–266 (1974)

Hillman, W.S.: Nonphotosynthetic light requirement in *Lemna minor* and its partial satisfaction by kinetin. Science *126*, 165–166 (1957)

Jamieson, W.D., Hutzinger, O.: Identification of simple naturally occurring indoles by mass spectrometry. Phytochemistry *9*, 2029–2036 (1970)

Jones, R.L., Varner, J.E.: The bioassay of gibberellins. Planta *72*, 155–161 (1967)

Kaldewey, H., Wakhloo, J.L., Weis, A., Jung, H.: The *Avena* geocurvature test. Planta *84*, 1–10 (1969)

Kende, H.: Kinetin-like factors in the root exudate of sunflowers. Proc. Natl. Acad. Sci. USA *53*, 1302–1307 (1965)

Knock, B.A., Smith, I.C., Wright, D.E., Ridley, R.G., Kelly, W.: Compound identification by computer matching of low resolution mass spectra. Anal. Chem. *42*, 1516–1520 (1970)

Köhler, D., Lang, A.: Evidence for substances in higher plants interfering with response of dwarf peas to gibberellin. Plant Physiol. *38*, 555–560 (1963)

Koshimizu, K., Fukui, H., Mitsui, T., Ogawa, Y.: Identity of lupin inhibitor with abscisin II and its biological activity on growth of rice seedlings. Agric. Biol. Chem. *30*, 941–943 (1966)

Lederberg, J., Sutherland, G.L., Buchanan, B.G., Feigenbaum, E.A., Robertson, A.V., Duffield, A.M., Djerassi, C.: Application of artificial intelligence for chemical inference. I. The number of possible organic compounds. Acyclic structures containing C, H, O, and N. J. Am. Chem. Soc. *91*, 2973–2976 (1969)

Lenton, J.R., Perry, V.M., Saunders, P.F.: The identification and quantitative analysis of abscisic acid in plant extracts by gas-liquid chromatography. Planta *96*, 271–280 (1971)

Lenton, J.R., Perry, V.M., Saunders, P.F.: Endogenous abscisic acid in relation to photoperiodically induced bud dormancy. Planta *106*, 13–22 (1972)

Letham, D.S.: Regulators of cell division in plant tissues. V. A comparison of the activities of zeatin and other cytokinins in five bioassays. Planta *74*, 228–242 (1967)

Letham, D.S.: A new cytokinin bioassay and the naturally occurring cytokinin complex. In: Biochemistry and physiology of plant growth substances. Wightman, F., Setterfield, G., (eds.), pp. 19–31. Ottawa: Runge Press 1968
Lieberman, M., Knegt, E.: Influence of ethylene on indole-3-acetic acid concentration in etiolated pea epicotyl tissue. Plant Physiol. *60*, 475–477 (1977)
Manos, P.J., Goldthwaite, J.: An improved cytokinin bioasay using cultured soybean hypocotyl sections. Plant Physiol. *57*, 894–897 (1976)
Martin, G.C., Dennis, F.G., MacMillan, J., Gaskin, P.: Hormones in pear seeds. 1. Levels of gibberellins, abscisic acid, phaseic acid, dihydrophaseic acid and two metabolites of dihydrophaseic acid in immature seeds of *Pyrus communis* L. J. Am. Soc. Hortic. Sci. *102*, 16–19 (1977)
McFadden, W.H.: Interfacing chromatography and mass spectrometry. J. Chromatogr. Sci. *17*, 2–17 (1979)
McFadden, W.H., Bradford, D.C., Gaines, D.E., Gower, J.L.: Applications of combined liquid chromatography/mass spectrometry. Int. Laboratory. pp. 55–64 October (1977)
McWha, J.A., Philipson, J.J., Hillman, J.R., Wilkins, M.B.: Molecular requirements for abscisic acid activity in two bioassay systems. Planta *109*, 327–336 (1973)
Milborrow, B.V.: The identification of (+)-abscisin II [(+)-dormin] in plants and measurement of its concentrations. Planta *76*, 93–113 (1967)
Miller, C.O.: Kinetin and kinetin-like compounds. In: Modern methods of plant analysis Linskens, H.F., Tracey, M.V. (eds.), Vol. VI, pp. 194–202. Berlin-Heidelberg-New York: Springer 1963
Moewus, F.: Der Kressewurzeltest, ein neuer quantitativer Wuchsstofftest. Biol. Zentralbl. *68*, 118–140 (1949)
Moffat, A.C., Smalldon, K.W., Brown, C.: Optimum use of paper, thin layer and gas liquid chromatography for the identification of basic drugs 1) Determination of effectiveness of a series of chromatographic steps. J. Chromatogr. *90*, 1–7 (1974)
Murakami, Y.: The microdrop method, a new rice seedling test for gibberellins and its use for testing extracts of rice and morning glory. Bot. Mag. *79*, 33–43 (1968)
Murashige, T., Skoog, F.: A revised medium for rapid growth and bioassays with tobacco tissue cultures. Physiol. Plant. *15*, 473–497 (1962)
Nicholls, P.B., Paleg, L.G.: A barley endosperm bioassay for gibberellins. Nature (London) *199*, 823–824 (1963)
Nitsch, J.P., Nitsch, C.: Studies on the growth of coleoptile and first internode sections: A new sensitive straight-growth test for auxins. Plant. Physiol. *31*, 94–111 (1956)
Ockerse, R., Galston, A.W.: Gibberellin-auxin interaction in pea stem elongation. Plant Physiol. *42*, 47–54 (1967)
Ogawa, Y.: Studies on the conditions for gibberellin assay using rice seedling. Plant Cell Physiol. *4*, 227–237 (1963)
Osborne, D.J., McCalla, D.R.: Rapid bioassay for kinetin and kinins using senescing leaf tissue. Plant Physiol. *36*, 219–221 (1961)
Pengelly, W., Meins, F.: A specific radioimmunoassay for nanogram quantities of the auxin, indole-3-acetic acid. Planta *136*, 173–180 (1977)
Phinney, B.O.: Growth response of single-gene dwarf mutants in maize to gibberellic acid. Proc. Natl. Acad. Sci. USA *42*, 185–189 (1956)
Powell, L.E., Tautvydas, K.J.: Chromatography of gibberellins on silica gel partition columns. Nature (London) *213*, 292–293 (1967)
Quebedaux, B., Sweetser, P.B., Rowell, J.C.: Abscisic acid levels in soybean reproductive structures during development. Plant Physiol. *58*, 363–366 (1976)
Railton, I.D., Reid, D.M., Gaskin, P., MacMillan, J.: Characterization of abscisic acid in chloroplasts of *Pisum sativum* cv. Alaska by combined gas chromatography-mass spectrometry. Planta *117*, 179–182 (1974)
Reeve, D.R., Crozier, A.: An assessment of gibberellin structure-activity relationships. J. Exp. Bot. *25*, 431–445 (1974)
Ridley, R.G.: Progress in the Mass Spectrometry Data Centre, 1966–70. In: Advances in mass spectrometry. Quayle, A. (ed.), Vol. V, pp. 307–313. Amsterdam: Elsevier Publishing Co. Ltd. 1971

Rivier, L., Pilet, P.-E.: Indolyl-3-acetic acid in cap and apex of maize roots: identification and quantification by mass spectrometry. Planta *120*, 107–112 (1974)

Rivier, L., Milon, H., Pilet, P.-E.: Gas chromatography-mass spectrometric determinations of abscisic acid levels in the cap and the apex of maize roots. Planta *134*, 23–27 (1977)

Ross, J.D., Bradbeer, J.W.: Studies in seed dormancy V. The content of endogenous gibberellins in seeds of *Corylus avellana* L. Planta *100*, 288–302 (1971)

Rouvray, D.H.: Isomer enumeration methods. Chem. Soc. Rev. **3**, 355–372 (1974)

Sakurai, N., Shibata, K., Kamisaka, S.: Stimulation of cucumber hypocotyl elongation by dihydroconiferyl alcohol. Interactions between dihydroconiferyl alcohol and auxin or gibberellin. Plant Cell Physiol. *15*, 709–716 (1974)

Scott, R.P.W.: Liquid chromatography detectors. J. Chromatogr. Library VII. Amsterdam: Elsevier Scientific Publishing Company 1977

Seeley, S.D., Powell, L.E.: Electron capture-gas chromatography for sensitive assay of abscisic acid. Anal. Biochem. *35*, 530–533 (1970)

Sivori, E.M., Sonvico, V., Fernandez, N.O.: Determination of abscisic acid following Paleg's method. Plant Cell Physiol. *12*, 993–996 (1971)

Stenhagen, E., Abrahamson, S., McLafferty, F.M.: Atlas of mass spectral data. New York: John Wiley and Sons 1969

Stoessl, A., Venis, M.A.: Determination of submicrogram levels of indole-3-acetic: a new, highly specific method. Anal. Biochem. *34*, 344–351 (1970)

Summons, R.E., MacLeod, J.K., Parker, C.W., Letham, D.S.: The occurrence of raphanatin as an endogenous cytokinin in radish seed. Identification and quantification by gas chromatographic-mass spectrometric analysis using deuterium-labelled standards. FEBS Lett. *82*, 211–214 (1977)

Sweetser, P.B., Vatvars, A.: High-performance liquid chromatographic analysis of abscisic acid in plant extracts. Anal. Biochem. *71*, 68–78 (1976)

Taylor, H.F., Burden, R.S.: Xanthoxin, a recently discovered plant growth inhibitor. Proc. R. Soc. London Ser. B *180*, 317–346 (1972)

Thimann, K.V., Schneider, C.L.: The relative activities of different auxins. Am. J. Bot. *26*, 328–333 (1939)

Thompson, A.G., Horgan, R., Heald, J.K.: A quantitative analysis of cytokinin using single-ion-current-monitoring. Planta *124*, 207–210 (1975)

Tillberg, E.: An abscisic acid-like substance in dry and soaked *Phaseolus vulgaris* seeds determined by the *Lemna* growth bioassay. Physiol. Plant. *34*, 192–195 (1975)

Tucker, D.J., Mansfield, T.A.: A simple bioassay for detecting "antitranspirant" activity of naturally occurring compounds such as abscisic acid. Planta *98*, 157–163 (1971)

Van Onckelen, H.A., Verbeek, R.: Isolation and assay of a cytokinin from barley. Phytochemistry *11*, 1677–1680 (1972)

Van Overbeek, J., Went, F.W.: Mechanism and quantitative application of the pea test. Bot. Gaz. *99*, 22–41 (1937)

Van Staden, J., Bornman, C.H.: *Spirodela* growth test: a possible bioassay for abscisic acid. J.S. Afr. Bot. *36*, 9–12 (1970)

Vogel, A.I.: A text-book of quantitative inorganic analysis. London: Longmans, Green and Co. Ltd., 1st Ed. 1939, 3rd Ed. 1961

Wangen, L.E., Woodward, W.S., Isenhour, T.L.: Small computer, magnetic tape oriented, rapid search system applied to mass spectrometry. Anal. Chem. *43*, 1605–1614 (1971)

Went, F.W.: Wuchsstoff und Wachstum. Recl. Trav. Bot. Neerl. *25*, 1–116 (1928)

Went, F.W.: On the pea test method for auxin, plant growth hormone. Proc. K. Ned. Akad. Wet. *37*, 547–555 (1934)

Whyte, P., Luckwill, L.C.: A sensitive bioassay for gibberellins based on retardation of leaf senescence in *Rumex obtusifolius* (L.). Nature (London) *210*, 1360 (1966)

Williams, P.M., Bradbeer, J.W., Gaskin, P., MacMillan, J.: Studies in seed dormancy VIII. The identification and determination of gibberellins A_1 and A_9 in seeds of *Corylus avellana* L. Planta *117*, 101–108 (1974)

Selected Bibliography

1. Jeffreys, H.: Scientific inference, 2nd. Ed. Cambridge: Cambridge University Press 1957
2. Smart, J.C.C.: Between science and philosophy. New York: Random House 1968
3. Calvin, M.: Chemical evolution. Oxford: Clarendon Press 1969
4. Cairns-Smith, A.G.: The life puzzle. Edinburgh: Oliver and Boyd 1971
5. Brillouin, L.: Science and information theory 2nd. Ed. New York: Academic Press 1957
6. Shannon, C.E., Weaver, W.: The mathematical theory of communication. Urbana: The University of Illinois Press 1949
7. Kaiser, H.: Quantitation in elemental analysis. Anal. Chem. *42* (*2*) 24A-41A (1971)

4 Biosynthesis and Metabolism of Plant Hormones

G. SEMBDNER, D. GROSS, H.-W. LIEBISCH, and G. SCHNEIDER

4.1 Biosynthetic Pathways

4.1.1 Auxins

Indole-3-acetic acid (IAA) (Structure 4.1) has been conclusively shown to be a natural hormone, ubiquitously present in higher plants. Evidence is gradually accumulating that there are indolic auxins other than IAA naturally occurring in several plants, i.e., chlorinated derivatives (see Sect. 4.1.1f), or auxin-like compounds which are non-indolic in chemical structure, such as phenylacetic acid (cf. SCHNEIDER and WIGHTMAN, 1974). The existence of a non-indolic "citrus auxin" (KHALIFAH et al., 1963), however, is doubtful after repeated re-investigation (GOLDSCHMIDT et al., 1971; TAKAHASHI et al., 1975). Research work on steroid-like plant hormones is still in course (VENDRIG, 1967a, b, 1971; HEFTMANN, 1971, 1975; KOPCEWICZ, 1971; KOPCEWICZ and ROGOZINSKA, 1972; GEUNS, 1974, 1977, 1978; GEUNS and VENDRIG, 1974). This section deals with both IAA as the principal auxin of higher plants and its chlorinated derivatives naturally occurring in the plant. Biosynthesis of IAA in microorganisms is not discussed in detail.

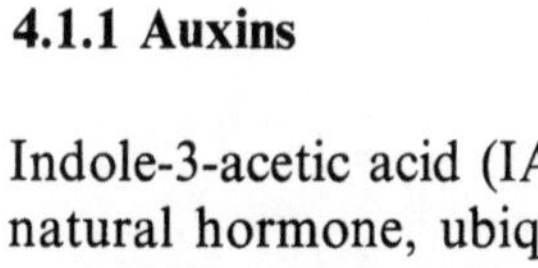

Structure 4.1. Indole-3-acetic acid

a) Tryptophan as Primary Precursor

The protein amino acid, L-tryptophan, has been suggested to be the main intermediate in the pathway of IAA biosynthesis both in microorganisms and higher plants. This suggestion is supported by the close chemical similarity of tryptophan and IAA, as well as their ubiquitous occurrence as natural constituents of higher plants.

Since THIMANN's studies, showing that cultures of the mould *Rhizopus suinus* form IAA when supplied with tryptophan (THIMANN, 1935), a number of metabolic experiments have been carried out both in vivo and in vitro with the aim of elucidating the biosynthetic pathway of IAA and, especially, of finding out the intermediary steps in the enzymatic conversion of tryptophan into IAA.

The formation of IAA from tryptophan has been studied both in intact plant tissue and in cell-free preparations: enzyme systems from spinach leaves (WILDMAN et al., 1947); pineapple leaves (GORDON and SANCHEZ, 1949); pea

Table 4.1. Incorporation of radioactive labelled tryptophan

Material	Reference
Cabbage seedlings	WIGHTMAN (1962), RIDDLE and MAZELIS (1965), KUTACEK and KEFELI (1970)
Pea seedlings	KUTACEK and KEFELI (1970)
Tomato shoots	WIGHTMAN (1964), GIBSON et al. (1972b), WIGHTMAN (1973)
Bean shoots	BLACK and HAMILTON (1971)
Bean roots	MITCHELL and DAVIES (1972)
Cucumber hypocotyls	SHERWIN and PURVES (1969)
Barley shoots	GIBSON et al. (1972b)
Wheat leaves (healthy and rust infected)	KIM and ROHRINGER (1969)
Oat coleoptiles	BLACK and HAMILTON (1971)
Maize coleoptiles	LIBBERT and SILHENGST (1970), KUTACEK and KEFELI (1970), HEERKLOSS and LIBBERT (1976a, b)
Maize endosperm	HALL and BANDURSKI (1978)
Lime fruit	KHALIFAH (1967)
Lens roots	PILET (1964a)
Tobacco petiole tissue	LIU et al. (1978a)
Cell-free extracts from:	
Tobacco bud tissue	PHELPS and SEQUEIRA (1967)
Pea shoot tips	MOORE and SHANER (1967, 1968), MOORE (1969)
Mung bean seedlings	WIGHTMAN and COHEN (1968)
Soybean callus tissue	BLACK and HAMILTON (1976)
Tobacco callus tissue	LIU et al. (1978a)

(LIBBERT, 1962); oat coleoptile tissue (LANTICAN and MUIR, 1967; MUIR and LANTICAN, 1968); dwarf pea tissue (MUIR and LANTICAN, 1968); shoot tips of *Phaseolus aureus* (GORDON, 1958; GORDON and PALEG, 1961), and pollen, pollen tubes, and pistils from flowers of tobacco (LUND, 1956). These and earlier investigations are reviewed by LARSEN (1951a), GORDON (1954, 1961), and more recently by LIBBERT et al. (1970a), and SCHNEIDER and WIGHTMAN (1974, 1978).

Moreover, tracer studies have provided considerable evidence that tryptophan is metabolized to IAA in higher plants. It was shown, in several laboratories, that radioactivity from exogenously supplied (^{14}C- or ^{3}H)-labelled tryptophan was incorporated into IAA. This was conclusively demonstrated for the first time by experiments in which (2-^{14}C)-DL-tryptophan was incubated with water melon tissue slices (DANNENBURG and LIVERMAN, 1957). The incorporation of radioactively labelled tryptophan into IAA was found both in several plant tissues and enzyme preparations (Table 4.1).

Further evidence for tryptophan being a precursor of IAA has been given by experiments in which ^{14}C-labelled indole and ^{3}H-serine or (^{14}C, ^{3}H) double-labelled tryptophan was fed to tips of pea seedlings, grown in sterile conditions (ERDMANN and SCHIEWER, 1971). The $^{14}C:^{3}H$ ratios of the supplied precursor and of IAA produced were found to be the same. These results indicate that indole is incorporated into IAA via tryptophan. Therefore, no bypass from indole to IAA exists in the plant material used. Similar double-labelling experiments with sterile-grown maize coleoptiles also show that there is no path for the biosynthesis of IAA from indole which does not involve tryptophan as an intermediate (HEERKLOSS and LIBBERT, 1976a, b). These findings are in accord with studies in which indole or indole-3-glycerol phosphate were incubated with crude or dialyzed enzyme extracts of pea seedlings (ERDMANN et al., 1969b). No formation of IAA could be detected either with, or without, the addition of serine-like co-substrates. Auxin activity of indole-3-glyceric acid and of its N-acetyl derivative apparently is based on their conversion into IAA (ERDMANN et al., 1969a).

These experiments confirm that all IAA synthesized from indole is produced on a pathway involving the synthesis of tryptophan as an intermediate. Thus, it can be stated that tryptophan is the primary precursor of IAA. However, chemical degradation of IAA biosynthesized from labelled tryptophan in order to localize the radio-carbon within the molecule has not yet been reported, due to the low amounts of IAA in plant material. It should also be mentioned that experiments have been done showing no elongation of oat coleoptiles supplied with tryptophan. This has led to the suggestion that tryptophan might not be a direct precursor of IAA (WINTER, 1966; THIMANN and GROCHOWSKA, 1968).

b) Pathways of IAA Formation from Tryptophan

The conversion of tryptophan into IAA involves a side-chain degradation by deamination, decarboxylation, and two oxidation steps. The exact pathway through which IAA is synthesized in higher plants is, however, still controversial. As discussed by LIBBERT et al. (1970a) and SCHNEIDER and WIGHTMAN (1974, 1978), IAA can be formed by different pathways (see Fig. 4.1). These reviews thoroughly cover the field of IAA biosynthesis, and only a few further results have been published subsequently. Therefore, the biosynthetic pathways of IAA formation is discussed only briefly in terms of the major intermediates.

c) Indole-3-Pyruvic Acid Pathway

Indole-3-Pyruvic Acid. Indole-3-pyruvic acid formed from tryptophan by transamination represents the first intermediate in the biosynthetic pathway of IAA. Transamination as the initial reaction is catalyzed by a widely distributed multispecific aminotransferase, reacting with the three aromatic amino acids tryptophan, phenylalanine, and tyrosine (WIGHTMAN and COHEN, 1968; WIGHTMAN, 1973; TRUELSEN, 1973). The identification of indole-3-pyruvic acid is difficult due to its instability during isolation and quantification and its relatively low amounts in plant tissue.

The identification of this α-keto acid has been established both by chromatography and chemical methods. Thus, the occurrence and formation of indole-

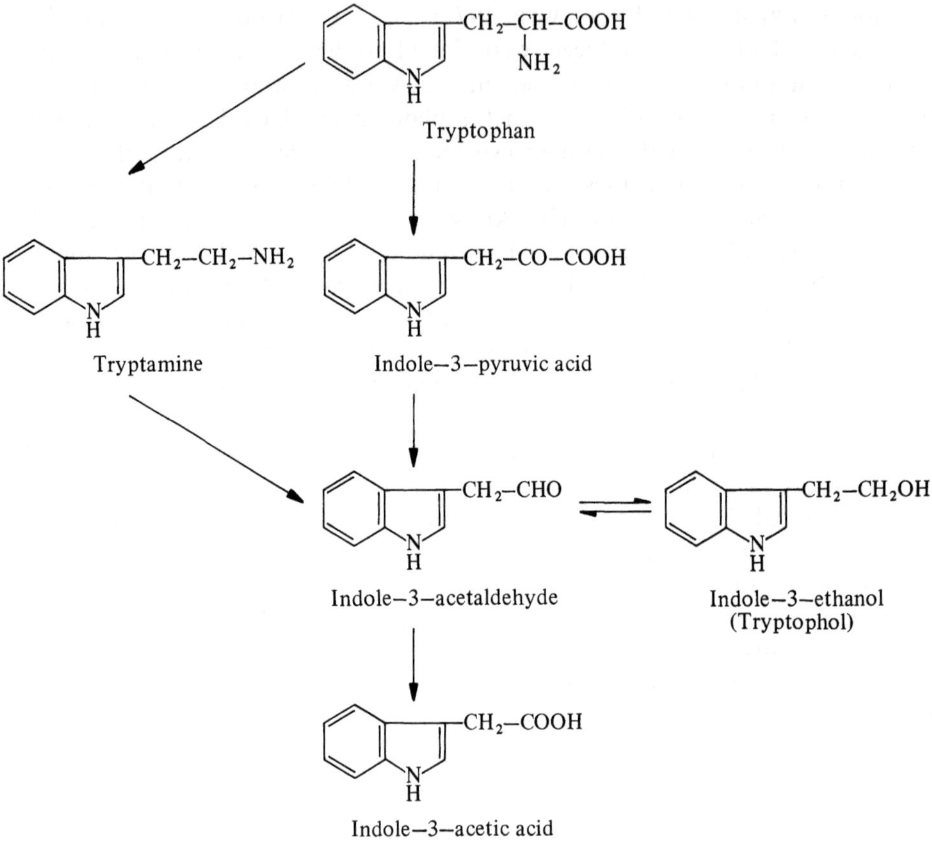

Fig. 4.1. Biosynthesis of IAA. Indole-3-pyruvic acid and tryptamine pathways

pyruvate from tryptophan have been reported in several plant tissues: homogenates of pea tissue (LIBBERT and BRUNN, 1961; LIBBERT et al., 1968; MOORE and SHANER, 1968); sterile cell-free preparations of mung bean seedlings (GORDON and PALEG, 1961; WIGHTMAN and COHEN, 1968; TRUELSEN, 1972); lime fruit (KHALIFAH, 1967); wheat leaves (KIM and ROHRINGER, 1969); sterile pea stem sections (LIBBERT et al., 1970b); barley and tomato shoots (GIBSON et al., 1972a), and tomato shoot extracts (GIBSON et al., 1972b; WIGHTMAN, 1973). A soluble enzyme capable of converting L-tryptophan into indole-3-pyruvic acid by transamination reacts only with the L-form of tryptophan, phenylalanine, and tyrosine (WIGHTMAN and COHEN, 1968; GIBSON et al., 1972a; TRUELSEN, 1972). Tryptophan aminotransferase activity was found to be present in different species of *Nicotiana* (LIU et al., 1978a).

Indole-3-Acetaldehyde and Indole-3-Ethanol. The next step in the biosynthetic pathway leading to IAA involves the decarboxylation of indole-3-pyruvic acid forming indole-3-acetaldehyde. This intermediate has been isolated from pea seedlings (RAJAGOPAL, 1967a, 1968a; MOORE and SHANER, 1968; LIBBERT et al., 1970b), mung bean seedlings (WIGHTMAN and COHEN, 1968), barley and tomato shoots (SCHNEIDER et al., 1972; WIGHTMAN, 1973), *Helianthus* seedlings (RAJAGOPAL, 1967a), and cucumber seedlings (PURVES and BROWN, 1978).

A number of experiments show that indole-3-acetaldehyde is derived from tryptophan via indole-3-pyruvic acid (WIGHTMAN and COHEN, 1968; GIBSON et al., 1972b; WIGHTMAN, 1973; for further citations see in LIBBERT et al., 1970a, and SCHNEIDER and WIGHTMAN, 1974, 1978). The presence of indole-3-pyruvic acid decarboxylase has been described in cell-free extracts from mung bean seedlings (WIGHTMAN and COHEN, 1968) and tomato extracts (GIBSON et al., 1972b).

The subsequent enzymatic conversion of indole-3-acetaldehyde into IAA is the final step of the pathway; it has been shown to occur both in plant tissues and tissue extracts (LARSEN, 1949, 1951b; CLARKE and MANN, 1957; GORDON and SANCHEZ, 1949; PHELPS and SEQUEIRA, 1967; RAJAGOPAL, 1967b, 1968a, b; WIGHTMAN and COHEN, 1968; LIBBERT et al., 1970c; GIBSON et al., 1972a).

An NAD-dependent indoleacetaldehyde dehydrogenase was shown to be present in the cytoplasmic supernatant preparations of mung bean seedlings (WIGHTMAN and COHEN, 1968) as well as in tomato shoots (WIGHTMAN, 1973). In *Avena* coleoptiles the activity of indole-3-acetaldehyde oxidase is not controlled by the phytochrome system (RAJAGOPAL and BULARD, 1975). A soluble NAD-independent enzyme, oxidizing indole-3-acetaldehyde to IAA, has been extracted and purified from oat coleoptiles; it is active only in the presence of oxygen (RAJAGOPAL, 1971). An indole-3-acetaldehyde oxidase, very similar to the *Avena* enzyme, has been isolated from *Nicotiana glauca*, *N. langsdorfii*, and their F_1 hybrid (LIU et al., 1978a). Extracts of light-grown cucumber seedlings also catalyze the oxidation of indole-3-acetaldehyde to IAA (BOWER et al., 1978). The enzyme has been characterized as a metalloflavoprotein acting without added cofactors. The indole-3-acetaldehyde is the major substrate in vivo. Besides this oxidation of indole-3-acetaldehyde to IAA its enzymic reduction to indole-3-ethanol (tryptophol) seems to be of great importance with respect to indole metabolism in higher plants. Indole-3-ethanol is assumed to be a storage product apparently involved in the regulation of IAA biosynthesis. This indolic compound occurs naturally in different higher plants (Chap. 1) and can also be formed in plant homogenates. For example, it has been reported to occur in pea (LIBBERT and BRUNN, 1961; RAJAGOPAL, 1967a, 1968a; LIBBERT et al., 1970b), sunflower (WINTER, 1966; RAJAGOPAL, 1967a), tomato (WIGHTMAN, 1964; GIBSON et al., 1972b; SCHNEIDER et al., 1972), and cucumber (RAYLE and PURVES, 1967a, 1968).

Indole-3-ethanol possesses auxin-like growth-promoting activity, as demonstrated with different plant material (LARSEN, 1947; PURVES et al., 1967; RAYLE and PURVES, 1968; EVANS and RAYLE, 1970). Its activity is apparently due to its conversion into IAA.

Experiments using tomato shoots (GIBSON et al., 1972a), pea stem sections (LIBBERT et al., 1970c) and tobacco callus tissue (LIU et al., 1978a) have shown that indole-3-ethanol is derived from the metabolism of tryptophan. Moreover, the conversion of tryptamine [see Sect. 4.1.1 d] into indole-3-ethanol has also been demonstrated in cucumber (SHERWIN and PURVES, 1969) as well as in etiolated pea seedlings (LIBBERT et al., 1970c; MAGNUS et al., 1973). In later studies indole-3-ethyl-β-D-glucopyranoside was isolated and identified as a new indolic metabolite.

The reversible conversion of indole-3-acetaldehyde into indole-3-ethanol has been studied by different methods. Infiltration of plant tissue with the aldehyde gives rise to both IAA and tryptophol (RAJAGOPAL, 1967b, 1968b). Cell-free preparations of aseptic, etiolated pea shoots, as well as of tissues and cell-free preparations of oat seedlings, possess activities to metabolize the aldehyde to IAA and tryptophol (RAJAGOPAL, 1968a). Moreover, the enzymatic reduction of indole-3-acetaldehyde to the corresponding alcohol has been studied in mung bean seedlings (WIGHTMAN and COHEN, 1968), tomato and barley shoots (GIBSON et al., 1972a, WIGHTMAN, 1973) and in some species of *Nicotiana* (LIU et al., 1978a). In cucumber seedlings, three distinct enzymes have been demonstrated to catalyze the reduction of indole-3-acetaldehyde with the concomitant oxidation of NADH and NADPH, respectively (BOWER et al., 1976; BROWN and PURVES, 1976, 1980; PURVES and BROWN, 1978). NADH-specific indole-3-acetaldehyde reductase was found to be present in the cytosol fractions obtained from hypocotyl and cotyledon tissue, whereas NADPH-specific indole-3-acetaldehyde reductase activity was associated with a microsomal fraction derived from these tissues. The enzymes show a strong substrate specificity for indole-3-acetaldehyde.

The subcellular and organelle compartmentation of the NADPH- and NADH-specific indole-3-acetaldehyde reductases indicates a separation of the terminal steps of IAA biosynthesis into at least two subcellular compartments.

Reconversion of indole-3-ethanol to IAA has been shown to occur rapidly in cucumber seedling shoots by means of tracer studies (RAYLE and PURVES, 1967b, 1968; SHERWIN and PURVES, 1969). An enzyme catalyzing the oxidation of indole-3-ethanol to indole-3-acetaldehyde by use of molecular oxygen was isolated from cucumber seedlings and partially purified (VICKERY and PURVES, 1972). Further characterization of this indole-3-ethanol oxidase shows that the enzyme may be a flavoprotein, which is inhibited by indole-3-acetaldehyde and IAA. Therefore, it is suggested that this enzyme may have a regulatory function in the IAA biosynthesis (PERCIVAL et al., 1973).

Summarizing the results on the indole-3-pyruvic acid pathway of IAA biosynthesis it can be stated that both the main intermediates (cf. Fig. 4.1) and the most important enzymes involved (tryptophanaminotransferase, indole-3-pyruvic acid decarboxylase, indole-3-acetaldehyde oxidase) are well established. It has been suggested that the enzymatically-catalyzed formation of indole-3-ethanol and its enzymatic reversion to indole-3-acetaldehyde play an important role in regulation of IAA biosynthesis. Besides these biosynthetic and enzymatic investigations little information is available on the regulation of IAA formation. This is especially true of the regulatory mechanisms. Some data are given by studies dealing with the influence of the developmental stage and the physiological conditions on the endogenous auxin level. However, in most cases direct correlation between these factors and IAA biosynthesis was not proved. Auxin level may also be controlled by IAA catabolism (see Sect. 4.2.1 d). Furthermore, a number of experiments have been published concerning the effects of other hormone groups on IAA in plants, e.g., inhibition of IAA biosynthesis by abscisic acid and its reversal by gibberellin (ANKER, 1975). These studies do not yet contribute much to our understanding of the regulation of IAA biosynthesis sensu strictu and thus will not be reviewed in detail.

Epiphytic Bacteria. The interaction between plants and epiphytic bacteria in the occurrence and biosynthesis of IAA has been studied extensively by LIBBERT and coworkers (LIBBERT et al., 1970a, and literature cited therein; LIBBERT et al., 1969a, b, 1970b, d, e; WICHNER and LIBBERT, 1968a, b; LIBBERT and MANTEUFFEL, 1970; SCHIEWER and ERDMANN, 1970; KUNERT and LIBBERT, 1972; MANTEUFFEL, 1973; MANTEUFFEL et al., 1973a, b, c). It was shown that the epiphytic microflora colonizing the plant surface contain many bacterial strains which are able to convert tryptophan into IAA. Furthermore, it has been demonstrated that epiphytic bacteria increase the IAA content within plant tissue; in general sterile plants contain less auxin than non-sterile ones. Thus, the amount of IAA formed from tryptophan can be affected by the presence of epiphytic bacteria.

There is, however, unequivocal evidence that higher plants are capable of biosynthesizing IAA by themselves. This is shown by estimation of IAA biosynthesis by sterile plant tissues, by the detection of postulated intermediates as natural plant constituents and by the isolation of enzymes catalyzing the conversion of tryptophan into IAA.

d) Tryptamine Pathway

Besides the indole-3-pyruvic acid pathway (Sect. 4.1.1c), a second route for the biosynthesis of IAA has been found in which tryptophan is decarboxylated to tryptamine followed by its conversion into IAA via indole-3-acetaldehyde which is thus a common intermediate of both the indole-3-pyruvic acid and the tryptamine pathway (Fig. 4.1).

Tryptamine was found to occur naturally in tomato and barley shoots (SCHNEIDER et al., 1972), and is apparently distributed in many plant families. A tryptamine forming L-tryptophan decarboxylase was shown to be present in hypocotyls of cucumber seedlings (SHERWIN, 1970) and in tomato and barley shoots (GIBSON et al., 1972a, b; WIGHTMAN, 1973) and in species of *Nicotiana*, examined by LIU et al. (1978a). The occurrence of this enzyme seems to be limited to certain plant species. The decarboxylation of ^{14}C-labelled tryptophan to ^{14}C-tryptamine and its conversion into ^{14}C-IAA has been shown to exist in tobacco (PHELPS and SEQUEIRA, 1967, 1968), cucumber (SHERWIN and PURVES, 1969), and barley and tomato (GIBSON et al., 1972a; WIGHTMAN, 1973). Tryptamine can be oxidized to indole-3-acetaldehyde by plant amino oxidases (CLARKE and MANN, 1957; THIMANN and GROCHOWSKA, 1968; PERCIVAL and PURVES, 1974). The role of tryptamine as a precursor of IAA has also been studied in *Avena* coleoptile tips (THIMANN and GROCHOWSKA, 1968) and pea seedlings (LIBBERT et al., 1968, 1970c; MAGNUS et al., 1973).

Experiments showing growth stimulation of pea stem sections by joint application of indole and ethanolamine led to the suggestion that these two substances, exogenously supplied, act as precursors of IAA formed via tryptamine

(ISHIGAMI and SUZUKI, 1970). In *Avena* coleoptile tissue and apical tissue of dwarf pea, an enzyme system has been isolated which converts tryptamine into IAA (LANTICAN and MUIR, 1967; MUIR and LANTICAN, 1968). The purified pea enzyme possesses a high substrate specificity; its molecular weight was determined to be 1.84×10^5 daltons. This amino oxidase has been shown to be specifically stimulated by phosphate (MCGOWAN and MUIR, 1971).

It is interesting that both the indole-3-pyruvic acid and tryptamine pathways can operate in barley and tomato shoots (GIBSON et al., 1972a). In tomato, the pathway via indole-3-pyruvic acid was shown to be the major route for the conversion of tryptophan into IAA (WIGHTMAN, 1973). The present data indicate that the indolepyruvic acid pathway is the predominant route of IAA biosynthesis in higher plants. In a number of species it is apparently the only one. The significance of some other postulated minor routes (see LIBBERT et al., 1970a; SCHNEIDER and WIGHTMAN, 1974) is not yet known.

e) Other Possible Pathways

On the basis of evidence obtained in metabolism studies and tracer experiments an indole-3-acetaldoxime → indole-3-acetonitrile pathway has been postulated. As discussed in detail by LIBBERT et al. (1970a) and SCHNEIDER and WIGHTMAN (1974), IAA can be formed from tryptophan via indole-3-acetaldoxime: tryptophan → indole-3-acetaldoxime → indole-3-acetonitrile → IAA.

The indoleacetaldoxime naturally occurring in plants (KINDL, 1968) can be metabolized to IAA by two routes. The first one is catalyzed by the action of a hydro-lyase leading to the formation of indole-3-acetonitrile which can be hydrolyzed to IAA by nitrilase. The nitrilase seems to be of limited distribution in the plant kingdom; it was found in Gramineae, Musaceae, and Cruciferae (MAHADEVAN and THIMANN, 1964; THIMANN and MAHADEVAN, 1964). The indole-3-acetaldoxime hydro-lyase has been detected in banana leaves (MAHADEVAN, 1963) and isolated and partially purified from the fungus *Gibberella* (KUMAR and MAHADEVAN, 1963; SHUKLA and MAHADEVAN, 1970).

In a second pathway indole-3-acetaldehyde is formed by hydrolysis of the oxime followed by oxidation to IAA or reduction to indole-3-ethanol as shown in several plant species (RAJAGOPAL and LARSEN, 1972) [cf. 4.1.1c]. After feeding of the oxime some species showed an increased indoleacetonitrile content. In cell-free preparations from oat coleoptiles the conversion of indolealdoxime to indoleacetaldehyde has been demonstrated (RAJAGOPAL and LARSEN, 1972).

Indole-3-acetaldoxime is known to be a precursor of indole glucosinolates such as glucobrassicin (Structure 4.2) widely distributed in Brassicaceae and related families (SCHRAUDOLF, 1965, 1968; KUTACEK and KEFELI, 1968). The biosynthetic route leading to glucobrassicin has been studied by means of labelled

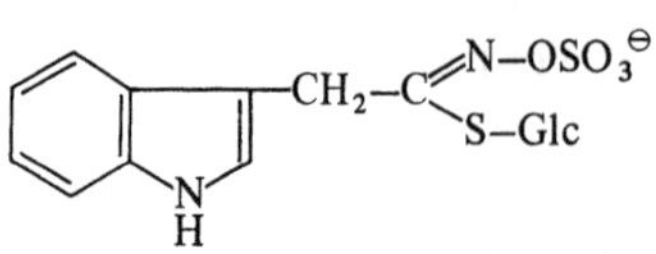

Structure 4.2. Glucobrassicin

tryptophan (SCHRAUDOLF and BERGMANN, 1965; SCHRAUDOLF, 1966; KUTACEK and KRALOVA, 1972 and literature cited therein). Glucobrassicin can be transformed to indoleacetonitrile, which might serve as a precursor of IAA. Thus, in Brassicaceae a special route of IAA formation from indole glucosinolates may exist. Its physiological significance, however, is not known and has been doubted (SCHRAUDOLF and WEBER, 1969).

f) Chlorinated Auxins

A new group of auxins was discovered when the methyl ester of 4-chloroindole-3-acetic acid was isolated from immature seeds of pea (GANDAR and NITSCH, 1964, 1967; MARUMO et al., 1968a, b; ENGVILD et al., 1978). The 4-chloroindole-3-acetic acid and its methyl ester have been also found in immature seeds of *Vicia faba* (HOFINGER and BÖTTGER, 1979). Biological activities of this chlorinated IAA, as well as of a number of synthesized IAA derivatives differing in substitution of the benzene ring, were studied using different bioassays (MARUMO et al., 1974; BÖTTGER et al., 1978). The native 4-chloroindole-3-acetic acid was shown to be more active than the major auxin IAA in the *Avena* straight growth bioassay. The incorporation of ^{36}Cl into 4-chloroindolylacetic acid and its methyl ester in immature seeds of pea has been demonstrated (ENGVILD, 1974, 1975), but the biosynthetic pathway has not yet been investigated. From the occurrence of both the chlorinated tryptophan derivatives (MARUMO and HATTORI, 1970; HATTORI and MARUMO, 1972) and chlorinated IAA in immature pea seed one can conclude that introduction of chlorine into the indole ring occurs before side chain degradation of tryptophan takes place.

4.1.2 Cytokinins

The biosynthetic pathway of cytokinins is closely related to the metabolism of RNA, especially of tRNA (SKOOG and ARMSTRONG, 1970; HALL, 1973; KAMINEK, 1975; BURROWS, 1975; LETHAM, 1978). This is deduced from the following facts: (a) in consequence of their adenine moiety, both RNA and cytokinins branch off from the same purine pool; (b) cytokinins have been shown to exist in nature as the base, the ribonucleoside, and the ribonucleotide (see Fig. 4.2); (c) tRNA's have been demonstrated to contain active cytokinins and cytokinin-like adenine derivatives always situated at the important position adjacent to the anticodon triplet (Fig. 5.1.3).

a) Biosynthesis of tRNA-Cytokinins

The isoprenoid side chain of i^6Ade (Fig. 4.2), present in many tRNA species (HALL, 1973; KAMINEK, 1975), was proved to originate from mevalonic acid and from 3-methylbut-2-enylpyrophosphate. These radioactive labelled precursors were fed to *Lactobacillus acidophilus* (PETERKOVSKY, 1968), yeast cultures or enzyme preparations from yeast and rat liver (FITTLER et al., 1968; KLINE et al., 1969), enzyme preparations from *Escherichia coli* (BARTZ and SÖLL, 1972;

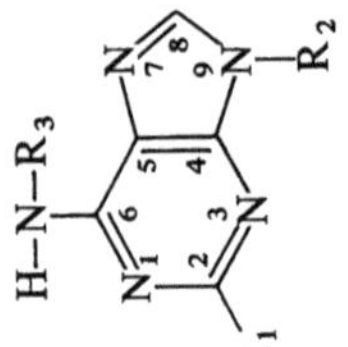

Substituents			Trivial name	Systematic name	Abbreviation
R_1	R_2	R_3^a			
H	H		N^6–(Δ^2–Isopentenyl) adenine	6–(3–Methylbut–2–enylamino)–purine	i^6Ade
H	Ribosyl	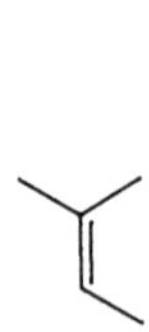	N^6–(Δ^2–Isopentenyl) adenosine	6–(3–Methylbut–2–enylamino)–9–β–D–ribofuranosylpurine	i^6A
H	H	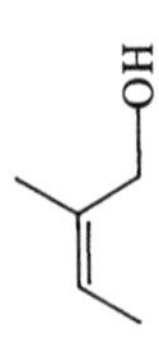	cis–Zeatin	6–(4–Hydroxy–3–methyl–cis–but–2–enylamino)–purine	c–io^6Ade
H	Ribosyl	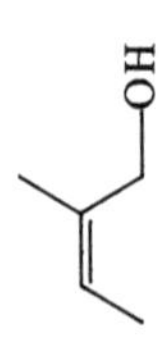	cis–Zeatin riboside	6–(4–Hydroxy–3–methyl–cis–but–2–enylamino)–9–β–D–ribofuranosylpurine	c–io^6A
H	H	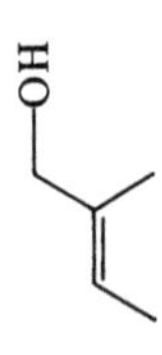	trans–Zeatin	6–(4–Hydroxy–3–methyl–trans–but–2–enylamino)–purine	t–io^6Ade
H	Ribosyl	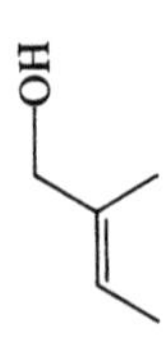	trans–Zeatin riboside	6–(4–Hydroxy–3–methyl–trans–but–2–enylamino)–9–β–D–ribofuranosyl–purine	t–io^6A

ROSENBAUM and GEFTER, 1972), as well as to tobacco tissue cultures or crude enzyme extracts therefrom (CHEN and HALL, 1969). The radioactive tRNA isolated contained the specific label in the side chain of the i^6Ade as well as of the zeatin (io^6Ade) moiety (MURAI et al., 1975). Adenine, simultaneously supplied, did not affect the specific label. Therefore, the N^6-alkylation has

H	H	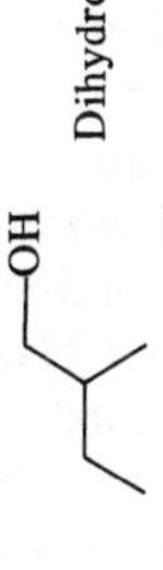	Dihydrozeatin	6–(4–Hydroxy–3–methylbutylamino)–purine	H_2–io^{6}Ade
H	Ribosyl	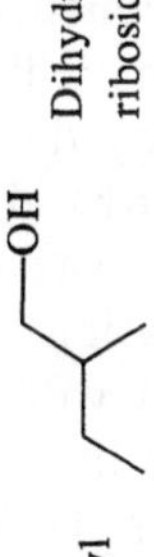	Dihydrozeatin riboside	6–(4–Hydroxy–3–methylbutylamino)–9–β–D–ribofuranosylpurine	H_2–io^{6}A
CH_3–S	H	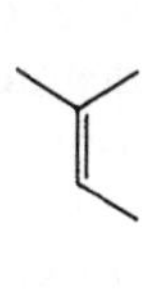		2–Methylthio–6–(3–methylbut–2–enylamino)–purine	ms^2–i^{6}Ade
CH_3–S	Ribosyl	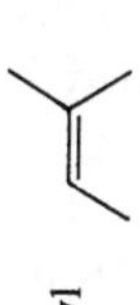		2–Methylthio–6–(3–methylbut–2–enylamino)–9–β–D–ribofuranosylpurine	ms^2–i^{6}A
CH_3–S	H	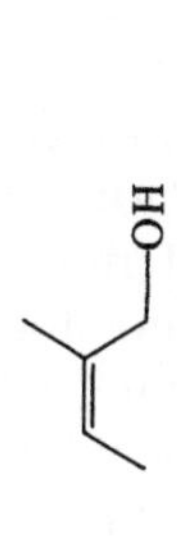		2–Methylthio–6–(4–hydroxy–3–methyl–cis–but–2–enylamino)–purine	ms^2–c–io^{6}Ade
CH_3–S	Ribosyl	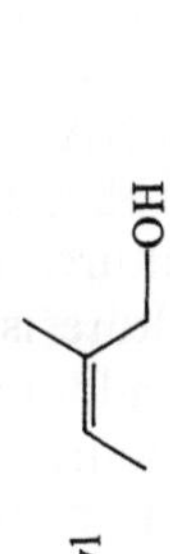		2–Methylthio–6–(4–hydroxy–3–methyl–cis–but–2–enylamino)–9–β–D–ribofuranosylpurine	ms^2–c–io^{6}A
CH_3–S	H	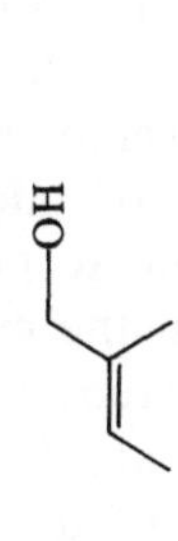		2–Methylthio–6–(4–hydroxy–3–methyl–trans–but–2–enylamino)–purine	ms^2–t–io^{6}Ade
CH_3–S	Ribosyl	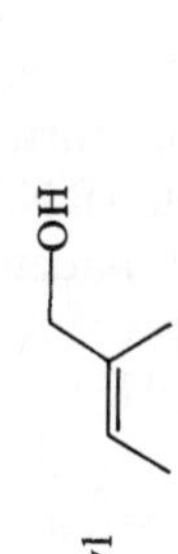		2–Methylthio–6–(4–hydroxy–3–methyl–trans–but–2–enylamino)–9–β–D–ribofuranosylpurine	ms^2–t–io^{6}A

[a] In all cases N^6 is linked to the C–1 of the isoprenoid side chain.

Fig. 4.2. Structures, names, and abbreviations of naturally occurring cytokinins

to occur on the adenine moiety fixed in the intact tRNA molecule (CHEN and HALL, 1969).

Other N^6-substitutions of adenine occurring at the tRNA level were assumed to lead to 6-methylaminopurine (HELGESON and LEONARD, 1966), 6-(2-hydroxybenzylamino)-9-β-D-ribofuranosyl-purine (HORGAN et al., 1975) as well as to the incorporation of the side chain of 6-benzylaminopurine into tRNA of soyabean callus cultures after being supplied with this artificial cytokinin (ELLIOT and MURRAY, 1972).

The specific oxidation of the isoprenoid side chain of tRNA-bound i^6Ade is assumed to result mainly in cis-zeatin (c-io^6Ade) detected both at the macromolecular level in microbes (HALL et al., 1967; BURROWS et al., 1970; PLAYTIS and LEONARD, 1971) and in free state in cultures of *Corynebacterium fascians* (SCARBROUGH et al., 1973). This cis-isomer of zeatin has been shown to be the main form present in tRNA especially of bacteria. Nevertheless, there is some evidence for the simultaneous occurrence of the trans-isomer in small amounts bound in the tRNA of plants, e.g., peas (VREMAN et al., 1972) or as the free nucleoside in *Agrobacterium tumefaciens* (CHAPMAN et al., 1976). Differences in the biosynthesis of cytokinins in cytoplasmic tRNA and chloroplast tRNA became apparent by qualitative and quantitative studies of the cytokinins in spinach tRNA's. Chloroplast tRNA, corresponding to bacterial (prokaryotic) tRNA, was found to contain i^6A and ms^2-c-io^6A but no zeatin ribosides (io^6A) whereas tRNA preparations from the whole leaf contain cis-zeatin riboside (c-io^6A) and trans-zeatin riboside (t-io^6A) as well as ms^2-c-io^6A and ms^2-t-io^6A (VREMAN et al., 1978).

The 2-methylthio-substitution leading to ms^2-i^6Ade and ms^2-io^6Ade has been found to occur apparently after N^6 substitution in tRNA, whereby the sulphur may originate from cysteine, whilst methionine serves as the methyl donor (GEFTER, 1969; GEFTER and RUSSELL, 1969; SKOOG and ARMSTRONG, 1970).

b) Biosynthesis of Free Cytokinins

The occurrence of cytokinin constituents in tRNA led to the conclusion that tRNA's act as direct intermediate or precursor of free cytokinins which could be released by the turnover of tRNA (CHEN et al., 1976a) or by degradation of moribund cell tRNA of microbes (ARMSTRONG et al., 1976b). However, by applying (^{14}C)-adenine to cultures of moss callus evidence was obtained that the i^6Ade present in the culture is not formed by degradation of tRNA (BEUTELMANN, 1973). Moreover, additional feeding experiments with tobacco cell cultures suggest that an alternative tRNA-free cytokinin pathway may be functional in plant tissue (CHEN et al., 1976b; CHEN and ECKERT, 1977). From a cytokinin autotrophic tobacco tissue culture an enzyme has been isolated that is capable of catalyzing the isoprenylation of 5′-AMP to 5′-i^6-AMP (CHEN and MELITZ, 1979). Recently, further evidence for the de novo synthesis was gained by comparing free and tRNA-bound cytokinins of poplar and lupin (BURROWS, 1978). It was found that the tRNA of both species contains the same cytokinins: i^6Ade, ms^2i^6Ade, c-io^6Ade, and ms^2-c-io^6Ade. The free cytokinins, however, were shown to be quite different, namely, the nucleoside of dihydrozeatin (H_2-io^6Ade) in lupin and 6-(2-hydroxybenzylamino)-purine in poplar. Furthermore, by using a cell-free system from *Dictyostelium discoideum* the AMP was shown to be a direct precursor of i^6Ade and its nucleoside (TAYA et al., 1978a).

The apparent existence of a tRNA-free pathway is confirmed by the fact that stereospecific hydroxylation of the free i^6Ade and its nucleoside to trans-zeatin (t-io^6Ade) and its riboside takes place in plant tissue, e.g., *Zea mays* (LETHAM, 1966a; MIURA and MILLER, 1969; MIURA and HALL, 1973) and in cytokinin-autotrophic tobacco callus (EINSET and SKOOG, 1973) as well as in cultures of microorganisms (MILLER, 1967; MIURA and HALL, 1973). In contrast to cis-hydroxylation found in tRNA biosynthesis (see above), trans-hydroxylation is considered to be the specific pathway to trans-zeatin which seems to be the major free plant cytokinin (PURSE et al., 1976).

The occasional occurrence of free cis-zeatin and its riboside is assumed to originate from the tRNA turnover (KAMINEK, 1975; ARMSTRONG et al., 1976b).

Biosynthetically, dihydrozeatin (KOSHIMIZU et al., 1967; PURSE et al., 1976) may be derived from the metabolic hydrogenation of trans-zeatin which has been observed when the latter compound was applied to bean axes (SONDHEIMER and TZOU, 1971) (see Sect. 4.3.2).

Naturally occurring free cytokinins substituted in positions other than, or additional to, N^6 were assumed to originate from tRNA metabolism (SKOOG and ARMSTRONG, 1970; ARMSTRONG et al., 1976b). Furthermore, the transformation of free cytokinins to ribosides or ribotides (see Sect. 4.4.2) either by adenosine phosphoribosyltransferase in vitro (CHEN and ECKERT, 1977) or by tobacco callus and sycamore tissue in vivo (LALOUE et al., 1974, 1977) may be biosynthetically relevant with respect to the incorporation of cytokinins into tRNA of tobacco callus (FOX and CHEN, 1967; WALKER et al., 1974; ARMSTRONG et al., 1976a) and *Agrobacterium tumefaciens* (HAHN et al., 1976).

4.1.3 Plant Hormones of Terpenoid Origin: General Terpenoid Pathway

Based on the number of carbon atoms and on the characteristic methyl branching, abscisic acid belongs to the group of sesquiterpenes, whereas the gibberellins are diterpenes. Therefore, these terpenoid phytohormones fit into the general scheme of terpene formation according to the biogenetic isoprene rule from isoprene units (RUZICKA, 1959; WENKERT, 1955).

Mevalonic acid has been shown to be a specific terpene precursor which is first converted into the isoprenoid C_5-compound isopentenyl pyrophosphate (Fig. 4.3). Isomerization, catalyzed by a specific isomerase, yields dimethylallyl pyrophosphate which acts as starter for chain elongation by reaction with isopentenyl pyrophosphate. The formation of the monoterpenoid geranyl pyrophosphate from dimethylallyl pyrophosphate and isopentenylpyrophosphate is catalyzed by a prenyltransferase. Sequential reaction of geranyl pyrophosphate with prenyltransferase yields farnesyl and geranylgeranyl pyrophosphates which act as direct precursors of sesquiterpenoids and diterpenoids, respectively.

The biologically active isomer of mevalonic acid [(+)-form] has the 3(R)-configuration indicated in Fig. 4.3 (EBERLE and ARIGONI, 1960). Only this enantiomer can be utilized by mevalonate kinase both in plants and animals. Moreover, mevalonic acid has three prochiral centres (C-2, C-4, and C-5). In chemical

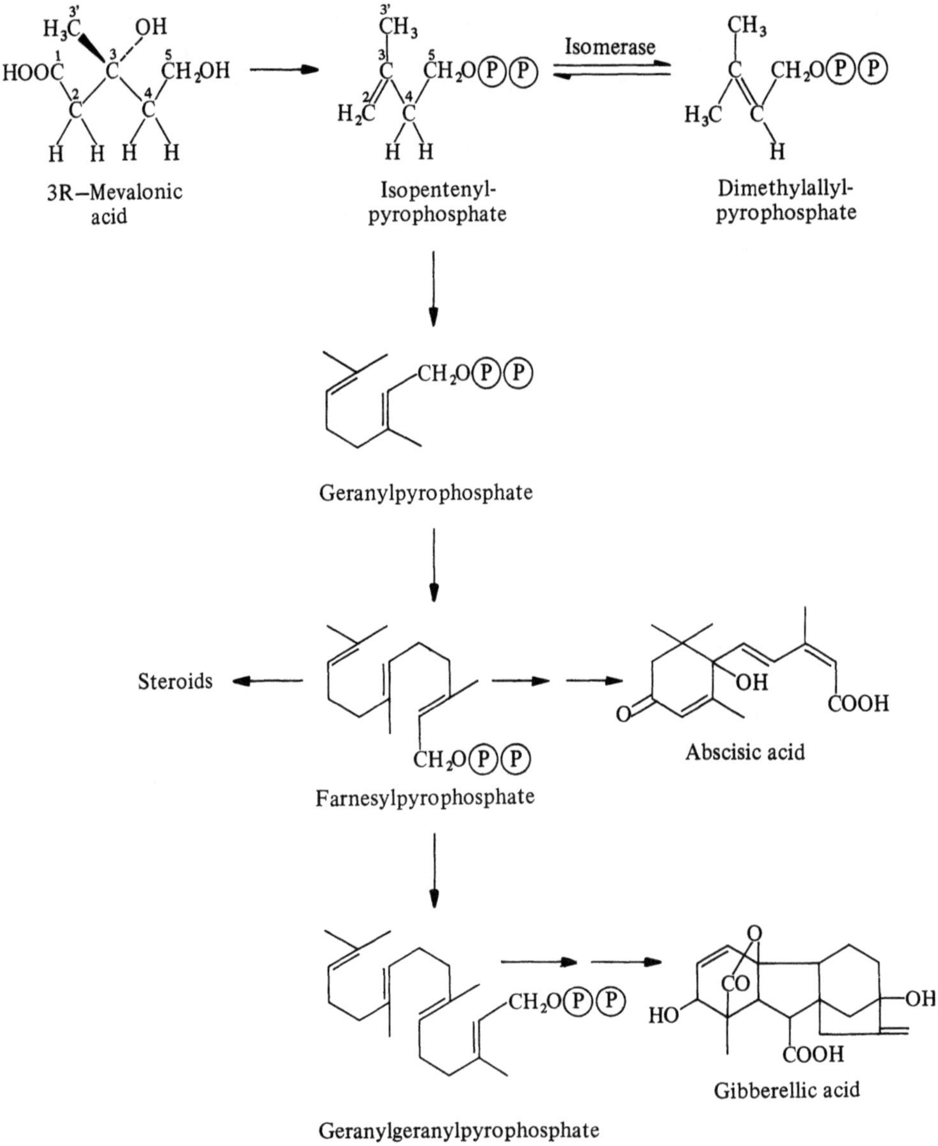

Fig. 4.3. Biosynthesis of sesquiterpenes and diterpenes from mevalonate

reactions these hydrogen atoms are equivalent and normally not distinguished. But, in enzyme-catalyzed reactions the two hydrogen atoms attached to each of these carbon atoms are stereospecifically different. For instance, studies using specific tritium labelling of mevalonic acid in position-4 have shown that one proton is lost stereospecifically from C-4 when the double bond is formed as an isopentenylpyrophosphate is linked to dimethylallylpyrophosphate. In isoprenoids with all trans double bonds (squalene, phytoene) the 4(S)-hydrogen is eliminated from C-4 and the 4(R)-hydrogen atom is retained (route a in Fig. 4.4) (POPJAK and CORNFORTH, 1966; GOODWIN, 1971; CORNFORTH et al.,

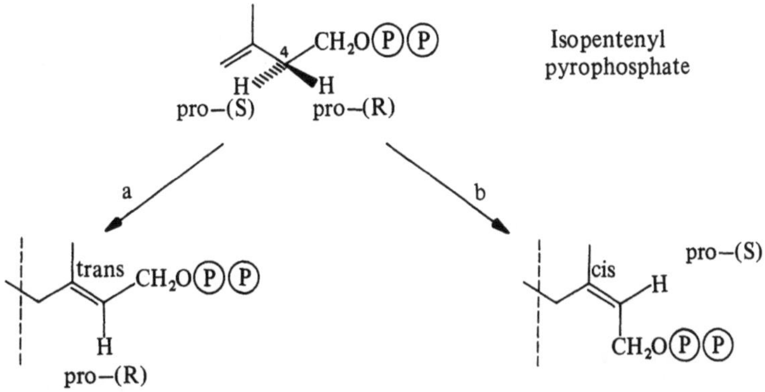

Fig. 4.4. Stereochemistry of the hydrogen elimination at C-4 of mevalonate during formation of a cis- or trans-double bond

1972). In terpenes with all cis double bonds (rubber) 4(R) is lost (route b in Fig. 4.4). In terpenes with cis and trans double bonds both types of hydrogen elimination are found. These results, obtained from studies on the biosynthesis of structurally different isoprenoids, are discussed in relation to experiments on the biosynthesis of abscisic acid and gibberellins which are described in the following sections.

4.1.4 Abscisic Acid

Since clarification of the structure (Chap. 1, Table 1.16) of abscisic acid (ABA) was accomplished in 1965 (OHKUMA et al., 1965), there have been an increasing number of publications dealing with chemical and biochemical aspects as well as the multiple physiological effects of ABA. These problems are discussed in a number of reviews (ADDICOTT and LYON, 1969; MILBORROW, 1969b, 1971, 1974a, b, 1978b; ADDICOTT, 1970; HARTMANN, 1970; WAREING and RYBACK, 1970; DÖRFFLING, 1971, 1972; GROSS, 1972; WAIN, 1975; BURDEN and TAYLOR, 1976, WAREING, 1978). The following covers studies on biosynthesis of ABA using radioactively labelled precursors such as mevalonic acid and other compounds considered to be possible intermediates.

With regard to the biosynthetic pathway of ABA two possibilities have been discussed in the past. The first one is that ABA is biosynthesized by a unique route from three C_5-units through a C_{15}-intermediate. In contrast to this direct synthesis of ABA the other pathway discussed envisages the photolytic splitting of a C_{40}-carotenoid such as violaxanthin to give a C_{15}-compound which is converted into ABA by oxidation.

a) Mevalonic Acid as Precursor of ABA

In 1969 it was first found that (2-^{14}C)-mevalonic acid is incorporated into ABA in ripening fruits of tomato and avocado (NODDLE and ROBINSON, 1969). Later, the radioactivity of mevalonic acid labelled with ^{14}C and/or ^{3}H was

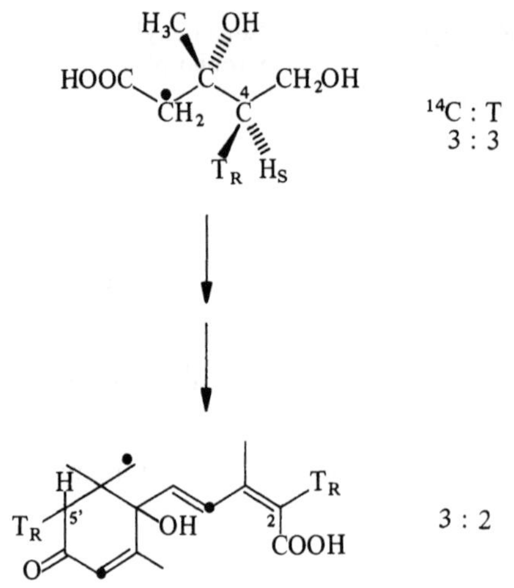

Fig. 4.5. Incorporation of [2-^{14}C,4(R)-$^{3}H_1$]-mevalonic acid into ABA. ● = ^{14}C; T = ^{3}H

shown to be incorporated into ABA on many occasions (ROBINSON and RYBACK, 1969; MILBORROW and NODDLE, 1970; MILBORROW, 1972b, 1974c, 1975a, b; MILBORROW and ROBINSON, 1973; LOVEYS et al., 1975; MILBORROW and MALLABY, 1975). No complete chemical degradation procedure for localization of ^{14}C in isolated ABA has been described, but it was established that the C_{15}-carbon skeleton of ABA is formed by three isoprene units which are derived equally from labelled mevalonic acid. Direct incorporation into ABA of the isoprenoid precursor, mevalonic acid, has been proved by application of tritiated mevalonic acid and analysis of the amount and position of tritium in the ABA isolated from plant tissue.

Interesting results were obtained from experiments using different stereospecifically labelled species of mevalonic acid. These studies have demonstrated the stereochemistry of the hydrogen elimination that occurs during the formation of the Δ^2-, Δ^4- and $\Delta^{2'}$-double bonds of ABA.

[2-^{14}C, 4(R)-4-$^{3}H_1$]-Mevalonic acid with a ^{14}C/^{3}H ratio of 3:3 was applied to avocado fruit (ROBINSON and RYBACK, 1969). The supplied precursor carried a C-4(R) tritium and a C-4(S) protium in order to establish whether the R or the S hydrogen atom is retained in the ABA molecule biosynthesized. The ^{14}C/^{3}H ratio of the ABA obtained was estimated to be 3:2. After removal of the C-5′ hydrogen atom by exchange with base it fell to 3:1. A contrasting experiment, using [2-^{14}C, 4(S)-$^{3}H_1$]-mevalonic acid, resulted in a complete loss of tritium. From these experimental data it was concluded that the 4(R)-hydrogen atom of mevalonic acid was retained at C-2 and C-5′ of ABA (Fig. 4.5). The second hydrogen atom at C-5′ presumably comes from the medium during cyclization.

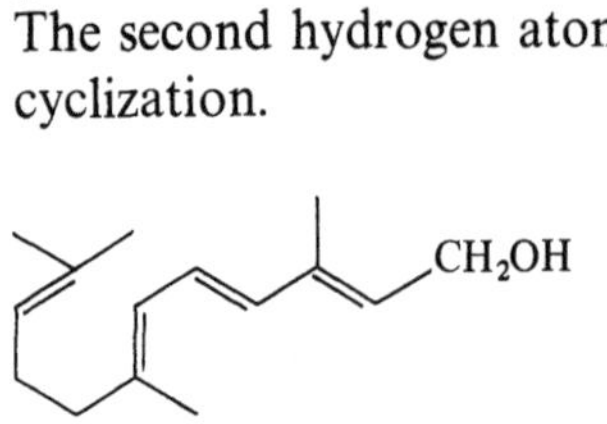

Structure 4.3 2,4,6-trans-Dehydrofarnesol

Fig. 4.6. Incorporation of [2-^{14}C,5(S)-5-^{3}H]-mevalonic acid into ABA. ● = ^{14}C; T = ^{3}H

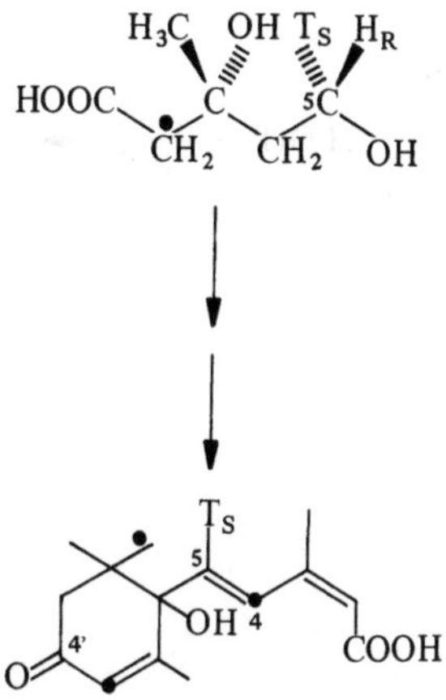

^{14}C : T
3 : 3

3 : 1

Transferring these findings to the results obtained in biosynthetic studies of other isoprenoids mentioned above (Sect. 4.1.3), the elimination of the 4(S)-hydrogen suggests that ABA arises from an intermediate having a trans-Δ^2-double bond, probably farnesyl pyrophosphate, which is converted into the cis-Δ^2-double bond of ABA by a subsequent isomerization reaction. The ABA from avocado fruit supplied with (2-^{14}C)-2,4,6-trans-dehydrofarnesol was found to contain ^{14}C, but it was not chemically degraded (MILBORROW, 1974b). The corresponding cis-Δ^2-isomer was not incorporated into ABA. To date, trans-2,4,6-dehydrofarnesol (Structure 4.3) is not known as a natural compound.

[2-^{14}C, 5(S)-5-^{3}H$_1$]-Mevalonic acid with an initial ^{14}C/^{3}H ratio of 3:3 applied to avocado fruit was incorporated into ABA to give an isotope ratio of 3:1 (MILBORROW, 1972b). Therefore, only one of three tritium atoms in the mevalonic acid units, participating in the ABA formation, was retained whereas the two other tritium atoms were removed, as could be expected. This experiment demonstrated that the C-5 hydrogen atom of ABA is derived from the 5(S)-hydrogen atom of mevalonate as indicated in Fig. 4.6. In further studies using [2-^{14}C, 2(R)-2-^{3}H$_1$]- and [2-^{14}C, 2(S)-2-^{3}H$_1$]-mevalonic acid it has been shown that the hydrogen atoms at C-3′ and C-4 of ABA are both derived from the pro-2(R)-hydrogen atom of mevalonate (Fig. 4.7) (MILBORROW, 1972a, 1974a, b). Thus, the hydrogens retained at C-4 and C-5 of ABA are the same as those at the analogous positions of carotenoids.

Moreover, MILBORROW (1975a) investigated which of the two methyl groups at C-6′ of ABA is derived from C-2 of mevalonate. ABA, biosynthesized from (2-^{14}C, 2-^{3}H)-mevalonate by avocado mesocarp, was isolated and applied to tomato shoots to be metabolized to phaseic acid (see Sect. 4.2.3). Formation of phaseic acid involves the hydroxylation of one of the geminal C-6′ methyl groups of ABA giving metabolite C. Its cyclization by forming the cis-oxymethylene bridge leads to phaseic acid. The ^{14}C/^{3}H ratio of applied ABA and isolated phaseic acid was found to be the same. Therefore, it was suggested that the pro-6′(S)-methyl group of ABA is derived from C-3′ of mevalonate. The 6′(R)-methyl group of ABA which is retained as the 6′-methyl group of

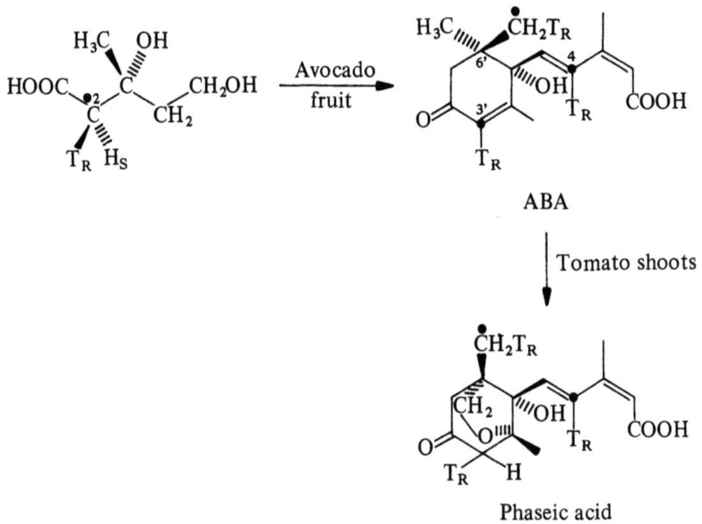

Fig. 4.7. Incorporation of [2-^{14}C,2(R)-2-$^{3}H_1$]-mevalonic acid into ABA and its conversion into phaseic acid. ● = ^{14}C; T = ^{3}H

phaseic acid corresponds to the C-2 of mevalonate (Fig. 4.7). This result was confirmed by Kuhn-Roth-oxidation of methyl phaseate to give (^{14}C,^{3}H) acetate.

Finally, it was demonstrated that the methyl groups attached to C-3, C-2′, and the pro-(S)-methyl group at C-6′ of ABA are derived from the methyl group at C-3 in each of the three isoprene residues which form the ABA molecule (MILBORROW, 1975b). This result was obtained by feeding of (3′-^{14}C, 3′-^{3}H)-mevalonate (^{14}C/^{3}H ratio 3:3) to avocado fruit. The corresponding ratio in the biosynthesized ABA was the same as in the supplied mevalonate.

b) Compounds with a Pre-Formed Carbon Skeleton of ABA

Although the specific incorporation of mevalonate into ABA has been intensively studied, only a few compounds possessing a pre-formed ABA skeleton have been shown to be converted into ABA (Fig. 4.8, 4.9).

(2-^{14}C)-trans, trans-ABA was not converted enzymatically into cis,trans-ABA by tomato plants and fruit and, thus, has to be excluded as a precursor of ABA (MILBORROW, 1970). Experiments using green tomato fruit have shown that (2-^{14}C)-1′,2′-epoxyionylidene acetic acid did not act as a precursor of ABA. Under physiological conditions, the 1′,4′-epidioxide of ABA is rearranged to give ABA in a rapid non-enzymic reaction (SONDHEIMER et al., 1969). The cis-1′,4′-diol of ABA, and to a lesser extent the trans-isomer, were transformed into ABA in bean axes (WALTON and SONDHEIMER, 1972b) and wheat leaves (MILBORROW, 1970), but this oxidation was also found to be a non-enzymic reaction proceeding spontaneously in aqueous solution (MILBORROW and MALLABY, 1975). Shoots and fruit of tomato, and avocado mesocarp supplied with racemic (2-^{14}C)-1′,2′-epoxyionylidene acetic acid metabolize it to ABA (MILBOR-

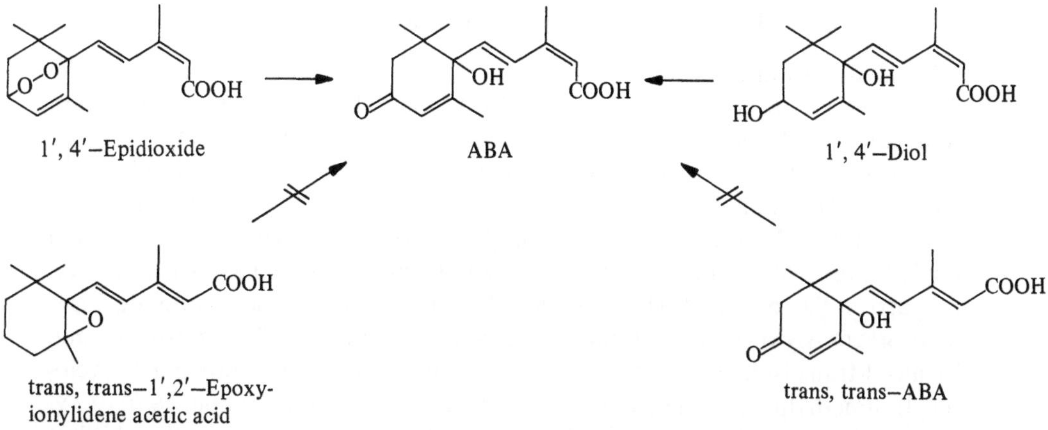

Fig. 4.8. Substances checked for incorporation into ABA

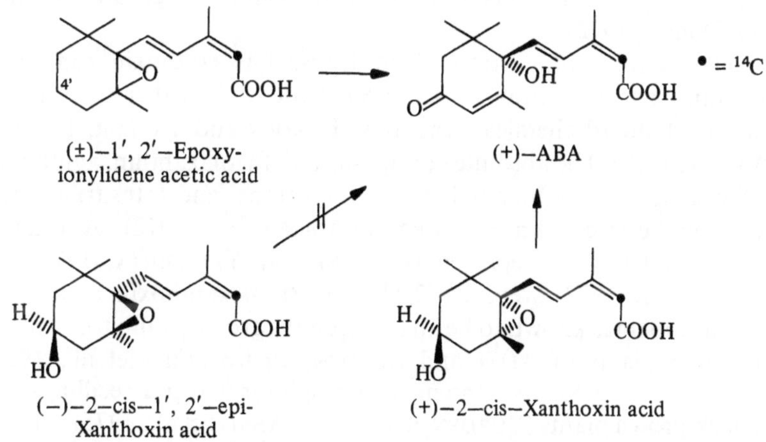

Fig. 4.9. Incorporation of 1′,2′-epoxyionylidene acetic acid and xanthoxin acid into ABA

ROW and NODDLE, 1970; MILBORROW and GARMSTON, 1973). By ^{18}O labelling, the oxygen of the epoxy group was shown to be used to form the tertiary hydroxyl group of ABA. The epoxide, however, does not appear to be a natural intermediate in the pathway from mevalonate to ABA. A metabolite isolated after feeding the cis-epoxide was identified as (–)-epi-1′(R)-2′(R)-4′(S)-2-cis-xanthoxin acid (MILBORROW and GARMSTON, 1973). It bears the 1′,2′-epoxy group at the same side of the ring as the 4′(S)-hydroxyl group. Further experiments have shown that (2-^{14}C)-(+)-cis-xanthoxin acid is incorporated into ABA. Results from these studies suggested that the hydroxylation occurs at C-4′ with both enantiomers of the epoxide, whereas subsequent conversion to ABA occurs only with the isomer having the hydroxyl and epoxide trans to each other. Probably, it is not a naturally occurring intermediate, since no traces of the acid could be found in nature.

c) Formation of ABA by Degradation of Carotenoids

ABA is structurally most similar to the end portion of certain carotenoids. This gave rise to the hypothesis that ABA could be formed by degradation of a carotenoid such as violaxanthin (TAYLOR and SMITH, 1967). Exposure of leaf xanthophyll epoxides like violaxanthin or neoxanthin to bright light led to the formation of a seed germination inhibitor (TAYLOR and SMITH, 1967; TAYLOR, 1968). As is shown in Fig. 4.10, photo-oxidation in vitro of violaxanthin yielded three main products which were isolated and identified as 4-(1′,2′-epoxy-4′-hydroxy-2′,6′,6′-trimethyl-1′-cyclohexyl)-trans-3-buten-2-one, loliolide isolated previously from several plant species, and a mixture of 2-cis-4-trans- and 2-trans-4-trans-isomers of 5-(1′,2′-epoxy-4′-hydroxy-2′,6′,6′-trimethyl-1′-cyclohexyl)-3-methylpentadienal, termed xanthoxin (BURDEN and TAYLOR, 1970; TAYLOR and BURDEN, 1970a, 1972). Moreover, the potent plant growth inhibitor xanthoxin was obtained by neutral zinc permanganate oxidation of violaxanthin in aqueous acetone (BURDEN and TAYLOR, 1970; TAYLOR and BURDEN, 1972) or by incubation of violaxanthin with soybean lipoxygenase and linoleate (FIRN and FRIEND, 1972).

The mixture of cis,trans- and trans,trans-xanthoxin can be oxidized to cis,-trans- and trans,trans-abscisic aldehyde and further to trans,trans-abscisic acid by standard chemical reactions (BURDEN and TAYLOR, 1970; TAYLOR and BURDEN, 1972). The absolute configuration of violaxanthin, xanthoxin, and ABA is the same. The (±)-methyl ethers of cis,trans- and trans,trans-xanthoxin have been synthesized from β-ionone (BURDEN et al., 1972). A total synthesis of xanthoxin has been reported (ORITANI and YAMASHITA, 1973); and TAYLOR and BURDEN (1973) prepared (2-^{14}C)-cis, trans-xanthoxin.

Xanthoxin is known to be an endogenous growth inhibitor, widely distributed in higher plants (TAYLOR and BURDEN, 1970b; FIRN et al., 1972; BÖTTGER, 1978; GOTO, 1978). Its content in red light-grown pea seedlings is higher than in dark-grown plants (BURDEN et al., 1971; ANSTIS et al., 1975). The light-induced

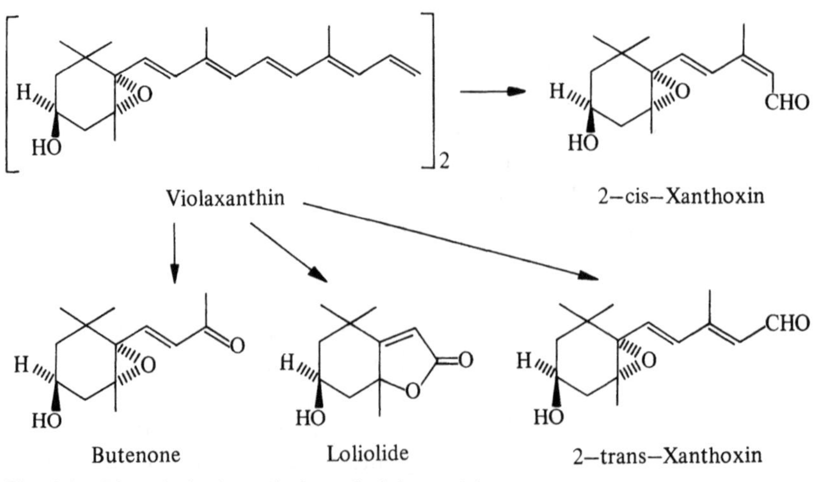

Fig. 4.10. Photolytic degradation of violaxanthin

growth-inhibiting substance in sunflower hypocotyl was also identified as xanthoxin (THOMPSON and BRUINSMA, 1977). Furthermore, the ABA and xanthoxin contents in spinach (ZEEVAART, 1974) and *Lolium temulentum* were studied (KING et al., 1977).

(2-^{14}C)-cis,trans-Xanthoxin, fed to shoots of tomato and dwarf bean, was converted into ABA and a second acidic metabolite which has been tentatively identified as phaseic acid (TAYLOR and BURDEN, 1973). This experiment indicates a viable biogenetic route to ABA by way of xanthophylls and xanthoxin, but no intermediates between xanthoxin and ABA have yet been isolated.

From all the experiments published on the biosynthesis of ABA it may be concluded that ABA in plants is mainly formed from mevalonic acid by the direct pathway. An alternative route via xanthophylls may be operating in plant tissue under certain circumstances, possibly varying with plant species, developmental stage, and environmental factors.

4.1.5 Gibberellins

a) General View

CROSS et al. (1956) first suggested the terpenoid origin of gibberellins. This view was supported by the incorporation of radioactively labelled acetate (BIRCH et al., 1959, 1960), mevalonate (BIRCH et al., 1959), and dimethylacrylic acid (REDEMANN and MEULI, 1959; ZWEIG and DEVAY, 1959) into GA_3 by cultures of the fungus *Gibberella fujikuroi*. A chemical degradation of the GA_3 molecule following acetate and mevalonate feedings gave a labelling pattern as outlined in Fig. 4.11 (BIRCH et al., 1959, 1960). This led to the following conclusions:

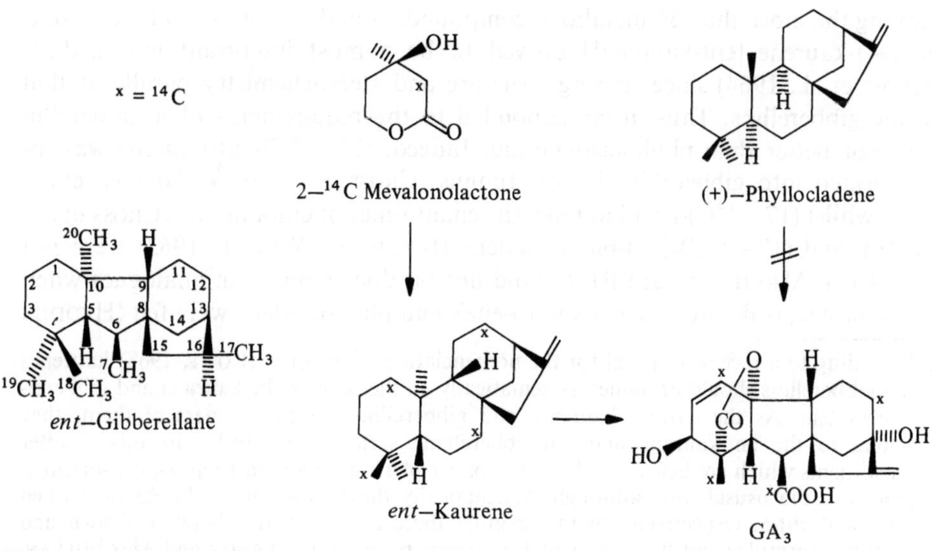

Fig. 4.11. Labelling pattern of GA_3 following application of (2-^{14}C)-mevalono lactone. Structures of ent-gibberellane and phyllocladene

(a) four mevalonate units are incorporated; (b) a phyllocladene-type tetracyclic diterpene should act as an intermediate which is susceptible for contraction of ring B; (c) at C-10 an angular methyl group has to be eliminated.

Due to its high capacity for gibberellin formation the fungus *Gibberella fujikuroi (Fusarium moniliforme)* has been widely used in biosynthetic studies. Aside from wild strains, such as ACC 917, several mutants have been introduced e.g., mutant B1-41a with a genetic block between ent-kaurenal and ent-kaurenoic acid (BEARDER et al., 1973a, 1974), mutant R-9 which lacks formation of GA_1 and GA_3 (BEARDER et al., 1973c), and mutant REC-193A, a slowly growing mutant with a low content of gibberellins (BEARDER et al., 1973b). These mutants can be used in preference to the wild type to investigate the conversion of non-labelled gibberellins or their precursors.

With improved analytical methods, higher plants also became accessible to biosynthetic experiments. A high degree of information is available from studies with enzyme preparations isolated from selected organs of higher plants.

The biosynthesis and metabolism of gibberellins have been reviewed on many occasions (CROSS et al., 1968b, c; LANG, 1970; MACMILLAN 1971, 1974a, b; MACMILLAN and PRYCE, 1973; WEST, 1973; TAKAHASHI, 1974; BARENDSE 1975; BEARDER and SPONSEL, 1977; GRAEBE and ROPERS, 1978; HEDDEN et al., 1978; RAPPAPORT and ADAMS, 1978).

Many features of the enzyme-mediated sequence leading from mevalonate to the diterpenes are known in detail, mainly as a result of the close similarity with mammalian terpene synthesis. This could be demonstrated by incorporating copalyl pyrophosphate (HANSON and WHITE, 1969a, c) and all-trans-geranylgeranyl pyrophosphate (HANSON and WHITE, 1969c; EVANS et al., 1970; HANSON and HAWKER, 1971) into fungal gibberellins as well as into GA_1 and GA_3 of wheat seedlings (SCHILLING et al., 1974, 1978).

In order to approach the problem of the cyclization step a thorough search for metabolites of terpenoid origin was carried out with cultures of the fungus. Among the more than 30 metabolic compounds found and structurally elucidated (−)-kaurene [ent-kaurene[1]] proved to be a most important intermediate (CROSS et al., 1964) since its ring structure and stereochemistry paralleled that of the gibberellins. Thus, it corresponded to the requirements of a gibberellin precusor better than phyllocladene did. Indeed, ($17-{}^{14}C$)-ent-kaurene was incorporated into gibberellins by the fungus (CROSS et al., 1964; GRAEBE et al., 1965) whilst ($17-{}^{14}C$)-(+)-kaurene, the enantiomer of ent-kaurene (CROSS et al., 1968a), and ($17-{}^{14}C$)-(+)-phyllocladene (BIRCH and WINTER, 1963) were not (Fig. 4.11). Also the mutant B1-41a did not produce gibberellin analogues when (+)-kaurene, isokaurene (ent-kaur-15-ene), and phyllocladene were fed (HEDDEN

1 According to a recent proposal for the nomenclature of terpenes (ROWE, 1968) kaurenes and gibberellins should be named systematically on the basis of the kaurane and gibberellane system. As the natural kaurenes and gibberellins are enantiomers of them, they belong to the ent-kaurane and ent-gibberellane series, respectively. In this chapter substituents which are below or above the paper plane are referred to as α- or β-oriented groups in the ususal way, although systematically these substituents should be termed ent-β and ent-α, respectively. In this manner there is a consensus between drawn and described formulae and this agrees with a former proposal (HARRISON and MACMILLAN, 1971)

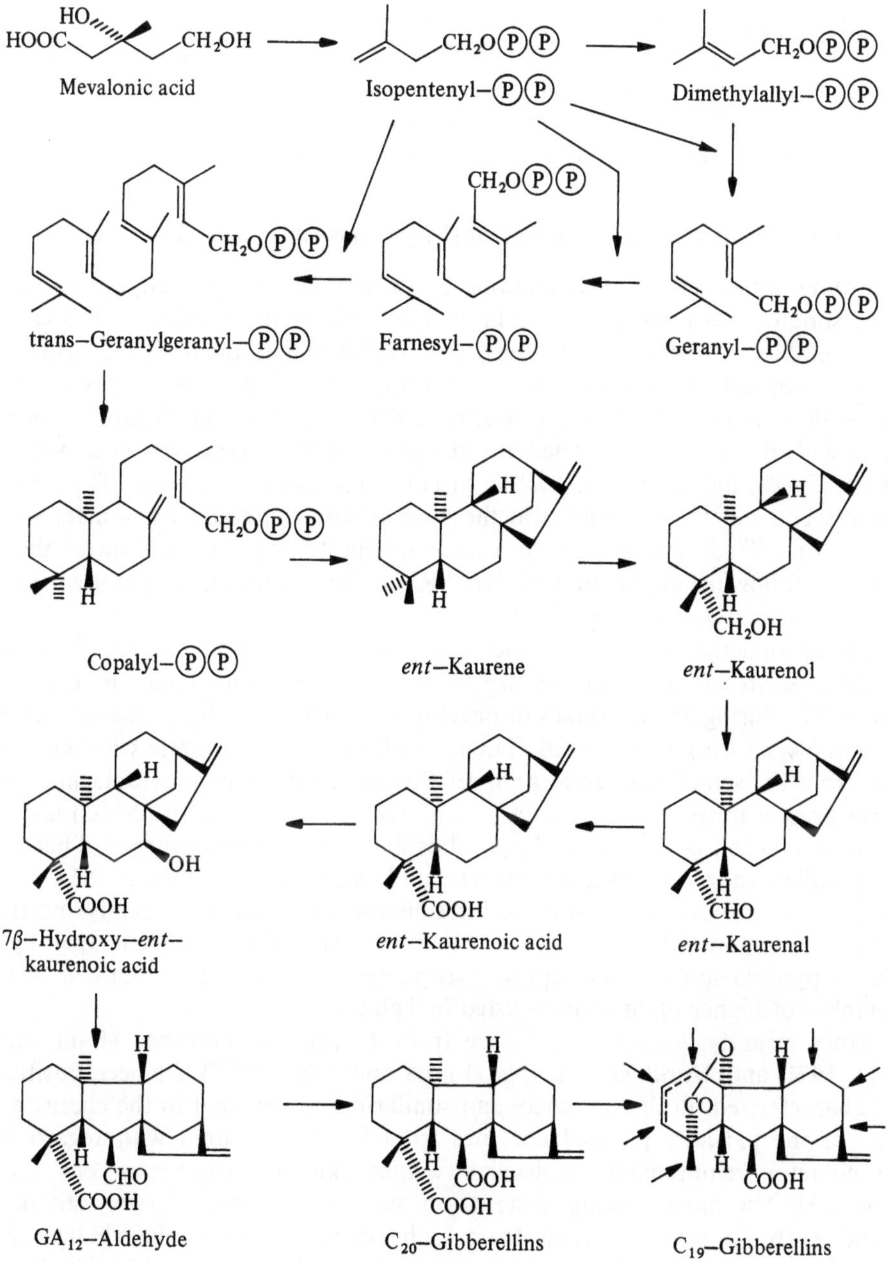

Fig. 4.12. Biosynthetic route leading to gibberellins → denotes known positions of oxygenation

et al., 1977). These results demonstrate both the high degree of stereospecificity in the biosynthetic route and the absence of (+)-gibberellins. ent-Kaurene, being an intermediate, one can divide the various biosynthetic steps on the way to gibberellins into seven stages. (a) formation of geranylgeranyl pyrophosphate from mevalonate (b) cyclization to give ent-kaurene (c) oxidative conversion of ent-kaurene, (d) contraction of ring B to give C_{20}-gibberellins, (e) oxidative removal of C_{20} to give C_{19}-gibberellins, (f) conversion of gibberellins by oxidative reactions ("metabolism", see Sect. 4.2.4), and (g) conjugation of gibberellins (see Sect. 4.3.4). Figure 4.12 summarizes the present understanding of intermediates and interconversions in the biosynthesis of gibberellins.

b) Enzymic Cyclization of Geranylgeranyl Pyrophosphate to ent-Kaurene

According to theoretical considerations, cyclization of geranylgeranyl pyrophosphate should pass through a bicyclic and a tricyclic intermediate (WENKERT, 1955; RUZICKA, 1959), and, in fact, (15-^{3}H)-labda-8,13-dien-15-yl pyrophosphate (=copalyl pyrophosphate) was specifically incorporated into GA_3 by *Gibberella* cultures (HANSON and WHITE, 1969a, c). However, feeding experiments with the tricyclic intermediate, pimara-8,14-diene (HANSON and WHITE, 1968a; CROSS and STEWART, 1968a, b) gave contradictory results. This compound seems not to be included in the biosynthetic route and its isomers with a $\Delta^{7(8)}$- or $\Delta^{8(9)}$-double bond were ruled out due to the ^{3}H-retention at these positions following application of ^{3}H, ^{14}C-labelled mevalonate (HANSON and WHITE, 1968a; HANSON et al., 1968).

The introduction of enzymatic methods made a breakthrough in the gibberellin field. Seeds of many species are known to accumulate high amounts of gibberellins during certain stages of development. This is so for immature seeds of *Marah macrocarpus* which led GRAEBE et al. (1965) to elaborate cell-free systems from the semi-liquid endosperm of this material. Soluble cytoplasmic enzymes present in the 105000 *g* supernatant, when incubated with (2-^{14}C)-mevalonate or labelled geranylgeranyl pyrophosphate, yielded ent-kaurene. This enzyme, called kaurene synthetase (UPPER and WEST, 1967), requires ATP and divalent ions as cofactors. Similar cell-free enzyme preparations, catalyzing the in vitro formation of labelled ent-kaurene from radioactive mevalonate, geranylgeranyl pyrophosphate, and copalyl pyrophosphate, have been obtained from a number of higher plant sources listed in Table 4.2.

More recently, kaurene synthetase from *Fusarium moniliforme* (FALL and WEST, 1971) and *Marah macrocarpus* (FROST and WEST, 1977) has been purified and characterized. Both differences and similarities are evident in the characteristics of the partially purified enzymes from the two sources with regard to the cofactor requirement, molecular weight, kinetic parameters etc. (see Table 4.3). The most striking observation was the existence of two activities, A and B, that are not separable from each other. It is uncertain whether the catalytic sites responsible for kaurene synthetase activities A and B, are identical, or separately located in a single protein, or even reside on separate proteins. Activity A converts geranylgeranyl pyrophosphate into the bicyclic copalyl pyrophosphate, and activity B continues the cyclization of the latter to give ent-kaurene. Doubly labelled (^{32}P,2-^{14}C)-geranylgeranyl pyrophosphate is converted by activity A into copalyl pyrophosphate with retention of the isotope ratio. Furthermore, on simultaneous incubation of fungal kaurene synthetase with (^{14}C)-geranylgeranyl pyrophosphate and ^{3}H-copalyl pyrophosphate the ent-kaurene formed was predominantly labelled with radio-carbon. These experi-

Table 4.2. Kaurene synthetase preparations from higher plant sources

Species	Plant tissue	Reference
Marah macrocarpus[a]	Endosperm, plastids, or mitochondria	GRAEBE et al. (1965), UPPER and WEST (1967), OSTER and WEST (1968), SIMCOX et al. (1975), FROST and WEST (1977), HEDDEN et al. (1977)
Marah oreganus	Endosperm	COOLBAUGH and HAMILTON (1976)
Pisum sativum	Immature fruits and seeds, shoot tips, or cotyledons	ANDERSON and MOORE (1967), GRAEBE (1968b), COOLBAUGH and MOORE (1971a, b), COOLBAUGH et al. (1973), ECKLUND and MOORE (1974), GOMEZ-NAVARETTE and MOORE (1978)
Phaseolus coccineus	Suspensor	CECCARELLI et al. (1979)
Ricinus communis	Seedlings	ROBINSON and WEST (1967, 1970a), SITTON and WEST (1975)
Robinia pseudoacacia	Seedling shoots	NOWAK and BROWN (1979)
Cucurbita maxima[b]	Endosperm	Graebe (1969)
Solanum lycopersicum	Germinating seeds, cell suspension cultures	YAFIN and SHECHTER (1975)
Triticum aestivum	Germinating seeds	SCHILLING et al. (1974)
Secale cereale	Germinating seeds	SCHILLING et al. (1974)
Zea mays	Coleoptiles	HEDDEN and PHINNEY (1979), MELLON and WEST (1979)
Gibberella fujikuroi (for comparison)	Mycelium	SHECHTER and WEST (1969), FALL and WEST (1971), EVANS and HANSON (1972), HEDDEN et al. (1977)

[a] Previously denoted as *Echinocystis macrocarpa*
[b] Previously denoted as *Cucurbita pepo*

ments demonstrate that the copalyl pyrophosphate, originating from the action of activity A, served as a better substrate for activity B than an intermediate out of the pool (FALL and WEST, 1971).

Seedlings of *Ricinus communis*, at special stages of development, yield cell-free enzyme preparations that are capable of converting mevalonate, geranylgeranyl pyrophosphate, and copalyl pyrophosphate into ent-kaurene and four additional diterpene hydrocarbons, ent-beyerene, ent-isopimaradiene [(+)-sandaracopimaradiene)], ent-13,16-cycloatisane [(−)-trachylobane] and casbene (SHECHTER and WEST, 1969; ROBINSON and WEST, 1970a, b). In comparing these structures (Fig. 4.13) with those of the postulated intermediates in the cyclization mechanism of kaurene synthetase (Fig. 4.14) it could be imagined that the diterpenes may originate from transient carbonium ions on the way to ent-kaurene. Purifica-

Table 4.3. Properties of kaurene synthetases isolated from *Fusarium* (FALL and WEST, 1971) and *Marah* (FROST and WEST, 1977)

	Fusarium moniliforme	*Marah macrocarpus*
M.W. (daltons)	≈460000	≈45000
Cofactor requirement	$Mg^{2+} > Co^{2+} > Ni^{2+}$	$Mg^{2+} > Mn^{2+} > Co^{2+}$
pH Optima		
Activity A	7.5	7.3
Activity B	6.9	6.9
Activity of growth retardants and inhibitors of terpenoid biosynthesis[a]		
Activity A	Amo 1618 > Carvadan > Phosfon D	SKF-525A > Phosfon D > Phosfon S > Q 64
Activity B	SKF-3301A > Q 64 > Q 58	Anchem[b] > Deoxycholate[b] > SKF-525A > SKF-3301A

[a] Structures are given in Fig. 4.18
[b] Not tested with kaurene synthetase isolated from *Fusarium*

tion of *Ricinus* kaurene synthetase, however, indicated that separate enzymic proteins participate in the cyclization reaction leading from geranylgeranyl pyrophosphate to each of the diterpene hydrocarbons with the exception of ent-kaurene and trachylobane where no evidence for a separation of the activities was observed. Kaurene synthetase seems to be a key enzyme with regard to the regulation by both internal factors and synthetic plant growth regulators [see Sect. 4.1.5f].

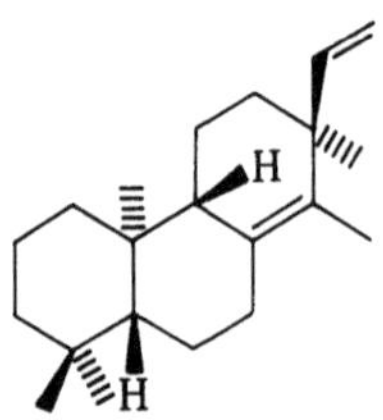

ent–Isopimaradiene
((+)–Sandaracopimaradiene)

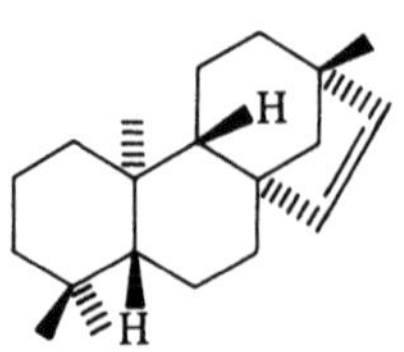

ent–Beyerene

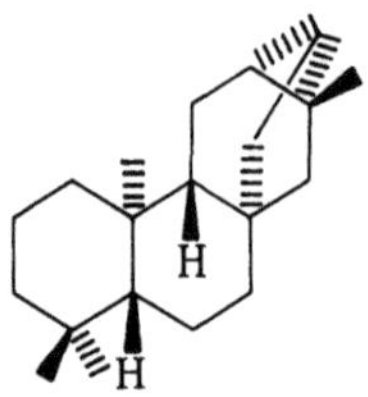

ent–13, 16–Cycloatisane
((–)–Trachylobane)

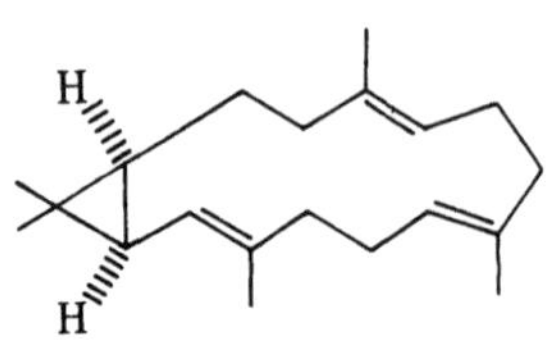

Casbene

Fig. 4.13. Diterpene hydrocarbons formed in cell-free extracts of *Ricinus communis* along with ent-kaurene

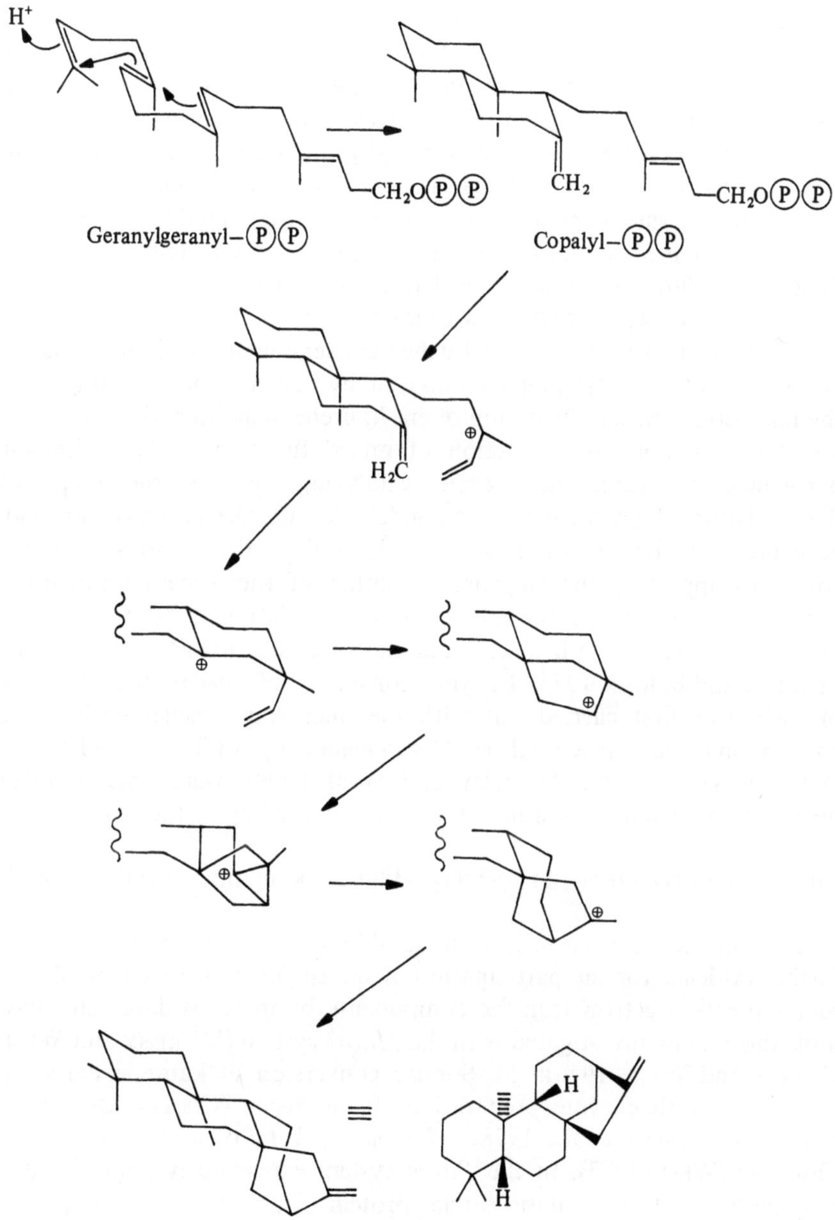

Fig. 4.14. Postulated reaction mechanism of ent-kaurene synthesis

c) Oxidation of ent-Kaurene

Based on the concept that further conversion of ent-kaurene must be oxidative in nature, several naturally occurring oxidation products of ent-kaurene have been applied in labelled form to cultures of the fungus. Among them, the C-19 functionalized compounds ent-kaurenol (GRAEBE et al., 1965; GALT, 1965;

VERBISCAR et al., 1967), ent-kaurenal (DENNIS and WEST, 1967), ent-kaurenoic acid (GEISSMAN et al., 1966), and 7β-hydroxy-ent-kaurenoic acid (HANSON and WHITE, 1969b; LEW and WEST, 1971; HANSON et al., 1972) were incorporated into gibberellins in good yield. These experiments supported the view of a stepwise formation of the α-oriented carboxyl group (see Fig. 4.12). The same pathway is operating in higher plants, as could be demonstrated by the incorporation of the intermediates mentioned above into gibberellins or gibberellin-like substances of *Brassica, Hordeum,* and *Pharbitis* species (BENNETT et al., 1966; STODDART, 1969; BARENDSE and KOK, 1971; TAKEBA and TAKIMOTO, 1971; FAULL et al., 1974; MURPHY and BRIGGS, 1975).

In contrast to its formation further conversion of ent-kaurene is catalyzed not by soluble enzyme preparations but by enzymes precipitating along with the microsomal pellet. Oxidation of ent-kaurene includes hydroxylations at C-19 and C-7, probably by the action of mixed function oxidases, known to be particulate in nature. The enzymes catalyzing the two other steps involving the oxidation of an alcohol to an aldehyde and the latter to an acid might be expected to be dehydrogenases or flavin-dependent oxidases. However, all enzymes supporting the stepwise oxidation of the α-methyl function into a carboxyl group as well as the subsequent hydroxylation at C-7 do have properties normally associated with mixed function oxidases (MURPHY and WEST, 1969; MURPHY and BRIGGS, 1975). Enzymic conversion of ent-kaurene into ent-kaurenoic acid was first carried out with the microsomal pellet derived from the endosperm of immature seeds of *Marah macrocarpus* (DENNIS and WEST, 1967). With the same system MURPHY and WEST (1969) were able to incorporate heavy oxygen from $^{18}O_2$ into ent-kaurenol according to the sequence

$$\text{ent-kaurene} + \text{NADPH} + \text{H}^+ + {}^{18}\text{O}_2 \rightarrow {}^{18}\text{O-ent-kaurenol} + \text{NADP}^+ + \text{H}_2{}^{18}\text{O}.$$

A photochemical action spectrum at 450 nm, and inhibition by CO, provided further evidence for the participation of mixed function oxidases. The cofactor requirements, electron transfer components of these oxidases etc. have been most thoroughly investigated with the *Marah* system (MURPHY and WEST, 1969; HASSON and WEST, 1976a, b). Similar conversion of kaurene derivatives was carried out with enzyme preparations from *Pisum* (COOLBAUGH and MOORE, 1971a, b; ROPERS et al., 1978), *Hordeum* (MURPHY and BRIGGS, 1975) and *Fusarium* (WEST, 1973). In the *Pisum* system exogenously applied ent-kaurene may be bound to a catalytic carrier protein of about 15×10^6 daltons prior to oxidation (MOORE et al., 1972). Similar observations were made with the *Marah* system (HASSON and WEST, 1976a). This behaviour parallels the participation of a "sterol carrier protein" included in the conversion of squalene into cholesterol by rat liver cells (RITTER and DEMPSEY, 1971; SCALLEN et al., 1971). On present knowledge this conversion from ent-kaurene to 7β-hydroxy-ent-kaurenoic acid is common in higher plants. Moreover, structural isomers of 7β-hydroxy-ent-kaurenoic acid have been reported to occur in certain species. Whereas 7β-hydroxy-ent-kaurenoic acid and ent-kaurenolides seem to be widely distributed, the two other compounds, shown in Fig. 4.15 are found in single species only. The role of steviol and xylopic acid as intermediates of gibberellin biosyn-

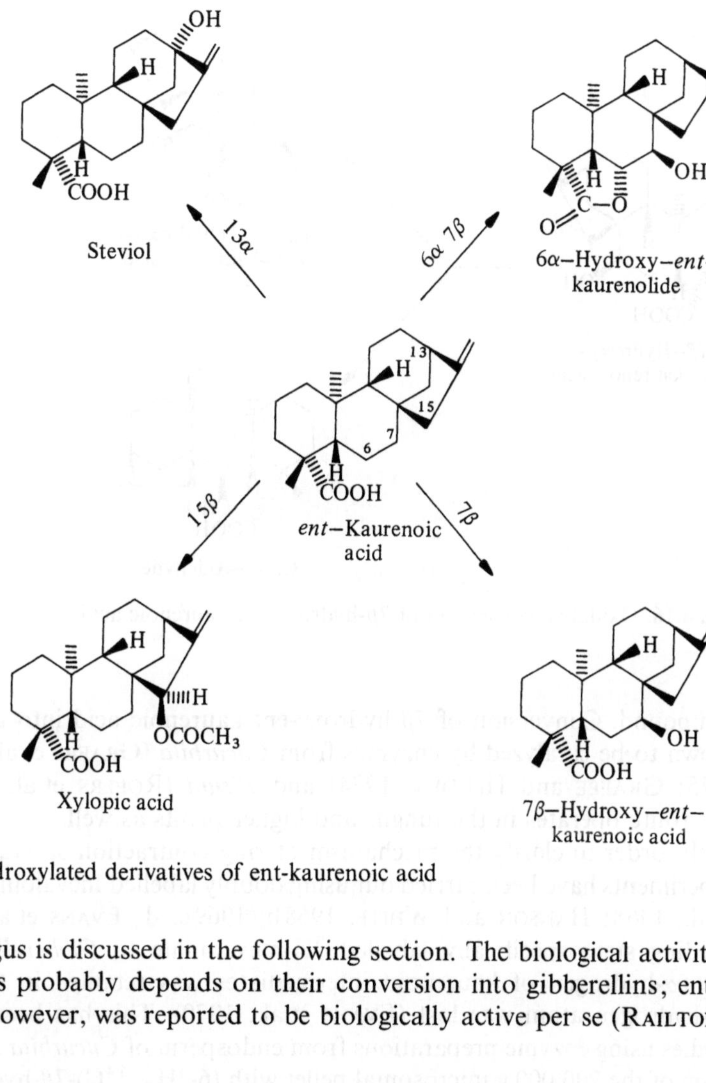

Fig. 4.15. Monohydroxylated derivatives of ent-kaurenoic acid

thesis in the fungus is discussed in the following section. The biological activity of ent-kaurenoids probably depends on their conversion into gibberellins; ent-kaurenoic acid, however, was reported to be biologically active per se (RAILTON et al., 1975).

d) Contraction of Ring B in ent-Kaurene

Feeding of (2-^{14}C)-mevalonate to cultures of *G. fujikuroi* gives labelled GA_3 with the COOH group attached to ring B being radioactive. Hence, this group originates from C-2 of mevalonate which was extruded as C-7 of the kaurene skeleton during ring contraction (BIRCH et al., 1959). This ring contraction step separates the gibberellins from all the other tetracyclic diterpenes.

Several B ring functionalized kaurenes have been tested as intermediates (CROSS et al., 1968b, 1970; JEFFERIES et al., 1970, 1974; EVANS et al., 1970; HANSON et al., 1972; WEST, 1973; BEARDER et al., 1975b). Only 7β-hydroxy-ent-kaurenol (JEFFERIES et al., 1974) and 7β-hydroxy-ent-kaurenoic acid (HANSON and WHITE, 1969b; LEW and WEST, 1971; HANSON et al., 1972) yielded suitable incorporation levels into gibberellins by the fungus. The latter represents the real intermediate, immediately preceding GA_{12}-aldehyde as the first gibberellin

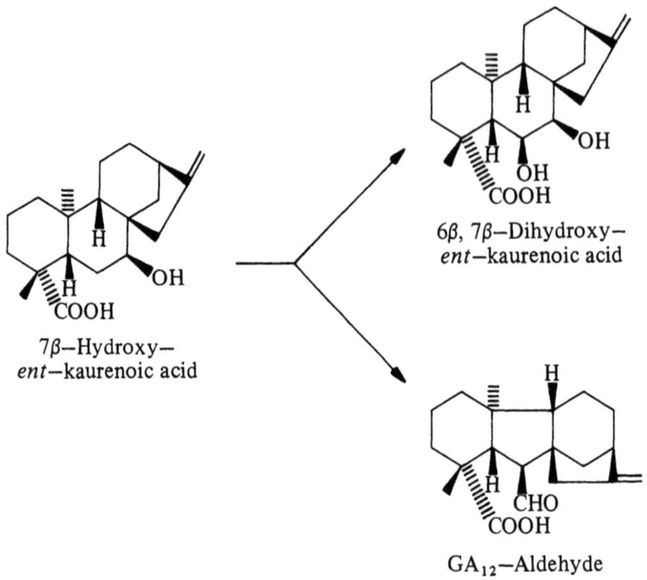

Fig. 4.16. Oxidative conversion of 7β-hydroxy-ent-kaurenoic acid

compound. Conversion of 7β-hydroxy-ent-kaurenoic acid into gibberellins was shown to be catalyzed by enzymes from *Cucurbita* (GRAEBE et al., 1972, 1974b, 1975; GRAEBE and HEDDEN, 1974) and *Pisum* (ROPERS et al., 1978). Hence, this route operates in the fungus and higher plants as well.

In order to clarify the mechanism of ring contraction several incorporation experiments have been carried out using doubly labelled mevalonic acid (HANSON et al., 1968; HANSON and WHITE, 1968b, 1969c, d; EVANS et al., 1970, 1973). Feeding stereospecifically labelled ^{3}H-mevalonate to *Gibberella* showed that the 6α-hydrogen of kaurenoid intermediates was retained in GA_3 whilst the 6β-hydrogen atom was lost (EVANS et al., 1970). This has been confirmed by studies using enzyme preparations from endosperm of *Cucurbita maxima*. Incubation of the 200,000 *g* microsomal pellet with (6-$^3H_2$, ^{14}C)-7β-hydroxy-ent-kaurenoic acid yielded GA_{12}-aldehyde and GA_{12} which both had lost half of the ^{3}H label (GRAEBE et al., 1975). These results conflict with the retention of two ^{3}H-atoms in GA_{12}-aldehyde obtained from application of (1-$^3H_2$,1-^{14}C)-geranyl phosphate to *Gibberella* cultures (HANSON and HAWKER, 1971; HANSON et al., 1972).

In the *Cucurbita* system, GA_{12}-aldehyde is formed along with 6β,7β-dihydroxy-ent-kaurenoic acid (Fig. 4.16). Kinetic studies established both metabolites to be produced simultaneously and at a rate indicating a monomolecular reaction (GRAEBE and HEDDEN, 1974).

The enzymes catalyzing the conversion of ent-kaurenoids into gibberellins possess a low substrate specificity, as was demonstrated by supplying several hydroxylated kaurenoids (Fig. 4.15) other than the true intermediates to *Gibberella* cultures. This was first shown with labelled steviol incorporated by wild-type cultures into 7β,13-dihydroxy-kaurenolide and probably GA_{20} (RUDDAT et al.,

1965b; HANSON and WHITE, 1968a). In cultures of the mutant B1-41a steviol, steviol acetate, and isosteviol (rerranged rings C/D) were metabolized to the corresponding derivatives of normal fungal gibberellins (MACMILLAN, 1974a; BEARDER et al., 1973d, 1976a). Desacetylxylopic acid gave the 15β-hydroxy-analogues of GA_{15}, GA_{24}, GA_{25}, and GA_9 (MACMILLAN, 1974a; BEARDER et al., 1975c).

Kaurenoids hydroxylated at 3β-position were efficiently converted into the usual 3β-hydroxylated gibberellins and 2β-hydroxylated ent-kaurenoids formed GA_3 by dehydration (LUNNON et al., 1977). On the contrary, JEFFERIES et al. (1974) found that substitution at ring A prevents formation of gibberellins. ent-Kaurenoids, dehydrogenated in ring A, yielded hitherto unknown gibberellins (COOK et al., 1971; BAKKER et al., 1972, 1974). ent-Kaur-6,16-diene gave GA_3 in low yield (CROSS et al., 1968b). On comparing the rates of incorporation of genuine and unnatural precursors into gibberellins the following observations were made: (a) 2-hydroxylation of ring A in kaurenoids reduced the ring contraction step, (b) methylation, or acetoxylation at the 7β- and 15β-positions, of ent-kaurenoic acid completely inhibited contraction of ring B, but not the 3β-hydroxylation, (c) 3β-hydroxylation is suppressed by rearrangement of rings C/D in precursors (MACMILLAN, 1974a; LUNNON et al., 1977).

Kaurenolides are not incorporated into gibberellins (CROSS et al., 1968b). On the contrary, ring B contraction took place with C/D modified analogues of ent-kaurene derivatives to give novel tetra- and pentacyclic analogues of gibberellins (BEARDER et al., 1979; HANSON et al., 1979; WADA et al., 1979).

e) Conversion of C_{20}-Gibberellins into C_{19}-Gibberellins

GA_{12}-aldehyde appears to be the first compound possessing the ent-gibberellane skeleton, but is still lacking the carboxyl group at C-6. When comparing the incorporation rates of various labelled precursors into gibberellins by *Gibberella* cultures it has been suggested that hydroxylation at C-3 occurs prior to the formation of the carboxyl group at C-6 (CROSS et al., 1968c; BEARDER et al., 1973b). According to the biogenetic theory during cyclization of geranylgeranyl pyrophosphate one hydrogen from the biological medium is introduced at C-3 (see Fig. 4.14) whereas the other one comes from the 4-pro-R hydrogen of mevalonic acid as could be demonstrated by feeding 4R(4-^{3}H,2-^{14}C)-mevalonic acid (HANSON et al., 1968). The latter hydrogen is substituted by the hydroxyl group. Direct hydroxylations usually proceed with retention of the configuration at the carbon atom. Evidence for this stereochemical reaction has been defined by the incorporation of (3α-^{3}H,17-^{14}C)-ent-kaurene into GA_3 with retention of the tritium label at the α-position (DAWSON et al., 1975); thus, a 3-keto intermediate is ruled out.

Apparently, at the stage of GA_{12}-aldehyde a "branching point" is reached leading to either the 3-non-hydroxylated gibberellins via GA_{12} or through GA_{14}-aldehyde to the 3-hydroxylated hormones. This view has been established in several ways. Application of labelled GA_{12}-7-alcohol to *Gibberella* strain ACC 917 gave the same ratio of 3-hydroxylated and non-hydroxylated gibberellins in radioactive form as in a feed with (2-^{14}C)-mevalonic acid, whereas from

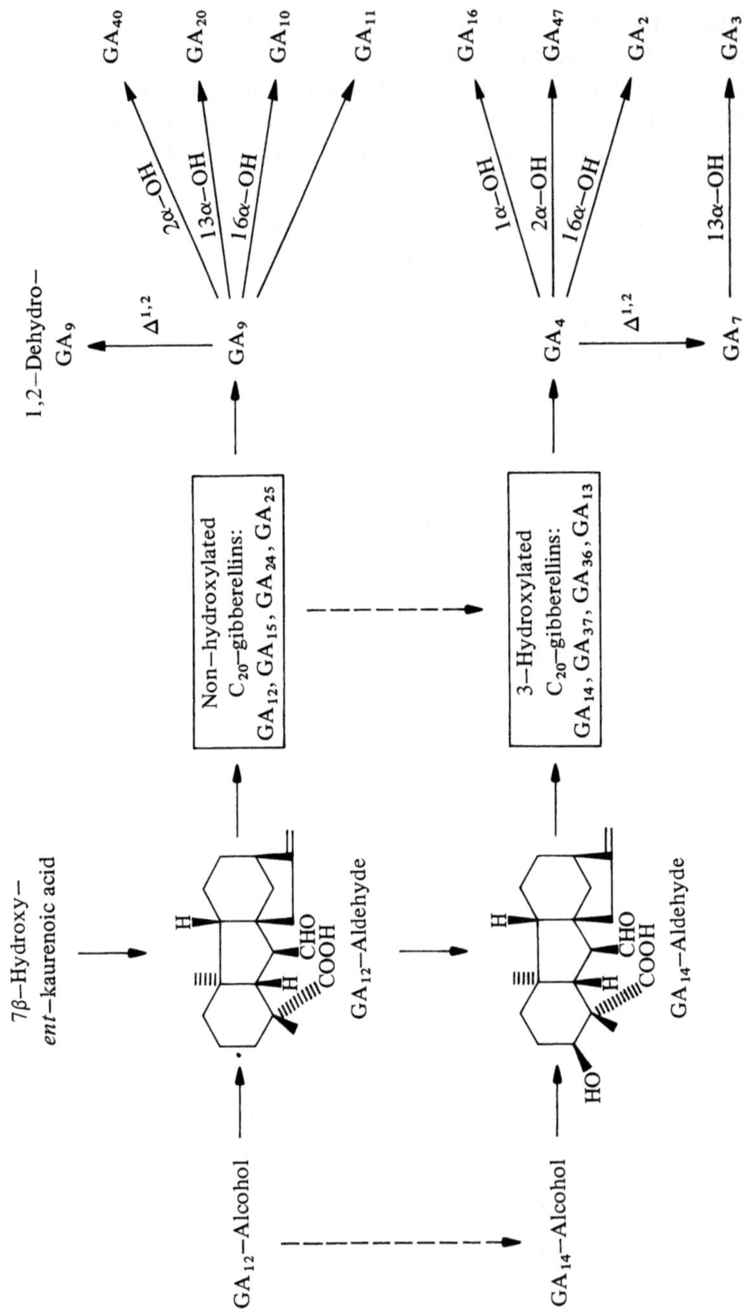

Fig. 4.17. Main pathways of gibberellin conversions in *Gibberella* cultures

GA_{12} only the non-hydroxylated hormones were derived (EVANS and HANSON, 1975). 3-Hydroxylation prior to oxidation of C-7 has been confirmed by using *G. fujikuroi* strain REC-193 A (BEARDER et al., 1973b) and the mutant B1-41a (HEDDEN et al., 1974; BEARDER et al., 1973a, 1975b).

GA_{12}-7-alcohol and GA_{14}-7-alcohol are not genuine metabolites of the fungus (HANSON and HAWKER, 1973) but they have been efficiently converted

Table 4.4. Hydroxylation pattern of several C_{20}-gibberellins of plant and fungal origin

Substituents at C-3	C-13	C-10				
		CH_3	CH_2OH	CHO	COOH	OH
H	H	GA_{12}	GA_{15}	GA_{24}	GA_{25}	GA_9
OH	H	GA_{14}	GA_{37}	GA_{36}	GA_{13}	GA_4
H	OH	GA_{53}	GA_{44}	GA_{19}	GA_{17}	GA_{20}
OH	OH	GA_{18}	GA_{38}	GA_{23}	GA_{28}	GA_1

into C_{20}- and C_{19}-gibberellins when fed exogenously (CROSS et al., 1968c; HEDDEN et al., 1974; EVANS and HANSON, 1975). Therefore, the conversion of radiocarbon labelled GA_{12}-alcohol into ^{14}C-GA_3 has been used to produce the latter on a preparative scale (HANSON and HAWKER, 1973). Its formation by reduction of aldehydes has been achieved with enzyme preparations from pea only (ROPERS et al., 1978). The role of gibberellin alcohols as transient metabolites merits further investigation.

Present knowledge of gibberellin interconversion by the fungus is outlined in Fig. 4.17. It seems noteworthy to mention that the type of gibberellins formed in the fungus is dependent on the pH value used in the culture (BEARDER et al., 1973a, 1975b). Furthermore, due to differences in uptake and transport of precursors some contradictory results have been obtained. Presuming GA_{12} and GA_{14} to be early gibberellins in the biosynthetic route, further conversion to the C_{19} hormones should be achieved by successive oxidative elimination of the C-20 group. According to Table 4.4 it seems reasonable (CROSS and NORTON, 1966) to eliminate C-20 via -CH_2OH, -CHO, and -COOH by decarboxylation of a β,γ-unsaturated intermediate. Unlike sterol demethylation, where unsaturated intermediates occur prior to removal of C-19 (corresponding to C-20 in gibberellins) as formic acid, the loss of C-20 in gibberellins requires rings A and B to be saturated. This was first shown by HANSON and WHITE (1969c) who found tritium label in positions 1, 5, and 9 of the gibberellane skeleton following application of stereospecifically labelled ^3H-mevalonic acid which precludes a double bond at these positions. The failure of GA_{13} to be converted into other gibberellins by the fungus over a 24-h period (CROSS et al., 1968c; HANSON and HAWKER, 1972), together with the lack of biological activity of hormones with a C-20 carboxyl group, makes it unlikely that these compounds are intermediates.

As a consequence, HANSON and WHITE (1968a, 1969c) postulated a biochemical equivalent of a Baeyer-Villiger-type oxidation of a C-20 aldehyde function, followed by solvolysis of the resultant ester. Several C_{20}-gibberellins with different oxidation levels at C-20 such as the acids GA_{13} and GA_{25}, the aldehydes GA_{36} and GA_{24}, and the lactonized alcohols GA_{15} and GA_{37} have been supplied to the mutant B1-41a of *Gibberella fujikuroi* (BEARDER et al., 1975b). None of them was incorporated into C_{19}-gibberellins, though all of them have been obtained along with C_{19}-hormones on feeding GA_{12} and GA_{14}, respectively.

A detailed analysis indicated GA_{36} was not an intermediate on the route from GA_{12}-aldehyde to GA_3 (BEARDER et al., 1975b). Results presented above and in Fig. 4.17 led to a consideration of the conversion of C_{20}-gibberellins as either C-7 aldehydes or 19,20-δ-lactones.

In a detailed study of the reaction mechanism and origin of the OH-group at C-10 the (19-^{18}O)-GA_{12} and its 7-alcohol were fed to the *Gibberella* mutant B1-41a (BEARDER et al., 1976c). In both cases all gibberellins isolated did not show a decrease in ^{18}O content with the exception of GA_{15}, formed from GA_{19} by lactonization in which one of the two ^{18}O atoms is lost. The complete retention of ^{18}O in the C_{19}-gibberellins rules out a Baeyer-Villiger-type oxidation, and also 19,20-δ-lactones as intermediates.

Because of the low incorporation rates of GA_{13}-anhydride (HANSON and HAWKER, 1972; BEARDER et al., 1975b) this compound can be ruled out as an intermediate. In another approach, biosynthetically labelled ent-kaurene was fed to cultures of *Gibberella*. During incorporation into C_{19}-gibberellins $^{14}CO_2$, derived from C-20, was produced. This led to the assumption of a peroxy acid as an intermediate in the formation of C_{19}-hormones (DOCKERILL et al., 1977).

Despite intensive investigations with numerous precursors and intermediates the reaction mechanism leading to the C_{19}-gibberellins remains unclear. Enzymatic reaction conditions, however, have been well documented (GRAEBE and HEDDEN, 1974; ROPERS et al., 1978).

Using the microsomal pellet from endosperm of immature seeds of *Cucurbita maxima*, ent-kaurene through successive oxidation yielded GA_{12}-aldehyde in the presence of oxygen and NADPH (GRAEBE et al., 1972; BOWEN et al., 1972). With the occurrence of GA_{12}-aldehyde the cell-free biosynthesis of an ent-gibberellane has been achieved for the first time. Further conversion of GA_{12}-aldehyde with the 15,000 *g* supernatant in the presence of Mn^{2+} ions yielded GA_{12} (GRAEBE et al., 1974b), but if manganese is omitted from the system of incubation with GA_{12}-aldehyde far more gibberellins were found, namely GA_{12}, GA_{13}, GA_{15}, GA_{24}, GA_{36}, GA_{37}, GA_{43}, and for the first time a C_{19}-gibberellin, GA_4, (GRAEBE et al., 1974a, b). As an intermediate one should expect GA_9, which has never been found in incubation mixtures, but which when fed, was efficiently converted into GA_4 (GRAEBE and HEDDEN, 1974).

On incubation of the low-speed supernatant of a *Gibberella* homogenate with (7-^{3}H,17-^{14}C)-GA_{12}-aldehyde, labelled GA_{14}-aldehyde and possibly GA_{13}-aldehyde were formed, confirming that 3-hydroxylation occurs at an early stage (DOCKERILL et al., 1977). On the contrary, a similar enzyme preparation obtained from immature pea seeds converted the GA_{12}-aldehyde into GA_{44} and the hitherto unknown 13-hydroxyGA_{12} (GA_{53}), thus indicating that 13-hydroxylation (ROPERS et al., 1978) occurs at an early stage in Leguminosae.

f) Predominant Steps in Regulation of Biosynthesis

Starting about 30 years ago, a number of growth retardants have been introduced, the effect of which usually can be counteracted by application of gibberellins. The most important growth retardants are listed in Fig. 4.18 (cf. CATHEY, 1964; NICKELL, 1978). On the basis of kinetic analysis, LOCKHART (1962) showed

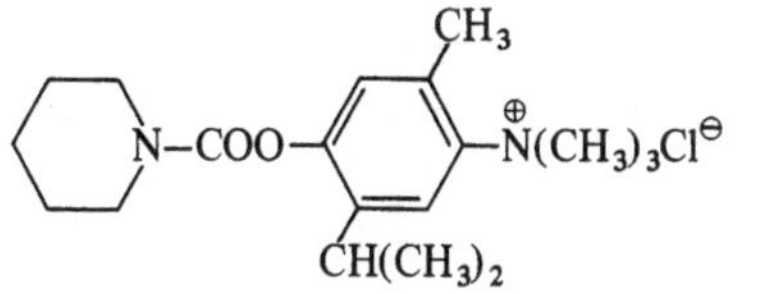

Amo–1618

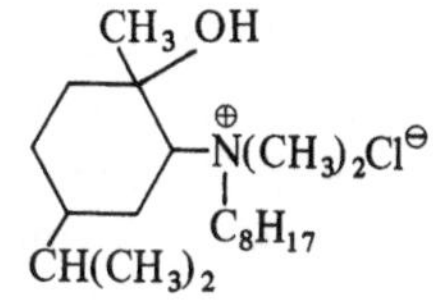

Q64

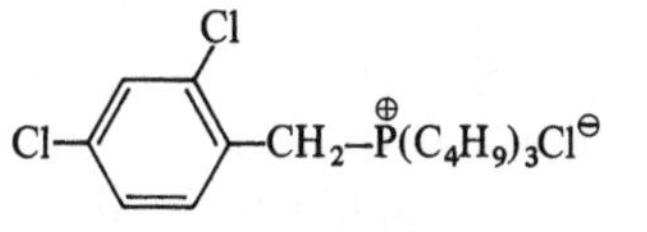

Phosfon D

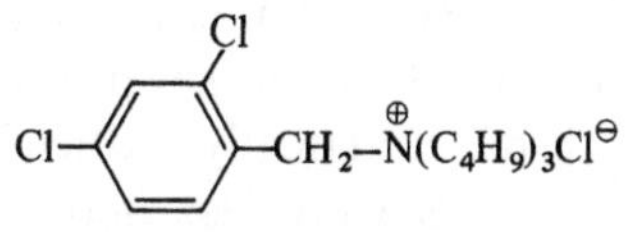

Phosfon S

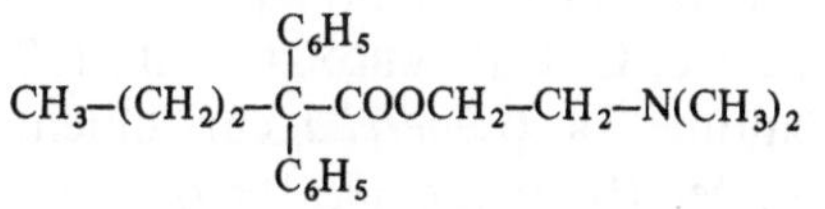

SKF–525A

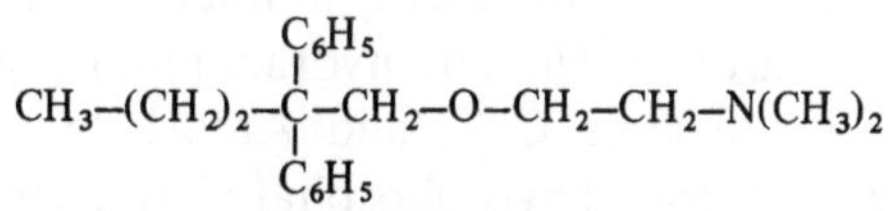

SKF–3301A

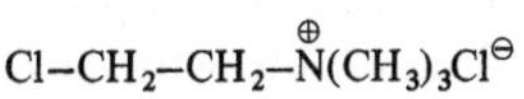

CCC, Cycocel,
Chlormequat

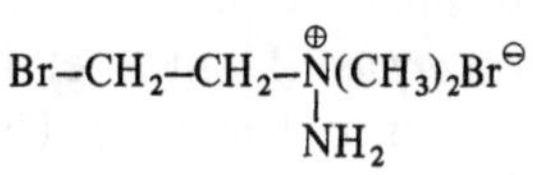

BMH

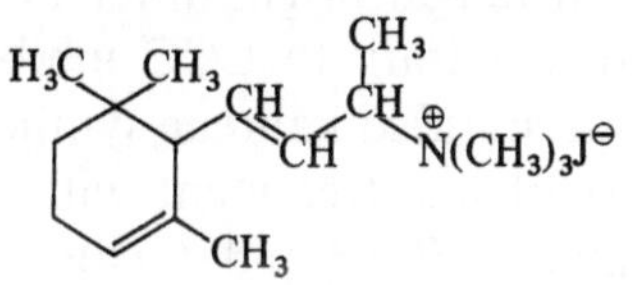

N, N, N–Trimethyl–1–methyl–(2′, 6′, 6′–trimethylcyclohex–2′–en–1′–yl)–prop–2–enylammonium iodide

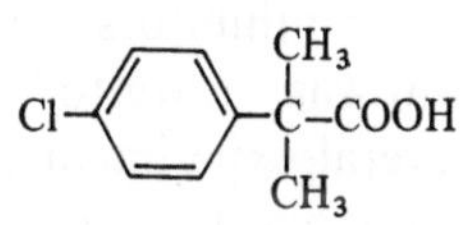

2–(4–Chlorophenyl)–isobutyric acid

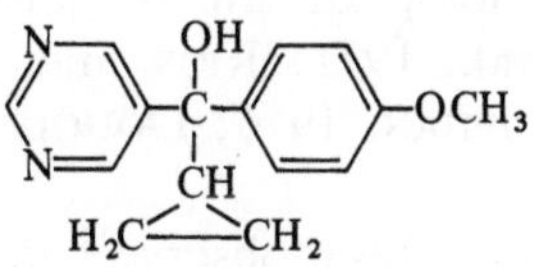

Ancymidol (EL–531)

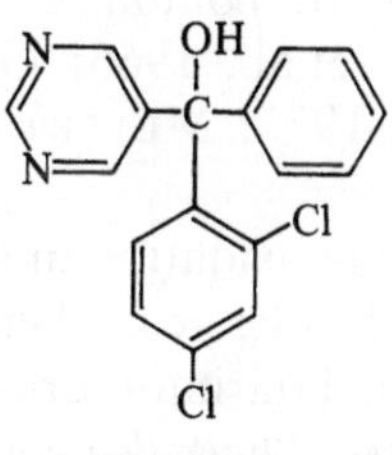

Triarimol

Fig. 4.18. Synthetic plant growth regulators inhibiting gibberellin biosynthesis

that CCC and Phosfon D interact with GA_3 in a competitive manner. In the meantime it became obvious that some growth retardants may act as inhibitors of the gibberellin biosynthesis. This could be demonstrated by using the fungus which, on treatment with Amo-1618 or CCC, produced reduced gibberellin levels (KENDE et al., 1963; NINNEMANN et al., 1964; RUDDAT, 1965a; HARADA and LANG, 1965; ZEEVAART and OSBORNE, 1965; ADAMIEC, 1966; BARNES et al., 1969; SMITH and SADRI, 1970; CROSS and MYERS, 1969; MERTZ and HENSON,

1967b). Appropriate experiments with higher plants supported the role of Amo-1618, CCC, and others to be inhibitors of the gibberellin biosynthesis (i.e., BALDEV et al., 1965; ZEEVAART, 1966; YOMO and INUMA, 1966; JONES and PHILLIPS, 1967; DALE and FELIPPE, 1968; WYLIE et al., 1970; KUO and PHARIS, 1975; ECKERT et al., 1978; SCHILLING et al., 1978). Also kaurene analogues (WADA, 1978a), N-substituted imidazoles (WADA, 1978b), some quarternary ammonium iodides of ABA analogues (HEDDEN et al., 1977; CHO et al., 1979; MUROFUSHI et al., 1979), and several chlorinated phenyl isobutyric acids (SEMBDNER et al., 1980) drastically reduced gibberellin levels in cultures of the fungus. Other retardants were less active or inactive in diminishing the rate of gibberellin formation, e.g., Phosfon S, Phosfon D (HARADA and LANG, 1965), SKF-7997 [=tris-(diethylaminoethyl)-phosphate trihydrochloride] (MOORE and ANDERSON, 1966), and hydrazonium analogues of CCC (SEMBDNER et al., 1972).

On feeding CCC and Amo-1618 to cultures of *Gibberella* enrichment of geranylgeranyl pyrophosphate has been observed (BARNES et al., 1969). Unambiguous evidence for the inhibition of kaurene synthetase by Amo-1618, Phosfon, and CCC was given for the first time by experiments with crude cell-free preparations of *Marah macrocarpus* (DENNIS et al., 1965) and later on with several enzyme preparations listed in Table 4.2. Using purified preparations of kaurene synthetase, isolated from *Gibberella* and *Marah*, the activity A has been found to be most susceptible to inhibition by the retardants (FALL and WEST, 1971; FROST and WEST, 1977). The enzymes of plant and fungal origin differ in their behaviour towards the inhibitors (see Table 4.3). Contrary to CCC inhibition in vivo, the retardant was shown to be inactive with purified kaurene synthetase. Furthermore, in several experiments using higher plants, treatment with CCC gave results contradictory to that mentioned above (REID and CARR, 1967; REID and CROZIER, 1970; HALEVY and SHILO, 1970; SNIR and KESSLER, 1975). Thus, one has to agree with CLELAND'S (1965) conclusion that there is more than one mode of action of CCC. Factors other than inhibition of gibberellin biosynthesis seem also to be combined with physiological effects caused by Amo-1618 (HEATHERBELL et al., 1966; GASPAR et al., 1971; RIOV and JAFFE, 1973a, b; CROZIER et al., 1973; MÜLLER and SCHUPHAN, 1976; DOUGLAS and PALEG, 1972, 1978a, b).

Inhibition of ent-kaurene oxidation in *Gibberella* has been observed on addition of N-substituted imidazoles, e.g., 1-geranylimidazole (WADA, 1978b). Ancymidol and the fungicide, Triarimol, drastically reduced the content of endogenous gibberellins in treated *Phaseolus vulgaris* plants (SHIVE and SISLER, 1976). The former has been found to inhibit oxidation of ent-kaurene in cell-free preparations of *Pisum* and *Marah* (COOLBAUGH and HAMILTON, 1976). A similar inhibition of the conversion of ent-kaurenol to ent-kaurenal in the *Marah* system was obtained with SKF-525A (DENNIS and WEST, 1967). The growth retardant B-995 (succinic acid-2,2-dimethylhydrazide) does not operate through inhibition of gibberellin biosynthesis (REED et al., 1965; KUO and PHARIS, 1975).

As mentioned above, the cyclization reaction catalyzed by kaurene synthetase has been shown to be the site of inhibition by synthetic plant growth retardants and furthermore it has been suggested to be the site of endogenous regulation of the biosynthetic pathways leading to gibberellins (see Sect. 4.1.5b). In order to compare ent-kaurene formation in rapidly differentiating and in non-differentiating tissues YAFIN and SHECHTER (1975) found only activity B of kaurene synthetase in cell suspension cultures of tomato and tobacco, whilst in germinating tomato seeds both activity A and B were operating.

In an investigation of the subcellular localization of kaurene synthetase in rapidly developing systems, the synthesis of ent-kaurene, from either geranylgeranyl pyrophosphate or copalylpyrophosphate, has been observed in plastids of *Marah macrocarpus*, whereas in organelle preparations of etiolated pea shoot tips and developing castor bean endosperm the activity A was missing (SIMCOX et al., 1975). At present it is impossible to determine at what developmental stage kaurene synthetase is expressed in the plastids. From the inhibition of kaurene formation in enzyme extracts of etiolated pea seedlings by chloramphenicol or cycloheximide it is tentatively concluded that the increase in kaurene synthetase activity during morphogenesis is due to photo-induction of the biosynthetic pathway from mevalonate to ent-kaurene (GOMEZ-NAVARETTE and MOORE, 1978).

A diminished formation of ent-kaurene has been demonstrated in homogenates derived from coleoptiles of the dwarf-5 mutant of *Zea mays* when compared with normal maize (HEDDEN and PHINNEY, 1979). Thus, the d_5-gene seems to control at least one of the two steps between geranylgeranyl pyrophosphate and the tetracyclic terpenes. In cell-free extracts from *Ricinus* seedlings infected with microbial pathogens an increased formation of casbene (cf. Sect. 4.1.5b) with concomitant decrease of ent-kaurene has been observed in comparison with non-infected material (SITTON and WEST, 1975; DUEBER et al., 1978). *Zea mays* seedlings responded to fungal infection with 50 to 100 fold increase in diterpene biosynthesis (MELLON and WEST, 1979).

For an internal regulation of energy-utilizing metabolic sequences the ratio of 5′-adenylates available (adenylate energy charge) is of great importance. Within the biosynthetic route leading from mevalonate to cyclic diterpenes pyrophosphomevalonate decarboxylase was the enzyme most subjected to regulation by adenylate energy charge (KNOTZ et al., 1977) whilst the activity of the other enzymes was not influenced.

4.1.6 Ethylene

Ethylene formation in plants has been known since 1934 (GANE, 1934), but biosynthesis of ethylene has been studied for less than two decades. Nevertheless, a number of reviews have already been published (JANSEN, 1965; MAPSON, 1969, 1970; PRATT and GOESCHL, 1969; SPENCER, 1969; YANG and BAUR, 1969; MAPSON and HULME, 1970; MCGLASSON, 1970; ABELES, 1972; YANG, 1974), and most of the work cited therein will not be referred to here. The following report deals with ethylene biosynthesis in higher plants only; ethylene biogenesis in fungi and bacteria differs with respect to the pathway and other aspects (cf. ABELES, 1972; CHOU and YANG, 1973; DASILVA et al., 1974; YANG, 1974; PRIMROSE, 1976). The way research in ethylene biosynthesis has progressed shows some outstanding peculiarities depending mainly on the following facts:

(a) Difficulties in detecting ethylene production, by means of disintegrated plant tissue and cell-free systems, led to usage of non-enzymic model systems. (b) Due to the simple chemical structure of ethylene, there are many potential precursors. (c) The biosynthetic route represents a very limited side-shoot of general metabolic pathways.

Substances which have been proposed as precursors of ethylene are methionine, linolenic acid, propanal, β-alanine, β-hydroxypropionic acid, acrylic acid, ethionine, ethanol, glycerol, organic acids like acetic acid, fumaric acid etc., as well as sucrose and glucose (ABELES, 1972; YANG, 1974). Among them, methionine is the only one indubitably established as a physiological precursor in ethylene biosynthesis. This may be true for linolenic acid, and possibly for some other substances, too, but these are questions which are still under investigation.

a) Ethylene Formation in Model Systems

The Cu^{2+}-ascorbate model system (LIEBERMAN and MAPSON, 1964) rapidly converts methionine into ethylene. In this conversion, methional (CH_3-S-CH_2-CH_2-CHO) appears to be an intermediate, and ethylene is derived from carbons-3 and -4 of methionine (LIEBERMAN et al., 1965, 1966). Catalytic reactions require reduction of Cu^{2+} to Cu^+ by ascorbic acid and re-oxidation of Cu^+ to Cu^{2+} by O_2, and include the formation of hydrogen peroxide and a superoxide anion radical necessary for the conversion of methional to ethylene (BEAUCHAMP and FRIDOVICH, 1970).

Linolenic acid also forms ethylene with the Cu-ascorbate system under certain circumstances (LIEBERMAN and MAPSON, 1964). Prior to catalytic degradation, linolenic acid is activated by the introduction of three conjugated double bonds, followed by some degree of oxidation and formation of peroxidized acids (cf. MAPSON, 1969; LIEBERMAN, 1975). Contrary to methionine, the conversion of linolenic acid yields not only ethylene but also ethane. Propanal, a decomposition product of peroxidized linolenic acid, was also shown to be an effective precursor of ethylene in the model system (LIEBERMAN and KUNISHI, 1967). Further studies concerning this "linolenate-propanal pathway" (cf. review of YANG, 1974) indicated that carbons-2 and -3 of propanal were converted into ethylene or ethane as a unit (YANG and BAUR, 1969). The reaction mechanism involves the formation of a $C_2H_5^{\cdot}$ radical as an intermediate, which is, in turn, either oxidized by Cu^{2+} to yield ethylene or is reduced by ascorbate to yield ethane.

Furthermore, a non-enzymic conversion of methionine to ethylene was observed in a photochemical system, based on flavin mononucleotide (FMN) (cf. review of MAPSON, 1969). Also in this system, methional serves as an intermediate and ethylene derives from methionine carbons-3 and -4. The reaction mechanism proposed by YANG et al. (1967) is mechanistically related to the decomposition of 2-chloroethane phosphonic acid (Ethrel, Ethephon) to ethylene (see YANG, 1974).

Model systems like those described may be considered as prototypes of in vivo processes in which enzymes play the role of catalysts (cf. MAPSON, 1970). Starting from the results obtained in model systems several authors have studied the effect of methionine and linolenic acid on ethylene production in plant tissues and cell-free systems.

b) Synthesis in Vivo – Physiological Pathways

Although completely disintegrated, post-climacteric apple pulp tissue does not form ethylene, slices of apple tissue continue to synthesize ethylene under appropriate conditions (cf. MAPSON, 1969). ^{14}C-Methionine, applied to apple tissue, is efficiently converted to ethylene, as first reported by LIEBERMAN et al. (1966). In addition to apple tissue, the role of methionine as an ethylene precursor was also demonstrated in banana tissue and in IAA-treated pea stem sections (BURG and CLAGETT, 1967; LIEBERMAN and KUNISHI, 1975), in tomato and cauliflower floret tissue (MAPSON et al., 1970), in avocado tissue (BAUR et al., 1971), in leaves of bean plants (ABELES, 1972), in IAA-treated mung bean sections (SAKAI and IMASEKI, 1972), and in senescent flower tissue of morning glory (KENDE and HANSON, 1976; HANSON and KENDE, 1976a), as well as in albedo tissue of mandarin fruit (HYODO, 1977). These results demonstrate thoroughly that ethylene, in higher plants, is derived from methionine. This conclusion is confirmed by the following facts: (a) L-methionine, the naturally occurring isomer, is converted into ethylene to a much higher extent than D-methionine (BAUR and YANG, 1969a). (b) Ethionine, which degrades to ethylene as efficiently as methionine in the model system, is not converted into ethylene in tissues (BAUR et al., 1971). (c) Specific radioactivity of ethylene corresponds to the amount of labelled methionine administered to apple tissue (OWENS et al., 1971; YANG and BAUR, 1972). (d) The ability of different plant tissues to convert methionine into ethylene parallels the ability to produce ethylene endogenously, as was shown with climacteric and preclimacteric avocado fruits (BAUR et al., 1971), IAA-treated and non-treated pea and mung bean stems (BURG and CLAGETT, 1967; SAKAI and IMASEKI, 1972) as well as with tissue induced to high ethylene production by wounding (ABELES and ABELES, 1972; HANSON and KENDE, 1976b). (e) Rhizobitoxine (Structure 4.4), an inhibitor of pyridoxal phosphate-dependent reactions and a specific inhibitor of ethylene production from methionine (see Sect. 4.1.6c), inhibits ethylene biosynthesis in higher plants but not in *Penicillium digitatum*, which forms ethylene from another source (OWENS et al., 1971).

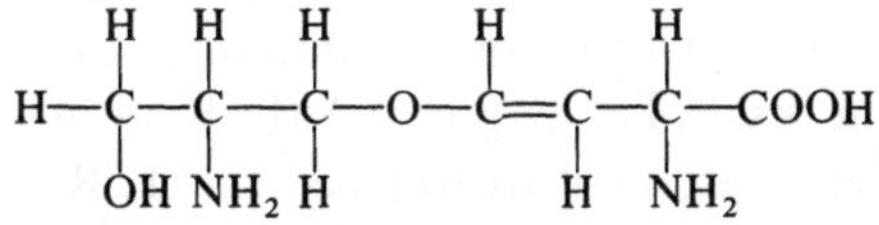

Structure 4.4. Rhizobitoxine (L-2-amino-4-(2′-amino-3′-hydroxypropoxy)-trans-3-butenoic acid)

To study substrate specificity, several radioactive derivatives of methionine have been applied to apple tissue (YANG and BAUR, 1972). The results indicate that the structural requirements for methionine as an ethylene precursor are highly specific (see YANG, 1974). Also, in apple tissue, just as in model systems, ethylene is derived from C-3 and C-4 of methionine; C-1 is converted into CO_2 and C-2 to formic acid, whereas sulphur and the methyl-carbon appear to be retained in the tissue (LIEBERMAN et al., 1966; BURG and CLAGETT, 1967).

Methionine metabolism and recycling of methionine sulphur during continuous ethylene synthesis was further studied with apple tissue (BAUR and YANG, 1972a, b; MURR and YANG, 1975b; ADAMS and YANG, 1977; MURR, 1977). A methionine sulphur cycle, proposed by BAUR and YANG (1972b), was based on the detection of S-methylcysteine as a major metabolite of methionine and the assumption that the fragment, produced from the methylthio group of methionine during ethylene formation, is methanethiol (cf. YANG, 1974). However, methanethiol has never been identified as a product of methionine during ethylene biosynthesis.

2,4-Dinitrophenol, an uncoupler of oxidative phosphorylation, has been shown to inhibit the conversion of methionine to ethylene in apple tissue, suggesting a requirement for ATP (Sect. 4.1.6c). Thus, S-adenosylmethionine (SAM) has been postulated to be an intermediate being degraded into C_2H_4, CO_2, HCOOH, NH_3, and 5′-methylthioadenosine (MTA) (MURR and YANG, 1975a, b). This proposal was confirmed recently (ADAMS and YANG, 1977). After feeding (methyl-^{14}C)- or (^{35}S)-methionine, labelled 5-methylthioribose (MTR) was identified as the predominant product, and MTA a minor one, in climacteric (ethylene producing) apple tissue but not in preclimacteric tissue. Inhibition of ethylene formation in climacteric tissue by L-2-amino-4-(2′-aminoethoxy)-trans-3-butenoic acid (aminoethoxyvinylglycine) (AVG), the ethoxy analogue of rhizobitoxine, a powerful inhibitor of pyridoxal-dependent reactions, gave a corresponding inhibition of MTR labelling. When fed (^{35}S)-MTA, radioactivity was efficiently incorporated into MTR and methionine. However, feeding of radioactive MTR, either ^{35}S or (^{35}S, methyl-^{3}H), gave incorporation into methionine but not MTA, indicating that the methylthio group of MTA is transferred as a unit into methionine via MTR.

Very recently the pathway has been completed by the identification of 1-aminocyclopropane-1-carboxylic acid (ACC) as intermediate precursor of ethylene (ADAMS and YANG, 1979; LÜRSSEN and NAUMANN, 1979; BOLLER et al., 1979). It was further shown that the conversion of methionine to ACC is strongly inhibited by AVG and that the conversion of ACC to ethylene requires oxygen. Furthermore, the role of SAM in ethylene biosynthesis has been confirmed by comparing the conversion of methionine and selenomethionine in ethylene-generating system (KONZE and KENDE, 1979b). From these results a revised scheme has been derived for the production of ethylene from methionine; the first step being the activation of methionine by ATP to give SAM. It is in turn fragmented to ACC and MTA, the latter being then hydrolyzed to MTR, which donates its methylthio group to a four-carbon acceptor to reform methionine. ACC is decomposed to ethylene and other products by mechanisms still under investigation. The mechanism, postulated for the formation and decomposition of ACC, is included in Fig. 4.19 which summarizes present knowledge on the biosynthesis of ethylene.

In tracer experiments, using (U-^{14}C)-L-methionine, (methyl-^{3}H)-L-methionine and ^{35}S-L-homocysteine thiolactone, the major soluble metabolite was identified as S-methylmethionine in morning glory flowers (HANSON and KENDE, 1976a, b). It acts as a methyl donor and, in senescent ethylene-producing flower tissue, the methyl group was utilized for methionine formation with homo-

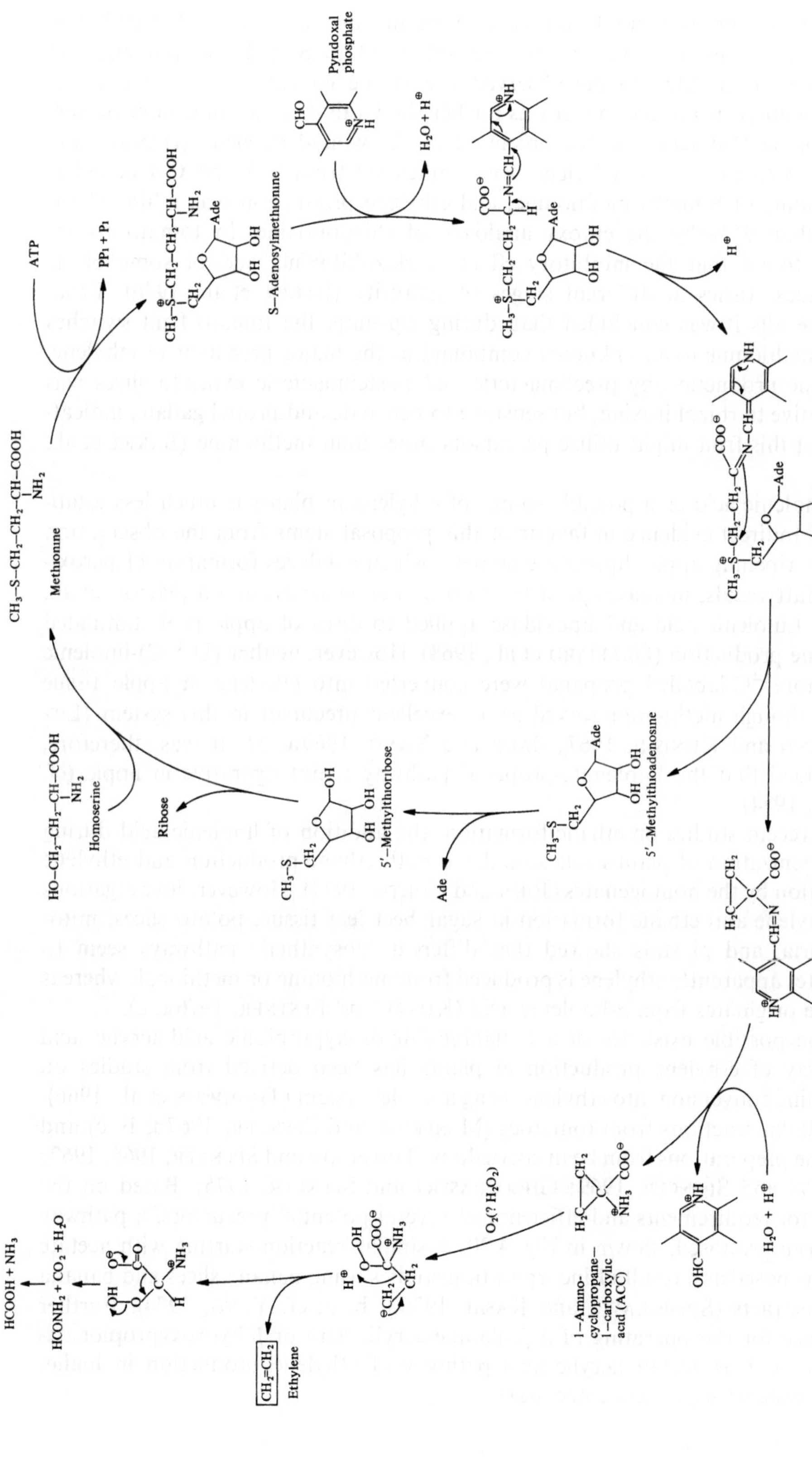

Fig. 4.19. Cycle for methionine metabolism related to ethylene biosynthesis and postulated mechanism for the conversion of S-adenosylmethionine to 1-aminocyclopropane-1-carboxylic acid. (According to ADAMS and YANG, 1977, 1979). The substituted pyridinecarboxaldehyde stands for pyridoxal phosphate

cysteine serving as methyl acceptor. From the two molecules of methionine produced in this reaction, one was remethylated to S-methylmethionine, and the other contributed to the observed rise in the content of free methionine used for ethylene production. It was further shown that in morning glory flowers methionine also serves as the precursor of the wound ethylene (HANSON and KENDE, 1976b). Wound ethylene was synthesized from C-3 and C-4 of either methionine or S-methylmethionine, and ethylene production was inhibited for more than 95% by the ethoxy analogue of rhizobitoxine. In tomato tissues it was found that the inhibitory effect of rhizobitoxine, and of some of its analogues, varies at different stages of maturity (BAKER et al., 1976). From these results it was concluded that, during ripening, the tomato fruit switches from methionine to an unknown compound as the major precursor of ethylene. Ethylene production by preclimacteric and postclimacteric avocado slices was insensitive to rhizobitoxine, but sensitive to benzoate and propyl gallate, indicating that this fruit might utilize precursors other than methionine (BAKER et al., 1976).

Linolenic acid as a possible source of ethylene in plants is much less established. Indirect evidence in favour of this proposal stems from the observation that, in ripening apple, lipoxidase activity, which catalyzes formation of peroxidized fatty acids, increases prior to the rise in ethylene evolution (MEIGH et al., 1967). Linolenic acid and lipoxidase applied to discs of apple peel stimulated ethylene production (GALLIARD et al., 1968). However, neither (U-^{14}C)-linolenic acid nor ^{14}C-labelled propanal were converted into ethylene in apple tissue slices, though methionine served as an excellent precursor in this system (LIEBERMANN and KUNISHI, 1967; BAUR and YANG, 1969a, b). It was, therefore, concluded that the linolenate-propanal pathway is not operative in apple (cf. YANG, 1974).

In recent studies on ethane formation, the addition of linolenic acid during homogenization of plant roots stimulated both ethane production and ethylene liberation by the homogenates (JOHN and CURTIS, 1977). However, investigations on ethylene and ethane formation in sugar beet leaf tissue, potato slices, mitochondria, and plastids showed that different biosynthetic pathways seem to operate; apparently ethylene is produced from methionine or methional, whereas ethane originates from α-linolenic acid (KONZE and ELSTNER, 1976a, c).

The possible existence of a β-alanine-β-hydroxypropionic acid-acrylic acid pathway of ethylene production in plants has been derived from studies on β-alanine conversion into ethylene using a model system (THOMPSON et al., 1966), subcellular fractions from tomatoes (MEHERIUK and SPENCER, 1967a, b, c) and enzyme preparations from bean cotyledons (THOMPSON and SPENCER, 1966, 1967; STINSON and SPENCER, 1969; GHOOPRASERT and SPENCER, 1975). Based on the co-factor requirements and efficiency of several potential precursors, a pathway has been proposed, shown in Fig. 4.20. A similar reaction starting with acetate was proposed as a result of incorporation studies using banana slices and banana pulp extracts (SHIMOKAWA and KASAI, 1970a, b, c; cf. YANG, 1974). Further evidence for the operating of a β-alanine-acrylic acid or β-hydroxypropionate-acrylic acid or acetate-acrylic acid pathway of ethylene production in higher plant systems seems to be necessary.

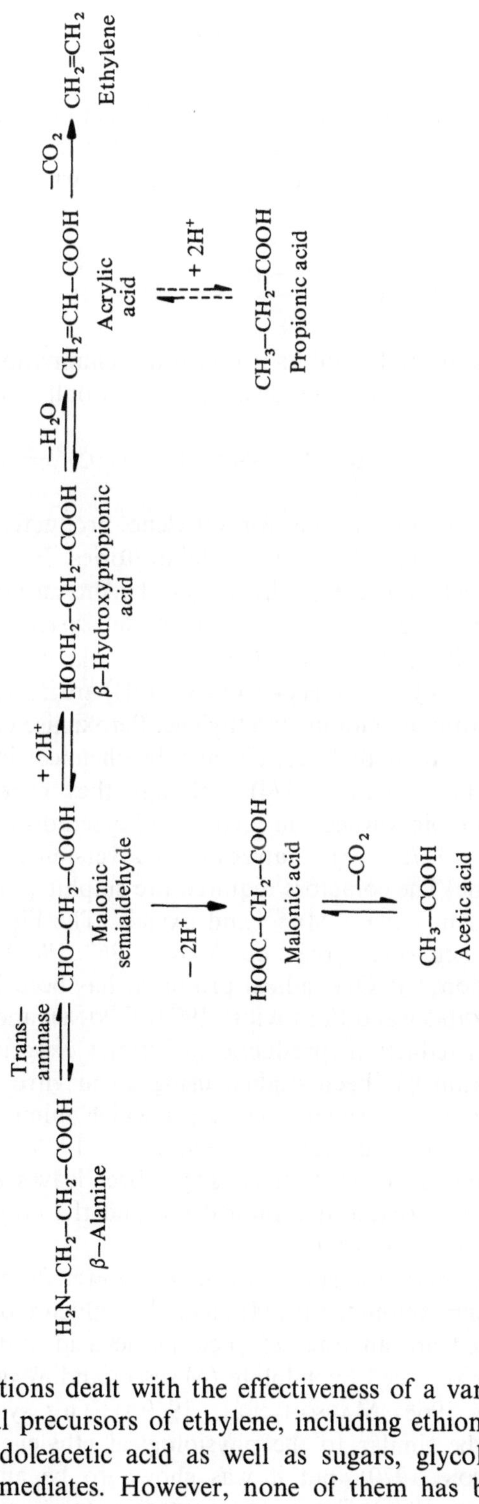

Fig. 4.20. Conversion of β-alanine to ethylene. (According to GHOOPRASERT and SPENCER, 1975)

A number of further investigations dealt with the effectiveness of a variety of different substances as potential precursors of ethylene, including ethionine, ethanol, ethane, fumaric acid, indoleacetic acid as well as sugars, glycolysis and tricarboxylic acid cycle intermediates. However, none of them has been shown to be a direct precursor of ethylene in vivo (cf. reviews of SPENCER, 1969; ABELES, 1972; YANG, 1974).

c) Enzymes – Reaction Mechanisms

As far as current knowledge of the biochemical mechanisms of olefine-forming reactions is concerned, two types of reaction mechanism for ethylene production can be deduced (cf. YANG, 1974). One of them involves the formation of a $CH_3-CH_2\cdot$ radical as an intermediate which is in turn oxidized to form ethylene:

$$R-CH_2-CH_3 \xrightarrow{-R^{\cdot}} \dot{C}H_2-CH_3 \xrightarrow{-H^{\cdot}} CH_2{=}CH_2$$

The second involves a concerted elimination process in which a precursor degrades into ethylene by a "push and pull" mechanism:

$$X-CH_2-CH_2-Y \longrightarrow X^- + CH_2{=}CH_2 + Y^+$$

Enzymatic systems for ethylene production from methionine have been sought extensively in several laboratories. However, the enzymatic production of ethylene outside the living cell, by the natural system, could not be demonstrated. Clues from enzymatic model systems indicate that either methional ($CH_3-S-CH_2-CH_2-CHO$) or α-keto-γ-methylthiobutyric acid, KMB ($CH_3-S-CH_2-CH_2-CO-COOH$) are intermediates in the sequence of reactions from methionine to ethylene. Peroxidase-catalyzed ethylene-producing systems are of both historical and biochemical interest (see reviews of MAPSON, 1969, 1970; YANG, 1974), although they were shown not to be involved in ethylene biosynthesis in vivo. The horseradish peroxidase system described by YANG (1967, 1968) catalyzes the conversion of either methional or KMB into ethylene. The cofactors required are sulphite, monophenol, Mn^{2+}, and oxygen; H_2O_2 can replace Mn^{2+} and oxygen. The biochemical mechanism accounting for this enzymatic reaction (YANG, 1967, 1969) is based on monovalent oxygen reduction; the OH radical produced has been identified as the oxidizing agent (BEAUCHAMP and FRIDOVICH, 1970; ELSTNER and KONZE, 1974a, b). Monovalent oxygen reduction (production of triplet oxygen) as a prerequisite for ethylene formation has been studied using an in vitro system (illuminated chloroplast lamellae) from sugar beet leaves which mimics the loss of compartmentation in the wounded leaf (ELSTNER et al., 1976). 3-Hydroxytyramine (dopamine) identified in extracts from sugar beet leaves stimulated both the production of the superoxide free radical ion and the ethylene formation from methional in the in vitro system.

An enzymatic system from cauliflower florets includes a transaminase, converting methionine to KMB, as well as glucose oxidase and peroxidase; cofactors required are an ester of p-coumaric acid and methane sulphinic acid which can be replaced by sulphite (MAPSON and WARDALE, 1967, 1968; MAPSON and MEAD, 1968; MAPSON et al., 1969a). This system was assumed to represent or to be similar to the physiological ethylene-forming system in plant tissue (MAPSON, 1970), but it was shown to be an artificial one (LIEBERMAN and KUNISHI, 1971). Further peroxidase systems catalyzing the conversion of KMB into ethylene have been developed by using indoleacetic acid, phenolic and sulphinic acids, as well as hydroperoxide-generating components (MAPSON and WARDALE, 1971, 1972). Among the phenolic compounds present in tomato fruits, p-coumaric acid and naringenin were capable of acting as phenolic substrates in these systems, other phenolic compounds being inhibitory. Both the endogenous naringenin concentration in the tomato peel and the ethylene production increased at the onset of the climacteric (WARDALE, 1973).

Generally, ethylene production in peroxidase-catalyzed model systems proved to be different from the natural ethylene pathway in a number of respects (cf. YANG, 1974), and the results available fail to support the proposal that KMB, methional, and peroxidase are involved in ethylene biosynthesis in vivo.

The role of oxidative metabolism in ethylene formation was first demonstrated by HANSEN (1942) showing that ethylene production did not occur when O_2 was removed. Several authors (see review of ABELES, 1972) have shown that conversion of methionine into ethylene has an absolute oxygen requirement, and the involvement of oxidative phosphorylation was demonstrated by inhibitor studies using respiratory uncoupler (BURG and THIMANN, 1960; SHIMOKAWA and KASAI, 1966; LIEBERMAN et al., 1966; SPENCER, 1969; BAUR et al., 1971; BURG, 1973; LAU et al., 1974; MURR and YANG, 1975a). From results obtained with 2,4-dinitrophenol, an uncoupler of oxidative phosphorylation, a strict requirement for ATP has been concluded (BURG, 1973; MURR and YANG, 1975a, b). In addition to this "normal" aerobic ethylene-producing system a quasi-anaerobic system was detected in embryonic axes and cotyledons excised from cocklebur seeds. Its activity was high in the dormant state and low in after-ripened axes (KATOH and ESASHI, 1975; ESASHI et al., 1976a, b).

Rhizobitoxine and some rhizobitoxine-like analogues were shown to inhibit ethylene production in a number of tissues, but rhizobitoxine is ineffective in the model systems (OWENS et al., 1971; LIEBERMAN et al., 1974; HANSON and KENDE, 1976b; KENDE and HANSON, 1976; ZIMMERMANN et al., 1977; BAKER et al., 1978). Accordingly, it was suggested that a pyridoxal phosphate-mediated enzyme reaction may be involved in conversion of methionine to ethylene; and further results obtained indicated that rhizobitoxine inhibits the very first step in the reaction of the enzyme system with methionine (LIEBERMAN, 1975). Another potent inhibitor of pyridoxal phosphate enzymes, L-canaline, also markedly inhibited the conversion of methionine into ethylene by plant tissue (MURR and YANG, 1975a) and an in vitro system (spinach chloroplasts lamellae) which converts D,L-methionine into ethylene, depending on pyridoxal phosphate, ferredoxin, NADP-ferredoxin reductase, and NADPH as electron donor (KONZE and ELSTNER, 1976a, b). This ferredoxin-reductase-system of chloroplasts can be substituted by xanthinoxidase, another non-heme iron-flavoprotein enzyme.

The recent results on intermediates of ethylene biosynthesis (see p. 320) have supplied an improved base for further investigations of enzymes and reaction mechanisms: formerly proposed involvement of pyridoxal phosphate-linked enzymes in ethylene biosynthesis could be confirmed more solidly by studying the action of the rhizobitoxine analogue AVG which interferes with pyridoxal-phosphate enzymes (RANDO, 1974). The only step blocked by AVG is the conversion of SAM to ACC (ADAMS and YANG, 1979; KONZE and KENDE, 1979a). Additionally, the activation of this enzymatic reaction by pyridoxal phosphate could be demonstrated and its mechanism was postulated (ADAMS and YANG, 1979; YU et al., 1979a; see Fig. 4.19). The possible activation of SAM by coenzyme A suggested by LÜRSSEN et al. (1979) awaits experimental evidence. Few data have been published concerning the mechanism of ACC conversion to ethylene. ADAMS and YANG (1979) proposed an involvement of hydrogen peroxide. This suggestion corresponds with the inhibitory action of catalase (KONZE and KENDE, 1979a).

The activity of the enzymes catalyzing the individual reactions from methionine to ethylene, as well as some of the enzyme characteristics, have been determined. KONZE and KENDE (1979b) investigated kinetic parameters of methionine adenosyltransferase in an ethylene-forming system from *Ipomoea* using methionine and selenomethionine as substrates. A soluble enzyme converting SAM to ACC has been isolated from tomato and investigated with respect to kinetics, substrate specificity, and inhibition by rhizobitoxine analogues (YU et al., 1979a; BOLLER et al., 1979).

ACC oxidation yielding ethylene was demonstrated using homogenates of etiolated pea seedlings (KONZE and KENDE, 1979a). This ethylene-synthesizing system appeared to be associated, at least in part, with a particulate fraction and could be saturated only at very high ACC concentrations.

d) Regulation of Ethylene Biosynthesis

Rates of ethylene production vary in different organs and tissues and are dependent on growth and developmental stages. Many publications deal with correlations of ethylene production to physiological processes such as germination, vegetative growth, flower formation, ripening of fruits, leaf senescence, abscission etc., or demonstrate effects of environmental conditions (light, temperature, minerals etc.), and stress factors (wounding, water, chemical effectors, etc.) as well as hormonal interrelationships in ethylene formation (cf. reviews of BURG, 1962; HANSEN, 1966; MAPSON, 1969; PRATT and GOESCHL, 1969; SPENCER, 1969; MAPSON and HULME, 1970; ABELES, 1972, 1973; YANG, 1974). However, little is known on the regulatory mechanisms of ethylene biosynthesis. Ethylene production in plant tissues can be governed by the availability of substrates, cofactors, and effectors, as well as the rate of biosynthesis of the ethylene-forming enzyme system and the rate of degradation of this system. There is good evidence that the total content of methionine, being the major if not the only precursor of ethylene, does not regulate the ethylene synthesis in ripening fruit tissues (BAUR et al., 1971; BAUR and YANG, 1972b) and in pea or bean seedling tissue, induced by auxin (BURG and CLAGETT, 1967; SAKAI and IMASEKI, 1972; LIEBERMAN and KUNISHI, 1975). These observations indicate that ethylene production in these tissues is controlled by factors other than the synthesis of methionine (cf. YANG, 1974). However, methionine metabolism is involved in the regulation of ethylene biosynthesis in senescent flower tissue of morning glory (HANSON and KENDE, 1976a; see below).

Knowledge of the enzyme system that catalyzes the conversion of methionine to ethylene has been improved only recently (see Sect. 4.1.6c). Studies using inhibitors of nucleic acid and protein synthesis, as well as some other chemical effectors, have proved that biosynthesis and turnover of the ethylene-forming enzyme system represent an important part of the regulatory system which may be affected in very different ways. Results published cannot yet be generalized and, therefore, will be used as selected examples in to the following discussion.

Regulation by Physiological Stages

Fruit Ripening. Major systems for ethylene production are the mature and ageing fruit tissues (BURG, 1962; HANSEN, 1966; PRATT and GOESCHL, 1969; SACHER, 1973; COOMBE, 1976). In most fleshy fruits an exponential rise in the rate of ethylene emanation precedes the onset of ripening as shown by several authors e.g., in banana (cf. MAPSON, 1969), apple (SFAKIOTAKIS and DILLEY, 1973), and a variety of other fruits (cf. SPENCER, 1969). In citrus, correlation of increased ethylene production with normal ripening processes was disputed (AHARONI, 1968; EAKS, 1970). Avocado fruits were shown to have a preclimacteric increase in ethylene production and, later, a second phase of climacteric ethylene production associated with ripening (ADATO and GAZIT, 1977b). Both in avocado and cherimoya fruit the increase in internal ethylene concentration does not precede the increase in CO_2 production (KOSIYACHINDA and YOUNG, 1975). In several fruit trees, the machinery for ethylene production apparently is present during premature stages, but the concentration of ethylene does not rise to a physiologically effective level until just before the onset of the climacteric (cf. PRATT and GOESCHL, 1969). Mechanisms limiting ethylene production in preclimacteric stages may vary from species to species (cf. MAPSON, 1970); of the factors involved, O_2 concentration and plant hormones seem to be important (cf. SACHER, 1973).

In apple fruit the rate of ethylene production rises when it is detached from the tree, indicating that an inhibitor from the tree may be effective. In avocado initiation of climacteric, ethylene is also suggested to be related to inhibitory factors arriving from leaves and regulated by the sink of growing seeds (ADATO and GAZIT, 1977c). Induction of climacteric ethylene evolution in fruit tissues needs the formation of enzymatic proteins since inhibitors of protein and RNA synthesis are effectively inhibitory (FRENKEL et al., 1968; MAPSON, 1969). Results on the efficiency of various inhibitors of DNA, RNA, and protein synthesis in postclimacteric apple tissue, as well as data on natural ageing of the ethylene-forming system, gave the conclusion that it exists in a dynamic state in which enzymes are produced and degraded simultaneously (LIEBERMAN and KUNISHI, 1975). Peach seeds after excision from the fruit develop, in the seed coat, capacity for high ethylene synthesis which is sensitive to inhibition by cycloheximide (JERIE and CHALMERS, 1976).

In another group, including certain kinds of banana, the fruit becomes more sensitive to ethylene at maturation and thus the endogenous level reaches an effective concentration (cf. PRATT and GOESCHL, 1969; MAPSON, 1970). The exponential rise in ethylene production in climacteric fruits is explained by an autocatalytic mechanism of ethylene synthesis (MAPSON, 1969; PRATT, 1974), a phenomenon well-known in flower senescence (BURG and DIJKMAN, 1967; NICHOLS, 1968; KENDE and BAUMGARTNER, 1974). However, in non-ripening stages of sycamore fig exogenously applied ethylene was shown to act as an auto-inhibitor of ethylene biosynthesis (ZERONI et al., 1976). Similar kinds of auto-inhibition have been described also for chilled avocado fruit (ZAUBERMAN and FUCHS, 1973) and preclimacteric banana fruit slices (VENDRELL and MCGLASSON, 1971). With respect to the mechanism of ethylene-induced ethylene synthesis, there is experimental support suggesting that ethylene affects the permeability of cell membranes such that enzymes and substrates are brought into contact (MAPSON, 1969; KENDE and BAUMGARTNER, 1974; HANSON and KENDE, 1975).

In contrast to the tree fruits mentioned, in several annual crops such as tomato and melon, the ethylene-forming system apparently is not developed or activated completely until the fruit reaches a critical physiological age. If the fruit is harvested too young, it is never capable of being fully activated (cf. PRATT and GOESCHL, 1969; ABELES, 1973). This may be explained by the assumption that during ripening the tomato fruit switches from methionine to another unknown pathway of ethylene biosynthesis (BAKER et al., 1976). Investigations using abnormally ripening tomato mutants such as "rin" and "nor" will open new possibilities in clarifying regulation of ethylene biosynthesis (cf. HERNER and SINK, 1973; MCGLASSON et al., 1975b; MIZRAHI et al., 1975). For interpretation of the results published by the authors, a divergence of mechanisms controlling ethylene synthesis in different tissues and possibly also different pathways might be assumed. Mechanisms of seed dormancy and germination in cocklebur involve two kinds of ethylene-producing systems, the normal aerobic and a quasi-anaerobic one, acting in the seed organs at different stages. Oxygen concentration may operate as a factor in controlling the activity of the two systems (ESASHI et al., 1976a, b). During germination of lettuce seeds ethylene production occurred at two distinctly different rates: a very low initial rate and a 100-fold increased rate at radicle emergence (DUNLAP and MORGAN, 1977).

Flower Development. Flowers exhibit a sharp increase in ethylene production either as a result of pollination or natural senescence, and respond to ethylene or propylene treatment by premature senescence and auto-catalytic ethylene production as was shown e.g., in carnation (SMITH et al., 1964; NICHOLS, 1966, 1968, 1971, 1977; MAYAK and DILLEY, 1976), orchid (BURG and DIJKMAN, 1967), rose (MAYAK et al., 1972), cotton (DURHAM and MORGAN, 1973), and morning glory (KENDE and BAUMGARTNER, 1974). In some genera the reproductive flower organs have been shown to produce more ethylene than the petals with the style apparently being a source of high ethylene production (HALL and FORSYTH, 1967; LIPE and MORGAN, 1973). In carnation ethylene surge after pollination or during natural senescence emanates from both gynaecium – especially the style – and petals (NICHOLS, 1977). In the protogynous annona flower a rise in ethylene production was found to precede the male stage. The highest rate of ethylene production occurs in the anthers, and is possibly induced by a signal (auxin?) towards the end of the female stage (BLUMENFELD, 1975).

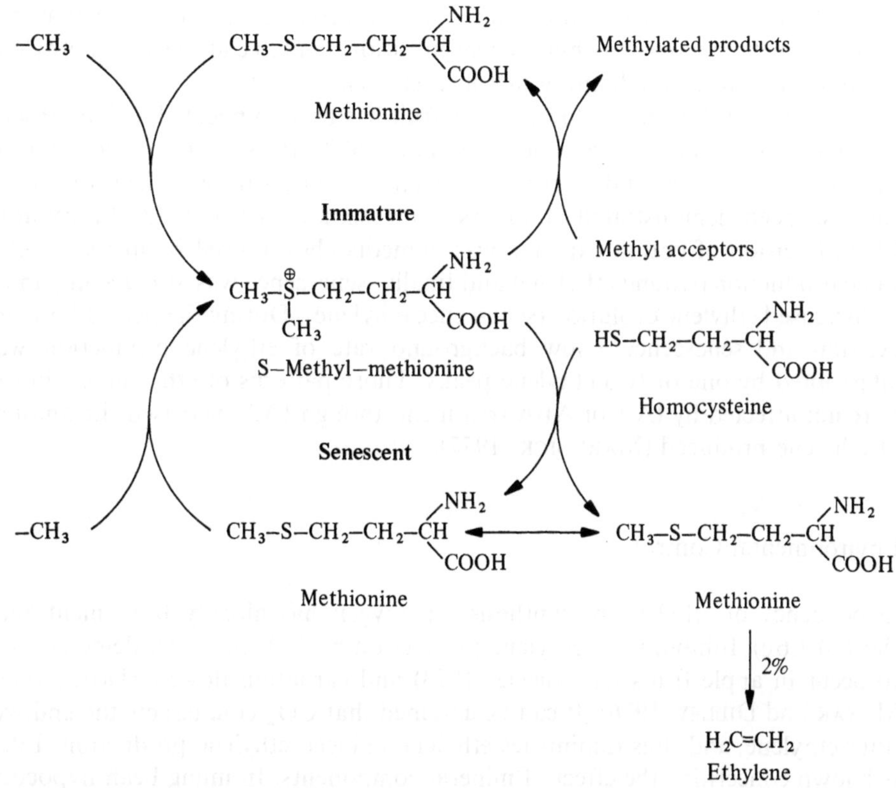

Fig. 4.21. Scheme of methionine metabolism in immature and senescent flower tissue of morning-glory according to HANSON and KENDE (1976a)

Regulation of ethylene biosynthesis during senescence of flower tissue in morning glory is closely related to methionine metabolism (KENDE and HANSON, 1976; HANSON and KENDE, 1976a). It was shown that an ethylene-generating system develops as an integral part of the ageing process in flower tissue; and the methionine pathway was proved to be the main, if not the only, route of ethylene biosynthesis. Summarizing their results, the authors proposed a scheme of ethylene production controlled by methionine metabolism (Fig. 4.21). In both immature and senescent flower tissue methionine is metabolized to S-methyl-methionine (SMM). In senescing tissue the free methionine content increases about tenfold, due to a change in SMM catabolism; SMM synthesis from methionine is not affected. In ageing tissue SMM donates a methyl group to homocysteine to yield two molecules of methionine; this reaction may be restricted in immature tissue. Ethylene formation depends on methionine derived from SMM and homocysteine, possibly because ethylene-generating and methionine-synthesizing systems are located in the same cellular compartment. During ageing, the supply of homocysteine to this system increases, either through direct stimulation of homocysteine formation or due to a breakdown in the compartmentation between stored homocysteine and the methionine-synthesizing system. As related to the total amount of free methionine in senescent flower tissue, only 2% is utilized for ethylene production, indicating that not the total level of methionine but its concentration available at the site of ethylene synthesis may be a rate-limiting factor in ethylene evolution.

Abscission. Ethylene produced by the fruit has been implicated in fruit abscission (BLANPIED, 1972; LIPE and MORGAN, 1972, 1973; JERIE, 1976; ADATO and GAZIT, 1977a, b) and increased ethylene evolution prior to leaf abscission has also been demonstrated (JACKSON and OSBORNE, 1970, 1972; MARYNICK, 1977). Excision of cotton explants was immediately followed by increased ethylene production (wound ethylene) and finally, senescence was also accompanied by increased ethylene evolution (senescence ethylene). During the period between excision and senescence a low background rate of ethylene production was interrupted by one or two ethylene peaks. Those patterns of ethylene evolution were not affected by IAA or ABA treatment, though IAA increased the amount of ethylene produced (MARYNICK, 1977).

Environmental Control

Dependency of ethylene biosynthesis on oxygen has already been mentioned (Sect. 4.1.6c). Inhibition of ethylene production by CO_2 has been demonstrated to occur in apple fruits (cf. ABELES, 1973) and carnation flowers (UOTA, 1969; MAYAK and DILLEY, 1976). It can be assumed that CO_2 counteracts the endogenous ethylene, and thus diminishes ethylene-induced ethylene production. Little is known concerning the effect of mineral components. In mung bean hypocotyl sections, the ethylene-producing system has a specific requirement for Ca^{2+} and kinetin (LAU and YANG, 1974, 1975, 1976a). Ca^{2+} could be substituted only by Sr^{2+} but not by other divalent cations. Combined application of Cu^{2+} plus Ca^{2+} gave a remarkable synergistic stimulation of ethylene production, due to Cu^{2+} enhanced uptake of Ca^{2+} into the bean hypocotyl tissue.

Varying effects of temperature on ethylene formation have been demonstrated in various tissues (ABELES, 1973). Generally, a direct correlation exists between temperature and rate of ethylene production, with Q_{10} values and optimum and maximum temperatures changing from tissue to tissue. Recently, optimum temperatures, Q_{10} values and calculated activation energies have been published for the ethylene-synthesizing system in apple and tomato tissues, with Arrhenius plots of ethylene synthesis showing discontinuities (MATTOO et al., 1977). During vernalization of radish and pea seedlings, the rate of ethylene production increased, reaching a maximum several days after starting low temperature treatment, and then decreased (SUGE, 1977). Temperature shock either by heat or low temperature, or drastic change of temperature at a physiologically sensitive stage, can induce rapid increase in ethylene production, thus resembling stress ethylene (see below).

Ethylene production can be regulated by light. In etiolated pea seedlings ethylene production decreased after a single dose of red light, and far-red irradiation reversed the effect of red light, indicating that the phytochrome system is involved (GOESCHL et al., 1967). In a number of studies the observation has been confirmed and extended to other systems, e.g., bean seedlings, sorghum seedlings, and rice coleoptiles (KANG et al., 1967; GOESCHL and PRATT, 1968; BURG and BURG, 1968; KANG and RAY, 1969; IMASEKI et al., 1971; KANG and BURG, 1972; CRAKER et al., 1973). Thus, the regulatory effect of phytochrome on ethylene production seems certain, but regulation mechanisms are unknown. A slight increase in ethylene production following illumination has been observed in oat seedlings (MEHERIUK and SPENCER, 1964), lettuce seeds (ABELES and LONSKI, 1969), cranberries (CRAKER, 1971), sorghum stem sections (CRAKER et al., 1971), and plant tissue cultures (LA RUE and GAMBORG, 1971), but the action spectrum has not been established.

Stress-Induced Ethylene Production. It is well known that ethylene production increases rapidly following trauma caused by phytotoxic chemicals, extreme temperature and drought, mechanical wounding, ionizing irradiation, disease, insect damage, and other injury. Stress ethylene is a product of injured but living cells since it decreases when damage is extreme enough to kill the tissue. This phenomenon has been studied extensively, and literature published up to 1972 was comprehensively reviewed by ABELES (1973).

A number of recent investigations deals with the evolution of ethylene by plants subjected to mechanical stress (wound ethylene). When plant tissues are cut, bruished, or pressed etc., the rate of ethylene production begins to rise, increasing by 10- to 100-fold in 1 to 10 h and then declines (HANSON and KENDE, 1976b; JACKSON and CAMPBELL, 1976; KENDE and HANSON, 1976; CHALMERS and FARAGHER, 1977; MARYNIK, 1977). The kinetics of wound ethylene induction and the rapid transmission of the wound signal have been studied in the apical meristematic region of etiolated pea seedlings (SALTVEIT and DILLEY, 1978a, b,c). The pathway of synthesis of wound ethylene in morning glory flower tissues was shown to be closely related, if not identical, to that of unwounded tissue (HANSON and KENDE, 1976a). Dependence of wound ethylene formation on tissue destruction degree has been demonstrated in discs of sugar beet leaves by means of point freezing, indicating that ethylene formation occurs in the non-decompartmentalized but physiologically perturbed cells surrounding the frozen leaf parts (ELSTNER and KONZE, 1976).

In host-parasite complexes leading to hypersensitivity, stimulation of ethylene production seems to be associated with the necrotic process (cf. MONTALBINI and ELSTNER, 1977), possibly resembling wound ethylene. In some other fungal diseases, however, higher amounts of ethylene were produced during the expression of susceptibility (DALY et al., 1970; GENTILE and MATTA, 1975; PEGG and CRONSHAW, 1976). Results obtained with bean varieties either susceptible or hypersensitive to *Uromyces phaseoli,* led the authors to conclude that stimulation of ethylene formation after mechanical wounding and at different stages after fungal infection constitute different processes. Differences may touch either trigger mechanisms or the pathway of ethylene formation (MONTALBINI and ELSTNER, 1977).

Water stress was found to increase ethylene production in detached avocado fruits (ADATO and GAZIT, 1974), detached orange leaves (BEN-YEHOSHUA and ALONI, 1974), *Vicia faba* plants (EL-BELTAGY and HALL, 1974), intact cotton petioles (MCMICHAEL et al., 1972), and young cotton bolls, either attached to the plant or detached (GUINN, 1976). Detailed studies on the interrelationship of water potential (ψ) in tissue and the level of ethylene have been performed using excised wheat leaves (WRIGHT, 1977).

Increased ethylene concentrations have been described also in flooded plants and immersed cuttings (KAWASE, 1972a, b, 1974; EL-BELTAGY and HALL, 1974; JACKSON and CAMPBELL, 1975a, b). The main cause for high endogenous ethylene level was shown to be the accumulation of ethylene in flooded portions, due to the blockade of ethylene escape by water (KAWASE, 1976).

In addition to the well-known stimulation of ethylene production by X-rays, UV irradiation has also been shown to induce stress ethylene in apple fruits (CHALMERS and FARAGHER, 1977).

When applied in appropriate concentrations, a large number of substances are known to induce stress ethylene formation. Those are metal ions (Cu^{2+}, Fe^{3+}, Hg^{2+}), KI, ozone, ascorbic acid and its derivatives, trichloroacetic acid, iodoacetic acid, monoiodoacetamide, $NaClO_3$, $NaHCN_2$, KOCN, sodium chloroacetate, ammonium thiocyanate, sodium ethylmercuri-thiosalicylate, 8-hydroxyquinoline sulphate, ethyl hydrogen 1-propyl phosphate, coumarin, herbicides, and fungal metabolites (see ABELES, 1973). Furthermore, stimulation of ethylene biosynthesis has been found by phenolics (catechine, 4-methylcatechine) (FUCHS, 1970), NH_4F (BANGERTH, 1975), and phytotoxic manganese concentrations (FOWLER and MORGAN, 1972). As already mentioned, Cu^{2+}-stimulated ethylene production has been studied recently in relation to the action of Ca^{2+} and kinetin (LAU and YANG, 1976a). The modes of action of the different chemicals in induction of stress ethylene are almost unknown; possibly some indirect effects are involved.

Cycloheximide, used commercially in *Citrus* fruit abscission, is known to induce stress ethylene (cf. ABELES, 1973). This effect has been further described in banana (MCGLASSON et al., 1971), olive (HARTMAN et al., 1972), mirabelle (BANGERTH, 1975), and apple (CHALMERS and FARAGHER, 1977). In contrast cycloheximide, being an inhibitor of protein synthesis, can also block ethylene production in maturing fruit (LIEBERMAN and KUNISHI, 1975), excised seed (JERIE and CHALMERS, 1976), as well as with auxin-induced and stress-induced ethylene formation (ABELES, 1973; LIEBERMAN and KUNISHI, 1975).

Though the data reported do not completely explain the mechanisms of induction and regulation of stress ethylene, some general conclusions can be drawn:

1. Stress factors, apparently causing different primary effects, change the properties of membranes and compartmentation without killing the cell.

2. Compartmentation changes may be responsible for induction of stress ethylene.

3. Stress ethylene is produced, at least after wounding and chemical injury, by the methionine pathway. However, involvement of other pathways cannot be excluded.

4. The stress ethylene-producing system needs continuous synthesis of enzyme protein.

Control by Other Hormones

Many publications deal with the effects of plant hormones, especially the effects of auxin, on ethylene production in plant tissues and results have already been summarized (ABELES, 1973; BURG, 1973; LIEBERMAN, 1975; SAKAI, 1975b; KONDO et al., 1975). Furthermore, special aspects of plant hormone interaction in ethylene production have been reviewed and studied recently in relation to senescence and fruit ripening (SACHER, 1973; LIEBERMAN et al., 1977) and water stress (HSIAO, 1973).

Auxin. Physiological evidence on the stimulation of ethylene production by auxin dates back to 1935; exact confirmation was first given in 1964 (ABELES and RUBINSTEIN, 1964a; MORGAN and HALL, 1964). Promotion of ethylene production by auxin is widespread; however, little or no auxin effect on ethylene production could be detected in seeds, mature fruit, and cell cultures as well as in relation to abscission (ABELES, 1973; LIEBERMAN and KUNISHI, 1975; MARYNICK, 1977). Induction of ethylene synthesis by auxin occurs after a lag phase of about 30–60 min (BURG, 1973), the lag period varying with the tissue (CHADWICK and BURG, 1967, 1970) and auxin concentration (BURG and BURG, 1969; SHINGO and IMASEKI, 1971). Dose-response curves obtained in different tissues were similar, showing no effect at 10^{-7} M and below, a threshold effect at 10^{-6} M and an increasing ethylene production (tenfold or higher) up to 10^{-3} M (cf. ABELES, 1973). The rates of ethylene production in response to auxin, however, may differ markedly, as was shown in coleoptile and mesocotyle segments of oat and maize (MALLOCH and OSBORNE, 1976). Auxin must be continuously present for ethylene to be produced (cf. ABELES, 1973; LAU and YUNG, 1974); and synthetic auxin-like compounds, such as NAA and 2,4-D are also known to be active (FOWLER and MORGAN, 1972; ABELES, 1973). Oxygen was shown to be essential for both the induction process and the synthesis of auxin-induced ethylene (IMASEKI et al., 1977).

The best-studied systems for auxin-induced ethylene production are the subhook stem section system of etiolated pea seedlings and the mungbean hypocotyl segment system. These systems produce ethylene by the methionine pathway (BURG and CLAGETT, 1967; SAKAI and IMASEKI, 1972); the pea system is specifically inhibited by rhizobitoxine (OWENS et al., 1971). Recent studies have shown that auxin stimulates enzymatic conversion of SAM to ACC, the same step inhibited by rhizobitoxine (JONES and KENDE, 1979; YU and YANG, 1979; YU et al., 1979b). Auxin-mediated ethylene formation can be blocked by inhibitors of RNA and protein synthesis, indicating that continuous synthesis of protein is required for high rates of ethylene production (ABELES, 1966b; KANG et al., 1971; SAKAI and IMASEKI, 1971; SHINGO and IMASEKI, 1971; STEEN and CHADWICK, 1973; IMASEKI et al., 1977). Detailed further studies, using many different inhibitors of DNA, RNA, and protein synthesis, indicated that auxin action might be placed at the level of nucleoplasmic RNA polymerase II (LIEBERMAN and KUNISHI, 1975).

Varying auxin effects on the ethylene production of whole fruits and tissue slices have been described in relation to ripening processes, for example in banana, apple, tomato, and avocado (cf. SACHER, 1973; LIEBERMAN, 1975; TINGWA and YOUNG, 1975; LIEBERMAN et al., 1977). In ripening pears ethylene production is stimulated by IAA, 2,4-D and the IAA oxidation product 3-methylene oxindole (FRENKEL and DYCK, 1973; FRENKEL, 1975). A synergistic activity in IAA-induced ethylene production by mung bean hypocotyl sections has been shown for 5-hydroxyindole-3-acetic acid and 5-hydroxydioxindole-3-acetic acid (SUZUKI et al., 1977). Lipids, stimulating respiration and auxin-induced cell elongation in pea stem sections, also initiate a period of ethylene formation (IWATA and STOWE, 1973).

Cytokinins are known to increase ethylene production in various plant species (ABELES, 1973; LIEBERMAN, 1975; ZIMMERMAN et al., 1977). Response to cytokinins varies in different tissues. Usually cytokinins cause a rather low (two- to fourfold) increase in ethylene production; however, in very young etiolated pea seedlings kinetin is more effective than IAA (FUCHS and LIEBERMAN, 1968). In contrast, ethylene production by tissue slices from pre-climacteric, climacteric, and post-climacteric apples, tomatoes, and avocados was significantly reduced by the cytokinin, isopentenyladenosine (LIEBERMAN et al., 1977). In cytokinin-mediated ethylene formation apparently the normal methionine pathway is involved (ZIMMERMAN et al., 1977). A number of publications deal with synergistic effects on ethylene production by cytokinins and auxin or cytokinins and Ca^{2+} (BURG and BURG, 1968a; FUCHS and LIEBERMAN, 1968; LAU and YANG, 1973, 1974, 1975, 1976b; LAU and YUNG, 1974; HRADILIK, 1975; IMASEKI et al., 1975; KONDO et al., 1975; LAU et al., 1977). Contrary to those results, obtained with mung bean hypocotyls (e.g., LAU et al., 1977), kinetin and Ca^{2+}, either alone or in combination, had no effect on the ethylene production by excised pea epicotyls (MONDAL, 1975).

Gibberellins are either inactive or have a small promotive effect on ethylene formation. Both plants showing a slight stimulation of ethylene production by gibberellins and those not responding to gibberellins have been described (ABELES, 1973; LIEBERMAN, 1975; MONDAL, 1975).

Abscisic acid has been shown to promote ethylene production in leaves and fruits, but this effect could not be correlated to abscisic acid-induced abscission (cf. ABELES, 1973). During cotton explant abscission ABA did not change the basic pattern of ethylene evolution (MARYNICK, 1977). Only slight or non-existent enhancement of ethylene production by ABA is referred to also in other publications (KONDO et al., 1975; LIEBERMAN, 1975; SAKAI, 1975b). In carnation, ABA-accelerated flower senescence is preceded by a stimulation of ethylene production (MAYAK and DILLEY, 1976). In other systems, e.g., cell suspension cultures, peanut seeds, and soybean seedlings, ABA can decrease ethylene production without complete inhibition (cf. ABELES, 1973). In the mung-bean hypocotyl system ABA inhibited IAA-induced ethylene production. Benzyladenine did not overcome ABA-inhibition and ABA did not inhibit the stimulative effect of benzyladenine on both endogenous and IAA-induced ethylene production (KONDO et al., 1975). In apple, tomato and avocado fruit slices, the ABA effect on ethylene production varied with stages of maturation and senescence (LIEBERMAN et al., 1977).

Effects of Synthetic Plant Growth Regulators, Herbicides, and Other Inhibitors

Plant growth retardants, like CCC, Alar, and Phosphon D, have been shown to reduce ethylene production in various plants. However, in some tissues, inhibitory effects could not be observed or a slight promotion of ethylene formation by CCC was found (cf. ABELES, 1973). Little is known on the mode of action of growth retardants in ethylene formation; possibly, indirect effects, e.g., lowering of endogenous auxin level, may be involved.

Stimulation of ethylene production by 2,4-D, and some related herbicides, might be due to their auxin-like activity (cf. ABELES, 1973). Other herbicides promoting ethylene production apparently induce chemical stress ethylene.

Inhibitors of Ethylene Biosynthesis. Studies using specific inhibitors of ethylene biosynthesis give insight into the pathway and its regulatory mechanisms. This has been well demonstrated by the detection and usage of rhizobitoxine, rhizobitoxine-like analogues, and L-canaline being potent inhibitors of pyridoxal phosphate enzymes (see Sect. 4.1.6c and literature cited therein). Inhibition of ethylene production by diethyldithiocarbamate suggested that a copper-containing enzyme may be involved (LIEBERMAN et al., 1966). Further information on ethylene biosynthesis is derived from inhibitory studies using oxidative phosphorylation uncoupler, especially 2,4-dinitrophenol (cf. Sect. 4.1.6c). Cyanide, which inhibits the cytochrome a/a_3 complex, is without effect on ethylene production in apple tissue, and iodoacetate which inhibits glyceraldehyde-3-phosphate dehydrogenase is also much less effective than 2,4-dinitrophenol in this system (BURG and THIMANN, 1960; SHIMOKAWA and KASAI, 1966).

Cobaltous ions (Co^{2+}) greatly inhibit ethylene production by vegetative and fruit tissues in the presence, or in the absence, of kinetin and auxin (LAU and YANG, 1974, 1976a, b; GROVER and PURVES, 1976). Both basal and induced ethylene production was significantly inhibited by Co^{2+} concentrations as low as 10 µM, and at concentrations of 0.1 mM or greater inhibition was strong or nearly complete. Tracer experiments showed that Co^{2+} inhibits the conversion of methionine to ethylene; the mode of action, however, is not yet clear. In mung bean hypocotyls, also Ni^{2+} and Hg^{2+} strongly inhibited ethylene production in the absence or presence of kinetin, while other metal ions (Zn^{2+}, Mg^{2+}, Sn^{2+}, and Ba^{2+}) had a slight or no inhibitory effect (LAU and YANG, 1976a).

Certain phenolic acids, such as 3,5-diiodo-4-hydroxybenzoic acid, have been shown to inhibit endogenous and auxin-induced ethylene production in cress seedling roots (ROBERT et al., 1975, 1976a, b). The effect does not depend on reduction of IAA level. Benzyl isothiocyanate inhibits ethylene production in discs of unaged papaya fruits but not of aged papaya, indicating that the induction of the ethylene-producing system is affected. Benzyl isothiocyanate may be formed by enzymatic hydrolysis of benzyl glucosinolate, ubiquitously present in papayas, and, thus, it is proposed as an endogenous regulator of ethylene production in papaya fruit (PATIL and TANG, 1974). N,N-dimethyl tryptophan, a plant growth-inhibiting substance isolated from seeds of *Abrus precatorius*, has been shown to inhibit IAA- and kinetin-stimulated ethylene production by lettuce seedlings as well as IAA-stimulated ethylene production by wheat coleoptile sections (ANDERSON et al., 1975). Acetylcholine chloride inhibited IAA-induced ethylene production in etiolated bean tissues, possibly by mimicking effects of red light (PARUPS, 1976). Proteinaceous inhibitors of auxin-induced ethylene production by mung bean hypocotyl segments have been isolated, partially purified, and biochemically characterized from mung bean seedlings (SAKAI and IMASEKI, 1973a, b), leguminous seeds (SAKAI, 1975), and marine algae (WATANABE and KONDO, 1976). The mung bean inhibitor apparently achieves its activity by reversible inactivation of an ethylene-synthesizing enzyme. Action of the inhibitory protein may be associated with the plasma membrane (ODAWARA et al., 1977).

4.2 Metabolism (Interconversion and Catabolism)

4.2.1 Auxins

Knowledge of the metabolism of plant auxins is restricted to the catabolism of IAA. In a number of reviews on this subject, results have been summarized and progress has been assessed during the last two decades (HEWITT, 1958; RAY, 1958; GALSTON and HILLMAN, 1961; HARE, 1964; SHANTZ, 1966; PILET and GASPAR, 1968; SCOTT, 1972; THIMANN, 1972; SCHNEIDER and WIGHTMAN, 1974).

a) Routes and Products of IAA Catabolism

Many publications deal with the metabolism of IAA, studied either in vitro or in vivo by incubation of radioactive labelled IAA, and both enzymatic and non-enzymatic degradation reactions have been described. IAA, in water solution, is degraded by acids, ionizing radiation, ultra-violet light, visible light in the presence of sensitizing pigment, and by oxygen or peroxide in the presence of suitable redox systems. Additionally, IAA degradation is effected by autoclaving and several other factors such as heavy metals, pH etc. (GALSTON and HILLMAN, 1961; DEVERALL, 1965; KENNEY et al., 1969; IVERSEN and AASHEIM, 1970; EPSTEIN and LAVEE, 1975). Direct involvement of H_2O_2 in the inactivation of IAA in plant tissue has been shown recently (SCHNEIDER and GÜNTHER, 1975; OMRAN, 1977). Non-enzymatic aerobic oxidation of IAA can be achieved through co-oxidation of sulphite to sulphate (YANG and SALEH, 1973; HORNG and YANG, 1975).

IAA, exogenously applied to plant tissues, is degraded with rates and degrees of destruction which vary considerably with the tissue and physiological conditions (TROXLER and HAMILTON, 1965; DAVIES, 1972; THIMANN, 1972). Commonly, decarboxylation of IAA takes place, as was frequently demonstrated by measuring $^{14}CO_2$ output after application of (1-^{14}C)-indoleacetic acid (FANG et al., 1959; RAY, 1962; DEVERALL and DALY, 1964; AASHEIM and IVERSEN, 1971; HAMILTON et al., 1976). Enzymatic IAA decarboxylation in *Avena* coleoptiles showed similar rates and kinetics in the dark and in the light. Photoactivated decarboxylation of IAA was found only in vitro by UV irradiation (without riboflavin) or blue light (plus riboflavin) (MENSCHICK et al., 1977). IAA-oxidase activities in phototropic stimulated plants did not differ in the illuminated and shaded sides (LIBBERT and BORMANN, 1966). In apple leaves a high rate of photolytic decarboxylation of (1-^{14}C)-IAA has been demonstrated in vivo (GRO-

CHOWSKA, 1974). Incubation of apple callus tissue in a medium containing ($1\text{-}^{14}C$)-IAA resulted in a high degree (90%) of decarboxylation by old non-growing tissue and a lower decarboxylation (20%) by young growing tissue. Decarboxylation apparently was enzymatic (EPSTEIN and LAVEE, 1975). In mature olive leaves a more intense decarboxylation of ($1\text{-}^{14}C$)-IAA occurred than in young ones. The rate of decarboxylation was dependent on the presence of peltate scales of the leaves. IAA-decomposing bacteria (cf. LIBBERT and RISCH, 1969; LIBBERT et al., 1970e) might have been involved but did not play a major role in decarboxylation of applied IAA by the olive leaf (EPSTEIN and LAVEE, 1977a).

Products and pathway of IAA catabolism have been thoroughly studied by use of horseradish peroxidase (HINMAN and LANG, 1965). Both in vitro and in vivo experiments have been carried out, in some cases with ($2\text{-}^{14}C$)-IAA (TROXLER and HAMILTON, 1965; MORITA et al., 1967; SCHNEIDER and WIGHTMAN, 1974; EPSTEIN et al., 1975; MINCHIN and HARMEY, 1975; HAMILTON et al., 1976; EPSTEIN and LAVEE, 1977b; and literature cited therein).

3-Hydroxymethyloxindole (3-OH-MeOx) and 3-methyleneoxindole (3-MeneOx) are major products of IAA oxidation by enzyme preparations (Fig. 4.22). Conversion of IAA to 3-OH-MeOx and/or 3-MeneOx has been demonstrated using horseradish peroxidase (RAY and THIMANN, 1956; HINMAN and LANG, 1965; MORITA et al., 1967), cell-free enzyme preparations from corn coleoptiles (HAGER and SCHMIDT, 1968), *Parthenocissus tricuspidata* crown gall tissue (HAMILTON et al., 1976) as well as extracts from, and intact, pea seedlings (TULI and MOYED, 1967; MOYED and TULI, 1968). Both compounds have been tentatively identified as IAA oxidation products in vivo in apple callus tissue (EPSTEIN et al., 1975). Dehydration of 3-OH-MeOx to 3-MeneOx can occur non-enzymatically but an enzyme catalyzing this conversion has been partially purified from wheat germ (BASU and TULI, 1972b).

3-Methyloxindole (3-MeOx) has been detected as a minor compound of IAA catabolism in *Parthenocissus* crown gall tissue (HAMILTON et al., 1976). In the fungus *Phycomyces blakesleeanus*, which was shown to have an IAA-degrading system, 3-MeOx has been chromatographically identified (HILGENBERG et al., 1976). Reduction of 3-MeneOx to 3-MeOx has been demonstrated both in vitro and in vivo by a specific 3-methyleneoxindole reductase from peas (STILL et al., 1965a, b; MOYED and WILLIAMSON, 1967a, b; TULI and MOYED, 1967, 1969; MOYED and TULI, 1968).

Several authors have described the biological activity of IAA oxidation products, especially of 3-MeneOx, and discussed the possibility that degradation products, rather than IAA itself, may be active in growth regulation (FUKUYAMA and MOYED, 1964; STILL et al., 1965a, b; MEUDT, 1967, 1972; HAGER and SCHMIDT, 1968; MOYED and TULI, 1968; BASU and TULI, 1972a; DEMOREST and STAHMANN, 1972). However, in a number of publications 3-MeneOx and 3-MeOx were proved to be without growth-regulatory activity in different systems and thus the former thesis can be abandoned (ANDERSEN et al., 1972; EVANS and RAY, 1972; THIMANN, 1972; RAY, 1974; HILGENBERG et al., 1976; HAMILTON et al., 1976). In fruits, 3-MeneOx may function as a senescence promotor (FRENKEL, 1972, 1975; FRENKEL et al., 1975).

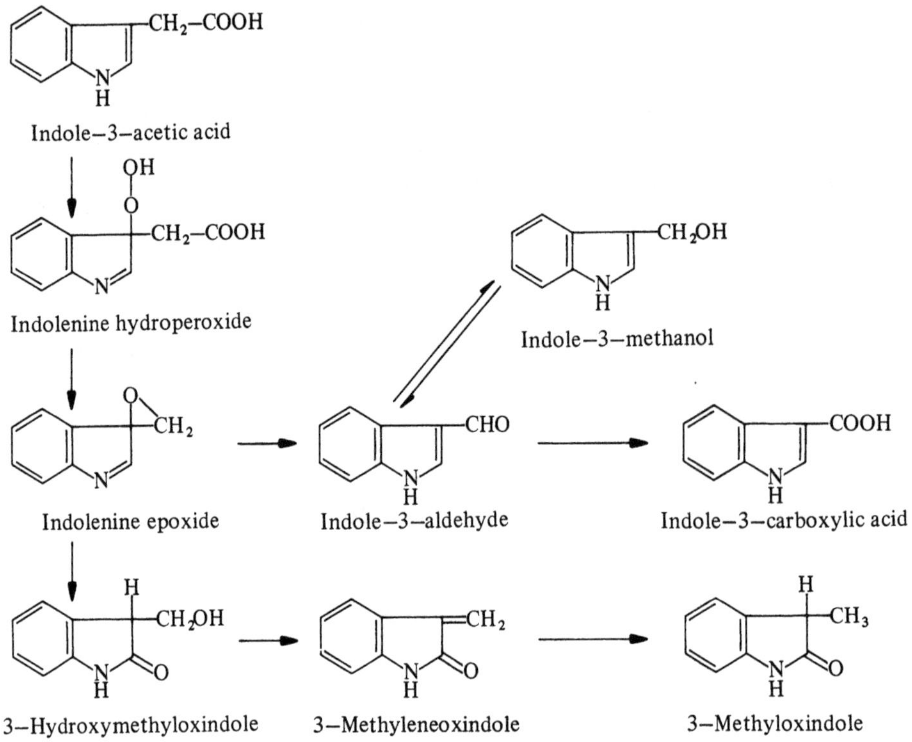

Fig. 4.22. Schematic pathway of oxidative IAA degradation

Indolealdehyde (IAld) is another product of IAA catabolism in plants, as was shown by (2-^{14}C)-IAA application to tomato and barley shoots (cf. SCHNEIDER and WIGHTMAN, 1974), pea stem sections (MAGNUS et al., 1971), pea roots (MORRIS et al., 1969), bean stems (DAVIES, 1972), and olive leaves (EPSTEIN and LAVEE, 1977b). IAld has also been demonstrated to be a native constituent of plants, and it can be produced from IAA by enzyme preparations from plant material (cf. SCHNEIDER and WIGHTMAN, 1974). IAld can be reversibly reduced to indole-3-methanol (MAGNUS et al., 1971).

Indole-3-carboxylic acid (ICA) has been detected in some plant systems, such as wheat and pea, as well as in epiphytic bacteria (SEELEY et al., 1956; CLARKE et al., 1959; WIGHTMAN, 1964; DULLAART, 1967; HOREGOTT, 1970; KUTAČEK and KEFELI, 1970; LIBBERT et al., 1970b, c, d, e; MAGNUS et al., 1971). In olive leaves ICA was shown to be the first and major metabolite of IAA (EPSTEIN and LAVEE, 1977b). ICA can be produced by oxidation of IAld (LIBBERT et al., 1970a, c; MAGNUS et al., 1971; SCHNEIDER and WIGHTMAN, 1974).

Recently some new products of peroxidase catalyzed oxidation of IAA have been identified, e.g. 3-acetoxyindole, 3-(indol-3-yl-methyl)-oxindole, and 2-(indol-3-ylmethyl)indolyl-3-acetic acid (SUZUKI and KAWARADA, 1978).

The pathways leading to these metabolites of IAA have not been elucidated in full detail, but several steps were proved by studies on the mechanism of

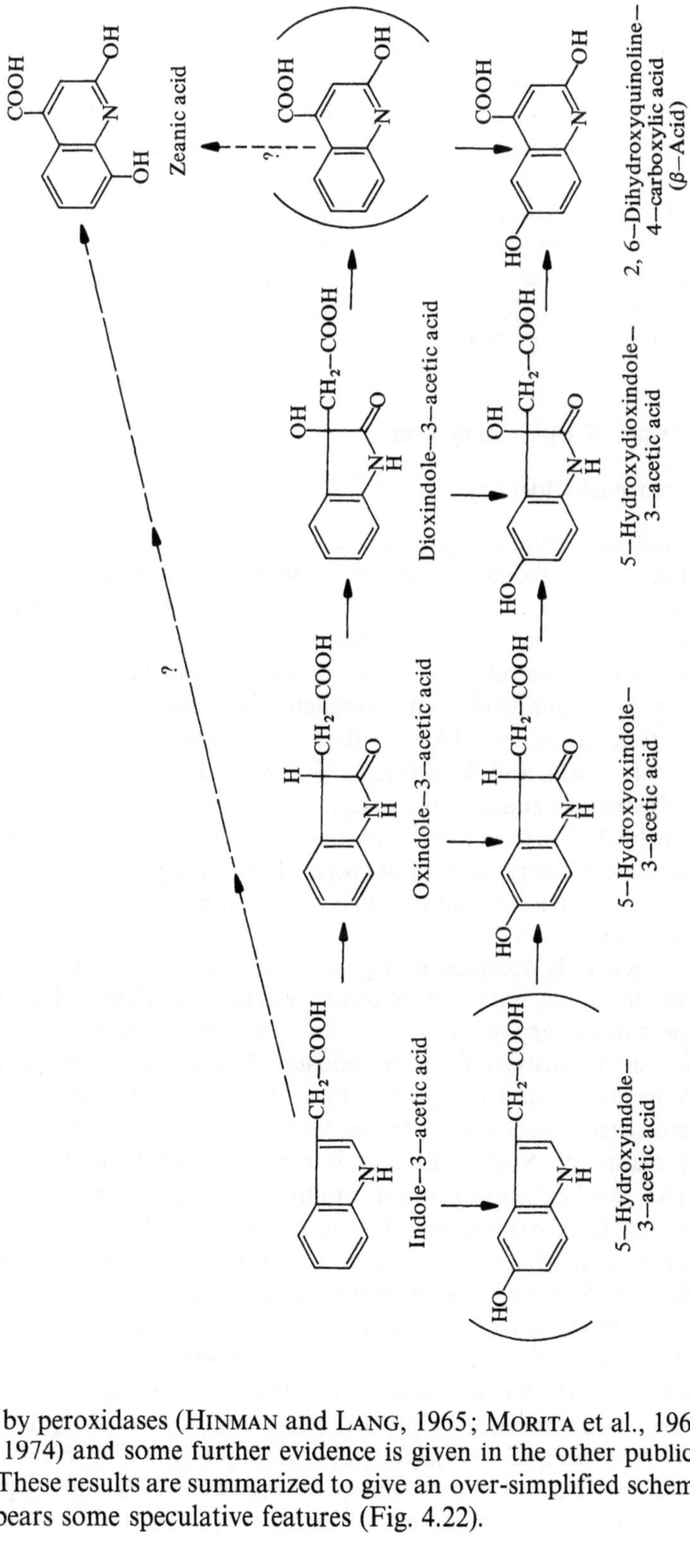

Fig. 4.23. Possible oxidation pathway of IAA to β-acid. (According to KINASHI et al., 1976)

IAA degradation by peroxidases (HINMAN and LANG, 1965; MORITA et al., 1967; RICARD and JOB, 1974) and some further evidence is given in the other publications referred to. These results are summarized to give an over-simplified scheme, which of course bears some speculative features (Fig. 4.22).

The β-acid (2,6-dihydroxy-quinoline-4-carboxylic acid) isolated from rice bran (SAHASHI, 1925, 1926, 1927; KINASHI et al., 1976) and zeanic acid isolated from corn steep liquor (MATSUSHIMA and ARIMA, 1973; MATSUSHIMA et al., 1973) may be considered to be oxidation products of IAA. Further indication of the possible existence of an IAA oxidation pathway other than that described above (Fig. 4.22) was given by the isolation of some oxindoles structurally related to IAA (KINASHI et al., 1976). The authors proposed a tentative scheme of IAA oxidation without decarboxylation leading to the β-acid (Fig. 4.23). Dioxindole-3-acetic acid, one of the oxindoles isolated from rice bran, was shown to be a major product of aerobic oxidation of IAA with bisulphite (HORNG and YANG, 1975).

b) Enzymes of IAA Catabolism

Occurrence and Identity

It is well established that plants contain enzymes capable of oxidizing IAA and that "IAA oxidases" are as universal as IAA (HARE, 1964; PILET and GASPAR, 1968; SCHNEIDER and WIGHTMAN, 1974). Information given in these reviews on the occurrence, isolation, and characterization of IAA-oxidizing enzymes cannot be repeated in full detail, but will be summarized with respect to some important aspects and completed by recent results. A number of recent publications deals with IAA oxidase from vascular plants (e.g., RICARD and JOB, 1974; EVANS and SCHMITT, 1975; NANDA et al., 1975; PUPPO and RIGAUD, 1975; HAMILTON et al., 1976; LEE, 1977; MENSCHICK et al., 1977; and literature cited below). Furthermore, IAA-degrading systems have been demonstrated in liverwort (SPAETH and MARAVOLO, 1973), fungi (HILGENBERG et al., 1976), in bacteroids (RIGAUD and PUPPO, 1975) and in bacteria isolated from root nodules (MENNES, 1973b).

It is generally accepted that peroxidase is responsible for some IAA oxidase activity, but it has not yet been clearly established whether IAA oxidase activity and peroxidase activity reside on the same molecule or whether a true IAA oxidase exists, distinct from peroxidase. Resolution of the problem, requiring rigorous purification of the enzymes, is complicated by the fact that plants contain several isoenzymes. Peroxidase isoenzymes have been demonstrated in many plants (cf. SCHNEIDER and WIGHTMAN, 1974; GOVE and HOYLE, 1975), and IAA oxidase also occurs in multiple molecular forms. So far, isoenzymes possessing IAA oxidase activity have been found in horseradish peroxidase (SEQUEIRA and MINEO, 1966), turnip, radish, and morning glory (ENDO, 1968), broad bean (SAHULKA, 1970), lupin (MENNES, 1973a, b), leaf extracts of tobacco (MEUDT, 1967), spinach (PENEL and GREPPIN, 1972), and birch (GOVE and HOYLE, 1975), extracts of wheat seedlings (MACHÁČKOVÁ and ZMRHAL, 1974), wheat coleoptiles (CHAPPET and DUBOUCHET, 1975), oat coleoptiles (GORDON and HENDERSON, 1973), morning glory (YONEDA and ENDO, 1969, 1970), mung bean (FRENKEL and HESS, 1974), pea (MACNICOL, 1966, 1973), cucumber (RETIG and RUDICH, 1972), and peppermint plants (BARZ, 1977), barley grains (MINCHIN and HARMEY, 1975), pear and blueberry (FRENKEL, 1972), tobacco callus (LEE,

1972) and cell suspension cultures (SHINISHI and NOGUCHI, 1975). The isoenzyme pattern varies from species to species and may depend on the plant tissue and its developmental stage.

Several studies have shown that electrophoretically separated proteins possess peroxidase activity but no IAA oxidase activity and the non-identity of IAA oxidases with peroxidases was concluded (ENDO, 1968; YONEDA and ENDO, 1969, 1970; SAHULKA, 1970; FRENKEL, 1972; PENEL and GREPPIN, 1972; GORDON and HENDERSON, 1973; CHAPPET and DUBOUCHET, 1975; MINCHIN and HARMEY, 1975). A number of authors, however, have reported that all peroxidase isoenzymes contained IAA oxidase activity as well (MCCUNE, 1961; MACNICOL, 1966; KAY et al., 1967; RETIG and RUDICH, 1972; MENNES, 1973a; SRIVASTAVA and VAN HUYSTEE, 1973; FRENKEL and HESS, 1974; MACHÁČKOVÁ and ZMRHAL, 1974; GOVE and HOYLE, 1975; SHINISHI and NOGUCHI, 1975). In enzyme preparations of pea roots (VAN DER MAST, 1969) and tobacco roots, and in the isoenzyme spectrum of horseradish peroxidase (SEQUEIRA and MINEO, 1966), single oxidase fractions have been found which did not show peroxidase activity. However, further studies on this subject could not confirm the presence of peroxidase-free IAA oxidase (HOYLE, 1972; GOVE and HOYLE, 1975). The present knowledge seems to be insufficient to answer conclusively the initial question whether IAA oxidase activity and peroxidase activity reside upon the same protein molecule, at least when the universality of this phenomenon in higher plants is also questioned. Interpretation and comparison of the results reported are difficult since the techniques differed considerably. Neither staining of gels nor wet assay of eluted isoenzymes are always reliable, and the presence of inhibitors, as well as special claims to co-factors and optimum reaction conditions, have to be considered (cf. GOVE and HOYLE, 1975).

It appears, then, that plants contain several peroxidase isoenzymes, some of which (or all?) are also able to oxidize IAA. The enzymes may differ in the rates of catalysis of peroxidation and IAA oxidation. Thus horseradish peroxidase, is generally high in peroxidase activity and low in oxidizing IAA, and turnip peroxidase, P_7, has both high peroxidase and high IAA oxidase activity (cf. SCHNEIDER and WIGHTMAN, 1974). Within a zymogramme, ratios of peroxidase and IAA oxidase activity were found to vary directly among isoenzymes, i.e., isoenzymes with high oxidase activity also possess high peroxidase activity and vice versa (GOVE and HOYLE, 1975). Other authors have proposed that one, or a few, of these isoenzymes may be relatively higher in oxidase than peroxidase activity (MACNICOL, 1966; KAY et al., 1967). A suggestion made by RICARD et al. (1972) implies the existence of a sequence of enzyme proteins having different ratios of acidic groups in the haem cleft, with an associated difference in activity of peroxide and oxygen activation. Another approach did not contribute much to the question of the sites of peroxidase and IAA oxidase activity. By removing the haem prosthetic group of horseradish peroxidase it was found that the apoenzyme still had IAA oxidase activity but had lost its activity towards peroxidase substrates such as guaiacol (SIEGEL and GALSTON, 1967; GALSTON et al., 1968; HOYLE, 1972). But other authors demonstrated that IAA degradation required both the haem group and the apoprotein of horseradish peroxidase (KU et al., 1970; LEE, 1977). Purified leghaemoglobin from root nodules of soybean showed IAA oxidase activity which was assumed to be related to the pseudoperoxidase activity known for this haemoprotein. The ferric form was found to be more active than ferrous leghaemoglobin (PUPPO and RIGAUD, 1975).

Mode of Action. Co-factors and Inhibitors

The oxygen-consuming IAA degradation by pure peroxidases, especially from horseradish and turnip in the absence of exogenous H_2O_2, has been studied since 1955 (KENTEN, 1955), but the mechanism of this reaction is not yet completely understood. Various reaction schemes were proposed (i.e., MAC LACHLAN and WAYGOOD, 1956; RAY, 1962; HINMAN and LANG, 1965; RICARD and NARI, 1966, 1967; FOX and PURVES, 1968; GELINAS, 1973; YAMAZAKI and YAMAZAKI, 1973; MACHÁČKOVÁ and ZMRHAL, 1974) and literature has been reviewed (RICARD and JOB, 1974; SCHNEIDER and WIGHTMAN, 1974). Recently, a thorough study of the reaction mechanisms of IAA degradation by various peroxidases employing stopped-flow and low-temperature spectroscopic techniques has been performed by RICARD and JOB (1974). The scheme developed by the authors (Fig. 4.24) gives an explanation of most of the experimental facts published. The main results may be summarized by the following general conclusions: (a) IAA is binding on peroxidase, and haem protein is reduced ($Fe_p^{3+} \rightarrow Fe_p^{2+}$, steps 1 and 2); (b) Reaction of IAA with peroxidase results in the formation of other spectroscopically distinct enzyme forms, called compound (Co) I, II, III; (c) Co III is formed by oxygenation of ferroperoxidase (steps 3–5), and Co II mostly originates from Co III by reduction (steps 9–11). Co II can also be formed by reaction of IAA with Co I (steps 16 and 18), and Co I arises from ferriperoxidase either in the presence of H_2O_2 or IAA peroxide derived from IAA free radical (steps 15 and 17); (d) Co II is a key intermediate in a cycle giving rise to 3-MeneOx as the major end-product (steps 12–14), including non-enzymatic transformation of IAA epoxide. It can also be formed by non-enzymatic oxidation of IAA free radical (IAA·) evolved at several reaction steps; (e) A second reaction cycle involving ferroperoxidase and Co III leads to IAld (steps 3–8); (f) The shift between these two cycles is induced by variations of IAA/peroxidase ratio and different experimental conditions (e.g., pH).

Recent studies on the mechanism of IAA oxidation by horseradish peroxidase yielded some further results suggesting, e.g. that the ferric enzyme is reduced by the IAA free radical and not by IAA itself (NAKAJIMA and YAMAZAKI, 1979).

The mode of action of IAA-oxidase systems other than horseradish is almost unknown. Wheat peroxidase action was proposed to be very similar to that of horseradish peroxidase (MACHÁČKOVÁ and ZMRHAL 1974, 1976). IAA decarboxylation in *Avena* coleoptiles (MENSCHICK et al., 1977) showed a time-dependency similar to IAA-oxidase activity measured in maize coleoptiles (EVANS and SCHMITT, 1975). The kinetics is consistent with a possible enzyme induction.

The activity of IAA oxidase can be altered by naturally occurring substances such as minerals, phenols, coumarins, and organic acids. Mn^{2+} has been found to be a cofactor for many IAA oxidase systems (GALSTON and HILLMAN, 1961;

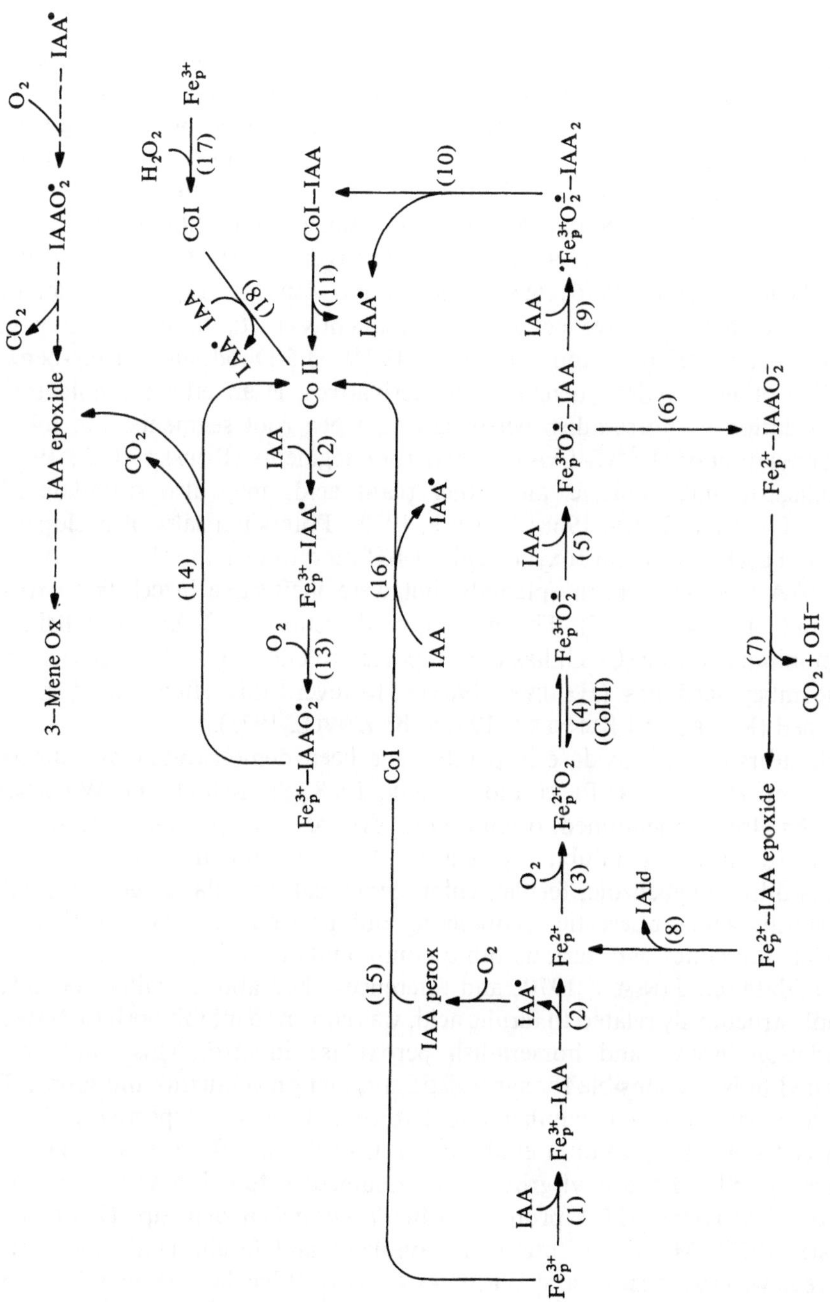

Fig. 4.24. Reaction scheme of peroxidase catalyzed IAA oxidation. (According to RICARD and JOB, 1974)

SCHNEIDER and WIGHTMAN, 1974). Manganese may act directly or via its effect on other natural cofactors and inhibitors in the system. Depending on concentration and the peculiarities of the system, Mn^{2+} can promote, have no effect upon, or inhibit the reaction (HOYLE and ROUTLEY, 1974). Sodium bisulphite at low concentrations enhanced and, at higher concentrations, inhibited IAA oxidation by enzyme preparation from liverwort (SPAETH and MARAVOLO, 1973).

Other types of effector of IAA oxidase systems are represented by some substituted phenols (GALSTON and HILLMAN, 1961). Numerous reports indicate that monophenols and m-diphenols act as cofactors or activators while o- and p-diphenols and polyphenols inhibit IAA oxidase activity or induce a lag period (GORTNER et al., 1958; PILET, 1964b; 1966; LEE and SKOOG, 1965; STONIER and YONEDA, 1967a; YONEDA and STONIER, 1967; GELINAS and POSTLETHWAIT, 1969; JANSSEN, 1970; VAN DER MAST, 1970; IMBERT and WILSON, 1972; RUNKOVA et al., 1972; GELINAS, 1973; GOVE and HOYLE, 1975; MACHÁČKOVÁ et al., 1975; HAMILTON et al., 1976; MACHÁČKOVÁ and ZMRHAL, 1976). 2,4-Dichlorophenol, the most effective cofactor, is frequently used, though it may give spurious results (HOYLE and ROUTLEY, 1974). 3,5-Diiodo-4-hydroxy-benzoic acid, like other p-hydroxybenzoic acid derivatives, is an effective cofactor of IAA oxidation by horseradish peroxidase and pea root segments, but inhibits decarboxylation of (1-^{14}C)-IAA by cress root segments (ROBERT et al., 1976a).

Malic, succinic, fumaric, and other plant acids may also stimulate IAA oxidase (LANE and KING, 1968; RUBERY, 1972). Potassium salts of cyclopentylacetic and cyclohexane carboxylic acids were found to increase the in vivo activity of IAA oxidase in bean epicotyls, but were ineffective in cell-free extracts (LOH and SEVERSON, 1975). The plant growth retardant 2-chloroethyltrimethylammonium chloride (CCC) has been reported to enhance IAA oxidase activity in cucumber seedlings (HALEVY, 1963). However, this effect could not be confirmed (KNYPL and RENNERT, 1967a, b; KNYPL, 1973).

Inhibitors of IAA oxidase in plants have been demonstrated by numerous authors (cf. HARE, 1964; PILET and GASPAR, 1968; SCHNEIDER and WIGHTMAN, 1974). As already mentioned, o- and p-dihydric phenols, polyphenols, and coumarins are known to inhibit, or retard, IAA oxidation in vitro and in vivo, e.g., 3,4-dihydroxybenzoic, ferulic, chlorogenic, caffeic, ellagic, gentisic, gallic, and vanillic acids, quercetin, scopoletin, and pelarginidin, etc. Furthermore, phenylene diamines and various other compounds have been shown to inhibit IAA oxidase (cf. JANSSEN, 1970; and references cited above). Dihydroconiferyl alcohol, structurally related to ferulic acid, was shown to inhibit both (5-^{3}H)-IAA degradation in vivo and horseradish peroxidase in vitro. This inhibition is suggested to be responsible for synergistic stimulation of auxin- and gibberellin-induced elongation of cucumber and lettuce hypocotyl, respectively, by this "cotyledon factor" (SAKURAI et al., 1975). Catechins, biflavans, and flavolans, possessing ortho-dihydroxyl groups, are assumed to function as natural inhibitors of IAA oxidase (IAA protectors) in *Prunus avium* stem tips (FEUCHT and NACHIT, 1977). Modes of action of scopoletin and ferulic acid have already been reviewed (SCHNEIDER and WIGHTMAN, 1974). It has been shown that scopoletin (SIROIS and MILLER, 1972; MILLER et al., 1975) and ferulic acid (GELINAS, 1973) are themselves oxidized by the IAA-horseradish peroxidase system. The inhibitors attack enzyme transformation, which might be a key factor for the lag phase induced (LEE and CHAPMAN, 1977), and can react preferentially with both horseradish peroxidase Co II and Co III (cf. Fig. 4.24). Quercetin known to inhibit auxin destruction by IAA-oxidase (SANO, 1971) has been shown not to be a simple inhibitor but rather a much better substrate than IAA (BARZ, 1977). To that group of phenolic inhibitors which produce only a temporary inhibition, since the inhibitors are metabolized in the reaction, belong also catechol, protocatechuic, and caffeic acid, as well as some 7-hydroxy-2,3-dihydrobenzofuran derivatives which are metabolites of the carbamate insecticide "carbofuran". In contrast, two keto compounds of this type were found to be stable in the IAA-horseradish peroxidase system, and, thus, to produce a persistent inhibition (LEE, 1976, 1977; LEE and CHAPMAN, 1977). A series of publications deals with high molecular weight "auxin protectors" from tobacco, Japanese morning glory, and several other dicotyledonous plants; these compounds induce a lag in oxidation of IAA by horseradish peroxidase, rather than affect the oxidation rate (tobacco protectors: PHIPPS, 1965, 1966; *Pharbitis* protectors: YONEDA and STONIER, 1966, 1967; STONIER and YONEDA, 1967a, b; STONIER et al., 1968a, b, 1970a, b; STONIER, 1972; NOVAK and GALSTON, 1971; STONIER and YANG, 1971, 1973). Auxin protectors are found in juvenile tissue, in wound tissue and in crown gall tissue (STONIER and YANG, 1971; STONIER, 1972), and gradients of protector activity diminishing from younger to older tissue have been reported (PHIPPS, 1966; YONEDA and STONIER, 1966, 1967; STONIER and YONEDA, 1967b). A thermostable, nondialyzable "auxin protector" from tobacco appears to be a protein-Fe-chlorogenic acid-rutin complex (PHIPPS, 1965, 1966). From Japanese morning glory auxin protectors A (MW 200,000), I (MW 8000), and II (MW 2000) have been isolated (STONIER and YONEDA, 1967b). Polymers or oligomers of phenols may be involved in the protectors (STONIER et al., 1970a), with an active site of a protector being probably an o-dihydroxyphenol (STONIER et al., 1970b; STONIER and YANG, 1971, 1973). Auxin protectors appear to be cellular poisers whose antioxidant properties inhibit not only IAA oxidation but also peroxidase-catalyzed oxidations, probably leading to lignin formation (STONIER and YONEDA, 1967a; STONIER et al., 1970a; STONIER and YANG, 1971). Furthermore, auxin protectors and low molecular weight o-dihydric phenols were shown to inhibit peroxidase-catalyzed oxidation of glutathione (STONIER and YANG, 1973). Considering that the juvenile state apparently is associated with the presence of antioxidants, such as high SH gradients, and in view of the importance of reduced glutathione and other SH compounds in cell division, the authors suggested that those substances could be part of a regulatory mechanism determining the juvenile state. This coincides with earlier results on the inhibition of peroxidase-catalyzed IAA oxidation by reduced glutathione, cysteine, NADH, and ascorbic acid (BETZ, 1963; LALORAYA et al., 1972).

c) Physiological Significance of IAA Catabolism

Considerable indirect and some direct evidence suggest that the IAA-oxidase system is involved in the control of endogenous auxin levels and, thus, participate in the regulation of physiological processes such as cell growth and differentiation, flowering, fruit ripening, senescence, abscission, apical dominance, geotropism, dwarfism, host-parasite interactions, etc. Investigations concerning the physiological significance of enzymatic IAA degradation have been reviewed in detail by GALSTON and HILLMAN (1961) and PILET and GASPAR (1968). The authors also discussed precautions and difficulties in interpretation of the results. A number of subsequent publications give some additional information but no fundamental advances. An inverse relationship between IAA-oxidase activity and endogenous IAA content has been demonstrated during abscission in bean seedlings (JAIN et al., 1969) and light-induced hook opening of pea seedlings by precise kinetic studies (DE GREEF et al., 1977). In developing cotton hairs, changes in IAA-oxidase and peroxidase activity showed that IAA catabolism was low during the elongation phase but very high in the period of secondary thickening (JASDANWALA et al., 1977). A consistent increase in IAA-oxidase isoenzymes has been demonstrated during ripening of fruits, suggesting that oxidative degradation may represent a mechanism for the inactivation of endogenous IAA in the fruit and thereby the promotion of ripening (FRENKEL, 1972, 1975; FRENKEL and HAARD, 1973). Hyperauxiny in root nodules could be correlated with low IAA-oxidase activity (MENNES, 1973b); but high IAA-oxidase activity in cabbage roots was suggested not to be essential for the regulation of the endogenous IAA concentration (RAA, 1971). Changes in IAA level in relation to pathogenesis of fungal diseases may be attributed to alteration in IAA-oxidase activity (cf. VIZAROVA, 1973). IAA degradation by epiphytic bacteria isolated from plants seems to be without physiological significance (LIBBERT and RISCH, 1969).

A number of further publications deal with correlations between endogenous IAA and physiological parameters, but give no evidence that changes in auxin level are due to enzymatic IAA degradation (i.e., DE YOE and ZAERR, 1976; JINDAL and HEMBERG, 1976; TELTSCHEROVA et al., 1976).

d) Control of IAA Catabolism in vivo

The regulatory role of phenolic substances in IAA oxidation in vivo is impressively suggested by the results reviewed earlier (Sect. 4.2.1b), though almost all workers used in vitro enzyme systems. Only a few deal with the action of natural or artificial effectors in plant tissue (e.g., SAKURAI et al., 1975; ROBERT et al., 1976a; LOH and SEVERSON, 1975). Manganese dependency of IAA-oxidase systems had also been referred to. Boron deficiency was found to increase IAA-oxidase activity in root tips of *Cucurbita pepo* (BOHNSACK and ALBERT, 1977).

Plant hormones have been shown to influence the level or activity of IAA-oxidase in vivo in a wide variety of tissues. Effects of auxins, gibberellins, and cytokinins on IAA-oxidase and peroxidase activity have already been summarized (PILET and GASPAR, 1968; SCHNEIDER and WIGHTMAN, 1974). The data demonstrate that these hormones can affect enzyme activity and isoenzyme pattern in a direct or indirect manner. Hormone effect is often concentration-dependent, and there may be interaction between hormones. These conclusions are completed by few recent results on effects of synthetic auxins (GROCHOWSKA and KARASZEWSKA, 1974; RYUGO and BREEN, 1974) and cytokinins (HRADILIK, 1976). Kinetin was shown to have profound effects on IAA-oxidase and peroxidase, differing with respect to both the hormone concentration and the isoenzyme pattern in subcellular fractions (LEE, 1974).

Ethylene promotes the activities of IAA-oxidase (HALL and MORGAN, 1964) and peroxidase in leaves (HERRERO and HALL, 1960), cotyledonary buds (HRADILIK, 1976), and storage organs (STAHMANN et al., 1966; GAHAGAN et al., 1968; IMASEKI et al., 1968; IMASEKI, 1970; BIRECKA et al., 1973). However, ethylene had little effect on the rate of metabolism of applied (2-^{14}C)-IAA in either maize or oat seedlings (MALLOCH and OSBORNE, 1976).

Abscisic acid inhibited peroxidase activity in lentil (FRIES, 1973), but not in sugar cane (GAYLER and GLASZIOU, 1969). Either inhibitory or stimulatory effects of ABA on IAA-oxidase and peroxidase activities in *Trigonella* seedlings have been observed, depending on both the organ and the hormone concentration as well as the reference basis used (MEGHA and LALORAYA, 1977). IAA oxidation, mediated by horseradish peroxidase, was enhanced by ABA at high concentration (RUNKOVA et al., 1972).

Action and interaction of hormones, natural inhibitors, and cofactors contribute to the regulation of IAA-oxidase activity. Nutritional conditions and physical factors can also control activity of this enzyme system, either directly or by influencing hormonal and other chemical effectors. Regulation mechanisms, however, are almost unknown. Only few recent publications deal with environmental stress affecting IAA-oxidase. Low temperature was shown to increase the enzyme activity in wheat (BOLDUC et al., 1970) and *Chrysanthemum* (TOMPSETT and SCHWABE, 1974). Water stress enhanced IAA-oxidase activity in etiolated pea seedlings and tomato leaves (DARBYSHIRE, 1971a, b). IAA-oxidase activity in light-grown wheat leaves, however, did not increase with water stress, but declined markedly (MILLS and TODD, 1973). Kinetic studies during phytochrome-mediated hook opening of etiolated pea seedlings demonstrated a rapid increase in peroxidase and IAA-oxidase in red-light pulse-treated hook regions compared with the dark control. The increase in peroxidase activity was much faster than that of IAA-oxidase. Concomitant with the rise in IAA-oxidase activity, a gradual decrease in endogenous IAA content was noticed during 2 h after the red light pulse (DE GREEF et al., 1977). Irradiation of intact oat seedlings with red, far-red, and blue light did not alter the IAA oxidase activity of coleoptile segments, but UV irradiation decreased the enzyme activity (RAJAGOPAL and BULARD, 1975). The kinetics and rate of (1-^{14}C)-IAA decarboxylation by oat coleoptiles did not change after being illuminated, suggesting that the IAA-degrading enzyme is not induced or activated by light (MENSCHICK et al., 1977). Electrical currents, applied directly to tomato plants, altered peroxidase activity in different tissues, e.g., increase in leaf and petiole and decrease in root (BRATTON and HENRY, 1977).

4.2.2 Cytokinins

Cytokinins undergo a number of metabolic reactions leading to either new cytokinins (interconversion), or degradation products, or conjugated forms. As compared to the auxins, gibberellins, and abscisic acid, the cytokinin metabolism shows some peculiarities resulting from the fact that cytokinins occur both in the free state and incorporated into tRNA, and that the tRNA-free cytokinins are present either as bases, nucleosides or nucleotides (see Sect. 4.3.2). Hence, metabolic changes can take place both with tRNA and free cytokinins and metabolic pathways can occur at the level of the base, riboside, or ribotide. Results published on the metabolism of natural cytokinins can be summarized as outlined in Fig. 4.25 (see also Fig. 4.31).

The natural cytokinin i^6A (for abbreviations see Fig. 4.2) in corn kernels is hydroxylated at C-4 to form ribosylzeatin (MIURA and HALL, 1973). This particular reaction was first noted in *Rhizopogon roseolus* (MIURA and MILLER, 1969). Hydroxylation of the 3-methylbut-2-enyl side chain occurred also in tobacco callus (EINSET and SKOOG, 1973) and in tobacco cell suspensions (LALOUE et al., 1977). However, it is not known if this reaction takes place at the i^6Ade, i^6A, or i^6Ade ribotide level. Transformation of i^6A into N^6-(3-hydroxy-3-methylbutylamino)-purine has been reported using marrow preparations and pea seedlings (MCLENNAN et al., 1968).

Zeatin and its riboside were shown to be converted into dihydrozeatin (H_2-io^6Ade) and its riboside (H_2-io^6A) in bean axes (SONDHEIMER and TZOU, 1971). Considering, furthermore, the natural occurrence of dihydrozeatin in lupin seeds (KOSHIMIZU et al., 1967) and in sycamore spring sap (PURSE et al., 1976), and of dihydrozeatin-O-glucoside in bean leaves (WANG et al., 1977), the following sequence can be suggested: i^6Ade → zeatin → dihydrozeatin. However, the ability to convert the 4-hydroxy-3-methylbut-2-enyl side chain of zeatin or its riboside to saturated analogues does not appear to be universal. Conversion of either zeatin or zeatin conjugates to dihydro analogues has been shown in *Populus alba* leaves (LETHAM et al., 1977) and root nodules of *Alnus glutinosa* (HENSON and WHEELER, 1977a, b, c). Radish seedlings, however, failed to convert zeatin into dihydrozeatin (PARKER and LETHAM, 1973; GORDON et al., 1974).

The isolation of hydroxylated derivatives of dihydrozeatin [6-(3,4-dihydroxy-3-methylbutylamino)-purine and 6-(2,3,4-trihydroxy-3-methylbutylamino)-purine] from immature maize kernels indicates the possible existence of some further steps in the transformation of zeatin (LETHAM, 1973). Another natural compound, isolated by LETHAM, is the 2-hydroxy-derivative of zeatin, which might be formed by oxidizing the purine ring. At present, oxidation at the purine ring system of natural cytokinins has been demonstrated in vitro only. i^6Ade and zeatin are oxidized by xanthine oxidase with the formation, first of the 8-hydroxy-, and then of the 2,8-dihydroxy-derivatives (CHHEDA and MITTELMAN, 1972; MCLENNAN and PATER, 1973; CHEN et al., 1975). These derivatives showed lower biological activity than i^6Ade and zeatin and it is still unknown whether this same reaction occurs in vivo. Ribosylzeatin and i^6A were not attacked by the enzyme system.

Formation of 2-methylthio-derivatives may be assumed to be a special metabolic pathway of natural cytokinins. Both the ms^2-i^6A and 2-methylthio-cis and trans-ribosylzeatin (ms^2-c-io^6A and ms^2-t-io^6A) have been isolated from tRNA species (HARADA et al., 1968; ARMSTRONG et al., 1969, BURROWS et al., 1970; HECHT et al., 1969; BABCOCK and MORRIS, 1970; VREMAN et al., 1972, 1974; TIMMAPPAYA and CHERAYIL, 1974; BURROWS, 1976). The ms^2-c-io^6A has been shown to occur also in the free state in *Corynebacterium fascians* (ARMSTRONG et al., 1976b). Whether the methylthio group is introduced into the

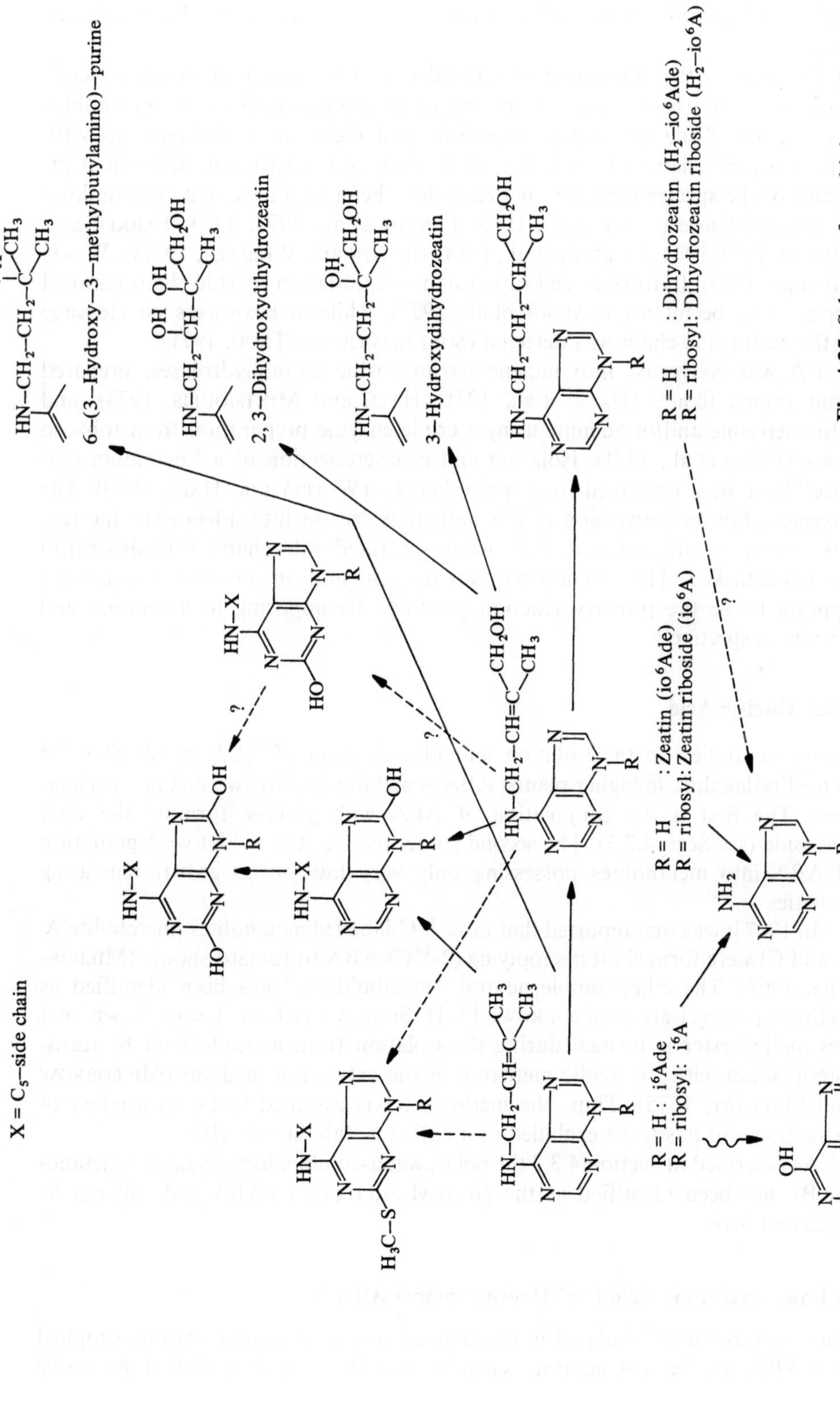

Fig. 4.25. Scheme of cytokinin metabolism

purine nucleus only at the tRNA level (see Sect. 4.1.2) cannot be decided at present.

Cleavage of the side chain of natural cytokinins has been observed both in vivo by plant tissue and in vitro by enzyme preparations. Common metabolites of zeatin and i^{6}Ade are adenine, adenosine, and adenosine 5′-phosphate (AMP). Amounts and kinetics of formation of the metabolites varied in different plant organs of the species used, i.e., maize, radish, bean seedlings, and root nodules of *Alnus glutinosa* (CHEN et al., 1968; PARKER et al., 1972, 1973; PARKER and LETHAM, 1973, 1974; GORDON et al., 1974; HENSON and WHEELER, 1977b; WAREING et al., 1977). In tobacco cell suspensions, degradation by side chain removal appeared to be intense (LALOUE et al., 1977), while in bean axes no cleavage of the zeatin side chain was detected (SONDHEIMER and TZOU, 1971).

i^{6}A was converted into inosine by adenosine aminohydrolases, prepared from animal tissues (HALL et al., 1971; HALL and MINTSIOULIS, 1973), and into adenosine and/or adenine using a crude enzyme preparation from tobacco tissue (PAČES et al., 1971). Isolation and characterization of a "cytokinin oxidase" have been described from maize kernels (WHITTY and HALL, 1974). The enzyme catalyzes conversion of i^{6}A and ribosylzeatin into adenosine; the free bases serve as substrates as well, while saturated side chains (dihydrozeatin) are not attacked. The reaction requires oxygen; and an unstable intermediate appears to be the primary reaction product, decomposing to adenosine and adenine respectively.

4.2.3 Abscisic Acid

Extensive studies on the isolation and identification of ABA metabolites led to the finding that, in higher plants, ABA is metabolized by two distinct mechanisms. The first is the conjugation of ABA with glucose forming the ester glucoside (see Sect. 4.3.3). The second route involves the oxidative degradation of ABA into metabolites possessing only very low or no growth-regulating activities.

In 1967 it was first reported that three ^{14}C-labelled metabolites (metabolite A, B, and C) were formed after supplying (2-^{14}C)-ABA to tomato shoots (MILBORROW, 1968). The ether-soluble neutral "metabolite A" has been identified as methyl (+)-abscisate (MILBORROW, 1971). Some years later it was shown that this methyl ester is formed during the isolation from avocado fruit by transesterification with the acidic methanol of the extraction medium (MILBORROW and MALLABY, 1975). Thus, the methyl ester is assumed to be an artefact of extraction and has to be excluded as a native metabolite of ABA.

As described in Section 4.3.3 the polar, water-soluble, hydrolyzable "metabolite B" has been identified as the glucosyl ester of (+)-ABA and will not be discussed here.

a) Hydroxyabscisic Acid ("6′-Hydroxymethyl ABA")

The "metabolite C" isolated in crystalline form from tomato shoots, supplied with ABA, has been structurally elucidated as the 6′-hydroxymethyl derivative

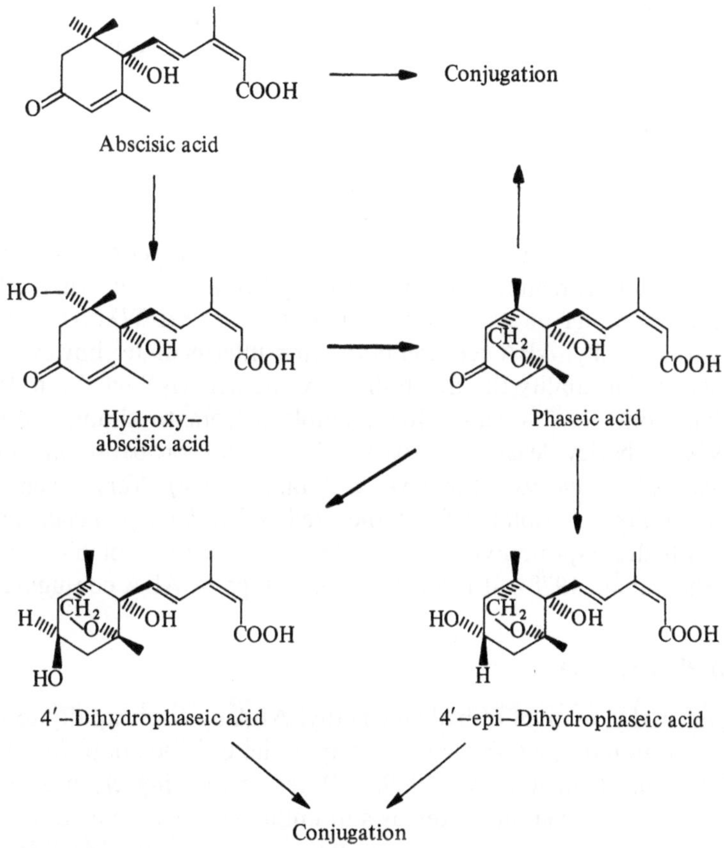

Fig. 4.26. Metabolism of abscisic acid

of (+)-ABA (see Fig. 4.26) (MILBORROW, 1969a, 1970). It is interesting that this 6′-hydroxymethyl ABA is derived biosynthetically only from the (+)-enantiomer of ABA (naturally occurring), as shown by application of (+)- and (−)-ABA to tomato shoots (MILBORROW, 1970). As indicated in Fig. 4.26, only the pro-6′(S)-methyl group of ABA, derived from C-3′ of mevalonate (see Sect. 4.1.4a) is hydroxylated (MILBORROW, 1975a). This result was obtained by experiments in which ABA, labelled in the 6′-methyl group, was supplied to the plant material. In isolated bean axes (+)-ABA, exogenously supplied, was metabolized much faster than (−)-ABA, supporting the conclusion that reversal in chirality is more important to metabolic conversion than to the biological activity of the molecule (SONDHEIMER et al., 1971).

The particulate enzyme catalyzing this ABA-hydroxylation has been isolated from cell-free extracts of *Echinocystis lobata* liquid endosperm; it has a requirement for O_2 and NADPH and possesses a high substrate specificity for (+)-ABA (GILLARD and WALTON, 1976). This is in agreement with previous feeding experiments using (2-^{14}C)-labelled ABA and some derivatives. It has been shown that trans, trans-ABA, methyl and ethyl abscisate, and the cis- and trans-1′,4′-diols of ABA were not metabolized to their respective analogues and are less active in growth inhibition than ABA (WALTON and SONDHEIMER, 1972b). By

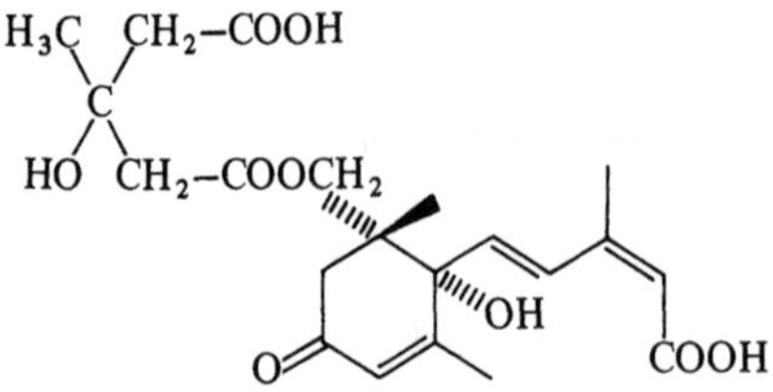

Structure 4.5. β-Hydroxy-β-methyl-glutaryl ester of 6′-hydroxymethyl ABA

acetylation of short term incubation mixtures, acetylated 6′-hydroxymethyl ABA was obtained, demonstrating that 6′-hydroxylation of ABA takes place at the very beginning of its metabolism (GILLARD and WALTON, 1976). Hence, 6′-hydroxymethyl ABA represents the first intermediate; however, due to its high chemical instability the free 6′-hydroxymethyl ABA has been detected in higher plants only in few cases, for example in tomato shoots (MILBORROW, 1968, 1969a), barley leaves (CUMMINS, 1973) and probably in root segments of *Phaseolus coccineus* (HARTUNG and BEHL, 1974). Very recently a novel ABA metabolite was isolated from the seeds of *Robinia pseudoacacia* and identified as β-hydroxy-β-methyl-glutaryl ester (Structure 4.5) of 6′-hydroxymethyl ABA (HIRAI et al., 1978). It represents a new type of ABA conjugate.

b) Phaseic Acid

The methyl ester of 6′-hydroxymethyl ABA cyclises rapidly to methyl phaseate, by addition of the 6′-hydroxyl group to the C-2′ position, forming a tetrahydrofuran ring (MILBORROW, 1969a). Whether the ring closure in the plant tissue is enzymatic or spontaneous is still unknown. Phaseic acid had been previously isolated from immature seeds of *Phaseolus coccineus* (MACMILLAN et al., 1960; MACMILLAN and PRYCE, 1968). The chemical structure originally proposed (MACMILLAN and PRYCE, 1968, 1969a, b) was revised (MILBORROW, 1969a) and, additionally, the absolute configuration has been confirmed by comparison with the NMR and IR spectra of methyl phaseate and the two epimers of methyl 4′-dihydrophaseate (MILBORROW, 1975c). Thus, the structure of phaseic acid is as shown in Fig. 4.26.

Phaseic acid possesses very low biological activity in different bioassay systems based on growth inhibition, abscission, or stomata closure. The compound was shown to be identical with the "metabolite M-1" rapidly formed in excised bean axes after incubation with (2-^{14}C)-ABA (WALTON and SONDHEIMER, 1972a; TINELLI et al., 1973; HARRISON and WALTON, 1975). In a number of further investigations dealing with the metabolism of exogenously applied ^{14}C-ABA, phaseic acid and its 4′-dihydro derivative as well, have been detected by means of chromatographic methods or combined gas chromatography-mass spectrometry (e.g. DATHE et al., 1978a; DEWDNEY and MCWHA, 1978; DURLEY et al., 1971; GASKIN et al., 1973; GILLARD and WALTON, 1976; HARRISON and WALTON, 1975; KRIEDEMANN et al., 1975; LOVEYS, 1977; LOVEYS et al., 1975; MARTIN et al., 1977a, b; SONDHEIMER et al., 1974; TINELLI et al., 1973; WALTON and SONDHEIMER, 1972a; WALTON et al., 1973, 1976; ZEEVAART, 1977; ZEEVAART and MILBORROW, 1976). Recent literature on detection of phaseic acid and dihydrophaseic acid in plants is rapidly increasing (see e.g. ADESOMOJU et al., 1980; KING, 1979;

MILBORROW and VAUGHAM, 1969; PHILLIPS and HOFMANN, 1979; TIETZ et al., 1979). In some further studies phaseic and 4′-dihydrophaseic acid have been tentatively identified as metabolites of ABA (e.g. DÖRFFLING et al., 1974; GROSS and SCHÜTTE, 1974, 1977). From a few experiments it can be concluded that phaseic acid appears to exert a direct effect on the photosynthetic apparatus of the leaf and to act as a specific endogenous inhibitor of photosynthesis (LOVEYS and KRIEDEMANN, 1974; KRIEDEMANN et al., 1975; but see SHARKEY and RASCHKE, 1980, cited in Chap. 1).

c) 4′-Dihydrophaseic Acid and Its 4′-Epimer

The "metabolite M-2" rapidly formed in bean axes from supplied (2-^{14}C)-ABA via phaseic acid and accumulated as a major ABA metabolite (WALTON and SONDHEIMER, 1972a) was identified as 4′-dihydrophaseic acid by means of chemical and spectroscopical methods (TINELLI et al., 1973). From NMR and IR data it was concluded that the 4′-dihydrophaseic acid isolated from beans (TINELLI et al., 1973) has the 4′-hydroxyl group linked to the cyclohexane ring opposite to the side of the oxymethylene bridge whereas the oxymethylene bridge and the 1′-hydroxyl group are on the same side of the cyclohexane ring (MILBORROW, 1975c). In various metabolic studies, using added ^{14}C-ABA, the 4′-dihydrophaseic acid has been detected in different plant material (Table 4.5). It has been reported that the concentration of native dihydrophaseic acid in mature beans is 100 times that of ABA (WALTON et al., 1973). Within 4 h, more than 40% of the radioactivity of ^{14}C-ABA taken up by bean roots was metabolized, and after 11 h only about 18% of the ABA remained unmetabolized, with more than 50% converted to dihydrophaseic acid (WALTON et al., 1976).

A soluble enzyme preparation, reducing phaseic acid to 4′-dihydrophaseic acid was obtained from cell-free extracts of *Echinocystis lobata* liquid endosperm; it requires a soluble co-factor in addition to NADP (GILLARD and WALTON, 1976). Using either the crude enzyme extract or both the particulate hydroxylating enzyme mentioned above and the phaseic acid-reducing system, the sequence indicated in Fig. 4.26 was verified. This is in full agreement with feeding experiments on plant tissues already mentioned.

Besides the major metabolites, phaseic acid and its 4′-dihydro derivative, the 4′-epimer of 4′-dihydrophaseic acid, designated as epi-4′-dihydrophaseic acid, was isolated from bean seedlings incubated with (2-^{14}C)- or (4′-^{18}O)-ABA (ZEEVAART and MILBORROW, 1976). It was shown to be a minor metabolite of ABA; its concentration, including the native conjugated form, amounted to only 1.2% of that of 4′-dihydrophaseic acid. Both epimers are formed separately from phaseic acid by enzymatic reduction of the 4′-keto group without loss of the 4′-oxygen.

Moreover, it was reported that, in dormant ash seed and excised bean leaves, a further compound, more polar than dihydrophaseic acid, is formed following application of (2-^{14}C)-ABA. It is derived apparently from dihydrophaseic acid (SONDHEIMER et al., 1974; HARRISON and WALTON, 1975). Both immature and mature seeds of *Pyrus communis* contain two compounds (methyl TMSi derivatives M^+ 382 and 456, respectively), probably being derivatives of phaseic acid (hydroxyphaseic acid or keto-4′-dihydrophaseic acid and a hydroxydihydrophaseic acid) (MARTIN et al., 1977a, b). Additional substances detected may be epimers or isomers differing in the position of hydroxyl groups.

In the methanolic extract of lettuce seeds, supplied with (2-^{14}C)-ABA, two metabolites have been found; one of them was chromatographically resolved into two components, possibly di- and trihydroxy derivatives of ABA (ROBERTSON and BERRIE, 1977). It is not known if one of these compounds is identical with an unidentified metabolite previously found in lettuce fruits after inhibition with ^{14}C-ABA (McWHA and HILLMAN, 1973). After feeding (2-^{14}C)-ABA to wheat plants phaseic acid, dihydrophaseic acid, glucosyl abscisate, and 6 unidentified acetone-soluble metabolites have been found (DEWDNEY and McWHA, 1978). All these results show that further metabolites differing from phaseic acid and its dihydro derivatives are formed during the metabolism of ABA. In apple seeds decarboxylation of (1-^{14}C)-ABA has been observed (RUDNICKI and CZAPSKI, 1974) and thus, the decarboxylation of ABA is postulated as a mode of degradation but, so far, no metabolites of this pathway have been found.

All these investigations demonstrate that ABA, applied to plant tissue, is normally metabolized rapidly. However, a few exceptions have been published: (2-^{14}C)-ABA applied to cotton seedlings (SHINDY et al., 1973), and to cucumber plants (FRIEDLÄNDER et al., 1976), was shown not to be transformed metabolically though uptake and translocation took place.

The biologically inactive 2-trans-isomer of ABA has been isolated from several plants, and it was conclusively shown to occur naturally. Nevertheless, the formation of 2-trans-ABA is assumed to be a non-metabolic change probably catalyzed by light. Furthermore, it is known that isomerization takes place in solvents during isolation procedure leading to an equilibrium mixture of ABA and 2-trans-ABA (MILBORROW, 1974a, b). Some recent publications (e.g., FUKUI et al., 1977; NAUMANN and SACHS, 1977; SHAYBANY et al., 1977; DATHE et al., 1978a) confirm the natural existence of the trans-isomer and give some further evidence that it may be a product of environmental stress factors such as irradiation, chilling, storage etc.

4.2.4 Gibberellins

a) Interconversion of C_{19}-Gibberellins in the Fungus

The main gibberellin, produced in cultures of *Gibberella fujikuroi*, is GA_3. Studying the rate at which labelled precursors were incorporated into fungal gibberellins an early increase in the GA_4/GA_7 fraction has been observed, followed by a decrease with concomitant accumulation of the radioactivity in the GA_1/GA_3 fraction (GEISSMAN et al., 1966; VERBISCAR et al., 1967). The intermediacy of GA_4 and GA_7 on the way to GA_3 was established by re-feeding experiments (VERBISCAR et al., 1967; PITEL et al., 1971; BEARDER et al., 1973a, 1975b).

This conversion includes dehydrogenation of GA_4 to give GA_7 then 13-hydroxylation to GA_3 (PITEL et al., 1971). The alternative pathway leading through GA_1 to GA_3 is a minor one (BEARDER et al., 1973c) and may be due to the non-specificity of the dehydrogenating enzyme (PITEL et al., 1971). Further metabolism (see Fig. 4.17) of GA_4 by direct hydroxylation leads to GA_1, GA_{16}, and GA_{47} (PITEL et al., 1971; MCINNES et al., 1973, 1977; BEARDER et al., 1973a, 1975b). In addition, GA_2 was found (MCINNES et al., 1977). Analogously GA_9 is transformed into GA_{10} (CROSS et al., 1964, 1968b). Using the mutant B1-41a, GA_9 is metabolized to GA_{20} and GA_{40} as main metabolites along with GA_{10}, GA_{11}, 1,2-dehydro-GA_9 and its lactone isomer, 16,17-dihydroxydihydro-GA_9, 3-epi-GA_4 and probably 12α-hydroxy-GA_9 (BEARDER et al., 1976b). GA_9 has never been incorporated into GA_3. Gibberellins A_1, A_3 and A_{16} seem to be "dead ends" of the metabolism that showed no further conversions in cultures of the fungus (PITEL et al., 1971; BEARDER et al., 1975b).

b) Metabolism of Gibberellins in Higher Plants

As in the fungus the further metabolism of gibberellins in higher plants is characterized by oxidative attack of the ent-gibberellane skeleton. The most favoured positions are C-2, C-3, and C-13. At present more than 40 different gibberellins have been isolated from plant sources and, hence, it seems unlikely that a single pathway is operating in all cases. The most common reactions are direct hydroxylations. A summary of the results of GA metabolism obtained by in vivo and in vitro experiments is given in Table 4.5, arranged according to the hydroxylation pattern.

Among the conversions listed, there are only a few occurring at the C_{20} level. Using cell-free systems from the endosperm of immature *Cucurbita* seeds several 3β-hydroxylated gibberellins have been obtained by incubation with GA_{12}-aldehyde, indicating that 3β-hydroxylation precedes the loss of C-20 (see Sect. 4.1.5e). In contrast, enzyme preparations present in the low-speed supernatant of pea seeds homogenates converted GA_{12}-aldehyde into 13-hydroxy GA_{12} (GA_{53}) and GA_{44} (ROPERS et al., 1978). This evidence led to the suggestion that 13-hydroxylation is an early metabolic step at least in Leguminosae. Further support for 13-hydroxylation at the level of C_{20}-gibberellins came from the conversion of GA_{14} into GA_{18} in pea shoots (DURLEY et al., 1974a). These findings are in agreement with the natural occurrence of 13-hydroxylated gibberellins in Leguminosae, whereas in Cucurbitaceae they seem to be absent (see TAKAHASHI, 1974; BEARDER and SPONSEL, 1977; SPONSEL et al., 1979).

Also, 2β-hydroxylation takes place with C_{20}-gibberellins, as demonstrated by the conversion of GA_{13} to GA_{43} by means of cell-free enzyme preparations from the endosperm of immature *Cucurbita* seeds (GRAEBE and HEDDEN, 1974; GRAEBE et al., 1974a). The reaction was carried out with the high speed supernatant, supplemented by iron ions (Fe^{2+} or Fe^{3+}) and a pyridine nucleotide (NADPH). In *Gibberella fujikuroi*, GA_{13} seems to be a "dead end" of gibberellin biosynthesis, whereas in *Cucurbita* it can be oxidized to GA_{43}, thus demonstrating differences between the metabolic pathways in the fungus and higher plants. Since GA_{43} is known to be an endogenous gibberellin of *Cucurbita* it may be concluded that this pathway is operating both in vitro and in vivo.

Table 4.5. Formation of hydroxyl groups during metabolism of gibberellins in higher plants

Hydroxyl group formed at	Plant species	Material	Reference
C−2β			
$GA_1 \rightarrow GA_8$	*Zea mays*	Shoot tips, apex, leaf disks, seedlings	RAPPAPORT et al. (1974), DAVIES and RAPPAPORT (1975a, b)
	Hordeum vulgare	Aleurone layers	NADEAU et al. (1972)
	Oryza sativa	Shoots	RAILTON et al. (1973)
	Phaseolus vulgaris	Germinating seeds, cell-free system	NADEAU and RAPPAPORT (1972), YAMANE et al. (1975, 1977), PATTERSON and RAPPAPORT (1974, 1975)
	Phaseolus coccineus	Seedlings	REEVE et al. (1975)
	Pisum sativum	Shoots	DURLEY et al. (1974a, b), RAPPAPORT et al. (1974)
	Nicotiana tabacum	Callus cultures	RAPPAPORT et al. (1974)
	Olea europaea	Leaves, callus cultures	RAPPAPORT et al. (1974)
	Cucumis sativus	Seedlings	RUDICH et al. (1976)
	Lactuca sativa	Hypocotyls	KUHN-SILK et al. (1976)
$GA_4 \rightarrow GA_{34}$	*Oryza sativa*	Seedlings	DURLEY and PHARIS (1973)
	Pinus attenuata	Pollen	KAMIENSKA et al. (1973, 1976)
	Pseudotsuga menziesii	Seedlings	WAMPLE et al. (1975)
$GA_9 \rightarrow GA_{51}$	*Pisum sativum*	Ripening seeds in vitro	FRYDMAN and MACMILLAN (1975), SPONSEL and MACMILLAN (1977)
$GA_{13} \rightarrow GA_{43}$	*Cucurbita maxima*	Cell-free system	GRAEBE and HEDDEN (1974), GRAEBE et al. (1974a)
$GA_{20} \rightarrow GA_{29}$	*Pisum sativum*	Seedlings, shoots, chloroplasts, ripening seeds in vitro	RAILTON et al. (1974a, b), RAILTON and REID (1974), FRYDMAN and MACMILLAN (1975), SPONSEL and MACMILLAN (1977, 1978), TAYLOR and RAILTON (1977), DURLEY et al. (1979)
	Phaseolus vulgaris	Immature seeds	YAMANE et al. (1977)
	Pharbitis nil	Seedlings	BARENDSE et al. (1977)
	Bryophyllum daigremontianum		DURLEY et al. (1975)

Table 4.5. (Continued)

Hydroxyl group formed at	Plant species	Material	Reference
C−3β			
$GA_5 \rightarrow GA_3$	*Pisum sativum*	Seedlings	DURLEY et al. (1973)
GA_{12} Aldehyde $\rightarrow\rightarrow GA_{37}$	*Cucurbita maxima*	Cell-free system	GRAEBE and HEDDEN (1974), GRAEBE et al. (1974a, b)
$GA_9 \rightarrow GA_4$	*Cucurbita maxima*	Cell-free system	GRAEBE and HEDDEN (1974)
$GA_{20} \rightarrow GA_1$	*Nicotiana tabacum*	Callus cultures	LANCE et al. (1976)
	Phaseolus vulgaris	Germinating seeds	YAMANE et al. (1975, 1977)
C−3α			
$GA_{20} \rightarrow$ 3-epi-GA_1	*Bryophyllum daigremontianum*		DURLEY et al. (1975)
C−12α			
$GA_9 \rightarrow$ 2,3-Dihydro GA_{31}	*Pisum sativum*	Shoots, ripening seeds in vitro	RAILTON (1974a), RAILTON et al. (1974a), FRYDMAN and MACMILLAN (1975), SPONSEL and MACMILLAN (1977)
C−13α			
$GA_4 \rightarrow GA_1$	*Oryza sativa*		DURLEY and PHARIS (1973)
	Phaseolus vulgaris	Immature Seeds	YAMANE et al. (1975, 1977)
	Lactuca sativa	Seedlings	DURLEY et al. (1976)
	Pinus attenuata	Pollen	KAMIENSKA et al. (1973, 1976)
$GA_9 \rightarrow GA_{20}$	*Pisum sativum*	Seedlings, shoots, ripening seeds in vitro	RAILTON (1974a, b), RAILTON et al. (1974a), FRYDMAN and MACMILLAN (1975), SPONSEL and MACMILLAN (1977)
	Lactuca sativa	Hypocotyls	NASH et al. (1978)
GA_{12} Aldehyde $\rightarrow\rightarrow GA_{44}$	*Pisum sativum*	Cell-free system	ROPERS et al. (1978)
$GA_{14} \rightarrow GA_{18}$	*Pisum sativum*	Etiolated shoots	DURLEY et al. (1974a)
C−16α			
$GA_9 \rightarrow GA_{10}$	*Pisum sativum*	Shoots	RAILTON et al. (1974a)
$GA_4 \rightarrow GA_2$	*Pinus attenuata*	Pollen	KAMIENSKA et al. (1973)
	Pseudotsuga menziesii	Seedlings	WAMPLE et al. (1975)

From Table 4.5 it may be concluded that hydroxylation in the 2β-position is a common feature of gibberellin metabolism in higher plants. Since 2β-hydroxylated gibberellins are biologically inactive, as compared with non-hydroxylated ones, this metabolic step might play a role in regulating gibberellin-mediated processes (PATTERSON and RAPPAPORT, 1974). The apparent differences in some metabolic pathways elucidated for the fungus, e.g., formation of biologically active 2α-hydroxy gibberellins, may be related to the fact that GA's are without physiological significance in *Gibberella*. For example, GA_1 is shown to be metabolized to GA_8 in at least nine species (see Table 4.5). In a similar way 2β-hydroxylation takes place with GA_4, GA_9, and GA_{20}, giving GA_{34}, GA_{51}, and GA_{29}, respectively (Table 4.5). This hydroxylating system has been best investigated for the conversion of GA_1 into GA_8. Several characteristics of this transformation indicate that a direct hydroxylation is operating with the exclusion of GA_3 as an intermediate. The requirement for a reduced cosubstrate, either NADPH or ascorbate, and the release of ^3H from (1,2-^3H)-GA_1 during GA_8 formation in cell-free systems of *Phaseolus vulgaris*, coincides with the mixed-function oxidase mechanism (PATTERSON and RAPPAPORT, 1974).

$$(1,2\text{-}^3H)\text{-}GA_1 + NADPH + H^+ + O_2 \xrightarrow{Fe^{2+}} (1\text{-}^3H)\text{-}GA_8 + NADP + {}^3HOH$$

This enzyme has been detected in the 95,000 *g* supernatant derived from homogenates of imbibed seeds of *Phaseolus vulgaris*; highest activities were obtained from cotyledons. So far as is known, its substrate specificity is restricted to GA_1; oxygen and Fe^{2+} are required, and one function of ascorbate is to maintain Fe in the reduced form. Carbon monoxide is known to compete reversibly with O_2 for binding to P-450. Inhibition of hydroxylation by CO, observed in the system of PATTERSON and RAPPAPORT (1975), was due to reducing O_2 levels, and, participation of cytochrome in this reaction was ruled out. It seems likely that the mechanism elucidated for 2β-hydroxylation of GA_1 may also operate in the other hydroxylation reactions. Lettuce hypocotyls rapidly convert GA_9, probably to GA_{20} by 13-hydroxylation (NASH et al., 1978). Simultaneous application of GA_9 and 2,2′-dipyridyl inhibited this reaction; but the addition of more $FeSO_4$ overcame the inhibiting action of the chelator.

Besides the enzymic investigations using the *Cucurbita* system, extensive studies on the metabolism of gibberellins were undertaken mainly with legumes (YAMANE et al., 1975, 1977; FRYDMAN and MACMILLAN, 1975; SPONSEL and MACMILLAN, 1977). These experiments demonstrated the conversion of the applied labelled gibberellins into more polar ones during plant development. Present knowledge on the main pathways of gibberellin metabolism is outlined in Fig. 4.27. It should be noted that both the rate, and the route of gibberellin interconversions may differ in the different plant organs as well as in relation to the developmental stage (see Sect. 4.4.4). Furthermore, artificial metabolites may occur, as could be demonstrated by the conversion of GA_9 into dihydro-GA_{31} and GA_{20} in pea seeds (FRYDMAN and MACMILLAN, 1975).

Some important features of gibberellin interconversion remain to be resolved, e.g., the nature of the direct C_{20}-precursor for plant gibberellins, the possible existence of independent ways to a single gibberellin (e.g., GA_1), and the forma-

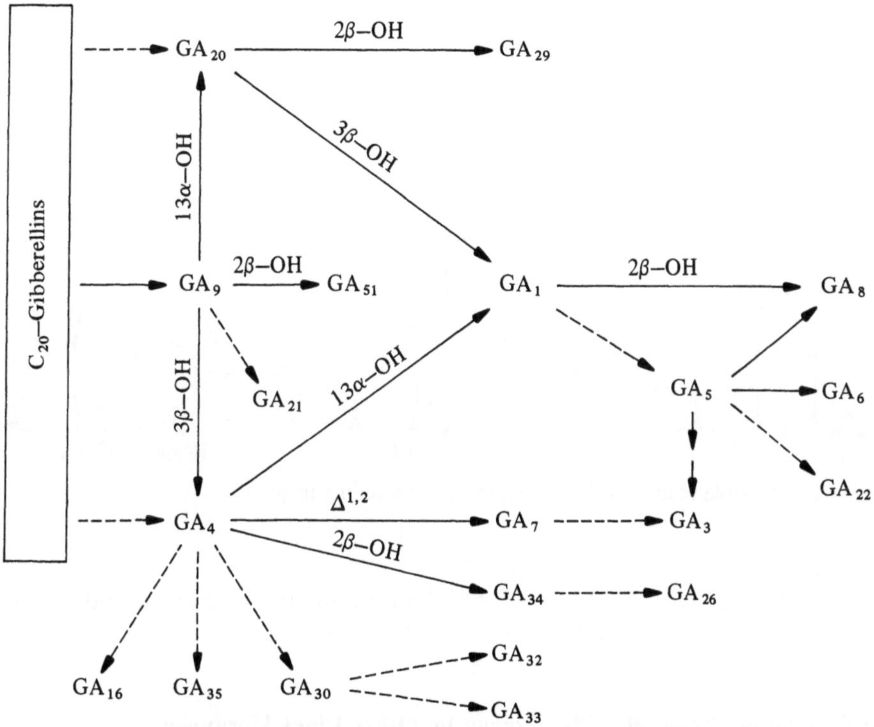

Fig. 4.27. Interconversion of C_{19}-gibberellins in higher plants. (→ Routes experimentally confirmed; --→ Proposed pathways)

tion of GA_3 which has never been obtained in labelled form following application of radioactive GA_4.

c) Catabolism of Gibberellins

In most cases metabolic experiments with labelled gibberellins led to the formation of unknown radioactive compounds, which, at least in part, may represent degradative reactions. Thus apart from conjugation, catabolic routes seem to occur as a means of reducing the biological activity of the gibberellins. Investigations on this topic have been directed mainly to *Pisum sativum*. Unlike other legumes conjugation in peas seems to play no predominant role in the deactivation of these hormones (FRYDMAN and MACMILLAN, 1975; SPONSEL and MACMILLAN, 1977). On feeding biosynthetically labelled GA_{29} RAILTON (1976) found unknown radioactive metabolites to a low extent. Exogenously applied GA_{29}, however, did not behave like the native hormone (SPONSEL and MACMILLAN, 1978) which disappears on ageing of pea seed. A more sophisticated method had to be elaborated in order to eliminate the errors caused by the apparent metabolic non-equivalence of endogenous and exogenously applied gibberellins. In comparing the fate of the natural GA_{29}, the GA_{29} formed from labelled GA_{20} and exogenously fed labelled GA_{29} a tentative route of gibberellin catabolism in peas (Fig. 4.28) was postulated.

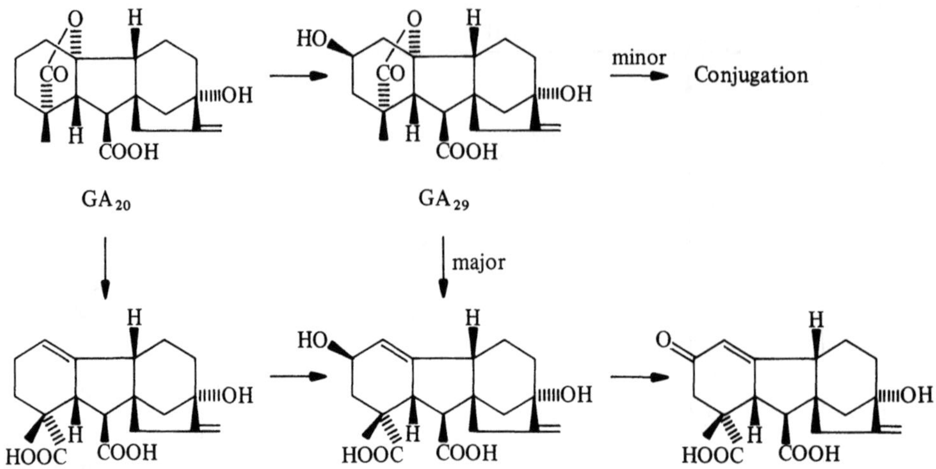

Fig. 4.28. Possible route for deactivation of gibberellins in peas

This scheme is supported by the isolation and structure elucidation of intermediates (SPONSEL and MACMILLAN, 1978; DURLEY et al., 1979).

d) Control of Gibberellin Metabolism by Other Plant Hormones and by Environmental Factors

Abscisic acid counteracts some gibberellin-promoted physiological responses and thus it has been thought to influence biosynthesis and metabolism of gibberellins. In the fungus, however, gibberellin biosynthesis was not inhibited following the addition of ABA (SMITH and SADRI, 1970; SEMBDNER et al., 1972).

With regard to the metabolism of single gibberellins the results of NADEAU et al. (1972) demonstrating that ABA enhanced uptake of ^{3}H-GA_1 into barley aleurone layers and increased its metabolism were confirmed by STOLP et al. (1973, 1977). Specific and reversible non-covalent binding of GA_1 in subcellular fractions of wheat endosperm, enriched with aleurone grains, is inhibited by ABA (JELSEMA et al., 1977). It has been postulated that the enhanced metabolism of GA_1 in the presence of ABA is due to the lower amount of bound gibberellin. In contrast, although ABA inhibited germination of lettuce seeds, it did not alter uptake of GA_4 and its conversion to GA_1 (DURLEY et al., 1976). In etiolated seedlings of dwarf pea, either wilting or application of ABA, led to a diminished rate of ^{3}H-GA_{20} metabolism to GA_{29} (TAYLOR and RAILTON, 1977).

Exogenously supplied cytokinins may exert some control on the gibberellin metabolism in tomato shoots (REID and RAILTON, 1974). Application of benzyladenine increased the metabolism of GA_9 and GA_{20} in seedlings of *Pisum sativum* (RAILTON, 1974b) and, likewise, the conversion of GA_4 in lettuce seedlings (DURLEY et al., 1976).

Ethylene-induced senescence in *Citrus* peel resulted in an increased formation of water-soluble gibberellins from exogenously applied ^{3}H-GA_3 (GOLDSCHMIDT and GALILY, 1974). This is in accordance with experiments using cucumber, in which the ethylene-releasing chemical, ethephon, increased ABA-like inhibitors parallel to a decrease in gibberellins (RUDICH et al., 1972). Animal hormones have also been tested, and KOPCEWICZ (1969) interpreted an increase of gibberellin-like content, obtained in oestrone-treated dwarf pea plants, in terms of enhanced biosynthesis.

Influence of light on biosynthesis and metabolism of gibberellins was proposed by BRIAN (1958). Later, light was shown to affect the levels of gibberellins and gibberellin-like substances in various plants either in consequence of illumination – especially red light treatment – or in response to photoperiod (reviewed by JONES, 1973).

Correlations between gibberellin content and photoperiod have been demonstrated for a number of plants, e.g., *Salix viminalis* (HOAD and BOWEN, 1968), *Trifolium silvestris* (STODDART and LANG, 1968), *Silene armeria* (CLELAND and ZEEVAART, 1970), *Spinacia oleracea* (ZEEVAART, 1971a), tobacco species (GRIGORIEVA et al., 1971) and *Bryophyllum daigremontianum* (DURLEY et al., 1975). In the photoperiod-sensitive G-2 line of pea plants GA_9 metabolism was inhibited by long-day conditions and was restored under short-day. In contrast, in the photoperiod-insensitive I-line, GA_9 metabolism was independent of the photoperiod, and the metabolites differed qualitatively from those in G-2 line (PROEBSTING et al., 1978).

In *Silene* plants labelled GA_5 was more rapidly metabolized under long-day than short-day conditions (VAN DEN ENDE and ZEEVAART, 1971). Increased metabolism of applied GA_4 has been observed in light-grown *Phaseolus coccineus* as compared to dark-grown plants (BOWN et al., 1975). The rate of GA_3 conversion to conjugates was also shown to be consistently higher in light- than in dark-grown *Pharbitis* seedlings. In contrast, metabolism of GA_{20} was enhanced in darkness and under short-day conditions in *Pisum sativum, Bryophyllum daigremontianum,* and *Nicotiana* cell cultures (RAILTON, 1974a, b; DURLEY et al., 1975; LANCE et al., 1976). As pea chloroplasts have been found to metabolize GA_9 (RAILTON and REID, 1974) they may thus represent a site of light-controlled interconversions of gibberellins.

Brief exposure of barley leaves to red light resulted in a rapid increase of the endogenous gibberellin levels followed by a decline (REID et al., 1968). A comparable increase of the amounts of gibberellins on red light treatment has been further reported for peas (KÖHLER, 1965, 1966a, 1970; NESKOVIC and KONJEVIC, 1974), *Pinus silvestris* (KOPCEWICZ and PORAZINSKI, 1973), apple embryos (SMOLENSKA and LEWAK, 1971), wheat leaves (BEEVERS et al., 1970; LOVEYS and WAREING, 1971; COOKE et al., 1975), homogenates of etiolated barley leaves (REID et al., 1972), and barley etioplasts (EVANS and SMITH, 1976). The effect is enhanced by a period of darkness after red light irradiation, but is reversed on exposure to far-red (KÖHLER, 1966a; REID et al., 1972; COOKE et al., 1975; EVANS and SMITH, 1976), indicating that phytochrome is involved in controlling the process. However, the mechanism of phytochrome action is unknown and, furthermore, it seems to be questionable whether the red light-enhanced hormone activity is due to de novo biosynthesis or metabolism, or altered membrane permeation. At least in part, release of gibberellins from organelles may contribute to the total increase, as could be demonstrated by increasing gibberellin levels on red light irradiation of intact etioplasts in vitro (COOKE and SAUNDERS, 1975; COOKE et al., 1975; COOKE and KENDRICK, 1976; EVANS and SMITH, 1976). Nevertheless, the origin and exact structural elucidation of the gibberellins formed on red light illumination remain obscure and merit further investigation. Any interpretation must take into account the observation that red light alters uptake and efflux of gibberellins in cytoplasmic compartments of etiolated pea stem sections (MERTZ, 1975). Furthermore, red light treatment may increase the turnover of gibberellins such as GA_9 in homogenates of etiolated barley leaves (REID et al., 1972).

In cultures of the fungus, *Fusarium moniliforme*, it is reported that light enhanced the incorporation of labelled acetate and leucine into the gibberellins (MERTZ and HENSON, 1967a, b; MERTZ, 1970). Changes in the rate of biosynthesis and metabolism in relation to the developmental stage of the plant has been reported several times, e.g., *Pharbitis* (BARENDSE, 1971; BARENDSE and LANG, 1972), *Pisum* (FRYDMAN and MACMILLAN, 1975; SPONSEL and MACMILLAN, 1977), *Phaseolus vulgaris* (YAMANE et al., 1975, 1977), *Phaseolus coccineus* (SEMBDNER et al., 1972), *Gladiolus* (HALEVY et al., 1974), *Tulipa* (AUNG and REES, 1974), *Oryza* (RAILTON, 1978), *Secale* (REID et al., 1974), *Picea* (LORENZI et al., 1975b, 1976; DUNBERG, 1976) or *Pseudotsuga* (WAMPLE et al., 1975). In some cases, however, unequivocal identification of the metabolites is needed. Qualitative differences in the metabolic pattern of isolated plant organs as opposed to intact seedlings further complicate the problem (NASH and CROZIER, 1975; KUHN-SILK and ERICKSON, 1976; ARIAS et al., 1976).

4.2.5 Ethylene

A number of publications which have already been reviewed by YANG (1974) deal with the incorporation and metabolization of ^{14}C- or ^{3}H-labelled ethylene by fruits (BUHLER et al., 1957; JANSEN, 1963, 1964, 1965, 1969) and vegetative tissues (HERRERO and HALL, 1959; HALL et al., 1967; SHIMOKAWA and KASAI, 1968; SHIMOKAWA et al., 1969). Generally, ethylene was incorporated into tissue only in extremely low amounts and the label was found in a variety of metabolically unrelated compounds, like benzene, toluene, organic acids, etc. In some cases no incorporation or metabolism of ethylene could be detected (BUHLER et al., 1957; BEHMER, 1958; CARLTON et al., 1961). By use of deuterated ethylene no exchange of hydrogen between the ethylene and the tissue was found (ABELES et al., 1972; BEYER, 1972). The former results obtained by using labelled ethylene should be considered with care because no adequate precautions were taken with regard to $^{14}C_2H_4$ purity or microbial contamination (cf. BEYER, 1975b). This is especially true of studies in which $^{14}C_2H_4$ was regenerated from mercury perchlorate, a reagent found to cause the formation of labelled impurities readily incorporated by plant tissue (JANSEN, 1965, 1969; BEYER, 1975c). Recent studies demonstrated that etiolated pea seedlings, grown under aseptic conditions, and cut carnations, treated with ultra-high purity $^{14}C_2H_4$, actively incorporated ethylene into ethanol-soluble tissue metabolites and oxidized the hormone to CO_2 (BEYER, 1975a, b, 1977; GIAQUINTA and BEYER,

1977). Propylene, like ethylene, is also oxidized to CO_2 and incorporated into polar metabolites by pea seedlings (BEYER, 1978). In morning glory flowers oxidation of $^{14}C_2H_4$ to $^{14}CO_2$ is very low 2 to 3 days prior to flower opening and increases dramatically about 1 day prior to full bloom (BEYER and SUNDIN, 1978). Prior to leaf abscission in cotton ethylene oxidation to CO_2 rapidly increased in the abscission zone (BEYER, 1979). In broad bean a dramatic metabolism of ethylene has also been found (JERIE and HALL, 1978; JERIE et al., 1978). In the whole fruit and particularly in segments of maturing cotyledons about 95% of the applied ethylene was metabolized within a short time, and the major metabolite was identified by combined GC-MS to be ethylene oxide. The authors concluded that, in this tissue, ethylene metabolism constitutes a mechanism for removing ethylene from the cell. Ethylene oxide is produced as a metabolite of ethylene also by cell free preparations from cotyledons of *Vicia faba* (DODDS et al., 1979). However, French bean cotyledons were shown not to metabolize ethylene in detectable amounts but to accumulate $^{14}C_2H_4$, apparently by compartmentation (JERIE et al., 1979). Recently ethylene glycol has been identified to be one of the stable end-products of ethylene metabolism in pea seedlings; ethylene oxide may be a precursor, and this oxidative process is assumed to be direct involved in the mechanism of ethylene action (BLOMSTROM and BEYER, 1980).

4.3 Conjugation

The term "conjugated plant hormones" is given to plant hormones which are metabolically bound to other low molecular weight compounds by covalent binding (SEMBDNER et al., 1974). Consequently, the term excludes adducts to cellular particles, associations with receptor proteins and other high-molecular complexes, which normally are called "bound plant hormones" (LANG, 1970). Non-identified polar fractions which possibly belong to the conjugates are also often termed "bound" hormones.

Conjugation represents part of the metabolism of plant hormones; but contrary to biosynthetic and metabolic routes, conjugation does not exclude reversibility, i.e., reconversion of the conjugates to the free hormone forms. Thus, conjugation may be involved in regulation of the level of active hormones by alteration of their physical, chemical, and biological properties. Much information has been published concerning the formation, natural occurrence, isolation, and chemical structures of plant hormone conjugates. However, the physiological role of conjugation needs further detailed investigation, requiring decisive analytical progress in this field.

4.3.1 Auxins

Recently conjugation of auxin (IAA) has assumed more and more importance, not least because of the findings that, in plants, the content of IAA conjugates or bound forms can be much higher than the amount of free IAA (BANDURSKI and SCHULZE, 1974, 1977; MERKYS et al., 1974; ZIMMERMANN et al., 1976). The physiological role of IAA conjugates has been discussed with regard to translocation processes, biosynthesis, storage, and mode of action of auxins (SCHNEIDER and WIGHTMAN, 1974; LOH and SEVERSON, 1975; BANDURSKI, 1977; HALL, 1977). All IAA conjugates, so far structurally elucidated, are exclusively derivatives of the IAA carboxyl group (SHANTZ, 1966; SCHNEIDER and WIGHTMAN, 1974; SEMBDNER, 1974). No evidence for N-glycosylation, well known for cytokinins (see Sect. 4.3.2), has been obtained for IAA. With respect to the conjugating moiety, the IAA conjugates can be divided into three groups: (a) Peptidyl IAA conjugates (the carboxylic group of IAA is linked to an amino-acid unit by a peptide bond). (b) Glycosyl IAA conjugates (the carboxylic group of IAA is linked to a sugar unit by a glycosidic or an ester bond). (c) myo-Inosityl IAA conjugates (the carboxylic group of IAA is linked to myo-inositol by an ester linkage).

Further derivatization of the conjugating moiety is not excluded. Glucobrassicin and neoglucobrassicin, being special IAA derivatives, the occurrence of which is restricted to a limited number of systematical families (KUTAČEK and KEFELI, 1968; SCHRAUDOLF, 1968), are briefly discussed in connection with the biosynthesis of IAA (see Sect. 4.1.1e).

a) Peptidyl IAA Conjugates

The first IAA conjugate to be identified was indole-3-acetyl-L-aspartic acid (IAAsp; 1a in Fig. 4.29). It was shown to be formed in pea seedlings after application of exogenous IAA (ANDREAE and GOOD, 1955). The IAAsp formation which can be induced by auxin application in all plants studied (see references in MOLLAN et al., 1972; EPSTEIN and LAVEE, 1977b), usually has a lag period of about 2 h (ANDREAE and VAN YSSELSTEIN, 1956, 1960; ZENK, 1962; SÜDI, 1969; GOREN and BUKOVAC, 1973). The full enzymatic activity reached after 6–8 h persists for a longer time (ANDREAE, 1964, 1967; ZENK, 1961b, 1964). L-Aspartic acid-N-acylase has been considered responsible for IAAsp formation (ZENK, 1962). The enzyme can be induced by auxins, other than IAA, and it is able to transform artificial auxins into asparate conjugates (ZENK, 1962; SCHRAUDOLF, 1971; VENIS, 1972; see also Chapter 5).

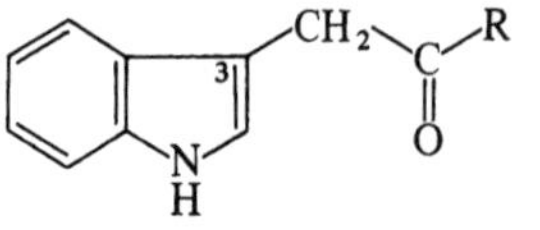

1a: R = Asp
1b: R = Gly
1c: R = Ala
1d: R = Val
1e: R = Glu
1f: R = Lys

Fig. 4.29. Amino acid conjugates of IAA

Ethylene was found to influence aspartate formation of auxins, as shown in cotton and *Coleus* (BEYER and MORGAN, 1970). However, no effect of ethylene on IAA- or NAA-induced aspartate formation could be found in detached primary leaves of cowpea (GOREN and BUKOVAC, 1973; GOREN et al., 1974). When a proteinaceous inhibitor of ethylene biosynthesis, isolated from mung bean, was used, a common enzymatic step of both ethylene and IAAsp formation could be established but any further metabolic interactions were not obtained (SAKAI and IMASEKI, 1973b). Kinetin applied together with IAA to mung bean tissue reduced IAAsp formation significantly (LAU et al., 1974).

IAAsp is not only a metabolite of exogenously applied IAA but is also probably a naturally occurring IAA conjugate widely distributed in plants (KLÄMBT, 1960; ROW et al., 1961; v. RAUSENDORFF-BARGEN, 1962; WEAVER and JACKSON, 1963; OLNEY, 1968; TILLBERG, 1974b. In addition, monomethyl-4-chloroindolyl-3-acetyl aspartate has been characterized (HATTORI and MARUMO, 1972) as an endogenous compound from immature seeds of *Pisum sativum*.

In addition to aspartate formation, the major conjugation reaction, other peptidyl-linked IAA conjugates were found to be formed after exogenous application of IAA, such as IAA-L-lysine in *Pseudomonas* (1f in Fig. 4.29) (HUTZINGER and KOSUGE, 1968) as well as IAA-glycine (1b), IAA-alanine (1c), IAA-valine (1d), and IAA-glutamic acid (1e) in *Parthenocissus* crown gall tissue (FEUNG et al., 1976). However these metabolites have not yet been shown to be endogenous compounds.

The chemical synthesis of IAAsp was performed by MOLLAN et al. (1972). Moreover, peptidyl IAA-conjugates of 20 different α-amino acids were chemically synthesized (FEUNG et al., 1975).

There are some indications in the literature that IAA can be linked by a peptide bond to high molecular weight units (SCHNEIDER and WIGHTMAN, 1974). From *Avena* seeds an IAA complex containing a glucoprotein of about 5,000–20,000 daltons could be isolated (PERCIVAL and BANDURSKI, 1976). IAA bound to proteins, RNA and DNA proteids has been partially characterized and is discussed with respect to the mode of action and binding site of the receptor (MERKYS et al., 1974; DAVIES, 1976; IHL, 1976).

Without doubt, IAAsp formation, being the major conjugation process in auxin metabolism, is of physiological significance. However, its physiological role is not yet clear in detail. IAAsp formation, and the balance of IAAsp with free IAA content within plant tissue, have been studied intensively with respect to growth processes. Light was shown to stimulate IAAsp formation in pea seedlings, and dwarf plants were found to contain up to three times the IAAsp content of normal plants (LANTICAN and MUIR, 1969; MUIR, 1972). Contrarily, in bean seedlings IAAsp formation has proved to be light-independent and is probably not involved in etiolation (TILLBERG, 1974a, b; FLINN, 1975). These findings agree with results obtained on rice seedlings in which only IAA uptake is phytochrome-controlled, whereas IAAsp content is not affected (SHERWIN and FURUYA, 1973). Some earlier reports considered IAAsp to be only an inactive storage product (MORRIS et al., 1969) or more likely a true detoxification product (ANDREAE and GOOD, 1955; ANDREAE and VAN YSSELSTEIN, 1956; ZENK, 1964; ANDREAE, 1967). Opposed to this suggestion, varying but significant auxin activity of IAAsp was found in some bioassay systems (THURMAN and STREET, 1962; TILLBERG, 1974a). Previously, IAAsp was reported to be more active than IAA during long-term experiments in stimulating cell division in tobacco tissue suspension, whereas IAA showed a higher short-term activity (REKOSLAVSKAYA and GAMBURG, 1976). This behaviour can be explained by increased stability of IAAsp compared with IAA to IAA oxidase and continuous hydrolytic release of IAA from the conjugate (HAMILTON et al., 1961; REKOSLAVSKAYA and GAMBURG, 1976; HALL, 1977; COHEN and BANDURSKI, 1978). The bioassay of 20 synthetic amino-acid conjugates of IAA showed that several of them, including IAAsp, are at least as active as IAA in oat coleoptile and soybean callus bioassay (FEUNG et al., 1977). From these results, the authors concluded that peptidyl IAA conjugates may be directly involved in controlling or balancing the physiological behaviour of the plant organs and tissues.

b) Glycosyl IAA Conjugates

Esterification of a carboxyl group by a sugar is a very common detoxification step in the metabolism of biologically active substances; therefore, sugar-containing IAA conjugates were predicted to occur. An IAA-arabinose complex was isolated from *Zea mays* but no structural details were given (STEWARD and SHANTZ, 1959). Rhamnose-bound IAA was isolated and partially characterized from *Peltophorium ferrugineum* (GANGULY et al., 1974).

Conjugation of IAA to glucose was first observed by KLÄMBT (1961) and ZENK (1961a, 1964) after feeding IAA to wheat coleoptiles and pea epicotyls respectively. On the basis of enzymatic hydrolysis, the conjugate was assumed to be indole-3-acetyl-1′-β-D-glucopyranosyl ester. Unlike IAAsp formation, the conversion of IAA into IAA glucosyl ester was shown to start without lag phase and increased up to about 12 h. In subsequent studies, IAA glucosyl ester was found to occur in stems of peas and beans (DAVIES, 1972), in barley leaves (SCHNEIDER, 1965), and oat coleoptiles (KEGLEVIC and POKORNY, 1969). From maize kernels three isomeric IAA glucosyl esters were isolated, and structurally elucidated, by mass spectra to be indole-3-acetyl-2′-D-glucopyranosyl ester, indole-3-acetyl-4′-D-glucopyranosyl ester, and indole-3-acetyl-6′-D-glucopyranosyl ester (Chap. 1, Table 1.2) (EHMANN, 1974). The same three isomers were formed in vitro using an enzyme extract from immature corn kernels. This in vitro reaction requires glucose CoASH and ATP whereby CoAS-IAA apparently acts as an intermediate (KOPCEWICZ et al., 1974). The speciality of the 2′, 4′ or 6′ binding site may be due to a pathway mechanistically differing from that leading to the 1′-binding normally observed. Furthermore, a high molecular weight sugar ester from *Zea mays* has been characterized as IAA-β-1→4 cellulosic glucan, the hydrolysis of which led to cellobiose, cellotriose, and cellotetrose with β-1→4 linkages (PISKORNIK and BANDURSKI, 1972). The correlation between IAA bound to glucan cell wall synthesis has been discussed (RAY, 1973; DAVIES, 1976).

With respect to the absence of a lag phase, the formation of IAA glucosyl ester is considered to act as a very rapid detoxification mechanism in vivo, especially during the lag phase of aspartate formation (ZENK, 1964). Using chemically synthesized IAA glucosyl ester, its biological activity was shown to be as high as IAA, leading to the conclusion that this conjugate may be active in growth induction (KEGLEVIC and POKORNY, 1969). However, having in mind that glucosyl esters may be easily hydrolyzed during bioassay, further detailed studies on the physiological role of glycosyl IAA conjugates seem to be necessary.

c) myo-Inositol IAA Conjugates

From kernels of *Zea mays*, a very rich source of IAA conjugates, a third group of IAA conjugates was isolated which always contain myo-inositol as the esterifying alcohol (LABARCA et al., 1965). Indole-3-acetyl-2′-myo-inositol (Chap. 1, Table 1.2) was the first representative of this group (NICHOLLS, 1967). The structure of this compound was established by NMR (NICHOLLS et al., 1971), confirming the axial configuration of the ester linkage. myo-Inositol is able to bind more than on IAA unit, as demonstrated by the isolation of di-(indole-3-acetyl)- and tri-(indole-3-acetyl)-myo-inositols (Chapt. 1, Table 1.2) from mature kernels of *Zea mays* (EHMANN and BANDURSKI, 1974). Moreover, a myo-inositol unit bound to IAA can be glycosylated additionally, whereby – in each case so far established – the C-5′ is linked to the glycosidic C-1 of the sugar moiety. Thus, 5′-(1-L-arabinopyranosyl)-2′-(indole-3-acetyl)-myo-inositol and 5′-(1-D-galactopyranosyl)-2′-(indole-3-acetyl)-myo-inositol (Chap. 1, Table 1.2) were isolated from maize (UEDA et al., 1970). As shown by GC-MS, at least two further isomers (see Chap. 1, Table 1.2) were present, the IAA residue of which might be linked to the equatorial hydroxyl groups at C-1′ or C-4′ (UEDA and BANDURSKI, 1974). In vitro experiments using crude enzyme preparations from immature maize kernels indicated that IAA is initially linked to the equatorial C-1′ or C-4′ of myo-inositol. Subsequent acyl migration leads to other isomers (KOPCEWICZ et al., 1974). Recently, IAA glucosyl ester has been shown to be the precursor of in vitro formation of IAA-myo-inositol (MICHALCZUK and BANDURSKI, 1980). The chemical synthesis of IAA-myo-inositol was performed recently (see BANDURSKI, 1977).

Concerning biological activity, IAA-myo-inositol was shown to be less active than IAA in coleoptile section tests, but more active in stimulating tissue culture growth (NICHOLLS, 1967). From results with horseradish peroxidase it was concluded that the esterification of IAA to myo-inositol prohibits oxidative degradation (COHEN and BANDURSKI, 1978). With respect to the physiological role of IAA-myo-inositol derivatives, it has been hypothesized that these conjugates might be storage forms which would be able to protect the IAA during desiccation of the kernels (UEDA and BANDURSKI, 1974). Applying IAA-myo-inositol to corn kernels, it could be shown that the transport to the shoot is about 1000 times the rate of IAA translocation. Therefore, IAA-myo-inositols are assumed to be functional as transport forms. Moreover, by finding that light is able to decrease IAA content and growth rate to the same extent, whereas the IAA ester content is increased, the myo-inositol conjugation may be a tool for expressing light control by means of esterifying IAA or releasing it from the conjugate (BANDURSKI, 1978; BANDURSKI et al., 1977).

4.3.2 Cytokinins

a) Metabolic Conjugation Pathways

Purines are known to form N(9)-ribosides and phosphorylated ribosides. Far more types of conjugation reaction have been found to operate in the metabolism of cytokinins (see HALL, 1973; SEMBDNER, 1974).

Among the naturally occurring cytokinin conjugates sugar derivatives dominate. With regard to the conjugating moiety and the linkage (Fig. 4.30) one can divide the cytokinin conjugates into five groups as outlined in Table 4.6. Zeatin riboside has been found most frequently, i.e., in sweet corn (LETHAM, 1966a, b; 1973), water melon seeds (PRAKASH and MAHESHWARI, 1970) chicory root (BUI-DANG-HA and NITSCH, 1970), poplar leaves (HEWETT and WAREING, 1973a, b), *Yucca* phloem exudate (VONK, 1974), coconut milk (VAN STADEN and DREWS, 1975), kohlrabi tubers (CONRAD, 1975), *Vinca rosea* crown gall tumour tissue (MILLER, 1975), pea root nodules (SYONO and TORREY, 1976) and even from a mycorrhizal fungus (MILLER, 1967).

First metabolic experiments were undertaken with *Phaseolus vulgaris*. Following application of (8-^{14}C)-zeatin to bean axes a rapid conjugation was observed, resulting in the formation of labelled zeatin riboside, zeatin riboside-5′-phosphate, and dihydrozeatin together with its riboside and riboside-5′-phosphate (SONDHEIMER and TZOU, 1971). Embryos of *Fraxinus americana* converted labelled zeatin within 12h to its riboside as well as to the mono-, di-, and triphosphoribosides (TZOU et al., 1973). Excised inflorescence stalks of *Yucca flaccida* transformed

Table 4.6. Types of cytokinin conjugates

	Type of conjugate	Conjugating agent linked to
1	Cytokinin-N-ribosides and Cytokinin-N-riboside-5′-phosphates	N-9
2	Cytokinin-N-glucosides	N-3, N-7 or N-9
3	Cytokinin-O-glucosides	O-4′
4	Ribosyl cytokinin glucosides	N-9, O-4′
5	Cytokinin amino acid conjugates	N-3, N-9

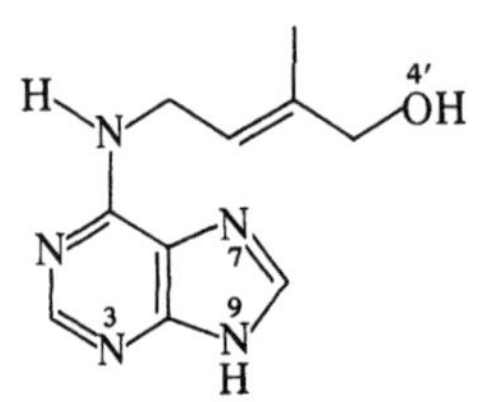

trans–Zeatin

Fig. 4.30. Possible sites for conjugating moieties

labelled zeatin into the riboside-5′-phosphates but when added to phloem sap only the riboside-5′-monophosphate was formed (VONK, 1976). Zeatin riboside and zeatin riboside-5′-monophosphate have been obtained after administration of zeatin to seedlings of *Zea mays* (PARKER et al., 1973; PARKER and LETHAM, 1974) and *Raphanus sativus* (PARKER and LETHAM, 1973; GORDON et al., 1974).

The riboside and its 5′-phosphates have also been detected in feeding experiments with labelled i^6Ade or i^6A (see Fig. 4.2). Cytokinin-requiring tobacco cells and cultures of *Acer pseudoplatanus* converted 77% and 60% respectively of the applied radioactivity into the three riboside-5′-phosphates (LALOUE et al., 1974, 1975, 1977).

In cytokinin ribosides and its phosphates the ribosyl moiety is always attached to N-9 of the cytokinin. Thus, this type of conjugation parallels the behaviour of other purines (e.g., adenine), and whether cytokinin ribotides could be incorporated directly into tRNA has been discussed (see Sect. 4.1.2.a).

Zeatin glucosides have been reported to occur in a series of higher plants, e.g., rice roots (YOSHIDA and ORITANI, 1972), pea roots (SHORT and TORREY, 1972), sweet corn (LETHAM, 1973), *Picea sitchensis* (LORENZI et al., 1975a), coconut milk (VAN STADEN, 1976a), leaves and honeydew of *Salix babylonica* (VAN STADEN, 1976b), radish seeds (SUMMONS et al., 1977). *Vinca rosea* crown gall tumor tissues contain glucosides of zeatin and ribosyl zeatin (MORRIS, 1977; PETERSON and MILLER, 1977). For some time there was confusion with regard to the structure of the cytokinin glucosides due to the fact that the ring size of the sugar and the site of the linkage have not been unequivocally identified. Cytokinin glucosides differ from each other in their biological activity and behaviour towards β-glucosidases. In order to establish the correct structure several metabolites were prepared synthetically and compared with the natural ones (COWLEY et al., 1975; DUKE et al., 1975). On present knowledge, all glucosides seem to be β-D-glucopyranosides with the sugar linked to N-3, N-7, or N-9 of the purine nucleus or to O-4 of the zeatin side chain. At least in de-rooted radish seedlings, glucosylation is not restricted to N^6-substituted adenines with strong cytokinin activity (LETHAM et al., 1978).

In the case of zeatin the N(7)-glucoside was first observed as metabolite in different organs of *Raphanus sativus* and denoted as raphanatin (Chap. 1, Table 1.15) (PARKER et al., 1972; PARKER and LETHAM, 1973; GORDON et al., 1974). De-rooted maize seedlings showed degradative reactions forming adenine, adenosine, and AMP in preference to zeatin-N(7)-glucoside. Roots, however, gave mainly zeatin riboside, zeatin riboside-5′-monophosphate, and zeatin-N(9)-glucoside (PARKER and LETHAM, 1974). The last was also formed as a main metabolite in cultured embryos (PARKER and LETHAM, 1974), and tissue cultures and intact seedlings of maize (PARKER et al., 1973). In tobacco cells i^6Ade-N(7)-glucoside occurred after application of labelled i^6Ade (LALOUE et al., 1975). In comparing the metabolic fate of cytokinins and their ribosides, 7-glucosylation was found to operate with the free bases only while the ribosides predominantly were phosphorylated at C-5′ (LALOUE et al., 1975, 1977).

Another glucoside, zeatin-O(4)-glucoside, has been obtained after application of labelled zeatin to de-rooted seedlings of *Lupinus angustifolius* (PARKER et al., 1975), cytokinin-requiring soybean callus (HORGAN, 1975), and leaves of *Populus alba*, which additionally, formed ribosyl dihydrozeatin-O-glucoside (LETHAM et al., 1977). WAREING et al. (1977) investigated the metabolic differences between roots and aerial parts of *Phaseolus vulgaris;* the results obtained parallel those of maize. Leaves of de-rooted and de-budded bean seedlings rapidly transformed (8-^{14}C)-zeatin into zeatin-O-glucoside which was subsequently degraded. Buds showed a similar conversion. On the contrary, rooted de-budded seedlings yielded mainly zeatin riboside-phosphates. The major cytokinin conjugate in the stem and petiole of bean seedlings has been identified as dihydrozeatin riboside. As minor metabolites dihydrozeatin-O-glucoside and the glucosides of zeatin riboside and dihydrozeatin riboside occurred in both rooted and de-rooted systems. Recently, the riboside and O-glucoside of dihydrozeatin have also been shown to be genuine cytokinin conjugates of *Phaseolus vulgaris* (WANG et al., 1977; WANG and HORGAN, 1978) and *Populus* species (DUKE et al., 1979).

Injection of (8-^{14}C)-zeatin into root nodules of *Alnus glutinosa* yielded the riboside and O-glucoside of zeatin together with dihydrozeatin-O-glucoside and several degradation products (HENSON, 1978b; HENSON and WHEELER, 1977b). It has been observed that the capacity for glucosylation of zeatin rises during expansion and maturation of leaves (HENSON, 1978a). N(3)-Glucosides have not yet been found as metabolites of naturally occurring cytokinins but of benzylaminopurine.

Additionally to the N(7)-, N(9)-, and O(4)-glucosides mentioned above, seedlings of *Lupinus angustifolius* form a zeatin amino-acid conjugate called lupinic acid (PARKER et al., 1975; see Sect. 4.3.2b). Its structure has been established as β-(zeatin-9-yl)-alanine (MACLEOD et al., 1975). In addition dihydrolupinic acid has been isolated as a minor metabolite of labelled zeatin in *Lupinus* seedlings (PARKER et al., 1978). Another type of amino-acid conjugate, called discadenine, has been isolated from the slime mould *Dictyostelium discoideum* (OBATA et al., 1973) and structurally elucidated as γ-(N^6-isopentenyladenin-3-yl)-α-amino butyric acid (ABE et al., 1976). Adenosine-5′-phosphate was shown to be the direct precursor of the mould cytokinins (TAYA et al., 1978a). Crude *Dictyostelium* enzyme preparations synthesized discadenine from i^{6}Ade and S-adenosylmethionine (TAYA et al., 1978b).

These results demonstrate that rapid conjugation of cytokinins takes place in plant tissue. The rate of conjugation and the type of conjugate formed in plants depend both on the plant organ and its physiological stage and on the environmental conditions (i.e., HEWETT and WAREING, 1973a, b, c; HENSON, 1978b). For example, red light treatment of poplar leaves resulted in a large and specific increase of N^6-(2-hydroxybenzylamino)-purine riboside (HEWETT and WAREING, 1973c). The various routes of cytokinin conjugation are summarized in the metabolic pathways outlined in Fig. 4.31.

Cytokinin-N(7)-glucosides have been previously considered as the biologically active cytokinins (FOX et al., 1973) following the observation that N^6-benzyladenine-N(7)-glucoside persisted in cytokinin-requiring tobacco cultures for a long time after all the labelled benzyladenine fed had disappeared. Recent experimental results (LALOUE, 1977) are in contrast with this assumption and favour the role of cytokinin ribosides as translocation forms and cytokinin glucosides as protected or storage forms (HEWETT and WAREING, 1973c; GORDON

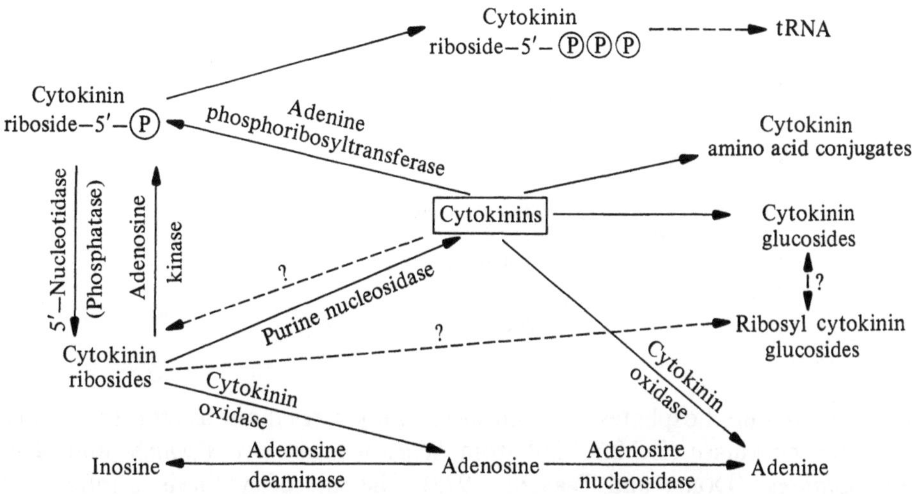

Fig. 4.31. Formation and degradation of naturally occurring cytokinin conjugates

et al., 1974; PARKER and LETHAM, 1973; LALOUE et al., 1975; DAVEY and VAN STADEN, 1976; LALOUE, 1977; GAWER et al., 1977). This is in agreement with the well-known occurrence of cytokinin ribosides in xylem sap and of cytokinin glucosides in coconut milk. The stability of at least the cytokinin-N(7)-glucosides depends on the protecting sugar moiety that renders the molecule insensitive to enzymatic degradation that occurs upon N(6)-side chain removal by cytokinin oxidases (WHITTY and HALL, 1974).

Cytokinin ribosides and glucosides show considerable biological activity in several bioassay systems. Usually the conjugates are less active than the free bases. As zeatin-O-glucoside in soybean callus is rapidly transformed to zeatin and its riboside (VAN STADEN and PAPAPHILIPPOU, 1977), biological activity may refer to the free bases, at least partially. Previously postulated "cytokinesine I" (WOOD et al., 1972) has been structurally elucidated as zeatin riboside (MILLER, 1975).

b) Enzymic Investigations

Kinetic studies of cytokinin conversions in plants are difficult to interpret because of interference by the activities of several cytokinin-metabolizing enzymes. In some cases enzyme preparations have been used to follow single steps of cytokinin metabolism. Most interest has been focussed on the formation of cytokinin ribosides. It has been demonstrated that purine nucleoside phosphorylase (EC 2.4.2.1) from *E. coli* catalyzed the in vitro transfer of a ribosyl or deoxyribosyl moiety from corresponding donors (e.g., ribosyl phosphate) to various N^6-substituted aminopurines including N^6-benzyladenine (SIVADJIAN et al., 1969). With regard to the biosynthesis of the cytokinin nucleotides the enzymes involved have been identified as either purine phosphoribosyl transferase (operating with free cytokinins as substrates) or nucleoside kinase (operating with exogenously supplied cytokinin ribosides). The formation of cytokinin

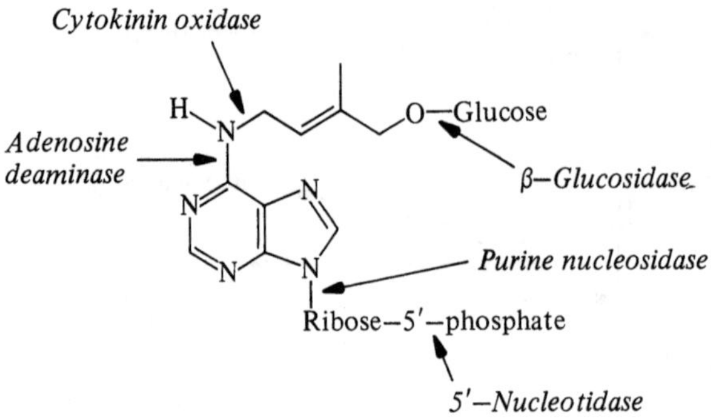

Fig. 4.32. Sites for the attack of cytokinin conjugate-degrading enzymes

riboside-5′-monophosphates has been successfully carried out using preparations of adenosine kinase (EC 2.7.1.20) from *Schizosaccharomyces pombe* and *Acer pseudoplatanus* (DOREE and TERRINE, 1973). The rate of synthesis is influenced by the nature of the substituent at N^6, as could be demonstrated with adenosine and several synthetic cytokinin ribosides as substrates. Adenosine kinase preparations of wheat germ and tobacco callus tissue converted labelled i^6A to the riboside-5′-monophosphate under conditions of adenosine phosphorylation (CHEN and ECKERT, 1977).

Metabolism of i^6A has also been studied in man (CHHEDA and MITTELMAN, 1972; DIVEKAR et al., 1974) demonstrating monophosphorylation and rapid degradation as well. i^6A was found to be a competitive inhibitor of adenosine kinase and other enzymes of the purine metabolism (DIVEKAR et al., 1974). Removal of the N^6 side chain can be achieved by the action of partially purified cytokinin oxidase from maize kernels (WHITTY and HALL, 1974) and tobacco tissues (PAČES et al., 1971). Mammalian cytokinin oxidase preparations seem to be accompanied by adenosine deaminase (EC 3.5.4.4) since incubation with cytokinin ribosides yielded inosine as product (CHASSY and SUHADOLNIK, 1967; TERRINE et al., 1969; HALL et al., 1971; HALL and MINTSIOULIS, 1973).

Zeatin-O-glucoside is susceptible to enzymic cleavage by β-glucosidases whilst N-glucosides are not. Removal of the ribose moiety is catalyzed by the action of a purine nucleosidase (cytokinin hydrolase). This enzyme has been detected in barley leaves and rape seedlings (PAČES, 1976; PAČES et al., 1977). Conversion of i^6A to i^6Ade was inhibited by the addition of trans-zeatin riboside (PAČES, 1976) as were other enzymic degradation reactions (KAMINEK and PAČES, 1977). An outline of the metabolic reactions of cytokinins and the enzymes involved is given in Figs. 4.31 and 4.32.

Investigations concerning the enzymatic synthesis of cytokinin glucosides led to the isolation of soluble enzymes from expanded radish cotyledons (ENTSCH and LETHAM, 1979; ENTSCH et al., 1979) and senescent leaves of *Ginkgo biloba* (VAN STADEN, 1979) that formed 7- and 9-glucosides on incubation with zeatin or benzyladenine. On the other hand MURAKOSHI et al. (1977) extended their work on the enzyme-catalyzed formation of heterocyclic β-substituted alanines, including purine amino acid conjugates. Using enzyme preparations from immature seeds or hypocotyls of *Lupinus luteus* and seedlings of water melon, the in vitro synthesis of lupinic acid was achieved starting with (3-^{14}C)-O-acetyl

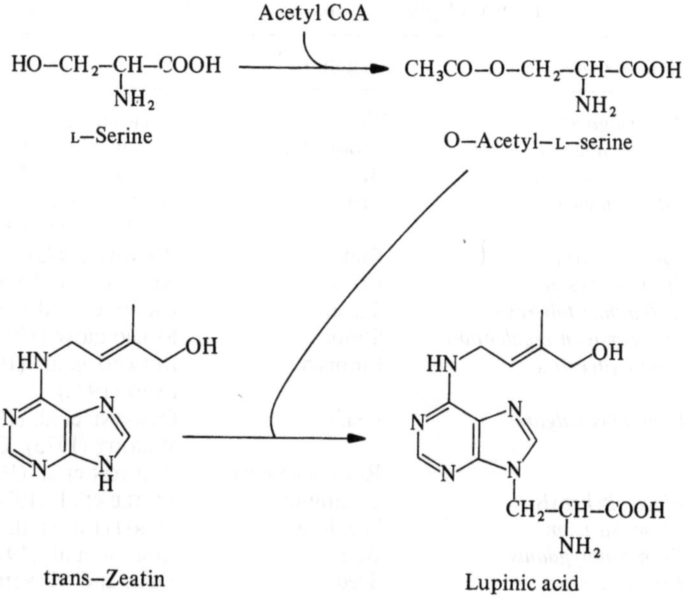

Fig. 4.33. Enzymic formation of lupinic acid

serine and zeatin (Fig. 4.33). Lupinic acid synthetase appears to be specific for the L-amino acid as a substrate. In a similar way discadenine is formed enzymatically (see Sect. 4.3.2a) from i^6Ade and S-adenosylmethionine (TAYA et al., 1978b).

4.3.3 Abscisic Acid

Conjugation of ABA has been shown to play an important role in the metabolism of this plant hormone. Reviews dealing with this topic have already been published (MILBORROW, 1974a, b; LEHMANN, 1980).

The formation of a polar, water-soluble, hydrolyzable "Metabolite B" was first detected in tomato shoots supplied with (2-^{14}C)-ABA (MILBORROW, 1968). In immature fruits of *Lupinus luteus* a naturally occurring ABA conjugate has been isolated (KOSHIMIZU et al., 1968) and structurally elucidated to be the β-D-glucopyranosyl ester of (+)-ABA (Chap. 1, Table 1.16). Synthesis of the ABA ester glucoside was achieved by combined chemical and enzymatical procedures (LEHMANN et al., 1975; LEHMANN and SCHÜTTE, 1977).

MILBORROW (1970) demonstrated the occurrence of glucosyl abscisate in the pseudocarp of *Rosa arvensis*. In his experiments, identification was carried out both by comparison of the IR and NMR spectra and by co-chromatography with an authentic sample of glucosyl abscisate or its O-tetraacetate. This, and the above-cited work of KOSHIMIZU et al. (1968), are the only two rigorous identifications. In other publications the glucosyl ester of ABA, isolated from plant material, has only been tentatively identified. In a number of publications a polar fraction, which can be hydrolyzed by alkaline to ABA and glucose,

Table 4.7. Occurrence of glucosyl abscisate in plants – partial characterization

Plant species	Organ/Tissue	Reference
Beta vulgaris	Shoot	MILBORROW (1978 a)
Betula pubescens	Wood, bark	DATHE et al. (1978 b)
Betula verrucosa	Bud	HARRISON and SAUNDERS (1975)
Citrus sinensis	Fruit	GOLDSCHMIDT et al. (1973), SWEETSER and VATVARS (1976)
Fagus sylvatica	Bud	WRIGHT (1975)
Fragaria vesca	Fruit	RUDNICKI and PIENIAZEK (1971)
Helianthus tuberosus	Tuber	CHARNAY and BONGEN-OTTOKO (1977)
Lycopersicon esculentum	Shoot	MILBORROW (1978 a)
Malus silvestris	Embryo	BULARD et al. (1974), BARTHE and BULARD (1978)
Phaseolus vulgaris	Leaf	OSBORNE et al. (1972), HIRON and WRIGHT (1973), GOTO (1978)
	Root segments	WALTON et al. (1976)
Picea sitchensis	Cambium	LITTLE et al. (1978)
Pisum sativum	Seedling	DÖRFFLING et al. (1974)
Prunus amygdalus	Bud	LESHEM et al. (1974)
Prunus persica	Seed	DIAZ and MARTIN (1972)
Ribes nigrum	Bud	WRIGHT (1975)
Spinacea oleracea	Plant	ZEEVAART (1971 b, 1974)
Taxus baccata	Embryo	LE PAGE-DEGIVRY (1973)
Triticum aestivum	Leaf	KING (1976), DEWDNEY and MCWHA (1978)
Vitis vinifera	Bud	DÜRING and ALLEWELDT (1973)

has been designated as "bound" or "conjugated" ABA or "ABA-glucose complex". In most cases it may represent the glucose ester of ABA. The reported occurrence, in higher plants, of glucosyl abscisate partially characterized as "conjugated ABA" is summarized in Table 4.7. According to recent results ABA may also be conjugated to lipids, "gluten-like" proteins, and carbohydrates other than glucose (DEWDNEY and MCWHA, 1978).

A number of studies have been published dealing with the changes in the levels of free ABA and its conjugate in plant organs as related to physiological processes. For example, conjugated ABA was found to increase progressively in embryos of peach seeds as stratification proceeded, whereas the free ABA content decreased (DIAZ and MARTIN, 1972). The authors postulated that the conjugate was a storage form. *Citrus* fruit peel contains relatively low amounts of free and bound ABA but, during natural and ethylene-induced senescence, both free ABA and its conjugate accumulate rapidly, attaining finally a 10:1 ratio of conjugated to free ABA (GOLDSCHMIDT et al., 1973). In buds of *Ribes nigrum* and *Fagus silvatica* the highest level of free ABA occurs in the autumn at about the onset of winter dormancy; the ABA content declines throughout the winter and reaches its lowest value before bud burst. In a parallel way the level of the conjugate increases throughout the autumn and winter (WRIGHT, 1975). Corresponding investigations were done with nodes of vine, carrying dormant buds, and with leaves (DÜRING and ALLEWELDT, 1973). Progress in dormancy break of *Prunus amygdalus* bud is related to a decrease in total free ABA and accompanied by a parallel increase in conjugated ABA (LESHEM et al., 1974). Both free and conjugated ABA were found in unripe and ripe strawberry fruit, where conjugated ABA presents a considerable portion of ABA-like compounds in young unripe fruit (RUDNICKI and PIENIAZEK, 1971). During a period of wilting of pea seedlings, the ABA level increased about 20 times and the level of conjugated ABA 7–10 times. After watering, both ABA and conjugate contents dropped back to the level of control plants within 24–48 h (DÖRFFLING et al., 1974). In leaves of dwarf bean the levels of both free and conjugated ABA were shown to rise during a warm-air-induced wilting cycle (HIRON and WRIGHT, 1973). From these and from some further physiological results, it may be concluded that conjugation of ABA is apparently involved in regulation of the physiologically active hormone level. The ABA conjugate possibly may function as a storage and/or translocation form; however, its physiological role needs further detailed investigation. Recently it has been shown that in shoots of tomato and *Beta vulgaris* the ABA glucosyl ester is not reconverted to free ABA during wilting and thus conjugation seems to be irreversible in the plants studied (MILBORROW, 1978a). An ABA glucosylating enzyme has been isolated from cell suspension cultures of *Macleaya microcarpa* (LEHMANN and SCHÜTTE, 1980).

A number of experiments on the formation of the glucosyl ester of ABA have been carried out by application of unlabelled or labelled ABA to different plants. When apple seedlings are supplied with ABA and trans-ABA water-soluble complexes are formed containing glucose and ABA or trans-ABA (POWELL and SEELEY, 1974). ABA injected to apples was mainly converted into conjugated ABA which remained at a high concentration throughout storage; however, the amount of free ABA was always higher than that in control fruits (WILLS et al., 1976). MILBORROW's studies (1970) on the metabolism of ABA have shown that the glucose ester of ^{14}C-ABA can be isolated from tomato shoots after supplying ^{14}C-ABA. In these experiments it was demonstrated that both (+)- and (−)-ABA are metabolized but in different manner. (−)-ABA is esterified with glucose more rapidly than the natural (+)-enantiomer. When trans-ABA, assumed to be a naturally occurring metabolite (see Sect. 4.2.3), was fed to tomato shoots it was converted into its glucose ester ten times faster than ABA, but it was not isomerized enzymatically to cis-ABA. ^{14}C-ABA administered to barley leaves gave two metabolites identified as glucosyl ester of ABA and metabolite C (CUMMINS, 1973). After application of (2-^{14}C)-ABA to barley and wheat seedlings, the glucosyl ester of ABA, as well as three acidic metabolites, were isolated; the amount of the ester increased with prolongation of incubation time (GROSS and SCHÜTTE, 1974, 1977). Similar results were obtained with shoots of *Agrostemma githago* and *Macleaya microcarpa*. Labelled ABA supplied to cell suspension cultures of *Agrostemma* was partly transformed into the glucosyl ester, whereas acidic metabolites could not be found in the cells or in the culture filtrate (GROSS and SCHÜTTE, 1977; HARTMANN, 1975).

4.3.4 Gibberellins

Although many references to the occurrence of conjugated gibberellins (formerly called "bound gibberellins" (McComb, 1961)) and their metabolic formation were published in the early 1960's (Murakami, 1961; Sembdner et al., 1964; see ref. Schreiber et al., 1970), it was not until 1967 that the first endogenously occurring GA conjugate was identified as an O-glucoside of GA_8 (Schreiber et al., 1967, 1970). Since then, the group of GA conjugates has grown rapidly by isolating further glucosides and other types of conjugates from higher plants and fungi, by metabolical conversion of free GA's into GA conjugates as well as by chemically synthesizing conjugated GA's (see Sembdner, 1974, 1980). The known GA conjugates can be divided into (a) hydroxyl-bound GA-O-conjugates (glucosyl and acyl derivatives), (b) carboxyl-bound GA-O-conjugates (glucosyl and alkyl ester), and (c) carboxyl-bound GA-N-conjugates (amino-acid derivatives).

All known conjugates, including those which have been isolated as metabolites of applied free GA's and those which have been chemically synthesized, are given in Table 4.8; see also Chap. 1).

a) Endogenous Gibberellin Conjugates

The most common conjugates are those in which glucose is bound to a hydroxyl group or the carboxyl group. The 3-O-glucosides of gibberellenic acid and the $\Delta^{1(10)}$-19,2-lactone isomer of GA_3, isolated from seed of *Pharbitis nil* may be artefacts (Yokota et al., 1976).

In addition to the glucosyl conjugates, the 3-O-acetyl derivatives of GA_3 and GA_1 have been isolated from cultures of *G. fujikuroi* (Schreiber et al., 1966; McInnes et al., 1977) and the n-propyl esters of GA_3, and possibly of GA_1, have been reported in *Cucumis sativus* (Hemphill et al., 1973).

Another type of GA conjugate has been isolated from *Pharbitis nil* and structurally elucidated as gibberethione (Chap. 1, Table 1.8), formerly called "pharbitic acid" (Yokota et al., 1974). This conjugate might be formed by joining mercaptopyruvic acid or cysteine to an α,β-unsaturated keto system of GA_3.

Despite some indications (McComb, 1961; Jones, 1964; Sembdner and Schreiber, 1965) no N-containing peptidyl conjugates have been isolated from natural sources. GA_{32} acetonide, occasionally found in GA preparations of *Prunus persica*, was concluded to be an artificial product, formed during the isolation procedure (Yamaguchi et al., 1975a, b).

b) Chemical Synthesis of Gibberellin Conjugates

After initial attempts (Schreiber et al., 1969), the chemical synthesis of the O(3)-glucopyranosides of GA_1, GA_3, GA_4 (Schneider, 1972; Schneider et al., 1974, 1977b) and the GA_8-O(2)-glucopyranoside (Schneider, 1980) has been realized. GA's containing a tertiary 13-hydroxyl group were found to be glucosylated in this position at least as easily as in position C-3, leading to the O(13)-β-D-glucopyranosides of GA_1, GA_3, GA_5 (SCHNEIDER et al., 1974, 1977b) and of GA_8 (SCHNEIDER, 1980), and the GA_3-O(3,13)-β-D-diglucoside (SCHNEIDER et al., 1977b). O(13)-Glucosides represent a new type of GA glucosyl conjugate not yet found to occur naturally. However, some indications are obtained by feeding non-polar GA's to plant material (see Sect. 4.3.4.c) that they may well exist. The amount of GA-O-glucosides formed is influenced by the position and stereochemistry of the hydroxyl group involved in glucosylation (SCHNEIDER, 1980). Using Wilzbach tritiation, uniformly labelled GA_3-O(3)-β-D-glucopyranoside has been prepared (SCHNEIDER, 1978a).

The synthesis of GA glucosyl esters has been realized for the naturally occurring β-D-glucopyranosyl esters of GA_1, GA_4, GA_{37} (HIRAGA et al., 1974c), and GA_9 (LORENZI et al., 1976). Additionally, the GA_3-O-β-D-glucopyranosyl ester (HIRAGA et al., 1974c; MIERSCH and LIEBISCH, 1974, 1980) and its α-D-isomer, not yet found naturally, have been synthesized (MIERSCH and LIEBISCH, 1980). Moreover, the synthesis of the β-D-glucopyranosyl ester of GA_5-O(13)-β-D-glucopyranoside has been performed (SCHNEIDER et al., 1977a).

A series of GA conjugates with amino acids linked to the carboxyl group by a peptide bond have been synthesized including oligo-peptidyl derivatives (ADAM et al., 1977). This type of conjugation has not yet been found to occur naturally. In addition, (^{14}C)-labelled amino acid GA conjugates have been prepared (LISCHEWSKI and ADAM, 1976). The structure of gibberethione (pharbitic acid) was confirmed by its chemical synthesis (YOKOTA et al., 1976). A survey on GA conjugates synthesized chemically is given in Table 4.8, column c.

c) Metabolism of Gibberellin Conjugates

Since GA conjugates, especially glucosyl derivatives, play an outstanding role in GA metabolism, many papers have dealt with the analysis of free and conjugated GA's in plants and their interconversion in the plant (see LANG, 1970). The transformation of free GA_3 into a glucosyl derivative was first reported to occur during incubation of a GA_3 solution with leaf discs of several plant species, resulting in the formation of GA_3-O-glucopyranoside (MURAKAMI, 1961). This conversion has also been observed after applying GA_3 to maturing runner beans (SEMBDNER et al., 1968; SEMBDNER, 1980) or dwarf kidney beans; in addition to GA_3-O(3)-glucoside, the O(3)-glucosides of gibberellenic acid, "iso-GA_3", and an unidentified substance were obtained (ASAKAWA et al., 1974a, b). In *Pharbitis nil,* transformation of labelled GA_3 to GA_3-glucoside was found in both seedlings and maturing seeds. The conversion seemed to be inducible by light (BARENDSE, 1974). The GA_3 glucoside formed is accumulated in the cotyledons of the seeds and hydrolyzed to free GA_3 during germination (BARENDSE and DE KLERK, 1975; BARENDSE and GILISSEN, 1977). In studying the glucosylation of GA's in maturing runner beans and other leguminous fruits, an UDP glucose requiring glucosyl transferase was isolated, which showed selectivity in accepting and glucosylating GA_3 in vitro (MÜLLER et al., 1974).

In vivo, however, glucosylation seems to be a commonly occurring process, proceeding at different biosynthetic or metabolic levels. This could be shown

Table 4.8. Survey on gibberellin conjugates

Conjugate	Isolated from plant		
	(a) Naturally	(b) Metabolically (after applic. of free GA)	(c) Chemically synthesized
GA_1-O(3)-β-D-glucopyranoside	YOKOTA et al. (1978)	NADEAU et al. (1972)[a], YAMANE et al. (1975, 1977) STODDART and JONES (1977)[a]	SCHNEIDER (1972, 1974), SCHNEIDER et al. (1977b)
GA_1-O(13)-β-D-glucopyranoside	–	–	SCHNEIDER et al. (1977b)
GA_1-O-β-D-glucopyranosyl ester	HIRAGA et al. (1972, 1974a, b)	YAMANE et al. (1975, 1977)[a], STODDART and JONES (1977)[a]	HIRAGA et al. (1974c)
GA_1-O(3)-acetate	McINNES et al. (1977)	–	–
GA_1-O-n-propyl ester	HEMPHILL et al. (1973)	–	–
N-GA_1-oyl-glycine	–	–	ADAM et al. (1977)
N-GA_1-oyl-L-proline	–	–	ADAM et al. (1977)
N-GA_1-oyl-glycylglycine	–	–	ADAM et al. (1977)
GA_3-O(3)-β-D-glucopyranoside	YOKOTA et al. (1969a, 1971b)	SEMBDNER et al. (1968, 1972), ASAKAWA et al. (1974b) BARENDSE (1974, 1975)[a], BARENDSE and DE KLERK (1975)[a]	SCHNEIDER (1974), SCHNEIDER et al. (1974, 1977b)
GA_3-O(13)-β-D-glucopyranoside	–	–	SCHNEIDER (1974), SCHNEIDER et al. 1974, 1977b)
GA_3-O(3,13)-di-β-D-glucopyranoside	–	–	SCHNEIDER et al. (1977b),
GA_3-O-β-D-glucopyranosyl ester	–	–	HIRAGA et al. (1974c) MIERSCH and LIEBISCH (1974, 1980)
GA_3-O-α-D-glucopyranosyl ester	–	–	MIERSCH and LIEBISCH (1980)
GA_3-O(3)-acetate	SCHREIBER et al. (1966) McINNES et al. (1977)	–	CROSS (1954)
GA_3-O-n-propyl ester	HEMPHILL et al. (1973)	–	HEMPHILL et al. (1973)
N-GA_3-oyl amino-acid conjugates (R = L-Ala, D-Ala, L-Val, L-Leu, L-Phe, L-Ser, L-Pro, L-Tyr, L-Asp, Gly, GlyGly)	–	–	ADAM et al. (1977)
GA_4-O(3)-β-D-glucopyranoside	–	–	SCHNEIDER et al. (1977b)
GA_4-O-β-D-glucopyranosyl ester	HIRAGA et al. (1972, 1974a, d)	TAKAHASHI et al. (1976), YAMANE et al. (1975, 1977)	HIRAGA et al. (1974c)

GA_5-O(13)-β-D-glucopyranoside	–	Takahashi et al. (1976)[a], Yamane et al. (1975, 1977)[a]	Schneider et al. (1977b)
GA_5-O-β-D-glucopyranosyl ester	–	Takahashi et al. (1976)[a], Yamane et al. (1975, 1977)[a]	Schneider et al. (1977a)
O(13)-β-D-glucopyranosyl-GA_5-O-β-D-glucopyranosyl ester	–	–	Schneider et al. (1977a)
GA_8-O(2)-β-D-glucopyranoside	Schreiber et al. (1967, 1970), Hiraga et al. (1974a, b, d), Tamura et al. (1968), Yokota et al. (1969b), Harada and Yokota (1970)	Dathe et al. (1978a)[a], Sembdner et al. (1968), Nadeau and Rappaport (1972), Takahashi et al. (1976), Yamane et al. (1975, 1977), Davies and Rappaport (1975a, b), Nadeau et al. (1972)	Schneider (1979)
GA_8-O(13)-β-D-glucopyranoside	–	–	Schneider (1979)
GA_9-O-β-D-glucopyranosyl ester	Lorenzi et al. (1976)	–	Lorenzi et al. (1976)
GA_{20}-O(13)-β-D-glucopyranoside	–	Takahashi et al. (1976)[a], Yamane et al. (1975, 1977)[a]	–
GA_{26}-O(2)-β-D-glucopyranoside	Yokota et al. (1969a, b, 1971b)	–	–
GA_{27}-O(2)-β-D-glucopyranoside	Yokota et al. (1969a, b, 1971b)	–	–
GA_{29}-O(2)-β-D-glucopyranoside	Yokota et al. (1970, 1971b)	Railton et al. (1974b), Barendse (1976)[a]	–
Dihydro-GA_{31}-O-glucoside	–	Sponsel and MacMillan (1976, 1977)[a]	–
GA_{34}-O-glucoside	–	Wample et al. (1975)[a]	–
GA_{35}-O(11)-β-D-glucopyranoside	Yamane et al. (1971)	–	–
GA_{37}-O-β-D-glucopyranosyl ester	Hiraga et al. (1972, 1974a, d)	–	Hiraga et al. (1974c)
GA_{38}-O-β-D-glucopyranosyl ester	Hiraga et al. (1972, 1974a, d)	–	–
Gibberethione	Yokota et al. (1974)	–	Yokota et al. (1976)

[a] Tentative identification

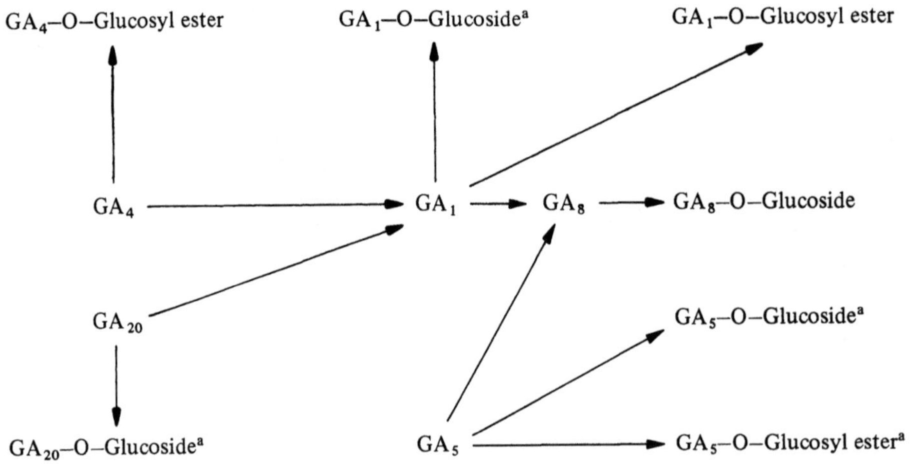

[a] Tentative identification

Fig. 4.34. Scheme of gibberellin glucosylation in bean seeds. (According to YAMANE et al., 1975)

by feeding free GA's to maturing French beans resulting in the formation of the O-glucosyl esters of GA_1, GA_4, and GA_5 as well as the O-glucosides of GA_1, GA_5, GA_8, and GA_{20} (see Fig. 4.34) (YAMANE et al., 1975). In early immature seeds the capability of glucosylation is strikingly decreased (YAMANE et al., 1977). In general, during seed ripening the content of free GA's was found to decrease to the same extent as the GA conjugate level increases, as shown by feeding ^3H-GA_8 to runner beans (SEMBDNER, 1979). During germination, a reconversion of GA conjugates into free GA's has been observed in *Phaseolus coccineus* (SEMBDNER et al., 1972; SEMBDNER, 1980), *Phaseolus vulgaris* (TAKAHASHI et al., 1976), and *Pharbitis nil* (BARENDSE and DE KLERK, 1975). However, germinating seeds and seedlings of *Phaseolus vulgaris* are also capable of converting free GA's into glucosides and glucosyl esters (NADEAU and RAPPAPORT, 1972; NASH and CROZIER, 1975). Metabolism of ^3H-GA_1 in seedlings of *Phaseolus vulgaris* (NADEAU and RAPPAPORT, 1972), in dwarf and normal maize plants (DAVIES and RAPPAPORT, 1975a, b) and in barley aleurone layers (NADEAU et al., 1972) always led to the GA_8 glucoside and, in barley aleurone cells, also to GA_1 glucoside (RAPPAPORT et al., 1974). Additionally, a polar amphoteric GA metabolite was isolated apparently containing a peptide bond (NADEAU and RAPPAPORT, 1974; STOLP et al., 1977). The conversion of ^3H-GA_1 to both GA_1-O-glucoside and GA_1-O-glucosyl ester was observed in excised lettuce hypocotyls (STODDART and JONES, 1977). In dwarf varieties of *Pisum sativum* the metabolism of free GA's was often shown to result in polar glucoside-like GA derivatives (MUSGRAVE and KENDE, 1970; RAILTON et al., 1974b). Thus, the application of GA_9 and GA_{20} to immature seeds led to dihydroGA_{31}-O-glucoside and a GA_{29} conjugate (FRYDMAN and MACMILLAN, 1975; SPONSEL and MACMILLAN, 1976, 1977). GA_4 fed to shoots of *Douglas fir* was assumed to be converted into GA_{34}-O-glucoside (WAMPLE et al., 1975).

The GA conjugates formed in the course of metabolical experiments are summarized in Table 4.8, column (b). But it should be pointed out that most of these results given were based only on chromatographic identification of the conjugates or on their hydrolytic cleavage, followed by identification of the GA moiety. Neither method, however, is able to clarify conclusively the position, or stereochemistry, of the linkage. Still more restrictions have to be considered in quantitative analysis of GA conjugates.

d) Enzymatic Hydrolysis and Biological Activity of Gibberellin Conjugates

From the beginning enzymatic hydrolysis of GA conjugates to the free GA's has been considered necessary for the observed biological activity, and function, of the conjugates. First indications were obtained from findings that Condurit B epoxide, a specific inhibitor of β-glucosidase, was able to suppress the biological activity of GA_8-O(2)-glucoside in dwarf pea bioassay (SEMBDNER et al., 1972, 1973). The hydrolysis rate of ^3H-GA_3-O-glucosylester was measured in different bioassay plants and found to parallel the biological activity observed (HIRAGA et al., 1974c; LIEBISCH, 1974).

Further studies, using in vitro systems of different enzyme preparations, showed a strong dependence of hydrolysis on conjugate type and stereochemistry. Thus, GA-O-β-D-glucosyl esters, as well as GA-O-α-D-glucosylesters, were found to be hydrolyzed readily by β-glucosidases and α-glucosidases, respectively (LIEBISCH, 1974). Enzymatic hydrolysis of a series of different GA-O-glucosides has been studied in vitro using both commercial enzymes and enzyme preparations from plant sources (KNÖFEL et al., 1974; MÜLLER et al., 1978). The rate of hydrolysis was found to be clearly influenced by the stereochemistry of the GA hydroxyl group linked to the sugar moiety (SCHNEIDER and SCHLIEMANN, 1979; SCHLIEMANN and SCHNEIDER, 1979). N-GA-oyl conjugates have been shown to be completely resistent to enzymatic hydrolysis (BORGMANN, 1977).

The biological activity of GA conjugates has been studied intensively using a series of bioassay systems such as dwarf pea, dwarf corn, dwarf rice, morning glory, lettuce hypocotyl, barley and wheat endosperm, and *Amaranthus* (SEMBDNER et al., 1968, 1972, 1973, 1974, 1976; YOKOTA et al., 1969b, 1971a; HARADA and YOKOTA, 1970; YAMANE et al., 1973; HIRAGA et al., 1974c; LIEBISCH, 1974; OGAWA and TAKAHASHI, 1974; HEMPHILL et al., 1973; LORENZI et al., 1976; BERNHARDT et al., 1979). The GA conjugates possess different relative activities, as compared to those of the corresponding free GA's depending on the chemical structure, the bioassay system and the application technique (via the shoot or the roots).

In summarizing all published results, it can be stated that conjugation leads to a partial or complete loss of biological activity. The potency of GA conjugates, as measured in different bioassays, parallels the hydrolytical cleavage and is, therefore, based upon the ability of plant enzymes and possibly of certain microbial activities (in non-sterile assays) to hydrolyze glucosidic, glucosyl, or alkyl ester, and amide linkages.

e) Biological Function of Gibberellin Conjugates

On the basis of their low biological activity, GA conjugates must be considered to be deactivated products of gibberellin metabolism (LANG, 1970; SEMBDNER, 1974, 1980; BARENDSE and DE KLERK, 1975). Glucosylation apparently leads to reversible deactivation, whereas GA-amino acid conjugates seem to be irreversibly deactivated. There is good evidence to conclude that GA conjugation is a useful means of regulating the physiologically active level of the GA content within plant tissue. Formation of GA conjugates takes place preferentially during seed ripening and reconversion is known to occur during seed germination. Hence, GA conjugates can be considered as depot or storage forms (HEMPHILL et al., 1973; LORENZI et al., 1976; SEMBDNER, 1980).

Dwarfism in *Zea mays* and *Pharbitis nil* has been discussed in connection with the ratio of free and conjugated GA's but no striking relationship could be derived from the results obtained (BARENDSE and LANG, 1972; DAVIES and RAPPAPORT, 1975a). Another possible function of GA glucosyl conjugates as favoured forms for long-distance translocation has been concluded from their occurrence in bleeding sap of trees (SEMBDNER et al., 1968; DATHE et al., 1978b). It must be added, however, that conjugation processes of gibberellins represent not in all plant species the main pathway of metabolism (SPONSEL et al., 1979).

4.4 Localization of Biosynthesis and Metabolism[1]

The wide field of biosynthesis and metabolic research has hitherto been directed mainly to the pathways including the identification and structural elucidation of precursors and intermediates, as well as the isolation of the enzymes involved. Far less attention has been given to the sites at which these processes take place. Systematic and detailed investigations on the sites of biosynthesis and metabolism in plant organs and tissues, as well as their cellular and subcellular location, are very rare. Contributions to this field have been made by the following research lines:

1. Occurrence of the endogenous hormones. However, correlations between location of hormones and sites of biosynthesis and/or metabolism are not cogent, as plant hormones can be translocated. Further requirements, which are very difficult to fulfill, are absolute identification and exact quantification of the hormones in plant tissues and cells.
2. Distribution pattern of the hormones studied in exudates, diffusates, and by some other physiological methods.
3. Localized detection of intermediates.
4. Incorporation of radioactive precursors by plant tissues in vivo and cell-free enzymic preparations in vitro.
5. Isolation of the enzymes from the tissue or fractionated cell organelles.

1 In addition, the reader is referred to the forthcoming volume 10 of this series, in particular to Chap. 6 (MATHYSSE and SCOTT)

Results obtained by these methods, and giving information on sites of biosynthesis and metabolism are quite heterogenous in the different plant hormone groups. Interpretation of the results is difficult and raises several questions, e.g., the mechanism of plant hormone transport and the ability of hormones and precursors to permeate through membranes.

With regard to the latter problem, information on the control of cell membrane permeability by phytochrome etc. becomes important. Sometimes data on localized hormone formation did not answer the question whether it resulted from de novo synthesis or by transformation of preformed precursors, or whether re-conversion of conjugated hormones led to active free forms.

4.4.1 Auxin

Since the early history of auxin research the coleoptile tip of cereals has been a site of IAA biosynthesis. This conception has been confirmed during studies on the biosynthetic pathway using radioactive precursors (cf. SCHNEIDER and WIGHTMAN, 1974, 1978; HEERKLOSS and LIBBERT, 1976a, b). At present, the view is widely accepted that the main sites of auxin synthesis are the meristemic tissues and young growing parts of plants, especially the shoot apex, buds, developing seeds, etc. Indications of the existence of an IAA-synthesizing system in these and other tissues of higher plants are given by studies on the (1) detection, identification, and quantification of native IAA (cf. SCHNEIDER and WIGHTMAN, 1974; McDOUGALL and HILLMAN, 1978; MOUSDALE et al., 1978), (2) occurrence of intermediate, (3) incorporation of radioactive labelled precursors, and (4) isolation of enzymes involved in the pathway. Results concerning (2)–(4) are reviewed in Section 4.1.1.

Systematic studies on the location of IAA biosynthesis are rare in recent literature. WIGHTMAN (1973), in studying the main site(s) of IAA synthesis in tomato shoots, found that, contrary to the generally accepted view, the formation of IAA is not confined to tissues at the shoot apex, but also occurs readily in all developing and mature leaves of a vegetative shoot system. The rapidly expanding leaves in the upper part of the shoot rather than the shoot tip tissues were the most active sites of IAA synthesis. The results were obtained by feeding radioactive tryptophan to excised whole shoots, shoot tips and leaves of different ages as well as to cell-free enzymic preparations of these tissues. Results, published by EDELBLUTH and KALDEWEY (1976), suggest that the shoot of *Narcissus* possesses two different sites of auxin production, namely, the apical region, represented either by the flower bud, the flower or the fruit, and the scape.

IAA biosynthesis has also been demonstrated to occur in plant tissue cultures (BLACK and HAMILTON, 1976). This discloses new possibilities in studying subcellular location of the IAA synthetic system. At present only little information is available on the compartmentation of single enzymes involved in the pathway. Thus, after subcellular fractionation of cucumber hypocotyl and coteyledon tissue the NADH-specific indole-3-acetaldehyde reductase was found only in the cytosol fractions, while a portion of the NADPH-specific indole-3-acetaldehyde reductase activity was associated with a microsomal fraction (BOWER et al., 1976).

Some further information is available on the localization of enzymes catalyzing IAA catabolism. Peroxidase and IAA-oxidase systems are present in various, if not all parts of higher plants. In plant organs IAA-oxidase activity normally shows an increasing gradient from young (growing) to old (non-growing) tissue, being highest in the old tissue (GALSTON and HILLMAN, 1961; cf. Sect. 4.2.1 c). However, there may be a high content of specific types of IAA-oxidase in some rapidly growing cells (LEE, 1971), and also, in the *Avena* coleoptile, IAA oxidase activity decreases along the length from apex to base (RAJAGOPAL and LARSEN, 1974). The distribution picture is complicated by the existence of natural inhibitors and cofactors which may themselves have distribution gradients (cf. Sect. 4.1.2b). Furthermore, IAA degradation in vivo and enzyme activity in homogenates are not automatically correlated. Cellular location of peroxidase has been reviewed in detail (SCHNEIDER and WIGHTMAN, 1974), showing that part of the peroxidase activity of the cell is soluble, but peroxidases are also associated with various organelles and subcellular compartments such as ribosomes, microsomes, lysosomal-like particles, plasmalemma, tonoplast, and other membranes. Isoperoxidases are assumed to differ in their subcellular distribution (MACNICOL, 1973; LEE, 1974). Intracellular location of IAA-oxidase activity has been much less studied. Binding of IAA-oxidase isozymes to the cell wall has been described (e.g., RIDGE and OSBORNE, 1970; DARIMONT et al., 1973), and localization of IAA-oxidase, distinct from peroxidase, has been demonstrated (LEE, 1974). The author demonstrated IAA oxidase and peroxidase in all subcellular fractions of tobacco callus cells, but both the IAA oxidase/peroxidase ratio and isoperoxidase composition were found to differ greatly in cytoplasmic and organelle-bound fractions, with IAA oxidase activities being particularly high in the plasma membrane fraction.

4.4.2 Cytokinins

Cytokinins and cytokinin-like substances have been isolated from numerous plant species and different plant parts. Seeds and fruits are known to contain high cytokinin levels (MILLER, 1965; LETHAM and WILLIAMS, 1969; BURROWS and CARR, 1970; KRASNUK et al., 1972; LETHAM, 1973; SHULMAN and LAVEE, 1976; DAVEY and VAN STADEN, 1978a, b; VAN STADEN and BUTTON, 1978). These findings do not unequivocally prove the seeds to be sites of cytokinin biosynthesis. However, further studies using *Zea mays* endosperm (MIURA and HALL, 1973), as well as fruits of tomato, tobacco, pea and bean (PETERSON and FLETCHER, 1973), indicate that seed and fruit tissues may be capable of producing cytokinins. The strongest evidence supporting this conception is the increase of cytokinin content, found in seeds of 10-day-old pea pods cultured independently from a root system (HAHN et al., 1974). Contradictory results were obtained using 5-day-old pea pods, showing the seeds to be a sink with a high metabolic activity (VAN STADEN and BUTTON, 1978).

A major site of cytokinin synthesis in higher plants is the root, especially the mitotically active root tip (cf. TORREY, 1976). This has been circumstantially documented by the detection of high cytokinin levels in roots (YOSHIDA and ORITANI, 1972; BEEVER and WOOLHOUSE, 1973), root tips (SHORT and TORREY, 1972) and in the xylem sap of various plants (KENDE and SITTON, 1967; KLÄMBT, 1968; KENDE, 1971; SKENE, 1972; BEEVER and WOOLHOUSE, 1973; HEWETT and WAREING, 1973a, b; DAVEY and VAN STADEN, 1976; VAN STADEN and DAVEY, 1976; HENSON and WHEELER, 1977d). Secretion of cytokinin into the culture medium has also been reported from roots of *Xanthium*, maize, tomato, and *Lupinus* (HENSON and WAREING, 1976; DAVEY and VAN STADEN, 1978a, b, c) as well as from excised tomato root tips, subcultured throughout eight passages (KODA and OKAZAWA, 1978). From this, and the findings that there are qualitative and quantititive differences in the endogenous cytokinins of roots, xylem sap, and shoot tissues (YOSHIDA et al., 1971; ENGELBRECHT, 1972; SKENE, 1972; YOSHIDA and ORITANI, 1972; GORDON et al., 1974; HENSON and WAREING, 1976; DAVEY and VAN STADEN 1978a, b, c), it can be generally stated that cytokinins originate mainly in the root system and are transported to the shoot via the transpiration stream in xylem.

Root nodules of *Phaseolus vulgaris* (PUPPO et al., 1974), *Pisum sativum* (SYONO and TORREY, 1976), *Vicia faba* (HENSON and WHEELER, 1976), and *Alnus glutinosa* (HENSON and WHEELER, 1977a, b), as well as leaf nodules of *Psychotria punctata* (EDWARDS and LAMOTTE, 1975), have also been shown to be sites of cytokinin formation. In pea root nodules it was shown that the nodule meristem is the source of typical "plant cytokinins", e.g., io^6Ade and its derivations, whereas i^6A and its derivations are found in nodules originating from the infecting bacteria *Rhizobium leguminosarum*. Cytokinin production seems to be higher in the meristematic tissue than in the infected nodule cells in which nitrogen is fixed (SYONO et al., 1976; SYONO and TORREY, 1976). The high level of cytokinins detected in nodules led to the suggestion that they might be a source of cytokinins for the plant (RODRIGUEZ-BARRUECA and DE CASTRO, 1973; HENSON and WHEELER, 1977a, d).

In addition to the root sites, described above, it has recently been demonstrated that cytokinin biosynthetic sites are also located in the shoot. Rootless tobacco plants, originated from regenerating tobacco callus in vitro, were capable of synthesizing i^6Ade and its derivatives from (8-^{14}C)-Ade (CHEN and PETSCHOW, 1978a). Tobacco callus, known to be most active in metabolizing cytokinins, may possibly also be capable of biosynthesizing them (EINSET and SKOOG, 1973; LALOUE et al., 1975, 1977). At the cellular level, differences of cytokinin biosynthesis in chloroplast tRNA and "normal leaf tRNA" of spinach was found (VREMAN et al., 1978; see Sect. 4.1.2a). The occurrence of cytokinin-like bases in tRNA is assumed to be restricted to only a few tRNA-species, e.g., $tRNA^{Leu}$, in higher eukaryotes (STRUXNESS et al., 1979).

Considering the metabolism of cytokinins, some processes seem to be restricted to distinct plant species or plant tissues. Thus, conversion of io^6-Ade to dihydro derivatives was obtained in bean axes, in *Populus alba*, and in root nodules of *Alnus glutinosa* (SONDHEIMER and TZOU, 1971; LETHAM et al., 1977; HENSON and WHEELER, 1977c) whereas radish seedlings were not active (PARKER and LETHAM, 1973; GORDON et al., 1974). Similarly, side chain splitting of cytokinins by cytokinin oxidase was found to occur in vivo in seedlings of maize, radish and bean, maize caryopses, tobacco cell cultures, as well as in vitro using enzyme preparations (CHEN et al., 1968; PARKER et al., 1972, 1973; PARKER and LETHAM, 1973, 1974; GORDON et al., 1974; WHITTY and HALL, 1974; LALOUE et al., 1977). However, it could not be observed in bean axes, rape seedlings, and barley leaves (SONDHEIMER and TZOU, 1971; PAČES, 1976; PAČES et al., 1977).

On the other hand, metabolic processes such as ribosylation and subsequent phosphorylation or glucosylation of cytokinins seem to be ubiquitous. Thus, root tissues are known to be most active in converting cytokinins into ribosides which are transported to the sink organs via the transpiration stream in xylem (SKENE, 1972; PARKER et al., 1972, 1973; HEWETT and WAREING, 1973a, b; GORDON et al., 1974; HENSON and WAREING, 1976; VAN STADEN, 1976b; HENSON and WHEELER, 1977c, d; DAVEY and VAN STADEN, 1978b, c; RAMINA, 1979). Furthermore, the formation of ribosides and ribotides from administered cytokinins has been observed in ash embryos (TZOU et al., 1973), bean axes (SONDHEIMER and TZOU, 1971), intact seedlings of maize and radish (PARKER et al., 1973; PARKER and LETHAM, 1973, 1974; GORDON et al., 1974), as well as in soybean callus cultures (VAN STADEN and PARAPHILIPPOU, 1977).

Favoured sites of glucosylation seem to be the leaves, especially senescing leaves, and the glucosides formed are considered to be storage forms translocated via the phloem into storage organs (PARKER et al., 1973; LANGILLE and FORSLINE, 1974; LETHAM et al., 1977; VAN STADEN, 1976b, 1977; HENSON, 1978a, b). Furthermore glucosylation of cytokinins at different positions has been reported in various other plant parts, such as radish cotyledons (PARKER et al., 1972), root tissues and de-rooted seedlings of radish (GORDON et al., 1974), bean axes (SONDHEIMER and TZOU, 1971) and soybean callus (HORGAN, 1975).

Reconversion of ribosyl derivatives to free cytokinins was found to dominate in mature leaves (PAČES, 1976, LETHAM et al., 1977). The hydrolysis of ribosides by using crude enzyme preparations from barley leaves or rape seedlings could also be achieved (PAČES, 1976; PAČES et al., 1977). The increase of cytokinin activity after wounding of potato tubers was assumed to occur by enzymatic conversion of storage forms of cytokinins (CONRAD and KÖHN, 1975).

Rates of cytokinin biosynthesis and metabolism have been shown to be related to the physiological stage (see above). Some effects of external factors have also been reported. Thus, red light influenced cytokinin conjugation in poplar leaves (HEWETT and WAREING, 1973c). Desiccation resulted in a decrease of cytokinin level in *Lactuca sativa*, possibly by inactivation which can be reversed by re-watering (ITAI and VAADIA, 1971; AHARONI et al., 1977). Abscisic acid and cycloheximide were found to curtail the conversion of io^{6}Ade into H_2-io^{6}Ade in bean axes while the total amount of zeatin metabolism was not affected (SONDHEIMER and TZOU, 1971). ABA was found to inhibit the conversion of kinetin into its nucleotide in germinating lettuce seeds (MIERNYK, 1979). Other inhibitors, as well as increased N, K, P nutrition, resulted in a decrease of cytokinin level in *Coleus* (BANKO and BOE, 1975).

However, systematic investigations on the regulation of cytokinin biosynthesis and metabolism have not yet been performed.

4.4.3 Abscisic Acid

Numerous reports deal with the occurrence and concentration of ABA in whole plants or plant parts (cf. reviews cited in Sect. 4.1.4), but very few experiments have been done to determine the cellular and intracellular location in relation to sites of synthesis and metabolism.

Most experiments on ABA biosynthesis have been carried out with avocado fruits, due to the high concentration of ABA (up to 10 mg/kg). Moreover, the incorporation of labelled mevalonate into ABA is more efficient than in any other plant tissue known at present. Mevalonate incorporation into ABA has also been demonstrated in cotyledons of mature avocado seed, avocado stem, green tomato fruit, embryo and endosperm of developing wheat grain, and in suspension cultures from grape pericarp tissue (see Sect. 4.1.4a and literature cited therein).

In 1974 it was found that (a) ABA is present in pea chloroplasts (RAILTON et al., 1974c) and (b) lysed chloroplasts, isolated from both ripening avocado and bean fruit and avocado leaves, are able to synthesize ABA from (2-^{14}C)-mevalonate (MILBORROW, 1974c). Therefore, it was suggested that ABA is biosynthesized within the chloroplasts. This assumption is in agreement with experiments on the intracellular location of ABA both in stressed and non-stressed spinach leaf tissue and in chloroplast preparations (LOVEYS, 1977). In non-stressed spinach leaves 97% of total ABA has been found in the chloroplasts. Mild stress resulted in a doubling of chloroplastic ABA, whereas leaf ABA rose by a factor of 11. Thus, the majority of ABA in stressed spinach leaf tissue is located outside the chloroplast. While the possibility of extra-chloroplastic ABA synthesis cannot be discounted, the data indicate that stress-induced ABA synthesis occurs in the mesophyll chloroplasts and that the ABA readily migrates from there to other parts.

Sites of ABA synthesis and metabolism in stressed and non-stressed *Ricinus communis* plants have been studied by analyzing the level of ABA and its two metabolites, phaseic acid and dihydrophaseic acid (ZEEVAART, 1977). It was found that in non-stressed plants ABA is synthesized and metabolized in mature leaves and both ABA and its two metabolites are translocated in the phloem to the shoot tips. Leaves in a very early stage of development are capable of synthesizing ABA in response to water stress. ABA was not metabolized by endosperm, but by aleurone layers of barley seeds, indicating that metabolism was tissue-specific (DASHEK et al., 1979).

4.4.4 Gibberellins

The literature reviewed by LANG (1970), JONES (1973) and RAPPAPORT and ADAMS (1978) strongly supports the view that both the root apical region and young leaves of apical buds are sites of GA biosynthesis in plant seedlings. This concept is based mainly on the occurrence of gibberellins in bleeding sap (xylem exudate) (CARR et al., 1964; PHILLIPS and JONES, 1964; JONES and

PHILLIPS, 1967; REID and CARR, 1967; SITTON et al., 1967; CARR and REID, 1968; JONES and LANG, 1968; SKENE, 1967; MURAKAMI, 1968), phloem exudates (KLUGE et al., 1964; HOAD and BOWEN, 1968) and in agar diffusates (JONES and PHILLIPS, 1966, 1967; STODDART and LANG, 1968; JONES and LANG, 1968). Applying inhibitors of gibberellin biosynthesis to the roots rapidly decreased the amounts of hormone collected in diffusates or exudates (JONES and PHILLIPS, 1967; REID and CARR, 1967), thus demonstrating de novo synthesis. Removal of root apices from *Phaseolus coccineus* seedlings resulted in decreased shoot growth and disappearance of GA_1 (CROZIER and REID, 1971). From these and other results CROZIER and REID concluded that shoots are the site of biosynthesis, whereas roots are a site of gibberellin conversion.

Another organ of GA biosynthesis and metabolism is the developing seed. Immature seeds and fruits are known to contain high amounts of gibberellins, with components and quantities varying with different physiological stages (cf. TAKAHASHI et al., 1976). Detailed studies on the occurrence (DURLEY et al., 1971; HIRAGA et al., 1974a) and metabolism (YAMANE et al., 1975, 1977) of gibberellins in *Phaseolus vulgaris* seeds supported the view that they are a site of metabolism. Investigations on the metabolism of GA_9 using *Pisum sativum* seeds cultivated in vitro prove this organ to be autonomous in metabolizing gibberellins (SPONSEL and MACMILLAN, 1977). During maturation of seeds the formation of gibberellin sugar conjugates has been observed in *Phaseolus coccineus* (SEMBDNER et al., 1964, 1972) and *Phaseolus vulgaris* (YAMANE et al., 1975, 1977). In general, maturing seeds seem to be the predominant site of conjugate formation (see Sect. 4.3.4c).

Inside the seed the sites of biosynthesis may be the cotyledons of the embryo (*Pisum sativum,* cf. COOLBAUGH and MOORE, 1971a, 1971b) or the endosperm (Cucurbitaceae, cf. UPPER and WEST, 1967; GRAEBE et al., 1974a, b). This was shown by studies using cell-free enzyme preparations from the different plant parts in order to follow biosynthetic steps starting from mevalonate (see Table 4.2). The suspensor, an embryonic organ in many plant species, is also discussed as a site of GA biosynthesis in *Phaseolus coccineus* (ALPI et al., 1975; CIONINI et al., 1976). Cell-free enzyme preparations from suspensors converted mevalonate into ent-kaurene, whereas an endosperm system did not (CECCARELLI et al., 1979). In cereal caryopses the scutellum proved to be a site of GA biosynthesis, probably from preformed precursors (*Hordeum vulgare:* BRIGGS, 1972). A de novo synthesis has been found to operate in embryonic axis of chilled *Corylus* seeds, whereas isolated cotyledons showed release of free gibberellins from "bound" forms only (ARIAS et al., 1976). Studies on the distribution of free gibberellins and GA_{16} conjugate in fruit parts of rye suggested that sites of biosynthesis and metabolism seem to be strictly related to the developmental stage (DATHE and SEMBDNER, 1978).

Recent investigations have shown significant levels of gibberellin-like activity associated with lysates of plastids from leaves of higher plants (STODDART, 1968; RAILTON and WAREING, 1973; RAILTON and REID, 1974; COOKE and SAUNDERS, 1975; COOKE and KENDRICK, 1976; BROWNING and SAUNDERS, 1977). Chloroplasts of *Solanum andigena* contained the same amounts of gibberellin-like substances as leaf extracts (RAILTON and WAREING, 1973) and treatment of isolated chloroplasts of *Triticum aestivum* with Triton X-100 yielded GA_4 and GA_9 in unusually high levels (BROWNING and SAUNDERS, 1977). The latter result is not however reproducible (P.F. SAUNDERS, personal communication).

These observations led to the investigation of plastids as a possible site of gibberellin biosynthesis. In an early attempt STODDART (1969) found that ent-kaurenoic acid was incorporated into a GA_1/GA_3-like compound using *Brassica* chloroplast preparations. The preceding steps are shown to operate in plastids also, as could be demonstrated by the conversion of geranylgeranyl pyrophosphate to ent-kaurene in proplastids of *Marah* endosperm (SIMCOX et al., 1975) and with enzyme extracts of pea chloroplasts (MOORE and COOLBAUGH, 1976). Oxidation of ent-kaurenol was shown to operate in plastids of developing *Hordeum distichon* grains (MURPHY and BRIGGS, 1975) and in microsomal fractions of *Marah* (HASSON and WEST, 1976a, b). The failure of mevalonate incorporation into kaurenoids and gibberellins may be due to the loss of mevalonate kinase activity during aqueous preparation procedures in the isolation of chloroplasts. The presence of geranylgeranyl pyrophosphate synthetase has been established for proplastids from castor bean endosperm (GREENE et al., 1975). Metabolism of GA_9 and GA_{20} in isolated pea chloroplast led to the suggestion that these organelles are the major site of gibberellin interconversion in shoots (RAILTON and REID, 1974).

Recent investigations provide further evidence for metabolic routes, specifically localized organelles. Protoplasts of barley leaves converted labelled GA_1 to GA_8, GA_8 glucoside and unknown metabolites, whereas vacuoles of the same plant source were incapable of further conversion of the rapidly formed GA_8 (RAPPAPORT and ADAMS, 1978).

4.4.5 Ethylene

Ethylene production has been proved to occur in all plant organs – roots, stems, leaves, buds, tubers, bulbs, flowers, fruits, and seeds, etc. Rates of ethylene production vary from organ to organ and are dependent on growth and developmental processes (see Sect. 4.1.6d). Within the same organ, the rates of ethylene production vary in different tissues (cf. ABELES, 1973). Knowledge of the localization of ethylene biosynthesis at the tissue level, or at the cellular level, is rather limited. In peach seeds, ethylene production is localized in the seed coat (JERIE and CHALMERS, 1976), and in avocado, the preclimacteric ethylene is also produced mainly by the seed coat (ADATO and GAZIT, 1977a, b). Climacteric tomato fruits showed the highest rates of ethylene production in the peel (WARDALE, 1973). Studies on proteinaceous inhibitors of ethylene production by mung bean hypocotyl segments suggested that auxin-induced ethylene production is localized in the epidermis (SAKAI and IMASEKI, 1973b). Subcellular localization of ethylene biosynthesis is almost a closed book. Several authors, observing a loss of ethylene-producing capacity by disintegration of the tissue, concluded that enzymes located in organelles are necessary for ethylene biosynthesis (cf. ABELES, 1973; ELSTNER and KONZE, 1976; KONZE and ELSTNER, 1976a, c). Ethylene-producing systems using subcellular fractions have been described, such as preparations of chloroplasts (PORUTSKII et al., 1962; PORUTSKII and MATKOVSKII, 1963; ELSTNER et al., 1976; KONZE and ELSTNER, 1976a, c) and mitochondria (SPENCER, 1969; SPENCER and MEHERIUK, 1963; RAM CHANDRA et al., 1963; MEHERIUK and SPENCER, 1964, 1967a, b; STINSON and SPENCER, 1970; KONZE and ELSTNER, 1976c).

However, there are several reasons to suspect that the evidence from these systems, and results obtained therefrom, do not reflect the in vivo situation. Involvement of cellular membranes, probably the plasma membrane, in regulating ethylene production, has been demonstrated recently (ODAWARA et al., 1977). Affecting membrane structures by application of phospholipase, lecithin, or detergents resulted in inhibition of ethylene production by mung bean tissue. A similar conclusion was derived from studies on the temperature-dependence of ethylene production in apple and tomato and the effect of Triton X-100, suggesting that the ethylene-synthesizing system is associated with a cell wall-cell membrane complex and with a phase change in the lipid component of the membranes (MATTOO et al., 1977). Further information was obtained from studies using protoplasts (MATTOO and LIEBERMAN, 1977a, b). Isolated protoplasts from apple tissue do not synthesize ethylene, whereas protoplasts, on regenerating their cell walls, regain the ability to synthesize ethylene. Ethylene biosynthesis was dependent on application of methionine and could be inhibited by a rhizobitoxine analogue and propyl gallate.

Information on the localization of ethylene metabolism is not yet available.

Acknowledgement. The authors wish to thank Mrs. C. KLOSE for the valuable assistance in preparing the manuscript.

References

Aasheim, T., Iversen, T.H.: Decarboxylation and transport of auxin in segments of sunflower and cabbage roots. II. Chromatographic study using IAA-1-C^{14} and IAA-5-H^3. Physiol. Plant. *24*, 325–329 (1971)

Abe, H., Uchiyama, H., Tanaka, Y., Saito, H.: Structure of discadenine, a spore germinating inhibitor from the cellular slime mold, *Dictyostelium discoideum*. Tetrahedron Lett. 3807–3810 (1976)

Abeles, A.L., Abeles, F.B.: Biochemical pathway of stress induced ethylene. Plant Physiol. *50*, 496–498 (1972)

Abeles, F.B.: Ethylene production from linolenic acid. Nature (London) *210*, 23–25 (1966a)

Abeles, F.B.: Auxin stimulation of ethylene evolution. Plant Physiol. *41*, 585–588 (1966b)

Abeles, F.B.: Ethylene production from linolenic acid. Nature (London) *210*, 23–25 (1969a)

Abeles, F.B.: Abscission: Role of cellulase. Plant Physiol. *44*, 447–452 (1969b)

Abeles, F.B.: Biosynthesis and mechanism of action of ethylene. Annu. Rev. Plant Physiol. *23*, 259–292 (1972)

Abeles, F.B.: Ethylene in plant biology. 302 pp. New York, London: Academic Press 1973

Abeles, F.B., Lonski, J.: Stimulation of lettuce seed germination by ethylene. Plant Physiol. *44*, 277–280 (1969)

Abeles, F.B., Rubinstein, B.: Regulation of ethylene evolution and leaf abscission by auxin. Plant Physiol. *39*, 963–969 (1964a)

Abeles, F.B., Rubinstein, B.: Cell-free ethylene evolution from etiolated pea seedlings. Biochim. Biophys. Acta *93*, 675–677 (1964b)

Abeles, F.B., Ruth, J.M., Forrence, L.E., Leather, G.R.: Mechanisms of hormone action. Use of deuterated ethylene to measure isotopic exchange with plant material and the biological effects of deuterated ethylene. Plant Physiol. *49*, 669–671 (1972)

Adam, G., Lischewski, M., Sych, F.-J., Ulrich, A.: Gibberelline XLVI. Synthese von Gibberellin-Aminosäurekonjugaten. Tetrahedron *33*, 95–100 (1977)

Adamiec, A.: The effect of plant growth regulators on strains of *Gibberella fujikuroi.* Acta Soc. Bot. Pol. *35*, 489–510 (1966)

Adams, D.O., Yang, S.F.: Methionine metabolism in apple tissue. Implication of S-adenosylmethionine as an intermediate in the conversion of methionine to ethylene. Plant Physiol. *60*, 892–896 (1977)

Adams, D.O., Yang, S.F.: Ethylene Biosynthesis: Identification of 1-aminocyclopropane-1-carboxylic acid as an intermediate in the conversion of methionine to ethylene. Proc. Natl. Acad. Sci. USA *76*, 170–174 (1979)

Adato, I., Gazit, S.: Water-deficit stress, ethylene production, and ripening in avocado fruits. Plant Physiol. *53*, 45–46 (1974)

Adato, I., Gazit, S.: Role of ethylene in avocado fruit development and ripening I. Fruit drop. J. Exp. Bot. *28*, 636–643 (1977a)

Adato, I., Gazit, S.: Role of ethylene in avocado fruit development and ripening. II. Ethylene production and respiration by harvested fruits. J. Exp. Bot. *28*, 644–649 (1977b)

Adato, I., Gazit, S.: Changes in the initiation of climacteric ethylene in harvested avocado fruits during their development. J. Sci. Food Agric. *28*, 240–242 (1977c)

Addicott, F.T.: Plant hormones in the control of abscission. Biol. Rev. *45*, 485–524 (1970)

Addicott, F.T., Lyon, J.L.: Physiology of abscisic acid and related substances. Annu. Rev. Plant. Physiol. *20*, 139–164 (1969)

Adesomoju, A.A., Okogun, J.I., Ekong, D.E.U., Gaskin, P.: GC-MS Identification of abscisic acid and abscisic acid metabolites in seed of *Vigna unguiculata*. Phytochemistry *19*, 223–225 (1980)

Aharoni, N., Blumenfeld, A., Richmond, A.E.: Hormonal activity in detached lettuce leaves as affected by leaf water content. Plant Physiol. *59*, 1169–1173 (1977)

Aharoni, Y.: Respiration of oranges and grapefruits harvested at different stages of development. Plant Physiol. *43*, 99–102 (1968)

Alpi, A., Tognoni, F., D'Amato, F.: Growth regulation levels in the embryo and suspensor of *Phaseolus coccineus* at two stages of development. Planta *127*, 153–162 (1975)

Andersen, A.S., Møller, J.B., Hansen, J.: 3-Methyleneoxindole and plant growth regulation. Physiol. Plant. *27*, 105–108 (1972)

Anderson, J.D., Moore, T.C.: Biosynthesis of (−)-kaurene in cell-free extracts of immature pea seeds. Plant Physiol. *42*, 1527–1534 (1967)

Anderson, J.D., Mandava, N., Garrett, S.: Inhibition of hormone-induced ethylene synthesis by the indole plant-growth inhibitor from *Abrus precatorius* seeds. Plant Cell Physiol. *16*, 1233–1236 (1975)

Andreae, W.A.: Auxin metabolism and root growth inhibition. In: Régulateurs naturels de la croissance végétale. Nitsch, J.P. (ed.), pp. 559–573. Paris: CNRS 1964

Andreae, W.A.: Uptake and metabolism of indoleacetic acid, naphthaleneacetic acid, and 2,4-dichlorophenoxyacetic acid by pea root segments in relation to growth inhibition during and after auxin application. Can. J. Bot. *45*, 737–753 (1967)

Andreae, W.A., Good, N.E.: The formation of indole-acetyl aspartic acid in pea seedlings. Plant Physiol. *30*, 380–382 (1955)

Andreae, W.A., van Ysselstein, M.W.H.: Studies on 3-indoleacetic acid metabolism. III. The uptake of 3-indoleacetic acid by pea epicotyls and its conversion to 3-indoleacetyl-aspartic acid. Plant Physiol. *31*, 235–240 (1956)

Andreae, W.A., van Ysselstein, M.W.H.: Studies on 3-indoleacetic acid metabolism. V. Effect of calcium ions on 3-indoleacetic acid uptake and metabolism by pea roots. Plant Physiol. *35*, 220–224 (1960)

Anker, L.: Auxin synthesis inhibition by abscisic acid, and its reversal by gibberellic acid. Acta Bot. Neerl. *24*, 339–347 (1975)

Anstis, P.J.P., Friend, J., Gardner, D.C.J.: The role of xanthoxin in the inhibition of pea seedling growth by red light. Phytochemistry *14*, 31–35 (1975)

Arias, I., Williams, P.W., Bradbeer, J.W.: Studies in seed dormancy. IX. The role of gibberellin biosynthesis and the release of bound gibberellin in the post-chilling accumulation of gibberellin in seeds of *Corylus avellana* L. Planta *131*, 135–139 (1976)

Armstrong, D.J., Burrows, W.J., Evans, P.K., Skoog, F.: Isolation of cytokinins from tRNA. Biochem. Biophys. Res. Commun. *37*, 451–456 (1969)

Armstrong, D.J., Murai, N., Taller, B.J., Skoog, F.: Incorporation of cytokinin N^6-benzyladenine into tobacco callus transfer ribonucleic acid and ribosomal ribonucleic acid preparations. Plant Physiol. *57*, 15–22 (1976a)

Armstrong, D.J., Scarbrough, E., Skoog, F., Cole, D.L., Leonard, N.J.: Cytokinins in *Corynebacterium fascians* cultures. Isolation and identification of 6-(4-hydroxy-3-methyl-cis-2-butenylamino)-2-methylthiopurine. Plant Physiol. *58*, 749–752 (1976b)

Asakawa, Y., Tamari, K., Inoue, K., Kaji, J.: Translocation and intracellular distribution of tritiated gibberellin A_3. Agric. Biol. Chem. *38*, 713–717 (1974a)

Asakawa, Y., Tamari, K., Shoji, A., Kaji, J.: Metabolic products of gibberellin A_3 and their interconversion in dwarf kidney bean plants. Agric. Biol. Chem. *38*, 719–725 (1974b)

Aung, L.H., Rees, A.R.: Changes in endogenous gibberellin levels in *Tulipa* bulblets during ontogeny. J. Exp. Bot. *25*, 745–751 (1974)

Babcock, D.F., Morris, R.O.: Quantitative measurement of isoprenoid nucleosides in transfer ribonucleic acid. Biochemistry *9*, 3701–3705 (1970)

Baker, J.E., Lieberman, M., Anderson, J.D.: Inhibition of ethylene production in fruit slices by a rhizobitoxine analog and free radical scavengers. Plant Physiol. *61*, 886–888 (1978)

Bakker, H.J., Jefferies, P.R., Knox, J.R.: Gibbane metabolites from ent-kaura-2,16-dien-19-ol. Tetrahedron Lett. 2723–2728 (1972)

Bakker, H.J., Cook, I.F., Jefferies, P.R., Knox, J.R.: Gibberellin metabolites from ent-kaura-2,16-dien-19-ol and its succinate in *Gibberella fujikuroi*. Tetrahedron *30*, 3631–3640 (1974)

Baldev, B., Lang, A., Agatep, A.O.: Gibberellin production in pea seeds developing in excised pods: Effect of growth retardant AMO-1618. Science *147*, 155–157 (1965)

Bandurski, R.S.: Chemistry and physiology of myoinositol esters of indole-3-acetic acid. In: Cyclitols and the phosphoinositides. Wells, W.W., Eisenberg, F. (eds.), pp. 35–54. New York, London: Pergamon Press 1978

Bandurski, R.S., Schulze, A.: Concentration of indole-3-acetic acid and its ester in *Avena* and *Zea*. Plant Physiol. *54*, 257–262 (1974)

Bandurski, R.S., Schulze, A.: Concentration of indole-3-acetic acid and its derivatives in plants. Plant Physiol. *60*, 211–213 (1977)

Bandurski, R.S., Schulze, A., Cohen, J.D.: Photo-regulation of the ratio of ester to free indole-3-acetic acid. Biochem. Biophys. Res. Commun. *79*, 1219–1223 (1977)

Bangerth, F.: Zur Wirkungsweise einiger Substanzen bei der Reduktion der Haltekräfte von Explantaten verschiedener Steinobstarten. Angew. Bot. *49*, 31–39 (1975)

Banko, T.J., Boe, A.A.: Effects of pH, temperature, nutrition, ethephon and chlormequat on endogenous cytokinin levels of *Coleus blumei* Benth. J. Am. Soc. Hortic. Sci. *100*, 168–172 (1975)

Barendse, G.W.M.: Formation of bound gibberellins in *Pharbitis nil*. Planta *99*, 290–301 (1971)

Barendse, G.W.M.: Accumulation and metabolism of radioactive gibberellic acid in seedlings of *Pharbitis nil* Chois. In: Plant growth substances 1973. pp. 332–341. Tokyo: Hirokawa 1974

Barendse, G.W.M.: Biosynthesis, metabolism, transport, and distribution of gibberellins. In: Gibberellin and plant growth. Krishnamoorthy, H.N. (ed.), pp. 65–89. New Delhi: Wiley 1975

Barendse, G.W.M.: The metabolism of gibberellins in *Pharbitis nil*. In: Abstr. 9th Int. Conf. Plant Growth Subst. 1976. p. 24. Lausanne 1976

Barendse, G.W.M., Gilissen, H.A.M.: The diffusion of gibberellins into agar and water during early germination of *Pharbitis nil* Choisy. Planta *137*, 169–175 (1977)

Barendse, G.W.M., Klerk de, G.J.M.: The metabolism of applied gibberellic acid in *Phar-*

bitis nil Choisy: Tentative identification of its sole metabolite as gibberellic acid glucoside and some of its properties. Planta *126*, 25–35 (1975)

Barendse, G.W.M., Kok, N.J.J.: Incorporation of ^{14}C-kaurene into the gibberellin of a higher plant (*Pharbitis nil.* Choisy). Plant Physiol. *48*, 476–479 (1971)

Barendse, G.W.M., Lang, A.: Comparison of endogenous gibberellins and of the fate of applied radioactive gibberellin A_1 in a normal and a dwarf strain of Japanese morning glory. Plant Physiol. *49*, 836–841 (1972)

Barendse, G.W.M., Klerk, de, G.J.M., Mierlo, J.v.: Metabolism of radioactive gibberellins in *Pharbitis nil* Choisy. In: Plant growth regulators. Kudrev, T., Ivanova, I., Karanov, E. (eds.), pp. 161–164. Sofia: Academy of Sciences 1977

Barnes, M., Light, E.N., Lang, A.: The action of plant growth retardants on terpenoid biosynthesis. Planta *88*, 172–182 1969)

Barthe, P., Bulard, C.: Bound and free abscisic acid levels in dormant and after-ripened embryos of *Pyrus malus* L. cv. Golden delicious. Z. Pflanzenphysiol. *90*, 201–208 (1978)

Bartz, L., Söll, D.: N^6-(Δ^2-isopentenyl)adenosine: biosynthesis in vitro in transfer RNA by an enzyme purified from *Escherichia coli.* Biochimie *54*, 31–39 (1972)

Barz, W.: Degradation of polyphenols in plants and plant cell suspension cultures. Physiol. Veg. *15*, 261–277 (1977)

Basu, P.A., Tuli, V.: Auxin activity of 3-methyleneoxindole in wheat. Plant Physiol. *50*, 499–502 (1972a)

Basu, P.S., Tuli, V.: Enzymatic dehydration of 3-hydroxymethyloxindole. Plant Physiol. *50*, 503–506 (1972b)

Basu, P.S., Tuli, V.: The binding of indole-3-acetic acid and 3-methyleneoxindole to plant macromolecules. Plant Physiol. *50*, 507–509 (1972c)

Baur, A.H., Yang, S.F.: Ethylene production from propanol. Plant Physiol. *44*, 189–192 (1969a)

Baur, A.H., Yang, S.F.: Precursors of ethylene. Plant Physiol. *44*, 1347–1349 (1969b)

Baur, A.H., Yang, S.F.: Formation of ethionine from homocysteine and of S-methylmethionine from methionine in apple tissue. Phytochemistry *11*, 2503–2505 (1972a)

Baur, A.H., Yang, S.F.: Methionine metabolism in apple tissue in relation to ethylene biosynthesis. Phytochemistry *11*, 3207–3214 (1972b)

Baur, A.H., Yang, S.F., Pratt, H.K., Biale, J.B.: Ethylene biosynthesis in fruit tissues. Plant Physiol. *47*, 696–699 (1971)

Bearder, J.R., Sponsel, V.M.: Selected topics in gibberellin metabolism. Biochem Soc. Trans. *5*, 569–582 (1977)

Bearder, J.R., Hedden, P., MacMillan, J., Wels, C.M.: Gibberellin biosynthesis in the mutant B1–41a of *Gibberella fujikuroi.* J.C.S. Chem. Commun. 777–778 (1973a)

Bearder, J.R., MacMillan, J., Phinney, B.O.: 3-Hydroxylation of gibberellin A_{12}-aldehyde in *Gibberella fujikuroi* strain REC-193A. Phytochemistry *12*, 2173–2179 (1973b)

Bearder, J.R., MacMillan, J., Phinney, B.O.: Conversion of gibberellin A_1 into gibberellin A_3 by the mutant R-9 of *Gibberella fujikuroi.* Phytochemistry *12*, 2655–2659 (1973c)

Bearder, J.R., MacMillan, J., Wels, C.M., Phinney, B.O.: Metabolism of steviol and its derivatives by *Gibberella fujikuroi,* mutant B1–41a. J.C.S. Chem. Commun. 778–779 (1973d)

Bearder, J.R., MacMillan, J., Wels, C.M., Chaffey, M.B., Phinney, B.O.: Position of the metabolic block for gibberellin biosynthesis in mutant B1–41a of *Gibberella fujikuroi.* Phytochemistry *13*, 911–917 (1974)

Bearder, J.R., MacMillan, J., Wels, C.M., Phinney, B.O.: The metabolism of steviol to 13-hydroxylated ent-gibberellanes and ent-kauranes. Phytochemistry *14*, 1741–1748 (1975a)

Bearder, J.R., MacMillan, J., Phinney, B.O.: Fungal products XIV. Metabolic pathway from ent-kaurenoic acid to the fungal gibberellins in mutant B1–41a of *Gibberella fujikuroi.* J. Chem. Soc. Perkin *I,* 721–726 (1975b)

Bearder, J.R., Dennis, F.G., MacMillan, J., Martin, G.C., Phinney, B.O.: A new gibberellin (A_{45}) from seed of *Pyrus communis* L. Tetrahedron Lett. 669–670 (1975c)

Bearder, J.R., Frydman, V.M., Gaskin, P., MacMillan, J., Wels, C.M., Phinney, B.O.: Fungal products. Part XVI. Conversion of isosteviol and steviol into gibberellin anal-

ogues by mutant B1–41a of *Gibberella fujikuroi* and the preparation of [^{3}H] gibberellin A_{20}. J. Chem. Soc. Perkin *I*, 173–178 (1976a)
Bearder, J.R., Frydman, V.M., Gaskin, P., Harvey, W.E., MacMillan, J., Phinney, B.O.: Fungal products. Part XVII. Microbiological hydroxylation of gibberellin A_9 and its methyl ester. J. Chem. Soc. Perkin *I*, 178–183 (1976b)
Bearder, J.R., MacMillan, J., Phinney, B.O.: Origin of the oxygen atoms in the lactone bridge of C_{19}-gibberellins. J.C.S. Chem. Commun. 834–835 (1976c)
Bearder, J.R., MacMillan, J., Matsuo, A., Phinney, B.O.: Conversion of trachylobanic acid into novel pentacyclic analogues of gibberellins by *Gibberella fujikuroi*, mutant B1–41a. J.C.S. Chem. Commun. *1979*, 649–650
Beauchamp, C., Fridovich, J.: A mechanism for the production of ethylene from methional. J. Biol. Chem. *245*, 4641–4646 (1970)
Beever, J.E., Woolhouse, H.W.: Increased cytokinin from root system of *Perilla frutescens* and flower and fruit development. Nature (London) New Biol. *246*, 31–32 (1973)
Beevers, L., Loveys, B., Pearson, J.A., Wareing, P.F.: Phytochrome and hormonal control of expansion and greening of etiolated wheat leaves. Planta *90*, 286–294 (1970)
Behmer, M.: Untersuchungen über den Austausch von Kohlendioxid und Äthylen bei lagernden Äpfeln. In: Mitt. Höheren Bundeslehr Versuchsanst. Wein Obstbau Klosterneuburg Gartenbau, Schoenbrunn Ser. B. *8*, 257–273 (1958)
Bennett, R.D., Ko, S.T., Heftmann, E.: Effect of photoperiodic floral induction on the metabolism of a gibberellin precursor, (−)-kaurene, in *Pharbitis nil*. Plant Physiol. *41*, 1360–1363 (1966)
Bennett, T.D., Lieber, E.R., Heftmann, E.: Biosynthesis of steviol from (−)-kaurene. Phytochemistry *6*, 1107–1110 (1967)
Ben-Yehoshua, S., Aloni, B.: Effect of water stress on ethylene production by detached leaves of Valencia orange (*Citrus sinensis* Osbeck). Plant Physiol. *53*, 863–865 (1974)
Bernhardt, D., Köhler, K.-H., Sembdner, G.: Die Aktivität von einigen freien und konjugierten Gibberellinen im *Triticum*-Halbkaryopsen- und *Amaranthus*-Gibberellintest. Biochem. Physiol. Pflanzen *174*, 607–615 (1979)
Betz, A.: Ascorbinsäure, NADH, Cystein und Glutathion hemmen den durch Peroxydase katalysierten oxydativen Abbau von β-Indolylessigsäure. Z. Bot. *51*, 424–433 (1963)
Beutelmann, P.: Untersuchungen zur Biosynthese eines Cytokinins in Calluszellen von Laubmoossporophyten. Planta *112*, 181–189 (1973)
Beyer, E.M., Jr.: Mechanism of ethylene action. Biological activity of deuterated ethylene and evidence against isotopic exchange and cis-trans-isomerization. Plant Physiol. *49*, 672–675 (1972)
Beyer, E.M., Jr.: Abscission: the initial effect of ethylene is in the leaf blade. Plant Physiol. *55*, 322–327 (1975a)
Beyer, E.M., Jr.: $^{14}C_2H_4$: Its incorporation and metabolism by pea seedlings under aseptic conditions. Plant Physiol. *56*, 273–278 (1975b)
Beyer, E.M., Jr.: Ethylene-C^{14} incorporation and metabolism in pea seedlings. Nature (London) *255*, 144–147 (1975c)
Beyer, E.M., Jr.: $^{14}C_2H_4$: Its incorporation and oxidation to $^{14}CO_2$ by cut carnations. Plant Physiol. *60*, 203–206 (1977)
Beyer, E.M., Jr.: Rapid metabolism of propylene by pea seedlings. Plant Physiol. *61*, 893–895 (1978)
Beyer, E.M., Jr.: [^{14}C] Ethylene metabolism during leaf abscission in cotton. Plant Physiol. *64*, 971–974 (1979)
Beyer, E.M., Jr., Morgan, P.W.: Effect of ethylene on the uptake, distribution and metabolism of indoleacetic acid-1-^{14}C and 2-^{14}C and naphthaleneacetic acid-1-^{14}C. Plant Physiol. *46*, 157–162 (1970)
Beyer, E.M., Jr., Sundin, O.: $^{14}C_2H_4$-metabolism in morning glory flowers. Plant Physiol. *61*, 896–899 (1978)
Bezemer-Sybrandy, S.M., Veldstra, H.: Investigations on cytokinins. IV. The metabolism of benzylamino purine in *Lemna minor*. Physiol. Plant. *25*, 1–7 (1971)
Birch, A.J., Winter, J.: A partial synthesis of ^{14}C-phyllocladene: some observations on the biosynthesis of gibberellic acid. J. Chem. Soc. 5547–5548 (1963)

Birch, A.J., Rickards, R.W., Smith, H., Harris, A., Whalley, W.B.: Studies in relation to biosynthesis. XXI. Rosenolactone and gibberellic acid. Tetrahedron *7*, 241–251 (1959)

Birch, A.J., Rickards, R.W., Smith, H., Winter, J., Turner, W.B.: The allogibberic-gibberic acid rearrangement. Chem. Ind. *1960*, 401–402 (1960)

Birecka, H., Briber, K.A., Catalfamo, J.L.: Comparative studies on tobacco pith and sweet potato root isoperoxidases in relation to injury, indoleacetic acid, and ethylene effects. Plant Physiol. *52*, 43–49 (1973)

Black, R.C., Hamilton, R.H.: Indoleacetic acid biosynthesis in *Avena* coleoptile tips and excised bean shoots. Plant Physiol. *48*, 603–606 (1971)

Black, R.C., Hamilton, H.: Indoleacetic acid synthesis in soybean cotyledon callus tissue. Plant Physiol. *57*, 437–439 (1976)

Blanpied, G.D.: A study of ethylene in apple, red raspberry and cherry. Plant Physiol. *49*, 627–630 (1972)

Blomstrom, D.C., Beyer, E.M., Jr.: Plants metabolise ethylene to ethylene glycol. Nature (London) *283*, 66–68 (1980)

Blumenfeld, A.: Ethylene and the annona flower. Plant Physiol. *55*, 265–269 (1975)

Bohnsack, C.W., Albert, L.S.: Early effects of boron deficiency on indoleacetic acid oxidase levels of squash root tips. Plant Physiol. *59*, 1047–1050 (1977)

Bolduc, R., Cherry, J., Blair, B.: Increase in indoleacetic acid oxidase activity of winter wheat by cold treatment and gibberellic acid. Plant Physiol. *45*, 461–464 (1970)

Boller, T., Herner, R.C., Kende, H.: Assay for and enzymatic formation of an ethylene precursor, 1-aminocyclopropane-1-carboxylic acid. Planta *143*, 293–303 (1979)

Borgmann, E.: Untersuchungen über die biologische Aktivität und den Stoffwechsel konjugierter Gibberelline. 107 pp. Diss. Martin-Luther-Univ. Halle-Wittenberg (1977)

Böttger, M.: The occurrence of cis,trans- and trans,trans-xanthoxin in pea roots. Z. Pflanzenphysiol. *86*, 265–268 (1978)

Böttger, M., Engvild, K.C., Soll, H.: Growth of *Avena* coleoptiles and pH drop of protoplast suspensions included by chlorinated indoleacetic acids. Planta *140*, 89–92 (1978)

Bowen, D.H., MacMillan, J., Graebe, J.E.: Determination of specific radioactivity of ^{14}C-compounds by mass spectroscopy. Phytochemistry *11*, 2253–2257 (1972)

Bower, P.J., Brown, H.M., Purves, W.K.: Auxin biogenesis. Subcellular compartmentation of indoleacetaldehyde reductases in cucumber seedlings. Plant Physiol. *57*, 850–854 (1976)

Bower, P.J., Brown, H.M., Purves, W.K.: Cucumber seedling indoleacetaldehyde oxidase. Plant Physiol. *61*, 107–110 (1978)

Bown, A.W., Reeve, D.R., Crozier, A.: The effect of light on the gibberellin metabolism and growth of *Phaseolus coccineus* seedlings. Planta *126*, 83–91 (1975)

Bratton, B.O., Henry, E.W.: Electrical stimulation and its effects on indoleacetic acid and peroxidase levels in tomato plants (*Lycopersicon esculentum*). J. Exp. Bot. *28*, 338–344 (1977)

Brian, P.W.: Role of gibberellin-like hormones in regulation of plant growth and flowering. Nature (London) *181*, 1122–1123 (1958)

Briggs, D.E.: Enzyme formation, cellular breakdown and the distribution of gibberellins in the endosperm of barley. Planta *108*, 351–358 (1972)

Brown, H.M., Purves, W.K.: Isolation and characterization of indoleacetaldehyde reductases from *Cucumis sativus*. J. Biol. Chem. *251*, 907–913 (1976)

Brown, H.M., Purves, W.K.: Indoleacetaldehyde reductase of Cucumis sativus L. Kinetic properties and role in auxin biosynthesis. Plant Physiol. *65*, 107–113 (1980)

Browning, G., Saunders, P.F.: Membrane localized gibberellins A_9 and A_4 in wheat chloroplasts. Nature (London) *265*, 375–377 (1977)

Buhler, D.R., Hansen, E., Wang, C.H.: Incorporation of ethylene into fruits. Nature (London) *179*, 48–49 (1957)

Bui-Dang-Ha, D., Nitsch, J.P.: Isolation of zeatin riboside from chicory root, Planta *95*, 119–126 (1970)

Bulard, C., Barthe, M.P., Garrello, G., Le Page-Degivry, M.-T.: Mise en évidence d'acide abscissique lié á β-D-glucopyranose dans les embryons dormants de *Pyrus malus* L., C. R. Acad. Sci. Sér. B *278*, 2145–2148 (1974)

Burden, R.S., Taylor, H.F.: The structure and chemical transformations of xanthoxin. Tetrahedron Lett. 4071–4074 (1970)

Burden, R.S., Taylor, H.F.: Xanthoxin and abscisic acid. Pure Appl. Chem. *47*, 203–209 (1976)

Burden, R.S., Firn, R.D., Hiron, R.W.P., Taylor, H.F., Wright, S.T.C.: Induction of plant growth inhibitor xanthoxin in seedlings by red light. Nature (London) New Biol. *234*, 95–96 (1971)

Burden, R.S., Dawson, G.W., Taylor, H.F.: Synthesis and plant growth inhibitory properties of (±)-O-methylxanthoxin. Phytochemistry *11*, 2295–2299 (1972)

Burg, S.P.: The physiology of ethylene formation. Annu. Rev. Plant Physiol. *13*, 265–302 (1962)

Burg, S.P.: Ethylene in plant growth. Proc. Natl. Acad. Sci. USA *70*, 591–597 (1973)

Burg, S.P., Burg, E.A.: Ethylene formation in pea seedlings, its relation to the inhibition of bud growth caused by indole-3-acetic acid. Plant Physiol. *43*, 1069–1074 (1968a)

Burg, S.P., Burg, E.A.: Auxin stimulated ethylene formation: its relationship to auxin inhibited growth root geotropism and other plant processes. In: Biochemistry and physiology of plant growth substances. Wightman, F., Setterfield, G. (eds.) pp. 1275–1294. Ottawa: Runge Press 1968b

Burg, S.P., Burg, E.A.: Qual. Plant Mater. Veg. *19*, 185 (1969)

Burg, S.P., Clagett, C.O.: Conversion of methionine to ethylene in vegetative tissues and fruits. Biochem. Biophys. Res. Commun. *27*, 125–130 (1967)

Burg, S.P., Dijkman, M.J.: Ethylene and auxin participation in pollen induced fading of *Vanda* orchid blossoms. Plant Physiol. *42*, 1648–1650 (1967)

Burg, S.P., Thimann, K.V.: Studies on the ethylene production of apple tissue. Plant Physiol. *35*, 24–35 (1960)

Burrows, W.J.: Mechanism of action of cytokinins. Curr. Adv. Plant Sci. *7*, 837–847 (1975)

Burrows, W.J.: Mode of action of N,N′-diphenylurea: the isolation and identification of the cytokinins in the transfer RNA from tobacco callus grown in the presence of N,N′-diphenylurea. Planta *130*, 313–316 (1976)

Burrows, W.J.: Evidence in support of biosynthesis de novo of free cytokinins. Planta *138*, 53–57 (1978)

Burrows, W.J., Carr, D.J.: Cytokinin content of pea seeds during their growth and development. Physiol. Plant. *23*, 1064–1070 (1970)

Burrows, W.J., Armstrong, D.J., Kaminek, M., Skoog, F., Bock, R.M., Hecht, S.M., Dammann, L.G., Leonard, N.J., Occolowitz, J.: Isolation and identification of four cytokinins from wheat germ transfer ribonucleic acid. Biochemistry *9*, 1867–1872 (1970)

Carlton, B.C., Peterson, C.E., Tolbert, N.E.: Effects of ethylene and oxygen on production of a bitter compound by carrot roots. Plant Physiol. *36*, 550–552 (1961)

Carr, D.J., Reid, D.M.: The physiological significance of the synthesis of hormones in roots and their export to the shoot system. In: Biochemistry and physiology of plant growth substances. Wightman, F., Setterfield, G. (eds.), pp. 1169–1185. Ottawa: Runge Press 1968

Carr, D.J., Reid, D.M., Skene, K.G.M.: The supply of gibberellins from the root to the shoot. Planta *63*, 382–392 (1964)

Cathey, H.M.: Physiology of growth retarding chemicals. Annu. Rev. Plant Physiol. *15*, 271–302 (1964)

Ceccarelli, N., Lorenzi, R., Alpi, A.: Kaurene and kaurenol biosynthesis in cell-free system of *Phaseolus coccineus* suspensor. Phytochemistry *18*, 1657–1658 (1979)

Chadwick, A.V., Burg, S.P.: An explanation of the inhibition of root growth by indole-3-acetic acid. Plant Physiol. *42*, 415–420 (1967)

Chadwick, A.V., Burg, S.P.: Regulation of root growth by auxin-ethylene interaction. Plant Physiol. *45*, 192–200 (1970)

Chalmers, D.J., Faragher, J.D.: Regulation of anthocyanin synthesis in apple skin. II. Involvement of ethylene. Aust. J. Plant Physiol. *4*, 123–131 (1977)

Chapman, R.W., Morris, R.O., Zaerr, J.R.: Occurrence of trans-ribosylzeatin in *Agrobacterium tunefaciens* tRNA. Nature (London) *262*, 153–154 (1976)

Chappet, A., Dubouchet, J.: Variations de l'activité des isoperoxydases du coléoptile de blé pendent l'auxesis. Physiol. Veg. *13*, 153–162 (1975)
Charnay, D., Bongen-Ottoko, B.: Acide abscissique et dormance chez le Topinambour. Physiol. Veg. *15*, 403–412 (1977)
Chassy, B.M., Suhadolnik, R.J.: Adenosine aminohydrolase. J. Biol. Chem. *242*, 3655–3658 (1967)
Chen, C.-M., Eckert, R.L.: Phosphorylation of cytokinin by adenosine kinase from wheat germ. Plant Physiol. *59*, 443–447 (1977)
Chen, C.-M., Hall, R.H.: Biosynthesis of N^6(Δ^2-isopentenyl)-adenosine in transfer ribonucleic acid of cultured tobacco pith tissue. Phytochemistry *8*, 1687–1695 (1969)
Chen, C.M., Melitz, D.K.: Cytokinin biosynthesis in a cell-free system from cytokinin-autotrophic tobacco tissue culture. FEBS Letters *107*, 15–20 (1979)
Chen, C.-M., Petschow, B.: Cytokinin biosynthesis in cultured rootless tobacco plants. Plant Physiol. *62*, 861–865 (1978a)
Chen, C.-M., Petschow, B.: Metabolism of cytokinin. Ribosylation of cytokinin bases by adenosine phosphorylase from wheat. Plant Physiol. *62*, 871–874 (1978b)
Chen, C.-M., Logan, D.M., McLennan, B., Hall, R.H.: Studies on the metabolism of a cytokinin, N^6-(Δ^2-isopentenyl)adenosine. Plant Physiol. *43*, Suppl. 18 (1968)
Chen, C.-M., Smith, O.O., Hartnell, G.F.: Biological activity of ribose-modified N^6-(Δ^2-isopentenyl)adenosine derivative. Can. J. Biochem. *52*, 1154–1161 (1974)
Chen, C.-M., Smith, O'Brien, C., McChesney, J.D.: Biosynthesis and cytokinin activity of 8-hydroxy and 2,8-dihydroxy derivatives of zeatin and N^6-(Δ^2-isopentenyl)adenine. Biochemistry *14*, 3088–3093 (1975)
Chen, C.-M., Eckert, R.L., McChesney, J.D.: N^6-(Δ^2-isopentenyl)adenosine from crown gall tumor tissue of *Vinca rosea*. Phytochemistry *15*, 1565 (1976a)
Chen, C.-M., Eckert, R.L., McChesney, J.D.: Evidence for the biosynthesis of transfer RNA-free cytokinin. FEBS Lett. *64*, 429–434 (1976b)
Chheda, G.B., Mittelman, A.: N^6-(Δ^2-isopentenyl)adenosine metabolism in man. Biochem. Pharmacol. *21*, 27–37 (1972)
Cho, K.Y., Sakurai, A., Kamiya, Y., Takahashi, N., Tamura, S.: Effects of the new plant growth retardants of quarternary ammonium iodides on gibberellin biosynthesis in *Gibberella fujikuroi*. Plant Cell Physiol. *20*, 75–81 (1979)
Chou, T.W., Yang, S.F.: The biosynthesis of ethylene in *Penicillium digitatum*. Arch. Biochem. Biophys. *157*, 73–82 (1973)
Cionini, P.G., Bennici, A., Alpi, A., D'Amato, F.: Suspensor, gibberellin, and in vitro development of *Phaseolus coccineus* embryos. Planta *131*, 115–117 (1976)
Clarke, A.J., Mann, P.J.G.: Oxydation of tryptamine to indole-3-acetaldehyde by plant amine oxidase. Biochem. J. *65*, 763–774 (1957)
Clarke, G., Dye, M.H., Wain, R.L.: Occurrence of 3-indolylacetic and 3-indolecarboxylic acids in tomato crown gall tissue extracts. Nature (London) *184*, 825–826 (1959)
Cleland, C.F., Zeevart, J.A.D.: Gibberellins in relation to flowering and stem elongation in long-day plant *Silene armeria*. Plant Physiol. *46*, 392–400 (1970)
Cleland, R.: Evidence on the site of action of growth retardants. Plant Cell Physiol. *6*, 7–15 (1965)
Cohen, J.D., Bandurski, R.S.: The bound auxins: Protection of indole-3-acetic acid from peroxidase-catalyzed oxidation. Planta *139*, 203–208 (1978)
Conrad, K.: Evidence for the origin of zeatin riboside-like cytokinin in grated kohlrabi tissue by several bioassays. Biochem. Physiol. Pflanz. *168*, 341–347 (1975)
Conrad, K., Köhn, B.: Zunahme von Cytokinin und Auxin in verwundeten Speichergeweben von *Solanum tuberosum*. Phytochemistry *14*, 325–328 (1975)
Cook, I.F., Jefferies, P.R., Knox, J.R.: A gibbane metabolite from (−)-kaura-2,16-dien-19-ol. Tetrahedron Lett. 2157–2160 (1971)
Cooke, R.J., Kendrick, R.E.: Phytochrome controlled gibberellin metabolism in chloroplast envelopes. Planta *131*, 303–307 (1976)
Cooke, R.J., Saunders, P.F.: Photocontrol of gibberellin levels as related to the unrolling of etiolated wheat leaves. Planta *126*, 151–160 (1975)
Cooke, R.J., Saunders, P.F., Kendrick, R.E.: Red light induced production of gibberellin-

like substances in homogenates of etiolated wheat leaves and in suspensions of intact etioplasts. Planta *124*, 319–328 (1975)
Coolbaugh, R.C., Hamilton, R.: Inhibition of ent-kaurene oxidation and growth by α-cyclopropyl-α-(p-methoxyphenyl)-5-pyrimidine methyl alcohol. Plant Physiol. *57*, 245–248 (1976)
Coolbaugh, R.C., Moore, T.C.: Apparent changes in rate of kaurene biosynthesis during development of pea seeds. Plant Physiol. *44*, 1364–1367 (1969)
Coolbaugh, R.C., Moore, T.C.: Localization of enzymes catalyzing kaurene biosynthesis in immature pea seeds. Phytochemistry *10*, 2395–2400 (1971 a)
Coolbaugh, R.C., Moore, T.C.: Metabolism of kaurene in cell-free extracts of immature pea seeds. Phytochemistry *10*, 2401–2412 (1971 b)
Coolbaugh, R.C., Moore, T.C., Barlow, S.A., Ecklund, P.R.: Biosynthesis of ent-kaurene in cell-free extract of *Pisum sativum* shoot tips. Phytochemistry *12*, 1613–1618 (1973)
Coombe, B.G.: The development of fleshy fruits. Annu. Rev. Plant Physiol. *27*, 507–528 (1976)
Cornforth, J.W., Clifford, K., Mallaby, R., Phillips, G.T.: Stereochemistry of isopentenyl pyrophosphate isomerase. Proc. R. Soc. London Ser. B *182*, 277–295 (1972)
Cowley, D.E., Jenkins, I.D., MacLeod, J.K., Summons, R.E., Letham, D.S., Wilson, M.M., Parker, C.W.: Structure and synthesis of unusual cytokinin metabolites. Tetrahedron Lett. 1015–1018 (1975)
Craker, L.E.: Post-harvest color promotion in cranberry with ethylene. Hort. Sci. *6*, 137–139 (1971)
Craker, L.E., Stanley, L.A., Starbuck, M.J.: Ethylene control of anthocyanine synthesis in *Sorghum*. Plant Physiol. *48*, 349–352 (1971)
Craker, L.E., Abeles, F.B., Shropshire, W.Jr.: Light-induced ethylene production in *Sorghum*. Plant Physiol. *51*, 1082–1083 (1973)
Cross, B.E.: Gibberellic acid. Part I. J. Chem. Soc. 4670–4676 (1954)
Cross, B.E., Myers, P.L.: The effect of plant growth retardants on the biosynthesis of diterpenes by *Gibberella fujikuroi*. Phytochemistry *8*, 79–83 (1969)
Cross, B.E., Norton, K.: The biosynthesis of gibberellic acid. Chem. Commun. 535–536 (1965)
Cross, B.E., Norton, K.: The role of gibberellins A_{13} and A_{14} in the biosynthesis of gibberellic acid. Tetrahedron Lett. 6003–6007 (1966)
Cross, B.E., Stewart, J.C.: (−)-Pimaradiene, a new precursor of the gibberellins. Tetrahedron Lett. 5195–5196 (1968 a)
Cross, B.E., Stewart, J.C.: Biosynthesis of gibberellic acid. A correction. Tetrahedron Lett. 6321–6322 (1968 b)
Cross, B.E., Grove, J.F., MacMillan, J., Mulholland, T.P.C.: Gibberellic acid. IV. Structures of gibberic and allo-gibberic acids and possible structures of gibberellic acid. Chem. Ind. 954–955 (1956)
Cross, B.E., Galt, R.H.B., Hanson, J.R.: The biosynthesis of gibberellins. Part I. (−)-Kaurene as a precursor of gibberellic acid. J. Chem. Soc. C, 295–300 (1964)
Cross, B.E., Norton, K., Stewart, J.C.: An attempt to find evidence for the existence of (+)-gibberellic acid. Phytochemistry *7*, 83–84 (1968 a)
Cross, B.E., Galt, R.H.B., Norton, K.: The biosynthesis of the gibberellins. II. Tetrahedron *24*, 231–237 (1968 b)
Cross, B.E., Norton, K., Stewart, J.C.: The biosynthesis of gibberellins. III. J. Chem. Soc. *C*, 1054–1063 (1968 c)
Cross, B.E., Stewart, J.C., Stoddart, J.L.: 6β,7β-Dihydroxykaurenoid acid: its biological activity and possible role in the biosynthesis of gibberellic acid. Phytochemistry *9*, 1065–1071 (1970)
Crozier, A., Reid, D.M.: Do roots synthesize gibberellins? Can. J. Bot. *49*, 967–975 (1971)
Crozier, A., Reid, D.M., Reeve, D.R.: Effects of Amo 1618 on growth, morphology, and gibberellin content of *Phaseolus coccineus* seedlings. J. Exp. Bot. *24* 923–934 (1973)
Cummins, W.R.: The metabolism of abscisic acid in relation to its reversible action on stomata in leaves of *Hordeum vulgare* L. Planta *114*, 159–167 (1973)

Dale, J.E.: Gibberellins and early growth in seedlings of *Phaseolus vulgaris*. Planta *89*, 155–164 (1969)
Dale, J.E., Felippe, G.M.: The gibberellin content and early seedling growth of plants of *Phaseolus vulgaris* treated with the growth retardant CCC. Planta *80*, 288–298 (1968)
Daly, J.M., Seevers, P.M., Ludden, P.: Studies on wheat stem rust resistance controlled at the Sr6 locus. III. Ethylene and disease reaction. Phytopathology *60*, 1648–1652 (1970)
Dannenburg, W.N., Liverman, J.L.: Conversion of tryptophan-2-^{14}C to indoleacetic acid by watermelon tissue slices. Plant Physiol. *32*, 263–269 (1957)
Darbyshire, B.: The effect of water stress on indoleacetic acid oxidase in pea plants. Plant Physiol. *47*, 65–67 (1971a)
Darbyshire, B.: Changes in indoleacetic acid oxidase activity associated with plant water potential. Physiol. Plant. *25*, 80–84 (1971b)
Darimont, E., Schwachhofer, K., Gaspar, T.: Isoperoxydases de hydroxyproline dans les parois cellulaires des racines de lentille. Biochim. Biophys. Acta *321*, 461–466 (1973)
Dashek, W.V., Singh, B.N., Walton, D.C.: Abscisic acid localization and metabolism in barley aleurone layers. Plant Physiol. *64*, 43–48 (1979)
Dasilva, E.J., Henriksson, E., Henriksson, L.E.: Ethylene production by fungi. Plant Sci. Lett. *2*, 63–66 (1974)
Dathe, W., Sembdner, G.: Distribution of gibberellins and abscisic acid in different fruit parts of rye (*Secale cereale* L.). Gibberellins LXVII. Biochem. Physiol. Pflanz. *173*, 440–447 (1978)
Dathe, W., Schneider, G., Sembdner, G.: Endogenous gibberellins and inhibitors in caryopses of rye. Phytochemistry *17*, 963–966 (1978a)
Dathe, W., Sembdner, G., Kefeli, V.I., Vlasov, P.V.: Gibberellins, ABA, and related inhibitors in branches and bleeding sap of birch (*Betula pubescens* Ehrh.). Biochem. Physiol. Pflanz. *173*, 238–248 (1978b)
Davey, J.E., van Staden, J.: Cytokinin translocation: Changes in zeatin and zeatin riboside levels in root exudate of tomato plants during their development. Planta *130*, 69–72 (1976)
Davey, J.E., van Staden, J.: Cytokinin activity in *Lupinus albus*. I. Distribution in vegetative and flowering plants. Physiol. Plant. *43*, 77–81 (1978a)
Davey, J.E., van Staden, J.: Cytokinin activity in *Lupinus albus*. II. Distribution in fruiting plants. Physiol. Plant. *43*, 82–86 (1978b)
Davey, J.E., van Staden, J.: Cytokinin activity in *Lupinus albus*. III. Distribution in fruits. Physiol. Plant. *43*, 87–93 (1978c)
Davies, P.J.: The fate of exogenously applied indoleacetic acid in light-grown stems. Physiol. Plant. *27*, 262–270 (1972)
Davies, P.J.: Bound auxin formation in growing stems. Plant Physiol. *57*, 197–202 (1976)
Davies, L.J., Rappaport, L.: Metabolism of tritiated gibberellins in d-5 dwarf maize. I. In excised tissues and intact dwarf and normal plants. Plant Physiol. *55*, 620–625 (1975a)
Davies, L.J., Rappaport, L.: Metabolism of tritiated gibberellins in d-5 dwarf maize. II. (^{3}H)-Gibberellin A_1, (^{3}H)-gibberellin A_3, and related compounds. Plant Physiol. *56*, 60–66 (1975b)
Dawson, R.M., Jefferies, P.R., Knox, J.R.: Cyclization and hydroxylation stereochemistry in the biosynthesis of gibberellic acid. Phytochemistry *14*, 2593–2597 (1975)
De Greef, J.A., van Hoof, R., Caubergs, R.: Light-induced changes in auxin metabolism during hook opening of etiolated bean seedlings. Biochem. Soc. Transact. *5*, 1049–1051 (1977)
Deleuze, G.G., McChesney, J.D., Fox, J.E.: Identification of a stable cytokinin metabolite. Biochem. Biophys. Res. Commun. *48*, 1426–1432 (1972)
Demorest, D.M., Stahmann, M.A.: The binding of the peroxidase oxidation products of indole-3-acetic acid to histone. Biochem. Biophys. Res. Commun. *47*, 227–233 (1972)
Dennis, D.T., West, C.A.: Biosynthesis of gibberellins. III. The conversion of (−)-kaurene to (−)-kauren-19-oic acid in endosperm of *Echinocystis macrocarpa* Greene. J. Biol. Chem. *242*, 3293–3300 (1967)
Dennis, D.T., Upper, C.D., West, C.A.: An enzymic site of inhibition of gibberellin biosyn-

thesis by Amo 1618 and other plant growth retardants. Plant Physiol. *40*, 948–952 (1965)
Deverall, B.J.: Apparently spontaneous decarboxylation of indolyl-3-acetic acid. Nature (London) *207*, 828–829 (1965)
Deverall, B.J., Daly, J.M.: Metabolism of indoleacetic acid in rust diseases. II. Metabolites of carboxyl-labelled indoleacetic acid in tissues. Plant Physiol. *39*, 1–9 (1964)
Dewdney, S.J., McWha, J.A.: The metabolism and transport of abscisic acid during grain fill in wheat. J. Exp. Bot. *29*, 1299–1308 (1978)
De Yoe, D.R., Zaerr, J.B.: Indole-3-acetic acid in *Douglas fir*. Plant Physiol. *58*, 299–303 (1976)
Diaz, D.H., Martin, G.C.: Peach seed dormancy in relation to endogenous inhibitors and applied growth substances. J. Am. Soc. Hortic. Sci. *97*, 651–654 (1972)
Divekar, A.Y., Slocum, H.K., Hakala, M.T.: N^6-(Δ^2-Isopentenyl)adenosine-5′-monophosphate: formation and effect of purine metabolism in cellular and enzymatic systems. Mol. Pharmacol. *10*, 529–543 (1974)
Dockerill, B., Hanson, J.R.: The fate of C-20 in C_{19} gibberellin biosynthesis. Phytochemistry *17*, 701–704 (1978)
Dockerill, B., Evans, R., Hanson, J.R.: Removal of C-20 in gibberellin biosynthesis. J.C.S. Chem. Commun. 919–921 (1977)
Dodds, J.H., Musa, S.K., Jerie, P.H., Hall, M.A.: Metabolism of ethylene to ethylene oxide by cell-free preparations from *Vicia faba* L. Plant Sci. Lett. *17*, 109–114 (1979)
Doree, M., Guern, J.: Short-term metabolism of some exogenous cytokinins in *Acer pseudoplatanus* cells. Biochim. Biophys. Acta *304*, 611–622 (1973)
Doree, M., Terrine, C.: Enzymatic synthesis of ribonucleoside-5′-phosphates from some N^6-substituted adenosines. Phytochemistry *12*, 1017–1023 (1973)
Doree, M., Terrine, C., Trapy, F.: Kinetic study of the metabolism of some exogenous cytokinins in *Acer pseudoplatanus*. C. R. Acad. Sci. Ser. D *275*, 1131–1134 (1972)
Dörffling, K.: Das Phytohormon Abscisinsäure. Biol. Rundsch. *9*, 129–143 (1971)
Dörffling, K.: Recent advances in abscisic acid research. In: Hormonal regulation in plant growth and development. Kaldewey, H., Vardar, Y. (eds.), pp. 281–295. Weinheim: Chemie 1972
Dörffling, K., Sonka, B., Tietz, D.: Variation and metabolism of abscisic acid in pea seedlings during and after water stress. Planta *121*, 57–66 (1974)
Douglas, T.J., Paleg, L.G.: Inhibition of sterol biosynthesis by isopropyl-4-dimethylamino-5-methylphenyl-1-piperidine carboxylate methylchloride in tobacco and rat liver preparations. Plant Physiol. *49*, 417–420 (1972)
Douglas, T.J., Paleg, L.G.: Amo 1618 and sterol biosynthesis in tissues and sub-cellular fractions of tobacco seedlings. Phytochemistry *17*, 705–712 (1978a)
Douglas, T.J., Paleg, L.G.: Amo 1618 effects on incorporation of ^{14}C-MVA and ^{14}C-acetate into sterols in *Nicotiana* and *Digitalis* seedlings and cell-free preparations from *Nicotiana*. Phytochemistry *17*, 713–718 (1978b)
Dueber, M.T., Adolf, W., West, C.A.: Biosynthesis of the diterpene phytoalexin casbene. Plant Physiol. *62*, 598–603 (1978)
Duke, C.D., Liepa, A.J., MacLeod, J.K., Letham, D.S., Parker, C.W.: Synthesis of raphanatin and its 6-benzylaminopurine analogue. J.C.S. Chem. Commun. 964–965 (1975)
Duke, C.D., Letham, D.S., Parker, C.W., McLeod, J.K., Summons, R.E.: The complex of O-glucosylzeatin derivatives formed in *Populus* species. Phytochemistry *18*, 819–824 (1979)
Dullaart, J.: Quantitative estimation of indoleacetic acid and indolecarboxylic acid in root nodules and roots of *Lupinus luteus* L. Acta Bot. Neerl. *16*, 222–230 (1967)
Dunberg, A.: Changes in gibberellin-like substances and indole-3-acetic acid in *Picea abies* during period of shoot elongation. Physiol. Plant. *38*, 186–190 (1976)
Dunlap, J.R., Morgan, P.W.: Characterization of ethylene/gibberellic acid control of germination in *Lactuca sativa* L. Plant Cell Physiol. *18*, 561–568 (1977)
Durham, J.I., Morgan, P.W.: Production of ethylene by cotton flower petals. Plant Physiol. *51*, Suppl. 30 (1973)

Düring, H., Alleweldt, G.: Der Jahresgang der Abscisinsäure in vegetativen Organen von Reben. Vitis *12*, 26–31 (1973)

Durley, R.C., Pharis, R.P.: Interconversion of gibberellin A_4 to gibberellins A_1 and A_{34} by dwarf rice, cultivar Tan-ginbozu. Planta *109*, 357–361 (1973)

Durley, R.C., MacMillan, J., Pryce, R.J.: Investigation of gibberellins and other growth substances in the seed of *Phaseolus multiflorus* and of *Phaseolus vulgaris* by gas chromatography and by gas chromatography – mass spectrometry. Phytochemistry *10*, 1891–1908 (1971)

Durley, R.C., Railton, I.D., Pharis, R.P.: Interconversion of gibberellin A_5 to gibberellin A_3 in seedlings of dwarf *Pisum sativum*. Phytochemistry *12*, 1609–1612 (1973)

Durley, R.C., Railton, I.D., Pharis, R.P.: Conversion of gibberellin A_{14} to other gibberellins by etiolated shoots of dwarf *Pisum sativum*. Phytochemistry *13*, 547–551 (1974a)

Durley, R.C., Railton, I.D., Pharis, R.P.: The metabolism of gibberellin A_1 and gibberellin A_{14} in seedlings of dwarf *Pisum sativum*. In: Plant growth substances 1973. pp. 285–293. Tokyo: Hirokawa 1974b

Durley, A.C., Pharis, R.P., Zeevaart, J.A.D.: Metabolism of ^3H gibberellin A_{20} by plants of *Bryophyllum daigremontianum* under long- and short-day conditions. Planta *126*, 139–149 (1975)

Durley, R.C., Bewley, J.D., Railton, I.D., Pharis, R.P.: Effect of light, abscisic acid, and ^6N-benzyladenine on the metabolism of ^3H gibberellin A_4 in seeds and seedlings of lettuce, cv. Grand Rapids. Plant Physiol. *57*, 699–703 (1976)

Durley, R.C., Sassa, T., Pharis, R.P.: Metabolism of tritiated gibberellin A_{20} in immature seeds of dwarf pea, cv. 'Meteor'. Plant Physiol. *64*, 214–219 (1979)

Eaks, L.I.: Respiratory response, ethylene production, and response to ethylene of citrus fruit during ontogeny. Plant Physiol. *45*, 334–338 (1970)

Eberle, M., Arigoni, D.: Absolute Konfiguration des Mevalolactons. Helv. Chim. Acta *43*, 1508–1513 (1960)

Eckert, H., Podlesak, W., Schilling, G.: Wirkung wachstumsretardierender Verbindungen auf Gibberellingehalte und -transport sowie auf das Sproßwachstum von *Triticum* und *Secale*. Biochem. Physiol. Pflanz. *172*, 553–565 (1978)

Ecklund, P.R., Moore, T.C.: Correlation of growth rate and de-etiolation with rate of ent-kaurene biosynthesis in pea. Plant Physiol. *53*, 5–10 (1974)

Edelbluth, E., Kaldewey, H.: Auxin in scapes, flower buds, flowers, and fruits of daffodil (*Narcissus pseudonarcissus* L.). Planta *131*, 285–291 (1976)

Edwards, E.J., Lamotte, C.E.: Evidence for cytokinin in bacterial leaf nodules of *Psychotria punctata* (Ribiaceae). Plant Physiol. *56*, 425–428 (1975)

Ehmann, A.: Isolation of 2-O(indole-3-acetyl)-D-glucopyranose, 4-0-(indole-3-acetyl)-D-glucopyranose, and 6-0-(indole-3-acetyl)-D-glucopyranose from kernels of *Zea mays* by gas liquid chromatography – mass spectrometry. Carbohydr. Res. *34*, 99–114 (1974)

Ehmann, A., Bandurski, R.S.: The isolation of di-0-(indole-3-acetyl)-myo-inositol and tri-0-(indole-3-acetyl)-myo-inositol from mature kernels of *Zea mays*. Carbohydr. Res. *36*, 1–12 (1974)

Einset, J.W., Skoog, F.: Biosynthesis of cytokinins in cytokinin-autotrophic tobacco callus. Proc. Natl. Acad. Sci. USA *70*, 658–660 (1973)

El-Beltagy, A.S., Hall, M.A.: Effect of water stress upon endogenous ethylene levels in *Vicia faba*. New Phytol. *73*, 47–60 (1974)

Elliot, D.C., Murray, A.W.: A quantitative limit for cytokinin incorporation into transfer ribonucleic acid by soyabean callus tissue. Biochem. J. *130*, 1157–1160 (1972)

Elstner, E.F., Konze, J.R.: Ethylene production by isolated chloroplasts. Z. Naturforsch. *29c*, 710–716 (1974a)

Elstner, E.F., Konze, J.R.: Light-dependent ethylene production by isolated chloroplasts. FEBS Lett. *45*, 18–21 (1974b)

Elstner, E.F., Konze, J.R.: Effect of point freezing on ethylene and ethane production by sugar beet leaf disks. Nature (London) *263*, 351–352 (1976)

Elstner, E.F., Konze, J.R., Selman, B.R., Stoffer, C.: Ethylene formation in sugar beet leaves. Evidence for the involvement of 3-hydroxytyramine and phenoloxidase after wounding. Plant Physiol. *58*, 163–168 (1976)

Endo, T.: Indoleacetate oxidase activity of horseradish and other plant peroxidase isoenzymes. Plant Cell Physiol. *9*, 333–341 (1968)
Engelbrecht, L.: Cytokinins in leaf-cuttings of *Phaseolus vulgaris* L. during their development. Biochem. Physiol. Pflanz. *163*, 335–343 (1972)
Engvild, K.C.: Chloroindolyl-3-acetic acid and its methylester. Incorporation of ^{36}Cl in immature seeds of pea and barley. Physiol. Plant. *32*, 84–88 (1974)
Engvild, K.C.: Natural chlorinated auxins labelled with radioactive chloride in immature seeds. Physiol. Plant. *34*, 286–287 (1975)
Engvild, K.C., Egsgaard, H., Larsen, E.: Gas chromatographic – mass spectrometric identification of 4-chloroindolyl-3-acetic acid methyl ester in immature green peas. Physiol. Plant. *42*, 365–368 (1978)
Entsch, B., Letham, D.S.: Enzymic glucosylation of the cytokinin, 6-benzylaminopurine. Plant Sci. Letters *14*, 205–212 (1979)
Entsch, B., Parker, C.W., Letham, D.S.: Enzymic glucosylation of the phytohormone zeatin. Proc. Austral. Biochem. Soc. *11*, 53 (1979)
Epstein, E., Lavee, S.: Uptake and fate of IAA in apple callus tissue using IAA-1-^{14}C. Plant Cell Physiol. *16*, 553–561 (1975)
Epstein, E., Lavee, S.: Uptake, translocation, and metabolism of IAA in the olive (*Olea europea*). I. Uptake and translocation of (1-^{14}C) IAA in detached "Manzanilla" olive leaves. J. Exp. Bot. *28*, 619–628 (1977a)
Epstein, E., Lavee, S.: Uptake, translocation, and metabolism of IAA in olive (*Olea europaea*). II. The fate of exogenously applied (2-^{14}C) IAA to detached "Manzanilla" olive leaves. J. Exp. Bot. *28*, 629–635 (1977b)
Epstein, E., Klein, I., Lavee, S.: Uptake and fate of IAA in apple callus tissues: Metabolism of IAA-2-^{14}C. Plant Cell Physiol. *16*, 305–311 (1975)
Erdmann, N., Schiewer, U.: Tryptophan-dependent indoleacetic acid biosynthesis from indole, demonstrated by double labelling experiments. Planta *97*, 135–141 (1971)
Erdmann, N., Libbert, E., Schiewer, U.: Auxinaktivität und Metabolismus von Indolyl-3-glycerinsäure und 3-(N-Acetylindolyl-3)-glycerinsäuremethylester. Flora (Jena) Abt. A *160*, 350–360 (1969a)
Erdmann, N., Schiewer, U., Libbert, E.: Untersuchungen zur Indol-3-essigsäure-Bildung aus Indol, Indol-3-glycerin und Indol-3-glycerinphosphat. Flora (Jena) Abt. A *160*, 500–511 (1969b)
Erichsen, V., Knoop, B., Bopp, M.: Uptake, transport and metabolism of cytokinin in moss protonema. Plant Cell Physiol. *19*, 839–850 (1978)
Esashi, Y., Ohhara, Y., Kotaki, K., Watanabe, K.: Two C_2H_4-producing systems in cocklebur seeds. Planta *129*, 23–26 (1976a)
Esashi, Y., Watanabe, K., Ohhara, Y., Katoh, H.: Dormancy and impotency of cocklebur seeds. V. Growth and ethylene production of axial segments in response to oxygen. Aust. J. Plant Physiol. *3*, 701–710 (1976b)
Evans, A., Smith, H.: Localization of phytochrome in etioplasts and its regulation in vitro of gibberellin levels. Proc. Natl. Acad. Sci. USA *73*, 138–142 (1976)
Evans, M.L., Ray, P.M.: Inactivity of 3-methyleneoxindole as mediator of auxin action on cell elongation. Plant Physiol. *52*, 186–189 (1972)
Evans, M.L., Rayle, D.L.: The timing of growth promotion and conversion to indole-3-acetic acid for auxin precursors. Plant Physiol. *45*, 240–243 (1970)
Evans, M.L., Schmitt, M.R.: The nature of spontaneous changes in growth rate in isolated coleoptile segments. Plant Physiol. *55*, 757–762 (1975)
Evans, R., Hanson, J.R.: The formation of (−)-kaurene in a cell-free system of *Gibberella fujikuroi*. J. Chem. Soc. Perkin *I*, 2382–2385 (1972)
Evans, R., Hanson, J.R.: Studies in terpenoid biosynthesis. Part. XIII. The biosynthetic relationship of the gibberellins in *Gibberella fujikuroi*. J. Chem. Soc. Perkin *I*, 663–666 (1975)
Evans, R., Hanson, J.R., White, A.F.: Studies in terpenoid biosynthesis, Part VI. The stereochemistry of some stages in tetracyclic diterpene synthesis. J. Chem. Soc. *C*, 2601–2603 (1970)
Evans, R., Hanson, J.R., Mulheirn, L.J.: Studies in terpenoid biosynthesis X. Incorporation

of 5(S)-5-^{3}H-mevalonic acid into gibberellic acid. J. Chem. Soc. Perkin *I*, 753–756 (1973)
Fall, R.R., West, C.A.: Purification and properties of kaurene synthetase from *Fusarium moniliforme*. J. Biol. Chem. *246*, 6913–6928 (1971)
Fang, S.C., Theisen, P., Butts, J.S.: Metabolic studies of applied indoleacetic acid-1-^{14}C in plant tissues as affected by light and 2,4-D treatment. Plant Physiol. *34*, 26–32 (1959)
Faull, K.F., Coombe, B.G., Paleg, L.G.: Biosynthesis of gibberellins in barley and dwarf rice seedlings. Aust. J. Plant Physiol. *1*, 199–210 (1974)
Feucht, W., Nachit, M.: Flavolans and growth-promoting catechins in young shoot tips of prunus species and hybrids. Physiol. Plant. *40*, 230–234 (1977)
Feung, C.S., Mumma, R.O., Hamilton, R.O.: Indole-3-acetic acid. Mass spectra and chromatographic properties of amino acid conjugates. J. Agric. Food Chem. *23*, 1120–1124 (1975)
Feung, C.S., Hamilton, R.H., Mumma, R.O.: Metabolism of indole-3-acetic acid. III. Identification of metabolites isolated from crown gall. Plant Physiol. *58*, 666–669 (1976)
Feung, C.S., Hamilton, R.H., Mumma, R.O.: Metabolism of indole-3-acetic acid. IV. Biological properties of amino acid conjugates. Plant Physiol. *59*, 91–93 (1977)
Firn, R.D., Friend, J.: Enzymatic production of the plant growth inhibitor, xanthoxin. Planta *103*, 263–266 (1972)
Firn, R.D., Burden, R.S., Taylor, H.F.: The detection and estimation of the growth inhibitor xanthoxin in plants. Planta *102*, 115–126 (1972)
Fittler, F., Kline, L.K., Hall, R.H.: N^6-(Δ^2-Isopentenyl)-adenosine: biosynthesis in vitro by an enzyme extract from yeast and rat liver. Biochem. Biophys. Res. Commun. *31*, 571–576 (1968)
Flinn, A.M.: Regulation of leaflet photosynthesis by developing fruit in the pea. Physiol. Plant. *31*, 275–278 (1975)
Fowler, J.L., Morgan, P.W.: The relationship of the peroxidative indoleacetic acid oxidase system to in vivo ethylene synthesis in cotton. Plant Physiol. *49*, 555–559 (1972)
Fox, J.E., Chen, C.M.: Characterization of labeled ribonucleic acid from tissue grown on ^{14}C-containing cytokinins. J. Biol. Chem. *242*, 4490–4494 (1967)
Fox, L.R., Purves, W.K.: The mechanism of peroxidase catalysis of 3-indoleacetic acid oxidation. In: Biochemistry and physiology of plant growth substances. Wightman, F., Setterfield, G. (eds.), pp. 301–309. Ottawa: Runge Press 1968
Fox, J.E., Sood, C.K., Buckwalter, B., McChesney, J.D.: The metabolism and biological activity of a 9-substituted cytokinin. Plant Physiol. *47*, 275–281 (1971)
Fox, J.E., Cornette, J., Deleuze, G., Dyson, W., Giersak, C., Niu, P., Zapata, J., McChesney, J.: The formation, isolation, and biological activity of a cytokinin 7-glucoside. Plant Physiol. *52*, 627–632 (1973)
Frenkel, C.: Involvement of peroxidase and indole-3-acetic acid oxidase isozymes from pear, tomato, and blueberry fruit in ripening. Plant Physiol. *49*, 757–763 (1972)
Frenkel, C.: Oxidative turnover of auxins in relation to the onset of ripening in bartlett pear. Plant Physiol. *55*, 480–484 (1975)
Frenkel, C., Dyck, R.: Auxin inhibition of ripening in bartlett pears. Plant Physiol. *51*, 6–9 (1973)
Frenkel, C., Haard, N.F.: Initiation of ripening in bartlett pear with an antiauxin alpha (p-chlorophenoxy)isobutyric acid. Plant Physiol. *52*, 380–384 (1973)
Frenkel, C., Hess, C.E.: Isozymic changes in relation to root initiation in mung bean. Can. J. Bot. *52*, 295–297 (1974)
Frenkel, C., Klein, I., Dilley, D.R.: Protein synthesis in relation to ripening of some fruits. Plant Physiol. *43*, 1146–1153 (1968)
Frenkel, C., Haddon, V.R., Smallheer, J.M.: Promotion of softening and ethylene synthesis in bartlett pears by 3-methylene oxindole. Plant Physiol. *56*, 647–649 (1975)
Friedländer, M., Atsmon, D., Galun, E.: Uptake, transport, and stability of (2-^{14}C)-abscisic acid in cucumber following foliar application. Plant Cell Physiol. *17*, 965–972 (1976)
Fries, D.: Effect of ABA and kinetin on the growth, peroxidase activity and the isoperoxidase spectrum of *Lens culinaria*. Med. Ann. Physiol. Veg. Univ. Brux. *17*, 83–104 (1973)

Frost, R.G., West, C.A.: Properties of kaurene synthetase from *Marah macrocarpus*. Plant Physiol. *59*, 22–29 (1977)

Frydman, V.M., MacMillan, J.: The metabolism of gibberellins A_9, A_{20}, and A_{29} in immature seeds of *Pisum sativum* cv. 'Progress No. 9'. Planta *125*, 181–195 (1975)

Fuchs, Y.: Ethylene production by citrus fruit peel. Stimulation by phenol derivatives. Plant Physiol. *45*, 533–534 (1970)

Fuchs, Y., Lieberman, M.: Effect of kinetin, IAA, and gibberellic acid on ethylene production and their interactions in growth of seedlings. Plant Physiol. *43*, 2029–2036 (1968)

Fujita, T., Masuda, I., Takao, S., Fujita, E.: Biosynthesis of natural products. Part 1. Incorporation of ent-kaur-16-ene and ent-kaur-16-en-15-one into enmein and oridonin. J. Chem. Soc. Perkin *I*, 2098–2102 (1976)

Fukui, H., Koshimizu, K., Usuda, S., Yamazaki, Y.: Isolation of plant growth regulators from seeds of *Cucurbita pepo* L. Agric. Biol. Chem. *41*, 175–180 (1977)

Fukuyama, T.T., Moyed, N.S.: Inhibition of cell growth by photooxidation products of indole-3-acetic acid. J. Biol. Chem. *239*, 2392–2397 (1964)

Gahagan, H.E., Holm, R.E., Abeles, F.B.: Effect of ethylene on peroxidase activity. Physiol. Plant. *21*, 1270–1279 (1968)

Galliard, T., Rhodes, M.J.C., Wooltorton, L.S.C., Hulme, A.C.: Metabolic changes in excised fruit tissue. III. The development of ethylene biosynthesis during the ageing of disks of apple peel. Phytochemistry *7*, 1465–1470 (1968)

Galston, A.W., Hillman, W.S.: The degradation of auxin. In: Encyclopedia of plant physiology. Ruhland, W. (ed.), Vol. XIV, pp. 647–670. Berlin-Heidelberg-New York: Springer 1961

Galston, A.W., Lavee, S., Siegel, B.Z.: The induction and repression of peroxidase isozymes by 3-indoleacetic acid. In: Biochemistry and physiology of plant growth substances. Wightman, F., Setterfield, G. (eds.), pp. 455–472. Ottawa: Runge Press 1968

Galt, R.H.B.: New metabolites of *Gibberella fujikuroi* IX. Gibberellin A_{13}. J. Chem. Soc. 3143–3151 (1965)

Gandar, J.C., Nitsch, C.: Extraction d'une substance de croissance a partir des jeunes graines de *Pisum sativum*. In: Régulateurs naturels de la croissance végétale. Nitsch, J.P. (ed.); Vol. 123, pp. 169–178. Paris: C.N.R.S. 1964

Gandar, J.C., Nitsch, C.: Isolement de l'ester methylique d'un acide chloro-3-indolylacetique a partir de graines immatures de pois, *Pisum sativum*. C. R. Acad. Sci. Ser. D *265*, 1795–1798 (1967)

Gane, R.: Production of ethylene by some ripening fruits. Nature (London) *134*, 1008 (1934)

Ganguly, T., Ganguly, S.N., Sircar, P.K., Sircar, S.M.: Rhamnose bound indole-3-acetic acid in the floral parts of *Peltophorum ferrugineum*. Physiol. Plant. *31*, 330–332 (1974)

Gaskin, P., MacMillan, J., Zeevaart, J.A.D.: Identification of GA_{20}, abscisic acid, and phaseic acid from flowering *Bryophyllum daigremontianum* by combined gas chromatography-mass spectrometry. Planta *111*, 347–352 (1973)

Gaspar, T., Verbeek, R., Khan, A.A.: Some effects of Amo 1618 on growth, peroxidase and α-amylase which cannot be easily explained by inhibition of gibberellin biosynthesis. Plant Physiol. *24*, 552–555 (1971)

Gawer, M., Laloue, M., Terrine, C., Guern, J.: Metabolism and biological significance of benzyladenine-7-glucoside. Plant Sci. Lett. *8*, 267–274 (1977)

Gayler, K.R., Glasziou, K.T.: Plant enzyme synthesis: hormonal regulation of invertase and peroxidase synthesis in sugar cane. Planta *84*, 185–194 (1969)

Gefter, M.L.: The in vitro synthesis of 2′-O-methylguanosine and 2-methylthio-^{6}N-(γ,γ-dimethylallyl)adenosine in transfer RNA of *Escherichia coli*. Biochem. Biophys. Research Commun. *36*, 435–441 (1969)

Gefter, M.L., Russell, R.L.: Role of modification in tyrosine transfer RNA: A modified base affecting ribosome binding. J. Mol. Biol. *39*, 145–157 (1969)

Geissman, T.A., Verbiscar, A.J., Phinney, B.O., Cragg, G.: Studies on the biosynthesis of gibberellins from (−)-kaurenoic acid in cultures of *Gibberella fujikuroi*. Phytochemistry *5*, 933–947 (1966)

Gelinas, D.A.: Proposed model for the peroxidase-catalyzed oxidation of indole-3-acetic acid in the presence of the inhibitor ferulic acid. Plant Physiol. *51*, 967–972 (1973)

Gelinas, D., Postlethwait, S.N.: IAA oxidase inhibitors from normal and mutant maize plants. Plant Physiol. *44*, 1553–1559 (1969)
Gentile, I.A., Matta, A.: Production of and some effects of ethylene in relation to *Fusarium* wilting of tomato. Physiol. Plant Pathol. *5*, 27–37 (1975)
Geuns, J.M.C.: Physiological activity of corticosteroids in etiolated mung bean plants. Z. Pflanzenphysiol. *74*, 42–51 (1974)
Geuns, J.M.C.: Structure requirements of corticosteroids for physiological activity in etiolated mung bean seedlings. Z. Pflanzenphysiol. *81*, 1–16 (1977)
Geuns, J.M.C.: Steroid hormones and plant growth and development. Phytochemistry *17*, 1–14 (1978)
Geuns, J.C.M., Vendrig, J.C.: Hormonal control of sterol biosynthesis in *Phaseolus aureus*. Phytochemistry *13*, 919–922 (1974)
Ghooprasert, P., Spencer, M.: Preparation and purification of an enzyme system for ethylene synthesis from acrylate. Physiol. Veg. *13*, 579–589 (1975)
Giaquinta, R., Beyer, E.M., Jr.: $^{14}C_2H_4$: distribution of ^{14}C-labeled tissue metabolites in pea seedlings. Plant Cell Physiol. *18*, 141–148 (1977)
Gibson, R.A., Barrett, G., Wightman, F.: Biosynthesis and metabolism of indol-3yl-acetic acid. III. Partial purification and properties of a tryptamine-forming L-tryptophan decarboxylase from tomato shoots. J. Exp. Bot. *23*, 775–786 (1972a)
Gibson, R.A., Schneider, E.A., Wightman, F.: Biosynthesis and metabolism of indolyl-3-acetic acid II. In vivo experiments with ^{14}C-labelled precursors of IAA in tomato and barley shoots. J. Exp. Bot. *23*, 381–399 (1972b)
Gillard, D.F., Walton, D.C.: Abscisic acid metabolism by a cell-free preparation from *Echinocystis lobata* liquid endosperm. Plant Physiol. *58*, 790–795 (1976)
Goeschl, J.D., Pratt, H.K.: Regulatory roles of ethylene in the etiolated growth habit of *Pisum sativum*. In: Biochemistry and physiology of plant growth substances. Wightman, F., Setterfield, G. (eds.), pp. 1229–1242. Ottawa: Runge Press 1968
Goeschl, J.D., Pratt, H.K., Bonner, B.: An effect of light on the production of ethylene and the growth of the plumula portion of etiolated pea seedlings. Plant Physiol. 42, 1077–1080 (1967)
Goldschmidt, E.E., Galily, D.: The fate of endogenous gibberellins and applied radioactive gibberellin A_3 during natural and ethylene-induced senescence in citrus peel. Plant Cell Physiol. *15*, 485–491 (1974)
Goldschmidt, E.E., Goren, R., Monselise, S.P., Takahashi, N., Igoshi, H., Yamaguchi, I., Hirose, K.: Auxin in citrus. A reappraisal. Science *174*, 1256–1257 (1971)
Goldschmidt, E.E., Goren, R., Even-Chen, Z., Bittner, S.: Increase in free and bound abscisic acid during natural and ethylene-induced senescence of citrus fruit peel. Plant Physiol. *51*, 879–882 (1973)
Gomez-Navarette, G., Moore, T.C.: Effects of protein synthesis inhibitors on ent-kaurene biosynthesis during morphogenesis of etiolated pea seedlings. Plant Physiol. *61*, 889–892 (1978)
Goodwin, T.W.: Biosynthesis of carotenoids and plant triterpenes. Biochem. J. *123*, 293–329 (1971)
Gordon, M.E., Letham, D.S., Parker, C.W.: The metabolism and translocation of zeatin in intact radish seedlings. Ann. Bot. (London) *38*, 809–825 (1974)
Gordon, S.A.: Occurrence, formation, and inactivation of auxins, Annu. Rev. Plant Physiol. *5*, 341–378 (1954)
Gordon, S.A.: Intracellular localization of the tryptophan-indoleacetate enzyme system. Plant Physiol. *33*, 23–27 (1958)
Gordon, S.A.: The biogenesis of auxin. In: Handbuch der Pflanzenphysiologie. Ruhland, W. (ed.), Vol. 14, pp. 620–646. Berlin-Göttingen-Heidelberg: Springer 1961
Gordon, S.A., Paleg, L.G.: Formation of auxin from tryptophan through action of polyphenols. Plant Physiol. *36*, 838–845 (1961)
Gordon, S.A., Sanchez, N.F.: The biosynthesis of auxin in the vegetative pineapple. II. The precursors of indoleacetic acid. Arch. Biochem. Biophys. *20*, 367–385 (1949)
Gordon, W.R., Henderson, J.H.M.: Isoperoxidases of (IAA oxidase) oxidase in oat coleoptiles. Can. J. Bot. *51*, 2047–2052 (1973)

Goren, R., Bukovac, M.J.: Mechanism of NAA conjugation. No effect of ethylene. Plant Physiol. *51*, 907–913 (1973)
Goren, R., Bukovac, M.J., Flore, J.A.: Mechanism of indole-3-acetic acid conjugation. No induction by ethylene. Plant Physiol. *53*, 164–166 (1974)
Gortner, W.A., Kent, M., Southerland, G.K.: Ferulic and p-coumaric acids in pineapple tissue as modifiers of pineapple indoleacetic acid oxidase. Nature (London) *181*, 630–631 (1958)
Goto, N.: Gibberellins and inhibitors in leaves of tall and dwarf beans. Physiol. Plant. *42*, 359–364 (1978)
Gove, J.P., Hoyle, M.C.: The isozymic similarity of indoleacetic acid oxidase to peroxidase in birch and horseradish. Plant Physiol. *56*, 684–687 (1975)
Graebe, J.E.: Isoprenoid biosynthesis in a cell-free system from pea shoots. Science *157*, 73 (1967)
Graebe, J.E.: Biosynthese isoprenoider Verbindungen in zellfreien Extrakten aus Erbsenpflanzen. Hoppe Seylers Z. Physiol. Chem. *349*, 6–7 (1968a)
Graebe, J.E.: Biosynthesis of kaurene, squalene, and phytoene from mevalonate-2-^{14}C in a cell-free system from pea fruits. Phytochemistry *7*, 2003–2020 (1968b)
Graebe, J.E.: The enzymic preparation of ^{14}C-kaurene. Planta *85*, 171–174 (1969)
Graebe, J.E., Hedden, P.: Biosynthesis of gibberellins in cell free system. In: Biochemistry and chemistry of plant growth regulators. Schreiber, K., Schütte, H.R., Sembdner, G. (eds.), pp. 1–16. Inst. Pl. Biochem. Acad. Sci. GDR, Halle: 1974
Graebe, J.E., Ropers, H.J.: Gibberellins. In: Phytohormones and related compounds – a comprehensive treatise. The Biochemistry of Phytohormones and Related Compounds. Letham, D.S., Goodwin, P.B., Higgins, T.J.V. (eds.), Vol. I. pp. 107–204. Amsterdam: Elsevier 1978
Graebe, J.E., Dennis, D.T., Upper, C.D., West, C.A.: Biosynthesis of gibberellins. I. The biosynthesis of (−)-kaurene, (−)-kaurenol, and trans-geranylgeraniol in endosperm-nucellus of *Echinocystis macrocarpa* Greene. J. Biol. Chem. *240*, 1847–1854 (1965)
Graebe, J.E., Bowen, D.H., MacMillan, J.: The conversion of mevalonic acid into gibberellin A_{12}-aldehyde in a cell-free system from *Cucurbita pepo*. Planta *102*, 261–271 (1972)
Graebe, J.E., Hedden, P., Gaskin, P., MacMillan, J.: The biosynthesis of a C_{19}-gibberellin from mevalonic acid in a cell-free system from higher plant. Planta *120*, 307–309 (1974a)
Graebe, J.E., Hedden, P., Gaskin, P., MacMillan, J.: Biosynthesis of gibberellins A_{12}, A_{15}, A_{24}, A_{36}, and A_{37} by a cell-free system from *Cucurbita maxima*. Phytochemistry *13*, 1433–1440 (1974b)
Graebe, J.E., Hedden, P., MacMillan, J.: The ring contraction step in gibberellin biosynthesis. J.C.S. Chem. Commun., 161–162 (1975)
Green, T.R., Dennis, D.T., West, C.A.: Compartmentation of isopentenylpyrophosphate isomerase and prenyl transferase in developing castor bean endosperm. Biochem. Biophys. Res. Commun. *64*, 976–982 (1975)
Grigorieva, N.Y., Kucherov, V.F., Loschnikova, N.V., Chailakhian, M.K.: Endogenous gibberellins and gibberellin-like substances in long-day and short-day species of tobacco plants. A possible correlation with photo-periodic response. Phytochemistry *10*, 509–517 (1971)
Grochowska, M.J.: Photolytic decarboxylation of carboxyl-carbon-14-labelled indol-3-yl-acetic acid in leaves of the apple tree. J. Exp. Bot. *25*, 638–645 (1974)
Grochowska, M.J., Karaszewska, A.: The effect of naphthalene-acetic acid sprays on the rate of decarboxylation of ^{14}C-labelled indoleacetic acid in apple leaves. In: Biochemistry and chemistry of plant growth regulators. Schreiber, K., Schütte, H.R., Sembdner, G. (eds.), pp. 345–352. Inst. Pl. Biochem. Acad. Sci. GDR, Halle (Saale): 1974
Gross, D.: Chemie und Biochemie der Abscisinsäure. Pharmazie *27*, 619–630 (1972)
Gross, D., Schütte, H.R.: On the metabolism of abscisic acid-2-^{14}C in wheat and barley. In: Biochemistry and chemistry of plant growth regulators. Schreiber, K., Schütte, H.R., Sembdner, G. (eds.), pp. 219–224. Inst. Pl. Biochem. Acad. Sci. GDR, Halle (Saale): 1974
Gross, D., Schütte, H.R.: On the metabolism of abscisic acid-2-^{14}C. In: Plant growth regulators. Kudrev, T., Ivanova, I., Karanov, E. (eds.), pp. 203–206. Bulg. Acad. Sci. Sofia: 1977

Grover, S., Purves, W.K.: Cobalt and plant development, interactions with ethylene in hypocotyl growth. Plant Physiol. *57*, 886–889 (1976)

Guern, J., Doree, M., Sadorge, P.: Transport, metabolism, and biological activity of some cytokinins. In: Biochemistry and physiology of plant growth substances. Wightman, F., Setterfield, G. (eds.), pp. 1155–1167. Ottawa: Runge Press 1968

Guinn, G.: Water deficit and ethylene evolution by young cotton bolls. Plant Physiol. *57*, 403–405 (1976)

Hager, A., Schmidt, R.: Auxintransport and Phototropismus. I. Die lichtbedingte Bildung eines Hemmstoffes für den Transport von Wuchsstoffen in Koleoptilen. Planta *83*, 347–371 (1968)

Hahn, H., De Zacks, R., Kende, H.: Cytokinin formation in pea seeds. Naturwissenschaften *61*, 170 (1974)

Hahn, H., Heftmann, E., Blumbach, M.: Cytokinins: Production and biogenesis of N^6-(Δ^2-iso-pentenyl)adenine in cultures of *Agrobacterium tumefaciens* strain B 6. Z. Pflanzenphysiol. *79*, 143–153 (1976)

Halevy, A.H.: Interaction of growth-retarding compounds and gibberellin on indoleacetic acid oxidase and peroxidase of cucumber seedlings. Plant Physiol. *38*, 731–737 (1963)

Halevy, A.H., Shilo, R.: Promotion of growth and flowering and increase in content of endogenous gibberellins in gladiolus plants treated with growth retardant CCC. Physiol. Plant. *23*, 820–827 (1970)

Halevy, A.H., Simchon, S., Shillo, R.: Changes in "free" and two forms of "bound" gibberellins in the various stages of dormancy of gladiolus corms. In: Plant growth substances 1973. pp. 64–74. Tokyo: Hirokawa 1974

Hall, I.V., Forsyth, F.R.: Production of ethylene by flowers following pollination and treatments with water and auxin. Can. J. Bot. *45*, 1163–1166 (1967)

Hall, P.L.: Movement of indole-3-acetic acid from the endosperm to the shoot of *Zea mays*. L. M.S. Thesis. Michigan State University, East Lansing (1977)

Hall, P.L., Bandurski, R.S.: Movement of indole-3-acetic acid and tryptophan-derived indole-3-acetic acid from the endosperm to the shoot of *Zea mays* L. Plant Physiol. *61*, 425–429 (1978)

Hall, R.H.: Cytokinins as a probe of developmental processes. Annu. Rev. Plant Physiol. *24*, 415–444 (1973)

Hall, R.H., Mintsioulis, G.: Enzymatic activity that catalyzes degradation of N^6-(Δ^2-isopentenyl)adenosine. J. Biochem. *73*, 739–748 (1973)

Hall, R.H., Csonka, L., David, H., McLennan, B.: Cytokinins in the soluble RNA of plant tissues. Science *156*, 69–71 (1967)

Hall, R.H., Alam, S.N., McLennan, B.D., Terrine, C., Guern, J.: N^6(Δ^2-Isopentenyl)adenosine. Can J. Biochem. *49*, 623–630 (1971)

Hall, W.C., Morgan, P.W.: Auxin-ethylene interrelationships. In: Régulateurs naturels de la croissance végétale. Nitsch, J.P. (ed.), pp. 727–745. Paris: C.N.R.S. 1964

Hall, W.C., Miller, C.S., Herrero, F.A.: Studies with ^{14}C-labeled ethylene. In: 4th Int. Conf. Plant Growth Regul. pp. 751–778. Ames: Iowa State University Press 1961

Hamilton, R.H., Bandurski, R.S., Grigsby, B.H.: Isolation of indole-3-acetic acid from corn kernels and etiolated corn seedlings. Plant Physiol. *36*, 354–359 (1961)

Hamilton, R.H., Meyer, H.E., Burke, R.E., Feung, C.S., Mumma, R.O.: Metabolism of indole-3-acetic acid. II. Oxindole pathway in *Parthenocissus tricuspidata* crown-gall tissue cultures. Plant Physiol. *58*, 77–81 (1976)

Hansen, E.: Quantitative study of ethylene production in relation to respiration of pears. Bot. Gaz. *103*, 543 (1942)

Hansen, E.: Postharvest physiology of fruits. Annu. Rev. Plant Physiol. *17*, 459–480 (1966)

Hanson, A.D., Kende, H.: Ethylene-enhanced ion and sucrose efflux in morning glory flower tissue. Plant Physiol. *55*, 663–669 (1975)

Hanson, A.D., Kende, H.: Methionine metabolism and ethylene biosynthesis in senescent flower tissue of morning glory. Plant Physiol. *57*, 528–537 (1976a)

Hanson, A.D., Kende, H.: Biosynthesis of wound ethylene in morning glory flower tissue. Plant Physiol. *57*, 538–541 (1976b)

Hanson, J.R., Hawker, J.: The ring contraction stage in gibberellin biosynthesis. J. Chem. Soc. *D*, 208 (1971)
Hanson, J.R., Hawker, J.: The formation of the C_{19}-gibberellins from gibberellin A_{13} anhydride. Tetrahedron Lett. 4299–4302 (1972)
Hanson, J.R., Hawker, J.: Preparation of (^{14}C) gibberellic acid. Phytochemistry *12*, 1073–1075 (1973)
Hanson, J.R., White, A.F.: The transformation of steviol by *Gibberella fujikuroi*. Tetrahedron *24*, 6291–6293 (1968a)
Hanson, J.R., White, A.F.: Aspects of gibberellin biosynthesis. Chem. Commun. 1689–1690 (1968b)
Hanson, J.R., White, A.F.: The incorporation of a labdadienol into the tetracyclic diterpenes. Chem. Commun. 103 (1969a)
Hanson, J.R., White, A.F.: The oxidative modification of the kauranoid ring B during gibberellin biosynthesis. Chem. Commun. 410–411 (1969b)
Hanson, J.R., White, A.F.: Studies in terpenoid biosynthesis. Part IV. Biosynthesis of the kaurenolides and gibberellic acid. J. Chem. Soc. *C*, 981–985 (1969c)
Hanson, J.R., White, A.F.: The stereochemistry of some stages in gibberellin biosynthesis. Chem. Commun. 1071–1072 (1969d)
Hanson, J.R., Hough, A., White, A.F.: The mevalonoid origin of some hydrogen atoms in the tetracyclic diterpenes. Chem. Commun. 467–468 (1968)
Hanson, J.R., Hawker, J., White, A.F.: Studies in terpenoid biosynthesis. IX. The sequence of oxidation on ring B in kaurene-gibberellin biosynthesis. J. Chem. Soc. Perkin *I*, 1892–1895 (1972)
Hanson, J.R., Sarah, F.Y., Fraga, B.M., Hernandez, M.G.: The microbial preparation of two 'atisagibberellins'. Phytochem. *18*, 1875–1876 (1979)
Harada, H., Lang, A.: Effect of some (2-chlorethyl)trimethyl-ammonium chloride analogs and other growth retardants on gibberellin biosynthesis in *Fusarium moniliforme*. Plant Physiol. *40*, 176–183 (1965)
Harada, H., Yokota, T.: Isolation of gibberellin A_8 glucoside from shoot apices of *Althea rosea*. Planta *92*, 100–104 (1970)
Harada, F., Gross, H.J., Kimura, F., Chang, S.H., Nishimura, S., Rajbhandary, U.L.: 2-Methylthio-N^6-(Δ^2-isopentenyl)adenosine: a component of *E. coli* tyrosine transfer RNA. Biochem. Biophys. Res. Commun. *33*, 299–306 (1968)
Hare, R.C.: Indoleacetic acid oxidase. Bot. Rev. *30*, 129–165 (1964)
Harrison, D.M., MacMillan, J.: Two new gibberellins, A_{24} and A_{25}, from *Gibberella fujikuroi*, their isolation, structure, and correlation with gibberellins A_{13} and A_{15}. J. Chem. Soc. *C*, 631–636 (1971)
Harrison, M.A., Saunders, P.F.: The abscisic acid content of dormant birch buds. Planta *123*, 291–298 (1975)
Harrison, M.A., Walton, D.C.: Abscisic acid metabolism in water-stressed bean leaves. Plant Physiol. *56*, 250–254 (1975)
Hartman, H.T., El-Hamady, M., Whisler, J.: Abscission induction in the olive by cycloheximide. J. Am. Soc. Hortic. Sci. *97*, 781–785 (1972)
Hartmann, S.: Untersuchungen zum Metabolismus von Abscisinsäure. Diplomarbeit, Martin-Luther-Univ. Halle-Wittenberg (1975)
Hartmann, W.: Knospenruhe und Abscisinsäure. Biol. Rundsch. *8*, 154–158 (1970)
Hartung, W., Behl, R.: Transport and metabolism of abscisic-2-^{14}C acid in root segments of *Phaseolus coccineus*. Planta *120*, 299–305 (1974)
Hasson, E.P., West, C.A.: Properties of the system for the mixed function oxidation of kaurene and kaurene derivatives in microsomes of the immature seed of *Marah macrocarpus*. Cofactor requirements. Plant Physiol. *58*, 473–478 (1976a)
Hasson, E.P., West, C.A.: Properties of the system for the mixed function oxidation of kaurene and kaurene derivatives in microsomes of the immature seed of *Marah macrocarpus*. Electron transfer components. Plant Physiol. *58*, 479–484 (1976b)
Hattori, H., Marumo, S.: Monomethyl-4-chloroindolyl-3-acetyl-L-aspartate and absence of indolyl-3-acetic acid in immature seeds of *Pisum sativum*. Planta *102*, 85–90 (1972)
Heatherbell, D.A., Howard, B.H., Wicken, A.J.: The effects of growth retardants on the

respiration and coupled phosphorylation of preparations from etiolated pea seedlings. Phytochemistry *5*, 635–642 (1966)
Hecht, S.M., Leonard, N.J., Burrows, W.J., Armstrong, B.J., Skoog, F., Occolowitz, J.: Cytokinin of wheat germ transfer RNA: 6-(4-hydroxy-3-methyl-2-butenylamino)-2-methylthio-9-β-D-ribofuranosylpurine. Science *166*, 1272–1274 (1969)
Hedden P., Phinney, B.O.: Kaurene biosynthesis in the dwarf-5 mutant of *Zea mays*. Plant Physiol. *57*, 86 (1976)
Hedden, P., Phinney, B.O.: Comparison of ent-kaurene and ent-isokaurene synthesis in cell-free systems from etiolated shoots of normal and dwarf-5 maize seedlings. Phytochemistry *18*, 1475–1479 (1979)
Hedden, P., MacMillan, J., Phinney, B.O.: Fungal products. Part XII. Gibberellin A_{14}-aldehyde, an intermediate in gibberellin biosynthesis in *Gibberella fujikuroi*. J. Chem. Soc. Perkin *I*, 587–592 (1974)
Hedden, P., Phinney, B.O., MacMillan, J., Sponsel, V.M.:Metabolism of kaurenoids by *Gibberella fujikuroi* in the presence of the plant growth retardant; N,N,N-trimethyl-1-methyl-(2′,6′,6′-trimethylcyclohex-2′-en-1′-yl)prop-2-enylammonium iodide. Phytochemistry *16*, 1913–1917 (1977)
Hedden, P., MacMillan, J., Phinney, B.O.: The metabolism of the gibberellins. Annu. Rev. Plant Physiol. *29*, 149–192 (1978)
Heerkloss, R., Libbert, E.: Auxinbiosynthese in vivo aus Indol und Tryptophan unter sterilen Bedingungen. Biol. Plant. *18*, 327–334 (1976a)
Heerkloss, R., Libbert, E.: On the question of β-indolyl-acetic acid synthesis from indole without a tryptophan intermediate. Planta *131*, 299–302 (1976b)
Heftmann, E.: Functions of sterols in plants. Lipids *6*, 128 (1971)
Heftmann, E.: Review. Functions of steroids in plants. Phytochemistry *14*, 891–901 (1975)
Helgeson, J.P., Leonard, N.J.: Cytokinins: identification of compounds isolated from *Corynebacterium fascians*. Proc. Natl. Acad. Sci. USA *56*, 60–63 (1966)
Hemphill, D.D., Baker, L.R., Sell, H.N.: Isolation of novel conjugated gibberellins from *Cucumis sativus* seed. Can. J. Biochem. *51*, 1647–1653 (1973)
Henson, I.E.: Cytokinins and their metabolism in leaves of *Alnus glutinosa* L. Gaertn.. Effects of leaf development. Z. Pflanzenphysiol. *86*, 363–369 (1978a)
Henson, I.E.: Types, formation, and metabolism of cytokinins in leaves of *Alnus glutinosa* (L.) Gaertn. J. Exp. Bot. *29*, 935–951 (1978b)
Henson, I.E., Wareing, P.F.: Cytokinins in *Xanthium strunarium* L.: Distribution in the plant and production in root system. J. Exp. Bot. *27*, 1268–1278 (1976)
Henson, I.E., Wheeler, C.T.: Hormones in plants bearing nitrogen fixing root nodules: The distribution of cytokinins in *Vicia faba* L. New Phytol. *76*, 433–439 (1976)
Henson, I.E., Wheeler, C.T.: Hormones in plants bearing nitrogen-fixing root nodules: Distribution and seasonal changes in levels of cytokinins in *Alnus glutinosa* (L.) Gaertn. J. Exp. Bot. *28*, 205–214 (1977a)
Henson, I.E., Wheeler, C.T.: Hormones in plants bearing nitrogen-fixing root nodules: Partial characterisation of cytokinins from root nodules of *Alnus glutinosa* (L.) Gaertn. J. Exp. Bot. *28*, 1076–1086 (1977b)
Henson, I.E., Wheeler, C.T.: Hormones in plants bearing nitrogen-fixing root nodules: Metabolism of [8-^{14}C]zeatin in root nodules of *Alnus glutinosa* (L.) Gaertn. J. Exp. Bot. *28*, 1087–1098 (1977c)
Henson, I.E., Wheeler, C.T.: Hormones in plants bearing nitrogen-fixing root nodules: Cytokinin transport from the root nodules of *Alnus glutinosa* (L.) Gaertn. J. Exp. Bot. *28*, 1099–1110 (1977d)
Herner, R.C., Sink, Jr., K.C.: Ethylene production and respiratory behavior of the *rin* tomato mutant. Plant Physiol. *52*, 38–42 (1973)
Herrero, F.A., Hall, W.C.: Preliminary results on absorption and metabolism of ethylene-C^{14} in plants. Proc. Assoc. South. Agric. Work. *56*, 224–225 (1959)
Herrero, F.A., Hall, W.C.: General effects of ethylene on enzyme systems in the cotton leaf. Physiol. Plant. *13*, 736–750 (1960)
Hewett, E.W., Wareing, P.F.: Cytokinins in *Populus robusta* Schneid.: a complex in leaves. Planta *112*, 225–233 (1973a)

Hewett, E.W., Wareing, P.F.: Cytokinins in *Populus x robusta:* Changes during chilling and bud burst. Physiol. Plant. *28*, 393–399 (1973b)
Hewett, E.W., Wareing, P.F.: Cytokinins in *Populus x robusta* (Schneid): Light effects on endogenous levels. Planta *114*, 119–129 (1973c)
Hewitt, E.J.: The role of mineral elements in the activity of plant enzyme systems. Oxidation of indoleacetic acid and analogous systems. In: Handbuch der Pflanzenphysiologie. Ruhland, W. (ed.), Vol. IV, pp. 451–454. Berlin-Göttingen-Heidelberg: Springer 1958
Hilgenberg, W., Hanke, R., Bürstell, H.: Tryptophanstoffwechsel bei *Phycomyces blakesleeanus* Bgff. III. Der Nachweis eines Abbauweges der Indol-3-essigsäure. Biol. Zentralbl. *95*, 703–712 (1976)
Hinman, R.L., Lang, J.: Peroxidase-catalyzed oxidation of indole-3-acetic acid. Biochemistry *4*, 144–158 (1965)
Hiraga, K., Yokota, T., Murofushi, N., Takahashi, N.: Isolation and characterization of a free gibberellin and glucosyl esters of gibberellins in mature seeds of *Phaseolus vulgaris*. Agric. Biol. Chem. *36*, 345–347 (1972)
Hiraga, K., Yokota, T., Murofushi, N., Takahashi, N.: Isolation and characterization of gibberellins in mature seeds of *Phaseolus vulgaris*. Agric. Biol. Chem. *38*, 2511–2520 (1974a)
Hiraga, K., Kawase, S., Yokota, T., Murofushi, N., Takahashi, N.: Plant growth substances in seed of *Phaseolus vulgaris*. II. Isolation and characterization of plant growth substances in immature seeds and etiolated seedlings of *Phaseolus vulgaris*. Agric. Biol. Chem. *38*, 2521–2527 (1974b)
Hiraga, K., Yamane, H., Takahashi, N.: Biological activity of some synthetic gibberellin glucosyl esters. Phytochemistry *13*, 2371–2376 (1974c)
Hiraga, K., Yokota, T., Murofushi, N., Takahashi, N.: Plant growth regulators in immature and mature seeds of *Phaseolus vulgaris*. In: Plant growth substances 1973, pp. 75–85. Tokyo: Hirokawa 1974d
Hirai, N., Fukui, H., Koshimizu, K.: A novel abscisic acid metabolite from seeds of *Robinia pseudoacacia*. Phytochemistry *17*, 1625–1627 (1978)
Hiron, R.W., Wright, S.T.C.: The role of endogenous abscisic acid in the response of plants to stress. J. Exp. Bot. *24*, 769–781 (1973)
Hoad, G.V., Bowen, M.R.: Evidence for gibberellin-like substances in phloem exudate of higher plants. Planta *82*, 22–32 (1968)
Hofinger, M., Böttger, M.: Identification by GC-MS of 4-chloro-indolylacetic acid and its methyl ester in immature *Vicia faba* seeds. Phytochemistry *18*, 653–654 (1979)
Horegott, H. von: The degradation of tryptophan in *Cordyceps militaris* strains in saprophytic cultures. Biochem. Physiol. Pflanz. *164*, 500–508 (1970)
Horgan, R.: A new cytokinin metabolite. Biochem. Biophys. Res. Commun. *65*, 358–363 (1975)
Horgan, R., Hewett, E.W., Horgan, J.M., Purse, J., Wareing, P.F.: A new cytokinin from *Populus x robusta*. Phytochemistry *14*, 1005–1008 (1975)
Horng, A.J., Yang, S.F.: Aerobic oxidation of indole-3-acetic acid with bisulfite. Phytochemistry *14*, 1425–1428 (1975)
Hoyle, M.C.: Indoleacetic acid oxidase: A dual catalytic enzyme? Plant Physiol. *50*, 15–18 (1972)
Hoyle, M.C., Routley, D.G.: Promotion of the IAA oxidase-peroxidase complex enzyme in yellow birch. In: Mechanisms of regulation of plant growth. Bieleska, R.L., Ferguson, A.R., Cresswell, M.M. (eds.), pp. 743–750. Wellington: The Royal Society of New Zealand 1974
Hradilik, J.: Estimation of ethylene produced by pea (*Pisum sativum* L.) plants inhibited with auxin and treated with cytokinin. Biochem. Physiol. Pflanz. *167*, 459–461 (1975)
Hradilik, J.: Effects of auxin, cytokinin, ethrel and cotyledon excision on the peroxidase activity in cotylar buds of decapitated pea (*Pisum sativum* L.) plants. Biol. Plant. *18*, 93–98 (1976)
Hsiao, T.C.: Plant responses to water stress. Annu. Rev. Plant Physiol. *24*, 519–570 (1973)
Hutzinger, O., Kosuge, T.: Microbial synthesis and degradation of indole-3-acetic acid.

III. The isolation and characterization of indole-3-acetyl-ε-L-lysin. Biochemistry *7*, 601–605 (1968)
Hyodo, H.: Ethylene production by albedo tissue of *Satsuma mandarin* (*Citrus unshiu* Marc.) fruit. Plant Physiol. *59*, 111–113 (1977)
Ihl, M.: Indole-acetic binding proteins in soybean cotyledons. Planta *131*, 223–228 (1976)
Imaseki, H.: Induction of peroxidase activity by ethylene in sweet potato. Plant Physiol. *46*, 172–174 (1970)
Imaseki, H., Uritani, I., Stahmann, M.A.: Production of ethylene by injured sweet potato root tissue. Plant Cell Physiol. *9*, 757–768 (1968)
Imaseki, H., Pjon, C.J., Furuya, M.: Phytochrome action in *Oryza sativa* L. IV. Red and far red reversible effect on the production of ethylene. Plant Physiol. *48*, 241–244 (1971)
Imaseki, H., Kondo, K., Watanabe, A.: Mechanism of cytokinin action on auxin-induced ethylene production. Plant Cell Physiol. *16*, 777–787 (1975)
Imaseki, H., Watanabe, A., Odawara, S.: Role of oxygen in auxin-induced ethylene production. Plant Cell Physiol. *18*, 577–586 (1977)
Imbert, M.P., Wilson, L.A.: Effects of chlorogenic and caffeic acids on IAA oxidase preparations from sweet potato roots. Phytochemistry *11*, 2671–2676 (1972)
Ishigami, T., Suzuki, Y.: The elongation promotion of pea stem sections by indole together with ethanolamine. Z. Pflanzenphysiol. *62*, 460–463 (1970)
Itai, C., Vaadia, Y.: Cytokinin activity in water-stressed shoots. Plant Physiol. *47*, 87–90 (1971)
Iversen, T.H., Aasheim, T.: Decarboxylation and transport of auxin in segments of sunflower and cabbage roots. Planta *93*, 354–362 (1970)
Iwata, T., Stowe, B.B.: Probing a membrane matrix regulating hormone action II. The kinetics of lipid-induced growth and ethylene production. Plant Physiol. *51*, 691–701 (1973)
Jackson, M.B., Campbell, D.J.: Movement of ethylene from roots to shoots, a factor in the responses of tomato plants to waterlogged soil conditions. New Phytol. *74*, 397–406 (1975a)
Jackson, M.B., Campbell, D.J.: Ethylene and waterlogging effects in tomato. Ann. Appl. Biol. *81*, 102–105 (1975b)
Jackson, M.B., Campbell, D.J.: Production of ethylene by excised segments of plant tissue prior the effect of wounding. Planta *129*, 273–274 (1976)
Jackson, M.B., Osborne, D.J.: Ethylene, the natural regulator of leaf abscission. Nature (London) *225*, 1019–1022 (1970)
Jackson, M.B., Osborne, D.J.: Abscisic acid, auxin, and ethylene in explant abscission. J. Exp. Bot. *23*, 849–862 (1972)
Jaffe, M.J.: On the molecular mode of action of the growth retardant Amo-1618. Plant Physiol. *47*, Suppl. 49 (1971)
Jain, M.L., Kadkade, P.G., van Huysse, P.: The effect of growth regulatory chemicals on abscission and IAA-oxidizing enzyme system of dwarf bean seedlings. Physiol. Plant. *22*, 1038–1042 (1969)
Jansen, E.F.: Metabolism of labeled ethylene in the avocado: Appearance of tritium in the methyl group of toluene. J. Biol. Chem. *238*, 1552–1555 (1963)
Jansen, E.F.: Metabolism of labeled ethylene in the avocado. II. Benzene and toluene from ethylene-^{14}C; benzene from ethylene-^{3}H. J. Biol. Chem. *239*, 1664–1667 (1964)
Jansen, E.F.: Ethylene and polyacetylenes. In: Plant biochemistry. Bonner, J., Varner, J.E. (eds.), pp. 641–644. New York, London: Academic Press 1965
Jansen, E.F.: The metabolism of ethylene in the avocado. In: Food science and technology. Leitch, J.M., (ed.), Vol. I, pp. 475–481. New York, London: Academic Press 1969
Janssen, M.G.H.: Inhibitors of pea root indole acetic oxidase. Acta Bot. Neerl. *19*, 108–111 (1970)
Jasdanwala, R.T., Singh, Y.D., Chinoy, J.J.: Auxin metabolism in developing cotton hairs. J. Exp. Bot. *28*, 1111–1116 (1977)
Jefferies, P.R., Knox, J.R., Ratajczak, T.: Metabolites from the succinate ester of (−)-kaurenol. Tetrahedron Lett. 3229–3231 (1970)

Jefferies, P.R., Knox, J.R., Ratajczak, T.: Metabolic transformations of some ent-kaurenes in *Gibberella fujikuroi*. Phytochemistry *13*, 1423–1431 (1974)
Jelsema, C.L., Ruddat, M., Moore, D.J., Williamson, F.A.: Specific binding of gibberellin A_1 to aleurone grain fraction from wheat endosperm. Plant Cell Physiol. *18*, 1009–1019 (1977)
Jerie, P.H.: The role of ethylene in abscission of cling peach fruit. Aust. J. Plant. Physiol. *3*, 747–754 (1976)
Jerie, P.H., Chalmers, D.J.: Some characteristics of ethylene production in peach (*Prunus persica* L.) seeds. Planta *132*, 13–17 (1976)
Jerie, P.H., Hall, M.A.: The identification of ethylene oxide as a major metabolite of ethylene in *Vicia faba* L. Proc. R. Soc. London Ser. B *200*, 87–94 (1978)
Jerie, P.H., Zeroni, M., Hall, M.A.: Aspects of the role of ethylene in fruit ripening. Acta Hortic. *80*, 325–332 (1978)
Jerie, P.H., Shaari, A.R., Hall, M.A.: The compartmentation of ethylene in developing cotyledons of *Phaseolus vulgaris* L. Planta *144*, 503–507 (1979)
Jindal, K.K., Hemberg, T.: Influence of gibberellic acid on growth and endogenous auxin level in epicotyl and hypocotyl tissues of normal and dwarf bean plants. Physiol. Plant. *38*, 78–82 (1976)
John, W.W., Curtis, R.W.: Isolation and identification of the precursor of ethane in *Phaseolus vulgaris* L. Plant Physiol. *59*, 521–522 (1977)
Jones, D.F.: Examination of the gibberellins of *Zea mays* and *Phaseolus multiflorus* using thin-layer chromatography. Nature (London) *202*, 1309–1310 (1964)
Jones, J.F., Kende, H.: Auxin-induced ethylene biosynthesis in subapical stem sections of etiolated seedlings of *Pisum sativum* L. Planta *146*, 649–656 (1979)
Jones, R.L.: Gibberellins: Their physiological role. Annu. Rev. Plant Physiol. *24*, 571–598 (1973)
Jones, R.L., Lang, A.: Extractable and diffusible gibberellins from light- and dark-grown pea seedlings. Plant Physiol. *43*, 629–634 (1968)
Jones, R.L., Phillips, I.D.J.: Organs of gibberellin synthesis in light-grown sunflower plants. Plant Physiol. *41*, 1381–1386 (1966)
Jones, R.L., Phillips, I.D.J.: Effects of CCC on the gibberellin content of excised sunflower organs. Planta *72*, 53–59 (1967)
Kamienska, A., Durley, R.C., Pharis, R.P.: Metabolism of gibberellin A_4, a native gibberellin of pine pollen. Plant Physiol. *51*, Suppl. 37 (1973)
Kamienska, A., Durley, R.C., Pharis, R.P.: Endogenous gibberellins in pine pollen. III. Conversion of 1,2-^{3}H-GA_4 to gibberellins A_1 and A_{34} in germinating pollen of *Pinus attenuata* Lemm. Plant Physiol. *58*, 68–70 (1976)
Kaminek, M.: Die Zytokinine und ihre Beziehungen zu den Transfer-Ribonukleinsäuren. Biol. Rundsch. *13*, 137–152 (1975)
Kaminek, M., Pačes, V.: Inhibition of enzymatic degradation of N^6-(Δ^2-isopentenyl)-adenosine by cis- and trans isomers of ribosylzeatin and their activity in the *Amaranthus* bioassay. In: Plant growth regulators. Kudrev, T., Ivanova, I., Karanov, E. (eds.), pp. 190–194. Bulg. Acad. Sci. Sofia: 1977
Kang, B.G., Burg, S.P.: Relation of phytochrome-enhanced geotropic sensitivity to ethylene production. Plant Physiol. *50*, 132–135 (1972)
Kang, B.G., Ray, R.M.: Role of growth regulators in the bean hypocotyl hook opening response. Planta *87*, 193–205 (1969)
Kang, B.G., Yocum, C.S., Burg, S.P., Ray, P.M.: Ethylene and carbon dioxide: mediation of hypocotyl hook-opening response. Science *156*, 958 (1967)
Kang, B.G., Newcomb, W., Burg, S.P.: Mechanism of auxin-induced ethylene production. Plant Physiol. *47*, 504–509 (1971)
Katoh, H., Esashi, Y.: Dormancy and impotency of cocklebur seeds. I. CO_2, C_2H_4, O_2 and high temperature. Plant Cell Physiol. *16*, 687–696 (1975)
Kawase, M.: Effect of flooding on ethylene concentration in horticultural plants. J. Am. Soc. Hortic. Sci. *97*, 584–588 (1972a)
Kawase, M.: Submersion increases ethylene and stimulates rooting in cuttings. Proc. Int. Plant Propag. Soc. *22*, 360–366 (1972b)

Kawase, M.: Role of ethylene in induction of flooding damage in sunflower. Physiol. Plant. *31*, 29–38 (1974)

Kawase, M.: Ethylene accumulation in flooded plants. Physiol. Plant. *36*, 236–241 (1976)

Kay, E., Shannon, L.M., Lew, J.Y.: Peroxidase isoenzymes from horseradish roots. II. Catalytic properties. J. Biol. Chem. *242*, 2470–2473 (1967)

Keglević, D., Pokorny, M.: The chemical synthesis of 1-0-(indol-3′-ylacetyl)-β-D-glucopyranose. Biochem. J. *114*, 827–832 (1969)

Kende, H.: Preparation of radioactive gibberellin A_1 and its metabolism in dwarf peas. Plant Physiol. *42*, 1612–1618 (1967)

Kende, H.: The cytokinins. Int. Rev. Cytol. *31*, 301–338 (1971)

Kende, H., Baumgartner, B.: Regulation of ageing in flowers of *Ipomoea tricolor* by ethylene. Planta *116*, 279–289 (1974)

Kende, H., Hanson, A.D.: Relationship between ethylene evolution and senescence in morning glory flower tissue. Plant Physiol. *57*, 523–527 (1976)

Kende, H., Sitton, D.: The physiological influence of kinetin and gibberellin-like root hormones. Ann. N.Y. Acad. Sci. *144*, 235–243 (1967)

Kende, H., Ninnemann, H., Lang, A.: Inhibition of gibberellic acid biosynthesis in *Fusarium moniliforme* by AMO 1618 and CCC. Naturwissenschaften *50*, 599–600 (1963)

Kenney, G., Südi, J., Blackman, G.E.: The uptake of growth substances. XII. Differential uptake of indole-3yl-acetic acid through the epidermal and cut surfaces of etiolated stem segments. J. Exp. Bot. *20*, 820–840 (1969)

Kenten, R.H.: Oxidation of indolyl-3-acetic acid by waxpod bean root sap and peroxidase systems. Biochem. J. *59*, 110–121 (1955)

Khalifah, R.A.: Metabolism of DL-tryptophan-3-^{14}C by the fruitlets of *Citrus aurantifolia*. Physiol. Plant. *20*, 355–360 (1967)

Khalifah, R.A., Lewis, L.N., Coggins, C.W. Jr.: New natural growth promoting substances in young citrus fruits. Science *142*, 399–400 (1963)

Kim, W.K., Rohringer, R.: Metabolism of aromatic compounds in healthy and rost-infected primary leaves of wheat. IV. Studies on the metabolism of tryptophan. Can. J. Bot. *47*, 1425–1433 (1969)

Kinashi, H., Suzuki, Y., Takeuchi, S., Kawarada, A.: Possible metabolic intermediates from IAA to β-acid in rice bran. Agric. Biol. Chem. *40*, 2465–2470 (1976)

Kindl, H.: Oxydasen und Oxygenasen in höheren Pflanzen. I. Über das Vorkommen von Indolyl-(3)-acetaldehydoxim und seine Bildung aus L-Tryptophan. Hoppe Seylers Z. Physiol. Chem. *349*, 519–520 (1968)

King, R.W.: Abscisic acid in developing wheat grains and its relationship to grain growth and maturation. Planta *132*, 43–51 (1976)

King, R.W.: Abscisic acid synthesis and metabolism in wheat ears. Austr. J. Plant Physiol. *6*, 99–108 (1979)

King, R.W., Evans, L.T., Firn, R.D.: Abscisic acid and xanthoxin contents in the long-day plant *Lolium temulentum* L. Aust. J. Plant Physiol. *4*, 217–223 (1977)

Klämbt, H.D.: Indol-3-acetylasparaginsäure, ein natürlich vorkommendes Indolderivat. Naturwissenschaften *47*, 398 (1960)

Klämbt, H.-D.: Wachstumsinduktion und Wuchsstoffmetabolismus im Weizenkoleoptilzylinder. II. Stoffwechselprodukte der Indol-3-essigsäure und der Benzoesäure. Planta *56*, 618–631 (1961)

Klämbt, H.-D.: Cytokinine aus *Helianthus annuus*. Planta *82*, 170–178 (1968)

Kline, L.K., Fittler, F., Hall, R.H.: N^6-(Δ^2-Isopentenyl) adenosine. Biosynthesis in transfer ribonucleic acid in vitro. Biochemistry *8*, 4361–4371 (1969)

Kluge, M., Reinhard, E., Ziegler, H.: Gibberellinaktivität von Siebröhrensäften. Naturwissenschaften *51*, 145–146 (1964)

Knöfel, H.D., Müller, P., Sembdner, G.: Studies on the enzymatical hydrolysis of gibberellin-O-glucosides. In: Biochemistry and chemistry of plant growth regulators. Schreiber, K., Schütte, H.-R., Sembdner, G. (eds.), pp. 121–124, Inst. Pl. Biochem. Acad. Sci. GDR, Halle 1974

Knotz, J., Coolbaugh, R.C., West, C.A.: Regulation of the biosynthesis of ent-kaurene from mevalonate in the endosperm of immature *Marah macrocarpus* seeds by adenylate energy charge. Plant Physiol. *60*, 81–85 (1977)

Knypl, J.S.: No effect of plant growth retarding compounds and growth stimulators on indole-3-acetic acid oxidase activity in greening cucumber cotyledons. Acta Soc. Bot. Pol. *XLII*, 441–452 (1973)

Knypl, J.S., Rennert, A.: Growth and indole-3-acetic acid oxidase in cucumber treated with coumarin, (2-chloroethyl)trimethylammonium chloride and (2,4-dichloro-benzyl)tributylphosphonium chloride. Flora (Jena) Abt. A *158*, 468–478 (1967a)

Knypl, J.S., Rennert, A.: Growth inhibitors and the indole-3-acetic acid oxidase contents of cucumber hypocotyl sections. Naturwissenschaften *54*, 544 (1967b)

Koda, Y., Okazawa, Y.: Cytokinin production by tomato root: Occurrence of cytokinins in staled medium of root culture. Physiol. Plant. *44*, 412–416 (1978)

Köhler, D.: Die Wirkung von schwachem Rotlicht und Chlorcholinchlorid auf den Gibberellingehalt normaler Erbsensämlinge und die Ursache der unterschiedlichen Empfindlichkeit von Zwerg- und Normalerbsensämlingen gegen ihr eigenes Gibberellin. Planta *67*, 44–54 (1965)

Köhler, D.: Die Abhängigkeit der Gibberellinproduktion von Normalerbsen vom Phytochromsystem. Planta *69*, 27–33 (1966a)

Köhler, D.: Veränderungen des Gibberellingehaltes von Salatsamen nach Belichtung. Planta *70*, 42–45 (1966b)

Köhler, D.: Die Wirkung des Rotlichtes auf das Wachstum von Erbsenkeimlingen und ihren Gehalt an gibberellinähnlichen Substanzen. Z. Pflanzenphysiol. *62*, 426–435 (1970)

Kondo, K., Watanabe, A., Imaseki, H.: Relationships in actions of indoleacetic acid, benzyladenine and abscisic acid in ethylene production. Plant Cell Physiol. *16*, 1001–1007 (1975)

Konze, J.R., Elstner, E.F.: Pyridoxalphosphate-dependent ethylene production from methionine by isolated chloroplasts. FEBS Lett. *66*, 8–11 (1976a)

Konze, J.R., Elstner, E.F.: Effect of point freezing on ethylene and ethane production by sugar beet leaf disks. Nature (London) *263*, 351–352 (1976b)

Konze, J.R., Elstner, E.F.: Ethylene- and ethane formation in leaf disks, plastids and mitochondria. Ber. Dtsch. Bot. Ges. *89*, 547–553 (1976c)

Konze, J.R., Kende, H.: Ethylene formation from 1-aminocyclopropane-1-carboxylic acid in homogenates of etiolated pea seedlings. Planta *146*, 293–301 (1979a)

Konze, J.R., Kende, H.: Interactions of methionine and selenomethionine with methionine adenosyltransferase and ethylene-generating systems. Plant Physiol. *63*, 507–510 (1979b)

Kopcewicz, J.: Effect of estrone on the content of endogenous gibberellins in the dwarf pea. Naturwissenschaften *56*, 334 (1969)

Kopcewicz, J.: Influence of steroidal hormones on flower sex expression in *Ecballium elaterium* (L.) A. Rich. Z. Pflanzenphysiol. *65*, 92–94 (1971)

Kopcewicz, J., Porazinski, J.: Effect of red light irradiation on metabolism of free and bound gibberellins in Scots pine (*Pinus silvestris*). Bull. Acad. Pol. Sci. Ser. Sci. Biol. *21*, 383–387 (1973)

Kopcewicz, J., Rogozinska, J.H.: Effect of estrogens and gibberellic acid on cytokinin and abscisic acid-like compound contents in pea. Experientia *28*, 1516–1517 (1972)

Kopcewicz, J., Ehmann, A., Bandurski, R.S.: Enzymatic esterification of indole-3-acetic acid to *myo* inositol and glucose. Plant Physiol. *54*, 846–851 (1974)

Koshimizu, K., Kusaki, T., Mitsui, T., Matsubara, S.: Isolation of a cytokinin, (−)-dihydrozeatin, from immature seeds of *Lupinus luteus*. Tetrahedron Lett. 1317–1320 (1967)

Koshimizu, K., Inui, M., Fukui, H., Mitsui, T.: Isolation of (+)-abscisyl-β-D-glucopyranoside from immature fruit of *Lupinus luteus*. Agric. Biol. Chem. *32*, 789–791 (1968)

Kosiyachinda, S., Young, R.E.: Ethylene production in relation to the initiation of respiratory climacteric in fruit. Plant Cell Physiol. *16*, 595–602 (1975)

Krasnuk, M., Witham, F.H., Tegley, J.R.: Cytokinins extracted from pinto bean fruit. Plant Physiol. *48*, 320–324 (1972)

Kriedemann, P.E., Loveys, B.R., Downton, W.J.S.: Internal control of stomatal physiology and photosynthesis. II. Photosynthetic responses to phaseic acid. Aust. J. Plant Physiol. *2*, 553–567 (1975)

Ku, H.S., Yang, S.F., Pratt, H.K.: Inactivity of apoperoxidase in indoleacetic acid oxidation and in ethylene formation. Plant Physiol. *45*, 358–359 (1970)

Kuhn-Silk, W., Erickson, R.O.: Compartmental analysis of gibberellin A_1 supplied to excised lettuce hypocotyls. Plant Physiol. *57*, Suppl. 444 (1976)

Kuhn-Silk, W., Stoddart, J.L., Jones, R.L.: GA metabolism during hormone induced growth. Plant Physiol. *57*, Suppl. 445 (1976)

Kumar, S.A., Mahadevan, S.: 3-Indoleacetaldoxime hydrolyase: A pyridoxal-5′-phosphate activated enzyme. Arch. Biochem. Biophys. *103*, 516–518 (1963)

Kunert, R., Libbert, E.: Beziehungen zwischen Pflanzen und epiphytischen Bakterien hinsichtlich ihres Auxinstoffwechsels. X. Die Exsudation von Aminosäuren und Kohlenhydraten durch Maissprosse als Ernährungsgrundlage für epiphytische Bakterien. Biochem. Physiol. Pflanz. *163*, 524–535 (1972)

Kuo, C.G., Pharis, R.P.: Effects of AMO-1618 and B-995 on growth and endogenous gibberellin content of *Cupressus arizonica* seedlings. Physiol. Plant. *34*, 288–292 (1975)

Kuraishi, S., Muir, R.M.: Mode of action of growth retarding chemicals. Plant Physiol. *38*, 19–24 (1963)

Kutáček, M., Kefeli, V.I.: The present knowledge of indole compounds in plants of the *Brassicaceae* family. In: Biochemistry and physiology of plant growth substances. Wightman, F., Setterfield, G. (eds.), pp. 127–155. Ottawa: Runge Press 1968

Kutáček, M., Kefeli, V.: Biogenesis of indole compounds from D- and L-tryptophan in segments of etiolated seedlings of cabbage, maize and pea. Biol. Plant. *12*, 145–158 (1970)

Kutáček, M., Kralova, M.: Biosynthesis of the glucobrassicin aglycon from ^{14}C and ^{15}N labelled L-tryptophan precursors. Biol. Plant. *14*, 279–285 (1972)

Labarca, C., Nichols, P.B., Bandurski, R.S.: A partial characterization of indoleacetylinositols from *Zea mays*. Biochem. Biophys. Res. Commun. *20*, 641–646 (1965)

Laloraya, M.M., Srivastav, H.N., Baksi, Nandita, Shah, Nandita, Pandya, K.J., Singh, V.D.: Role of growth regulators, flavonoid pigments and peroxidase in stomata and trichome differentiation. In: Biology of the land plants. Puri, V., Murty, Y.S., Gupta, P.K., Banerji, D. (eds.), pp. 143–159. Meerut (India): Sarita Praheskan 1972

Laloue, M.: Cytokinins: 7-Glucosylation is not a prerequisite of the expression of their biological activity. Planta *134*, 273–275 (1977)

Laloue, M., Hall, R.H.: Cytokinins in *Rhizopogon roseolus*, secretion of N-(9-β-D-ribofuranosyl-9H)purine-6-ylcarbamoyl) threonine into the culture medium. Plant Physiol. *51*, 559–562 (1973)

Laloue, M., Terrine, C., Gawer, M.: Cytokinins: formation of the nucleoside-5′-triphosphate in tobacco and *Acer* cells. FEBS Lett. *46*, 45–50 (1974)

Laloue, M., Gawer, M., Terrine, C.: Modalités de l'utilisation des cytokinines exogènes par les cellules de tabac cultivés en milieu liquide agité. Physiol. Veg. *13*, 781–796 (1975)

Laloue, M., Terrine, C., Guern, J.: Cytokinins: metabolism and biological activity of N^6-(Δ^2-isopentenyl)adenosine and N^6-(Δ^2-isopentenyl)adenine in tobacco cells and callus. Plant Physiol. *59*, 478–483 (1977)

Lance, B., Durley, R.C., Reid, D.M., Thorpe, T.A., Pharis, R.P.: Metabolism of [^{3}H]Gibberellin A_{20} in light- and dark-grown tobacco callus cultures. Plant Physiol. *58*, 387–392 (1976)

Lane, H.C., King, E.E.: Stimulation of indoleacetic acid oxidase and inhibition of catalase in cotton extracts by plant acids. Plant Physiol. *43*, 1699–1702 (1968)

Lang, A.: Gibberellins: Structure and metabolism. Annu. Rev. Plant Physiol. *21*, 537–570 (1970)

Langille, A.R., Forsline, P.L.: Influence of temperature and photoperiod on cytokinin pools in the potato, *Solanum tuberosum* L. Plant Sci. Lett. *2*, 189–191 (1974)

Lantican, B.P., Muir, R.M.: Isolation and properties of the enzyme system forming indoleacetic acid. Plant Physiol. *42*, 1158–1160 (1967)

Lantican, B.P., Muir, R.M.: Auxin physiology of dwarfism in *Pisum sativum*. Physiol. Plant. *22*, 412–423 (1969)

Larsen, P.: Neutral auxins. Nature (London) *159*, 842–843 (1947)

Larsen, P.: Conversion of indoleacetaldehyde to indoleacetic acid in excised coleoptiles and in coleoptile juice. Am. J. Bot. *36*, 32–41 (1949)

Larsen, P.: Formation, occurrence and inactivation of growth substances. Annu. Rev. Plant Physiol. *2*, 169–198 (1951 a)

Larsen, P.: Enzymatic conversion of indoleacetaldehyde and naphthalene acetaldehyde to auxins. Plant Physiol. *26*, 697–707 (1951 b)

La Rue, T.A.G., Gamborg, O.L.: Ethylene production by plant cell cultures. Variation in production during growing cycle and in different plant species. Plant Physiol. *48*, 394–398 (1971)

Lau, O.L., Yang, S.F.: Mechanism of a synergistic effect of kinetin on auxin-induced ethylene production suppression of auxin conjugation. Plant Physiol. *51*, 1011–1014 (1973)

Lau, O.L., Yang, S.F.: Synergistic effect of calcium and kinetin on ethylene production by mung bean hypocotyl. Planta *118*, 1–6 (1974)

Lau, O.L., Yang, S.F.: Interaction of kinetin and calcium in relation to their effect on stimulation of ethylene production. Plant Physiol. *55*, 738–740 (1975)

Lau, O.L., Yang, S.F.: Stimulation of ethylene production in the mung bean hypocotyls by cupric ion, calcium ion, and kinetin. Plant Physiol. *57*, 88–92 (1976 a)

Lau, O.L., Yang, S.F.: Inhibition of ethylene production by cobaltous ion. Plant Physiol. *58*, 114–117 (1976 b)

Lau, O.L., Yung, K.H.: Synergistic effect of kinetin on IAA-induced ethylene production. Plant Cell Physiol. *15*, 29–35 (1974)

Lau, O.L., Murr, D.P., Yang, S.F.: Effect of the 2,4-dinitrophenol on auxin-induced ethylene production and auxin conjugation by mung bean tissue. Plant Physiol. *54*, 182–185 (1974)

Lau, O.L., John, W.W., Yang, S.F.: Effect of different cytokinins on ethylene production by mung bean hypocotyls in the presence of indole-3-acetic acid or calcium ion. Physiol. Plant. *39*, 1–3 (1977)

Lee, T.T.: Promotion of indoleacetic acid oxidase isoenzymes in tobacco callus cultures by indoleacetic acid. Plant Physiol. *48*, 56–59 (1971)

Lee, T.T.: Changes in indoleacetic acid oxidase isoenzymes in tobacco. Tissues after treatment with 2,4-dichlorophenoxyacetic acid. Plant Physiol. *49*, 957–960 (1972)

Lee, T.T.: Cytokinin control in subcellular localization of indoleacetic acid oxidase and peroxidase. Phytochemistry *13*, 2445–2454 (1974)

Lee, T.T.: Insecticide-plant interaction: carbofuran effect on indole-3-acetic acid metabolism and plant growth. Life Sci. *18*, 205–210 (1976)

Lee, T.T.: Role of phenolic inhibitors in peroxidase-mediated degradation of indole-3-acetic acid. Plant Physiol. *59*, 372–375 (1977)

Lee, T.T., Chapman, R.A.: Inhibition of enzymic oxidation of indole-3-acetic acid by metabolites of the insecticide carbofuran. Phytochemistry *16*, 35–39 (1977)

Lee, T.T., Skoog, F.: Effects of hydroxybenzoic acids on indoleacetic acid inactivation by tobacco callus extracts. Physiol. Plant. *18*, 577–585 (1965)

Lehmann, H.: Abscisic acid conjugation. In: Plant growth and differentiation. Kefeli, V.J. (ed.). Moskwa: Nauka 1980 in press

Lehmann, H., Schütte, H.R.: Zur Darstellung O-substituierter Zuckerester. J. Prakt. Chem. *319*, 117–122 (1977)

Lehmann, H., Schütte, H.R.: Purification and characterization of an abscisic acid glucosylating enzyme from cell suspension cultures of *Macleaya microcarpa*. Biochem. Physiol. Pflanz. *96*, 277–280 (1980)

Lehmann, H., Miersch, O., Schütte, H.R.: Synthese von Abscisyl-β-D-glucopyranosid. Z. Chem. *15*, 443 (1975)

Leopold, A.C.: Antagonism of some gibberellin actions by a substituted pyrimidine. Plant Physiol. *48*, 537–540 (1971)

Le Page-Degivry, M.-T.: Intervention d'un inhibiteur lié dans la dormance embryonnaire de *Taxus baccata* L. Co. R. Acad. Sci. Ser. D, *277*, 177–180 (1973)
Leshem, Y., Philosoph, S., Wurzburger, J.: Glucosylation of free trans-trans-abscisic acid as a contributing factor in bud dormancy break. Biochem. Biophys. Res. Commun. *57*, 526–531 (1974)
Letham, D.S.: Purification and probable identity of a new cytokinin in sweet corn extracts. Life Sci. *5*, 551–554 (1966a)
Letham, D.S.: Isolation and probable identity of a third cytokinin in sweet corn extracts. Life Sci. *5*, 1999–2004 (1966b)
Letham, D.S.: Cytokinins from *Zea mays*. Phytochemistry *12*, 2445–2455 (1973)
Letham, D.S.: Cytokinins. In: Phytohormones and related compounds – A comprehensive treatise. Letham, D.S., Higgins, J., Goodwin, P.Z. (eds.), Vol. I, pp. 205–293. Amsterdam: Elsevier 1978
Letham, D.S., Williams, M.W.: Regulators of cell division in plant tissue. VIII. The cytokinins of apple fruit. Physiol. Plant. *22*, 925–936 (1969)
Letham, D.S., Wilson, M.M., Parker, C.W., Jenkins, I.D., MacLeod, J.K., Summons, R.E.: Regulators of cell division in plant tissues. XXIII. The identity of an unusual metabolite of 6-benzylaminopurine. Biochem. Biophys. Acta *399*, 61–70 (1975)
Letham, D.S., Parker, C.W., Duke, C.C., Summons, R.E., MacLeod, J.K.: O-Glucosyl zeatin and related compounds – a new group of cytokinin metabolites. Ann. Bot. (London) *41*, 261–263 (1977)
Letham, D.S., Summons, R.E., Entsch, B., Gollnow, B.I., Parker, C.W., MacLeod, J.K.: Glucosylation of cytokinin analogues. Phytochemistry *17*, 2053–2057 (1978)
Lew, F.T., West, C.A.: (−)-Kaur-16-en-7β-ol-19-oic acid, an intermediate in gibberellin biosynthesis. Phytochemistry *10*, 2065–2076 (1971)
Libbert, E.: Bildung von Indol-3-essigsäure in vivo aus Tryptophan durch Transaminierung. Naturwissenschaften *49*, 423 (1962)
Libbert, E., Bormann, H.: Die Aktivität der IES-Oxydase in phototropisch gereizten Pflanzen. Flora (Jena) Abt. A, *157*, 68–79 (1966)
Libbert, E., Brunn, K.: Nachweis von Indol-3-brenztraubensäure und Indol-3-äthanol (Tryptophol) bei der enzymatischen Auxinbildung aus Tryptophan in vitro. Naturwissenschaften *48*, 741 (1961)
Libbert, E., Manteuffel, R.: Interactions between plants and epiphytic bacteria regarding their auxin metabolism. VII. The influence of the epiphytic bacteria on the amount of diffusible auxin from corn coleoptiles. Physiol. Plant. *23*, 93–98 (1970)
Libbert, E., Risch, H.: Interactions between plants and epiphytic bacteria regarding their auxin metabolism. V. Isolation and identification of the IAA-producing and -destroying bacteria from pea plants. Physiol. Plant. *22*, 51–58 (1969)
Libbert, E., Silhengst, P.: Interactions between plants and epiphytic bacteria regarding their auxin metabolism. VIII. Transfer of ^{14}C-indoleacetic acid from epiphytic bacteria to corn coleoptiles. Physiol. Plant. *23*, 480–487 (1970)
Libbert, E., Wichner, S., Duerst, E., Kaiser, W., Kunert, R., Manicki, A., Manteuffel, R., Riecke, E., Schröder, R.: Auxin content and auxin synthesis in sterile and non-sterile plants, with special regard to the influence of epiphytic bacteria. In: Biochemistry and physiology of plant growth substances. Wightman, F., Setterfield, G. (eds.), pp. 213–230. Ottawa: Runge Press 1968
Libbert, E., Drawert, A., Schröder, R.: Pathways of IAA production from tryptophan by plants and by their epiphytic bacteria: a comparison. I. IAA formation by sterile pea sections in vivo as influenced by IAA oxidase inhibitors and by transaminase coenzyme. Physiol. Plant. *22*, 1217–1225 (1969a)
Libbert, E., Kaiser, W., Kunert, R.: Interaction between plants and epiphytic bacteria regarding their auxin metabolism. VI. The influence of the epiphytic bacteria on the content of extractable auxin in the plant. Physiol. Plant. *22*, 432–439 (1969b)
Libbert, E., Erdmann, N., Schiewer, U.: Auxinbiosynthese. Biol. Rundsch. *8*, 369–390 (1970a)
Libbert, E., Fischer, E., Drawert, A., Schröder, R.: Pathways of IAA production from

tryptophan by plants and by their epiphytic bacteria: A comparison. II. Establishment of the tryptophan metabolites, effects of a native inhibitor. Physiol. Plant. *23*, 278–286 (1970b)

Libbert, E., Schröder, R., Drawert, A., Fischer, E.: Pathways of IAA production from tryptophan by plants and by their epiphytic bacteria: A comparison. III. Metabolism of tryptamine, indoleacetaldehyde, indoleethanol and indoleacetamide. Effects of a native inhibitor. Physiol. Plant. *23*, 287–293 (1970c)

Libbert, E., Schröder, R., Drawert, A.: Sites and mode of action of a native inhibitor from *Pisum sativum* affecting the biogenesis of auxin. Biochem. Physiol. Pflanz. *161*, 310–319 (1970d)

Libbert, E., Manteuffel, R., Siegel, E.: Interactions between plants and epiphytic bacteria regarding their auxin metabolism. IX. The influence of the epiphytic bacteria on the auxin production from tryptophan applied to corn shoots and coleoptiles. Physiol. Plant. *23*, 784–791 (1970e)

Lieberman, M.: Biosynthesis and regulatory control of ethylene in fruit ripening. A review. Physiol. Veg. *13*, 489–499 (1975)

Lieberman, M., Kunishi, A.T.: Propanol may be a precursor of ethylene in metabolism. Science *158*, 938 (1967)

Lieberman, M., Kunishi, A.T.: An evaluation of 4-S-methyl-2-keto-butyric acid as an intermediate in the biosynthesis of ethylene. Plant Physiol. *47*, 576–580 (1971)

Lieberman, M., Kunishi, A.T.: Ethylene-forming systems in etiolated pea seedling and apple tissue. Plant Physiol. *55*, 1074–1078 (1975)

Lieberman, M., Mapson, L.W.: Genesis and biogenesis of ethylene. Nature (London) *204*, 343–345 (1964)

Lieberman, M., Kunishi, A.T., Mapson, L.W., Wardale, D.A.: Ethylene production from methionine. Biochem. J. *97*, 449–459 (1965)

Lieberman, M., Kunishi, A.T., Mapson, L.W., Wardale, D.A.: Stimulation of ethylene production in apple tissue slices by methionine. Plant Physiol. *41*, 376–382 (1966)

Lieberman, M., Kunishi, A.T., Owens, L.D.: Specific inhibitors of ethylene production as retardants of the ripening process in fruits. In: Facteurs et régulation de la maturation des fruits. Colloq. Int. C.N.R.S. No. 238. Ulrich, R. (ed.), pp. 161–170. Paris: C.N.R.S. 1974

Lieberman, M., Baker, J.E., Sloger, M.: Influence of plant hormones on ethylene production in apple, tomato, and avocado slices during maturation and senescence. Plant Physiol. *60*, 214–217 (1977)

Liebisch, H.-W.: Uptake, translocation and metabolism of labelled GA_3 glucosyl ester. In: Biochemistry and chemistry of plant growth regulators. Schreiber, K., Schütte, H.-R., Sembdner, G. (eds.), pp. 109–113. Inst. Pl. Biochem. Acad. Sci. GDR, Halle 1974

Lipe, J.A., Morgan, P.W.: Ethylene: Role in fruit abscission and dehiscence processes. Plant Physiol. *50*, 759–764 (1972)

Lipe, J.A., Morgan, P.W.: Ethylene, a regulator of young fruit abscission. Plant Physiol. *51*, 949–953 (1973)

Lischewski, M., Adam, G.: Darstellung von Gibberellin-A_3-oyl-2-^{14}C-glycin. Z. Chem. *16*, 357 (1976)

Little, C.H.A., Heald, J.K., Browning, G.: Identification and measurement of indoleacetic and abscisic acids in the cambial region of *Picea sitchensis* (Bong) Carr. by combined gas chromatography-mass spectrometry. Planta *139*, 133–139 (1978)

Liu, S.-T., Katz, C.D., Knight, C.A.: Indole-3-acetic acid synthesis in tumorous and nontumorous species of *Nicotiana*. Plant Physiol. *61*, 743–747 (1978a)

Liu, W.T., Pool, R., Wenkert, W., Kriechmann, P.S.: Changes in photosynthesis, stomatal resistance and abscisic acid of *Vitis iabruscana* though drought and irrigation cycles. Am. J. Enol. Vitic. *29*, 239–246 (1978b)

Lockhart, J.A.: Kinetic studies of certain anti-gibberellins. Plant Physiol. *37*, 759–764 (1962)

Loh, J.W.C., Severson, J.G., Jr.: Stimulation of indoleacetic acid oxidase of bean plants by naphthenates. Phytochemistry *14*, 1265–1267 (1975)

Lorenzi, R., Horgan, R., Wareing, P.F.: Cytokinins in *Picea sitchensis:* Identification and relation to growth. Biochem. Physiol. Pflanz. *168*, 333–339 (1975a)
Lorenzi, R., Horgan, R., Heald, J.K.: Gibberellins in *Picea sitchensis* Carriere: Seasonal variation and partial characterization. Planta *126*, 75–82 (1975b)
Lorenzi, R., Horgan, R., Heald, J.K.: Gibberellin A_9 glucosyl ester in needles of *Picea sitchensis*. Phytochemistry *15*, 789–790 (1976)
Loveys, B.R.: The intracellular location of abscisic acid in stressed and non-stressed leaf tissue. Physiol. Plant. *40*, 6–10 (1977)
Loveys, B.R., Kriedemann, P.E.: Internal control of stomatal physiology and photosynthesis. I. Stomatal regulation and associated changes in endogenous levels of abscisic and phaseic acid. Aust. J. Plant Physiol. *1*, 407–415 (1974)
Loveys, B.R., Wareing, P.F.: The red light controlled production of gibberellin in etiolated wheat leaves. Planta *98*, 109–116 (1971)
Loveys, B.R., Brien, C.J., Kriedemann, P.E.: Biosynthesis of abscisic acid under osmotic stress. Studies based on a dual labeling technique. Physiol. Plant. *33*, 166–170 (1975)
Lund, H.A.: The biosynthesis of indoleacetic acid in the styles and ovaries of tobacco preliminary to the setting of fruit. Plant Physiol. *31*, 334–339 (1956)
Lunnon, M.W., MacMillan, J., Phinney, B.O.: Fungal products. 20. Transformations of 2- and 3-hydroxylated kaurenoids by *Gibberella fujikuroi*. J. Chem. Soc. Perkin *I*, 2308–2316 (1977)
Lürssen, K., Naumann, K.: 1-Aminocyclopropane-1-carboxylic acid – a new intermediate of ethylene biosynthesis. Naturwissenschaften *66*, 264 (1979)
Lürssen, K., Naumann, K., Schröder, R.: 1-Aminocyclopropane-1-carboxylic acid – an intermediate of the ethylene biosynthesis in higher plants. Z. Pflanzenphysiol. *92*, 285–294 (1979)
Macháčková, J., Zmrhal, Z.: Oxidation of indole acetic acid by wheat peroxidase. In: Biochemistry and chemistry of plant growth regulators. Schreiber, K., Schütte, H.R., Sembdner, G. (eds.), pp. 353–358. Inst. Pl. Biochem. Acad. Sci. GDR Halle (Saale): 1974
Macháčková, J., Zmrhal, Z.: Comparison of the effect of some phenolic compounds on wheat coleoptile section growth with their effect on IAA-oxidase activity. Biol. Plant. *18*, 147–151 (1976)
Macháčková, J., Gančeva, K., Zmrhal, Z.: The role of peroxidase in the metabolism of indole-3-acetic acid and phenols in wheat. Phytochemistry *14*, 1251–1254 (1975)
MacLachlan, G.A., Waygood, E.R.: Kinetics of the enzymically catalyzed oxidation of indoleacetic acid. Can. J. Biochem. Physiol. *34*, 1233–1250 (1956)
MacLeod, J.E., Summons, R.E., Parker, C.W., Letham, D.S.: Lupinic acid, a purinyl amino acid and novel metabolite of zeatin. J. Chem. Soc. 809–810 (1975)
MacMillan, J.: Diterpenes – The Gibberellins. In: Aspects of terpenoid chemistry and biochemistry. Goodwin, T.W. (ed.), pp. 158–180. New York, London: Academic Press 1971
MacMillan, J.: Metabolic processes related to gibberellin biosynthesis in mutants of *Gibberella fujikuroi.* In: Biochemistry and chemistry of plant growth regulators. Schreiber, K., Schütte, H.R., Sembdner, G. (eds.), pp. 33–49. Inst. Pl. Biochem. Acad. Sci. GDR Halle: 1974a
MacMillan, J.: Recent aspects of the chemistry and biosynthesis of the gibberellins. In: Recent advances in phytochemistry. Runeckles, V.C., Sondheimer, E., Walton, D.C. (eds.), Vol. VII, pp. 1–19. New York, London: Academic Press 1974b
MacMillan, J.: Some aspects of gibberellin metabolism in higher plants. In: Plant growth regulation. Pilet, P.-E. (ed.), pp. 129–138. Berlin-Heidelberg-New York: Springer 1977
MacMillan, J., Pryce, R.J.: Phaseic acid, a putative relative of abscisic acid, from seed of *Phaseolus multiflorus.* Chem. Commun. 124–126 (1968)
MacMillan, J., Pryce, R.J.: Plant hormones. – IX. Phaseic acid, a relative of abscisic acid from seed of *Phaseolus multiflorus*. Possible structures. Tetrahedron *25*, 5893 (1969a)
MacMillan, J., Pryce, R.J.: Plant hormones – X. The constitution of phaseic acid; a relative of abscisic acid from *Phaseolus multiflorus*. An interpretation of the mass spectrum of phaseic acid and a probable structure. Tetrahedron *25*, 5903–5914 (1969b)

MacMillan, J., Pryce, R.J.: The gibberellins. In: Phytochemistry. Miller, L.P. (ed.), Vol. III, pp. 283–326. New York: Van Nostrand-Reinhold 1973
MacMillan, J., Seaton, J.C., Suter, P.J.: Plant hormones – I. Isolation of gibberellin A_1 and gibberellin A_5 from *Phaseolus multiflorus*. Tetrahedron *11*, 60–66 (1960)
MacNicol, P.K.: Peroxidases of the Alaska pea (*Pisum sativum* L.) Enzymatic properties and distribution within the plant. Arch. Biochem. Biophys. *117*, 347–356 (1966)
MacNicol, P.K.: Isoperoxidase pattern and internode length genotype in *Pisum*. Phytochemistry *12*, 1273–1279 (1973)
Magnus, V., Iskrič, S., Kveder, S.: Indole-3-methanol – a metabolite of indole-3-acetic acid in pea seedlings. Planta *97*, 116–125 (1971)
Magnus, V., Iskrič, S., Kveder, S.: The formation of tryptophol glucoside in the tryptamine metabolism of pea seedlings. Planta *110*, 57–62 (1973)
Mahadevan, S.: Conversion of 3-indoleacetaldoxime to 3-indoleacetonitrile by plants. Arch. Biochem. Biophys. *100*, 557–558 (1963)
Mahadevan, S., Thimann, K.V.: Nitrilase. II. Substrate specificity and possible mode of action. Arch. Biochem. Biophys. *107*, 62–68 (1964)
Malloch, K.R., Osborne, D.J.: Auxin and ethylene control of growth in seedlings of *Zea mays* L. and *Avena sativa* L. J. Exp. Bot. *27*, 992–1003 (1976)
Manteuffel, R.: Beziehungen zwischen Pflanzen und epiphytischen Bakterien hinsichtlich ihres Auxinstoffwechsels. XV. Bakterielle Besiedlungsdichte und Bakterienausbreitung auf Pflanzenoberflächen unter Laboratoriumsbedingungen. Biochem. Physiol. Pflanz. *164*, 454–457 (1973)
Manteuffel, R., Siegl, E., Kunert, R., Libbert, E.: Beziehungen zwischen Pflanzen und epiphytischen Bakterien hinsichtlich ihres Auxinstoffwechsels. XIII. Bakterielle Auxinsynthese in Pflanzenexsudaten. Biochem. Physiol. Pflanz. *164*, 303–314 (1973a)
Manteuffel, R., Siegl, E., Kunert, R., Libbert, E.: Beziehungen zwischen Pflanzen und epiphytischen Bakterien hinsichtlich ihres Auxinstoffwechsels. XIV. Der Einfluß der epiphytischen Bakterien auf das Wachstum, den Geotropismus und die Adventivwurzelbildung der Pflanzen. Biochem. Physiol. Pflanz. *164*, 338–348 (1973b)
Manteuffel, R., Siegl, E., Libbert, E.: Beziehungen zwischen Pflanzen und epiphytischen Bakterien hinsichtlich ihres Auxinstoffwechsels. XI. Der Einfluß von IES-bildenden, IES-abbauenden und IES-neutralen Bakterien auf den Auxingehalt der Maispflanze im Vergleich zur Wirkung von Zucker-, Aminosäuren- oder Vitaminapplikationen. Biochem. Physiol. Pflanz. *163*, 586–595 (1973c)
Mapson, L.W.: Biogenesis of ethylene. Biol. Rev. *44*, 155–187 (1969)
Mapson, L.W.: Biosynthesis of ethylene and the ripening of fruit. Endeavour *29*, 29–33 (1970)
Mapson, L.W., Hulme, A.C.: The biosynthesis, physiological effects and mode of action of ethylene. Prog. Phytochem. *2*, 343 (1970)
Mapson, L.W., Mead, A.: Biosynthesis of ethylene. Dual nature of cofactor required for the enzymic production of ethylene from methional. Biochem. J. *108*, 875–881 (1968)
Mapson, L.W., Wardale, D.A.: Biosynthesis of ethylene. Formation of ethylene from methional by a cell-free enzyme system from cauliflower florets. Biochem. J. *102*, 574–585 (1967)
Mapson, L.W., Wardale, D.A.: Biosynthesis of ethylene. Enzymes involved in its formation from methional. Biochem. J. *107*, 433–442 (1968)
Mapson, L.W., Wardale, D.A.: Enzymes involved in the synthesis of ethylene from methionine, or its derivatives, in tomatoes. Phytochemistry *10*, 29–39 (1971)
Mapson, L.W., Wardale, D.A.: Role of indolyl-3-acetic acid in the formation of ethylene from 4-methylmercapto-2-oxo butyric acid by peroxidase. Phytochemistry *11*, 1371–1387 (1972)
Mapson, L.W., Self, R., Wardale, D.A.: Biosynthesis of ethylene. Methanesulphinic acid as cofactor in the enzymic formation of ethylene from methional. Biochem. J. *111*, 413–418 (1969a)
Mapson, L.W., March, J.F., Wardale, D.A.: Biosynthesis of ethylene. 4-Methylmercapto-2-oxo butyric acid: an intermediate in the formation from methionine. Biochem. J. *115*, 653–661 (1969b)

Mapson, L.W., March, J.F., Rhodes, M.J.C., Wooltorton, L.S.C.: A comparative study of the ability of methionine or linolenic acid to act as precursors of ethylene in plant tissues. Biochem. J. *117*, 473–479 (1970)

Martin, G.C., Dennis, F.G., Gaskin, P., MacMillan, J.: Identification of gibberellins A_{17}, A_{25}, A_{45}, abscisic acid, phaseic acid, and dihydrophaseic acid in seeds of *Pyrus communis*. Phytochemistry *16*, 605–607 (1977a)

Martin, G.C., Dennis, F.G., Jr., MacMillan, J., Gaskin, P.: Hormones in pear seeds. I. Levels of gibberellins, abscisic acid, phaseic acid, dihydrophaseic acid, and two metabolites of dihydrophaseic acid in immature seeds of *Pyrus communis* L. J. Am. Soc. Hortic. Sci. *102*, 16–19 (1977b)

Marumo, S., Hattori, H.: Isolation of D-4-chlorotryptophan derivatives as auxin-related metabolites from immature seeds of *Pisum sativum*. Planta *90*, 208–211 (1970)

Marumo, S., Abe, H., Hattori, H., Munakata, K.: Isolation of a novel auxin, methyl 4-chloroindoleacetate from immature seeds of *Pisum sativum*. Agric. Biol. Chem. *32*, 117–118 (1968a)

Marumo, S., Hattori, H., Abe, H., Munakata, K.: Isolation of 4-chloroindolyl-3-acetic acid from immature seeds of *Pisum sativum*. Nature (London) *219*, 959–960 (1968b)

Marumo, S., Hattori, H., Yamamoto, A.: Biological activity of 4-chloroindolyl-3-acetic acid. In: Plant growth substances 1973, pp. 419–428. Tokyo: Hirokawa 1974

Marynik, M.: Patterns of ethylene and carbon dioxide evolution during cotton explant abscission. Plant Physiol. *59*, 484–489 (1977)

Matsushima, H., Arima, K.: Physiological activities of zeanic acid, a new plant-growth promotor from corn steep liquor. Agric. Biol. Chem. *37*, 1873–1880 (1973)

Matsushima, H., Fukumi, H., Arima, K.: Isolation of zeanic acid, a natural plant growth regulator from corn steep liquor and its chemical structure. Agric. Biol. Chem. *37*, 1865–1871 (1973)

Mattoo, A.K., Lieberman, M.: Evidence that the ethylene-synthesizing enzyme system in plants is associated with a cell wall-cell membrane complex. Fed. Proc. *36*, 703 (1977a)

Mattoo, A.K., Lieberman, M.: Localization of the ethylene-synthesizing system in apple tissue. Plant Physiol. *60*, 794–799 (1977b)

Mattoo, A.K., Bauer, J.E., Chalutz, E., Lieberman, M.: Effect of temperature on the ethylene-synthesizing systems in apple, tomato and *Penicillium digitatum*. Plant Cell Physiol. *18*, 715–719 (1977)

Mayak, S., Dilley, D.R.: Regulation of senescence in carnation (*Dianthus caryophyllus*). Effect of abscisic acid and carbon dioxide on ethylene production. Plant Physiol. *58*, 663–665 (1976)

Mayak, S., Halevy, A.H., Katz, M.: Correlative changes in phytohormones in relation to senescence processes in rose petals. Physiol. Plant. *27*, 1–4 (1972)

McCalla, D.R., Moore, J.D., Osborne, D.J.: The metabolism of a kinin, benzyladenine. Biochim. Biophys. Acta *55*, 522–528 (1962)

McComb, A.J.: "Bound" gibberellin in mature runner bean seeds. Nature (London) *192*, 575–576 (1961)

McCune, D.C.: Multiple peroxidases of corn. Ann. N.Y. Acad. Sci. *94*, 723–730 (1961)

McDougall, J., Hillman, J.R.: Analysis of indole-3-acetic acid using GC-MS techniques. In: Isolation of plant growth substances. Hillman, J.R. (ed.), pp. 1–25. Cambridge, London, New York, Melbourne: Cambridge University Press 1978

McGlasson, W.B.: The ethylene factor. In: The biochemistry of fruits and their products. Hulme, A.C. (ed.), Vol. I, pp. 475–519. New York, London: Academic Press 1970

McGlasson, W.B., Palmer, J.K., Vendrell, M., Brady, C.J.: Metabolic studies with banana fruit slices. II. Effects of inhibitors on respiration, ethylene production and ripening. Aust. J. Biol. Sci. *24*, 1103–1112 (1971)

McGlasson, W.B., Dostal, H.C., Tigchelaar, E.C.: Comparison of propylene-induced responses of immature fruit of normal and *rin* mutant tomatoes. Plant Physiol. *55*, 218–222 (1975a)

McGlasson, W.B., Poovaiah, B.W., Dostal, H.C.: Ethylene production and respiration

in ageing leaf segments and in disks of fruit tissue of normal and mutant tomatoes. Plant Physiol. *56*, 547–549 (1975b)
McGowan, R.E., Muir, R.M.: Purification and properties of amine oxidase from epicotyls of *Pisum sativum*. Plant Physiol. *47*, 644–648 (1971)
McInnes, A.G., Smith, D.G., Arsenault, G.P., Vining, L.C.: Biosynthesis of gibberellins in *Gibberella fujikuroi*. Gibberellin A_{16}. Can. J. Biochem. *51*, 1470–1474 (1973)
McInnes, A.G., Smith, D.G., Durley, R.C., Pharis, R.P., Arsenault, G.P., MacMillan, J., Gaskin, P., Vining, L.C.: Biosynthesis of gibberellins in *Gibberella fujikuroi*. Gibberellin A_{47}. Can. J. Biochem. *55*, 728–735 (1977)
McLennan, B.D., Pater, A.: Evidence for the oxidation of N^6-(Δ^2-isopentenyl)adenine and N^6-(3-hydroxy-3-methylbutyl) adenine by xanthine oxidase. Can. J. Biochem. *51*, 1123–1126 (1973)
McLennan, B.D., Logan, D.M., Hall, R.H.: Enzymic degradation of N^6(Δ^2-isopentenyl) adenosine. Proc. Am. Assoc. Cancer Res. *9*, 47 (1968)
McMichael, B.L., Jordan, W.R., Powell, R.D.: An effect of water stress on ethylene production by intact cotton petioles. Plant Physiol. *49*, 658–660 (1972)
McWha, J.A., Hillmann, J.R.: Uptake and metabolism of 2-[^{14}C]-abscisic acid in lettuce fruits var. Great Lakes. Planta *110*, 345–351 (1973)
Megha, B.M., Laloraya, M.M.: Effect of abscisic acid on growth, IAA oxidase, peroxidase and ascorbate oxidizing systems in *Trigonella foenum graecum* L. seedlings. Biochem. Physiol. Pflanz. *171*, 269–277 (1977)
Meheriuk, M., Spencer, M.S.: Effects of nitrogen, and respiratory inhibitors on ethylene production by a sub-cellular fraction from tomatoes. Nature (London) *204*, 43–45 (1964)
Meheriuk, M., Spencer, M.S.: Studies on ethylene production by a cellular fraction from ripening tomatoes. I. Effects of several substrates, cofactors and cations. Phytochemistry *6*, 535–543 (1967a)
Meheriuk, M., Spencer, M.S.: Studies on ethylene production by a subcellular fraction from ripening tomatoes. II. Effects of several inhibitors. Phytochemistry *6*, 545–549 (1967b)
Meheriuk, M., Spencer, M.S.: Studies on ethylene production by a subcellular fraction from ripening tomatoes. III. Effects of addition of β-alanine and cofactors for decarboxylation. Phytochemistry *6*, 551–558 (1967c)
Meigh, D.F., Jones, J.D., Hulme, A.C.: The respiration climacteric in the apple. Production of ethylene and fatty acids in fruit attached to and detached from the tree. Phytochemistry *6*, 1507–1515 (1967)
Mellon, J.E., West, C.A.: Diterpene biosynthesis in maize seedlings in response to fungal infection. Plant Physiol. *64*, 406–410 (1979)
Mennes, A.M.: The indole-3-acetic acid oxidase of *Lupinus luteus* L. I. A qualitative comparison of the activity of this enzyme in root nodules and roots. Acta Bot. Neerl. *22*, 694–705 (1973a)
Mennes, A.M.: The indole-3-acetic oxidase of *Lupinus luteus* L. II. A quantitative comparison of the activity of this enzyme in root nodules and roots. Acta Bot. Neerl. *22*, 706–729 (1973b)
Menschick, R., Hild, V., Hager, A.: Decarboxylierung von Indolylessigsäure in Zusammenhang mit dem Phototropismus in *Avena*-Koleoptilen. Planta *133*, 223–228 (1977)
Merkys, A., Anisimouviene, N., Putrimas, A.: Comparative study of the IAA content in systematically different plants. Biochem. Physiol. Pflanz. *165*, 67–81 (1974)
Mertz, D.: Light-stimulated incorporation of leucine into the gibberellin of *Gibberella fujikuroi*. Plant Cell Physiol. *11*, 273–279 (1970)
Mertz, D.: Effect of red light on uptake and efflux of gibberellin. Plant Physiol. *56* Suppl. 76 (1975)
Mertz, D., Henson, W.: Light-stimulated biosynthesis of the gibberellins by *Fusarium moniliforme*. Nature (London) *214*, 844–846 (1967a)
Mertz, D., Henson, W.: The effect of the plant growth retardants AMO 1618 and CCC on gibberellin production in *Fusarium moniliforme:* Light-stimulated biosynthesis of gibberellins. Physiol. Plant. *20*, 187–199 (1967b)

Meudt, W.J.: Studies on the oxidation of indole-3-acetic acid by peroxidase enzymes. Ann. N.Y. Acad. Sci. *144*, 118–128 (1967)

Meudt, W.J.: Electrophoretic isolation and growth activity of indole-3-acetic acid oxidation products. In: Plant growth substances 1970. Carr, D.J. (ed.), pp. 110–116. Berlin-Heidelberg-New York: Springer 1972

Michalczuk, L., Bandurski, R.S.: UDP-Glucose: indoleacetic acid glucosyl transferase and indoleacetyl-glucose: myo-inositol indoleacetic transferase. Biochem. Biophys. Res. Commun. *93*, 588–592 (1980)

Miernyk, J.A.: Abscisic acid inhibition of kinetin nucleoside formation in germinating lettuce seeds. Physiol. Plant. *45*, 63–66 (1979)

Miernyk, J.A., Blaydes, D.F.: Short-term metabolism of radioactive kinetin during lettuce seed germination. Physiol. Plant. *39*, 4–8 (1977)

Miersch, O., Liebisch, H.-W.: Utilization of enzyme preparations in the course of synthesis of radioactive labelled gibberellin A_3 and some derivatives. In: Biochemistry and chemistry of plant growth regulators. Schreiber, K., Schütte, H.-R., Sembdner, G. (eds.), pp. 125–131. Inst. Pl. Biochem. Acad. Sci. GDR, Halle 1974

Miersch, O., Liebisch, H.-W.: Synthese von Gibberellin-glucopyranosylestern. Z. Chem. (1980) in press

Milborrow, B.V.: Identification and measurement of (+)-abscisic acid [(+)-abscisin II, (+)-dormin] in plants. In: Biochemistry and physiology of plant growth substances. Wightman, F., Setterfield, G. (eds.), pp. 1531–1545. Ottawa: Runge Press 1968

Milborrow, B.V.: Identification of "metabolite C" from abscisic acid and a new structure for phaseic acid. Chem. Commun. 966–967 (1969a)

Milborrow, B.V.: The occurrence and function of abscisic acid in plants. Sci. Prog. (London) *57*, 533–558 (1969b)

Milborrow, B.V.: The metabolism of abscisic acid. J. Exp. Bot. *21*, 17–29 (1970)

Milborrow, B.V.: Abscisic acid. In: Aspects of terpenoid chemistry and biochemistry. Goodwin, T.W. (ed.), pp. 137–151, New York, London: Academic Press 1971

Milborrow, B.V.: Stereochemical aspects of the formation of double bonds in abscisic acid. Biochem. J. *128*, 1135–1146 (1972a)

Milborrow, B.V.: The biosynthesis and degradation of abscisic acid. In: Plant growth substances 1970. Carr, D.J. (ed.), pp. 281–290. Berlin-Heidelberg-New York: Springer 1972b

Milborrow, B.V.: The chemistry and physiology of abscisic acid. Annu. Rev. Plant Physiol. *25*, 259–307 (1974a)

Milborrow, B.V.: Chemistry and biochemistry of abscisic acid. In: The chemistry and biochemistry of plant hormones. Rec. Adv. Phytochem. Runeckles, V.C., Sondheimer, E., Walton, D.C. (eds.), Vol. VII, pp. 57–91. New York, London: Academic Press 1974b

Milborrow, B.V.: Biosynthesis of abscisic acid by a cell-free system. Phytochemistry *13*, 131–136 (1974c)

Milborrow, B.V.: The stereochemistry of cyclization in abscisic acid. Phytochemistry *14*, 123–128 (1975a)

Milborrow, B.V.: The origin of the methyl groups of abscisic acid. Phytochemistry *14*, 2403–2405 (1975b)

Milborrow, B.V.: The absolute configuration of phaseic and dihydrophaseic acids. Phytochemistry *14*, 1045–1053 (1975c)

Milborrow, B.V.: The stability of conjugated abscisic acid during wilting. J. Exp. Bot. *29*, 1059–1066 (1978a)

Milborrow, B.V.: Abscisic acid. In: Phytohormones and related compounds – A comprehensive treatise. Letham, D.S., Goodwin, P.G., Higgins, T.J.V. (eds.), Vol. 1, pp. 295–347. Elsevier: Biomedical Press 1978b

Milborrow, B.V., Garmston, M.: Formation of (−)-1′,2′-epi-2-cis-xanthoxin acid from a precursor of abscisic acid. Phytochemistry *12*, 1597–1608 (1973)

Milborrow, B.V., Mallaby, R.: Occurrence of methyl (+)-abscisate as an artefact of extraction. J. Exp. Bot. *26*, 741–748 (1975)

Milborrow, B.V., Noddle, R.C.: Conversion of 5-(1,2-epoxy-2,6,6-trimethylcyclohexyl)-3-

methylpenta-cis-2-trans-4-dienoic acid into abscisic acid in plants. Biochem. J. *119*, 727–734 (1970)
Milborrow, B.V., Robinson, D.R.: Factors affecting the biosynthesis of abscisic acid. J. Exp. Bot. *24*, 537–548 (1973)
Milborrow, B.V., Vaugham, G.: The long term metabolism of (±)-(2-^{14}C) abscisic acid by apple seeds. J. Exp. Bot. *30*, 983–995 (1979)
Miller, C.O.: Evidence for the natural occurrence of zeatin and derivatives: Compounds from maize which promote cell division. Proc. Natl. Acad. Sci. USA *54*, 1052–1058 (1965)
Miller, C.O.: Zeatin and zeatinriboside from a mycorrhizal fungus. Science *157*, 1055–1057 (1967)
Miller, C.O.: Cell-division factors from *Vinca rosea* L. crown gall tumor tissue. Proc. Natl. Acad. Sci. USA *72*, 1883–1886 (1975)
Miller, R.W., Sirois, J.C., Morita, H.: Reaction of coumarins with horseradish peroxidase. Plant Physiol. *55*, 35–41 (1975)
Mills, V.M., Todd, G.W.: Effects of water stress on the indoleacetic acid oxidase activity in wheat leaves. Plant Physiol. *51*, 1145–1146 (1973)
Minchin, A., Harmey, M.A.: The metabolism of indoleacetic acid by barley grains. Planta *122*, 245–254 (1975)
Mirai, N., Fukui, H., Koshimizu, K.: A novel abscisic acid metabolite from seeds of *Robinia pseudoacacia*. Phytochemistry *17*, 1625–1627 (1978)
Mitchell, E.K., Davies, P.J.: Indoleacetic acid synthesis in sterile roots of *Phaseolus coccineus*. Plant Cell Physiol. *13*, 1135–1138 (1972)
Miura, G.A., Hall, R.H.: Trans-ribosylzeatin. Its biosynthesis in *Zea mays* endosperm and the mycorrhizal fungus *Rhizopogon roseolus*. Plant Physiol. *51*, 563–569 (1973)
Miura, G.A., Miller, C.O.: 6-(γγ-Dimethylallylamino)purine as a precursor of zeatin. Plant Physiol. *44*, 372–376 (1969)
Mizrahi, Y., Dostal, H.C., McGlasson, W.B., Cherry, J.H.: Effects of abscisic acid and benzyladenine on fruits of normal and *rin* mutant tomatoes. Plant Physiol. *56*, 544–546 (1975)
Mollan, R.C., Donnelly, D.M.X., Harmey, M.A.: Synthesis of indole-3-acetylaspartic acid. Phytochemistry *11*, 1485–1488 (1972)
Mondal, M.H.: Effects of gibberellic acid, calcium, kinetin, and ethylene on growth and cell wall composition of pea epicotyls. Plant Physiol. *56*, 622–625 (1975)
Montalbini, P., Elstner, E.F.: Ethylene evolution by rust-infected, detached bean (*Phaseolus vulgaris* L.) leaves susceptible and hypersensitive to *Uromyces phaseoli* (Pers.) Wint. Planta *135*, 301–306 (1977)
Moore, T.C.: Comparative net biosynthesis of indoleacetic acid from tryptophan in cell-free extracts of different parts of *Pisum sativum* plants. Phytochemistry *8*, 1109–1120 (1969)
Moore, T.C., Anderson, J.D.: Inhibition of the growth of peas by tris-(2-diethylaminoethyl)-phosphate trihydrochloride. Plant Physiol. *41*, 238–243 (1966)
Moore, T.C., Coolbaugh, R.C.: Conversion of geranylgeranyl pyrophosphate to ent-kaurene in extracts of sonicated chloroplasts. Phytochemistry *15*, 1241–1247 (1976)
Moore, T.C., Shaner, C.A.: Biosynthesis of indoleacetic acid from tryptophan-^{14}C in cell-free extracts of pea shoot tips. Plant Physiol. *42*, 1787–1796 (1967)
Moore, T.C., Shaner, C.A.: Synthesis of indoleacetic acid from tryptophan via indolepyruvic acid in cell-free extracts of pea seedlings. Arch. Biochem. Biophys. *127*, 613–621 (1968)
Moore, T.C., Barlow, S.A., Coolbaugh, R.C.: Participation of non-catalytic carrier protein in the metabolism of kaurene in cell-free extracts of pea seeds. Phytochemistry *11*, 3225–3233 (1972)
Morgan, P.W., Hall, W.C.: Accelerated release of ethylene by cotton following application of indolyl-3-acetic acid. Nature (London) *201*, 99 (1964)
Morita, Y., Kominato, Y., Shimizu, K.: Studies on phytoperoxidase. Part 19. Some further aspects of oxidation of indole-3-acetic acid by peroxidase. Mem. Res. Inst. Food Sci. Kyoto Univ. *28*, 1–17 (1967)
Morris, D.A., Briant, R.E., Thomson, P.C.: The transport and metabolism of ^{14}C-labelled indoleacetic acid in intact pea seedlings. Planta *89*, 178–197 (1969)

Morris, R.O.: Mass spectroscopic identification of cytokinins. Glucosyl zeatin and glucosyl ribosylzeatin from *Vinca rosea* crown gall. Plant Physiol. *59*, 1029–1033 (1977)
Mousdale, D.M.A., Butcher, D.N., Powell, R.G.: Spectrophotofluorimetric methods of determining indole-3-acetic acid. In: Isolation of plant growth substances. Hillman, J.R. (ed.), pp. 27–39. Cambridge, London, New York, Melbourne: Cambridge University Press 1978
Moyed, H.S., Tuli, V.: The oxindole pathway of 3-indoleacetic acid metabolism and the action of auxins. In: Biochemistry and physiology of plant growth substances. Wightman, F., Setterfield, G. (eds.), pp. 289–300. Ottawa: Runge Press 1968
Moyed, H.S., Williamson, V.: 3-Methyleneoxindole reductase of peas. Plant Physiol. *42*, 510–514 (1967a)
Moyed, H.S., Williamson, V.: Multiple 3-methyleneoxindole reductases of peas. Differential inhibition by synthetic auxins. J. Biol. Chem. *242*, 1075–1077 (1967b)
Muir, R.M.: The control of growth by the synthesis of IAA and its conjugation. In: Plant growth substances 1970. Carr, D.J. (ed.), pp. 96–101. Berlin-Heidelberg-New York: Springer 1972
Muir, R.M., Lantican, B.P.: Purification and properties of the enzyme system forming indoleacetic acid. In: Biochemistry and physiology of plant growth substances. Wightman, F., Setterfield, G. (eds.), pp. 259–272. Ottawa: Runge Press 1968
Müller, H., Schuphan, W.: Biochemische Untersuchungen über den Wirkungsmechanismus des Wachstumsregulators CCC in Tomatenpflanzen (L. esculentum): 1. Wirkung von CCC auf den Mevalonsäurestoffwechsel unter besonderer Berücksichtigung der Gibberellin- und Sterinbiosynthese. Qualitas Plant. *26*, 263–282 (1976)
Müller, P., Knöfel, D., Sembdner, G.: Studies on the enzymatical synthesis of gibberellin-O-glucosides. In: Biochemistry and chemistry of plant growth regulators. Schreiber, K., Schütte, H.-R., Sembdner, G. (eds.), pp. 115–119. Inst. Pl. Biochem. Acad. Sci. GDR, Halle 1974
Müller, P., Knöfel, H.D., Liebisch, H.-W., Miersch, O., Sembdner, G.: Untersuchungen zur Spaltung von Gibberellinglucosiden. Biochem. Physiol. Pflanz. *173*, 396–409 (1978)
Murai, N., Armstrong, D.J., Skoog, F.: Incorporation of mevalonic acid into ribosylzeatin in tobacco callus RNA preparations. Plant Physiol. *55*, 853–858 (1975)
Murakami, Y.: Formation of gibberellin A_3 glucoside in plant tissues. Bot. Mag. *74*, 424–425 (1961)
Murakami, Y.: Gibberellin-like substances in roots of *Oryza sativa*, *Pharbitis nil*, and *Ipomoea batatas*, and the site of their synthesis in the plant. Bot. Mag. *81*, 334–343 (1968)
Murakoshi, I., Ikegami, F., Ookawa, N., Haginiwa, J., Letham, D.S.: Enzymatic synthesis of lupinic acid, a novel metabolite of zeatin in higher plants. Chem. Pharm. Bull. *25*, 520–522 (1977)
Murofushi, N., Nagura, S., Takahashi, N.: Metabolism of steviol by *Gibberella fujikuroi* in the presence of plant growth retardant. Agric. Biol. Chem. *43*, 1159–1161 (1979)
Murphy, G.J.P., Briggs, D.E.: Gibberellin estimation and biosynthesis in germinating *Hordeum distichon*. Phytochemistry *12*, 1299–1309 (1973a)
Murphy, G.J.P., Briggs, D.E.: ent-Kaurene, occurrence and metabolism in *Hordeum distichon*. Phytochemistry *12*, 2597–2605 (1973b)
Murphy, G.J.P., Briggs, D.E.: Metabolism of ent-kaurenol-17-^{14}C, ent-kaurenal-17-^{14}C and ent-kaurenoic acid-17^{14}C by germinating *Hordeum distichon* grains. Phytochemistry *14*, 429–433 (1975)
Murphy, P.J., West, C.A.: The role of mixed function oxidases in kaurene metabolism in *Echinocystis macrocarpa* Greene endosperm. Arch. Biochem. Biophys. *133*, 395–407 (1969)
Murr, D.P.: Methionine metabolism in apple tissue. Experientia *33*, 1559–1561 (1977)
Murr, D.P., Yang, S.F.: Inhibition of in vivo conversion of methionine to ethylene by L-canaline and 2,4-dinitrophenol. Plant Physiol. *55*, 73–82 (1975a)
Murr, D.P., Yang, S.F.: Conversion of 5′-methylthioadenosine to methionine by apple tissue. Phytochemistry *14*, 1291–1292 (1975b)
Musgrave, A., Kende, H.: Radioactive gibberellin A_5 and its metabolism in dwarf peas. Plant Physiol. *45*, 56–61 (1970)

Musgrave, A., Kays, S.E., Kende, H.: Uptake and metabolism of radioactive gibberellins by aleuron layers. Planta *102*, 1–10 (1972)
Nadeau, R., Rappaport, L.: Metabolism of gibberellin A_1 in germinating bean seeds. Phytochemistry *11*, 1611–1616 (1972)
Nadeau, R., Rappaport, L.: An amphoteric conjugate of [^{3}H]-gibberellin A_1 from barley aleurone layers. Plant Physiol. *54*, 809–812 (1974)
Nadeau, R., Rappaport, L., Stolp, C.L.: Uptake and metabolism of ^{3}H-gibberellin A_1 by barley aleurone layers: Response to abscisic acid. Planta *107*, 315–324 (1972)
Nakajima, R., Yamazaki, I.: The mechanism of indole-3-acetic acid oxidation by horseradish peroxidases. J. Biol. Chem. *254*, 872–878 (1979)
Nanda, K.K., Gurumurti, K., Chibbar, R.N.: Evidence for the allosteric nature of IAA oxidase system in *Phaseolus mungo* hypocotyls. Experientia *31*, 635–637 (1975)
Nash, L.J., Crozier, A.: Translocation and metabolism of [^{3}H]-gibberellins by light grown *Phaseolus coccineus* seedlings. Planta *127*, 221–231 (1975)
Nash, L.J., Jones, R.L., Stoddart, J.L.: Gibberellin metabolism in excised lettuce hypocotyls: Response to GA_9 and the conversion of [^{3}H]GA_9. Planta *140*, 143–150 (1978)
Naumann, W.D., Sachs, M.: Variations in abscisic acid content of strawberry plants during winter time in Israel. Gartenbauwissenschaft *40*, 37–39 (1977)
Neskovic, M., Konjevic, R.: The non-reversible effects of red and far red light on the content of gibberellin-like substances in pea internodes. J. Exp. Bot. *25*, 733–739 (1974)
Newhall, W.F.: Correlation of pseudocholinesterase inhibition and plant growth retardation by quaternary ammonium derivatives of (+)-limonene. Nature (London) *223*, 965–966 (1969)
Nicholls, P.B.: Isolation of indole-3-acetyl-2-O-myoinositol from *Zea mays*. Planta *72*, 258–264 (1967)
Nicholls, P.B., Ong, B.L., Tate, M.E.: Assignment of the ester linkage of 2-O-indoleacetyl-*myo*-inositol isolated from *Zea mays*. Phytochemistry *10*, 2207–2209 (1971)
Nichols, R.: Ethylene production during senescence of flowers. J. Hortic. Sci. *41*, 279–290 (1966)
Nichols, R.: The response of carnation (*Dianthus caryophyllus*) to ethylene. J. Hortic. Sci. *43*, 335–349 (1968)
Nichols, R.: Induction of flower senescence and gynaecium development in the carnation (*Dianthus caryophyllus*) by ethylene and 2-chloroethylphosphonic acid. J. Hortic. Sci. *46*, 323–332 (1971)
Nichols, R.: Sites of ethylene production in the pollinated and unpollinated senescing carnation (*Dianthus caryophyllus*) inflorescence. Planta *135*, 155–159 (1977)
Nickell, L.G.: Plant growth regulators. Controlling biological behavior with chemicals. Chem. Eng. News *9*, 18–29 (1978)
Ninnemann, H., Zeevaart, J.A.D., Kende, H., Lang, A.: The plant growth retardant CCC as inhibitor of gibberellin biosynthesis in *Fusarium moniliforme*. Planta *61*, 229–235 (1964)
Noddle, R.C., Robinson, D.R.: Biosynthesis of abscisic acid: Incorporation of radioactivity from 2-^{14}C-mevalonic acid by intact fruit. Biochem. J. *112*, 547–548 (1969)
Nowak, J., Brown, G.N.: Free and bound gibberellin activities and ent-kaurene synthesis during induction of cold hardiness in black locust seedlings. Physiol. Plant. *45*, 11–16 (1979)
Novak, J., Galston, A.W.: Studies on auxin protector substances, IAA-oxidase and peroxidase in cotyledons of *Pharbitis nil*. Plant Cell Physiol. *12*, 931–940 (1971)
Obata, Y., Abe, H., Tanaka, Y., Yanagisawa, K., Uchiyama, M.: Isolation of a spore germination inhibitor from a cellular slime mold *Dictyostelium discoideum*. Agric. Biol. Chem. *37*, 1989–1990 (1973)
Odawara, S., Watanabe, A., Imaseki, H.: Involvement of cellular membrane in regulation of ethylene production. Plant Cell Physiol. *18*, 569–575 (1977)
Ogawa, Y., Takahashi, N.: Comparative biological effectiveness of gibberellin A_3 glucoside and gibberellin A_3. Bull. Fac. Agric. Mich. Univ. 261–267 (1974)
Ohkuma, K., Addicott, F.T., Smith, O.C., Thiessen, W.E.: The structure of abscissin II. Tetrahedron Lett. 2529–2535 (1965)

Olney, H.O.: Growth substances from *Veratrum tenuipetalum*. Plant Physiol. *43*, 293–302 (1968)

Omran, R.G.: The direct involvement of hydrogen peroxide in indoleacetic acid inactivation. Biochem. Biophys. Res. Commun. *78*, 970–976 (1977)

Oritani, T., Yamashita, K.: Synthesis of ($\pm$)-xanthoxins. Agric. Biol. Chem. *37*, 1215–1217 (1973)

Osborne, D.J., Jackson, M.B., Milborrow, B.V.: Physiological properties of abscission accelerator from senescent leaves. Nature (London) New Biol. *240*, 98–101 (1972)

Oster, M.O., West, C.A.: Biosynthesis of trans-geranylgeranyl pyrophosphate in endosperm of *Echinocystis macrocarpa* Greene. Arch. Biochem. Biophys. *127*, 112–123 (1968)

Owens, L.D., Lieberman, M., Kunishi, A.: Inhibition of ethylene production by rhizobitoxine. Plant Physiol. *48*, 1–4 (1971)

Pačes, V.: Metabolism of cytokinins in barley leaves. Biochem. Biophys. Res. Commun. *72*, 830–839 (1976)

Pačes, V., Werstiuk, E., Hall, R.H.: Conversion of N^6-(Δ^2-isopentenyl) adenosine to adenosine by enzyme activity in tobacco tissue. Plant Physiol. *48*, 775–778 (1971)

Pačes, V., Rosenberg, I., Kaminek, M., Holy, A.: Metabolism of cytokinins in rape seedlings. Coll. Czech. Chem. Commun. *42*, 2452–2458 (1977)

Parker, C.W., Letham, D.S.: Regulators of cell division in plant tissues. XVI. Metabolism of zeatin by radish cotyledons and hypocotyls. Planta *114*, 199–218 (1973)

Parker, C.W., Letham, D.S.: Regulators of cell division in plant tissues. XVIII. Metabolism of zeatin in *Zea mays* seedlings. Planta *115*, 337–344 (1974)

Parker, C.W., Letham, D.S., Cowley, D.E., McLeod, J.K.: Raphanatin, an unusual purine derivative and a metabolite of zeatin. Biochem. Biophys. Res. Commun. *49*, 460–466 (1972)

Parker, C.W., Wilson, M.M., Letham, D.S., Cowley, D.E., McLeod, J.K.: The glucosylation of cytokinins. Biochem. Biophys. Res. Commun. *55*, 1370–1376 (1973)

Parker, C.W., Letham, D.S., Wilson, M.M., Jenkins, I.D., McLeod, J.K., Summons, R.E.: The identity of two new cytokinin metabolites. Ann. Bot. (London) *39*, 375–376 (1975)

Parker, C.W., Letham, D.S., Gollnow, B.I., Summons, R.E., Duke, C.C., MacLeod, J.K.: Regulators of cell division in plant tissues. XXV. Planta *142*, 239–251 (1978)

Parups, E.V.: Acetylcholine and synthesis of ethylene in etiolated bean tissues. Physiol. Plant. *36*, 154–156 (1976)

Patil, S.S., Tang, C.S.: Inhibition of ethylene evolution in papaya pulp tissue by benzyl isothiocyanate. Plant Physiol. *53*, 585–588 (1974)

Patterson, R.J., Rappaport, L.: The conversion of gibberellin A_1 to gibberellin A_8 by a cell-free enzyme system. Planta *119*, 183–191 (1974)

Patterson, R.J., Rappaport, L.: Characterization of an enzyme from *Phaseolus vulgaris* seeds which hydroxylates GA_1 to GA_8. Phytochemistry *14*, 363–368 (1975)

Pegg, G.F., Cronshaw, D.K.: Ethylene production in tomato plants infected with *Verticillium albo-atrum*. Physiol. Plant Pathol. *8*, 279–295 (1976)

Penel, C., Greppin, H.: Evolution de l'activité auxines-oxidasique et peroxydasique lors de l'induction photoperiodique et de la sexualité de l'épinard. Plant Cell Physiol. *13*, 151–156 (1972)

Percival, F.W., Bandurski, R.S.: Esters of indole-3-acetic acid from *Avena* seeds. Plant Physiol. *58*, 60–67 (1976)

Percival, F.W., Purves, W.K.: Multiple amine oxidases in cucumber seedlings. Plant Physiol. *54*, 601–607 (1974)

Percival, F.W., Purves, W.K., Vickery, L.E.: Indole-3-ethanol oxidase: kinetics, inhibition, and regulation by auxins. Plant Physiol. *51*, 739–743 (1973)

Peterkovsky, A.: The incorporation of mevalonic acid into the N^6-(Δ^2-isopentenyl)adenosine of transfer ribonucleic acid in *Lactobacillus acidophilus*. Biochemistry *7*, 472–482 (1968)

Peterson, C.A., Fletcher, R.A.: Formation of fruits on rootless plants. Can. J. Bot. *51*, 1899–2005 (1973)

Peterson, J.V., Miller, C.O.: Glucosyl zeatin and glucosyl ribosylzeatin from *Vinca rosea* L. crown gall tumor tissue. Plant Physiol. *59*, 1026–1028 (1977)

Pethe-Sadorge, P., Signor, Y., Guern, J.: Sur la synthèse des nucléosides-5′-monophosphates de cytokinines par les cellules d'*Acer pseudoplatanus*. C. R. Acad. Sci. Ser. D *275*, 2493–2496 (1977)

Petrids, C., Verbeek, R., Massart, L.: Detection in barley of precursors in the gibberellin (GA_3) biosynthesis. Naturwissenschaften *53*, 331–332 (1966)

Phelps, R.H., Sequeira, L.: Synthesis of indoleacetic acid via tryptamine by a cell-free system from tobacco terminal buds. Plant Physiol. *42*, 1161–1163 (1967)

Phelps, R.H., Sequeira, L.: Auxin biosynthesis in a host-parasite complex. In: Biochemistry and physiology of plant growth substances. Wightman, F., Setterfield, G. (eds.), pp. 197–212. Ottawa: Runge Press 1968

Phillips, I.D.J., Hofmann, A.: Abscisic acid, abscisic acid-esters and phaseic acid in vegetative terminal buds of *Acer pseudoplatanus* during emergence from winter dormancy. Planta *146*, 591–595 (1979)

Phillips, I.D.J., Jones, R.L.: Gibberellin-like activity in bleeding-sap of root systems of *Helianthus annuus* detected by a new dwarf pea epicotyl assay and other methods. Planta *63*, 269–278 (1964)

Phipps, J.: La plante adulte de tabac: mise en évidence et répartition du système auxin-oxidasique. C.R. Acad. Sci. *261*, 3864–3867 (1965)

Phipps, J.: Le catabolisme auxinique chez le tabac: ses modalités dans la plante saine et parasitée par le virus de la mosaïque. Toulouse: Imprimerie du Commerce 1966

Pilet, P.-E.: Tryptophan treatment and endogenous auxin in the root. In: Régulateurs naturels de la croissance végétale. Nitsch. J.P. (ed.), Vol. 123, pp. 543–558. Paris: C.N.R.S. 1964a

Pilet, P.-E.: Effect of chlorogenic acid on the auxin catabolism and the auxin content of root tissues. Phytochemistry *3*, 617–621 (1964b)

Pilet, P.-E.: Auxin oxidase feedback effect. C. R. Acad. Sci. Ser. D *263*, 864–867 (1966)

Pilet, P.-E., Gaspar, T.: Catabolisme auxinique. Monogr. Physiol. Veg. 148 pp Paris: Masson 1968

Piskornik, Z., Bandurski, R.S.: Purification and partial characterization of a glucan containing indole-3-acetic acid. Plant Physiol. *51*, 176–182 (1972)

Pitel, D.W., Vining, L.C., Arsenault, G.P.: Biosynthesis of gibberellins in *Gibberella fujikuroi*. The sequence after gibberellin A_4. Can. J. Biochem. *49*, 194–200 (1971)

Playtis, A.J., Leonard, N.J.: The synthesis of ribosyl-cis-zeatin and thin layer chromatographic separation of the cis and trans isomers of ribosylzeatin. Biochem. Biophys. Res. Commun. *45*, 1–5 (1971)

Popjak, G., Cornforth, J.W.: Substrate stereochemistry in squalene biosynthesis. Biochem. J. *101*, 553–568 (1966)

Porutzkii, G.V., Matkovskii, K.I.: Izn. Inst. Biol. Metod. Popov. Bulg. Akad. Nauk *13*, 147 (1963)

Porutzkii, G.V., Luchko, A.S., Matkovskii, K.I.: Sov. Plant Physiol. *9*, 382 (1962)

Powell, L.E., Seeley, S.D.: Metabolism of abscisic acid to a water soluble complex in apple. J. Am. Soc. Hortic. Sci. *99*, 439–441 (1974)

Prakash, R., Maheshwari, S.C.: Studies on cytokinins in water melon seeds. Physiol. Plant. *23*, 792–799 (1970)

Pratt, H.K.,: The role of ethylene in fruit ripening. In: Facteurs et regulation de la maturation des fruits. Ulrich, R. (ed.), pp. 153–160. Colloq. Int. C.N.R.S. No. 238. Paris: C.N.R.S. 1974

Pratt, H.K., Goeschl, J.D.: Physiological roles of ethylene in plants. Annu. Rev. Plant Physiol. *20*, 541–584 (1969)

Primrose, S.B.: Formation of ethylene by *Escherichia coli*. J. Gen. Microbiol. *95*, 159–165 (1976)

Proebsting, W.M., Davies, P.J., Marx, G.A.: Photoperiod-induced changes in gibberellin metabolism in relation to apical growth and senescence in genetic lines of peas (*Pisum sativum* L.). Planta *141*, 231–238 (1978)

Puppo, A., Rigaud, J.: Indole-3-acetic acid (IAA) oxidation by leghemoglobin from soybean nodules. Physiol. Plant. *35*, 181–185 (1975)

Puppo, A., Rigaud, J., Barthe, P.: Sur la presence de cytokinines dans les nodules de *Phaseolus vulgaris* L. C. R. Acad. Sci. Ser. D *279*, 2029–2032 (1974)

Purse, J.G., Horgan, R., Horgan, J.M., Wareing, P.F.: Cytokinins of sycamore spring sap. Planta *132*, 1–8 (1976)

Purves, W.K., Brown, H.M.: Indoleacetaldehyde in cucumber seedlings. Plant Physiol. *61*, 104–106 (1978)

Purves, W.K., Rayle, D.L., Johnson, K.D.: Actions and interactions of growth factors on cucumber hypocotyl segments. Ann. N.Y. Acad. Sci. *144*, 169–179 (1967)

Raa, J.: Degradation of indol-3-yl-acetic acid in homogenates and segments of cabbage roots. Physiol. Plant. *24*, 494–505 (1971)

Railton, I.D.: Studies on gibberellins in shoots of light-grown peas. II. The metabolism of tritiated gibberellin A_9 and gibberellin A_{20} by light- and dark-grown shoots of dwarf *Pisum sativum* cv. 'Meteor'. Plant Sci. Lett. *3*, 207–212 (1974a)

Railton, I.D.: Effect of N^6-benzyladenine on the rate of turnover of ^3H-GA_{20} by shoots of dwarf *Pisum sativum*. Planta *120*, 197–200 (1974b)

Railton, I.D.: The preparation of 2,3-^3H-GA_{29} and its metabolism by etiolated seedlings and germinating seeds of dwarf *Pisum sativum* (Meteor). J. S. Afr. Bot. *42*, 147–156 (1976)

Railton, I.D.: Control of gibberellin A_1 levels by 2β-hydroxylation during growth of dwarf *Oryza sativa* L. var. Tan-ginbozu. S. Afr. J. Sci. *74*, 191 (1978)

Railton, I.D., Reid, D.M.: Studies on gibberellins in shoots of light grown peas. III. Interconversion of ^3H-GA_9 and ^3H-GA_{20} to other gibberellins by an in vitro system derived from chloroplasts of *Pisum sativum*. Plant Sci. Lett. *3*, 303–308 (1974)

Railton, I.D., Wareing, P.F.: Effects of daylength on endogenous gibberellins in *Solanum andigena*. II. Metabolism of gibberellin A_1 by potato shoots. Physiol. Plant. *28*, 127–131 (1973)

Railton, I.D., Durley, R.C., Pharis, R.P.: Interconversion of gibberellin A_1 to gibberellin A_8 by shoots of dwarf *Oryza sativa*. Phytochemistry *12*, 2351–2352 (1973)

Railton, I.D., Durley, R.C., Pharis, R.P.: Metabolism of tritiated gibberellin A_9 by shoots of dark-grown dwarf pea, cv. Meteor. Plant Physiol. *54*, 6–12 (1974a)

Railton, I.D., Murofushi, N., Durley, R.C., Pharis, R.P.: Interconversion of gibberellin A_{20} to gibberellin A_{29} by etiolated seedlings and germinating seeds of dwarf pea, *Pisum sativum*. Phytochemistry *13*, 793–796 (1974b)

Railton, I.D., Reid, D.M., Gaskin, P., MacMillan, J.: Characterization of abscisic acid in chloroplasts of *Pisum sativum* cv. 'Alaska' by combined gas-chromatography and mass spectrometry. Planta *117*, 179–182 (1974c)

Railton, I.D., Durley, R.C., Pharis, R.P.: A lack of correlation between the biological activity and rate of metabolism of ent-^3H-17-kaurenoic acid by seedlings of dwarf rice cv. Tan-ginbozu. Plant Cell Physiol. *16*, 943–951 (1975)

Rajagopal, R.: Occurrence of indole acetaldehyde and tryptophol in the extracts of etiolated shoots of *Pisum* and *Helianthus* seedlings. Physiol. Plant. *20*, 655–660 (1967a)

Rajagopal, R.: Metabolism of indole-3-acetaldehyde. I. Distribution of indoleacetic acid and tryptophol forming activities in plants. Physiol. Plant. *20*, 982–990 (1967b)

Rajagopal, R.: Occurrence and metabolism of indoleacetaldehyde in certain higher plant tissues under aseptic conditions. Physiol. Plant. *21*, 378–385 (1968a)

Rajagopal, R.: Metabolism of indole-acetaldehyde. II. On dismutation. Physiol. Plant. *21*, 1076–1096 (1968b)

Rajagopal, R.: Metabolism of indole-3-acetaldehyde. III. Some characteristics of the aldehyde oxydase of *Avena* coleoptiles. Physiol. Plant. *24*, 272–281 (1971)

Rajagopal, R., Bulard, C.: Auxin balance in the *Avena* coleoptile. Effect of light quality. Physiol. Veg. *13*, 1–11 (1975)

Rajagopal, R., Larsen, P.: Metabolism of indole-3-acetaldoxime in plants. Planta *103*, 45–54 (1972)

Rajagopal, R., Larsen, P.: Auxin balance in the Avena coleoptile. Physiol. Plant. *31*, 119–124 (1974)

Ram Chandra, G., Spencer, M., Meheriuk, M.: Evolution of ethylene by subcellular particles from tomatoes as influenced by components of the system. Nature *199*, 767–769 (1963)

Ramina, A.: Aspects of [8-^{14}C]-benzylaminopurine metabolism in *Phaseolus vulgaris*. Plant Physiol. *63*, 298–300 (1979)

Rando, R.R.: Chemistry and enzymology of K_{cat} inhibitors. Science *185*, 320–324 (1974)
Rappaport, L., Adams, D.: Gibberellins: synthesis, compartmentation and physiological process. Philos. Trans. R. Soc. London Ser. B *284*, 521–539 (1978)
Rappaport, L., Hsu, A., Thompson, R., Yang, S.F.: Fate of ^{14}C-gibberellin A_3 in plant tissues. Ann. N.Y. Acad. Sci. *144*, 211–218 (1967)
Rappaport, L., Davies, L., Lavee, S., Nadeau, R., Patterson, R., Stolp, C.F.: Significance of metabolism of ^{3}H-GA_1 for plant regulators. In: Plant growth substances 1973, pp. 314–324. Tokyo: Hirokawa 1974
Raussendorff-Bargen, G. v.: Indolderivate im Apfel. Planta *58*, 471–482 (1962)
Ray, P.M.: Destruction of auxin. Annu. Rev. Plant Physiol. *9*, 81–118 (1958)
Ray, P.M.: Destruction of indoleacetic acid. IV. Kinetics of enzymic oxidation. Arch. Biochem. Biophys. *96*, 199–209 (1962)
Ray, P.M.: Regulation of β-glucan synthetase activity by auxin in pea stem tissue. II. Metabolic requirements. Plant Physiol. *51*, 609–614 (1973)
Ray, P.M.: The chemistry and biochemistry of plant hormones. Rec. Adv. Phytochem. *6*, 93–122 (1974)
Ray, P.M., Thimann, K.V.: The destruction of indoleacetic acid. I. Action of an enzyme from *Omphalia flavida*. Arch. Biochem. Biophys. *64*, 175 (1956)
Rayle, D.L., Purves, W.K.: Isolation and identification of indole-3-ethanol (tryptophol) from cucumber seedlings. Plant Physiol. *42*, 520–524 (1967a)
Rayle, D.L., Purves, W.K.: Conversion of indole-3-ethanol to indole-3-acetic acid in cucumber seedling shoots. Plant Physiol. *42*, 1091–1093 (1967b)
Rayle, D.L., Purves, W.K.: Studies on 3-indoleethanol in higher plants. In: Biochemistry and physiology of plant growth substances. Wightman, F., Setterfield, G. (eds.), pp. 153–161, Ottawa: Runge Press 1968
Redemann, C.T., Meuli, L.J.: Senecionate as a precursor of gibberellic acid in *Fusarium moniliforme*. Naturwissenschaften *46*, 382–383 (1959)
Reed, D.J., Moore, T.C., Anderson, J.D.: Plant growth retardant B-995: A possible mode of action. Science *148*, 1469–1471 (1965)
Reeve, D.R., Crozier, A., Durley, R.C., Reid, D.M., Pharis, R.P.: Metabolism of ^{3}H-gibberellin A_1 and ^{3}H-gibberellin A_4 by *Phaseolus coccineus* seedlings. Plant Physiol. *55*, 42–44 (1975)
Reid, D.M., Carr, D.J.: Effects of dwarfing compound, CCC, on the production and export of gibberellin-like substances by root systems. Planta *73*, 1–11 (1967)
Reid, D.M., Crozier, A.: CCC-induced increase of gibberellin levels in pea seedlings. Planta *94*, 95–106 (1970)
Reid, D.M., Railton, I.D.: The influence of benzyladenine on the growth and gibberellin content of waterlogged tomato plant. Plant Sci. Lett. *2*, 151–156 (1974)
Reid, D.M., Clements, J.B., Carr, D.J.: Red light induction of gibberellin synthesis in leaves. Nature (London) *217*, 580–582 (1968)
Reid, D.M., Tuing, M.S., Durley, R.C., Railton, I.D.: Red-light-enhanced conversion of ^{3}H-GA_9 into other gibberellin-like substances in homogenates of etiolated barley leaves. Planta *108*, 67–75 (1972)
Reid, D.M., Pharis, R.P., Roberts, D.W.A.: Effects of four temperature regimes on the gibberellin content of winter wheat cv. Kharkov. Physiol. Plant. *30*, 53–57 (1974)
Reid, W.W.: Effect of SKF 7997 and SKF 525 on diterpene and sterol biosynthesis in *Gibberella fujikuroi* from 2-^{14}C mevalonate. Biochem. J. *113*, 37p–38p (1969)
Rekoslavskaya, N.I., Gamburg, K.Z.: On the biological activity of IAA conjugates. Biochem. Physiol. Pflanz. *169*, 299–303 (1976)
Retig, N., Rudich, J.: Peroxidase and IAA oxidase activity and isoenzyme patterns in cucumber plants, as affected by sex expression and ethephon. Physiol. Plant. *27*, 156–160 (1972)
Ricard, J., Job, D.: Reaction mechanisms of indole-3-acetate degradation by peroxidases. A stopped-flow and low-temperature spectroscopic study. Eur. J. Biochem. *44*, 359–374 (1974)
Ricard, J., Nari, J.: Contribution à l'étude des mécanismes de la dégradation de l'acide β-indolylacétique par la peroxydase de raifort. Biochem. Biophys. Acta *113*, 57–70 (1966)

Ricard, J., Nari, J.: The formation and reactivity of peroxidase compound III. Biochem. Biophys. Acta *132*, 321–329 (1967)

Ricard, J., Mazza, G., Williams, R.J.P.: Oxidation-reduction potentials and ionization states of two turnip peroxidases. Eur. J. Biochem. *28*, 566–578 (1972)

Riddle, V., M., Mazelis, M.: Conversion of tryptophan to indoleacetamide and further conversion to indoleacetic acid by plant preparations. Plant Physiol. *40*, 481–484 (1965)

Ridge, I., Osborne, D.J.: Hydroxyproline and peroxidases in cell walls of *Pisum sativum*: regulation by ethylene. J. Exp. Bot. *21*, 843–856 (1970)

Rigaud, J., Puppo, A.: Indole-3-acetic acid catabolism by soybean bacteroids. J. Gen. Microbiol. *88*, 223–228 (1975)

Riov, J., Jaffe, M.J.: A cholinesterase from bean roots and its inhibition by plant growth retardants. Experientia *29*, 264–265 (1973a)

Riov, J., Jaffe, M.J.: Cholinesterases from plant tissues. II. Inhibition of bean cholinesterase by 2-isopropyl-4-dimethylamino-5-methylphenyl-1-piperidine carboxylate methylchloride (Amo-1618). Plant Physiol. *52*, 233–235 (1973b)

Ritter, M.C., Dempsey, M.E.: Specificity and role in cholesterol biosynthesis of a squalene and sterol carrier protein. J. Biol. Chem. *246*, 1536–1539 (1971)

Robert, M.L., Taylor, H.F., Wain, R.L.: Ethylene production by cress roots and excised cress root segments and its inhibition by 3.5-diiodo-4-hydroxybenzoic acid. Planta *126*, 273–284 (1975)

Robert, M.L., Taylor, H.F., Wain, R.L.: The effects of 3,5-diiodo-4-hydroxybenzoic acid on the oxidation of IAA and auxin-induced ethylene production by cress root segments. Planta *129*, 53–57 (1976a)

Robert, M.L., Taylor, H.F., Wain, R.L.: The effect of certain phenolic acids on the growth and ethylene production of cress seedling roots. Planta *132*, 95–96 (1976b)

Robertson, J., Berrie, A.M.M.: Abscisic acid and the germination of thermodormant lettuce fruits (*Lactuca sativa* cv. Grand Rapids). The fate of isotopically labelled abscisic acid. Physiol. Plant. *39*, 51–59 (1977)

Robinson, D.R., Ryback, G.: Incorporation of tritium from [(4R)-4-^{3}H] mevalonate into abscisic acid. Biochem. J. *113*, 895–897 (1969)

Robinson, D.R., West, C.A.: Biosynthesis of (−)-kaurene and other diterpenes in germinating castor bean seeds. Fed. Proc. *26*, 454 (1967)

Robinson, D.R., West, C.A.: Biosynthesis of cyclic diterpenes in extracts from seedlings of *Ricinus communis* L. I. Identification of diterpene hydrocarbons formed from mevalonate. Biochemistry *9*, 70–79 (1970a)

Robinson, D.R., West, C.A.: Biosynthesis of cyclic diterpenes in extracts from seedlings of *Ricinus communis* L. II. Conversion of geranylgeranyl pyrophosphate into diterpene hydrocarbons and partial purification of the cyclization enzyme. Biochemistry *9*, 80–89 (1970b)

Rodriguez-Barrueco, C., Bermudez de Castro, F.: Cytokinin-induced pseudonodules on *Alnus glutinosa*. Physiol. Plant. *29*, 277–280 (1973)

Ropers, H.J., Graebe, J.E., Gaskin, P., MacMillan, J.: Gibberellin biosynthesis in a cell-free system from immature seeds of *Pisum sativum*. Biochem. Biophys. Res. Commun. *80*, 690–697 (1978)

Rosenbaum, N., Gefter, M.Z.: Δ^{2}-Isopentenylpyrophosphate transfer ribonucleic acid Δ^{2}-isopentenyltransferase from *Escherichia coli*: Purification and property of the enzyme. J. Biol. Chem. *247*, 5676–5680 (1972)

Row, V.V., Sanford, W.W., Hitchcock, A.E.: Indole-3-acetyl-D,L-aspartic acid as a naturally occurring indole compound in tomato seedlings. Contrib. Boyce Thompson Inst. *21*, 1–10 (1961)

Rowe, J.R.: The common and systematic nomenclature of cyclic diterpenes. Proposal IUPAC Commission of Organic Nomenclature, 3rd. rev. ed., 1968

Rubery, P.H.: Studies on indole acetic acid oxidation by liquid medium from crown gall tissue culture cells: the role of malic acid and other related compounds. Biochem. Biophys. Acta *261*, 21–34 (1972)

Ruddat, M.: Inhibition of the biosynthesis of steviol by a growth retardant. Nature (London) *211*, 971 (1966)

Ruddat, M., Heftmann, E., Lang, A.: Chemical evidence for the mode of action of Amo 1618, a plant growth retardant. Naturwissenschaften *52*, 267 (1965a)
Ruddat, M., Heftmann, E., Lang, A.: Conversion of steviol to a gibberellin-like compound by *Fusarium moniliforme*. Arch. Biochem. Biophys. *111*, 187–190 (1965b)
Rudich, J., Halevy, A.H., Kedar, N.: The level of phytohormones in monoecious and gynoecious cucumbers as effected by photoperiod and etephon. Plant Physiol. *50*, 585–590 (1972)
Rudich, J., Sell, H.M., Baker, L.R.: Transport and metabolism of ^{3}H-gibberellin A_1 in dioecious cucumber seedlings. Plant Physiol. *57*, 734–737 (1976)
Rudnicki, R., Czapski, J.: The uptake and degradation of 1-^{14}C-abscisic acid by apple seeds during stratification. Ann. Bot. (London) *38*, 189–192 (1974)
Rudnicki, R., Pieniazek, J.: Free and bound abscisic acid in developing and ripe strawberries. Bull. Acad. Pol. Sci. Ser. Sci. Biol. *19*, 421–423 (1971)
Runkova, L.V., Lis, E.K., Tomaszewski, M., Antoszewski, R.: Function of phenolic substances in the degradation system of IAA in strawberries. Biol. Plant. *14*, 71–81 (1972)
Ruzicka, L.: History of the isoprene rule. Proc. Chem. Soc. 341–360 (1959)
Ryugo, K., Breen, P.J.: Indoleacetic acid metabolism in cuttings of plum (*Prunus scerasifera* x *P. munsoniana* cv. Marianna 2624). J. Am. Soc. Hortic. Sci. *99*, 247–251 (1974)
Sacher, J.A.: Senescence and postharvest physiology. Annu. Rev. Plant Physiol. *24*, 197–224 (1973)
Sahashi, Y.: Über das Vorkommen von Di-hydroxy-chinolin-carbonsäure (β-Säure von U. Suzuki) in der Reiskleie. Biochem. Z. *159*, 221–234 (1925)
Sahashi, Y.: Über die Konstitution der durch Hydrolyse von Rohoryzanin entstehenden β-Säure (Dioxychinolincarbonsäure). II. Biochem. Z. *168*, 69–72 (1926)
Sahashi, Y.: Synthese der durch Hydrolyse des Roh-Oryzanins entstehenden β-Säure (2,6-Dioxychinolin-4-carbonsäure). Biochem. Z. *189*, 208–215 (1927)
Sahulka, J.: Electrophoretic study on peroxidase, indoleacetic acid oxidase, and o-diphenol oxidase fractions in extracts from different growth zones of *Vicia faba* L. roots. Biol. Plant. *12*, 191–198 (1970)
Sakai, S.: Distribution of the proteinaceous inhibitor of ethylene synthesis in leguminous seeds. Plant Cell Physiol. *16*, 529–532 (1975a)
Sakai, S.: Interaction of plant growth regulators in auxin induced ethylene production. Okayama Univ. Ohara Inst. Agric. Biol. Ber. Ohara Daigaku Nogyo Seibutsu Kenkyujo *16*, 121–128 (1975b)
Sakai, S., Imaseki, H.: Auxin-induced ethylene production by mung bean hypocotyl segments. Plant Cell Physiol. *12*, 349–359 (1971)
Sakai, S., Imaseki, H.: Ethylene biosynthesis: methionine as an in vivo precursor of ethylene in auxin treated mungbean hypocotyl segments. Planta *105*, 165–173 (1972)
Sakai, S., Imaseki, H.: A proteinaceous inhibitor of ethylene biosynthesis by etiolated mung bean hypocotyl sections. Planta *113*, 115–128 (1973a)
Sakai, S., Imaseki, H.: Properties of the proteinaceous inhibitor of ethylene synthesis: action on ethylene production and indoleacetylaspartate formation. Plant Cell Physiol. *14*, 881–892 (1973b)
Sakurai, N., Shibata, K., Kamisaka, S.: Stimulation of auxin-induced elongation of cucumber hypocotyl sections by dihydroconiferyl alcohol. Dihydroconiferyl alcohol inhibits indole-3-acetic acid degradation in vivo and in vitro. Plant Cell Physiol. *16*, 845–855 (1975)
Saltveit, M.E., Jr., Dilley, D.R.: Rapidly induced wound ethylene from excised segments of etiolated *Pisum sativum* L., cv. Alaska. I. Characterization of the response. Plant Physiol. *61*, 447–450 (1978a)
Saltveit, M.E., Jr., Dilley, D.R.: Rapidly induced wound ethylene from excised segments of etiolated *Pisum sativum* L., cv. Alaska. II. Oxygen and temperature dependency. Plant Physiol. *61*, 675–679 (1978b)
Saltveit, M.E., Jr., Dilley, D.R.: Rapidly induced wound ethylene from excised segments of etiolated *Pisum sativum* L., cv. Alaska. III. Induction and transmission of the response. Plant Physiol. *62*, 710–712 (1978c)
Sano, H.: Studies on peroxidase isolated from etiolated Alaska pea seedlings. II. Effect

of quercetin on the oxidation of indole-3-acetic acid. Biochim. Biophys. Acta *227*, 565–575 (1971)
Scallen, T.J., Schuster, M.W., Dhar, A.K.: Evidence for a noncatalytic carrier protein in cholesterol biosynthesis. J. Biol. Chem. *246*, 224–230 (1971)
Scarbrough, E., Armstrong, D.J., Skoog, F., Frihart, C.R., Leonard, N.J.: Isolation of cis-zeatin from *Corynebacterium fascians* cultures. Proc. Natl. Acad. Sci. USA *70*, 3825–3829 (1973)
Scharf, P.: Der Einfluß der 2.4-Dichlor-phenoxy-essigsäure auf den Indol-3-essigsäure-Stoffwechsel von *Phaseolus vulgaris* L. Diss. Pädag. Hochschule Potsdam 1969
Schiewer, U., Erdmann, N.: Bakterielle Auxinbildung in unsterilen Sproßhomogenaten nach Fraktionierung an Sephadex. Biochem. Physiol. Pflanz. *161*, 593–597 (1970)
Schilling, G., Podlesak, W., Eckert, H.: The action of some plant growth inhibitors in wheat studied by cell-free systems. In: Biochemistry and chemistry of plant growth regulators. Schreiber, K., Schütte, H.-R., Sembdner, G. (eds.), pp. 17–32. Inst. Pl. Biochem. Acad. Sci. GDR, Halle 1974
Schilling, G., Eckert, H., Podlesak, W.: Biosynthese und Metabolismus von GA_1 und GA_3 bei *Triticum aestivum*, untersucht mit Hilfe wachstumsretardierender Verbindungen. Biochem. Physiol. Pflanz. *172*, 567–579 (1978)
Schliemann, W., Schneider, G.: Untersuchungen zur enzymatischen Hydrolyse von Gibberellin-O-glucosiden. II. Hydrolysegeschwindigkeiten von Gibberellin-O(13)-glucosiden. Biochem. Physiol. Pflanz. *174*, 738–745 (1979)
Schneider, E.A.: Studies on the biosynthesis and degradation of 3-indoleacetic acid and gramine in barley shoots, 237 pp. Ph. D. thesis. Carleton University, Ottawa: 1965
Schneider, E.A., Wightman, F.: Metabolism of auxin in higher plants. Annu. Rev. Plant Physiol. *25*, 487–513 (1974)
Schneider, E.A., Wightman, F.: Auxins. In: Phytohormones and related compounds – A comprehensive treatise. The Biochemistry of phytohormones and related compounds. Letham, D.S., Goodwin, P.B., Higgins, T.J.V. (eds.), Vol. I. pp. 29–105. Amsterdam: Elsevier 1978
Schneider, E.A., Gibson, R.A., Wightman, F.: Biosynthesis and metabolism of indole-3-yl-acetic acid. I. The native indoles of barley and tomato shoots. J. Exp. Bot. *23*, 152–170 (1972)
Schneider, G.: Partialsynthese von Gibberellin-A_1-O(3)-β-D-glucopyranosid. Tetrahedron Lett. 4053–4054 (1972)
Schneider, G.: Studies on the synthesis of gibberellin-O-glucosides. In: Biochemistry and chemistry of plant growth regulators. Schreiber, K., Schütte, H.-R., Sembdner, G. (eds.), pp. 157–160, Inst. Pl. Biochem. Acad. Sci. GDR, Halle 1974
Schneider, G.: Synthesen radioaktiv markierter Verbindungen, Darstellung von GA_3-O(3)-β-D-glucopyranosid-U-^{3}H. Z. Chem. *18*, 217 (1978a)
Schneider, G.: Studies on the glucosylation of gibberellins. Symp. Papers, 11th IUPAC Int. Symp. Chem. Nat. Products, Varna 1978, Vol. I 259–260, Sofia: Bulg. Acad. Sci. (1978b)
Schneider, G., Schliemann, W.: Untersuchungen zur enzymatischen Hydrolyse von Gibberellin-O-glucosiden. I. Hydrolysegeschwindigkeiten von Gibberellin-O(2)- und -O(3)-glucosiden. Biochem. Physiol. Pflanz. *174*, 746–751 (1979)
Schneider, G., Sembdner, G., Schreiber, K.: Zur Synthese von Gibberellin A_3-β-D-glucopyranosiden. Z. Chem. *14*, 474–475 (1974)
Schneider, G., Miersch, O., Liebisch, H.-W.: Synthese von O-β-D-glucopyranosyl-gibberellin-O-β-D-glucopyranosylestern. Tetrahedron Lett. 405–406 (1977a)
Schneider, G., Sembdner, G., Schreiber, K.: Synthese von O(3)- und O(13)-glucosylierten Gibberellinen. Tetrahedron *33*, 1391–1397 (1977b)
Schneider, J., Günther, G.: Effect of diquat on the level and the enzymatic oxidation of indole-3-acetic acid [IAA], in etiolated pea plants. Biochem. Physiol. Pflanz. *167*, 151–158 (1975)
Schraudolf, H.: Zur Verbreitung von Glucobrassicin und Neoglucobrassicin in höheren Pflanzen. Experientia *21*, 520–522 (1965)
Schraudolf, H.: Der Stoffwechsel von Indolderivaten in *Sinapis alba* L. I. Synthese und

Umsetzung von L-Tryptophan in etiolierten Hypocotylsegmenten nach Applikation von Indol-2-^{14}C. Phytochemistry *5*, 83–90 (1966)
Schraudolf, H.: Untersuchungen zur Verbreitung von Indolylglucosinolaten in Cruciferen. Experientia *24*, 434 (1968)
Schraudolf, H.: Wachstum. Fortschr. Bot. *33*, 121–140 (1971)
Schraudolf, H., Bergmann, F.: Der Stoffwechsel von Indolderivaten in *Sinapis alba* L. II. Untersuchungen zur Biogenese und Umsetzung von Indolglucosinolaten mit Hilfe von ringmarkiertem C^{14}-Tryptophan und S^{35}-Sulfat. Planta *67*, 75–95 (1965)
Schraudolf, H., Weber, H.: IAN-Bildung aus Glucobrassicin: pH-Abhängigkeit und wachstumsphysiologische Bedeutung. Planta *88*, 136–143 (1969)
Schreiber, K., Schneider, G., Sembdner, G., Focke, I.: Gibberelline. X. Isolierung von O(2)-Acetyl-Gibberellinsäure als Stoffwechselprodukt von *Fusarium moniliforme* Sheld. Phytochemistry *5*, 1221–1225 (1966)
Schreiber, K., Weiland, J., Sembdner, G.: Isolierung und Struktur eines Gibberellinglucosides. Tetrahedron Lett. 4285–4288 (1967)
Schreiber, K., Weiland, J., Sembdner, G.: Gibberelline XV. Synthese von O(2)-β-D-Glucopyranosyl-gibberellin-A_3-methylester. Tetrahedron *25*, 5541–5545 (1969)
Schreiber, K., Weiland, J., Sembdner, G.: Isolierung von Gibberellin-A_8-O(3)-β-D-glucopyranosid aus Früchten von *Phaseolus coccineus*. Phytochemistry *8*, 189–198 (1970)
Scott, T.K.: Auxins and roots. Annu. Rev. Plant Physiol. *23*, 235–258 (1972)
Seeley, R.C., Fawcett, R.L., Wain, R.L., Wightman, F.: In: The chemistry and mode of action of plant growth substances. Wain, R.L., Wightman, F. (eds.), pp. 234–247. London: Butterworths Scientific Publ. 1956
Sembdner, G.: Conjugation of plant hormones. In: Biochemistry and chemistry of plant growth regulators. Schreiber, K., Schütte, H.-R., Sembdner, G. (eds.), pp. 283–302, Inst. Pl. Biochem. Acad. Sci. GDR, Halle 1974
Sembdner, G.: Conjugated gibberellins. In: Plant growth and differentiation. Kefeli, V.I. (ed.), Moskau: Nauka 1980 in press
Sembdner, G., Schreiber, K.: Gibberelline IV. Über die Gibberelline von *Nicotiana tabacum* L. Phytochemistry *4*, 49–56 (1965)
Sembdner, G., Schneider, G., Weiland, J., Schreiber, K.: Über ein gebundenes Gibberellin aus *Phaseolus coccineus* L. Experientia *20*, 89–90 (1964)
Sembdner, G., Weiland, J., Aurich, O., Schreiber, K.: Isolation, structure and metabolism of a gibberellin glucoside. In: Plant growth regulators. S.C.I. Monogr. Vol. 31, pp. 70–86. London 1968
Sembdner, G., Weiland, J., Schneider, G., Schreiber, K., Focke, I.: Recent advances in the metabolism of gibberellins. In: Plant growth substances 1970. Carr, D.J. (ed.), pp. 145–150. Berlin-Heidelberg-New York: Springer 1972
Sembdner, G., Schulze, C., Adam, G., Voigt, B., Hung, P.D., Weiland, J., Schreiber, K.: Biologische Aktivität von Gibberellin-Derivaten. In: Wirkungsmechanismen von Herbiziden und synthetischen Wachstumsregulatoren, Part. 10. Barth, A., Jacob, F., Feyerabend, G. (eds.), pp. 206–212. Halle: Wissenschaftliche Beiträge der Martin-Luther-Universität Halle 1973
Sembdner, G., Adam, G., Lischewski, M., Sych, F.J., Schulze, C., Knöfel, D., Müller, P., Schneider, G., Liebisch, H.-W., Schreiber, K.: Biological activity and metabolism of conjugated gibberellins. In: Plant growth substances 1973. pp. 349–355. Tokyo: Hirokawa 1974
Sembdner, G., Borgmann, E., Schneider, G., Liebisch, H.-W., Miersch, O., Adam, G., Lischewski, M., Schreiber, K.: Biological activity of some conjugated gibberellins. Planta *132*, 249–257 (1976)
Sembdner, G., Bergner, C., Royl, B., Liebisch, H.-W.: Untersuchungen zur Wirkungsweise von Phenoxyisobuttersäuren als Wachstumsregulatoren. In: Wirkungsmechanismen von Herbiziden und synthetischen Wachstumsregulatoren. Schütte, H.R. (ed.), pp. 308–315, Jena: VEB Fischer 1979
Sequeira, L., Mineo, L.: Partial purification and kinetics of indoleacetic acid oxidase from tobacco roots. Plant Physiol. *41*, 1200–1208 (1966)
Sfakiotakis, E.M., Dilley, D.R.: Induction of autocatalytic ethylene production in apple

fruits by propylene in relation to maturity and oxygen. Am. Soc. Hortic. Sci. *98*, 504–508 (1973)
Shantz, E.M.: Chemistry of naturally-occurring growth-regulating substances. Annu. Rev. Plant Physiol. *17*, 409–438 (1966)
Shaybany, B., Weinbaum, S.A., Martin, G.C.: Identification of ABA stereoisomers in French prune seeds and association of ABA with ethylene-enhanced prune abscission. J. Am. Soc. Hortic. Sci. *102*, 501–503 (1977)
Shechter, I., West, C.A.: Biosynthesis of gibberellins. IV. Biosynthesis of cyclic diterpenes from transgeranylgeranyl pyrophosphate. J. Biol. Chem. *244*, 3200–3209 (1969)
Sherwin, J.E.: A tryptophan decarboxylase from cucumber seedlings. Plant Cell Physiol. *11*, 865–872 (1970)
Sherwin, J.E., Furuya, M.: A red-far red reversible effect on uptake of exogenous indoleacetic acid in etiolated rice coleoptiles. Plant. Physiol. *51*, 295–298 (1973)
Sherwin, J.E., Purves, W.K.: Tryptophan as an auxin precursor in cucumber seedlings. Plant Physiol. *44*, 1303–1309 (1969)
Shimokawa, K., Kasai, Z.: Biogenesis of ethylene in apple tissue. I. Formation of ethylene from glucose, acetate, pyruvate, and acetaldehyde in apple tissue. Plant Cell Physiol. *7*, 1–9 (1966)
Shimokawa, K., Kasai, Z.: A possible incorporation of ethylene into RNA in Japanese morning glory seedlings. Agric. Biol. Chem. *32*, 680–682 (1968)
Shimokawa, K., Kasai, Z.: The role of β-hydroxypropionate in ethylene biosynthesis. I. Ethylene formation from acetate-2, C-14 and fumarate-2,3-C-14 in banana fruits. Agric. Biol. Chem. *34*, 1633–1639 (1970a)
Shimokawa, K., Kasai, Z.: The role of β-hydroxypropionate in ethylene biosynthesis. II. Ethylene formation from propionate-2-^{14}C in banana pulp slices and homogenates. Agric. Biol. Chem. *34*, 1640–1645 (1970b)
Shimokawa, K., Kasai, Z.: Ethylene formation from acrylic acid by a banana pulp extract. Agric. Biol. Chem. *34*, 1646–1651 (1970c)
Shimokawa, K., Yokoyama, K., Kasai, Z.: Fixation of ethylene-^{14}C by Japanese morning glory seedling (*Pharbitis nil* Chois.). Mem. Res. Inst. Food Sci. Kyoto Univ. *30*, 1–7 (1969)
Shindy, W.W., Asmundson, C.M., Smith, O.F., Kumamoto, J.: Absorption and distribution of high specific radioactivity 2-^{14}C-ABA in cotton seedlings. Plant Physiol. *52*, 443–447 (1973)
Shingo, S., Imaseki, H.: Auxin-induced ethylene production by mungbean hypocotyl segments. Plant Cell Physiol. *12*, 349–359 (1971)
Shinshi, H., Noguchi, M.: Relationships between peroxidase, IAA oxidase and polyphenol oxidase. Phytochemistry *14*, 1255–1258 (1975)
Shive, J.B., Sisler, H.D.: Effects of ancymidol (a growth retardant) and triarimol (a fungicide) on the growth, sterols, and gibberellins of *Phaseolus vulgaris* (L.). Plant Physiol. *57*, 640–644 (1976)
Short, K.C., Torrey, J.G.: Cytokinins in seedling roots of pea. Plant Physiol. *49*, 155–166 (1972)
Shukla, P.S., Mahadevan, S.: Indoleacetaldoxime hydro-lyase (4.2.1.2a). III. Further studies on the nature and mode of action of the enzyme. Arch. Biochem. Biophys. *137*, 166–174 (1970)
Shulman, Y., Lavee, S.: Endogenous cytokinins in maturing manzanillo olive fruits. Plant Physiol. *57*, 490–492 (1975)
Siebert, K., Clagett, C.O.: Formic acid from carbon 2 of methionine in ethylene production in apple tissue. Plant Physiol. *44*, Suppl. 30 (1969)
Siegel, B.Z., Galston, A.W.: Indoleacetic acid oxidase activity of apoperoxidase. Science *157*, 1557–1559 (1967)
Silk, W.K., Jones, R.L., Stoddart, J.L.: Growth and gibberellin A_1 metabolism in excised lettuce hypocotyls. Plant Physiol. *59*, 211–216 (1977)
Simcox, P.D., Dennis, D.T., West, C.A.: Kaurene synthetase from plastids of developing plant tissues. Biochem. Biophys. Res. Commun. *66*, 166–172 (1975)
Sinska, I., Lewak, S.: Is the gibberellin A_4 biosynthesis involved in the removal of dormancy in apple seeds. Plant Sci. Lett. *9*, 163–170 (1977)

Sinska, I., Lewak, S., Gaskin, P., MacMillan, J.: Reinvestigation of apple-seed gibberellins. Planta *114*, 359–364 (1973)

Sirois, J.C., Miller, R.W.: The mechanism of the scopoletin-induced inhibition of the peroxidase catalyzed degradation of indole-3-acetate. Plant Physiol. *49*, 1012–1018 (1972)

Sitton, D., West, C.A.: Casbene: an antifungal diterpene produced in cell-free extracts of *Ricinus communis* seedlings. Phytochemistry *14*, 1921–1925 (1975)

Sitton, D., Richmond, A., Vaadia, Y.: On the synthesis of gibberellins in roots. Phytochemistry *6*, 1101–1105 (1967)

Sivadjian, A., Sadorge, P., Gawer, M., Terrine, C., Guern, J.: Enzymic synthesis of some N^6-substituted adenine nucleosides. Physiol. Vég. *7*, 31–42 (1969)

Skene, K.G.M.: Gibberellin-like substances in root exudate of *Vitis vinifera*. Planta *74*, 250–262 (1967)

Skene, K.G.M.: Cytokinins in the xylem sap of grape vine canes: Changes in activity during cold storage. Planta *104*, 89–92 (1972)

Skoog, F., Armstrong, D.J.: Cytokinins. Annu. Rev. Plant Physiol. *21*, 359–383 (1970)

Smith, O.E., Sadri, H.A.: Effects of abscisic acid on gibberellin biosynthesis in *Fusarium moniliforme*. Plant Cell Physiol. *11*, 345–348 (1970)

Smith, W.H., Meigh, D.F., Parker, J.C.: Effect of damage and fungal infection on the production of ethylene by carnations. Nature (London) *204*, 92–93 (1964)

Smolenska, G., Lewak, S.: Gibberellins and the photosensitivity of isolated embryos from non-stratified apple seeds. Planta *99*, 144–151 (1971)

Snir, I., Kessler, B.: The influence of the growth retardant CCC on endogenous gibberellin in cucumber seedlings. Planta *125*, 73–75 (1975)

Sondheimer, E., Tzou, D.S.: The metabolism of hormones during seed germination and dormancy. II. The metabolism of 8-^{14}C-zeatin in bean axes. Plant Physiol. *47*, 516–519 (1971)

Sondheimer, E., Michniewicz, B.M., Powell, L.E.: Biological and chemical properties of the epidioxide isomer of abscisic acid and its rearrangement products. Plant Physiol. *44*, 205–209 (1969)

Sondheimer, E., Galson, E.C., Chang, Y.P., Walton, D.C.: Asymmetry, its importance to the action and metabolism of abscisic acid. Science *174*, 829–831 (1971)

Sondheimer, E., Galson, E.C., Tinelli, E., Walton, D.C.: The metabolism of hormones during seed germination and dormancy. IV. The metabolism of (S)-2-^{14}C-abscisic acid in ash seed. Plant Physiol. *54*, 803–808 (1974)

Spaeth, S.C., Maravolo, N.C.: Partial purification and characterization of an auxin-degrading enzyme in the liverwort *Marchantia polymorpha* L. Bot. Gaz. *134*, 274–282 (1973)

Spector, C., Phinney, B.O.: Gibberellin biosynthesis. Genetic studies in *Gibberella fujikuroi*. Physiol. Plant. *21*, 127–136 (1968)

Spencer, M.: Ethylene in nature. Fortschr. Chem. Org. Naturst. *27*, 31–80 (1969)

Spencer, M., Meheriuk, M.: Influence of temperature and ageing on ethylene production by a sub-cellular fraction from tomatoes. Nature (London) *199*, 1077–1078 (1963)

Sponsel, V.M., MacMillan, J.: The metabolism of gibberellins in immature seeds of *Pisum sativum* cv. Progress No. 9. In: Abstr. 9th. Int. Conf. Plant Growth Subst. 1976. pp. 366–368. Lausanne 1976

Sponsel, V.M., MacMillan, J.: Further studies on the metabolism of gibberellins (GAs) A_9, A_{20}, and A_{29} in immature seeds of *Pisum sativum* cv. Progress No. 9. Planta *135*, 129–130 (1977)

Sponsel, V.M., MacMillan, J.: Metabolism of gibberellin A_{29} in seeds of *Pisum sativum* cv. Progress No. 9; Use of (^{2}H) and (^{3}H) GAs and the identification of a new GA catabolite. Planta *144*, 69–78 (1978)

Sponsel, V.M., Gaskin, P., MacMillan, J.: The identification of gibberellins in immature seeds of *Vicia faba*, and some chemotaxonomic considerations. Planta *146*, 101–105 (1979)

Srivastava, O.P., van Huystee, R.B.: Evidence for close association of peroxidase, polyphenol oxidase, and IAA (indole-3-acetic acid) oxidase isoenzymes of peanut suspension culture medium. Can. J. Bot. *51*, 2207–2215 (1973)

Stahmann, M.A., Clare, B.G., Woodbury, W.: Increased disease resistance and enzyme activity induced by ethylene and ethylene production by black rot infected sweet potato tissue. Plant Physiol. *41*, 1505–1512 (1966)

Steen, D.A., Chadwick, A.V.: Effects of cycloheximide on indoleacetic acid-induced ethylene production in pea root tips. Plant Physiol. *52*, 171–173 (1973)

Stekoll, M., West, C.A.: Purification and properties of an elicitor of castor bean phytoalexin from culture filtrates of the fungus *Rhizopogon stolonifer*. Plant Physiol. *61*, 38–45 (1978)

Steward, F.C., Shantz, E.M.: The chemical regulation of growth (Some substances and extracts which induce growth and morphogenesis). Annu. Rev. Plant Physiol. *10*, 379–404 (1959)

Still, C.C., Fukuyama, T.T., Moyed, H.S.: Inhibitory oxidation products of indole-3-acetic acid. J. Biol. Chem. *240*, 2612–2618 (1965a)

Still, C.C., Oliver, C., Moyed, H.S.: Inhibitory oxidation products of IAA: enzymatic formation and detoxification by pea seedlings. Science *149*, 1249–1251 (1965b)

Stinson, R.A., Spencer, M.S.: *β*-Alanine as an ethylene precursor. Investigation towards preparation, and properties of a soluble enzyme system from a subcellular fraction of bean cotyledons. Plant Physiol. *44*, 1217–1226 (1969)

Stinson, R.A., Spencer, M.: Respiratory control, oxidative phosphorylation, respiration, rate of ATP hydrolysis, and ethylene evolution in subcellular particulate fractions from cotyledons of germinating seedlings. Can. J. Biochem. *48*, 541–546 (1970)

Stoddart, J.L.: The association of gibberellin-like activity with the chloroplast fraction of leaf homogenates. Planta *81*, 106–112 (1968)

Stoddart, J.L.: Incorporation of kaurenoic acid into gibberellins by chloroplast preparations of *Brassica oleracea*. Phytochemistry *8*, 831–837 (1969)

Stoddart, J.L., Jones, R.L.: Gibberellin metabolism in excised lettuce hypocotyls. Evidence for the formation of gibberellin A_1 glucosyl conjugates. Planta *136*, 261–269 (1977)

Stoddart, J.L., Lang, A.: The effect of daylength on gibberellin synthesis in leaves of red clover (*Trifolium pratense* L.). In: Biochemistry and physiology of plant growth substances. Wightman, F., Setterfield, G. (eds.), pp. 1371–1381. Ottawa: Runge Press 1968

Stoddart, J.L., Breidenbach, W., Nadeau, R., Rappaport, L.: Selective binding of ^{3}H-gibberellin A_1 by protein fractions from dwarf pea epicotyls. Proc. Natl. Acad. Sci. USA *71*, 3255–3259 (1974)

Stolp, C.F., Nadeau, R., Rappaport, L.: Effect of abscisic acid on uptake and metabolism of ^{3}H-gibberellin A_1 and ^{3}H-pseudogibberellin A_1 by barley half-seeds. Plant Physiol. *52*, 546–548 (1973)

Stolp, C.F., Nadeau, R., Rappaport, L.: Abscisic acid and the accumulation, biological activity and metabolism of four derivatives of [^{3}H]-gibberellin A_1 in barley aleurone layers. Plant Cell Physiol. *18*, 721–728 (1977)

Stonier, T.: The role of auxin protectors in autonomous growth. In: Les cultures de tissus de plantes. Proc. 2nd Int. Conf. Plant Tissue Cult., Strasbourg 1970. Hirth, M.L., Morel, G. (eds.), No. 193, pp. 423–435. Paris: C.N.R.S. 1972

Stonier, T., Yang, H.: Studies on auxin protectors. X. Protector levels and lignification in sunflower crown gall tissue. Physiol. Plant. *25*, 474–481 (1971)

Stonier, T., Yang, H.: Studies on auxin protectors. XI. Inhibition of peroxidase-catalyzed oxidation of glutathione by auxin protectors and o-dihydroxyphenols. Plant Physiol. *51*, 391–395 (1973)

Stonier, T., Yoneda, Y.: Stem internode elongation in the Japanese morning glory (*Pharbitis nil* Choisy) in relation to an inhibitor system of auxin destruction. Physiol. Plant. *20*, 13–19 (1967a)

Stonier, T., Yoneda, Y.: Auxin destruction and growth in Japanese morning glory stems. Ann. N.Y. Acad. Sci. *144*, 129–135 (1967b)

Stonier, T., Rodriguez-Tormes, F., Yoneda, Y.: Studies on auxin protectors. IV. The effect of manganese on auxin protector-I of the Japanese morning glory. Plant Physiol. *43*, 69–72 (1968a)

Stonier, T., Yoneda, Y., Rodriguez-Tormes, F.: Studies on auxin protectors. V. On the mechanism of IAA protection by protector-I of the Japanese morning glory. Plant Physiol. *43*, 1141–1145 (1968b)

Stonier, T., Hudek, J., Vande-Stouwe, R., Yang, H.: Studies of auxin protectors. VIII. Evidence that auxin protectors act as cellular poisers. Physiol. Plant. *23*, 775–783 (1970a)

Stonier, T., Singer, R.W., Yang, H.: Studies on auxin protectors. IX. Inactivation of certain protectors by polyphenol oxidase. Plant Physiol. *46*, 454–457 (1970b)

Struxness, L.A., Armstrong, D.J., Gillam, I., Tener, G.M., Burrow, W.J., Skoog, F.: Distribution of cytokinin in active ribonucleosides in wheat germ t-RNA species. Plant Physiol. *63*, 35–41 (1979)

Südi, J.: Increase in the capacity of pea tissue to form acylaspartic acids especially induced by auxins. New Phytol. *65*, 9–21 (1969)

Suge, H.: Changes in ethylene production of vernalized plants. Plant Cell Physiol. *18*, 1167–1171 (1977)

Summons, R.E., MacLeod, J.K., Parker, C.W., Letham, D.S.: The occurrence of raphanatin as an endogenous cytokinin in radish seed. FEBS Lett. *82*, 211–214 (1977)

Suzuki, Y., Kawarada, A.: Products of peroxidase catalyzed oxidation of indolyl-3-acetic acid. Agric. Biol. Chem. *42*, 1315–1321 (1978)

Suzuki, Y., Kinashi, H., Takeuchi, S., Kawarada, A.: (+)-5-Hydroxy-dioxindole-3-acetic acid, a synergist from rice bran of auxin-induced ethylene production in plant tissue. Phytochemistry *16*, 635–637 (1977)

Sweetser, P.B., Vatvars, A.: High-performance liquid chromatographic analysis of abscisic acid in plant extracts. Anal. Biochem. *71*, 68–78 (1976)

Syono, K., Torrey, J.G.: Identification of cytokinins of root nodules of the garden pea *Pisum sativum* L. Plant Physiol. *57*, 602–606 (1976)

Syono, K., Newcomb, W., Torrey, J.G.: Cytokinin production in relation to the development of pea root nodules. J. Bot. *54*, 2155–2162 (1976)

Takahashi, N.: Recent progress in the chemistry of gibberellins. In: Plant growth substances 1973, pp. 228–240. Tokyo: Hirokawa 1974

Takahashi, N., Yamaguchi, I., Kono, T., Igoshi, M., Hirose, K., Suzuki, K.: Characterization of plant growth substances in *Citrus unshiu* and their change in fruit development. Plant Cell Physiol. *16*, 1101–1111 (1975)

Takahashi, N., Murofushi, N., Yamane, H.: Metabolism of gibberellins in maturing and germinating bean seeds. In: Abstr. 9th Int. Conf. Plant Growth Subst. 1976. p. 383. Lausanne 1976

Takeba, G., Takimoto, A.: Metabolism of (−)-kauren-^{3}H and (−)-kaurenol-^{3}H during the photoperiodic flora induction in *Pharbitis nil*. Plant Cell Physiol. *12*, 81–88 (1971)

Taya, Y., Tanaka, Y., Nishimura, S.: 5′-AMP is a direct precursor of cytokinin in *Dictyostelium discoideum*. Nature (London) *271*, 545–547 (1978a)

Taya, Y., Tanaka, Y., Nishimura, S.: Cell-free biosynthesis of discadenine, a spore germination inhibitor of *Dictyostelium discoideum*. FEBS Lett. *89*, 326–328 (1978b)

Taylor, C.M., Railton, I.D.: The influence of wilting and abscisic acid application on gibberellin interconversion in etiolated seedlings of dwarf *Pisum sativum* cv. Meteor. Plant Sci. Lett. *9*, 317–322 (1977)

Taylor, H.F.: Carotenoids as possible precursors of abscisic acid in plants. In: Plant growth regulators S.C.I. Monogr. Vol. 31, pp. 22–35. London 1968

Taylor, H.F., Burden, R.S.: Identification of plant growth inhibitors produced by photolysis of violaxanthin. Phytochemistry *9*, 2217–2223 (1970a)

Taylor, H.F., Burden, R.S.: Xanthoxin, a new naturally occurring plant growth inhibitor. Nature (London) *227*, 302–304 (1970b)

Taylor, H.F., Burden, R.S.: Xanthoxin, a recently discovered plant growth inhibitor. Proc. R. Soc. London Ser. B *180*, 317–346 (1972)

Taylor, H.F., Burden, R.S.: Preparation and metabolism of 2-^{14}C-cis,trans-xanthoxin. J. Exp. Bot. *24*, 873–880 (1973)

Taylor, H.F., Smith, T.A.: Production of plant growth inhibitors from xanthophylls: a possible source of dormin. Nature (London) *215*, 1513–1514 (1967)

Teltscherová, L., Pavlová, L., Pleskotová, D.: Changes in the content of endogenous auxins in apical buds of *Chenopodium rubrum* L. induced with respect to the endogenous rhythm in capacity to flower. Biol. Plant. *19*, 205–211 (1976)

Terrine, C., Sadorge, P., Gawer, M., Guern, J.: Etude de la dégradation par voie enzymatique des analogues N^6-substitués de l'adénosine. I. Action de l'adénosine aminohydrolase in vitro. Physiol. Vég. *7*, 425–435 (1969)

Thimann, K.V.: On the plant growth hormone produced by *Rhizopus suinus*. J. Biol. Chem. *109*, 279–291 (1935)
Thimann, K.V.: The natural plant hormones. In: Plant physiology – A treatise. Steward, F.C. (ed.), Vol. VIb, pp. 1–359. New York, London: Academic Press 1972
Thimann, K.V., Grochowska, M.: The role of tryptophan and tryptamine as IAA precursors. In: Biochemistry and physiology of plant growth substances. Wightman, F., Setterfield, G. (eds.), pp. 231–242. Ottawa: Runge Press 1968
Thimann, K.V., Mahadevan, S.: Nitrilase. I. Occurrence, preparation and general properties of the enzyme. Arch. Biochem. Biophys. *105*, 133–141 (1964)
Thomas, T.H., Khan, A.A., O'Toole, D.F.: The location of cytokinins and gibberellins in wheat seeds. Physiol. Plant. *42*, 61–66 (1978)
Thompson, A.G., Bruinsma, J.: Xanthoxin: A growth inhibitor in light-grown sunflower seedlings, *Helianthus annuus* L. J. Exp. Bot. *28*, 804–810 (1977)
Thompson, J.E., Spencer, M.: Preparation and properties of an enzyme system for ethylene production. Nature (London) *210*, 595–597 (1966)
Thompson, J.E., Spencer, M.S.: Ethylene production from β-alanine by an enzyme powder. Can. J. Biochem. *45*, 563–572 (1967)
Thompson, J.E., Tribe, T.A., Spencer, M.S.: A non-enzymatic model for ethylene production from β-alanine. Can. J. Biochem. *44*, 389–391 (1966)
Thurman, D.A., Street, H.E.: Metabolism of some indole auxins in excised tomato roots. J. Exp. Bot. *13*, 369–377 (1962)
Tietz, D., Doerffling, K., Woehrle, D., Erzleben, J., Liemann, F.: Identification by combined gas chromatography-mass spectrometry of phaseic acid and dihydrophaseic acid and characterization of further abscisic acid metabolites in pea seedlings. Planta *147*, 168–173 (1979)
Tillberg, E.: Lèvels of indol-3yl-acetic acid and acid inhibitors in green and etiolated bean seedlings (*Phaseolus vulgaris*). Physiol. Plant. *31*, 106–111 (1974a)
Tillberg, E.: Occurrence of endogenous indole-3-acetyl-aspartic acid in light- and dark-grown bean seedlings (*Phaseolus vulgaris*). Physiol. Plant. *31*, 271–274 (1974b)
Timmappaya, G., Cherayil, J.D.: Unique presence of 2-methylthioribosylzeatin in ribonucleic acid of the bacterium *Pseudomonas aeruginosa*. Biochem. Biophys. Res. Commun. *60*, 665–672 (1974)
Tinelli, E.T., Sondheimer, E., Walton, D.C., Gaskin, P., MacMillan, J.: Metabolites of 2-^{14}C-abscisic acid. Tetrahedron Lett. 139–140 (1973)
Tingwa, P.O., Young, R.E.: The effect of indole-3-acetic acid and other growth regulators on the ripening of avocado fruits. Plant Physiol. *55*, 937–940 (1975)
Tompsett, P.B., Schwabe, W.W.: Growth hormone changes in *Chrysanthemum marifolium*. Effects of environmental factors controlling flowering. Ann. Bot. *38*, 269–285 (1974)
Torrey, J.G.: Root hormones and plant growth. Annu. Rev. Plant Physiol. *27*, 435–459 (1976)
Troxler, R.F., Hamilton, R.H.: Metabolism of indole-3-acetic acid by geranium stem callus cultures. Plant Physiol. *40*, 400–405 (1965)
Truelsen, T.A.: Indole-3-pyruvic acid as an intermediate in the conversion of tryptophan to indole-3-acetic acid. I. Some characteristics of tryptophan transaminase from mung bean seedlings. Physiol. Plant. *26*, 289–295 (1972)
Truelsen, T.A.: Indole-3-pyruvic acid as an intermediate in the conversion of tryptophan to indole-3-acetic acid. II. Distribution of tryptophan transaminase activity in plants. Physiol. Plant. *28*, 67–70 (1973)
Tuli, V., Moyed, H.S.: Inhibitory oxidation products of indole-3-acetic acid: 3-hydroxymethyloxindole and 3-methyleneoxindole as plant metabolites. Plant Physiol. *42*, 425–430 (1967)
Tuli, V., Moyed, H.S.: The role of 3-methyleneoxindole in auxin action J. Biol. Chem. *244*, 4916–4920 (1969)
Tzou, D.S., Galson, E.C., Sondheimer, E.: The metabolism of hormones during seed germination and release from dormancy. III. The effects and metabolism of zeatin in dormant and nondormant ash embryos. Plant Physiol. *51*, 894–897 (1973)
Ueda, M., Bandurski, R.S.: Structure of indole-3-acetic acid myoinositol esters and pentamethyl myoinositols. Phytochemistry *13*, 243–253 (1974)

Ueda, M., Ehmann, A., Bandurski, R.S.: Gas-liquid chromatographic analysis of indole-3-acetic acid myoinositol esters in maize kernels. Plant Physiol. *46*, 715–719 (1970)

Uota, M.: Carbon dioxide suppression of ethylene-induced sleepiness of carnation blooms. J. Am. Soc. Hortic. Sci. *94*, 598–601 (1969)

Upper, C.D., West, C.A.: Biosynthesis of gibberellins. II. Enzymic cyclization of geranylgeranyl pyrophosphate to kaurene. J. Biol. Chem. *242*, 3285–3292 (1967)

Upper, C.D., West, C.A.: Enzymic synthesis of (−)-kaurene and related diterpenes. In: Methods in enzymology. Clayton, R.B. (ed.), Vol. XV, pp. 481–490. New York, London: Academic Press 1969

van den Ende, H., Zeevaart, J.A.D.: Influence of daylength on gibberellin metabolism and stem growth in *Silene armeria*. Planta *98*, 164–176 (1971)

van der Mast, C.A.: Separation of IAA degrading enzymes from pea roots on columns of polyvinylpyrrolidone. Acta Bot. Neerl. *18*, 620–626 (1969)

van der Mast, C.A.: The inhibiting effect of caffeic acid on the enzymatic degradation of IAA is exerted via the non-IAA degrading peroxidases. Acta Bot. Neerl. *19*, 659–664 (1970)

van Staden, J.: The identification of zeatin glucoside from coconut milk. Physiol. Plant. *36*, 123–126 (1976a)

van Staden, J.: Occurrence of a cytokinin glucoside in the leaves and in the honeydew of *Salix babylonica*. Physiol. Plant. *36*, 225–228 (1976b)

van Staden, J.: Seasonal changes in the cytokinin content of the leaves of *Salix babylonica*. Physiol. Plant. *40*, 296–299 (1977)

van Staden, J.: The metabolism of zeatin by cell-free plant extracts. Z. Pflanzenphysiol. *92*, 399–405 (1979)

van Staden, J., Button, J.: The cytokinin content of as especially cultured pea fruits. Z. Pflanzenphysiol. *87*, 129–135 (1978)

van Staden, J., Davey, J.E.: Cytokinin translocation in the xylem sap of herbaceous plants. Z. Pflanzenphysiol. *77*, 377–382 (1976)

van Staden, J., Drews, S.E.: Identification of zeatin and zeatin riboside in coconut milk. Physiol. Plant. *34*, 106–109 (1975)

van Staden, J., Papaphilippou, A.P.: Biological activity of O-β-D-glucopyranosylzeatin. Plant Physiol. *60*, 649–650 (1977)

Vendrell, M., McGlasson, W.B.: Inhibition of ethylene production in banana fruit tissue by ethylene treatment. Austr. J. Biol. Sci. *24*, 885–895 (1971)

Vendrig, J.C.: Sterol-like compounds with auxin activity in human urine. Z. Pflanzenphysiol. *57*, 113–127 (1967a)

Vendrig, J.C.: Steroid derivatives as native auxins in *Coleus*. Ann. N.Y. Acad. Sci. *144*, 81 (1967b)

Vendrig, J.C.: Further research on the phytohormones in urine. Z. Pflanzenphysiol. *64*, 297–308 (1971)

Venis, M.A.: Auxin-induced conjugation systems in peas. Plant Physiol. *49*, 24–27 (1972)

Verbiscar, A.J., Cragg, G., Geissman, T.A., Phinney, B.O.: Studies on the biosynthesis of gibberellins. II. The biosynthesis of gibberellins from (−)kaurenol, and the conversion of gibberellins ^{14}C-GA_4 and ^{14}C-GA_7 into ^{14}C-GA_3 by *Gibberella fujikuroi*. Phytochemistry *6*, 807–814 (1967)

Vickery, L.E., Purves, W.K.: Isolation of indole-3-ethanol oxidase from cucumber seedlings. Plant Physiol. *49*, 716–721 (1972)

Vizarova, G.: Thin-layer chromatography of auxin in barley cultivar 'Professor Schiemann" infected by powdery mildew. Biol. Plant. *15*, 270–273 (1973)

Vonk, C.R.: Studies on phloem exudation from *Yucca flaccida* HEW. XIII. Evidence for the occurrence of a cytokinin nucleoside in the exudate. Acta Bot. Neerl. *23*, 541–548 (1974)

Vonk, C.R.: Studies on phloem exudation from *Yucca flaccida* HEW. XIV. Metabolism of 8-^{14}C-zeatin in an excised inflorescence stalk in phloem exudate and in flower sap. Acta Bot. Neerl. *25*, 153–166 (1976)

Vreman, H.J., Skoog, F., Frihart, C.R., Leonhard, N.J.: Cytokinins in *Pisum* transfer ribonucleic acid. Plant Physiol. *49*, 848–851 (1972)

Vreman, H.J., Schmitz, R.Y., Skoog, F., Playtis, A.J., Frihart, C.R., Leonard, N.J.: Synthesis of 2-methylthio-cis- and trans-ribosylzeatin and their isolation from *Pisum* tRNA. Phytochemistry *13*, 31–37 (1974)

Vreman, H.J., Thomas, R., Corse, J., Swaminathan, S., Murai, N.: Cytokinins in tRNA obtained from *Spinacia oleracea* L. leaves and isolated chloroplasts. Plant Physiol. *61*, 296–306 (1978)

Wada, K.: Inhibition of gibberellin biosynthesis by geraniol derivatives and 17-nor-azakauranes. Agric. Biol. Chem. *42*, 787–791 (1978a)

Wada, K.: New gibberellin biosynthesis inhibitors, 1-N-decyl- and 1-geranylimidazole: Inhibitors of (−)-kaurene 19-oxidation. Agric. Biol. Chem. *42*, 2411–2413 (1978b)

Wada, K., Imai, T., Shibata, K.: Microbial production of unnatural gibberellins from (−)-kaurene derivatives in *Gibberella fujikuroi*. Agric. Biol. Chem. *43*, 1157–1158 (1979)

Wain, R.L.: Some developments in research on plant growth inhibitors. Proc. R. Soc. London Ser. B *191*, 335–352 (1975)

Walker, G.C., Leonard, N.J., Armstrong, D.J., Murai, N., Skoog, F.: The mode of incorporation of 6-benzylaminopurine into tobacco callus transfer ribonucleic acid. A double labeling determination. Plant Physiol. *54*, 737–743 (1974)

Walton, D.C., Sondheimer, E.: Metabolism of $2\text{-}^{14}C\text{-}(\pm)$-abscisic acid in excised bean axes. Plant Physiol. *49*, 285–289 (1972a)

Walton, D.C., Sondheimer, E.: Activity and metabolism of $^{14}C\text{-}(\pm)$-abscisic acid and derivatives. Plant Physiol. *49*, 290–292 (1972b)

Walton, D.C., Dorn, B., Fey, J.: The isolation of an abscisic-acid metabolite, 4′-dihydrophaseic acid, from non-imbibed *Phaseolus vulgaris* seed. Planta *112*, 87–90 (1973)

Walton, D.C., Harrison, M.A., Cote, P.: The effects of water stress on abscisic acid levels and metabolism in roots of *Phaseolus vulgaris* L. and other plants. Planta *131*, 141–144 (1976)

Wample, R.L., Durley, R.C., Pharis, R.P.: Metabolism of gibberellin A_4 by vegetative shoots of *Douglas fir* at three stages of ontogeny. Physiol. Plant. *35*, 273–278 (1975)

Wang, T.L., Horgan, R.: Dihydrozeatin riboside, a minor cytokinin from the leaves of *Phaseolus vulgaris* L. Planta *140*, 151–153 (1978)

Wang, T.L., Thompson, A.G., Horgan, R.: A cytokinin glucoside from the leaves of *Phaseolus vulgaris* L. Planta *135*, 285–288 (1977)

Wardale, D.A.: Effect of phenolic compounds in *Lycopersicon esculentum* on the synthesis of ethylene. Phytochemistry *12*, 1523–1530 (1973)

Wareing, P.F.: Abscisic acid as a natural growth regulator. Philos. Trans. R. Soc. London Ser. B *284*, 483–498 (1978)

Wareing, P.F., Ryback, G.: Abscisinsäure, eine neu entdeckte, das pflanzliche Wachstum regulierende Substanz. Endeavour *29*, 84–88 (1970)

Wareing, P.F., Horgan, R., Henson, J.E., Davis, W.: Cytokinin relations in the whole plant. In: Plant growth regulation. Pilet, P.E. (ed.), pp. 147–153, Berlin-Heidelberg-New York: Springer 1977

Watanabe, T., Kondo, N.: Isolation of a proteinaceous inhibitor of ethylene biosynthesis from marine algae. Agric. Biol. Chem. *40*, 1877–1878 (1976)

Weaver, G.M., Jackson, H.O.: Genetic differences in leaf abscission and the activity of naturally occurring growth regulators in peach. Can. J. Bot. *41*, 1405–1418 (1963)

Wenkert, E.: Structural and biogenetic relationships in the diterpene series. Chem. Ind. *1955*, 282–284 (1955)

West, C.A.: Biosynthesis of gibberellins. In: Biosynthesis and its control in plants. Milborrow, B.V. (ed.), pp. 143–169. New York, London: Academic Press 1973

Whitty, C.D., Hall, R.H.: A cytokinin oxidase in *Zea mays*. Can. J. Biochem. *52*, 789–799 (1974)

Wichner, S., Libbert, E.: Interactions between plants and epiphytic bacteria regarding their auxin metabolism. I. Detection of IAA-producing epiphytic bacteria and their role in long duration experiments on tryptophan metabolism in plant homogenates. Physiol. Plant. *21*, 227–241 (1968a)

Wichner, S., Libbert, E.: Interactions between plants and epiphytic bacteria regarding their auxin metabolism. II. Influence of IAA-producing epiphytic bacteria on short-term

IAA production from tryptophan in plant homogenates. Physiol. Plant. *21*, 500–509 (1968b)

Wightman, F.: Metabolism and biosynthesis of 3-indoleacetic acid and related indole compounds in plants. Can. J. Bot. *40*, 689–718 (1962)

Wightman, F.: Pathways of tryptophan metabolism in tomato plants. In: Regulateurs naturels de la croissance végétale. Nitsch, J.P. (ed.), Vol. 123, pp. 191–212. Paris: C.N.R.S. 1964

Wightman, F.: Biosynthesis of auxins in tomato shoots. Biochem. Soc. Symp. *38*, 247–275 (1973)

Wightman, F., Cohen, D.: Intermediary steps in the enzymatic conversion of tryptophan to IAA in cell-free systems from mung bean seedlings. In: Biochemistry and physiology of plant substances. Wightman, F., Setterfield, G. (eds.), pp. 273–288. Ottawa: Runge Press 1968

Wildman, S.G., Ferri, M.G., Bonner, J.: The enzymatic conversion of tryptophan to auxin by spinach leaves. Arch. Biochem. Biophys. *13*, 131–134 (1947)

Wills, R.B.H., Scott, K.J., Franklin, M.J.: Abscisic acid and the development of storage break down in apples. Phytochemistry *15*, 1817–1818 (1976)

Wilson, M.M., Gordon, M.E., Letham, D.S., Parker, C.W.: Regulators of cell division in plant tissues. XIX. The metabolism of 6-benzylaminopurine in radish cotyledons and seedlings. J. Exp. Bot. *25*, 725–732 (1974)

Winter, A.: A hypothetical route for the biogenesis of IAA. Planta *71*, 229–239 (1966)

Wood, H.N., Lin, H.C., Braun, A.C.: The inhibition of plant and animal adenosine 3,5-cyclic monophosphate phosphodiesterases by a cell-division promoting substance from tissues of higher plant species. Proc. Natl. Acad. Sci. USA *69*, 403–406 (1972)

Woolley, D.J., Wareing, P.F.: The role of roots, cytokinins and apical dominance in the control of lateral shoot form in *Solanum andigena*. Planta *105*, 33–42 (1972)

Wright, S.T.C.: Seasonal changes in the levels of free and bound abscisic acid in black currant (*Ribes nigrum*) buds and beech (*Fagus sylvatica*) buds. J. Exp. Bot. *26*, 161–174 (1975)

Wright, S.T.C.: The relationship between leaf water potential (ψ leaf) and the levels of abscisic acid and ethylene in excised wheat leaves. Planta *134*, 183–189 (1977)

Wylie, A.W., Ryuge, K., Sachs, R.M.: Effects of growth retardants on biosynthesis of gibberellin precursors in root tips of peas, *Pisum sativum* L. J. Am. Soc. Hortic. Sci. *95*, 627–630 (1970)

Yafin, Y., Shechter, I.: Comparison between biosynthesis of ent-kaurene in germinating tomato seeds and cell suspension cultures of tomato and tobacco. Plant Physiol. *56*, 671–675 (1975)

Yamaguchi, I., Yokota, T., Murofushi, N., Takahashi, N., Ogawa, Y.: Isolation of gibberellins A_5, A_{32}, A_{32} acetonide and (+)-abscisic acid from *Prunus persica*. Agric. Biol. Chem. *39*, 2399–2403 (1975a)

Yamaguchi, I., Yokota, T., Murofushi, N., Takahashi, N.: Structure elucidation of gibberellin A_{32} and its acetonide. Agric. Biol. Chem. *39*, 2405–2410 (1975b)

Yamane, H., Yamaguchi, I., Murofushi, N., Takahashi, N.: Isolation and structure of gibberellin A_{35} and its glucoside from immature seed of *Cytisus scoparius*. Agric. Biol. Chem. *35*, 1144–1146 (1971)

Yamane, H., Yamaguchi, I., Yokota, T., Murofushi, N., Takahashi, H., Katsumi, M.: Biological activities of new gibberellins A_{30}-A_{31} and A_{35}-glucoside. Phytochemistry *12*, 255–261 (1973)

Yamane, H., Yamaguchi, I., Murofushi, N., Takahashi, N.: Isolation and structures of gibberellin A_{35} and its glucoside from immature seed of *Cytisus scoparius*. Agric. Biol. Chem. *38*, 649–655 (1974)

Yamane, H., Murofushi, N., Takahashi, N.: Metabolism of gibberellins in maturing and germinating bean seeds. Phytochemistry *14*, 1195–1200 (1975)

Yamane, H., Murofushi, N., Osada, H., Takahashi, N.: Metabolism of gibberellins in early immature bean seeds. Phytochemistry *16*, 831–835 (1977)

Yamazaki, H., Yamazaki, I.: The reaction between indole-3-acetic acid and horseradish peroxidase. Arch. Biochem. Biophys. *154*, 147–159 (1973)

Yang, S.F.: Ethylene formation from methional by horseradish peroxidase. Arch. Biochem. Biophys. *122*, 401–407 (1967)

Yang, S.F.: Biosynthesis of ethylene. In: Biochemistry and physiology of plant growth substances. Wightman, F., Setterfield, G. (eds.), pp. 1217–1228. Ottawa: Runge Press 1968

Yang, S.F.: Further studies on ethylene formation from α-keto-γ-methylthiobutyric acid or β-methylthiopropionaldehyde by peroxidase in the presence of sulfite and oxygen. J. Biol. Chem. *244*, 4360–4365 (1969)

Yang, S.F.: The biochemistry of ethylene: Biogenesis and metabolism. In: The chemistry and biochemistry of plant hormones. Rec. Adv. Phytochem. Runeckles, V.C., Sondheimer, E., Walton, D.C. (eds.), Vol. VII, pp. 131–178. New York, London: Academic Press 1974

Yang, S.F., Baur, A.H.: Pathways of ethylene biosynthesis. Qual. Plant. Mater. Veg. *19*, 201–220 (1969)

Yang, S.F., Baur, A.H.: Biosynthesis of ethylene in fruit tissues. In: Plant growth substances 1970. Carr, D.J. (ed.), pp. 510–517. Berlin-Heidelberg-New York: Springer 1972

Yang, S.F., Saleh, M.A.: Destruction of indole-3-acetic acid during the aerobic oxidation of sulfite. Phytochemistry *12*, 1463–1466 (1973)

Yang, S.F., Ku, H.S., Pratt, H.K.: Photochemical production of ethylene from methionine and its analogues in the presence of flavin mononucleotide. J. Biol. Chem. *242*, 5274–5280 (1967)

Yokota, T., Takahashi, N., Murofushi, N., Tamura, S.: Structures of new gibberellin glucosides in immature seeds of *Pharbitis nil.* Tetrahedron Lett. 2081–2084 (1969a)

Yokota, T., Takahashi, N., Murofushi, N., Tamura, S.: Isolation of gibberellin A_{26} and A_{27} and their glucosides from immature seeds of *Pharbitis nil.* Planta *87*, 180–184 (1969b)

Yokota, T., Murofushi, N., Takahashi, N.: Structure of new gibberellin glucoside in immature seeds of *Pharbitis nil.* Tetrahedron Lett. 1489–1491 (1970)

Yokota, T., Murofushi, N., Takahashi, N., Katsumi, M.: Biological activities of gibberellins and their glucosides in *Pharbitis nil.* Phytochemistry *10*, 2943–2949 (1971a)

Yokota, T., Murofushi, N., Takahashi, N., Tamura, S.: Gibberellins in immature seeds of *Pharbitis nil.* Part III. Isolation and structures of gibberellin glucosides. Agric. Biol. Chem. *35*, 583–595 (1971b)

Yokota, T., Yamazaki, S., Takahashi, N., Iitaki, Y.: Structure of pharbitic acid, a gibberellin related diterpenoid. Tetrahedron Lett. 2957–2960 (1974)

Yokota, T., Yamane, H., Takahashi, N.: The synthesis of gibberethiones, gibberellin-related diterpenoids. Agric. Biol. Chem. *40*, 2507–2508 (1976)

Yokota, T., Kobayashi, S., Yamane, H., and Takahashi, N.: Isolation of a novel gibberellin glucoside, 3-O-β-D-glucopyranosyl-gibberellin A_1 from *Dolichos lablab* seed. Agric. Biol. Chem. *42*, 1811–1812 (1978)

Yomo, H., Iinuma, H.: Production of gibberellin-like substance in the embryo of barley during germination. Planta *71*, 113–118 (1966)

Yoneda, Y., Endo, T.: Effect of low concentration of hydrogen peroxide on indoleacetate oxidase zymogram in *Pharbitis nil.* Plant Cell Physiol. *10*, 235–237 (1969)

Yoneda, Y., Endo, T.: Peroxidase isoenzymes and their indoleacetate oxidase activity in the Japanese morning glory, *Pharbitis nil.* Plant Cell Physiol. *11*, 503–506 (1970)

Yoneda, Y., Stonier, T.: Elongation of stem internodes in the Japanese morning glory (*Pharbitis nil*) in relation to auxin destruction. Physiol. Plant. *19*, 977–981 (1966)

Yoneda, Y., Stonier, T.: Distribution of three auxin protector substances in seeds and shoots of the Japanese morning glory (*Pharbitis nil*). Plant Physiol. *42*, 1017–1020 (1967)

Yoshida, R., Oritani, T.: Cytokinin glucoside in roots of the rice plant. Plant Cell Physiol. *13*, 337–343 (1972)

Yoshida, R., Oritani, T., Nishi, A.: Kinetin-like factors in the root exudate of rice plants. Plant Cell Physiol. *12*, 89–94 (1971)

Yu, Y.B., Yang, S.F.: Auxin-induced ethylene production and its inhibition by aminoethoxyvinylglycine and cobalt ion. Plant Physiol. *64*, 1074–1077 (1979)

Yu, Y.B, Adams, D.O., Yang, S.F.: Regulation of auxin-induced ethylene production in mung bean hypocotyls. Role of 1-aminocyclopropane-1-carboxylic acid. Plant Physiol. *63*, 589–590 (1979a)

Yu, Y.B., Adams, D.O., Yang, S.F.: 1-Aminocyclopropane-carboxylate synthase, a key enzyme in ethylene biosynthesis. Arch. Biochem. Biophys. *198*, 280–286 (1979b)

Zauberman, G., Fuchs, Y.: Ripening processes in avocados stored in ethylene atmosphere in cold storage. J. Am. Soc. Hortic. Sci. *98*, 477–480 (1973)

Zeevaart, J.A.D.: Reduction of the gibberellin content of *Pharbitis* seeds by CCC and after-effects in the progeny. Plant Physiol. *41*, 856–862 (1966)

Zeevaart, J.A.D.: Effects of photoperiod on growth rate and endogenous gibberellins in long-day rosette plant spinach. Plant Physiol. *47*, 821–827 (1971a)

Zeevaart, J.A.D.: (+)-Abscisic acid content of spinach in relation to photoperiod and water stress. Plant Physiol. *48*, 86–90 (1971b)

Zeevaart, J.A.D.: Levels of (+)-abscisic acid and xanthoxin in spinach under different environmental conditions. Plant Physiol. *53*, 644–648 (1974)

Zeevaart, J.A.D.: Sites of abscisic acid synthesis and metabolism in *Ricinus communis* L. Plant Physiol. *59*, 788–791 (1977)

Zeevaart, J.A.D., Milborrow, B.V.: Metabolism of abscisic acid and the occurrence of epi-dihydrophaseic acid in *Phaseolus vulgaris*. Phytochemistry *15*, 493–500 (1976)

Zeevaart, J.A.D., Osborne, H.D.: Comparative effects of some Amo 1618 analogs on gibberellin production in *Fusarium moniliforme* and on growth in higher plants. Planta *66*, 320–330 (1965)

Zenk, M.H.: 1-(Indole-3-acetyl)-β-D-glucose, a new compound in the metabolism of indole-3-acetic acid in plants. Nature (London) *191*, 493–494 (1961a)

Zenk, M.H.: Darstellung und Eigenschaften aktivierter Verbindungen der Indol-3-essigsäure. Naturwissenschaften *48*, 455–456 (1961b)

Zenk, M.H.: Aufnahme und Stoffwechsel von α-Naphthylessigsäure durch Erbsenepikotyle. Planta *58*, 75–94 (1962)

Zenk, M.H.: Isolation, biosynthesis and function of indoleacetic acid conjugates. In: Régulateurs naturels de la croissance végétale. Nitsch, J.P. (ed.), pp. 241–249. Paris: C.N.R.S. 1964

Zeroni, M., Galil, T., Ben-Yehoshua, S.: Autoinhibition of ethylene formation in nonripening stages of the fruit of sycamore fig (*Ficus sycomorus* L.). Plant Physiol. *57*, 647–650 (1976)

Zimmerman, R.H., Lieberman, M., Broome, O.C.: Inhibitory effect of a rhizobitoxine analog on bud growth after release from dormancy. Plant Physiol. *59*, 158–160 (1977)

Zimmermann, H., Siegert, C., Karl, R.: Veränderungen des Wuchsstoffes Indolyl-3-essigsäure bei der Keimung von *Avena sativa* L. Z. Pflanzenphysiol. *80*, 225–235 (1976)

Zweig, G., Devay, J.E.: On the biosynthesis of gibberellins from carbon-^{14}C-substrates by *Fusarium moniliforme*. Mycologia *51*, 877 (1959)

5 Molecular and Subcellular Aspects of Hormone Action

J.L. STODDART and M.A. VENIS

5.1 Introduction

Throughout this chapter the terms hormone and growth regulator will be used interchangeably. The designation "hormone" was originally restricted by definition to compounds that are produced naturally and exert their action in tissues or cellular localities removed from the point of synthesis. Both terms are, however, widely used in the current literature on plant growth regulation and development to cover both natural and synthetic compounds.

5.1.1 Receptor Characteristics

Hormones in general, whether animal or plant, exert physiological or biochemical effects that are specific to their particular class. The active molecules must, therefore, be recognized by the cell and distinguished in some way from other compounds. There are several ways in which this selectivity could be achieved, but it has long been suspected that specific hormone-binding proteins (receptors) may fulfil this function. In animals the concept has received convincing experimental confirmation during the last 15 years and efforts to characterize similar systems in plants are now intensifying.

In a candidate receptor system, binding of the hormone to the macromolecule should satisfy certain minimum criteria: – a) Binding should be of high affinity. The saturation range of the receptor should be compatible with the range of hormone concentrations known to be active physiologically.

b) Binding specificity for different hormone analogues should be in approximate accordance with the relative biological activities of the compounds (but factors other than intrinsic site affinity can contribute towards overall activity).

c) Binding should be of finite capacity and should be reversible, although there may be considerable differences in the rates of association and dissociation of the complex.

d) Binding should lead to a hormone-specific response.

The last criterion – that of the receptor-dependent response – is always the most difficult to establish. In the absence of known receptor mutants it will prove particularly difficult to establish for plant hormones.

5.1.2 General Methodology of Receptor Studies

a) Analysis of Binding Data

From the law of mass action governing the equilibrium:

$$L+R \rightleftharpoons L-R$$

where L = ligand, R = Receptor, L−R = ligand−receptor complex, it can be shown that:

$$B=\frac{n K_a F}{1+K_a F}$$

$$=\frac{n F}{K_D+F} \qquad \text{EDSALL and WYMAN (1958)} \qquad (1)$$

where B = mol of ligand bound
F = concentration of free or unbound ligand
K_a = association constant
K_D = dissociation constant $=\frac{1}{K_a}$
n = number of binding sites

The most commonly used linear transformations of this equation for graphic representation of binding data are:

$$\frac{1}{B}=\frac{K_D}{nF}+\frac{1}{n} \qquad \text{Double reciprocal plot } \left(\frac{1}{B} \text{ vs. } \frac{1}{F}\right)$$

or

$$\frac{B}{F}=\frac{n}{K_D}-\frac{B}{K_D} \qquad \text{Scatchard plot } \left(\frac{B}{F} \text{ vs. B}\right)$$

The binding constants K_D and n can be obtained from the intercepts and slopes of the appropriate plots (Fig. 5.1). Deviations from simple linear plots can arise in a variety of circumstances, e.g., with cooperative interactions or the presence of multiple classes of binding sites, and there are numerous methods for extracting the binding constants in such situations. An example of such a method for the case of two sets of sites has been described by HUNSTON (1975): for a brief treatment of cooperative interactions, see DEMEYTS (1976).

b) Acquisition of Binding Data

Estimation of the binding constants K_D and n necessitates the measurement, under equilibrium conditions, of B, the amount of hormone bound to the receptor, and F, the concentration of unbound hormone, at a series of total hormone concentrations. The hormone must normally be available radio-labelled, and at a specific activity sufficiently great to permit binding measure-

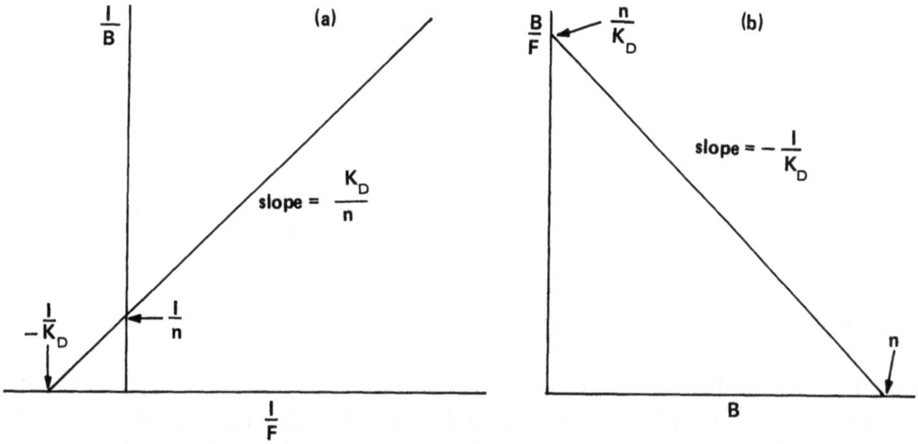

Fig. 5.1 a, b. Linear transformations of binding data. **a** Double reciprocal plot. **b** Scatchard plot

ments at concentrations well below receptor saturation. Usually, a fixed concentration of labelled hormone is used, together with increasing concentrations of the same, unlabelled hormone. The affinity of analogues not available in radioactive form can, under appropriate conditions, be estimated from their ability to displace the labelled hormone from the receptor sites.

For soluble receptors, the method of equilibrium dialysis is often used. The macromolecular receptor is placed on one side of a semi-permeable membrane (e.g., inside a dialysis bag) and dialysed against labelled hormone. At equilibrium, F is obtained from the label concentration in the "outside" solution, while B is derived from the difference in radioactivity (inside)–(outside). Dialysis half-cells are available commercially and for multiple assays they are more convenient than the original bag technique.

Bound and free hormone can be separated by gel filtration or by ultracentrifugation. Binding will be underestimated (or even impossible to determine) unless the rate of dissociation of the hormone-receptor complex is sufficiently slow. Both techniques can, however, be applied under equilibrium conditions by prerunning and eluting the gel filtration column with a fixed concentration of labelled hormone (HUMMEL and DREYER, 1962), or by centrifuging the receptor-hormone mixture through a gradient containing an appropriate uniform concentration of the radioactive hormone. Variations of ultrafiltration techniques have been used in ligand-binding studies. The method described by PAULUS (1966) is suitable for multiple samples and can detect low levels of binding.

In cases where dissociation of hormone from the receptor is slow, or when conditions can be manipulated to make it so, the use of charcoal to adsorb unbound hormone provides a convenient, rapid, and sensitive method of assay. To the labelled hormone-receptor mixture, charcoal or dextran-coated charcoal is added and after a brief period of mixing and incubation in the cold, the charcoal is removed by centrifugation. Receptor-bound hormone is then determined by counting aliquots of the supernatant. This method has been extensively applied in steroid receptor work (CLARK, J.H. et al., 1976).

Receptors in particulate preparations, such as membrane fractions, are conveniently studied by pelleting the particles from a suspension containing radioactive hormone. This is an equilibrium method and the amount of bound hormone is obtained by measurement of the radio-label in the pellet.

5.2 Auxins

5.2.1 Structure-Activity Relationships

Numerous attempts have been made to encompass the diverse groups of chemicals exhibiting auxin activity (see Chap. 1 and Fig. 5.2) within a unified structure-activity framework. The historical development of the different concepts has been ably discussed in the first series of the *Encyclopedia* (JÖNSSEN, 1961) and elsewhere (AUDUS, 1972). Not surprisingly, early ideas received progressive modification in the light of the discovery of new active chemicals, though later theories have often been in some respects more restrictive than their predecessors. In some of the earliest and most prescient discussions of structure-activity rules, VELDSTRA (1944a, b) proposed that the structures of the then-known auxins suggested, amongst other features, interaction of the molecules at a membrane surface by non-covalent forces. Over thirty years later, these key proposals – action at a membrane, and non-covalent interaction – are wholly in accord with current thinking on auxin action. Chemical reactivity theories – notably the nucleophilic attack (MUIR et al., 1949) and nucleophilic displacement (HANSCH et al., 1951) hypotheses, which proposed covalent linkages of the $-COOH$ group and an ortho-position on the auxin ring with $-NH_2$ and $-SH$ groups respectively of a protein – have never received satisfactory experimental support. There is every reason to believe that physiological activity resides in the auxin molecule per se, rather than in some product of auxin metabolism. Thus, while TULI and MOYED (1969) have claimed that the reactive intermediate in IAA action is an oxidation product, 3-methyleneoxindole (3-Menox), the activity of synthetic auxins such as NAA and 2,4-D could only be accommodated with difficulty (MOYED and WILLIAMSON, 1967), and other laboratories have failed to detect any auxin activity for 3-Menox (e.g., EVANS and RAY, 1973; HAMILTON et al., 1976).

No single structure-activity theory yet proposed can be regarded as totally satisfactory. One of the more recent proposals is the charge-separation theory of PORTER and THIMANN (1959, 1965) which suggests that active auxin molecules are characterized by the presence of a fractional positive charge situated at a distance of about 0.55 nm from the negatively charged carboxyl group (Fig. 5.3). This theory is not without its anomalies, but it does appear to accommodate most of the different classes of auxins, including the aliphatic dithiocarbamates (KERK et al., 1955), which were difficult to deal with under any preceding hypothesis (see discussions by THIMANN, 1969 and AUDUS, 1972). FARRIMOND et al. (1978) have re-evaluated the charge distribution in IAA and several other auxin molecules. They conclude that the charge separation concept is valid, but that

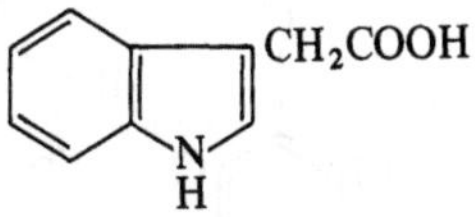

IAA
Indole–3–acetic acid

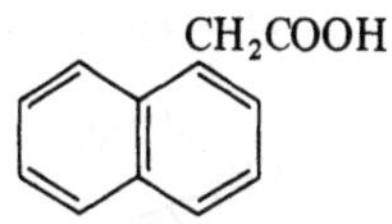

NAA
Naphthalene–1–acetic acid

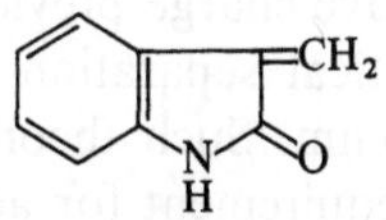

I–Menox
3–Methylene oxindole

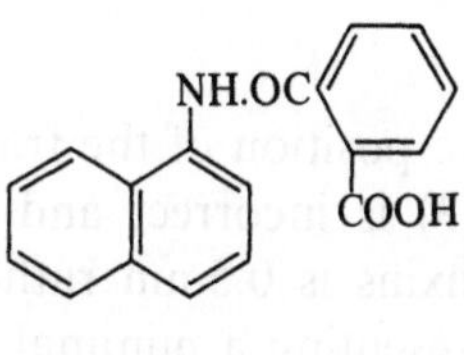

NPA
1–N–Naphthylphthal-amic acid

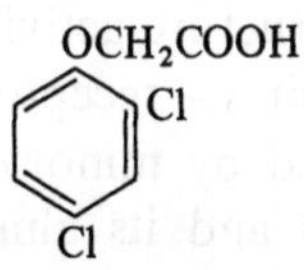

2,4–D
2,4–Dichlorophenoxy-acetic acid

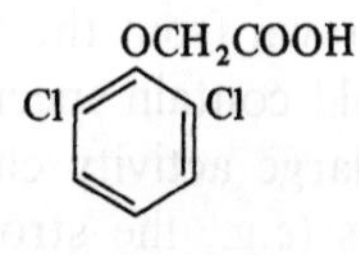

2,6–D
2,6–Dichloro-phenoxyacetic acid

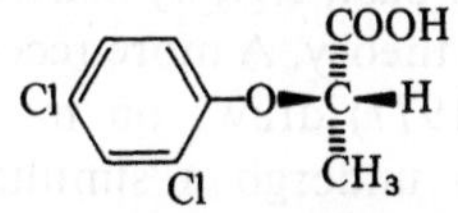

D(+)–2,4–DP

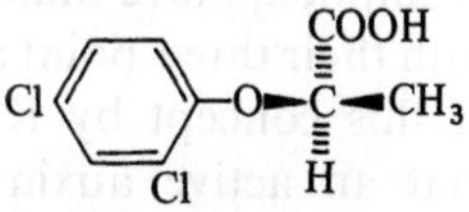

L(–)–2,4–DP

2,4–Dichlorophenoxy-propionates

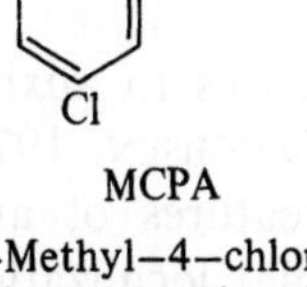

MCPA
2–Methyl–4–chloro-phenoxyacetic acid

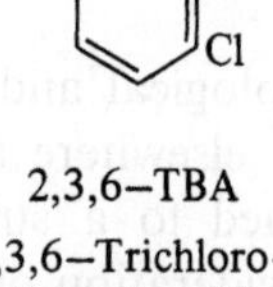

2,3,6–TBA
2,3,6–Trichloro-benzoic acid

Fig. 5.2. Structures of some auxin analogues and related compounds referred to in Section 5.2

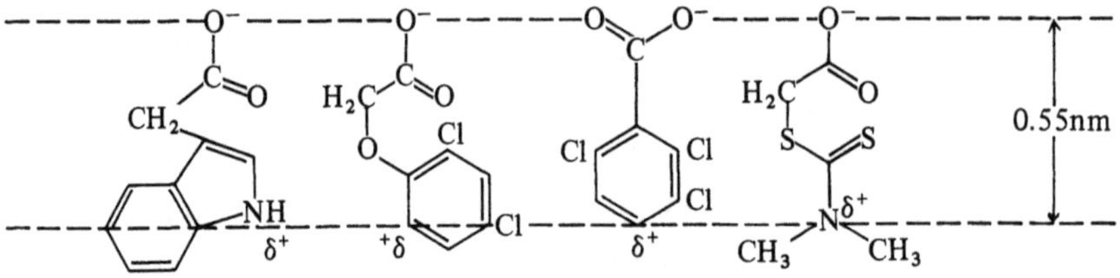

Fig. 5.3. Porter-Thimann charge-separation model for auxin activity. (After THIMANN, 1969)

the position of the fractional positive charge previously assigned for IAA (Fig. 5.3) is incorrect and that the critical separation distance common to active auxins is 0.5 nm rather than 0.55 nm. Such theories can only be regarded as presenting a minimal structural requirement for activity, since they are unable to reconcile the contrasting activities of enantiomeric auxins having identical charge-separation distances, e.g., the active D(+)- and inactive L(−)-phenoxypropionates (SMITH et al., 1952).

From the standpoint of the present chapter, the important conclusion to be drawn from the large body of structure-activity literature is that plant cells should contain specific recognition sites – receptors – for auxins. In particular, the large activity changes occasioned by minor structural variations within a series (e.g., the strong auxin, 2,4-D and its almost inactive analogue 2,6-D), and the differing behaviour of auxin optical isomers, alluded to above, imply a configurational recognition site with a precision that can probably only be accommodated in a protein molecule. The concept of an appropriate molecular "shape" for correct fit of the auxin molecule in a receptor site has been stated, or implied, more than once in the past, e.g., by SMITH et al. (1952) in connection with their three-point attachment theory. A more recent and explicit re-statement of this concept by KAETHNER (1977) draws on many earlier ideas to suggest that an active auxin is able to undergo a simultaneous re-orientation with a receptor site of defined shape and charge distribution. KATEKAR (1979) has also formulated structure-activity proposals based on postulates as to the nature of the auxin receptor site.

5.2.2 Potential Location of Auxin Receptors

Physiological and biochemical responses to auxins are dealt with in Chapter 6 and elsewhere (e.g., RAY, 1974; JACOBSEN, 1977). The present discussion is confined to a summary of those features of auxin action most relevant to a consideration of the likely subcellular localization of receptors.

During the 1960s in particular, a large body of evidence was accumulated to show that auxins are able to alter the amount and type of RNA and protein synthesis in treated tissues, to increase the activity and/or amount of RNA polymerases (particularly the nucleolar enzyme), and to enhance selectively the activity of other enzymes in a manner suggesting enzyme induction. In the case of cellulase, an enzyme that may be involved in cell-wall loosening during auxin-induced growth, the detailed studies of MACLACHLAN and co-workers have resulted in convincing evidence for the de novo synthesis of cellulase mRNA in auxin-treated pea epicotyls (VERMA et al., 1975). Investigations of this kind led to widespread acceptance of a transcriptional regulatory role for auxins.

This "gene activation hypothesis" has been challenged, largely as a result of the development of sensitive techniques for the continuous monitoring of the early events of auxin-induced growth. These investigations made it clear that the lag period preceding auxin-induced growth was in the range of only 8–15 min, and under some conditions could be much shorter (see Chapter 6 and EVANS, 1974, for a detailed review of rapid auxin responses). It was argued, probably correctly, that this early growth must precede any changes in RNA or protein synthesis. Coupled with these observations has been the renewed interest in H^+-induced growth (Chapter 6), particularly following the realization that the cuticle may present a considerable barrier to the passage of protons. Using "peeled" sections, optimum H^+-induced extension growth was observed at pH 5 (RAYLE, 1973) compared with the previously determined pH 3 optimum for sections with intact cuticles. In addition, such sections were found to excrete H^+ in response to auxin treatment, acidification of the medium being detectable in 20–30 min (CLELAND, 1973; RAYLE, 1973). These observations provided support for the "proton pump hypothesis" of HAGER et al. (1971), who had suggested, based on similarities between rapid H^+-induced and auxin-induced growth, that auxins activate a plasma membrane ATPase, thereby causing an outward pumping of protons from the cytoplasm to the cell wall. The resulting acidification of the wall is considered to increase cell-wall plasticity by means that remain the subject of debate.

While the gene activation and proton pump hypotheses represent very divergent viewpoints, they are not mutually exclusive. Thus, although the initial growth response to auxin is almost certainly too rapid to be dependent on new RNA and protein synthesis, any sustained response does require new protein synthesis, and over longer periods auxins can and do selectively alter macromolecular synthesis. Equally, it would now seem that for some tissues the growth data are not readily interpretable in terms of the H^+ pump theory (e.g., VANDERHOEF et al., 1977) and that acid-induced growth can account for only a part of the overall growth response to auxin. The observations of PENNY et al. (1972) and VANDERHOEF and STAHL (1975) may help to resolve the different viewpoints. Both groups of workers noted that if growth *rate* rather than total growth of excised sections is measured with time, then two phases in the auxin response are discernable. These phases can be distinguished by several criteria (e.g., the second, but not the first, response is eliminated by cycloheximide) and the normal dual-phase growth rate plot has been interpreted (VANDERHOEF and STAHL, 1975) as two overlapping growth responses, beginning about 12 min and 40 min after auxin application. The growth rate-time curve for acidic solutions showed only a single peak, resembling the first auxin response, thereby reinforcing earlier indications that H^+-induced growth mimics only a portion of the auxin response (RAYLE, 1973).

It would seem, therefore, that the overall growth curve observed following auxin application may represent the summation of a rapid response, frequently detectable within minutes and which is mimicked by low pH, and a second response that is blocked by cycloheximide and whose timing (35–45 min after auxin treatment) is consistent with a requirement for macromolecular synthesis. The timing of the first response and of other early auxin responses (EVANS, 1974), together with more recent direct observations of rapid auxin-induced changes in plasma membrane structure (MORRÉ and BRACKER, 1976; HELGERSON et al., 1976) are suggestive of an interaction with membrane-bound receptors. The second auxin growth response and longer-term effects on enzyme synthesis may reflect selective transcriptional changes which, by analogy with steroid hormones for example (see Sect. 5.4.2), can be expected to require the intervention of "soluble" (nuclear/cytoplasmic) receptors for the hormone.

Before discussing evidence for the existence of these different classes of auxin receptor, two points require brief mention. First, the proton pump model of auxin action requires activation of a membrane-bound ATPase. Such activation was first claimed by KASAMO and YAMAKI (1974), but for reasons discussed elsewhere (VENIS, 1977a) the data are less than convincing. More recently, ERDEI et al. (1979) have reported up to 50% stimulation of a microsomal ATPase of rice roots by IAA or 2,4-D (10^{-10}–10^{-8} M), both in vivo and in vitro. It is not yet clear whether this is an auxin-specific response, as no physiologically inactive analogues were tested. The fungal toxin fusicoccin (a diterpene glucoside, structure see p. 534), which promotes growth and H^+ secretion in a manner similar though not identical to that of auxin, has also been reported (BEFFAGNA et al., 1977) to stimulate (in vivo and in vitro) a plasmalemma K^+-ATPase in maize coleoptiles and spinach leaves by about 30%. The effects await confirmation, but they were obtained in a large number of experiments, using physiologically effective concentrations (1–10 μM), and an inactive analogue of fusicoccin was without effect.

The second point is that any attempt to investigate the association of plant hormone receptors with plasmalemma is hindered by the absence of any unambigous marker for the plasma membrane in plants. As noted by QUAIL and BROWNING (1977), all the commonly used criteria (e.g., K^+-ATPase, glucan synthetase II, high sterol:phospholipid ratio) are themselves correlates of the phosphotungstate-chromate staining procedure, the specificity of which is open to question. Attempts to apply surface labelling techniques to excised tissue sections have been complicated by evidence of spurious labelling (QUAIL and BROWNING, 1977 and references therein). Recently, ANDERSON and RAY (1978) claim to have overcome this problem by using an enzyme substrate as the label. Radioactive UDP-glucose supplied to pea stem slices was incorporated into a 1,3-glucan associated with membranes whose isopycnic distribution was similar to that of the putative plasma membrane marker K^+-ATPase. Another approach, currently being investigated in several laboratories, would be to apply general surface labelling procedures to isolated plant protoplasts. GALBRAITH and NORTHCOTE (1977) have described labelling of soybean protoplasts with diazotized sulphanilic acid. When the lysed protoplasts were fractionated on isopycnic gradients, the distribution of radioactivity and enzyme markers suggested an association of Mg^{2+}-ATPase and perhaps acid phosphodiesterase with the plasma membrane. Notwithstanding these promising but preliminary developments, reference in subsequent sections to "plasma membrane fractions" or some similar phrase should be regarded as shorthand for "fractions enriched in marker activities that are generally regarded as criteria of plasma membrane content."

5.2.3 Nuclear and Cytoplasmic Receptors

a) Auxin Mediator Proteins

The investigations discussed under this heading describe preparations that mediate an auxin reponse, but for which actual auxin binding has not been demonstrated.

MATTHYSSE and PHILLIPS (1969) reported that RNA synthesis by tobacco or soybean nuclei was stimulated by addition of the auxin, 2,4-D, provided that 2,4-D was also included in the nuclear isolation medium. If the nuclei were sedimented from a medium lacking 2,4-D, then RNA synthesis by the nuclei was unaffected by subsequent addition of the auxin, but auxin sensitivity could be restored by adding back the supernatant fraction obtained from the nuclear centrifugation. It was inferred that, in the absence of 2,4-D, some factor required for the auxin response was lost from the nuclei. This factor, apparently protein, could be partially purified either from the nuclear lysates or from post-chromatin supernatant of pea buds. The factor was also active, in the presence of 2,4-D, in stimulating RNA synthesis by exogenous polymerase with pea bud chromatin as template, but not with purified DNA. It was claimed that this effect was auxin-specific and that the stimulation was maintained when saturating levels of *E. coli* polymerase were added, suggesting that the chromatin template had been de-repressed. However, no data were presented in support of these statements. Factors prepared from different tissues of peas showed somewhat greater activity on chromatin derived from the same tissue (MATTHYSSE, 1970). No further elaboration of this system has been described, although a purification procedure similar to that of MATTHYSSE and PHILLIPS (1969) has been used in the isolation of a receptor from coconuts, discussed later.

Treatment of soybean hypocotyls with 2,4-D for 4–24 h before harvesting causes a large increase in the activity of chromatin-bound RNA polymerase (O'BRIEN et al., 1968), the increase being due to enhancement of the nucleolar enzyme, RNA polymerase I (HARDIN and CHERRY, 1972). A protein fraction prepared from the cotyledons stimulated polymerase from control, but not from 2,4-D-treated tissue (HARDIN et al., 1970). It was suggested that this protein might mediate the action of the hormone in vivo, so that polymerase from auxin-treated tissue, having been already "activated" by the hormone-receptor complex, would be unresponsive to in vitro addition of the factor. The degree of stimulation by the factor in the cell-free system (32%) was very much lower than that obtained following treatment with 2,4-D in vivo (over 200%).

A comparable situation has been described for lentil root nucleolar RNA polymerase, the activity of which is doubled in tissue treated with IAA (TEISSERE

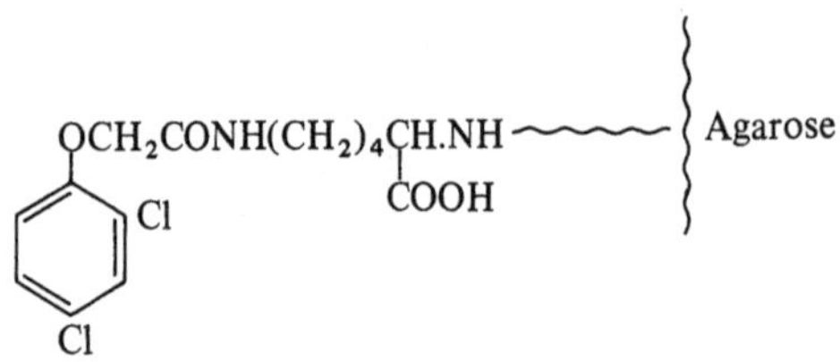

Fig. 5.4. 2,4-D-lysyl affinity adsorbent

et al., 1973). Fractionation of the non-histone chromosomal proteins yielded four fractions that stimulated activity of the polymerase (TEISSERE et al., 1975), two of which were studied in more detail and appeared to be initiation factors. It was claimed that the "level" of one of the factors, γ, was doubled in auxin-treated tissue compared with control tissue. It is possible that γ could bear some relationship to the soybean factor of HARDIN et al. (1970), although the soybean protein was derived from the high-speed supernatant fraction, rather than the chromosomal pellet.

Attempts have been made to isolate auxin receptors by affinity chromatography, using the ε-L-lysine derivatives of 2,4-D or IAA coupled (Fig. 5.4) to cyanogen bromide-activated agarose (VENIS, 1971). When crude supernatants prepared from pea or maize shoots were passed over such columns, small amounts of protein were retained and could be eluted by various means, typically with 1 M NaCl followed by 2 mM KOH. The fractions obtained were tested for their effects on DNA-dependent RNA synthesis, supported by *E. coli* polymerase. Fractions eluted by NaCl were invariably inhibitory, while KOH fractions promoted RNA synthesis by 40–200% in different preparations. Stimulation was not dependent on the addition of auxin to the reaction mixture. Activity of the KOH fraction was unchanged by dialysis, but completely destroyed by freezing or by brief heating. Various trivial explanations for factor activity were investigated and eliminated. From time-course experiments and the effects of rifampicin it was suggested that the factor was acting on RNA chain initiation. However, a recent report (KESSLER and HARTMANN, 1977) that rifampicin only *appears* to act as an inhibitor of initiation because of its much reduced affinity for free polymerase, relative to polymerase attached to DNA, renders this conclusion less certain.

To account for the lack of an auxin requirement for stimulation of RNA synthesis by the affinity column protein, it was suggested (VENIS, 1971) either that passage through the affinity column may transform the factor to an active configuration in which further contact with auxin is not required or, that in vivo, auxin may simply permit activity to be expressed by transporting an inherently active regulatory protein from the cytoplasm to the nucleus (HARDIN et al., 1970). Alternatively, a regulatory subunit might be stripped off the protein during passage through the column, as reported for affinity chromatography of cyclic AMP-dependent protein kinase (WILCHEK et al., 1971).

Binding of labelled auxins by the active fractions could not be convincingly demonstrated by equilibrium dialysis. Again, it was possible that passage through the column alters the configuration of the factor to one in which auxin is not bound, either by removal of a binding subunit or through inactivation of the binding site under the somewhat harsh (2 mM KOH) conditions of elution.

In more recent years, the extensive application of affinity chromatography in numerous areas has pointed up many of its limitations and would suggest that it is more advantageous to apply the technique at a point where some biological activity, e.g., binding, is already demonstrable, rather than to fish around optimistically in a crude extract. Nevertheless the auxin affinity columns described have found application in other laboratories. BISWAS and co-workers (BISWAS et al., 1975) have reported that auxin receptors from coconuts (discussed in the next section) can be isolated by chromatography on the IAA-lysine column. RIZZO et al. (1977), using the 2,4-D-lysine column and preparation methods identical to those previously outlined (VENIS, 1971), have obtained a similar though more active factor from soybean hypocotyls. DNA-dependent RNA synthesis with *E. coli* polymerase was stimulated two- to seven-fold by the KOH-eluted fraction. Stimulation was also observed, in the presence of 0.1 μM 2,4-D, using soybean polymerase I, but not with the nucleoplasmic enzyme polymerase II. The effect (25–80% stimulation) was much smaller than that obtained with the bacterial enzyme, perhaps because of the impure nature of the soybean enzyme. It may be that the factor is involved in mediating the previously discussed enhancement of RNA polymerase activity following treatment of soybeans with 2,4-D (O'BRIEN et al., 1968).

b) Auxin-Binding Proteins

Using a procedure broadly similar to that of MATTHYSSE and PHILLIPS (1969), BISWAS and co-workers have outlined the preparation from young coconut endosperm nuclei of a receptor protein, n-IRP (MONDAL et al., 1972; BISWAS et al., 1975; ROY and BISWAS, 1977). In a completely homologous system, containing nucleoplasmic RNA polymerase, DNA, and initiation factor, RNA synthesis was doubled by n-IRP, provided that IAA (1 μM) was also present. Most of the enhanced RNA synthesis was the result of increased chain initiation. Hybridization and gel electrophoretic analysis of reaction products suggested that new kinds of RNA were produced in the stimulated reaction. n-IRP itself binds to (^{3}H)-DNA, as judged by retention of the label on membrane filters, but binding is further enhanced by IAA, at an optimum concentration of 0.1 μM. IRP is apparently a single polypeptide of 100,000 daltons. Equilibrium dialysis data, performed at one ligand concentration only, have been interpreted as showing that IRP binds IAA and 2,4-D, but not the inactive compound benzoic acid. From a fuller kinetic analysis of IAA binding, the dissociation constant, K_D, was estimated at 7.5 μM with 0.5 binding sites per protein molecule. Since the binding protein was thought to be homogeneous, the latter figure was taken to indicate either a requirement for two IRP molecules to bind one molecule of IAA, or more probably, partial inactivation of IRP during purification.

The most recent report (ROY and BISWAS, 1977) describes purification of n-IRP on the IAA-lysine affinity column, using 1 M KSCN elution in place of 2 mM KOH. Additionally, this paper deals with the similar affinity purification and properties of a chromosomal IAA receptor protein, c-IRP, derived from coconut chromatin. Binding was assayed by a dextran-charcoal method, and Scatchard analysis indicated two sets of IAA-binding sites, with K_D values of 58 nM and 8.15 μM. The latter is referred to as "non-specific binding", though the dissociation constant is in fact very similar to that of n-IRP (7.5 μM). The methods description of c-IRP isolation states that the crude material was passed through CM-cellulose (which would remove any n-IRP present; MONDAL et al., 1972) prior to affinity chromatography. However, the legend to the affinity column profile indicates that the total non-histone protein fraction obtained after chromatin solubilization and re-association was used. Furthermore, the column eluates were assayed for binding at 5 μM IAA, which would be expected to detect only the lower affinity binding site. It is also unclear as to whether the Scatchard analysis was carried out using crude or purified c-IRP. For all these reasons, it is difficult to decide whether the lower-affinity c-IRP site represents contamination by n-IRP. The ^{14}C-IAA-n-IRP complex was reported to bind to coconut chromatin and to cause a twofold increase in chromatin transcription by *E. coli* RNA polymerase. RNA synthesis supported by chick erythrocyte chromatin was unaffected. The model proposed, by analogy with steroid receptors (see Sect. 5.4.2), is that the high affinity c-IRP in chromatin acts as an acceptor site for the n-IRP-IAA complex to trigger specific gene transcription.

Ostensibly, the reported properties of the coconut receptors make them the best candidates for genuine auxin receptors in existence. It is therefore unfortunate that many of the claims made for n-IRP have only been outlined in conference proceedings (BISWAS et al., 1975) and that experimental details elsewhere are also often unsatisfactory. For example, nowhere is it possible to deduce how much starting material is required to produce a given quantity of either receptor. The auxin-specificity of c-IRP binding has not been examined at all, and the data for n-IRP (single point figures with only two compounds other than IAA) are inadequate. Nor does it appear that any compound other than IAA has been tested in relation to the reported effects on RNA synthesis. Additional criticisms of the data have been made by KENDE and GARDNER (1976). It is also unfortunate, of course, that immature coconut endosperm is such an inaccessible source material. Publication of adequate experimental details might, nevertheless, render possible independent evaluation of the data, or at least permit the application of similar techniques to more readily available plant material.

Charcoal or dextran-charcoal techniques have also been applied in three other reports on IAA binding in soluble systems:

i) LIKHOLAT et al. (1974) found that IAA added to the homogenization and wash media enhanced RNA synthesis by wheat coleoptile chromatin, but was without effect if added only to the in vitro reaction mixture. This situation is comparable to that found by MATTHYSSE and PHILLIPS (1969) for nuclei. Binding of IAA-^{14}C by the cytosol fraction was detected after addition of dextran-coated charcoal to remove the unbound auxin and was about five times higher in extracts from 36-h-old, rather than from 72-h-old, plants. These and other data on stimulation of chromatin RNA synthesis by membrane fractions incubated with IAA, were interpreted in terms of a transition of the auxin receptors from a "soluble" to a membrane-bound state during passage of the cells from a dividing (36-h) to an elongating (72-h) phase. The membrane effects on RNA synthesis are somewhat similar to those described in an earlier report (discussed in Sect. 5.2.4.b) by HARDIN et al. (1972). Auxin binding by the cytosol fraction was saturable upon addition of unlabelled IAA, but no attempt to estimate the binding parameters was made.

ii) OOSTROM et al. (1975) used a dextran-charcoal method to detect high affinity auxin binding in the cytosol of tobacco pith. Using IAA-^{3}H of high specific activity, binding sites with a K_D of 10 nM were found in extracts from cultured pith explants, whereas extracts derived from freshly excised pith showed only low affinity binding. The concentration of binding sites was extremely low (ca. 0.01 p mol/g) suggesting losses or inactivation during extraction. As yet, no information on binding specificity is available.

iii) IHL (1976) has reported binding of IAA to a 100,000 *g* supernatant fraction from soybean cotyledons, using either gel filtration or charcoal adsorption to remove free IAA. Unfortunately, most of the data were obtained by feeding IAA-^{14}C in vivo and there is no evidence to indicate that the radio-label associated with the macromolecular fraction is in fact IAA, or that the binding is reversible in vitro. Similarly, binding saturation by unlabelled IAA or other analogues was determined by feeding tissue 10 μM IAA-^{14}C with or without a large excess (2.75 mM) of the cold compound. The observed reduction in macromolecular binding could therefore represent, amongst other things, competition for uptake.

Finally in this section, mention should be made of the report by WARDROP and POLYA (1977) of a soluble, high-affinity auxin-binding protein from bean leaves. The assay procedure consisted of incubating a crude protein fraction with IAA-^{14}C or NAA and then measuring radio-label in the precipitate obtained at 90% ammonium sulphate saturation. The results described could be repeated when this assay method was followed, but a more stringent assay avoiding protein precipitation (equilibrium dialysis in the absence of ammonium sulphate) resulted in no saturable binding whatsoever (VENIS, unpublished data). Equilibrium dialysis in the presence of ammonium sulphate yielded apparently saturable binding, as with the pellet assay. It was felt, therefore, that the binding observed by WARDROP and POLYA (1977) was an artefact arising from protein precipitation by ammonium sulphate.

5.2.4 Membrane-Bound Receptors

a) The Corn Coleoptile System

The publication in 1971 and 1972 of two papers by HERTEL and co-workers (LEMBI et al., 1971; HERTEL et al., 1972) on binding of growth regulators to corn coleoptile membranes provided a welcome stimulus to research on plant hormone receptors. The corn coleoptile system has proved to be convenient and reproducible and, although binding to membranes from other plants has been reported, corn coleoptiles remain, at the time of writing, the source material of choice.

i) General Characterization of the Binding Sites. The first paper in this field (LEMBI et al., 1971) dealt with binding of naphthylphthalamic acid (NPA Fig. 5.2), a potent synthetic inhibitor of polar auxin transport. Membrane preparations from corn coleoptiles were fractionated on sucrose gradients and their ability to bind NPA-^{3}H was assayed by a pelleting technique. Binding of NPA was correlated with plasma membrane content of the fractions as determined by the phosphotungstate-chromate staining method. A similar conclusion has been reached more recently (NORMAND et al., 1975; RAY, 1977) by correlation of binding with glucan synthetase II on sucrose gradients. The K_D for binding of NPA to the main class of sites is 0.22 μM, though a smaller fraction of even higher affinity sites ($K_D = 13$ nM) was also detected (THOMSON, 1972). The kinetics of the association and dissociation reactions have been studied by TRILLMICH and MICHALKE (1979). The NPA sites are apparently distinct from auxin-binding sites, since IAA, NAA, and 2,4-D did not compete (THOMSON, 1972). Neither was competition observed with TIBA, another transport inhibitor, nor with GA_3 nor abscisic acid. The affinity of NPA for the bulk of the binding sites ($K_D = 0.22$ μM) is well correlated with the saturation kinetics of its effect on auxin transport (THOMSON et al., 1973). The binding sites have recently been solubilized from the membranes using the detergent Triton X-100 (GARDNER and SUSSMAN, 1979).

Several NPA derivatives were found to displace NPA-^{3}H from the binding sites, in a manner that was reasonably compatible with their relative activities as transport inhibitors (THOMSON and LEOPOLD, 1974). In addition, various morphactins, another group of synthetic auxin transport inhibitors, also displaced NPA-^{3}H. The interaction of one of the morphactins was established as competitive by double reciprocal analysis, with a K_i value (0.29 μM) about the same as the K_D for NPA.

The function of these binding sites, which recognize two groups of unnatural growth regulators but not any of the known plant hormones tested, remains a mystery. It is possible, as suggested by THOMSON (1972), that their presence signifies an, as yet, unidentified natural hormone.

Using procedures similar to those employed with NPA, it was possible to detect saturable binding of NAA-^{14}C and IAA-^{3}H to heterogeneous membrane preparations from corn coleoptiles (HERTEL et al., 1972). Radioactive NAA in the pellets was displaceable by unlabelled auxins such as IAA and 2,4-D, but not by the inactive compound, benzoic acid. Abscisic acid, GA_3, and NPA also failed to interact with the binding sites. The dissociation constants were estimated as 1–2 μM for NAA and 3–4 μM for IAA. Binding activity was destroyed by heating or by treatment with sodium lauryl sulphate, and could be largely separated from nuclear and mitochondrial markers. The same system was investigated further by NORMAND et al. (1975), who obtained a K_D for NAA of 1.66 μM and showed that maximum NAA binding was associated with membranes that band on sucrose gradients at a lower density than the NPA-binding peak. Changes in the original procedure, in particular prior separation of the membranes from the supernatant by pelleting and re-suspen-

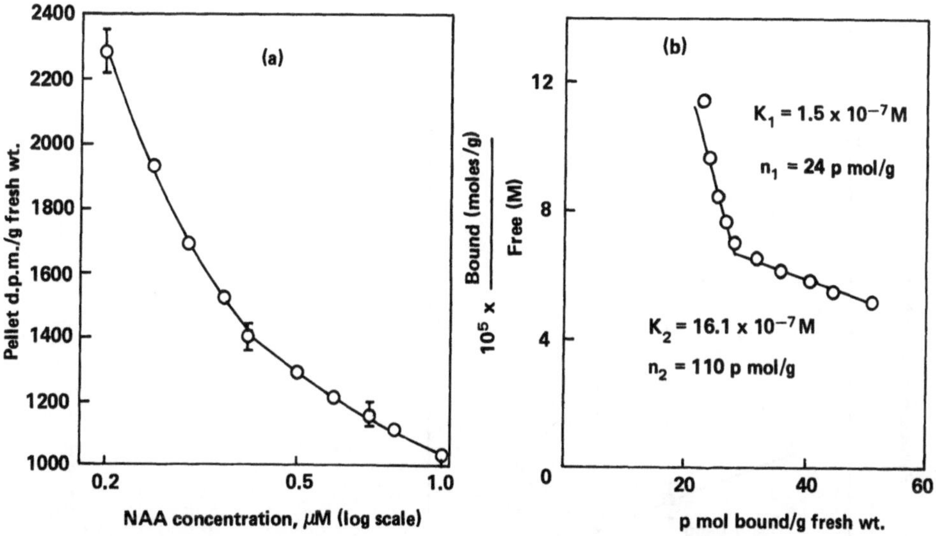

Fig. 5.5a, b. Binding of NAA-^{14}C by corn membranes. **a** displacement of NAA-^{14}C in membrane pellets by unlabelled NAA. **b** Scatchard-Hunston analysis of the binding data. (BATT et al., 1976)

sion, yielded improved binding data and a revised K_D for NAA of 0.5 μM (HERTEL, 1974).

In the original studies of HERTEL et al. (1972), displacement by unlabelled NAA of NAA-^{14}C binding to the membrane pellets was examined over an extended concentration range (0.2–200 μM). When binding was investigated in detail over a more restricted concentration range, 0.2–1 μM (BATT et al., 1976), Scatchard analysis of the data (Fig. 5.5) yielded a biphasic plot, suggesting the presence within the membrane preparation of at least two sets of high affinity binding sites for NAA, with dissociation constants of 0.15 μM (site 1) and 1.6 μM (site 2). IAA-^{14}C also appeared to bind to two sets of sites ($K_1 = 1.7$ μM; $K_2 = 5.8$ μM), whose concentrations were in good agreement with those deduced for NAA (n_1, n_2, in Fig. 5.5). The binding specificity of each class of sites was examined by double reciprocal analysis of complete NAA-binding curves performed in the presence or absence of a fixed concentration of a given auxin analogue. From a large number of such experiments with a range of auxin analogues it was found that all compounds tested, including inactive analogues such as benzoic acid, were able to compete with NAA for site 1 binding. Site 2 on the other hand exhibited binding specificity compatible with the expected properties of an auxin receptor as defined previously by HERTEL et al. (1972), in that only active auxins, anti-auxins or transport inhibitors were able to compete with NAA for the binding sites.

The discrete nature of the two sets of NAA-binding sites was further suggested by experiments in which substantial resolution of site 1 and site 2 binding was achieved (BATT and VENIS, 1976). By differential centrifugation of the membrane preparation it was found that the 4,000–10,000 *g* fraction contained a single population of NAA-binding sites (linear Scatchard plot)

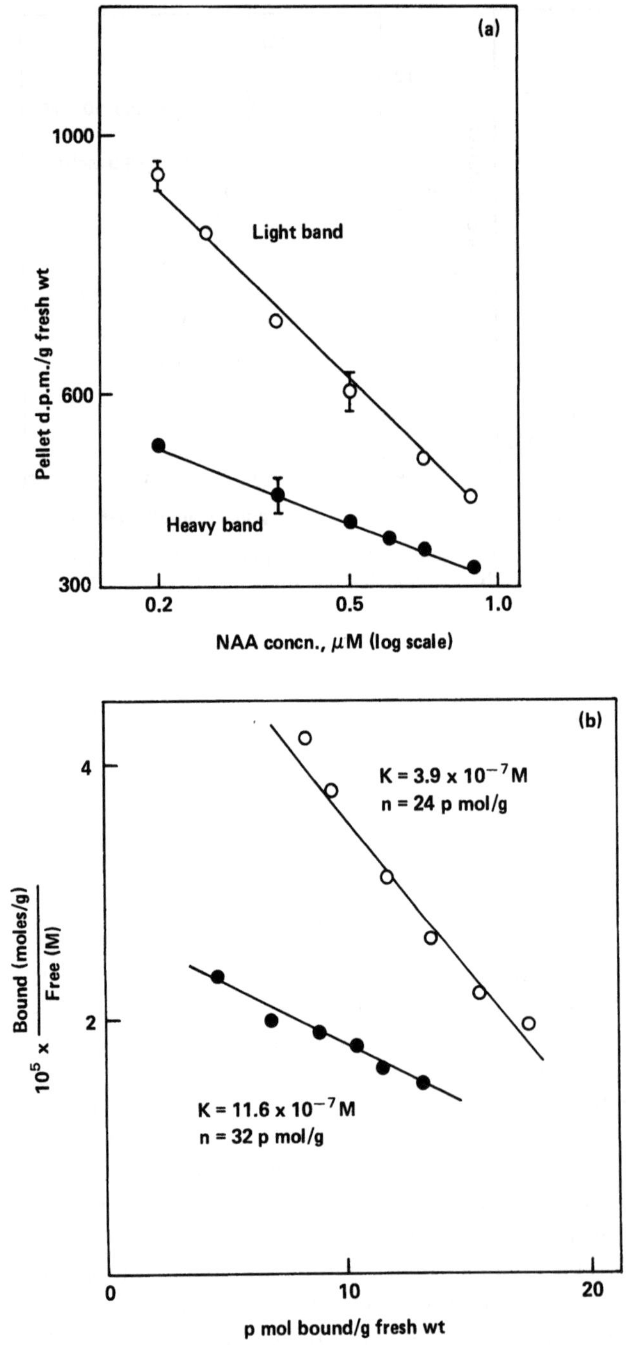

Fig. 5.6a, b. Binding of NAA-^{14}C by corn membrane fractions separated on sucrose gradients. **a** pellet radioactivity as a function of NAA concentration. **b** Scatchard analysis of the binding data. (BATT and VENIS, 1976)

with a $K_D = 1.6\,\mu M$, characteristic of site 2, the auxin-specific site. The 10,000–38,000 *g* fraction, on the other hand, still yielded a biphasic Scatchard plot and therefore appeared to contain both sets of binding sites. The total membrane preparation (4,000–38,000 *g*) could be fractionated on discontinuous sucrose gradients into two bands. The kinetics of NAA-^{14}C binding by these bands (Fig. 5.6) suggested that one distinct set of binding sites predominated in each band. The dissociation constants, $0.38\,\mu M$ for the light band, and $1.2\,\mu M$ for the heavy band, were in reasonable agreement with those determined previously for site 1 and site 2 respectively in unfractionated membrane preparations, implying that substantial resolution of the binding sites had been obtained. The binding specificities of the two membrane bands were also compatible with this conclusion. By analysis of binding activities and enzymic and chemical markers in membrane fractions from multi-step sucrose gradients, it was suggested that site 1 is located in endoplasmic reticulum (ER) or Golgi membranes, while site 2 is associated with plasma membrane fractions.

RAY et al. (1977a) could distinguish only a single kinetic class of binding site, with a K_D of 0.5–$0.7\,\mu M$ for NAA. From the distribution of binding activity on continuous sucrose gradients, RAY (1977) concluded that the bulk of the NAA-binding sites are associated with ER, while a minority of the sites may be located on Golgi and/or plasma membranes. More recently, however, DOHRMANN et al. (1978) have reported evidence for *three* types of auxin-binding sites in the membranes, based on the differential affinities of sucrose gradient fractions for different auxin analogues. These have been termed site I (associated with ER), site II (associated with tonoplast membranes and distinguished by low affinity for the weak auxin, phenylacetic acid) and site III, found in plasma membrane fractions and characterized by preferential binding of 2,4-D. Site I and site II could be distinguished kinetically, either by assaying in the presence (site II) or absence (site I) of phenylacetic acid or, as described by BATT and VENIS (1976), by assaying membrane fractions resolved on sucrose gradients. The affinities of NAA for sites I and II were in reasonable agreement with the values for the sites designated by BATT et al. (1976) as site 1 and site 2 (Table 5.1).

Site III has been studied in greater detail in zucchini hypocotyls (JACOBS and HERTEL, 1978). IAA binds to this site with a K_D of $1.5\,\mu M$ and curiously, the binding is enhanced by TIBA and NPA, inhibitors of auxin transport. From the effects of these compounds, the apparent site localization in the plasma membrane and from the correlation between transport inhibition and binding competition by auxin analogues, it was suggested that site III may represent an auxin transport site. The major anomaly is NAA, which strongly inhibits IAA transport, but competes very poorly for site III binding. In addition, the 2-NAA isomer binds with very much greater affinity than would be expected from its activity as a transport inhibitor.

Corn coleoptiles contain a low molecular weight modifier of auxin binding, termed supernatant factor (SF), which reduces the binding affinity of the membrane sites for NAA (RAY et al., 1977a, b). Removal of SF is largely responsible for the improved NAA binding obtained by washing the membrane preparations. It was reported that the affinity for certain physiologically inactive ligands

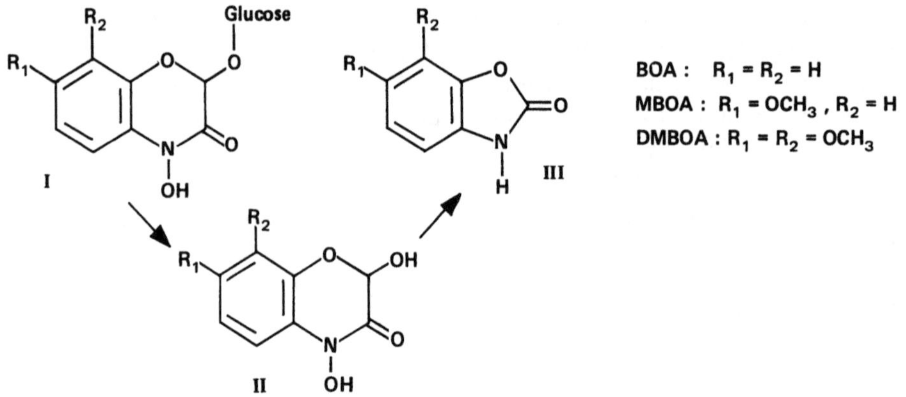

Fig. 5.7. Benzoxazolinones and precursors in corn

is reduced to an even greater extent, while the binding affinity for some auxin analogues, e.g., 2,4-D, is actually increased by SF. The effect of SF appears to be mainly on site I (DOHRMANN et al., 1978) and it has been suggested that it may function as a natural regulator of auxin activity. SF has now been identified (Venis and Watson, 1978) as a mixture of 6-methoxy-2-benzoxazolinone (MBOA) and 6,7-dimethoxy-2-benzoxazolinone (DMBOA, Fig. 5.7). The parent compound 2-benzoxazolinone (BOA) was also found, but in much smaller amount. These are known natural products in corn and in certain other Gramineae, and have been implicated in resistance of cereals to fungi, insects and chloro-s-triazine herbicides (KLUN et al., 1970 and references therein). In the intact plant the compounds occur as glucosides (I), from which the benzoxazinones (II) are released enzymically upon cell damage, e.g., by pathogenic attack or by homogenization. On warming in dilute aqueous solution the aglucones undergo ring contraction to produce the benzoxazolinones III (Fig. 5.7).

MBOA and DMBOA were isolated by following inhibition of NAA-receptor binding during purification, a convenient measure of SF activity. Both compounds are active in such assays (Fig. 5.8), DMBOA being about 50 times more active than MBOA; BOA is only weakly active. The compounds were also found to inhibit auxin-induced growth of oat coleoptiles and the relative activity of BOA, MBOA, and DMBOA in such assays (Fig. 5.9) correlates well with their inhibitory action on auxin-receptor binding (Fig. 5.8). In view of the very restricted species distribution of benzoxazolinone precursors (see references in VENIS and WATSON, 1978), the compounds cannot have a general role in auxin physiology. Nevertheless, the good correlation between the activity of the benzoxazolinones in inhibiting on the one hand auxin-induced growth (Fig. 5.9) and on the other hand auxin-binding site interaction (Fig. 5.8) is significant in providing the first direct evidence that these binding sites represent physiological receptor sites for auxin action.

Corn coleoptile membranes are also able to bind radioactive fusicoccin with high affinity (DOHRMANN et al., 1977). The K_D was estimated as 2 nM, though

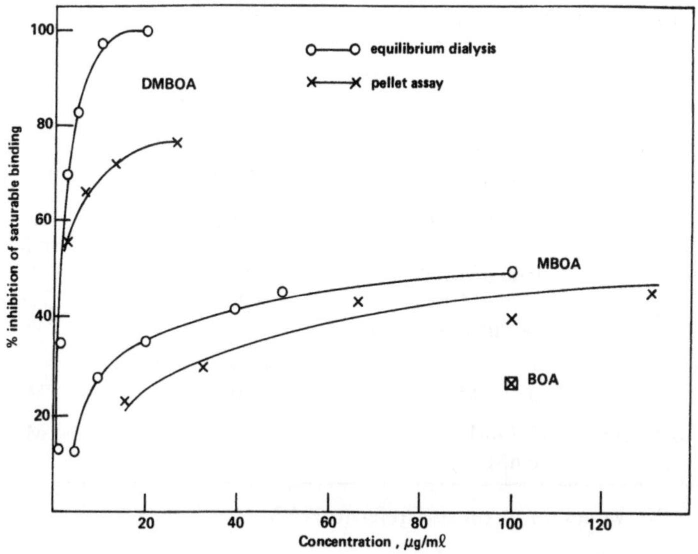

Fig. 5.8. Benzoxazolinone inhibition of NAA binding to membrane-bound (pellet assay) or solubilized (equilibrium dialysis; see Sect. 5.2.4.a. iii, receptors) in corn

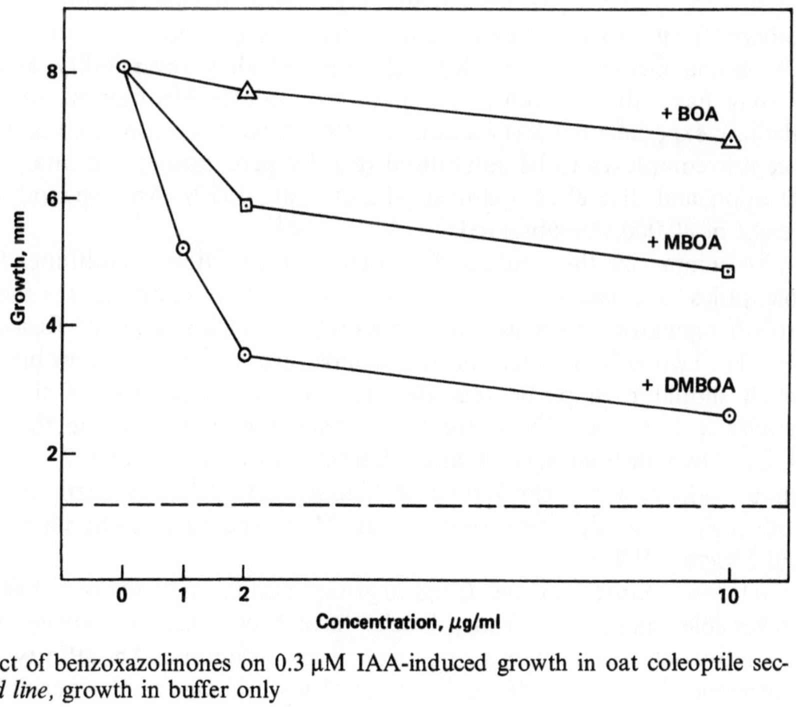

Fig. 5.9. Effect of benzoxazolinones on 0.3 µM IAA-induced growth in oat coleoptile sections. *Dashed line,* growth in buffer only

Table 5.1. Binding sites in corn coleoptile membranes

Ligands	Site nomenclature and/or K_D[a]	No. of binding sites, pmol/g	Tentative location
Auxins	0.15 μM (site 1)	25–30	ER/Golgi
	1.60 μM (site 2)	100–120	PM (plasma membrane)
	0.5–0.7 μM	30–50	Mainly ER; some in Golgi and/or PM
	0.40 μM (site I)	40	ER
	1.30 μM (site II)	20	Tonoplast
	>5 μM (site III)	40	PM
NPA	13 nM	~10	
	0.22 μM	~90	PM
Fusicoccin	0.7 nM		PM
	6 nM		

[a] K_D values for auxin sites refer to NAA

a more recent account (BALLIO, 1979) suggests that there are two sets of binding sites with affinities of 0.7 nM and 6 nM. Neither NAA nor NPA was able to compete for the fusicoccin-binding sites. An unusual feature of the binding was that it appeared to be a slow, temperature-dependent process, requiring incubation for 1 h at 26 °C. This is comparable with the behavior of some animal hormone receptors (see Sect. 5.4), but is in marked contrast to the very rapid equilibration of the auxin-binding sites (RAY et al., 1977a). On isopycnic sucrose gradients, fusicoccin binding coincided closely with NPA-binding activity, from which it was concluded that the fusicoccin sites are located in the plasma membrane. The high affinity and slow reversibility of fusicoccin binding have allowed confirmation of the cellular localization from in vivo labelling experiments (BEFFAGNA et al., 1979) and have also enabled fusicoccin-receptor complexes to be solubilized (e.g. by perchlorate) and analyzed by gel filtration and disc electrophoresis (PESCI et al., 1979). An apparent molecular weight of 80,000 was observed.

To summarize the findings discussed in this section, membranes from corn coleoptiles have been shown to carry distinct sets of binding sites for the synthetic growth regulators NPA and morphactins, for auxins, and for fusicoccin (Table 5.1). Two different terminologies have arisen for the auxin-binding sites, which should perhaps be regarded as a healthy reflection of the degree of activity in this area. The current consensus would seem to be that there are at least two distinct sets of auxin-binding sites (sites 1 and 2 or I and II), while evidence for a third type of binding, site III, has been presented for both corn coleoptiles (DOHRMANN et al., 1978) and zucchini hypocotyls (JACOBS and HERTEL, 1978).

ii) Active-Site Chemistry. General group-modifying reagents and site-directed irreversible inhibitors (affinity labels) have been used extensively in probing the active sites of enzymes and cholinergic receptors. An affinity label is a compound that bears sufficient structural resemblance to the normal substrate

Other properties	Reference
Less specific than site 2	BATT et al. (1976)
Auxin-specific	BATT et al. (1976)
Affinity modulated by SF	RAY et al. (1977a, b)
Less specific than site II	DOHRMANN et al. (1978)
Low affinity for PAA	DOHRMANN et al. (1978)
K_D 2,4-D = 5 μM	DOHRMANN et al. (1978)
	THOMSON (1972)
Also binds morphactins	THOMSON and LEOPOLD (1974)
Binding is time- and temp.-dependent	DOHRMANN et al. (1977)
	BALLIO (1979)

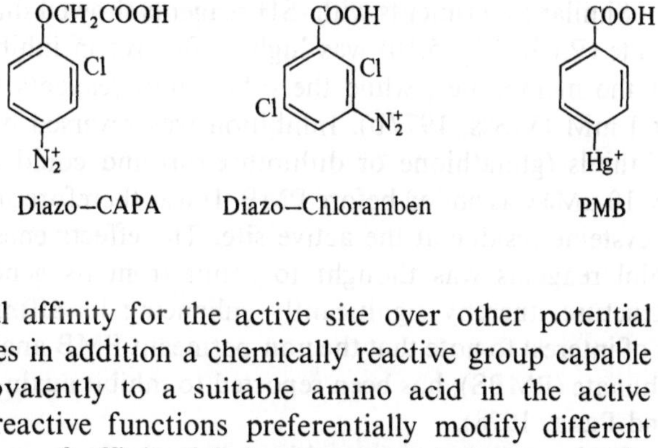

Diazo–CAPA Diazo–Chloramben PMB

Fig. 5.10. Site-directed inhibitors of auxin binding

or ligand to have preferential affinity for the active site over other potential binding sites, but which carries in addition a chemically reactive group capable of attaching the molecule covalently to a suitable amino acid in the active site environment. Different reactive functions preferentially modify different amino acids and hence the use of affinity labels can enable inferences to be drawn concerning the residues in the region of the active site.

Using corn coleoptile membranes, two compounds have been investigated as potential auxin affinity labels (VENIS, 1977b). When membranes were treated at various pH values with the diazonium salts (Fig. 5.10) of 2-chloro-4-amino-phenoxyacetic acid (CAPA) or of 2,5-dichloro-3-aminobenzoic acid (Chloramben), then pelleted from the reaction medium and resuspended at pH 5.5, their ability to bind NAA-^{14}C was impaired in comparison with control membranes incubated similarly but without diazonium salt. Diazo-CAPA was effective only at pH 8–9, while inhibition by diazo-Chloramben was largely independent of reaction pH over the range pH 6–9. Because modification by diazo-Chloramben could be carried out at a pH close to the optimum NAA-binding value, pH 5.5 (BATT et al., 1976; RAY et al., 1977a), it was possible to show that prior addition of unlabelled NAA to the membranes protected the binding sites against inhibition by this diazonium analogue. This ligand protection experiment suggested that the compound did indeed react at the active auxin-binding site. From the different pH dependences of reaction, it was proposed that diazo-Chloramben modified a histidine residue, while diazo-CAPA reacted with tyrosine or

Table 5.2. Suggested amino-acid environment of auxin receptor sites in corn membranes

Observation	Inference	Reference
1. pH Dependence of auxin binding	Histidine (-COO′-binding) Aspartate/glutamate (δ^+-binding)	VENIS (1977b)
2. Diazo-Chloramben inhibition at pH 6	Histidine	VENIS (1977b)
3. Diazo-CAPA inhibition at pH 8–9	Tyrosine/lysine	VENIS (1977b)
4. Inhibition by PMB	Cysteine	VENIS (1977b)
5. Inhibition by DTE	-S-S-	RAY et al. (1977a)

lysine. Arguments based on the pH profile of NAA binding also suggested the presence of a histidine (possibly a second residue) at the binding site, and in addition an aspartate or glutamate residue.

Similar experiments with -SH reagents showed that 50–100 μM p-mercuribenzoate (PMB, Fig. 5.10) was highly effective in inhibiting auxin-binding activity of the membranes, while the other thiol reagents tested were ineffective even at 1 mM (VENIS, 1977b). Inhibition was reversed by subsequent brief addition of thiols (glutathione or dithiothreitol) and could be greatly reduced if NAA (5–10 μM) was added before PMB. It was therefore suggested that PMB modified a cysteine residue at the active site. The effectiveness of PMB relative to other thiol reagents was thought to result from its general similarity to an auxin structure, thereby significantly enhancing its affinity for the binding site. It is of interest to note that the non-permeant PMB analogue, p-mercuribenzenesulphonate (PMBS), has been reported to inhibit auxin-induced growth (GOLDBERG and PRAT, 1978).

Prolonged treatment with reducing agents, e.g., dithioerythritol (DTE), 90 min at 0 °C, was found to inhibit subsequent auxin binding by the membranes (RAY et al., 1977a). Such inhibition was largely reversible by oxidants and could be partly prevented by the prior addition of 100 μM NAA, while the physiologically inactive benzoic acid afforded little or no protection. The presence of a disulphide group at the binding site was therefore inferred. More recently, however, CROSS and BRIGGS (1978) could not confirm DTE inhibition of auxin binding in solubilized extracts.

The indirect information obtained so far on the possible amino acid environment of the active auxin-binding site is summarized in Table 5.2.

iii) Solubilization and Purification of Auxin-Binding Sites. Auxin-binding activity in corn membranes can be readily solubilized by Triton X-100 (BATT

→

Fig. 5.11 a, b. Purification of auxin-binding proteins solubilized from corn membranes. **a** DEAE-Bio-Gel. **b** binding peak from **a** on Sephadex G100. Binding of NAA-^{14}C assayed by equilibrium dialysis. ——A_{280}; ----binding. *BB* binding buffer. *DBB* five fold dilution of same. (VENIS, 1977c)

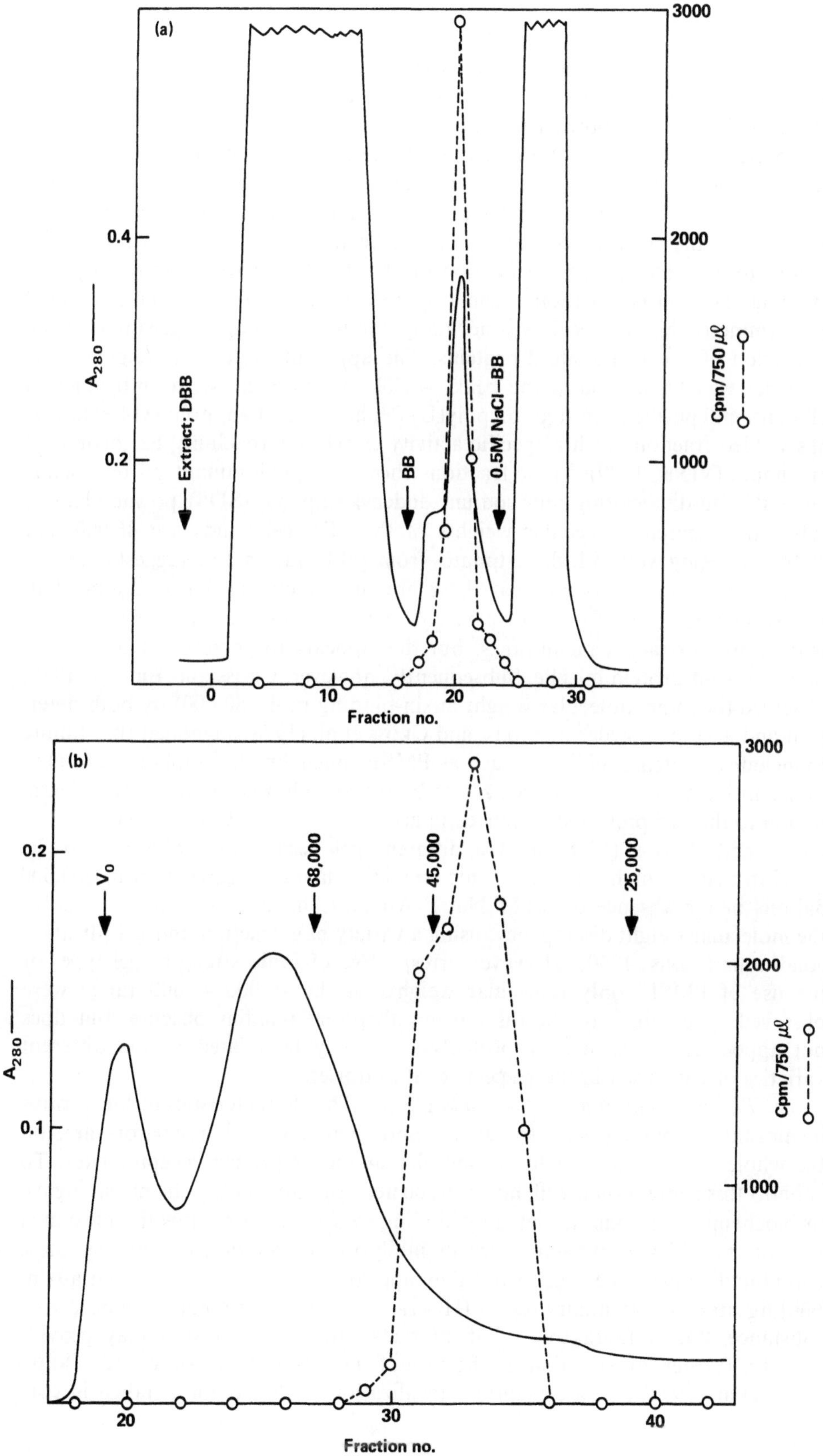
(a)
3000
2000
1000
0.4
0.2
A280
Cpm/750 µl
Extract; DBB
BB
0.5M NaCl–BB
0
10
20
30
Fraction no.
(b)
3000
2000
1000
0.2
0.1
A280
Cpm/750 µl
V0
68,000
45,000
25,000
20
30
40
Fraction no.

et al., 1976; RAY et al., 1977a; CROSS et al., 1978), but such extracts are not amenable to extensive purification. However, it has been found that the use of a modified acetone powder technique enables the binding sites to be obtained in a buffer-soluble form without recourse to detergent treatment (VENIS, 1977c). Kinetic analysis suggested the presence of two sets of binding sites in these extracts, with dissociation constants very similar to those reported for site 1 and site 2 in the membrane-bound state (see Sect. 5.2.4.a.i). The binding species appear to be proteins and purification of about 100-fold was obtained by ion-exchange and gel permeation chromatography (Fig. 5.11). On Sephadex G100 an asymmetric binding peak was normally obtained, perhaps suggesting partial resolution of site 1 and site 2 proteins. The apparent molecular weight at the shoulder was 47,300 and at the peak, 40,300. More recently, the introduction of additional purification steps on poly(U)-Sepharose and aminohexyl-Sepharose has yielded fractions with a specific activity closely approaching the theoretical maximum (VENIS, 1979). These fractions show two predominant protein bands on both non-dissociating and sodium dodecyl sulphate (SDS) polyacrylamide gels. The apparent molecular weights of the SDS gel bands are 46,000 and 40,000, agreeing well with the estimates from gel filtration and suggesting, since no subunit bands are seen, that these are monomeric proteins. CROSS et al. (1978) have reported a molecular weight of 90,000 for the auxin-binding sites from Triton-extracted membranes, but this appears to correspond to the size of a detergent-protein micelle. Subsequently, however, CROSS and BRIGGS (1978) observed the same molecular weight auxin-binding peak (80,000) by both detergent and acetone powder methods, and CROSS et al. (1978) suggested that failure to include a protease inhibitor, such as PMSF (phenylmethylsulphonylfluoride), in the initial membrane homogenization buffer could have caused receptor degradation to the size previously reported in acetone powder extracts (VENIS, 1977c). CROSS and BRIGGS (1978) used a different gel permeation column (Bio-Gel A 1.5 m), run at higher pH and ionic strength, claiming aggregation to 200,000 daltons in the absence of 0.1 M NaCl. Attempts have been made to resolve the molecular weight discrepancy, using a variety of extraction and gel filtration conditions (VENIS, 1980). However, irrespective of ionic strength, gel type, or the use of PMSF, only molecular weights in the 40,000–45,000 range were observed. The source of the discrepancy therefore remains obscure, but does not appear to be caused by proteolysis. It may be related to the different varieties of corn used in the respective laboratories.

iv) Physiological Roles of the Binding Sites. The characteristics of the various membraneous binding sites for auxins and related growth regulators are, on the whole, consistent with the proposal that they represent receptor sites. To support this contention it will, however, be necessary to identify the physiological or biochemical response initiated by binding to the receptor. Thus the presumed role of the NPA-morphactin sites in inhibition of auxin transport by these compounds must be indirect, since they appear to be quite distinct from auxin-binding sites. A preliminary report (HERTEL et al., 1976) of a naturally occurring substance, "Kartoffelfaktor", that interacts with the NPA sites may provide clues for further investigation. In the case of fusicoccin it was suggested (DOHRMANN et al., 1977) that a cell-surface localization of the receptor, taken in conjunction with the reported activation of a plasma membrane ATPase (BEFFAGNA et al., 1977; Sect. 5.2.2) would be compatible with the very rapid effects of this compound on H^+ extrusion and cell extension. Fusicoccin-binding and ATP-ase activities could be resolved after perchlorate solubilization (TOGNOLI et al., 1979), but from the similar sensitivity to inhibitors of the two activities in the particulate preparations, it was suggested that they are normally integrated in the membrane in an interactive multi-unit complex. The somewhat slower effects of auxins on extension growth and proton extrusion were considered by RAY (1977) to be consistent with their being initiated by combination of auxin with the ER-located sites rather than with the plasma membrane. Alternatively, it has been suggested (VENIS, 1977a) that the ER sites (site 1) might represent a "pro-receptor" which undergoes a maturation process to a final auxin-specific form (site 2) in the plasma membrane. Another possibility is that auxin binds initially to site 2 which functions as a primary discriminator (auxin/not auxin?) and perhaps mediates some auxin-specific response at the plasma membrane; site 1, which has higher binding affinity, could then accept auxin from site 2 and possibly transmit the auxin signal in some way to the nucleus. On this model, therefore, site 1 would not necessarily have to have the same degree of specificity as site 2. The auxin-binding sites solubilized from corn membranes do not themselves appear to have ATPase activity (CROSS et al., 1978; VENIS, 1980) and auxin stimulation of ATPase in the intact membranes has not been observed. On the other hand, auxin activation of a microsomal ATPase in rice roots has been reported (Sect. 5.2.2). If the effects can be confirmed and shown to be auxin-specific, and if auxin-binding activity can be demonstrated in the same system, it may become feasible to initiate studies on the possible functional inter-dependence of hormone-binding sites and ATP-ase, as in the case of fusicoccin. Meanwhile, the most direct evidence of a physiological role for the auxin-binding sites is the correlation between the effects of benzoxazolinones or auxin binding and on auxin-induced growth (Sect. 5.2.4.a.i). In addition the ability of PMB to block the binding sites and of its non-permeant analogue, PMBS, to inhibit auxin-induced growth is also suggestive (Sect. 5.2.4.a.ii). Site III has been suggested to function as an auxin transport site (JACOBS and HERTEL, 1978), but the evidence available at present is very indirect.

b) Membrane-Bound Sites in Other Tissues

Binding of IAA to particulate fractions from buds of light-grown peas (JABLONOVIC and NOODEN, 1974) and from mung bean hypocotyls (KASAMO and YAMAKI, 1976) has been noted. In both reports, saturable, relative to total, binding was low, no kinetic data were presented, and information on analogue competition was inadequate for any assessment of binding specificity. More recently, auxin binding has been described in a particulate preparation from tobacco pith callus (VREUGDENHIL et al., 1979). The affinities for different auxins were broadly comparable to the corn membrane system, but binding differed in being time- and temperature-dependent (cf. fusicoccin, Sect. 5.2.4.a.i). In this and in other respects, particulate binding in tobacco callus also differed from the soluble IAA-binding component described previously in this tissue by the same group (see Sect. 5.2.3.b).

DÖLLSTÄDT et al. (1976) have described saturable binding of IAA and certain substituted phenoxyacetic acids to particulate preparations from pea epicotyls and roots. The dissociation constants for IAA and MCPA were tentatively estimated at 1 μM or lower, with 5×10^4 IAA-binding sites per cell. In experiments examining displacement of IAA-^{14}C or MCPA-^{14}C from the binding sites by unlabelled compounds, the order of effectiveness of the phenoxyacetic acids was in general agreement with their relative auxin activities. Interestingly, MCPA could displace IAA-^{14}C, but IAA did not compete for binding of MCPA-^{14}C. It was therefore suggested that phenoxyacetic acid may bind to a site on the receptor that is distinct from the auxin-binding site and affect IAA binding allosterically. The possibility of distinct binding sites for phenoxyacetic acids could be of interest in relation to the herbicidal activity of these compounds. Alternatively this type of binding could be related to site III binding (Sect. 5.2.4.a.i) described by DOHRMANN et al. (1978) and by JACOBS and HERTEL (1978).

A model proposed by HARDIN et al. (1972) suggests that both rapid and long-term auxin responses could be mediated by a common auxin receptor in the plasma membrane. The basis of this suggestion was their finding that plasma membrane fractions stimulated soybean RNA polymerase and that the active factor could be released from the membranes by incubation with IAA or 2,4-D, but not by the inactive analogue 3,5-dichlorophenoxyacetic acid. This report bears a superficial resemblance to some of the findings of LIKHOLAT et al. (1974) for wheat coleoptiles (see Sect. 5.2.3.ii). The soybean factor has been characterized only in so far as showing that it is certainly not protein, is ethanol-soluble and, judging by its chromatographic properties, is of low molecular weight (CLARK J.E. et al., 1976). It is hard to visualize such a substance as a receptor in the generally accepted sense, and it would seem more appropriate to regard it as a "stimulatory factor".

5.3 Cytokinins

5.3.1 Structure-Activity Relationships

High activity in cell division assays is associated with an adenine molecule containing an intact purine ring and an N^6-substituent (Chap. 1.5 and Fig. 5.12). There are, of course, exceptions to this generalization such as N,N′-diphenylurea and the 8-azapurines, but these have greatly reduced potencies when compared with analogous purine compounds (SKOOG and ARMSTRONG, 1970). The same authors have also reported that the exchange of C and N atoms in the 7 and 8-positions reduces activity by two orders of magnitude and that the substitution of O and S for N in the N^6-position reduces effectiveness in the tobacco callus assay by up to 90%. However, the effect of such structural changes

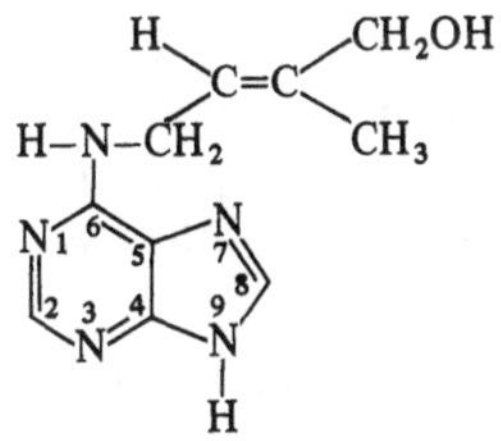

(1)

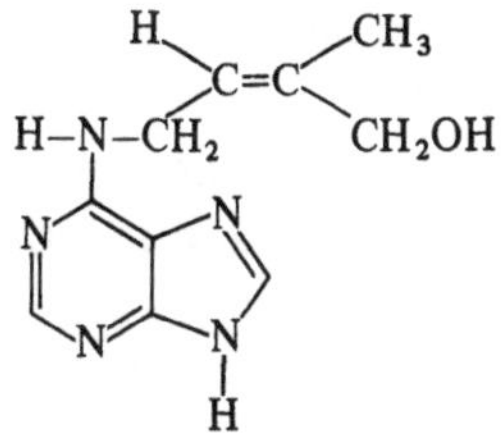

(2)

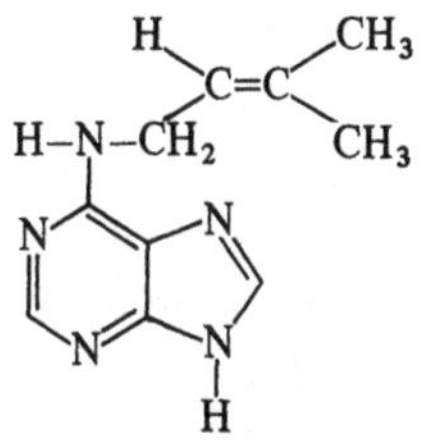

(3)

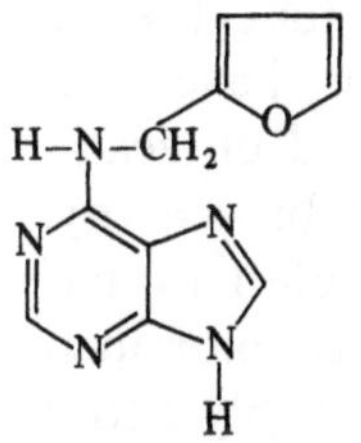

(4)

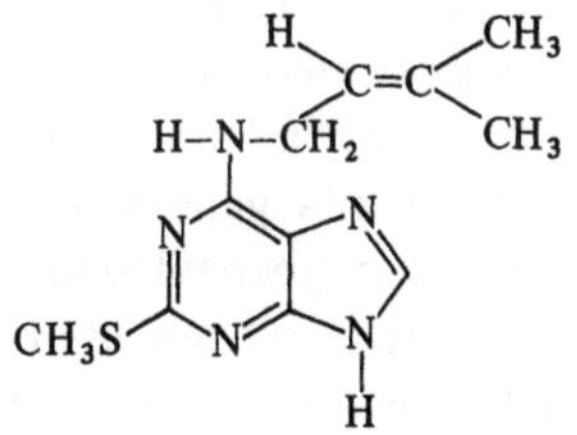

(5)

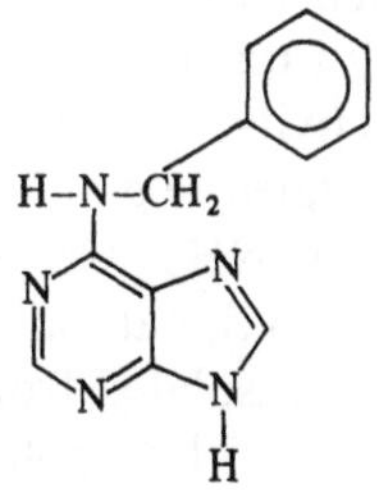

(6)

1. *trans*-Zeatin
2. *cis*-Zeatin
3. N^6-(Δ^2-Isopentenyl) adenine
4. N^6-Furfuryladenine (kinetin)
5. 2-Methylthio-N^6-(Δ^2-isopentenyl) adenine
6. N^6-Benzyladenine

Fig. 5.12. Structures of some natural and synthetic cytokinins referred to in Section 5.3

is a function of the bioassay system in which they are evaluated and it appears, for example, that a purine nucleus is not essential for effective retardation of senescence in excised leaf segments. Compounds based on other ring systems, exemplified by pyrimidine derivatives and benzimidazole, can be highly effective.

The presence of an N^6 side chain is of considerable significance for the level of activity in callus bioassays. This subject is extensively discussed by SKOOG and ARMSTRONG (1970) who indicate that both the length of the side chain, with an optimum of about five carbon atoms, and the degree of unsaturation are important for high biological activity. Dihydrozeatin, with no side-chain double bond, is only one-tenth as active as zeatin. However, there appears to be no specific requirement for critical reactive groups at fixed points in the side chain and the influence of this structural feature on the level of biological effectiveness is believed to be conditioned by the general molecular dimensions and the charge or polarity. The stereochemistry of the double bond in the side chain of zeatin is important. Both cis and trans forms of zeatin occur naturally but the cis form is appreciably less active in promoting the growth of tobacco callus tissue (HALL and SRIVASTAVA, 1968). The large terminal ring structures, characteristic of synthetic cytokinins such as N^6-furfuryladenine and N^6-benzyladenine, indicate that relatively massive molecular dimensions may be advantageous in this region of the molecule and SKOOG et al. (1967) have established a hierarchy of effectiveness for various terminal substituents (benzyl > furfuryl = phenyl > cyclohexyl). The acceptable variability in the composition of the side chain suggests that it may not interact directly with the primary cellular site of action but may exert a directing effect on the orientation of the molecule during such an interaction.

Additional substitutions at other sites on the purine ring can have variable effects on biological activity. Substitution in the 1-position usually leads to loss of activity in N^6-adenine derivatives (SKOOG et al., 1967) and there is evidence to suggest that substitution at the 3-position may have a similarly deleterious effect. An unsubstituted 1-position is regarded as being critical for cytokinin activity. Natural compounds, substituted in the 2- and 6-positions, have been isolated, e.g., (2-methylthio)isopentenyladenine, but structures based on zeatin with NH_2, CH_3S, or OH substituents generally exhibit reduced bio-activity. A novel, naturally occurring, o-hydroxybenzyladenine (Chap. 1, Table 1.15) has been isolated from leaves of poplar (HORGAN et al., 1975) and this compound showed low potency as a cell-division factor. It was, however, highly active in leaf expansion and senescence assays. On the other hand, 2-halogenation of zeatin enhanced activity in callus assays (SKOOG and ARMSTRONG, 1970) and these workers also noted that a 9-methyl substituent had a similar effect. The latter observation is supported by KENDE and TAVARES (1968) who found that 9-methyl-N^6-benzyladenine was more active than the N^6-mono-substituted compound in the soybean callus assay.

Of the mono-substituted adenines, only those with substituents in the N^6-position exhibit cytokinin activity.

5.3.2 Cytokinins in Transfer RNA (tRNA)

N^6-(Δ^2-Isopentenyl)adenosine (i^6A) or its hydroxylated derivative, zeatin, have been widely detected as constituents of plant tRNA whilst 2-methylthio-isopentenyladenosine (ms^2-i^6A) has been shown to occur in both plants and bacteria (see Chap. 1, Tables 1.10–1.13). Both cis- and trans-zeatin have been isolated from pea shoot tRNA but it is generally considered that the cis form normally predominates (HALL, 1973).

The incorporated cytokinin is located at the 3′ end of the anticodon sequence (Fig. 5.13) in a position which is believed to exert an influence on the configuration of that part of the tRNA molecule. When yeast tRNA is fractionated it is found that not all species contain a cytokinin base; the distribution of i^6A being restricted to those forms containing an anticodon corresponding to an mRNA codon commencing with the letter U (uridine) and thus to tRNA's for cysteine, leucine, phenylalanine, serine, tryptophan, and tyrosine (ARMSTRONG et al., 1969; PETERKOFSKY and JESENSKY, 1969). Species of tRNA responding to codons commencing with the letter A (adenine) contain the compound N-[9-(β-D-ribofuranosyl-9H)purin-6-ylcarbamoyl] threonine (Ad-CO-thr, Chap. 1, Fig. 1.7) in the same position and there are indications that this compound also possesses cytokinin activity (see HALL, 1973).

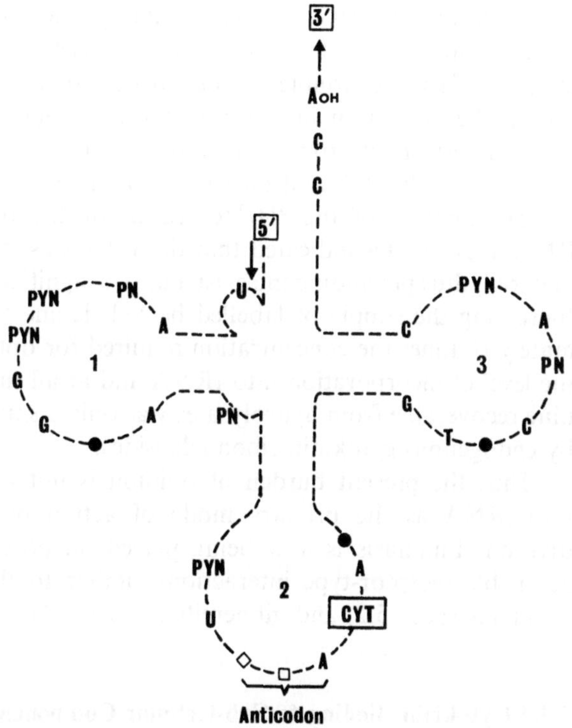

Fig. 5.13. Schematic clover-leaf structure for transfer RNA (tRNA) showing position of cytokinin residue. *A* adenine, *C*, cytidine, *G* guanine, *T* thymidine, *U* uridine, *PN* purine nucleoside, *PYN* pyrimidine nucleoside, *CYT* cytokinin (i^6A)

There is, therefore, an implied function for cytokinins as essential constituents of tRNA molecules with a role in the association of the anticodon loop with the ribosome tRNA or mRNA binding sites. Can such a mechanism explain the observed physiological effects of applied cytokinins in various plant growth systems? An essential feature for activity of any cytokinin is the possession of a suitable side chain and this feature is also known to be necessary for the proper association of the tRNA anticodon loop with the ribosome in the presence of mRNA (FITTLER and HALL, 1966). It is now clear (Chap. 4.1.2.b) that i^6A is not incorporated directly into tRNA but rather is initially present as adenosine to which the Δ^2-isopentenyl side chain is transferred via isopentenyl pyrophosphate (CHEN and HALL, 1969; ROSENBAUM and GEFTER, 1972), although there may be exceptions to this rule (e.g., HORGAN et al., 1975). There cannot, therefore, be any suggestion that the rate of incorporation of cytokinins into tRNA bears a direct relationship to the pool of free species present in the tissue.

One of the more compelling arguments against a direct relationship between free and tRNA cytokinin is based upon tissue cultures which require the addition of free cytokinins for the maintenance of continuous growth. In the presence of the synthetic cytokinin, benzyladenine (BA, Fig. 5.12), the callus generates the full range of natural cytokinin tRNA inclusions but also incorporates a small proportion of the BA as the riboside (BURROWS et al., 1971). Evidently, the cytokinin inclusions are derived from both endogenous and exogenous supplies and the response to BA cannot be wholly explained in terms of its incorporation into tRNA. The possibility that endogenous cytokinins may act as side-chain donors for adenosine in tRNA has been ruled out by WALKER et al. (1974) using double-labelled benzyladenine (^3H in the phenyl ring and ^{14}C at the 8-position in the purine ring). They found that, in tobacco callus tissue, a very small proportion of the applied cytokinin was incorporated intact into tRNA, about 1 molecule of benzyladenine per 10,000 tRNA molecules. An examination of the $^3H/^{14}C$ ratio for the benzyladenine recovered from RNA hydrolysates indicated that the value was the same as that in the starting material, thus precluding the existence of a significant degree of transbenzylation. Increasing the supply of labelled benzyladenine to a level of 0.4 μM, approximately 10 times the concentration required for optimal growth, did not increase the level of incorporation into tRNA and at all times the amount of benzyladenine recoverable from hydrolysates was only a small fraction of that contributed by endogenous cytokinin ribonucleosides.

Thus the present burden of opinion is not in favour of the incorporation into tRNA as the primary mode of action of cytokinins in regulating cell division. Emphasis is now being placed on possible allosteric mechanisms or reversible receptor-type interactions similar to those discussed in the context of auxins (Sect. 5.2) and gibberellins (Sect. 5.4).

5.3.3 Cytokinin Binding to Sub-Cellular Components

It has been shown that cytokinins can increase the rate of incorporation of labelled amino acids into proteins synthesized by isolated organelles such as nuclei, mitochondria, and plastids (e.g., BHATTACHARYYA and ROY, 1969; DAVIES and COCKING, 1967; TAKEGAMI and YOSHIDA, 1977). These responses occurred after a lag period of only a few minutes, suggesting that there could be a direct influence on elements of the translation apparatus.

The first indication that cytokinin responses might be effected via a protein interaction were provided by MATTHYSSE and ABRAMS (1970). These authors reported the recovery of a mediator protein, from pea chromatin preparations, which stimulated RNA synthesis in an in vitro system containing *E. coli* polymerase and homologous pea DNA. Unfortunately, the nature and properties of this factor were not elucidated further.

BERRIDGE et al. (1970a, b) have studied interactions between cytokinins and ribosomes isolated from leaves and dwarf pea shoots of Chinese cabbage (*Brassica pekinensis*). They found that, when such preparations were incubated with 8-^{14}C-kinetin or [G-^3H]benzyladenine, a significant proportion of the radioactivity remained associated with the ribosomes after gel-filtration on Sephadex G-200. The binding showed a temperature dependence with 1.34 molecules of cytokinin bound per ribosome at 4 °C and 0.5 molecules at 20 °C when the cytokinin was applied at a concentration of 23 μM. The binding was shown to be exchangeable by equilibrium dialysis and increased with cytokinin concentration up to a value of 20 μg/ml (92 μM) when four molecules of benzyladenine were bound per ribosome. This property was reflected in the observation that binding in G-200 eluates could only be observed when the column had been

Table 5.3. Comparison of the activities of various purine derivatives in promoting expansion or retarding senescence of Chinese cabbage leaf discs and correlation of these activities with binding to ribosome preparations. (Data from BERRIDGE et al., 1970b)

Compound	Disc expansion[a]		Chlorophyll retention[b]	Molecules bound per ribosome
	23 μM	2.3 μM	2.0 μM	
Adenosine	1.7	0.5	4	0.34
Adenine	5.8	0.0	0	0.10
3-Benzyladenine	8.9	6.8	0	0.51
9-Benzyladenine	9.3	7.1	3	0.53
6-Morpholinopurine	17.0	4.6	7	0.10
Kinetin	23.8	18.7	36	1.00±0.05
6-Benzylaminopurine	29.0	23.4	52	1.35±0.05

[a] % Increase in fresh weight over control
[b] Expressed as a retardation index (LETHAM, 1967)

equilibrated with radioactive cytokinin prior to the application of the sample. Studies involving detergent extraction and density-gradient centrifugation suggested that, in these preparations, the binding was predominantly to ribosomes with a sedimentation coefficient of 83S. Affinity was not reduced by washing the preparations with 0.5 M ammonium chloride.

Specificity of interaction was examined using a range of substituted adenines and the binding values were compared with the ability of the test compound to promote expansion or retard senescence in discs of Chinese cabbage leaf (Table 5.3). At a concentration of 2 μM there was a very close correlation between cytokinin activity and ribosomal binding but at higher application levels (23 μM), 6-morpholinopurine promoted expansion at a rate comparable with that given by kinetin and thus did not conform to the general pattern of correlation.

Similar studies on the association between cytokinins and higher plant ribosomes were conducted by FOX and ERION (1975) using preparations derived from freshly milled wheat germ or tobacco callus cultures. In this instance binding was determined either by means of equilibrium dialysis or by pelleting previously incubated ribosomes at 150,000 *g*. Substantial binding was observed; the wheat preparations gave values around 1.24 mol of cytokinin per ribosome, and the tobacco preparations gave values of 2.26 mol per ribosome, at 4.7 μM. These values compared with 0.34 and 0.20 obtained respectively with rat liver and *E. coli* ribosome preparations.

The kinetics of binding were examined by competitive equilibrium dialysis using methylene-^{14}C-N^{6}-benzyladenine of varying specific activity. These data were plotted by the SCATCHARD (1949) procedure to yield biphasic plots of the relationship between the amount of bound cytokinin and the ratio of bound to free ligand (Sect. 5.1.2). FOX and ERION (1975) noted that the data for the ribosomal binding of benzyladenine, in contrast to the findings of BERRIDGE et al. (1970a, b), gave clear indications of the existence of two classes of binding sites with K_D values of 0.6 μM for the high affinity site and 0.1 mM for the low affinity association.

The binding capacity of the ribosomes was lowered by washing with 0.5 M KCl and this appeared to be due to the selective removal of high affinity sites. Low affinity, non-specific binding was unaffected by salt washing. Material, thus dissolved, was also effective in binding cytokinins giving a Scatchard plot consistent with the sole presence of high affinity sites (FOX and ERION, 1977). Purification of this cytokinin binding fraction (designated CBF-1) by ammonium sulphate precipitation and gel-filtration suggested a mean molecular weight of 93,000 daltons. It was calculated that CBF-1 occurred in wheat germ in the ratio of one binding unit per ribosome and competitive dialysis studies indicated affinities which were in direct relationship to the cytokinin activities of the tested compounds. The presence of a 9-ribosyl substituent reduced binding to one-tenth of the level observed with the free base. Crude CBF-1 preparations could be inactivated by boiling or by trypsin digestion but were insensitive to ribonuclease, properties consistent with identification as a protein.

Free CBF-1, and a second cytokinin binding factor with a molecular weight of approximately 30,000 daltons (CBF-2), were detected in the cytosol. The free CBF-1 activity was 2 or 3 times greater than that encountered in the ribosomal fraction. SDS polyacrylamide gel electrophoresis of ribosomal CBF-1, after preliminary purification on phosphocellulose, ion-exchange and gel-filtration columns, indicated the presence of three subunits with molecular weights of 40,000, 50,000 and 54,000 daltons (ERION and FOX, 1977). Subsequent studies have shown that the intact protein is not required for binding activity and that cytokinins associate with one, or more, of the sub-units, as yet unspecified (PRATT and FOX, 1978).

A binding protein with similar properties, isolated from wheat germ by affinity chromatography on an agarose column substituted with a kinetin riboside derivative, has been described by MOORE (1977). In the presence of mercaptoethanol this material had an apparent molecular weight of 185,000 daltons as measured by gel filtration but, in the absence of the thiol reagent, peaks were observed at 37,000, 56,000 and 91,000 daltons (MOORE, personal communication). These values show some correspondence to those determined by ERION and FOX and the form with the apparent molecular weight of 185,000 daltons is tentatively explained by MOORE as being an oligomer, composed of two 37,000 and two 56,000 sub-units.

POLYA and DAVIS (1978) have also described the properties of a purified soluble protein, isolated from wheat germ. It has a K_D of 0.2 μM for kinetin and an apparent molecular weight of 180,000 daltons. On gel filtration it co-chromatographs exactly with a single peak of carbohydrate and protein and is displaced from a concanavalin A – Sepharose 4B column by α-methyl glucoside.

The binding proteins isolated by these groups show high affinity kinetics and good specificity for active cytokinins but, surprisingly, none bound zeatin and this anomaly remains to be explained.

Affinity chromatography was also employed by TAKEGAMI and YOSHIDA (1975) in their studies on binding proteins from tobacco leaves and by KLÄMBT

(1977) for maize shoots and wheat germ. The former workers used a column consisting of benzyladenine linked to CNBr-activated agarose and the active fraction was eluted by 0.1 N KOH. This procedure resulted in a 40-fold enhancement of binding capacity per unit protein when compared with the crude extract and there was reasonable specificity for active cytokinin structures. However, IAA and tryptophan were also bound by this extract. This material differed from that described by ERION and FOX by virtue of its low molecular weight of the order of 6,000 daltons. It is significant that this factor has been shown to bind specifically to 40 S ribosomal sub-units in vitro and can be subsequently released by washing with 0.5 M KCl (TAKEGAMI and YOSHIDA, 1977), behaviour resembling that observed by FOX and ERION for CBF-1. However, the large molecular weight difference seems to preclude a common identity. It is more probable that the binding component of TAKEGAMI and YOSHIDA relates to the membrane-located glycopeptide site in *Achlya* which bound cytokinins, IAA, tryptophan, and Ca^{2+} (LÉJOHN, 1975). Four binding components isolated by KLÄMBT (1977) from maize shoots using an i^6A riboside-agarose column have not yet been sized or kinetically characterized. The molecular weight discrepancies between the binding components so far described must be resolved before their inter-relationships can be properly assessed.

A note of caution in relation to the observed kinetics of cytokinin-binding proteins has been sounded by KENDE and GARDNER (1976), who point out that the K_D for the high affinity site described by FOX and ERION, at 0.6 μM, is equal to, or exceeds, the concentration at which many physiological responses are saturated. There may, however, be other components in the intact cell which modify the response to a given concentration of cytokinin and the importance of this observation remains to be assessed.

What is the significance of cytokinin binding to isolated ribosomes? BERRIDGE et al. (1970a) suggest that the reversible binding may "stabilize" ribosomes, alluding possibly to the association between the large and small ribosomal sub-units. There is, nevertheless, at least an equal possibility that the interaction may relate to ribosomal properties associated with other facets of protein synthesis. The same authors were unable to detect any effects of cytokinins on in vitro protein synthesis in organelle preparations from Chinese cabbage or in reconstituted systems containing ribosomes, tRNA, and supernatant factors from the same species. Preparations from etiolated pea seedlings gave similar negative results, but these could be due to a loss of the soluble binding component during ribosome isolation.

At present we must also face the possibility that the binding may not have regulatory implications and is a manifestation of a general affinity of ribosomal proteins for nucleotides. This alternative is supported by the extensive and detailed studies on the ribosomal assembly process which show that the constituent proteins are added to the 16 S ribosomal RNA strand in a specific order and that these proteins have sites capable of recognizing and interacting with defined polynucleotide sequences (e.g., COX and BONANOU, 1969; MIZUSHIMA and NOMURA, 1970). If a part of these sites, or others of a similar nature, remains unmasked after assembly it is conceivable that they might constitute a general binding potential on the ribosome for nucleotide structures, including cytokinins. Alternatively, the effect may be due to association with unoccupied tRNA binding sites on the large ribosomal sub-unit. In this context it is relevant to note that selective iodination of the side chain of i^6A, incorporated in tRNA, destroys the ability of that tRNA to attach to the ribosomal binding site (FITTLER and HALL, 1966).

We need to know whether both parts of the ribosome have cytokinin binding capacity and it would also be informative to conduct competitive binding experiments for cytokinin using a range of polynucleotides rather than cytokinin analogues. Knowledge of the degree to which components of the ribosomal assembly and tRNA binding sites are salt-soluble would also be valuable.

Binding of benzyladenine and kinetin, but not zeatin, to isolated pea nuclei has been reported (BERRIDGE et al., 1970b) with affinities in the region of 1 nM/mg DNA but there was no correlated stimulation of RNA synthesis when a complete mixture of nucleoside triphosphates was added to the system. This finding is at variance with the data of MATTHYSSE and ABRAMS (1970), who were able to demonstrate a stimulation of RNA synthesis by cytokinins in isolated pea and soybean callus nuclei as well as in in vitro systems based upon pea chromatin and *E. coli* polymerase.

SUSSMAN and KENDE (1975) and GARDNER et al. (1975) have described binding of benzyladenine to fractions from tobacco callus and protonemata of *Funaria hygrometrica*. Further studies on the tobacco system, based upon cell cultures, have indicated that two classes of sites are present in an 80,000 *g* pelletable fraction (SUSSMAN and KENDE, 1978). The major component had a low affinity for cytokinins (K_D 7.7×10^{-6} M) and no specificity for biologically active structures. A second, high affinity site (K_D 1.4×10^{-7} M) was detected and this was able to discriminate between a series of halogenated benzyladenine derivatives in a manner which correlated with their biological effectiveness. This site was also heat labile. The sub-cellular location of these sites remains to be determined.

It seems well established that cytokinin-binding protein moieties exist in a number of plant tissues and that some of these have kinetic properties consistent with a receptor function. The high affinity sites show good specificity for active cytokinin structures but we cannot yet determine whether the interaction bears a relationship to growth response. The difficulty of looking at such sites against a substantial background of non-specific binding will, undoubtedly, be greatly eased by the recent development of photo-affinity labelled cytokinins (SUSSMAN and KENDE, 1977).

5.4 Gibberellins

5.4.1 Structure-Activity Relationships

At present there are 57 known gibberellins (GA's) of natural origin and their structures are given in Chap. 1, Table 1.7. This number does not include the myriad of structural variants and functional derivatives which have been prepared during researches into the biosynthesis and biological potency of the GA's. All naturally-occurring forms are characterised by the common feature of an intact ent-gibberellane ring system (Chap. 1, Fig. 1.4).

Partial coverage of the biological implications of changes in structure, in relation to various bioassay systems, has been provided by BRIAN et al. (1967) and CROZIER et al. (1970). Conclusions about structural requirements for activity vary according to the plant system in which the compound is tested and structures which are highly active in one system may have low or zero activity in others. The availability of radioactive GA's has uncovered a major complicating factor in the determination of biological activity, namely that GA's are rapidly metabolized in many plant tissues, making it difficult to assess whether it is the compound supplied or one of its metabolites which is eliciting the growth response. This situation is particularly well illustrated in lettuce hypocotyls. Assays with intact seedlings indicate that GA_9 and GA_{20} are both highly active in promoting hypocotyl elongation. However, when labelled GA_9 is fed to excised hypocotyls it is rapidly hydroxylated, probably at the 13-position to form GA_{20}, and the conversion is virtually complete when GA_9 is supplied at concentrations below 1.0 μM (NASH et al., 1978). Growth occurs in the excised system in response to added GA_9. When the interconversion of GA_9 is blocked by a specific hydroxylation inhibitor, 2,2′-dipyridyl, growth ceases, and can be restored by adding GA_1 or GA_3 (but not GA_{20}) to the medium. This observation suggests that further hydroxylation of the tentatively identified GA_{20} is required to provide the active molecular species or that concurrent pathways (e.g., via GA_4) may be operating. Thus, although investigation of the biological activity of GA_9 by intact plant bioassay superficially suggests that the structure per se is biologically active, closer examination reveals that hydroxylation at the 13-position and probably other sites is essential before any growth stimulation can occur (NASH et al., 1978). A similar situation can be envisaged where an active structure could be inactivated by hydroxylation at a crucial site during uptake (see later). Data on structural requirements for biological action should, therefore, be evaluated in the light of the metabolic events which can occur in the tissue prior to the occurrence of the primary interaction.

The range of metabolic transformations which have been observed varies from tissue to tissue. For example, relatively little metabolism occurs when labelled GA's are applied to barley aleurone layers. CROZIER et al. (1970) have postulated that bioassay systems which exhibit a response to a wide range of GA's (e.g., Tan-ginbozu dwarf rice) are capable of catalyzing extensive metabolic transformations and that tissues with high specificity are incapable of carrying out significant metabolic changes. Against such a background it is difficult, at present, to make general statements about structure activity relationships.

For a GA to act it may be assumed that the GA molecule must fit a sub-cellular site where the topology is related to the molecular dimensions of the hormone and that the association is stabilized and orientated, by the specific interaction between charged groups on the two components. An ionizable carboxyl function at the 7-position seems to be a ubiquitous feature of active GA's and largely determines their partitioning behaviour at various pH

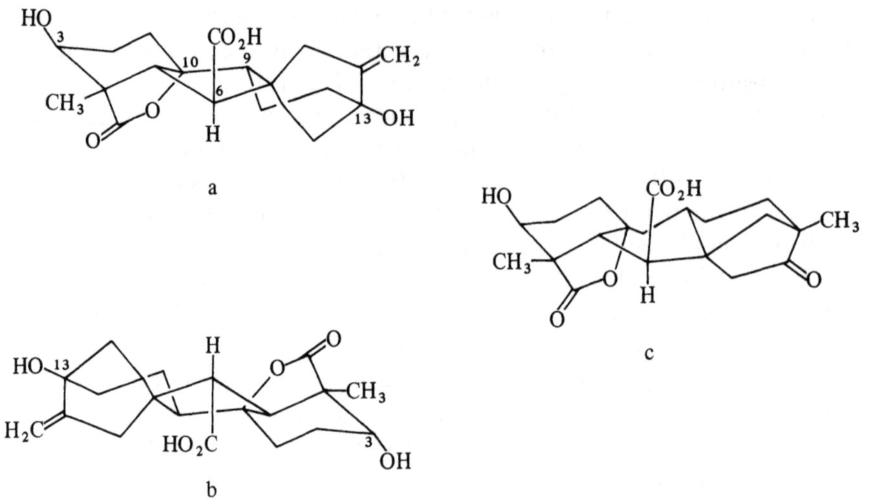

Fig. 5.14a–c. Orientation of the lactone function and D-ring when viewed in the plane of the A, B, and C rings. **a** gibberellin A_1. **b** gibberellin A_1 rotated through 180°. **c** gibberellin A_1 derivative with D-ring inversion (biologically inactive)

values. Masking of the group by methylation drastically reduces biological activity (HIRAGA et al., 1974) suggesting that the carboxyl may interact with a positively charged entity at the site of action. Highly active GA's possess a 19,10-lactone function and the stereochemical implication of this, in terms of molecular symmetry, are depicted in Fig. 5.14. When a molecular model is viewed in the approximate plane of the A, B, and C rings, (Fig. 5.14.a), it can be seen that the D-ring and lactone function are projected on opposite sides of the molecule with 3- and 13-substituents diametrically opposed. Rotation through 180°, on an axis through the 6-carbon and bisecting the 9,10-bond, results in a basically similar configuration with hydroxyl groups in the 3- and 13-positions occupying the same spatial positions (Fig. 5.14.b). This approximate two-fold axis of symmetry has been discussed by FRYDMAN and MACMILLAN (1975) in relation to the equivalence of the 2β- and 12α-positions in hydroxylation reactions. BRIAN et al. (1967) have described a number of rearrangement products in which ring D has the opposite stereochemical configuration to that normally encountered. These compounds were found to be virtually devoid of biological activity and it is evident that they do not exhibit the described rotational symmetry (Fig. 5.14.c). The same is true of non-lactonic gibberellins ($GA_{12,13,14,17,19,23,24,25,28,36,39,41,42}$) which generally have limited potency in bioassays and may require conversion to lactonic forms before acting. Thus, there is at least a superficial case to suggest that the overall shape, conferred on the GA molecule by the configuration of the lactone bridge and D-ring, is important in determining the "fit" of the hormone at the active site.

The positions at which the molecule is hydroxylated are also important determinants of the level of biological activity. Very high activity is generally associated with 3β-hydroxylation and this substitution is most effective when the 13-position is similarly substituted. Hydroxylation at the 13-position alone results in moderate activity. Such statements have, however, to be qualified by the knowledge that other structural requirements for high activity, such as the lactone function, must also be present and that activity depends on the type of bioassay. Highly potent negative effects are exerted by hydroxylation at the 2β-position, as exemplified by the conversion of GA_1 to GA_8 which results in the complete loss of biological activity (Chap. 4.2.4.b). The ability to catalyze this conversion seems to be a widespread property of plant tissues (e.g., NADEAU and RAPPAPORT, 1972; NADEAU et al., 1972; STODDART et al., 1974; DAVIES and RAPPAPORT, 1975; FRYDMAN and MACMILLAN, 1975). A 2β-hydroxylating enzyme present only in the soluble sub-cellular fraction of *Phaseolus vulgaris* seeds, has been studied by PATTERSON et al. (1975). This, therefore, provides an enzymatic mechanism for regulating the pool of active GA in tissues where 2β-hydroxylation can be applied as an inactivation step. In the presence of a soluble enzyme capable of a one-step irreversible inactivation, it can be assumed that only those GA molecules attached to the active site(s) or protected by compartmentalization would survive in an unsubstituted state. If the GA response requires a reversible association between the growth regulator and a finite number of action sites, then the response can only be sensitive to the rate of synthesis when a means exists for the continuous removal of excess biologically active molecular species from the available pool. Hydroxylation at carbon-2 provides a simple mechanism for this purpose. Short-term demands for GA's at levels exceeding the maximum biosynthetic rate might then be met by release from protected compartments, possibly exemplified by organelles bounded by lipid membranes.

A detailed discussion of 2-hydroxylation has been provided by SPONSEL et al. (1977). They considered the effects of hydroxylation in the 2-, 3-, and 13-positions in a series of structures including GA_{40}, GA_{43}, GA_{46}, GA_{47}, and GA_{51}. 2β-Hydroxylation of GA_9 and GA_{13}, to yield GA_{51} and GA_{43} respectively, resulted in complete inactivation. In contrast, 2α-hydroxylation of GA_9, to produce GA_{40}, diminished, but did not eliminate, biological activity and the same was true of the analogous conversion of GA_4 to GA_{47}.

The orientation of hydroxyl groups is also of critical importance at other molecular sites. It has been shown, for example, that the 3-hydroxyl in GA_1 must be in the β-orientation and that conversion to the α-epimer (CROSS et al., 1961) is accompanied by an almost total loss of biological potency (BRIAN et al., 1967). The 3α-hydroxy stereochemistry also confers immunity to further hydroxylation in the 2-position (PATTERSON et al., 1975), suggesting that the β-stereochemistry of the 3-OH is essential for successful interaction with growth regulatory sites and metabolizing enzymes.

There has been continuing interest in the idea of "long-lasting" GA derivatives with substitutions at critical positions which either prevent metabolic deactivation or stabilize the association with the active site. Replacement of a hydrogen atom by fluorine in a carbon-hydrogen bond produces only a small-dimensional increase but markedly enhances electronegativity and hydrogen-bonding potential (KIRK and COHEN, 1971). Fluorination of steroids at significant sites has resulted in enhancement of hormonal effectiveness (TAYLOR and KENT, 1965) and a number of investigators have explored the possibility that similar effects might be obtained with GA's (BATESON and CROSS, 1972; BANKS and CROSS, 1975; STODDART, 1972; JONES, 1976). An 18-fluoro-substitution in GA_9 and GA_{12} resulted in reduced potency in a range of bioassay systems with the exception of the lettuce hypocotyl system where 18-fluorinated GA_9 was approximately 5 times as active as GA_9 at concentrations up to 1.0 μM. Interestingly, 18-fluoro-GA_3 acted as a competitive inhibitor of GA_3 action in the same assay, halving the response when present as 30% of the available GA.

Replacement of the 3- and 13-hydroxyl groups in GA_3 by fluorine reduced activity by an order of magnitude in the dwarf rice assay and by three orders of magnitude in the barley α-amylase assay; however, the compound was almost completely inactive in the lettuce hypocotyl test. Replacement of the 13-hydroxyl in GA_5 and 13-fluoro-GA_9 had slightly promotive effects in all bioassays. Competition studies with the 3,13-difluoro derivative of GA_3 again indicated strong inhibitory effects when applied in conjunction with GA_3 itself. Thus, fluorination does not appear promising as a means of prolonging the bio-activity of GA's, but studies along similar lines using methyl substituents may be more productive (MACMILLAN, personal communication).

In general the effects of substitution on biological activity vary with the site of addition and some of the results obtained differ from expectation. There are, however, indications that enhancement of activity can occur and that inhibitory structures can also be produced. The latter offer the possibility that low dose dwarfing agents, acting as competitive inhibitors of endogenous GA at the site of action, might be feasible. Similarly, substitutions which interfere with normal metabolic turnover processes may permit the extension of the period of effectiveness of exogenously applied GA.

The present consensus of opinion is that neither the O-glucosyl ethers nor the glucosyl esters are biologically active per se. Those glucosyl conjugates which do show activity in conventional bio-assays are believed to owe their activity to hydrolytic cleavage to the free GA either by plant enzymes or by microbial contamination under non-sterile conditions of bio-assay. The topics are discussed in more detail in Chapter 4 where literature references can be found.

The function of GA conjugates is still an open question. It has been suggested that they are storage or depot forms of the free GA's. More specifically it has been proposed that the polar, water-soluble conjugates may act as transportable forms which are hydrolyzed to the active free GA in the sink tissue. Currently there is little direct evidence to support this idea. Indeed, conjugate GA's do not appear to be turned over at an appreciable rate (MUSGRAVE and KENDE, 1970). Furthermore simple hydrolysis of glucosyl ethers of 2β-hydroxylated GA's such as GA_8 and GA_{29}, would only yield biologically inactive GA's. Time-course studies with excised lettuce hypocotyls (STODDART and JONES, 1977) show that conjugation, in this system, occurs on a much longer time scale than does GA-action on elongation. In experiments where growth is stopped by the addition of a hydroxylation inhibitor, conjugates accumulate at an accelerated rate (NASH et al., 1978). This observation suggests that conversion to a conjugate is not necessary before the GA can exert its action.

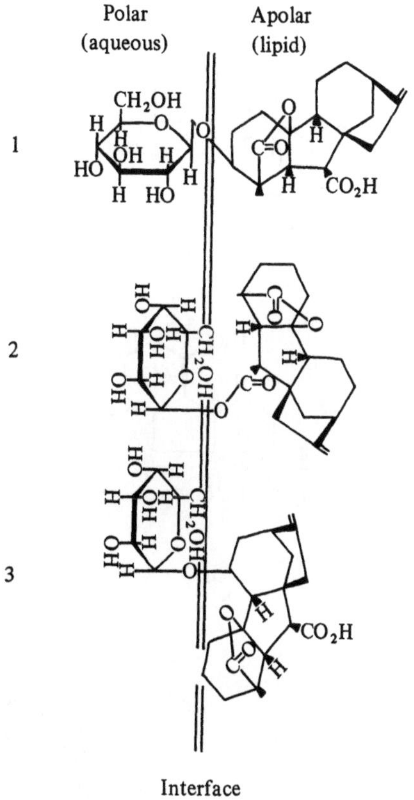

Fig. 5.15. Limitations imposed by glycosylation on the orientation of a gibberellin at an aqueous/lipid interface. *1* A-ring ether conjugation; *2* Ester conjugation; *3* C-ring ether conjugation

In seeking a cellular function for GA conjugates it should be borne in mind that the attachment of a glycosyl group alters the molecular dimensions in different regions of the molecule. This can place interesting limitations on the behaviour of the GA portion of the association at aqueous/lipid interfaces, for example at the surface of a cellular membrane. Some possible orientations, thus imposed, are depicted in Fig. 5.15 which also illustrates the limitations which the polar sugar portion could impose on the penetration of the GA-molecule into the apolar zone. If GA's exert their primary action at a site within, or bounded by, a lipid membrane it can also be envisaged that the sugar moiety might stabilize the GA in the "wrong" orientation or prevent sufficient penetration to the site of action. The same mechanism may operate to compartmentalize a pool of "unwanted" GA's in, for example, the vacuole.

5.4.2 Transcriptional Control: Analogies with Animal Steroid Hormones

Gibberellins, plant sterols, and animal steroid hormones have close structural relationships and all are products of the terpenoid pathway. They share common early biosynthetic steps to the C_{15}-sesquiterpene, farnesyl pyrophosphate (FPP) (see Chap. 4.1.3 and Fig. 4.3). The structures of some plant and animal steroids are shown in Fig. 5.16, together with ent-kaurene and GA_3 for comparison. The biochemistry of plant steroids has been reviewed by HEFTMANN (1963).

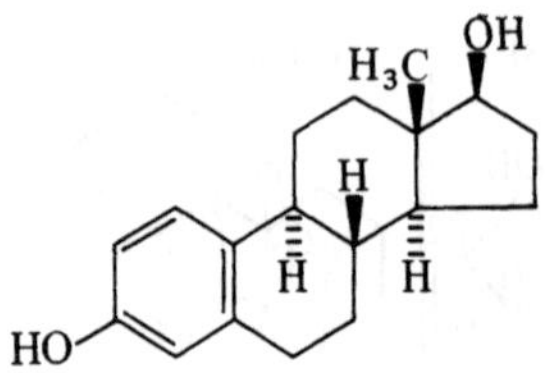

1.

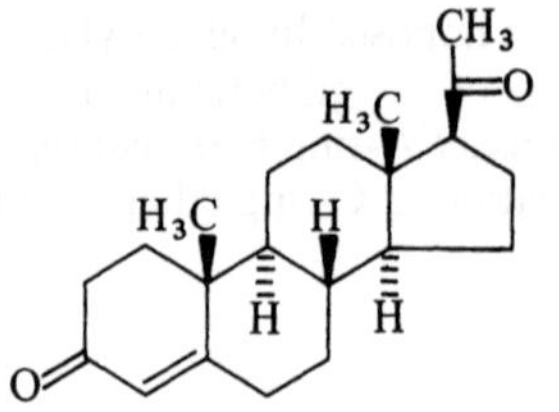

2.

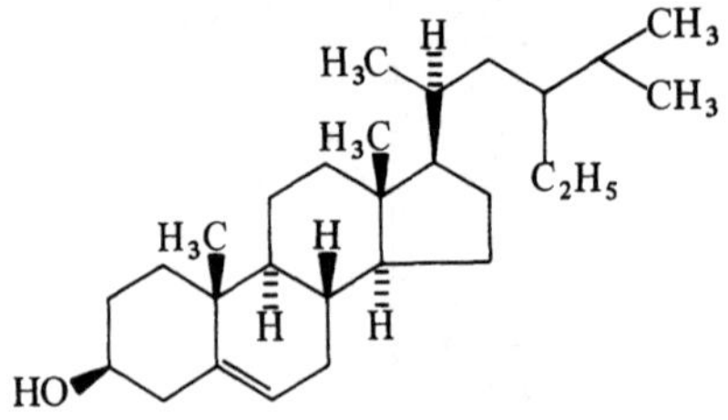

3.

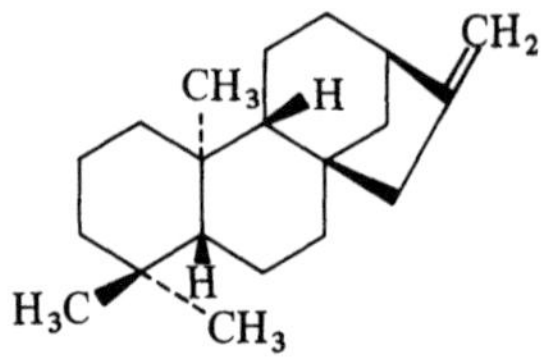

4.

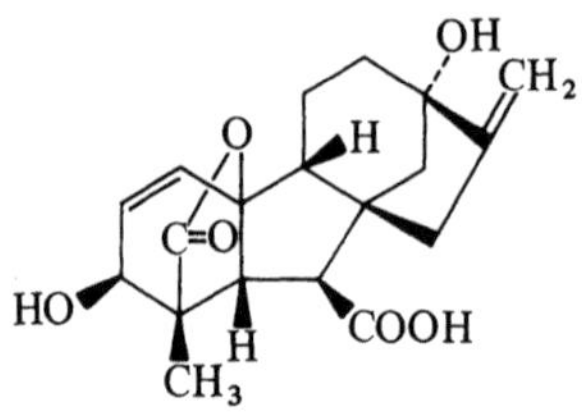

5.

Fig. 5.16. Structures of selected steroids, ent-kaurene, and gibberellin A_3. *1* oestradiol. *2* progesterone. *3* β-sitosterol, *4* ent-kaurene. *5* gibberellin A_3

a) Steroid Hormones

The structural similarities have prompted suggestions that the GA's may have a mode of action which resembles that of animal steroid hormones. In order to examine this possibility it is necessary to outline briefly current views on steroid hormone action.

Specific receptors are the key elements in the animal system. They are soluble cytoplasmic proteins with a sedimentation coefficient of approximately 4S (ca 90,000 molecular weight) and are recoverable as an 8S dimeric complex in cell extracts. There is some discussion as to whether the 4S or 8S form is the normal configuration in intact cells (e.g., O'MALLEY and MEANS, 1974). Receptor proteins are present in all cells but are at a higher concentration in target tissues such as the uterus or oviduct. Binding of steroid and protein is a highly structure-specific process, discriminating in favour of active steroid configurations in the presence of a large excess of structurally similar, but biologically inactive, analogues. Activating mechanisms also exist, exemplified

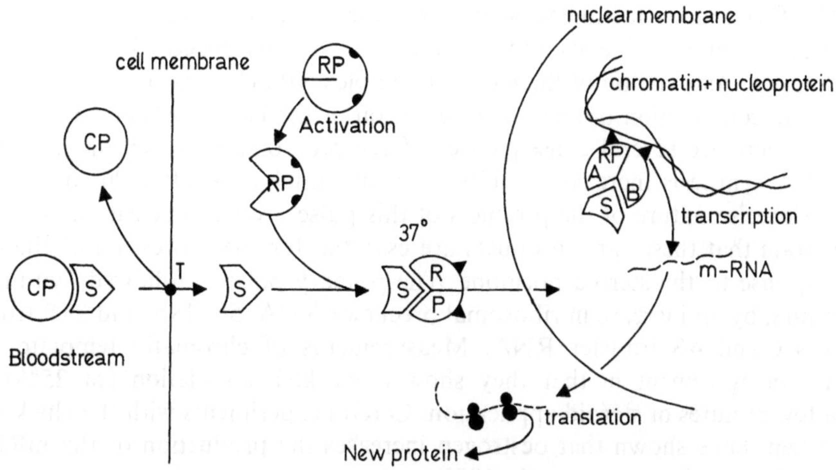

Fig. 5.17. Schematic summary of the action mechanism of an animal steroid hormone (e.g., oestradiol in a uterine cell). *A*, *B* receptor subunit identification; *CP* carrier protein; *RP* receptor protein; *S* steroid hormone; *T* transport system

by the observed enhancement of progesterone binding in tissues previously exposed to oestrogen. The affinity of the receptor for steroid is characteristically high with dissociation constants in the 1 nM to 100 pM range. Therefore, the characterization of steroid receptor mechanisms in animal cells was consequent upon the availability of hormones of high specific radioactivity (100 Ci/mmol) which allowed the detection of the interaction. Hormone-protein complexes were characterized by gel-filtration and sucrose density-gradient centrifugation. Affinity constants were derived by competitive binding experiments in an equilibrium dialysis system. Detailed references to methodology can be found in GORELL et al. (1972), O'MALLEY and MEANS (1974), SICA et al. (1973), and YAMAMOTO and ALBERTS (1976).

Uterine tissue fed with (^{3}H)-17β-oestradiol shows binding in both cytoplasm and nucleus. It is considered that binding in the two sites is not a reflection of separate primary events but rather that it arises as a result of translocation of the cytoplasmic steroid-protein complex to the nucleus (JENSEN et al. 1968). It is thought that the translocation process is initiated by a change in the conformation of the receptor protein occurring after association with the hormone. This is termed the 37° transformation and is reflected in a change in the sedimentation value of the individual sub-units from 4S to 5S. The passage of the complex across the nuclear membrane has not been described, but there have been concentrated studies on the events occuring within the nucleus after entry of the complex. For detailed consideration the reader is directed to O'MALLEY and MEANS (1974) but the essential elements can be summarized as follows (Fig. 5.17). The receptor is thought to consist of two sub-units (A and B), one of which is complexed with the steroid hormone. Sub-unit A is believed to attach directly to the DNA strand in the ratio of one receptor unit per 10^6 DNA nucleotide pairs. The association, in the case of progesterone and purified chick DNA, has a K_D value of about 0.3 nM. The second sub-unit

(B) forms an attachment with a specific fraction of chromatin non-histone protein but the kinetics and stoichiometry of this process have not yet been detailed. Association of the receptor complex with chromatin results in a change in the composition or steric conformation of the latter component.

There are several consequences of receptor binding to chromatin. Nuclear RNA synthesis increases rapidly with an initial peak after 20 to 30 min but the precise nature of the products of this pulse are still unclear. It is, however, certain that these early products are essential for the expression of the cellular response to the steroid hormone. The primary pulse is followed, within a few hours, by an increase in ribosomal precursor RNA, 5S, 18S, and 28S ribosomal RNA and 4S transfer RNA. Measurements of chromatin template activity are in agreement in that they show a marked stimulation (ca 25%) within a few minutes of steroid application. Careful experiments with the chick oviduct system have shown that oestrogen increases the production of the mRNA for ovalbumin (Comstock et al., 1972).

A summary of the postulated response scheme for steroid hormones is depicted in Fig. 5.17. There is still considerable uncertainty about many of the details but the overall picture is reasonably clear. The characteristic and vital features are the existence of receptor proteins in the cytosol, the migration of complexes to the nucleus, two-point association with chromatin, and the formation of more, or new, RNA species which are, in turn, transported out of the nucleus for translation.

b) Gibberellins

To what extent can the steroid system, or elements of it, be applied to the action of the GA's in plant tissues? Is it valid to attempt to do so? In addressing the first query it must be recognized at the outset that there is really no firm evidence from plants which would allow any sort of meaningful overall conclusion. However, various workers have investigated gibberellin-tissue interactions in a way which allows us to consider obliquely facets of the question.

Firstly, the existence or otherwise of specific receptor proteins for GA's in the plant cell cytoplasm has been investigated. The search for such receptors was made possible by the synthesis of GA's with high specific radioactivity. For example (1,2-^{3}H)-GA_1 at 40–50 Ci/mmol has been prepared (Pitel and Vining, 1970; Nadeau and Rappaport, 1974) and has been used to probe the existence of specific receptor proteins in dark-grown pea epicotyls. Epicotyl sections, fed with labelled GA_1 or GA_5, concentrated radioactivity in the growing hook region of the epicotyl and irradiation with red light, which decreases sensitivity to exogenous GA, greatly reduces the extent of accumulation (Musgrave et al., 1969).

Using ^{3}H-GA_1 of very high specific activity, Stoddart et al. (1977) detected associations between GA_1 and soluble macromolecular components in extracts of the epicotyl hook. After feeding the ^{3}H-GA_1 to intact shoots a 20,000 g supernatant was prepared and subjected to gel filtration on Sephadex G-200 in a manner directly analogous to that employed in steroid hormone studies. The type of profile which was obtained is illustrated in Fig. 5.18. Binding

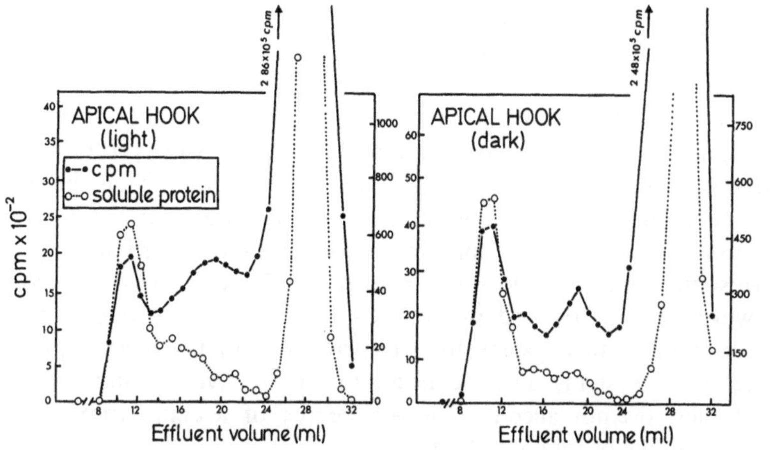

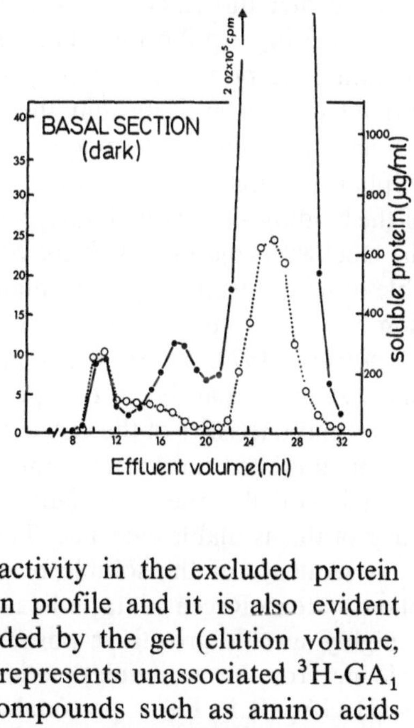

Fig. 5.18. Sephadex G-200 elution profiles from various homogenates of dwarf pea epicotyls. Columns loaded with 20,000 *g* supernatants from dark-grown material fed with [^{3}H] gibberellin A_1. (Redrawn from STODDART et al., 1974)

of GA_1 is indicated by the presence of radioactivity in the excluded protein peak (elution volume, 8–13 ml) on the column profile and it is also evident that some binding occurs to components retarded by the gel (elution volume, 13–22 ml). The very large peak of radioactivity represents unassociated ^3H-GA_1 which coincides with low molecular weight compounds such as amino acids and peptides. When ^3H-GA_1 uptake proceeded in the light a reduction occurred in the total bound radioactivity and an even greater reduction was obtained when non-GA-responsive basal sections of hypocotyls were examined. Using similar methods, KONJEVIĆ et al. (1976) have reported binding of ^{14}C-GA_3 to epicotyl extract components excluded by Sephadex G-10.

Are such associations significant? As anyone who has fractionated extracts of plants fed with isotopically labelled compounds will testify, there is a tendency for non-specific binding to proteins and cellular components to occur, an effect which can be exaggerated by the use of very high specific activity compounds. Potentially significant events must, therefore, be viewed against this background. The ability of the binding process to differentiate between biologically active and inactive structures has been tested in pea epicotyls by supplying ^3H-GA_1

in the presence of inactive, or less active structural analogues. Gibberellin A_8 and 3-epi-GA_1, biologically inactive compounds, did not reduce the overall ^3H-GA_1 binding level and were not recovered from any of the macromolecular associations. On the other hand 17-keto-GA_1, which has biological activity in the intact dwarf pea assay, competed with ^3H-GA_1 for the binding sites and was recoverable from the macromolecular fraction. Thus the structure specificity and the correlation of the level of binding with the GA sensitivity of the tissue would suggest that the associations observed in pea extracts may have some significance in relation to the biological activity of that compound. However, as KENDE and GARDNER (1976) have pointed out, the kinetics of GA binding in this system do not correlate with the dose-response curves for intact plant systems and the levels of radioactivity found in a bound form in extracts do not account for the degree of accumulation observed in the growing zones of intact sections.

A further difficulty is raised by the failure to demonstrate binding effects in vitro using equilibrium dialysis in a manner analogous to that employed in animal receptor studies. Competitive dialysis studies on ^3H-GA_1 associations formed in vivo indicate that the radioactivity is not readily exchangeable although, surprisingly, the majority of the GA_1 can be displaced by ammonium sulphate precipitation (STODDART, 1975). Competition for uptake and for loading of the binding sites with $^3HGA_1$, can be applied in vivo by progressively diluting the label with cold GA, with the result that labelled complexes become undetectable at concentrations greater than 1.0 mM. Cold GA_8, in the same concentration range, does not compete with ^3H-GA_1 for the occupation of binding sites (STODDART, 1974, 1975). Whilst the initial gel-filtration profiles are suggestive of a system comparable to the specific steroid hormone receptor, an examination of the characteristics of the GA-associations indicates that this conclusion cannot be contemplated until much more kinetic and physical data are available.

It is equally true that plant cytosol receptors cannot be ruled out on the basis of the available evidence. Transfer of radioactivity to the nuclear fraction, a consequence of the soluble receptor system, has not been observed for any plant system. Recent data indicate that in the lettuce hypocotyl, a tissue with a highly specific growth response to some GA's, there is no autoradiographic evidence for the accumulation in the nuclear region of radioactivity supplied as ^3H-GA_1 (R.L. JONES, personal communication). Nevertheless, there is a body of evidence to suggest that GA's influence the rate of transcription in some systems. Nuclei isolated from light-grown pea internodes showed an increase of 60% to 90% in the rate of incorporation of labelled RNA precursor when GA_3 was present throughout the extraction procedure and this RNA, when examined by nearest-neighbour frequency analysis and methylated albumin - kieselguhr (MAK) column chromatography, was shown to differ from control RNA both in chain length and base sequence (JOHRI and VARNER, 1968). An effect on RNA synthesis was also obtained when GA_3 was added to nuclei after isolation but this occurred over a much longer time course (16–18 h) and was considered not to be related to the mechanism of the primary action. The fact that responses are obtained with isolated nuclei, albeit in the presence of cytosol proteins during the early stages of separation, is not supportive of the idea that soluble receptors are obligately involved in the transfer of GA into the nucleus. This does not preclude specific associations with macromolecules as a means of protecting active GA's from metabolism during passage through the cytoplasm and across the nuclear envelope, nor does it rule out the possibility that, in plants, specific receptors may be intranuclear.

Support for the preceding findings has been given by McCOMB et al. (1970) who found that GA_3-treated plants showed marked increases in RNA polymerase activity and that these changes preceded the enhancement of growth rate. Addition of *E. coli* polymerase to pea chromatin isolated at various points on the response curve did not indicate that GA increased the template activity, suggesting that the RNA response was quantitative rather than qualitative. Changes in template activity may have been below the sensitivity threshold of the technique but there are precedents, drawn from soybean hypocotyls treated with 2,4-dichlorophenoxyacetic acid, for increases in polymerase activity without change in available template (O'BRIEN et al., 1968).

Probably the best evidence in favour of GA influencing events at the transcriptional level has been obtained with the barley aleurone system. JACOBSEN and ZWAR (1974a, b) and HO and VARNER (1974) have found that GA_3 stimulates the incorporation of ribonucleosides into a wide range of poly(A)-containing RNA's and that the effect can be detected within 4 h of exposure to GA_3. Total and poly(A)-containing RNA from treated and control aleurone layers were translated in vitro in a wheat germ cell-free system. Total RNA gave rise to a spectrum of protein products varying in size from 8,000 to 70,000 daltons. Only two polypeptides were noticeably increased by GA_3 treatment and these had molecular weights of 35,000 and 45,000 daltons (HIGGINS et al., 1976). The poly(A)-RNA fraction had a higher template activity (10×) but produced the same range of protein products in the wheat germ system.

Immunoprecipitation with monospecific antibody prepared against barley α-amylase yielded a single polypeptide which had similar electrophoretic properties to authentic α-amylase. Time-course studies of the rate of α-amylase synthesis in vivo and in vitro showed an increase with time in the presence of GA_3 and the proportion of the total protein made in the cell-free system which was precipitable by anti-amylase showed an increase which followed the same time course (Fig. 5.19). These data provide the strongest evidence to date to support a transcriptional role for GA in the barley aleurone cell expressed as an increase in the mRNA for α-amylase. The authors point out that several mechanisms exist whereby the response could be achieved e.g., (i) a decreased rate of mRNA degradation; (ii) an enhancement of the translational capacity of mRNA for α-amylase; (iii) synthesis of new mRNA molecules.

On the basis of the existing data it is not possible to choose between these mechanisms. It is also evident that, because of the relatively long lag period between GA addition and maximal rate of response, we cannot be certain that the observed change is a primary response to GA. Other cellular interactions may have occurred prior to the step influencing transcription.

JONES and CHEN (1976) have applied immunohistochemical techniques to sections of barley aleurone layers and have shown that α-amylase-specific immunofluorescence is localized in the perinuclear region. They have also noted

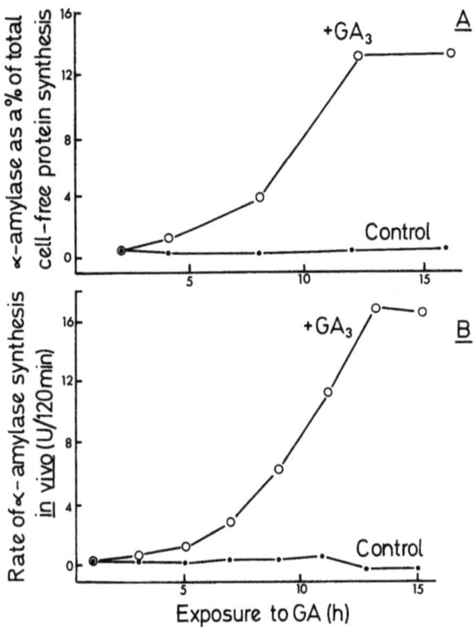

Fig. 5.19. Time course of α-amylase production during cell-free translation of total RNA isolated from gibberellin A_3-treated and control barley aleurone layers. **A** Time course of α-amylase production during cell-free translation of total RNA isolated from gibberellin A_3-treated and control aleurone layers. **B** Rates of α-amylase synthesis by intact gibberellin A_3-treated and control aleurone layers. (Redrawn from HIGGINS et al., 1976)

that GA-stimulated proliferation of endoplasmic reticulum tends to be similarly concentrated. These data give support to the case for GA-stimulated α-amylase synthesis being dependent upon RNA export from the nucleus, but provide no evidence that this is a primary response.

Overall, therefore, it is conjectural whether a steroid type of mechanism exists in plants. Cytosol receptors may, or may not, be present and there is little evidence for accumulation of hormone in the nucleus. Both aspects require considerable further study. The experiments with isolated nuclei from peas are suggestive of a direct response of the organelle, expressed as a stimulation of long-term RNA synthesis, invoking in turn the suggestion that a form of GA receptor system exists within the nucleus itself. On the other hand, shorter-term stimulation of transcription appears to require the presence of extra-nuclear components and, in this instance, soluble receptors or feedback effects from an extra-nuclear primary action site may be involved.

The general area of hormonal control of RNA metabolism in plants has been reviewed by JACOBSEN (1977).

5.4.3 Effects on Membrane Organization

Evidence relating to control at the post-transcriptional level as a possible mode of action for GA emanates mainly from studies conducted with the barley aleurone layer. Cytological studies, notably those by JONES (1969a, b) have indicated that, in the presence of GA_3, endoplasmic reticulum (ER) and rough endoplasmic reticulum (RER) become more prominent in electron micrographs of the treated cells. There has been some discussion as to whether these observations reflect a net increase in the total amount of membrane material or merely represent a change in the organization of a pre-existing membrane pool. The same question can be raised in relation to the ribosomal component of RER (e.g., JACOBSEN, 1977).

JOHNSON and KENDE (1971) have reported that phosphorylcholine-cytidyl transferase (PCT) and phosphorylcholine-glyceride transferase (PGT) increase two- to four fold in activity during the period between 2 and 12 h after GA_3 treatment, suggesting that the synthesis of lecithin (a major phospholipid membrane component) is stimulated. The pathway is shown below.

$$\textit{Choline} \xrightarrow[\text{Choline kinase}]{\text{ATP}} \textit{Phosphorylcholine} \xrightarrow[\text{PCT}]{\text{Cytidine triphosphate}} \textit{CDP-choline} \xrightarrow[\text{PGT}]{\text{Diglyceride}} \textit{Lecithin}$$

Similarly, EVINS and VARNER (1971) found that ^{14}C-choline was incorporated into a partially purified ER preparation at a significantly higher rate in the presence of GA_3.

Polysome formation (i.e., the aggregation of ribosomes) also appears to be stimulated in the presence of GA_3 and the number of ribosomes per cell is increased (EVINS and VARNER, 1972). Enhancement of polysome formation by GA_3 was negated by the addition of abscisic acid.

These data point to a role for GA_3 in the regulation of polysome and RER formation, either as a consequence of transcription or by a direct effect on the organelles themselves. The effects of various inhibitors on polysome formation were examined by EVINS and VARNER (1972). Fluorouracil, an inhibitor of ribosomal RNA synthesis, depressed the GA_3-induced increase in ribosome numbers but had no effect on polysome frequency. In contrast, actinomycin-D inhibited both ribosome appearance and conversion to polysomes with the inference that stimulation by GA_3 of these processes was dependent upon mRNA synthesis.

An interesting facet of the relationship between GA_3 and RER formation has been described by ARMSTRONG and JONES (1973). These authors studied the effect of water stress, induced by 0.6 M solutions of polyethylene glycol, on the synthesis of α-amylase by GA_3-treated aleurone layers. This treatment caused a dramatic reduction in the binding of ribosomes to ER when assessed under the electron microscope, and density gradient studies indicating a marked suppression of polysome formation confirmed the effect. Incorporation studies with ^3H-puromycin showed that the synthetic capacity of the individual ribosomes was not impaired by the presence of the polyethylene glycol and the total RNA content of the cells was only marginally depressed. On removal of the osmotic stress there was a re-association of RER components and the aleurone layers recovered their α-amylase synthetic capacity in a linear fashion over the course of 4–5 h. These results demonstrate the close dependence of enzyme synthesis on the binding of polysomes to ER, but suggest that GA_3 is not directly involved in the attachment of the polysomes to the membrane. There have been suggestions that animal steroid hormones may regulate ribosome attachment. SUNSHINE et al. (1971) and BLYTH et al. (1971) have produced evidence to indicate the existence of binding sites on rat liver membranes with a high affinity for steroid hormones and postulated that these were adjacent to ribosome attachment sites. Occupation of the steroid site was considered to be permissive for ribosome binding.

It seems most probable that the observations on barley aleurone cells, correlating increased membrane and RER formation with the presence of GA_3, are reflections of a series of secondary events consequent upon an increase in RNA production. The possibility that RER formation may be directly influenced, either by an action of GA_3 in potentiating ribosome binding or in facilitating attachment of mRNA to the ribosome, remains unsupported and cannot, therefore, be considered as a serious alternative at this time. However, the recent observation by JONES (1978) that EDTA-treated (i.e., smooth) ER from GA-incubated aleurone layers has a buoyant density of 1.11 g/cm^{-3} compared to a value of 1.13 g/cm^{-3} for control layers opens the door on a new aspect of membrane organization and suggests that the growth regulator may exert an effect on lipid/protein relationships in ER.

5.4.4 Action in Artificial Membrane Systems

The possibility that the biological effects of GA's may be the consequence of an interaction with cellular membranes has already been mentioned in Section 5.4.3 and the idea of a membrane-located receptor site is one which is frequently advanced. Interaction of GA_3 with plant membrane components has been directly investigated by WOOD and PALEG (1972, 1974), using artificial membrane micelles derived from crude soybean lecithin (phospholipid).

The models are based upon the ability of phospholipids to become arranged in self-ordered concentric membrane structures (liposomes) when dispersed in aqueous media (e.g., BANGHAM et al., 1965). To produce the liposomes crude soybean lecithin, deposited initially as a dried film on the wall of a flask, is shaken vigorously with aqueous buffer containing a tracer compound (sugar, dye, etc.). The agitation causes the formation of membranous structures enclosing tracer compound at the concentration present in the original solution. The liposomes can be freed from residual buffer and from untrapped tracer compound by gel-filtration. In addition to lecithin the membranes also contain a sterol and dicetyl phosphate, the latter acting as a negatively charged agent to ensure separation of the individual membrane layers. The identity of the sterol does not seem to be critical, but the greatest entrapment capacity was observed with β-sitosterol.

Liposomes made in this manner had a glucose leakage rate of between 20% and 30% when incubated in buffer at 35 °C and this efflux tended to decline with increasing incubation time (WOOD and PALEG, 1972). In the presence of GA_3 a concentration-dependent enhancement of leakage rate was observed (Fig. 5.20) saturating at a value of approximately 1.2 mM and with a lower threshold at 25 μM. The lecithin used for the liposome preparations contained high impurity levels and consisted of 75% polar lipids (of which about 50% were phospholipids) and 25% non-polar material as well as sterols, free fatty acids, and glycerides. Leakage rates and GA_3 responsiveness were also assessed in liposomes prepared from purified phospholipid fractions. In these studies

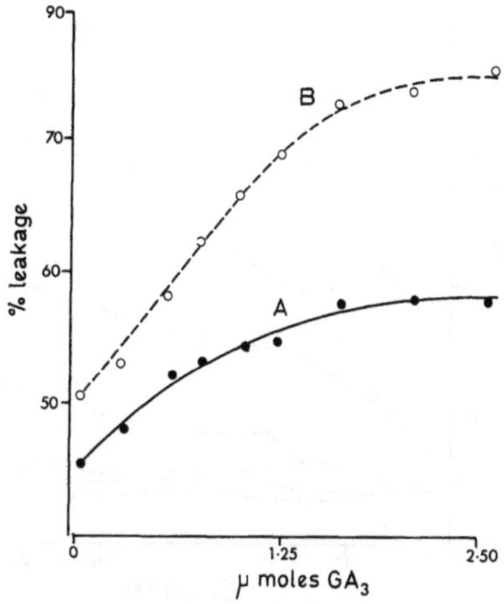

Fig. 5.20. Effect of gibberellin A_3 on the permeability of liposomes to glucose. *A* liposomes prepared from lecithin, cholesterol, and dicetyl phosphate. *B* liposomes prepared from lecithin, β-sitosterol, and dicetyl phosphate. (Data from WOOD and PALEG, 1972)

all liposomes contained 20% cholesterol, 20% dicetyl phosphate, 50% purified phosphatidyl choline, and the remaining 10% consisted of various purified phospholipid fractions. The behaviour of the preparations was compared with that achieved with crude lecithin. The glucose entrapment properties were markedly enhanced in all preparations containing phosphatidyl serine and at optimal β-sitosterol levels the glucose content of the micelles approached three times the value found in crude lecithin liposomes. However, none of the preparations containing phosphatidyl serine showed any marked responsiveness to GA_3 and the largest increase in leakage rate (9.1%) was obtained with phosphatidyl inositol addition. Even this increase was minute when compared to the 65% enhancement obtained when GA_3 was added to crude phosphatidyl choline preparations.

It is possible that the GA_3 response in the crude lecithin liposomes relates to some unidentified component not present in the purified phospholipid preparations; an explanation which WOOD and PALEG (1972) link with the observation that such liposomes have high initial leakage rates which may mask or preclude a subsequent hormonal response. The possibility that a non-phospholipid impurity may be responsible for the observed effects on permeability is dismissed on the grounds that steroids do elicit a leakage response. More disturbing is the observation that GA_8 (biologically inactive in most assay systems) is as potent as GA_3 in accelerating glucose leakage. A response was also obtained with indole-3-acetic acid, but kinetin was ineffective.

Plant lipids undergo a thermal transition between the liquid-crystalline and gel states at a temperature which is related to the degree of saturation of the membrane fatty acids (e.g., RAISON, 1973), a process which results in modified kinetics for membrane-associated enzymes. WOOD and PALEG (1974) examined the effect of GA_3 on thermal transitions occurring in their artificial membrane systems (Fig. 5.21). In the absence of the hormone an abrupt decrease in leakage rate occurred at a temperature of 25 °C and the loss of glucose remained steady at 5% at all temperatures below that value. A similar discontinuity was seen in the presence of GA_3 but here the transition occurred at about

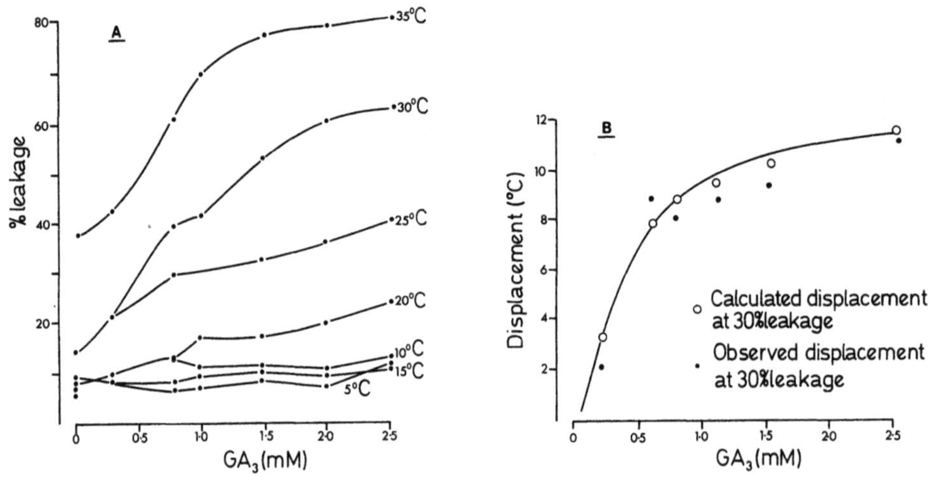

Fig. 5.21. Effect of temperature on the rate of glucose leakage from liposomes exposed to increasing gibberellin concentrations. **A** Leakage curves for selected temperatures plotted against gibberellin concentrations. **B** Interactions of temperature and gibberellin concentration determined at a constant glucose leakage rate of 30%. Based on an Arrhenius transformation of the information in **A**. (Redrawn from WOOD and PALEG, 1974)

15 °C. The rate of leakage in the presence of GA_3 was greater both above and below the transition point (Fig. 5.21 A).

When the data were plotted according to the Arrhenius relationship (rate of leakage vs. $^1/T^0$ K) essentially parallel slopes were obtained for the entire concentration range from 0.25 to 2.5 mM GA_3 with a progressive displacement of the line with increasing temperature. The displacement of the phase change could be assessed by adopting a standard leakage rate for each GA_3 concentration and the relationship obtained is depicted in Fig. 5.21 B. The curve is biphasic with a higher rate of response up to approximately 1.0 mM and a reduced slope above that concentration. The maximum change of transition temperature observed in these studies was of the order of 12 °C.

Effects on permeability elicited by steroid hormones in similar preparations have been described by BANGHAM et al. (1965) and WEISSMANN et al. (1965). In animal systems pharmacologically active steroids have been shown to affect the membranes of lysosomes, erythrocytes, and mitochondria, causing visible disruption of the organelle surface in many instances. Steroids not possessing an oxygen function in the 11 or 17 position (e.g., deoxycorticosterone or progesterone) were particularly potent in this respect (DE DUVE et al., 1966; BLECHER and WHITE, 1960). These effects were distinguished from those caused by Triton X-100 and digitonin by electron microscope studies which indicated that the surfactants caused complete disruption of the liposome micelles whilst the steroids did not. WOOD and PALEG (1974) were also able to demonstrate that diethyl stilboesterol caused an alteration in both the leakage rate and the thermal transition point in their liposome preparations.

MUDD and KLEINSCHMIDT (1970) have shown that the 35-carbon polyene antibiotic, filipin, isolated from *Streptomyces griseus*, will greatly enhance the efflux of betacyanin from discs of red beet and that this process occurs without any apparent ultrastructural change in either the tonoplast or plasmalemma. The compound also depresses fungal spore germination but its influence can be reversed by the addition of cholesterol to the medium (GOTTLIEB et al., 1960). Studies with lipid mono- and bilayers (KINSKY et al., 1968) have shown that the leakage rate is only affected if cholesterol has previously been incorporated into the membrane.

There is evidence, therefore, that GA_3 and other compounds can affect the permeability of liposomal membranes without necessarily causing the type of total disruption characteristic of detergents. What are the possible molecular bases for such an action? Phospholipid/cholesterol structures tend to have a highly ordered arrangement based upon the regularity of bonding or charge interactions between arrays of similar molecules. WILLMER (1961) has suggested that the insertion of dissimilar molecules (such as steroids or gibberellins) into such a matrix would result in discontinuities. These could have effects on the permeability properties of the membrane. This hypothesis has been examined for steroids by various workers (e.g., TAYLOR and HAYDON, 1965) without definitive results; mainly because of the analytical problems posed by the very small amounts of hormone actually incorporated into the membrane. There is evidence, however, derived from electron micrographs (WEISSMANN et al., 1965), to indicate that liposomes treated with deoxycorticosterone differ in conformation from those in control preparations.

KENDE and GARDNER (1976) have drawn attention to the parallels between plant responses to GA's and the olfactory response in animals. In both instances the effect is evident over several decades of concentration and also diverges from Michaelis-Menten saturation kinetics at the higher end of the range. The olfactory process can be explained in terms of the stimulatory compound partitioning into an epithelial membrane (MOZELL, 1970; MOZELL and JAGODOWICZ, 1973) and the response occurs when a molecule desorbs from the membrane leaving a "hole" or ionophore through which ions can pass to initiate nervous impulses. There is, thus, a clear parallel with the changes in solute flow observed in the lecithin liposome/GA_3 systems.

What is the nature of the association of GA_3 with the liposomal membrane? WOOD et al. (1974) have studied this problem using proton magnetic resonance spectroscopy. Solutions of phosphatidyl choline (lecithin) in deuterochloroform, examined at a frequency of 60 MHz, showed identifiable resonances for the N,N,N-trimethylamino group and olefinic protons. Complex envelopes of signals associated with -$(CH_2)_n$- and -CH-O-protons were also observed. In the presence of GA_3 there was a proportional shift of the trimethylamino group resonance towards higher magnetic field. Smaller resonance band-broadening responses were also detected for other features of the molecule.

The authors interpreted these changes as being a consequence of an electrostatic attraction between the cationic trimethylamine group of lecithin and the carboxyl group of GA_3:

$$\text{lecithin} + GA_3 \leftrightharpoons \text{complex}$$

It is rightly emphasized that the observed shifts occurred in a deuterochloroform environment and that the data should not be immediately extrapolated to the aqueous state. They are, nevertheless, suggestive of an association of GA_3 with a basic membrane component in the liposome.

The implication of the carboxyl group does raise some serious points in relation to the specificity of the GA effects on liposome permeability. In their original publication WOOD and PALEG (1972) showed that GA_3 and indole-3-acetic acid were active in altering permeability whilst kinetin was not, and in 1974 the same authors showed that the methyl ester of GA_3 did not cause a shift in the temperature of phase transition in liposomes. These data, together with the proton magnetic resonance studies, argue strongly in favour of a carboxyl group being the over-riding feature required for activity as a regulator of permeability. It is argued that the activity of non-carboxylic steroids such as diethylstilboestrol points to features other than, or in addition to, the carboxyl group as having importance, but it is difficult to assess, in such instances, what the effect of gross variations in the basic structure of the two molecules might be. If the capture of a "foreign" molecule within the membrane matrix is the important event initiating increased permeability, the actual nature of the chemical bond may not be critical and a multiplicity of chemical interactions could occur in the presence of compounds with suitable and available functional groups. It would be informative to study the effectiveness of compounds such as abscisic acid and, possibly more significantly, the efficacy of a range of carboxyl-containing moieties such as organic acids and the herbicidal phenol carboxylic acids.

The specificity of phospholipid binding to IAA and other auxins has been examined by VEEN (1974). He confirmed the observation of WEIGL (1969) that lecithin and IAA bind in a molar ratio of 1.0:0.8 but showed that this did not hold for other synthetic auxins and analogues, as exemplified by naphthalene-1-acetic acid. Biologically inactive analogues of NAA (e.g., naphthalene-2-acetic acid) also showed an association with lecithin suggesting that the process was not directly related to the biological action of the auxins.

Gibberellin responses in vivo are highly structure- and stereo-specific processes and the molecular features which affect biological activity have already been discussed in Section 5.4.1. This means that it is comparatively easy to test the specificity of an in vivo GA response by considering the effectiveness of a range of structural variants of known potency. To date there is only one such piece of information for the lecithin liposome system, namely a comparison of GA_1 (highly active) and GA_8 (inactive). Both were equally effective in increasing the rate of glucose leakage and this would seem to mitigate against the use of the liposome system as a general model for in vivo GA action. Indeed, it is possible that effects on liposomes may be more related to uptake and penetration phenomena than they are to interactions with a primary growth-regulating site. Much more data is required before any useful conclusions can be drawn.

Overall, the idea that GA action is concerned with membrane properties and that the primary action site is also membrane-located has many attractions. It fits well with the kinetic data for response, provides possible explanations for the diversity of type and timing of response, suggests why intensive efforts to isolate soluble receptor proteins may have failed, and maintains a reasonable analogy with aspects of steroid hormone action in animal systems. Active GA's have a strong affinity for lipid-containing sub-cellular structures, less active structures and conjugates are much more polar. The highly active, but polar, GA_{32} is a notable exception to this generalization.

These observations suggest that cellular membranes will continue to be investigated as a primary site of GA action, although we may still be ultimately concerned with proteinaceous sites located within a membrane matrix.

5.4.5 Changes in Cell Extensibility

Recent studies on GA action in lettuce hypocotyl sections, where response is entirely confined to cell elongation, have indicated that an increase in cell extensibility occurs on exposure to the growth regulator (e.g., SILK and JONES, 1975; STUART and JONES, 1977; STUART et al., 1977).

In contrast to auxin effects, there is no accompanying proton efflux (STUART and JONES, 1978) and wall acidification was not detected. The nature of the wall-softening process and its relationship to auxin effects on the turnover of cell-wall glycoprotein linkages remains to be explored.

However, the distribution of ^3H-GA_1 amongst sub-cellular fraction from lettuce (*Lactuca sativa* L.) hypocotyl sections has been examined by differential centrifugation (STODDART, 1979a). Material pelleting at 2000 *g* (2KP) accounted for 2–5% of the total tissue radioactivity and no other significant interactions were detected. After rigorous washing procedures, the 2KP fraction was found to consist mainly of cell-wall material and the radioactivity remained associated during sucrose density-gradient centrifugation. 2KP labelling increased linearly with time and was only partially removed by chasing for up to 40 h in the presence of unlabelled GA_1. A linear plot was obtained for the relationship between external concentration and the log of 2KP GA_1 content with a maximum percentage incorporation at 10^{-5} mol l^{-1}. Incorporation of radioactivity was reduced by chasing with unlabelled GA_1 or GA_9, but not by GA_8.

Labelling of the 2KP fraction increased uniformly with temperature between 15° and 30 °C but showed little change below 15 °C. Growth rate and 2KP labelling were highly correlated (r=0.989).

The incorporated radioactivity was stable in 0.1 M buffer at pH 3 and pH 9 and also in 1 M salt, organic solvents, protease, or cellulase. Both 1 M potassium hydroxide (KOH) and a quarternary ammonium hydroxide tissue solubilizer effected an 80% release of incorporated ^3H. No radioactivity was released during polyacrylamide gel electrophoresis (STODDART, 1979b).

Gel chromatography of KOH digests indicated the presence of labelled compounds with a higher molecular weight than GA_1. This material remained near to the origin during high-voltage paper electrophoresis. Calculations based upon KOH extraction of 2KP material suggested that 5% to 20% of the total uptake of ^3H-GA_1 could be recovered from this fraction.

The relationship between protein synthesis and incorporation into the 2KP fraction was also investigated (STODDART and WILLIAMS, 1979). Concentrations of L-2-(4-methyl-2,6-dinitroanilino)-N-methylpropionamide (MDMP) between

10^{-7} M and 10^{-4} M caused increasing inhibition of growth, 2KP labelling and incorporation of ^{14}C leucine into soluble protein. Transfer to MDMP early or late in the course of GA response caused reductions in both growth and incorporation into the 2KP fraction. Exposure to the inhibitor had more effect at 4 h than at 20 h. The proportions of alkali-soluble and insoluble radioactivity in the 2KP fraction were also altered by this treatment, resulting in a higher proportion of KOH-soluble ^3H samples treated with MDMP at 4 h.

Whilst it is not yet possible to say whether the close relationship between growth and 2KP incorporations is causal or consequential, it must nevertheless be considered as a possible mode of action in a system where the ultimate response to GA appears to be related to wall-softening.

5.5 Abscisic Acid

The brevity of this section reflects the dearth of direct information on ABA receptors. The biological activities of a large number of ABA analogues have been reviewed by MILBORROW (1974); more recently, BITTNER et al. (1977) have presented data on 36 aromatic analogues. The structure-activity pattern is often complicated by susceptibility to metabolism during the lengthy bioassays frequently used. Nevertheless, the large activity changes resulting from minor structural variations argue the need for a precise molecular fit in some receptor site. Thus isomerization of Δ^2-cis-ABA to Δ^2-trans-ABA virtually eliminates activity, while in a rapid bioassay system the naturally occurring (+)-optical isomer is very much more active than the (−)-enantiomer (CUMMINS and SONDHEIMER, 1973).

PEARSON and WAREING (1969) found that 1 μM ABA inhibited RNA polymerase of radish cotyledon chromatin. The effect was observed only if ABA was included in the initial grinding medium, not if it was simply added to the assay mixture, suggesting perhaps the loss of an essential mediator protein during chromatin preparation (cf. MATTHYSSE and PHILLIPS, 1969; Sect. 5.2.3.a). This observation does not seem to have been pursued further. One report of actual ABA binding has recently appeared. HOCKING et al. (1978) found that ^3H-ABA bound in a saturable manner to membraneous preparations from bean leaves. On sucrose gradients, maximum binding was obtained in fractions thought to be enriched in plasma membrane. Scatchard analysis suggested two classes of binding sites, the higher affinity site having a $K_D = 35$ nM. The data are preliminary in nature and no information on binding specificity is available. Nevertheless, it is to be hoped that this report will stimulate further work on possible ABA receptor systems.

5.6 Ethylene

Ethylene (C_2H_4) is the only known gaseous plant growth regulator. It is moderately water-soluble, one volume dissolving in four volumes of water at 0 °C and in nine volumes at 25 °C. These properties allow it to pass rapidly between tissues with the minimum of hindrance in either the gaseous or liquid phase. It has been observed to influence a plethora of plant responses and functions including cell-division, stem growth, flowering, fruit ripening, leaf abscission, leaf senescence, epicotyl curvature, response to drought stress, and seed germination. For descriptions and discussion of such responses reference should be made to Volume 10 of this Encyclopedia and to the numerous available review articles (e.g., PRATT and GOESCHL, 1969; ABELES, 1972).

5.6.1 Structure-Activity Relationships

BURG and BURG (1967) have discussed the structural features of the ethylene molecule which are necessary for biological action, measured as effects on the growth and curvature of etiolated dwarf pea stem sections. Activity requires the presence of an unsaturated bond adjacent to the terminal carbon atom and potency shows an inverse relationship to molecular size. The terminal carbon must be electrophilic and substitutions at this position which cause electron delocalisation reduce biological effectiveness (e.g., insertion of fluorine). The inactivity of acetonitrile, contrasted with the high potency of methyl acetylene, suggests that N cannot be substituted for the C-atom at the terminus of the double bond.

5.6.2 Aspects of Molecular Action

In seeking an action mechanism to unify the range of responses there has been a tendency to consider general effects on basic cellular components and membranes have been a prime site of interest. This aspect was explored in an artificial system by MEHARD et al. (1970) who studied effects on the surface tension and conductivity of films of lipid (lecithin or cholesterol), protein (bovine serum albumin or cytochrome c) or mixtures of the two. A range of gases comprising ethylene, propylene, propane, l-butene, and butane were tested and all gave reversible decreases in the surface tension of the thin films. The response increased with increasing molecular size, the smallest changes being induced by ethylene and the largest by butane. No changes in the conductivity of a lecithin-cholesterol bilayer were induced by any of the treatments but chloroform vapour was highly effective in this respect. Thus, in such membrane systems ethylene had no specific action which could be invoked either to support theories of a similar modification in natural cellular membranes or to indicate an analogy between membrane-based anaesthetic effects of ethylene in animals and the growth regulatory properties in plants. The authors are, however, careful to point out that such results do not preclude low-level ethylene interactions in membranes, exemplified by binding to a metal group in a key enzyme, which might have secondary effects on permeability. The possibility of ethylene binding to a metal at the site of action has been examined by BEYER (1972) using a series of deuterated ethylenes (C_2D_4, cis-$C_2H_2D_2$, trans-$C_2H_2D_2$).

It has been shown that deuteration increases the stability of ethylene-silver ion complexes (ATKINSON et al., 1967) and, therefore, if biological action involves a metal ion interaction, deuteration of the applied ethylene would be expected to enhance the response. In studies on the growth of etiolated pea epicotyl sections C_2H_4 and C_2D_4 were found to be identical in their effectiveness.

The deuterated compounds were also used to probe the possibility that the primary interaction with tissue might involve breaking of carbon-hydrogen bonds in the growth regulator. There are indications that deuteration interferes with reactions involving such bond cleavages (WIBERG, 1955) and, where the event constitutes a rate-limiting step, the reaction rate can be several times slower than with protonated compounds. BEYER argued that the identical effectiveness of ethylene and tetradeuteroethylene in the pea system indicates that C-H bonds are not broken during primary action, assuming that this process constitutes the rate-limiting step. Similarly, application of asymmetrically deuterated ethylenes revealed no change in cis-trans isomerization during action.

On the basis of such evidence it can be concluded that ethylene does not undergo any molecular change whilst exerting its action in the cell, provided that biological action is not associated with incorporation of ethylene into cellular components and that the scale of interaction is not below the threshold detection limits of the methods used.

The possibility that ethylene acts via an incorporation mechanism has not been extensively studied and it is possible, for example, that this regulator may act by entering into associations with membrane proteins in a manner which limits dynamic changes between conformational states, thus biasing the properties of the membrane in which the protein is located.

In a series of recent papers, HALL and co-workers have made significant contributions towards an understanding of the molecular fate of ethylene in plants, and have described for the first time a system that has at least some of the properties expected of an ethylene receptor. It was found that many plant species have mechanisms for compartmenting ethylene (JERIE et al., 1978; JERIE, SHAARI and HALL, 1979). Subsequent work with cell-free systems showed that in *Vicia* this phenomenon was a reflection of oxidation to ethylene oxide (DODDS et al., 1979), while in *Phaseolus,* compartmentation was due to the presence of high affinity binding sites for ethylene, with a K_D (calculated at infinite site dilution) of 0.1 nM (BENGOCHEA et al., 1980a). The ability of structural analogues to compete with ethylene for the binding sites closely parallelled their relative physiological effectiveness (BENGOCHEA et al., 1980b). The investigations were facilitated by the very slow rates of association and dissociation of the ethylene-binding site complex, but this behaviour raises questions as to whether the system functions as a true receptor or whether it has some other role. It has been suggested (BEYER, 1975; BEYER and BLOMSTROM, 1980) that ethylene metabolism and ethylene action may be integrally related. The similarity between the affinities of ethylene for the binding sites in *Phaseolus* and for the oxidation system in *Vicia* have led BENGOCHEA et al. (1980a) to propose that the two activities could represent different manifestations of a common system involved in the regulation of a plant's response to ethylene.

5.7 Concluding Remarks

No single plant hormone "receptor" system yet described satisfies completely all the criteria outlined in Section 5.1.1. Auxin receptors have received the most attention experimentally, but even here the only instance where a binding-dependent biological response has been noted (alteration of RNA synthesis) is the coconut receptor system of BISWAS and co-workers. As already discussed in Section 5.2.3b, the auxin-specificity of this system has not been adequately examined, nor does the nature of the source material lend itself to general experimentation. Corn coleoptiles are a rich source of binding sites (Sect. 5.2.4) and those for auxins have been studied in some detail. The properties of the auxin sites are generally compatible with the view that some of them at least may be receptors, though their relationship to the physiology of auxin action remains to be established. This will not be an easy task, but the fairly rapid development of this system in the last five years and its general availability to any laboratory are grounds for optimism.

In general it is reasonable to consider that evolution proceeds in a conservative manner and that similar types of compound act via similar mechanisms in plants and animals. Overall metabolism provides many examples to support this precept. However, when dealing with hormones it is prudent to consider the dangers of squeezing plants into the animal mould and to be aware that unique or subtly different action pathways may be involved.

References

Abeles, F.B.: Biosynthesis and mechanism of action of ethylene. Annu. Rev. Plant Physiol. *23*, 259–292 (1972)

Anderson, R.L., Ray, P.M.: Labelling of the plasma membrane of pea cells by a surface-localised glucan synthetase. Plant Physiol. *61*, 723–730 (1978)

Armstrong, D.J., Burrows, W.J., Skoog, F., Roy, K.L., Söll, D.: Cytokinins: distribution in transfer RNA species of *Escherichia coli*. Proc. Natl. Acad. Sci. USA *63*, 834–841 (1969)

Armstrong, J.E., Jones, R.L.: Osmotic regulation of α-amylase synthesis and polyribosome formation in aleurone cells of barley. J. Cell. Biol. *59*, 444–455 (1973)

Atkinson, J.G., Russell, A.A., Stuart, R.S.: Gas chromatographic studies of isotopically labelled ethylenes. Can. J. Chem. *45*, 1963–1969 (1967)

Audus, L.J.: Plant growth substances, Vol. I. London: Leonard Hill 1972

Ballio, A.: Chemistry and plant growth regulator activity of fusicoccin derivatives and analogues. In: Advances in pesticide science. Proc. 4th Int. Congr. Pesticide Chem. Geissbühler, H. (ed.) Part II, pp. 366–372 Oxford: Pergamon Press 1979

Bangham, A.D., Standish, M.M., Weissmann, G.: The action of steroids and streptolysin S on the permeability of phospholipid structures to cations. J. Mol. Biol. *13*, 253–259 (1965)

Banks, R.E., Cross, B.E.: Bridgehead fluorinations: Preparation of 7-fluorogibberellins. Chem. Ind. *2*, 90 (1975)

Bateson, J.H., Cross, B.E.: The production of fluorogibberellic acid and fluorogibberellin A_9 by *Gibberella fujikuroi*. J. Chem. Soc. Chem. Comm. 649–655 (1972)

Batt, S., Venis, M.A.: Separation and localization of two classes of auxin binding sites in corn coleoptile membranes. Planta *130*, 15–21 (1976)
Batt, S., Wilkins, M.B., Venis, M.A.: Auxin binding to corn coleoptile membranes: kinetics and specificity. Planta *130*, 7–13 (1976)
Beffagna, N., Cocucci, S., Marré, E.: Stimulating effect of fusicoccin on K-activated ATPase in plasmalemma preparations from higher plant tissues. Plant Sci. Lett. *8*, 91–98 (1977)
Beffagna, N., Pesci, P., Tognoli, L., Marré, E.: Distribution of fusicoccin bound *in vivo* among subcellular fractions from maize coleoptiles. Plant Sci. Lett. *15*, 323–330 (1979)
Bengochea, T., Dodds, J.H., Evans, D.E., Jerie, P.H., Niepel, B., Shari, A.R., Hall, M.A.: Studies on ethylene binding by cell-free preparations from cotyledons of *Phaseolus vulgaris* L.: Separation and characterisation. Planta, *148*, 397–406 (1980a)
Bengochea, T., Acaster, M.A., Dodds, J.H., Evans, D.E., Jerie, P.H., Hall, M.A.: Studies on ethylene binding by cell-free preparations from cotyledons of *Phaseolus vulgaris* L.: Effects of structural analogues of ethylene and of inhibitors. Planta, *148*, 407–411 (1980b)
Berridge, M.V., Ralph, R.K., Letham, D.S.: The binding of cytokinin to plant ribosomes. Biochem. J. *119*, 75–84 (1970a)
Berridge, M.V., Ralph, R.K., Letham, D.S.: On the significance of cytokinin binding to plant ribosomes. In: Plant growth substances 1970. Carr, D.J. (ed.), pp. 248–255. Berlin-Heidelberg-New York: Springer 1970b
Beyer, E.M.: Mechanism of ethylene action. Biological activity of deuterated ethylene and evidence against isotopic exchange and cis-trans isomerisation. Plant Physiol. *49*, 672–675 (1972)
Beyer, E.M.: ^{14}C-Ethylene incorporation and metabolism in pea seedlings. Nature (London) *255*, 144–147 (1975)
Beyer, E.M., Blomstrom, D.C.: Ethylene metabolism and its possible physiological role in plants. In: Proc. 10th Int. Conf. Plant Growth Substances, Madison, Wisconsin, July 1979. Berlin-Heidelberg-New York: Springer, in press, 1980
Bhattacharyya, J., Roy, S.C.: Growth promoters and the synthesis of protein in plant mitochondria. I. Effect of kinetin on the incorporation of amino acids into mitochondrial protein. Biochem. Biophys. Res. Commun. *35*, 606–610 (1969)
Biswas, B.B., Ganguly, A., Das, A., Roy, P.: Action of indoleacetic acid: a plant growth hormone on transcription. In: Regulation of growth and differentiated function in eukaryote cells. Talwar, G.P. (ed.), pp. 461–477. New York: Raven Press 1975
Bittner, S., Gorodetsky, M., Har-Paz, I., Mizrahi, Y., Richmond, A.E.: Synthesis and biological effects of aromatic analogs of abscisic acid. Phytochemistry *16*, 1143–1151 (1977)
Blecher, A., White, A.: Alterations produced by steroids in adenosine triphosphatase activity and volume of lymphosarcoma and liver mitochondria. J. Biol. Chem. *235*, 3404–3412 (1960)
Blyth, C.A., Freedman, R.B., Rabin, B.R.: Sex specific binding to microsomal membranes of rat liver. Nature (London) New Biol. *230*, 137–139 (1971)
Brian, P.W., Grove, J.F., Mulholland, T.P.C.: Relationships between structure and growth-promoting activity of the gibberellins and some allied compounds in four test systems. Phytochemistry *6*, 1475–1499 (1967)
Burg, S.P., Burg, E.A.: Molecular requirements for the biological activity of ethylene. Plant Physiol. *42*, 144–152 (1967)
Burrows, W.J., Skoog, F., Leonard, N.J.: Isolation and identification of cytokinins located in the transfer ribonucleic acid of tobacco callus grown in the presence of 6-benzylamino-purine. Biochemistry *10*, 2189–2194 (1971)
Chen, C.M., Hall, R.H.: Biosynthesis of N^6-(Δ^2-isopentenyl) adenosine in the transfer ribonucleic acid of cultured tobacco pith tissue. Phytochemistry *8*, 1687–1695 (1969)
Clark, J.E., Morre, D.J., Cherry, J.H., Yunghans, W.N.: Enhancement of RNA polymerase activity by non-protein components from plasma membranes of soybean hypocotyl. Plant. Sci. Lett. *7*, 233–238 (1976)
Clark, J.H., Peck, E.J., Schrader, W.T., O'Malley, B.W.: Estrogen and progesterone recep-

tors: methods for characterization, quantification and purification. In: Methods in cancer research. Busch, H. (ed.), Vol. XII, pp. 367–417. New York, London: Academic Press 1976
Cleland, R.: Auxin-induced hydrogen ion excretion from *Avena* coleoptiles. Proc. Natl. Acad. Sci. USA *70*, 3092–3093 (1973)
Comstock, J.P., Rosenfeld, G.C., O'Malley, B.W., Means, A.R.: Estrogen-induced changes in translation and specific messenger RNA levels during oviduct differentiation. Proc. Natl. Acad. Sci. USA *69*, 2377–2380 (1972)
Cox, R.A., Bonanou, S.A.: A possible structure for the rabbit reticulocyte ribosome. Biochem. J. *114*, 769–774 (1969)
Cross, B.E., Grove, J.F., Morrison, A.: Gibberellic acid. Part XVIII. Some rearrangements of ring A. J. Chem. Soc. 2498–2515 (1961)
Cross, J.W., Briggs, W.R.: Properties of a solubilized auxin-binding protein from coleoptiles and primary leaves of *Zea mays*. Plant Physiol. *62*, 152–157 (1978)
Cross, J.W., Briggs, W.R., Dohrmann, U.C., Ray, P.M.: Auxin receptors of maize coleoptile membranes do not have ATPase activity. Plant Physiol. *61*, 581–584 (1978)
Crozier, A., Kuo, C.C., Durley, R.C., Pharis, R.P.: The biological activities of 26 gibberellins in nine plant bioassays. Can. J. Bot. *48*, 867–877 (1970)
Cummins, W.R., Sondheimer, E.: Activity of the asymmetric isomers of abscisic acid in a rapid bioassay. Planta *111*, 365–369 (1973)
Davies, J.W., Cocking, E.C.: Protein synthesis in tomato fruit locule tissue: Incorporation of amino-acids into protein by aseptic cell-free systems. Biochem. J. *104*, 23–33 (1967)
Davies, L.J., Rappaport, L.: Metabolism of tritiated gibberellins in d-5 dwarf maize. Plant Physiol. *55*, 620–625 (1975)
DeDuve, C., Wattiaux, R.: Functions of lysosomes. Annu. Rev. Physiol. *28*, 435–492 (1966)
DeMeyts, P.: Insulin and growth hormone receptor in human cultured lymphocytes and peripheral blood monocytes. In: Methods in receptor research. Part 1, Blecher, M. (ed.), pp. 301–383. New York, Basel: Marcel Dekker 1976
Dodds, J.H., Musa, S.K., Jerie, P.H., Hall, M.A.: Metabolism of ethylene to ethylene oxide by cell-free preparations from *Vicia faba*. Pl. Sci. Lett., *17*, 109–114 (1979)
Dohrmann, U., Hertel, R., Pesci, P., Cocucci, S.M., Marrè, E., Randazzo, G., Ballio, A.: Localization of "in vitro" binding of the fungal toxin fusicoccin to plasma-membrane rich fractions from corn coleoptiles. Plant Sci. Lett *9*, 291–299 (1977)
Dohrmann, U., Hertel, R., Kowalik, H.: Properties of auxin binding sites in different subcellular fractions from maize coleoptiles. Planta *140*, 97–106 (1978)
Dollstädt, R., Hirschberg, K., Winkler, E., Hubner, G.: Bindung von Indolylessigsäure und Phenoxyessigsäure an Fraktionen aus Epikotylen und Wurzeln von *Pisum sativum* L. Planta *130*, 105–111 (1976)
Edsall, J.T., Wyman, J.: Biophysical chemistry, Vol. I. New York, London: Academic Press 1958
Erdei, L., Toth, I., Zsoldos, F.: Hormonal regulation of Ca^{2+}-stimulated K^+ influx and Ca^{2+}, K^+-ATPase in rice roots: *in vivo* and *in vitro* effects of auxins and reconstitution of the ATPase. Physiol. Plant. *45*, 448–452 (1979)
Erion, J.L., Fox, J.E.: Purification and properties of a cytokinin binding protein isolated from wheat germ ribosomes. Plant Physiol. Suppl. *59*, 83 (1977)
Evans, M.L.: Rapid responses to plant hormones. Annu. Rev. Plant. Physiol. *25*, 195–224 (1974)
Evans, M.L., Ray, P.M.: Inactivity of 3-methyleneoxindole as mediator of auxin action on cell elongation. Plant Physiol. *52*, 186–189 (1973)
Evins, W.H., Varner, J.E.: Hormone controlled synthesis of endoplasmic reticulum in barley aleurone cells. Proc. Natl. Acad. Sci. USA *68*, 1631–1633 (1971)
Evins, W.H., Varner, J.E.: Hormonal control of polyribosome formation in barley aleurone layers. Plant Physiol. *49*, 348–352 (1972)
Farrimond, J.A., Elliott, M.C., Clack, D.W.: Charge Separation as a component of the structural requirements for hormone activity. Nature (London) *274*, 401–402 (1978)

Fittler, F., Hall, R.H.: Selective modification of yeast seryl-tRNA and its effect on the acceptance and binding functions. Biochem. Biophys. Res. Commun. *25*, 441–446 (1966)
Fittler, F., Kline, L.K., Hall, R.H.: $N^6(\Delta^2$-isopentenyl)adenosine: biosynthesis *in vitro* by an enzyme extract from yeast and rat liver. Biochem. Biophys. Res. Commun. *31*, 571–576 (1968)
Fox, J.E., Erion, J.L.: A cytokinin binding protein from higher plant ribosomes. Biochem. Biophys. Res. Commun. *64*, 694–700 (1975)
Fox, J.E., Erion, J.L.: Cytokinin binding proteins in higher plants. In: Plant growth regulation. Pilet, P.-E. (ed.), pp. 139–146. Berlin-Heidelberg-New York: Springer 1977
Frydman, V.M., MacMillan, J.: The metabolism of gibberellins A_9, A_{20} and A_{29} in immature seeds of *Pisum sativum* cv. Progress No. 9. Planta *125*, 181–195 (1975)
Galbraith, D.W., Northcote, D.H.: The isolation of plasma membrane from protoplasts of soybean suspension cultures. J. Cell. Sci. *24*, 295–310 (1977)
Gardner, G., Sussman, M.R.: Solubilization and characterization of the receptor for the auxin transport inhibitor N-1-naphthylphthalamic acid. Abstracts, 10th Int. Conf. Plant Growth Substances, Madison, Wisconsin, July 1979, p. 4 Berlin, Heidelberg, New York: Springer 1980 in press
Gardner, G., Sussman, M.R., Kende, H.: Cytokinin binding to particulate fractions from moss protonemata. Plant Physiol. Suppl. *56*, 28 (1975)
Goldberg, R., Prat, R.: Spontaneous and auxin-induced growth in mung bean hypocotyl segments. Abstr. Fed. Eur. Soc. Plant Physiol. Meet. pp. 216–217. Edinburgh 1978
Gorell, T.A., Gilbert, L.I., Siddall, J.B.: Binding proteins for an ecdysone metabolite in the crustacean hepatopancreas. Proc. Natl. Acad. Sci. USA *69*, 812–815 (1972)
Gottlieb, D., Carter, H.E., Wu, L.C., Slonecker, J.H.: Inhibition of fungi by filipin and its antagonism by sterols. Phytopathology *50*, 594–603 (1960)
Hager, A., Menzel, H., Krauss, A.: Versuche und Hypothese zur Primärwirkung des Auxins beim Streckungswachstum. Planta *100*, 47–75 (1971)
Hall, R.H.: Cytokinins as a probe of developmental processes. Annu. Rev. Plant. Physiol. *24*, 415–444 (1973)
Hall, R.H., Strivastava, B.I.S.: Cytokinin activity of compounds obtained from soluble RNA. Life Sci. *7(II)* 7–13 (1968)
Hamilton, R.H., Meyer, H.E., Burke, R.E., Feung, C.S., Mumma, R.D.: Metabolism of indole-3-acetic acid. II. Oxindole pathway in *Parthenocissus tricuspidata* crown-gall tissue cultures. Plant Physiol. *58*, 77–81 (1976)
Hansch, C., Muir, R.M., Metzenberg, R.L.: Further evidence for a chemical reaction between plant growth regulators and a plant substrate. Plant Physiol. *26*, 812–821 (1951)
Hardin, J.W., Cherry, J.G.: Solubilization and partial characterization of soybean chromatin-bound RNA polymerase. Biochem. Biophys. Res. Commun. *48*, 299–306 (1972)
Hardin, J.W., O'Brien, T.J., Cherry, J.H.: Stimulation of chromatin bound RNA polymerase activity by a soluble factor. Biochim. Biophys. Acta *224*, 667–670 (1970)
Hardin, J.W., Cherry, J.H., Morré, D.J., Lembi, C.A.: Enhancement of soybean RNA polymerase activity by a factor released by auxin from isolated fractions enriched in plasma membranes. Proc. Natl. Acad. Sci. USA *69*, 3146–3150 (1972)
Heftmann, E.: Biochemistry of plant steroids. Annu. Rev. Plant. Physiol. *14*, 225–248 (1963)
Helgerson, S.L., Cramer, W.A., Morré, D.J.: Evidence for an increase in microviscosity of plasma membranes from soybean hypocotyls induced by the plant hormone, indole-3-acetic acid. Plant Physiol. *58*, 548–551 (1976)
Hertel, R.: Auxin transport and *in vitro* auxin binding. In: Membrane transport in plants. Zimmerman, U., Dainty, J. (eds.), pp. 457–461. Berlin-Heidelberg-New York: Springer 1974
Hertel, R., Thomson, K-St., Russo, V.E.A.: In vitro auxin binding to particulate cell fractions from corn coleoptiles. Planta *107*, 325–340 (1972)
Hertel, R., Dohrmann, U., Jesaitis, A.J., Peterson, W.: Various receptor sites in membrane fractions from corn coleoptiles: in vitro binding of auxins, riboflavin and a naturally occurring analog of morphactin. Abstr. 9th. Int. Conf. Plant Growth Subst. Pilet, P.-E. (ed.), pp. 140a–140c. Lausanne 1976

Higgins, T.J.V., Zwar, J.A., Jacobsen, J.V.: Gibberellic acid enhances the level of translatable mRNA for α-amylase in barley aleurone layers. Nature (London) *260*, 166–169 (1976)

Hiraga, K., Yamane, H., Takahashi, N.: Biological activity of some synthetic gibberellin glucosyl esters. Phytochemistry *13*, 2371–2376 (1974)

Ho, D.T.H., Varner, J.E.: Hormonal control of messenger ribonucleic acid metabolism in barley aleurone layers. Proc. Natl. Acad. Sci. USA *71*, 4783–4786 (1974)

Hocking, T.J., Clapham, J., Cattell, K.J.: Abscisic acid binding to subcellular fractions from leaves of *Vicia faba*. Planta *138*, 303–304 (1978)

Horgan, R., Hewett, E.W., Horgan, J.M., Purse, J., Wareing, P.F.: A new cytokinin from *Populus* × *robusta*. Phytochemistry *14*, 1005–1008 (1975)

Hummel, J.P., Dreyer, W.J.: Measurement of protein-binding phenomena by gel filtration. Biochim. Biophys. Acta *63*, 530–532 (1962)

Hunston, D.L.: Two techniques for evaluating small molecule-macromolecule binding in complex systems. Anal. Biochem. *63*, 99–109 (1975)

Ihl, M.: Indole-acetic acid binding proteins in soybean cotyledon. Planta *131*, 223–228 (1976)

Jablonovic, M., Nooden, L.D.: Changes in competable IAA binding in relation to bud development in pea seedlings. Plant Cell Physiol. *15*, 687–692 (1974)

Jacobs, M., Hertel, R.: Auxin binding to subcellular fractions from *Cucurbita* hypocotyls: in vitro evidence for an auxin transport carrier. Planta *142*, 1–10 (1978)

Jacobsen, J.V.: Regulation of ribonucleic acid metabolism by plant hormones. Annu. Rev. Plant. Physiol. *28*, 537–564 (1977)

Jacobsen, J.V., Zwar, J.A.: Gibberellic acid causes increased synthesis of RNA which contains poly (A) in barley aleurone tissue. Proc. Natl. Acad. Sci. USA *71*, 3290–3293 (1974a)

Jacobsen, J.V., Zwar, J.A.: Gibberellic acid and RNA synthesis in barley aleurone layers: Metabolism of rRNA and tRNA and of RNA containing polyadenylic acid sequences. Aust. J. Plant Physiol. *1*, 343–349 (1974b)

Jensen, E.V., Suzuki, T., Kawashima, W., Stumpf, E., Jungblut, P.W., DeSombre, E.R.: A two-step mechanism for the interaction of estradiol with rat uterus. Proc. Natl. Acad. Sci. USA *59*, 632–639 (1968)

Jerie, P.H., Shaari, A.R., Zeroni, M., Hall, M.A.: The partition coefficient of $^{14}C_2H_4$ in plant tissue as a screening test for metabolism or compartmentation of ethylene. New Phytol. *81*, 499–504 (1978)

Jerie, P.H., Shaari, A.R., Hall, M.A.: The compartmentation of ethylene in developing cotyledons of *Phaseolus vulgaris* L. Planta *144*, 503–507 (1979)

Johnson, K.D., Kende, H.: Hormonal control of lecithin synthesis in barley aleurone cells: regulation of the CDP-choline pathway by gibberellin. Proc. Natl. Acad. Sci. USA *68*, 2674–2677 (1971)

Johri, M.M., Varner, J.E.: Enhancement of RNA synthesis in isolated pea nuclei by gibberellic acid. Proc. Natl. Acad. Sci. USA *59*, 260–276 (1968)

Jones, R.L.: Gibberellic acid and the fine structure of barley aleurone cells. II Changes during the synthesis and secretion of α-amylase. Planta *88*, 73–86 (1969a)

Jones, R.L.: The effect of ultracentrifugation on fine structure and α-amylase production in barley aleurone cells. Plant Physiol. *44*, 1428–1438 (1969b)

Jones, R.L.: Gibberellic acid and the endoplasmic reticulum of barley aleurone layers. Plant Physiol. Suppl. *61*, 110 (1978)

Jones, R.L., Chen, R.F.: Immunohistological localisation of α-amylase in barley aleurone cells. J. Cell Sci. *20*, 183–198 (1976)

Jones, T.W.A.: Biological activities of fluorogibberellins and interactions with unsubstituted gibberellins. Phytochemistry *15*, 1825–1827 (1976)

Jönssen, A.: Chemical structure and growth activity of auxins and antiauxins. In: Encyclopedia of plant physiology. Ruhland W. (ed.), Vol. XIV, pp. 959–1006. Berlin-Göttingen-Heidelberg: Springer 1961

Kaethner, T.M.: Conformational change theory for auxin structure-activity relationships. Nature (London) *267*, 19–23 (1977)

Kasamo, K., Yamaki, T.: Effect of auxin on Mg^{++}-activated and inhibited ATPases from mung bean hypocotyls. Plant Cell Physiol. *15*, 965–970 (1974)
Kasamo, K., Yamaki, T.: In vitro binding of IAA to plasma membrane-rich fractions containing Mg^{++}-activated ATPase from mung bean hypocotyls. Plant Cell Physiol. *17*, 149–164 (1976)
Katekar, G.F.: Auxins: on the nature of the receptor site and molecular requirements for auxin activity. Phytochemistry *18*, 223–233 (1979)
Kaufman, P.B.: Effect of gibberellic acid on plasticity and elasticity of *Avena* stem segments. Plant Physiol. *56*, 757–760 (1975)
Kende, H., Gardner, G.: Hormone binding in plants. Annu. Rev. Plant. Physiol. *27*, 267–290 (1976)
Kende, H., Tavares, J.E.: On the significance of cytokinin incorporation into RNA. Plant Physiol. *43*, 1244–1248 (1968)
Kerk, G.J.M. van der, Raalte, M.H. van, Sijpesteijn, A.K., Veen, R. van der: A new type of plant growth regulating substance. Nature (London) *176*, 308–310 (1955)
Kessler, C., Hartmann, G.R.: The two effects of rifampicin on the RNA polymerase reaction. Biochem. Biophys. Res. Commun. *74*, 50–56 (1977)
Kinsky, S.C., Haxby, J., Kinsky, C.B., Demel, R.A., van Deenen, L.L.M.: Effect of cholesterol incorporation on the sensitivity of liposomes to the polyene antibiotic filipin. Biochem. Biophys. Acta *152*, 174–185 (1968)
Kirk, K.L., Cohen, L.A.: Photochemical decomposition of diazonium fluoroborates. Application to the synthesis of ring-fluorinated imidazoles. J. Am. Chem. Soc. *93*, 3060–3061 (1971)
Klämbt, D.: Cytokinin and cell metabolism. In: Plant growth regulation. Pilet, P.-E. (ed.), pp. 154–160. Berlin-Heidelberg-New York: Springer 1977
Klun, J.A., Tipton, C.L., Robinson, J.F., Ostrem, D.L., Beroza, M.: Isolation and identification of 6,7-dimethoxy-2-benzoxazolinone from dried tissues of *Zea mays* (L.) and evidence of its cyclic hydroxamic acid precursor. J. Agric. Food Chem. *18*, 663–665 (1970)
Konjević, R., Grubisić, D., Marković, R., Petrović, J.: Gibberellic acid binding proteins from pea stems. Planta *131*, 125–128 (1976)
LéJohn, H.B.: A rapid and sensitive binding system for N^6-substituted adenines, and some urea and thiourea derivatives, that show cytokinin activity in cell division tests. Can. J. Biochem. *53*, 768–775 (1975)
Lembi, C.A., Morre, D.J., Thomson, K.-St., Hertel, R.: N-1-naphthylphthalamic-acid-binding activity of a plasma membrane-rich fraction from maize coleoptiles. Planta *99*, 37–45 (1971)
Letham, D.S.: Regulators of cell division in plant tissues. V. A comparison of zeatin and other cytokinins in five bioassays. Planta *74*, 228–242 (1967)
Likholat, T.V., Pospelov, V.A., Morozova, T.M., Salganik, R.I.: Capacity of wheat coleoptile cells for specific binding of auxin and effect of the hormone on template activity of chromatin in seedlings of different ages. Sov. Plant Physiol. *21*, 779–784 (1974)
Matthysse, A.G.: Organ specificity of hormone receptor-chromatin interactions. Biochim. Biophys. Acta *199*, 519–521 (1970)
Matthysse, A.G., Abrams, M.: A factor mediating interaction of kinins with the genetic material. Biochim. Biophys. Acta *199*, 511–518 (1970)
Matthysse, A.G., Phillips, C.: A protein intermediary in the interaction of a hormone with the genome. Proc. Natl. Acad. Sci. USA *63*, 897–903 (1969)
McComb, A.J., McComb, J.A., Duda, C.T.: Increased ribonucleic acid polymerase activity associated with chromatin from internodes of dwarf pea plants treated with gibberellic acid. Plant Physiol. *46*, 221–223 (1970)
Mehard, C.W., Lyons, J.M., Kumamoto, J.: Utilisation of model membranes in a test for the mechanism of ethylene action. J. Membr. Biol. *3*, 173–179 (1970)
Milborrow, B.V.: The chemistry and physiology of abscisic acid. Annu. Rev. Plant. Physiol. *25*, 259–307 (1974)
Mizushima, S., Nomura, M.: Assembly mapping of 30S ribosomal proteins from *E. coli*. Nature (London) *226*, 1214–1218 (1970)
Mondal, H., Mandal, R.K., Biswas, B.B.: RNA stimulated by indole acetic acid. Nature (London) New Biol. *240*, 111–113 (1972)

Moore, H.: Kinetin binding protein from wheat germ. Plant Physiol. Suppl. *59*, 17 (1977)
Morré, D.J., Bracker, C.E.: Ultrastructural alteration of plant plasma membranes induced by auxin and calcium ions. Plant Physiol. *58*, 544–547 (1976)
Moyed, H.S., Williamson, V.: Multiple 3-methyleneoxindole reductases of peas. Differential inhibition by synthetic auxins. J. Biol. Chem. *242*, 1075–1077 (1967)
Mozell, M.M.: Evidence for a chromatographic model of olfaction. J. Gen. Physiol. 56:46–63 (1970)
Mozell, M.M., Jagodowicz, M.: Chromatographic separation of odorants by the nose: Retention times measured across in vivo olfactory mucosa. Science *181*, 1247–1248 (1973)
Mudd, J.B., Kleinschmidt, M.G.: Effect of filipin on the permeability of red beet and potato tuber discs. Plant Physiol. *45*, 517–518 (1970)
Muir, R.M., Hansch, C., Gallup, A.H.: Growth regulation by organic compounds. Plant Physiol. *24*, 359–366 (1949)
Musgrave, A., Kende, H.: Radioactive gibberellin A_5 and its metabolism in dwarf peas. Plant Physiol. *45*, 56–61 (1970)
Musgrave, A., Kays, S.E., Kende, H.: In vivo binding of radioactive gibberellins in dwarf pea shoots. Planta *89*, 165–177 (1969)
Nadeau, R., Rappaport, L.: Metabolism of gibberellin A_1 in germinating bean seeds. Phytochemistry *11*, 1611–1616 (1972)
Nadeau, R., Rappaport, L.: The synthesis of [^{3}H] gibberellin A_3 and [^{3}H] gibberellin A_1 by the palladium-catalysed actions of carrier-free tritium on gibberellin A_3. Phytochemistry *13*, 1537–1545 (1974)
Nadeau, R., Rappaport, L., Stolp, C.F.: Uptake and metabolism of [^{3}H] gibberellin A_1 by barley aleurone layers: response to abscisic acid. Planta *107*, 315–324 (1972)
Nash, L.J., Jones, R.L., Stoddart, J.L.: Gibberellin metabolism in excised lettuce hypocotyls: Response to GA_9 and the conversion of [^{3}H]-GA_9. Planta *140*, 143–150 (1978)
Normand, G., Hartman, M.A., Schuber, F., Benveniste, P.: Charactérisation de membranes de coléoptiles de maïs fixant l'auxine et l'acide N-naphtyl phtalamique. Physiol. Vég. *13*, 743–761 (1975)
O'Brien, T.J., Jarvis, B.C., Cherry, J.H., Hanson, J.B.: Enhancement by 2,4-D of chromatin RNA polymerase in soybean hypocotyl tissue. Biochim. Biophys. Acta *169*, 35–43 (1968)
O'Malley, B., Means, A.R.: Female steroid hormones and target cell nuclei. Science *183*, 610–620 (1974)
Oostrom, H., Van Loopik-Detmers, M.A., Libbenga, K.R.: A high affinity receptor for indoleacetic acid in cultured tobacco pith explants. FEBS Lett. *59*, 194–197 (1975)
Patterson, R., Rappaport, L., Breidenbach, R.W.: Characterisation of an enzyme from *Phaseolus vulgaris* seeds which hydroxylates GA_1 to GA_8. Phytochemistry *14*, 363–368 (1975)
Paulus, H.: A rapid and sensitive method for measuring the binding of radioactive ligands to proteins. Anal. Biochem. *32*, 91–100 (1969)
Pearson, J.A., Wareing, P.F.: Effect of abscisic acid on activity of chromatin. Nature (London) *221*, 672–673 (1969)
Penny, D., Penny, P., Monro, J., Bailey, R.W.: Cell elongation and auxin action in lupin hypocotyls. In: Plant growth substances 1970. Carr, D.J. (ed.), pp. 52–61. Berlin-Heidelberg-New York: Springer 1972
Pesci, P., Tognoli, L., Beffagna, N., Marrè, E.: Solubilization and partial purification of a fusicoccin-receptor complex from maize microsomes. Plant Sci. Lett. *15*, 313–322 (1979)
Peterkofsky, A., Jesensky, C.: The localisation of N^6-(Δ^2-isopentenyl) adenosine among acceptor species of transfer ribonucleic acid in *Lactobacillus acidophilus*. Biochemistry *8*, 3798–3809 (1969)
Pitel, D.W., Vining, L.C.: Preparation of gibberellin A_1-3,4-^{3}H. Can. J. Biochem. *48*, 259–263 (1970)
Polya, G.M., Davis, A.W.: Properties of a high-affinity cytokinin-binding protein from wheat germ. Planta *139*, 139–147 (1978)
Porter, W.L., Thimann, K.V.: Molecular and functional complementarity of auxins and phosphatides. Abstr. 9th. Int. Bot. Congr. Vol. II, pp. 305–306. University of Toronto Press 1959

Porter, W.L., Thimann, K.V.: Molecular requirements for auxin action. I. Halogenated indoles and indoleacetic acid. Phytochemistry *4*, 229–243 (1965)
Pratt, H.M., Fox, J.E.: Subunit structure and amino-acid composition of a cytokinin-binding protein. Plant Physiol. Suppl. *61*, 10 (1978)
Pratt, H.K., Goeschl, J.D.: Physiological roles of ethylene in plants. Annu. Rev. Plant Physiol. *20*, 541–584 (1969)
Quail, P.H., Browning, A.: Failure of lactoperoxidase to iodinate specifically the plasma membrane of *Cucurbita* tissue segments. Plant Physiol. *59*, 759–766 (1977)
Raison, J.K.: Temperature-induced phase changes in membrane lipids and their influence on metabolic regulation. In: Rate control of biological processes. Soc. Exp. Biol. Symp. XXVII. Davies, D.D. (ed.), pp. 485–512. Cambridge: University Press 1973
Ray, P.M.: The biochemistry of the action of indoleacetic acid on plant growth. In: Runeckles, V.C., Sondheimer, F. and Walton, D.C. (eds.), Rec. Adv. Phytochem. *7*, 93–123 (1974)
Ray, P.M.: Auxin binding sites of maize coleoptiles are localized on membranes of the endoplasmic reticulum. Plant Physiol. *59*, 594–599 (1977)
Ray, P.M., Dohrmann, V., Hertel, R.: Characterization of naphthaleneacetic acid binding to receptor sites on cellular membranes of maize coleoptiles tissue. Plant Physiol. *59*, 357–364 (1977a)
Ray, P.M., Dohrmann, U., Hertel, R.: Specificity of auxin-binding sites on maize coleoptile membranes as possible receptor sites for auxin action. Plant Physiol. *60*, 585–591 (1977b)
Rayle, D.L.: Auxin-induced H^+-ion secretion in *Avena* coleoptiles and its implications. Planta *114*, 63–73 (1973)
Rizzo, P.J., Pederson, K., Cherry, J.H.: Stimulation of transcription by a soluble factor isolated from soybean hypocotyl by 2,4-D affinity chromatography. Plant. Sci. Lett. *8*, 205–211 (1977)
Rosenbaum, N., Gefter, M.L.: Δ^2 isopentenylpyrophosphate: Transfer ribonucleic acid Δ^2 isopentenyltransferase from *Escherichia coli*. Purification and properties of the enzyme. J. Biol. Chem. *247*, 5675–5680 (1972)
Roy, P., Biswas, B.B.: A receptor protein for indoleacetic acid from plant chromatin and its role in transcription. Biochem. Biophys. Res. Commun. *74*, 1597–1606 (1977)
Scatchard, G.: The attractions of proteins for small molecules and ions. Ann. N.Y. Acad. Sci. *51*, 660–672 (1949)
Sica, V., Nola, E., Parikh, I., Puca, G.A., Cautrecasas, P.: Purification of oestradiol receptors by affinity chromatography. Nature (London) New Biol. *244*, 36–39 (1973)
Silk, W.K., Jones, R.L.: Gibberellin response in lettuce hypocotyl sections. Plant Physiol. *56*, 267–272 (1975)
Skoog, F., Armstrong, D.J.: Cytokinins. Annu. Rev. Plant Physiol. *21*, 359–384 (1970)
Skoog, F., Hamzi, H.Q., Szweykowska, A.M., Leonard, N.J., Carraway, K.L., Fujii, T., Helgeson, J., Loeppky, R.N.: Cytokinins: Structure/activity relationships. Phytochemistry *6*, 1169–1192 (1967)
Smith, M.S., Wain, R.L., Wightman, F.: Studies of plant growth-regulating substances V. Steric factors in relation to mode of action of certain aryloxyalkylcarboxylic acids. Ann. Appl. Biol. *39*, 295–307 (1952)
Sponsel (née Frydman), V.M., Hoad, G.V., Beeley, L.J.: The biological activities of some new gibberellins (GAs) in six plant bioassays. Planta *135*, 143–147 (1977)
Stoddart, J.L.: The biological activity of fluorogibberellins. Planta *107*, 81–88 (1972)
Stoddart, J.L.: Gibberellin receptors. Rep. Welsh Pl. Breed. Sta. 62–63 (1974)
Stoddart, J.L.: Characterisation of high-molecular-weight [^{3}H] gibberellin A_1 complexes. Rep. Welsh Plant Breed Stn. 80 (1975)
Stoddart, J.L.: Interaction of [^{3}H] gibberellin A_1 with a sub-cellular fraction from lettuce (*Lactuca sativa* L.) hypocotyls. I. Kinetics of labelling. Planta *146*, 353–361 (1979a)
Stoddart, J.L.: Interaction of [^{3}H] gibberellin A_1 with a sub-cellular fraction from lettuce (*Lactuca sativa* L.) hypocotyls. II. Stability and properties of the association. Planta *146*, 363–8 (1979b)
Stoddart, J.L., Jones, R.L.: Gibberellin metabolism in excised lettuce hypocotyls: Evidence for the formation of gibberellin A_1 glucosyl conjugates. Planta *136*, 261–269 (1977)

Stoddart, J.L., Williams, P.D.: Interaction of [^{3}H] gibberellin A_1 with a sub-cellular fraction from lettuce (*Lactuca sativa* L.) hypocotyls. III. Requirement for protein synthesis. Planta *147*, 264–268 (1979)

Stoddart, J.L., Briedenbach, W., Nadau, R., Rappaport, L.: Selective binding of [^{3}H] gibberellin A_1 by protein fractions from dwarf pea epicotyls. Proc. Natl. Acad. Sci. USA *71*, 3255–3259 (1974)

Stuart, D.A., Jones, R.L.: The roles of extensibility and turgor in gibberellin- and dark-stimulated growth. Plant Physiol. *59*, 61–68 (1977)

Stuart, D.A., Jones, R.L.: The relationship between proton efflux and gibberellin-stimulated growth in hypocotyl sections. Planta *61*, 180–183 (1978)

Stuart, D.A., Durnam, D.J., Jones, R.L.: Cell elongation and cell division in elongating lettuce hypocotyl sections. Planta *135*, 249–255 (1977)

Sunshine, G.H., Williams, D.J., Rabin, B.R.: Sex specific role in binding of ribosomes to endoplasmic reticulum. Nature (London) New Biol. *230*, 133–134 (1971)

Sussman, M.R., Kende, H.: Cytokinin binding to particulate fractions from tobacco callus. Plant Physiol. Suppl. *56*, 28 (1975)

Sussman, M.R., Kende, H.: The synthesis and biological properties of 8-azido-N^6-benzyl-adenine, a potential photoaffinity reagent for cytokinins. Planta *137*, 91–96 (1977)

Sussman, M.R., Kende, H.: *In vitro* cytokinin binding to a particulate fraction of tobacco cells. Planta *140*, 251–259 (1978)

Takegami, T., Yoshida, K.: Isolation and purification of a cytokinin binding protein from tobacco leaves by affinity column chromatography. Biochem. Biophys. Res. Commun. *67*, 782–789 (1975)

Takegami, T., Yoshida, K.: Specific interaction of cytokinin binding protein with 40S ribosomal subunits in the presence of cytokinin *in vitro*. Plant Cell Physiol. *18*, 337–345 (1977)

Taylor, J.L., Haydon, D.A.: The interaction of progesterone with lipid films at the air-water interface. Biochim. Biophys. Acta *94*, 488–493 (1965)

Taylor, N.F., Kent, P.W.: Structural aspects of monofluoro-steroids. In: Advances in fluorine chemistry. Stacey, M., Tatlow, J.C., Sharpe, A.C. (eds.), Vol. IV, pp. 113–141. London: Butterworths 1965

Teissere, M., Penon, P., Ricard, J.: Hormonal control of chromatin availability and of the activity of purified RNA polymerases in higher plants. FEBS Lett. *30*, 65–70 (1973)

Teissere, M., Penon, P., Van Huystee, R.B., Azou, Y., Ricard, J.: Hormonal control of transcription in higher plants. Biochim. Biophys. Acta *402*, 391–402 (1975)

Thimann, K.V.: The auxins. In: Physiology of plant growth and development. Wilkins, M.B. (ed.), pp. 3–45. London: McGraw-Hill 1969

Thomson, K.S.: The binding of 1-N-naphthylphthalamic acid (NPA), an inhibitor of auxin transport, to particulate fractions of corn coleoptiles. In: Hormonal regulation in plant growth and development. Kaldeway, H., Vardar, Y. (eds.), pp. 83–88. Weinheim: Chemie 1972

Thomson, K.S., Leopold, A.C.: In vitro binding of morphactins and 1-N-naphthylphthalamic acid in corn coleoptiles, and their affects on auxin transport. Planta *115*, 259–270 (1974)

Thomson, K.S., Hertel, R., Muller, S., Tavares, J.E.: 1-N-naphthylphthalamic acid and 2,3,5-triiodobenzoic acid. In vitro binding to particulate cell fractions and action on auxin transport on corn coleoptiles. Planta *109*, 337–352 (1973)

Tognoli, L., Beffagna, N., Pesci, P., Marrè, E.: On the relationship between ATPase and fusicoccin binding capacity of crude and partially-purified microsomal preparations from maize coleoptiles. Plant Sci. Lett. *16*, 1–14 (1979)

Trillmich, K., Michalke, W.: Kinetic characterization of N-1-naphthylphthalamic acid binding sites from maize coleoptile homogenates. Planta *145*, 119–127 (1979)

Tuli, V., Moyed, H.S.: The role of 3-methyleneoxindole in auxin action. J. Biol. Chem. *244*, 4916–4920 (1969)

Vanderhoef, L.N., Stahl, C.A.: Separation of two responses to auxin by means of cytokinin inhibition. Proc. Natl. Acad. Sci. USA *72*, 1822–1825 (1975)

Vanderhoef, L.N., Lu, T-Y.S., Williams, C.A.: Comparison of auxin-induced and acid-induced elongation in soybean hypocotyl. Plant Physiol. *59*, 1004–1007 (1977)

Veen, H.: Specificity of phospholipid binding to indole acetic acid and other auxins. Z. Naturforsch. *29c*, 39–41 (1974)

Veldstra, H.: Researches on plant growth substances IV. Relation between chemical structure and physiological activity I. Enzymologia *11*, 97–136 (1944a)

Veldstra, H.: Researches on plant growth substances IV. Relation between chemical structure and physiological activity II. Contemplations on place and mechanism of the action of the growth substances. Enzymologia *11*, 137–163 (1944b)

Venis, M.A.: Stimulation of RNA transcription from pea and corn DNA by protein retained on Sepharose coupled to 2,4-dichlorophenoxyacetic acid. Proc. Natl. Acad. Sci USA *68*, 1824–1827 (1971)

Venis, M.A.: Receptors for plant hormones. Adv. Bot. Res. *5*, 53–88 (1977a)

Venis, M.A.: Affinity labels of auxin binding sites in corn coleoptile membranes. Planta *134*, 145–149 (1977b)

Venis, M.A.: Solubilisation and partial purification of auxin-binding sites of corn membranes. Nature (London) *66*, 268–269 (1977c)

Venis, M.A.: Receptors for plant growth regulators. In: Advances in pesticide science. Proc. 4th Int. Congr. Pesticide Chem. Geissbühler, H. (ed.) Part II, pp. 487–493 Oxford: Pergamon Press 1979

Venis, M.A., Watson, P.J.: Naturally occurring modifiers of auxin-receptor interaction in corn: identification as benzoxazolinones. Planta *142*, 103–107 (1978)

Venis, M.A.: Purification and properties of membrane-bound auxin receptors in corn. In: Proc. 10th Int. Conf. Plant Growth Substances, Madison, Wisconsin, July 1979. Berlin, Heidelberg, New York: Springer 1980, in press

Verma, D.P.S., MacLachlan, G.A., Byrne, H., Ewings, D.: Regulation and in vitro translation of messenger ribonucleic acid for cellulase from auxin-treated pea epicotyls. J. Biol. Chem. *250*, 1019–1026 (1975)

Vreugdenhil, D., Burgers, A., Libbenga, K.R.: A particle-bound auxin receptor from tobacco pith callus. Plant Sci. Lett. *16*, 115–121 (1979)

Walker, G.C., Leonard, N.J., Armstrong, D.J., Murai, N., Skoog, F.: The mode of incorporation of 6-benzylaminopurine into tobacco callus transfer ribonucleic acid. Plant Physiol. *54*, 737–743 (1974)

Wardrop, A.J., Polya, G.M.: Properties of a soluble auxin-binding protein from dwarf bean seedlings. Plant Sci. Lett. *8*, 155–163 (1977)

Weigl, J.: Einbau von Auxin in gequollene Lecithin-Lamellen. Z. Naturforsch. *24 B*, 365–366 (1969)

Weissmann, G., Sessa, G., Weissmann, S.: Effects of steroids and Triton X-100 on glucose filled phospholipid/cholesterol structures. Nature (London) *208*, 649–651 (1965)

Wiberg, K.B.: The deuterium isotope effect. Chem. Rev. *55*, 713–743 (1955)

Wilchek, M., Salomon, Y., Lowe, M., Selinger, Z.: Conversion of protein kinase to a cyclic AMP independent form by affinity chromatography on N^6-caproyl 3′,5′-cyclic adenosine monophosphate-Sepharose. Biochem. Biophys. Res. Commun. *45*, 1177–1184 (1971)

Willmer, E.N.: Steroids and cell surfaces. Biol. Rev. Cambridge Philos. Soc. *36*, 368–398 (1961)

Wood, A., Paleg, L.G.: The influence of gibberellin acid on the permeability of model membrane systems. Plant Physiol. *50*, 103–108 (1972)

Wood, A., Paleg, L.G.: Alteration of liposomal membrane fluidity by gibberellic acid. Aust. J. Plant Physiol. *1*, 31–40 (1974)

Wood, A., Paleg, L.G., Spotswood, T.M.: Hormone-phospholipid interaction: a possible hormonal mechanism of action in the control of membrane permeability. Aust. J. Plant Physiol. *1*, 167–169 (1974)

Yamamoto, K.R., Alberts, B.M.: Steroid receptors: Elements for modulation of eukaryotic transcription. Annu. Rev. Biochem. 721–746 (1976)

6 Molecular Effects of Hormone Treatment on Tissue

M. Zeroni and M.A. Hall

6.1 Introduction

Exogenous application of hormones is one of the most common methods in research into the mode of action of such substances. The rationale behind such approaches is based mainly on the idea of replacement of the endogenous, naturally occurring hormone by an exogenous hormone, the level of which may be controlled and its effect monitored. This is fairly easily accomplished in animal systems, where the gland synthesizing a particular hormone can be dissected out and the target tissue monitored after application of the exogenous hormone via the blood stream. This is rarely easy and often impossible to achieve with plant material because in most instances any given hormone may be produced in more than one location within the plant. Furthermore, the same organ or cell which produces the hormone may also be the target.

Even where there is a relatively clear distinction between the site of production of the hormone and that of its action, the very removal of the source of the hormone may induce major changes in plant processes, many of which may be quite unrelated to those directly controlled by the hormone. This may result either in the action of the hormone being masked or, alternatively, may lead to spurious findings owing to effects of the applied hormone on secondary or tertiary processes. Thus, removal of all or part of a major source of cytokinins – the roots – may affect the supply of water, mineral salts, and amino acids to the upper part of the plant and hence modify growth and development to such an extent that the result of exogenous application of cytokinins in lieu of those normally derived from the root may in no way reflect their real role.

Another difficulty with plant tissue is that at any given time each cell is likely to contain representatives of all the known hormones at varying concentrations. These hormones will interact differently with each other and with applied hormone at different times and at the same time in different processes.

In spite of these difficulties, exogenous application of hormones to plant tissues is a powerful tool in hormone research which has its own merits. It may be argued that if an endogenous hormone has a certain effect on a process, additional hormone should enhance or promote the effect, provided that the endogenous hormone is not present at an optimal concentration. This must therefore be assessed initially by determining the effect of a range of concentrations of the hormone on the process under study.

Equally, if it is known that a certain process occurs at a particular stage of development and it is thought to be hormone-dependent, an attempt may be made to induce it somewhat earlier by hormone treatment. The systems based on these criteria, however, make the often unjustified assumption that the tissue is "ready" for the change to occur and "waits" for the appropriate hormone concentration to be achieved (see Sect. 6.5).

In this chapter we shall attempt to evaluate some of the main systems, their strengths and their weaknesses, with a view to determining whether the results obtained, and the conclusions drawn, may serve as bases for hypotheses on hormonal action in growth and development.

However suitable for such studies a biological system may appear to be, some general principles must be borne in mind. It may seem facile to enumerate these in a review at this level, but experience shows that a large number of reports do not take them into account, either in their entirety or in part.

Thus, whatever the external concentration of the applied hormone may be, the internal concentration of the same hormone will be different and will depend on the rate of penetration of the hormone into the cells and tissues, the rate of metabolism of the applied hormone (which may be different from that of the endogenous hormone) and the concentration and distribution of the endogenous hormone already present in the tissue. The presence or absence of other substances such as ions or sugars (see Sect. 6.5) may change or even invert the natural action of a hormone. Very often the effects of synthetic hormones on plant tissues may be different from those of their natural equivalents in the same tissue because of differences in distribution, metabolism, specificity, etc. The duration of hormone application may also be critical since a particular process may be affected differentially in time.

One last point must be made in this connection and it is perhaps the most important. Many workers fail to distinguish between direct and indirect effects of growth regulators. While this is not always easy to ascertain, it is a feature which should be given more prominence in the interpretation of data. In the past, a great deal of effort has been expended on systems where a distinction has not been made between processes permissive for a phenomenon and those controlling it.

Thus, it has now been established that hormones rarely determine the way in which a cell or tissue may respond, but rather that in most cases their role is more likely to be that of inducing certain responses the nature of which is predetermined by factors intrinsic to the cell – in other words the "programming" of the cell (Wareing, 1971; Hall, 1976). Most of the systems which have been worked on fall into this category and the result has been that much of the work on "mechanism of action" of growth regulators has looked at the processes necessary for a developmental phenomenon to occur – e.g., the production of cell wall hydrolases in abscission – rather than the effects which bring about such changes. Only relatively recently have attempts been made to look at hormone-binding sites (Chap. 5 of this volume) and the processes controlled by hormones in the short term.

This does not mean that such work on "permissive" processes is valueless, since it has provided much information on cellular processes in general and given leads as to where to look for the primary processes controlled by hormones. But it does mean that out of the colossal mass of experimental data collected in the last twenty years only a small part has given us information on basic modes of hormone action.

We know that probably all plant processes are affected by growth regulators and moreover someone has probably investigated all of them at one time or another. Hence, a full treatment of all this work would not only require a whole volume but would leave the reader little wiser after he had read it than before.

Because of these considerations we have been rather selective in those areas we have chosen to cover in this chapter. We have also chosen on the whole to group the text into a consideration of developmental systems rather than individual metabolic processes such as effects on nucleic acid and protein synthesis, since it is the connection between these latter processes in the mediation of a particular response which appears to us more important than the effect on processes considered in isolation.

We have also in the main concentrated on those systems which have been most extensively investigated and/or which seem to be most susceptible to an early solution.

6.2 Effects of Hormones on Processes Involved in Growth

In general, plant cells are bounded by, and their size limited by, a cell wall. It follows therefore that for a cell to enlarge there must be an increase in the area of the cell wall. The phenomenon of irreversible cell enlargement, which, when integrated, represents the overall growth of the plant, has long been a subject of interest to botanists, more especially so since the early work of Darwin and Went showing that phototropic curvature in coleoptiles is the result of unequal growth and furthermore that this growth is controlled by the differential distribution of a plant hormone, namely IAA.

It appears that the control of cell extension in most systems is vested in IAA and most work has been performed using natural or synthetic auxins. Nevertheless, all the other groups of growth regulators have at some time or other been implicated in the process. Thus, it is well established that application of gibberellins to intact plants or excised sections of plants may result in remarkable promotions of extension growth. Equally, in appropriate systems such as cotyledon tissue, cytokinins may induce increases in cell size. Abscisic acid will normally inhibit overall extension growth, although at high concentrations extension may be transitorily promoted as a consequence of the low pH (see below). The effects of ethylene on growth are varied. In some plants, ethylene promotes extension growth and it is noteworthy that such plants are often hydrophytes or live in environments where waterlogging is an important factor e.g., *Callitriche* (Musgrave et al., 1972; Musgrave and Walters, 1974). This may be related to the finding by a number of workers (Dowdell et al., 1972; Smith and Dowdell, 1974) that ethylene levels may be relatively high in such environments; alternatively, outward diffusion of ethylene from a plant growing in water may be much reduced, resulting in increased endogenous levels.

On the other hand, ethylene inhibits cell elongation in both the roots and shoots of a wide range of species. Like cytokinins it may induce isodiametric cell expansion in the presence of IAA (BURG and BURG, 1966) and indeed, the observation that IAA at high concentration may induce isodiametric expansion has been attributed to the induction of increased production of ethylene by auxin treatment (BURG and BURG, 1966).

It is important to note that the kinetics of growth promotion or inhibition, where they have been studied, are not identical in the case of all growth regulators, nor indeed for the same growth regulator in different situations. The following section therefore will deal primarily with cell extension as controlled by auxin, for which most information is available. Having established the general phenomena associated with auxin-induced growth, we will return to a consideration of the mediation of cell expansion by other growth regulators.

6.2.1 Effects of Auxins on Cell Extensibility

It has been shown that the rate of cell elongation may be expressed in terms of a single equation of the form (RAY et al., 1972)

$$v = L\varphi \cdot \frac{\Delta\pi - Y}{L + \varphi} \tag{1}$$

where v represents the relative change in cell volume (dv/dt); φ is a yielding compliance (usually termed extensibility); Y is a yield threshold; $\Delta\pi$ the difference in osmotic potentials of cell and medium; and L the permeability of the cell to water (the hydraulic conductivity).

If the hydraulic conductivity is large relative to the extensibility then the equation reduces to

$$v = \varphi(\Delta\pi - Y). \tag{2}$$

This also implies that $\Delta\psi$ (the difference in water potential between medium and cell) in the equation $\Delta\psi = \Delta\pi - P$ is negligible and hence $\Delta\pi = P$ (the turgor pressure). This gives a third equation

$$v = \varphi(P - Y). \tag{3}$$

This latter is the equation most familiar to physiologists and may be useful as such, but it must be borne in mind that this is a special case and that at least in oat coleoptiles $\Delta\psi$ is by no means negligible (RAY and RUESINK, 1963) and Eq. (1) must be used. The quantification of the components of φ is a complex rheological problem, a detailed discussion of which is not appropriate here, and the reader is referred to various articles on the subject (CLELAND, 1971; GREEN, 1972); suffice to say that it comprises components of elastic and plastic extensibility (EEi and PEi) which represent the ability of the wall to be extended elastically (EEi) and the potential of the wall to undergo plastic extension (PEi), at any one moment. In addition we must include the continuous extensibility (WEc) which is the potential of the wall for irreversible extension over a particular time interval (CLELAND, 1971).

In general it is the parameter PEi which is normally measured, by means of the Instron technique (CLELAND, 1965). Briefly, the technique consists of fixing a piece of the tissue under study between two clamps and extending it at a constant rate, the force across the material being measured continuously. When a particular stress (force on the material per unit area) has been attained, the clamps are returned to their original position and the material re-extended to the same force. The first extension provides a compliance value [the ratio of strain (extension/initial length) to stress] of total extensibility and the second a compliance of elastic extensibility. From the difference between these two values a figure for compliance of plastic extensibility (DP) may be obtained (CLELAND, 1971). It is generally accepted that DP is a reasonable approximation to PEi. Using these techniques and earlier variants it appears that EEi is not related directly to extension, although treatment of coleoptile tissue with auxin does lead to increases in this parameter (MASUDA, 1969; MORRÉ and EISINGER, 1968) which may, moreover, be very rapid. More importantly it has been shown that there *is* a good correlation between DP and the rate of cell elongation (CLELAND, 1967; LOCKHART, 1967). Thus, treatment of *Avena* coleoptiles with auxin leads to a threefold increase in DP over the first 90 min, after which a constant level is maintained for some time (CLELAND, 1967). It should nevertheless be pointed out that DP and PEi are not necessarily identical and that DP may represent a measure of the past rate of wall-loosening (CLELAND, 1971).

WEc has been extensively measured by CLELAND and his co-workers, who have shown a close relationship between increases in this parameter and growth stimulation in response to auxin treatment.

It is clear from the foregoing discussion that auxin may radically affect the value of φ [Eq. (1)]. A possible relationship between auxin action and turgor pressure is still a matter of some controversy. Early work indicated that auxins did not influence cell extension by increasing turgor pressure, and the fact that, in some tissues, a drop of turgor pressure eventuates during rapid extension seemed to prove the point. At that time, however, the involvement of yield threshold (Y) was not appreciated and the inclusion of this factor has cast doubt on the interpretation of this work.

Using external osmotica to modify the growth rate, CLELAND (1959) showed that auxin increased the growth rate at any pressure in excess of Y but did not affect Y during steady-state growth. On the other hand RAY (1969) has suggested that auxin may reduce Y, although his conclusions have been challenged (CLELAND, 1971). Nevertheless, in *Nitella* cells it has been found that Y may be a variable (GREEN, 1968; GREEN et al., 1971). Again, the arguments concerning the processes are beyond the scope of this article (but see CLELAND, 1971). Nevertheless, it is important to establish whether a turgor pressure in excess of Y is required not merely for extension itself but for the "loosening" of the wall prior to extension. In other words, is such a pressure necessary for the biochemical modifications of the wall which lead to "loosening" (see also Sect. 6.3)? Much of the argument regarding this aspect has revolved around the demonstration of so-called "stored growth". Originally, it appeared that, if growth was inhibited by incubating tissue in hypertonic solutions of mannitol in the presence or absence of auxin and then returning to water, the auxin pre-treated sections extended more than controls, suggesting that a potential for growth could build up during periods when the turgor pressure was below Y (KOBAYASHI et al., 1956; THIMANN, 1954). Not all this early work has been confirmed, however (ADAMSON and ADAMSON, 1957; RAY, 1961), and the increase in extensibility, shown by sections treated with auxin in the presence of mannitol, is not necessarily a potential for extra extension (CLELAND, 1971). Although this problem remains unresolved it seems likely that auxin-induced "loosening" steps only occur when the wall is extended or under tension. The implications of this proposal will be discussed below.

Whatever the final outcome of the controversies set out above it is clear that treatment with auxin leads to a change in the properties of the cell wall and it is in relation to this long-standing piece of information that most work has been orientated. Namely, how does auxin bring this about?

It is well established that cell extension does not involve a mere physical stretching of the cell wall, in that extension is sensitive to temperature and may be slowed down by metabolic inhibitors. Physical wall extension on the other hand is unaffected by such factors. Clearly, therefore, some biochemical modification of the wall must precede (and possibly follow) extension itself, although, as indicated above, the biochemical event(s) may well be affected by turgor pressure which also drives extension.

The hypotheses which have been proposed to account for auxin-induced wall loosening are legion and reflect not only a progression in thought and concepts but also the relative popularity of particular areas of research. The confusion has been compounded by an almost total ignorance of the chemical structure of plant cell walls, a problem which has only just recently begun to be resolved.

Early work stressed the possibility of direct effect of auxin on the cell wall itself. Thus BENNET-CLARK (1956) proposed that IAA chelated calcium ions in the cell wall, thus permitting increased "slippage" between polyuronide components of the wall, which, it was proposed, were held together by ionic linkages via calcium. This theory was subsequently abandoned when it was shown that auxin treatment did not, in fact, lead to changes in the distribution of calcium in the wall (CLELAND, 1960a). A variant on this theme suggested that auxin prevented the de-esterification of carboxyl groups in polyuronides via pectin-methylesterase (SCHRANK, 1956) – leading to effects similar to those proposed by BENNET-CLARK. It was shown, however, that ethionine could competitively inhibit methylesterification without at the same time affecting extension (CLELAND, 1960b).

Two main strands run through subsequent work: firstly, the events in the cell wall itself which lead to changes in its properties, and secondly, the complexity of the events leading up to that change in properties.

a) Kinetic Studies on Cell Extension

Although measurement of the kinetics of auxin-induced growth has a venerable history (see EVANS, 1974) it is only in the last ten years that a vastly increased number of papers have appeared in the literature. The now classical paper of EVANS and RAY (1969) can probably claim the credit for stimulating this renewed interest. It is generally observed that after addition of auxin there is a "lag" or "latent" period before growth rate increases. This latent period varies between about 6 and 15 min (see Table 6.1). Subsequent to the latent period, growth rate increases to a maximum over a period of about 10 to 20 min. In many cases, but not in all (DURAND and ZENK, 1972), the growth rate then decreases somewhat, reaching a minimum some 10–20 min after the first maximum. Growth rate may then stabilize (BARKLEY and EVANS, 1970; PENNY, 1970) or rise to a second maximum only slightly lower than the first (PHILIPSON et al., 1973a; PENNY, 1970) (Fig. 6.1). It should be noted that some workers have reported that during the initial latent period there may actually be a transient reduction in growth rate (BARKLEY and EVANS, 1970; RAYLE et al., 1970a) but it is not clear whether this is a real effect or a systematic error (PENNY and PENNY, 1978).

These relatively simple observations have had the effect of concentrating the minds of workers in the field wonderfully indicating, as they do, that, if the key to the role of auxin in inducing increased growth is to be found, it must involve a process or processes occurring within the latent period.

The length of the latent period and the occurrence of various processes within it are clearly most important. Table 6.1 shows the latent period for various auxin effects in elongating tissue; these will be referred to where appro-

Table 6.1. Latent periods in the effect of auxin on cellular processes

Phenomenon affected	Material	Latent period (min)	Reference
Metabolism			
Stimulation of CO_2 uptake	*Avena* coleoptile	"immediate"	YAMAKI (1954)
Release of xyloglucan from cell wall	Pea stem	10–15	LABAVITCH and RAY (1974b)
Stimulation of RNA synthesis	*Avena* coleoptile	10	ESNAULT (1965) MASUDA and KAMISAKA (1969)
		40	MASUDA et al. (1967)
Stimulation of protein synthesis	Pea stem	60	TREWAVAS (1968a)
Stimulation of cell wall synthesis	Pea stem	60	NEUMANN (1971)
Stimulation of respiration	Pea stem Wheat coleoptile	120 120	MARRÈ and FORTI (1958) ROWAN et al. (1972)
Increased enzyme activity			
RNA polymerase	Soybean plasmamembrane	1	MORRÉ and CHERRY (1977)
$\beta 1 \rightarrow 3$ Glucanase	*Avena* coleoptile	10	MASUDA and YAMAMOTO (1970)

Table 6.1 (continued)

Phenomenon affected	Material	Latent period (min)	Reference
Glucan synthetase "Xylan" synthetase "Galactan" synthetase	Pea stem	15	RAY (1973a)
β-Galactosidase	*Avena* coleoptile	60	JOHNSON et al. (1974)
Cellulase	Pea stem	360	FAN and MACLACHLAN (1966)
β1→3 Glucanase, β1→6 Glucanase, Hemicellulase	Barley coleoptile, Pea stem	180	TANIMOTO and MASUDA (1968)
RNA polymerase	Soybean	1	O'BRIEN et al. (1968a, b)
Electrophysiology			
Change in membrane potential	Bean root	1	JENKINSON (1962)
	Avena coleoptile	7–8	CLELAND and LOMAX (1977)
Growth and extensibility			
Stimulation of growth	Sunflower	6	UHRSTRÖM (1969)
	Pea, etiolated	6–9	BARKLEY and EVANS (1970)
		21	JACOBS and RAY (1976)
	Pea, green	11–14 (32)	MURAYAMA and UEDA (1973) PENNY (1969)
	Avena coleoptile	12	EVANS and RAY (1969) RAY and RUESINK (1962) PHILIPSON et al. (1973b)
	Soybean	12	VANDERHOEF and STAHL (1975)
	Lupin	13–19	PENNY (1970)
	Maize coleoptile	21	JACOBS and RAY (1976)
Increased wall extensibility	Pea stem	2–3	BURSTROM et al. (1970)
	Sunflower hypocotyl	4	UHRSTRÖM (1969)
	Avena coleoptile	20	CLELAND (1967)
Protein secretion			
	Soybean	∞	VANDERHOEF et al. (1977)
	Sunflower	60	ILAN (1973)
	Avena coleoptile	>37	PENNY et al. (1975)
	Avena coleoptile	15–20	CLELAND (1973)
	Avena coleoptile	10–14	CLELAND (1976)
	Maize coleoptile	12	JACOBS and RAY (1976)
	Pea, etiolated	12	JACOBS and RAY (1976)

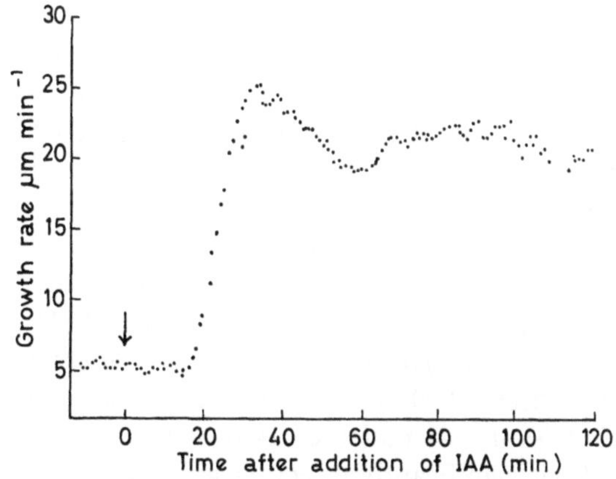

Fig. 6.1. Effect of auxin on the growth rate of light-grown pea epicotyl segments. Segments were soaked for 110 min in 20 mM tris-maleate buffer pH 6.1 and 30 µM indole-3-acetic acid was then added (at *arrow*). (PENNY et al., 1972)

priate. However, the period itself is a variable. As might be expected, the latent period varies from tissue to tissue but of even more significance are observations showing that it may be increased or reduced by appropriate means in the same system. Some of these differences are undoubtedly due to systematic variations, but equally some are not. The concentration of auxin employed may have a marked effect on the latent period. Thus, RAYLE et al. (1970a) reported that the lag period was unaffected by changes in auxin concentration between 5 and 1,000 µM. On the other hand NISSL and ZENK (1969) showed that in *Avena* the latent period decreased with increasing IAA concentration over the range 10 nM to 5 mM. On the whole, however, above a minimum concentration of ca. 0.6 µM auxin, most workers do not observe effects of auxin concentration on the latent period.

It is generally agreed that increased temperature leads to a shortening of the latent period (EVANS and RAY, 1969; PENNY et al., 1972; MURAYAMA and UEDA, 1973; PHILIPSON et al., 1973a, b; NISSL and ZENK, 1969). DURAND and ZENK (1972) in *Avena* and MURAYAMA and UEDA (1973) with pea have demonstrated that a combination of high temperature and high auxin concentration may virtually eliminate the latent period. It is possible, however, that the low pH of the auxin solution is responsible for this effect (PHILIPSON et al., 1973a, b). The situation is further complicated by the observation of BARKLEY and LEOPOLD (1973) that if auxin is supplied to green pea stem segments simultaneously with a drop in the pH of the incubation medium from 6.3 to 3.0 the latent period is reduced from 10–15 min to 1 min. It should be noted that low pH itself may not induce growth in this tissue (BARKLEY and LEOPOLD, 1973 and Sect. 6.2.1c).

b) Effects on Nucleic Acids and Proteins

It is now accepted that continued RNA and protein synthesis are required if growth is to be sustained (NOODEN and THIMANN, 1963; KEY, 1964; CHERRY, 1974). These findings, combined with others which demonstrated that auxin could stimulate synthesis of protein and RNA in elongating tissue (KEY, 1964;

KEY and INGLE, 1968), led to the proposition of the so-called gene activation hypothesis which suggested that auxin derepressed certain genes, the expression of which was necessary for the extension process. Further studies indicated that synthesis of specific proteins and RNA species was promoted by auxin in tissue capable of elongation (PATTERSON and TREWAVAS, 1967; DAVIES et al., 1968; TREWAVAS, 1968a). Most of the newly synthesized RNA appears to be ribosomal (KEY and SHANNON, 1964; DAVIES et al., 1968; TREWAVAS, 1968a) although effects on mRNA (MASUDA and TANIMOTO, 1967; MASUDA et al., 1967; MASUDA, 1968a; MIASSOD et al., 1971) and tRNA (ESNAULT, 1968) have been claimed. However, a series of papers by KEY and his co-workers (KEY and INGLE, 1964; KEY, 1966; KEY et al., 1967) using "specific" inhibitors of RNA synthesis indicated that synthesis of rRNA could be greatly inhibited without significantly affecting growth. Inhibition of DNA-like RNA synthesis with actinomycin D did, however, impair the growth response.

More recent work on this field has not in general been very conclusive. Thus, THOMPSON and CLELAND (1971) and CLELAND et al. (1972) could not observe any change in hybridizable RNA in pea stem in response to auxin treatment, suggesting that no new species of RNA were formed. O'BRIEN et al. (1968a) demonstrated a promotory effect of auxin on RNA polymerase in soybean chromatin, but the effect took a considerable time to develop. MATTHYSE and PHILLIPS (1969) showed an effect of auxin in vitro upon an RNA polymerase system. However, other workers using similar systems (HARDIN et al., 1970) have been unable to reproduce these results. More recently, HARDIN et al. (1972) and MORRÉ and CHERRY (1977), using isolated plasmalemma vesicles treated with auxin, have shown release of a transcription factor that stimulates RNA polymerase in vitro (see also Sect. 6.5). The effect is very rapid.

The kinetic studies on auxin-induced cell extension described above soon cast doubt on the role of many of the phenomena set out above in the induction of elongation, both because of calculations such as those of EVANS and RAY (1969) showing that auxin-induced growth could only occur via gene activation if the half life of the auxin-induced RNA and protein was about 2–3 min and also because only rarely (see Table 6.1) could promotion of RNA or protein synthesis be observed within the latent period. This led to the proposal that many of the changes were "growth-permissive" or growth-induced rather than growth-causative, or were related to continued growth rather than its induction. This coincided with newer views on the role of hormones in growth and development, suggesting that in most cases growth regulators "switch on" predetermined developmental pathways rather than initiating them per se (e.g., WAREING, 1971).

The number of enzymes in elongating tissue, the synthesis of which is promoted by auxin, is legion and includes both those degrading and those synthesizing components of the wall, e.g., 1→4-glucanase (FAN and MACLACHLAN, 1966, 1967a, b), 1→3-glucanase (MASUDA, 1968b), pectinesterase (DATKO and MACLACHLAN, 1968), and 1→3 and 1→4β-glucan synthetase (HALL and ORDIN, 1968; ABDUL-BAKI and RAY, 1971). The significance of these findings was also cast in doubt by the kinetic experiments on auxin-induced growth. On the other hand it was shown (PENNY and GALSTON, 1966; NELSON et al., 1969) that, whereas inhibitors of protein synthesis could inhibit growth immediately or nearly so, inhibitors of RNA synthesis did not. This led to the idea of the existence of an unstable protein whose synthesis was dependent on auxin and which was a prerequisite for the induction of growth. Much work has since been expended on this "growth-limiting protein" (GLP). The experiments have normally employed cycloheximide as an inhibitor of protein synthesis; in general, treatment of stem or coleoptile tissue with this substance results in an inhibition of auxin-induced growth (EVANS and RAY, 1969; PENNY, 1971; POPE and BLACK, 1972; MURAYAMA and UEDA, 1973) but growth still occurs and, in general, the latent period is unaffected. These studies and others have led to the proposition that the GLP has a short half-life – of the order of 12 min – a suggestion which raises a number of problems, not the least of which is that such a half-life is at least an order of magnitude smaller than any so far observed for defined proteins in eukaryotes. Explanations for this observation range from suggesting that the GLP is a structural protein and that its apparently short half-life is a reflection of its incorporation into some cell fraction (EVANS and HOKANSON, 1969; POPE and BLACK, 1972) or that it undergoes a conformational change from an active to inactive form (PENNY and PENNY, 1978). Recent work by VANDERHOEF et al. (1976) suggests that the first and second maxima of growth (Sect. 6.2.1.a) possess GLP's with half-lives of 28 and 11 min respectively, although the authors are at pains to point out that the fact that a protein is necessary for growth does not imply that it has anything to do with auxin action.

This field is still very active but it is appropriate to point out that the calculations of half-lives of proteins is a difficult undertaking at best, and especially so when the nature of the protein is unknown. Equally, the use of inhibitors is fraught with difficulty and although the workers in the field are aware of the limitations of such techniques (see review by PENNY and PENNY, 1978) there still remains a major element of doubt as to the specificity of action of inhibitors. These reservations are borne out by some recent work of BLACK and DHEIDAH (1976) indicating that some of the effects of cycloheximide may be attributable to its effect on respiration, especially if kinetic studies are carried out in continuous recording vessels of the type used by EVANS and RAY (1969).

c) The Acid Growth Effect

It has long been known that treatment of tissue at low pH may stimulate elongation (BONNER, 1934). In recent years this phenomenon has been studied in much greater detail (RAYLE et al., 1970b; RAYLE and CLELAND, 1970; see also reviews by RAYLE and ZENK, 1973, and EVANS, 1974) especially since the demonstration of auxin-induced proton pumps in elongating tissue (see Sect. 6.3). The effect of acid on the growth response is very short-lived – about 30 min – and in contrast to the effect of auxin it is not invariably affected by metabolic inhibitors (EVANS et al., 1971; RAYLE and CLELAND, 1972). Treatment of frozen-thawed tissue sections at low pH also induces increases in wall extensibility, an effect not observable with auxin. The latent period for response to low pH is a minute or less, and the rate of elongation induced is comparable

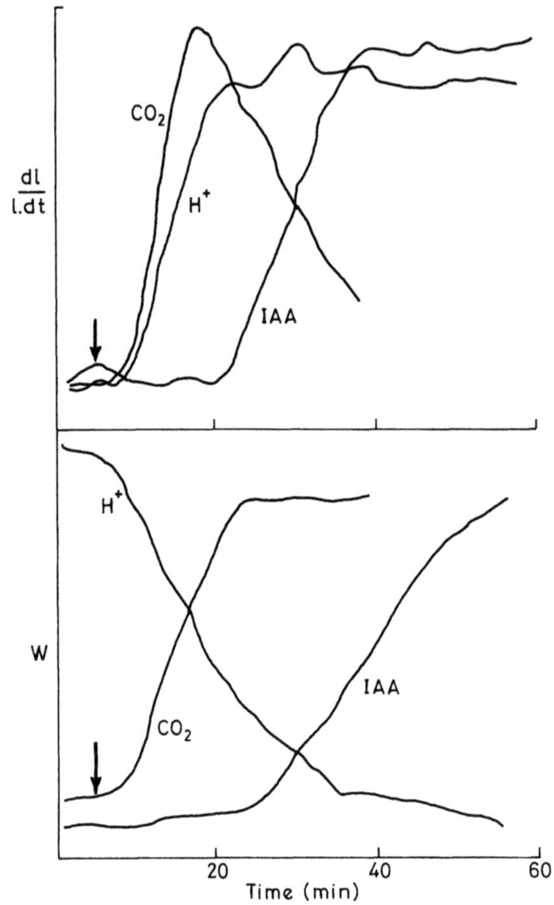

Fig. 6.2. Relative elongation rate ($dl/(l \cdot dt)$) and radius (w) of lupin hypocotyl segments. At *arrow* segments were transferred from 1 mM phosphate buffer pH 6.5 to the experimental treatments which were: IAA, 30 μM indole-3-acetic acid in 1 mM phosphate buffer pH 6.5; H^+, 10 mM citrate phosphate buffer, pH 4.0; CO_2, saturated in 1 mM phosphate buffer. Segments were taken from 4-day-old lupin seedlings grown in continuous light. (PERLEY et al., 1975)

to that shown by treatment with optimal auxin concentrations (BARKLEY and LEOPOLD, 1973; EVANS et al., 1971; HAGER et al., 1971; RAYLE and CLELAND, 1972) (Fig. 6.2). Further, the acid and auxin responses have similar Q_{10} values and similar effects on wall extensibility (RAYLE and CLELAND, 1970). Despite these similarities and others (see PENNY and PENNY, 1978), and the demonstration of auxin-induced proton pumps, the concept first proposed by HAGER et al. (1971), that auxin acts on cell extension via activation of a proton pump, is still very controversial. EVANS (1974) and PENNY and PENNY (1978) have set out the criteria which need to be fulfilled if the acid model of auxin action is valid, namely that (i) proton efflux must be detectable in the latent period (ii) all tissues responsive to auxin must also be responsive to acid (iii) there must be polysaccharide-modifying enzymes in the cell which exhibit increased activity at low pH and/or possibly acid-labile linkages capable of modification at pH 4–5 and/or pH-dependent changes in wall matrix conformation (iv) the properties of auxin- and acid-promoted growth must be similar.

In general, condition (iv) is fulfilled although PERLEY et al. (1975) found that, in lupin, whereas auxin-induced growth is accompanied by an increase

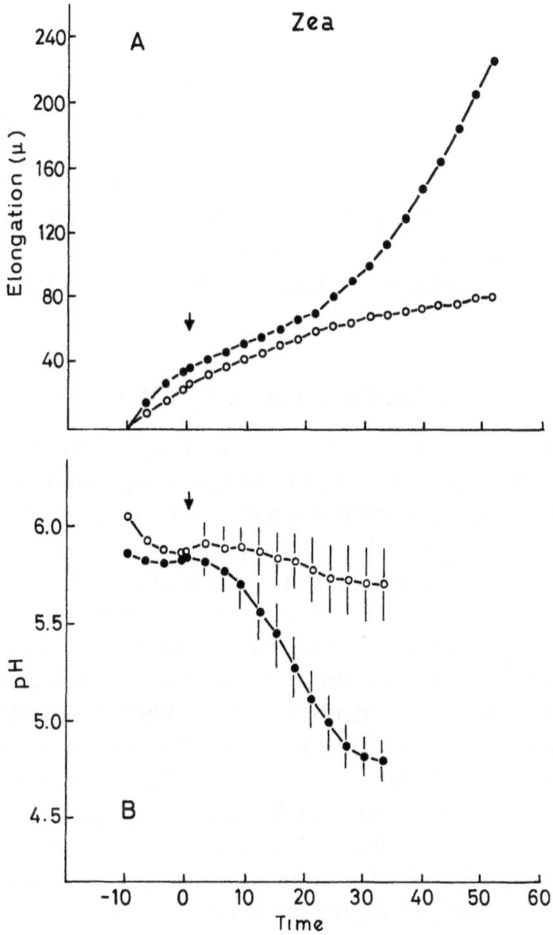

Fig. 6.3 A, B. Comparison of auxin-induced decrease in free space pH (as measured by a pH microelectrode inserted into the tissue) with auxin-induced increase in elongation rate in *Zea* coleoptiles. 20 μM IAA (●) or 1 mM phosphate-citrate buffer pH 6 (○) added at *arrow*. (JACOBS and RAY, 1976)

in stem thickness, a decrease is observed with acid (Fig. 6.2). Condition (ii) is not universally fulfilled since it has been reported that green pea stem tissue does not respond to acid, even though it does to auxin (BARKLEY and LEOPOLD, 1973). Equally, KAUFMAN et al. (1969) have shown that in *Avena* internodes acid-induced growth is observable even though the tissue is unresponsive to auxin. It should be noted that the failure of BARKLEY and LEOPOLD to obtain a pH response in green pea tissue may be a matter of penetration of H^+, since if slits are made in the epidermis of such tissue an acid growth effect can be demonstrated.

Condition (i) is probably the most controversial. Reports range from situations where auxin-induced growth is not accompanied by a pH drop (e.g., PENNY et al., 1975; POPE et al., 1975; VANDERHOEF et al., 1977) through situations where a drop in pH could only be detected after growth had commenced (CLELAND, 1973) to observations by CLELAND (1976) and JACOBS and RAY (1976) of a decrease in pH prior to growth induction (Fig. 6.3). At least some of this conflict can probably be attributed to differences of technique but almost certainly not all. It is perhaps significant that the different types of observation [and this applies to some extent to condition (ii)] are often correlated with the type of tissue, namely a lack of response with green, dicotyledonous stems and a positive response with etiolated, usually monocotyledonous tissue. Furthermore, the biphasic response to auxin is apparently more readily demonstrated in dicotyledons than monocotyledons (see PENNY and PENNY, 1978, for comparisons). This suggests that, in some tissues at least, auxin has an effect on the induction of growth other than, but possibly complemented by, that afforded by acidification (see Sect. 6.2).

The evidence relating to conditions (iii) and (iv) will be dealt with in the next section.

d) Effect on Metabolism of the Cell Wall

The cell wall consists of three components, namely cellulose microfibrils, the matrix and water. The mechanical properties of this system depend on a number of factors the nature of which has been set out by NORTHCOTE (1972) as follows; (a) the length and cross-sectional area of the microfibrillar component, (b) the relative amounts of fibre and matrix, (c) the arrangement of the fibres in the matrix and (d) the interfacial interactions between individual components of the matrix and between fibre and matrix. These criteria imply that a change in wall properties can be brought about by a change in either the microfibrillar or matrix components or both. While it is certainly true that the orientation of cellulose microfibrils determines the shape of cells and contributes largely to the tensile properties of the wall, no convincing evidence has been adduced to show that changes in the degree of polymerization (DP) of the polysaccharide are directly involved in the mediation of auxin-induced extension. Thus, although the activity of $\beta 1 \rightarrow 4$ glucan synthetase is enhanced by auxin treatment (HALL and ORDIN, 1968), this increase does not occur if the tissue is prevented from extending (SMITH, E.J., 1976). Equally, since the $\beta 1 \rightarrow 4$ glucanases isolated from higher plants do not appear to be able to degrade crystalline cellulose, it is difficult to understand how a decrease in the DP of microfibrillar cellulose could be brought about. WONG et al. (1977a) have suggested that the role of higher plant cellulases lies in the control of cellulose biosynthesis. It thus seems probable that the changes in the properties of primary cell walls leading to increased extensibility are the result of changes in the connections between and within components of the matrix and between the latter and the microfibrillar component. Hypotheses over the years have tended to fall into two categories, namely, those suggesting that auxin-induced growth occurs via a stimulation of polysaccharide biosynthesis, and those proposing a critical role for polysaccharide degradation.

It has long been established that auxin-induced growth can occur without *net* polysaccharide synthesis and vice versa (BENNET-CLARK, 1956; RAY, 1961, 1962). Thus if there is a role for polysaccharide synthesis in the mediation of auxin-induced growth it must involve the synthesis of specific polysaccharides or differences in the mode of incorporation of polysaccharides into the wall (BAKER and RAY, 1965). It is not altogether clear how polysaccharide synthesis would affect extensibility in the short term but it is possible that incorporation of polyuronides could achieve this. However, the latent periods for auxin-induced synthesis of polysaccharide constituents are, in general, longer than those for growth (BAKER and RAY, 1965) and the auxin-induced increases in the activity of polysaccharide synthesizing enzymes so far observed may be prevented if the tissue is prevented from extending. However, RAY (1973a) has shown that, in pea, the latent period for the increase in the activity of β-glucan synthetase by auxin is between 10 and 15 min. This increase is dependent on the presence of sucrose and is not inhibited by cycloheximide or actinomycin D. Respiratory inhibitors such as KCN, DNP, azide, and CCCP inhibit the activation, as do pretreatments with hypertonic mannitol solutions. Other synthetases appear to be involved since similar results were obtained using UDP xylose and UDP galactose as glycosyl donors. RAY proposes that the effect is due to the activation of previously existent enzyme possibly via phosphorylation of the enzyme in a way analogous to the activation of glycogen synthetase. A role for cellular transport phenomena is suggested by the results with mannitol. Work by VAN DER WOUDE et al. (1972) showing an apparent in vitro effect of auxin on β-glucan synthetases from onion may be related to RAY's findings, although the latter could not obtain his effect in vitro. While RAY's work constitutes the demonstration of the most rapid effect yet shown of auxin on a polysaccharide synthetase, the timing of the response precludes a role for it in the causation of elongation.

By far the most attention has been focussed on the role of polysaccharide breakdown, either directly, by work on the enzymes involved, or indirectly, by studying polysaccharide turnover. The situation is far from satisfactory, however, largely because the detailed structure of wall polysaccharides other than cellulose was, until recently, rather obscure and because the polysaccharides themselves were not available for use as substrates. Thus, workers have tended to look at enzymes, the significance of which in terms of wall biochemistry is doubtful to say the least. With some notable exceptions the enzymes studied have been glycosidases [e.g., β-galactosidase] (JOHNSON et al., 1974). While such enzymes (or, as they may be, exoglycanases) may be important in modifying wall structure, it seems more likely that endoglycanases or transglycosylases are involved. The former are more likely because endoenzymes bring about more rapid reductions in DP and hence changes in mechanical properties than exoenzymes; the latter are more likely because their action is reversible and because their action would be consistent with the concept of bond breakage followed by re-formation which has been suggested as the way in which changes in the rheological properties of the wall are brought about (CLELAND, 1971). The studies of MASUDA and his co-workers on $\beta 1 \rightarrow 3$ glucanase deserve special mention in this context both because the effect of auxin upon the enzyme is rapid (MASUDA and YAMAMOTO, 1970) and because treatment of coleoptile tissue with the enzyme may lead to elongation in the absence of auxin (MASUDA, 1968b), although not all workers have been able to repeat these latter results (CLELAND, 1968). However, for reasons set out below it seems possible that the observed effects on enzyme activity may be related more to wound-induced callose formation than to the induction of extension. NEVINS and his co-workers (LOESCHER and NEVINS, 1972, 1973; NEVINS, 1975a, b) have proposed a rather similar hypothesis to that of MASUDA, namely that auxin-induced growth is

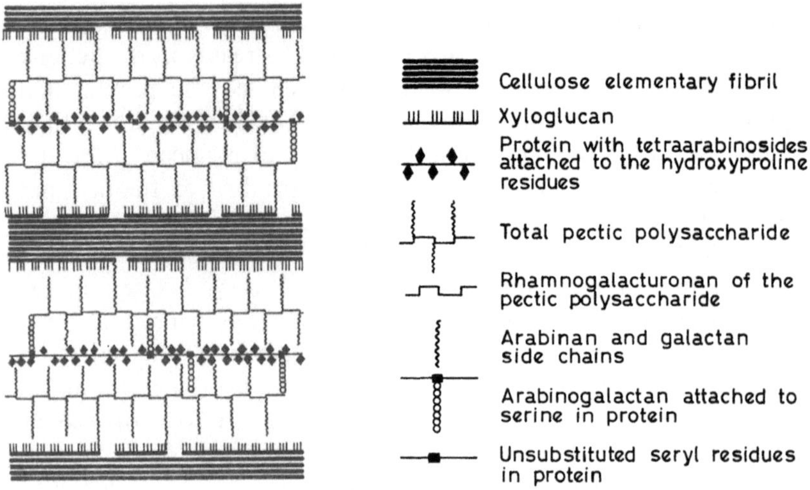

Fig. 6.4. Proposed structure for sycamore callus cell wall. (KEEGSTRA et al., 1973)

related to a loss or turnover of non-cellulosic glucan, a conclusion supported by the work of KATZ and ORDIN (1967a). The same reservations apply to this work as to that of MASUDA mentioned above. In addition, since some of NEVINS' conclusions depend on the use of the polysaccharidase inhibitor, nojirimycin, the fact that the latter inhibits auxin-induced proton extrusion (CLELAND, personal communication) must be taken into account.

MACLACHLAN and YOUNG (1962) developed the gravimetric technique for studying cell wall turnover, and many workers have used this or its variants in subsequent years – in particular the studies of BAKER and RAY (1965), and of KATZ and ORDIN (1967b) (for a list of studies on turnover see LAMPORT, 1970).

Rather similar considerations apply to this work on polysaccharide turnover as those set out above for enzymes. Unless turnover of a particular sugar can be attributed to a particular polysaccharide the object of the experiment, namely the identification of the enzyme involved, is unlikely to be resolved. On the other hand, the very fact that this work did show that turnover of some wall fractions is stimulated by auxin was valuable in itself and permitted some of the more recent advances.

The turning point in these studies arose from the work of ALBERSHEIM and his group who, in a series of papers (TALMADGE et al., 1973; KEEGSTRA et al., 1973) described the development of new techniques for the study of cell wall structures and, using these techniques, were able to propose a model for the cell wall of cultivated *Acer* cells. Recently, DARVILL et al. (1977, 1978) have produced a tentative model for the cell wall of *Zea mays* coleoptiles. These structures are shown in Figs. 6.4 and 6.5.

These discoveries have been exploited by other workers. Thus, LABAVITCH and RAY (1974a, b) and GILKES and HALL (1977) have shown that in pea epicotyl – the cell wall of which appears to resemble that of *Acer* rather closely

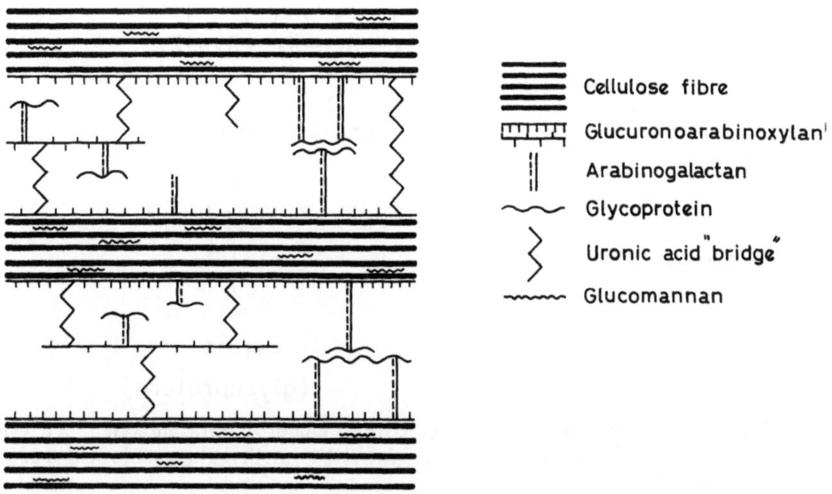

Fig. 6.5. Proposed structure for maize coleoptile cell wall. (DARVILL et al., 1977)

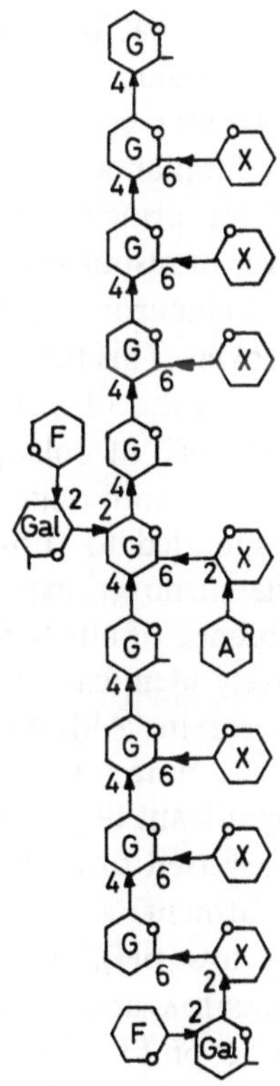

Fig. 6.6. Proposed structure for sycamore callus xyloglucan. *X* xylose; *G* glucose; *A* arabinose; *Gal* galactose; *F* fucose. (KEEGSTRA et al., 1973)

(GILKES and HALL, 1977) – xyloglucan (Fig. 6.6) is the only polysaccharide component of the wall to exhibit substantial turnover specifically in response to auxin treatment (although a number of components turn over in both control and auxin treatments). The effect was observed within 15 min of auxin addition in the work of LABAVITCH and RAY. Moreover, JACOBS and RAY (1975) have

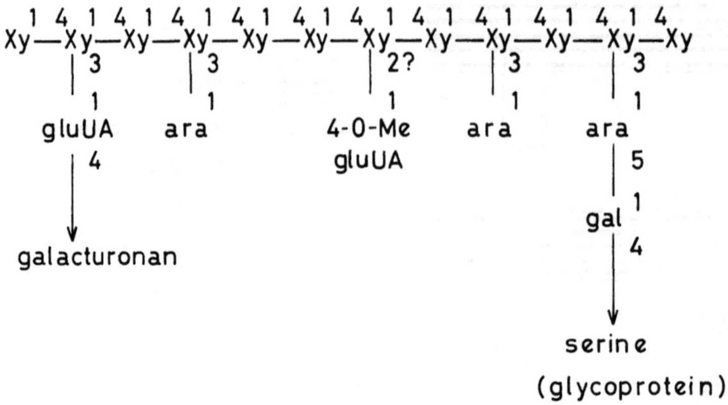

Fig. 6.7. Proposed structure for maize coleoptile glucurono-arabinoxylan. *Xy* xylose; *ara* arabinose; *gluUA* glucuronic acid; *4-O-Me gluUA* 4-O-methyl glucuronic acid; *gal* galactose. (DARVILL et al., 1977)

shown that treatment of pea epicotyl sections at low pH results in a release of xyloglucan, an observation confirmed by KHODARY (1977) using cell sheets (DARVILL et al., 1977) derived from the same tissue.

Similar work with *Zea mays* (DARVILL et al., 1977, 1978) showed that part of the principal matrix component, a glucuronoarabinoxylan (Fig. 6.7), turns over in response to auxin treatment such that interconnections between individual glucuronoarabinxylan chains and between these and cell-wall protein are broken. This turnover occurred even if extension was inhibited by simultaneous incubation of auxin-treated sections in mannitol. In similar experiments using "sheets" of coleoptile tissue two or three cells thick, it was shown that the mono- and oligo-saccharides released during auxin-stimulated growth corresponded to those which would have been expected from consideration of the turnover experiments. Moreover, incubation of tissue at pH 4.5 in the absence of auxin resulted in a release of mono- and oligo-saccharides, qualitatively identical with and quantitatively similar to that observed in the presence of auxin. Although the sugar release experiments were not performed over short time intervals, calculations from the turnover experiments indicate that significant turnover of some linkages was occurring shortly after auxin addition. Increased turnover of $\beta 1 \rightarrow 3$-linked glucose was observed in response to auxin treatment, as was release of $\beta 1 \rightarrow 3$ glucan (this process was also promoted by low pH). At first sight this seems to lend support to the hypotheses of MASUDA and NEVINS referred to above. However, it was observed that the effect on $\beta 1 \rightarrow 3$ glucan synthesis was restricted to the extremities of auxin-treated coleoptiles, making it unlikely that the effect is related to growth of the whole coleoptiles. The work on *Pisum sativum* and *Zea mays* is summarized in Tables 6.2 and 6.3.

It is also appropriate at this juncture to mention the possible role of cell-wall glycoprotein. It is now well established that such glycoproteins are structurally dissimilar to those occurring within the protoplasm, containing, as they do, a high proportion of hydroxyproline (LAMPORT, 1969) to which are

Table 6.2. Summary of glycosidic linkages in the cell walls of etiolated pea epicotyl tissue affected by pH or auxin treatments. Data from polysaccharide release or turnover experiments. The number of asterisks indicates the magnitude of turnover or release; dashes indicate no effect. Linkages other than those shown exhibit turnover but this was not correlated with specific treatments. [Summarized from LABAVITCH and RAY (1974b); JACOBS and RAY (1975); GILKES and HALL (1977); KHODARY and HALL (unpublished)]

LINKAGE	IAA(M)			pH	
	10^{-7}	10^{-5}	10^{-4}	4.5	6.0
t-ara	*	**	***	***	*
3-ara	–	**	**	**	–
5-ara	*	*	**	**	
t-xyl	*	**	***	***	*
2-xyl	*	**	***	***	*
t-man	–	–	*	*	–
4,6-glu	*	**	***	***	*

Abbreviations: ara = arabinose, xyl = xylose, man = mannose, glu = glucose, t = terminal

Table 6.3. Summary of glycosidic linkages in the cell walls of *Zea mays* L. coleoptile tissue affected by auxin or pH treatments. Data from polysaccharide release or turnover experiments. The number of asterisks indicates the magnitude of turnover or release; dashes indicate no effect. Linkages other than those shown exhibit turnover but this was not correlated with specific treatments. [Summarized from Darvill et al. (1977, 1978)]

LINKAGE	IAA(M)			pH	
	10^{-7}	10^{-5}	10^{-3}	4.5	6.0
t-ara	*	***	*	***	*
3,4-xyl	*	***	*	***	*
t-gal	–	**	–	**	*
4-gal	–	**	–	**	–
3-glu	*	***	**	***	*
t-gal UA	*	***	–	***	–
4-gal UA					
t-glu UA	*	***	–	***	–
4-glu UA	**	***	–	***	–

Abbreviations: ara = arabinose, xyl = xylose, gal = galactose, glu = glucose, UA = uquinoic acid, t = terminal

attached short oligoarabinosides (LAMPORT and MILLER, 1971). It also appears that the polysaccharide components of the matrix are covalently linked to this protein, possibly via a galactose-serine linkage (KEEGSTRA et al., 1973). GIESEN and KLÄMBT (1969) observed that auxin enhanced the incorporation of proline into cell-wall protein although WINTER et al. (1970) could not find a correlation between auxin-induced growth and hydroxyproline content in the wall. Moreover, CLELAND and KARLSNES (1967) and SADAVA and CHRISPEELS (1973) present evidence suggesting that there is an increase in hydroxyproline residues in cell walls during the *cessation* of elongation. However, while the possible role of cell protein in auxin-induced extension must remain open, it may be significant that the galactose-serine bond is acid-labile (LAMPORT, 1970) and appeared to be cleaved in response to auxin and acid treatment in the work of DARVILL et al. (1977, 1978).

6.2.2 Other Growth Regulators

Although the effects of other growth regulators on cell extension have been studied in a number of systems, compared to auxin, information on the basis of these effects – especially on the cell wall – is, to say the least, sparse.

a) Gibberellins

The dramatic effects of gibberellins on stem extension are well known but relatively few measurements have been made on the kinetics of the process. In general in those instances where it has been measured (WARNER and LEOPOLD, 1971; MCCOMB and BROUGHTON, 1972; ROSE, 1974) the latent period is of the order of 30 min or more, significantly longer than is usually the case with auxin.

A number of workers have observed interactions between auxins and gibberellins in cell extension. Thus, extended pretreatment (0.5–8 h) with GA_3 enhances the subsequent response of cucumber hypocotyl segments to auxin (KAZAMA and KATSUMI, 1974).

There have been few measurements of changes in mechanical properties in response to treatment with gibberellins. As with auxins, treatment of tissue capable of extending in response to gibberellins in general leads to increases in plasticity (NAKAMURA et al., 1975), although this is not always the case (YODA and ASHIDA, 1961). CLELAND et al. (1968) demonstrated differing effects of auxin and gibberellins on the mechanical properties of cucumber hypocotyl cell walls and suggest that gibberellins may exercise their effects by changing the osmotic potential of the cell sap. ADAMS et al. (1975), working with *Avena* stem internodes, have shown a marked increase in plasticity but not elasticity 1 to 2 h after treatment with GA_3. This corresponded well with the timing of GA_3-induced growth and wall synthesis. The effect was blocked by cycloheximide.

Some recent work (SILK and JONES, 1975; STUART and JONES, 1977, 1978) does have a significant bearing on the possible role of proton pumps in elongation. Using lettuce hypocotyl segments, it has been shown that growth in response to GA_3 proceeds in a manner similar to auxin but that the increased growth rate is not accompanied by acidification of the growth medium. On the other hand HEBARD et al. (1976), working with *Avena* stem segments, have correlated acidification with elongation in response to GA_3. The difference between control and GA_3 treatments was not large, however; moreover, a brief (10-min), preincubation with GA_3 actually depressed acidification relative to controls. Recent work on GA-binding and cell wall extension appears on p. 497.

b) Cytokinins

Cytokinins in general inhibit endogenous or auxin-induced growth in stem tissue (BIRMINGHAM and MACLACHLAN, 1972; HASHIMOTO 1961; KATSUMI, 1962) although not in all cases (VANDERHOEF et al., 1973; VANDERHOEF and STAHL, 1975). When etiolated soybean hypocotyl segments are treated simultaneously with auxin and isopentenyl adenine only the first peak of the type of growth rate curve shown in Fig. 6.1 appears (VANDERHOEF and STAHL, 1975). It has frequently been observed that simultaneous addition of auxin and cytokinin inhibits extension but not fresh weight increase since isodiametric extension is promoted (BIRMINGHAM and MACLACHLAN, 1972; KATSUMI, 1962; HASHIMOTO, 1961). This effect is analogous to that produced by ethylene (see below) and a number of other substances. The basis of the effect is unclear, although GILKES and HALL (1977) have observed that the auxin-induced increase in xyloglucan turnover in etiolated pea epicotyls is severely inhibited by treatment with kinetin.

The effects of cytokinins on cell expansion in cotyledon tissue is well established and MARRÈ and his co-workers (MARRÈ et al., 1974a, b) have demonstrated that this effect is accompanied by a decrease in the pH of the bathing medium and an increase in transmembrane electropotential. The effect can be mimicked by treatment with the fungal toxin fusicoccin (see Sect. 6.5.2.f). They suggest that this implies a central role for proton pumps in cell enlargement generally. On the other hand, despite this coincidence between proton extrusion and growth in response to cytokinin treatment, the bulk of the evidence on cytokinin effects at the subcellular level indicates an involvement at the level of nucleic acids and proteins.

Thus ROYCHOUDHURY et al. (1965) showed that kinetin increased both RNA levels within, and RNA release from, coconut milk nuclei although incubation was for several hours. Since then, other workers have shown effects of cytokinins on protein synthesis (BHATTACHARYYA and ROY, 1969; KLÄMBT, 1974). Other work has suggested specific binding of cytokinin to ribosomes (FOX and ERION, 1975) and acceleration of protein synthesis in vitro (KLÄMBT, 1976), although in the last-mentioned work the effect was rather small. More recently KLÄMBT (1977), using an in vitro system from wheat germ which included poly(U), demonstrated an 80%–100% stimulation of phenylalanine incorporation into protein by cytokinin.

The demonstration that cytokinins are present in tRNA and are positioned (p. 473) next to the anticodon (e.g., HALL et al., 1967) led many workers to suggest that herein lay the key to cytokinin activity; either via incorporation of cytokinin into tRNA or via side chain donation from a cytokinin to the appropriate purine in tRNA. The importance of the positioning of the cytokinin next to the anticodon was demonstrated by work showing that blocking the cytokinin in seryl tRNA with iodine interfered with codon-anticodon interaction (FITTLER and HALL, 1966).

However, a series of papers (BURROWS et al., 1971; SKOOG et al., 1973; ARMSTRONG et al., 1976) make the general proposition outlined above highly unlikely. Thus, it was shown that cytokinin-dependent callus cultures were still capable of producing tRNA which contained cytokinin and that certain synthetic compounds with high cytokinin activity displayed structural features which made it highly unlikely that they would be incorporated into tRNA. Moreover, radioactive cytokinins were shown to be incorporated into rRNA more readily than into tRNA. These considerations have been extensively reviewed (KENDE, 1971; KENDE and GARDNER, 1976) (see also p. 473).

c) Ethylene

In most instances ethylene at physiological concentrations inhibits extension in stems (BURG and BURG, 1966) and roots (HALL et al., 1977). There are, however, a number of instances where ethylene increases the growth rate, for example in *Callitriche* (MUSGRAVE et al., 1972), in *Helianthus* petioles (PALMER, 1975) and in stems (KU et al., 1969; SUGE et al., 1971) and roots (JOHN, 1977; SMITH and ROBERTSON, 1971) of rice. The kinetics of these effects have only rarely been documented. In *Helianthus* the latent period prior to increased extension is of the order of 1 h. For inhibitory effects the latent period varies between 6 min (WARNER and LEOPOLD, 1971) and 20 min (HALL et al., 1977). Some recent work by OSBORNE (1977) has shown that in the water plants *Regnellidium diphyllum* and *Ranunculus sceleratus* growth may be enhanced by both auxin and ethylene. The effect of either growth regulator does not seem to be mediated by effects on levels of the other. Since it is unlikely that the mechanism of extension in the same tissue is different for the two growth regulators, it seems probable that these tissues possess separate receptors operating the same basic systems.

Ethylene effects are often related to auxin. The work of BURG and BURG (1968) suggested that in pea epicotyl extension, the inhibitory effects of high auxin concentrations could be attributed to auxin-induced ethylene production. As with cytokinins the inhibition of elongation was associated with an increase in isodiametric expansion. RIDGE (1973) has suggested that this transverse cell expansion is a direct result of cellulose microfibril reorientation and DATKO and MACLACHLAN (1968) have correlated auxin-induced swelling with the promotion of cellulase activity in pea epicotyls. On the other hand, using the same system, RIDGE and OSBORNE (1969) failed to demonstrate a similar promotion during ethylene-induced enlargement. RIDGE and OSBORNE (1971) have suggested that ethylene affects elongation by promoting an increase in hydroxyproline residues in the cell wall (pp. 528–529 and 564).

d) Abscisic Acid

It is well established that abscisic acid normally inhibits endogenous or auxin-induced cell extension (WARNER and LEOPOLD, 1971). The length of the latent period varies widely depending on the tissue used and the pretreatments applied, but it may be as short as 5 min (WARNER and LEOPOLD, 1971; REHM and CLINE, 1973). It has also been shown that ABA inhibits auxin-induced H^+ secretion (RAYLE and JOHNSON, 1975; RAYLE, 1973). Whether this is the sole reason for its effects on extension generally is somewhat doubtful, however, especially in the light of its well-documented effects on nucleic acid and protein metabolism (see Sect. 6.5).

e) 3-Methyleneoxindole

Considerable interest was generated when TULI and MOYED (1969) proposed that IAA was not itself active in promoting cell elongation but that this role devolved on 3-methyleneoxindole (Structure 6.1) which is formed from IAA via 3-hydroxymethyloxindole by peroxidase-catalyzed oxygenase oxidation (see Chap. 4.2.1). The basis for this belief was derived from experiments which in-

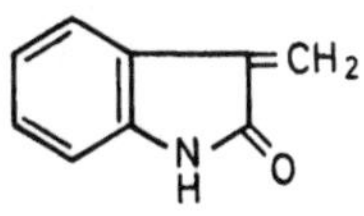

Structure 6.1. 3-Methylene-2-oxindole

dicated that 3-methyleneoxindole promoted growth of pea epicotyl segments as much as IAA but at a concentration an order of magnitude lower; further, chlorogenic acid, which inhibits the conversion of IAA to 3-methyloxindole, was reported to block IAA but not 3-methyleneoxindole effects on growth. It was suggested that the mechanism by which 3-methyleneoxindole brought about its effects lay in part in its function as a sulphydryl reagent and results were adduced showing that it could reverse feedback inhibition of amino-acid biosynthesis (Tuli and Moyed, 1966). The effects of synthetic auxins such as 2,4-D and NAA – which are not of course subject to the same oxidation processes as IAA – were explained by results indicating that they reduced the activity of specific pyridine nucleotide dependent 3-methyleneoxindole reductases which convert 3-methyleneoxindole to 3-methyloxindole (which appeared inactive in growth tests). Thus 3-methyleneoxindole derived from IAA oxidation would build up in the tissue (Moyed and Williamson, 1967).

However, neither Andersen et al. (1972) nor Evans and Ray (1973) could verify the results of Moyed's group either in terms of the growth response of peas or *Avena* or the effects of chlorogenic acid. Equally, in other work the only effect of either 3-hydroxymethyloxindole or 3-methyleneoxindole on growth of *Avena* coleoptiles which could be demonstrated was one of inhibition (Hall et al., 1971).

Whatever the reasons, the 3-methyleneoxindole hypothesis appears to be invalid.

f) Fusicoccin

Although not a growth regulator from higher plants, fusicoccin (Structure 6.2), a toxin extracted from *Fusicoccum amygdali* (Ballio et al., 1964; Graniti, 1964), has been used so frequently in recent studies on cell extension that it merits mention here (see also Sect. 6.3 and p. 452). Fusicoccin induces marked changes in ion transport phenomena and growth in a wide range of types and organs of eukaryotes. Most of these effects are much greater than those obtained by natural hormones in the same situations (see Marrè, 1977). The latent period for the effect of fusicoccin on elongation is much shorter than for auxin (Dohrmann et al., 1974) and the effect of the toxin is associated with a rapid reduction in the pH of the medium. The possible mechanism of this effect will be discussed below but it is appropriate to point out that results with this substance have been used to support the "acid-growth" hypothesis not only for auxin-induced growth but also for cytokinins and possibly other growth regulators (e.g., Lado

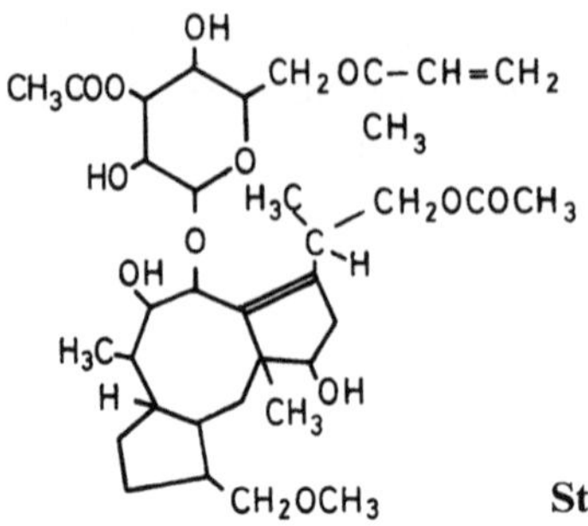

Structure 6.2. Fusicoccin

et al., 1973; MARRÈ et al., 1974a, b, c; MARRÈ, 1977; YAMAGATA et al., 1975). However, while the evidence is compelling, much more work needs to be done on this system.

6.2.3 Conclusion

The last ten years have seen considerable renewed interest in the effects of growth regulators on cell extension, which has, at least in part, led to a better understanding of the processes involved. It is clear that all growth regulators can affect cell expansion, most being able to either promote or inhibit the process. What is not clear is whether the final step is identical for all growth regulators – or even for the same growth regulator in different tissues. Early work on the "acid-growth" hypothesis seemed to point to a certain universality, but later work appears to have cast considerable doubt on this. On the other hand, the apparent conflict between experimental data may not be as serious as it seems, and in this connection it is appropriate to reassess the criteria set out in Sect. 6.2.1.c. As noted above (Sect. 6.2.1.d) criteria (iii) and (iv) appear to be satisfied at least in *Zea* and *Pisum*. We have pointed out that conditions (i) and (ii) are not universally fulfilled but we need to ask not only whether there is an explanation for this but also whether the criteria are necessarily valid. Thus, regarding condition (i), even if the activity of a proton pump is increased by auxin this may not manifest itself in a decrease in the pH of the surrounding medium within the latent period for growth. Many workers have failed to appreciate that the pH of the external medium in which sections are incubated is necessarily different from the pH of their cell walls, both because of the diffusion gradient set up by the proton pump, because of cuticular barriers to such diffusion, and because of Donnan effects in the walls themselves. Cuticular barriers may also have another effect by allowing a rapid decrease in cell-wall pH which in turn may have effects on proton extrusion via feedback mechanisms before a change in external pH is detected. The work of CLELAND (1976) and JACOBS and RAY (1976) uses techniques designed to overcome such errors and it is perhaps significant that, here, increased proton extrusion in response to auxin treatment is demonstrable within the latent period (Fig. 6.3).

It is also important to point out that even if the proton pump hypothesis is valid, conditions (ii) and (iii) might not be fulfilled. Thus, if the activity of a proton pump is increased this will result in a *rise* in cytoplasmic pH, which will in turn affect processes such as CO_2 fixation into malate by PEP carboxylase thus lowering the cell's osmotic potential. In an acid treatment, the converse would more likely be the case. Equally, the distribution of H^+ in the cell would be different in the two treatments, not only in the direction of the gradients of H^+ but also in the absolute concentrations of H^+ in different parts of the tissue. Hence differences between an auxin treatment and an acid treatment would not be unexpected.

Perhaps the most important consideration in relation to this controversy relates more to an attitude of mind than a conflict of data. Thus, the control of cell extension has been studied for so long that workers in the field have begun to lose sight of certain basic truths about the process they are studying. Hence, one refers to the "induction" or "re-induction" of growth by auxin. We have often done so in this chapter. Neither of these terms is really appropriate. What, in fact, usually happens is that when a piece of plant tissue is excised for growth studies, there is a *deceleration* of growth from one steady state to another and this is followed by an *acceleration* to a new steady state after the addition of growth regulator. This distinction is not merely a matter of semantics since induction or re-induction imply the initiation of processes not previously extant or their re-initiation, whereas acceleration does not. Clearly, in normal development there is a transition from the meristematic to the elongating phase where true "induction" of growth is involved but this is not the case in most systems described in the literature where we are dealing with a population of cells which are all growing at the time of excision. The situation is even more complicated if we consider the work of WRIGHT (1968) on coleoptile tissue which shows that during normal ontogeny growth is affected by several growth regulators in sequence.

The most important lesson to be drawn from the observations set out above is that if, as is certain beyond reasonable doubt, a large number of different processes are involved in growth, then these processes do not necessarily change in step during the acceleration or deceleration phases. Equally, some, or possibly all, of these processes are affected by auxin or other growth regulators. To some extent this makes the determination of the "primary" mode of action of growth regulators in elongating systems somewhat academic since we may only be looking at the process which is most limiting at a given time. Such an elucidation only becomes important if the other processes involved in growth are consequent upon the "primary" process.

If, therefore, we view growth as a process involving the cooperation of a number of systems, most if not all of the experimental evidence on the subject may be reconciled. Thus, a number of workers have shown that auxin effects can be separated from acid effects (see Sect. 6.2.1.c). Clearly, if auxin alters the activity of a number of processes in elongating systems – including proton extrusion – although not necessarily simultaneously – then it is hardly surprising that not all the effects of auxin are mirrored by the effects of low pH.

We can extrapolate these ideas somewhat further by considering the changes in the cell wall. Thus, if the concept of breakage of glycosidic linkages followed by re-formation at a different site is a characteristic of cell-wall extension,

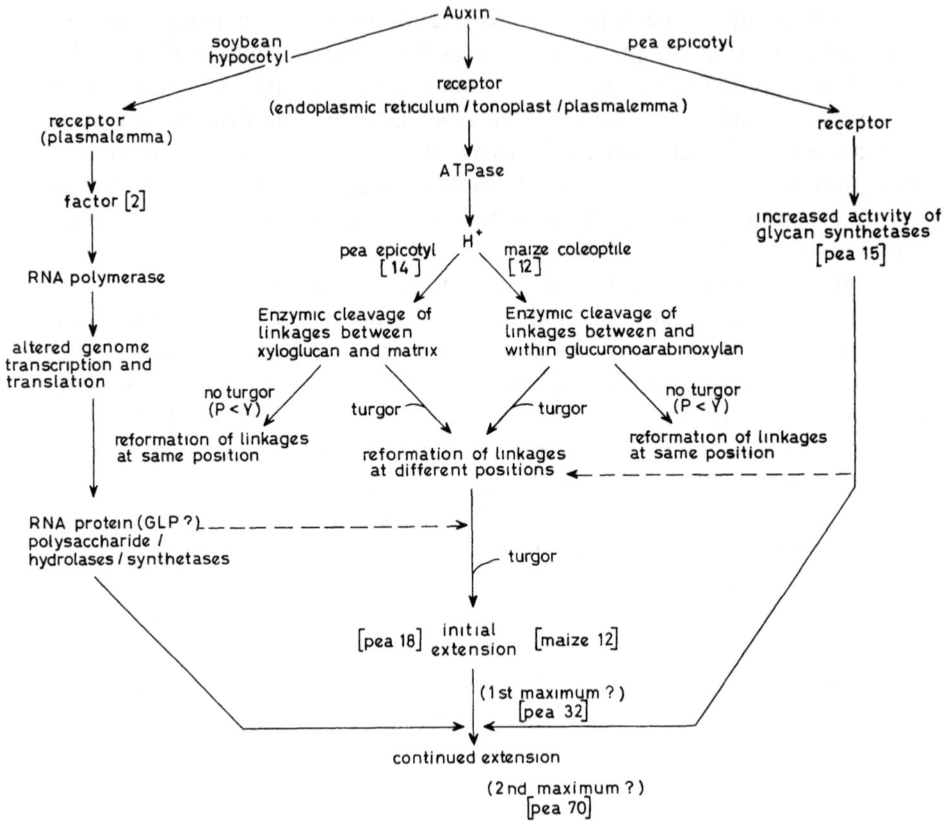

Fig. 6.8. Hypothetical summary of events leading to extension. Figures in square brackets indicate time in minutes for event to take place after addition of auxin (where known). (See text and PENNY, D. et al., 1972; RAY, 1973b; LABAVITCH and RAY, 1974a, b; JACOBS and RAY, 1976; CLELAND, 1976; MORRÉ and CHERRY, 1977; GILKES and HALL, 1977; DARVILL et al., 1977, 1978)

then it is logical to suppose that for such a process to function over a prolonged period, new sites must be continuously provided. Polysaccharide and/or wall protein synthesis presumably fulfils this role and it is interesting in this connection to refer again to the work of RAY (1973a, b) showing that the activity of polysaccharide synthetases is enhanced by auxin, although at a time somewhat later than the acceleration of elongation. The effect is not apparently related to the increased proton pump activity since it is not reproduced by changes in intra- or extracellular pH. It is also interesting to note that the pattern of polysaccharide deposition differs depending on whether growth is occurring or not. BAKER and RAY (1965) showed that incorporation of polysaccharides into the wall associated with growth is intussusceptional, i.e., within the wall, whereas in the absence of growth incorporation is by apposition to the inner surface of the wall.

It seems that these phenomena represent one of the other processes affected by auxin and involved in growth which we referred to earlier and one moreover that may explain the "two-phase" response to auxin described in Section 6.2.1.a (Fig. 6.1). Thus the fall-off in growth rate between the two "peaks" may represent the lag between the acceleration of the proton pump and the increase in activity of the polysaccharide synthetases. Equally, since net polysaccharide synthesis maintains wall thickness during growth we might expect a thinning of the wall at low pH where elongation is promoted but polysaccharide synthesis is not (cf. Fig. 6.2).

Lastly, if processes other than the acceleration of a proton pump are affected independently by auxin but are equally necessary for continued growth, we have an explanation of the short duration of the acid growth effect.

An attempt to combine some of the results of a number of workers on the role of auxins in cell extension is shown in Fig. 6.8.

6.3 Effects of Growth Regulators on Ion Transport and Regulation of Membrane Properties

The concept that the mode of action of growth regulators may, in at least some instances, be exercised via effects on changes in the properties of cell membranes is not a new one but has only recently returned to the fore, not least because of the demonstration of membrane binding sites for hormones (see Chap. 5), the demonstration of auxin-induced proton pumps and the role of ABA in the stomatal mechanism. However, with the exception of the work referred to, the numbers of hard facts are currently outweighed by the numbers of hypotheses.

Rayle and Cleland (1970) and Hager et al. (1971) first proposed that hydrogen ions act as a second messenger between the primary site of auxin action and the site of wall loosening. Since these sites are in the protoplasm and cell wall respectively such a proposition, if valid, necessitated the existence of a mechanism for proton transport between these sites. Cleland (1973) demonstrated the existence of auxin-induced acidification in *Avena* coleoptiles and his work was followed by a number of other similar reports (e.g., Marrè et al., 1973a, b, in peas and Ilan, 1973, in *Helianthus*). The possible relationship of these effects to auxin-induced cell extension is discussed above (Sect. 6.2), here, the nature of the phenomenon will be considered.

The demonstration that auxins may induce plant sections to acidify the medium in which they are incubated does not itself necessarily signify the existence of a proton pump in the true sense. Penny et al. (1975) and Penny and Penny (1978) have listed a number of alternative explanations for the phenomenon including OH^- and HCO_3^- exchange, increased OH^- and K^+ symport, increased OH^- import, and mechanisms involving changes in the transport and metabolism of cell wall constituents. Such possibilities have by no means all been eliminated in all instances and the matter is still somewhat controversial. For example, Sloane and Sadava (1975) attribute auxin-induced acidification in pea stem segments to effects on CO_2 production by the plant tissue.

Notwithstanding these reservations it is clear that proton pumps are ubiquitous in pro- and eukaryotic systems and, although some of the systems which have been studied in higher plants and which will be discussed below have not always been shown to be proton pumps in the strict sense, nevertheless it is appropriate to discuss such systems in the general framework of proton pumps.

RAVEN and SMITH (1977) have recently reviewed the characteristics, functions and regulation of active proton extrusion. They define active proton transport as a "net movement of H^+ across a membrane in the opposite direction to that predicted from prevailing passive driving forces of H^+ concentration gradient and electrical potential difference". This movement clearly requires an energy input which can be either a redox process or ATP use. Although classical proton pumps are normally considered to involve pumping of protons alone it has been suggested (see below) that electroneutral K^+/H^+ exchange or electrogenic K^+/H^+ exchange may occur. As much of the work has been performed simultaneously with the fungal toxin fusicoccin and with auxins the effects will be discussed in parallel.

The nature of the proton pumps induced by auxin is a matter of some controversy. MARRÈ et al. (1974c) showed that in pea internode tissue both fusicoccin- and auxin-induced proton extrusion appear to be coupled to K^+ uptake. K^+ and to a lesser extent Na^+ stimulate proton extrusion, whereas Cs^+ or Li^+ have little activity or may inhibit. The H^+/K^+ ratio was about 0.9. Similar figures were obtained by HASCHKE and LÜTTGE (1973). Fusicoccin causes hyperpolarization of the membrane potential in a number of systems (COCUCCI et al., 1976; MARRÈ et al., 1974b; PITMAN et al., 1975) and the same effect has been demonstrated for auxin in oat coleoptiles (CLELAND and LOMAX, 1977; PRINS et al., 1976).

These data imply that the pump is either a simple electrogenic proton pump coupled with passive K^+ uptake or involves electrogenic K^+/H^+ exchange. Which process in fact operates for auxin and fusicoccin is unclear. MARRÈ and his co-workers interpret their results in terms of electrogenic K^+/H^+ exchange, but their data are better in this regard for fusicoccin than for auxin. Similar conclusions have been reached by PITMAN et al. (1977) in barley roots. CLELAND and LOMAX (1977) on the other hand believe that whereas fusicoccin-induced proton secretion is due to electrogenic K^+/H^+ exchange (although results with oat protoplasts indicate that the system does not have an absolute requirement for K^+), auxin-induced proton secretion is due to a simple electrogenic proton pump.

Notwithstanding the work of SLOANE and SADAVA (1975) (see above) some of the earliest work (LADO et al., 1972) seems to exclude the possibility that, in pea internode segments, the pH drop in the incubation media is due to CO_2 production. The situation here is further complicated by results indicating that CO_2 fixation may be involved in the process overall (HASCHKE and LÜTTGE, 1977). It also appears that in peas at least the pH drop cannot be explained by changes in the rate of extrusion of weak organic acids or other substances buffering in the pH range between 4.5 and 6.5 (MARRÈ et al., 1973b).

That the effects of both auxin and fusicoccin are closely related to respiration is demonstrated by the fact that inhibitors of terminal oxidation such as carbon monoxide and of oxidative phosphorylation such as DNP and CCCP inhibit auxin- and fusicoccin-stimulated acidification (CLELAND, 1973; MARRÈ et al., 1973b). It is of interest to note that whilst auxin does not stimulate acidification by fully grown pea internodes, fusicoccin does (MARRÈ et al., 1973a). Proton pump stimulation is tissue-specific for natural and synthetic growth regulators, responding to auxins in elongating tissue such as coleoptiles or dicotyledonous stem segments; antiauxins such as p-chlorophenoxyisobutyric acid are without effect (MARRÈ et al., 1973a). In cotyledon enlargement, only cytokinins are effective (MARRÈ et al., 1974a, b) and abscisic acid appears to be most effective in the inhibition of proton extrusion associated with stomatal opening (see Sect. 6.3.1). In seeds, ABA inhibits proton extrusion whilst GA_3 promotes it (LADO et al., 1974, 1975). It should be noted, however, that fusicoccin stimulates proton extrusion in all these systems. Whilst fusicoccin stimulates proton efflux in roots of barley (PITMAN et al., 1975) and pea, bean, and maize (LADO et al., 1976), IAA is without effect. Neither can ABA reverse the fusicoccin effect (PITMAN et al., 1975).

As with stomata, CO_2 fixation and malate accumulation appear to play a role (see Sect. 6.3.1). Thus HASCHKE and LÜTTGE (1977) have shown that in *Avena* coleoptiles auxin-induced proton extrusion is followed by increased CO_2 fixation into malate. They suggest that this is a homeostatic response to outward pumping of protons and not a direct effect of auxin. Indeed the activity of PEP-carboxylase in *Avena* is unaffected by auxin treatment (BOWEN et al., 1976). HASCHKE and LÜTTGE further propose that the newly formed malate, as well as contributing to the control of cytoplasmic pH, also contributes to the maintenance of water potential during growth.

So far, the precise site(s) at which auxin and other growth regulators have their effect on proton extrusion and even the exact mechanism of the extrusion itself are unclear. BEFFAGNA et al. (1977) have shown a stimulatory effect of fusicoccin on a plasmalemma-bound ATPase in preparations from maize coleoptiles and spinach leaves. CLELAND and LOMAX (1977) have not, however, been able to demonstrate a similar phenomenon in *Avena*. On the basis of auxin-binding studies RAY (1977) and DOHRMANN et al. (1978) have proposed that auxin-activated proton pumps may be located on the endoplasmic reticulum and the tonoplast.

Other recent studies have considerable bearing on the relationship between auxin and ionic relations. Thus, ZIMMERMANN and his co-workers have elaborated a theory on a turgor-sensing mechanism in plant cell membranes (for a full discussion see ZIMMERMANN et al., 1977). Briefly, it has been shown that the thickness of cell membranes depends on turgor, cell-wall stretching and the magnitude of the membrane potential (COSTER et al., 1976). In turn, membrane resistance (presumed to be a measure of salt permeability) has a dependence on pressure, exhibiting a maximum value (termed the critical pressure) below which net ion uptake predominates and above which efflux predominates. This effect apparently applies both to the plasmalemma and the tonoplast (ZIMMERMANN et al., 1976). In *Valonia*, addition of auxin raises the critical pressure substantially, hence affecting the point at which ion-flux equilibrium is attained

(ZIMMERMANN et al., 1976). This finding is corroborated by the work of MORRÉ and BRACKER (1976) and HELGERSON et al. (1976) showing that in in situ and isolated plasmalemma from *Glycine max*, auxin treatment leads to ultrastructural changes, rapid decreases in membrane thickness and in microviscosity of the hydrocarbon regions of the membrane. It must be borne in mind that the elastic modulus of *Valonia* cell walls does not change in response to auxin. In angiosperms, the elastic modulus of the cell wall may vary in response to auxin treatment, which in turn will affect membrane thickness and hence resistance. In the general context of hormonal effects on permeability, KANG and BURG (1971) stimulated a good deal of controversy by claiming that auxin increased the permeability of pea stem segments to water. However, work in three separate laboratories (DOWLER et al., 1974) has been unsuccessful in repeating this finding, which must therefore be considered unproven at this time. In an extension of the studies described above, ZIMMERMANN and STEUDLE (1977) and ZIMMERMANN et al. (1977) have shown that the location of IAA in artificial lipid bilayers may be altered by KCl such that, at low salt concentrations (1 mM), IAA is absorbed onto the surface and in the polar head region, whereas at higher concentrations (10 mM) IAA is absorbed into the hydrocarbon region. To what extent studies with such systems can explain the effects of IAA on membrane structures and properties described earlier remains to be seen, but clearly such re-location of auxin could have profound implications for auxin-binding studies and any interaction with membrane-bound systems such as ATPase.

Because ethylene partitions more readily into lipid than into water it has often been proposed that the mode of action of ethylene might lie in effects on membrane properties. This supposition appeared to be supported by studies on senescence and fruit ripening where obvious changes in permeability occur and these processes may be promoted by ethylene. However, ABELES (1972) has pointed out that such changes are more likely the result, rather than the cause, of senescence although this is not necessarily so in all cases (SACHER, 1973). Equally, a number of workers have been unable to detect any effect of ethylene on membrane permeability (BURG, 1964; SACHER and SALMINEN, 1969; BURG, 1968). In recent studies, KENDE and HANSON (1977), working with flowers of *Ipomoea tricolor*, have come to the conclusion that although ethylene does not trigger flower senescence it does affect cellular compartmentation via effects on membrane permeability. It is proposed that this changed compartmentation leads to further ethylene biosynthesis, thus explaining the autocatalytic effects of the gas on its own synthesis.

It is interesting to note that ethylene affects ATPase in a number of systems (MADEIKYTE and TURKOVA, 1965; STEWART and FREEBAIRN, 1969). Although this work was done in relation to effects of ethylene on respiration, it may have a bearing on membrane permeability. Certainly MEHARD and LYONS (1970) and KU and LEOPOLD (1970) have shown that ethylene may affect mitochondrial membrane permeability, although the effect is not specific.

Ethylene also appears to affect secretory processes in that it has been shown to increase the rate of release of α-amylase from barley half seeds (JONES, 1968), of peroxidase from peas (RIDGE and OSBORNE, 1970), of cellulase in the abscission zone (ABELES and LEATHER, 1971) and of cellulase from fungal mycelia (MADKOUR, 1977).

Observations consistent with the effects of ethylene on secretory processes have been provided by recent studies with fractions derived from *Phaseolus vulgaris* cotyledons which indicate that specific binding sites for ethylene are present on the endomembrane system (HALL et al., 1980).

6.3.1 Hormones and Stomata

A large number of workers have shown that both natural and synthetic growth regulators may affect stomatal movement. FERRI and LEX (1948) and BRADBURY and ENNIS (1952) studied effects of foliar and soil applications of 2-naphthoxyacetic acid and 2,4-D on intact *Nasturtium* and kidney bean plants respectively and observed decreased transpiration rates and increases in stomatal resistance. While JOHANSEN (1954) was unable to demonstrate any effect of IAA upon stomatal resistance in *Sinapis alba* when the growth regulator was applied to the leaves, on the other hand soil applications did result in stomatal closure. However, if treated leaves were excised, the stomata re-opened even if the petioles were dipped into 0.6 mM IAA. ZELITCH (1961, 1963) was unable to induce stomatal closure in tobacco leaf discs with IAA, although 2,4-D and 2-naphthoxyacetic acid were effective in this system. On the other hand ALLERUP (1964) demonstrated that supplying IAA, 2(3-indole)butyric acid or naphthalene-1-acetic acid (NAA) to barley seedlings previously grown in nutrient solution, induced rapid, albeit transient, increases in transpiration rate. TAL and IMBER (1971) also observed that pre-treatment of tomato plants with 2,4-D increased transpiration rate per unit leaf area. MANSFIELD (1967) tested the effects of a number of auxins on stomata of detached *Xanthium* leaves and found that whereas the effect of 2-naphthoxyacetic acid and NAA on stomatal closure could be reversed by flushing the leaves with CO_2-free air the effect of 2,4-D could not be so reversed. He proposed therefore that the effect of 2,4-D was directly on the guard cells, whereas that of the other two growth regulators was mediated via an effect on CO_2 production or utilization. TAL et al. (1974) showed that tomato plants treated with IAA during development exhibited excessive stomatal opening. They correlated this with other findings on the "wilty" mutant of tomato (TAL and IMBER, 1970) whose stomata resist closure and which has higher endogenous concentrations of IAA than normal varieties. The most recent work is that of NOWAKOWSKI and LUBANSKA (1975) who demonstrated that treatment of spring wheat with IAA increased transpiration rates.

It is unclear whether the work described can be ascribed to "direct" or "indirect" effects of auxin. However, the fact that, in general, auxins are more effective if applied to the root rather than directly or through cut petioles may indicate that auxins affect root permeability and hence water supply which will in turn affect stomatal opening.

LIVNE and VAADIA (1965) found increased rates of transpiration in barley leaves inserted in solutions of kinetin. Similar findings were reported by MEIDNER (1967). KEMP et al. (1957) observed that kinetin-treated tomato plants wilted more rapidly than controls. It is apparent, however, that such effects are not necessarily of general occurrence. Thus, LUKE and FREEMAN (1968) reported that excised leaves of *Phaseolus lunatus, Helianthus annuus* and *Acer saccharinum* did not exhibit increased transpiration when treated with kinetin. Similar results have been reported by TUCKER and MANSFIELD (1971) and HORTON (1971) with *Commelina communis* and *Vicia faba* respectively.

Several explanations have been advanced to account for the lack of response in some systems. Thus, LIVNE and VAADIA (1965) found that kinetin affects stomatal aperture in fully expanded barley leaves but not in growing leaves and suggested that the difference lay in differences in endogenous levels of cytokinins. PALLAS and BOX (1970) suggest that the effect of kinetin may be related to changes in the solute content of epidermal cells. DAS et al. (1976) have shown in epidermal strips of *Commelina benghalensis* and *Tridax procumbens* that kinetin does not affect stomatal aperture whereas benzyladenine does, thus implying that different responses between species may be the result of specificity for particular cytokinins. On the other hand, RASCHKE (1975) suggests that kinetin (and presumably cytokinins generally) may act indirectly on stomata by preventing leaf ageing. This would be in accord with the work of LIVNE and VAADIA (1965) mentioned above.

There are few reports of effects of gibberellins on stomata. LIVNE and VAADIA (1965) observed that the rate of transpiration of barley leaves was increased by GA_3 but this was later shown to be true only for freshly excised tissue (LIVNE and VAADIA, 1972). HALEVY and KESSLER (1963) showed that treatment of bean plants with CCC and Phosphon D led to an increase in drought tolerance and LIVNE and VAADIA (1972) have interpreted this as evidence for the involvement of gibberellins in stomatal regulation.

The situation with ethylene is somewhat contradictory. PALLAGHY and RASCHKE (1972) showed that stomatal aperture in *Zea mays* and *Pisum sativum* was unaffected by ethylene treatment. Similar results have been reported by EL-BELTAGY and HALL (1974) in *Vicia faba*. BROWNING (1974) showed that the synthetic growth regulator, ethephon (2-chloroethane phosphonic acid), which releases ethylene, reduced transpiration and stomatal opening in *Coffea arabica*. On the other hand VITAGLIANO (1975) showed that ethephon treatment of *Olea europaea* stimulates stomatal opening. The results with ethephon should be viewed with some caution however, since on degradation, this compound yields not only ethylene, but also phosphate and chloride, either or both of which could conceivable affect stomata.

By far the most comprehensive data on hormones and stomata relates to effects of abscisic acid. WRIGHT (1969) first showed that water stress led to increases in inhibitor levels in wheat and subsequently, this inhibitor was shown to be abscisic acid (WRIGHT and HIRON, 1969). This work has been confirmed many times subsequently. LITTLE and EIDT (1968) demonstrated that abscisic acid could induce stomatal closure in woody species; similarly, JONES and MANSFIELD (1970) and CUMMINS et al. (1971) were able to demonstrate that ABA fed through leaf petioles induced rapid stomatal closure, that the effect was rapidly reversed when ABA was removed and that the effect was not related to an increased CO_2 concentration in the intercellular spaces. Only very low concentrations of abscisic acid are necessary to induce stomatal closure. Thus, KRIEDEMANN et al. (1972) showed that ABA induced closure within three minutes of application and that this could occur at concentrations as low as 8.9 pmol cm^{-2} leaf area. Similar results have been reported for woody angiosperms (DAVIES and KOZLOWSKI, 1975). The only exception to the general observation that ABA induces stomatal closure is that of LANCASTER et al. (1977) with yellow lupin leaves, which appear to be insensitive to the growth regulator.

TAL and IMBER (1970, 1972) working on the "wilty" mutant of tomato "flacca" which has low endogenous levels of abscisic acid and whose stomata resist closure, showed that the stomata could be induced to close by treatment with ABA. This effect does not, however, appear to be analogous to the "rapid" effects of ABA since the stomata respond only slowly in "flacca" and repeated application of the growth regulator is necessary.

There appears to be an antagonism between abscisic acid and cytokinins in that MITTELHAUSER and VAN STEVENINCK (1969) observed that increased stomatal opening induced by kinetin in wheat leaves could be partially reversed by treatment with ABA. On the other hand reduction in stomatal aperture induced by applied ABA cannot be readily reversed by cytokinin treatment (HORTON, 1971; MIZRAHI et al., 1971).

It appears in general that the effects of abscisic acid upon stomatal opening do, at least in part, reflect the role of this growth regulator in the natural control of the stomatal mechanism. As pointed out above, however, results with other growth regulators are varied and somewhat contradictory. Moreover, many of the effects reported for other growth regulators are the result of long-term exposures, suggesting that explanation should be sought in overall effects on growth and development rather than specific effects on the stomatal mechanism itself. This is particularly noticeable in the case of cytokinins where if the treatment is limited to a few hours such that senescence does not interfere, the effects of the growth regulator are in general very small.

The explanation of the effects of cytokinins and abscisic acid must be considered in relation to the mechanism of stomatal movement itself. It is inappropriate to discuss here in detail the proposed mechanism and the reader is referred to excellent review articles on this subject (HSIAO, 1976; RASCHKE, 1975, 1977). Briefly, however it is established that potassium salts are utilized by the plant to change the water potential of guard cells such that an inflow of water occurs resulting in higher turgor and hence stomatal opening. Depending on the system the K^+ may or may not be paired with Cl^-. The K^+ that are not paired by Cl^- are neutralized by organic anions, mainly malate. HORTON and MORAN (1971) and MANSFIELD and JONES (1971) showed a reduction in accumulation of K^+ in guard cells of *Vicia faba* and *Commelina communis* treated with ABA. The effect of abscisic acid is strongly related to effects of CO_2. Thus, RASCHKE (1974) has shown that stomata insensitive to CO_2 can be sensitized by abscisic acid; equally, the responses of stomata to abscisic acid may be enhanced by CO_2. GRANITI (1964) reported that plants treated with fusicoccin wilt easily. Similarly, the induced stomatal opening observed in similar plants is associated with increased K^+ content in the guard cells. Furthermore, fusicoccin can overcome the effects of CO_2 and abscisic acid on stomatal closure

(SQUIRE and MANSFIELD, 1972, 1974). RASCHKE (1977) reports that fusicoccin accelerates proton excretion from epidermal strips of *Vicia faba* and *Commelina communis*. This work suggests a central role for K^+/H^+ exchange in the stomatal mechanism (see also Sect. 6.3).

RASCHKE (1977) proposes that ABA prevents stomatal opening and causes closure in more or less the same way as CO_2, namely by accelerating a build-up of H^+ and malate in the cytoplasm probably by inhibiting acid removal via K^+/H^+ exchange or active expulsion of H^+. He also suggests that ABA detaches from its site of action relatively readily when the medium around the cells contains little of the growth regulator[1].

6.3.2 Hormone-Directed Transport

The mobilizing effects of growth regulators are now well established (e.g., MOTHES et al., 1959; SETH and WAREING, 1967) but the mechanism whereby application of growth regulators to, for example, decapitated internodes of pea and bean seedlings brings about an attraction of metabolites, presumably via the phloem, is unknown.

While the effect may merely reflect an increase in metabolism brought about by the growth regulator, thus creating a "sink", some recent work has raised another possibility. Thus, MALEK and BAKER (1977) have shown that loading of ^{14}C-sugars from the petiole of *Ricinus communis* is stimulated by K^+ and H^+. These workers also showed that the pH of a solution perfusing the hollow petiole fell but that this drop was greater in the presence of K^+ and less in the presence of sugars. It is proposed that proton co-transport of sugars from the free space is driven by a linked proton efflux/potassium influx pump. This type of co-transport is well documented in microorganisms (TANNER et al., 1977). Preliminary experiments indicate that the process is affected by growth regulators in *Ricinus* (BAKER, personal communication) and herein may lie the mode of action of hormones in affecting translocation.

6.4 Effects of Hormones on Non-Growing Systems

6.4.1 The Aleurone Layer System

The effects of growth regulators on the mobilization of food reserves in seeds deserves a special mention in this account for several reasons. Firstly for its intrinsic interest, but also because it exemplifies a system which is clearly "pre-programmed" to respond in a particular way to appropriate internal and external signals; moreover, the systems involved appear to implicate all the known groups of plant growth regulators. Last, but not least, this is a system which, with

[1] The reader is referred also to Vol. 7 of this Encyclopedia (Plant Movements, eds. W. HAUPT and M.E. FEINLEIB) Chapter 4.1 by K. RASCHKE

the exception of growth processes, has probably been more intensively investigated than any other.

The aleurone is a peripheral layer of cells around the endosperm of gramináceous seeds. Its cells are rich in proteins, and possess conspicuous nuclei. On germination, the aleurone tissue is active for a short period but then quickly deteriorates along with the rest of the endosperm (ESAU, 1967).

PALEG (1960) and YOMO (1961) showed independently that treatment of barley seeds with gibberellic acid (GA_3) caused an increase in amylolytic enzyme activity leading to a release of reducing sugars from the endosperm starch. It was shown that the GA_3 treatment simulated the effect of the embryo on the endosperm, thus leading to the conclusion that the embryo is the synthesizing organ for GA and the aleurone layer is the target tissue of the hormone in this system (VARNER and CHANDRA, 1964). The aleurone layers of rice (OGAWA, 1966) and wheat (PHILLIPS and PALEG, 1972) behave in a similar fashion to that of barley.

It is now well established that GA_3-treatment induces de novo synthesis of α-amylase (FILNER and VARNER, 1967; VARNER and JOHRI, 1968; PALEG, 1960; FILNER et al., 1969) when the hormone is applied in concentrations ranging from 10^{-8} to 10^{-6} M, with an optimum at 10^{-7} M (CHANDRA and DUYNSTEE, 1968). JACOBSEN et al. (1970) found that GA_3 induces the production of four α-amylases, all synthesized de novo.

There is a lag period of 6 to 8 h between the application of the GA_3 and the synthesis of the enzyme (Fig. 6.9) and the continuous presence of GA_3 is required throughout the lag period. GA_3 cannot be dispensed with after the initiation of the response and is required during the period of amylase synthesis (Fig. 6.10). CHANDRA and DUYNSTEE (1968) found that supplying GA_3-treated aleurone with actinomycin D (an inhibitor of RNA synthesis which prevents transcription) inhibits the synthesis of α-amylase. The later the treatment with actinomycin D during the lag period the smaller becomes its effect on α-amylase synthesis, and when applied after the lag period it has little effect on levels of the enzyme. This result suggests that sufficient quantity of the RNA involved in α-amylase synthesis, whose synthesis is inhibited by actinomycin-D, is produced during the lag period. On the other hand VARNER and JOHRI (1968) suggested that continuous RNA synthesis is required during the production of the enzyme. CHANDRA and DUYNSTEE (1968) have further found that a prolonged (30 h) treatment with very low concentrations of GA_3 (lower than 10^{-12} M) which do not induce α-amylase synthesis *do* enhance a higher incorporation of precursors into salt-soluble RNA. Treatment with somewhat higher concentrations of GA_3 (10^{-12}–10^{-10} M) which still do not induce α-amylase synthesis *do* enhance the incorporation of precursors into all RNA fractions. Application of 10^{-8} M GA_3 allowed an increase in α-amylase synthesis but decreased the specific activity of RNA fractions after 30 h. A 30 min treatment with 10^{-6} M GA_3 induced soluble and ribosomal RNA formation that was not found in the control. Subsequently, the control began forming new soluble and ribosomal RNA and by 90 min all nucleic acid fractions of both control and GA_3-treated aleurones were labelled. The DNA:RNA fraction labelled only in the RNA, suggesting that DNA synthesis did not occur. After $2^1/_2$ h of GA_3 treatment a new heavy

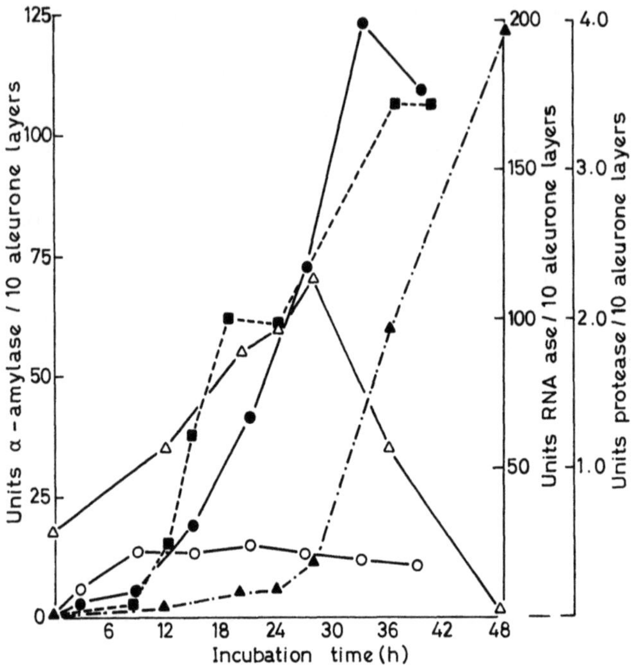

Fig. 6.9. Time course of synthesis and release of α-amylase, protease, and ribonuclease from barley aleurone layers during incubation in gibberellic acid. α-amylase: in aleurone layers (○), in medium (●); protease: in medium (■); ribonuclease: in aleurone layers (△), in medium (▲). (Modified from VARNER and CHANDRA, 1964; JACOBSEN and VARNER, 1967; CHRISPEELS and VARNER, 1967a)

ribosomal RNA fraction appeared. This fraction did not appear in the control and actinomycin D prevented its appearance without affecting other RNA fractions significantly. Thus, the authors suggested that GA-induced α-amylase synthesis is dependent upon RNA synthesis as a whole and on the synthesis of a new heavy ribosomal RNA. HO and VARNER (1974) found that an RNA, which contains poly(adenylic acid [poly(A)-RNA] and may be messenger RNA, is present in aleurone layers and is synthesized there during the incubation of the tissue with or without GA_3. However, addition of GA_3 enhanced the rate of synthesis of poly (A)-RNA within 3–4 h and this reached a maximum at a level 50% to 60% above the control 10–12 h after treatment. Cordicepin (an inhibitor of RNA synthesis) inhibited production of α-amylase only if added 12 h or less after GA_3 application but inhibited total RNA production as well as poly (A)-RNA synthesis whenever added. This last result suggests that α-amylase formation depends on stable messenger RNA formed during the first 12h after GA_3 application. On the other hand, ZWAR and JACOBSEN (1972) found that RNA, polydispersed on acrylamide gels and sedimenting between 5S and 14S was stimulated by 300% after 16h treatment of aleurone layers with GA_3. There was no effect after 4 h and very little effect after 8 h. Ribosomal and transfer RNA were not affected by the treatment. The time course of α-amylase synthesis paralleled the appearance of this "GA-RNA". Actinomycin D inhib-

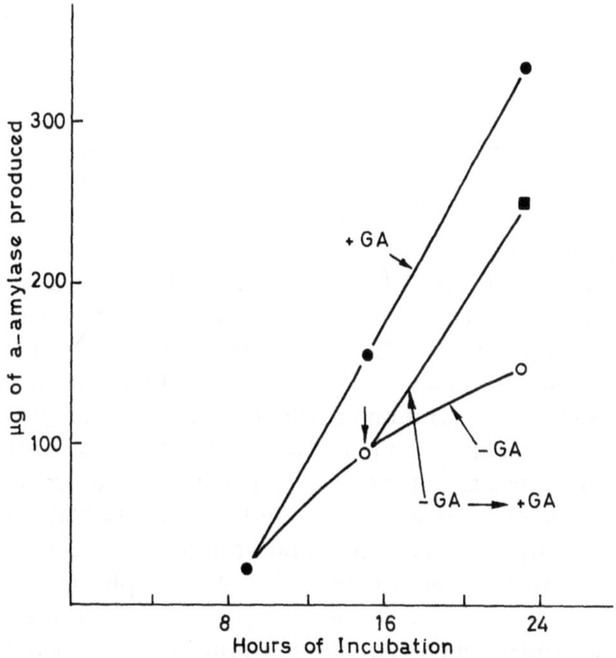

Fig. 6.10. Effect of gibberellic acid (GA_3) added and removed at different times upon production of α-amylase by barley aleurone layers. All aleurone layers incubated for 7 h in 0.5 µM GA_3, then either in GA_3 for a further 16 h (●), or GA_3 removed by four consecutive 30-min rinses (○), or GA_3 added back at 15 h (■). (CHRISPEELS and VARNER, 1967a)

ited α-amylase and "GA-RNA" to much the same degree but fluorouracil, which halved the incorporation of label into ribosomal RNA, had no effect on either "GA-RNA" or α-amylase. ABA, when applied together with GA, eliminated the synthesis of both "GA-RNA" and α-amylase. This work suggests that, after 16 h, a "GA-RNA", which may well be a messenger RNA, is produced in response to GA treatment. This "GA-RNA" comprises only 1% of the total RNA and its appearance may be dependent on other RNA changes previously reviewed.

EVINS (1971) noticed that GA_3 treatment (1 µM) of aleurone layers enhanced the formation of polyribosomes within 2 to 4 h and increased the synthesis of ribosomes. These newly formed polyribosomes were shown to be responsible for the synthesis of α-amylase induced by GA_3. The addition of ABA (0.25 pM) to this system prevented the GA_3-induced increases in the percentage of polyribosomes, the formation of monosomes and the synthesis of α-amylase (EVINS and VARNER, 1972). The addition of ABA to aleurone cells already producing α-amylase inhibited further synthesis of the enzyme and decreased the percentage of polysomes but did not change the number of ribosomes per cell. Changes in α-amylase synthesis caused by midcourse removal of GA_3 and its re-addition later on or by treatment with 5-fluorouracil and actinomycin-D correlated with changes in the percentage of polyribosomes.

It is well established, however, that the aleurone system is not a one hormone–one enzyme system. The activity of many other enzymes is influenced by GA_3 application (JACOBSEN and VARNER, 1967) and other hormones appear to be involved in α-amylase synthesis (KHAN, 1971; JACOBSEN, 1973). Equally, other systems in the aleurone are activated by other hormones (see also below). CHRISPEELS and VARNER (1967a, b, c) have found that application of ABA inhibits the GA_3-induced formation of α-amylase, without significantly affecting amino-acid incorporation into total protein and only slightly depressing RNA synthesis. VARNER and JOHRI (1968) confirmed this observation and found a lag period of 4 h between the time of ABA application and the complete inhibition of enzyme production. HO and VARNER (1976) showed that 5 μM ABA is enough to inhibit GA_3-induced α-amylase formation if added simultaneously with GA_3. A higher concentration of ABA (10–25 μM) is required to reduce the α-amylase activity if added 12 h after GA_3. ABA (up to 50 μM) neither forms a complex with α-amylase nor reduces its activity, but rather, inhibits its synthesis. ABA did not seem to have any effect on the stability of the mRNA responsible for α-amylase formation. Application of cordicepin (an inhibitor of RNA synthesis) with, or immediately after, ABA application caused α-amylase synthesis to be sustained or restored respectively. This suggests that the ABA effect is dependent upon continuous synthesis of short-lived RNA.

JACOBSEN (1973) found that inhibition of GA_3-induced α-amylase synthesis by ABA can be relieved only partially by additional GA_3 alone, and hence it is not certain whether ABA is a competitive inhibitor of GA_3 or not. Although ethylene has no effect on α-amylase synthesis per se, it promotes the release of the enzyme from the aleurone cells to the surrounding medium (JONES, 1968), and also partially relieves ABA inhibition when applied at concentrations between 4.5×10^{-10} and 4.5×10^{-7} M (JACOBSEN, 1973). Furthermore, when applied together, additional GA_3 and ethylene almost eliminated the ABA inhibition. Both the length of the lag phase and the rate of α-amylase synthesis during the linear phase are affected by all three hormones: ABA extended the lag phase and decreased the rate of synthesis, while GA_3 and ethylene shortened the lag phase and increased the rate of synthesis of the enzyme. One last point in this connection is that the effect of ABA is highly specific in that it does not affect substrate induction of nitrate reductase in this system even when amylase activity is inhibited by about 90% (FERRARI and VARNER, 1969).

Cytokinins do not affect α-amylase production when applied alone (KHAN, 1969; KHAN and WATERS, 1969) but treatment with kinetin (0.5–50 μM) or ^{6}N-benzyl-adenine (0.25–200 μM) overcomes the inhibition by ABA of GA_3-induced synthesis (KHAN and DOWNING, 1968). KHAN (1971) has suggested that the production of α-amylase depends upon three hormones; ABA, cytokinin, and gibberellin, ABA having a "preventive" role and cytokinins a "permissive" role. However, such a definition seems unnecessarily restrictive when applied to the metabolic events under review here, although it may be valid in relation to the original sense in which it was used, namely the effects of applied hormones on the breaking of seed dormancy (see also below).

The work described above has added much to our knowledge of the mode of action of hormones in the aleurone system in particular and in plants in general, but many further questions are raised. For example in what order do ABA, ethylene, cytokinins, and gibberellins appear in the aleurone layer and what controls their appearance?

As DNA synthesis appears not to be involved in the system as described, is the gene responsible for α-amylase synthesis already "activated" and if so why does it not produce mRNA for α-amylase before GA_3 is added or in the presence of ABA? Is GA_3 the substance transported from the embryo to the aleurone layer or does something else arrive at the aleurone layer carrying the message for GA_3 synthesis? DUFFUS and DUFFUS (1969) and GALSKY and LIPPINCOTT (1969) found that treatment of isolated aleurone layers with 3,5-cAMP and 3,5-cGMP simulates the effect of gibberellin on α-amylase production. The appearance of α-amylase activity following this treatment was delayed by approximately 12 h relative to the effect of GA_3. Inhibitors of DNA synthesis (sarcomycin and mitomycin) prevented enzyme synthesis induced by the cyclic mononucleotides but not by GA_3 (KESSLER and KAPLAN, 1973). Inhibitors of gibberellin biosynthesis, (2-chloroethyl) trimethylammonium chloride (CCC) and 2-isopropyl-4-dimethylamino-5-methylphenyl-1-piperidine carboxylate methyl chloride (AMO-1618), reduced cyclomononucleotide-induced formation of α-amylase progressively with increasing concentration (1–100 μM) (KESSLER, 1973). Enhanced thymidine incorporation into DNA was observed after 3,5-cAMP treatment of aleurone layers; this effect could be inhibited by sarcomycin and mitomycin C. KESSLER (1973) has suggested that treatment with 3,5-cAMP or 3,5-cGMP involves first a stage of DNA synthesis followed by the formation of gibberellin which results in α-amylase synthesis. The authors have their own doubts whether cAMP penetrates the cell as such or whether it is first metabolized but in any case these results raise the question as to whether studies on GA_3 treatment of aleurone layers reveal only a part of the natural processes occurring in the germinating seed.

Changes other than those connected with RNA synthesis have been observed in aleurone layers during the lag period between GA_3 application and α-amylase synthesis. GA_3 treatment increased the rate of synthesis of endoplasmic reticulum in aleurone layers eightfold within 4 h of application. This coincided with an increase in the number of ribosomes which could be isolated as polysomes (JONES, 1969a, b; EVINS and VARNER, 1971). The authors suggested that the polysomes were attached to the endoplasmic reticulum. Rough endoplasmic reticulum was formed and this vesiculated mainly in the basal regions of the aleurone cells facing the endosperm. The smooth vesicles thus formed apparently become appressed to the plasma membrane and may be involved in protein secretion (VIGIL and RUDDAT, 1973). Increases in the rate of release of α-amylase into the medium and its appearance in cell homogenates correlated directly with the formation and subsequent vesiculation of the rough endoplasmic reticulum. When actinomycin D was added to aleurone simultaneously with GA_3 the cells exhibited large areas of disarranged segments of fragmented rough endoplasmic reticulum. The reduction in α-amylase synthesis and release relative to controls without actinomycin D was 45% and 63% respectively in similar experiments (VIGIL and RUDDAT, 1973). In other studies on membrane synthesis it was found (JOHNSON and KENDE, 1971) that the enzymes of the cytidine diphosphate choline pathway, which is involved in lecithin biosynthesis required in the production of rough endoplasmic reticulum, are under GA_3 control. The first enzyme of the pathway – choline kinase – is found in the soluble protein fraction, and its activity is not affected by prior treatments of aleurone with GA_3 (1 μM), even after 22 h. The second and third enzymes of the pathway – phosphorylcholine-cytidyl transferase (PC-C) and phosphorylcholine-glyceride transferase (PC-G) respectively – are associated with the particulate fractions and their activities are increased three- to four-fold during the lag phase of gibberellin-induced α-amylase synthesis (0 to 8 h). The effect was evident 2 h after gibberellin treatment. Inhibitors that block α-amylase formation (cycloheximide at 10 μg/ml and actinomycin D at 50 μg/ml) also inhibit the stimulation of these membrane-bound enzymes by the hormone. ABA (1 μM), which did not affect the activity of PC-C and PC-G in vitro severely inhibited the in vivo promotion of enzyme activity by GA_3. BEN-TAL and VARNER (1974) also showed that the increase in activity of PC-G transferase could not be prevented within 4 h by a mixture of amino-acid analogues, nor by cordicepin, although it was shown that the inhibitors had penetrated the cells in effective concentrations by that time. These workers therefore concluded that neither RNA nor protein synthesis are required for this GA_3 effect and that GA_3 promotes, and ABA inhibits, the activation of these enzymes which are already present in the aleurone cells, presumably as zymogens, or within compartments.

More direct evidence on the involvement of GA_3 in the increase in membrane synthesis was provided by KOEHLER and VARNER (1973), who found that treatment of aleurone layers with 1 μg/ml GA_3 increased, by three- to five-fold, orthophosphate incorporation into phospholipids soluble in a chloroform-methanol mixture. The effect was measurable 4 to 6 h after GA_3 treatment and reached its maximum after 8 to 12 h. The enhancement was inhibited by cycloheximide, 6-methylpurine and ABA. The inhibition by ABA was concentration-dependent between 10^{-10} and 10^{-6} M. The increase in incorporation occurred throughout the subcellular fraction and was not restricted to a specific structure or organelle. Moreover, it was shown that there was a proportional increase in incorporation into all phospholipids separable by thin-layer chromatography. Thus, it may be that general de novo membrane synthesis and especially production of rough endoplasmic reticulum is controlled by GA_3 and is required for the subsequent production of hydrolases.

COLLINS et al. (1972) found that treatment of aleurone layers with 100 μg/ml GA_3 did not affect the overall levels of nucleotides in the cells though a transient increase in ^{32}P incorporation into nucleotides was observed. This effect was maximal 30 to 60 min after GA_3 application and then declined. By 180 min after GA_3 treatment, incorporation of ^{32}P was lower than in controls. The nucleoside triphosphates exhibited the greatest changes, particularly cytidine triphosphate. The absolute increases in the specific radioactivity of the four nucleoside triphosphates were approximately equal after 30 min of GA_3 treatment, though their initial levels were quite different. COLLINS et al. (1972) argue that these findings do not support the hypothesis that the effect of GA_3 is restricted to nucleic acid synthesis since if this process alone was stimulated, the increases in the specific activity of the four nucleosides would be expected to be inversely proportional to their initial levels, assuming approximately equimolar incorporation of nucleotides into RNA. Thus, they have suggested that GA_3 brings about an alteration and an increase in membrane synthesis and that the new membranes, with their attached polyribosomes, act as participants in the initiation of the synthesis of extracellular hydrolytic enzymes. A somewhat divergent view of effects on membrane activities is taken by LAIDMAN and his co-workers using wheat aleurone. They indicate that phospholipid synthesis and formation of endoplasmic reticulum is independent of the embryo and that this process occurs merely as a consequence of imbibition. On the other hand they too find that the vesiculation of the endoplasmic reticulum described above is GA_3-dependent (LAIDMAN et al., 1974).

Treatment of cereal half seeds or isolated aleurone layers with GA_3 greatly enhances the activity of a variety of other enzymes such as protease (JACOBSEN and VARNER, 1967), ribonuclease (VARNER and JOHRI, 1968) (Fig. 6.9), β-amylase (JACOBSEN et al., 1970), phosphatase and β1→3-glucanase (TAIZ and JONES, 1970; JONES, 1971), polyphenol oxidase (mono-phenolase) (TANEJA and SACHER, 1974), β1→4-xylanase, β-xylopyranosidase and β-arabinofuranosidase (TAIZ and HONIGMAN, 1976) the enzymes of fatty acid β-oxidation and the enzymes of the glyoxylate cycle (DOIG and LAIDMAN, 1972) and some isoperoxidases (TAO and KHAN, 1975; JACOBSEN and VARNER, 1967). Most of the enzymes mentioned above are formed in the aleurone cells but exert their hydrolytic action outside the cytoplasm of those cells; either in the endosperm cells (α-amylase, β-amylase, protease, ribonuclease, peroxidases) or in the aleurone cell walls (β1→3-glucanase, β-xylanase, β-xylopyranosidase, and α-arabinofuranosidase) which are degraded during the germination process (JONES, 1969b; BRIGGS, 1973). The effect of GA_3 upon the activity of these enzymes may be expressed in three ways: (i) by increasing the rate of enzyme synthesis; (ii) by influencing the rate of enzyme secretion from the cell and (iii) by affecting either the stability of the enzyme or its performance. In some cases, GA_3 affects both the rate of synthesis of an enzyme and its secretion. This is the case for α-amylase (CHRISPEELS and VARNER, 1967a; VIGIL and RUDDAT, 1973), and protease (VARNER and JOHRI, 1968; JACOBSEN and VARNER, 1967). In other cases GA_3 only affects enzyme secretion, for example β1→3-glucanase (JONES, 1971; TAIZ and JONES, 1970; BENNETT and CHRISPEELS, 1972), phosphatases (JONES, 1971), β-amylase (JACOBSEN et al., 1970) and ribonuclease (CHRISPEELS and VARNER, 1967a; BENNETT and CHRISPEELS, 1972). In this category β1→3-glucanase and ribonuclease are synthesized de novo during the imbibition of the half seeds and application of GA_3 adds very little to their rate of synthesis (BENNETT and CHRISPEELS, 1972) though it controls their secretion (JONES, 1971; CHRISPEELS and VARNER, 1967a). In the release of acid phosphatase GA_3 exerts its effect on the cell walls rather than on cell membranes (ASHFORD and JACOBSEN, 1974). It was found that the enzyme passes through the plasma membrane and is trapped in the periplasmic space even in the absence of GA_3. The release of the enzyme into the medium outside the cell wall is, however, GA_3-dependent. This effect is presumably related to GA_3 effects on the cell-wall-degrading enzymes referred to above. In the third category GA affects only the performance of the enzymes but neither their release not their synthesis. Enzymes in this category are the enzymes of the cytidine diphosphate-choline pathway PC-C and PC-G referred to above (JOHNSON and KENDE, 1971; BEN-TAL and VARNER, 1974) and some peroxidase isoenzymes (TAO and KHAN, 1975).

Lastly, we need to consider those other processes in aleurone tissue unaffected by GA's and apparently under the control of other hormones. Thus, a factor emanating from the endosperm and which can be replaced by cytokinin initiates the mobilization of part of the triglyceride reserve and promotes the retention of macronutrient mineral ions in the aleurone tissue (TAVENER and LAIDMAN, 1972a). Equally, a factor or factors emanating from the embryo induces neutral lipase activity and the mobilization of another part of the triglyceride reserve. This effect can be mimicked by a combination of IAA and glutamine (TAVENER and LAIDMAN, 1972b).

While phytase is present in the dry grain and only requires hydration to bring about its activation (EASTWOOD and LAIDMAN, 1971a) a factor from the embryo which can be replaced by glutamine or certain other nitrogenous compounds causes a doubling in activity. This appears to be an activation effect on an inactive form of the enzyme rather than an effect on de novo synthesis (EASTWOOD and LAIDMAN, 1971a). Connected with this effect is the effect of cytokinin on the retention of macronutrient ions referred to above. In wheat seed, during the first 24 h of germination there is a decline in the capacity of aleurone layers to release ions (mainly K^+, Mg^{2+} and phosphate); this is followed by a recovery phase. EASTWOOD and LAIDMAN (1971b) using aleurone, embryo and endosperm tissue alone and in combination established that a cytokinin from the endosperm and GA_3 from the embryo act sequentially to control the mobilization of mineral reserves from the aleurone tissue.

While the work described above has related to cereal aleurone tissue there are clear analogues with other seeds. Thus in fenugreek (*Trigonella foenum-graecum* L.) and other legumes the pattern of breakdown of the seed reserves shows a close parallelism to the situation in cereals. However, in these leguminous seeds the reserve polysaccharide is a galactomannan and the initiation of its breakdown is not dependent on factors from the embryo. Actinomycin D has little effect on the mobilization of the reserves and in contrast to cereal aleurone many ribosomes are present prior to imbibition. GRANT REID and MEIER (1972) suggest that the apparent lack of embryonic control of reserve mobilization reflects the less important role of the endosperm in leguminous seeds; however the work of LAIDMAN and his group referred to above should encourage caution here since the possible involvement of hormonal factors from the endosperm cannot be ruled out.

Surprisingly, rather less can be said in conclusion to all this work than might be expected. Clearly, the aleurone system involves the induction, activation, or modification of a multiplicity of processes; equally, these processes must take place in a well-defined sequence. All types of hormone appear to be involved and they appear to intervene at a number of different levels of organization and metabolism, sometimes alone, sometimes in combination. It is also apparent for these reasons that the tissue is "pre-programmed" to respond in a particular way.

The likelihood of the existence of a single "master reaction" for each growth regulator seems highly improbable, although the possibility that the coordination of several processes by a single hormone is brought about by the activities of "integrator" genes as proposed by BRITTEN and DAVIDSON (1969) cannot be excluded. It seems at least as likely that specific hormone receptors are required for most of the stages affected.

6.5 Effects of Hormones on Differentiating Systems

Unlike the situation pertaining to hormonal effects on growth or on non-growing systems which we have discussed previously, it is impossible to single out a system or systems to typify what is known about effects of hormones at the subcellular level, prior to and during differentiation.

The meaning of differentiation is itself difficult to define. In this chapter we will use the term in the sense of those changes occurring in a cell or a tissue starting at the multipotential stage up to the point of assuming the structure typical of the final role of the cell or the tissue; the growth phase of this progression has already been dealt with above and will only be referred to briefly. An example of such a process would be the changes occurring in a meristematic cell, one product of the division of which is to become a xylem element. Many difficulties arise in research on this type of phenomenon. It is usually impossible to define which of the growing meristematic cells will re-divide and retain its meristematic nature and which will continue growing and eventually differentiate. It is also difficult to separate the differentiating cells from each other in order to study the differences between cells differentiating in different directions into different tissues.

In order to overcome these and other difficulties the method of organ, tissue, and cell culture was devised. It was soon observed that hormones, especially auxins and cytokinins, are essential for the success of the culture, and that the level of hormones and their ratio in the medium may influence the appearance of the culture and affect the differentiation of the cultured cells (OSWALD et al., 1977; UCHIMIYA and MURASHIGE, 1976). Several culture techniques have been developed, all leading to dividing, multipotential cells, sometimes fairly uniform and even synchronized, in which processes connected with cell division and differentiation may be studied.

Two main problems exist in the application of such culture techniques and in the interpretation of the results obtained:

1. The method is highly artificial. The results of the differentiation obtained by experimental manipulations with tissue explants never duplicate the complex organisation of tissues characteristic of the intact plant (TORREY et al., 1971). The time and place of the initial cell divisions, in comparison with the surrounding tissues, is important to all the developmentally connected processes which follow. This led TORREY et al. (1971) to the conclusion that "series of unalterable events are set into motion at the time the cell 'originates'

and cannot be stopped short of full cellular differentiation". Thus, a change in the meristem's organization, or in the time of cell divisions, may affect the whole sequence of processes as well as the resulting differentiation.

2. The method is based on dedifferentiation of already specialized cells which are supposed to become totipotent again by excision from their natural surrounding tissues and culturing them on the appropriate medium. The sole factor determining the future development of the tissue is supposed to be the composition of the medium. In practice, the ingredients of the medium are *not* the sole factors affecting the differentiation of the cells. The type of tissue from which the cells or explants were excised, its age, the species used, and other factors all affect the behaviour of the culture. Thus, WAREING (1971) stated that a distinction should be made between the effects of plant hormones on meristematic cells, cells about to undergo maturation, and mature cells, because they differ in their state of determination. A number of apparent discrepancies between different studies can be explained on this basis and will be discussed later in this chapter. (For example compare HALL and ORDIN, 1968 with RAY, 1973a, b.)

A different problem altogether is that relating to the type of hormone used and its concentration. In many studies synthetic hormones (such as 2,4-D, NAA, ethephon, kinetin, etc.) have been used, without taking several precautions necessary. It must be shown that the same results can be obtained with natural and artificial hormones and that there is not a basic difference in their effects (RITZERT and TURIN, 1970; LEE, 1972b; MILLER, 1969). Some precautions should be taken in relation to the concentration of the hormone used, especially when it is an artificial one. Thus, different hormone levels were required to obtain the same result when 2,4-D and NAA were used (MATSUMOTO et al., 1973; LEE, 1971). Several research workers have used very high hormone concentrations, sometimes so high that they are on the verge of poisoning the tissue (HARDIN et al., 1970 – 4.5×10^{-3} M 2,4-D; VERMA et al., 1975 – 4.5×10^{-3} M 2,4-D daily for five days; PUECH et al., 1976 – 500 ppm ethephon; BIRECKA et al., 1976 – 2.2×10^{-6} M ethylene). Many such studies will be cited in this section and the reader's attention is called to this.

Most studies agree that DNA synthesis takes place prior to differentiation and that hormones affect this synthesis (ROBERTS, 1976). FOSKET (1970) found that treatment of cultivated *Coleus* stem segments with 2.9×10^{-7} M IAA induced xylem differentiation which could be detected after 3 days and DNA synthesis which began 2 days after treatment. Thus the auxin-induced DNA synthesis occurs either together with or just before the commencement of differentiation. 5-Fluorodeoxyuridine (FUDR), which blocks the activity of thymidylate synthetase [thus blocking DNA synthesis in the nucleus as well as in other organelles (MATTHYSSE and TORREY, 1967)], consistently blocked DNA synthesis and xylogenesis, but only if added to the medium during the first three days of culture. When added later on, after the peak period of DNA synthesis had passed, it had no effect. KAMISAKA and MASUDA (1970) also found that 2,4-D (4.5×10^{-6} M) and GA_3 (3×10^{-5} M) induced thymidine incorporation into DNA and that FUDR inhibited it.

HOLM and KEY (1971) support this view and add that in rootless basal tissue of soybean hypocotyls, 2,4-D at 10^{-4} M induced an approximately nine-fold increase in DNA synthesis. Chromatin extracted from the tissues of both control and auxin-treated explants had the same RNA synthetic capacity as chromatin from intact seedlings. FUDR blocked the DNA synthesis induced by auxin and the enhancement of chromatin activity was inhibited by 70%, but this level was still threefold higher than the control. Thus, the authors concluded that auxin influenced the synthetic activity of existing DNA templates, and more of the existing genome is available for transcription after auxin treatment. TIMMIS and INGLE (1975) support this conclusion by showing that the cellular rRNA content is positively regulated but that maximum gene utilization is not normally employed. The degree of utilization responds to the total genome rather than to the specific rRNA dosage. The mechanisms of control of this regulation appear to be at the transcriptional level.

TORREY and FOSKET (1970) showed that the cortex of pea root segments cultured on auxin-containing medium without cytokinin produced a callus in which tracheary elements never appeared. When kinetin was added to the medium, a callus was formed in which the majority of the cells were stimulated to divide and had synthesized DNA on the third day of culture, prior to cell division. These cells were polyploid. After 5–7 days of culture those cells which had recently undergone division formed tracheary elements. Thus the authors concluded that auxin plus kinetin treatment brings about DNA synthesis which results in polyploid cells which in turn divide to form tracheary elements. PHILLIPS and TORREY (1973) subsequently showed that the cortical cells, cultured on medium containing IAA and kinetin, exhibited endo-reduplication (NAGL, 1976) and continued DNA synthesis without entering the mitosis period of the cell cycle beginning 24 h after excision. Cyto-differentiation evidently occurred sometime between successive DNA cycles. Mitoses commenced at approximately 48 h after excision and were all tetraploid at first, but as the experiment progressed the proportion of 4n cells decreased and the octaploid population increased. Tracheary elements commenced differentiating after 7 days. These workers also found that the distribution of cells with different "C values" (2C=1n) in cortical cells in the tissue culture was as follows: 2C=3%; 4C=57%; 8C=39% and 16C=1%. Most of the differentiating tracheary elements were either tetraploid or octaploid but cytodifferentiation occurred in approximately 3% of the diploid cells as well. LIBBENGA and TORREY (1973) and SIMPSON and TORREY (1977) also confirmed that in this system both auxin (IAA or 2,4-D) and cytokinin (kinetin) are required to initiate the sequence of DNA synthesis, endo-reduplication, cell division, and tracheary element differentiation. On the other hand FOSKET and SHORT (1973) found in a cytokinin-dependent variety of cultured soybean that, although the increase in cell number was a function of exogenous zeatin concentration, the amount of tritiated thymidine incorporated into DNA was unrelated to the rate of cell division; the highest level of DNA synthesis was found on the "minus cytokinin" medium. The authors concluded that the cells became polyploid by endo-reduplication in the absence of cytokinin and that cytokinin was acting in the initiation of cytokinesis.

This view is supported by several other studies. LEFFLER et al. (1971) found that chromatin isolated from soybean hypocotyls treated with 2,4-D at 4.5×10^{-4}–4.5×10^{-3} M prior to extraction supported a higher level of DNA synthesis than did chromatin from untreated plants. JOUANNEAU and TANDEAU DE MARSAC (1973) found that, although cytokinin induced synchronous divisions in suspension cultures of tobacco, it was not essential to mitosis and DNA synthesis proceeded normally in the presence of auxin, with or without cytokinin, for at least the time required for one completion of the cell cycle. YEOMAN (1974) described a burst of DNA synthesis during the S period of the cell cycle in Jerusalem artichoke pieces treated with 2,4-D. KOVOOR and MELET (1972) stated that the DNA produced after treatment of Jerusalem artichoke tuber explants with 5×10^{-6} M NAA was different in its re-association kinetics to that of resting (control) explants. WARDELL (1975) found that deprivation of tobacco stem segments of IAA by moving them from light to dark, defoliating them, or inverting defoliated segments resulted in a decrease in thymidine incorporation into DNA. For segments given all three treatments IAA (1.1×10^{-5} M) restored the rate of thymidine incorporation within 60 min with a lag time of approximately 25 min. This tends to show that DNA synthesis is the first effect of auxin, before RNA and protein synthesis are affected.

NAGL and RUCKER (1972) observed DNA amplification during histogenesis and morphogenesis of *Cymbidium* protocorms. 2,4-D treatment caused an increase in the number of chromatids per chromosome (polyteny) in differentiating tracheary elements. The same treatment produced some giant nuclei in differentiating xylem elements which exhibited polytenic structures at approximately the 128C level and led to mis-differentiation. NAGL et al. (1972) explored more deeply the problem of nucleus organization as related to DNA synthesis under hormonal treatment. They produced evidence for a close relationship between cyto-differentiation and changes at the level of the cell cycle such as a diminution in the rate of division, endo-mitotic short-cut of the mitotic cycle and extra replication of specific genes. Treatment of *Cymbidium* with 10^{-7} M IAA doubled the mitotic index and greatly enhanced thymidine incorporation into DNA in the meristem. GA_3 (10^{-4} M) had the same effect while 2,4-D (10^{-6} M) inhibited the mitotic index tenfold and the incorporation of thymidine into DNA sixfold. In the differentiating region (that is, after differentiation had commenced) IAA and GA_3 slightly enhanced DNA synthesis in storage cells, but IAA had no effect and GA_3 inhibited DNA synthesis 54-fold in assimilating tissue. 2,4-D had no effect on DNA synthesis in storage cells, but doubled that occurring in the assimilating cells. The authors concluded that 2,4-D inhibited DNA replication in the mitotic cycle but not in the endo-mitotic short-cut and stimulated extra replication of heterochromatin DNA. GA_3 stimulated DNA replication in mitosis, slightly stimulated the endomitotic cycle but reduced DNA amplification significantly. Effects of IAA are similar to those of GA_3 but DNA amplification seems not to be affected. All molecular effects were followed by morphological changes, but whereas IAA or GA_3 treatments did not alter the direction of differentiation markedly, 2,4-D did. In other tissues, e.g., root meristems of three species of *Albium,* NAGL (1972) was able to demonstrate a selective blockage of the cell cycle in G1, G2, and the mitotic stage and between karyokinesis and cytokinesis by treatments with colchicine, kinetin, ABA, and IAA at appropriate concentrations, combinations, and durations. Thus he suggested that any step of the cell cycle is independently controlled both by a specific balance of growth regulators and by specific synthesis of nucleic acids and further that the various forms of endopolyploidization and DNA amplification are also regulated by the hormonal balance available to a cell (NAGL, 1976). In connection with this hypothesis NISHI et al. (1977) found that when a strain of carrot cell culture was deprived of auxins, all cells stopped dividing and were arrested at the G1 phase of the cell cycle. When returned to auxin-containing medium, a burst of cell division occurred after DNA synthesis. Cytokinin-dependent cell suspension cultures of *Acer* and *Nicotiana* could be induced to become synchronous in respect to cell cycle by excluding kinetin from the culture medium and adding it again after the culture stopped growing (YEOMAN, 1974). In this way it was found that 2,4-D treatment induced a burst of protein and RNA (mainly rRNA) synthesis which accompanied DNA synthesis during the S period, which in turn induced a new increase in rRNA dependent on the newly synthesized DNA, leading to the mitotic period. Histones were synthesized during the S period together with DNA. Changes in the pattern of protein distribution occurred as the cell progressed towards division. Enzymes were synthesized periodically only once per cycle and not all at the same time.

Ethylene treatment reduces or inhibits DNA synthesis and cell division during the first steps of growth and differentiation (APELBAUM et al., 1974). SATO et al. (1976) showed that ethylene, if present continuously at a concentration of 4.5×10^{-10} M, reduced by 50% the rate of DNA synthesis induced by cutting of discs from potato tuber. It was shown that a process, necessary for DNA synthesis and susceptible to ethylene action, began about 6 h after cutting and continued for a limited period of approximately 12 h. As this wound-induced DNA synthesis required biphasic synthesis of protein, i.e., during the 6 h after cutting and 6 h just prior to DNA synthesis, and was dependent upon prior synthesis of RNA, the effect of ethylene on the synthesis of RNA and proteins was checked, but no change was observed in either process. It was noticed, however, that ethylene decreased the number of cells which entered the S phase of the cell cycle, at which stage DNA synthesis takes place. On the other hand, HOLM et al. (1970) found that elongating sections of soybean seedlings treated with ethylene showed a small increase in RNA content. Moreover, chromatin extracted from such sections after treatment showed a 35%–60% increase in its capacity for RNA synthesis. Auxin treatment of similar sections caused marked accumulation of RNA and DNA and chromatin isolated from auxin-treated tissue showed an eight- to tenfold increase in its capacity to synthesize RNA. Ethylene reduced the auxin-enhanced nucleic acid synthesis. Nearest-neighbour analysis of the RNA produced by chromatin after ethylene or auxin treatment showed that the RNA was different in each treatment and different from the controls. Different results were obtained in other experiments with soybean seedlings. Thus, KEY et al. (1966) found that 2,4-D (5×10^{-4} M–10^{-2} M) blocked nucleic acid and protein synthesis in the apical zone but induced the synthesis of DNA and RNA in other parts of the seedling, RNA synthesis preceding that of DNA. There was preferential synthesis of rRNA in response to 2,4-D treatment and the RNA/DNA ratio increased, while the protein to RNA ratio decreased. CHEN et al. (1975) reported that nuclei isolated from control and 2,4-D treated (2.5×10^{-3} M) mature soybean hypocotyl tissue were similar in size but the nucleoli were significantly larger in treated plants. The DNA content per nucleus was the same (4 ± 1 picogram) in control and auxin-treated tissue but the ratio DNA: RNA: protein was 1:3.1:11 for control and 1:5.4:21.7 for treated plants respectively. While in the control the level of RNA polymerase II activity was 50%–60% higher than that of RNA polymerase I, the nuclei of auxin-treated tissue contained about 2.5 times more RNA polymerase I activity than controls. The RNA polymerase III reported from animals was not detected.

O'BRIEN et al. (1968a, b) found that pretreatment of soybean hypocotyls with 5×10^{-3} M 2,4-D doubled the activity of RNA polymerase bound to the chromatin. The RNA products formed by control and 2,4-D-induced chromatin systems were found to be different in respect to elution profiles on methylated albumin-kieselguhr (M.A.K.) columns. In a cell-free system the 2,4-D pretreatment enhanced amino-acid incorporation into protein twofold. The authors concluded that 2,4-D affected the rate of endogenous RNA polymerase only but did not make new templates available. On the other hand HOLM et al. (1970), while working on the same system with similar treatments, concluded that 2,4-D affected template availability rather than RNA polymerase activity. MCCOMB et al. (1970) found that GA_3 increased the level of RNA polymerase associated with chromatin without a detectable increase in the amount of DNA template available. YANAGISHIMA and SHIMODA (1973) found that there were two different kinds of RNA necessary for auxin to exert its action on expansion of cells of Jerusalem artichoke and yeasts; one necessary to render the cells responsive to auxin and the other whose synthesis is induced by auxin. The first type of RNA was extracted from yeasts and oat coleoptiles and transformed yeast cells and Jerusalem artichoke tuber tissue from the original "potentially auxin-responsive" state to the "actual auxin-responsive" state. GA_3 was found to induce the synthesis of this type of RNA or to cause its accumulation, although GA_3 does not cause expansion growth by itself.

KEY and INGLE (1968) also found that 2,4-D treatment enhanced the synthesis of total RNA but this time in growing as well as in maturing sections of the soybean hypocotyl. The treatment resulted in accumulation of soluble and ribosomal RNA in the cells and in AMP-rich RNA synthesis as well. The authors showed that the growth of the cells was solely dependent on continuous synthesis of AMP-rich RNA and that soluble or ribosomal RNA was not required. They also found that radioactivity from ^{14}C-labelled IAA and 2,4-D was incorporated into all fractions of nucleic acid and that the incorporation is markedly inhibited by actinomycin D. The authors concluded that the incorporation resulted largely from utilization of products of auxin catabolism rather than direct attachment of the auxin molecule to RNA.

KEY and VANDERHOEF (1973) came to the conclusion that, in tobacco pith culture, different IAA concentrations which induce different reactions in the tissue also induce different nucleic acid contents of the cells. Thus low concentrations of IAA which induced cell division but little cell enlargement increased the amount of DNA per cell, while treatment with high concentrations of IAA caused mainly cell enlargement and increased the RNA content of the cells. 5-Fluorouracil inhibited accumulation of tRNA and rRNA but not the amount of AMP-rich RNA. THOMPSON and CLELAND (1971) found that in pea stem sections no change in hybridizable RNA could be detected in response to auxin over a 2–24 h period regardless of the auxin (IAA or 2,4-D) concentration used – optimal or supraoptimal for growth. However, when large doses of 2,4-D were applied to intact pea seedlings, definite changes in the hybridizable RNA were detected both 8 and 24 h after treatment. It must be noted that in intact seedlings high concentrations of 2,4-D cause changes in the direction of growth and in the pattern of differentiation. BIRMINGHAM and MACLACHLAN (1972) suggested that auxins caused RNA synthesis and polysome aggregation together with RNAase synthesis, two-thirds of the RNAase being kept apart from the polysomes by compartmentalization. N^6-Benzyladenine suppressed the synthesis of RNAase. Thus, the equilibrium between these hormones, together with GA, decided the direction of growth prior to differentiation. PILET and BRAUN (1970) found that IAA treatment of whole *Lens* seedlings caused an increase in RNA content and a decrease in RNAase activity. Addition of IAA to excised roots inhibited both the decrease in RNA levels and the increase in RNAase activity. DOVE (1971) was able to reduce the increase in RNAase activity after excision of tomato leaflets by treatment with either 10^{-6} M IAA or 9×10^{-6} M kinetin: IAA being the more effective. Treatment with IAA plus kinetin was less effective than treatment with either of the hormones alone in reducing the rise in RNAase activity. N^6-Benzyladenine at 4.5×10^{-6} M or higher or $2.2–4.5 \times 10^{-5}$ M kinetin increased RNAase activity. A list of hormones which are able to suppress the rise in RNAase activity after excision include IAA (wheat coleoptile, pea green stem internodes, tomato leaflets, lentil roots), kinetin (oat leaves, tobacco leaves, tomato leaflets), 8-aza-adenine (tobacco leaves), p-fluorophenylanine (tobacco leaves), NAA (*Rhoeo discolor* leaf sections), and others (DOVE, 1972). ABA enhances the increase in RNAase activity in many of these systems. DAVIES (1976) confirmed that auxins (IAA or NAA, 0.5% w/w in lanolin) caused a decrease in monosomes and an increase in polysomes, especially membrane-bound polysomes. GA_3 or benzyladenine were ineffective. These membrane-bound polysomes may, in DAVIES' opinion, be involved in increased utilization of mRNA as a template for enzyme synthesis. The connection between auxin treatment and membranes may be correlated with the work on membrane-bound auxin receptors (BATT et al., 1976; BATT and VENIS, 1976; DOLLSTADT et al., 1976) (see Chap. 5). HARDIN et al. (1970) in work mentioned above found that a proteinaceous factor isolated from untreated soybean etiolated seedlings was able to increase in vitro RNA synthesis by chromatin-bound RNA polymerase. The same factor failed to increase RNA synthesis by chromatin extracted from seedlings previously treated with 4.5×10^{-3} 2,4-D. The activity of chromatin from 2.4-D treated plants without the factor was higher than that of the chromatin of control plants with the added factor. The authors argued that the 2,4-D had already "turned on" the chromatin by means of the factor so that an additional supply had no effect. By further fractionation the authors were able to separate the factor into a high molecular weight fraction which was active in the endogenous test (soybean RNA-synthesizing medium) only and a low molecular weight fraction that was active in the *E. coli* polymerase assay. The authors suggested that auxins or auxin-like hormones interact with a receptor which enables them to move across the nuclear membrane where a modification of the auxin-receptor complex may occur. The complex, modified or not, interacts with RNA polymerase to allow transcription of hormone-specific genes, not previously expressed. VENIS (1971) also found a protein fraction in pea stem tissue which could be distinguished by affinity chromatography when bound to 2,4-D. This fraction when added to an in vitro RNA-synthesizing medium (containing DNA and *E. coli* RNA polymerase) enhanced RNA synthesis without additional auxin. HARDIN and CHERRY (1972) showed that isolated plasma membrane fractions could act in the same way as the proteinaceous factor and HARDIN et al. (1972) found that when untreated plasma membranes were centrifuged from the suspension the supernatant solution lost its ability to enhance the activity of solubilized mRNA polymerase from soybean chromatin. If the membranes were treated with 2,4-D, the supernatant solution retained the ability to stimulate the activity of the RNA polymerase. DAVIES (1973) suggested that auxin interacted with cell membranes to produce a protein factor which selectively promoted the production of mRNA specific for RNA polymerase. If this is true the process may bear some relation to replication of specific genes during the cell cycle and to the production of specific enzymes, thus affecting the pattern of differentiation. JOHRI and VARNER (1968) found that although GA_3 had no effect on the rate of synthesis of RNA in vitro when supplied to pea sections via the incubation medium it *was* able to enhance the rate of RNA synthesis when it was included in the medium used to extract nuclei. Nearest-neighbour analysis of the RNA produced by the control and the GA_3-treated nuclei showed that the GA_3 caused qualitative changes in the RNA as well as an increase in the rate of RNA synthesis. The authors interpreted these results by assuming that some factor or factors present in the nucleus or cytoplasm is involved in RNA synthesis. GA_3 could prevent the loss of the factor or be involved in transporting the factor from the cytoplasm into the nucleus. It may be involved in the activity of RNA polymerase.

Kinetin removes a limitation that prevents the synthesis of RNA and genome expression. ABA, on the other hand, causes depression or inhibition of RNA and protein synthesis (RIJVEN and PARKASH, 1971; WYEN et al., 1972). The response may be specific, as some enzymes respond to these treatments [nitrate reductase and the "relative purine specific endo-nuclease" (hydrolyses RNA)] while other enzymes do not respond ("sugar non-specific endo-nuclease I" and "alkaline phosphodiesterase"). It was shown (SUSSEX et al., 1975) that a cytokinin (N^6-benzyladenine) can completely counteract ABA inhibition of growth and RNA synthesis. Zeatin failed to counteract inhibition of RNA synthesis by ABA in the same system – excised embryonic bean axes (WALTON et al., 1970) – although it partially reversed the inhibition of growth. As the effect of ABA on ATP pool size and O_2 consumption was marginal (3% and 6% respectively) it was concluded (WALBOT et al., 1975) that ABA acts directly on RNA synthesis and not through a control of the precursors and that it affects growth through its effect on RNA synthesis. However, as cytokinins affect specific enzymes and ABA inhibits DNA, rRNA, and tRNA synthesis (but probably not mRNA synthesis, at least in this system) it seems that these two hormones affect growth and differentiation in two different ways.

Many studies indicate that cytokinin treatment is not a prerequisite for the induction or enhancement of protein synthesis in tobacco cell culture (JOUANNEAU and PÉAUD-LENOEL, 1967; JOUANNEAU, 1968, 1970) or in cultured tissue of cytokinin-dependent soybean (FOSKET and SHORT, 1973). On the other hand, some workers suggest that plant cells contain sufficient cytokinin to keep protein synthesis going for a few days. Thus MAAS and KLÄMBT (1977) found a 35% rise in protein synthesis after treatment with 1.9×10^{-6} M kinetin following cytokinin starvation of tobacco cells (in this experiment a 27% standard deviation was recorded). This rise in protein synthesis persisted even in the presence of actinomycin D and the authors concluded that kinetin-controlled protein synthesis is independent of transcription and is accelerated either by enhancement of the specific activity of polysomes or by a shortening of the time necessary for polypeptide synthesis.

Auxins accelerate protein synthesis in a variety of systems, via the mechanism of enhanced transcription and translation (MAAS and KLÄMBT, 1977; SIMPSON and TORREY, 1977) and isolated chromatin produced more DNA polymerase and RNA polymerase if the plant cells were treated with 2,4-D before chromatin extraction (LEFFLER et al., 1971). GA_3 caused a rise in amino-acid production and in the incorporation of amino-acids into proteins, especially in the shoot and root meristems and in the endosperm cells of charlock seedlings (EDWARDS, 1976).

In order to demonstrate hormonal effects on differentiation one should be able to show changes in enzyme pattern, and especially of enzymes related to the direction of differentiation which the tissue will follow. ASHTON (1976) came to the conclusion (1) that the response of proteolytic enzymes to hormones is not a primary response, (2) that hormones are not specific to a single enzyme and the specific hormones required to induce de novo synthesis of such enzymes vary from species to species, e.g., GA_3 for barley and cytokinin for squash and (3) that a degree of control of synthesis of a dipeptidase in squash cotyledons is affected through four hormones acting in concert. This last conclusion by ASHTON may indicate that differentiation processes are controlled by different interactions between several hormones. Thus, each hormone, at different concentrations and in concert with other hormones at specific concentrations, may affect the synthesis or activation of an enzyme. It is also possible that one hormone, when present in different tissues or in the same tissue during different stages of its development may act via different mechanisms.

KAUFMAN et al. (1973) found that GA_3 increased invertase activity in excised *Avena* stem sections, the products providing materials for secondary wall synthesis. At the same time the products played significant roles in the regulation of invertase. CHERRY (1968) found that GA_3 stimulated, and IAA inhibited, increased invertase activity in sugar beet slices which had been washed in water and supplied immediately with the hormones after the tissue had been cut. However, if GA_3 was given for an 8-h period during the first day, the later it was given the smaller was the stimulation on the first day and the greater was the inhibition on the second day. ABA promoted the development of invertase in the first and second days of washing to about the same extent as GA_3. GLASZIOU et al. (1968) used slices of sugar-cane stems taken from the zone of rapid cell elongation at the base of expanding internodes. Invertase and peroxidase showed a switch-over from transcription-limited synthesis to translation-limited synthesis. During the transcription-limited synthesis GA_3 at a high concentration (5×10^{-3} M) scarcely affected invertase but severely repressed peroxidase synthesis, whereas neither enzyme was affected during the period when translation was limited. Low concentrations of GA_3 (10^{-7} M) markedly increased invertase levels at the transcription-limited stage. NAA depressed enzyme synthesis when transcription was limited and had no effect when translation was limited. GA_3 had no effect on α-amylase or acid phosphatase levels throughout the experiment. Thus, in *Avena* stems and sugar beet slices (mature tissues), GA_3 increased invertase synthesis while, in growing tissues of sugar-cane, it affected the enzyme only in the second stage and when given at a low concentration. RICARDO (1976) showed that the activity of acid-invertase in intact carrot roots declined with age; fructose, GA_3, and kinetin had no effect on this decline. When portions of the roots were excised and incubated in water the enzyme activity rose markedly; GA_3 stimulated this increase and fructose and kinetin inhibited it. Fructose was most effective in young, non-tuberous roots while the effects of GA_3 and kinetin were most marked in mature tuberous roots.

Other enzymes involved in sugar metabolism show the same pattern but other hormones and different concentrations are active. In dwarf pea internodes GA_3 stimulated amylase, β-fructosidase and starch phophorylase, and caused cell elongation, cell division and secondary cell-wall synthesis. Treatment with glucose and glucose derivatives mimicked the morphological effects of GA_3, and the authors (BROUGHTON and MCCOMB, 1971) concluded that the overall effect of GA_3 was to provide more substrate for cell metabolism and wall formation. On the other hand KATSUMI (1970) found that in dwarf maize seedlings the elongation of the leaf sheath and the length of the first leaf were proportional to α-amylase activity. IAA promoted the growth of the leaf sheath but not the α-amylase activity and GA_3 promoted both growth and α-amylase activity in the first leaf but the content of reducing sugar declined. Thus, he concluded that the processes of growth and α-amylase activity and the effect of GA_3 on them are independent of one another. HUFF and ROSS (1975) found that light treatment elevated the activity of amylase, while zeatin treatment reduced this elevation. In the dark, zeatin promoted growth but increased neither sugar levels nor amylase activity. LOVELL and MOORE (1970) subsequently showed that N^6-benzyladenine reduced the movement of ^{14}C-labelled assimilates from mustard cotyledons to the base of the petioles by reducing the sink strength of the latter.

CHRISPEELS (1976) suggested that hormones caused a modulation of the biosynthesis and subsequent secretion of macromolecules involved in cell-wall building. UCHIMIYA and MURASHIGE (1976) found that NAA (up to 3.5×10^{-6} M) was essential for wall regeneration on isolated protoplasts of tobacco, higher concentrations being less effective or inhibitory. Kinetin (4.6×10^{-6} M or higher) was inhibitory and lower concentrations had no effect. MACLACHLAN et al. (1968) found that 0.5% IAA in lanolin fed to decapitated dark-grown seedlings of Alaska pea, maize, sunflower, and bean induced both the synthesis and activity of cellulase. RNA and protein synthesis were required for the synthesis of the enzyme but neither DNA synthesis nor cell division were required. Polyribosomal preparations that contained cellulase activity were isolated from IAA-treated and untreated pea epicotyls and provided in vitro with ingredients necessary for protein synthesis. Polyribosomes from untreated tissue were capable of protein synthesis in vitro but no cellulase was found, while the "IAA-treated" preparation produced cellulase within 15 min. By using sophisticated purification and characterization methods BYRNE et al. (1975) found that 2,4-D caused the formation of two forms of cellulase, one was buffer-soluble (BS) and the other buffer-insoluble (BI). The molecular weight of the BS was approximately 20,000 and that of BI approximately 70,000; both were produced in equal amounts after auxin treatment. BI was found to be localized mainly at the inner surface of the cell wall in close association with microfibrils and BS was localized mainly within the distended endoplasmic reticulum. The Golgi complex and plasma membrane appeared to be completely devoid of any cellulase; the cells showed distinct large vacuoles and were at least partially differentiated (BAL et al., 1976). The two enzymes had different kinetic constants, different amino-acid compositions, different mobility in electrophoretic and chromatographic systems and different sedimentation behaviour and immunological properties. On the other hand both enzymes showed true $\beta 1 \rightarrow 4$ endoglucanase properties, acting specifically on β-1,4-linkages (thus being β-1,4 glucan 4-glucanohydrolases, EC 3.2.1.4). Despite all the differences mentioned above the authors claimed that the active sites of the two isozymes were the same (WONG et al., 1977b). When poly (A)-containing RNA was fractionated from the polysomes of 2,4-D-treated pea epicotyls and translated in a wheat embryo cell-free system the translation products included only the BS cellulase and never BI cellulase (VERMA et al., 1975). The authors concluded that in spite of the differences in amino-acid composition, and in other properties, BI cellulase is modified from the smaller BS cellulase during its secretion from the cell to the wall. Thus, auxin treatment changed the course of development of partially differentiated cells (showing as lateral expansion and a swollen region) by causing the synthesis of an enzyme capable of promoting cell "loosening" and enhancing cellulose synthesis by generating primer chain ends within the continuous microfibrillar network of the wall. HOGETSU et al. (1974a, b) found that in azuki bean (*Azukia angularis* = *Vigna angularis*) epicotyl sections, coumarin inhibited the incorporation of glucose into cellulose whether auxin was present or not, but did not affect microtubules associated with the wall. Colchicine on the other hand broke up the microtubules but did not affect cellulose synthesis. Both colchicine and coumarin reversed the promoting effect of GA_3 and the inhibitory effect of kinetin on IAA-induced growth of the sections, but did not affect the IAA-induced growth itself. This may seem in contrast to the other above-mentioned results, but NANCE (1973) and MONDAL (1975) found that 10^{-5} M kinetin, while inhibiting the elongation of sections of pea epicotyls and promoting lateral expansion, caused the formation of cell wall rich in pectic uronic acids. MONDAL (1975) added that Ca^{2+} and ethylene had the same effect as kinetin while GA caused elongation accompanied by the development of walls comparatively low in pectic substances. VERMA et al. (1976) showed that cells of *Acer pseudoplatanus* grown in suspension culture with 4.5×10^{-7} M 2,4-D and 4.4×10^{-6} M 6-benzylaminopurine (BA) removed myoinositol from the medium at a rapid rate and used it for pectin biosynthesis. Raising the BA concentration in the medium tenfold caused a drastic reduction in myoinositol uptake, but cell growth and monomer composition of pectic polysaccharides were not affected. A tenfold increase in 2,4-D concentration in the medium instead of BA had little or no effect on myoinositol uptake but reduced the utilization of myoinositol for pectin synthesis.

Ethylene is also involved in affecting the quality of the cell wall and the direction of growth during differentiation (see also Sect. 6.2.2.c). RIDGE and OSBORNE (1970, 1971) showed that exogenous ethylene induced parallel increases in hydroxyproline content and peroxidase activity in the covalently bound cell-wall protein of pea shoots. They assumed that the real effect of ethylene was an increase in the hydroxyproline-rich peroxidases in the cell walls. This enhanced level of hydroxyproline possibly led to an increased cross-linking by glycosidic bonds of hydroxyproline-rich proteins with polysaccharides in the cell wall and this has implications for wall plasticity and cell growth (OSBORNE et al., 1972; see also Sect. 6.2.2.c). RIDGE (1973) added that ethylene affects the re-orientation of cellulosic fibrils during cell wall growth. ROBERTS and BABA (1968) suggested that exogenous proline may stimulate xylogenesis (and thus secondary wall formation) in cultured explants of *Coleus*. Taking into account the decisive role of peroxidases in the pathway of lignin formation (MAZELIS, 1962; ROBERTS, 1976) ethylene may be responsible for the shift from growth and deposition of the primary wall to the production and deposition of lignin.

Lignin is an aromatic polymer of polyhydroxyphenylpropanoid subunits. The pathway of lignin biosynthesis includes the phenylalanine and tyrosine pools. PAL (phenylalanine ammonia-lyase) converts phenylalanine to cinnamic acid, whereas L-tyrosine ammonia-lyase produces p-coumaric acid from tyrosine. The cinnamyl alcohols, oxidized by peroxidases and other oxidative enzymes, lead to polymerization of coniferyl, sinapyl and p-coumaryl alcohols, glucosides of which may furnish lignin precursors. The lignin molecules are formed chiefly by dehydrogenative polymerization of 1-(4-hydroxyphenyl) allyl alcohol, i.e., p-hydroxycinnamic alcohol and its phenyl-substituted mono- and di-methoxylated derivatives (BERLYN, 1970; ROBERTS, 1976).

GA_3 seems to stimulate lignin synthesis when applied to tobacco tissue culture, but only when used at low concentration (3×10^{-7} M). High concentrations (3×10^{-5} M) had no effect (LI et al., 1970). ABA reduced lignin content in this system but GA_3 overcame the inhibiting effect of ABA. GA_3 and ABA had the same effects on other phenols. TORREY et al. (1971) stated that a change in the auxin-cytokinin ratio may be involved in lignification. SHAH et al. (1976) showed that while maximum growth of cultivated *Cassia* tissue occurred on a medium containing 9×10^{-6} M 2,4-D and 1.9×10^{-6} M kinetin, maximum polyphenol synthesis was observed in tissues growing more slowly on 9×10^{-7} M 2,4-D. In the presence of 9×10^{-6} M 2,4-D a higher kinetin concentration in the medium (up to 1.9×10^{-5} M) gave higher polyphenol concentrations. Without 2,4-D in the medium, sub-optimal and supra-optimal kinetin concentrations suppressed polyphenol accumulation. The richer the tissue was with polyphenols the higher became the peroxidase activity in the tissue, a maximum being recorded at 9×10^{-7} M 2,4-D. RANJEVA et al. (1975) added that enzymes of aromatic metabolism occurring in *Petunia* calluses can be divided into two groups according to their response to hormones: group A included PAL, cinnamate hydroxylase, p-coumarate hydroxylase, and the enzymes activating phenylpropanoid units and group B included chalcone-flavonone isomerase (involved in the synthesis of flavonoids) and coniferyl alcohol dehydrogenase (leading to monomers of lignin). In the presence of both 4.5×10^{-6} M 2,4-D and 8.9×10^{-7} M BA, group A enzymes have two peaks of activity at the fourth and eighteenth days. In the presence of BA alone only the first peak appears and in the presence of 2,4-D alone only the second peak is observed. Coniferyl alcohol dehydrogenase has a requirement for BA and chalcone-flavonone isomerase is delayed but not inhibited in the absence of one hormone. Thus, enzymes involved in the synthesis and activation of phenylpropanoid units seem to act coordinately and the common pathway leading to activated cinnamic acids and specific metabolic steps of lignin and flavonoid synthesis are regulated in a different way. Ethylene treatment (10^{-6} M) was found to increase PAL synthesis caused by excision of gherkin (*Cucumis sativus* L.) hypocotyls (ENGELSMA and VAN BRUGGEN, 1971). Actinomycin D and cycloheximide effectively inhibited the ethylene-induced increase in PAL activity (CHALUTZ, 1973).

The peroxidase activity of tobacco pith cells increases as they become older and larger and further from the apex. As pith cells are excised and cultivated the total activity of peroxidases rises sharply within 24 h, due to the appearance of new isoenzymes. This increase is favoured by kinetin (9×10^{-7} M) and inhibited by IAA (10^{-4} M). IAA represses the appearance of the new isoenzymes and subsequently induces the appearance of a new isoenzyme (LAVEE and GALSTON, 1968). LEE (1971, 1972a) showed that IAA and 2,4-D may affect the synthesis of peroxidase isoenzymes serving as IAA oxidases but later on (LEE, 1972b) found that IAA, kinetin and GA_3 affect the appearance of many isoenzymes, only two of which had the same characteristics as IAA oxidases. In these experiments the isoenzymes were resolved into three groups, namely slow (P1–4), medium (P5–7) and fast (P8–11) migrating peroxidases. The appearance of the fast-migrating isoenzymes required both kinetin and IAA, kinetin becoming inhibitory at 5×10^{-6} M. GA_3 further increased the content of all peroxidases when IAA and kinetin were present at optimal concentrations but GA_3 could not replace IAA nor overcome the inhibitory effect of 5×10^{-6} M kinetin. ABA, cycloheximide and actinomycin D inhibited the formation of the fast-migrating isoperoxidases. 2,4-D at low concentrations (10^{-7}–10^{-6} M) promoted the activity of a fast-migrating isoperoxidase and at high concentrations (10^{-5}–10^{-4} M) inhibited it, but caused an increase in other peroxidases of lower mobility (see also RITZERT and TURIN, 1970). GASPAR et al. (1973) also found that IAA, kinetin, and ABA at different concentrations influenced the appearance of new isoenzymes in lentil embryonic axes. Ethylene also induced the activity and synthesis of new isoenzymes of peroxidase in injured or sliced tissues (IMASEKI, 1970; BIRECKA et al., 1973, 1976).

Other combinations of hormone treatment in different sorts of tissue may direct the pathway towards flavonoids or anthocyanin formation. Thus, MILLER

(1969) found that either 2,4-D or a combination of NAA and kinetin may lead to deoxyisoflavone synthesis in soybean callus tissue, the kinetin being involved in RNA or protein synthesis (MILLER, 1972). Anthocyanin formation induced by light is probably controlled by ethylene which, given early in the lag phase, induced a higher anthocyanin content and when given after the lag phase inhibited anthocyanin formation (CRAKER et al., 1971). When given during the dark period before exposure to light, ethylene increased the rate of anthocyanin synthesis when the tissue was placed in the light (CRAKER and WETHERBEE, 1973). ABA and ethephon acted synergistically to enhance the induction of both total phenolics and anthocyanin in leaves and fruit of grapevine. The magnitude of this interaction depended on the age of the tissues (PIRIE and MULLINS, 1976). In *Populus* cell suspension culture, kinetin inhibited and NAA stimulated anthocyanin production (MATSUMOTO et al., 1973) while, in the common fig fruit, ethephon first caused a sharp decline in chlorophylls, β-carotene, lutein, violaxanthin, and neoxanthin, and only after a lag period of four days did it affect anthocyanin accumulation.

The interactions between hormones and effects of different concentrations of hormones referred to in this section may reflect something more than differences between individual species or between the existing hormone status of tissues. It has long been observed in the induction of vascular tissues in callus (e.g., WETMORE et al., 1964) that the concentration of auxin at the point where the tracheary elements are formed is much lower (ca. 2.9×10^{-7} M) than the concentration for elongation (ca. 5×10^{-5} M). Since cell elongation occurs below the apical meristem prior to lignification and secondary thickening, the observed concentration dependence in callus would appear to mimic the situation in a normal meristem where one would expect, and indeed it has been shown, that there is a gradient of auxin concentration from the apex downwards. In other words the concentration dependence of such events in artificial systems may reflect the situation as regards the auxin status in the normal growing apex, a status which is controlled not only by rates of auxin biosynthesis but also by other constraints within the plant such as polar auxin transport.

Although little biochemical data is available, the induction of vascular tissue by growth regulators also provides us with an interesting example of interactions between hormones and other metabolites. Thus, while the initiation of a meristem in callus is vested largely in growth regulators, WETMORE and RIER (1963) showed that the *type* of vascular tissue formed depended on sucrose concentration, high and low sucrose concentrations (11.6×10^{-2} M and 5.8×10^{-2} M) favouring the formation of phloem and xylem respectively. Furthermore, JEFFS and NORTHCOTE (1967) demonstrated that the requirement was specific for sucrose and not merely for carbohydrate, in that of fifteen carbohydrates tested only sucrose, trehalose, and maltose induced organized nodules of meristematic tissue and of the remainder cellobiose, lactose, and glucose gave xylem but no organized nodules. As regards specificity it is perhaps significant that, of the carbohydrates tested, only sucrose, maltose, and trehalose have α-glucosyl residues at their non-reducing ends.

6.6 Conclusions

It must be apparent to the reader, even in relation to the very restricted field covered by this chapter, that although a massive body of data has been accumulated over the last twenty or so years, nevertheless a real understanding of the mode of action of endogenous growth regulators at the subcellular level is still beyond our grasp. On the other hand it would be too pessimistic to state that we have not advanced at all. Indeed, in the case of one or two systems we appear to be approaching something resembling an explanation. Thus, if part of the mode of action of auxin in promoting growth is in fact via the increased activity of a proton pump and if the specific binding sites currently being characterized (Chap. 5 of this volume) from growing tissue such as *Zea* are those which control the increased proton efflux, then the establishment of a connection between the two may well not be far off. The same might be said to a greater or lesser extent about each of the other groups of hormones in specific systems which have been noted in the text. It would seem therefore that much future work is likely to be concentrated on determining how a change in the properties of a receptor, brought about by interaction with a hormone, is translated into a primary metabolic response.

This latter statement begs several questions, however. As we pointed out in the introduction, a distinction must be drawn between effects of hormones on tissues and organs the subsequent development of which is determined – for example, most growth systems and the aleurone system – and those systems where the hormones themselves appear to determine the pattern of development. That is, a differentiation must be made between situations where the hormone is merely a messenger and those where it is both messenger and message.

In the case of tissues or cells whose pattern of development is determined and which merely await the hormonal message it is not conceptually difficult to marry what we know about such systems to hypotheses on modes of hormone action. Thus, the competence of such systems to respond to a given stimulus by a hormone or hormones presumably would reflect the presence of the appropriate receptors to interact with the hormone(s) and subsequently mediate the appropriate metabolic events. The question as to whether there are only a few such types of receptor in a given system, or many, is unknown, but even if the former is the case, the amplification of the response could readily be achieved by a process of "cascade" regulation where a number of subordinate responses are activated by the primary processes (e.g., Britten and Davidson, 1969). On the other hand, in those instances where hormones may determine the pattern of differentiation we have a more difficult problem. Wareing (1971) points out that the numbers of such instances must be small if only because these processes must involve the switching of large sectors of the genome whereas only five classes of endogenous hormone have been identified so far, although there is no prima facie reason why sites with different affinities for a given hormone, together with interactions between receptors for different hormones, might not increase the number of possible permutations. Such examples relate to instances where hormones effect a switch in development in meristematic tissues, for example the regeneration of roots and shoots in callus in response to auxin/cytokinin ratio and the initiation of cambial activity in callus.

The fact that only a few such cases exist in no way diminishes the problem, for if we accept the reasoning set out above about systems which are already determined, then we have the "hen and egg" situation where the hormone in a determined system interacts with receptors whose distribution has in the first instance (i.e., in the meristem or in its formation) been determined by the same hormone or other hormones. Moreover, in, for example, the case of hormonal initiation of root and shoot meristems in callus the effect is that of a "trigger" since there is no need for the continued presence of the hormone for further growth – that is, a stable change has been induced.

The question which needs to be resolved is whether the nature of the primary action of the hormone is different in kind when it is acting as a determining factor than when it is initiating a predetermined response. At this stage in the state of the field we can only speculate, since it is possible to construct consistent hypotheses to fit either case.

References

Abdul-Baki, A.A., Ray, P.M.: Regulation by auxin of carbohydrate metabolism in cell wall synthesis by pea stem tissue. Plant Physiol. *47*, 537–544 (1971)

Abeles, F.B.: Biosynthesis and mechanism of action of ethylene. Annu. Rev. Plant Phys. *23*, 259–592 (1972)

Abeles, F.B., Leather, G.R.: Abscission: Control of cellulase secretion by ethylene. Planta *97*, 87–91 (1971)

Adams, P.A., Montague, M.J., Tepfer, M., Rayle, D.L., Ikuma, H., Kaufman, P.B.: Effect of gibberellic acid on the plasticity and elasticity of *Avena* stem segments. Plant Physiol. *56*, 757–760 (1975)

Adamson, H., Adamson, D.: The lag period in auxin-induced expansion of storage tissue and coleoptiles. Aust. J. Biol. Sci. *10*, 435–442 (1957)

Allerup, J.: Induced transpiration changes: Effects of some growth substances added to the root medium. Physiol. Plant *17*, 899–908 (1964)

Andersen, A.S., Moller, I., Hansen, J.: 3-Methyleneoxindole and plant growth regulation. Physiol. Plant. *27*, 105–108 (1972)

Apelbaum, A., Sfakiotakis, E., Dilley, D.R.: Reduction in extractable desoxyribonucleic acid polymerase activity in *Pisum sativum* seedlings by ethylene. Plant Physiol. *54*, 125–128 (1974)

Armstrong, D.J., Murai, N., Taller, B.J., Skoog, F.: Incorporation of cytokinin N^6-Benzyladenine into tobacco callus transfer ribonucleic acid and ribosomal ribonucleic acid preparations. Plant Physiol. *57*, 15–22 (1976)

Ashford, A.E., Jacobsen, J.V.: Cytochemical localization of phosphatase in barley aleurone cells: The pathway of gibberellic-acid-induced enzyme release. Planta *120*, 81–105 (1974)

Ashton, F.M.: Mobilization of storage proteins of seeds. Annu. Rev. Plant Physiol. *27*, 95–117 (1976)

Baker, D.B., Ray, P.M.: Relation between the effects of auxin on cell wall synthesis and cell elongation. Plant Physiol. *40*, 360–367 (1965)

Bal, A.K., Verma, D.P.S., Byrne, H., Maclachlan, G.A.: Subcellular localization of cellulases in auxin treated pea. J. Cell Biol. *69*, 97–105 (1976)

Ballio, A., Chain, E.B., De Leo, P., Erlanger, B.F., Mauri, M., Totolo, A.: Fusicoccin:

a new wilting toxin produced by *Fusicoccum amygdali* Del. Nature (London) *203*, 297 (1964)

Barkley, G.M., Evans, M.L.: Timing of the auxin responses in etiolated pea stem sections. Plant Physiol. *45*, 143–147 (1970)

Barkley, G.M., Leopold, A.C.: Comparative effects of hydrogen ions, carbon dioxide, and auxin on pea stem segment elongation. Plant Physiol. *52*, 76–78 (1973)

Batt, S., Venis, M.A.: Separation and localization of two classes of auxin binding sites in corn coleoptile membranes. Planta *130*, 15–21 (1976)

Batt, S., Wilkins, M.B., Venis, M.A.: Auxin binding to corn coleoptile membranes: kinetics and specificity. Planta *130*, 7–13 (1976)

Beffagna, N., Cocucci, S., Marrè, E.: Stimulating effect of fusicoccin on K-activated ATPase in plasmalemma preparations from higher plant tissues. Plant Sci. Lett. *8*, 91–98 (1977)

Bennet-Clark, T.A.: Salt accumulation and mode of action of auxin. A preliminary hypothesis. In: The chemistry and mode of action of plant growth substances. Wain, R.L., Wightman, F. (eds.), pp. 284–291. London: Butterworth 1956

Bennett, P.A., Chrispeels, M.J.: De novo synthesis of ribonuclease and β-1,3-glucanase by aleurone cells of barley. Plant Physiol. *49*, 445–447 (1972)

Ben-Tal, Y., Varner, J.E.: An early response to gibberellic acid not requiring protein synthesis. Plant Physiol. *54*, 813–816 (1974)

Berlyn, G.P.: Ultrastructural and molecular concepts of cell wall formation. Wood Fiber *2*, 196–227 (1970)

Bhattacharya, J., Roy, S.C.: Growth promoters and the synthesis of protein in plant mitochondrial protein. Biochem. Biophys. Res. Commun. *35*, 606–610 (1969)

Birecka, H., Briber, K.A., Catalfamo, J.L.: Comparative studies on tobacco pith and sweet potato root isoperoxidases in relation to injury, indoleacetic acid and ethylene effects. Plant Physiol. *52*, 43–49 (1973)

Birecka, H., Catalfamo, J.L., Urban, P.: Cell isoperoxidases in sweet potato plants in relation to mechanical injury and ethylene. Plant Physiol. *57*, 74–79 (1976)

Birmingham, B.C., MacLachlan, G.A.: Generation and suppression of microsomal ribonuclease activity after treatment with auxin and cytokinin. Plant Physiol. *49*, 371–375 (1972)

Black, M., Dheidah, M.: The growth physiology of coleoptiles in continuous-record vessels: Anomalous kinetics of cycloheximide action on auxin-controlled growth. Physiol. Plant *37*, 89–95 (1976)

Bonner, J.: The relation of hydrogen ions to the growth rate of the *Avena* coleoptile. Protoplasma *21*, 406–423 (1934)

Bowen, A., Dymock, J., Hill, B.: Auxin and dark carbon dioxide fixation in *Avena sativa* coleoptile sections. Plant Physiol. *57*, Suppl. 19 (1976)

Bradbury, D., Ennis, W.B.: Stomatal closure in kidney bean treated with ammonium 2,4-dichlorophenoxyacetate. Am. J. Bot. *39*, 324–328 (1952)

Briggs, D.E.: Biosynthesis and its control in plants. Milborrow, B.V. (ed.), pp. 219–277. New York, London: Academic Press 1973

Britten, R.J., Davidson, E.H.: Gene regulation for higher cells: A theory. Science *165*, 349–357 (1969)

Broughton, W.J., McComb, A.J.: Changes in the pattern of enzyme development in gibberellin-treated pea internodes. Ann. Bot. *35*, 213–228 (1971)

Browning, G.: 2-Chlorethanephosphoric acid reduces transpiration and stomatal opening in *Coffea arabica*. Planta *121*, 175–179 (1974)

Burg, S.P.: Studies on the formation and function of ethylene gas in (banana) plant tissue. Colloq. Int. C.N.R.S. *123*, 719–725 (1964)

Burg, S.P.: Ethylene, plant senescence and abscission. Plant Physiol. *43*, 1503–1511 (1968)

Burg, S.P., Burg, E.A.: The interaction between auxin and ethylene and its role in plant growth. Proc. Natl. Acad. Sci. USA *55*, 262–269 (1966)

Burg, S.P., Burg, E.A.: Auxin stimulated ethylene formation; its relationship to auxin inhibited growth, root geotropism and other plant processes. In: Biochemistry and physiology of plant growth substances. Wightman, F., Setterfield, G. (eds.), pp. 1275–1294. Ottawa: Runge Press 1968

Burrows, W.J., Skoog, F., Leonard, N.J.: Isolation and identification of cytokinins located in the transfer ribonucleic acid of tobacco callus grown in the presence of 6-benzylaminopurine. Biochem. Am. Chem. Soc. *10*, 2189–2193 (1971)

Burström, H.G., Uhrström, I., Olausson, B.: Influence of auxin on Young's modulus in stem and roots in *Pisum* and the theory of changing the modulus in tissue. Physiol. Plant *23*, 1223–1233 (1970)

Byrne, H., Christou, N.V., Verma, D.P.S., MacLachlan, G.A.: Purification and characterization of two cellulases from auxin-treated pea epicotyls. J. Biol. Chem. *250*, 1012–1018 (1975)

Chalutz, E.: Ethylene-induced phenylalanine ammonia-lyase activity in carrot roots. Plant Physiol. *51*, 1033–1036 (1973)

Chandra, G.R., Duynstee, E.E.: Hormonal regulation of nucleic acid metabolism in aleurone cells. In: Biochemistry and physiology of plant growth substances. Wightman, F., Setterfield, G. (eds.), pp. 793–814. Ottawa: Runge Press 1968

Chen, Y.M., Lin, C.Y., Chang, H., Guilfoyle, T.J., Key, J.L.: Isolation and properties of nuclei from control and auxin-treated soybean hypocotyl. Plant Physiol. *56*, 78–82 (1975)

Cherry, J.H.: Regulation of invertase in washed sugar beet tissue. In: Biochemistry and physiology of plant growth substances. Wightman, F., Setterfield, G. (eds.), pp. 417–431. Ottawa: Runge Press 1968

Cherry, J.H.: Regulation of RNA polymerase activity by a plasma membrane factor. In: Mechanisms of regulation of plant growth. Bieleski, R.L., Ferguson, A.R., Cresswell, M.H. (eds.), Vol. 12, pp. 751–757. Wellington: Bull. R. Soc. N. Z. 1974

Chrispeels, M.J.: Biosynthesis, intracellular transport and secretion of extracellular macromolecules. Annu. Rev. Plant Physiol. *27*, 19–38 (1976)

Chrispeels, M.J., Varner, J.E.: Gibberellic acid-enhanced synthesis and release of α-amylase and ribonuclease by isolated barley aleurone layers. Plant Physiol. *42*, 398–406 (1967a)

Chrispeels, M.J., Varner, J.E.: Hormonal control of enzyme synthesis: On the mode of action of gibberellic acid and abscisin in aleurone layers of barley. Plant Physiol. *42*, 1008–1016 (1967b)

Chrispeels, M.J., Varner, J.E.: Inhibition of gibberellic acid-induced formation of α-amylase by absiscin II. Nature (London) *212*, 1066–1067 (1967c)

Cleland, R.E.: Effect of osmotic concentration on auxin-action and on irreversible and reversible expansion of the *Avena* coleoptile. Physiol. Plant. *12*, 809–825 (1959)

Cleland, R.E.: Effect of auxin upon loss of calcium from cell walls. Plant Physiol. *35*, 581–584 (1960a)

Cleland, R.E.: Auxin-induced methylation in maize. Nature (London) *185*, 44 (1960b)

Cleland, R.E.: Auxin-induced cell wall loosening in the presence of Actinomycin D. Plant Physiol. *40*, 595–600 (1965)

Cleland, R.E.: Extensibility of isolated cell walls: Measurement and changes during cell elongation. Planta *74*, 197–209 (1967)

Cleland, R.E.: Auxin and wall extensibility: reversibility of the auxin-induced wall-loosening process. Science *160*, 192–193 (1968)

Cleland, R.E.: Cell wall extension. Annu. Rev. Plant Physiol. *22*, 197–222 (1971)

Cleland, R.E.: Auxin-induced hydrogen ion excretion from *Avena* coleoptiles. Proc. Natl. Acad. Sci. USA *70*, 3092–3097 (1973)

Cleland, R.E.: Kinetics of hormone-induced H^+ excretion. Plant Physiol. *58*, 210–213 (1976)

Cleland, R.E., Karlsnes, A.: A possible role for hydroxyproline-containing proteins in the cessation of cell elongation. Plant Physiol. *42*, 669–671 (1967)

Cleland, R.E., Lomax, T.: Hormonal control of H^+-excretion from oat cells. In: Regulation of cell membrane activities in plants. Marrè, E., Ciferri, O. (eds.), pp. 161–172. Amsterdam, Oxford, New York: North Holland 1977

Cleland, R.E., Thompson, M.L., Rayle, D.L., Purves, W.K.: Differences in effects of gibberellins and auxin on wall extensibility of cucumber hypocotyls. Nature (London) *219*, 510–511 (1968)

Cleland, R.E., Thompson, W.F., Haughton, P.M., Rayle, D.L.: Macromolecule synthesis and wall extensibility in relation to the mechanism of auxin-induced cell elongation. In: Plant growth substances 1970. Carr, D.J. (ed.), pp. 1–8. Berlin-Heidelberg-New York: Springer 1972

Cocucci, M., Marrè, E., Ballarin-Denti, A., Scachi, A.: Characteristics of fusicoccin-induced changes of trans-membrane potential and ion uptake in maize root segments. Plant Sci. Lett. *6*, 143–156 (1976)

Collins, G.G., Jenner, C.F., Paleg, L.G.: The metabolism of soluble nucleotides in wheat aleurone layers treated with gibberellic acid. Plant Physiol. *49*, 404–410 (1972)

Coster, H.G.L., Steudle, E., Zimmermann, U.: Turgor pressure sensing in plant cell membranes. Plant Physiol. *58*, 636–43 (1976)

Craker, L.E., Wetherbee, P.J.: Ethylene, light and anthocyanin synthesis. Plant Physiol. *51*, 436–438 (1973)

Craker, L.E., Standley, L.A., Starbuck, M.J.: Ethylene control of anthocyanin synthesis in sorghum. Plant Physiol. *48*, 349–352 (1971)

Cummins, W.R., Kende, H., Raschke, K.: Specificity and reversibility of the rapid stomatal response to abscisic acid. Planta *99*, 347–351 (1971)

Darvill, A.G.: Studies on maize cell walls in relation to extension growth. Ph.D. thesis. University of Wales (1976)

Darvill, A.G., Smith, C.J., Hall, M.A.: Auxin induced proton release, cell wall structure and elongation growth: A hypothesis. In: Regulation of cell membrane activities in plants. Marrè, E., Ciferri, O. (eds.), pp. 275–282. North Holland: Elsevier 1977

Darvill, A.G., Smith, C.J., Hall, M.A.: Cell wall structure and elongation growth in *Zea mays* coleoptile tissue. New Phytol. *80*, 503–516 (1978)

Das, V.S.R., Rao, I.M., Raghavendra, A.S.: Reversal of abscisic acid induced stomatal closure by benzyl adenine. New Phytol. *76*, 449–452 (1976)

Datko, A.H., Maclachlan, G.A.: Indoleacetic acid and the synthesis of glucanases and pectic enzymes. Plant Physiol. *43*, 735–742 (1968)

Davies, D.D., Patterson, B.D., Trewavas, A.J.: Studies on the mechanism of action of indoleacetic acid. In: Plant growth regulators. Soc. Chem. Ind. Monogr. *31*, 208–233 (1968)

Davies, E.: Polyribosomes from peas. VI. Auxin-stimulated recruitment of free monosomes into membrane-bound polysomes. Plant Physiol. *57*, 516–518 (1976)

Davies, P.J.: Current theories on the mode of action of auxin. Bot. Rev. *39*, 139–171 (1973)

Davies, W.J., Kozlowski, T.T.: Effects of applied abscisic acid and plant water stress on transpiration of woody angiosperms. For. Sci. *21 (2)*, 191–196 (1975)

Doig, R.I., Laidman, D.L.: The occurrence of an induced glyoxylate cycle in wheat aleurone tissue. Biochem. J. *128*, 88 (1972)

Dohrmann, U., Johnson, K.D., Rayle, D.L.: Comparison of auxin and fusicoccin: growth, H^+ secretion, glycosidase activation, and binding to membrane vesicles in *Avena* coleoptiles. Plant Physiol. Suppl. *53* 43 (1974)

Dohrmann, U., Hertel, R., Kowalik, H.: Properties of auxin binding sites in different subcellular fractions from maize coleoptiles. Planta *140*, 97–106 (1978)

Dollstadt, R., Hirschberg, K., Winkler, E., Hubner, G.: Bindung von Indolylessigsäure und Phenoxyessigsäure an Fraktionen aus Epikotylen und Wurzeln von *Pisum sativum*. Planta *130*, 105–111 (1976)

Dove, L.D.: Short term responses and chemical control of ribonuclease activity in tomato leaflets. New Phytol. *70*, 397–401 (1971)

Dove, L.D.: Environmental and chemical control of RNA breakdown in leaves. In: Nucleic acid and proteins in higher plants. Farkas, G.L. (ed.), Vol. XIII, pp. 299–307. Symp. Biol. Hung. Budapest: Akademiai Kiade 1972

Dowdell, R.J., Smith, K.A., Crees, R., Restall, S.W.F.: Field studies of ethylene in the soil atmosphere – equipment and preliminary results. Soil Biol. Biochem. *4*, 325–331 (1972)

Dowler, M.J., Rayle, D.L., Cande, W.Z., Ray, P.M., Durand, H., Zenk, M.H.: Auxin does not alter the permeability of pea segments to tritium labelled water. Plant Physiol. *53*, 229–32 (1974)

Duffus, C.M., Duffus, J.H.: A possible role for cyclic AMP in gibberellic acid triggered release of α-amylase in barley endosperm slices. Experientia *25*, 581 (1969)

Durand, H., Zenk, M.H.: Initial kinetics of auxin-induced cell elongation in coleoptiles. In: Plant growth substances 1970. Carr, D.J. (ed.), pp 62–67. Berlin-Heidelberg-New York: Springer 1972

Eastwood, D., Laidman, D.L.: The mobilization of macronutrient elements in the germinating wheat grain. Phytochemistry *10*, 1275–1284 (1971a)

Eastwood, D., Laidman, D.L.: The hormonal control of inorganic ion release from wheat aleurone layers. Phytochemistry *10*, 1459–1467 (1971b)

Edwards, M.: Dormancy in seeds of charlock (*Sinapis arvensis* L.). Early effects of gibberellic acid on the synthesis of amino acids and proteins. Plant Physiol. *58*, 626–630 (1976)

El-Beltagy, A.S., Hall, M.A.: Effect of water stress upon endogenous ethylene levels in *Vicia faba*. New Phytol. *73*, 47–60 (1974)

Engelsma, G., Bruggen, J.M.H. van: Ethylene production and enzyme induction in excised plant tissues. Plant Physiol. *48*, 94–96 (1971)

Esau, K.: Plant anatomy. New York: Wiley 1967

Esnault, R.: Recherches sur le mode d'action de l'acide β-indolyl-acetique en relation avec le metabolisme des acides nucléiques. Bull. L. Soc. Fr. Physiol. Veg. *11*, 55–67 (1965)

Esnault, R.: Etude de l'action de l'acide β-indolyl-acetique sur le metabolisme de l'ARN de segments de coleoptiles d'avoine. Bull. Soc. Chim. Biol. *50*, 1887–1913 (1968)

Evans, M.L.: Rapid responses to plant hormones. Annu. Rev. Plant Physiol. *25*, 195–223 (1974)

Evans, M.L., Hokanson, R.: Timing of the response of coleoptiles to the application and withdrawal of various auxins. Planta *85*, 85–95 (1969)

Evans, M.L., Ray, P.M.: Timing of the auxin response in coleoptiles and its implications regarding auxin action. J. Gen. Physiol. *53*, 1–20 (1969)

Evans, M.L., Ray, P.M.: Inactivity of 3-methyleneoxindole as mediator of auxin action on cell elongation. Plant Physiol. *52*, 186–189 (1973)

Evans, M.L., Ray, P.M., Reinhold, L.: Induction of coleoptile elongation by carbon dioxide. Plant Physiol. *47*, 335–341 (1971)

Evins, W.H.: Enhancement of polyribosome formation and induction of tryptophan-rich proteins by gibberellic acid. Biochemistry *10*, 4295–4303 (1971)

Evins, W.H., Varner, J.E.: Hormone controlled synthesis of endoplasmic reticulum in barley aleurone cells. Proc. Natl. Acad. Sci. USA *68*, 1631–1633 (1971)

Evins, W.H., Varner, J.E.: Hormonal control of polyribosome formation in barley aleurone layers. Plant Physiol. *49*, 348–352 (1972)

Fan, D.F., Maclachlan, G.A.: Control of cellulase activity by indole acetic acid. Can. J. Bot. *44*, 1025–1034 (1966)

Fan, D.F., Maclachlan, G.A.: Studies on the regulation of cellulase activity and growth in excised pea epicotyl sections. Can. J. Bot. *45*, 1837–1844 (1967a)

Fan, D.F., Maclachlan, G.A.: Massive synthesis of ribonucleic acid and cellulase in pea epicotyl in response to indoleacetic acid, with and without concurrent cell division. Plant Physiol. *42*, 1114–1122 (1967b)

Ferrari, T.E., Varner, J.E.: Substrate induction of nitrate reductase in barley aleurone layers. Plant Physiol. *44*, 85–88 (1969)

Ferri, M.G., Lex, A.: Stomatal behaviour as influenced by treatment with β-naphthoxyacetic acid. Contrib. Boyce Thompson Inst. *15*, 283–290 (1948)

Filner, P., Varner, J.E.: A test for de novo synthesis of enzymes: Density labelling with H_2O^{18} of barley α-amylase induced by gibberellic acid. Proc. Natl. Acad. Sci. USA *58*, 1520–1526 (1967)

Filner, P., Wray, J.L., Varner, J.E.: Enzyme induction in higher plants. Science *165*, 358–367 (1969)

Fittler, F., Hall, R.H.: Selective modification of yeast seryl-t-RNA and its effect on the acceptance and binding functions. Biochem. Biophys. Res. Commun. *25*, 441–446 (1966)

Fosket, D.E.: The time course of xylem differentiation and its relation to deoxyribonucleic acid synthesis in cultured *Coleus* stem segments. Plant Physiol. *46*, 64–68 (1970)

Fosket, D.E., Short, K.C.: The role of cytokinin in the regulation of growth, DNA synthesis and cell proliferation in cultured soybean tissue (*Glycine max* var. Biloxi). Physiol. Plant. *28*, 14–23 (1973)
Fox, J.E., Erion, J.L.: A cytokinin binding protein from higher plant ribosomes. Biochem. Biophys. Res. Commun. *64*, 694–700 (1975)
Galsky, A.G., Lippincott, J.A.: Promotion and inhibition of α-amylase production in barley endosperm by cyclic 3′, 5′-adenosine monophosphate and adenosine diphosphate. Plant Cell Physiol. *10*, 607–620 (1969)
Galston, A.W., Lavee, S., Siegel, B.Z.: The induction and repression of peroxidase isoenzymes by 3-indoleacetic acid. In: Biochemistry and physiology of plant growth substances. Wightman, F., Setterfield, G. (eds.), pp. 455–472. Ottawa: Runge Press 1968
Gaspar, T., Khan, A.A., Fries, D.: Hormonal control of isoperoxidases in lentil embryonic axis. Plant Physiol. *51*, 146–149 (1973)
Giesen, M., Klämbt, D.: Untersuchungen zur Beeinflussung des Prolin-Einbaues in Proteine durch Auxin und Hydroxyprolin in Weizenkoleoptilzylindern. Planta *85*, 73–84 (1969)
Gilkes, N.R., Hall, M.A.: The hormonal control of cell wall turnover in *Pisum sativum* L. New Phytol. *78*, 1–15 (1977)
Glasziou, K.T., Gayler, K.R., Waldron, J.C.: Effects of auxin and gibberellic acid on the regulation of enzyme synthesis in sugar cane stem tissue. In: Biochemistry and physiology of plant growth substances. Wightman, F., Setterfield, G. (eds.), pp. 433–442. Ottawa; Runge Press 1968
Graniti, A.: The role of toxins in the pathogenesis of infections by *Fusicoccum amygdali* Del. on Almond and Peach. In: Host-parasite relations in plant pathology. Kiraly, Z., Ubriszy, G. (eds.), pp. 211–217. Budapest: Magyar Tudományas Akademia 1964
Grant Reid, J.S., Meier, H.: The function of the aleurone layer during galactomannan mobilisation in germinating seeds of fenugreek (*Trigonella foenum-graecum* L.), crimson clover (*Trifolium incarnatum* L.) and lucerne (*Medicago sativa* L.): A correlative biochemical and ultrastructural study. Planta *106*, 44–60 (1972)
Green, P.B.: Growth physics in Nitella: a method for continuous in vivo analysis of extensibility based on a micro-manometer technique for turgor pressure. Plant Physiol. *43*, 1169–84 (1968)
Green, P.B.: On the biophysical control of growth rate in *Avena* coleoptile sections. In: Plant growth substances 1970. Carr, D.J. (ed.), pp. 9–16. Berlin-Heidelberg-New York: Springer 1972
Green, P.B., Erickson, R.O., Buggy, J.: Metabolic and physical control of cell elongation rate. In vivo studies in *Nitella*. Plant Physiol. *47*, 423–430 (1971)
Hager, A., Menzel, H., Kraus, A.: Experiments and hypothesis concerning the primary action of auxin in elongation growth. Planta *100*, 47–75 (1971)
Halevy, A.H., Kessler, B.: Increased tolerance of bean plants to soil drought by means of growth retarding substances. Nature (London) *197*, 310–311 (1963)
Hall, M.A.: Hormones and differentiation in plants. In: Developmental biology of plants and animals. Graham, C.F., Wareing, P.F. (eds.), pp. 216–231. Blackwell Scientific Publications 1976
Hall, M.A., Ordin, L.: Auxin-induced control of cellulose synthetase activity in *Avena* coleoptile sections. In: Biochemistry and physiology of plant growth substances. Wightman, F., Setterfield, G. (eds.), pp. 659–671. Ottawa: Runge Press 1968
Hall, M.A., Brown, R.L., Ordin, L.: Inhibitory products of the action of peroxyacetyl nitrate upon indole-3-acetic acid. Phytochemistry *10*, 1233–1238 (1971)
Hall, M.A., Kapuya, J.A., Sivakumaran, S., John, A.: The role of ethylene in the response of plants to stress. Pestic. Sci. *18*, 217–223 (1977)
Hall, M.A., Acaster, M.A., Dodds, J.H., Evans, D.E., Jones, J.F., Jerie, P.H., Mutumba, G.C., Niepel, B., Shaari, A.R.: Ethylene and seeds. Proc. 10th. Int. Conf. Plant Growth Subst. 1980 in press
Hall, R.H., Cxonka, L., David, H., McLennon, B.: Cytokinins in the soluble RNA of plant tissue. Science *156*, 69–71 (1967)
Hardin, J.W., Cherry, J.H.: Solubilization and partial characterization of soybean chromatin-bound RNA polymerase. Biochem. Biophys. Res. Commun. *48*, 299–306 (1972)

Hardin, J., Cherry, J.H., Morré, D.J., Lembi, C.A.: Enhancement of RNA polymerase activity by a factor released by auxin from plasma membrane. Proc. Natl. Acad. Sci. USA *69*, 3146–3150 (1972)

Hardin, J.W., O'Brien, T.J., Cherry, J.H.: Stimulation of chromatin-bound RNA polymerase activity by a soluble factor. Biochem. Biophys. Acta *224*, 667–670 (1970)

Haschke, H.P., Lüttge, U.: A beta-indoleacetic acid (IAA) dependant K^+-H^+ exchange mechanism as related to growth of *Avena* coleoptiles. Z. Naturforsch. *28c*, 555–558 (1973)

Haschke, H.P., Lüttge, U.: Auxin action on K^+-H^+-exchange and growth, $^{14}CO_2$ fixation and malate accumulation in *Avena* coleoptile segments. In: Regulation of cell membrane activities in plants. Marrè, E., Ciferri, O. (eds.), pp. 243–250. North Holland: Elsevier 1977

Hashimoto, T.: Synergistic effect of indoleacetic acid and kinetin on primary thickening of pea stem segments. Bot. Mag. *74*, 110–117 (1961)

Hebard, F.V., Amatangelo, S.J., Dayanandan, P., Kaufman, P.B.: Studies on acidification of the media by *Avena* stem segments in the presence and absence of gibberellic acid. Plant Physiol. *58*, 671–675 (1976)

Helgerson, S.L., Cramer, W.A., Morré, D.J.: Evidence for an increase in microviscosity of plasma membranes from soyabean hypocotyls induced by the plant hormone, indole-3-acetic acid. Plant Physiol. *58*, 548–551 (1976)

Ho, D.T.-H., Varner, J.E.: Hormonal control of messenger ribonucleic acid metabolism in barley aleurone layers. Proc. Natl. Acad. Sci. USA *71*, 4783–4786 (1974)

Ho, D.T.-H., Varner, J.E.: Response of barley aleurone layers to abscisic acid. Plant Physiol. *57*, 175–178 (1976)

Hogetsu, T., Shibaoka, H., Shimokoriyama, M.: Involvement of cellulose synthesis in actions of gibberellin and kinetin on cell expansion. Gibberellin-coumarin and kinetin-coumarin interactions on stem elongation. Plant Cell Physiol. *15*, 265–272 (1974a)

Hogetsu, T., Shibaoka, H., Shimokoriyama, M.: Involvement of cellulose synthesis in actions of gibberellin and kinetin on cell expansion. 2,6-Dichlorobenzonitrate as a new cellulose-synthesis inhibitor. Plant Cell Physiol. *15*, 389–393 (1974b)

Holm, R.E., Key, J.L.: Inhibition of auxin-induced deoxyribonucleic acid synthesis and chromatin activity by 5-fluorodeoxyuridine in soybean hypocotyl. Plant Physiol. *47*, 606–608 (1971)

Holm, R.E., O'Brien, T.J., Key, J.L., Cherry, J.H.: The influence of auxin and ethylene on chromatin-directed ribonucleic acid synthesis in soybean hypocotyl. Plant Physiol. *45*, 41–45 (1970)

Horton, R.F.: Stomatal opening: the role of abscisic acid. Can. J. Bot. *49*, 583–585 (1971)

Horton, R.F., Moran, L.: Abscisic acid inhibition of potassium influx into stomatal guard cells. Z. Pflanzenphysiol. *66*, 193–196 (1972)

Hsiao, T.C.: Stomatal ion transport. In: Transport in plants, Part B. Lüttge, U., Pitman, M.G. (eds.), Vol II, pp. 195–221. Berlin-Heidelberg-New York: Springer 1976

Huff, A.K., Ross, C.W.: Promotion of radish cotyledons enlargement and reducing sugar content by zeatin and red light. Plant Physiol. *56*, 429–433 (1975)

Imaseki, H.: Induction of peroxidase activity by ethylene in sweet potato. Plant Physiol. *46*, 172–174 (1970)

Ilan, I.: On auxin-induced pH drop and on the improbability of its involvement in the primary mechanisms of auxin-induced growth promotion. Physiol. Plant *28*, 146–148 (1973)

Jacobs, M., Ray, P.M.: Promotion of xyloglucan metabolism by acid pH. Plant Physiol. *56*, 373–376 (1975)

Jacobs, M., Ray, P.M.: Rapid auxin-induced decrease in free space pH and its relationship to auxin-induced growth in maize and pea. Plant Physiol. *58*, 203–209 (1976)

Jacobsen, J.V.: Interaction between gibberellic acid, ethylene and abscisic acid in control of amylase synthesis in barley aleurone layers. Plant Physiol. *51*, 198–202 (1973)

Jacobsen, J.V., Varner, J.E.: Gibberellic acid-induced synthesis of protease by isolated aleurone layers of barley. Plant Physiol. *42*, 1596–1600 (1976)

Jacobsen, J.V., Scandalios, J.G., Varner, J.E.: Multiple forms of amylase induced by gibberellic acid in isolated barley aleurone layers. Plant Physiol. *45*, 367–371 (1970)

Jeffs, R.A., Northcote, D.M.: The influence of indole-3-ylacetic acid and sugar on the pattern of induced differentiation in plant tissue culture. J. Cell. Sci. *2*, 77–88 (1967)

Jenkinson, I.A.: Bioelectric oscillations of bean roots. Further evidence for a feed back oscillator. III. Excitation and and inhibition of oscillations by osmotic pressure, auxins and antiauxins. Aust. J. Biol. Sci. *15*, 114 (1962)

Johansen, S.: Effect of indole-acetic acid on stomata and photosynthesis. Physiol. Plant. *7*, 531–537 (1954)

John, A.: The effect of ethylene on the growth or roots of *Hordeum vulgaris* and *Oryza sativa*. Ph.D. thesis. University of Wales (1977)

Johnson, K.D., Kende, H.: Hormonal control of lecithin synthesis in barley aleurone cells: Regulation of the CDP-choline pathway by gibberellin. Proc. Natl. Acad. Sci. USA *68*, 2674–2677 (1971)

Johnson, K.D., Daniels, D., Dowler, M.J., Rayle, D.L.: Activation of *Avena* coleoptile cell wall glycosidases by hydrogen ions and auxin. Plant Physiol. *53*, 224–228 (1974)

Johri, M.M., Varner, J.E.: Enhancement of RNA synthesis in isolated pea nuclei by gibberellic acid. Proc. Natl. Acad. Sci. USA *59*, 269–276 (1968)

Jones, R.J., Mansfield, T.A.: Suppression of stomatal opening in leaves treated with abscisic acid. J. Exp. Bot. *21*, 714–719 (1970)

Jones, R.L.: Ethylene-enhanced release of α-amylase from barley aleurone cells. Plant Physiol. *43*, 442–444 (1968)

Jones, R.L.: Gibberellic acid and fine structure of barley aleurone cells. I. Changes during the lag phase of α-amylase synthesis. Planta *87*, 119–133 (1969a)

Jones, R.L.: Gibberellic acid and the fine structure of barley aleurone cells. II. Changes during the synthesis and secretion of α-amylase. Planta *88*, 73–86 (1969b)

Jones, R.L.: Gibberellic acid-enhanced release of β-1,3-glucanase from barley aleurone cells. Plant Physiol. *47*, 412–416 (1971)

Jouanneau, J.P.: Relations entre l'activité de la kinetine et l'activité de synthèse des proteines dans des cultures de cellules de tabac. C. R. Acad. Sci. *267*, 320–323 (1968)

Jouanneau, J.P.: Renouvellement des proteins et effect specifique de la kinetine sur des cultures de cellules de tabac. Physiol. Plant. *23*, 232–244 (1970)

Jouanneau, J.P., Péaud-Lenoel, C.: Croissance et synthèse des proteins de suspension cellulaires de tabac sensibles a la kinetine. Physiol. Plant. *20*, 834–850 (1967)

Jouanneau, J.P., Tandeau de Marsac, N.: Stepwise effects of cytokinin activity and DNA synthesis upon mitotic cycle events in partially synchronized tobacco cells. Exp. Cell. Res. *77*, 167–174 (1973)

Kamisaka, S., Masuda, Y.: Auxin induced growth of *Helianthus* tuber tissue. V. Role of DNA synthesis in hormonal control of cell growth. Physiol. Plant. *23*, 343–350 (1970)

Kang, B.G., Burg, S.P.: Rapid change in water efflux induced by auxins. Proc. Natl. Acad. Sci. USA *68*, 1730–1733 (1971)

Katsumi, M.: Physiological effects of kinetin on the thickening of etiolated pea stem sections. Physiol. Plant. *15*, 115–121 (1962)

Katsumi, M.: Effect of gibberellin A_3 on the elongation, α-amylase activity, reducing sugar content and oxygen-uptake of the leaf sheath of dwarf maize seedlings. Physiol. Plant. *23*, 1077–1084 (1970)

Katz, M., Ordin, L.: A cell wall polysaccharide-hydrolyzing enzyme system in *Avena sativa* L. coleoptiles. Biochim. Biophys. Acta *141*, 126–134 (1967a)

Katz, M., Ordin, L.: Metabolic turnover in cell wall constituents of *Avena sativa* L. coleoptile sections. Biochim. Biophys. Acta *141*, 118–125 (1967b)

Kaufman, P.B., Petering, L., Adams, P.A.: Regulation of growth and cellular differentiation in developing *Avena* internodes by gibberellic acid and indole-3-acetic acid. Am. J. Bot. *56*, 918–27 (1969)

Kaufman, P.B., Ghosheh, N.S., La Croix, J.D., Soni, S.L., Ikuma, H.: Regulation of invertase levels in *Avena* stem segments by gibberellic acid, sucrose, glucose and fructose. Plant Physiol. *52*, 221–228 (1973)

Kazama, H., Katsumi, M.: Auxin-gibberellin relationships in their effects on hypocotyl elongation of light-grown cucumber seedlings. II. Effect of GA_3-pretreatment on IAA-induced elongation. Plant Cell Physiol. *15*, 307–314 (1974)

Keegstra, K., Talmadge, K.W., Bauer, W.D., Albersheim, P.: The structure of plant cell walls. III. A model of suspension – cultured sycamore cell walls based on the interconnections of macromolecular components. Plant Physiol. *51*, 188–196 (1973)
Kemp, H.T., Fuller, R.G., Davidson, R.S.: Inhibition of plant growth by root drench application of kinetin. Science *126*, 1182 (1957)
Kende, H.: The cytokinins. Int. Rev. Cytol. *31*, 301–37 (1971)
Kende, H., Gardner, G.: Hormone binding in plants. Annu. Rev. Plant Physiol. *27*, 267–90 (1976)
Kende, H., Hanson, A.D.: On the role of ethylene in ageing. In: Plant growth regulation. Pilet, P.E. (ed.), pp. 173–180. Berlin-Heidelberg-New York: Springer 1977
Key, J.L.: Ribonucleic acid and protein synthesis as essential processes for cell elongation. Plant Physiol. *39*, 365–370 (1964)
Key, J.L.: Effect of purine and pyrimidine analogues on growth and RNA metabolism in the soybean hypocotyl. The selective action of 5-fluorouracil. Plant Physiol. *41*, 1257–1264 (1966)
Key, J.L., Ingle, J.: Requirement for the synthesis of DNA-like RNA for growth of excised plant tissue. Proc. Natl. Acad. Sci. USA *52*, 1382–1388 (1964)
Key, J.L., Ingle, J.: RNA metabolism in response to auxin. In: Biochemistry and physiology of plant growth substances. Wightman, F., Setterfield, G. (eds.), pp. 711–722. Ottawa: Runge Press 1968
Key, J.L., Shannon, J.C.: Enhancement by auxin of RNA synthesis in excised soybean hypocotyl tissue. Plant Physiol. *39*, 360–364 (1964)
Key, J.L., Vanderhoef, L.N.: Plant hormones and developmental regulation: role of transcription and translation. In: Developmental regulation: Aspects of cell differentiation. Coward, S.J. (ed.), pp. 49–83. New York, London: Academic Press 1973
Key, J.L., Lin, C.Y., Gifford, E.M. Jr., Dengler, R.: Relations of 2,4-D-induced growth aberrations to changes in nucleic acid metabolism in soybean seedlings. Bot. Gaz. *127*, 87–94 (1966)
Key, J.L., Barnett, N.M., Lin, C.Y.: RNA and protein biosynthesis and the regulation of cell elongation by auxin. Ann. N.Y. Acad. Sci. *144*, 49–62 (1967)
Kessler, B., Kaplan, B.: Cyclic purine mononucleotides: Induction of gibberellin biosynthesis in barley endosperm. Physiol. Plant. *27* (3) 424–431 (1972)
Khan, A.A.: Cytokinin-inhibitor antagonism in the hormonal control of α-amylase synthesis and growth in barley seed. Physiol. Plant. *22*, 94–103 (1969)
Khan, A.A.: Cytokinins: permissive role in seed germination. Science *171*, 853–859 (1971)
Khan, A.A., Downing, R.D.: Cytokinin reversal of abscisic acid inhibition of growth and α-amylase synthesis in barley seed. Physiol. Plant. *21*, 1301–1307 (1968)
Khan, A.A., Waters, E.C.: On the hormonal control of post-harvest dormancy and germination in barley seeds. Life Sci. *8*, Part II 729–736 (1969)
Khodary, S.: Studies on the turnover of cell wall polysaccharides in relation to growth in *Pisum sativum* L. Ph.D. thesis. University of Wales (1977)
Klämbt, D.: Einfluß von Auxin und Cytokinin auf die RNA Synthese in sterilem Tabakgewebe. Planta *118*, 7–16 (1974)
Klämbt, D.: Cytokinin effects on protein synthesis of in vitro systems of higher plants. Plant Cell Physiol. *17*, 73–76 (1976)
Klämbt, D.: Cytokinin and Cell Metabolism. In: Plant growth regulation. Pilet, P.E. (ed.), pp. 154–160. Berlin-Heidelberg-New York: Springer 1977
Kobayashi, S., Hatakeyama, I., Ashida, J.: Effect of auxin upon the water uptake of *Avena* coleoptiles. Bot. Mag. *69*, 16–23 (1956)
Koehler, D.E., Varner, J.E.: Hormonal control of orthophosphate incorporation into phospholipids of barley aleurone layers. Plant Physiol. *52*, 208–214 (1973)
Kovoor, A., Melet, D.: Rapidly reassociating DNA in Jerusalem artichoke rhizomes and auxin-treated explants. In: Nucleic acids and proteins in higher plants. Symp. Biol. Hung. Farkas, G.L. (ed.), Vol. XIII, pp. 29–32. Budapest: Academiai Kiado 1972
Kriedmann, P.E., Loveys, B.R., Fuller, G.L., Leopold, A.C.: Abscisic acid and stomatal regulation. Plant Physiol. *49*, 842–847 (1972)
Ku, H.S., Leopold, A.C.: Mitochondrial responses to ethylene and other hydrocarbons. Plant Physiol. *46*, 842–844 (1970)

Ku, H.S., Suge, H., Rappapart, L., Pratt, H.K.: Stimulation of rice coleoptile growth by ethylene. Planta *90*, 333–339 (1969)

Labavitch, J.M., Ray, P.M.: Turnover of cell wall polysaccharides in elongating pea segments. Plant Physiol. *53*, 669–673 (1974a)

Labavitch, J.M., Ray, P.M.: Relationship between promotion of xyloglucan metabolism and induction of elongation by indoleacetic acid. Plant Physiol. *54*, 499–502 (1974b)

Lado, P., Caldogno, F.R., Colombo, R., Marrè, E.: Regulation of pH in the plant cell wall and cell enlargement. II. Auxin-induced decrease in pH of the medium of incubation of pea internode segments. Rend. Accad. Naz. Lincei. *53*, 583–588 (1972)

Lado, P., Rasi-Caldogno, F., Pennachioni, A., Marrè, E.: Mechanism of the growth-promoting action of fusicoccin. Interaction with auxin and effects of inhibitors of respiration and protein synthesis. Planta *110*, 311–320 (1973)

Lado, P., Rasi-Caldogno, F., Colombo, R.: Promoting effect of fusicoccin on seed germination. Physiol. Plant. *31*, 149–152 (1974)

Lado, P., Rasi-Caldogno, F., Colombo, R.: Acidification of the medium associated with normal and fusicoccin-induced seed germination. Physiol. Plant. *34*, 359–364 (1975)

Lado, P., De Michelis, M.I., Cerana, R., Marrè, E.: Fusicoccin-induced, K^+-stimulated proton secretion and acid-induced growth of apical root segments. Plant Sci. Lett. *6*, 5–20 (1976)

Laidman, D.L., Colborne, A.J., Doig, R.I., Varty, K.: The multiplicity of induction systems in wheat aleurone tissue. In: Mechanisms of regulation of plant growth. Bieleski, R.L., Ferguson, A.R., Cresswell, M.M. (eds.), Vol. 12. Wellington: Bull. R. Soc. N. Z. 1974

Lamport, D.T.A.: The isolation and partial characterisation of hydroxyproline-rich glycopeptides obtained by enzymic degradation of primary cell walls. Biochem. Am. Chem. Soc. *8*, 1155–1163 (1969)

Lamport, D.T.A.: Cell wall metabolism. Annu. Rev. Plant. Physiol. *21*, 235–270 (1970)

Lamport, D.T.A., Miller, D.H.: Hydroxyproline arabinosides in the plant kingdom. Plant Physiol. *48*, 454–456 (1971)

Lancaster, J.E., Mann, J.A., Porter, N.G.: Ineffectiveness of abscisic acid in stomatal closure of yellow lupin, *Lupinus luteus* var. Weiko III. J. Exp. Bot. *28*, 184–191 (1977)

Lavee, S., Galston, A.W.: Hormonal control of peroxidase activity in cultured *Pelargonium* pith. Am. J. Bot. *55* (8) 890–893 (1968)

Lee, T.T.: Promotion of indoleacetic acid oxidase isoenzymes in tobacco callus cultures by indoleacetic acid. Plant Physiol. *48*, 56–59 (1971)

Lee, T.T.: Changes in indoleacetic acid oxidase isoenzymes in tobacco tissues after treatment with 2,4-dichlorophenoxyacetic acid. Plant Physiol. *49*, 957–960 (1972a)

Lee, T.T.: Interaction of cytokinin, auxin and gibberellin on peroxidase isoenzymes in tobacco tissue cultured in vitro. Can. J. Bot. *50*, 2471–2477 (1972b)

Leffler, H.R., O'Brien, T.J., Glover, D.V., Cherry, J.H.: Enhanced desoxyribonucleic acid polymerase activity of chromatin from soybean hypocotyls treated with 2,4-dichlorophenoxyacetic acid. Plant Physiol. *48*, 43–45 (1971)

Li, H.C., Rice, E.L., Rohrbaugh, L.M., Wender, S.H.: Effects of abscisic acid on phenolic content and lignin biosynthesis in tobacco tissue culture. Physiol. Plant. *23*, 928–936 (1970)

Libbenga, K.R., Torrey, J.G.: Hormone-induced endoreduplication prior to mitosis in cultured pea root cortex cells. Am. J. Bot. *60*, 293–299 (1973)

Little, C.H.A., Eidt, D.C.: Effect of abscisic acid on budbreak and transpiration in woody species. Nature (London) *220*, 498–499 (1968)

Livne, A., Vaadia, Y.: Stimulation of transpiration rate in barley leaves by kinetin and gibberellic acid. Physiol. Plant. *18*, 658–664 (1965)

Livne, A., Vaadia, Y.: Water deficits and hormone relations. In: Water deficits and plant growth. Kozlowski, T.T. (ed.), Vol. III, pp. 255–275. New York, London: Academic Press 1972

Lockhart, J.A.: An analysis of irreversible plant cell elongation. J. Theoret. Biol. *8*, 264–275 (1965)

Lockhart, J.A.: Physical nature of irreversible deformation of plant cells. Plant Physiol. *42*, 1545–1552 (1967)

Loescher, W.H., Nevins, D.J.: Auxin-induced changes in *Avena* coleoptile cell wall composition. Plant Physiol. *50*, 556–563 (1972)

Loescher, W.H., Nevins, D.J.: Turgor-dependant changes in *Avena* coleoptile cell wall composition. Plant Physiol. *52*, 248–251 (1973)

Lovell, P., Moore, K.: The effect of benzylaminopurine on growth and ^{14}carbon translocation in excised mustard cotyledons. Physiol. Plant. *23*, 179–186 (1970)

Luke, H.H., Freeman, T.E.: Stimulation of transpiration by cytokinins. Nature (London) *217*, 873–874 (1968)

Maas, H., Klämbt, D.: Cytokinin effect on protein synthesis in vivo in higher plants. Planta *133*, 117–120 (1977)

MacLachlan, G.A., Young, M.: Breakdown and synthesis of cell walls during growth. Nature (London) *195*, 1319–1321 (1962)

MacLachlan, G.A., Davies, E., Fan, D.F.: Induction of cellulase by 3-indoleacetic acid. In: Biochemistry and physiology of plant growth substances. Wightman, F., Setterfield, G. (eds.), pp. 443–453. Ottawa: Runge Press 1968

Madeikyte, E., Turkova, N.S.: Variations in growth direction and in nucleotide phosphate metabolism in plants under the influence of some physiologically active substances. I. Variations of the activity of adenosine-triphosphate at leaf epinasty and stem curvature in tomato and sunflower. Liet. TSR Mokslu Akad. Darb. Ser. C 37–45 (1965)

Madkour, M.A.: Studies on the role of ethylene in the control of growth and development in *Sclerotium rolfsii* Sacc., *Lycopersicon esculentum* Mill. and *Vicia faba* L. Ph.D. thesis. University of Wales (1977)

Malek, F., Baker, D.A.: Proton co-transport of sugars in phloem loading. Planta *135*, 297–299 (1977)

Mansfield, T.A.: Stomatal behaviour following treatment with auxin-like substances and phenylmercuric acetate. New Phytol. *66*, 325–330 (1967)

Mansfield, T.A., Jones, R.J.: Effects of abscisic acid on potassium uptake and starch content of stomatal guard cells. Planta *101*, 147–158 (1971)

Marrè, E.: Physiologic implications of the hormonal control of ion transport in plants. In: Plant growth regulation. Pilet, P.E. (ed.), pp. 54–66. Berlin-Heidelberg-New York: Springer 1977

Marrè, E., Forti, G.: Metabolic responses to auxin. III. The effects of auxin on ATP levels as related to the induced respiration increase. Physiol. Plant. *11*, 36–47 (1958)

Marrè, E., Lado, P., Rasi-Caldogno, F., Colombo, R.: Correlation between cell enlargement in pea internode segments and decrease in the pH of the medium of incubation. I. Effects of fusicoccin, natural and synthetic auxins and mannitol. Plant Sci. Lett. *1*, 179–184 (1973a)

Marrè, E., Lado, P., Rasi-Caldogno, F., Colombo, R.: Correlation between cell enlargement in pea internode segments and decrease in the pH of the medium of incubation II. Effects of inhibitors of respiration, oxidative phosphorylation and protein synthesis. Plant Sci. Lett. *1*, 185–192 (1973b)

Marrè, E., Colombo, R., Lado, P., Rasi-Caldogno, F.: Correlation between proton extrusion and stimulation of cell enlargement. III. Effects of fusicoccin and cytokinins on leaf fragments and isolated cotyledons. Plant Sci. Lett. *2*, 139–150 (1974a)

Marrè, E., Lado, P., Ferroni, A., Ballarin-Denti, A.: Transmembrane potential increase induced by auxin, benzyladenine and fusicoccin. Correlation with proton extrusion and cell enlargement. Plant Sci. Lett. *2*, 257–265 (1974b)

Marrè, E., Lado, P., Rasi-Caldogno, F., Colombo, R., DeMichelis, M.I.: Evidence for the coupling of proton extrusion to K^+ uptake in pea internode segments treated with fusicoccin or auxin. Plant Sci. Lett. *3*, 365–379 (1974c)

Masuda, Y.: Requirement of RNA biosynthesis for the auxin induced elongation of oat coleoptile and pea stem cells. In: Biochemistry and physiology of plant growth substances. Wightman, F., Setterfield, G. (eds.), pp. 699–710. Ottawa: Runge Press 1968a

Masuda, Y.: Role of cell-wall degrading enzymes in cell-wall loosening in oat coleoptiles. Planta *83*, 171–184 (1968b)

Masuda, Y.: Auxin-induced cell expansion in relation to cell wall extensibility. Plant Cell Physiol. *10*, 1–9 (1969)

Masuda, Y., Kamisaka, S.: Rapid stimulation of RNA biosynthesis by auxin. Plant Cell Physiol. *10*, 79–86 (1969)
Masuda, Y., Tanimoto, E.: Effect of auxin and antiauxin on the growth and RNA synthesis of etiolated pea internode. Plant Cell Physiol. *8*, 459–465 (1967)
Masuda, Y., Yamamoto, R.: Effect of auxin on $\beta 1 \rightarrow 3$ glucanase activity in *Avena* coleoptile. Dev. Growth Differ. *11*, 287–296 (1970)
Masuda, Y., Tanimoto, E., Wada, S.: Auxin stimulated RNA synthesis in oat coleoptile cells. Physiol. Plant. *20*, 713–719 (1967)
Matsumoto, T., Nishida, K., Noguchi, M., Tamaki, E.: Some factors affecting the anthocyanin formation by *Populus* cells in suspension culture. Agric. Biol. Chem. *37*, 561–567 (1973)
Matthysse, A.G., Phillips, C.: A protein intermediary in the interaction of a hormone with the genome. Proc. Natl. Acad. Sci. USA *63*, 897–930 (1969)
Matthysse, A.G., Torrey, J.G.: Nutritional requirements for polyploid mitoses in cultured pea root segments. Physiol. Plant. *20*, 661–672 (1967)
Mazelis, M.: The pyridoxal phosphate-dependent oxidative decarboxylation of methionine by peroxidase. J. Biol. Chem. *237*, 104–108 (1962)
McComb, A.J., Broughton, W.J.: Metabolic changes in internodes of dwarf pea plants treated with gibberellic acid. In: Plant growth substances 1970. Carr, D.J. (ed.), pp. 407–413. Berlin-Heidelberg-New York: Springer 1972
McComb, A.J., McComb, J.A., Duda, C.T.: Increased ribonucleic acid polymerase activity associated with chromatin from internodes of dwarf pea plants treated with gibberellic acid. Plant Physiol. *46*, 221–223 (1970)
Mehard, C.W., Lyons, J.M.: A lack of specificity for ethylene induced mitochondrial changes. Plant Physiol. *46*, 36–39 (1970)
Meidner, H.: The effect of kinetin on stomatal opening and the rate of intake of carbon dioxide in mature primary leaves of barley. J. Exp. Bot. *18*, 556–561 (1967)
Miassod, R., Penon, P., Teissière, M., Ricard, J., Cecchini, J-P.: Contribution à l'étude d'acides ribonucleiques à marquage rapide de la racine de lentille. Isolement et characterisation, action d'une hormone sur leur synthèse et sur celle de peroxydases. Biochem. Biophys. Acta *224*, 423–440 (1971)
Miller, C.O.: Control of deoxyisoflavone synthesis in soybean tissue. Planta *87*, 26–35 (1969)
Miller, C.O.: Modification of the cytokinin promotion of deoxysoflavone synthesis in soybean tissue. Plant Physiol. *49*, 310–313 (1972)
Mittelhauser, C.T., Van Steveninck, R.F.M.: Stomatal closure and inhibition of transpiration induced by RS-ABA. Nature (London) *221*, 281–282 (1969)
Mizrahi, Y., Blumenfeld, A., Bittner, S., Richmond, A.E.: Abscisic acid and cytokinin contents of leaves in relation to salinity and relative humidity. Plant Physiol. *48*, 752–755 (1971)
Momotani, Y., Kato, J.: Effect of gibberellin A_3 on in vivo and in vitro induction of α-amylase isozymes. In: Plant growth substances 1970. Carr, D.J. (ed.), pp. 352–355. Berlin-Heidelberg-New York: Springer 1972
Mondal, M.H.: Effect of gibberellic acid, calcium, kinetin and ethylene on growth and cell wall composition of pea epicotyls. Plant Physiol. *56*, 622–625 (1975)
Morré, D.J., Bracker, C.E.: Ultrastructural alteration of plant plasma membranes induced by auxin and calcium ions. Plant Physiol. *58*, 544–547 (1976)
Morré, D.J., Cherry, J.H.: Auxin hormone-plasma membrane interactions. In: Plant growth regulation. Pilet, P.E. (ed.), pp. 35–43. Berlin-Heidelberg-New York: Springer 1977
Morré, D.J., Eisinger, W.R.: Cell wall extensibility: its control by auxin and relationship to cell elongation. In: Biochemistry and physiology of plant growth substances. Wightman, F., Setterfield, G. (eds.), pp. 625–646. Ottawa: Runge Press 1968
Mothes, K., Engelbrecht, L., Kulaeva, O.N.: Über die Wirkung des Kinetins auf Stickstoffverteilung und Eiweißsynthese in isolierten Blättern. Flora (Jena) *147*, 445–464 (1959)
Moyed, H.S., Williamson, V.: Multiple 3-methyleneoxindole reductases of peas. J. Biol. Chem. *242*, 1075–1077 (1967)
Murayama, K., Ueda, K.: Short-term responses of *Pisum* stem segments to indole-3-acetic acid. Plant Cell Physiol. *14*, 973–979 (1973)

Musgrave, A., Walters, J.: Ethylene and buoyancy control rachis elongation of the semi-aquatic form *Regnillidium diphyluum*. Planta *121*, 51–56 (1974)
Musgrave, A., Jackson, M.B., Ling, E.: *Callitriche* stem elongation is controlled by ethylene and gibberellin. Nature (London) New Biol. *238*, 93–96 (1972)
Nagl, W.: Selective inhibition of cell cycle stages in the *Allium* root meristem by colchicine and growth regulators. Am. J. Bot. *59*, 346–359 (1972)
Nagl, W.: Nuclear organisation. Annu. Rev. Plant. Physiol. *27*, 39–69 (1976)
Nagl, W., Rucker, W.: Beziehungen zwischen Morphogenese und nuclearem DNS-Gehalt bei aseptischen Kulturen von *Cymbidium* nach Wuchsstoffbehandlung. Z. Pflanzenphysiol. *67*, 120–134 (1972)
Nagl, W., Hendon, J., Rücker, W.: DNA amplification in *Cymbidium* protocorms in vitro, as it relates to cytodifferentiation and hormone treatment. Cell Differ. *1*, 229–237 (1972)
Nakamura, T., Sekine, S., Arai, K., Takahashi, N.: Effects of gibberellic acid and indole-3-acetic acid on stress relation properties of pea hook cell wall. Plant Cell Physiol. *16*, 127–128 (1975)
Nance, J.F.: Effect of calcium and kinetin on growth and cell wall composition of pea epicotyls. Plant Physiol. *51*, 312–317 (1973)
Nelson, H., Ilan, I., Reinhold, L.: The effect of inhibitors of protein and RNA synthesis on auxin induced growth. Isr. J. Bot. *18*, 129–134 (1969)
Neumann, P.M.: Possible involvement of a glycerophosphate compound in auxin induced growth. Planta *99*, 56–62 (1971)
Nevins, D.J.: The effects of nojirimycin on plant growth and its implications concerning a role for exo-β-glucanases in auxin-induced cell expansion. Plant Cell Physiol. *16*, 347–356 (1975a)
Nevins, D.J.: The in vitro stimulation of IAA-induced modification of *Avena* cell wall polysaccharides by an exo-glucanase. Plant Cell Physiol. *16*, 495–503 (1975b)
Nishi, A., Kato, K., Takahashi, M., Yoshida, R.: Partial synchronization of carrot cell culture by auxin deprivation. Physiol. Plant. *39*, 9–12 (1977)
Nissl, D., Zenk, M.H.: Evidence against induction of protein synthesis during auxin-induced initial elongation of *Avena* coleoptiles. Planta *89*, 323–341 (1969)
Nooden, L.D., Thimann, K.V.: Evidence for a requirement for protein synthesis for auxin-induced cell enlargement. Proc. Natl. Acad. Sci. USA *50*, 194–200 (1963)
Northcote, D.H.: Chemistry of the plant cell wall. Annu. Rev. Plant Physiol. *23*, 113–132 (1972)
Nowakowski, W., Lubanska, G.: Wplyw IAA i GA_3 na natezenie fotosyntexy, oddychania i transpiracji siewek pszenicy w glebie. Acta Agro. Bot. *28*, 79–87 (1975)
O'Brien, T.J., Jarvis, B.C., Cherry, J.H., Hanson, J.B.: Enhancement by 2,4-dichlorophenoxyacetic acid of chromatin RNA polymerase in soybean hypocotyl tissue. Biochim. Biophys. Acta *169*, 35–43 (1968a)
O'Brien, T.J., Jarvis, B.C., Cherry, J.H., Hanson, J.B.: The effect of 2,4-dichlorophenoxyacetic acid on RNA synthesis by soybean hypocotyl chromatin. In: Biochemistry and physiology of plant growth substances. Wightman, F., Setterfield, G. (eds.), pp. 747–759. Ottawa: Runge Press 1968b
Ogawa, Y.: Effects of various factors on the increase of α-amylase activity in rice endosperm induced by gibberellin A_3. Plant Cell Physiol. *7*, 509–517 (1966)
Osborne, D.J.: Auxin and ethylene and the control of cell growth. Identification of three classes of target cells. In: Plant growth regulation. Pilet, P.E. (ed.), pp. 161–171. Berlin-Heidelberg-New York: Springer 1977
Osborne, D.J., Ridge, I., Sargent, J.A.: Ethylene and the growth of plant cells: role of peroxidase and hydroxyproline-rich proteins. In: Plant growth substances 1970. Carr, D.J. (ed.), pp. 534–542. Berlin-Heidelberg-New York: Springer 1972
Oswald, T.H., Smith, A.E., Phillips, D.V.: Callus and plantlet regeneration from cell cultures of ladino clover and soybean. Physiol. Plant. *39*, 127–134 (1977)
Paleg, L.G.: Physiological effects of gibberellic acid. I. On carbohydrate metabolism and amylase activity of barley endosperm. Plant Physiol. *35*, 293–299 (1960)
Pallas, J.E. Jr., Box, J.E. Jr. Explanation for the stomatal response of excised leaves to kinetin. Nature (London) *227*, 87–88 (1970)

Pallaghy, C.K., Raschke, K.: No stomatal response to ethylene. Plant Physiol. *49*, 275–276 (1972)

Palmer, J.: Temperature sensitivity of the latent phase in ethylene-induced elongation. Plant Physiol. *55*, 581–582 (1975)

Patterson, B.D., Trewaves, A.J.: Changes in the pattern of protein synthesis induced by 3-indolylacetic acid. Plant Physiol. *42*, 1081–1086 (1967)

Penny, D., Penny, P., Monro, J., Bailey, R.W.: Cell elongation and auxin action in lupin hypocotyls. In: Plant growth substances 1970. Carr, D.J. (ed.), pp. 52–61. Berlin-Heidelberg-New York: Springer 1972

Penny, P.: The rates of response of excised stem segments to auxins. North Am. J. Bot. *7*, 290–301 (1969)

Penny, P.: Auxin action and cell elongation: a rational approach. Ph.D. thesis. Massey University, New Zealand (1970)

Penny, P.: Growth-limiting proteins in relation to auxin-induced elongation in lupin hypocotyl. Plant Physiol. *48*, 720–723 (1971)

Penny, P., Galston, A.W.: The kinetics of inhibition of auxin-induced growth in green pea stem segments by Actinomycin-D and other substances. Am. J. Bot. *53*, 1–7 (1966)

Penny, P., Penny, D.: Rapid responses to phytohormones. In: Plant hormones and related compounds. Vol. II. Letham, D.S., Goodwin, P.B., Higgins, T.J. (eds.), pp. 537–587. North Holland: Elsevier 1978

Penny, P., Penny, D., Marshall, D., Heyes, J.K.: Early responses of excised stem segments to auxins. J. Exp. Bot. *23*, 23–36 (1972)

Penny, P., Dunlop, J., Perley, J.E., Penny, D.: pH and auxin-induced growth: a casual relationship? Plant Sci. Lett. *4*, 35–40 (1975)

Perley, J.E., Penny, D., Penny, P.: A difference between auxin-induced and hydrogen-ion-induced growth. Plant Sci. Lett. *4*, 133–136 (1975)

Phillips, M.L., Paleg, L.G.: The isolated aleurone layer. In: Proc. 7th Int. Conf. Plant Growth Subst. 1970. Carr, D. (ed.), pp. 396–406. Berlin-Heidelberg-New York: Springer 1972

Phillips, R., Torrey, J.G.: DNA synthesis, cell division and specific cell differentiation in cultured pea root cortical explants. Dev. Biol. *31*, 336–347 (1973)

Phillips, R., Torrey, J.G.: DNA levels in differentiating tracheary elements. Dev. Biol. *39*, 322–325 (1974)

Philipson, J.J., Hillman, J.R., Wilkins, M.B.: Studies on the action of abscisic acid on IAA-induced rapid growth of *Avena* coleoptile segments. Planta *114*, 87–93 (1973a)

Philipson, J.J., Hillman, J.R., Wilkins, M.B.: The effects of temperature and IAA concentration on the latent period for IAA-induced growth of *Avena* coleoptile segments. Planta *114*, 323–329 (1973b)

Pilet, P.E., Braun, R.: Ribonuclease activity and auxin effects in the lens root. Physiol. Plant *23*, 245–250 (1970)

Pirie, A., Mullins, M.G.: Changes in anthocyanin and phenolic content of grapevine leaf and fruit tissues treated with sucrose, nitrate and abscisic acid. Plant Physiol. *58*, 468–472 (1976)

Pitman, M.G., Schaefer, N., Wildes, R.A.: Relation between permeability to potassium and sodium ions and fusicoccin-stimulated hydrogen-ion efflux in barley roots. Planta *126*, 61–73 (1975)

Pitman, M.G., Anderson, W.P., Schaefer, N.: H^+ ion transport in plant roots. In: Regulation of cell membrane activities in plants. Marrè, E., Ciferri, O. (eds.), pp. 147–160. North Holland: Elsevier 1977

Pope, D., Black, M.: The effect of indole-3-acetic acid on coleoptile extension growth in the absence of protein synthesis. Planta *102*, 26–36 (1972)

Pope, D.G., Osborne, M., Abrams, J.: The proton pump theory of IAA-action a reappraisal. Plant Physiol. *56*, Suppl. Pap. No. 2 (1975)

Prins, H.B.A., Harper, J.R., Higinbotham, N., Cleland, R.E.: Kinetics of rapid hormone-induced changes in the membrane potential of oat coleoptile cells. Plant Physiol. Suppl. *57*, 3 (1976)

Puech, A.A., Reberz, C.A., Crane, J.C.: Pigment changes associated with application of

ethephon (2-chloroethyl phosphoric acid) to fig (*Ficus carica* L.) fruits. Plant Physiol. *57*, 504–509 (1976)
Ranjeva, R., Boudet, A.M., Harada, H., Marigo, G.: Phenolic metabolism in petunia tissues. I. Characteristic responses of enzymes involved in different steps of polyphenol synthesis to different hormonal influences. Biochim. Biophys. Acta *399*, 23–30 (1975)
Raschke, K.: Abscisic acid sensitizes stomata to CO_2 in leaves of *Xanthium strumarium* L. In: Plant growth substances 1973. Proc. 7th Int. Conf. Plant Growth Subst. pp. 1151–1158. Tokyo: Hirokawa 1974
Raschke, K.: Stomatal action. Annu. Rev. Plant Phys. *26*, 309–340 (1975)
Raschke, K.: The stomatal turgor mechanism and its reponses to CO_2 and abscisic acid: Observations and a hypothesis. In: Regulation of cell membrane activities in plants. Marrè, E., Ciferri, O. (eds.), pp. 173–183. North Holland: Elsevier 1977
Raven, J.A., Smith, F.A.: Characteristics, functions and regulation of active proton extrusion. In: Regulation of cell membrane activities in plants. Marrè, E., Cifferi, O. (eds.), pp. 25–40. North Holland: Elsevier 1977
Ray, P.M.: Hormonal regulation of plant cell growth. In: Control mechanisms in cellular processes. Bonner, D.M. (ed.), pp. 185–212. New York: Ronald 1961
Ray, P.M.: Cell wall synthesis and cell elongation in oat coleoptile tissue. Am. J. Bot. *49*, 928–939 (1962)
Ray, P.M.: The action of auxin on cell enlargement in plants. In: Communication in development. Lang, A. (ed.), Dev. Biol. Suppl. Vol. III, pp. 172–205. New York, London: Academic Press 1969
Ray, P.M.: Regulation of β-glucan synthetase activity by auxin in pea stem tissue. I. Kinetic aspects. Plant Physiol. *51*, 601–608 (1973a)
Ray, P.M.: Regulation of β-glucan synthetase activity, by auxin in pea stem tissue. II. Metabolic requirements. Plant Physiol. *51*, 609–614 (1973b)
Ray, P.M.: Auxin binding sites of maize coleoptiles are localised on membranes of the endoplasmic reticulum. Plant Physiol. *59*, 594–599 (1977)
Ray, P.M., Ruesink, A.W.: Kinetic experiments on the nature of the growth mechanism in oat coleoptile cells. Dev. Biol. *4*, 377–397 (1962)
Ray, P.M., Ruesink, A.W.: Osmotic behaviour of oat coleoptile tissue in relation to growth. J. Gen. Physiol. *47*, 83–101 (1963)
Ray, P.M., Green, P.B., Cleland, R.: Role of turgor in plant cell growth. Nature (London) *239*, 163–164 (1972)
Ray, P.M., Dohrmann, U., Hertel, R.: Characterization of naphthalenacetic acid binding to receptor sites on cellular membranes of maize coleoptile tissue. Plant Physiol. *59*, 357–364 (1977)
Rayle, D.L.: Auxin-induced hydrogen-ion secretion in *Avena* coleoptiles and its implications. Planta *114*, 63–73 (1973)
Rayle, D.L., Cleland, R.: Enhancement of wall loosening and elongation by acid solutions. Plant Physiol. *46*, 250–253 (1970)
Rayle, D.L., Cleland, R.: The in vitro acid-growth response: relation to in vivo growth responses and auxin action. Planta *104*, 282–296 (1972)
Rayle, D.L., Johnson, K.D.: Direct evidence that auxin-induced growth is related to hydrogen ion secretion. Plant Physiol. Suppl. *51*, 2 (1975)
Rayle, D., Zenk, M.H.: Cell extension growth: Some recent advances. Biochem. Soc. Symp. *38*, 235–246 (1973)
Rayle, D.L., Evans, M.L., Hertel, R.: Action of auxin on cell elongation. Proc. Natl. Acad. Sci. USA *65*, 184–191 (1970a)
Rayle, D.L., Haughton, P.M., Cleland, R.: An in vitro system that stimulates plant cell extension growth. Proc. Natl. Acad. Sci. USA *67*, 1814–1817 (1970b)
Rehm, M.M., Cline, M.G.: Rapid growth inhibition of *Avena* coleoptile segments by abscisic acid. Plant Physiol. *51*, 93–96 (1973)
Ricardo, C.P.P.: Effects of sugars, gibberellic acid and kinetin on acid invertase of developing carrot roots. Phytochemistry *15*, 615–617 (1976)
Ridge, I.: The control of cell shape and rate of cell expansion by ethylene: effects on microfibrillar orientation and cell wall extensibility in etiolated peas. Acta Bot. Neerl. *22*, 144–158 (1973)

Ridge, I., Osborne, D.J.: Cell growth and cellulases: regulation by ethylene and indole-3-acetic acid in shoots of *Pisum sativum*. Nature (London) *223*, 318–319 (1969)

Ridge, I., Osborne, D.J.: Hydroxyproline and peroxidases in cell walls of *Pisum sativum:* regulation by ethylene. J. Exp. Bot. *21*, 843–856 (1970)

Ridge, I., Osborne, D.J.: Role of peroxidase when hydroxyproline-rich protein in plant cell walls is increased by ethylene. Nature (London) New Biol. *229*, 205–208 (1971)

Rijven, A.H.G.C., Parkash, V.: Action of kinetin on cotyledons of fenugreek. Plant Physiol. *47*, 56–64 (1971)

Ritzert, R.W., Turin, B.A.: Formation of peroxidases in response to indole-3-acetic acid in cultured tobacco cells. Phytochemistry *9*, 1701–1705 (1970)

Roberts, L.W.: Cytodifferentiation in plants; Xylogenesis as a model system: Cambridge, London, New York, Melbourne: Cambridge University Press 1976

Roberts, L.W., Baba, S.: Effect of proline on wound vessel member formation. Plant Cell Physiol. *9*, 353–360 (1968)

Rose, R.J.: Differential effect of cycloheximide on short term gibberellin and auxin growth kinetics of gamma coleoptiles. Plant Sci. Lett. *2*, 233–237 (1974)

Rowan, K.S., Gillbank, L.R., Spring, A.H.: Effects of IAA and cyanide on the growth and respiration of coleoptile sections from *Triticum*. In: Plant growth substances 1970. Carr, D.J. (ed.), pp. 76–81. Berlin-Heidelberg-New York: Springer 1972

Roychoudhury, R., Datta, A., Sen, P.: The mechanism of action of plant growth substances: The role of nuclear DNA in growth substance action. Biochim. Biophys. Acta *107*, 346–351 (1965)

Sacher, J.A.: Senescence and postharvest physiology. Annu. Rev. Plant Phys. *24*, 197–224 (1973)

Sacher, J.A., Salminen, S.O.: Comparative studies of effects of auxin and ethylene on permeability and synthesis of RNA and protein. Plant Physiol. *44*, 1371–1377 (1969)

Sadava, D., Chrispeels, M.J.: Hydroxyproline-rich cell wall protein (Extensin): Role in the cessation of elongation in excised pea epicotyls. Dev. Biol. *30*, 49–55 (1973)

Sato, T., Watanabe, A., Imaseki, H.: Effect of ethylene on DNA synthesis in potato tuber discs. Plant Cell Physiol. *17*, 1255–1262 (1976)

Schrank, A.R.: Ethionine inhibition of elongation and geotropic curvature of *Avena* coleoptiles. Arch. Biochim. Biophys. *61*, 348–355 (1956)

Seth, A.K., Wareing, P.F.: Hormone-directed transport of metabolites and its possible role in plant senescence. J. Exp. Bot. *18*, 65–77 (1967)

Shah, R.R., Subbaiah, K.V., Mehta, A.R.: Hormonal effect on polyphenol accumulation in *Cassia* tissues cultured in vitro. Can. J. Bot. *54*, 1240–1245 (1976)

Silk, W.K., Jones, R.L.: Gibberellin response in lettuce hypocotyl sections. Plant Physiol. *56*, 267–272 (1975)

Simpson, S.F., Torrey, J.G.: Hormonal control of deoxyribonucleic acid and protein syntheses in pea root cortical explants. Plant Physiol. *59*, 4–9 (1977)

Skoog, F., Miller, C.O.: Chemical regulation of growth and organ formation in plant tissues cultured in vitro. Symp. Soc. Exp. Biol. *11*, 118–131 (1957)

Skoog, F., Schmitz, R.Y., Bock, R.M., Hecht, S.M.: Cytokinin antagonists: Synthesis and physiological effects of 7-substituted 3-methylpyrazolo (4,3-d) pyrimidines. Phytochemistry *12*, 25–37 (1973)

Sloane, M., Sadava, D.: Auxin, carbon dioxide and hydrogen ions. Nature (London) *257*, 68 (1975)

Smith, C.J.: Studies on cell wall structure and metabolism in *Zea mays* L. Ph.D. thesis. University of Wales (1976)

Smith, E.J.: The hormonal control of β-glucan metabolism in cell walls of *Zea mays* and *Avena sativa*. Ph.D. thesis. University of Wales (1976)

Smith, K.A., Dowdell, R.J.: Field studies of the soil atmosphere. II. Occurrence of nitrous oxide. J. Soil Sci. *25*, 231–238 (1974)

Smith, K.A., Robertson, P.D. Effect of ethylene on root extension of cereals. Nature (Lond.) *234*, 148–149 (1971)

Spencer, F.S., Ziola, B., Maclochlan, G.A.: Particulate glucan synthetase activity: dependence on acceptor, activator and plant growth hormones. Can. J. Biol. Chem. *49*, 1326–1332 (1971)

Squire, G.R., Mansfield, T.A.: Studies of the mechanism of action of fusicoccin, the fungal toxin that induces wilting and its interaction with abscisic acid. Planta *105*, 71–78 (1972)
Squire, G.R., Mansfield, T.A.: The action of fusicoccin on stomatal guard cells and subsidiary cells. New Phytol. *73*, 433–440 (1974)
Stewart, E.R., Freebairn, H.T.: Ethylene, seed germination and epinasty. Plant Physiol. *44*, 955–958 (1969)
Stuart, D.A., Jones, R.L.: The roles of extensibility and turgour in gibberellin and dark stimulated growth. Plant Physiol. *59*, 61–68 (1977)
Stuart, D.A., Jones, R.L.: The role of acidification in gibberellic acid- and fusicoccin-induced elongation growth in lettuce *hypocotyl* sections. Planta *142*, 135–146 (1978)
Suge, H., Katsura, N., Inada, K.: Ethylene-light relationship in the growth of the rice coleoptile. Planta *101*, 365–368 (1971)
Sussex, I., Clutter, M., Walbot, V.: Benzyladenine reversal of abscisic acid inhibition of growth and RNA synthesis in germinating bean axes. Plant Physiol. *56*, 575–578 (1975)
Taiz, L., Honigman, W.A.: Production of cell wall hydrolyzing enzymes by barley aleurone layers in response to gibberellic acid. Plant Physiol. *58*, 380–386 (1976)
Taiz, L., Jones, R.L.: Gibberellic acid, β-1,3-glucanase and the cell wall of barley aleurone layers. Planta *92*, 73–84 (1970)
Tal, M., Imber, D.: Abnormal stomatal behaviour and hormonal imbalance in *flacca*, a wilty mutant of tomato. II. Auxin and abscisic acid activity. Plant Physiol. *46*, 373–376 (1970)
Tal, M., Imber, D.: Abnormal stomatal behaviour and hormonal imbalance in *flacca*, a wilty mutant of tomato. III. Hormonal effects on the water status in the plant. Plant Physiol. *47*, 849–850 (1971)
Tal, M., Imber, D.: The effect of abscisic acid on stomatal behaviour in *flacca*, a wilty mutant of tomato, in darkness. New Phytol. *71*, 81–84 (1972)
Tal, M., Imber, D., Gardi, I.: Abnormal stomatal behaviour and hormonal imbalance in *flacca*, a wilty mutant of tomato. IV. Effect of abscisic acid and auxin on stomatal behaviour and peroxidase activity. J. Exp. Bot. *25*, 51–60 (1974)
Talmadge, K.W., Keegstra, K., Bouer, W.D., Albersheim, P.: The structure of plant cell walls. I. The macromolecular components of the walls of suspension-cultured sycamore cells with a detailed analysis of the pectic polysaccharides. Plant Physiol. *51*, 158–173 (1973)
Taneja, S.R., Sacher, R.C.: Stimulation of polyphenol oxidase (monophenolase) activity in wheat endosperm by gibberellic acid, cycloheximide and actinomycin D. Planta *116*, 133–142 (1974)
Tanimoto, E., Masuda, Y.: Effect of auxin on cell wall degrading enzymes. Physiol. Plant. *21*, 820–826 (1968)
Tanimoto, E., Masuda, Y.: Auxin-induced synthesis of TB-RNA in pea stems. Plant Cell Physiol. *10*, 485–489 (1969)
Tanner, W., Komor, E., Fenzl, F., Decker, M.: Sugar-proton cotransport systems. In: Regulation of cell membrane activities in plants. Marrè, E., Ciferri, O. (eds.), pp. 79–90. Amsterdam: Elsevier/North Holland Biomedical Press 1977
Tao, K.L., Khan, A.A.: Occurrence of some enzymes in starchy endosperm and hormonal regulation of isoperoxidase in aleurone of wheat. Plant Physiol. *56*, 797–800 (1975)
Tautvydas, K.I., Galston, A.W.: Binding of indoleacetic acid to isolated pea nuclei. In: Plant growth substances 1970. Carr, D.J. (ed.), pp. 256–264. Berlin-Heidelberg-New York: Springer 1972
Tavener, R.J.A., Laidman, D.L.: The induction of triglyceride metabolism in the germinating wheat grain. Phytochemistry *11*, 981–987 (1972a)
Tavener, R.J.A., Laidman, D.L.: The induction of lipase activity in the germinating wheat grain. Phytochemistry *11*, 989–997 (1972b)
Thimann, K.V.: The physiology of growth in plant tissues. Am. Sci. *42*, 589–606 (1954)
Thompson, W.F., Cleland, R.: Auxin and ribonucleic acid synthesis in pea stem studied by deoxyribonucleic acid-ribonucleic acid hybridization. Plant Physiol. *48*, 663–670 (1971)
Timmis, J., Ingle, J.: Quantitative regulation of gene activity in plants. Plant Physiol. *56*, 255–258 (1975)

Torrey, J.G., Fosket, D.E.: Cell division in relation to cytodifferentiation in cultured pea root segments. Am. J. Bot. *57*, 1072–1080 (1970)
Torrey, J.G., Fosket, D.E., Hepler, P.K.: Xylem formation: a paradigm of cytodifferentiation in higher plants. Am. Sci. *59*, 338–352 (1971)
Trewavas, A.J.: Effect of IAA on RNA and protein synthesis. Arch. Biochem. Biophys. *123*, 324–335 (1968a)
Trewavas, A.J.: The effect of 3-indolylacetic acid on the levels of polysomes in etiolated pea tissue. Phytochemistry *7*, 673–681 (1968b)
Tucker, D.J., Mansfield, T.A.: A simple bioassay for detecting "Antitranspirant" activity of naturally occurring compounds such as abscisic acid. Planta *98*, 157–163 (1971)
Tuli, V., Moyed, H.S.: Desensitization of regulatory enzymes by a metabolite of plant auxin. J. Biol. Chem. *241*, 4564–4566 (1966)
Tuli, V., Moyed, H.S.: The oxindole pathway of 3-indoleacetic acid metabolism and the action of auxins. In: Biochemistry and physiology of plant growth substances. Wightman, F., Setterfield, G. (eds.), pp. 289–300. Ottawa: Runge Press 1969
Uchimiya, H., Murashige, T.: Influence of the nutrient medium on the recovery of dividing cells from tobacco protoplasts. Plant Physiol. *57*, 424–429 (1976)
Uhrström, I.: Time effects of auxin and calcium on growth and elastic modulus in hypocotyls. Physiol. Plant *22*, 271–287 (1969)
Vanderhoef, L.N., Stahl, C.A.: Separation of two responses to auxin by means of cytokinin inhibition. Proc. Natl. Acad. Sci. USA *72*, 1822–1825 (1975)
Vanderhoef, L.N., Stahl, C.A., Siegel, N., Ziegler, R.: The inhibition by cytokinin of auxin-promoted elongation in excised soya bean hypocotyl. Physiol. Plant. *29*, 22–27 (1973)
Vanderhoef, L.N., Stahl, C.A., Williams, C.A., Brinkman, K.A., Greenfield, J.C.: Additional evidence for separable responses to auxin in soya bean hypocotyl. Plant Physiol. *57*, 817–819 (1976)
Vanderhoef, L.N., Findley, J.S., Burke, J.J., Blizzard, W.E.: Auxin has no effect on modification of external pH by soybean hypocotyl cells. Plant Physiol. *59*, 1000–1003 (1977)
Van der Woude, W.J., Lembi, C.A., Morré, D.J.: Auxin (2,4D) stimulation in vivo and in vitro of polysaccharide synthesis in plasma membrane fractions isolated from onion stems. Biochem. Biophys. Res. Commun. *46*, 245–253 (1972)
Varner, J.E., Chandra, G.R.: Hormonal control of enzyme synthesis in barley endosperm. Proc. Natl. Acad. Sci. USA *52*, 100–106 (1964)
Varner, J.E., Johri, M.M.: Hormonal control of enzyme synthesis. In: Biochemistry and physiology of plant growth substances, Wightman, F., Setterfield, G. (eds.), pp. 793–814. Ottawa: Runge Press 1968
Venis, M.A.: Stimulation of RNA transcription from pea and corn DNA by protein retained on sepharose coupled to 2,4-dichlorophenoxyacetic acid. Proc. Natl. Acad. Sci. USA *68*, 1824–1827 (1971)
Verma, D.C., Tavares, J., Loewus, F.A.: Effect of benzyladenine, 2,4-dichlorophenoxyacetic acid and *β*-glucose on myo-inositol metabolism in *Acer pseudoplatanus* L. cell growth in suspension culture. Plant Physiol. *57*, 241–244 (1976)
Verma, D.P.S., Maclachlan, G.A., Byrne, H., Ewings, D.: Regulation and in vitro translation of messenger ribonucleic acid for cellulase from auxin-treated pea epicotyls. J. Biol. Chem. *250*, 1019–1026 (1975)
Vigil, E.L., Ruddat, M.: Effect of gibberellic acid and actinomycin D on the formation and distribution of rough endoplasmic reticulum in barley aleurone cells. Plant Physiol. *51*, 549–558 (1973)
Vitagliano, C.: Effects of ethephon on stomata, ethylene evolution and abscission in olive (*Olea europaea* L.) c.v. coratina. J. Am. Soc. Hortic. Sci. *100*, 482–485 (1975)
Walbot, V., Clutter, M., Sussex, I.: Effect of abscisic acid on growth, RNA metabolism and respiration in germinating bean axes. Plant Physiol. *56*, 570–574 (1975)
Walton, D.C., Soofi, G.S., Sondheimer, E.: The effect of abscisic acid on growth and nucleic acid synthesis in excised embryonic bean axes. Plant Physiol. *45*, 37–40 (1970)
Wardell, W.L.: Rapid initiation of thymidine incorporation into deoxyribonucleic acid in vegetative tobacus stem segments treated with indole-3-acetic acid. Plant Physiol. *56*, 171–176 (1975)
Wareing, P.F.: Some aspects of differentiation in plants. In: Control mechanisms of growth

and differentiation. Davies, D.D., Balls, M.J. (eds.), pp. 323–344. London: Cambridge University Press 1971
Warner, H.L., Leopold, A.C.: Timing of growth regulator responses in peas. Biochem. Biophys. Res. Commun. *44*, 989–994 (1971)
Wetmore, R.H., Rier, J.P.: Experimental induction of vascular tissues in callus of angiosperms. Am. J. Bot. *50*, 418–430 (1963)
Wetmore, R.H., De Maggio, M., Rier, J.P.: Contemporary outlook on the differentiation of vascular tissues. Phytomorphology *14*, 203–217 (1964)
Winter, H., Spaander, S., Wiersema, P.K.: The relationship between cellulose synthesis and an unidentified glucose-complex in pea stem segments. Acta. Bot. Neerl. *19*, 525–532 (1970)
Wong, Y-S., Fincher, G.B., Maclachlan, G.A.: Cellulase can enhance β-glucan syntheses. Science *195*, 679–681 (1977a)
Wong, Y-S., Fincher, G.B., Maclachlan, G.A.: Kinetic properties and substrate specificities of two cellulases from auxin-treated pea epicotyls. J. Biol. Chem. *252*, 1402–1407 (1977b)
Wright, S.T.C.: Multiple and sequential roles of plant growth regulators. In: Biochemistry and physiology of plant growth substances. Wightman, F., Setterfield, G. (eds.), pp. 521–542. Ottawa: Runge Press 1968
Wright, S.T.C.: An increase in the "Inhibitor β" content of detached wheat leaves following a period of wilting. Planta *86*, 10–20 (1969)
Wright, S.T.C., Hiron, R.W.P.: (+)-Abscisic acid, the growth inhibitor induced in detached wheat leaves by a period of wilting. Nature (London) *224*, 710–720 (1969)
Wyen, N.V., Uolvardy, J., Erdei, S., Farkas, G.L.: Hormonal control of nuclease level in *Avena* leaf tissues. In: Nucleic acids and proteins in higher plants. Symp. Biol. Hung. Farkas, G.L. (ed.), Vol. XIII, pp. 293–297. Budapest: Academai Kiads 1972
Yamagata, Y., Yamamoto, R., Masuda, Y.: Auxin and hydrogen ion actions on light-grown pea epicotyl segments. II. Effects of hydrogen ions on extension of the isolated epidermis. Plant Cell Physiol. *15*, 833–841 (1975)
Yamaki, T.: Sci. Pap. Coll. Gen. Educ. Univ. Tokyo *4*, 127–154 (1954)
Yanagishima, N., Shimoda, C.: Auxin and yeast. Bot. Rev. *39*, 1–14 1973
Yeoman, M.M.: Division synchrony in cultured cells. In: Tissue culture and plant science. Street H.E. (ed.), pp. 1–19. New York-London: Academic Press 1974
Yoda, S., Ashida, J.: Effect of auxin on the osmotic value in pea stem sections. Nature (London) *192*, 577–578 (1961)
Yomo, H.: Studies on the amylase activating substance V. Purification of the amylase activating substance in barley malt. Hakko Kyokaishi *18*, 603–606 (1960)
Yomo, H.: Studies on the amylase activating substance V. Purification of the amylase activating substance in barley malt. Hakko Kyokaishi *18*, 603–606 (1960). Cited in: Chem. Abst. *55*, 26–45 (1961)
Zelitch, I.: Biochemical control for stomatal opening in leaves. Proc. Natl. Acad. Sci. USA *47*, 1423–1433 (1961)
Zelitch, I.: The control and mechanisms of stomatal movement. Can. Agric. Exp. Stan. New Haven Bull. *664*, 18–36 (1963)
Zimmermann, U., Steudle, E.: Action of indoleacetic acid on membrane structure and transport. In: Regulation of cell membrane activities in plants. Marrè, E., Ciferri, O. (eds.), pp. 231–242. North Holland: Elsevier 1977
Zimmermann, U., Steudle, E., Lelkes, P.I.: Turgor pressure regulation in *Valonia utricularis*: Effect of cell wall elasticity and auxin. Plant Physiol. *58*, 608–613 (1976)
Zimmermann, U., Ashcroft, R.G., Coster, H.G.L., Smith, J.R.: The molecular organisation of bimolecular lipid membranes: The effect of KCl on the location of indole acetic acid in the membrane. Biochim. Biophys. Acta *469*, 23–32 (1977a)
Zimmermann, U., Beckers, F., Steudle, E.: Turgor sensing in plant cells by the electromechanical properties of the membrane. In: Transmembrane ionic exchange in plants. Thellier, M., Dainty, J. (eds.). CNRS 1977b
Zwar, J.A., Jacobsen, J.V.: A correlation between a ribonucleic acid fraction selectively labelled in the presence of gibberellic acid and amylase synthesis in barley aleurone layers. Plant Physiol. *49*, 1000–1006 (1972)

Author Index

Page numbers in *italics* refer to the references

Aasheim T, Iversen TH 336, *390*
Aasheim T, see Iversen TH 336, *411*
Abdul-Baki AA, Ray PM 520, *568*
Abdullaev ND, see Tarasov VA 77, *108*
Abe H, Uchiyama M, Sato R 13, 21, *80*, 133, 142, *190*
Abe H, Uchiyama H, Tanaka Y, Saito H 370, *390*
Abe H, see Marumo S 16, 20, *98*, 115, 130, 133, 138, 142, *197*, 289, *421*
Abe H, see Obata Y 370, *426*
Abe H, see Tanaka Y 43, *108*
Abeles AL, Abeles FB 319, *390*
Abeles FB 10, *80*, 317, 318, 319, 323, 325, 326, 328, 330, 331, 332, 333, 334, 335, 389, *390*, 499, *501*, 540, *568*
Abeles FB, Leather GR 541, *568*
Abeles FB, Lonski J 331, *390*
Abeles FB, Rubinstein B 11, *80*, 333, *390*
Abeles FB, Ruth JM, Forrence LE, Leather GR 362, *390*
Abeles FB, see Abeles AL 319, *390*
Abeles FB, see Craker LE 331, *398*
Abeles FB, see Gahagan HE 347, *404*
Abrahamson S, see Stenhagen E 250, *279*
Abrams J, see Pope DG 523, *581*
Abrams M, see Matthysse AG 474, 478, *506*
Acaster MA, see Bengochea T 500, *502*
Acaster MA, see Hall MA 541, *573*
Adam G, Lischewski M, Sych F-J, Ulrich A 377, 378, *391*
Adam G, see Lischewski M 377, *418*
Adam G, see Sembdner G 363, 376, 381, *434*
Adamiec A 315, *391*
Adams DO, Yang SF 320, 321, 325, 326, *391*
Adams DO, see Yu Y 326, 333, *443*, *444*
Adams G, see Rappaport L 302, 387, 389, *430*
Adams PA, Montague MJ, Tepfer M, Rayle DL, Ikuma H, Kaufman PB 530, *568*
Adams PA, see Kaufman PB 523, *575*
Adamson D, see Adamson H 516, *568*
Adamson H, Adamson D 516, *568*
Adato I, Gazit S 327, 330, 331, 389, *391*
Addicott FT 295, *391*
Addicott FT, Carns HR, Cornforth JW, Lyon JL, Milborrow BV, Ohkuma K, Ryback G, Smith OE, Thiessen WE, Wareing PF 48, *80*
Addicott FT, Carns HR, Lyon JL, Smith OE, McMeans JL 244, *276*
Addicott FT, Lyon JL 52, *80*, 295, *391*
Addicott FT, see Davis LA 52, *86*, 131, 174, *192*, 244, *276*
Addicott FT, see Geissman TA 77, *89*
Addicott FT, see Ohkuma K 48, 49, *101*, 131, 132, 143, 154, 155, *198*, 295, *426*
Adesomoju AA 23, 24, 25, 26, 27, 29, *80*
Adesomoju AA, Okogun JI, Ekong DEU, Gaskin P 352, *391*
Adesomoju AA, Okugun JI, Ekong DEU, Gaskin P 50, 51, *81*
Adolf W, see Dueber MT 317, *400*
Aducci P, Crosetti G, Federico R, Ballio A 6
Agatep AO, see Baldev B 316, *392*
Aharoni N, Blumenfeld A, Richmond AE 386, *391*
Aharoni Y 327, *391*
Aizawa A, see Kato T 54, 56, 67, 78, *94*
Ajami A, see Cleland C 61, *84*
Akamine EK 11, *81*
Akinosi N, see Nakagawa S 186, *197*
Alam M, see Berrie AMM *82*
Alam SN, Hall RH 173, *190*
Alam SN, see Dyson WH 44, 45, *87*
Alam SN, see Hall RH 350, 372, *407*
Albersheim P, see Keegstra K 526, 527, 529, *576*
Albersheim P, see Talmadge KW 526, *584*
Albert LS, see Bohnsack CW 346, *395*
Alberts BM, see Yamamoto KR 485, *510*
Albone KS, see Rivier L 4, *7*
Albrecht J, see Van Sumere CF 62, *109*
Allen JRF, Greenway AM, Baker DA 13, *81*, 186, *190*
Allerup J 541, *568*
Alleweldt G, see Düring H 374, 375, *401*
Almeida MS, see Evans LS 76, *88*

Almeida S, see Lynn DG 76, *97*
Aloni B, see Ben-Yehoshua S 331, *394*
Alpi A, Lorenzi R, Cionini PG, Bennicia A, D'Amato F 3, 5, *6*
Alpi A, Tognoni F, D'Amato F 388, *391*
Alpi A, see Ceccarelli N 5, *6*, 305, 388, *396*
Alpi A, see Cionini PG 388, *397*
Altree-Williams S, Howden MEH, Keegan JT, Malcolm HDR, Wyllie SG 60, *81*
Alves LM, Ruddat M 27, *81*
Amatangelo SJ, see Hebard FV 530, *574*
Andersen AS, Møller JB, Hansen J 337, *391*, 533, *568*
Andersen CR, see Carnes MG 176, 178, 180, *191*
Anderson JD, Mandava N, Garrett S 335, *391*
Anderson JD, Moore TC 305, *391*
Anderson JD, see Baker JE 322, 325, 328, *392*
Anderson JD, see Mandava N 76, *97*
Anderson JD, see Moore TC 316, *424*
Anderson JD, see Reed DJ 316, *430*
Anderson RL, Ray PM 452, *501*
Anderson WP, see Pitman MG 538, *581*
Andersson B, Haggstrom N, Andersson K 49, 50, *81*
Andersson K, see Andersson B 49, 50, *81*
Ando T, Tsukamoto Y 79, *81*
Andreae WA 365, *391*
Andreae WA, Good NE 364, 365, *391*
Andreae WA, van Ysselstein MWH 364, 365, *391*
Anetai M, see Takasugu M 51, 53, 54, *108*
Anker L 287, *391*
Anstis PJP, Friend J, Gardner DCJ 55, *81*, 300, *391*
Anthony DS, see Brook JL 169, *191*
Antoszewski R, Rudnicki R 168, *190*
Antoszewski R, see Runkova LV 344, 347, *432*
Anzai T, see Shibaoka H 77, *106*
Aoki H, see Crozier A 138, *192*
Aoyama T, see Murofushi N 27, *100*, 113, 114, 142, 161, *197*
Aoyama T, see Takahashi N 184, *200*
Apelbaum A, Sfakiotakis E, Dilley DR 557, *568*
Arai K, see Nakamura T 530, *580*
Arakawa H, see Nakazaki M 78, *100*
Arias I, Williams PW, Bradbeer JW 362, 388, *392*
Arigoni D, see Eberle M 293, *401*
Arima K, see Matsushima H 60, 75, *98*, 340, *421*
Armstrong DJ, Burrows WJ, Evans PK, Skoog F 139, 140, *190*, 348, *392*
Armstrong DJ, Burrows WJ, Skoog F, Roy KL, Söll D 472, *501*
Armstrong DJ, Evans PK, Burrows WJ, Skoog F, Petit J-F, Dahl JL, Steward T, Stominger JL, Leonard NJ, Hecht SM, Occolowitz J 43, *81*
Armstrong DJ, Murai N, Taller BJ, Skoog F 293, *392*, 531, *568*
Armstrong DJ, Scarbrough E, Skoog F, Cole DL, Leonard NJ 40, 43, *81*, 292, 293, 348, *392*
Armstrong DJ, see Burrows WJ 41, 43, 44, 45, *84*, 137, 168, *191*, 292, 348, *396*
Armstrong DJ, see Hecht SM 348, *409*
Armstrong DJ, see Murai N 290, *425*
Armstrong DJ, see Scarbrough E 40, *105*, 292, *433*
Armstrong DJ, see Skoog F 40, 44, *106*, 289, 292, 293, *436*, 470, 471, 472, *508*
Armstrong DJ, see Struxness LA 385, *438*
Armstrong DJ, see Walker GC 293, *441*, 474, *510*
Armstrong JE, Jones RL 491, *501*
Arora SK, Bates RB, Chou PCC, Sanchez WE, Brown KS, Galbraith MN 78, *81*
Arpino PJ, Guiochon G *276*
Arrigoni O, see De Leo P 75, *86*
Arsenault GP, see McInnes AG 26, *98*, 355, 376, 378, *422*
Arsenault GP, see Pitel DW 134, 136, 161, *198*, 354, 355, *428*
Asahira T, see Hayashi H 76, *91*
Asakawa Y, Tamari K, Inoue K, Kaji J 377, *392*
Asakawa Y, Tamari K, Shoji A, Kaji J 125, *190*, 377, 378, *392*
Asen S, Cathey HM, Stuart NW 70, *81*
Ashcroft RG, see Zimmermann U 539, 540, *586*
Ashford AE, Jacobsen JV 551, *568*
Ashida J, see Kobayashi S 516, *576*
Ashida J, see Yoda S 530, *586*
Ashton FM 561, *568*
Asmundson CM, see Shindy WW 354, *435*
Aso K, see Yabuta T 22, *111*
Assante G, Merlini L, Nasini G 52, *81*
Atkinson JG, Russell AA, Stuart RS 500, *501*
Atkinson MR, Eckermann G 71, *81*
Atmon D, see Friedländer M 354, *403*
Atsumi S, Kuraishi S, Hayashi T 120, *190*
Audus LJ 448, *501*
Audus LJ, Thresh R 241, *276*
Augier H 9, *81*
Aung LH, Rees AR 362, *392*
Aurich O, see Sembdner G 377, 378, 379, 381, 382, *434*
Avers CJ, Goodwin RH 67, *81*

Avery GS, see Berger J 13, *82*
Azou Y, see Teisseve M 454, *509*

Baba S, see Roberts LW 564, *583*
Babcock DF, Morris RO 41, *81*, 173, *190*, 348, *392*
Baeu RN, see Roy BN 61, *104*
Bagni N 76, *81*
Bagni N, Serafini Fracassini D 67, *81*
Bagni N, see Bertossi F 76, *82*
Baguley BC, see Staehelin M 43, *106*
Bailey RW, see Penny D 451, *507*, 536, *581*
Baker DA, see Allen JRF 13, *81*, 186, *190*
Baker DA, see Malek F 544, *578*
Baker DB, Ray PM 524, 525, 526, 536, 544, *568*
Baker JE, Lieberman M, Anderson JD 322, 325, 328, *392*
Baker JE, see Lieberman M 333, 334, *418*
Baker LR, see Hemphill DD 23, 24, 25, 36, *92*, 376, 378, 381, 382, *409*
Baker LR, see Rudich J 356, *432*
Bakker HJ, Cook IF, Jefferies PR, Knox JR 311, *392*
Bakker HJ, Jefferies PR, Knox JR 311, *392*
Baksi, see Laloraya MM 345, *415*
Bal AK, Verma DPS, Byrne H, Maclachlan GA 563, *568*
Balboa-Zavala O, Dennis FG 49, *81*
Baldev B, Lang A, Agatep AO 316, *392*
Ballarin-Denti A, see Cocucci M 538, *571*
Ballarin-Denti A, see Marré E 531, 534, 537, 539, *578*
Ballio A 464, 465, *501*
Ballio A, Chain EB, De Leo P, Erlanger BF, Mauri M, Totolo A 533, *568*
Ballio A, see Aducci P 6
Ballio A, see Dohrmann U 462, 465, 468, *503*
Bandurski RS 364, 367, 368, *392*
Bandurski RS, Schulze A 13, 20, 21, *81*, 264, 265, 267, *276*, 364, *392*
Bandurski RS, Schulze A, Cohen JD 21, *81*, 368, *392*
Bandurski RS, see Cohen JD 366, 367, *397*
Bandurski RS, see Ehmann A 15, *87*, 113, 120, 122, 123, 184, *192*, 367, *401*
Bandurski RS, see Hall PL 282, *407*
Bandurski RS, see Hamilton RH 115, *194*, 366, *407*
Bandurski RS, see Kopcewicz J 115, *195*, 366, 367, *414*
Bandurski RS, see Labarca C 113, 138, *195*, 367, *415*
Bandurski RS, see Michalczuk L 367, *423*
Bandurski RS, see Percival FW 13, 21, *102*, 365, *427*
Bandurski RS, see Piskornik Z 21, *102*, 113, 120, 130, 138, *198*, 366, *428*
Bandurski RS, see Ueda M 14, 15, *109*, 113, 115, 123, 170, *200*, 367, *439*, *440*
Bangerth F 332, *392*
Bangham AD, Standish MM Weissmann G 492, 494, *501*
Banko TJ, Boe AA 386, *399*
Banks RE, Cross BE 482, *501*
Barendse GWM 302, 362, 377, 378, 379, *392*
Barendse GWM, Gilissen HAM 377, *392*
Barendse GWM, Klerk de GJM 377, 378, 380, 382, *392*
Barendse GWM, Klerk de GJM, Mierlo JV 356, *393*
Barendse GWM, Kok NJJ 308, *393*
Barendse GWM, Lang A 362, 382, *393*
Barkley GM, Evans ML 517, 518, *569*
Barkley GM, Leopold AC 519, 522, 523, *569*
Barlow HWB, see Hancock RC 241, *277*
Barlow SA, see Coolbaugh RC 305, *398*
Barlow SA, see Moore TC 308, *424*
Barnes MF, Light EN, Lang A 315, 316, *393*
Barnett NM, see Key JL 520, *576*
Barrett G, see Gibson RA 284, 285, 286, 287, 288, *405*
Barthe MP, see Bulard C 374, *395*
Barthe P, Bulard C 374, *393*
Barthe P, see Puppo A 385, *428*
Barton DHR, see Simonson J 54, 78, *106*
Bartz L, Söll D 289, *393*
Barz W 340, 344, *393*
Baskin JM, Ludlow CJ, Harris TM, Wolf FT 67, *81*
Basu PA, Tuli V 337, *393*
Bates RB, see Arora SK 78, *81*
Bateson JH, Cross BE 482, *501*
Batt S, Venis MA 459, 460, 461, *502*, 559, *569*
Batt S, Wilkins MB, Venis MA 459, 461, 465, 466, 468, *502*, 559, *569*
Bauer JE, see Mattoo AK 330, 390, *421*
Bauer L, see Beutelmann P 44, *82*
Bauer WD, see Keegstra K 526, 527, 529, *576*
Baumgartner B, see Kende H 328, *413*
Baur AH, Yang SF 319, 320, 322, 326, *393*
Baur AH, Yang SF, Pratt HK, Biale JB 319, 325, 326, *393*
Baur AH, see Yang SF 317, 318, 319, *443*
Bawdekar AS, see Paul A 77, *102*
Beale MH, Gaskin PS, Kirkwood PS, MacMillan J 190, *190*
Bearder JR, Bleckschmidt S, Gaskin P, Graebe JE, MacMillan J 3, *6*
Bearder JR, Dennis FG, MacMillan J, Martin GC, Phinney BO 311, *393*

Bearder JR, Frydman VM, Gaskin P, Harvey WE, MacMillan J, Phinney BO 355, *394*
Bearder JR, Frydman VM, Gaskin P, MacMillan J, Wels CM, Phinney BO 311, *393*
Bearder JR, Hedden P, MacMillan J, Wels CM 302, 312, 313, 354, 355, *393*
Bearder JR, MacMillan J 3, *6*, 26, 30, 31, 34, *82*
Bearder JR, MacMillan J, Matsuo A, Phinney BO 311, *394*
Bearder JR, MacMillan J, Phinney BO 302, 309, 311, 312, 313, 314, 354, 355, *393*, *394*
Bearder JR, MacMillan J, Wels CM, Chaffey MB, Phinney BO 302, *393*
Bearder JR, MacMillan J, Wels CM, Phinney BO 33, *82*, 311, *393*
Bearder JR, Sponsel VM 302, 355, *393*
Beauchamp C, Fridovich J 318, 324, *394*
Beauchesne G, Goutarel R 40, *82*
Becker H, Kempf T 35, *82*
Beckers F, see Zimmermann U 539, 540, *586*
Beeley LJ, Gaskin P, MacMillan J 24, 25, 28, 31, 32, 50, 51, *82*
Beeley LJ, MacMillan J 32, *82*
Beeley LJ, see Sponsel VM 481, *508*
Beever JE, Woodlhouse HW 385, *394*
Beevers L, Loveys B, Pearson JA, Wareing PF 361, *394*
Beffagna N, Cocucci S, Marré E 469, *502*, 539, *569*
Beffagna N, Pesci P, Tognoli L, Marré E 452, 464, *502*
Beffagna N, see Pesci P 464, *507*
Beffagna N, see Tognoli L 469, *509*
Behl R, see Hartung W 352, *408*
Behmer M 362, *394*
Bell EA 56, 76, *82*
Bell EA, see Wilson MF 76, *111*
Bengochea T, Acaster MA, Dodds JH, Evans DE, Jerie PH, Hall MA 500, *502*
Bengochea T, Dodds JH, Evans DE, Jerie PH, Niepel B, Shari AR, Hall MA 500, *502*
Bennet-Clark TA 516, 524, *569*
Bennet-Clark TA, Kefford NP 52, *82*
Bennett PA, Chrispeels MJ 551, *569*
Bennett RD, Ko ST, Heftmann E 308, *394*
Bennett TD, Lieber ER, Heftmann E *394*
Bennici A, see Cionini PG 388, *397*
Bennicia A, see Alpi A 3, 5, *6*
Ben-Tal Y, Varner JE 550, 552, *569*
Bentley JA 20, 21, *82*
Bentley JA, Farrar KR, Housley S, Smith GF, Taylor WC 120, *191*
Benveniste P, see Normand G 458, *507*
Ben-Yehoshua S, Aloni B 331, *394*
Ben-Yehoshua S, see Zeroni M 328, *444*
Bercht CA, Samrak HM, Lousberg RJJC, Theuns H, Salemink CA 53, 54, *82*
Berger J, Avery GS 13, *82*
Bergmann F, see Schraudolf H 289, *434*
Bergmann L, Rennenberg H 80, *82*
Bergner C, see Sembdner G 316, 376, *434*
Bergren WR, see Haagen-Smit AJ 13, *90*, 113, *193*
Berlyn GP 564, *569*
Bermudez de Castro F, see Rodriguez-Barrueco C 385, *431*
Bernhardt D, Köhler K-H, Sembdner G 381, *394*
Beroza M, see Klun JA 462, *506*
Berridge MV, Ralph RK, Letham DS 474, 475, 477, 478, *502*
Berrie AMM, Don R, Buller D, Alam M, Parker W *82*
Berrie AMM, Parker W, Knights BA, Hendrie MR 69, *82*
Berrie AMM, Robertson J 50, *82*, 272, *276*
Berrie AMM, see Robertson J 354, *431*
Bertossi F, Bagni N, Moruzzi G, Caldarera CM 76, *82*
Betz A 345, *394*
Beutelmann P 292, *394*
Beutelmann P, Bauer L 44, *82*
Bewley JD, see Durley RC 357, 360, *401*
Beyer EM 499, 500, *502*
Beyer EM, Blomstrom DC 500, *502*
Beyer EM Jr 362, 363, *394*
Beyer EM Jr, Morgan PW 365, *394*
Beyer EM Jr, Sundin O 363, *394*
Beyer EM Jr, see Blomstrom DC 363, *395*
Beyer EM Jr, see Giaquinta R 362, *405*
Beynon JE, Williams AE 234, *276*
Bezemer-Sybrandy SM, Veldstra H *394*
Bhakuni DS, Joshi PP, Uprety H, Kapil RS 53, *82*
Bhakuni DS, see Satish S 53, *105*
Bhattacharyya J, Roy SC 474, *502*, 531, *569*
Bhattacharyya SC, see Mathur SB 77, *98*
Bhattacharyya SC, see Paul A 77, *102*
Biale JB 11, *82*
Biale JB, see Baur AH 319, 325, 326, *393*
Biale JB, see Pratt HK 11, *102*
Bialek K, Bielinska-Czarnecka M, Gaskin P, MacMillan J 49, *82*
Biddington NL, Thomas TH 141, *191*, 242, *276*

Bielinska-Czarnecka M, see Bialek K 49, *82*
Bielski RL 117, *191*
Bieman K, Tsunakawa S, Sonnenbichler J, Feldman H, Dütting D, Zachau HG 40, 43, *82*
Biggs RH, see Brook JL 169, *191*
Bigot C 242, *276*
Binks R, MacMillan J, Pryce RJ 184, *191*
Birch AJ, Rickards RW, Smith H, Harris A, Whalley WB 301, 309, *395*
Birch AJ, Rickards RW, Smith H, Winter J, Turner WB 301, *395*
Birch AJ, Winter J 302, *394*
Birecka H, Briber KA, Catalfamo JL 347, *395*, 554, 565, *569*
Birecka H, Catalfamo JL, Urban P 554, 565, *569*
Birmingham BC, Maclachlan GA 530, 531, 559, *569*
Biswas BB, Ganguly A, Das A, Roy P 455, 456, 501, *502*
Biswas BB, see Mondal H 455, 456, *506*
Biswas BB, see Roy P 455, *508*
Bittner S, Gazit S, Blumenfeld A 79, *82*
Bittner S, Gorodetsky M, Har-Paz I, Mizrahi Y, Richmond AE 498, *502*
Bittner S, see Goldschmidt EE 49, *90*, 117, *193*, 374, *405*
Bittner S, see Mizrahi Y 543, *579*
Black M, Dheidah M 521, *569*
Black M, see Pope D 521, *581*
Black RC, Hamilton RH 282, 383, *395*
Blackman GE, see Kenney G 336, *413*
Blair B, see Bolduc R 347, *395*
Blair C, see Henze H 235, *277*
Blanpied GD 330, *395*
Blaydes DF, see Miernyk JA *423*
Blecher A, White A 494, *502*
Bleckschmidt S, see Bearder JR 3, *6*
Blizzard WE, see Vanderhoef LN 518, 523, *585*
Blomstrom DC, Beyer EM Jr 363, *395*
Blomstrom DC, see Beyer EM 500, *502*
Blumbach M, see Hahn H 293, *407*
Blumenfeld A 328, *395*
Blumenfeld A, see Aharoni N 386, *391*
Blumenfeld A, see Bittner S 79, *82*
Blumenfeld A, see Mizrahi Y 543, *579*
Blyth CA, Freedman RB, Rabin BR 491, *502*
Bock RM, see Burrows WJ 41, 43, 44, 45, *84*, 292, 348, *396*
Bock RM, see Skoog F 531, *583*
Bock RM, see Swaminathan S 42, 43, 44, 45, *107*
Bode HR 77, *82*
Boe AA, see Banko TJ 386, *392*
Bohnsack CW, Albert LS 346, *395*
Bolduc R, Cherry J, Blair B 347, *395*
Boller T, Herner RC, Kende H 320, 326, *395*
Bonamy PA, Dennis FG 50, *83*
Bonanou SA, see Cox RA 477, *503*
Bonde EK, see Geissman TA 77, *89*
Bonga JM, see Little CHA 49, *96*, 143, *196*
Bongen-Ottoko B, see Charnay D 374, *397*
Bonner B, see Goeschl JD 330, *405*
Bonner J 56, 76, *83*, 241, *276*, 521, *569*
Bonner J, Galston AW 59, *83*
Bonner J, see Gray R 74, *90*
Bonner J, see Wildman SG 281, *442*
Bonner WD, see Thimann KV 69, *108*
Bopp M, see Erichsen V *402*
Borgmann E 381, *395*
Borgmann E, see Sembdner G 381, *434*
Bormann H, see Libbert E 336, *417*
Börner H 57, *83*
Bornman CH, see Van Staden J 244, *279*
Borowski D, see Dörffling K 55, *87*
Borrow A, Brian PW, Chester VE, Curtis PJ, Hemming HG, Henehan C, Jeffreys EG, Lloyd PB, Nixon GLF, Radley M 22, *83*
Borthwick HA, see Toole EH 57, *109*
Bose TK, see Roy BN 61, *104*
Botjes JO 11, *83*
Böttger M 55, *83*, 300, *395*
Böttger M, Engvild KC, Soll H 289, *395*
Böttger M, see Dörffling K 55, *87*
Böttger M, see Hofinger M 16, *92*, 184, *194*, 289, *410*
Boudet AM, see Ranjeva R 565, *582*
Bouer WD, see Talmadge KW 526, *584*
Bough WA, Gander JE 71, *83*
Bourdoux P, Vandervorst D, Hootelé C 18, *83*
Bowen A, Dymock J, Hill B 539, *569*
Bowen DH, Crozier A, MacMillan J, Reid DM 23, 24, 27, *83*, 142, 183, *191*, 244, *276*
Bowen DH, MacMillan J, Graebe JE 314, *395*
Bowen DH, see Crozier A 24, *86*
Bowen DH, see Graebe JE 310, 314, *406*
Bowen MR, see Hoad GV 361, 388, *410*
Bowen NR, see Lenton RJ 142, 173, *196*
Bower PJ, Brown HM, Purves WK 285, 286, 383, *395*
Bown AW, Reeve DR, Crozier A 361, *395*

Box JE Jr, see Pallas JE Jr 542, *580*
Boyle JTA, see Burrows WJ 43, *84*, 137, 168, *191*
Boysen Jensen P 12, *83*
Bracker CE, see Morré DJ 452, *507*, 540, *579*
Bradbeer JW, see Arias I 362, 388, *392*
Bradbeer JW, see Ross JD 268, *279*
Bradbeer JW, see Williams PM 23, 25, 49, *111*, 268, *279*
Bradbury D, Ennis WB 541, *569*
Bradford DC, see McFadden WH 262, *278*
Brady CJ, see McGlason WB 332, *421*
Brandes H, Kende H 242, *276*
Bratton BO, Henry EW 347, *395*
Braun AC, see Wood HN 371, *442*
Braun R, see Pilet PE 559, *581*
Breen PJ, see Ryugo K 346, *432*
Breidenbach RW, see Patterson R 481, *507*
Breidenbach W, see Stoddart JL *437*, 481, 486, 487, *509*
Brenner M, see Ciha AJ 272, *276*
Brenner ML, see Carnes MG 176, 178, 180, *191*
Brenner ML, see Ciha AJ 50, *84*, 181, *191*
Brenner MW, see Niederl JB 11, *101*
Brian PW 34, *83*, 361, *395*
Brian PW, Grove JF, MacMillan J 33, *83*
Brian PW, Grove JF, Mulholland TPC 33, 35, *83*, 479, 480, 481, *502*
Brian PW, Hemming HG, Lowe D 243, *276*
Brian PW, see Borrow A 22, *83*
Briant RE, see Morris DA 338, 365, *424*
Briber KA, see Birecka H 347, *395*, 554, 565, *569*
Bridges IG, Hillman JR, Wilkins MB 13, *83*, 184, *191*, 264, *276*
Brien CJ, see Loveys BR 296, 352, *419*
Briggs DE 35, *83*, 388, *395*, 551, *569*
Briggs DE, see Murphy GJP 308, 389, *425*
Briggs WR, see Cross JW 466, 468, 469, *503*
Brillouin L *280*
Brinkman KA, see Vanderhoef LN 521, *585*
Britten RJ, Davidson EH 553, 567, *569*
Brook JL, Biggs RH, StJohn PA, Anthony DS 169, *191*
Broome OC, see Zimmerman 325, 334, *444*
Broughton WJ, McComb AJ 562, *569*
Broughton WJ, see McComb AJ 530, *579*
Brown C, see Moffat AC 256, *278*
Brown GN, see Nowak J 305, *426*
Brown HM, Purves WK 286, *395*
Brown HM, see Bower PJ 285, 286, 383, *395*
Brown HM, see Purves WK 285, 286, *429*
Brown JC, Cross BE, Hanson JR 25, *83*
Brown KS, Sanchez WE 56, 78, *83*
Brown KS, see Arora SK 78, *81*
Brown RL, see Hall MA 533, *573*
Browning A, see Quail PH 452, *508*
Browning G 542, *569*
Browning G, Hoad GV, Gaskin P 49, *83*
Browning G, Saunders PF 24, 25, *83*, 116, *191*, 268, *276*, 388, 389, *395*
Browning G, see Little CHA 13, 50, *97*, 374, *418*
Bruce MI, Zwar JA, Kefford NP 40, *83*
Bruggen JMH van, see Engelsma G 565, *572*
Bruinsma J, Patil SS 74, *83*
Bruinsma J, see Thompson AG 55, *108*, 301, *439*
Brun WA, see Ciha AJ 50, *84*, 181, *191*, 272, *276*
Brunn K, see Libbert E 284, 285, *417*
Bryan RF, see Kupchan MS 80, *96*
Buchanan BG, see Lederberg J 235, 239, *277*
Buckwalter B, see Fox JE *403*
Buggy J, see Green PB 515, *573*
Buhler DR, Hansen E, Wang CH 362, *395*
Bui-Dang-Ha D, Nitsch JP 41, *83*, 368, *395*
Bukolova TP, see Komissarenko NF 67, *95*
Bukovac MJ, Yuda E, Murofushi N, Takahashi N 29, *83*
Bukovac MJ, see Davison RM 50, *86*
Bukovac MJ, see Goren R 364, 365, 369, 370, 371, *406*
Bukovac MJ, see Hopping ME 13, 42, *92*
Bukovac MJ, see Rudnicki R 60, *104*
Bulard C, Barthe MP, Garrello G, Le Page-Degivry M-T 374, *395*
Bulard C, see Barthe P 374, *393*
Bulard C, see Le Page-Degivry MT 49, *96*
Bulard C, see Rajagopal R 285, 347, *429*
Buller D, see Berrie AMM *82*
Buller DC, Parker W, Reid JSG 79, *84*
Bürcky K 35, *83*
Burden RS, Dawson GW, Taylor HF 300, *396*
Burden RS, Firn RD, Hiron RWP, Taylor HF, Wright STC 300, *396*
Burden RS, Taylor HF 129, *191*, 295, 300, *396*
Burden RS, see Firn RD 55, *88*, 117, 129, 175, *193*, 300, *403*
Burden RS, see Taylor HF 55, *108*, 129, 132, 174, *200*, 244, *279*, 300, 301, *438*

Burden RS, see Valio IMF 52, 53, *109*
Burden RS, see Wilkins H 75, *111*
Burg EA, see Burg SP 331, 333, 334, *396*, 499, *502*, 514, 532, *569*
Burg SP 10, 11, *84*, 325, 326, 327, 333, *396*, 540, *569*
Burg SP, Burg EA 331, 333, 334, *396*, 499, *502*, 514, 532, *569*
Burg SP, Clagett CO 319, 326, 333, *396*
Burg SP, Dijkman MJ 328, *396*
Burg SP, Stolwijk JAJ 12, *84*
Burg SP, Thimann KV 11, *84*, 325, 335, *396*
Burg SP, see Chadwick AV 333, *396*
Burg SP, see Kang BG 331, 333, *412*, 540, *575*
Burgers A, see Vreugdenhil D 469, *510*
Burke JJ, see Vanderhoef LN 518, 523, *585*
Burke RE, see Hamilton RH 336, 337, 340, 344, *407*, 448, *504*
Burrows WJ 41, 43, 44, *84*, 289, 292, 348, *396*
Burrows WJ, Armstrong DJ, Kaminek M, Skoog F, Bock RM, Hecht SM, Dammann LG, Leonard NJ, Occolowitz J 292, 348, *396*
Burrows WJ, Armstrong DJ, Kaminek M, Skoog F, Bock RM, Hecht SM, Dammann LG, Leonard NJ, Occolowitz JL 41, 43, 44, 45, *84*
Burrows WJ, Armstrong DJ, Skoog F, Hecht SM, Boyle JTA, Leonard NJ, Occolowitz J 43, *84*, 137, 168, *191*
Burrows WJ, Carr DJ 384, *396*
Burrows WJ, Skoog F, Leonard NJ 41, 43, 44, *84*, 139, *191*, 473, *502*, 531, *570*
Burrows WJ, see Armstrong DJ 43, *81*, 139, 140, *190*, 348, *392*, 472, *501*
Burrows WJ, see Hecht SM 348, *409*
Burrows WJ, see Struxness LA 385, *438*
Bürstell H, see Hilgenberg W 337, 340, *410*
Burstrom H, see Waller GR 75, *110*
Burstrom HG, Uhrström I, Olausson B 518, *570*
Buston HW, Roy SK 71, *84*
Buta JG, Flippen JL, Lusby WR 78, *84*
Buta JG, Worley JF 75, *84*
Butcher DN, see Mousdale DMA 383, *425*
Butler GW, see Conn EE 74, *84*
Button J, see Staden J van 384, *440*
Butts JS, see Fang SC 336, *403*
Byrd DJ, Kochen W, Idzko D, Knorr E 157, 160, *191*
Byrne H, Christou NV, Verma DPS, Maclachlan GA 563, *570*
Byrne H, see Bal AK 563, *568*
Byrne H, see Verma DPS 451, *510*, 554, *585*

Cairns-Smith AG *280*
Caldarera CM, see Bertossi F 76, *82*
Caldogno FR, see Lado P 538, *577*
Calero A, see Dominguez XA 74, *87*
Calvin M *280*
Campbell DJ, see Jackson MB 331, 332, *411*
Cande WZ, see Dowler MJ 540, *571*
Caplin SM, Steward FC 242, *276*
Cardenas E, see Dominguez XA 77, *87*
Carlton BC, Peterson CE, Tolbert NE 362, *396*
Carnes MG, Brenner ML, Andersen CR 176, 178, 180, *191*
Carns HR, see Addicott FT 48, *80*, 244, *276*
Carns HR, see Liu W-C 48, *97*
Carr DJ, Reid DM 388, *396*
Carr DJ, Reid DM, Skene KGM 387, *396*
Carr DJ, see Burrows WJ 384, *396*
Carr DJ, see Meinhenett R 79, *98*
Carr DJ, see Reid DM 316, 361, 388, *430*
Carraway KL, see Skoog F 471, 472, *508*
Carroll JE, see Thomas TH 167, 180, *200*
Carter HE, see Gottlieb D 495, *504*
Caruso IL, Smith RG, Smith LM, Cheng T-Y 186, *191*
Caruso JL, Smith RG, Smith LM, Cheng T-Y, Daves GD 13, *84*
Catalfamo JL, see Birecka H 347, *395*, 554, 565, *569*
Cathey HM 314, *396*
Cathey HM, see Asen S 70, *81*
Cattell KJ, see Hocking TJ 498, *505*
Caubergs R, see De Greef JA 346, 347, *399*
Cautrecasas P, see Sica V 485, *508*
Cavell BD, MacMillan J, Pryce RJ, Sheppard AC 161, 163, 171, 172, *191*
Ceccarelli N, Lorenzi R, Alpi A 5, *6*, 305, 388, *396*
Cecchini J-P, see Miassod R 520, *579*
Cerena R, see Lado P 539, *577*
Chadwick AV, Burg SP 333, *396*
Chadwick AV, see Steen DA 333, *437*
Chaffey MB, see Bearder JR 302, *393*
Chailakhian MK, see Grigorieva NY 361, *406*
Chain EB, see Ballio A 533, *568*
Chakraborty DF, see Roy S 74, *104*
Chakraborty DP, see Chowdhury BK 18, *84*
Challice JS 178, 179, *191*
Chalmers DJ, Faragher JD 331, 332, *396*

Chalmers DJ, see Jerie PH 327, 332, 389, *412*
Chalutz E 565, *570*
Chalutz E, see Mattoo AK 330, 390, *421*
Chan BG, see Kuhnle JA 80, *95*
Chandra GR, Duynstee EE 545, *570*
Chandra GR, see Varner JE 545, 546, *585*
Chandrasekaran S, see Corey EJ 38, *85*
Chang H, see Chen YM 557, *570*
Chang SH, see Harada F 43, *91*, 316, 348, *408*
Chang YP, see Sondheimer E 351, *436*
Chapman RA, see Lee TT 344, 345, *416*
Chapman RW, Morris RO, Zaerr JB 40, 43, 45, *84*, 292, *396*
Chappet A, Dubouchet J 341, *397*
Charnay D, Bongen-Ottoko B 374, *397*
Chassy BM, Suhadolnik RJ 372, *397*
Chen C-M, Eckert RL 292, 293, 372, *397*
Chen C-M, Eckert RL, McChesney JD 292, *397*
Chen C-M, Hall RH 166, *191*, 290, 291, *397*, 473, *502*
Chen C-M, Logan DM, McLennan B, Hall RH 350, 386, *397*
Chen C-M, Melitz DK 292, *397*
Chen C-M, Petschow B 385, *397*
Chen C-M, Smith OO, O'Brien C, McChesney JD 348, *397*
Chen C-M, Smith OO, Hartnell GF *397*
Chen C-M, see Dyson WH 44, 45, *87*
Chen C-M, see Fox JE 293, *403*
Chen RF, see Jones RL 488, 489, *505*
Chen YM, Lin CY, Chang H, Guilfoyle TJ, Key JL 557, *570*
Cheng T-Y, see Caruso JL 13, *84*, 186, *191*
Cherayil JD, Lipsett MN 40, *84*
Cherayil JD, see Timmappaya G 348, *439*
Cherry J, see Bolduc R 347, *395*
Cherry JG, see Hardin JW 453, 454, *504*
Cherry JH 519, 561, *570*
Cherry JH, see Clark JE 447, 470, *502*
Cherry JH, see Hardin JW 457, 470, *504*, 520, 554, 559, 560, *573*, *574*
Cherry JH, see Holm RE 557, 558, *574*
Cherry JH, see Leffler HR 555, 561, *577*
Cherry JH, see Mizrahi Y 328, *424*
Cherry JH, see Morré DJ 517, 520, 536, *579*
Cherry JH, see O'Brien TJ 453, 455, 489, *507*, 518, 520, 558, *580*
Cherry JH, see Rizzo PJ 455, *508*
Chester VE, see Borrow A 22, *83*
Chhabra BR, see Kalsi PS 78, *93*
Chheda GB, Mittelman A 348, 372, *397*
Chheda GB, see Dyson WH 44, 45, *87*
Chibbar RN, see Nanda KK 340, *426*
Chinoy JJ, see Jasdanwala RT 346, *411*
Cho KY, Sakurai A, Kamiya Y, Takahashi N, Tamura S 316, *397*
Chou C-H, see Muller CH 57, *100*
Chou PCC, see Arora SK 78, *81*
Chou TW, Yang SF 317, *397*
Chowdhury BK, Chakraborty DP 18, *84*
Chrispeels MJ 562, *570*
Chrispeels MJ, Varner JE 244, *276*, 546, 547, 548, 551, *570*
Chrispeels MJ, see Bennett PA 551, *569*
Chrispeels MJ, see Sadava D 529, *583*
Christou NV, see Byrne H 563, *570*
Chuang T-T, see Wang TSC 59, 60, *110*
CIAT Annual Report 34, *84*
Ciha AJ, Brenner M, Brun WA 272, *276*
Ciha AJ, Brenner ML, Brun WA 50, *84*, 181, *191*
Cionini PG, Bennici A, Alpi A, D'Amato F 388, *397*
Cionini PG, see Alpi A 3, 5, *6*
Clack DW, see Farrimond JA 448, *503*
Clagett CO, see Burg SP 319, 326, 333, *396*
Clagett CO, see Siebert K *435*
Clapham J, see Hocking TJ 498, *505*
Clardy J, see Springer JP 78, *106*
Clare BG, see Stahmann MA 347, *436*
Clark JE, Morré DJ, Cherry JH, Yunghans WN 447, 470, *502*
Clark JH, Peck EJ, Schrader WT, O'Malley BW 447, *502*
Clarke AJ, Mann PJG 285, 287, *397*
Clarke G, Dye MH, Wain RL 338, *397*
Cleland C, Ajami A 61, *84*
Cleland CF, Zeevaart JAD 361, *397*
Cleland R 316, *397*, 451, *503*
Cleland R, see Ray PM 514, *582*
Cleland R, see Rayle DL 521, 522, 537, *582*
Cleland R, see Thompson WF 520, 559, *584*
Cleland RE 514, 515, 516, 518, 523, 525, 526, 530, 534, 536, 537, 539, *570*
Cleland RE, Karlsnes A 529, *570*
Cleland RE, Lomax T 518, 538, 539, *570*
Cleland RE, Thompson ML, Rayle DL, Purves WK 530, *570*

Cleland RE, Thompson WF, Haughton PM, Rayle DL 520, *570*
Cleland RE, see Prins HBA 538, *581*
Clements JB, see Reid DM 361, *430*
Clifford K, see Cornforth JW 294, *398*
Clifford MN 140, *191*
Cline MG, see Rehm MM 532, *582*
Clum HH, see Sharp PB 35, *105*
Clutter M, see Sussex I 560, *584*
Clutter M, see Walbot V 560, *585*
Coccucci S, see Beffagna N 539, *569*
Cocking EC, see Davies JW 474, *503*
Cocucci M, Marré E, Ballarin-Denti A, Scachi A 538, *571*
Cocucci S, see Beffagna N 469, *502*
Cocucci SM, see Dohrmann U 462, 465, 468, *503*
Coggins CW Jr, see Khalifah RA 281, *413*
Coggins CW, see Jones WW 26, 49, *93*
Coggon P, see Cook CE 77, *85*
Cohen A, see Lavie O 74, *96*
Cohen D, see Wightman F 282, 283, 284, 285, 286, *442*
Cohen H, see Hedden P 4, *6*
Cohen JD, Bandurski RS 366, 367, *397*
Cohen JD, see Bandurski RS 21, *81*, 368, *392*
Cohen LA, see Kirk KL 481, *506*
Cohen N, see Marshall JA 77, *97*
Coke LB, Stuart KL, Whittle YG 56, *84*
Coke LB, see Stuart KL 56, *107*
Colborne AJ, see Laidman DL 551, 552, *577*
Cole DL, see Armstrong DJ 40, 43, *81*, 292, 293, 348, *392*
Cole RJ, see Springer JP 78, *106*
Collins GG, Jenner CF, Paleg LG 550, *571*
Colombo R, see Lado P 538, 539, *577*
Colombo R, see Marré E 531, 534, 537, 538, 539, *578*
Comman I *84*
Comstock JP, Rosenfeld GC, O'Malley BW, Means AR 486, *503*
Conn EE, Butler GW 74, *84*
Conrad K 368, *397*
Conrad K, Köhn B 386, *397*
Conway HF, see Stodola FH 22, 23, *107*
Cook CE, Whichard LP, Turner B, Wall ME, Egley GH 77, *85*
Cook CE, Whichard LP, Wall ME, Egley GH, Coggon P, Luhan PA, McPhail AT 77, *85*
Cook CE, see Wall ME 75, *110*
Cook IF, see Bakker HJ 311, *392*
Cook JC Jr, see Grove MD 3, *6*
Cook JF, Jefferies PR, Knox JR 311, *397*
Cooke RJ, Kendrick RE 362, 388, *397*
Cooke RJ, Saunders PF 362, 388, *397*
Cooke RJ, Saunders PF, Kendrick RE 361, 362, *397*
Coolbaugh RC, Hamilton R 305, 316, *398*
Coolbaugh RC, Moore TC 305, 308, 388, *398*
Coolbaugh RC, Moore TC, Barlow SA, Ecklund PR 305, *398*
Coolbaugh RC, see Knotz J 317, *413*
Coolbaugh RC, see Moore TC 308, 389, *424*
Coombe BG 29, *85*, 327, *398*
Coombe BG, Mander LN, Paleg LG, Turner JV 35, *85*
Coombe BG, see Faull KF 308, *403*
Cooper R, Gottlieb HE, Lavie D, Levy EC 74, *85*
Cooper R, Levy EC, Lavie D 74, *85*
Corcoran MR 58, *85*
Corcoran MR, Geissman TA, Phinney BO 58, *85*
Corcoran MR, see Jacobson A 58, *93*
Corcuera LJ, see Woodward MD 75, *111*
Corey EJ, Danheiser RL, Chandrasekaran S, Keck GE, Gopolan B, Larsen SD, Siret P, Gras J-L 38, *85*
Cornette J, see Fox JE 370, *403*
Cornforth JW, Clifford K, Mallaby R, Phillips GT 294, *398*
Cornforth JW, Draber W, Milborrow BV, Ryback G 48, *85*
Cornforth JW, Milborrow BV, Ryback G 48, 49, *85*, 190, *191*
Cornforth JW, Milborrow BV, Ryback G, Rothwell K, Wain RL 48, *85*
Cornforth JW, Milborrow BV, Ryback G, Wareing PF 48, 49, *85*
Cornforth JW, see Addicott FT 48, *80*
Cornforth JW, see Popjak G 294, *428*
Cornman I 71, *85*
Corse J, see Kuhnle JA 80, *95*
Corse J, see Vreman HJ 42, 43, 44, *109*, 128, *200*, 292, 385, *441*
Cory HJ, see Hashizume T 42, 45, *91*, 186, *194*
Coster HGL, Steudle E, Zimmermann U 539, *571*
Coster HGL, see Zimmermann U 539, 540, *586*
Cote P, see Walton DC 352, 353, 374, *441*
Cottenie J, see Van Sumere CF 57, 58, 67, 74, *109*
Court WA, see Kupchan MS 80, *96*
Cousins HH 10, *85*
Cowley DE, Jenkins ID, MacLeod JK, Summons RE, Letham DS, Wilson MM, Parker CW 369, *398*
Cowley DE, see Parker CW 45, *101*, 128, *198*, 369, 386, *427*
Cox RA, Bonanou SA 477, *503*

Cox RH, see Springer JP 78, *106*
Crafts CB, Miller CO 40, *85*
Cragg G, see Geissman TA 308, 354, *404*
Cragg G, see Verbiscar AJ 308, 354, *440*
Craker LE 331, *398*
Craker LE, Abeles FB, Shropshire W Jr 331, *398*
Craker LE, Stanley LA, Starbuck MJ 331, *398*, 566, *571*
Craker LE, Wetherbee PJ 566, *571*
Cramer WA, see Helgerson SL 452, *504*, 540, *574*
Crane FL, see Gibson MS 11, *89*
Crane JC, see Puech AA 554, *581*
Crees R, see Dowdell RJ 513, *571*
Crocker W, Hitchcock AE, Zimmermann PW 11, *85*
Crocker W, Knight LI 10, *85*
Crocker W, see Knight LI 10, *94*
Crocker W, see Zimmerman PW 10, *112*
Cronshaw DK, see Pegg GF 331, *427*
Crosby DG, Vlitos AJ 79, *85*
Crosetti G, see Aducci P 6
Cross BE 26, *85*, 378, *398*
Cross BE, Galt RHB, Hanson JR 25, *85*, 131, *192*, 302, 355, *398*
Cross BE, Galt RHB, Norton K 302, 309, 311, 355, *398*
Cross BE, Grove JF, MacMillan J, Mulholland TPC 301, *398*
Cross BE, Grove JF, Morrison A 481, *503*
Cross BE, Myers PL 315, *398*
Cross BE, Norton K 26, *85*, 313, *398*
Cross BE, Norton K, Stewart JC 302, 311, 313, *398*
Cross BE, Stewart JC 304, *398*
Cross BE, Stewart JC, Stoddart JL 35, *85*, 309, *398*
Cross BE, see Banks RE 482, *501*
Cross BE, see Bateson JH 482, *501*
Cross BE, see Brown JC 25, *83*
Cross BE, see Curtis PJ 22, *86*
Cross JW, Briggs WR 466, 468, *503*
Cross JW, Briggs WR, Dohrmann UC, Ray PM 468, 469, *503*
Crow WD, Nicholls W, Sterns M 80, *85*
Crow WD, Osawa T, Paton DM, Willing RR 74, *86*
Crow WD, Osawa T, Platz KM, Southerland DS 80, *85*
Crozier A, Aoki H, Pharis RP 138, 143, 176, *192*
Crozier A, Bowen DH, MacMillan J, Reid DM, Most BH 24, *86*
Crozier A, Kuo CC, Durley RC, Pharis RP 479, *503*
Crozier A, Reeve DR 134, 143, 177, 178, 181, *192*
Crozier A, Reid DM 388, *398*
Crozier A, Reid DM, Reeve DR 316, *398*
Crozier A, see Bowen DH 23, 24, 27, *83*, 142, 183, *191*, 244, *276*
Crozier A, see Bown AW 361, *395*
Crozier A, see Durley RC 134, 135, 176, *192*
Crozier A, see Nash LJ 362, 380, *426*
Crozier A, see Reeve DR 134, 138, 176, 178, *199*, 243, *278*, 356, *430*
Crozier A, see Reid DM 316, *430*
Csonka L, see Hall RH 41, *90*, 137, *194*, 292, 362, *407*, 531, *573*
Cummins WR 352, 375, *398*
Cummins WR, Kende H, Raschke K 244, *276*, 542, *571*
Cummins WR, Sondheimer E 498, *503*
Curtis PJ, Cross BE 22, *86*
Curtis PJ, see Borrow A 22, *83*
Curtis RW, see John WW 322, *412*
Cutler HG, Reid WW, Délétang J 78, *86*
Cutler HG, see Springer JP 78, *106*
Czapski J, see Rudnicki R 354, *432*

Dahl, JL, see Armstrong DJ 43, *81*
Dale JE *399*
Dale JE, Felippe GM 316, *399*
Dalessandro G, see De Leo P 75, *86*
Dalvi PR, Singh B, Salunkhe DK 77, *86*
Daly JM, Seevers PM, Ludden P 331, *399*
Daly JM, see Deverall BJ 336, *400*
D'Amato F, see Alpi A 3, 5, *6*, 388, *389*
D'Amato F, see Cionini PG 388, *397*
Dammann LG, see Burrows WJ 41, 43, 44, 45, *84*, 292, 348, *396*
Dandliker WB, see Haagen-Smit AJ 13, *90*, 113, *194*
Danheiser RL, see Corey EJ 38, *85*
Daniels D, see Johnson KD 518, 525, *575*
Dannenburg WN, Liverman JL 282, *399*
Darbyshire B 347, *399*
Darimont E, Schwachhofer K, Gaspar T 384, *399*
Darvill AG *571*
Darvill AG, Smith CJ, Hall MA 526, 527, 528, 529, 530, 536, *571*
Darwin C, Darwin F 12, *86*
Darwin F, see Darwin C 12, *86*
Das A, see Biswas BB 455, 456, 501, *502*
Das VSR, Rao LM, Raghavendra AS 542, *571*
Dashek WV, Singh BN, Walton DC 387, *399*
Dasilva EJ, Henriksson E, Henriksson LE 317, *399*
Dathe W, Schneider G, Sembdner G 25, 26, 28, 50, 51, *86*, 352, 354, 379, *399*

Dathe W, Sembdner G 388, *399*
Dathe W, Sembdner G, Kefeli VI, Vlasov PV 49, *86*, 374, 382, *399*
Datko AH, Maclachlan GA 520, 532, *571*
Datta A, see Roychoudhury R 531, *583*
Dauphin B, Teller G, Durand B 42, 44, *86*
Dauphin B, Teller G, Durrand B 186, *192*
Dauphin B, Teller G, Dward B 42, 44, *86*
Daves GD, see Caruso JL 13, *84*
Davey JE, Van Staden J 371, 384, 385, 386, *399*
Davey JE, see Staden J van 385, *440*
David H, see Hall RH 41, *90*, 137, *194*, 292, 362, *407*, 531, *573*
Davidson EH, see Britten RJ 553, 567, *569*
Davidson RS, see Kemp HT 541, *576*
Davies D, Patterson BD, Trewavas AJ 520, *571*
Davies E 559, *571*
Davies E, see Maclachlan GA 562, *578*
Davies JW, Cocking EC 474, *503*
Davies L, see Rappaport L 356, 380, *430*
Davies LJ, Rappaport L 356, 379, 380, 382, *399*, 481, *503*
Davies PJ 336, 338, 365, 366, *399*, 560, *571*
Davies PJ, Kozlowski TT 543, *571*
Davies PJ, see Mitchell EK 282, *424*
Davies PJ, see Proebsting WM 361, *428*
Davies WJ, see Fenton R 56, *88*
Davis AW, see Polya GM 476, *507*
Davis LA, Heinz DE, Addicott FT 174, *192*
Davis LA, Lyon JL, Addicott FT 52, *86*, 131, *192*, 244, *276*
Davis M, Pettet M, Scanlon DB, Ferrito V 52, 74, *86*
Davis W, see Wareing PF 45, *110*, 350, 370, *441*
Davison RM, Rudnicki RM, Bukovac MJ 50, *86*
Dawson GW, see Burden RS 300, *396*
Dawson RM, Jefferies PR, Knox JR 311, *399*
Dayanandan P, see Hebard FV 530, *574*
De Greef J, see Van Sumere CF 57, 58, 67, 74, *109*
De Leo P, Dalessandro G, De Santis A, Arrigoni O 75, *86*
De Leo P, see Ballio A 533, *568*
De Maggio M, see Wetmore RH 566, *586*
De Michelis MI, see Lado P 539, *577*
De Michelis MI, see Marré E 534, 537, *578*
De Pooter H, see Van Sumere CF 62, *109*
de Ropp RS, see Hall RH 39, *90*
De Santis A, see De Leo P 75, *86*
De Zacks R, see Hahn H 384, *407*
Decker M, see Tanner W 544, *584*
Dedio W, Zalik S 170, *192*
Dedonder A, see Van Sumere CF 62, *109*
DeDuve C, Wattiaux R 494, *503*
Deenen LLM van, see Kinsky SC 495, *506*
DeGreef J, DeProft M 175, *192*
DeGreef JA, Van Hoof R, Caubergs R 346, 347, *399*
Dekhuijen HM, Gevers ECT 117, 128, *192*
Dekhuijzen HM 45, *86*
Délétang J, see Cutler HG 78, *86*
Deleuze G, see Fox JE 370, *403*
Deleuze GG, McChesney JD, Fox JE *399*
Demel RA, see Kinsky SC 495, *506*
DeMeyts P 446, *503*
Demorest DM, Stahmann MA 337, *399*
Dempsey ME, see Ritter MC 308, *431*
Dengler R, see Key JL 557, *576*
Dennis DT, Upper CD, West CA 316, *399*
Dennis DT, West CA 308, 316, *399*
Dennis DT, see Graebe JE 302, 304, 305, 307, *406*
Dennis DT, see Green TR 310, 389, *406*
Dennis DT, see Simcox PD 305, 317, 389, *435*
Dennis FG Jr, see Martin GC 184, 186, *196*, *197*, 352, 354, *421*
Dennis FG, see Balboa-Zavala O 49, *81*
Dennis FG, see Bonamy PA 50, *83*
Dennis FG, see Martin GC 26, 28, 31, 35, 49, 51, *97*, 268, *278*, 352, 354, *421*
Dennis FG, see Mielke EA 50, *99*
Dennis PG, see Bearder JR 311, *393*
Denny FE 10, *86*
DeProft M, see DeGreef J 175, *192*
Desmos EK, Woolwine M, Wilson RH, McMillan C 61, *86*
DeSombre ER, see Jensen EV 485, *505*
Deuel P, see Geissman TA 77, *89*
Deuel PG, Geisman TA 77, *86*
Dev S 77, *86*
Devay JE, see Taylor PA 45, *108*
Devay JE, see Zweig G 304, *444*
Deverall BJ 336, *400*
Deverall BJ, Daly JM 336, *400*
Devon TK, Scott AI 57, *86*
DeVries JE, see Fukui HN 115, *193*
Dewar MJS 75, *86*
Dewdney SJ, McWha JA 352, 354, 374, *400*
Dey B, see Sircar PK 23, *106*

DeYoe DR, Zaerr JB 13, *86*, 133, 138, *192*, 346, *400*
Dhar AK, see Scallen TJ 308, *433*
Dheida M, see Black M 521, *569*
Diaz DH, Martin GC 374, *400*
Diaz E, see Romo J 77, *104*
Dijkman MJ, see Burg SP 328, *396*
Dilley DR, see Apelbaum A 557, *568*
Dilley DR, see Dayak S 328, 330, 334, *421*
Dilley DR, see Frenkel C 327, *403*
Dilley DR, see Saltveit ME Jr 331, *432*
Dilley DR, see Sfakiotakis EM 327, *434*
Divekar AY, Slocum HK, Hakala MT 372, *400*
Djerassi C, see Lederberg J 235, 239, *277*
Dockerill B, Evans R, Hanson JR 314, *400*
Dockerill B, Hanson JR *400*
Dodds JH, see Bengochea T 500, *502*
Dodds JH, see Hall MA 541, *573*
Doerffling K, see Tietz D 353, *439*
Dohrmann U, Hertel R, Kowalik H 461, 462, 464, 465, 470, *503*, 539, *571*
Dohrmann U, Hertel R, Pesci P, Cocucci SM, Marre E, Randazzo G, Ballio A 462, 465, 468, *503*
Dohrmann U, Johnson KD, Rayle DL 533, *571*
Dohrmann U, see Hertel R 468, *504*
Dohrmann U, see Ray PM 461, 464, 465, 466, 468, *508*, *582*
Dohrmann UC, see Cross JW 468, 469, *503*
Doig RI, Laidman DL 551, *571*
Doig RI, see Laidman DL 551, 552, *577*
Dollstädt R, Hirschberg K, Winkler E, Hubner G 470, *503*, 559, *571*
Dominguez Jr XA, see Dominguez XA 74, *87*
Dominguez XA, Cardenas E 77, *87*
Dominguez XA, Martinez C, Calero A, Dominguez Jr XA, Hinojosa M, Zamudio A, Zabel V, Smith WB, Watson WH 74, *87*
Dominguez XA, Merijanian A, Gonzalez BL 78, *87*
Dominguez XA, Merijanian A, Gonzalez BL, Zamudio A, Salazar AL 78, *87*
Dominguez XA, Perez FM, Leyter L 77, *87*
Dominguez XA, see Rojas Garciduenas M 77, 78, *104*
Don R, see Berrie AMM *82*
Don R, see Willmer CM 79, *111*
Donnelly DMX, see Mollan RC 21, *100*, 364, 365, *424*
Doods JH, Musa SK, Jerie PH, Hall MH 363, *400*, 500, *503*
Doree M, Guern J *400*
Doree M, Terrine C 372, *400*
Doree M, Terrine C, Trapy F *400*
Doree M, see Guern J *404*
Dörffling K 295, *400*
Dörffling K, Böttger M, Martin D, Schmidt V, Borowski D 55, *87*
Dörffling K, Sonka B, Tietz D 353, 374, 375, *400*
Dorn B, see Walton DC 51, *110*, 131, *201*, 352, 353, *441*
Dostal HC, see McGlason WB 328, *421*
Dostal HC, see Mizrahi Y 328, *424*
Dougherty TM, Schepartz AI 139, 140 *192*
Dougherty TM, see Lerner J 139, *196*
Douglas TJ, Paleg LG 316, *400*
Dove LD 559, *571*
Dowdell RJ, Smith KA, Crees R, Restall SWF 513, *571*
Dowdell RJ, see Smith KA 513, *583*
Dowler MJ, Rayle DL, Cande WZ, Ray PM, Durand H, Zenk MH 540, *571*
Dowler MJ, see Johnson KD 518, 525, *575*
Downing RD, see Khan AA 548, *576*
Downton WJS, see Kriedemann PE 51, 52, *95*, 352, 353, *414*
Draber W, see Cornforth JW 48, *85*
Drawert A, see Libbert E 284, 285, 286, 287, 338, *417*, *418*
Drewes SE, see Van Staden J 41, 42, *109*, 368, *440*
Dreyer WJ, see Hummel JP 447, *505*
Dubouchet J, see Chappet A 341, *397*
Duda CT, see McComb aJ 489, *506*, 558, *579*
Dueber MT, Adolf W, West CA 317, *400*
Duerst E, see Libbert E 284, 287, *417*
Duffield AM, see Lederberg J 235, 239, *277*
Duffus CM, Duffus JH 549, *572*
Duffus JH, see Duffus CM 549, *572*
Duke CC, Letham DS, Parker CW, McLeod JK, Summons RE *400*
Duke CC, Liepa AJ, MacLeod JK, Letham DS, Parker CW 47, *87*, 369, 370, *400*
Duke CC, see Letham DS 348, 370, 385, 386, *417*
Duke CC, see Parker CW 370, *427*
Duke CC, see Summons RE 4, *7*
Duke WB, see Putnam AR 57, *103*
Dullaart J 338, *400*
Dunberg A 362, *400*
Dunlap RJ, Morgan PW 328, *400*
Dunlop J, see Penny P 518, 523, 537, *581*
Dunn DB, Smith JD, Spahr PF 45, *87*
Durand B, see Dauphin B 42, 44, *86*
Durand H, Zenk MH 517, 519, *572*

Durand H, see Dowler MJ 540, *571*
Durham JI, Morgan PW 328, *400*
Düring H, Alleweldt G 374, 375, *401*
Durley RC, Bewley JD, Railton ID, Pharis RP 357, 360, *401*
Durley RC, Crozier A, Pharis RP, McLaughlin GE 134, 135, 176, *192*
Durley RC, Kannangara T, Simpson G 272, *276*
Durley RC, MacMillan J, Pryce RJ 23, 24, 25, 26, 27, 35, 51, *87*, 116, 130, *192*, 352, 388, *401*
Durley RC, Pharis RP 124, 125, *192*, 356, 357, *401*
Durley RC, Pharis RP, Zeevaart JAD 356, 357, 361, *401*
Durley RC, Railton ID, Pharis RP 169, *192*, 355, 356, 357, *401*
Durley RC, Sassa T, Pharis RP 356, 360, *401*
Durley RC, see Crozier A 479, *503*
Durley RC, see Glenn TL 140, *193*
Durley RC, see Kamienska A 23, 24, 25, *93*, 356, 357, *412*
Durley RC, see Kaufman PB 23, *94*
Durley RC, see Lance B 357, 361, *415*
Durley RC, see McInnes AG 355, 376, 378, *422*
Durley RC, see Railton ID 309, 356, 357, 379, 380, 387, *429*
Durley RC, see Reeve DR 356, *430*
Durley RC, see Reid DM 361, 362, *430*
Durley RC, see Wample RL 356, 357, 362, 379, 380, *441*
Durnam DJ, see Stuart DA 497, *509*
Durrand B, see Dauphin B 186, *192*
Dutky SR, see Mandava N 76, *97*
Dutky SR, see Thompson MJ 3, *8*
Dütting D, see Biemann K 40, 43, *82*
Dütting D, see Zachau HG 43, *112*
Duynstee EE, see Chandra GR 545, *570*
Dward B, see Dauphin B 42, 44, *86*
Dyck R, see Frenkel C 333, *403*
Dye MH, see Clarke G 338, *397*
Dymock J, see Bowen A 539 *569*
Dyson W, see Fox JE 370, *403*
Dyson WH, Chen C-M, Alam SN, Hall RH, Hong CI, Chheda GB 44, 45, *87*
Dyson WH, Hall RH 41, 44, *87*, 128, 139, 173, *192*

Eagles CF, see Wareing PF 48, *110*
Eaks LI 327, *401*
Eastwood D, Laidman DL 552, *572*
Eberle M, Arigoni D 293, *401*
Eckermann G, see Atkinson MR 71, *81*
Eckert H, Podlesak W, Schilling G 316, *401*
Eckert H, Schilling G, Podlesak W, Franke P 23, *87*
Eckert H, see Schilling G 302, 305, 316, *433*
Eckert RL, see Chen C-M 292, 293, 372, *397*
Ecklund RP, Moore TC 305, *401*
Ecklund PR, see Coolbaugh RC 305, *398*
Edelbluth E, Kaldewey H 383, *401*
Edsall JT, Wyman J 446, *503*
Edwards JE, Lamotte CE 385, *401*
Edwards M 561, *572*
Eeuwens CJ, Gaskin P, MacMillan J 27, *87*
Egley GH, see Cook CE 77, *85*
Egley GH, see Moreland DE 56, 60, 65, 76, *100*
Eglinton G, see MacMillan J 182, 183, *196*
Egsgaard H, see Engvild KC 16, 20, *88*, 289, *402*
Ehmann A 15, *87*, 113, 115, 123, 160, 184, *192*, 366, *401*
Ehmann A, Bandurski RS 15, *87*, 113, 120, 122, 123, 184, *192*, 367, *401*
Ehmann A, see Kopcewicz J 115, *195*, 366, 367, *414*
Ehmann A, see Ueda M 14, 15, *109*, 170, *200*, 367, *440*
Eichholzer JV, see Summons RE 4, *7*
Eidt DC, see Little CHA 542, *577*
Einset JW, Skoog F 293, 348, 385, *401*
Einset JW, Skoog FK 40, *87*
Einset JW, Swaminathan S, Skoog F 41, *87*, 139, *193*
Eisinger WR, see Morré DJ 515, *579*
Ekong DEU, see Adesomoju AA 50, 51, *81*, 352, *391*
El Tannir-Lomba J, see Sironval G 74, *106*
El-Beltagy AS, Hall MA 331, 332, *401*, 542, *572*
El-Hamady M, see Hartmann HT 332, *408*
Eliasson L, Strömquist L-H, Tillberg E 267, *277*
Eliasson L, see Steen I 138, *199*
Elliot DC, Murray AW 292, *401*
Elliot MC 18, *87*
Elliot MC, Greenwood MS 13, *87*
Elliott MC, Stowe BB 19, *87*, 114, 115, 123, 124, 130, 138, *193*
Elliott MC, see Farrimond JA 448, *503*
Elmer OH 11, *87*
El-Olemy MM, Reisch J 74, *88*
Elson GW, Jones DF, MacMillan J, Suter PJ 162, 163, *193*
Elstner EF, Konze JR 324, 331, 389, 390, *401*
Elstner EF, Konze JR, Selman BR, Stoffer C 324, 390, *401*
Elstner EF, see Konze JR 322, 325, 389, 390, *414*

Elstner EF, see Montalbini P 331, *424*
Embleton TW, see Jones WW 26, 49, *93*
Ende H van den, Zeevaart JAD 361, *440*
Endo M, see Nakanishi K 35, *100*
Endo T 340, 341, *402*
Endo T, see Yoneda Y 340, 341, *443*
Engelbrecht L 385, *402*
Engelbrecht L, see Mothes K 544, *579*
Engelsma G, Bruggen JMH van 565, *572*
Engvild KC 289, *402*
Engvild KC, Egsgaard H, Larsen E 16, 20, *88*, 289, *402*
Engvild KC, see Böttger M 289, *395*
Ennis WB, see Bradbury D 541, *569*
Entsch B, Letham DS 372, *402*
Entsch B, Parker CW, Letham DS 372, *402*
Entsch B, see Letham DS 369, *417*
Entsch B, see Summons RE 4, *7*
Enzie JV, see Jones MB 74, *93*
Epstein E, Klein I, Lavee S 337, *402*
Epstein E, Lavee S 336, 337, 338, 364, *402*
Erdei S, see Wyen NV 560, *586*
Erdel L, Toth I, Zsoldos F 452, *503*
Erdmann N, Libbert E, Schiewer U 283, *402*
Erdmann N, Schiewer U 283, *402*
Erdmann N, Schiewer U, Libbert E 283, *402*
Erdmann N, see Libbert E 282, 283, 285, 287, 288, 338, *417*
Erdmann N, see Schiewer U 287, *433*
Erichsen V, Knoop B, Bopp M *402*
Erickson RO, Rosen GV 71, *88*
Erickson RO, see Green PB 515, *573*
Erickson RO, see Kuhn-Silk W 362, *415*
Erion JL, Fox JE 475, 476, 477, *503*
Erion JL, see Fox JE 475, 476, 477, *504*, 531, *573*
Erlanger BF, see Ballio A 533, *568*
Erxleben H, see Kögl F 12, *95*
Erzieben J, see Tietz D 353, *439*
Esashi Y, Ohhara Y, Kotaki K, Watanabe K 325, 328, *402*
Esashi Y, Watanabe K, Ohhara Y, Katoh H 325, 328, *402*
Esashi Y, see Katoh H 325, *412*
Esau K 545, *572*
Esnault R 517, 520, *572*
Ettlinger MG, Kjaer A 75, *88*
Evans A, Smith H 361, 362, *402*
Evans DE, see Bengochea T 500, *502*
Evans DE, see Hall MA 541, *573*
Evans LS, Almeida MS, Lynn DG, Nakanishi K 76, *88*
Evans LS, see Lynn DG 76, *97*
Evans LT, see King RW 50, 55, *94*, 301, *413*
Evans ML 451, 452, *503*, 516, 521, 533, *572*
Evans ML, Hokanson R 521, *572*
Evans ML, Ray PM 337, *402*, 448, *503*, 517, 518, 519, 520, 521, 533, *572*
Evans ML, Ray PM, Reinhold L 521, 522, *572*
Evans ML, Rayle DL 285, *402*
Evans ML, Schmitt MR 340, 342, *402*
Evans ML, see Barkley GM 517, 518, *569*
Evans ML, see Rayle DL 517, 519, *582*
Evans PK, see Armstrong DJ 43, *81*, 139, 140, *190*, 348, *392*
Evans R, Hanson JR 305, 312, 313, *402*
Evans R, Hanson JR, Mulheirn LJ 310, *402*
Evans R, Hanson JR, White AF 302, 309, 310, *402*
Evans R, see Dockerill B 314, *400*
Evenari M 57, 75, *88*
Evenari M, see Lavie O 74, *96*
Evenari M, see Mayer AM 67, 69, *98*
Even-Chen Z, see Goldschmidt EE 49, *90*, 117, *193*, 374, *405*
Evins WH 547, *572*
Evins WH, Varner JE 491, *503*, 547, 549, *572*
Ewings D, see Verma DPS 451, *510*, 554, *585*

Falkner FC, Sweetman BJ, Watson JT 186, *193*
Fall RR, West CA 304, 305, 306, 316, *403*
Falt RHB, see Cross BE 131, *192*
Fan DF, Maclachlan GA 518, 520, *572*
Fan DF, see Maclachlan GA 562, *578*
Fang SC, Theisen P, Butts JS 336, *403*
Faragher JD, see Chalmers DJ 331, 332, *396*
Farkas GL, see Wyen NV 560, *586*
Farrar KR, see Bentley JA 120, *191*
Farrimond JA, Elliott MC, Clack DW 448, *503*
Faull KF, Coombe BG, Paleg LG 308, *403*
Fawcett RL, see Seeley RC 338, *434*
Feeny PP, see Whittaker RH 57, *111*
Feigenbaum EA, see Lederberg J 235, 239, *277*
Feldman H, see Biemann K 40, 43, *82*
Feldman H, see Zachau HG 43, *112*
Felippe GM, see Dale JE 316, *399*

Fenton R, Davies WJ, Mansfield TA 56, *88*
Fenton R, see Wellburn AR 53, 56, *110*
Fenzl F, see Tanner W 544, *584*
Fernandez NO, see Sivori EM 244, *279*
Fernel DI, see Stodola FH 22, 23, *107*
Ferrari TE, Varner JE 548, *572*
Ferri MG, Lex A 541, *572*
Ferri MG, see Wildman SG 281, *442*
Ferrito V, see Davis M 52, 74, *86*
Ferroni A, see Marré E 531, 534, 537, 539, *578*
Feucht W, Nachit M 344, *403*
Feung CS, Hamilton RH, Mumma RO 365, 366, *403*
Feung CS, Mumma RO, Hamilton RO 365, *403*
Feung CS, see Hamilton RH 336, 337, 340, 344, *407*, 448, *504*
Fey J, see Walton DC 51, *110*, 131, *201*, 352, 353, *441*
Filmer AAE, see Meigh DF 74, *98*
Filner P, Varner JE 545, *572*
Filner P, Wray JL, Varner JE 545, *572*
Fincher GB, see Wong Y-S 524, 563, *586*
Findley JS, see Vanderhoef LN 518, 523, *585*
Firn RD, Burden RS, Taylor HF 55, *88*, 117, 129, 175, *193*, 300, *403*
Firn RD, Friend J 55, *88*, 300, *403*
Firn RD, see Burden RS 300, *396*
Firn RD, see King RW 50, 55, *94*, 301, *413*
Fischer E, see Libbert E 284, 285, 286, 287, 338, *417*, *418*
Fisher DF, see Miller EV 11, *99*
Fittler F, Hall RH 473, 478, *504*, 531, *572*
Fittler F, Kline LK, Hall RH 289, *403*, *504*
Fittler F, see Kline LK 289, *413*
Fletcher RA, McCullagh D 242, *277*
Fletcher RA, see Peterson CA 384, *427*
Flinn AM 365, *403*
Flippen JL, see Buta JG 78, *84*
Flippen-Anderson JL, see Grove MD 3, *6*
Flippen-Anderson JL, see Thompson MJ 3, *8*
Flore JA, see Goren R 365, 369, 370, 371, *406*
Focke I, see Schreiber K 35, 36, *105*, 376, 378, *434*
Focke I, see Sembdner G 316, 360, 362, 378, 380, 381, 388, *434*
Forrence LE, see Abeles FB 362, *390*
Forsline PL, see Langille AR 386, *415*
Forsyth FR, see Hall IV 328 *407*
Forti G, see Marré E 517, *578*
Fosket DE 554, *572*
Fosket DE, Short KC 555, 561, *573*
Fosket JE, see Torrey JG 553, 555, 564, *585*
Fowden L 76, *88*
Fowden L, Lewis D, Tristram H 76, *88*
Fowden L, Richmond MH 76, *88*
Fowden L, see Smith IK 76, *106*
Fowler JL, Morgan PW 332, 333, *403*
Fox JE 45, *88*
Fox JE, Chen C-M 293, *403*
Fox JE, Cornette J, Deleuze G, Dyson W, Giersak G, Niu P, Zapata J, McChesney J 370, *403*
Fox JE, Erion JL 475, 476, 477, *504*, 531, *573*
Fox JE, Sood CK, Buckwalter B, McChesney JD *403*
Fox JE, see Deleuze GG *399*
Fox JE, see Erion JL 475, 476, 477, *503*
Fox JE, see Pratt HM 476, *508*
Fox LR, Purves WK 342, *403*
Fraga BM, see Hanson JR 311, *408*
Fragata M 75, *88*
Francis AW 235, *277*
Franke P, see Eckert H 23, *87*
Frankland B, Wareing PF 243, *277*
Franklin MJ, see Wills RBH 375, *442*
Frederico R, see Aducci P 6
Freebairn HT, see Stewart ER 540, *584*
Freedman RB, see Blyth CA 491, *502*
Freeman TE, see Luke HH 542, *578*
Frei RW, see Hutzinger O 159, *194*
Frenkel C 333, 337, 340, 341, 346, *403*
Frenkel C, Dyck R 333, *403*
Frenkel C, Haard NF 346, *403*
Frenkel C, Haddon VR, Smallheer JM 337, *403*
Frenkel C, Hess CE 340, 341, *403*
Frenkel C, Klein I, Dilley DR 327, *403*
Fridovich J, see Beauchamp C 318, 324, *394*
Friedländer M, Atmon D, Galun E 354, *403*
Friend J, see Anstis PJP 55, *81*, 300, *391*
Friend J, see Firn RD 55, *88*, 300, *403*
Fries D 347, *403*
Fries D, see Gaspar T 565, *573*
Frigerio A, see Gordon AE 186, *193*
Frihart CR, see Scarbrough E 40, *105*, 292, *433*
Frihart CR, see Vreman HJ 41, 43, 44, *109*, 139, *201*, 292, 348, *440*, *441*
Frost RG, West CA 304, 305, 306, 316, *404*
Frydman VM, Gaskin P, MacMillan J 25, 26, 27, 29, 30, 31, 35, 49, 51, *88*, 186, *193*, 268, *277*

Frydman VM, MacMillan J 27, 29, *88*, 170, *193*, 356, 357, 358, 359, 362, 380, *404*, 480, 481, *504*
Frydman VM, see Bearder JR 311, 355, *393*, *394*
Fuchs S, Haimovich J, Fuchs Y 245, 246, *277*
Fuchs S, see Fuchs Y 245, *277*
Fuchs Y 332, *404*
Fuchs Y, Gertman E 245, 246, *277*
Fuchs Y, Lieberman M 334, *404*
Fuchs Y, Mayak S, Fuchs S 245, *277*
Fuchs Y, see Fuchs S 245, 246, *277*
Fuchs Y, see Zauberman G 328, *444*
Fuganti C, see Piozzi F 75, *102*
Fujihiro H, see Okamoto T *198*
Fujii D, see Hedden P 4, *6*
Fujii T, Ogawa N 46, *88*
Fujii T, see Leonard NJ 40, *96*, 125, 127, *196*
Fujii T, see Skoog F 471, 472, *508*
Fujita E, Node M 35, *88*
Fujita E, Uchida I, Fujita T 78, *88*
Fujita E, see Fujita T *404*
Fujita K, see Kato T 54, 56, 67, 78, *94*
Fujita K, see Yamaguchi I 31, *112*
Fujita T, Masuda I, Takao S, Fujita E *404*
Fujita T, see Fujita E 78, *88*
Fukami H, see Ina K 54, *92*
Fukinbara T, see Kagawa T 134, 161, 162, *195*
Fukui H, Ishii H, Koshimizu K, Katsumi M, Ogawa Y, Mitsui T 27, 28, 30, 31, *88*. 134, *193*
Fukui H, Koshimizu K, Mitsui T 29, *88*, 124, 134, *193*
Fukui H, Koshimizu K, Nemori R 32, *89*
Fukui H, Koshimizu K, Usuda S, Yamazaki Y 30, 32, 50, 53, 54, 79, *88*, *89*, 137, *193*, 354, *404*
Fukui H, Nemori R, Koshimizu K, Yamazuki Y 30, 32, *88*
Fukui H, see Hirai N 50, 52, *92*, 352, *410*
Fukui H, see Koshimizu K 27, 28, 48, 49, 50, *95*, 117, 128, 131, 132, 134, 138, *195*, 244, *277*, 373, *414*
Fukui H, see Mirai N *424*
Fukui HN, DeVries JE, Wittwer SH, Sell HM 115, *193*
Fukumi H, see Matsushima H 60, 75, *98*, 340, *421*
Fukuyama TT, Moyed NS 337, *404*
Fukuyama TT, see Still CC 337, *437*
Fuller GL, see Kriedémann PE 543, *576*
Fuller RG, see Kemp HT 541, *576*
Funamizu M, see Shibuya T 49, *106*
Funke GL 77, *89*
Furuga T, see Nishio M 13, *101*
Furuya M, Galston AW, Stowe BB 64, *89*
Furuya M, see Imaseki H 331, *411*
Furuya M, see Sherwin JE 365, *435*
Furuya M, see Yamane H 3, *8*

Gabe EJ, Neidle S, Rogers D, Nordman CE 77, *89*
Gahagan HE, Holm RE, Abeles FB 347, *404*
Gaines DE, see McFadden WH 262, *278*
Galbraith DW, Northcote DH 452, *504*
Galbraith MN, Horn DHS 53, 54, *89*
Galbraith MN, Horn DHS, Ito S, Kodama M, Sasse JM 78, *89*
Galbraith MN, see Arora SK 78, *81*
Galil T, see Zeroni M 328, *444*
Galily D, see Goldschmidt EE 360, *405*
Galliard T, Grey TC 175, *193*
Galliard T, Rhodes MJC, Wooltorton LSC, Hulme AC 322, *404*
Gallup AH, see Muir RM 448, *507*
Galsky AG, Lippincott JA 549, *573*
Galson EC, see Sondheimer E 351, 352, 353, *436*
Galson EC, see Tzou DS 368, 386, *439*
Galston AW 64, *89*
Galston AW, Hillman WS 336, 342, 344, 384, *404*
Galston AW, Lavee S, Siegel BZ 341, *404*, *573*
Galston AW, see Bonner J 59, *83*
Galston AW, see Furuya M 64, *89*
Galston AW, see Lavee S 565, *577*
Galston AW, see Nowak J 345, *426*
Galston AW, see Ockerse R 241, *278*
Galston AW, see Penny P 520, *581*
Galston AW, see Siegel BZ 341, *435*
Galston AW, see Tautvydas KI *584*
Galt RHB 26, *89*, 307, *404*
Galt RHB, see Cross BE 25, *85*, 302, 309, 311, 355, *398*
Galun E, see Friedländer M 354, *403*
Gamborg OL, see La Rue TAG 331, *416*
Gamburg KZ, see Rekoslavskaya NI 366, *430*
Gančeva K, see Macháčková J 344, *419*
Gandar JC, Nitsch C 16, 20, *89*, 289, *404*
Gander JE, see Bough WA 71, *83*
Gane R 11, *89*, 317, *404*
Ganguly A, see Biswas BB 455, 456, 501, *502*
Ganguly SN, Ganguly T, Sircar SM 25, 26, *89*
Ganguly SN, Sircar SM 23, 24, 25, *89*
Ganguly SN, see Ganguly T 21, *89*, 366, *404*

Ganguly SN, see Gaskin P 36, 51, *89*
Ganguly SN, see Sanyal T 49, *104*
Ganguly SN, see Sircar PK 23, *106*
Ganguly T, Ganguly SN, Sircar PK, Sircar SM 21, *89*, 366, *404*
Ganguly T, see Ganguly SN 25, 26, *89*
Garcia M, see Maçaira LA 77, *97*
Gardi I, see Tal M 541, *584*
Gardner DCJ, see Anstis PJP 55, *81*, 300, *391*
Gardner G, Sussman MR 458, *504*
Gardner G, Sussman MR, Kende H 478, *504*
Gardner G, see Kende H 456, 477, 488, 495, *506*, 531, *576*
Garmston M, see Milborrow BV 299, *423*
Garrello G, see Bulard C 374, *395*
Garrett S, see Anderson JD 335, *391*
Garshtya LY, see Komissarenko NF 67, *95*
Gaskin P, Kirkwood PS, Lenton JR, MacMillan J, Radley ME 3, *6*, 13, 26, 27, 28, 31, 33, *89*
Gaskin P, MacMillan J 36, 49, 51, *89*, 117, 173, 181, 184, *193*
Gaskin P, MacMillan J, Ganguly SN, Sanyal T, Sircar PK, Sircar SM 36, 51, *89*
Gaskin P, MacMillan J, Zeevaart JAD 27, 49, *89*, 352, *404*
Gaskin P, see Adesomoju AA 50, 51, *81*, 352, *391*
Gaskin P, see Bearder JR 3, *6*, 311, 355, *393*, *394*
Gaskin P, see Beeley LJ 24, 25, 28, 31, 32, 50, 51, *82*
Gaskin P, see Bialek K 49, *82*
Gaskin P, see Browning G 49, *83*
Gaskin P, see Eeuwens CJ 27, *87*
Gaskin P, see Frydman VM 25, 26, 27, 29, 30, 31, 35, 49, 51, *88*, 186, *193*, 268, *277*
Gaskin P, see Graebe JE 31, *90*, 310, 314, 355, 356, 357, 388, *406*
Gaskin P, see Hedden P 4, *6*
Gaskin P, see Martin GC 26, 28, 31, 35, 49, 51, *97*, 184, 186, *196*, *197*, 268, *278*, 352, 354, *421*
Gaskin P, see McInnes AG 355, 376, 378, *422*
Gaskin P, see Most BH 49, *100*
Gaskin P, see Railton ID 49, *103*, 272, *278*, 387, *429*
Gaskin P, see Rivier L 4, *7*
Gaskin P, see Ropers AJ 308, 310, 313, 314, 355, 357, *431*
Gaskin P, see Sinska I 24, 25, *106*, *436*
Gaskin P, see Sponsel VM 4, 5, *7*, 26, 27, 29, 31, 33, 50, 51, *106*, 355, 382, *436*
Gaskin P, see Tinelli ET 51, *108*, 352, 353, *439*
Gaskin P, see Williams PM 268, *279*
Gaskin PS, see Beale MH 190, *190*
Gaspar R, see Darimont E 384, *399*
Gaspar T, Khan AA, Fries D 565, *573*
Gaspar T, Verbeek R, Khan AA 316, *404*
Gaspar T, see Pilet PE 336, 340, 344, 345, 346, *428*
Gawer M, Laloue M, Terrine C, Guern J 371, *404*
Gawer M, see Laloue M 293, 369, 371, 385, *415*
Gawer M, see Sivadjian A 371, *436*
Gawer M, see Terrine C 372, *438*
Gayler KR, Glasziou KT 347, *404*
Gayler KR, see Glasziou KT 561, *573*
Gazit S, see Adato I 327, 330, 331, 389, *391*
Gazit S, see Bittner S 79, *82*
Gefter ML 292, *404*
Gefter ML, Russell RL 292, *404*
Gefter ML, see Rosenbaum N 290, *431*, 473, *508*
Geissman TA 77, *89*
Geissman TA, Deuel P, Bonde EK, Addicott FT 77, *89*
Geissman TA, Verbiscar AJ, Phinney BO, Cragg G 308, 354, *404*
Geissman TA, see Corcoran MR 58, *85*
Geisman TA, see Deuel PG 77, *86*
Geissman TA, see Verbiscar AJ 308, 354, *440*
Geissman TA, see Winters TE 77, *111*
Gelinas DA 342, 344, *404*
Gelinas DA, Postlethwait SN 344, *405*
Gentile IA, Matta A 331, *405*
Geriotti G, see Piozzi F 75, *102*
Gertman E, see Fuchs Y 245, 246, *277*
Gesto MDV, see Mendez J 60, *98*
Gesto MDV, see Sancheza Tames R 60, *104*
Geuns JCM, Vendrig JC 281, *405*
Geuns JMC 78, *89*, 281, *405*
Gevers ECT, see Dekhuijen HM 117, 128, *192*
Ghooprasert P, Spencer M 322, 323, *405*
Ghosheh NS, see Kaufman PB 23, *94*, 561, *575*
Giaquinta R, Beyer EM Jr 362, *405*
Gibson MS, Crane FL 11, *89*
Gibson RA, Barrett G, Wightman F 284, 285, 286, 287, 288, *405*
Gibson RA, Schneider EA, Wightman F 282, 284, 285, 287, *405*
Gibson RA, see Schneider EA 18, 20, *105*, 285, 287, *433*
Giddings JC 254, *277*
Giersak G, see Fox JE 370, *403*
Giesen M, Klämbt D 529, *573*
Gifford EM Jr, see Key JL 557, *576*
Gilbert LI, see Gorell TA 485, *504*
Giliberti P, see Niederwieser A 116, 124, *197*

Gilissen HAM, see Barendse GWM 377, *392*
Gilkes NR, Hall MA 526, 527, 529, 531, 536, *573*
Gillam I, see Struxness LA 385, *438*
Gillard DF, Walton DC 351, 352, 353, *405*
Gillbank LR, see Rowan KS 517, *583*
Gilmore CJ, see Kupchan MS 80, *96*
Gimmler H, see Hartung W 5, *6*
Ginsburg T, see Staehelin M 43, *106*
Ginzburg B 175, *193*
Girardin JP 10, *89*
Glasziou KT, Gayler KR, Waldron JC 561, *573*
Glasziou KT, see Gayler KR 347, *404*
Glenn TL, Kuo CC, Durley RC, Pharis RP 140, *193*
Glover DV, see Leffler HR 555, 561, *577*
Gmelin R 114, 115, 123, 124, 130, *193*
Gmelin R, Virtanen AI 19, 20, *89*, *90*
Gmelin R, see Schlüter M 115, *199*
Godfrey JE, Waters JM 78, *90*
Goeschl JD, Pratt HK 331, *405*
Goeschl JD, Pratt HK, Bonner B 330, *405*
Goeschl JD, see Pratt HK 10, *102*, 317, 326, 327, 328, *428*, 499, *508*
Goldberg R, Prat R 466, *504*
Goldschmidt EE, Galily D 360, *405*
Goldschmidt EE, Goren R, Even-Chen Z, Bittner S 49, *90*, 117, *193*, 374, *405*
Goldschmidt EE, Goren R, Monselise SP, Takahashi N, Igoshi H, Yamaguchi I, Hirose K 281, *405*
Goldschmidt EE, Monselise SP 128, *193*
Goldschmidt EE, Monselise SP, Goren R 157, *193*
Goldthwaite J, see Manos PJ 242, *278*
Gollnov BI, see Letham DS 369, *417*
Gollnov BI, see Parker CW 370, *427*
Gollnow BI, see Summons RE 4, *7*
Gomez-Navarette G, Moore TC 305, 317, *405*
Gonchorov *235*
Gonzalez BL, see Dominguez XA 78, *87*
Good NE, see Andreae WA 364, 365, *391*
Goodwin RH, Taves C 67, *90*
Goodwin RH, see Avers CJ 67, *81*
Goodwin RH, see Pollock BM 67, *102*
Goodwin TW 294, *405*
Gopolan B, see Corey EJ 38, *85*
Gordon AE, Frigerio A 186, *193*
Gordon ME, Letham DS, Parker CW 348, 350, 369, 385, 386, *405*
Gordon ME, see Wilson MM *442*
Gordon SA 20, *90*, 282, *405*
Gordon SA, Paleg LG 282, 284, *405*
Gordon SA, Sanchez NF 281, 285, *405*
Gordon WR, Henderson JHM 340, 341, *405*
Gorell TA, Gilbert LI, Siddall JB 485, *504*
Goren R, Bukovac MJ 364, 365, *406*
Goren R, Bukovac MJ, Flore JA 365, 369, 370, 371, *406*
Goren R, Tomer E 67, *90*
Goren R, see Goldschmidt EE 49, *90*, 117, 157, *193*, 281, 374, *405*
Gorodetsky M, see Bittner S 498, *502*
Gortner WA, Kent M, Sutherland GK 344, *406*
Goto N 300, 374, *406*
Gottlieb D, Carter HE, Wu LC, Slonecker JH 495, *504*
Gottlieb HE, see Cooper R 74, *85*
Goutarel R, see Beauchesne G 40, *82*
Gove JP, Hoyle MC 340, 341, 344, *406*
Gower JL, see McFadden WH 262, *278*
Graebe JE 305, *406*
Graebe JE, Bowen DH, MacMillan J 310, 314, *406*
Graebe JE, Dennis DT, Upper CD, West CA 302, 304, 305, 307, *406*
Graebe JE, Hedden P 310, 314, 355, 356, 357, *406*
Graebe JE, Hedden P, Gaskin P, MacMillan J 31, *90*, 310, 314, 355, 356, 357, 388, *406*
Graebe JE, Hedden P, MacMillan J *406*
Graebe JE, Hedden P, Rademacher W 4, *6*
Graebe JE, Ropers HJ 34, *90*, 302, *406*
Graebe JE, see Bearder JR 3, *6*
Graebe JE, see Bowen DH 314, *395*
Graebe JE, see Hedden P 4, *6*
Graebe JE, see Rademacher W 4, 7, 24, 34, *103*
Graebe JE, see Ropers AJ 308, 310, 313, 314, 355, 357, *431*
Graniti A 533, 543, *573*
Grant Reid JS, Meier H 552, *573*
Gras J-L, see Corey EJ 38, *85*
Gray R, Bonner J 74, *90*
Green PB 514, 515, *573*
Green PB, Erickson RO, Buggy J 515, *573*
Green PB, see Ray PM 514, *582*
Green TR, Dennis DT, West CA 310, 389, *406*
Greene S, see Pollock BM 67, *102*
Greenfield JC, see Vanderhoef LN 521, *585*
Greenway AM, see Allen JRF 13, *81*, 186, *190*
Greenwood MS, Shaw S, Hillman JR, Ritchie A, Wilkins MB 13, *90*
Greenwood MS, see Elliot MC 13, *87*

Greppin H, see Penel C 340, 341, *427*
Grey TC, see Galliard T 175, *193*
Grigorieva NY, Kucherov VF, Loschmikova NV, Chailakhian MK 361, *406*
Grigsby BH, see Hamilton RH 115, *194*, 366, *407*
Grochowska M, see Thimann KV 283, 287, *439*
Grochowska MJ 336, 337, *406*
Grochowska MJ, Karaszewska A 346, *406*
Gross D 23, 56, *90*, 295, *406*
Gross D, Schütte HR 353, 375, *406*
Gross D, see Hartmann S *408*
Gross G, see Sembdner G 162, *199*
Gross HJ, see Harada F 43, *91*, 316, 348, *408*
Grotch SL 252, *277*
Grove JF 23, 33, *90*, 116, *193*
Grove JF, Jeffs PW, Mulholland TPC 23, *90*
Grove JF, MacMillan J, Mulholland TPC, Turner RB 24, *90*
Grove JF, see Brian PW 33, 35, *83*, 479, 480, 481, *502*
Grove JF, see Cross BE 301, *398*, 481, *503*
Grove MD, Spenser GF, Rohwedder WK, Mandava N, Worley JF, Warthen JD Jr, Steffens GL, Flippen-Anderson JL, Cook JC Jr 3, *6*
Grover S, Purves WK 335, *407*
Grubisić D, see Konjević R 487, *506*
Grümmer G 57, *90*
Grunwald C, Lockard RG 170, 171, *193*
Guern J, Doree M, Sadorge P *407*
Guern J, see Doree M *400*
Guern J, see Gawer M 371, *404*
Guern J, see Hall RH 350, 372, *407*
Guern J, see Laloue M 293, 348, 350, 369, 385, 386, *415*
Guern J, see Pethe-Sadorge P *427*
Guern J, see Sivadjian A 371 *436*
Guern J, see Terrine C 372, *438*
Guha R, see Roy S 74, *104*
Guilfoyle TJ, see Chen YM 557, *570*
Guinn G 331, *407*
Guiochon G, see Arpino PJ *276*
Gummons RE, see McLeod JK 4, *7*
Günther G, see Schneider J 336, *433*
Gurumurti K, see Nanda KK 340, *426*
Gurvich IA, see Zaretkii VI 184, *201*
Gutterman Y, see Lavie O 74, *96*

Haagen-Smit AJ, Dandliker WB, Wittwer SH, Murneek AE 13, *90*, 113, *194*
Haagen-Smit AJ, Leach WD, Bergren WR 13, *90*, 113, *193*
Haagen-Smit AJ, see Kögl F 12, 13, *95*
Haard NF, see Frenkel C 346, *403*
Haberlandt G 38, 39, *90*
Haddon VR, see Frenkel C 337, *403*
Haddon WF 4, *6*
Haddon WF, see Heftmann E 176, *194*
Hager A, Menzel H, Krauss A 451, *504*, 522, 537, *573*
Hager A, Schmidt R 337, *407*
Hager A, see Menschick R 336, 340, 342, 347, *422*
Haggstrom N, see Andersson B 49, 50, *81*
Haginiwa J, see Murakoshi I 372, *425*
Hahn H 180, *194*
Hahn H, De Zacks R, Kende H 384, *407*
Hahn H, Heftmann E, Blumbach M 293, *407*
Haimovich J, see Fuchs S 245, 246, *277*
Haines BL, see Muller CH 76, *100*
Hakala MT, see Divekar AY 372, *400*
Halevy AH 344, *407*
Halevy AH, Kessler B 542, *573*
Halevy AH, Shilo R 316, *407*
Halevy AH, Simchon S, Shilo R 362, *407*
Halevy AH, see Mayak S 328, *421*
Halevy AH, see Rudich J 361, *432*
Hall IV, Forsyth FR 328, *407*
Hall MA 512, *573*
Hall MA, Acaster MA, Dodds JH, Evans DE, Jones JF, Jerie PN, Mutumba GC, Niepel B, Shaari AR 541, *573*
Hall MA, Brown RL, Ordin L 533, *573*
Hall MA, Kapuya JA, Sivakumaran S, John A 532, *573*
Hall MA, Ordin L 520, 524, 554, *573*
Hall MA, see Bengochea T 500, *502*
Hall MA, see Darvill AG 526, 527, 528, 529, 530, 536, *571*
Hall MA, see Doods JH 363, *400*
Hall MA, see El-Beltagy AS 331, 332, *401*, 542, *572*
Hall MA, see Gilkes NR 526, 527, 529, 531, 536, *573*
Hall MA, see Jerie PH 12, *93*, 500. *505*
Hall MA, see Perie PH 363, *412*
Hall MA, see Sivakumaran S 50, *106*
Hall MH, see Jerie PH 500, *503*
Hall PL 364, 366, *407*
Hall PL, Bandurski RS 282, *407*
Hall RH 43, 44, *90*, 137, *194*, 289, 368, 372, *407*, 472, *504*
Hall RH, Alam SN, McLennan BD, Terrine C, Guern J 350, 372, *407*
Hall RH, Csonka L, David H, McLemman B 137, *194*
Hall RH, Csonka L, David H, McLennan B 41, *90*, 292, 362, *407*, 531, *573*

Hall RH, de Ropp RS 39, *90*
Hall RH, Mintsioulis G 350, 372, *407*
Hall RH, Robins MJ, Stasiuk L, Thedford R 43, *90*
Hall RH, Strivastava BIS 471, 500, *504*
Hall RH, see Alam SN 173, *190*
Hall RH, see Chen C-M 166, *191*, 290, 291, 350, 386, *397*, 473, *502*
Hall RH, see Dyson WH 41, 44, 45, *87*, 128, 139, 173, *192*
Hall RH, see Fittler F 289, *403*, 473, 478, *504*, 531, *572*
Hall RH, see Kline LK 289, *413*
Hall RH, see Laloue M *415*
Hall RH, see McLennan BD 348, *422*
Hall RH, see Miura GA 293, 348, 384, *424*
Hall RH, see Pačes V 350, 372, *427*
Hall RH, see Robins MJ 43, *103*
Hall RH, see Whitty CD 350, 371, 372, 386, *441*
Hall RS, see Robins MJ 137, 165, *199*
Hall SM, Medlow GC 13, *91*
Hall WC, Millers CS, Herrero FA *407*
Hall WC, Morgan PW 347, *407*
Hall WC, see Herrero FA 347, 362, *409*
Hall WC, see Morgan PW 333, *424*
Haltiwanger RC, see Kupchan MS 80, *96*
Hamilton R, see Coolbaugh RC 305, 316, *398*
Hamilton RH, Bandurski RS, Grigsby BH 115, *194*, 366, *407*
Hamilton RH, Meyer HE, Burke RE, Feung CS, Mumma RO 336, 337, 340, 344, *407*, 448, *504*
Hamilton RH, see Black RC 282, 383, *395*
Hamilton RH, see Feung CS 365, 366, *403*
Hamilton RH, see Troxler RF 336, 337, *439*
Hamilton RO, see Feung CS 365, *403*
Hammond RK, see Rudnicki R 60, *104*
Hamzi HQ, see Skoog F 471, 472, *508*
Hancock RC, Barlow HWB, Lacey HJ 241, *277*
Hanke R, see Hilgenberg W 337, 340, *410*
Hansch C, Muir RM, Metzenberg RL 448, *504*
Hansch C, see Muir RM 448, *507*
Hansen E 325, 326, 327, *407*
Hansen E, see Buhler DR 362, *395*
Hansen J, see Andersen AS 337, *391*, 533, *568*
Hanson AD, Kende H 319, 320, 322, 325, 326, 328, 329, 331, *407*
Hanson AD, see Kende H 319, 325, 329, 331, *413*, 540, *576*
Hanson JB, see O'Brien TJ 453, 455, 489, *507*, 518, 520, 558, *580*
Hanson JR 25, 26, 33, *91*, 116, *194*
Hanson JR, Hawker J 302, 310, 312, 313, 314, *408*
Hanson JR, Hawker J, White AF 308, 309, 310, *408*
Hanson JR, Hough A, White AF 304, 310, 311, *408*
Hanson JR, Sarah FY, Fraga BM, Hernandez MG 311, *408*
Hanson JR, White AF 302, 304, 308, 309, 310, 311, 313, *408*
Hanson JR, see Brown JC 25, *83*
Hanson JR, see Cross BE 25, *85*, 131, *192*, 302, 355, *398*
Hanson JR, see Dockerill B 314, *400*
Hanson JR, see Evans R 302, 305, 309, 310, 312, 313, *402*
Harada F, Gross HJ, Kimura F, Chang SH, Nishimura S, Rajbhandary UL 43, *91*, 316, 348, *408*
Harada H, Lang A 315, 316, *408*
Harada H, Nitsch JP 23, 25, *91*
Harada H, Yokota T 36, *91*, 131, *194*, 379, 381, *408*
Harada H, see Ranjeva R 565, *582*
Harada J, Nakayama H 58, *91*
Harborne JB 9, 57, *91*
Hardin JW, Cherry JG 453, *504*
Hardin JW, Cherry JH 559, *573*
Hardin JW, Cherry JH, Morré DJ, Lembi CA 457, 470, *504*, 520, 560, *574*
Hardin JW, O'Brien TJ, Cherry JH 453, 454, *504*, 520, 554, 559, *574*
Hare RC 336, 340, 344, *408*
Harmey MA, see Minchin A 337, 340, 341, *424*
Harmey MA, see Mollan RC 21, *100*, 364, 365, *424*
Har-Paz I, see Bittner S 498, *502*
Harper JR, see Prins HBA 538, *581*
Harris A, see Birch AJ 301, 309, *395*
Harris TM, see Baskin JM 67, *81*
Harrison DM, MacMillan J 28, *91*, 302, *408*
Harrison MA, Saunders PF 271, *277*, 374, *408*
Harrison MA, see Walton DC 352, 353, 374, *441*
Harrison MA, Walton DC 352, 353, *408*
Hartman MA, see Normand G 458, *507*
Hartmann GR, see Kessler C 454, *506*
Hartmann HT, El-Hamady M, Whisler J 332, *408*
Hartmann S, Gross D, Schütte HR *408*
Hartmann W 295, 375, *408*
Hartnell GF, see Chen C-M *397*
Hartsuck JA, Lipscomb WN 23, 33, *91*
Hartung W, Behl R 352, *408*

Hartung W, Gimmler H, Heilmann B, Kaiser G 5, *6*
Harvey RB 10, *91*
Harvey WE, see Bearder JR 355, *394*
Haschke HP, Lüttge U 538, 539, *574*
Hasegawa K, Hashimoto T 74, *91*
Hasegawa K, see Hashimoto T 74, *91*
Hashimoto T 530, 531, *574*
Hashimoto T, Hasegawa K, Kawarada A 74, *91*
Hashimoto T, Hasegawa K, Yamaguchi H, Saito M, Ishimoto S 74, *91*
Hashimoto T, Ikai T, Tamura S 49, *91*, 132, 137, *194*
Hashimoto T, Tajima M 74, *91*
Hashimoto T, see Hasegawa K 74, *91*
Hashizume T, Kimura K, Sugiyama T 41, *91*
Hashizume T, Sugiyama T, Imura M, Cory HJ, Scott MF, McCloskey JA 42, 45, *91*, 186, *194*
Hasson EP, West CA 308, 389, *408*
Hatakeyama I, see Kobayashi S 516, *576*
Hattori H, Marumo S 20, *91*, 138, 142, *194*, 289, 365, *408*
Hattori H, see Marumo S 16, 20, *98*, 115, 130, 133, 138, 142, *197*, 289, *421*
Haughton PM, see Cleland RE 520, *570*
Haughton PM, see Rayle DL 521, *582*
Hawker J, see Hanson JR 302, 308, 309, 310, 312, 313, 314, *408*
Haxby J, see Kinsky SC 495, *506*
Hayashi H, Koshimizu K, Asahira T, Matsubara S 76, *91*
Hayashi T, see Atsumi S 120, *190*
Hayashi Y, Sakan T 78, *91*
Hayashi Y, Yokai J, Watanabe Y, Sakan T 78, *91*
Hayashi Y, see Sakan T 53, *104*
Haydon DA, see Taylor JL 495, *509*
Heacock RA, see Hutzinger O 159, *194*
Heald JK, see Little CHA 13, 50, *97*, 374, *418*
Heald JK, see Lorenzi R 25, 36, *97*, 125, *196*, 362, 377, 379, 381, 382, *419*
Heald JK, see Thompson AG 45, *108*, 269, *279*
Heatherbell DA, Horward BH, Wicken AJ 316, *408*
Hebard FV, Amatangelo SJ, Dayanandan P, Kaufman PB 530, *574*
Hecht SM, Leonard NJ, Burrows WJ, Armstrong DJ, Skoog F, Occolowitz J 348, *409*
Hecht SM, see Armstrong BJ 43, *81*
Hecht SM, see Burrows WJ 41, 43, 44, 45, *84*, 137, 168, *191*, 292, 348, *396*
Hecht SM, see Skoog F 531, *583*
Hedden P 2, *6*
Hedden P, Graebe JE *6*
Hedden P, MacMillan J, Phinney BO 302, 312, 313, *409*
Hedden P, Phinney BO 5, *6*, 305, 317, *409*
Hedden P, Phinney BO, Heupel R, Fujii D, Cohen H, Gaskin P, MacMillan J, Graebe JE 4, *6*
Hedden P, Phinney BO, MacMillan J, Sponsel VM 302, 303, 305, 316, *409*
Hedden P, see Bearder JR 302, 312, 313, 354, 355, *393*
Hedden P, see Graebe JE 4, *6*, 31, *90*, 310, 314, 355, 356, 357, 388, *406*
Heerkloss R, Libbert E 282, 283, 383, *409*
Heftmann E 281, *409*, 483, *504*
Heftmann E, Saunders GA, Haddon WF 176, *194*
Heftmann E, see Bennett RD 308, *394*
Heftmann E, see Bennett TD *394*
Heftmann E, see Hahn H 293, *407*
Heftmann E, see Ruddat M 310, *432*
Heilmann B, see Hartung W 5, *6*
Heinz DE, see Davis LA 174, *192*
Helgerson SL, Cramer WA, Morré DJ 452, *504*, 540, *574*
Helgeson J, see Skoog F 471, 472, *508*
Helgeson JP, Leonard NJ 43, *91*, 291, *409*
Helgeson JP, see Woodward MD 75, *111*
Helgeson P, see Upper CD 173, 174, 184, *200*
Hemberg T 48, 56, *91, 92*
Hemberg T, Westlin PE 127, *194*
Hemberg T, see Jindal KK 346, *412*
Hemingway RJ, see Kupchan MS 77, *96*
Hemingway RJ, see Sequeira L 77, *105*
Hemming HG, see Borrow A 22, *83*
Hemming HG, see Brian PW 243, *276*
Hemphill DD, Baker LR, Sell HM 23, 24, 25, 36, *92*, 376, 378, 381, 382, *409*
Henbest HB, Jones ERH, Smith GF 17, *92*, 115, 130, *194*
Henderson JHM, see Gordon WR 340, 341, *405*
Hendon J, see Nagl W 556, *580*
Hendricks SB, see Toole EH 57, *109*
Hendrie MR, see Berrie AMM 69, *82*
Henehan C, see Borrow A 22, *83*
Henrick CA, see Katsumi M 35, *94*
Henriksson E, see Dasilva EJ 317, *399*
Henriksson LE, see Dasilva EJ 317, *399*
Henry EW, see Bratton BO 347, *395*
Henson IE 370, 386, *409*
Henson IE, Wareing PF 385, 386, *409*

Henson IE, Wheeler CT 348, 350, 370, 385, 386, *409*
Henson IE, see Wareing PF 45, *110*
Henson JE, see Wareing PF 350, 370, *441*
Henson W, see Mertz D 315, 362, *422*
Henze H, Blair C 235, *277*
Hepler PK, see Torrey JG 553, 564, *585*
Herald JK, see Thompson AG 185, *200*
Hernandez MG, see Hanson JR 311, *408*
Herner RC, Sink KC Jr 328, *409*
Herner RC, see Boller T 320, 326, *395*
Herout V, see Vokáč K 77, *109*
Herrero FA, Hall WC 347, 362, *409*
Herrero FA, see Hall WC *407*
Hertel R 459, *504*
Hertel R, Dohrmann U, Jesaitis AJ, Peterson W 468, *504*
Hertel R, Thomson K-S, Russo VEA 457, 458, 459, *504*
Hertel R, see Dohrmann U 461, 462, 464, 465, 468, 470, *503*, 539, *571*
Hertel R, see Jacobs M 461, 464, 469, 470, *505*
Hertel R, see Lembi CA 457, 458, *506*
Hertel R, see Ray PM 461, 464, 465, 466, 468, *508*, *582*
Hertel R, see Rayle DL 517, *582*
Hertel R, see Thomson KS 458, *509*
Herz W, Watanabe H, Miyazaki M, Kishida Y 77, *92*
Hess CE, see Frenkel C 340, 341, *403*
Heupel R, see Hedden P 4, *6*
Hewett EW, see Horgan R 41, 45, 47, *92*, 139, 173, *194*, 291, *410*, 472, 473, *505*
Hewett EW, Wareing PF 139, *194*, 368, 370, 385, 386, *409*, *410*
Hewitt EJ 336, *410*
Heyes JK, see Penny P 519, *581*
Hietala PK, see Virtanen AE 75, *109*
Higgins TJV, Zwar JA, Jacobsen JV 489, 490, *505*
Higinbotham N, see Prins HBA 538, *581*
Hild V, see Menschick R 336, 340, 342, 347, *422*
Hilgenberg W, Hanke R, Bürstell H 337, 340, *410*
Hill B, see Bowen A 539, *569*
Hillman JR 113, *194*
Hillman JR, Young I, Knights BA 49, *92*, 184, *194*, 272, *277*
Hillman JR, see Bridges IC 13, *83*
Hillman JR, see Bridges IG 13, *83*, 184, *191*, 264, *276*
Hillman JR, see Greenwood MS 13, *90*
Hillman JR, see Itai C 52, *93*
Hillman JR, see McDougall J 13, *98*, 383, *421*
Hillman JR, see McWha JA 50, *98*, 244, *278*, 354, *422*
Hillman JR, see Philipson JJ 517, 518, 519, *581*
Hillman JR, see White JC 13, *111*
Hillman WS 242, *277*
Hillman WS, see Galston AW 336, 342, 344, 345, 384, *404*
Hinman RI, Lang J 337, 339, 342, *410*
Hinojosa M, see Dominguez XA 74, *87*
Hiraga K, Kawabe S, Yokota T, Murofushi N, Takahashi N 23, 24, 25, 30, 35, 36, 49, *92*, 116, 131, 134, *194*, 378, 379, *410*
Hiraga K, Yamane H, Takahashi N 164, 171, 173, *194*, 377, 378, 379, 381, *410*, 480, *505*
Hiraga K, Yokota T, Murofushi N, Takahashi N 124, 126, 131, 147, 148, 149, *194*, 378, 379, 388, *410*
Hiraga K, see Yokota T 184, *201*
Hirai N, Fukui H, Koshimizu K 50, 52, *92*, 352, *410*
Hiremath SV, see Mathur SB 77, *98*
Hiron RW, Wright STC 374, 375, *410*
Hiron RWP, see Burden RS 300, *396*
Hiron RWP, see Wright STC 52, *111*, 542, *586*
Hirose K, see Goldschmidt EE 281, *405*
Hirose K, see Igoshi M 115, 130, 133, 142, 144, *195*
Hirose K, see Takahashi N 14, 20, *108*, 115, 145, 188, 189, *200*, 281, *438*
Hirschberg K, see Dollstädt R 470, *503*, 559, *571*
Hishada S, see Nakagawa S 186, *197*
Hitchcock AE, see Crocker W 11, *85*
Hitchcock AE, see Row VV 365, *431*
Ho DT-H, Varner JE 489, *505*, 546, 548, *574*
Hoad GV 24, 25, 26, 27, 31, 49, *92*
Hoad GV, Bowen MR 361, 388, *410*
Hoad GV, Phinney BO, Sponsel VM, MacMillan J 5, *6*
Hoad GV, see Browning G 49, *83*
Hoad GV, see Sponsel VM 481, *508*
Hocking TJ, Clapham J, Cattell KJ 498, *505*
Hodges R, Porte AL 53, *92*
Hofinger M, Böttger M 16, *92*, 184, *194*, 289, *410*
Hofmann A, see Phillips IDJ 353, *428*
Hogetsu T, Shibaoka H, Shimokoriyama M 563, *574*
Hokanson R, see Evans ML 521, *572*
Holm RE, Key JL 554, *574*
Holm RE, O'Brien TJ, Key JL, Cherry JH 557, 558, *574*
Holm RE, see Gahagan HE 347, *404*
Holub M, Samek Z, Poplawski J 53, *92*
Holy A, see Pačes V 372, 386, *427*

Hong CI, see Dyson WH 44, 45, *87*
Honigman WA, see Taiz L 551, *584*
Hootelé C, see Bourdoux P 18, *83*
Hopping ME, Bukovac MJ 13, *92*
Hopping ME, Young H, Bukovac MJ 42, *92*
Horegott H von 338, *410*
Horgan JM, see Horgan R 41, 45, 47, *92*, 139, 173, *194*, 291, *410*, 472, 473, *505*
Horgan JM, see Purse JG 41, 46, *103*, 293, 348, *429*
Horgan R 45, *92*, 117, 127, *194*, 369, 386, *410*
Horgan R, Hewett EW, Horgan JM, Purse J, Wareing PF 45, 47, *92*, 291, *410*, 472, 473, *505*
Horgan R, Hewett EW, Purse JG, Horgan JM, Wareing PF 41, *92*, 139, 173, *194*
Horgan R, Hewett EW, Purse JG, Wareing PF 139, 173, *194*
Horgan R, see Lorenzi R 25, 36, *97*, 125, *196*, 362, 369, 377, 379, 381, 382, *419*
Horgan R, see Purse JG 41, 46, *103*, 293, 348, *429*
Horgan R, see Thompson AG 45, *108*, 185, *200*, 269, *279*
Horgan R, see Wang TL 47, *110*, 127, 139, 173, 184, *201*, 348, 370, *441*
Horgan R, see Wareing PF 45, *110*, 350, 370, *441*
Hori S 22, *92*
Horn DHS, see Galbraith MN 53, 54, 78, *89*
Horng AJ, Yang SF 336, 340, *410*
Horton RF 542, 543, *574*
Horton RF, Moran L 543, *574*
Horward BH, see Heatherbell DA 316, *408*
Hoshino H, see Takahashi N 184, *200*
Hostettmann K 4, *7*
Hough A, see Hanson JR 304, 310, 311, *408*
Housley S, see Bentley JA 120, *191*
Howden MEH, see Altree-Williams S 60, *81*
Howell L, see Thomas TH 167, 180, *200*
Hoyle MC 341, *410*
Hoyle MC, Routley DG 343, 344, *410*
Hoyle MC, see Gove JP 340, 341, 344, *406*
Hradilik J 334, 346, 347, *410*
Hsiao TC 333, *410*, 543, *574*
Hsu A, see Rappaport L *430*
Huber J, see Noma M 3, *7*
Hubner G, see Dollstädt R 470, *503*, 559, *571*
Hudek J, see Stonier T 345, *437*
Huelin FE 11, *92*
Huelin FE, Kennett BH 12, *92*
Huff AK, Ross CW 562, *574*
Hulme AC, see Galliard T 322, *404*
Hulme AC, see Mapson LW 10, *97*, 317, 326, *420*
Hulme AC, see Meigh DF 322, *422*
Hummel JP, Dreyer WJ 447, *505*
Hung PD, see Sembdner G 381, *434*
Hunston DL 446, *505*
Hutzinger O, Heacock RA, MacNeill JD, Frei RW 159, *194*
Hutzinger O, Jamieson WD 184, *194*
Hutzinger O, Kosuge T 365, *410*
Hutzinger O, see Jamieson WD 184, *195*, 264, *277*
Hutzinger O, see Raj RK 116, 123, 133, 159, 161, *198*
Huystee RB van, see Srivastava OP 341, *436*
Huystee RB van, see Teissere M 454, *509*
Hwang Y-S, Matsui M 12, *92*
Hyeon SB, see Isoe S 53, *93*
Hyeon SB, see Sakan T 53, *104*
Hyodo H 319, *411*

Ichihara K, Kawai T, Kaji M, Noda M 79, *92*
Ichihara K, Kawai T, Noda M 79, *92*
Idzko D, see Byrd DJ 157, 160, *191*
Igarashi H, see Yabuta T 22, *111*
Igoshi H, see Goldschmidt EE 281, *405*
Igoshi M, Yamaguchi I, Takahashi N, Hirose K 115, 130, 133, 142, 144, *195*
Igoshi M, see Takahashi N 14, 20, *108*, 115, 145, 188, 189, *200*, 281, *438*
Ihl M 365, *411*, *505*
Iino Y, Tanaka A, Yamashita K 77, *92*
Iinuma H, see Yomo H 316, *443*
Iitaka Y, see Yokota T 13, 14, 36, *112*, 379, *443*
Ikai T, see Hashimoto T 49, *91*, 132, 137, *194*
Ikegami F, see Murakoshi I 372, *425*
Ikegami S, see Komoto N 49, 79, *95*, 131, 132, *195*
Ikekawa N, Kagawa T, Sumiki Y 170, *195*
Ikuma H, see Adams PA 530, *568*
Ikuma H, see Kaufman PB 561, *575*
Ilan I 518, 537, *574*
Ilan I, see Nelson H 520, *580*
Imai T, see Wada K 311, *441*
Imaseki H 347, *411*, 565, *574*
Imaseki H, Kondo K, Watanabe A 334, *411*
Imaseki H, Pjon CJ, Furuya M 331, *411*
Imaseki H, Uritani I, Stahmann MA 347, *411*
Imaseki H, Watanabe A, Odawara S 333, *411*
Imaseki H, see Kondo K 333, 334, *414*
Imaseki H, see Odawara S 336, 390, *426*
Imaseki H, see Sakai S 319, 326, 333, 335, 365, 389, *432*
Imaseki H, see Sato T 557, *583*
Imaseki H, see Shingo S 333, *435*
Imber D, see Tal M 541, 543, *584*

Imbert MP, Wilson LA 344, *411*
Imura M, see Hashizume T 42, 45, *91*, 186, *194*
Ina K, Sakato Y, Fukami H 54, *92*
Inada K, see Suge H 532, *584*
Ingle J, see Key JL 520, 558, *576*
Ingle J, see Timmis J 555, *584*
Inoue K, see Asakawa Y 377, *392*
Inui M, see Koshimizu K 27, 50, *95*, 117, 128, 138, *195*, 373, *414*
Iriuchijima S, see Murofushi N 27, *100*, 113, 114, 142, 161, *197*
Iriuchijima S, see Shibaoka H 77, *106*
Iriuchijima S, Tamura S 77, *93*
Isenberg FMR, see Thomas TH 167, 180, *200*
Isenhour TL, see Wangen LE 250, 251, 253, *279*
Ishida K, see Yamakawa T 4, *8*
Ishigami T, Suzuki Y 288, *411*
Ishii H, see Fukui H 27, 28, 30, 31, *88*, 134, *193*
Ishii H, see Koshimizu K 27, 28, *95*
Ishii T, see Nishio M 13, *101*
Ishikura H, see Nishimura S 43, *101*
Ishimoto S, see Hashimoto T 74, *91*
Iskrič S, see Magnus V 286, 287, 338, *420*
Isoe S, Hyeon SB, Katsumura S, Sakan T 53, *93*
Isoe S, see Sakan T 53, *104*
Isogai S, see Komoto N 49, *79, 95*
Isogai Y, Komoda Y, Okamoto T 49, 65, *93*
Isogai Y, Nomoto S, Noma T, Okamoto T 21, *93*
Isogai Y, Okamoto T, Koizumi T 14, 18, *93*, 130, *195*
Isogai Y, Okamoto T, Komoda Y 49, *93*, 131, 132, 137, *195*
Isogai Y, see Komoda Y 27, 80, *95*
Isogai Y, see Okamoto T 13, 20, 21, 75, *101*, 130, *198*
Itai C, Vaadia Y 386, *411*
Itai C, Weyers JDB, Hillman JR, Meidner H, Willmer C 52, *93*
Ito S, Kodama M 78, *93*
Ito S, see Galbraith MN 78, *89*
Itoh K, see Murofushi N 33, *100*
IUPAC-IUB Commission on Biochemical Nomenclature 40, *93*
Iversen TH, Aasheim T 336, *411*
Iversen TH, see Aasheim T 336, *390*
Iwamura J, see Komai K 77, *95*
Iwata T, Stowe BB 334, *411*

Jablonovic M, Nooden LD 469, *505*
Jablonski J, Skoog F 39, *93*
Jackson HO, see Weaver GM 365, *441*
Jackson MB, Campbell DJ 331, 332, *411*
Jackson MB, Osborne DJ 330, *411*
Jackson MB, see Musgrave A 513, 532, *580*
Jackson MB, see Osborne DM 374, *427*
Jackson RW, see Stodola FH 22, 23, *107*
Jacobs M, Hertel R 461, 464, 469, 470, *505*
Jacobs M, Ray PM 518, 523, 527, 529, 534, 536, *574*
Jacobsen JV 450, 490, 491, *505*, 548, *574*
Jacobsen JV, Scandalios JG, Varner JE 545, 551, *574*
Jacobsen JV, Varner JE 546, 548, 551, *574*
Jacobsen JV, Zwar JA 489, *505*
Jacobsen JV, see Ashford AE 551, *568*
Jacobsen JV, see Higgins TJV 489, 490, *505*
Jacobsen JV, see Zwar JA 539, 540, 546, *586*
Jacobson A, Corcoran MR 59, *93*
Jaffe MJ *411*
Jaffe MJ, see Riov J 316, *431*
Jagodowicz M, see Mozell MM 495, *507*
Jain ML, Kadkade PG, van Huysse P 346, *411*
Jamieson WD, Hutzinger O 184, *195*, 264, *277*
Jamieson WD, see Hutzinger O 184, *194*
Jänicke S, see Schneider G 171, 172, 183, *199*
Jansen EF 317, 362, *411*
Janssen MGH 344, *411*
Jarvis BC, see O'Brien TJ 453, 455, 489, *507*, 518, 520, 558, *580*
Jasdanwala RT, Singh YD, Chinoy JJ 346, *411*
Jaworski EG, see Mann JD 116, 143, *196*
Jefferies PR, Knox JR, Ratajczak T 309, 311, *411, 412*
Jefferies PR, see Bakker HJ 311, *392*
Jefferies PR, see Cook JF 311, *397*
Jefferies PR, see Dawson RM 311, *399*
Jefferies PR, see Katsumi M 35, *94*
Jeffreys EG, see Borrow A 22, *83*
Jeffreys H *280*
Jeffs PW, see Grove JF 23, *90*
Jeffs RA, Northcote DM 566, *575*
Jelsema CL, Ruddat M, Moore DJ, Williamson FA 360, *412*
Jenkins ID, see Cowley DE 369, *398*
Jenkins ID, see Letham DS *417*
Jenkins ID, see Parker CW 369, 370, *427*
Jenkins PA, Shepherd KR 49, *93*
Jenkinson IA 518, *575*
Jenner CF, see Collins GG 550, *571*
Jensen EV, Suzuki T, Kawashima W, Strumpf E, Jungblut PW, DeSombre ER 485, *505*

Jerie PH 330, *412*
Jerie PH, Chalmers DJ 327, 332, 389, *412*
Jerie PH, Hall MA 12, *93*, 363, *412*
Jerie PH, see Bengochea T 500, *502*
Jerie PH, see Dodds JH 363, *400*, 500, *503*
Jerie PH, Shaari AR, Hall MA 363, *412*, 500, *505*
Jerie PH, Shaari AR, Zeroni M, Hall MA 500, *505*
Jerie PH, Zeroni M, Hall MA 363, *412*
Jerie PN, see Hall MA 541, *573*
Jesaitis AJ, see Hertel R 468, *504*
Jesensky C, see Peterkofsky A 472, *507*
Jindal KK, Hemberg T 346, *412*
Job D, see Ricard J 339, 340, 342, 343, *430*
Johansen S 541, *575*
John A 532, *575*
John A, see Hall MA 532, *573*
John WW, Curtis RW 322, *412*
John WW, see Lau OL 334, *416*
Johnson AW 77, *93*
Johnson KD, Daniels D, Dowler MJ, Rayle DL 518, 525, *575*
Johnson KD, Kende H 491 *505*
Johnson KD, Kende HL 549, 552, *575*
Johnson KD, see Dohrmann U 533, *571*
Johnson KD, see Purves WK 285, *429*
Johnson KD, see Rayle DL 532, *582*
Johri MM, Varner JE 488, *505*, 560, *575*
Johri MM, see Varner JE 545, 548, 551, *585*
Jones DA 35, *93*
Jones DF 114, 116, *195*, 361, 376, 387, *412*
Jones DF, see Elson GW 162, 163, *193*
Jones ERH, Taylor WC 18, *93*, 115, *195*
Jones ERH, see Henbest HB 17, *92*, 115, 130, *194*
Jones JD, see Meigh DF 322, *422*
Jones JF, Kende H 333, *412*
Jones JF, see Hall MA 541, *573*
Jones KC 74, *93*, 161, *195*
Jones KC, West CA *93*
Jones MB, Enzie JV 74, *93*
Jones MG, Metzger JD, Zeevaart JAD 177, *195*
Jones MG, Zeevaart JAD 4, *7*
Jones RJ, Mansfield TA 542, *575*
Jones RJ, see Mansfield TA 543, *578*
Jones RL *412*, 490, 492, *505*, 540, 548, 549, 551, *575*
Jones RL, Chen RF 488, 489, *505*
Jones RL, Lang A 388, *412*
Jones RL, Phillips LDJ 316, 387, 388, *412*
Jones RL, Varner JE 243, *277*
Jones RL, see Armstrong JE 491, *501*
Jones RL, see Kuhn-Silk W 356, *415*
Jones RL, see Nash LJ 357, 358, *426*, 479, 482, *507*
Jones RL, see Phillips IDJ 387, *428*
Jones RL, see Silk WK *435*, 497, *508*, 530, *583*
Jones RL, see Stoddart JL 378, 380, *437*, 482, 488, *508*
Jones RL, see Stuart DA 497, *509*, 530, *584*
Jones RL, see Taiz L 551, *584*
Jones TWA 482, *505*
Jones WW, Coggins CW, Embleton TW 26, 49, *93*
Jönssen A 448, *505*
Jordan WR, see McMichael BL 331, *422*
Joseph-Nathan P, see Romo J 77, *104*
Joshi PP, see Bhakuni DS 53, *82*
Joshi RS, see Paul A 77, *102*
Jouanneau JP 561, *575*
Jouanneau JP, Péaud-Lenoel C 561, *575*
Jouanneau JP, Tandeau de Marsac N 556, *575*
Jung H, see Kaldewey H 241, *277*
Jungblut PW, see Jensen EV 485, *505*

Kabuto C, see Kato T 78, *94*
Kabuto C, see Tsunakawa M 78, *109*
Kadkade PG, see Jain ML 346, *411*
Kadyrov CS, see Kefeli VI 56, *94*, 117, *195*
Kaethner TM 450, *505*
Kagawa T, Fukinbara T, Sumiki Y 134, 161, 162, *195*
Kagawa T, see Ikekawa N 170, *195*
Kainuma K, see Tamura S 35, *108*
Kaiser G, see Hartung W 5, *6*
Kaiser H *280*
Kaiser W, see Libbert E 284, 287, *417*
Kaiss-Chapman RW, Morris RO 40, *93*
Kaiya S, see Kobayashi A 79, *94*
Kaji J, see Asakawa Y 125, *190*, 377, 378, *392*
Kaji M, see Ichihara K 79, *92*
Kaldewey H, Wakhloo JL, Weis A, Jung H 241, *277*
Kaldewey H, see Edelbluth E 383, *401*
Kalkarni GH, see Paul A 77, *102*
Kalsi PS, Chhabra BR, Singh OS 78, *93*
Kalsi PS, Singh OS, Chhabra BR 78, *93*
Kalsi PS, Vij VK, Singh OS, Wadia MS 77, *93*
Kamienska A, Durley RC, Pharis RP 23, 24, 25, *93*, 356, 357, *412*
Kaminek M 289, 293, *412*
Kaminek M, Paces V 372, *412*

Kaminek M, see Burrows WJ 41, 43, 44, 45, *84*, 292, 348, *396*
Kaminek M, see Pačes V 372, 386, *427*
Kamisaka S, Masuda Y 554, *575*
Kaminska S, see Sakurai N 62, *104*
Kamisaka S, see Sakurai N 241, *279*, 344, 346, *432*
Kamisaka S, see Shibata K 61, *106*
Kamiya K, see Nishikawa N 77, *101*
Kamiya Y, see Cho KY 316, *397*
Kanchan SD 77, *93*
Kang BG, Burg SP 331, *412*, 540, *575*
Kang BG, Newcomb W, Burg SP 333, *412*
Kang BG, Ray RM 331, *412*
Kang BG, Yocum CS, Burg SP, Ray PM 331, *412*
Kannangara T, see Durley RC 272, *276*
Kapil RS, see Bhakuni DS 53, *82*
Kaplan B, see Kessler B 549, *576*
Kapuya JA, see Hall MA 532, *573*
Karaszewska A, see Grochowska MJ 346, *406*
Karim A, see Kupchan MS 77, 80, *96*
Karl R, see Zimmermann H 364, *444*
Karlsnes A, see Cleland RE 529, *570*
Karrer W 57, *94*
Kasai Z, see Shimokawa K 175, *199*, 322, 325, 335, 362, *435*
Kasamo K, Yamaki T 452, 469, *506*
Kasymov SZ, see Tarasov VA 77, *108*
Katekar GF 450, *506*
Kato J, Katsumi M, Tamura S, Sakurai A 35, *94*
Kato J, see Momotani Y *579*
Kato J, see Murofushi N 27, 28, *100*, 113, 114, 134, 142, 161, *197*
Kato K, see Nishi A 557, *580*
Kato S, see Yamakawa T 4, *8*
Kato T, Kabuto C, Sasaki N, Tsunagawa M, Aizawa A, Fujita K, Kato Y, Kitahara Y 78, *94*
Kato T, Kobayashi M, Sasaki N, Kitahara Y, Takahashi N 67, *94*
Kato T, Tsunakawa M, Sasaki N, Aizawa H, Fujita K, Kitahara Y, Takahashi N 54, 56, 78, *94*
Kato T, see Kitahara Y 80, *94*
Kato T, see Yanagawa H 80, *112*
Kato Y, see Kato T 78, *94*
Katoh H, Esashi Y 325, *412*
Katoh H, see Esashi Y 325, 328, *402*
Katsui N, see Takasugu M 51, 53, 54, *108*
Katsumi M 530, 531, 562, *575*
Katsumi M, Phinney BO, Jefferies PR, Henrick CA 35, *94*
Katsumi M, see Fukui H 27, 28, 30, 31, *88*, 134, *193*
Katsumi M, see Kato J 35, *94*
Katsumi M, see Kazama H 530, *575*
Katsumi M, see Yamane H 381, *442*
Katsumi M, see Yokota T 381, *443*
Katsumura S, see Isoe S 53, *93*
Katsura N, see Suge H 532, *584*
Katz CD, see Liu S-T 282, 284, 285, 286, 287, *418*
Katz M, Ordin L 526, *575*
Katz M, see Mayak S 328, *421*
Kaufman PB *506*
Kaufman PB, Ghosheh NS, La Croix JD, Soni SL, Ikuma H 561, *575*
Kaufman PB, Ghosheh NS, Nakosteen L, Pharis RP, Durley RC, Morf W 23, *94*
Kaufman PB, Petering L, Adams PA 523, *575*
Kaufman PB, see Adams PA 530, *568*
Kaufman PB, see Hebard FV 530, *574*
Kawabe S, see Hiraga K 23, 24, 25, 30, 35, 36, 49, *92*, 116, 131, 134, *194*, 378, 379, *410*
Kawai H, see Koyama S 42, *95*
Kawai T, see Ichihara K 79, *92*
Kawarada A, see Hashimoto T 74, *91*
Kawarada A, see Kinashi H 339, 340, *413*
Kawarada A, Sumiki Y 23, 34, *94*
Kawarada A, see Suzuki Y 13, *107*, 334, 338, *438*
Kawarada A, see Takahashi N 23, 33, *107*, 131, *200*
Kawarada A, see Takeuchi S 75, *108*
Kawase M 332, *412, 413*
Kawashima W, see Jensen EV 485, *505*
Kay E, Shannon LM, Lew JY 341, *413*
Kays SE, see Musgrave A *426*, 486, *507*
Kazama H, Katsumi M 530, *575*
Keck GE, see Corey EJ 38, *85*
Kedar N, see Rudich J 361, *432*
Keegan JT, see Altree-Williams S 60, *81*
Keegstra K, Talmadge KW, Bauer WD, Albersheim P 526, 527, 529, *576*
Keegstra K, see Talmadge KW 526, *584*
Kefeli VI 59, *94*
Kefeli VI, Kadyrov CS 56, *94*, 117, *195*
Kefeli VI, Kutacek M 59, *94*
Kefeli VI, see Dathe W 49 *86*, 374, 382, *399*
Kefeli VI, see Kutáček M 20, *96*, 282, 288, 338, 364, *415*
Kefford NP, see Bennet-Clark TA 52, *82*
Kefford NP, see Bruce MI 40, *83*

Keglević D, Pokorny M 115, *195*, 366, 367, *413*
Keith B, Srivastava L 5, *7*
Keitt GW, see Sharp PB 35, *105*
Kelkar GR, see Mathur SB 77, *98*
Kellar GR, see Paul A 77, *102*
Kelly JN, see Niederl JB 11, *101*
Kelly W, see Knock BA 252, *277*
Kemp HT, Fuller RG, Davidson RS 541, *576*
Kemp JD, see Upper CD 173, 174, *200*
Kempf T, see Becker H 35, *82*
Kende H 40, 45, *94*, 242, *277*, 385, *413*, 531, *576*
Kende H, Baumgartner B 328, *413*
Kende H, Gardner G 456, 477, 488, 495, *506*, 531, *576*
Kende H, Hanson AD 319, 325, 329, 331, *413*, 540, *576*
Kende H, Ninnemann H, Lang A 315, *413*
Kende H, Sitton D 385, *413*
Kende H, Tavares JE 472, *506*
Kende H, see Boller T 320, 326, *395*
Kende H, see Brandes H 242, *276*
Kende H, see Cummins WR 244, *276*, 542, *571*
Kende H, see Gardner G 478, *504*
Kende H, see Hahn H 384, *407*
Kende H, see Hanson AD 319, 320, 322, 325, 326, 328, 329, 331, *407*
Kende H, see Johnson KD 491, *505*
Kende H, see Jones JF 333, *412*
Kende H, see Konze JR 320, 325, 326, *414*
Kende H, see Musgrave A 380, *425*, *426*, 482, 486, *507*
Kende H, see Ninnemann H 315, *426*
Kende H, see Sussman MR 478, *509*
Kende HL, see Johnson KD 549, 552, *575*
Kendrick RE, see Cooke RJ 361, 362, 388, *397*
Kennett BH, see Huelin FE 12, *92*
Kenney G, Südi J, Blackman GE 336, *413*
Kent M, see Gortner WA 344, *406*
Kent PW, see Taylor NF 481, *509*
Kenten RH 342, *413*
Kerk GJM van der, Raalte MH van, Sjipestejin AK, Veen R van der 448, *506*
Kessler B, Kaplan B 549, *576*
Kessler B, see Halevy AH 542, *573*
Kessler B, see Snir T 316, *436*
Kessler C *549*
Kessler C, Hartmann GR 454, *506*
Key JL 519, 520, *576*
Key JL, Barnett NM, Lin CY 520, *576*
Key JL, Ingle J 520, 558, *576*
Key JL, Lin CY, Gifford EM Jr, Dengler R 557, *576*
Key JL, Shannon JC 520, *576*
Key JL, Vanderhoef LN 558, *576*
Key JL, see Chen YM 557, *570*
Key JL, see Holm RE 554, 557, 558, *574*
Khalifah RA 282, 284, *413*
Khalifah RA, Lewis LN, Coggins CW Jr 281, *413*
Khan AA 77, *94*, 548, *576*
Khan AA, Downing RD 548, *576*
Khan AA, see Gaspar T 316, *404*, 565, *573*
Khan AA, see Tao KL 551, 552, *584*
Khan AA, see Thomas TH *439*
Khan AA, Waters EC 548, *576*
Khodary S 528, *576*
Khurana JP, Makeshwari SC 61, *94*
Kienzle F, Mayer H, Minder RE, Thommer H 49, 53, 55, *94*
Kikuta M, see Nakamura T 62, *100*
Kim WK, Rohringer R 282, 284, *413*
Kimura F, see Harada F 43, *91*, 316, 348, *408*
Kimura K, see Hashizume T 41, *91*
Kimura Y, Takesako K, Takahashi Y, Tamura S 60, *94*
Kinashi H, see Suzuki Y 13, *107*, 334, *438*
Kinashi H, Suzuki Y, Takeuchi S, Kawarada A 339, 340, *413*
Kindl H 18, *94*
Kindl H von 288, *413*
King EE, see Lane HC 344, *415*
King RW 50, *94*, 352, 374, *413*
King RW, Evans LT, Firn RD 50, 55, *94*, 301, *413*
Kinsky CB, see Kinsky SC 495, *506*
Kinsky SC, Haxby J, Kinsky CB, Demel RA, Deenen LLM van 495, *506*
Kint J, see Van Sumere CF 57, 58, 67, 74, *109*
Kirk KL, Cohen LA 481, *506*
Kirkwood PS, see Beale MH 190
Kirkwood PS, see Gaskin P 3, *6*, 13, 26, 27, 28, 31, 33, *89*
Kishida Y, see Herz W 77, *92*
Kiss G, Neukom H 19, *94*
Kitahara Y, Yanagawa II, Kato T, Takahashi N 80, *94*
Kitahara Y, see Kato T 54, 56, 67, 78, *94*
Kitahara Y, see Shibuya T 49, *106*
Kitahara Y, see Yanagawa H 80, *112*
Kitamura H, see Takahashi N 13, 23, 24, 33, *107*, *108*, 131, 134, *200*
Kjaer A 74, *94*
Kjaer A, see Ettlinger MG 75, *88*
Klämbt D 41, *94*, 476, 477, *506*, 531, *576*
Klämbt D, Thies G, Skoog F 43, *94*

Klämbt D, see Giesen M 529, *573*
Klämbt D, see Maas H 561, *578*
Klämbt HD 115, *195*, 365, 366, 385, *413*
Klein I, see Epstein E 337, *402*
Klein I, see Frenkel C 327, *403*
Klein I, see Swanson BT 49, *107*
Klein WH, see Leopold AC 72, *96*
Kleinschmidt MG, see Mudd JB 494, *507*
Klerk de GJM, see Barendse GWM 356, 377, 378, 380, 382, *392*
Kline LK, Fittler F, Hall RH 289, *413*
Kline LK, see Fittler F 289, *403*
Kluge M, Reinhard E, Ziegler H 388, *413*
Klun JA, Tipton CL, Robinson JF, Ostrem DL, Beroza M 462, *506*
Knegt E, see Lieberman M 266, *278*
Knight CA, see Liu T-S 282, 284, 285, 287, *418*
Knight LI, Crocker W 10, *94*
Knight LI, see Crocker W 10, *85*
Knights BA, see Berrie AMM 69, *82*
Knights BA, see Hillman JR 49, *92*, 184, *194*, 272, *277*
Knock BA, Smith IC, Wright DE, Ridley RG, Kelly W 252, *277*
Knöfel D, see Müller P 377, *425*
Knöfel D, see Sembdner G 363, 376, 381, *434*
Knöfel HD, Müller P, Sembdner G 381, *413*
Knöfel HD, see Müller P 381, *425*
Knoop B, see Erichsen V *402*
Knorr E, see Byrd DJ 157, 160, *191*
Knotz J, Coolbaugh RC, West CA 317, *413*
Knox JR, see Bakker HJ 311, *392*
Knox JR, see Cook JF 311, *397*
Knox JR, see Dawson RM 311, *399*
Knox JR, see Jefferies PR 309, 311, *411*, *412*
Knypl JS 344, *414*
Knypl JS, Rennert A 344, *414*
Ko ST, see Bennett RD 308, *394*
Kobayashi A, Kaiya S, Yamashita K 79, *94*
Kobayashi M, see Kato T 67, *94*
Kobayashi S, Hatakeyama I, Ashida J 516, *576*
Kobayashi S, see Yokota T 36, *112*, 378, *443*
Kochen W, see Byrd DJ 157, 160, *191*
Kochert G 9, *95*
Koda Y, Okazawa Y 385, *414*
Kodama M, see Galbraith MN 78, *89*
Kodama M, see Ito S 78, *93*
Koehler DE, Varner JE 550, *576*
Koeppe DE 74, *95*
Koeppe DE, Miller RJ 65, *95*
Kögl F, Erxleben H 12, *95*
Kögl F, Erxleben H, Haagen-Smit AJ 12, 13, *95*
Kögl F, Haagen-Smit AJ, Erxleben H 12, 13, *95*
Kögl F, Kostermanns DGFR 13, *95*
Köhler D 361, *414*
Köhler D, Lang A 243, *277*
Köhler K-H, see Bernhardt D 381, *394*
Köhn B, see Conrad K 386, *397*
Koizumi T, see Isogai Y 14, 18, *93*, 130, *195*
Koizumi T, see Okamoto T 13, 20, 21, 75, *101*, 130, *198*
Kok NJJ, see Barendse GWM 308, *393*
Komai K, Iwamura J, Ueki K 77, *95*
Komanine A, see Mizuno K 42, *99*
Kominato Y, see Morita Y 337, 339, *424*
Komissarenko NF, Zhamba GE, Garshtya LY, Bukolova TP 67, *95*
Komoda T, see Yamakawa T 4, *8*
Komoda Y, Isogai Y 27, 80, *95*
Komoda Y, Isogai Y, Okamoto T 27, *95*
Komoda Y, see Isogai Y 49, 65, *93*, 131, 132, 137, *195*
Komoda Y, see Kupchan MS 80, *96*
Komor E, see Tanner W 544, *584*
Komoto N, Ikegami S, Tamura S 131, 132, *195*
Komoto N, Isogai S, Tamura S 49, 79, *95*
Komoto N, Noma M, Ikegami S, Tamura S 49, 79, *95*
Kondo K, Watanabe A, Imaseki H 333, 334, *414*
Kondo K, see Imaseki H 334, *411*
Kondo N, see Watanabe T 336, *441*
Konishi TA 22, *95*
Konjević R, Grubisić D, Marković R, Petrović J 487, *506*
Konjevic R, see Neskovic M 361, *426*
Kono T, see Takahashi N 14, 20, *108*, 115, 145, 188, 189, *200*, 281, *438*
Kono Y, see Takeuchi S 75, *108*
Konze JR, Elstner EF 322, 325, 389, 390, *414*
Konze JR, Kende H 320, 325, 326, *414*
Konze JR, see Elstner EF 324, 331, 389, 390, *401*
Kopcewicz J 281, 361, *414*
Kopcewicz J, Ehmann A, Bandurski RS 115, *195*, 366, 367, *414*
Kopcewicz J, Porazinski J 361, *414*
Kopcewicz J, Rogozinska JH 281, *414*
Koreeda M, see Weiss G 53, 54, *110*
Koshimizu K, Fukui H, Kusaki T, Mitsui T, Ogawa Y 131, 132, *195*

Koshimizu K, Fukui H, Kusaki T, Ogawa Y, Mitsui T 27, *95*, 134, *195*
Koshimizu K, Fukui H, Mitsui T, Ogawa Y 48, 49, *95*, 244, *277*
Koshimizu K, Inui M, Fukui H, Mitsui T 27, 50, *95*, 117, 128, 138, *195*, 373, *414*
Koshimizu K, Ishii H, Fukui H, Mitsui T 27, 28, *95*
Koshimizu K, Kusaki T, Mitsui T, Matsubara S 293, 348, *414*
Koshimizu K, Matsubara S, Kusaki T, Mitsui T 46, *95*, 131, 143, 166, *195*
Koshimizu K, see Fukui H 27, 28, 29, 30, 31, 32, 50, 53, 54, 79, *88*, *89*, 124, 134, 137, *193*, 354, *404*
Koshimizu K, see Hayashi H 76, *91*
Koshimizu K, see Hirai N 50, 52, *92*, 352, *410*
Koshimizu K, see Mirai N *424*
Kosiyadincha S, Young RE 327, *414*
Kostermanns DGFR, see Kögl F 13, *95*
Kosuge T, see Hutzinger O 365, *410*
Kosuge T, see Taylor PA 45, *108*
Kotaki K, see Esashi Y 325, 328, *402*
Köves E, see Varga M 59, *109*
Kovoor A, Melet D 556, *576*
Kowalik H, see Dohrmann U 461, 462, 464, 465, 470, *503*, 539, *571*
Koyama S, Kawai H 42, *95*
Kozlowski TT, see Davies WJ 543, *571*
Kralova M, see Kutáček M 289, *415*
Krasnuk M, Witham FH, Tegley JR 41, 46, *95*, 384, *414*
Krasnuk M, see Tegley JR 128, *200*
Krauss A, see Hager A 451, *504*, 522, 537, *573*
Krausz JP 34, *95*
Kriechmann PS, see Liu WT *418*
Kriedemann PE, Loveys BR, Downton WJS 51, 52, *95*, 352, 353, *414*
Kriedémann PE, Loveys BR, Fuller GL, Leopold AC 543, *576*
Kriedemann PE, see Loveys BR 49, 51, *97*, 296, 352, 353, *419*
Krikorian AD, see Steward FC 56, 80, *107*
Ku HS, Leopold AC 540, *576*
Ku HS, Suge H, Rappaport L, Pratt HK 532, *577*
Ku HS, Yang SF, Pratt HK 341, *415*
Ku HS, see Yang SF 318, *443*
Kubo L, Muira L, Pettei MJ, Lee YW, Pilkiewicz F, Nakanishi K 78, *95*
Kubota T, see Shibata K 61, *106*
Kucherov VF, see Grigorieva NY 361, *406*
Kucherov VF, see Zaretkii VI 184, *201*
Kuhnle JA, Corse J, Chan BG 80, *95*
Kuhn-Silk W, Erickson RO 362, *415*
Kuhn-Silk W, Stoddart JL, Jones RL 356, *415*
Kulaeva ON, see Mothes K 544, *579*
Kulkarni GH, see Mathur SB 77, *98*
Kumamoto J, see Mehard CW 499, *506*
Kumamoto J, see Shindy WW 354, *435*
Kumar SA, Mahadevan S 288, *415*
Kunert R, Libbert E 287, *415*
Kunert R, see Libbert E 284, 287, *417*
Kunert R, see Manteuffel R 287, *420*
Kunishi A, see Owens LD 319, 325, 333, *427*
Kunishi AT, see Lieberman M 318, 319, 322, 324, 325, 326, 327, 332, 333, 335, *418*
Kuo CC, see Crozier A 479, *503*
Kuo CC, see Glenn TL 140, *193*
Kuo CG, Pharis RP 316, *415*
Kupchan MS, Hemingway RJ, Wasner D, Karim A 77, *96*
Kupchan MS, Komoda Y, Court WA, Thomas GJ, Smith RM, Karim A, Gilmore CJ, Haltiwanger RC, Bryan RF 80, *96*
Kupchan SM, see Sequeira L 77, *105*
Kurahashi O, see Yamakawa T 4, *8*
Kuraishi S 45, *96*
Kuraishi S, Muir RM *415*
Kuraishi S, see Atsumi S 120, *190*
Kurogochi S, Murofushi N, Ota Y, Takahashi N 27, 50, *96*, 186, 187, *195*
Kurosawa E 22, *96*
Kusaki T, see Koshimizu K 27, 46, *95*, 131, 132, 134, 143, 166, *195*, 293, 348, *414*
Kutáček M, Kefeli VI 20, *96*, 282, 288, 338, 364, *415*
Kutáček M, Kralova M 289, *415*
Kutacek M, see Kefeli VI 59, *94*
Kveder S, see Magnus V 286, 287, 338, *420*

La Croix JD, see Kaufman PB 561, *575*
La France R, see Little CHA 49, *96*, 143, *196*
La Rue TAG, Gamborg OL 331, *416*
Labarca C, Nicholls PB, Bandurski RS 113, 138, *195*, 367, *415*
Labavitch JM, Ray PM 517, 526, 527, 529, 536, *577*
Laboraya MM, Srivastav HN, Baksi, Nanadita, Shah, Nanadita, Pandya KJ, Singh VD 345, *415*
Lacey HJ, see Hancock RC 241, *277*
Lado P, Caldogno FR, Colombo R, Marré E 538, *577*
Lado P, De Michelis MI, Cerena R, Marré E 539, *577*
Lado P, Rasi-Caldogno F, Colombo R 539, *577*

Lado P, Rasi-Caldogno F, Pennachioni A, Marré E 533, 534, *577*
Lado P, see Marré E 531, 534, 537, 538, 539, *578*
Laidman DL, Colborne AJ, Doig RI, Varty K 551, 552, *577*
Laidman DL, see Doig RI 551, *571*
Laidman DL, see Eastwood D 552, *572*
Laidman DL, see Tavener RJA 552, *584*
Laloraya MM, see Megha BM 347, *422*
Laloue M 370, 371, *415*
Laloue M, Gawer M, Terrine C 369, 371, *415*
Laloue M, Hall RH *415*
Laloue M, Terrine C, Gawer M 293, 369, 385, *415*
Laloue M, Terrine C, Guern J 293, 348, 350, 369, 385, 386, *415*
Laloue M, see Gawer M 371, *404*
Lammi CJ, Lerner J 139, *196*
Lamotte CE, see Edwards JE 385, *401*
Lamport DTA 526, 528, 530, *577*
Lamport DTA, Miller DH 529, *577*
Lancaster JE, Mann JA, Porter NG 543, *577*
Lance B, Durley RC, Reid DM, Thorpe TA, Pharis RP 357, 361, *415*
Lane HC, King EE 344, *415*
Lang A 302, 363, 377, 382, 387, *415*
Lang A, see Baldev B 316, *392*
Lang A, see Barendse GWM 362, 382, *393*
Lang A, see Barnes MF 315, 316, *393*
Lang A, see Harada H 315, 316, *408*
Lang A, see Jones RL 388, *412*
Lang A, see Kende H 315, *413*
Lang A, see Köhler D 243, *277*
Lang A, see Ninnemann H 315, *426*
Lang A, see Ruddat M 35, *104*, 310, *432*
Lang A, see Stoddart JL 361, 388, *437*
Lang J, see Hinman RL 337, 339, 342, *410*
Langer P, Michajlowskij N 115, *196*
Langford CE, see Stodola FH 22, 23, *107*
Langille AR, Forsline PL 386, *415*
Langille AR, see Mauk CS 42, *98*
Lantican BP, Muir RM 282, 288, 365, *415*, *416*
Lantican BP, see Muir RM 282, 288, *425*
Larsen E, see Engvild KC 16, 20, *88*, 289, *402*
Larsen P 20, *96*, 282, 285, *416*
Larsen P, see Rajagopal R 288, 384, *429*
Larsen SD, see Corey EJ 38, *85*
Lau OL, John WW, Yang SF 334, *416*
Lau OL, Murr DP, Yang SF 325, 365, *416*
Lau OL, Yang SF 330, 332, 334, 335, *416*
Lau OL, Yung KH 333, 334, *416*
Lavee S, Galston AW 565, *577*
Lavee S, see Epstein E 336, 337, 338, 364, *402*
Lavee S, see Galston AW 341, *404*, *573*
Lavee S, see Rappaport L 356, 380, *430*
Lavee S, see Shulman Y 384, *435*
Lavender DP, see Webber JE 50, *110*
Laver ML, see Webber JE 50, *110*
Lavie D, see Cooper R 74, *85*
Lavie O, Levy DC, Cohen A, Evenari M, Gutterman Y 74, *96*
Le Page-Degivry M-T 374, *417*
Le Page-Degivry M-T, Bulard C, Milborrow RV 49, *96*
Le Page-Degivry M-T, see Bulard C 374, *395*
Le Poidevin N 79, *96*
Leach WD, see Haagen-Smit AJ 13, *90*, 113, *193*
Leather GR, see Abeles FB 362, *390*, 541, *568*
Leavitt RA, see Ries SK *103*
Lederberg J, Sutherland GL, Buchanan BG, Feigenbaum EA, Robertson AV, Duffield AM, Djerassi C 235, 239, *277*
Lee TT 340, 341, 345, 346, 384, *416*, 554, 565, *577*
Lee TT, Chapman RA 344, 345, *416*
Lee TT, Skoog F 344, *416*
Lee YW, see Kubo L 78, *95*
Leffler HR, O'Brien TJ, Glover DV, Cherry JH 555, 561, *577*
Lehmann H 373, *416*
Lehmann H, Miersch O, Schütte HR 373, *416*
Lehmann H, Schütte HR 373, 375, *416*
LéJohn HB 477, *506*
Lelkes PI, see Zimmermann U 539, 540, *586*
Lembi CA, Morre DJ, Thomson K-S, Hertel R 457, 458, *506*
Lembi CA, see Hardin JW 457, 470, *504*, 520, 560, *574*
Lembi CA, see Van der Woude WJ 525, *585*
Lenton JR 2, *7*
Lenton JR, Bowen NR, Saunders PF 142, 173, *196*
Lenton JR, Perry VM, Saunders PF 48, 49, *96*, 117, *196*, 271, *277*
Lenton JR, see Gaskin P 3, *6*, 13, 26, 27, 28, 31, 33, *89*
Leonard NJ, Fujii T 40, *96*, 125, 127, *196*
Leonard NJ, Playtis AJ, Skoog F, Schmitz RY 41, *96*, 167, *196*
Leonard NJ, see Armstrong DJ 40, 43, *81*, 292, 293, 348, *392*

Leonard NJ, see Burrows WJ 41, 43, 44, 45, *84*, 137, 139, 168, *191*, 292, 348, *396*, 473, *502*, 531, *570*
Leonard NJ, see Hecht SM 348, *409*
Leonard NJ, see Helgeson JP 43, *91*, 291, *409*
Leonard NJ, see Playtis AJ 41, *102*, 167, *198*, 292, *428*
Leonard NJ, see Scarbrough E 40, *105*, 292, *433*
Leonard NJ, see Scopes DIC 39, *105*
Leonard NJ, see Skoog F 471, 472, *508*
Leonard NJ, see Vreman HJ 41, 43, 44, *109*, 139, *201*, 292, 348, *440*, *441*
Leonard NJ, see Walker GC 293, *441*, 474, *510*
Leopold AC *416*
Leopold AC, Scott FI, Klein WH, Ranstad E 72, *96*
Leopold AC, see Barkley GM 519, 522, 523, *569*
Leopold AC, see Kriedémann PE 543, *576*
Leopold AC, see Ku HS 540, *576*
Leopold AC, see Thomson KS 458, 465, *509*
Leopold HC, see Warner HL 523, 530, 532, *586*
Lerner J, Dougherty TM, Schepartz AI 139, *196*
Lerner J, see Lammi CJ 139, *196*
Lesham Y, Philosoph S, Wurzburger J 374, 375, *417*
Letcher RM 74, *96*
Letham DS 9, 40, 41, 45, 56, 62, 79, *96*, 113, 127, 128, 131, 143, 149, 150, 151, 152, 154, *196*, 242, *277*, *278*, 289, 293, 348, 368, 369, 384, *417*, 475, *506*
Letham DS, Miller CO 40, 41, *96*
Letham DS, Parker CW, Duke CC, Summons RE, MacLeod JK 348, 370, 385, 386, *417*
Letham DS, Shannon JC, MacDonald IRC 40, 41, *96*, 127, *196*
Letham DS, Summons RE, Entsch B, Gollnov BI, Parker CW, MacLeod JK 369, *417*
Letham DS, Williams MW 384, *417*
Letham DS, Wilson MM, Parker CW, Jenkins ID, MacLeod JK, Summons RE *417*
Letham DS, see Berridge MV 474, 475, 477, 478, *502*
Letham DS, see Cowley DE 369, *398*
Letham DS, see Duke CC 47, *87*, 369, 370, *400*
Letham DS, see Entsch B 372, *402*
Letham DS, see Gordon ME 348, 350, 369, 385, 386, *405*
Letham DS, see MacLeod JE 370, *419*
Letham DS, see McLeod JK 4, *7*
Letham DS, see Murakoshi I 372, *425*
Letham DS, see Parker CW 45, *101*, 128, *198*, 348, 350, 369, 370, 371, 385, 386, *427*
Letham DS, see Summons RE 4, *7*, 45, 47, *107*, 186, *199*, 269, 270, *279*, 369, *438*
Letham DS, see Wilson MM *442*
Levy DC, see Lavie O 74, *96*
Levy EC, see Cooper R 74, *85*
Lew FT, West CA 308, 309, *417*
Lew JY, see Kay E 341, *413*
Lewak S, see Sinska I 24, 25, *106*, *435*
Lewak S, see Smolenska G 361, *436*
Lewis D, see Fowden L 76, *88*
Lewis LN, see Khalifah RA 281, *413*
Lex A, see Ferri MG 541, *572*
Leyter L, see Dominguez XA 77, *87*
Li HC, Rice EL, Rohrbaugh LM, Wender SH 564, *577*
Libbenga KR, see Oostrom H 457, *507*
Libbenga KR, see Vreugdenhil D 469, *510*
Libbenga KR, Torrey JG 555, *577*
Libbert E 282, *417*
Libbert E, Bormann H 336, *417*
Libbert E, Brunn K 284, 285, *417*
Libbert E, Drawert A, Schröder R 287, *417*
Libbert E, Erdmann N, Schiewer U 282, 283, 285, 287, 288, 338, *417*
Libbert E, Fischer E, Drawert A, Schröder R 284, 285, 287, 338, *417*
Libbert E, Kaiser W, Kunert R 287, *417*
Libbert E, Manteuffel R 287, *417*
Libbert E, Manteuffel R, Siegel E 337, 338, *418*
Libbert E, Risch H 337, 346, *417*
Libbert E, Schröder R, Drawert A 287, 338, *418*
Libbert E, Schröder R, Drawert A, Fischer E 285, 286, 287, 338, *418*
Libbert E, Silhengst P 282, *417*
Libbert E, Wichner S, Duerst E, Kaiser W, Kunert R, Manicki A, Manteuffel R, Riecke E, Schröder R 284, 287, *417*
Libbert E, see Erdmann N 283, *402*
Libbert E, see Heerkloss R 282, 283, 383, *409*
Libbert E, see Kunert R 287, *415*
Libbert E, see Manteuffel R 287, *420*
Libbert E, see Wichner S 287, *441*
Lichtenthaler HK, see Thiess DE 79, *108*
Lieber ER, see Bennett TD *394*
Lieberman M 318, 325, 333, 334, *418*
Lieberman M, Baker JE, Sloger M 333, 334, *418*
Lieberman M, Knegt E 266, *278*

Lieberman M, Kunishi AT 318, 319, 322, 324, 326, 327, 332, 333, *418*A
Lieberman M, Kunishi AT, Mapson LW, Wardale DA 318, 319, 325, 335, *418*
Lieberman M, Kunishi AT, Owens LD 325, *418*
Lieberman M, Mapson LW 318, *418*
Lieberman M, see Baker JE 322, 325, 328, *392*
Lieberman M, see Fuchs Y 334, *404*
Lieberman M, see Mattoo AK 330, 390, *421*
Lieberman M, see Owens LD 319, 325, 333, *427*
Lieberman M, see Zimmerman RH 325, 334, *444*
Liebisch H-W 381, *418*
Liebisch H-W, see Miersch O 377, 378, *423*
Liebisch H-W, see Müller P 381, *425*
Liebisch H-W, see Schneider G 377, 379, *433*
Liebisch H-W, see Sembdner G 316, 363, 376, 381, *434*
Liemann F, see Tietz D 353, *439*
Liepa AJ, see Duke CC 47, *87*, 369, 370, *400*
Light EN, see Barnes MF 315, 316, *393*
Likholat TV, Pospelov VA, Morozova TM, Salganik RI 456, 470, *506*
Lim HC, see Wood HN 371, *442*
Lin CY, see Chen YM 557, *570*
Lin CY, see Key JL 520, 557, *576*
Linehan DJ, see Vaughan D 58, *109*
Ling E, see Musgrave A 513, 532, *580*
Lipe JA, Morgan PW 328, 330, *418*
Lippincott JA, see Galsky AG 549, *573*
Lipscomb WN, see Hartsuck JA 23, 33, *91*
Lipsett MN, see Cherayil JD 40, *84*
Lis EK, see Runkova LV 344, 347, *432*
Lischewski M, Adam G 377, *418*
Lischewski M, see Adam G 377, 378, *391*
Lischewski M, see Sembdner G 363, 376, 381, *434*
Little CHA, Eidt DC 542, *577*
Little CHA, Heald JK, Browning G 13, 50, *97*, 374, *418*
Little CHA, Strunz GM, La France R, Bonga JM 49, *96*, 143, *196*
Liu T-S, Katz CD, Knight CA 282, 284, 285, 286, 287, *418*
Liu W-C, Carns HR 48, *97*
Liu WT, Pool R, Wenkert W, Kriechmann PS *418*
Liverman JL, see Dannenburg WN 282, *399*
Livne A, Vaadia Y 541, 542, *577*
Lloyd PB, see Borrow A 22, *83*
Lockard RG, see Grunwald C 170, 171, *193*
Lockhart JA 314, *418*, 515, *577*
Loeppky RN, see Skoog F 471, 472, *508*
Loescher WJ, Nevins DJ 525, *578*
Loewus FA, see Verma DC 563, 564, *585*
Logan DM, see Chen C-M 350, 386, *397*
Logan DM, see McLennan BD 348, *422*
Loh JWC, Severson JG Jr 344, 346, 364, *418*
Lomax T, see Cleland RE 518, 538, 539, *570*
Lona F 34, *97*
Lonski J, see Abeles FB 331, *390*
Loopik-Detmers MA van, see Oostrom H 457, *507*
Lorenzi R, Horgan R, Heald JK 36, *97*, 125, *196*, 362, 377, 379, 381, 382, *419*
Lorenzi R, Horgan R, Wareing PF 25, 36, *97*, 369, *419*
Lorenzi R, Saunders PF, Heald JK, Horgan R 25, *97*
Lorenzi R, see Alpi A 3, 5, *6*
Lorenzi R, see Ceccarelli N 5, *6*, 305, 388, *396*
Loschmikova NV, see Grigorieva NY 361, *406*
Lousberg RJJC, see Bercht CA 53, 54, *82*
Lovak EJ, see Sengupta SK 49, *105*
Lovell P, Moore K 562, *578*
Loveys B, see Beevers L 361, *394*
Loveys BR 352, 387, *419*
Loveys BR, Brien CJ, Kriedemann PE 296, 352, *419*
Loveys BR, Kriedemann PE 49, 51, *97*, 353, *419*
Loveys BR, see Kriedemann PE 51, 52, *95*, 352, 353, *414*, 543, *576*
Loveys BR, Wareing PF 361, *419*
Lowe D, see Brian PW 243, *276*
Lowe M, see Wilchek M 454, *510*
Lu T-YS, see Vanderhoef LN 451, *510*
Lubanska G, see Nowakowski W 541, *580*
Luchko AS, see Porutzkii GV 390, *428*
Luckwill LC, see Whyte P 243, *279*
Ludden P, see Daly JM 331, *399*
Ludlow CJ, see Baskin JM 67, *81*
Luhan PA, see Cook CE 77, *85*
Luke HH, Freeman TE 542, *578*
Lund HA 282, *419*
Lunnon MW, MacMillan J, Phinney BO 311, *419*
Lürssen K, Naumann K 320, *419*
Lürssen K, Naumann K, Schröder R 326, *419*
Lusby W, see Thompson MJ 3, *8*
Lusby WR, see Buta JG 78, *84*

Lüttge U, see Haschke HP 538, 539, *574*
Lynn DG, Nakanishi K, Patt SL, Occolowitz JL, Almeida S, Evans LS 76, *97*
Lynn DG, see Evans LS 76, *88*
Lyon JL, see Addicott FT 48, 52, *80*, 244, *276*, 295, *391*
Lyon JL, see Davis LA 52, *86*, 131, *192*, 244, *276*
Lyon JL, see Ohkuma K 48, 49, *101*, 131, 132, 143, 154, 155, *198*
Lyons JM, McGlasson WB, Pratt HK 175, *196*
Lyons JM, see Mehard CW 499, *506*, 540, *579*

Maas H, Klämbt D 561, *578*
Maçaira LA, Garcia M, Rabi JA 77, *97*
MacDonald IRC, see Letham DS 40, 41, *96*
Macháčková J, Gančeva K, Zmrhal Z 344, *419*
Macháčková J, Zmrhal Z 340, 341, 342, 344, *419*
Maclachlan GA, Davies E, Fan DF 562, *578*
Maclachlan GA, see Bal AK 563, *568*
Maclachlan GA, see Birmingham BC 530, 531, 559, *569*
Maclachlan GA, see Byrne H 563, *570*
Maclachlan GA, see Datko AH 520, 532, *571*
Maclachlan GA, see Fan DF 518, 520, *572*
Maclachlan GA, see Spencer FS *583*
Maclachlan GA, see Verma DPS 451, *510*, 554, *585*
Maclachlan GA, see Wong Y-S 524, 563, *586*
Maclachlan GA, Waygood ER 342, *419*
Maclachlan GA, Young M 526, *578*
MacLeod JE, Summons RE, Parker CW, Letham DS 370, *419*
MacLeod JK, see Cowley DE 369, *398*
MacLeod JK, see Duke CC 47, *87*, 369, 370, *400*
MacLeod JK, see Letham DS 348, 369, 370, 385, 386, *417*
MacLeod JK, see Parker CW 45, *101*, 128, *198*
MacLeod JK, see Summons RE 45, 47, *107*, 186, *199*, 269, 270, *279*, 369, *438*
MacMillan J 3, 7, 116, *196*, 302, 311, *419*
MacMillan J, Pryce RJ 51, 52, *97*, 302, 352, *419*, *420*
MacMillan J, Pryce RJ, Eglinton G, McCormick A 182, 183, *196*
MacMillan J, Seaton JC, Suter PJ 23, 24, 25, 35, *97*, 116, 131, 132, 146, 160, 161, *196*, 352, *420*
MacMillan J, Suter PJ 34, *97*, 134, *196*
MacMillan J, Takahashi N 34, *97*
MacMillan J, Wels CM 32, *97*, 136, 137, 170, 176, *196*
MacMillan J, see Beale MH 190, *190*
MacMillan J, see Bearder JR 3, *6*, 26, 30, 31, 33, 34, *82*, 302, 309, 311, 312, 313, 314, 354, 355, *393*, *394*
MacMillan J, see Beeley LJ 24, 25, 28, 31, 50, 51, *82*
MacMillan J, see Bialek K 49, *82*
MacMillan J, see Binks R 184, *191*
MacMillan J, see Bowen DH 23, 24, 27, *83*, 142, 183, *191*, 244, *276*, 314, *395*
MacMillan J, see Brian PW 33, *83*
MacMillan J, see Cavell BD 161, 163, 171, 172, *191*
MacMillan J, see Cross BE 301, *398*
MacMillan J, see Crozier A 24, *86*
MacMillan J, see Durley RC 23, 24, 25, 26, 27, 35, 51, *87*, 116, 130, *192*, 352, 388, *401*
MacMillan J, see Eeuwens CJ 27, *87*
MacMillan J, see Elson GW 162, 163, *193*
MacMillan J, see Frydman VM 25, 26, 27, 29, 30, 31, 35, 49, 51, *88*, 170, 186, *193*, 268, *277*, 356, 357, 358, 359, 362, 380, *404*, 480, 481, *504*
MacMillan J, see Gaskin P 3, *6*, 13, 26, 27, 28, 31, 33, 36, 49, *89*, 117, 173, 181, 184, *193*, 352, *404*
MacMillan J, see Graebe JE 31, *90*, 310, 314, 355, 356, 357, 388, *406*
MacMillan J, see Grove JF 24, *90*
MacMillan J, see Harrison DM 28, *91*, 302, *408*
MacMillan J, see Hedden P 4, *6*, 302, 303, 305, 312, 313, 316, *409*
MacMillan J, see Hoad GV 5, *6*
MacMillan J, see Lunnon MW 311, *419*
MacMillan J, see Martin GC 26, 28, 31, 35, 49, *97*, 184, 186, *196*, *197*, 268, *278*, 352, 354, *421*
MacMillan J, see McInnes AG 355, 376, 378, *422*
MacMillan J, see Most BH 49, *100*
MacMillan J, see Pryce RJ *198*
MacMillan J, see Railton ID 49, *103*, 272, *278*, 387, *429*
MacMillan J, see Rivier L 4, *7*
MacMillan J, see Ropers AJ 308, 310, 313, 314, 355, *431*
MacMillan J, see Sinska I 24, 25, *106*, *436*
MacMillan J, see Sponsel VM 4, 5, *7*, 26, 27, 29, 31, 32, 33, 50, 51, *106*, 189, 190, *199*, 355, 356, 357, 358, 359, 360, 362, 379, 380, 382, 388, *436*
MacMillan J, see Tinelli ET 51, *108*, 352, 353, *439*
MacMillan J, see Williams PM 268, *279*
MacNeill JD, see Hutzinger O 159, *194*
MacNicol PK 340, 341, 384, *420*
Madeikyte E, Turkova NS 540, *578*
Madkour MA 541, *578*
Magnus V, Iskrič S, Kveder S 286, 287, 338, *420*

Mahadevan S 288, *420*
Mahadevan S, see Kumar SA 288, *415*
Mahadevan S, see Shukla PS 288, *435*
Mahadevan S, Thimann KV 288, *420*
Mahadevan S, see Thimann KV 288, *439*
Makeshwari SC, see Khurana JP 61, *94*
Malcolm HDR, see Altree-Williams S 60, *81*
Malek F, Baker DA 544, *578*
Mallaby R, see Cornforth JW 294, *398*
Mallaby R, see Milborrow BV 117, 118, *197*, 296, 298, 350, *423*
Malloch KR, Osborne DJ 333, 347, *420*
Mandava N, Anderson JD, Dutky SR 76, *97*
Mandava N, see Anderson JD 335, *391*
Mandava N, see Grove MD 3, *6*
Mandava N, see Mitchell JW 3, *7*
Mandava N, see Thompson MJ 3, *8*
Mander LN, see Coombe BG 35, *85*
Manicki A, see Libbert E 284, 287, *417*
Mann JA, see Lancaster JE 543, *577*
Mann JD, Jaworski EG 116, 143, *196*
Mann PJG, see Clarke AJ 285, 287, *397*
Manos PJ, Goldthwaite J 242, *278*
Mansfield TA 541, *578*
Mansfield TA, Jones RJ 543, *578*
Mansfield TA, Wellburn AR, Moreira TJS 56, *97*
Mansfield TA, see Fenton R 56, *88*
Mansfield TA, see Jones RJ 542, *575*
Mansfield TA, see Ogunkanmi AB 56, *101*
Mansfield TA, see Squire GR 544, *584*
Mansfield TA, see Tucker DJ 244, *279*, 542, *585*
Mansfield TA, see Wellburn AR 53, 56, *110*
Manske RHF 53, *97*
Manteuffel R 287, *420*
Manteuffel R, Siegl E, Kunert R, Libbert E 287, *420*
Manteuffel R, Siegl E, Libbert E 287, *420*
Manteuffel R, see Libbert E 287, 337, 338, *417*, *418*
Mapson LW 317, 318, 319, 324, 326, 327, 328, *420*
Mapson LW, Hulme AC 10, *97*, 317, 326, *420*
Mapson LW, March JF, Rhodes MJC, Wooltorton LSC 319, *421*
Mapson LW, March JF, Wardale DA *420*
Mapson LW, Mead A 324, *420*
Mapson LW, Self R, Wardale DA 324, *420*
Mapson LW, Wardale DA 324, 325, *420*
Mapson LW, see Lieberman M 318, 319, 325, 335, *418*
Maravolo NC, see Spaeth SC 340, 343, *436*
March JF, see Mapson LW 319, *420*, *421*
Marigo G, see Ranjeva R 565, *581*
Marković R, see Konjević R 487, *506*
Marré E 533, 534, *578*
Marré E, Colombo R, Lado P, Rasi-Caldogno F 531, 534, 539, *578*
Marré E, Forti G 517, *578*
Marré E, Lado P, Ferroni A, Ballarin-Denti A 531, 534, 537, 539, *578*
Marré E, Lado P, Rasi-Caldogno F, Colombo R 537, 538, 539, *578*
Marré E, Lado P, Rasi-Caldogno F, Colombo R, De Michelis MI 534, 537, *578*
Marré E, see Beffagna N 452, 464, 469, *502*, 539, *569*
Marré E, see Cocucci M 538, *571*
Marré E, see Dohrmann U 462, 465, 468, *503*
Marré E, see Lado P 533, 534, 538, 539, *577*
Marré E, see Pesci P 464, *507*
Marré E, see Tognoli L 469, *509*
Marré E, see Tonzig S 80, *109*
Marshall D, see Penny P 519, *581*
Marshall JA, Cohen N 77, *97*
Martin D, see Dörffling K 55, *87*
Martin DMG, Reese CB 127, *196*
Martin GC, Dennis FG, Gaskin P, MacMillan J 26, 28, 31, 35, 49, 51, *97*, 352, 354, *421*
Martin GC, Dennis FG Jr, Gaskin P, MacMillan J 184, *197*
Martin GC, Dennis FG Jr, MacMillan J, Gaskin P 186, *196*, 352, 354, *421*
Martin GC, Dennis FG, MacMillan J, Gaskin P 35, *97*, 268, *278*
Martin GC, see Bearder JB 311, *393*
Martin GC, see Diaz DH 374, *400*
Martin GC, see Reed W 29, *103*
Martin GC, see Shaybany B 50, *105*, 354, *435*
Martinez C, see Dominguez XA 74, *87*
Martinez ML, see Wolf FT 56, 62, *111*
Marumo S, Abe H, Hattori H, Munakata K 16, 20, *98*, 130, *197*, 289, *421*
Marumo S, Hattori H 16, *98*, 142, *197*, 289, *421*
Marumo S, Hattori H, Abe H, Munakata K 16, 20, *98*, 115, 130, 133, 138, 142, *197*, 289, *421*
Marumo S, see Hattori H 20, *91*, 138, 142, *194*, 289, 365, *408*
Marumo S, see Takahashi N 117, *200*
Marx GA, see Proebsting WM 361, *428*
Marx JN, White EH 77, *98*

Marynik M 330, 331, 333, 334, *421*
Masamune T, see Takasugu M 51, 53, 54, *108*
Massart L, see Petrids C *428*
Mast CA van der 341, 344, *440*
Masuda I, see Fujita T *404*
Masuda Y 55, *98*, 515, 520, 525, 526, 528, *578*
Masuda Y, Kamisaka S 517, *579*
Masuda Y, Tanimoto E 520, *579*
Masuda Y, Tanimoto E, Wadà S 517, 520, *579*
Masuda Y, Yamamoto R 517, 525, *579*
Masuda Y, see Kamisaka S 554, *575*
Masuda Y, see Tanimoto E 518, *584*
Masuda Y, see Yamagata Y 534, *586*
Mathur SB, Hiremath SV, Kulkarni GH, Kelkar GR, Bhattacharyya SC 77, *98*
Matkovskii KI, see Porutzkii GV 390, *428*
Matsubara S, see Hayashi H 76, *91*
Matsubara S, see Koshimizu K 46, *95*, 131, 143, 166, *195*, 293, 348, *414*
Matsuhima H, Arima K 75, *98*
Matsui H, see Nakagawa S 186, *197*
Matsui M, see Hwang Y-S 12, *92*
Matsumoto T, Nishida K, Noguchi M, Tamaki E 554, 566, *579*
Matsuo A, see Bearder JR 311, *394*
Matsushima H, Arima K 340, *421*
Matsushima H, Fukumi H, Arima K 60, 75, *98*, 340, *421*
Matta A, see Gentile IA 331, *405*
Matthysse AG 453, *506*
Matthysse AG, Abrams M 474, 478, *506*
Matthysse AG, Phillips C 453, 455, 456, 498, *506*, 520, *579*
Matthysse AG, Torrey JG 554, *579*
Mattoo AK, Bauer JE, Chalutz E, Lieberman M 330, 390, *421*
Mattoo AK, Lieberman M 390, *421*
Mauk CS, Langille AR 42, *98*
Mauri M, see Ballio A 533, *568*
Mayak S, Dilley DR 328, 330, 334, *421*
Mayak S, Halevy AH, Katz M 328, *421*
Mayak S, see Fuchs Y 245, *277*
Mayer AM, Evenari M 67, 69, *98*
Mayer AM, Poljakoff-Mayber A 57, 67, *98*
Mayer H, see Kienzle F 49, 53, 55, *94*
Mazelis M 564, *579*
Mazelis M, see Riddle VM 282, *431*
Mazza G, see Ricard J 341, *431*
McCalla DR, Moore JD, Osborne DJ *421*
McCalla DR, see Osborne DJ 242, *278*
McCalla TM, see Schuman GE 79, *105*
McCapra F, McPhail AT, Scott AI, Sim GA, Young DW 23, 33, *98*
McChesney J, see Fox JE 370, *403*
McChesney JD, see Chen C-M 292, 348, *397*
McChesney JD, see Deleuze GG *399*
McChesney JD, see Fox JE *403*
McCloskey JA, Nishimura S 43, 44, 45, *98*
McCloskey JA, see Hashizume T 42, 45, *91*, 186, *194*
McCloskey JA, see Von Minden DL 184, *200*
McClure JW 58, *98*
McComb AJ 376, *421*
McComb AJ, Broughton WJ 530, *579*
McComb AJ, McComb JA, Duda CT 489, *506*, 558, *579*
McComb AJ, see Broughton WJ 562, *569*
McComb JA, see McComb AJ 489, *506*, 558, *579*
McCormick A, see MacMillan J 182, 183, *196*
McCormick A, see Pryce RJ *198*
McCullagh D, see Fletcher RA 242, *277*
McCune DC 341, *421*
McDonald IRC, see Letham DS 127, *196*
McDougall J, Hillman JR 13, *98*, 383, *421*
McFadden WH 262, *278*
McFadden WH, Bradford DC, Gaines DE, Gower JL 262, *278*
McGlason WB 317, *421*
McGlason WB, Dostal HC, Tigchelaar EC *421*
McGlason WB, Palmer JK, Vendrell M, Brady CJ 332, *421*
McGlason WB, Poovaiah BW, Dostal HC 328, *421*
McGlasson WB, see Lyons JM 175, *196*
McGlasson WB, see Mizrahi Y 328, *424*
McGlasson WB, see Vendrell M 328, *440*
McGlasson WG, Pratt HK 11, *98*
McGowan RE, Muir RM 288, *422*
McInnes AG, Smith DG, Arsenault GP, Vining LC 26, *98*, 355, *422*
McInnes AG, Smith DG, Durley RC, Pharis PP, Arsenault GP, MacMillan J, Gaskin P, Vining LC 355, 376, 378, *422*
McKeon W, see Nakanishi K 35, *100*
McLafferty FM, see Stenhagen E 250, *279*
McLaughlin GE, see Durley RC 134, 135, 176, *192*
McLemman B, see Hall RH 137, *194*
McLennan B, see Chen C-M 350, 386, *397*
McLennan B, see Hall RH 41, *90*, 292, 362, *407*

McLennan BD, Logan DM, Hall RH 348, *422*
McLennan BD, Pater A 348, *422*
McLennan BD, see Hall RH 350, 372, *407*
McLennon B, see Hall RH 531, *573*
McLeod JK, Gummons RE, Letham DS 4, *7*
McLeod JK, see Duke CC *400*
McLeod JK, see Parker CW 5, *7*, 369, 370, 386, *427*
McLeod JK, see Summons RE 4, *7*
McMeans JL, see Addicott FT 244, *276*
McMichael BL, Jordan WR, Powell RD 331, *422*
McMillan C, see Desmos EK 61, *86*
McPhail AT, see Cook CE 77, *85*
McPhail AT, see McCapra F 23, 33, *98*
McPhail AT, see Wall ME 75, *110*
McQuillin FJ 77, *98*
McWha JA, Hillman JR 50, *98*, 354, *422*
McWha JA, Philipson JJ, Hillman JR, Wilkins MB 244, *278*
McWha JA, see Dewdney SJ 352, 354, 374, *400*
Mead A, see Mapson LW 324, *420*
Means AR, see Comstock JP 486, *503*
Means AR, see O'Malley B 484, 485, *507*
Medlow GC, see Hall SM 13, *91*
Medlow GC, see White JC 13, *111*
Megha BM, Laloraya MM 347, *422*
Mehard CW, Lyons JM 540, *579*
Mehard CW, Lyons JM, Kumamoto J 499, *506*
Meheriuk M, Spencer M 331, 390, *422*
Meheriuk M, see Ram Chandra G 390, *429*
Meheriuk M, see Spencer M 390, *436*
Meheriuk M, Spencer MS 322, *422*
Meidner H 541, *579*
Meidner H, see Itai C 52, *93*
Meier H, see Grant Reid JS 552, *573*
Meigh DF 12, *98*
Meigh DF, Filmer AAE, Self R 74, *98*
Meigh DF, Jones JD, Hulme AC 322, *422*
Meigh DF, see Smith WH 328, *436*
Meinhenett R, Carr DJ 79, *98*
Meins F, see Pengelly W 246, *278*
Melet D, see Kovoor A 556, *576*
Melitz DK, see Chen C-M 292, *397*
Mellon JE, West CA 305, 317, *422*
Menary RC, see Van Staden J 42, *109*
Mendez J 50, *98*
Mendez J, Gesto MDV, Vasquez A, Vieitez E, Secare E 60, *98*
Mennes AM 340, 341, 346, *422*
Menschick R, Hild V, Hager A 336, 340, 342, 347, *422*
Menzel H, see Hager A 451, *504*, 522, 537, *573*
Merijanian A, see Dominguez XA 78, *87*
Merkys A 364, 365, *422*
Merlini L, see Assante G 52, *81*
Mertz D 362, *422*
Mertz D, Henson W 315, 362, *422*
Metha AR, see Shah RR 564, *583*
Metzenberg RL, see Hansch C 448, *504*
Metzger JD, Zeevaart JAD 4, *7*, 26, 27, 29, 31, 33, *98*
Metzger JD, see Jones MG 177, *195*
Meudt WJ 340, *423*
Meuli LJ, see Redemann CT 301, *430*
Meyer HE, see Hamilton RH 336, 337, 340, 344, *407*, 448, *504*
Mhaeshwari SC, see Prakash R 368, *428*
Miassod R, Penon P, Teissiére M, Ricard J, Cecchini J-P 520, *579*
Michajlowskij N, see Langer P 115, *196*
Michalczuk L, Bandurski RS 367, *423*
Michalke W, see Trillmich K 458, *509*
Michener HD 11, *98*
Michniewicz BM, see Sondheimer E 298, *436*
Mielke EA, Dennis FG 50, *99*
Mierlo JV, see Barendse GWM 356, *393*
Miernyk JA 386, *423*
Miernyk JA, Blaydes DF *423*
Miersch O, Liebisch H-W 377, 378, *423*
Miersch O, see Lehmann H 373, *416*
Miersch O, see Müller P 381, *425*
Miersch O, see Schneider G 377, 379, *433*
Miersch O, see Sembdner G 381, *434*
Milborrow BV 48, 49, 50, 51, 52, 54, *99*, 117, 118, 189, 190, *197*, 270, 271, *278*, 295, 296, 297, 298, 350, 351, 352, 353, 354, 373, 374, 375, 387, *423*, 498, *506*
Milborrow BV, Garmston M 299, *423*
Milborrow BV, Mallaby R 117, 118, *197*, 296, 298, 350, *423*
Milborrow BV, Noddle RC 168, *197*, 296, 298, 299, *423*
Milborrow BV, Purse JG, Wightman F 21, *99*
Milborrow BV, Robinson DR 296, *424*
Milborrow BV, Vaugham G *353*
Milborrow BV, see Addicott FT 48, *80*
Milborrow BV, see Cornforth JW 48, 49, *85*, 190, *191*
Milborrow BV, see Le Page-Degivry MT 49, *96*
Milborrow BV, see Osborne DM 374, *427*

Milborrow BV, see Zeevaart JAD 51, 54, *112*, 117, 118, *201*, 352, 353, *444*
Miller CO 40, 42, 45, *99*, 117, 127, 128, 131, 142, 143, *197*, 242, *278*, 293, 368, 371, 384, *424*, 554, 565, 566, *579*
Miller CO, Skoog F, Okomura FS, von Saltza MH, Strong FM 39, 40, *99*
Miller CO, Skoog F, Okumura MH, von Saltza MH, Strong FM 39, *99*
Miller CO, Skoog F, von Saltza MH, Strong FM 39, *99*
Miller CO, see Crafts CB 40, *85*
Miller CO, see Letham DS 40, 41, *96*
Miller CO, see Miura GA 293, 348, *424*
Miller CO, see Peterson JB 42, 47, *102*, 139, 142, 153, 154, *198*
Miller CO, see Peterson JV 369, *427*
Miller CO, see Skoog F 40, *106*, *583*
Miller CS, see Hall WC *407*
Miller DH, see Lamport DTA 529, *577*
Miller EV, Winston JR, Fisher DF 11, *99*
Miller RJ, see Koeppe DE 65, *95*
Miller RW, Sirois JC, Morita H 344, *424*
Miller RW, see Sirois JC 344, *436*
Mills VM, Todd GW 347, *424*
Milon H, see Rivier L 50, *103*, 272, *279*
Milstein LM, see Zaretkii VI 184, *201*
Minchin A, Harmey MA 337, 340, 341, *424*
Minco L, see Sequeira L 340, 341, *434*
Minder RE, see Kienzle F 49, 53, 55, *94*
Minoda Y, see Yamakawa T 4, *8*
Mintsioulis G, see Hall RH 350, 372, *407*
Mirai N, Fukui H, Koshimizu K *424*
Mitchell ED, Talbert NE 75, *99*
Mitchell EK, Davies PJ 282, *424*
Mitchell JC, see Rodriguez E 77, *104*
Mitchell JW, Mandava N, Worley JF, Plummer JR, Smith MV 3, *7*
Mitsuhashi M, Shibaoka H 78, *99*, 114, *197*
Mitsuhashi M, see Shibaoka H 77, *106*
Mitsui T, see Fukui H 27, 28, 29, 30, 31, *88*, 124, 134, *193*
Mitsui T, see Koshimizu K 27, 28, 46, 48, 49, 50, *95*, 117, 128, 131, 132, 134, 138, 143, 166, *195*, 244, *277*, 293, 348, 373, *414*
Mittelhauser CT, Steveninck RFM van 543, *579*
Mittelheuser CJ, van Steveninck REM 52, *99*
Mittelman A, see Chheda GB 348, 372, *397*
Miura GA, Hall RH 293, 348, 384, *424*
Miura GA, Miller CO 293, 348, *424*
Miyamoto M, see Yamaguchi I 31, *112*
Miyazaki M, see Herz W 77, *92*
Mizrahi Y, Blumenfeld A, Bittner S, Richmond AE 543, *579*
Mizrahi Y, Dostal HC, McGlasson WB, Cherry JH 328, *424*
Mizrahi Y, see Bittner S 498, *502*
Mizuno K, Komanine A 42, *99*
Mizushima S, Nomura M 477, *506*
Møller JB, see Andersen AS 337, *391*, 533, *568*
Moewus F 241, *278*
Moewus F, Schader E 73, *99*
Moffat AC, Smalldon KW, Brown C 256, *278*
Mojé W 57, 73, *100*
Molisch H 51, *100*
Mollan RC, Donnelly DMX, Harmey MA 21, *100*, 364, 365, *424*
Momotani Y, Kato J *579*
Monaco TJ, see Moreland DE 56, 60, 65, 76, *100*
Mondal H, Mondal RK, Biswas BB, 455, 456, *506*
Mondal MH 334, *424*, 563, *579*
Mondal RK, see Mondal H 455, 456, *506*
Mondelli R, see Piozzi F 75, *102*
Monro J, see Penny D 451, *507*, 521, 522, 536, *581*
Monselise SP, see Goldschmidt EE 128, 157, *193*, 281, *405*
Montague MJ, see Adams PA 530, *568*
Montalbini P, Elstner EF 331, *424*
Moore DJ, see Jelsema CL 360, *412*
Moore H 476, *507*
Moore JD, see McCalla DR *421*
Moore K, see Lovell P 562, *578*
Moore TC 282, *424*
Moore TC, Anderson JD 316, *424*
Moore TC, Barlow SA, Coolbaugh RC 308, *424*
Moore TC, Coolbaugh RC 389, *424*
Moore TC, Shaner CA 282, 284, *424*
Moore TC, see Anderson JD 305, *391*
Moore TC, see Coolbaugh RC 305, 308, 388, *398*
Moore TC, see Ecklund RP 305, *401*
Moore TC, see Gomez-Navarette G 305, 317, *405*
Moore TC, see Reed DJ 316, *430*
Moran L, see Horton RF 543, *574*
Moreira TJS, see Mansfield TA 56, *97*
Moreland DE, Egley GH, Worsham AD, Monaco TJ 56, 60, 65, 76, *100*

Morf W, see Kaufman PB 23, *94*
Morgan PW, Hall WC 333, *424*
Morgan PW, see Beyer EM Jr 365, *394*
Morgan PW, see Dunlap RJ 328, *400*
Morgan PW, see Durham JI 328, *400*
Morgan PW, see Fowler JL 332, 333, *403*
Morgan PW, see Hall WC 347, *407*
Morgan PW, see Lipe JA 328, 330, *418*
Morimoto H, Sanno Y, Oshio H 77, *100*
Morita H, see Miller RW 344, *424*
Morito Y, Kominato Y, Shimizu K 337, 339, *424*
Morozova TM, see Likholat TV 456, 470, *506*
Morré DJ, Bracker CE 452, *507*, 540, *579*
Morré DJ, Cherry JH 517, 520, 536, *579*
Morré DJ, Eisinger WR 515, *579*
Morré DJ, see Clark JE 447, 470, *502*
Morré DJ, see Hardin JW 457, 470, *504*, 520, 560, *574*
Morré DJ, see Helgerson SL 452, *504*, 540, *574*
Morré DJ, see Lembi CA 457, 458, *506*
Morré DJ, see Van der Woude WJ 525, *585*
Morris DA, Briant RE, Thomson PC 338, 365, *424*
Morris RO 4, *7*, 47, *100*, 154, 173, 184, *197*, 369, *425*
Morris RO, see Babcock DF 41, *81*, 173, *190*, 348, *392*
Morris RO, see Chapman RW 40, 43, 45, *84*, 292, *396*
Morris RO, see Kaiss-Chapman RW 40, *93*
Morris RO, see Zaerr JB 4, *8*
Morrison A, see Cross BE 481, *503*
Morufushi N, see Yokota T 116, 124, 130, 131, 134, 142, 164, *201*
Moruzzi G, see Bertossi F 76, *82*
Mosettig E, see Ruddat M 35, *104*
Most BH, Gaskin P, MacMillan J 49, *100*
Most BH, Williams JC, Parker KJ 173, *197*
Most BH, see Crozier A 24, *86*
Mothes K, Engelbrecht L, Kulaeva ON 544, *579*
Mousdale DMA, Butcher DN, Powell RG 383, *425*
Moyed HS, Tuli V 337, *425*
Moyed HS, Williamson V 337, *425*, 448, *507*, 533, *579*
Moyed HS, see Still CC 337, *437*
Moyed HS, see Tuli V 337, *439*, 448, *509*, 532, 533, *585*
Moyed NS, see Fukuyama TT 337, *404*
Mozell MM 495, *507*
Mozell MM, Jagodowicz M 495, *507*
Mudd JB, Kleinschmidt MG 494, *507*
Muir RM 365, *425*
Muir RM, Hansch C, Gallup AH 448, *507*
Muir RM, Lantican BP 282, 288, *425*
Muir RM, see Hansch C 448, *504*
Muir RM, see Kuraishi S *415*
Muir RM, see Lantican BP 282, 288, 365, *415*, *416*
Muir RM, see McGowan RE 288, *422*
Muira L, see Kubo L 78, *95*
Mulheirn LJ, see Evans R 310, *402*
Mulholland TPC, see Brian PW 33, 35, *83*, 479, 480, 481, *502*
Mulholland TPC, see Cross BE 301, *398*
Mulholland TPC, see Grove JF 23, 24, *90*
Muller CH 57, 76, *100*
Muller CH, Chou C-H 57, *100*
Muller CH, Muller WH, Haines BL 76, *100*
Müller H, Schuphan W 316, *425*
Müller P, Knöfel D, Sembdner G 377, *425*
Müller P, Knöfel HD, Liebisch H-W, Miersch O, Sembdner G 381, *425*
Müller P, see Knöfel HD 381, *413*
Müller P, see Sembdner G 363, 376, 381, *434*
Muller S, see Thomson KS 458, *509*
Muller WH, see Muller CH 76, *100*
Mullins MG, see Pirie A 566, *581*
Mumma RO, see Feung CS 365, 366, *403*
Mumma RO, see Hamilton RH 336, 337, 340, 344, *407*, 448, *504*
Munakata K, see Marumo S 16, 20, *98*, 115, 130, 133, 138, 142, *197*, 289, *421*
Murai F, see Sakan T 53, *104*
Murai N, Armstrong DJ, Skoog F 290, *425*
Murai N, see Armstrong DJ 293, *392*, 531, *568*
Murai N, see Vreman HJ 42, 43, 44, *109*, 292, 385, *441*
Murai N, see Walker GC 293, *441*, 474, *510*
Murakami Y 35, *100*, 243, *278*, 376, 377, 388, *425*
Murakoshi I, Ikegami F, Ookawa N, Haginiwa J, Letham DS 372, *425*
Muramo F, see Yamazaki S 78, *112*
Muramo S, Hattori H, Yamamoto A 289, *421*
Murashige T, Skoog F 242, *278*
Murashige T, see Uchimiya H 553, 562, *585*
Murayama K, Ueda K 518, 519, 521, *579*
Murneek AE, see Haagen-Smit AJ 13, *90*
Murofushi N, Iriuchijima S, Takahashi N, Tamura S, Kato J, Wada Y, Watanabe E, Aoyama T 27, *100*, 113, 114, 142, 161, *197*
Murofushi N, Nagura S, Takahashi N 316, *425*

Murofushi N, Sugimoto M, Itoh K, Takahashi N 27, 33, *100*
Murofushi N, Takahashi N, Yokota T, Kato J, Shiotani Y, Tamura S 28, *100*, 134, *197*
Murofushi N, Takahashi N, Yokota T, Tamura S 27, 28, *100*, 131, *197*
Murofushi N, Yokota T, Watanabe A, Takahashi N 25, 26, 27, 29, 30, 36, *100*, 142, *197*
Murofushi N, see Bukovac MJ 29, *83*
Murofushi N, see Hiraga K 23, 24, 25, 30, 35, 36, 49, *92*, 116, 124, 126, 131, 134, 147, 148, 149, *194*, 378, 379, 388, *410*
Murofushi N, see Kurogochi S 27, 50, *96*, 186, 187, *195*
Murofushi N, see Nakagawa S 186, *197*
Murofushi N, see Railton ID 356, 357, 379, 380, 387, *429*
Murofushi N, see Takahashi N 184, *200*, 378, 379, 380, 388, *438*
Murofushi N, see Yamaguchi I 24, 29, 31, 50, *111, 112*, 114, 116, 124, *201*, 376, *442*
Murofushi N, see Yamane H 23, 24, 25, 26, 27, 29, 30, 31, 36, *112*, 124, 131, 142, 170, *201*, 356, 357, 358, 362, 378, 379, 380, 381, 388, *442*
Murofushi N, see Yokota T 25, 27, 28, 29, 36, *112*, 378, 379, 381, *443*
Murphy GJP, Briggs DE 308, 389, *425*
Murphy PJ, West CA 308, *425*
Murr DP 320, *425*
Murr DP, Yang SF 320, 325, *425*
Murr DP, see Lau OL 325, 365, *416*
Murray AW, see Elliot DC 292, *401*
Musa SK, see Doods JH 363, *400*, 500, *503*
Musgrave A, Jackson MB, Ling E 513, 532, *580*
Musgrave A, Kays SE, Kende H *426*, 486, *507*
Musgrave A, Kende H 380, *425*, 482, *507*
Musgrave A, Walters J 513, *580*
Mutumba GC, see Hall MA 541, *573*
Myers PL, see Cross BE 315, *398*

Nachit M, see Feucht W 344, *403*
Nadeau R, Rappaport L 356, 379, 380, *426*, 481, 486, *507*
Nadeau R, Rappaport L, Stolp CF 356, 360, 378, 379, 380, *426*, 481, *507*
Nadeau R, see Rappaport L 356, 380, *430*
Nadeau R, see Stoddart JL *437*, 481, 486, 487, *509*
Nadeau R, see Stolp CF 360, 380, *437*
Näf U, see Nakanishi K 35, *100*
Näf U, see Sharp PB 35, *105*
Nagao M, see Tamura S 14, 17, *108*
Nagl W 555, 556, 557, *580*
Nagl W, Hendon J, Rücker W 556, *580*
Nagl W, Rucker W 556, *580*
Nagl W, see Przbyllok T 13, *103*
Nagura S, see Murofushi N 316, *425*
Nakagawa S, Matsui H, Yuda E, Murofushi N, Takahashi N, Akinosi N, Hishada S 186, *197*
Nakajima R, Yamazaki I 342, *426*
Nakamura T, Kikuta M, Watanabe M, Takahashi N 62, *100*
Nakamura T, Sekine S, Arai K, Takahashi N 530, *580*
Nakanishi K, Endo M, Näf U 35, *100*
Nakanishi K, Endo M, Näf U, McKeon W, Walker R 35, *100*
Nakanishi K, see Evans LS 76, *88*
Nakanishi K, see Kubo L 78, *95*
Nakanishi K, see Lynn DG 76, *97*
Nakanishi K, see Weiss G 53, 54, *110*
Nakata T, see Tanaka A 77, *108*
Nakayama H, see Harada J 58, *91*
Nakayama M, see Sana IS 76, *104*
Nakayama M, see Takeuchi S 75, *108*
Nakazaki M, Arakawa H 78, *100*
Nakosteen L, see Kaufman PB 23, *94*
Nanadita, see Laboraya MM 345, *415*
Nance JF 563, *580*
Nanda KK, Gurumurti K, Chibbar RN 340, *426*
Nari J, see Ricard J 342, *430, 431*
Nash LJ, Crozier A 362, 380, *426*
Nash LJ, Jones RL, Stoddart JL 357, 358, *426*, 479, 482, *507*
Nash LJ, see Reeve DR 134, 178, *199*
Nasini G, see Assante G 52, *81*
National Academy of Sciences 57, *100*
Naumann K, see Lürssen K 320, 326, *419*
Naumann WD, Sachs M 354, *426*
Neely PM, see Phinney BO 34, *102*
Neidle S, see Gabe EJ 77, *89*
Neljubov D 10, *101*
Nelson GEN, see Stodola FH 23, *107*, 133, *199*
Nelson H, Ilan I, Reinhold L 520, *580*
Nemori R, see Fukui H 30, 32, *88, 89*
Neskovic M, Konjevic R 361, *426*
Neukom H, see Kiss G 19, *94*
Neumann PM 517, *580*
Nevins DJ 525, 526, 528, *580*
Nevins DJ, see Loescher WH 525, *578*
Newcomb W, see Kang BG 333, *412*
Newcomb W, see Syono K 385, *438*

Newhall WF *426*
Nichimura S, see Harada F 316, 348, *408*
Nicholls PB 14, *101*, 113, 130, *197*, 367, *426*
Nicholls PB, Ong BL, Tate ME 14, *101*, 367, *426*
Nicholls PB, Paleg LG 243, *278*
Nicholls PB, see Labarca C 113, 138, *195*, 367, *415*
Nicholls W, see Crow WD 80, *85*
Nichols R 328, *426*
Nickell LG 314, *426*
Niederl JB, Brenner MW 11, *101*
Niederl JB, Brenner MW, Kelly JN 11, *101*
Niederwieser A, Giliberti P 116, 124, *197*
Nielsen N 13, *101*
Niepel B, see Bengochea T 500, *502*
Niepel B, see Hall MA 541, *573*
Ninnemann H, Zeevaart JAD, Kende H, Lang A 315, *426*
Ninnemann H, see Kende H 315, *413*
Nishi A, Kato K, Takahashi M, Yoshida R 557, *580*
Nishi A, see Yoshida R 385, *443*
Nishida K, see Matsumoto T 554, 566, *579*
Nishikawa N, Kamiya K, Takabatake A, Oshio H, Tomiie Y, Nitta I 77, *101*
Nishimura S 43, 44, 45, *101*
Nishimura S, Yamada Y, Ishikura H 43, *101*
Nishimura S, see Harada F 43, *91*
Nishimura S, see McCloskey JA 43, 44, 45, *98*
Nishimura S, see Taya Y 292, 370, 373, *438*
Nishmura S, see Tanaka Y 43, *108*
Nishio M, Zushi S, Ishii T, Furuga T, Syono K 13, *101*
Nissl D, Zenk MH 519, *580*
Nitsch C, see Gandar JC 16, 20, *89*, 289, *404*
Nitsch C, see Nitsch JP 64, *101*, 241, 244, *278*
Nitsch JP 166, 167, *197*
Nitsch JP, Nitsch C 64, *101*, 241, 244, *278*
Nitsch JP, see Bui-Dang-Ha D 41, *83*, 368, *395*
Nitsch JP, see Harada H 23, 25, *91*
Nitta I, see Nishikawa N 77, *101*
Niu P, see Fox JE 370, *403*
Nixon GLF, see Borrow A 22, *83*
Noda M, see Ichihara K 79, *92*
Noddle RC, Robinson DR 295, *426*
Noddle RC, see Milborrow BV 168, *197*, 296, 298, 299, *423*
Node M, see Fujita E 35, *88*
Noguchi M, see Matsumoto T 554, 566, *579*
Noguchi M, see Shinshi H 340, 341, *435*
Nola E, see Sica V 485, *508*
Noma M, Huber J, Pharis RP 3, *7*
Noma M, see Komoto N 49, 79, *95*
Noma T, see Isogai Y 21, *93*
Nomoto M, Tamura S 14, 17, *101*, 130, *198*
Nomoto M, see Tamura S 14, 17, *108*
Nomoto S, see Isogai Y 21, *93*
Nomura M, see Mizushima S 477, *506*
Nooden LD, Thimann KV 519, *580*
Nooden LD, see Jablonovic M 469, *505*
Nordman CE, see Gabe EJ 77, *89*
Normand G, Hartman MA, Schuber F, Benveniste P 458, *507*
Northcote DH 524, *580*
Northcote DH, see Galbraith DW 452, *504*
Northcote DM, see Jeffs RA 566, *575*
Norton K, see Cross BE 26, *85*, 302, 309, 311, 312, 313, 355, *398*
Nowacki EK, see Waller GR 75, *110*
Nowak J, Brown GN 305, *426*
Nowak J, Galston AW 345, *426*
Nowakowski W, Lubanska G 541, *580*
Nurneek AE, see Haagen-Smit AJ 113, *194*

Obata Y, Abe H, Tanaka Y, Yanagisawa K, Uchiyama M 370, *426*
Obreier JB, see Stowe BB 74, *107*
O'Brien C, see Chen C-M 348, *397*
O'Brien TJ, Jarvis BC, Cherry JH, Hanson JB 453, 455, 489, *507*, 518, 520, 558, *580*
O'Brien TJ, see Hardin JW 453, 454, *504*, 520, 554, 559, *574*
O'Brien TJ, see Holm RE 557, 558, *574*
O'Brien TJ, see Leffler HR 555, 561, *577*
Occolowitz J, see Armstrong DJ 43, *81*
Occolowitz J, see Burrows WJ 43, *84*, 137, 168, *191*, 292, 348, *396*
Occolowitz J, see Hecht SM 348, *409*
Occolowitz JL, see Burrows WJ 41, 43, 44, 45, *84*
Occolowitz JL, see Lynn DG 76, *97*
Ockerse R, Galston AW 241, *278*
Odawara S, see Imaseki H 333, *411*
Odawara S, Watanabe A, Imaseki H 336, 390, *426*
Ogawa N, see Fujii T 46, *88*
Ogawa Y 244, *278*, 545, *580*
Ogawa Y, Takahashi N 381, *426*
Ogawa Y, see Fukui H 27, 28, 30, 31, *88*, 134, *193*
Ogawa Y, see Koshimizu K 27, 48, 49, *95*, 131, 132, 134, *195*, 244, *277*
Ogawa Y, see Yamaguchi I 24, 29, 50, *111*, 114, 116, 124, *201*, 376, *442*
Ogunkanmi AB, see Wellburn AR 53, 56, *110*

Ogunkanmi AB, Wellburn AR, Mansfield TA 56, *101*
Ohashi Y, see Osawa T 77, *101*
Ohba A, see Tsunakawa M 78, *109*
Ohhara Y, see Esashi Y 325, 328, *402*
Ohkuma K, Addicott FT, Smith OE, Thiessen WE 48, 49, *101*, 295, *426*
Ohkuma K, Lyon JL, Addicott FT, Smith OE 48, 49, *101*, 131, 132, 143, 154, 155, *198*
Ohkuma K, see Addicott FT 48, *80*
Okabayashi M, see Yokota T 13, 14, 18, *112*, 115, 120, 130, 138, *201*
Okamoto T, Isogai Y, Koizumi T 13, 20, 21, 75, *101*, 130, *198*
Okamoto T, Isogai Y, Koizumi T, Fujishiro H, Sato Y *198*
Okamoto T, see Isogai Y 14, 18, 21, 65, *93*, 130, 131, 132, 137, *195*
Okamoto T, see Komoda Y 27, *95*
Okamoto T, Torii Y, Isogai Y *101*
Okazawa Y, see Koda Y 385, *414*
Okogun JI, see Adesomoju AA 352, *391*
Okomura FS, see Miller CO 39, 40, *99*
Okugun JI, see Adesomoju AA 50, 51, *81*
Okumura MH, see Miller CO 39, *99*
Olausson B, see Burstrom HG 518, *570*
Oliver C, see Still CC 337, *437*
Olney HO 365, *427*
O'Malley B, Means AR 484, 485, *507*
O'Malley BW, see Clark JH 447, *502*
O'Malley BW, see Comstock JP 486, *503*
Omran RG 336, *427*
Ong BL, see Nicholls PB 14, *101*, 367, *426*
Ookawa N, see Murakoshi I 372, *425*
Oostrom H, Loopik-Detmers MA van, Libbenga KR 457, *507*
Ordin L, see Hall MA 520, 524, 533, 554, *573*
Ordin L, see Katz M 526, *575*
Oritani T, Yamashita K 55, *101*, 300, *427*
Oritani T, see Yoshida R 128, *201*, 369, 385, *443*
Osada H, see Yamane H 23, 24, 25, 26, 27, 29, 30, 31, *112*, 170, *201*, 356, 357, 358, 362, 378, 379, 380, 388, *442*
Osawa T, Suzuki A, Tamura S 77, *101*
Osawa T, Suzuki A, Tamura S, Ohashi Y, Sasada Y 77, *101*
Osawa T, Taylor D, Suzuki A, Tamura S 77, *101*
Osawa T, see Crow WD 74, 80, *85*, 86
Osborne DJ 532, *580*
Osborne DJ, McCalla DR 242, *278*
Osborne DJ, Ridge I, Sargent JA 564, *580*
Osborne DJ, see Jackson MB 330, *411*
Osborne DJ, see Malloch KR 333, 347, *420*
Osborne DJ, see McCalla DR *421*
Osborne DJ, see Ridge I 384, *431*, 532, 540, 564, *583*
Osborne DJ, see Ward TM 11, *110*
Osborne DM, Jackson MB, Milborrow BV 374, *427*
Osborne DS 10, 11, *101*
Osborne HD, see Zeevaart JAD 315, *444*
Osborne J, see Ward TM 175, *201*
Osborne M, see Pope DG 523, *581*
Oshio H, see Morimoto H 77, *100*
Oshio H, see Nishikawa N 77, *101*
Oster MO, West CA 305, *427*
Ostrem DL, see Klun JA 462, *506*
Oswald TH, Smith AE, Phillips DV 553, *580*
Ota Y, see Kurogochi S 27, 50, *96*, 186, 187, *195*
Ota Y, see Sana IS 76, *104*
Ota Y, see Takeuchi S 75, *108*
Otake N, see Takahashi N 117, *200*
O'Toole DF, see Thomas TH *439*
Overland L 75, *101*
Owens LD, Lieberman M, Kunishi A 319, 325, 333, *427*
Owens LD, see Lieberman M 325, *418*

Pačes V 372, 386, *427*
Pačes V, Rosenberg I, Kaminek M, Holy A 372, 386, *427*
Pačes V, Werstiuk E, Hall RH 350, 372, *427*
Paces V, see Kaminek M 372, *412*
Paleg LG 545, *580*
Paleg LG, see Collins GG 550, *571*
Paleg LG, see Coombe BG 35, *85*
Paleg LG, see Douglas TJ 316, *400*
Paleg LG, see Faull KF 308, *403*
Paleg LG, see Gordon SA 282, 284, *405*
Paleg LG, see Nicholls PB 243, *278*
Paleg LG, see Phillips ML 545, *581*
Paleg LG, see Wood A 492, 493, 494, 495, 496, *510*
Pallaghy CK, Raschke K 542, *581*
Pallas JE Jr, Box JE Jr 542, *580*
Palmer J 532, *581*
Palmer JK, see McGlason WB 332, *421*
Palmer KH, see Wall ME 75, *110*
Pandya KJ, see Laboraya MM 345, *415*
Papaphilippou AP, see Staden J van 371, 386, *440*

Papernaja IB, see Zaretkii VI 184, *201*
Parikh I, see Sica V 485, *508*
Parkash V, see Rijven AHGC 560, *583*
Parker CW, Letham DS 348, 350, 369, 371, 385, 386, *427*
Parker CW, Letham DS, Cowley DE, MacLeod JK 45, *101*, 128, *198*, 369, 386, *427*
Parker CW, Letham DS, Gollnov BI, Summons RE, Duke CC, McLeod JK 370, *427*
Parker CW, Letham DS, Wilson MM, Jenkins ID, McLeod JK, Summons RE 369, 370, *427*
Parker CW, McLeod JK, Summons RE 5, *7*
Parker CW, Wilson M, Letham DS, Cowley DE, MacLeod JK 128, *198*
Parker CW, Wilson MM, Letham DS, Cowely DE, McLeold JK 369, 386, *427*
Parker CW, see Cowley DE 369, *398*
Parker CW, see Duke CC 47, *87*, 369, 370, *400*
Parker CW, see Entsch B 372, *402*
Parker CW, see Gordon ME 348, 350, 369, 385, 386, *405*
Parker CW, see Letham DS 348, 369, 370, 385, 386, *417*
Parker CW, see MacLeod JE 370, *419*
Parker CW, see Summons RE 4, *7*, 45, 47, *107*, 186, *199*, 269, 270, *279*, 369, *438*
Parker CW, see Wilson MM *442*
Parker JC, see Smith WH 328, *436*
Parker KJ, see Most BH 173, *197*
Parker W, see Berrie AMM 69, *82*
Parker W, see Buller DC 79, *84*
Parker W, see Willmer CM 79, *111*
Parups EV 335, *427*
Patchett AA, see Steward FC 76, *107*
Pater A, see McLennan BD 348, *422*
Patil SS 9, *102*
Patil SS, Tang CS 335, *427*
Patil SS, see Bruinsma J 74, *83*
Paton DM, see Crow WD 74, *86*
Patt SL, see Lynn DG 76, *97*
Patterson BD, Trewaves AJ 520, *581*
Patterson BD, see Davies DD 520, *571*
Patterson R, Rappaport L, Breidenbach RW 481, *507*
Patterson R, see Rappaport L 356, 380, *430*
Patterson RJ, Rappaport L 356, 358, *427*
Paul A, Bawdekar AS, Joshi RS, Kalkarni GH, Rao AS, Kellar GR, Bhattacharyya SC 77, *102*
Paulus H 447, *507*
Pavlová L, see Teltscherová L 346, *438*
Pé I, see Van Sumere CF 62, *109*
Pearson JA, Wareing PF 498, *507*
Pearson JA, see Beevers L 361, *394*
Péaud-Lenoel C, see Jouanneau JP 561, *575*
Peck EJ, see Clark JH 447, *502*
Pedersen M 43, *102*
Pederson K, see Rizzo PJ 455, *508*
Pegg GF 12, 22, 38, 52, *102*
Pegg GF, Cronshaw DK 331, *427*
Pendergrass A, see Thomas TH 167, 180, *200*
Penel C, Greppin H 340, 341, *427*
Pengelly W, Meins F 246, *278*
Pennachioni A, see Lado P 533, 534, *577*
Penny D, Penny P, Monro J, Bailey RW 451, *507*, 536, *581*
Penny D, see Penny P 517, 518, 519, 523, 524, 537, *581*
Penny D, see Perley JE 522, *581*
Penny P 517, 518, 521, *581*
Penny P, Dunlop J, Perley JE, Penny D 518, 523, 537, *581*
Penny P, Galston AW 520, *581*
Penny P, Penny D 517, 524, 537, *581*
Penny P, Penny D, Marshall D, Heyes JK 519, *581*
Penny P, see Penny D 451, *507*, 521, 522, 536, *581*
Penny P, see Perley JE 522, *581*
Penon P, see Miassod R 520, *579*
Penon P, see Teissere M 453, 454, *509*
Percival FW, Bandurski RS 13, 21, *102*, 365, *427*
Percival FW, Purves WK 287, *427*
Percival FW, Purves WK, Vikkery LE 286, 427
Perez FM, see Dominguez XA 77, *87*
Perley JE, Penny D, Penny P 522, *581*
Perley JE, see Penny P 518, 523, 537, *581*
Perry VM, see Lenton JR 48, 49, *96*, 117, *196*, 271, *277*
Pesci P, Tognoli L, Beffagna N, Marré E 464, *507*
Pesci P, see Beffagna N 452, 464, *502*
Pesci P, see Dohrmann U 462, 465, 468, *503*
Pesci P, see Tognoli L 469, *509*
Petering L, see Kaufman PB 523, *575*
Peterkofsky A, Jesensky C 472, *507*
Peterkovsky A 289, *427*
Peterson CA, Fletcher RA 384, *427*
Peterson CE, see Carlton BC 362, *396*
Peterson JB, Miller CO 42, 47, *102*, 139, 142, 153, 154, *198*
Peterson JV, Miller CO 369, *427*
Peterson W, see Hertel R 468, *504*
Pethe-Sadorge P, Signor Y, Guern J *427*

Petit J-F, see Armstrong DJ 43, *81*
Petrids C, Verbeek R, Massart L *428*
Petrović J, see Konjević R 487, *506*
Petschow B, see Chen C-M 385, *397*
Pettei MJ, see Kubo L 78, *95*
Pettet M, see Davis M 52, 74, *86*
Pharis PP, see McInnes AG 355, 376, 378, *422*
Pharis RP, see Crozier A 138, *192*, 479, *503*
Pharis RP, see Durley RC 124, 125, 134, 135, 169, 176, *192*, 355, 356, 357, 360, 361, *401*
Pharis RP, see Glenn TL 140, *193*
Pharis RP, see Kamienska A 23, 24, 25, *93*, 356, 357, *412*
Pharis RP, see Kaufman PB 23, *94*
Pharis RP, see Kuo CG 316, *415*
Pharis RP, see Lance B 357, 361, *415*
Pharis RP, see Noma M 3, *7*
Pharis RP, see Railton ID 309, 356, 357, 379, 380, 387, *429*
Pharis RP, see Reeve DR 356, *430*
Pharis RP, see Reid DM 362, *430*
Pharis RP, see Wample RL 356, 357, 362, 379, 380, *441*
Phelps RH, Sequeira L 282, 285, 287, *428*
Philipson JJ, Hillman JR, Wilkins MB 517, 518, 519, *581*
Philipson JJ, see McWha JA 244, *278*
Phillips C, see Matthysse AG 453, 455, 456, 498, *506*, 520, *579*
Phillips DV, see Oswald TH 553, *580*
Phillips GT, see Cornforth JW 294, *398*
Phillips IDJ 65, *102*
Phillips IDJ, Hofmann A 353, *428*
Phillips IDJ, Jones RL 387, *428*
Phillips LDJ, see Jones RL 316, 387, 388, *412*
Phillips ML, Paleg LG 545, *581*
Phillips R, Torrey JG 555, *581*
Philosoph S, see Lesham Y 374, 375, *417*
Phinney BO 3, *7*, 243, *278*
Phinney BO, West Ca, Ritzel M, Neely PM 34, *102*
Phinney BO, see Bearder JR 33, *82*, 302, 309, 311, 312, 313, 314, 354, 355, *393*, *394*
Phinney BO, see Corcoran MR 58, *85*
Phinney BO, see Geissman TA 308, 354, *404*
Phinney BO, see Hedden P 4, 5, *6*, 302, 303, 305, 312, 313, 316, 317, *409*
Phinney BO, see Hoad GV 5, *6*
Phinney BO, see Katsumi M 35, *94*
Phinney BO, see Lunnon MW 311, *419*
Phinney BO, see Spector C *436*
Phinney BO, see Verbiscar AJ 308, 354, *440*
Phinney BO, see West CA 23, 24, 34, *110*
Phipps J 345, *428*
Pieniazek J, see Rudnicki R 117, *199*, 374, 375, *432*
Pierce AE 169, *198*
Piironen E, Virtanen AI 20, *102*
Pilet PE 282, 344, *428*
Pilet PE, Braun R 559, *581*
Pilet PE, Gaspar T 336, 340, 344, 345, 346, *428*
Pilet P-E, see Rivier L 13, 50, *103*, 186, *199*, 266, 272, *279*
Pilkiewicz F, see Kubo L 78, *95*
Piozzi F, Fuganti C, Mondelli R, Geriotti G 75, *102*
Pirie A, Mullins MG 566, *581*
Piskornik Z, Bandurski RS 21, *102*, 113, 120, 130, 138, *198*, 366, *428*
Pitel DW, Vining LC 486, *507*
Pitel DW, Vining LC, Arsenault GP 134, 136, 161, *198*, 354, 355, *428*
Pitman MG, Anderson WP, Schaefer N 538, *581*
Pitman MG, Schaefer N, Wildes RA 538, 539, *581*
Pjon CJ, see Imaseki H 331, *411*
Platz KM, see Crow WD 80, *85*
Playtis AJ, Leonard NJ 41, *102*, 167, *198*, 292, *428*
Playtis AJ, see Leonard NJ 41, *96*, 167, *196*
Playtis AJ, see Vreman HJ 43, *109*, 348, *441*
Pleskotová D, see Teltscherová L 346, *438*
Plummer JR, see Mitchell JW 3, *7*
Podlesak W, see Eckert H 23, *87*, 316, *401*
Podlesak W, see Schilling G 302, 305, 316, *433*
Poison J, see Pousset J-L 53, 56, *102*
Pokorny M, see Keglević D 115, *195*, 366, 367, *413*
Poljakoff-Mayber A, see Mayer AM 57, 67, *98*
Pollard JK, Shantz EM, Steward FC 80, *102*
Pollard JK, see Steward FC 76, *107*
Pollock BM, Goodwin RH, Greene S 67, *102*
Polya GM, Davis AW 476, *507*
Polya GM, see Wardrop AJ 457, *510*
Pool R, see Liu WT *418*
Pool RM, Powell LE 178, *198*
Poovaiah BW, see McGlason WB 328, *421*
Pope D, Black M 521, *581*
Pope DG, Osborne M, Abrams J 523, *581*
Popjak G, Cornforth JW 294, *428*
Poplawski J, see Holub M 53, *92*
Poppleton BJ 78, *102*
Porazinski J, see Kopcewicz J 361, *414*

Porte AL, see Hodges R 53, *92*
Porter NG, van Steveninck PFM 48, 102
Porter NG, see Lancaster JE 543, *577*
Porter WL, Thimann KV 448, *507*, *508*
Porutzkii GV, Luchko AS, Matkovskii KI 390, *428*
Pospelov VA, see Likholat TV 456, 470, *506*
Postlethwait SN, see Gelinas DA 344, *405*
Pousset J-L, Poison J 53, 56, *102*
Powell JC, see Quebedeaux B 50, *103*
Powell LE 120, 121, 132, 133, 143, *198*
Powell LE, Seeley SD 375, *428*
Powell LE, Tautvydas KJ 134, 135, *198*, 241, *278*
Powell LE, see Pool RM 178, *198*
Powell LE, see Seeley SD 169, 174, *199*, 271, *279*
Powell LE, see Sondheimer E 298, *436*
Powell LE, see Swartz HJ 169, *200*
Powell RD, see McMichael BL 331, *422*
Powell RG, see Mousdale DMA 383, *425*
Prakash R, Maheshwari SC 368, *428*
Prat R, see Goldberg R 466, *504*
Pratt HK 327, *428*
Pratt HK, Goeschl JD 10, *102*, 317, 326, 327, 328, *428*, 498, *508*
Pratt HK, Young RE, Biale JB 11, *102*
Pratt HK, see Baur AH 319, 325, 326, *393*
Pratt HK, see Goeschl JD 330, 331, *405*
Pratt HK, see Ku HS 341, *415*, 532, *577*
Pratt HK, see Lyons JM 175, *196*
Pratt HK, see McGlasson WG 11, *98*
Pratt HK, see Yang SF 318, *443*
Pratt HM, Fox JE 476, *508*
Pridham JB 59, *102*
Primose SB 317, *428*
Prins HBA, Harper JR, Higinbotham N, Cleland RE 538, *581*
Procházka Ž, Šanda V 17, 18, 19, *102*, 115, 142, 143, *198*
Proebsting WM, Davies PJ, Marx GA 361, *428*
Proutzkii GV, Matkovskii KI 390, *428*
Pryce RJ 52, 61, *103*
Pryce RJ, MacMillan J, McCormick A *198*
Pryce RJ, see Binks R 184, *191*
Pryce RJ, see Cavell BD 161, 163, 171, 172, *191*
Pryce RJ, see Durley RC 23, 24, 25, 26, 27, 35, 51, *87*, 116, 130, *192*, 352, 388, *401*
Pryce RJ, see MacMillan J 51, 52, *97*, 182, 183, *196*, 302, 352, *419*, *420*
Przbyllok T, Nagl W 13, *103*
Puca GA, see Sica V 485, *508*
Puech AA, Reberz CA, Crane JC 554, *581*
Puppo A, Rigaud J 340, 342, *428*
Puppo A, Rigaud J, Barthe P 385, *428*
Puppo A, see Rigaud J 340, *431*
Purse J, see Horgan R 291, *410*, 472, 473, *505*
Purse JG, Horgan R, Horgan JM, Wareing PF 41, 46, *103*, 293, 348, *429*
Purse JG, see Horgan R 41, 45, 47, *92*, 139, *173*, *194*
Purse JG, see Milborrow BV 21, *99*
Purves WK, Brown HM 285, *429*
Purves WK, Rayle DL, Johnson KD 285, 286, *429*
Purves WK, see Bower PJ 285, 286, 383, *395*
Purves WK, see Brown HM 286, *395*
Purves WK, see Cleland RE 530, *570*
Purves WK, see Fox LR 342, *403*
Purves WK, see Grover S 335, *407*
Purves WK, see Percival FW 286, 287, *427*
Purves WK, see Rayle DL 18, *103*, 124, 130, 138, 145, *198*, 285, 286, *430*
Purves WK, see Sherwin JE 282, 285, 286, 287, *435*
Purves WK, see Vickery LE 286, *440*
Putnam AR, Duke WB 57, *103*

Quail PH, Browning A 452, *508*
Quebedeaux B, Sweetser PB, Powell JC 50, *103*, 272, *278*

Raa J 346, *429*
Raalte MH van, see Kerk GJM van der 448, *506*
Rabi JA, see Maçaira LA 77, *97*
Rabin BR, see Blyth CA 491, *502*
Rabin BR, see Sunshine GH 491, *509*
Rademacher W 4, *7*
Rademacher W, Graebe JE 4, *7*, 24, 34, *103*
Rademacher W, see Graebe JE 4, *6*
Radley M 34, *103*
Radley M, see Borrow A 22, *83*
Radley ME, see Gaskin P 3, *6*, 13, 26, 27, 28, 31, 33, *89*
Raghavendra AS, see Das VSR 542, *571*
Railton ID 5, *7*, 159, *198*, 357, 359, 360, 361, 362, *429*
Railton ID, Durley RC, Pharis RP 309, 356, 357, *429*
Railton ID, Murofushi N, Durley RC, Pharis RP 356, 357, 379, 380, 387, *429*
Railton ID, Reid DM 356, 361, 388, 389, *429*
Railton ID, Reid DM, Gaskin P, MacMillan J 49, *103*, 272, *278*, 387, *429*
Railton ID, Wareing PF 388, *429*

Railton ID, see Durley RC 169, *192*, 355, 356, 357, 360, *401*
Railton ID, see Reid DM 360, 361, 362, *430*
Railton ID, see Taylor CM 356, 360, *438*
Raison JK 493, *508*
Raj RK, Hutzinger O 116, 123, 133, 159, 161, *198*
Rajagopal R 284, 285, 286, *429*
Rajagopal R, Bulard C 285, 347, *429*
Rajagopal R, Larsen P 288, 384, *429*
Rajbhandary UL, see Harada F 43, *91*, 316, 348, *408*
Ralph RK, see Berridge MV 474, 475, 477, 478, *502*
Ram Chandra G, Spencer M, Meheriuk M 390, *429*
Ramina A 386, *429*
Ramstad E 71, *103*
Randazzo G, see Dohrmann U 462, 465, 468, *503*
Rando RR 325, *430*
Ranjeva R, Boudet AM, Harada H, Marigo G 565, *582*
Ranstad E, see Leopold AC 72, *96*
Rao AS, see Paul A 77, *102*
Rao LM, see Das VSR 542, *571*
Rapp A, Ziegler A 49, *103*
Rappaport L, Adams G 302, 387, 389, *430*
Rappaport L, Davies L, Lavee S, Nadeau R, Patterson R, Stolp CF 356, 380, *430*
Rappaport L, Hsu A, Thompson R, Yang SF *430*
Rappaport L, see Davies LJ 356, 379, 380, 382, *399*, 481, *503*
Rappaport L, see Ku HS 532, *577*
Rappaport L, see Nadeau R 356, 360, 378, 379, 380, *426*, 481, 486, *507*
Rappaport L, see Patterson R 481, *507*
Rappaport L, see Patterson RJ 356, 358, *427*
Rappaport L, see Redeman CT 79, *103*
Rappaport L, see Stoddart JL *437*, 481, 486, 487, *509*
Rappaport L, see Stolp CF 360, 380, *437*
Raschke K 52, *103*, 542, 543, 544, *582*
Raschke K, see Cummins WR 244, *276*, 542, *571*
Raschke K, see Pallaghy CK 542, *581*
Raschke K, see Sharkey T 52, 54, *105*
Rasi-Caldogno F, see Lado P 533, 534, 539, *577*
Rasi-Caldogno F, see Marré E 531, 534, 537, 538, 539, *578*
Ratajczak T, see Jefferies PR 309, 311, *411*, *412*
Raussendorff-Bargen G von 365, *430*
Rauthan BS, see Wightman F 21, *111*
Raven JA, Smith FA 538, *582*
Ray PM 336, 337, 342, 366, *430*, 450, 458, 461, 469, *508*, 515, 516, 518, 524, 525, 536, 539, 554, *582*
Ray PM, Dohrmann U, Hertel R 461, 464, 465, 466, 468, *508*, *582*
Ray PM, Green PB, Cleland R 514, *582*
Ray PM, Ruesink AW 514, 518, *582*
Ray PM, Thimann KV 337, *430*
Ray PM, see Abdul-Baki AA 520, *568*
Ray PM, see Anderson RL 452, *501*
Ray PM, see Baker DB 524, 525, 526, 536, 544, *568*
Ray PM, see Cross JW 468, 469, *503*
Ray PM, see Dowler MJ 540, *571*
Ray PM, see Evans ML 337, *402*, 448, *503*, 517, 518, 519, 520, 521, 522, 533, *572*
Ray PM, see Jacobs M 518, 523, 527, 529, 534, 536, *574*
Ray PM, see Kang BG 331, *412*
Ray PM, see Labavitch JM 517, 526, 527, 529, 536, *577*
Ray RM, see Kang BG 331, *412*
Rayle D, Zenk MH 521, *582*
Rayle DL 451, *508*, 532, *582*
Rayle DL, Cleland R 521, 522, 537, *582*
Rayle DL, Evans ML, Hertel R 517, 519, *582*
Rayle DL, Haughton PM, Cleland R 521, *582*
Rayle DL, Johnson KD 532, *582*
Rayle DL, Purves WK 18, *103*, 124, 130, 138, 145, *198*, 285, 286, *430*
Rayle DL, see Adams PA 530, *568*
Rayle DL, see Cleland RE 520, 530, *570*
Rayle DL, see Dohrmann U 533, *571*
Rayle DL, see Dowler MJ 540, *571*
Rayle DL, see Evans ML 285, *402*
Rayle DL, see Johnson KD 518, 525, *575*
Rayle DL, see Purves WK 285, *429*
Reberz CA, see Puech AA 554, *581*
Redeman CT, Rappaport L, Thompson RH 79, *103*
Redemann CT, Meuli LJ 301, *430*
Redemann CT, Wittwer SH, Sell HM 115, 130, *198*
Reed DJ, Moore TC, Anderson JD 316, *430*
Reed W, Martin GC 29, *103*
Rees AR, see Aung LH 362, *392*
Reese CB, see Martin DMG 127, *196*
Reeve DR, Crozier A 138, 176, 178, *199*, 243, *278*
Reeve DR, Crozier A, Durley RC, Reid DM, Pharis RP 356, *430*
Reeve DR, Yokota T, Nash LJ, Crozier A 134, 178, *199*
Reeve DR, see Bown AW 361, *395*

Reeve DR, see Crozier A 134, 143, 176, 177, 178, 181, *192*, 316, *398*
Rehm MM, Cline MG 532, *582*
Rehm S 75, *103*
Reid DM, Carr DJ 316, 388, *430*
Reid DM, Clements JB, Carr DJ 361, *430*
Reid DM, Crozier A 316, *430*
Reid DM, Pharis RP, Roberts DWA 362, *430*
Reid DM, Railton ID 316, 360, *430*
Reid DM, Tuing MS, Durley RC, Railton ID 361, 362, *430*
Reid DM, see Bowen DH 23, 24, 27, *83*, 142, 183, *191*, 244, *276*
Reid DM, see Carr DJ 387, 388, *396*
Reid DM, see Crozier A 24, *86*, 316, 388, *398*
Reid DM, see Lance B 357, 361, *415*
Reid DM, see Railton ID 49, *103*, 272, *278*, 356, 361, 387, 388, 389, *429*
Reid DM, see Reeve DR 356, *430*
Reid JSG, see Buller DC 79, *84*
Reid WW *430*
Reid WW, see Cutler HG 78, *86*
Reinhard E, see Kluge M 388, *413*
Reinhold L, see Evans ML 521, 522, *572*
Reinhold L, see Nelson H 520, *580*
Reisch J, see El-Olemy MM 74, *88*
Rekoslavskaya NI, Gamburg KZ 366, *430*
Rennenberg H, see Bergmann L 80, *82*
Rennert A, see Knypl JS 344, *414*
Restall SWF, see Dowdell RJ 513, *571*
Retig N, Rudich J 340, 341, *430*
Reynolds T 58, 78, *103*
Rhodes MJC, see Galliard T 322, *404*
Rhodes MJC, see Mapson LW 319, *421*
Ricard J, Job D 339, 340, 342, 343, *430*
Ricard J, Mazza G, Williams RJP 341, *431*
Ricard J, Nari J 342, *430*, *431*
Ricard J, see Miassod R 520, *579*
Ricard J, see Teissere M 453, 454, *509*
Ricardo CPP 562, *582*
Rice EL 56, 57, 65, 74, 76, *103*
Rice EL, see Li HC 564, *577*
Richman TL, see Ries SK 79, *103*
Richmond A, see Sitton D 388, *436*
Richmond AE, see Aharoni N 386, *391*
Richmond AE, see Bittner S 498, *502*
Richmond AE, see Mizrahi Y 543, *579*
Richmond MH, see Fowden L 76, *88*
Rickards RW, see Birch AJ 301, 309, *395*
Riddle VM, Mazelis M 282, *431*
Ridge I 532, 564, *582*
Ridge I, Osborne DJ 384, *431*, 532, 540, 564, *583*
Ridge I, see Osborne DJ 564, *580*
Ridley RG 253, *278*
Ridley RG, see Knock BA 252, *277*
Riecke E, see Libbert E 284, 287, *417*
Rier JP, see Wetmore RH 566, *586*
Ries SK, Richman TL, Wert VF 79, *103*
Ries SK, Wert V 79, *103*
Ries SK, Wert V, Sweeley LC, Leavitt RA *103*
Rigaud J, Puppo A 340, *431*
Rigaud J, see Puppo A 340, 342, 385, *428*
Rijven AHGC, Parkash V 560, *583*
Riov J, Jaffe MJ 316, *431*
Risch H, see Libbert E 337, 346, *417*
Ritchie A, see Greenwood MS 13, *90*
Ritter MC, Dempsey ME 308, *431*
Ritzel M, see Phinney BO 34, *102*
Ritzert RW, Turin BA 554, 565, *583*
Rivier L, Gaskin P, Albone KS, MacMillan J 4, *7*
Rivier L, Milon H, Pilet P-E 50, *103*, 272, *279*
Rivier L, Pilet P-E 13, *103*, 186, *199*, 266, *279*
Rizzo PJ, Pederson K, Cherry JH 455, *508*
Robbins WE, see Thompson MJ 3, *8*
Robert ML, Taylor HF, Wain RL 335, 344, 346, *431*
Roberts DWA, see Reid DM 362, *430*
Roberts JA, see Ward TM 11, *110*, 175, *201*
Roberts LW 554, 564, *583*
Roberts LW, Baba S 564, *583*
Robertson AV, see Lederberg J 235, 239, *277*
Robertson J, Berrie AMM 354, *431*
Robertson J, see Berrie AMM 50, *82*, 272, *276*
Robertson PD, see Smith KA 532, *583*
Robins MJ, Hall RH, Thedford R 43, *103*
Robins MJ, Hall RS, Thedford R 137, 165, *199*
Robins MJ, see Hall RH 43, *90*
Robinson DR, Ryback G 296, *431*
Robinson DR, West CA 305, *431*
Robinson DR, see Milborrow BV 296, *424*
Robinson DR, see Noddle RC 295, *426*
Robinson JF, see Klun JA 462, *506*
Robinson PM, Wareing PF 48. *104*
Robinson PM, see Wareing PF 48, *110*
Robinson T 57, 75, *104*

Rocha RF, see Valio IMF 35, *109*
Rodighiero G 67, *104*
Rodriguez AG 10, *104*
Rodriguez E, Towers GHN, Mitchell JC 77, *104*
Rodriguez-Barrueco C, Bermudez de Castro F 385, *431*
Rodriguez-Tormes F, see Stonier T 345, *437*
Rogers D, see Gabe EJ 77, *89*
Rogers MN, see Sengupta SK 49, *105*
Rogg H, see Staehelin M 43, *106*
Rogozinska JH, see Kopcewicz J 281, *414*
Rohrbaugh LM, see Li HC 564, *577*
Rohringer R, see Kim WK 282, 284, *413*
Rohwedder WK, see Grove MD 3, *6*
Rojas Garciduenas M, Dominguez XA 77, 78, *104*
Romo de Vivar A, see Romo J 77, *104*
Romo J, Romo de Vivar A, Trevino R, Joseph-Nathan P, Diaz E 77, *104*
Roper KB, see Stodola FH 22, 23, *107*
Ropers AJ, Graebe JE, Gaskin P, MacMillan J 308, 310, 313, 314, 355, 357, *431*
Ropers HJ, see Graebe JE 34, *90*, 302, *406*
Rosa JT 10, *104*
Rose RJ 530, *583*
Rosen GV, see Erickson RO 71, *88*
Rosenbaum N, Gefter ML 290, *431*, 473, *508*
Rosenberg I, see Pačes V 372, 386, *427*
Rosenfeld GC, see Comstock JP 486, *503*
Ross CW, see Huff AK 562, *574*
Ross JD, Bradbeer JW 268, *279*
Ross JD, see Williams PM 23, 25, 49, *111*
Rothwell K, Wain RL 48, *104*
Rothwell K, see Cornforth JW 48, *85*
Routley DG, see Hoyle MC 343, 344, *410*
Rouvray DH 234, 235, *279*
Row VV, Sanford WW, Hitchcock AE 365, *431*
Rowan KS, Gillbank LR, Spring AH 517, *583*
Rowe JR 302, *431*
Rowe JW 34, *104*
Rowell JC, see Quebedaux B 272, *278*
Roy BN, Roychondhury N, Bose TK, Baeu RN 61, *104*
Roy KL, see Armstrong DJ 472, *501*
Roy P, Biswas BB 455, *508*
Roy P, see Biswas BB 455, 456, 501, *502*
Roy S, Guha R, Chakraborty DF 74, *104*
Roy SC, see Bhattacharyya J 474, *502*, 531, *569*
Roy SK, see Buston HW 71, *84*
Roychondhury N, see Roy BN 61, *104*
Roychoudhury R, Datta A, Sen P 531, *583*
Royl B, see Sembdner G 316, 376, *434*
Rubery PH 344, *431*
Rubinstein B, see Abeles FB 11, *80*, 333, *390*
Rucker W, see Nagl W 556, *580*
Ruddat M 315, *431*
Ruddat M, Heftmann E, Lang A 310, *432*
Ruddat M, Lang A, Mosettig E 35, *104*
Ruddat M, see Alves LM 27, *81*
Ruddat M, see Jelsema CL 360, *412*
Ruddat M, see Vigil EL 549, 551, *585*
Rudich J, Halevy AH, Kedar N 361, *432*
Rudich J, Sell HM, Baker LR 356, *432*
Rudich J, see Retig N 340, 341, *430*
Rudnicki R, Czapski J 354, *432*
Rudnicki R, Hammond RK, Bukovac MJ 60, *104*
Rudnicki R, Pieniazek J 117, *199*, 374, 375, *432*
Rudnicki R, see Antoszewski R 168, *190*
Rudnicki RM, see Davison RM 50, *86*
Rudolph K 9, *104*
Ruesink AW, see Ray PM 514, 518, *582*
Runkova LV, Lis EK, Tomaszewski M, Antoszewski R 344, 347, *432*
Russell AA, see Atkinson JG 500, *501*
Russell RL, see Gefter ML 292, *404*
Russo VEA, see Hertel R 457, 458, 459, *504*
Ruth JM, see Abeles FB 362, *390*
Ruzicka L 293, 304, *432*
Ryback G 48, *104*
Ryback G, see Addicott FT 48, *80*
Ryback G, see Cornforth JW 48, 49, *85*, 190, *191*
Ryback G, see Robinson DR 296, *431*
Ryback G, see Wareing PF 295, *441*
Ryuge K, see Wylie AW 316, *442*
Ryugo K, Breen PJ 346, *432*

Sacher JA 327, 333, *432*, 540, *583*
Sacher JA, Salminen SO 540, *583*
Sacher RC, see Taneja SR 551, *584*
Sachs M, see Naumann WD 354, *426*
Sachs RM, see Wylie AW 316, *442*
Sadava D, Chrispeels MJ 529, *583*
Sadava D, see Sloane M 537, 538, *583*
Sadorge P, see Guern J *407*
Sadorge P, see Sivadjian A 371, *436*
Sadorge P, see Terrine C 372, *438*
Sadri HA, see Smith OE 315, 360, *436*

Safir D, see Winters TE 77, *111*
Sahashi Y 75, *104*, 340, *432*
Sahulka J 340, 341, *432*
Saito H, see Abe H 370, *390*
Saito M, see Hashimoto T 74, *91*
Saito Y, see Yamazaki S 78, *112*
Sakai S 333, 334, 335, *432*
Sakai S, Imaseki H 319, 326, 333, 335, 365, 389, *432*
Sakan T, Murai F, Isoe S, Hyeon SB, Hayashi Y 53, *104*
Sakan T, see Hayashi Y 78, *91*
Sakan T, see Isoe S 53, *93*
Sakato Y, see Ina K 54, *92*
Sakurai A, see ChO KY 316, *397*
Sakurai A, see Kato J 35, *94*
Sakurai A, see Tamura S 35, *108*
Sakurai N, Shibata K, Kaminska S 62, *104*
Sakurai N, Shibata K, Kamisaka S 241, *279*, 344, 346, *432*
Salazar AL, see Dominguez XA 78, *87*
Saleh MA, see Yang SF 336, *443*
Salemink CA, see Bercht CA 52, 54, *82*
Salganik RI, see Likholat TV 456, 470, *506*
Salminen SO, see Sacher JA 540, *583*
Salomon Y, see Wilchek M 454, *510*
Saltveit ME Jr, Dilley DR 331, *432*
Salunkhe DK, see Dalvi PR 77, *86*
Samek Z, see Holub M 53, *92*
Samek Z, see Vokáč K 77, *109*
Samrak HM, see Bercht CA 53, 54, *82*
Sana IS, Nakayama M, Ota Y 76, *104*
Sana IS, Ota Y 76, *104*
Sanchez NF, see Gordon SA 281, 285, *405*
Sanchez WE, see Arora SK 78, *81*
Sanchez WE, see Brown KS 56, 78, *83*
Sancheza Tames R, Gesto MDV, Vieitez E 60, *104*
Šanda V, see Procházka Ž 17, 18, 19, *102*, 115, 142, 143, *198*
Sanford WW, see Row VV 365, *431*
Sanno Y, see Morimoto H 77, *100*
Sano H 344, *432*
Santos A, see Wanner H 79, *110*
Sanyal T, Ganguly SN, Sircar PK, Sircar SM 49, *104*
Sanyal T, see Gaskin P 36, 51, *89*
Sanyal T, see Sircar PK 23, *106*
Sarah FY, see Hanson JR 311, *408*
Sargent JA, see Osborne DJ 564, *580*
Sasada Y, see Osawa T 77, *101*
Sasaki N, see Kato T 56, 67, 78, *94*
Sasaki N, see Tsunakawa M 78, *109*
Sasaki S, see Takahashi N 184, *200*
Sassa RP, see Durley RC 356, 360, *401*
Sasse JM, see Galbraith MN 78, *89*
Satish S, Bhakuni DS 53, *105*
Sato R, see Abe H 13, 21, *80*, 133, 142, *190*
Sato T, Watanabe A, Imaseki H 557, *583*
Sato Y, see Okamoto T *198*
Sato Y, see Yanagawa H 80, *112*
Saunders GA, see Heftmann E 176, *194*
Saunders PF 116, 117, *199*
Saunders PF, see Browning G 24, 25, *83*, 116, *191*, 268, *276*, 388, 389, *395*
Saunders PF, see Cooke RJ 361, 362, 388, *397*
Saunders PF, see Harrison MA 271, *277*, 374, *408*
Saunders PF, see Lenton JR 48, 49, *96*, 117, 142, 173, *196*, 271, *277*
Saunders PF, see Lorenzi R 25, *97*
Scachi A, see Cocucci M 538, *571*
Scallen TJ, Schuster MW, Dhar AK 308, *433*
Scandalios JG, see Jacobsen JV 545, 551, *574*
Scanlon DB, see Davis M 52, 74, *86*
Scarbrough E, Armstrong DJ, Skoog F, Frihart CR, Leonard NJ 40, *105*, 292, *433*
Scarbrough E, see Armstrong DJ 40, 43, *81*, 292, 293, 348, *392*
Scatchard G 475, *508*
Schader E, see Moewus F 73, *99*
Schaefer N, see Pitman MG 538, 539, *581*
Scharf P *433*
Scheider EA, see Gibson RA 282, 284, 285, 287, *405*
Schepartz AI, see Dougherty TM 139, 140, *192*
Schepartz AI, see Lerner J 139, *196*
Schiewer U, Erdmann N 287, *433*
Schiewer U, see Erdmann N 283, *402*
Schiewer U, see Libbert E 282, 283, 285, 287, 288, 338, *417*
Schilling G, Eckert H, Podlesak W 302, 316, *433*
Schilling G, Podlesak W, Ekkert H 302, 305, *433*
Schilling G, see Eckert H 23, *87*, 316, *401*
Schliemann W, Schneider G 381, *433*
Schliemann W, see Schneider G 381, *433*
Schlüter M, Gmelin R 115, *199*
Schmidt CJ, see Upper CD 173, 174, 184, *200*
Schmidt R, see Hager A 337, *407*
Schmidt V, see Dörffling K 55, *87*

Schmitt MR, see Evans ML 340, 342, *402*
Schmitz RY, see Leonard NJ 41, *96*, 167, *196*
Schmitz RY, see Scopes DIC 39, *105*
Schmitz RY, see Skoog F 531, *583*
Schmitz RY, see Vreman HJ 43, *109*, 348, *441*
Schneider CL, see Thimann KV 241, 279
Schneider EA 378, 379, *433*
Schneider EA, Gibson RA, Wightman F 18, 20, *105*, 285, 287, *433*
Schneider EA, Wightman F 20, 21, 62, *105*, 115, *199*, 281, 282, 283, 285, 288, 336, 337, 338, 340, 341, 342, 343, 344, 346, 364, 365, 383, 384, *433*
Schneider G 376, 377, *433*
Schneider G, Jänicke S, Sembdner G 171, 172, 183, *199*
Schneider G, Miersch O, Liebisch H-W 377, 379, *433*
Schneider G, Schliemann W 381, *433*
Schneider G, Sembdner G, Schreiber K 376, 377, 378, 379, *433*
Schneider G, see Dathe W 25, 26, 28, 50, 51, *86*, 352, 354, 379, *399*
Schneider G, see Schliemann W 381, *433*
Schneider G, see Schreiber K 376, 378, *434*
Schneider G, see Sembdner G 316, 360, 362, 363, 376, 378, 379, 380, 381, 388, *434*
Schneider J, Günther G 336, *433*
Schneider K, see Schreiber K 35, 36, *105*
Schnoes HK, see Woodward MD 75, *111*
Schrader WT, see Clark JH 447, *502*
Schrank AR 516, *583*
Schraudolf H 20, *105*, 288, 289, 364, *433*, *434*
Schraudolf H, Bergmann F 289, *434*
Schraudolf H, Weber H 289, *434*
Schreiber K 56, *105*
Schreiber K, Schneider G, Sembdner C, Focke I 376, 378, *434*
Schreiber K, Schneider K, Sembdner G, Focke I 35, 36, *105*
Schreiber K, Weiland J, Sembdner G 36, *105*, 376, 379, *434*
Schreiber K, see Schneider G 376, 377, 378, 379, *433*
Schreiber K, see Sembdner G 316, 360, 362, 363, 376, 377, 380, 381, 382, 388, *434*
Schrieber K, see Sembdner G 162, *199*
Schröder R, see Libbert E 284, 285, 286, 287, 338, *417*, *418*
Schröder R, see Lürssen K 326, *419*
Schuber F, see Normand G 458, *507*
Schulze A, see Bandurski RS 13, 20, 21, *81*, 264, 265, 267, *276*, 364, 368, *392*
Schulze C, see Sembdner G 363, 376, 381, *434*
Schuman GE, McCalla TM 79, *105*
Schuphan W, see Müller H 316, *425*
Schuster MW, see Scallen TJ 308, *433*
Schütte HR, see Gross D 353, 375, *406*
Schütte HR, see Hartmann S *408*
Schütte HR, see Lehmann H 373, 375, *416*
Schwabe WW, see Tompsett PB 347, *439*
Schwabe WW, see Valio IMF 52, 53, *109*
Schwachhofer K, see Darimont E 384, *399*
Schwaer C 77, *105*
Scopes DIC Zarnack U, Leonard NJ, Schmitz RY, Skoog F 39, *105*
Scott AI, see Devon TK 57, *86*
Scott AI, see McCapra F 23, 33, *98*
Scott FI, see Leopold AC 72, *96*
Scott KJ, see Wills RBH 375, *442*
Scott MF, see Hashizume T 42, 45, *91*, 186, *194*
Scott RPW 262, *279*
Scott TK 115, *199*, 336, *434*
Seaton JC, see MacMillan J 23, 24, 25, 35, *97*, 116, 131, 132, 146, 160, 161, *196*, 352, *420*
Secare E, see Mendez J 60, *98*
Seeley RC, Fawcett RL, Wain RL, Wightman F 338, *434*
Seeley SD, Powell LE 169, 174, *199*, 271, *279*
Seeley SD, see Powell LE 375, *428*
Seevers PM, see Daly JM 331, *399*
Seigler DS 74, *105*
Sekine S, see Nakamura T 530, *580*
Self R, see Mapson LW 324, *420*
Self R, see Meigh DF 74, *98*
Self R, see Ward TM 11, *110*, 175, *201*
Selinger Z, see Wilchek M 454, *510*
Sell HM, see Fukui HN 115, *193*
Sell HM, see Hemphill DD 23, 24, 25, 36, *92*, 376, 378, 381, 382, *409*
Sell HM, see Redemann CT 115, 130, *198*
Sell HM, see Rudich J 356, *432*
Sell HM, see Weller LE 115, 157, 159, *201*
Selman BR, see Elstner EF 324, 390, *401*
Sembdner G 21, 35, *105*, 364, 368, 377, 380, 382, *434*
Sembdner G, Adam G, Lischewski M, Sych FJ, Schulze C, Knöfel D, Müller P, Schneider G, Liebisch H-W, Schreiber K 363, 376, 381, *434*
Sembdner G, Bergner C, Royl B, Liebisch H-W 316, 376, *434*

Sembdner G, Borgmann E, Schneider G, Liebisch H-W, Miersch O, Adam G, Lischewski M, Schreiber K 381, *434*
Sembdner G, Gross G, Schrieber K 162, *199*
Sembdner G, Schneider G, Weiland J, Schreiber K 376, 388, *434*
Sembdner G, Schreiber K 376, *434*
Sembdner G, Schulze C, Adam G, Voigt B, Hung PD, Weiland J, Schreiber K 381, *434*
Sembdner G, Weiland J, Aurich O, Schreiber K 377, 378, 379, 381, 382, 388, *434*
Sembdner G, Weiland J, Schneider G, Schreiber K, Focke I 316, 360, 362, 378, 380, 381, 388, *434*
Sembdner G, see Bernhardt D 381, *394*
Sembdner G, see Dathe W 25, 26, 28, 49, 50, 51, *86*, 352, 354, 374, 379, 382, 388, *399*
Sembdner G, see Knöfel HD 381, *413*
Sembdner G, see Müller P 377, 381, *425*
Sembdner G, see Schneider G 171, 172, 183, *199*, 376, 377, 378, 379, *433*
Sembdner G, see Schreiber K 35, 36, *105*, 376, 378, 379, *434*
Sen P, see Roychoudhury R 531, *583*
Sengupta SK, Rogers MN, Lovak EJ 49, *105*
Sequeira L 115, *199*
Sequeira L, Hemingway RJ, Kupchan SM 77, *105*
Sequeira L, Minco L 340, 341, *434*
Sequeira L, see Phelps RH 282, 285, 287, *428*
Serafini Fracassini D, see Bagni N 67, *81*
Serebryakov EP, see Zaretkii VI 184, *201*
Sessa G, see Weissmann G 494, 495, *510*
Seta Y, see Takahashi N 13, 23, 24, 33, *107*, *108*, 131, 134, *200*
Seth AK, Wareing PF 544, *583*
Severson JG Jr, see Loh JWC 344, 346, 364, *418*
Sfakiiotakis E, see Apelbaum A 557, *568*
Sfakiotakis EM, Dilley DR 327, *434*
Shaari AR, see Hall MA 541, *573*
Shaari AR, see Jerie PH 363, *412*, 500, *505*
Shah, see Laboraya MM 345, *415*
Shah RR, Subbaiah KV, Mehta AR 564, *583*
Shaner CA, see Moore TC 282, 284, *424*
Shannon CE, Weaver W *280*
Shannon JC, see Key JL 520, *576*
Shannon JC, see Letham DS 40, 41, *96*, 127, *196*
Shannon LM, see Kay E 341, *413*
Shantz EM 336, 364, *435*
Shantz EM, see Pollard JK 80, *102*
Shantz EM, Steward FC 40, *105*
Shantz EM, see Steward FC 366, *437*
Shari AR, see Bengochea T 500, *502*
Sharkey T, Raschke K 52, 54, *105*
Sharp PB, Keitt GW, Clum HH, Näf U 35, *105*
Shaw S, Smallwood BM, Wilson DV *105*
Shaw S, see Greenwood MS 13, *90*
Shaybany B, Martin GC 50, *105*
Shaybany B, Weinbaum SA, Martin GC 50, *105*, 354, *435*
Shechter I, West Ca 305, *435*
Shechter I, see Yafin Y 305, 317, *442*
Sheldrake AR 20, 45, *105*, 113, 120, *199*
Shepherd KR, see Jenkins PA 49, *93*
Sheppard AC, see Cavell BD 161, 163, 171, 172, *191*
Sherwin JE 287, *435*
Sherwin JE, Furuya M 365, *435*
Sherwin JE, Purves WK 282, 285, 286, 287, *435*
Shibaoka H 77, *106*
Shibaoka H, Anzai T, Mitsuhashi M, Shimokoriyama M 77, *106*
Shibaoka H, Mitsuhashi M, Shimokoriyama M 77, *106*
Shibaoka H, Shimokoriyama M, Iriuchijima S, Tamura S 77, *106*
Shibaoka H, see Hogetsu T 563, *574*
Shibaoka H, see Mitsuhashi M 78, *99*, 114, *197*
Shibata K, Kubota T, Kamisaka S 61, *106*
Shibata K, see Sakurai N 62, *104*, 241, *279*, 344, 346, *432*
Shibata K, see Wada K 311, *441*
Shibuya T, Funamizu M, Kitahara Y 49, *106*
Shilo R, see Halevy AH 316, 362, *407*
Shimizu K, see Morita Y 337, 339, *424*
Shimoda C, see Yanagishima N 558, *586*
Shimokawa K, Kasai Z 175, *199*, 322, 325, 335, 362, *435*
Shimokawa K, Yokoyama K, Kasai Z 362, *435*
Shimokoriyama M, see Hogetsu T 563, *574*
Shimokoriyama M, see Shibaoka H 77, *106*
Shimura I, see Yokota T 13, 14, 18, *112*, 115, 120, 130, 138, *201*
Shindy WW, Asmundson CM, Smith OF, Kumamoto J 354, *435*
Shindy WW, Smith OE 13, 14, 18, 23, 24, 25, 26, 42, 44, 46, 49, *106*
Shingo S, Imaseki H 333, *435*
Shinshi H, Noguchi M 340, 341, *435*
Shiotani Y, see Murofushi N 28, *100*, 134, *197*
Shive JB, Sisler HD 316, *435*

Shoji A, see Asakawa Y 125, *190*, 377, 378, *392*
Short KC, see Fosket DE 555, 561, *573*
Short KC, Torrey JG 369, 385, *435*
Shropshire W Jr, see Craker LE 331, *398*
Shukla PS, Mahadevan S 288, *435*
Shulman Y, Lavee S 384, *435*
Sica V, Nola E, Parikh I, Puca GA, Cautrecasas P 485, *508*
Siddall JB, see Gorell TA 485, *504*
Sidyakin GP, see Tarasov VA 77, *108*
Siebert K, Clagett CO *435*
Siegel BZ, Galston AW 341, *435*
Siegel BZ, see Galston AW 341, *404*, *573*
Siegel E, see Libbert E 337, 338, *418*
Siegel N, see Vanderhoef LN 531, *585*
Siegel SM 71, *106*
Siegert C, see Zimmermann H 364, *444*
Siegl E, see Manteuffel R 287, *420*
Sievers AF, True RH 10, *106*
Signor Y, see Pethe-Sadorge P *427*
Silhengst P, see Libbert E 282, *417*
Silk WK, Jones RL 497, *508*, 530, *583*
Silk WK, Jones RL, Stoddart JL *435*
Sim GA, see McCapra F 23, 33, *98*
Sim GA, see Wall ME 75, *110*
Simchon S, see Halevy AH 362, *407*
Simcox PD, Dennis DT, West CA 305, 317, 389, *435*
Simolin AV, see Zaretkii VI 184, *201*
Simonson J, Barton DHR 54, 78, *106*
Simpson G, see Durley RC 272, *276*
Simpson GM, Wain RL 55, *106*
Simpson SF, Torrey JG 555, 561, *583*
Sindy WW, Smith OE 184, *199*
Singer RW, see Stornier T 345, *438*
Singh B, see Dalvi PR 77, *86*
Singh BN, see Dashek WV 387, *399*
Singh OS, see Kalsi PS 77, *93*
Singh VD, see Laboraya MM 345, *415*
Singh YD, see Jasdanwala RT 346, *411*
Sink KC Jr, see Herner RC 328, *409*
Sinska I, Lewak S *435*
Sinska I, Lewak S, Gaskin P, MacMillan J 24, 25, *106*, *436*
Sircar PK, Dey B, Sanyal T, Ganguly SN, Sircar SM 23, *106*
Sircar PK, see Ganguly T 21, *89*, 366, *404*
Sircar PK, see Gaskin P 36, 51, *89*
Sircar PK, see Sanyal T 49, *104*
Sircar SM, see Ganguly SN 23, 24, 25, 26, *89*
Sircar SM, see Ganguly T 21, *89*, 366, *404*
Sircar SM, see Gaskin P 36, 51, *89*
Sircar SM, see Sircar PK 23, *106*
Sircat SM, see Sanyal T 49, *104*
Siret P, see Corey EJ 38, *85*
Sirois JC, Miller RW 344, *436*
Sirois JC, see Miller RW 344, *424*
Sironval G, El Tannir-Lomba J 74, *106*
Sisler HD, see Shive JB 316, *435*
Sitton D, Richmond A, Vaadia Y 388, *436*
Sitton D, West CA 305, 317, *436*
Sitton D, see Kende H 385, *413*
Sivadjian A, Sadorge P, Gawer M, Terrine C, Guern J 371, *436*
Sivakumaran S, Hall MA 50, *106*
Sivakumaran S, see Hall MA 532, *573*
Sivori EM, Sonvico V, Fernandez NO 244, *279*
Sjipestejin AK, see Kerk GJM van der 448, *506*
Skene KGM 385, 386, 388, *436*
Skene KGM, see Carr DJ 387, *396*
Skoog F 2, *7*, 39, 40, 43, 44, *106*
Skoog F, Armstrong DJ 40, 44, *106*, 289, 292, 293, *436*, 470, 471, 472, *508*
Skoog F, Hamzi HQ, Szweykowska AM, Leonard NJ, Carraway KL, Fujii T, Helgeson J, Loeppky RN 471, 472, *508*
Skoog F, Miller CO *583*
Skoog F, Schmitz RY, Bock RM, Hecht SM 531, *583*
Skoog F, Strong FM, Miller CO 40, *106*
Skoog F, Tsui C 39, *106*
Skoog F, see Armstrong DJ 40, 43, *81*, 139, 140, *190*, 292, 293, 348, *392*, *501*, 531, *568*
Skoog F, see Burrows WJ 41, 43, 44, 45, *84*, 137, 139, 168, *191*, 292, 348, *396*, 473, *502*, 531, *570*
Skoog F, see Einset JW 41, *87*, 139, *193*, 293, 348, 385, *401*
Skoog F, see Hecht SM 348, *409*
Skoog F, see Jablonski J 39, *93*
Skoog F, see Klämbt D 43, *94*
Skoog F, see Lee TT 344, *416*
Skoog F, see Leonard NJ 41, *96*, 167, *196*
Skoog F, see Miller CO 39, 40, *99*
Skoog F, see Murai N 290, *425*
Skoog F, see Murashige T 242, *278*
Skoog F, see Scarbrough E 40, *105*, 292, *433*

Skoog F, see Scopes DIC 39, *105*
Skoog F, see Struxness LA 385, *438*
Skoog F, see Vreman HJ 41, 43, 44, *109*, 139, *201*, 292, 348, *440*, *441*
Skoog F, see Walker GC 293, *441*, 474, *510*
Skoog FK, see Einset JW 40, *87*
Sloane M, Sadava D 537, 538, *583*
Slocum HK, see Divekar AY 372, *400*
Sloger M, see Lieberman M 333, 334, *418*
Slonecker JH, see Gottlieb D 495, *504*
Smalldon KW, see Moffat AC 256, *278*
Smallheer JM, see Frenkel C 337, *403*
Smallwood BM, see Shaw S *105*
Smart JCC *280*
Smith, see Chen C-M 348, *397*
Smith AE, see Oswald TH 553, *580*
Smith CC, see Ohkuma K 295, *426*
Smith CJ *583*
Smith CJ, see Darvill AG 526, 527, 528, 529, 530, 536, *571*
Smith DG, see McInnes AG 26, *98*, 355, 376, 378, *422*
Smith EJ 524, *583*
Smith FA, see Raven JA 538, *582*
Smith GF, see Bentley JA 120, *191*
Smith GF, see Henbest HB 17, *92*, 115, 130, *194*
Smith H, see Birch AJ 301, 309, *395*
Smith H, see Evans A 361, 362, *402*
Smith IC, see Knock BA 252, *277*
Smith IK, Fowden L 76, *106*
Smith JD, see Dunn DB 45, *87*
Smith JR, see Zimmermann U 539, 540, *586*
Smith KA, Dowdell RJ 513, *583*
Smith KA, Robertson PD 532, *583*
Smith KA, see Dowdell RJ 513, *571*
Smith LM, see Caruso IL 186, *191*
Smith LM, see Caruso JL 13, *84*
Smith MS, Wain RL, Wightman F 450, *508*
Smith MV, see Mitchell JW 3, *7*
Smith OE, Sadri HA 315, 360, *436*
Smith OE, see Addicott FT 48, *80*, 244, *276*
Smith OE, see Ohkuma K 48, 49, *101*, 131, 132, 143, 154, 155, *198*
Smith OE, see Shindy WW 13, 14, 18, 23, 24, 25, 26, 42, 44, 46, 49, *106*
Smith OE, see Sindy WW 184, *199*
Smith OF, see Shindy WW 354, *435*
Smith OO, see Chen C-M *397*
Smith RG, see Caruso IL 186, *191*
Smith RG, see Caruso JL 13, *84*
Smith RM, see Kupchan MS 80, *96*
Smith TA 18, *106*
Smith TA, see Taylor HF 55, *108*, 300, *438*
Smith WB, see Dominguez XA 74, *87*
Smith WH, Meigh DF, Parker JC 328, *436*
Smolenska G, Lewak S 361, *436*
Snir I, Kessler B 316, *436*
Söding H 12, *106*
Sohns VE, see Stodola FH 22, 23, *107*
Söll D, see Armstrong DJ 472, *501*
Söll D, see Bartz L 289, *393*
Soll H, see Böttger M 289, *395*
Sondheimer E, Galson EC, Chang YP, Walton DC 351, *436*
Sondheimer E, Galston EC, Tinelli E, Walton DC 352, 353, *436*
Sondheimer E, Michniewicz BM, Powell LE 298, *436*
Sondheimer E, Tzou DS 293, 348, 350, 368, 385, 386, *436*
Sondheimer E, see Cummins WR 498, *503*
Sondheimer E, see Tinelli ET 51, *108*
Sondheimer E, see Tzou DS 368, 386, *439*
Sondheimer E, see Walton DC 298, 351, 352, 353, *441*, 560, *585*
Sondheimer R, see Tinelli ET 352, 353, *439*
Soni SL, see Kaufman PB 561, *575*
Sonka B, see Dörffling K 353, 374, 375, *400*
Sonnenbichler J, see Biemann K 40, 43, *82*
Sonvico V, see Sivori EM 244, *279*
Sood CK, see Fox JE *403*
Soofi GS, see Walton DC 560, *585*
Šorm F, see Vokáč K 77, *109*
Southerland DS, see Crow WD 80, *85*
Southerland GK, see Gortner WA 344, *406*
Spaander S, see Winter H 529, *586*
Spaeth SC, Maravolo NC 340, 343, *436*
Spahr PF, see Dunn DB 45, *87*
Spaulding DW, see Steffens GL 79, *107*
Spector C, Phinney BO *436*
Spence DJ, see Stodola FH 23, *107*, 133, *199*
Spencer FS, Ziola B, Maclachlan GA *583*
Spencer M 317, 323, 325, 326, 327, 390, *436*
Spencer M, Meheriuk M 390, *436*
Spencer M, see Ghooprasert P 322, 323, *405*
Spencer M, see Meheriuk M 331, 390, *422*
Spencer M, see Ram Chandra G 390, *429*

Spencer MS, see Meheriuk M 322, *422*
Spencer MS, see Stinson RA 322, 390, *437*
Spencer MS, see Thompson JE 322, *439*
Spenser GF, see Grove MD 3, *6*
Sponsel VM, Gaskin P, MacMillan J 4, 5, *7*, 26, 27, 29, 31, 33, 50, 51, *106*, 355, 382, *436*
Sponsel VM, Hoad GV, Beeley LJ 481, *508*
Sponsel VM, MacMillan J 32, *106*, 189, 190, *199*, 356, 357, 358, 359, 360, 362, 379, 380, 388, *436*
Sponsel VM, see Bearder JR 302, 355, *393*
Sponsel VM, see Hedden P 302, 303, 305, 316, *409*
Sponsel VM, see Hoad GV 5, *6*
Spotswood TM, see Wood A 495, *510*
Spring AH, see Rowan KS 517, *583*
Springer JP, Clardy J, Cox RH, Cutler HG, Cole RJ 78, *106*
Squire GR, Mansfield TA 544, *584*
Srivastav HN, see Laboraya MM 345, *415*
Srivastava BIS 114, 115, *199*
Srivastava L, see Keith B 6, *7*
Srivastava OP, Huystee RB van 341, *436*
Staden J van 369, 372, 386, *440*
Staden J van, Button J 384, *440*
Staden J van, Davey JE 385, *440*
Staden J van, Papaphilippou AP 371, 386, *440*
Staehelin M, Rogg H, Baguley BC, Ginsburg T, Wehrli W 43, *106*
Stahl CA, see Vanderhoef LN 451, *509*, 518, 521, 531, *585*
Stahmann MA, Clare BG, Woodbury W 347, *436*
Stahmann MA, see Demorest DM 337, *399*
Stahmann MA, see Imaseki H 347, *411*
Standish MM, see Bangham AD 492, 494, *501*
Stanley LA, see Craker LE 331, *398*, 566, *571*
Starbuck MJ, see Craker LE 331, *398*, 566, *571*
Stasiuk L, see Hall RH 43, *90*
Steele KD, see Zabkiewicz JA 13, *112*
Steen DA, Chadwick AV 333, *437*
Steen I, Eliasson L 138, *199*
Steffens GL, see Grove MD 3, *6*
Steffens GL, Tso LC, Spaulding DW 79, *107*
Stekoll M, West CA *437*
Stenhagen E, Abrahamson S, McLafferty FM 250, *279*
Stenlid G 64, 65, *107*
Sterns M 80, *107*
Sterns M, see Crow WD 80, *85*
Steudle E, see Coster HGL 539, *571*
Steudle E, see Zimmermann U 539, 540, *586*
Steveninck RFM van, see Mittelhauser CT 543, *579*
Steward FC, Krikorian AD 56, 80, *107*
Steward FC, Pollard JK, Patchett AA, Witkop B 76, *107*
Steward FC, Shantz EM 366, *437*
Steward FC, see Caplin SM 242, *276*
Steward FC, see Pollard JK 80, *102*
Steward FC, see Shantz EM 40, *105*
Steward T, see Armstrong DJ 43, *81*
Stewart ER, Freebairn HT 540, *584*
Stewart JC, see Cross BE 35, *85*, 302, 304, 309, 311, 312, *398*
Still CC, Fukuyama TT, Moyed HS 337, *437*
Still CC, Oliver C, Moyed HS 337, *437*
Stinson RA, Spencer MS 322, 390, *437*
StJohn PA, see Brook JL 169, *191*
Stocker H, see Wanner H 79, *110*
Stoddart JL 308, 388, 389, *437*, 482, 497, *508*
Stoddart JL, Breidenbach W, Nadeau R, Rappaport L *437*, 481, 486, 487, *509*
Stoddart JL, Jones RL 378, 380, *437*, 482, 488, *508*
Stoddart JL, Lang A 361, 388, *437*
Stoddart JL, Williams PD 5, *7*, 497, *509*
Stoddart JL, see Cross BE 35, *85*, 309, *398*
Stoddart JL, see Kuhn-Silk W 356, *415*
Stoddart JL, see Nash LJ 357, 358, *426*, 479, 482, *507*
Stoddart JL, see Silk WK *435*
Stodola FH 22, *107*
Stodola FH, Nelson GEN, Spence DJ 23, *107*, 133, *199*
Stodola FH, Roper KB, Fernel DI, Conway HF, Sohns VE, Langford CE, Jackson RW 22, 23, *107*
Stoessl A, Venis MA 206, 266, *279*
Stoffer C, see Elstner EF 324, 390, *401*
Stolp CF, Nadeau R, Rappaport L 360, 380, *437*
Stolp CF, see Nadeau R 356, 360, 378, 379, 380, *426*, 481, *507*
Stolp CF, see Rappaport L 356, 380, *430*
Stolwijk JAJ, see Burg SP 12, *84*
Stominger JL, see Armstrong DJ 43, *81*
Stonier T 345, *437*
Stonier T, Hudek J, Vande-Stouwe R, Yang H 345, *437*
Stonier T, Rodriguez-Tormes F, Yoneda Y 345, *437*
Stonier T, Singer RW, Yang H 345, *438*

Stonier T, Yang H 345, *437*
Stonier T, Yoneda Y 344, 345, *437*
Stonier T, Yoneda Y, Rodriguez-Tormes F 345, *437*
Stonier T, see Yoneda Y 344, 345, *443*
Stowe BB 21, *107*
Stowe BB, Obreier JB 74, *107*
Stowe BB, Thimann KV 157, 158, 159, *199*
Stowe BB, Yamaki T 22, *107*
Stowe BB, see Elliot MC 19, *87*
Stowe BB, see Elliott MC 114, 115, 123, 124, 130, 138, *193*
Stowe BB, see Furuya M 64, *89*
Stowe BB, see Iwata T 334, *411*
Street HE, see Thurman DA 366, *439*
Strivastava BIS, see Hall RH 471, 500, *504*
Strobel GA 9, *107*
Strömquist L-H, see Eliasson L 267, *277*
Strong FM, see Miller CO 39, 40, *99*
Strong FM, see Skoog F 40, *106*
Strumpf E, see Jensen EV 485, *505*
Strunz GM, see Little CHA 49, *96*, 143, *196*
Struxness LA, Armstrong DJ, Gillam I, Tener GM, Burrows WJ, Skoog F 385, *438*
Stuart DA, Durnam DJ, Jones RL 497, *509*
Stuart DA, Jones RL 497, *509*, 530, *584*
Stuart KL, Coke LB 56, *107*
Stuart KL, Woo-Ming RB 53, *107*
Stuart KL, see Coke LB 56, *84*
Stuart NW, see Asen S 70, *81*
Stuart RS, see Atkinson JG 500, *501*
Subbaiah KV, see Shah RR 564, *583*
Südi J 364, *438*
Südi J, see Kenney G 336, *413*
Suge H 330, *438*
Suge H, Katsura N, Inada K 532, *584*
Suge H, see Ku HS 532, *577*
Sugimoto M, see Murofushi N 33, *100*
Sugiyama T, see Hashizume T 41, 42, 45, *91*, 186, *194*
Suhadolnik RJ, see Chassy BM 372, *397*
Sumiki Y, see Ikekawa N 170, *195*
Sumiki Y, see Kagawa T 134, 161, 162, *195*
Sumiki Y, see Kawarada A 23, 34, *94*
Sumiki Y, see Takahashi N 13, 23, 24, 33, *107*, *108*, 131, 134, *200*
Sumiki Y, see Yabuta T 22, *111*
Summons RE, Duke CC, Eichholzer JV, Entsch B, Letham DS, McLeod JK, Parker CW 4, *7*
Summons RE, Entsch B, Letham DS, Gollnow BI, McLeod JK 4, *7*
Summons RE, Entsch B, Parker CW, Letham DS 4, *7*
Summons RE, MacLeod JK, Parker CW, Letham DS 45, 47, *107*, 186, *199*, 269, 270, *279*, 369, *438*
Summons RE, see Cowley DE 369, *398*
Summons RE, see Duke CC *400*
Summons RE, see Letham DS 348, 369, 370, 385, 386, *417*
Summons RE, see MacLeod JE 370, *419*
Summons RE, see Parker CW 5, *7*, 369, 370, *427*
Sundin O, see Beyer EM Jr 363, *394*
Sunshine GH, Williams DJ, Rabin BR 491, *509*
Sussex I, Clutter M, Walbot V 560, *584*
Sussex I, see Walbot V 560, *585*
Sussman MR, Kende H 478, *509*
Sussman MR, see Gardner G 458, 478, *504*
Suter PJ, see Elson GW 162, 163, *193*
Suter PJ, see MacMillan J 23, 24, 25, 34, 35, *97*, 116, 131, 132, 134, 146, 160, 161, *196*, 352, *420*
Sutherland GL, see Lederberg J 235, 239, *277*
Suzuki A, see Osawa T 77, *101*
Suzuki K, see Takahashi N 14, 20, *108*, 115, 145, 188, 189, *200*, 281, *438*
Suzuki T, see Jensen EV 485, *505*
Suzuki Y, Kawarada A 338, *438*
Suzuki Y, Kinashi H, Takeuchi S, Kawarada A 13, *107*, 334, *438*
Suzuki Y, see Ishigami T 288, *411*
Suzuki Y, see Kinashi H 339, 340, *413*
Swain T 57, *107*
Swaminathan S, Bock RM 42, 43, 44, 45, *107*
Swaminathan S, see Einset JW 41, *87*, 139, *193*
Swaminathan S, see Vreman HJ 42, 43, 44, *109*, 292, 385, *441*
Swanson BT, Wilkins HF, Weiser CF, Klein I 49, *107*
Swartz HJ, Powell LE 169, *200*
Sweeley LC, see Ries SK *103*
Sweetman BJ, see Falkner FC 186, *193*
Sweetser PB, see Quebedeaux B 50, *103*, 272, *278*
Sweetser PB, Vatvars A 118, 139, 181, *200*, 272, *279*, 374, *438*
Sych F-J, see Adam G 377, 378, *391*
Sych FJ, see Sembdner G 363, 376, 381, *434*
Syono K, Newcomb W, Torrey JG 385, *438*
Syono K, see Nishio M 13, *101*
Syono K, Torrey JG 368, 385, *438*
Szweykowska AM, see Skoog F 471, 472, *508*

Taiz L, Honigman WA 551, *584*
Taiz L, Jones RL 551, *584*
Tajima M, see Hashimoto T 74, *91*
Takabatake A, see Nishikawa N 72, *101*
Takahashi M, see Nishi A 557, *580*
Takahashi M, see Yokota T 13, 14, 18, *112*
Takahashi N 302, 355, *438*
Takahashi N, Kitamura H, Kawarada A, Seta Y, Takai M, Tamura S, Sumiki Y 23, 33, *107*, 131, *200*
Takahashi N, Marumo S, Otake N 117, *200*
Takahashi N, Murofushi N, Tamura S, Wasada N, Hoshino H, Tsuchiya T, Sasaki S, Aoyama T, Watanabe E 184, *200*
Takahashi N, Murofushi N, Yamane H 378, 379, 380, 388, *438*
Takahashi N, Seta Y, Kitamura H, Sumiki Y 13, 24, 33, *108*, 134, *200*
Takahashi N, Yamaguchi I, Kono T, Igoshi M, Hirose K, Suzuki K 14, 20, *108*, 115, 145, 188, 189, *200*, 281, *438*
Takahashi N, see Bukovac MJ 29, *83*
Takahashi N, see Cho KY 316, *397*
Takahashi N, see Goldschmidt EE 281, *405*
Takahashi N, see Hiraga K 23, 24, 25, 30, 35, 36, 49, *92*, 116, 124, 126, 131, 134, 147, 148, 149, 164, 171, 173, *194*, 377, 378, 379, 381, 388, *410*, 480, *505*
Takahashi N, see Igoshi M 115, 130, 133, 142, 144, *195*
Takahashi N, see Kato T 54, 56, 67, 78, *94*
Takahashi N, see Kitahara Y 80, *94*
Takahashi N, see Kurogochi S 27, 50, *96*, 186, 187, *195*
Takahashi N, see MacMillan J 34, *97*
Takahashi N, see Murofushi N 25, 26, 27, 28, 29, 30, 33, 36, *100*, 113, 114, 131, 134, 142, 161, *197*, 316, *425*
Takahashi N, see Nakagawa S 186, *197*
Takahashi N, see Nakamura T 62, *100*, 530, *580*
Takahashi N, see Ogawa Y 381, *426*
Takahashi N, see Watanabe N 27, 42, 44, 50, *110*, 137, 173, 186, *201*
Takahashi N, see Yamaguchi I 24, 29, 31, 50, *111*, *112*, 114, 116, 124, 177, *201*, 376, *442*
Takahashi N, see Yamane H 3, *8*, 23, 24, 25, 26, 27, 28, 29, 30, 31, 36, *117*, 124, 131, 142, 170, *201*, 356, 357, 358, 362, 378, 379, 380, 381, 388, *442*
Takahashi N, see Yanagawa H 80, *112*
Takahashi N, see Yokota T 25, 27, 28, 29, 36, *112*, 115, 116, 120, 124, 130, 131, 134, 138, 142, 164, 184, *201*, 376, 377, 378, 379, 381, *443*
Takahashi Y, see Kimura Y 60, *94*
Takai M, see Takahashi N 23, 33, *107*, 131, *200*
Takai M, see Tamura S 35, *108*
Takao S, see Fujita T *404*
Takasugu M, Anetai M, Katsui N, Masamune T 51, 53, 54, *108*
Takeba G, Takimoto A 308, *438*
Takegami T, Yoshida K 474, 476, 477, *509*
Takesako K, see Kimura Y 60, *94*
Takeuchi S, Kono Y, Kawarada A, Ota Y, Nakayama M 75, *108*
Takeuchi S, see Kinashi H 339, 340, *413*
Takeuchi S, see Suzuki Y 13, *107*, 334, *438*
Takimoto A, see Takeba G 308, *438*
Tal M, Imber D 541, 543, *584*
Tal M, Imber D, Gardi I 541, *584*
Talbert NE, see Mitchell ED 75, *99*
Taller BJ, see Armstrong DJ 293, *392*, 531, *568*
Talmadge KW, Keegstra K, Bouer WD, Albersheim P 526, *584*
Talmadge KW, see Keegstra K 526, 527, 529, *576*
Tamaki E, see Matsumoto T 554, 566, *579*
Tamari K, see Asakawa Y 125, *190*, 377, 378, *392*
Tamari K, see Yabuta T 22, *111*
Tamura N, see Murofushi N 113, 114, 131, 134, 142, 161, *197*
Tamura S 9, *108*
Tamura S, Nomoto M, Nagao M 14, 17, *108*
Tamura S, Sakurai A, Kainuma K, Takai M 35, *108*
Tamura S, see Cho KY 316, *397*
Tamura S, see Hashimoto T 49, *91*, 132, 137, *194*
Tamura S, see Iriuchijima S 77, *93*
Tamura S, see Kato J 35, *94*
Tamura S, see Kimura Y 60, *94*
Tamura S, see Komoto N 49, 79, *95*, 131, 132, *195*
Tamura S, see Murofushi N 27, 28, *100*
Tamura S, see Nomoto M 14, 17, *101*, 130, *198*
Tamura S, see Osawa T 77, *101*
Tamura S, see Shibaoka H 77, *106*
Tamura S, see Takahashi N 23, 33, *107*, 131, 184, *200*
Tamura S, see Yamazaki S 78, *112*
Tamura S, see Yokota T 25, 27, 28, 29, 36, *112*, 116, 124, 130, 131, 134, 142, 164, *201*, 378, 379, 381, *443*
Tamura T, see Yabuta T 22, *111*
Tanaka A, Nakata T, Yamashita K 77, *108*
Tanaka A, see Iino Y 77, *92*

Tanaka Y, Abe H, Uchiyama M, Taya Y, Nishimura S 43, *108*
Tanaka Y, see Abe H 370, *390*
Tanaka Y, see Obata Y 370, *426*
Tanaka Y, see Taya Y 292, 370, 373, *438*
Tandeau de Marsac N, see Jouanneau JP 556, *575*
Taneja SR, Sacher RC 551, *584*
Taneko K, see Yamane H 3, *8*
Tang CS, see Patil SS 335, *427*
Tanimoto E, Masuda Y 518, *584*
Tanimoto E, see Masuda Y 517, 520, *579*
Tanner W, Komor E, Fenzl F, Decker M 544, *584*
Tao KL, Khan AA 551, 552, *584*
Tarasov VA, Abdullaev ND, Kasymov SZ, Sidyakin GP 77, *108*
Tate ME, see Nicholls PB 14, *101*, 367, *426*
Tautvydas KI, Galston AW *584*
Tautvydas KJ, see Powell LE 134, 135, *198*, 241, *278*
Tavares J, see Verma DC 563, 564, *585*
Tavares JE, see Kende H 472, *506*
Tavares JE, see Thomson KS 458, *509*
Tavener RJA, Laidman DL 552, *584*
Taves C, see Goodwin RH 67, *90*
Taya Y, Tanaka Y, Nishimura S 292, 370, 373, *438*
Taya Y, see Tanaka Y 43, *108*
Taylor CM, Railton ID 356, 360, *438*
Taylor D, see Osawa T 77, *101*
Taylor HF 300, *438*
Taylor HF, Burden RS 55, *108*, 129, 132, 174, *200*, 244, *279*, 300, *301*, *438*
Taylor HF, Smith TA 55, *108*, 300, *438*
Taylor HF, see Burden RS 129, *191*, 295, 300, *396*
Taylor HF, see Firn RD 55, *88*, 117, 129, 175, *193*, 300, *403*
Taylor HF, see Robert ML 335, 344, 346, *431*
Taylor IEP, Wilkinson AJ 34, *108*
Taylor JL, Haydon DA 495, *509*
Taylor NF, Kent PW 481, *509*
Taylor PA, Kosuge T, Devay JE 45, *108*
Taylor WC, see Bentley JA 120, *191*
Taylor WC, see Jones ERH 18, *93*, 115, *195*
Tegley JR, Witham FH, Krasnuk M 128, *200*
Tegley JR, see Krasnuk M 41, 46, *95*, 384, *414*
Teissere M, Penon P, Huystee RB van, Azou Y, Ricard J 454, *509*
Teissere M, Penon P, Ricard J 453, 454, *509*
Teissiére M, see Miassod R 520, *579*
Teller G, see Dauphin B 42, 44, *86*, 186, *192*
Teltscherová L, Pavlová L, Pleskotová D 346, *438*
Tener GM, see Struxness LA 385, *438*
Tepfer M, see Adams PA 530, *568*
Terrine C, Sadorge P, Gawer M, Guern J 372, *438*
Terrine C, see Doree M 372, *400*
Terrine C, see Gawer M 371, *404*
Terrine C, see Hall RH 350, 372, *407*
Terrine C, see Laloue M 293, 348, 350, 369, 371, 385, 386, *415*
Terrine C, see Sivadjian A 371, *436*
Thakur ML 60, *108*
Thedford R, see Hall RH 43, *90*
Thedford R, see Robins MJ 43, *103*, 137, 165, *199*
Theisen P, see Fang SC 336, *403*
Theuns H, see Bercht CA 53, 54, *82*
Thies G, see Klämbt D 43, *94*
Thiess DE, Lichtenthaler HK 79, *108*
Thiessen WE, see Addicott FT 48, *80*
Thiessen WE, see Ohkuma K 48, 49, *101*, 295, *426*
Thimann KV 13, 56, 62, 64, *108*, 143, *200*, 281, 336, 337, *439*, 448, *509*, 516, *584*
Thimann KV, Bonner WD 69, *108*
Thimann KV, Grochowska M 283, 287, *439*
Thimann KV, Mahadevan S 288, *439*
Thimann KV, Schneider CL 241, *279*
Thimann KV, see Burg SP 11, *84*, 325, 335, *396*
Thimann KV, see Mahadevan S 288, *420*
Thimann KV, see Nooden LD 519, *580*
Thimann KV, see Porter WL 448, *507*, *508*
Thimann KV, see Ray PM 337, *430*
Thimann KV, see Stowe BB 157, 158, 159, *199*
Thimann KV, see Went FW 11, *110*
Thomas GJ, see Kupchan MS 80, *96*
Thomas R, see Vreman HJ 42, 43, 44, *109*, 292, 385, *441*
Thomas TH, Carroll JE, Isenberg FMR, Pendergrass A, Howell L 167, 180, *200*
Thomas TH, Khan AA, O'Toole DF *439*
Thomas TH, see Biddington NL 141, *191*, 242, *276*
Thommer H, see Kienzle F 49, 53, 55, *94*
Thompson AG, Bruinsma J 55, *108*, 301, *439*
Thompson AG, Horgan R, Heald JK 45, *108*, 269, *279*

Thompson AG, Horgan R, Herald JK 185, *200*
Thompson AG, see Wang TL 47, *110*, 127, 139, 173, 184, *201*, 348, 370, *441*
Thompson JE, Spencer MS 322, *439*
Thompson JE, Tribe TA, Spencer MS 322, *439*
Thompson MJ, Mandava N, Flippen-Anderson JL, Worley JF, Dutky SR, Robbins WE, Lusby W 3, *8*
Thompson ML, see Cleland RE 530, *570*
Thompson R, see Rappaport L *430*
Thompson RH 74, *108*
Thompson RH, see Redeman CT 79, *103*
Thompson WF, Cleland R 520, 559, *584*
Thompson WF, see Cleland RE 520, *570*
Thomson KS 458, 465, *509*
Thomson KS, Hertel R, Muller S, Tavares JE 458, *509*
Thomson KS, Leopold AC 458, 465, *509*
Thomson K-S, see Hertel R 457, 458, 459, *504*
Thomson K-S, see Lembi CA 457, 458, *506*
Thomson PC, see Morris DA 338, 365, *424*
Thorpe TA, see Lance B 357, 361, *415*
Thresh R, see Audus LJ 241, *276*
Thurman DA, Street HE 366, *439*
Tietz D, Doerffling K, Woehrle D, Erzieben J, Liemann F 353, *439*
Tietz D, see Dörffling K 353, 374, 375, *400*
Tifchelaar EC, see McGlason WB *421*
Tilford RH, see Wolf FT 56, 62, *111*
Tillberg E 21, *108*, 244, *279*, 365, 366, *439*
Tillberg E, see Eliasson L 267, *277*
Timmappaya G, Cherayil JD 348, *439*
Timmis J, Ingle J 555, *584*
Tinelli E, see Sondheimer E 352, 353, *436*
Tinelli ET, Sondheimer E, Walton DC, Gaskin P, MacMillan J 51, *108*, 352, 353, *439*
Tingwa PO, Young RE 333, *439*
Tipton CL, see Klun JA 462, *506*
Todd GW, see Mills VM 347, *424*
Tognoli L, Beffagna N, Pesci P, Marré E 469, *509*
Tognoli L, see Beffagna N 452, 464, *502*
Tognoli L, see Pesci P 464, *507*
Tognoni F, see Alpi A 388, *391*
Tolbert NE, see Carlton BC 362, *396*
Tomaszewski M, see Runkova LV 344, 347, *432*
Tomer E, see Goren R 67, *90*
Tomiie Y, see Nishikawa N 77, *101*
Tompsett PB, Schwabe WW 347, *439*
Tonzig S, Marré E 80, *109*
Toole EH, Hendricks SB, Borthwick HA, Toole VK 57, *109*
Toole VK, see Toole EH 57, *109*
Torrey JG 384, *439*
Torrey JG, Fosket DE 555, *585*
Torrey JG, Fosket DE, Hepler PK 553, 564, *585*
Torrey JG, see Libbenga KR 555, *577*
Torrey JG, see Matthysse AG 554, *579*
Torrey JG, see Phillips R 555, *581*
Torrey JG, see Short KC 369, 385, *435*
Torrey JG, see Simpson SF 555, 561, *583*
Torrey JG, see Syono K 368, 385, *438*
Torri Y, see Okamoto T *101*
Toth I, see Erdel L 452, *503*
Totolo A, see Ballio A 533, *568*
Towers GHN, see Rodriguez E 77, *104*
Trapy F, see Doree M *400*
Trevino R, see Romo J 77, *104*
Trewavas AJ 517, 520, *585*
Trewavas AJ, see Davies DD 520, *571*
Trewaves AJ, see Patterson BD 520, *581*
Tribe TA, see Thompson JE 322, *439*
Trillmich K, Michalke W 458, *509*
Tristram H, see Fowden L 76, *88*
Troxler RF, Hamilton RH 336, 337, *439*
True RH, see Sievers AF 10, *106*
Truelsen TA 283, 284, *439*
Tso LC, see Steffens GL 79, *107*
Tsuchiya T, see Takahashi N 184, *200*
Tsui C, see Skoog F 39, *106*
Tsukamoto Y, see Ando T 79, *81*
Tsunagawa M, see Kato T 54, 56, 78, *94*
Tsunakawa M, Ohba A, Sasaki N, Kabuto C 78, *109*
Tsunakawa S, see Biemann K 40, 43, *82*
Tucker DJ, Mansfield TA 244, *279*, 542, *585*
Tuing MS, see Reid DM 361, 362, *430*
Tuli V, Moyed HS 337, *439*, 448, *509*, 532, 533, *585*
Tuli V, see Basu PA 337, *393*
Tuli V, see Moyed HS 337, *425*
Turin BA, see Ritzert RW 554, 565, *583*
Turkova NS, see Madeikyte E 540, *578*
Turner B, see Cook CE 77, *85*
Turner JV, see Coombe BG 35, *85*
Turner RB, see Grove JF 24, *90*
Turner WB, see Birch AJ 301, *395*
Tzou DS, Galson EC, Sondheimer E 368, 386, *439*

Tzou DS, see Sondheimer E 293, 348, 350, 368, 385, 386, *436*

Uchida I, see Fujita E 78, *88*
Uchimiya H, Murashige T 553, 562, *585*
Uchiyama H, see Abe H 370, *390*
Uchiyama M, see Abe H 13, 21, *80*, 133, 142, *190*
Uchiyama M, see Obata Y 370, *426*
Uchiyama M, see Tanaka Y 43, *108*
Ueda K, see Murayama K 518, 519, 521, *579*
Ueda M, Bandurski RS 14, 15, *109*, 113, 115, 123, *200*, 367, *439*
Ueda M, Ehmann A, Bandurski RS 14, 15, *109*, 170, *200*, 367, *440*
Ueki K, see Komai K 77, *95*
Uhrström I 518, *585*
Uhrström I, see Burstrom HG 518, *570*
Ulrich A, see Adam G 377, 378, *390*
Umeya K, see Yokota T 13, 14, 18, *112*, 115, 120, 130, 138, *201*
Uolvardy J, see Wyen NV 560, *586*
Uota M 330, *440*
Upper CD, Helgeson JP, Schmidt CJ 184, *200*
Upper CD, Helgeson P, Kemp JD, Schmidt CJ 173, 174, *200*
Upper CD, West CA 304, 305, *440*
Upper CD, see Dennis DT 316, *399*
Upper CD, see Graebe JE 302, 304, 305, 307, *406*
Upper CD, see Woodward MD 75, *111*
Uprety H, see Bhakuni DS 53, *82*
Urban P, see Birecka H 554, 565, *569*
Uritani I, see Imaseki H 347, *411*
Usuda S, see Fukui H 30, 32, 50, 53, 54, 79, *88*, *89*, 137, *411*, 354, *404*

Vaadia Y, see Itai C 386, *411*
Vaadia Y, see Livne A 541, 542, *577*
Vaadia Y, see Sitton D 388, *436*
Valio IMF, Burden RS, Schwabe WW 52, 53, *109*
Valio IMF, Rocha RF 35, *109*
Valio IMF, Schwabe WW 52, 53, *109*
Van der Woude WJ, Lembi CA, Morré DJ 525, *585*
Van Hoof R, see De Greef JA 346, 347, *399*
van Huysse P, see Jain ML 346, *411*
Van Onckelen HA, Verbeek R 242, *279*
Van Overbeek J, Went FW 241, *279*
Van Rompuy LLL, Zeevaart JAD 78, *109*
Van Staden J, Bornman CH 244, *279*
Van Staden J, Drewes SE 41, 42, *109*, 368, *440*
Van Staden J, Menary RC 42, *109*
Van Staden J, see Davey JE 371, 384, 385, 386, *399*
van Steveninck PFM, see Porter NG 48, *102*
Van Steveninck RFM 48, *109*
Van Steveninck RFM, see Mittelheuser CJ 52, *99*
Van Sumere CF 57, 58, *109*
Van Sumere CF, Albrecht J, Dedonder A, De Pooter H, Pé I 62, *109*
Van Sumere CF, Cottenie J, De Greef J, Kint J 57, 58, 67, 74, *109*
Van Ysselstein MHW, see Andreae WA 364, 365, *391*
Vanderhoef LN, Findley JS, Burke JJ, Blizzard WE 518, 523, *585*
Vanderhoef LN, Lu T-YS, Williams CAy 451, *510*
Vanderhoef LN, Stahl CA 451, *509*, 518, 531, *585*
Vanderhoef LN, Stahl CA, Siegel N, Ziegler R 531, *585*
Vanderhoef LN, Stahl CA, Williams CA, Brinkman KA, Greenfield JC 521, *585*
Vanderhoef LN, see Key JL 558, *576*
Vandervorst D, see Bourdoux P 18, *83*
Vande-Stouwe R, see Stornier T 345, *437*
Varga M, Köves E 59, *109*
Varner JE, Chandra GR 545, 546, *585*
Varner JE, Johri MM 545, 548, 551, *585*
Varner JE, see Ben-Tal Y 550, 552, *569*
Varner JE, see Chrispeels MJ 244, *276*, 546, 547, 548, 551, *570*
Varner JE, see Evins WH 491, *503*, 547, 549, *572*
Varner JE, see Ferrari TE 548, *572*
Varner JE, see Filner P 545, *572*
Varner JE, see Ho DT-H 489, *505*, 546, 548, *574*
Varner JE, see Jacobsen JV 545, 546, 548, 551, *574*
Varner JE, see Johri MM 488, *505*, 560, *575*
Varner JE, see Jones RL 243, *277*
Varner JE, see Koehler DE 550, *576*
Varty K, see Laidman DL 551, 552, *577*
Vasquez A, see Mendez J 60, *98*
Vatvars A, see Sweetser PB 118, 139, 181, *200*, 272, *279*, 374, *438*
Vaugham G, see Milborrow BV *353*
Vaughan D, Linehan DJ 58, *109*
Vazquez A 69, *109*
Veen H 496, *510*
Veen R van der, see Kerk GJM van der 448, *506*
Veldstra H 448, *510*
Veldstra H, see Bezemer-Sybrandy SM *394*
Vendrell M, McGlasson WB 328, *440*
Vendrell M, see McGlason WB 332, *421*

Vendrig JC 281, *440*
Vendrig JC, see Geuns JCM 281, *405*
Venis MA 364, *440*, 452, 454, 455, 465, 466, 468, 469, *510*, 559, *585*
Venis MA, Watson PJ 75, *109*, 462, *510*
Venis MA, see Batt S 459, 460, 461, 465, 466, 468, *502*, 559, *569*
Venis MA, see Stoessl A 206, 266, *279*
Verbeek R, see Gaspar T 316, *404*
Verbeek R, see Petrids C *428*
Verbeek R, see Van Onckelen HA 242, *279*
Verbiscar AJ, Cragg G, Geissman TA, Phinney BO 308, 354, *440*
Verbiscar AJ, see Geissman TA 308, 354, *404*
Verma DC, Tavares J, Loewus FA 563, 564, *585*
Verma DPS, Maclachlan GA, Byrne H, Ewings D 451, *510*, 554, *585*
Verma DPS, see Bal AK 563, *568*
Verma DPS, see Byrne H 563, *570*
Vickery LE, Purves WK 286, *440*
Vickery LE, see Percival FW 286, *427*
Vieitez E, see Mendez J 60, *98*
Vieitez E, see Sancheza Tames R 60, *104*
Vigil EL, Ruddat M 549, 551, *585*
Vij VK, see Kalsi PS 77, *93*
Vining LC 135, *200*
Vining LC, see McInnes AG 26, *98*, 355, 376, 378, *422*
Vining LC, see Pitel DW 134, 136, 161, *198*, 354, 355, *428*, 486, *507*
Virtanen AE, Hietala PK, Wahlroos O 75, *109*
Virtanen AI 20, *109*
Virtanen AI, see Gmelin R 19, 20, *89*, *90*
Virtanen AI, see Piironen E 20, *102*
Vitagliano C 542, *585*
Vizarova G 346, *440*
Vlasov PV, see Dathe W 49, *86*, 374, 382, *399*
Vliegenhart JA, Vliegenhart JFG 12, *109*
Vliegenhart JFG, see Vliegenhart JA 12, *109*
Vlitos AJ, see Crosby DG 79, *85*
Vogel AI 203, 204, *279*
Voigt B, see Sembdner G 381, *434*
Vokáč K, Samek Z, Herout V, Šorm F 77, *109*
Von Minden DL, McCloskey JA 184, *200*
Von Saltza MH, see Miller CO 39, 40, *99*
Vonk CR 368, 369, *440*
Vreman HJ, Corse J 128, *200*
Vreman HJ, Schmitz RY, Skoog F, Playtis AJ, Frihart CR, Leonard NJ 43, *109*, 348, *441*
Vreman HJ, Skoog F, Frihart CR, Leonard NJ 41, 43, 44, *109*, 139, *201*, 292, 348, *440*
Vreman HJ, Thomas R, Corse J, Swaminathan S, Murai N 42, 43, 44, *109*, 292, 385, *441*
Vreugdenhil D, Burgers A, Libbenga KR 469, *510*

Wada K 316, *441*
Wada K, Imai T, Shibata K 311, *441*
Wadà S, see Masuda Y 517, 520, *579*
Wada T 53, *110*
Wada Y, see Murofushi N 27, *100*, 113, 114, 142, 161, *197*
Wadia MS, see Kalsi PS 77, *93*
Wahlroos O, see Virtanen AE 75, *109*
Wain RL 295, *441*
Wain RL, see Clarke G 338, *397*
Wain RL, see Cornforth JW 48, *85*
Wain RL, see Robert ML 335, 344, 346, *431*
Wain RL, see Rothwell K 48, *104*
Wain RL, see Seeley RC 338, *434*
Wain RL, see Simpson GM 55, *106*
Wain RL, see Smith MS 450, *508*
Wain RL. see Wilkins H 75, *111*
Wakhloo JL, see Kaldewey H 241, *277*
Walbot V, Clutter M, Sussex I 560, *585*
Walbot V, see Sussex I 560, *584*
Waldron JC, see Glasziou KT 561, *573*
Walker GC, Leonard NJ, Armstrong DJ, Murai N, Skoog F 293, *441*, 474, *510*
Walker R. see Nakanishi K 35, *100*
Wall ME, Wani MC, Cook CE, Palmer KH, McPhail AT, Sim GA 75, *110*
Wall ME, see Cook CE 77, *85*
Waller GR, Burstrom H 75, *110*
Waller GR, Nowacki EK 75, *110*
Walters J, see Musgrave A 513, *580*
Walton DC 3, *8*
Walton DC, Dorn B, Fey J 51, *110*, 131, *201*, 352, 353, *441*
Walton DC, Harrison MA, Cote P 352, 353, 374, *441*
Walton DC, Soofi GS, Sondheimer E 560, *585*
Walton DC, see Dashek WV 387, *399*
Walton DC, see Gillard DF 351, 352, 353, *405*
Walton DC, see Harrison MA 352, 353, *408*
Walton DC, see Sondheimer E 351, 352, 353, *436*
Walton DC, see Tinelli ET 51, *108*, 352, 353, *439*
Walton DC, Sondheimer E 298, 351, 352, 353, *441*
Wample RL, Durley RC, Pharis RP 356, 357, 362, 379, 380, *441*
Wang CH, see Buhler DR 362, *395*

Wang TL, Horgan R 47, *110*, 370, *441*
Wang TL, Thompson AG, Horgan R 47, *110*, 127, 139, 173, 184, *201*, 348, 370, *441*
Wang TSC, Yang T-K, Chuang T-T 59, 60, *110*
Wangen LE, Woodward WS, Isenhour TL 250, 251, 252, *279*
Wani MC, see Wall ME 75, *110*
Wanner H, Santos A, Stocker H 79, *110*
Ward TM, Wright M, Roberts JA, Self R, Osborne DJ 11, *110*
Ward TM, Wright M, Roberts JA, Self R, Osborne J 175, *201*
Wardale DA 325, 389, *441*
Wardale DA, see Lieberman M 318, 319, 325, 335, *418*
Wardale DA, see Mapson LW 324, 325, *420*
Wardell WL 556, *585*
Wardrop AJ, Polya GM 457, *510*
Wareing PF 48, 52, 57, *110*, 295, *441*, 512, 520, 554, 567, *585*
Wareing PF, Eagles CF, Robinson PM 48, *110*
Wareing PF, Horgan R, Henson IE, Davis W 45, *110*
Wareing PF, Horgan R, Henson JE, Davis W 350, 370, *441*
Wareing PF, Ryback G 295, *441*
Wareing PF, see Addicott FT 48, *80*
Wareing PF, see Beevers L 361, *394*
Wareing PF, see Cornforth JW 48, 49, *85*
Wareing PF, see Frankland B 243, *277*
Wareing PF, see Henson IE 385, 386, *409*
Wareing PF, see Hewett EW 139, *194*, 368, 370, 385, 386, *409, 410*
Wareing PF, see Horgan R 41, 45, 47, *92*, 139, 173, *194*, 291, *410*, 472, 473, *505*
Wareing PF, see Lorenzi R 25, 36, *97*, 369, *419*
Wareing PF, see Loveys BR 361, *419*
Wareing PF, see Pearson JA 498, *507*
Wareing PF, see Purse JG 41, 46, *103*, 293, 348, *429*
Wareing PF, see Railton ID 388, *429*
Wareing PF, see Robinson PM 48, *104*
Wareing PF, see Seth AK 544, *583*
Wareing PF, see Woolley DJ *442*
Warner HL, Leopold HC 523, 530, 532, *586*
Warthen JD Jr, see Grove MD 3, *6*
Wasada N, see Takahashi N 184, *200*
Wasner D, see Kupchan MS 77, *96*
Watanabe A, see Imaseki H 333, 334, *411*
Watanabe A, see Kondo K 333, 334, *414*
Watanabe A, see Murofushi N 142, *197*
Watanabe A, see Odawara S 336, 390, *426*
Watanabe A, see Sato T 557, *583*
Watanabe E, see Murofushi N 25, 26, 27, 29, 30, 36, *100*, 113, 114, 142, 161, *197*
Watanabe E, see Takahashi N 184, *200*
Watanabe H, see Herz W 77, *92*
Watanabe K, see Esashi Y 325, 328, *402*
Watanabe M, see Nakamura T 62, *100*
Watanabe N, Yokota T, Takahashi N 27, 42, 44, 50, *110*, 137, 173, 186, *201*
Watanabe T, Kondo N 336, *441*
Watanabe Y, see Hayashi Y 78, *91*
Waters EC, see Khan AA 548, *576*
Waters JM, see Godfrey JE 78, *90*
Watson JT, see Falkner FC 186, *193*
Watson PJ, see Venis MA 75, *109*, 462, *510*
Watson WH, see Dominguez XA 74, *87*
Wattiaux R, see DeDuve C 494, *504*
Waygood ER, see Maclachlan GA 342, *419*
Weaver GM, Jackson HO 365, *441*
Weaver W, see Shannon CE *280*
Webber JE, Laver ML, Zaerr JB, Lavender DP 50, *110*
Weber H, see Schraudolf H 289, *434*
Wehrli W, see Staehelin M 43, *106*
Weigl J 496, *510*
Weiland J, see Schreiber K 36, *105*, 376, 379, *434*
Weiland J, see Sembdner G 316, 360, 362, 376, 377, 378, 379, 380, 381, 382, 388, *434*
Weinbaum SA, see Shaybany B 50, *105*, 354, *435*
Weis A, see Kaldewey H 241, *277*
Weiser CF, see Swanson BT 49, *107*
Weiss G, Koreeda M, Nakanishi K 53, 54, *110*
Weissmann G, Sessa G, Weissmann S 494, 495, *510*
Weissmann G, see Bangham AD 492, 494, *501*
Weissmann S, see Weissmann G 494, 495, *510*
Wellburn AR, Ogunkanmi AB, Fenton R, Mansfield TA 53, 56, *110*
Wellburn AR, see Mansfield TA 56, *97*
Wellburn AR, see Ogunkanmi AB 56, *101*
Weller LE, Wittwer SH, Sell HM 115, 157, 159, *201*
Wels CM, see Bearder JR 33, *82*, 302, 311, 312, 313, 354, 355, *393*
Wels CM, see MacMillan J 32, 97, 136, 137, 170, 176, *196*
Wender SH, see Li HC 564, *577*

Wenkert E 293, 304, *441*
Wenkert W, see Liu WT *418*
Went FW 12, 57, *110*, 241, *279*
Went FW, Thimann KV 11, *110*
Went FW, see Van Overbeek J 241, *279*
Werstiuk E, see Pačes V 350, 372, *427*
Wert V, see Ries SK 79, *103*
Wert VF, see Ries SK 79, *103*
West CA 23, 24, *110*, 302, 308, 309, *441*
West CA, Phinney BO 23, 24, 34, *110*
West CA, see Dennis DT 308, 316, *399*
West CA, see Dueber MT 317, *400*
West CA, see Fall RR 304, 305, 306, 316, *403*
West CA, see Frost RG 304, 305, 306, 316, *404*
West CA, see Graebe JE 302, 304, 305, 307, *406*
West CA, see Green TR 310, 389, *406*
West CA, see Hasson EP 308, 389, *408*
West CA, see Jones KC *93*
West CA, see Knotz J 317, *413*
West CA, see Lew FT 308, 309, *417*
West CA, see Mellon JE 305, 317, *422*
West CA, see Murphy PJ 308, *425*
West CA, see Oster MO 305, *427*
West CA, see Phinney BO 34, *102*
West CA, see Robinson DR 305, *431*
West CA, see Shechter I 305, *435*
West CA, see Simcox PD 305, 317, 389, *435*
West CA, see Sitton D 305, 317, *436*
West CA, see Stekoll M *437*
West CA, see Upper CD 304, 305, *440*
Westlin PE, see Hemberg T 127, *194*
Wetherbee PJ, see Craker LE 566, *571*
Wetmore RH, De Maggio M, Rier JP 566, *586*
Wetmore RH, Rier JP 566, *586*
Weyers JDB, see Itai C 52, *93*
Whalley WB, see Birch AJ 301, 309, *395*
Wheeler AW 21, 76, *110*, *111*
Wheeler CT, see Henson IE 348, 350, 370, 385, 386, *409*
Whichard LP, see Cook CE 77, *85*
Whisler J, see Hartmann HT 332, *408*
White A, see Blecher A 494, *502*
White AF, see Evans R 302, 309, 310, *402*
White AF, see Hanson JR 302, 304, 308, 309, 310, 311, 313, *408*
White EH, see Marx JN 77, *98*
White JC, Medlow GC, Hillman JR, Wilkins MB 13, *111*
Whittaker RH 57, 74, *111*
Whittaker RH, Feeny PP 57, *111*
Whittle YG, see Coke LB 56, *84*
Whitty CD, Hall RH 350, 371, 372, 386, *441*
Whyte P, Luckwill LC 243, *279*
Wiberg KB 500, *510*
Wichner S, Libbert E 287, *441*
Wichner S, see Libbert E 284, 287, *417*
Wicken AJ, see Heatherbell DA 316, *408*
Wiersema PK, see Winter H 529, *586*
Wiesner J 38, *111*
Wightman F 21, *111*, 282, 283, 284, 285, 286, 287, 288, 338, 383, *442*
Wightman F, Cohen D 282, 283, 284, 285, 286, *442*
Wightman F, Rauthan BS 21, *111*
Wightman F, see Gibson RA 282, 284, 285, 286, 287, 288, *405*
Wightman F, see Milborrow BV 21, *99*
Wightman F, see Schneider EA 18, 20, 21, 62, *105*, 115, *199*, 281, 282, 283, 285, 287, 288, 336, 337, 338, 340, 341, 342, 343, 344, 346, 364, 365, 383, 384, *433*
Wightman F, see Seeley RC 338, *434*
Wightman F, see Smith MS 450, *508*
Wilchek M, Salomon Y, Lowe M, Selinger Z 454, *510*
Wildes RA, see Pitman MG 538, 539, *581*
Wildman SG, Ferri MG, Bonner J 281, *442*
Wilkins H, Burden RS, Wain RL 75, *111*
Wilkins HF, see Swanson BT 49, *107*
Wilkins MB, see Batt S 459, 461, 465, 466, 468, *502*, 559, *569*
Wilkins MB, see Bridges IG 13, *83*, 184, *191*, 264, *276*
Wilkins MB, see Greenwood MS 13, *90*
Wilkins MB, see McWha JA 244, *278*
Wilkins MB, see Philipson JJ 517, 518, 519, *581*
Wilkins MB, see White JC 13, *111*
Wilkinson AJ, see Taylor IEP 34, *108*
Williams AE, see Beynon JE 234, *276*
Williams CA, see Vanderhoef LN 451, *510*, 521, *585*
Williams DJ, see Sunshine GH 491, *509*
Williams JC, see Most BH 173, *197*
Williams MW, see Letham DS 384, *417*
Williams PD, see Stoddart JL 5, 7, 497, *509*
Williams PM, Bradbeer JW, Gaskin P, MacMillan J 268, *279*
Williams PM, Ross JD, Bradbeer JW 23, 25, 49, *111*

Williams PW, see Arias I 362, 388, *392*
Williams RJP, see Ricard J 341, *431*
Williamson FA, see Jelsema CL 360, *412*
Williamson V, see Moyed HS 337, *425*, 448, *507*, 533, *579*
Willing RR, see Crow WD 74, *86*
Willmer C, see Itai C 52, *93*
Willmer CM, Don R, Parker W 79, *111*
Willmer EN 495, *510*
Wills RBH, Scott KJ, Franklin MJ 375, *442*
Wilson DV, see Shaw S *105*
Wilson LA, see Imbert MP 344, *411*
Wilson MF, Bell EA 76, *111*
Wilson MM, Gordon ME, Letham DS, Parker CW *442*
Wilson MM, see Cowley DE 369, *398*
Wilson MM, see Letham DS *417*
Wilson MM, see Parker CW 128, *198*, 369, 370, 386, *427*
Wilson RH, see Desmos EK 61, *86*
Windholz M 54, 57, 76, *111*
Winkler E, see Dollstädt R 470, *503*, 559, *571*
Winston JR, see Miller EV 11, *99*
Winter A 283, 285, *442*
Winter H, Spaander S, Wiersema PK 529, *586*
Winter J, see Birch AJ 301, 302, *394*, *395*
Winters TE, Geissman TA, Safir D 77, *111*
Witham FH, see Krasnuk M 41, 46, *95*, 384, *414*
Witham FH, see Tegley JR 128, *200*
Witkop B, see Steward FC 76, *107*
Wittwer SH, see Fukui HN 115, *193*
Wittwer SH, see Haagen-Smit AJ 13, *90*, 113, *194*
Wittwer SH, see Redemann CT 115, 130, *198*
Wittwer SH, see Weller LE 115, 157, 159, *201*
Woehrle D, see Tietz D 353, *439*
Wolf FT, Tilford RH, Martinez ML 56, 62, *111*
Wolf FT, see Baskin JM 67, *81*
Wöllenweber HW 22, *111*
Wong Y-S, Fincher GB, Maclachlan GA 524, 563, *586*
Wood A, Paleg LG 492, 493, 494, 496, *510*
Wood A, Paleg LG, Spotswood TM 495, *510*
Wood HN, Lim HC, Braun AC 371, *442*
Woodbury W, see Stahmann MA 347, *436*
Woodlhouse HW, see Beever JE 385, *394*
Woodward MD, Corcuera LJ, Schnoes HK, Helgeson JP, Upper CD 75, *111*
Woodward WS, see Wangen LE 250, 251, 253, *279*
Woolley DJ, Wareing PF *442*
Wooltorton LSC, see Galliard T 322, *404*
Wooltorton LSC, see Mapson LW 319, *421*
Woolwine M, see Desmos EK 61, *86*
Woo-Ming RB, see Stuart KL 53, *107*
Worley JF, see Buta JG 75, *84*
Worley JF, see Grove MD 3, *6*
Worley JF, see Mitchell JW 3, *7*
Worley JF, see Thompson MJ 3, *8*
Worsham AD, see Moreland DE 56, 60, 65, 76, *100*
Wray JL, see Filner P 545, *572*
Wright DE, see Knock BA 252, *277*
Wright M, see Ward TM 11, *110*, 175, *201*
Wright STC 52, 55, *111*, 332, 374, *442*, 535, 542, *586*
Wright STC, Hiron RWP 52, *111*, 542, *586*
Wright STC, see Burden RS 300, *396*
Wright STC, see Hiron RW 374, 375, *410*
Wu LC, see Gottlieb D 495, *504*
Wulfson NS, see Zaretkii VI 184, *201*
Wurzburger J, see Lesham Y 374, 375, *417*
Wyen NV, Uolvardy J, Erdei S, Farkas GL 560, *586*
Wylie AW, Ryuge K, Sachs RM 316, *442*
Wyllie SG, see Altree-Williams S 60, *81*
Wyman J, see Edsall JT 446, *503*

Yabuta T, Sumiki Y 22, *111*
Yabuta T, Sumiki Y, Aso K, Tamura T, Igarashi H, Tamari K 22, *111*
Yafin Y, Shechter I 305, 317, *442*
Yamada Y, see Nishimura S 43, *101*
Yamagata Y, Yamamoto R, Masuda Y 534, *586*
Yamaguchi H, see Hashimoto T 74, *91*
Yamaguchi I, Miyamoto M, Yamane H, Murofushi N, Takahashi N, Fujita K 31, *112*
Yamaguchi I, Yokota T, Murofushi N, Takahashi N 29, *111*, 376, *442*
Yamaguchi I, Yokota T, Murofushi N, Takahashi N, Ogawa Y 24, 29, 50, *111*, 114, 116, 124, *201*, 376, *442*
Yamaguchi I, Yokota T, Yoshida S, Takahashi N 177, *201*
Yamaguchi I, see Goldschmidt EE 281, *405*
Yamaguchi I, see Igoshi M 115, 130, 133, 142, 144, *195*
Yamaguchi I, see Takahashi N 14, 20, *108*, 115, 145, 188, 189, *200*, 281, *438*
Yamaguchi I, see Yamane H 30, 36, *112*, 124, 131, 142, *201*, 379, 381, *442*
Yamakawa T, Kurahashi O, Ishida K, Kato S, Komoda T, Minoda Y 4, *8*
Yamaki T 517, *586*
Yamaki T, see Kasamo K 452, 469, *506*

Yamaki T, see Stowe BB 22, *107*
Yamamoto A, see Marumo S 289, *421*
Yamamoto KR, Alberts BM 485, *510*
Yamamoto R, see Masuda Y 517, 525, *579*
Yamamoto R, see Yamagata Y 534, *586*
Yamane H, Murofushi N, Osada H, Takahashi N 23, 24, 25, 26, 27, 29, 30, 31, *112*, 170, *201*, 356, 357, 358, 362, 378, 379, 380, 388, *442*
Yamane H, Murofushi N, Takahashi N 356, 357, 358, 362, 378, 379, 380, 388, *442*
Yamane H, Takahashi N, Taneko K, Furuya M 3, *8*
Yamane H, Yamaguchi I, Murofushi N, Takahashi N 30, 36, *112*, 124, 131, 142, *201*, 379, *442*
Yamane H, Yamaguchi I, Yokota T, Murofushi N, Takahashi N, Katsumi M 381, *442*
Yamane H, see Hiraga K 164, 171, 173, *194*, 377, 378, 379, 381, *410*, 480, *505*
Yamane H, see Takahashi N 378, 379, 380, 388, *438*
Yamane H, see Yamaguchi I 31, *112*
Yamane H, see Yokota T 36, *112*, 184, *201*, 376, 377, 378, 379, *443*
Yamashita K, see Iino Y 77, *92*
Yamashita K, see Kobayashi A 79, *94*
Yamashita K, see Oritani T 55, *101*, 300, *427*
Yamashita K, see Tanaka A 77, *108*
Yamazaka S, see Yokota T 13, 14, 36, *112*
Yamazaki H, Yamazaki I 342, *442*
Yamazaki I, see Nakajima R 342, *426*
Yamazaki I, see Yamazaki H 342, *442*
Yamazaki S, see Yokota T 376, 379, *443*
Yamazaki S, Tamura S, Muramo F, Saito Y 78, *112*
Yamazaki Y, see Fukui H 30, 32, 50, 53, 54, 79, *88*, *89*, 137, *193*, 354, *404*
Yanagawa H, Kato T, Kitahara Y 80, *112*
Yanagawa H, Kato T, Kitahara Y, Takahashi N, Sato Y 80, *112*
Yanagawa H, see Kitahara Y 80, *94*
Yanagisawa K, see Obata Y 370, *426*
Yanagishima N, Shimoda C 558, *586*
Yang H, see Stonier T 345, 437, *438*
Yang SF 10, 11, *112*, 317, 318, 319, 320, 322, 323, 324, 325, 326, 362, *443*
Yang SF, Baur AH 317, 318, 319, *443*
Yang SF, Ku HS, Pratt HK 318, *443*
Yang SF, Saleh MA 336, *443*
Yang SF, see Adams DO 320, 321, 325, 326, *391*
Yang SF, see Baur AH 319, 320, 322, 325, 326, *393*
Yang SF, see Chou TW 317, *397*
Yang SF, see Horng AJ 336, 340, *410*
Yang SF, see Ku HS 341, *415*
Yang SF, see Lau OL 325, 330, 332, 334, 335, 365, *416*
Yang SF, see Murr DP 320, 325, *425*
Yang SF, see Rappaport L *430*
Yang SF, see Yu Y 326, 333, *443*, *444*
Yang SF, see Yu YB 333, *443*
Yang T-K, see Wang TSC 59, 60, *110*
Yeoman MM 556, 557, *586*
Yocum CS, see Kang BG 331, *412*
Yoda S, Ashida J 530, *586*
Yokai J, see Hayashi Y 78, *91*
Yokota T, Hiraga K, Yamane H, Takahashi N 184, *201*
Yokota T, Kobayashi S, Yamane H, Takahashi N 36, *112*, 378, *443*
Yokota T, Murofushi N, Takahashi N 379, *443*
Yokota T, Murofushi N, Takahashi N, Katsumi M 381, *443*
Yokota T, Murofushi N, Takahashi N, Tamura S 25, 27, 28, 29, 36, *112*, 116, 124, 130, 131, 134, 142, 164, *201*, 378, 379, *443*
Yokota T, Okabayashi M, Takahashi M, Shimura I, Umeya K 13, 14, 18, *112*
Yokota T, Okabayashi M, Takahashi N, Shimura I, Umeya K 115, 120, 130, 138, *201*
Yokota T, Takahashi N, Murofushi N, Tamura S 378, 379, 381, *443*
Yokota T, Yamane H, Takahashi N 376, 377, 379, *443*
Yokota T, Yamazaka S, Takahashi N, Iitaka Y 13, 14, 36, *112*, 376, 379, *443*
Yokota T, see Harada H 36, *91*, 131, *194*, 379, 381, *408*
Yokota T, see Hiraga K 23, 24, 25, 30, 35, 36, 49, *92*, 116, 124, 126, 131, 134, 147, 148, 149, *194*, 378, 379, 388, *410*
Yokota T, see Murofushi N 25, 26, 27, 28, 29, 30, 36, *100*, 131, 134, 142, *197*
Yokota T, see Reeve DR 134, 178, *199*
Yokota T, see Watanabe N 27, 42, 44, 50, *110*, 137, 173, 186, *201*
Yokota T, see Yamaguchi I 24, 29, 50, *111*, 114, 116, 124, 177, *201*, 376, *442*
Yokota T, see Yamnae H 381, *442*
Yokoyama K, see Shimokawa K 362, *435*
Yomo H 545, *586*
Yomo H, Iinuma H 316, *443*
Yoneda Y, Endo T 340, 341, *443*
Yoneda Y, Stonier T 344, 345, *443*

Yoneda Y, see Stonier T 344, 345, *437*
Yoshida K, see Takegami T 474, 476, 477, *509*
Yoshida R, Oritani T 128, *201*, 369, 385, *443*
Yoshida R, Oritani T, Nishi A 385, *443*
Yoshida R, see Nishi A 557, *580*
Yoshida S, see Yamaguchi I 177, *201*
Young DW, see McCapra F 23, 33, *98*
Young H 42, *112*, 186, *201*
Young H, see Hopping ME 42, *92*
Young I, see Hillman JR 49, *92*, 184, *194*, 272, *277*
Young M, see Maclachlan GA 526, *578*
Young RE, see Kosiyadincha S 327, *414*
Young RE, see Pratt HK 11, *102*
Young RE, see Tingwa PO 333, *439*
Yu Y, Adams Do, Yang SF 326, 333, *443*, *444*
Yu YB, Yang SF 333, *443*
Yuda E, see Bukovac MJ 29, *83*
Yuda E, see Nakagawa S 186, *197*
Yung KH, see Lau OL 333, 334, *416*
Yunghans WN, see Clark JE 447, 470, *502*

Zabel V, see Dominguez XA 74, *87*
Zabkiewicz JA, Steele KD 13, *112*
Zachau HG, Dütting D, Feldman H 43, *112*
Zachau HG, see Biemann K 40, 43, *82*
Zaerr JB, Morris RO 4, *8*
Zaerr JB, see Chapman RW 40, 43, 45, *84*, 292, *396*
Zaerr JB, see DeYoe DR 13, *86*, 133, 138, *192*, 346, *400*
Zaerr JB, see Webber JE 50, *110*
Zalik S, see Dedio W 170, *192*
Zamudio A, see Dominguez XA 74, 78, *87*
Zapata J, see Fox JE 370, *403*
Zaretkii VI, Wulfson NS, Papernaĵa IB, Gurvich IA, Kucherov VF, Milstein LM, Serebryakov EP, Simolin AV 184, *201*
Zarnack U, see Scopes DIC 39, *105*
Zauberman G, Fuchs Y 328, *444*
Zeevaart JAD 50, 51, 55, *112*, 131, 175, *201*, 301, 316, 352, 361, 374, 387, *444*
Zeevaart JAD, Milborrow BV 51, 54, *112*, 117, 118, *201*, 352, 353, *444*
Zeevaart JAD, Osborne HD 315, *444*
Zeevaart JAD, see Cleland CF 361, *397*
Zeevaart JAD, see Durley RC 356, 357, 361, *401*
Zeevaart JAD, see Ende H van den 361, *440*
Zeevaart JAD, see Gaskin P 27, 49, *89*, 352, *404*
Zeevaart JAD, see Jones MG 4, *7*, 177, *195*
Zeevaart JAD, see Metzger JD 4, *7*, 26, 27, 29, 31, 33, *98*
Zeevaart JAD, see Ninnemann H 315, *426*
Zeevaart JAD, see Van Rompuy LLL 78, *109*
Zelitch I 541, *586*
Zenk MH 21, *112*, 115, 123, *201*, 364, 365, 366, 367, *444*
Zenk MH, see Dowler MJ 540, *571*
Zenk MH, see Durand H 517, 519, *572*
Zenk MH, see Nissl D 519, *580*
Zenk MH, see Rayle D 521, *582*
Zeroni M, Galil T, Ben-Yehoshua S 328, *444*
Zeroni M, see Jerie PH 363, *412*, 500, *505*
Zhamba GE, see Komissarenko NF 67, *95*
Ziegler A, see Rapp A 49, *103*
Ziegler H, see Kluge M 388, *413*
Ziegler R, see Vanderhoef LN 531, *585*
Zimmerman PW, Crocker W 10, *112*
Zimmerman RH, Lieberman M, Broome OC 325, 334, *444*
Zimmermann H, Siegert C, Karl R 364, *444*
Zimmermann PW, see Crocker W 11, *85*
Zimmermann U, Ashcroft RG, Coster HGL, Smith JR 539, 540, *586*
Zimmermann U, Beckers F, Steudle E 539, 540, *586*
Zimmermann U, Steudle E 539, 540, *586*
Zimmermann U, Steudle E, Lelkes PI 539, 540, *586*
Zimmermann U, see Coster HGL 539, *571*
Ziola B, see Spencer FS *583*
Zmrhal Z, see Macháčková J 340, 341, 342, 344, *419*
Zsoldos F, see Erdel L 452, *503*
Zushi S, see Nishio M 13, *101*
Zwar JA, Jacobsen JV 539, 540, 546, *586*
Zwar JA, see Bruce MI 40, *83*
Zwar JA, see Higgins TJV 489, 490, *505*
Zwar JA, see Jacobsen JV 489, *505*
Zweig G, Devay JE 301, *444*

Subject Index

(Page numbers in bold face refer to structures)

abbreviations
  general XV
  cytokinins 290–291
abienol **69**, 78
abrine,
  paper chromatography 158
abscisic acid (and related compounds) **49**, **53**, **54**
  see also abscisic acid biosynthesis
  adsorption chromatography 130–132, 154–155
  aldehyde 300
  α-amylase synthesis and 547–548
  anthocyanin and 566
  artefacts 117–118
  auxin-induced growth and 532
  bioassays 244
  cell division and 556
  cell extension and 513
  concentration in plants 52, 387
  conjugates (see also glucopyranosyl abscisate) 117–118, 374–375
  cytokinin metabolism and 386
  differentiation and 560
  discovery 48
  DNA synthesis and 560
  electron capture 169, 273
  ethylene production and 334
  fungal production 52
  gas liquid chromatography 131, 173–175, 271
  GC-MS 181, 184, 190, 272
  gel permeation chromatography 139, 240
  gibberellin metabolism and 360
  β-D-glucopyranosyl ester **50**, 117, 138, 373–375
  high pressure liquid chromatography 181, 272
  6′-hydroxylation 351–353
  IAA-oxidase and 347
  IAA peroxidase and 347
  inhibition of IAA biosynthesis 287
  invertase and 561
  isolation from cotton 134–135
  (−)-isomer 351, 387
  latent period and 532
  mass fragmentometry 272
  membrane receptors 498
  metabolic pathway 351
  metabolism and stress 387
  metabolism, sites of 5, 387
  metabolite C 350–352, **351**
  metabolite, unidentified 353–354
  occurrence 49–50
  optical rotatory dispersion 190, 270
  paper chromatography 154–155
  partition chromatography 134–135, 137–139
  phospholipid synthesis and 550
  polysomes and 491
  proton pump and 532, 539
  PVP chromatography 140, 142
  quantitative analysis 168, 190, 270–273
  receptors and hormone structure 498
  ribonuclease and 559
  RNA synthesis and 498, 560
  solvent partitioning 128–129, 154
  stomatal aperture and 542–543
  sublimation 143
  thin layer chromatography 168
  trans-abscisic acid 48, 117, 298, 354
  translocation 387
  UV fluorimetry 168
  UV isomerisation 271
  UV spectrometry 270
abscisic acid biosynthesis
  2,4,6-*trans*-dehydrofarnesol and 296–297
  from ionylidene acetic acid derivatives 298–299
  from mevalonic acids 295–298
  from violaxanthin 300
  2-*trans*-isomer and 298
  loliolide and 300
  origin of hydrogens 296–298
  phaseic acid 297–298
  sites of 387
  xanthoxin and 300
  xanthoxin acids and 299
absinthin 77
acanthotoxin **60**, 74
ACC, see 1-aminocyclopropane-1-carboxylic acid
accuracy (inaccuracy)
  by successive approximation 215, 217, 221
  definition 203, 205

accuracy (inaccuracy)
  detectors and 206, 211, 215
  hormone concentration and 211–212
  impurities and 208
  information theory and 226, 236, 239
  reference standard and 206
  verification in practice 236ff.
3-acetoxyindole,
  IAA catabolite 338
acetylcholine chloride 335
acetylenes
  allelopathy 79
  dehydromatricaria esters **72**, 79
  growth inhibition 79
N^{12}-acetyl-N^{12}-ethyltryptamine,
  paper chromatography 158
N-acetyl indole,
  partition column chromatography 132
N-acetylindoxyl,
  paper chromatography 158
3-acetyl-6-methoxybenzaldehyde **60**, 74
N^{12}-acetyltryptamine,
  paper chromatography 158
achilline **66**, 77
β-acid (2,6-dihydroxyquinoline-4-carboxylic acid) **339**
  IAA catabolite **339**, 340
  occurrence in rice 340
acrylic acid 318, 322–323
actinomycin D
  α-amylase synthesis and 545–546, 549–550
  ethylene, PAL and 565
  inhibition of GA_3 effects 550
  inhibition of RNA synthesis 520
  isoperoxidase formation and 565
  membrane-bound enzymes and 550
  nucleic acid synthesis and 558
  polisomes and 401
  protein synthesis and 561
  RNA synthesis inhibitor 545
adenine
  catabolite of cytokinins 350, 369
  gas liquid chromatography 174
  occurrence 45
  ribosome binding and biological activity 475
  thin-layer-chromatography 167
adenosine
  aminohydrolase 350
  catabolite of cytokinins 350, 369
  gas liquid chromatography 174
  ribosome binding and biological activity 475
  thin layer chromatography 167
adenosine 3′,5′-cyclic monophosphate (cAMP)
  α-amylase synthesis and 549
  DNA synthesis and GA_3 549
adenosine deaminase 371–372
adenosine kinase 371–372
adenosine-5′-monophosphate (AMP) 292
  catabolite of cytokinins 350, 369
adenosine nucleosidase 371
adenosine 2′(3′)-phosphate, paper chromatography 166
adenosine triphosphatase
  effect of ethylene 540
  effect of fusicoccin 539
S-adenosylmethionine **321**, 333, 370
  ethylene biosynthesis 320–321, 325–326
adsorption chromatography
  abscisic acid 131–132, 154
  auxins 130, 144–145
  cytokinins 131
  gibberellins 130–131, 146–149
  phaseic acid 130–131
  xanthoxin 131
aesculetin **58**, 67
affinity chromatography 454–455, 476–477
β-alanine 318, 322–323
alantolactone **66**, 77
Alar (B-995) 316, 334
alcohols (alipathic),
  as growth promoters 79
aleurone layers 544–553
  abscisic acid and 547–548, 551
  α-amylase synthesis 491
  endoplasmic reticulum 549–551
  enzymes of 545–551
  ethylene and 548
  GA_3 effects 489–490, 545–552
  ribosomes 547
  RNA synthesis 545–546, 548–549
alkaloids,
  effects on plants 75
allelopathy
  absinthin and *Artemesia absinthium* 77
  acetylenic acids and *Solidago altissima* 79
  3-acetyl-6-methoxybenzaldehyde and *Encelia farinosa* 74
  camphor and cineole and *Salvia leucophylla* 76
  gramine and barley 75
  juglone and *Juglans nigra* 74
  Podocarpus lactones 56
allylindole,
  paper chromatography 158
allylisothiocyanate **62**, 74
1-aminocyclopropane-1-carboxylic acid (ACC) 320–321, 325–326, 332
aminoethoxyvinylglycine (AVG) **319**, 322, 325–326, 335
AMO 1618 **315**
  α-amylase synthesis and 549
  GA biosynthesis inhibitor 316
amygdalin **62**, 74

α-amylase synthesis
abscisic acid and 547–548
cAMP and 549
6-benzyladenine and 548
cytokinins and 548–549
DNA synthesis and 549
ethylene and 540, 548
fluorouracil and 547
GA_3 and 489–490, 545–551, 562
cGMP and 549
inhibitors and 545–547, 549–550
kinetin and 548–549
lecithin synthesis 550
RNA synthesis and 546–547, 549
angelicin **59**, 67
antheridiogen 35, **38**
anthocyanin synthesis, hormone control of 565–566
anti-transpirants 79
β-arabinofuranosidase, effect of GA_3 551
5-O-β-L-arabinopyrosyl-2-O-indole-3-acetyl-myo-inositol **14**, 367
artefacts of isolation methods 114–120
abscisic acid 117–118
auxins 115
cytokinins 117, 128
gibberellins 115–116, 376
ascorbic acid **73**, 80
ascorbigen **19**
artefact of extraction 115
countercurrent distribution 142
asparagusic acid **73**, 80
dihydro-derivative **73**, 80
S-acetyl **73**, 80
S-oxide **73**, 80
L-aspartic acid-N-acylase 364
auxin(s) (abbreviations and numbering XV, 13)
see also auxin biosynthesis and indole-3-acetic acid
S-adenosylmethionine metabolism and 333
adsorption chromatography 130, 144
affinity chromatography 455
as artefacts 20, 115
ATP-ase activity and 451–452, 469
bioassays 241
biological activity and charge separation 448–449
biological activity and three-point theory 450
carbon dioxide uptake and 517
catabolism, sites of 384
cell division and 555–557
cell extension hypotheses 536–537
cell membrane and 452, 539–540
cellulase and 518
cellulase mRNA 451
cell wall enzymes and 520, 524–528
cell wall extension and 513–519, 522, 528, 530, 532, 534–537
cell wall proteins and 528–530
"citrus" 281
counter current distribution 142, 144
cytokinins, interaction with 530–531
differentiation and 569
discovery as plant hormones 12
distillation and sublimation 143
DNA synthesis and 555–557
enzyme activity and 450–455, 516–518, 520, 524–525, 536, 559
ethylene production and 333
galactosidase and 518
gas liquid chromatography 169–170
GC-MS 144, 184, 186, 264, 266
gel permeation chromatography 138–139
gene activation hypothesis 451
gibberellins, interaction with 530
glucanases and 517–518, 520, 524–525
glycan synthetases and 518, 520, 524, 536
glycosyl synthetases and 525
growth acceleration and 535
growth induction by 517–524, 535
hemicellulase and 518
high pressure liquid chromatography 176–177
initiation factors and 453–455
ion exchange chromatography 123–124
isolation from *Citrus* 144–145
latent period 517–519, 522–523
lecithin complexation 496
mass spectrometry 184, 266
membrane potential and 518
metal complexes 143
non-indole 21
occurrence 114–116
paper chromatography 144, 157–160
partition chromatography 132–133, 144–145
pectin esterase and 516, 520
picrate complexes 143
polysaccharide metabolism and 524–527
polysaccharide synthetases and 536
protectors 345
protein mediators and 453–455, 470
protein synthesis and secretion 517–521
proton pump and 451, 521–524, 534–538, 537–539
PVP chromatography 140
receptors from coconut 455–456
respiration and 517
ribonuclease synthesis and 559
RNA polymerase and 450–455, 517–518, 520, 536
RNA synthesis and 517–521, 557
selected ion monitoring 186–189, 266
solvent partitioning, 120–123, 144

auxin(s) (abbreviations and numbering XV, 13)
soya bean factor and 453–454, 470
stomal aperture 541
thin layer chromatography 144–145, 157–161
transcriptional regulation and 451
trimethylsilylation 170
xyloglucan and 517
auxin biosynthesis
4-chloroindole-3-acetic acid 289
epiphytic bacteria 287
indole-3-acetaldehyde decarboxylase 285
indole-3-acetaldehyde oxidase 285
indole-3-acetaldehyde reductase 286
indole-3-acetaldoxime hydro-lyase 288
indole-3-acetonitrile nitrilase 288
indole-3-ethanol oxidase 286
indole-3-pyruvic acid decarboxylase 285
inhibition by ABA 287
glucobrassin 288
pathways 284
sites of 583–584
tryptamineamino oxidase 288
tryptophanaminotransferase 283–284, 286
L-tryptophan decarboxylase 287
auxin receptors – see also corn coleoptiles
affinity chromatography 454–455
affinity labelling 464–466
auxin transport 461
binding sites and transport 458, 468
dithioerythritol and 466
membrane sites in pea and mung bean 469–470
properties 464–465
protein binding 455–457
sites 458–461, 464–468
structure-activity relationships and 450
supernatant factor 462
synthetic hormones and 461
transition, soluble to membrane 457
8-aza-adenine,
ribonuclease and 559
azetidine-2-carboxylic acid **63**, 76

B-995 (Alar) (succinic acid-2,2-dimethylhydrazide) 316, 334
basatasins **61**, 74
benzaldehyde 73
from mandelonitrile 74
benzoxazinone **462**
6,7-dimethoxy- **462**
0-glucoside **462**
6-methoxy- **462**
precursors of benzoxazolinones 462
2-benzoxazolinone (BOA) **63**, 75, **462**
6,7-dimethoxy- **462**
IAA-induced growth and 462–463
6-methoxy- **462**
NAA-binding and 462–463
supernatant factor and 462
3-benzyladenine, ribosome binding 475
6-benzyladenine **471**
affinity column chromatography 477
α-amylase synthesis and 548
binding sites 478
enzymes of flavonoid synthesis 565
enzymes of lignin synthesis and 565
ethylene formation and 334
gas liquid chromatography 174
gel permeation chromatography 139–140
gibberellin metabolism and 360
high pressure liquid chromatography 179
9-methyl- 472
pectin biosynthesis and 564
PVP chromatography 141
ribonuclease and 559
ribosome binding and 475–476, 559
tRNA cytokinins and 473–474
RNA synthesis and 560
stomatal aperture and 542
thin layer chromatography 167
translocation and 562
9-benzyladenine, ribosome binding 475
6-benzyladenine-N(7)-β-D-glucopyranoside 370
6-benzyladenine riboside
gas chromatography 174
gel permeation chromatography 139
high pressure liquid chromatography 179
PVP chromatography 141
thin layer chromatography 167
6-benzyladenosine – see 6-benzyladenine riboside
6-benzylaminopurine – see 6-benzyladenine
6-benzylamino-9-β-D-ribofuranosylpurine – see 6-benzyladenine riboside
benzyl glucosinolate 335
benzyl isothiocyanate 335
betacyanin,
efflux from beet discs 494–495
betaine **64**, 76
betalains 64
ent-beyerene 305, **306**
binary digit (bit) 227
binding constants 446
double reciprocal plots 446–447
Scatchard plots 446–447
bioassay methods
abscisic acid 244
auxins 241
cytokinins 242
gibberellins 241–245
linear response range 241–244
quantitation 242–245
sensitivity 241–244

bound hormones, definition 363
brassinolide
 structure 3
 synthetic isomers 3
bufotenin,
 thin layer chromatography 160

caffeic acid **58**, 63
camphor **65**, 76
campothecin **63**, 75
L-canaline, ethylene biosynthesis and 325, 335
canavanine **64**, 76
capric acid **71**, 79
6-(1-carboxy-2-hydroxy-propylamino)-9-ribofuranosylpurine **46**
 isolation from *Zea mays* 150–151
casbene 305, **306**, 317
CCC (Cycocel, Chlormequat) **315**, 316, 334
 α-amylase synthesis and 549
 drought tolerance and 542
 ethylene formation and 334
 GA action and 542
 GA_3 and amylase synthesis 549
 GA biosynthesis and 316
cell division
 abscisic acid and 556
 auxin and 555–557
 cytokinins and 555–556
 2,4-D and 556
 DNA synthesis and 554–558
 ethylene and 557
 GA_3 and 562
 RNA synthesis and 557–559
cell extensibility 514–515
cell extension and elongation – see also cell wall and under individual hormones
 abscisic acid and 513, 532
 auxin and 513–514, 516–519, 521–524, 528–529, 534–537
 calcium chelation and 516
 cytokinins and 513, 531
 equation 514
 ethylene and 513–514, 532
 extensibility 514–516
 fusicoccin and 533
 GA_1 and 497–498
 GA_3 and 513, 530, 563
 gibberellin and auxin interaction 530
 inhibitors 516
 ion transport and 537, 562
 latent period 517–519
 3-methyleneoxindole and 532–533
 pectin methylesterase and 516
 protein content and 528–530
 proton pump and 522–523, 531, 534–537
 turgor pressure 515–516
 yield threshold 514–516
cell membrane – see membrane, plant-cell
cellulase
 cellulose biosynthesis and 524, 532
 auxin and 518
 ethylene and 540
 induction by IAA 562
cell wall – see also cell extension
 auxin and 514–516, 520, 524–529, 562
 composition and structure 516, 524–525, 528–529, 532, 564
 cytokinin and 562–563
 ethylene and 564
 gibberellin and 497–498, 530, 562
 growth and hormones 562–563
 lignin synthesis 564–565
 model structures 526–527
 pectin methylesterase and 516
 polysaccharide metabolism 520, 524–526, 528–529, 535–537, 555–556, 563–564
 proteins 528–530
chalcone-flavonone isomerase 565
chelidonic acid **60**, 71
chloramben 465
 diazonium salt, and affinity labelling **465**
chloramphenicol 317
2-chloro-4-aminophenoxyacetic acid (CAPA), affinity labelling 465
chlorochrymonin **65**, 77
2-chloroethanephosphonic acid (Ethrel, Ethepon)
 anthocyanin induction 566
 chlorophyll and pigment reduction 566
 decomposition to ethylene 318
 ethylene release by 542
 stomatal opening and 542
2-chlorethyltrimethylammonium chloride – see CCC
chlorogenic acid **58**
 IAA conversion and 533
4-chloroindole-3-acetic acid **16**
 biosynthesis 289
 counter current distribution 142
 GC-MS identification 184
 gel permeation chromatography 138
 methyl ester **16**, 20
 monomethylaspartate **16**, 138, 142
 occurrence 16, 20, 289
 partition chromatography 133
 solvent partitioning 120
5α-cholestane,
 gas liquid chromatography 172
choline 491
choline kinase 491
chromatin template activity
 abscisic acid and 499
 2,4-dichlorophenoxyacetic acid and 453, 489
 gibberellin A_3 and 489

chromatin template activity
IAA-n-IRP complex and 456
IAA and membrane fractions 456–457
c-IRP and transcription 456
mediator protein and 474
oestradiol and 485–486
steroids and 486
supernatant factor and 453
chromatographic analysis
detectors 262
information content 255
peak capacity 254
performance parameters 257
chromosomal receptor protein (c-IRP) 455–456
affinity chromatography 456
IAA-binding and 456
chrysartemin A and B **65**, 77
cineole **65**, 76
cinnamate hydroxylase 565
cinnamic acid **58**
colchicine **35**, 75
cell division and 556
cell wall synthesis and 562
compounds G_1, G_2, and G_3 **73**, 80
Condurit B epoxide 381
confidence limit,
definition 204
coniferyl alcohol dehydrogenase 565
conjugate hormones, definition 363
copalylpyrophosphate (labda-8,13-dien-15-yl pyrophosphate) 302, **303**, 304–305, 307, 317
cordicepin
α-amylase synthesis and 546–548
PC-C and 550
PC-G and 550
RNA synthesis and 546
corn coleoptile receptors 457–469
affinity labelling 465
binding sites 458–461, 464, 468
coumarate hydroxylase 565
p-coumaric acid **58**, 63, 562
coumarin(s) **58**
cell wall synthesis and 562
gibberellin and 69
effects on plants 67, 69–70, 72
countercurrent distribution
auxins 142
gibberellins 142–143
cucurbic acid **72**, 79
cucurbitacin B **70**, 78
ent-13,16-cycloatisane (trachylobane) 305, **306**
cycloheximide
α-amylase synthesis and 550
cytokinin metabolism and 386
ethylene synthesis and 327, 332
isoperoxidase formation and 565
kaurene synthetase and 317
lecithin biosynthesis and 550
protein synthesis and 521
respiration and 521
α-cyperone **66**, 77
cysteine,
cytokinin biosynthesis 292
cytidine diphosphocholine (CDP-choline) 491
cytidine triphosphate 491
cytokinesine I,
identity with zeatin riboside 371
cytokinesis,
cytokinin and 555
cytokinin(s) (abbreviations, nomenclature, numbering 40, **290–291**)
adsorption chromatography 131
N^6-alkylation in tRNA 289–291
amphoteric nature 127
α-amylase synthesis and 548–549
auxin induced growth and 530–531
binding factor-1 and 2 476
bioassay 242, 471
biological activity and structure 470–472
biosynthesis – see cytokinin biosynthesis
cell division and 555–557
cell growth and 513, 531
conjugates – see cytokinin conjugates
conjugation and red light 386
differentiation and 568
discovery 39
DNA synthesis and 555, 557
enzyme degradation in extraction 117
function in tRNA 472–474
gas liquid chromatography 171–174
GC-MS 4, 184, 268–270, 273
gel permeation chromatography 139
glucosylation, in vitro 372–373
high pressure liquid chromatography 178–180
ion exchange chromatography 127–128, 149–150
isolation from *Vinca rosea* 153–154
isolation from *Zea mays* 128, 150–154
mediator protein 474
metabolic pathways 349
metabolism, abscisic acid and 386
metabolism and cycloheximide 386
metabolism and stress 386
metal complexes 143, 151
minerals in aleurone and 552
occurrence, free 45–47, 186, 292
occurrence in tRNA 4–5, 289–292, 472–474, 531
occurrence in *Zea mays* 40, 46
oxidase 350, 371, 372
paper chromatography 151–152, 165–168
partition chromatography 137
permethylation 4, 184

picrates 143, 154
protein synthesis and 474, 477–478, 531
proton pump and 531, 539
PVP chromatography 141, 153–154
quantitation by bio-assay 242
quantition by mass fragmentometry 269, 273
quantitation by SIM 4, 186
quantitation in *Populus* 269
quantitation in *Raphanus* 269
ribosome binding and function 477–478
RNA release and 531
side chain cleavage 350
sites of metabolism 385–386
solvent partitioning 125–128, 149–151
stomata and 541–543
thin layer chromatography 151–152, 165–168
triglycerides in aleurone and 552
trimethylsilylation 171, 184
xanthine oxidase oxidation 348
xyloglucan turnover and 531
cytokinin biosynthesis
from adenosine 292
from 3-methylbut-2-enylpyrophosphate 289
from mevalonic acid 289
in tRNA 292
isopentenyladenine in tRNA 290
isoprenylation of 5′-AMP 292
non tRNA 292
origin of 2-methylthio group 292
side chain oxidation in tRNA 292
sites 384–385
zeatin from AMP 292
cytokinin conjugates
bio-activity 371
enzymatic degradation, sites of 372
function 370–371
gas liquid chromatography 173
ion exchange chromatography 128
metabolic pathways 371
solvent partitioning 127, 165–167
stability 371
structural types **368**
cytokinin receptors
binding constants and biological activity 477
binding factor 1 and 2 476
binding sites 478
binding specificity 475–477
corn coleoptile membranes 457–469
membrane-glycopeptide 477
ribosome binding 474–475
ribosomes and protein assembly 477–478
soluble proteins 476–477
cytokinin ribosides
enzymatic hydrolysis 117
gel permeation chromatography 140
high pressure liquid chromatography 178
paper chromatography 165–168
thin layer chromatography 165–168
cytokinin ribotides
adsorption chromatography 131
barium salt 143
enzyme hydrolysis 117
isolation 150
paper and thin layer chromatography 166
cytosine,
PVP chromatography 167

daphnetin **58**, 67
dehydrocostus lactone **66**, 77
2,4,6-*trans*-dehydrofarnesol 296–297
dehydromatricaria esters **72**, 79
delcosine **62**, 75
deoxyribonucleic acid – see DNA
11-desoxycorticosterone **69**, 78
6-(1,2-dicarboxyethyl-amino-9-β-D-ribofuranosylpurine) **46, 349**
isolation from *Zea mays* 150–151
2,5-dichloro-3-aminobenzoic acid (Chloramben) **465**
diazonium salt and affinity labelling 465
2,4-dichlorophenoxy acetic acid (2,4-D) **449**
affinity chromatography and 454
biological activity 448, 450
cellulase synthesis in ribosomes 562
chromatin template activity and 453, 489
deoxy-*iso*flavone synthesis and 566
differentiation and 562
DNA polymerase and 561
DNA synthesis and 554–558
enzymes of flavonoid synthesis and 565
enzymes of lignin synthesis and 565
ethylene biosynthesis and 333, 335
membrane binding site III and 461
nuclear receptor protein (n-IRP) and 455
pectin biosynthesis and 564
peroxidase and 565
polyphenol synthesis and 564–565
polyteny and 556
RNA polymerase I and 453, 455, 558–561
RNA polymerase II and 455
RNA synthesis and 453, 557–559
soya bean protein and 453, 455
stomatal closure and 541
supernatant factor and 462
transpiration rate and 541
xylem differentiation and 556, 559
2,6-dichlorophenoxyacetic acid **449**, 450
D(+)-2,4-dichlorophenoxypropionic acid |D(+)-2,4-DP| **449**
L(−)-2,4-dichlorophenoxypropionic acid |L(−)-2,4-DP| **449**

tris-(diethylaminoethyl)phosphate trihydrochloride 316
diethylthiocarbamate 335
differentiation 567–568
 abscisic acid and 560
 auxin and 566, 569
 cytokinin and 555, 568
 2,4-D and 562
 ethylene and 557, 564
 GA_3 and 556
 sucrose and 566
 xylem and 554–556, 559–560
diglyceride 491
dihydroconiferyl alcohol **57**, 61–62
2,3-dihydrogibberellin A_{31}
 GA_9 metabolite 357–358
 glucoside 380
dihydrolupinic acid (β-(dihydrozeatin-9-yl)alanine) 370
4′-dihydrophaseic acid **51, 351**
 adsorption chromatography 131
 as artefact 117
 epimer **51, 351,** 352
 identity with metabolite M-2 353
 methyl ester as artefact 117
 occurrence 51, 352
 solvent partitioning 128
 translocation 386
4′-*epi*dihydrophaseic acid
 solvent partitioning 128
3,4-dihydrovomifoliol **54**
2′,3′-dihydroxy-dihydrozeatin **46, 349**
2,8-dihydroxy-N^6-(Δ^2-isopentenyl)adenine 348, **349**
6β,7β-dihydroxy-*ent*-kaurenoic acid 33, **38**
 biosynthesis 310
7β,13-dihydroxykaurenolide 310
2,6-dihydroxy quinoline-4-carboxylic acid (β-acid) **339**
 IAA-catabolite **339**, 340
 occurrence in rice 340
2,8-dihydroxyquinoline-4-carboxylic acid – see zeanic acid
2,8-dihydroxy-*trans*-zeatin 348, **349**
dihydrozeatin **46, 291**
 adsorption chromatography 131
 9-β-alanyl- 370
 biological activity 471
 biosynthesis 348
 2′,3′-dihydroxy- 348, **349**
 gas liquid chromatography 174
 glucosylation 370
 high pressure liquid chromatography 180
 3′-hydroxy- 348, **349**
 hydroxylation enzymatic 348
 metal complexes 143
 occurrence 46, 292–293
 paper chromatography 165–166
 picrate 143
 PVP chromatography 165–166
 ribosylation 370
 thin layer chromatography 167
dihydrozeatin-O-β-D-glucopyranoside **47**, 348
dihydrozeatin riboside 45, **47,** 348
dihydrozeatin-riboside-O(4)-β-D-glucopyranoside, a zeatin metabolite 369–370
β-(dihydrozeatin-9-yl)alanine 370
di-O-(indole-3-acetyl)myo-inositol **15,** 21
 see also IAA myo-inositol esters
3,5-di-iodo-4-hydroxybenzoic acid,
 ethylene biosynthesis and 335
6,7-dimethoxy-2-benzoxazolinone (DMBOA)
 inhibition of IAA-induced growth 462, 463
 inhibition of NAA binding 462–463
 supernatant factor, constituent of 462
dimethylacrylic acid 301
dimethylallylpyrophosphate 293, **294**
N^6,N^6-dimethyladenine
 paper chromatography 165
 thin layer chromatography 167
N^6, N^6-dimethyladenosine
 paper chromatography 165
 thin layer chromatography 167
3H-7,7-dimethyl-7,8,9-trihydropyrimido-(2,1-i) purine,
 paper chromatography 166
Nw-dimethyltryptamine,
 ion exchange chromatography 123
N,N-dimethyl-L-tryptophan **64,** 76, 335
2,4-dinitrophenol,
 ethylene biosynthesis and 320, 325, 335
dioxindole-3-acetic acid
 IAA-catabolite **339**
 occurrence 340
diphenylurea **39,** 40
discadenine[γ-(N^6-isopentenyladenin-3-yl)-α-aminobutyric acid],
 enzymic formation 370, 373
dithiocarbamates 448, **450**
dithioerythritol (DTE),
 auxin membrane binding and 466
DNA synthesis
 abscisic acid and 560
 aleurone and 549 ff.
 auxins and 556–557
 cell division and 555–556
 2,4-D and 554–558, 561
 ethylene and 557
 FUDR and 554
 gibberellin and 549, 556
 IAA and 554–556
 kinetin and 555–557
 mitomycin and 549
 NAA and 556

sarcomycin and 549
docosanol **71**, 79
dopamine(3-hydroxytrytamine) 324
dormancy,
induction by basatasins 74
duva-4,8,13-triene-1,3-diol **69**, 78

electrochemical detectors, performance parameters 249
electron capture detector
abscisic acid analysis 169, 273
performance parameters 249
endoplasmic reticulum
auxin receptor site 1 461
GA_3 and 490–492, 549–551
ethane,
from linolenic acid 318, 322
Ethepon – see 2-chloroethanephosphonic acid
ethionine 319
Ethrel – see 2-chloroethanephosphonic acid
ethylene – see also ethylene biosynthesis
α- and β-amylase and 540, 548
anthocyanin synthesis and 566
ATPase and 540
auxin-induced growth and 532
biological activity and metal complexes 499–500
biological activity and structure 499
biosynthetic pathway 321, 322
cell division and 557
cell growth and 513–514, 532
cell membrane and 540–541
cellulase and 532, 540–541
cell wall composition and 564
cell wall hydroxyproline and 532
cell wall synthesis and 563
compartmentation and binding 500
differentiation and 557, 564
discovery as plant hormone 10
DNA synthesis and 537
formation, non-enzymic 318, 324
GA_3 metabolism and 360
gas liquid chromatography 175
IAA aspartate formation and 365
latent period and 532
lipid bilayers and 499
metabolism 362–363, 500
mitochondria and 540
oxidation and binding 500
oxide 12, 363, 500
PAL activity and 565
peroxidase and 347, 540
quantities in plants 11
receptor and binding 500
stomatal aperture and 542
ethylene biosynthesis
abscisic acid and 334
abscission and 330
acetate-acrylic acid pathway 322–323
acetylcholine chloride and 335
β-ananine-acrylic acid pathway 322–323
autocatalysis 327–328
auto-inhibition 328
auxins and 328, 333
benzyladenine and 334
benzyl *iso*thiocyanate and 335
climacteric fruit and 327
cyanide and 335
cycloheximide and 332
cytokinins and methionine pathway 334
2,4-D and 333, 335
diethylthiocarbamate 335
3,5-di-iodo-4-hydroxybenzoic acid and 335
N,N-dimethyltryptophan and 335
2,4-dinitrophenol and 335
enzymology 324–326
ferridoxin and 325
from acrylic acid 322–323
from S-adenosylmethionine 320–321, 325
from β-alanine 322–323
from 1-aminocyclopropane-1-carboxylic acid 320–321, 325–326
from β-hydroxypropionic acid 322–323
from methionine 319–322, 325–326
fungal 319
fungal disease and 331
fruit ripening and 327
L-homocysteine thiolactone 320
homocysteine and 320–322
homoserine and 321
5-hydroxydioxindole-3-acetic acid and 334
5-hydroxyindole-3-acetic acid and 334
β-hydroxypropionate-acrylic acid pathway 322–323
inhibitors 319–320, 322, 325, 330, 335–336
isopentenyladenosine and 334
kinetin and 334
linolenate-propanol pathway 322
lipids and 334
membrane permeability and 328
metal ions and 330, 332, 335
methionine concentration and 326
methionine pathway 333
methionine transaminase and 324
3-methyleneoxindole and 333
S-methylmethionine and 320, 322, 329
5′-methylthioadenosine and 320–321
5′-methylthioribose and 320–321
NAA and 333
necrosis and 331
oxidative phosphorylation and 325
perioxidase and 324–325
plant growth retardants and 334–335
pollination and 328

ethylene biosynthesis
propyl gallate and 390
pyridoxal phosphate and 335
regulation by light 330–331
senescence and 328–329
sites of 389–390
stress-induced 331–332
synthetic auxins and 333
wound ethylene 322
xanthoxidase and 325
X-rays and 332
ethylene glycol, metabilite of ethylene 363
ethylene oxide 12, 363, 500
metabolite of ethylene 363
ethyl indole-3-carboxylate,
paper chromatography 159
N^{12}-ethyltryptamine,
paper chromatography 158
eugarzasadone (teucvin) **68,** 78

trans-farnesol **53**
farnesol pyrophosphate 293–294, 297
fatty acids
capric **71**, 79
9,10-dihydroxy-12-octadecenoic **72,** 79
fumaric **71,** 79
growth effects 79
linoleic **71**, 79
linolenic **71**, 79
oleic **71,** 79
palmitic **71,** 79
phaseolic **72,** 79
stearoylglycerol **72,** 79
ferridoxin 325
ferulic acid **58**
filipin 494–495
flavin mononucleotide 318
flavonoids
biosynthesis and hormone control of 565–566
effects on plant growth 64–65
flowering
gallic acid and 61
salicyclic acid and 61
α-trocopherol and 74
5-fluorodeoxyuridine (FUDR),
DNA synthesis and 534–535
p-fluorophenyladenine,
ribonuclease and 559
fluorouracil
α-amylase synthesis and 547
polysomes, GA_3 and 491
RNA accumulation and 558
fructosidase,
GA_3 and 562
FUDR and DNA synthesis 554
fumaric acid **71,** 79
6-furfurylaminopurine – see kinetin
6-furfurylamino-9-β-D-ribofuranosylpurine – see kinetin riboside
fusicoccin **534**
ATPase activity and 452, 539
binding and ATPase activity 469
binding sites 6, 462, 464
cell extension and 533
latent period 533
membrane potential and 538
proton-pump and 452, 523–524, 538
stomatal aperture and 543–544

5-O-β-D-galactopyranosyl-2-O-indole-acetyl-myoinositol **15**
β-galactosidase,
auxin and 518
gallic acid **57**
flowering inhibitor 61
propyl ester 390
gas chromatography-mass spectrometry
abscisic acid 184
auxins 184, 264, 266
capillary columns and 260
cytokinins 4, 184, 269–270
giberellins 4, 181–184, 244, 268–269, 273
information content and identification by 252
isotope dilution analysis 4, 264–265, 269–270
gas liquid chromatography (GLC)
abscisic acid 131, 142, 173–175
abscisic acid, quantitation 271
auxins 169–170
cytokinins 171–174
derivatisation 4, 168–169, 171, 173
detectors and sensitivity 169–170, 174, 259
ethylene 175
gibberellins 169–173
gibberellin quantitation 268
indoles, retention times 170
information content 259
isolation by 169
nucleosides 174
performance parameters 259
xanthoxin 131, 174
gel permeation chromatography
information content and 234, 240
of bean seed extract 235
of plant hormones 138–139
gene activation,
auxins and 451, 520
gentisic acid **57**
geranylgeranyl pyrophosphate
cyclisation in plastids 389
enzymatic cyclisation 304–306
gibberellin precursor 302–303
inhibitors and 316

kaurene synthetase and 317
proplastids, occurrence in 389
terpenoid precursor 293
l-geranylimidazole 316
geranylpyrophosphate 293–294, 310
Gibberella fujikuroi 302, 309–313
mutants and biosynthesis 302, 311–314
terpenoid metabolites 137
gibberellenic acid
artefact from GA_3 116
high pressure liquid chromatography 178
thin layer chromatography 163
gibberellenic acid 3-O-β-D-glucopyranosyl ether
artefact from GA_3 glucoside 116, 376
high pressure liquid chromatography 177
paper and thin layer chromatography 164
gibberellin(s) (nomenclature and numbering, XVII, 34–35, 38)
see also gibberellin biosynthesis, gibberellin conjugates, gibberellin receptors, and individual gibberellins.
adsorption chromatography 130–131, 146–149
artefacts of isolation procedures 116, 376
auxin interaction 530
bioassay 241–245
biological activity and carboxyl group 479–480, 496
biological activity and hydroxylation 4, 281, 480–481
biological activity and metabolism 4, 479
biological activity and molecular shape 480
biological activity and plant specificity 479
carboxyl group and membrane permeability 496
catabolic deactivation 359–360
cell extension and 5, 497, 513, 530
chemotaxonomy 4, 35, 37
conjugation and plant growth 482
countercurrent distribution 142–143
discovery 22, 34
droplet counter current chromatography 3
gas liquid chromatography 170–173
GC-MS 4, 181–184, 186–189, 244, 268–269, 273
gel permeation chromatography 138–139
(+)-gibberellins 303
glucosylation 377, 380–381
high pressure liquid chromatography 176–178
hydroxylation in plants 355–359
hydroxylation pattern 313
ion exchange chromatography 125
isolation from *Phaseolus* 146–150
latent period and 530
levels in plants 35
lipid structures and affinity for 497
liposomes and 496
metabolic pathways 303, 307, 310, 312, 359–360, 380
metabolism, higher plants 354–360
metabolism and abscisic acid 360
metabolism and light 361–362
occurrence 23–33
olfactory response analogy 495
osmotic potential and 530
paper chromatography 160–162
partition chromatography 133–137, 146–149, 241–243
partition coefficients 125
phytochrome and 361–362
proton pump and 530, 539
pseudo-rotational symmetry 480
PVP chromatography 140–141
quantitative analysis 186, 241–245, 267–269, 273
stability during extraction 116
stomatal aperture and 542
structures **23–33**
synergist from lettuce cotyledons 62
thin layer chromatography 147–149, 160–165, 241–243
transcriptional control and 486–490
vacuoles and metabolism 389
wall plasticity and 530
gibberellin A_1 **23**
abscisic acid and metabolism of 360
3-O-acetate 36, 376, 378
adsorption chromatography 130, 146–147, 148–150
benzyl ester 176
binding in pea epicotyl 5, 486–488
biosynthesis, fungal 355
biosynthesis in wheat seedlings 302
cell wall extension and 497
cell wall labelling 5, 478–479
dihydro- 134
3-epimer 357, 488
GA_4 metabolite 357
GA_{20} metabolite 357
gas liquid chromatography 170–173
GC-MS 182
gel permeation chromatography 240
β-D-glucopyranosyl ester 36, 116, 147–148, 164, 173, 377–378, 380–381
3-O-β-D-glucopyranosyl ether 36, 172, 376, 378, 380–381
13-O-β-D-glucopyranosyl ether 377–378
glycyl amide 378
glycylglycine amide 378
high pressure liquid chromatography 176–178
2β-hydroxylation in vivo 356, 358–359, 481

gibberellin A_1
isolation from *Phaseolus* 146, 149–150
liposomes and leakage rate 496
metabolism and ABA 360
metabolism in dwarf rice 5
metabolism to GA_3 355
methyl ester 131, 162, 172, 182
p-nitrophenyl ester, mass spectrum 176
17-norketone 488
nuclei of lettuce hypocotyl and 488
occurrence 3, 4, 23
paper chromatography 161–162
partition coefficients 125, 149–150
proline amide 378
propyl ester 36, 376, 378
protoplasts and metabolism 389
quantitative analysis 244–245, 268
solvent partitioning 124, 146–147
thin layer chromatography 147–149
gibberellin A_2 **23**
biosynthesis, fungal 312
GA_4 metabolite 357
gas liquid chromatography 170–172
methyl ester 131, 162, 172
occurrence 23
partition chromatography 134–137
partition coefficients 125
thin layer chromatography 162–163
gibberellin A_3 **23, 484**
3-O-acetate 36, 376, 378
adsorption chromatography 130
D-alanyl amide 378
L-alanyl amide 378
aleurone enzymes and 489–491, 549–551
aleurone layer and 545–551
α-amylase synthesis and 489–491, 545–551, 562
β-amylase and 551
L-aspartyl amide 378
benzyl ester 176
binding in pea epicotyl extracts 487
biosynthesis fungal 301, 309, 311–314, 354–355
biosynthesis, plants 302, 359
cell division and growth 530, 562–563
cell wall synthesis and 530, 562
chromatin template activity and 489
conversion to gibberellenic acid 116
counter current distribution 143
differentiation and 556
3,13-difluoro-derivative and biological activity 482
DNA synthesis and 549, 556
endoplasmic reticulum and 490–492, 549–551
enzyme stimulation 551, 561–562
ethylene and metabolism of 360
18-fluoro- and biological activity 482
GA_5 metabolite 356
gas liquid chromatography 170–173, 177
gel permeation chromatography 138
3, 13-O-β-D-glucopyranosyl diether 377–378
β-D-glucopyranosyl ester 377–378
3-O-β-D-glucopyranosyl ether 36, 116, 134, 164, 172–173, 183–184, 376, 378
13-O-β-D-glucopyranosyl ether 377–378
L-glycyl amide 378
L-glycylglycyl amide 378
growth retardants and 316
high pressure liquid chromatography 176–178
IAA induced growth and 563
invertase and 561
lactone isomer (iso GA_3) 116, 134–136
lactone isomer-3-O-β-D-glucopyranosyl ether 376
lecithin biosynthesis and 491, 550
lecithin complex 495–496
L-leucyl amide 378
lignin synthesis and 564
liposomes and leakage rate 492–493
liposome thermal transitions and 493–496
membrane organisation and 492, 495
membrane synthesis and 491, 549–551
metabolism and ethylene 360
metabolism and light 361
methyl ester 131, 162, 172
mineral reserves and 552
NMR of liposomes and 495
nucleotide synthesis and 550
occurrence 23
paper chromatography 162
partition chromatography 133–137
partition coefficients 125
PC-C(PCT) activity and 491, 550
PC-G(PGT) activity and 491, 550
L-phenylalanyl amide 378
phospholipid synthesis and 550
polysome formation and 491
L-prolyl amide 378
n-propyl ester 36, 376, 378
protein synthesis and 561
proton pump and 530
ribonuclease and 551
ribosomes and 490–492, 547
mRNA for α-amylase and 489
RNA polymerase 558, 560
RNA synthesis 488–489, 545–547, 558, 560
L-seryl amide 378
solvent partitioning 124–125
tetrahydro- **36**
thin layer chromatography 161–165
transpiration and 542
L-tyrosyl amide 378
L-valyl amide 378

gibberellin A_4 **24**
adsorption chromatography 130
benzyl ester 176
biological activity and 2-substitution 5
biosynthesis, *Cucurbita* system 4, 314
biosynthesis, fungal 312
conversion to GA_3 354–355
GA_9 metabolite 314, 357
gas liquid chromatography 170–173
GC-MS 186
gel permeation chromatography 138
β-D-glucopyranosyl ester 36, 124, 147, 164, 173, 376, 378, 380
3-O-β-D-glucopyranosyl ether 36, 376, 378
high pressure liquid chromatography 176–178
hydroxylations in plants 356–359
metabolism and abscisic acid 360
metabolism and 6-benzyladenine 360
metabolism, fungal 312, 354–355
methyl ester 162, 172
occurrence 24
paper chromatography 162–163
partition chromatography 134–137
partition coefficients 125
solvent partitioning 124–125
thin-layer chromatography 161–163
gibberellin A_5 **24**
adsorption chromatography 130, 146
benzyl ester 176
fluoro-derivative and biological activity 482
gas liquid chromatography 170–172, 177
GC-MS 182
O-β-D-glucopyranosyl ester 377, 379–380
13-O-β-D-glucopyranosyl ether 377, 379–380
high pressure liquid chromatography 176–178
isolation from *Phaseolus* seed 146
metabolism 356, 359
metabolism and light 361
methyl ester 162, 172, 182
p-nitrophenyl ester 176
occurrence 4, 24
paper chromatography 162
partition chromatography 134–137
partition coefficients 125
solvent partitioning 146
thin layer chromatography 162–163
gibberellin A_6 **24**
adsorption chromatography 130, 146
artefact from 116
biosynthesis 359
gas liquid chromatography 172
isolation from *Phaseolus* seed 146
methyl ester 162, 172
occurrence 24
paper chromatography 160–162
partition chromatography 134–135
partition coefficients 125
solvent partitioning 124–125, 146
thin layer chromatography 161–163
gibberellin A_7 **25**
benzyl ester 176
biosynthesis, fungal 312
gas liquid chromatography 170–172
gel permeation chromatography 138
19,2-lactone isomer 134–136
metabolism, fungal 354
metabolism, plant 359
methyl ester 162, 172
occurrence 25
paper chromatography 161–162
partition chromatography 134–137
partition coefficients 125
solvent partitioning 124–125
thin layer chromatography 161–165
gibberellin A_8 **25**
adsorption chromatography 130, 149
artefact of 116
"conjugate" 36
GA_1 binding and 488
GA_1 metabolite 356, 358–359
gas liquid chromatography 170–172
2-O-β-D-glucopyranosyl ether 36, 124, 134, 147–150, 164, 172, 183–184, 379–380, 482
13-O-β-D-glucopyranosyl ether 379
high pressure liquid chromatography 176–178
isolation from *Phaseolus* seed 146, 149–150
liposomes and 493
methyl ester 162, 172
occurrence 4, 25
paper chromatography 161–162
partition chromatography 134–137, 149–150
partition coefficients 125
receptor binding 488
solvent partitioning 124, 146
thin layer chromatography 161–163
vacuoles and metabolism 389
gibberellin A_9 **25**
benzyl ester 176
biosynthesis, fungal 312
1,2-dehydro- 312
13-fluoro- and biological activity 482
18-fluoro- and biological activity 482
gas liquid chromatography 170–172
β-D-glucopyranosyl ester 36, 125, 377, 379
high pressure liquid chromatography 176–178
12α-hydroxy- 357
15β-hydroxy- 311
metabolism and biological activity 479, 481
metabolism and light 361

gibberellin A_9
metabolism and 6-benzyladenine 360
metabolism, fungal 312, 355
metabolism, plants 356–359, 479
methyl ester 3, 162, 172
occurrence 25
paper chromatography 162
partition chromatography 134–137
partition coefficients 125
quantitative analysis 186, 268
solvent partitioning 124–125
2-substitution and biological activity 5
thin layer chromatography 162–163
gibberellin A_{10} **25**
GA_9 metabolite 312, 357
gas liquid chromatography 170–172
methyl ester 172
occurrence 25
partition chromatography 134–135
partition coefficients 125
thin layer chromatography 163
gibberellin A_{11} **25**
biosynthesis, fungal 312
gas liquid chromatography 170–172
methyl ester 172
occurrence 25
partition chromatography 134–135
partition coefficients 125
thin layer chromatography 163
gibberellin A_{12} **26**
7-alcohol 137, 311–314
7-aldehyde – see gibberellin A_{12}-aldehyde
biosynthesis 310–312, 314
gas liquid chromatography 170–172
high pressure liquid chromatography 176–178
methyl ester 172
occurrence 26
partition chromatography 134–137
partition coefficients 125
solvent partitioning 124–125
thin layer chromatography 163
gibberellin A_{12}-aldehyde 309, **310**, 311
biosynthesis, *Cucurbita* system 310, 314
biosynthesis, fungal 310, 312, 314
biosynthesis, *Pisum* system 310
high pressure liquid chromatography 178
3β-hydroxylation 355, 357
13-hydroxylation 355, 357
partition chromatography 136–137
gibberellin A_{13} **26**
7-aldehyde 314
anhydride 314
benzyl ester 176
biosynthesis, *Cucurbita* system 314
biosynthesis, fungal 312
gas liquid chromatography 170–172
2β-hydroxylation 355–356, 481
methyl ester 172
occurrence 26
non-conversion to GA_4 4
partition chromatography 134–137
partition coefficients 125
thin layer chromatography 163
gibberellin A_{14} **26**
7-alcohol 136–137, 312
7-aldehyde 136–137, 311–312, 314
benzyl ester 176
biosynthesis 311–312, 314
gas liquid chromatography 170–172
high pressure liquid chromatography 176–178
metabolism 355, 357
methyl ester 172
occurrence 26
partition chromatography 134–137
partition coefficients 125
thin layer chromatography 163
gibberellin A_{15} **26**
benzyl ester 176
biosynthesis, *Cucurbita* system 314
biosynthesis, fungal 312, 314
gas liquid chromatography 170–172
high pressure liquid chromatography 176–178
15β-hydroxy- 311
methyl ester 172
occurrence 26
partition chromatography 134–137
thin layer chromatography 163
gibberellin A_{16} **26**
biosynthesis 312
metabolism 359
occurrence 26
partition chromatography 134–137
partition coefficients 125
gibberellin A_{17} **26**
"conjugate" 36
gas liquid chromatography 177
GC-MS 186
methyl ester 36
occurrence 4, 26
partition chromatography 134–137
partition coefficients 125
quantitative analysis 186, 268
thin layer chromatography 165
gibberellin A_{18} **27**
GA_{14} metabolite 355, 357
gas liquid chromatography 170–172
methyl ester 172
occurrence 27
partition chromatography 134–135
partition coefficients 125
thin layer chromatography 163

gibberellin A_{19} **27**
gas liquid chromatography 170–172
high pressure liquid chromatography 177
methyl ester 172, 186
occurrence 4, 27
partition chromatography 134–135
partition coefficients 125
quantitation by SIM 186–187
thin layer chromatography 161–163
gibberellin A_{20} **27**
adsorption chromatography 130
benzyl ester 176, 251–252
biosynthesis, fungal 312
"conjugate" 36
from steviol 310
GA_9 metabolite 357–359
gas liquid chromatography 177
GC-MS 182
13-O-β-D-glucopyranoside 379–380
high pressure liquid chromatography 176–178
metabolism 356–357, 359–360, 479
metabolism and ABA 360
metabolism and 6-benzyladenine 360
metabolism and light 361
methyl ester 182
occurrence 4, 27
partition chromatography 134–137
quantitative analysis 186, 251–252, 268
gibberellin A_{21} **28**, 359
occurrence 28
partition chromatography 134–135
solvent partitioning 124–125
gibberellin A_{22} **28**, 359
occurrence 28
partition chromatography 134–135
partition coefficients 125
gibberellin A_{23} **28**
occurrence 28
partition chromatography 134–135
partition coefficients 125
solvent partitioning 124–125
gibberellin A_{24} **28**
biosynthesis, *Cucurbita* system 314
biosynthesis, fungal 312
15β-hydroxy- 311
occurrence 28
partition chromatography 134–137
partition coefficients 225
gibberellin A_{25} **28**
biosynthesis 312
occurrence 28
partition chromatography 134–137
partition coefficients 125
gibberellin A_{26} **28**
2-O-β-D-glucopyranosyl ether 36, 134, 164, 378
metabolism 359
occurrence 28
partition chromatography 134–135
partition coefficients 125
gibberellin A_{27} **29**
adsorption chromatography 130
occurrence 29
2-O-β-D-glucopyranosyl ether 36, 134, 164, 378
partition chromatography 134–135
partition coefficients 125
gibberellin A_{28} **29**
"conjugate" 36
occurrence 29
partition chromatography 134–137
partition coefficients 125
solvent partitioning 124–125
gibberellin A_{29} **29**
gas liquid chromatography 177
2-O-β-D-glucopyranosyl ether 36, 134, 164, 378, 390, 482
GA_{20} metabolite 356, 359–360
occurrence 4, 29
partition chromatography 134–135
partition coefficients 125
quantitative analysis 186, 268
gibberellin A_{30} **29**
metabolism 359
occurrence 29
partition chromatography 134–135
thin layer chromatography 165
gibberellin A_{31} **29**
2,3-dihydro derivative – see dihydrogibberellin A_{31}
occurrence 29
gibberellin A_{32} **29**
acetonide artefact 116
metabolism 359
occurrence 29
solvent partitioning 124
thin layer chromatography 165
gibberellin A_{33} **30**
metabolism 359
occurrence 30
partition chromatography 134–135
gibberellin A_{34} **30**
GA_4 metabolite 356, 358–359
glucoside 378, 380
occurrence 30
partition chromatography 134–135
thin layer chromatography 165
gibberellin A_{35} **30**
11-O-β-D-glucopyranoside 36, 378
metabolism 359
occurrence 30
gibberellin A_{36} **30**
biosynthesis, *Cucurbita* system 314

gibberellin A_{36}
biosynthesis, fungal 312, 314
GA_4 precursor 4
occurrence 30
partition chromatography 136–137
gibberellin A_{37} **30**
adsorption chromatography 130
biosynthesis, *Cucurbita* system 314, 357
biosynthesis, fungal 312
β-D-glucopyranosyl ester 36, 124, 147–148, 164, 173, 377, 379
occurrence 30
partition chromatography 136–137
gibberellin A_{38} **30**
β-D-glucopyranosyl ester 36, 124, 147–148, 164, 173, 378
occurrence 30
gibberellin A_{39} **30**
occurrence 30
gibberellin A_{40} **31**
biosynthesis, fungal 312
2-hydroxylation and biological activity 481
occurrence 31
thin layer chromatography 165
gibberellin A_{41} **31**
occurrence 31
gibberellin A_{42} **31**
occurrence 31
gibberellin A_{43} **31**
biological activity 481
biosynthesis, *Cucurbita* system 4, 314
GA_{13} metabolite 355–356
occurrence 31
gibberellin A_{44} **31**
biosynthesis, *Pisum* system 314, 355, 357
occurrence 4, 31
gibberellin A_{45} **31**
GC-MS 184
occurrence 31
gibberellin A_{46} **32**
biological activity 481
occurrence 32
gibberellin A_{47} **32**
biological activity 481
biosynthesis 312
occurrence 32
gibberellin A_{48} **32**
occurrence 32
partition chromatography 137
gibberellin A_{49} **32**
occurrence 32
partition chromatography 137
gibberellin A_{50} **32**
occurrence 32
gibberellin A_{51} **32**
biological activity 481
GA_9 metabolite 356, 358–359
occurrence 32
gibberellin A_{52} **32**
occurrence 32
gibberellin A_{53} **33**
biosynthesis, *Pisum* system 314, 355
occurrence 4, 33
gibberellin A_{54} **33**
occurrence 33
gibberellin A_{55} **33**
occurrence 33
gibberellin A_{56} **33**
occurrence 33
gibberellin A_{57} **33**
occurrence 33
gibberellin A_{58} (12α-hydroxy gibberellin A_4), from seed of *Cucurbita maxima* 3
gibberellin A_{59} (2,3-dehydro gibberellin A_{21}), from seed of *Canavalia gladiata* 3
gibberellin A_{60} (1β-hydroxy GA_{20}), from wheat germ 3
gibberellin biosynthesis
abscisic acid and 360
adenylate energy charge 317
Baeyer-Villiger oxidation 313–314
cell free enzyme systems 4, 304, 308, 310, 314, 316, 355
chloroplasts and 388–389
enzyme-substrate specificity 310–311
from acetate 301
from copalyl pyrophosphate 302, 304
(=labda-8,13-dienyl pyrophosphate)
from dimethyl acrylic acid 301
from GA_{12}-aldehyde 311–312
from GA_{14} 7-aldehyde 311–312
from geranylgeranyl pyrophosphate 302–303
from 7β-hydroxy *ent*-kaurenoic acid 308–310
from *ent*-kaurenal 308
from *ent*-kaurene 302–303, 308
from *ent*-kaurenoic acid 308
from *ent*-kaurenol 307
from mevalonic acid 301, 304, 309, 311
G. fujikuroi cultures 301, 309–311, 313–314, 316, 354–355
G. fujikuroi mutants 302, 311–314
3β-hydroxylation mechanism 311
inhibitors 315–316
kaurene synthetase 304–307
loss of carbon-20 4, 313–314
oestrone and 361
origin of lactone 314
pathways 303, 307, 310, 312, 359–360
pea seed 358
pea seedlings 355
peroxy-acid intermediate 314
pH and fungus 313
pyrophosphomevalonate decarboxylase 317

regulation by kaurene synthetase 316–317
ring B contraction 310
sites of 387–389
Sphaceloma manihoticola 34
suspensors of bean seed 5
wheat seedlings 302
gibberellin conjugates – see also individual gibberellins
adsorption chromatography 130–131, 147–148
biological activity 381
formation in vivo 378–381
function 382, 482–483
gas liquid chromatography 172–173
GC-MS 4, 183–184
germination and 381
high pressure liquid chromatography 177
hydrolysis and biological activity 381
hydrolysis during purification 116
ion exchange chromatography 125
isolation from *Phaseolus* 147–148
natural 378–379
occurrence 36
orientation at interface 483
paper chromatography 164
partition chromatography 134, 147–148, 150
peptide 377–378
solvent partitioning 124, 147
synthetic 378–379
thin layer chromatography 147–148, 164
gibberellin glucosides – see gibberellin conjugates and individual gibberellins
gibberellin glucosyl esters – see gibberellin conjugates and individual gibberellins
gibberellin glucosyl ethers – see gibberellin conjugates and individual gibberellins
gibberellin methyl esters
adsorption chromatography 130–131
gas liquid chromatography 170–172
GC-MS 182–184
natural occurrence 3, 36
thin layer chromatography 162
gibberellin receptors
binding and biological activity 488
membrane sites 492
pea epicotyls 5, 486–488
pea nucleus 488–489
steroid hormone analogy 484–486
gibberic acid,
high pressure liquid chromatography 178
*allo*gibberic acid,
high pressure liquid chromatography 178
*allo*gibberic acid-O(13)-glucoside,
gas liquid chromatography 172
gibberthione (pharbitic acid) **36**, 376, 379
$\beta 1 \rightarrow 3$ glucan,
auxin-induced turnover 528
glucan synthetases
auxins and 518, 520, 524, 536
plasma membrane marker 458
$\beta 1 \rightarrow 3$-glucanase
auxin effect on 517–518, 520, 524–525
elongation by 525
GA_3 effect on 551
$\beta 1 \rightarrow 4$-glucanase
auxin effect on 520
degradation of cellulose 524
$\beta 1 \rightarrow 6$ glucanase,
auxin effect on 518
glucinosolates – see glucobrassins
glucobrassins **19**
adsorption chromatography 130
biosynthesis 288–289
conversion to indole-3-acetonitrile 289
extraction 115
hydrolysis by myrosinase 20, 115
ion exchange chromatography 124
occurrence 20, 114
*neo*glucobrassin **19**
sulphoglucobrassin **19**
β-D-glucopyranosylabscisate **50**
formation and function 375
identity with metabolite B 373
occurrence 50
partial identification 374
purification by partition chromatography 138
6-(4-O-β-D-glucopyranosyl-3-methyl-*trans*-but-2-enylamino)purine – see zeatin-O-β-D-glucopyranoside
β-glucosidase
gibberellin glucosides and 381
zeatin glucosides and 372
glucosylribosylzeatin **47**
GC-MS 184
isolation from *Vinca* gall 153–154
occurrence 45, 47
glucosyl zeatin – see zeatin-O-β-D-glucopyranoside
glucuronoarabinoxylan **528**
auxin-induced turnover 528
cell-wall extension and 528
glutathione **73**, 80
glyceraldehyde-3-phosphate dehydrogenase 335
glycoproteins,
cell wall structure and 528–529
glycosidase,
effect on cell wall 525
Golgi membranes,
receptor site (I) and 461
gramine **62**
allelopathy and 75
colour reactions 158

gramine
ion exchange chromatography 123
paper chromatography 158
grandiflorenic acid 35, **38**
growth limiting protein 521
growth promoters (non-hormonal)
aliphatic alcohols 79
dihydroasparagusic acid 80
GA synergists 62
2-hydroxycinchoninic acid 75
myo-inositol 80
nicotinamide 75–76
polyamines 75
unsaturated ketones 77–78
guanine
gas liquid chromatography 174, 177
thin layer chromatography 167
guanosine
gas liquid chromatography 174
guanosine 3′,5′-cyclic monophosphate (cGMP)
α-amylase synthesis and 549
DNA synthesis and 549

harringtonolide **70**, 78
heliangine **65**, 77
helminthosporol 35, **38**
hemicellulase,
effect of auxin 518
heraclenol **59**, 67
high pressure liquid chromatography
abscisic acid 181, 292
auxins 176–177
CI-MS and 4
cytokinins 178–180
gibberellenic acid glucosyl ester 177
gibberellin 176–178
gibberellin benzyl esters 176
gibberellin glucosides 177
performance parameters 259
polarity range 261
quantitative 272
zeatin metabolites 4
histidine,
NAA-binding sites and 464
homo-arginine **64**, 76
homocysteine 320–322, **329**
homocysteine thiolactone 320
homoserine **321**
hormone concentration 203
and accuracy of estimation 212
hydrangeol **59**, 69
hydrogen cyanide,
from cyanogenic glucosides 74
6′-hydroxyabscisic acid (metabolite C) 52, **54**, 350–351, 375
β-hydroxy-β-methylglutaryl ester 52, **54**, **352**
p-hydroxybenzoic acid **57**
6-(2-hydroxybenzyl)adenine **47**
biological activity 472
PVP chromatography 141
6-(0-hydroxybenzyl)adenosine
gel permation chromatography 139
light-induced formation 370
occurrence 45, 292
PVP chromatography 141
quantitative analysis 269
3′-hydroxy-dihydrozeatin **46**, 349
isolation from *Zea mays* 151
5-hydroxyindole,
thin layer chromatography 160
5-hydroxyindole-3-acetic acid **339**
ethylene production and 234
ion exchange chromatography 123
paper chromatography 158
thin layer chromatography 158
7-hydroxyindole-3-acetic acid
paper chromatography 158
thin layer chromatography 160–161
8-hydroxy-N^6-(Δ^2-isopentenyl)adenine 348, **349**
7β-hydroxy-*ent*-kaurenoic acid **309**
biosynthesis 308–309
enzymatic oxidation 310
partition chromatography 136–137
ring B contraction 310, 312
7β-hydroxykaurenolide,
biosynthesis 4, **309**
6-(3-hydroxy-3-methylbutylamino)-purine 348, **349**
6-(4-hydroxy-3-methylbutylamine)purine – see dihydrozeatin
6-(4-hydroxy-3-methyl-*cis*-but-2-enylamino)purine – see *cis*-zeatin
6-(4-hydroxy-3-methyl-*trans*-but-2-enylamino) purine – see *trans*-zeatin
6-(4-hydroxy-3-methyl-*cis*-but-2-enylamino)-9-β-D-ribofuranosylpurine – see *cis*-zeatin riboside
6-(4-hydroxy-3-methyl-*trans*-but-2-enylamino)-9-β-D-ribofuranosylpurine – see *trans*-zeatin riboside
6-(4-hydroxy-3-methylbutylamino)-9-β-D-ribofuranosylpurine – see dihydrozeatin riboside
β-hydroxy-β-methylglutaryl-6′-hydroxyabscisic acid 52, **54**, **352**
3-hydroxymethyloxindole **20**
biological activity 533
glucobrassin metabolite 20
IAA catabolite 337, **338**
5-hydroxyoxindole-3-acetic acid **339**
p-hydroxyphenylacetic acid 21, **22**
hydroxyproline
cell wall proteins and 528–529
ethylene and 532
β-hydroxypropionic acid 322, **323**

2-hydroxy-*trans*-zeatin **46**, 348, **349**
isolation from *Zea mays* 151
8-hydroxy-*trans*-zeatin 348, **349**
3-hydroxytyramine (dopamine) 324
hypaphorine,
paper chromatography 158

identification
and information theory 227, 231
indican,
thin layer chromatography 160
indole(s) – see also individual compounds
gas liquid chromatography 170
IAA precursor 282–283
ion exchange chromatography 123–124
paper chromatography 157–160
partition column chromatography 132, 134
solvent partitioning 120–123
thin layer chromatography 159–161
indole-3-acetaldehyde **18**
countercurrent distribution 142
dehydrogenase 285
from indole-3-pyruvic acid 284–285
gas liquid chromatography 170
gel permeation chromatography 138
IAA precursor 285
isolation as dimethyl acetal 115
metabolism, 288
occurrence 284–285
oxidase 285–286
paper chromatography 158–160
partition chromatography 132
reductases 283, 286
solvent partitioning 120
thin layer chromatography 159–161
indole-3-acetaldoxime acetoxime) **18**
hydro-lyase 288
IAA precursor 288
metabolism 288
indole-3-acetamide **14**
adsorption chromatography 130, 144
as artefact 115
gas liquid chromatography 170
high pressure liquid chromatography 177
ion exchange chromatography 123
isolation from *Citrus unshiu* 144
occurrence 14
paper chromatography 144, 157–159
partition chromatography 144
solvent partitioning 120, 144
thin layer chromatography 144, 159–161
indole-3-acetic acid (IAA) **13** – see also auxin
adsorption chromatography 130, 144
affinity chromatography 454
α-amylase and 562
artefacts 115–116, 120
aspartate **21**, **364**, 365–366
benzoxazolines and 462–463
binding sites 455–456, 470
binding to proteins 365, 455–456, 470
bioassays 241
biosynthesis – see auxin biosynthesis
biosynthetic pathways 284
calcium chelation and 516
catabolism and pathways 338–339, 346
cell division and 556, 558
cell membrane and 540
cellulase and 562–563
cell wall glycosides and 529
countercurrent distribution 142
cytoplasmic receptor protein 456
decarboxylation 336–337, 347
differentiation and 554–556, 563, 566
distillation and sublimation 116, 144
DNA complex 365
DNA synthesis 554–556, 558
effects on cellular processes 517–519
elongation rate and 522–523
enzymes in aleurone 552, 562
epiphytic bacteria 287
esters – see named IAA esters
gas liquid chromatography 144, 169–171
GC-MS and trimethylsilyation 184, 186–189, 264, 266
gel permeation chromatography 138–139, 240
high pressure liquid chromatography 176
indole-α-pyrone fluorescence assay 266–267, 273
invertase and 561
ion exchange chromatography 123
isolation from *Citrus unshiu* 144
latent period 517–519
levels in plants 20
lipase and 562
liposome leakage rate and 493
membrane binding sites 458
3-methyleneoxindole precursor 533
nucleic acid synthesis and 558
occurrence in plants 13, 20, 169, 184–189
oxidation, non-enzymic 336
oxidative degradation 337–339, 342
paper chromatography 144, 157–160
partition chromatography 132–133, 144
peptide conjugates 364–366
peroxidase and 337, 339, 340–347, 384, 565
proton pump and 522–523
PVP chromatography 140
quantitative analysis 186, 241, 246, 251–252, 264–267, 273
radio-immunoassay 246
receptor proteins – see auxin receptors
ribonuclease 559
RNA complex 365

indole-3-acetic acid (IAA)
  RNA synthesis and 456, 558–559
  root permeability and 541
  solvent partitioning 120–122, 129, 144
  stability 4, 336
  stomatal closure and 541
  thin layer chromatography 144, 159–161
  transpiration and 541
  triglycerides in aleurone and 552
  turgor pressure and 515
indole-3-acetic acid ethyl ester **14**
  artefact 115
  gas liquid chromatography 169–170
  ion exchange chromatography 123
  occurrence 14
  paper chromatography 158–159
  partition chromatography 132
  solvent partitioning 120–123
  thin layer chromatography 159–161
indole-3-acetic acid glycosyl esters 14, **15**, 21, 113, 115
  adsorption chromatography 130
  ion exchange chromatography 123
  metabolism 286, 366–367
  solvent partitioning 120–123
indole-3-acetic acid methyl ester **14**
  adsorption chromatography 130
  ion exchange chromatography 123
  isolation 145
  occurrence 14
  paper chromatography 158
  solvent partitioning 120
indole-3-acetic acid myoinositol esters **15**, 21
  adsorption chromatography 130
  bioactivity and function 367
  gel permeation chromatography 138
  metabolism 365–367
  solvent partitioning 120–123
indole-3-acetic acid oxidase (IAA oxidase) 336, 340–342
  fungal disease and 346
  inhibitors 62–64
  kinetin and 346
  light and 336, 347
  localisation 384
  stress and 347
  substrate levels and 346
indole acetone,
  high pressure liquid chromatography 177
indole-3-acetonitrile **17**
  artefact 115
  countercurrent distribution 142
  from glucobrassicin 20, 288
  gas liquid chromatography 170
  high pressure liquid chromatography 177
  IAA precursor 288
  ion exchange chromatography 123
  1-methoxy- **17**
  4-methoxy- **17**
  nitrilase 288
  partition chromatography 132
  picrate 143
  solvent partitioning 120
  thin layer chromatography 158–160
indole acetoxime – see indole-3-acetaldoxime
indole-3-acetyl-alanine **365**, 366
indole-3-acetyl-L-aspartic acid **21**, **364**, 365–366
  L-aspartic acid-N-acylase 364
  biosynthesis 365
  ethylene and 365
  function of 365–366
  occurrence 365
indole-3-acetyl-β-1→4 cellulosic glucan 21, 366
indole-3-acetyl-1′-β-D-glucopyranosyl ester **21**, 366–367 – see also IAA glucosyl esters
indole-3-acetyl-2′-D-glucopyranosyl ester **15**, 366–367
indole-3-acetyl-4′-D-glucopyranosyl ester **15**, 366–367
indole-3-acetyl-6′-D-glucopyranosyl ester **15**, 366–367
indole-3-acetyl-1-β-glucose **21**
indole-3-acetyl-glutamic acid **365**, 366
indole-3-acetyl-glycine **365**, 366
indole-3-acetyl-L-lysine **365**, 366
  ion exchange chromatography 123
indole-3-acetyl-2′-myoinositol **14** – see also IAA myoinositol esters
  5′-(1-L-arabinopyranosyl)- **14**, 367
  5′-(1-D-galactopyranosyl **15**, 367
  precursors 365
Nδ-(indole-3-acetyl)-L-ornithione,
  ion exchange chromatography 123
indole-3-acetyl-valine 364, **365**
indole-3-acrylic acid
  gas liquid chromatography 171
  ion exchange chromatography 123
  partition chromatography 133
  solvent partitioning 120
  thin layer chromatography 160–161
indole-3-butyric acid
  gas liquid chromatography 171
  high pressure liquid chromatography 177
  paper chromatography 158–159
  partition chromatography 132–133
  solvent partitioning 120
  thin layer chromatography 161
  transpiration rate and 541
indole-3-carboxaldehyde **18**
  as artefact 20, 115
  counter current distribution 142
  gas liquid chromatography 170
  gel permeation chromatography 138

high pressure liquid chromatography 177
IAA catabolite **338**
ion exchange chromatography 123
paper chromatography 157–158
solvent partitioning 120
thin layer chromatography 161
indole-2-carboxylic acid,
gas liquid chromatography 171
indole-3-carboxylic acid **18**
artefact 20, 115
counter current distribution 142
gas liquid chromatography 171
high pressure liquid chromatography 177
ion exchange chromatography 123
paper chromatography 158–159
partition chromatography 132–133
solvent partitioning 120
thin layer chromatography 160–161
indole-5-carboxylic acid,
gas liquid chromatography 171
indole-3-ethanol (trypthophol) **18**
biosynthesis 285, 288
gas liquid chromatography 170
gel permeation chromatography 138, 145
IAA precursor 286
ion exchange chromatography 123
isolation from *Cucurbita* 145
occurrence 18, 283
oxidase 286
paper chromatography 158–159
partition chromatography 132
solvent partitioning 120
thin layer chromatography 145, 160–161
indole-3-ethyl-β-D-glucopyranoside – see also
IAA glycosylesters
IAA metabolite 286
indole glucosinolate,
from indole-3-acetaldoxime 288–289
indole-3-glyceraldehyde
auxin activity 283
N-acetate 283
indole-3-glycerol phosphate 283
indole-3-glyoxylic acid,
gas liquid chromatography 171
indole-3-hydroxymethyl **20**
indole-3-lactic acid
gas liquid chromatography 171
high pressure liquid chromatography 177
ion exchange chromatography 123
paper chromatography 158
partition chromatography 132
solvent partitioning 120
thin layer chromatography 160
indole-3-methanol,
IAA catabolite **338**
indole-1-methoxy-3-methylglucosinolate **19**
indole-3-methylglucosinolate **19**
indole-3-methyl-isothiocyanate **20**
artefact from glucobrassin 20
indolenine epoxide,
IAA catabolite **338**
indolenine hydroperoxide,
IAA catabolite **338**
indole-1-propionic acid,
gas liquid chromatography 171
indole-3-propionic acid
gas liquid chromatography 171
high pressure liquid chromatography 177
ion exchange chromatography 123
paper chromatography 157–159
partition chromatography 133
solvent partitioning 120
thin layer chromatography 160–161
indole-3-pyruvic acid
artefact from IAA 120
decarboxylase 285–286
from tryptophan 254, 283
high pressure liquid chromatography 177
indentification 283
ion exchange chromatography 123
occurrence 284
paper chromatography 158–160
partition chromatography 132
solvent partitioning 120–123
thin layer chromatography 160–161
indole-1-sulpho-3-methylglucosinolate **19**
indole-3-thioacrylic acid,
paper chromatography 158
indole-α-pyrone assay
analysis of indole-3-acetic acid 266–267, 273
hypothetical errors 206
internal standards error 208
successive approximation method and
221–222
N(β-3-indolylether)pyrrolidine,
paper chromatography 158
2-(indol-3-yl-methyl)indole-3-acetic acid,
IAA catabolite 338
3-(indol-3-yl-methyl)oxindole,
IAA catabolite 338
information content
accuracy and 226, 236, 239
binary digit (bit) 227
degree of correlation 228
distribution 229
equation 227
establishing indentity 227, 231–236, 238
example 234
gel permeation chromatography and 204,
235, 240
minimum for identification 235, 252
null hypothesis 231–233
of chemical structure 232
of chromatographic procedures 255

information content
of physicochemical methods 248–249
open ended system and 235, 239–240
precision and 227
sample purity and 229, 231, 239
sample size and 232
selectivity and sensitivity 254
structure generating algorithms and 239
infrared spectrometry, performance parameters 248
inhibitor(s)
abscisic acid 513, 548, 550, 560
actinomycin D 549–550, 558, 565
alcohols 79
aldehydes 73, 79
amino acids 76
ammonia 75
aromatic acids 58–59
AVG 320, 322, 325
betalains 64
caffeic acid 63, 345
L-canaline 325
CAPA 465
chloramben 465
cinnamic acid **58**
cordicepin 546, 550
coumarin 58, 64, 72, 563
cycloheximide 327, 521, 525, 550, 565
2,4-D 335, 557
diethyldithiocarbomate 335
diterpenoid 78
DTE 466
ethylene 514, 533, 557
fatty acids 79
flavonoids 58, 64
fluorouracil 547, 558
FUDR 554–555
grandinol **60**, 74
humic acids 58
hydrogen cyanide 74
hydroxybenzoic acid 59
hydroxycinnammic acid 59
inhibitor β 52, 60
isothiocyanates 74
ketones 79
lactones 67, 70, 77–78
lignans 74
MBOA and DMBOA 462
MDMP 497
mitomycin 549
NAA 461
of α-amylase synthesis 548–549
of auxin binding 466
of callus growth 80
of cell division 75–76
of NDA synthesis 549, 554, 557–558, 560
of dormancy 74
of ethylene biosynthesis 319–322, 325, 330, 335–336
of extension 533
of germination 61, 67, 73, 78–79, 300
of gibberellin activity 69, 75, 79
of gibberellin biosynthesis 315–316
of growth 462, 531
of IAA binding 466
of IAA oxidase 62–64, 344–345
of IAA peroxidase 344–345
of kaurene oxidation 316
of kaurene synthetase 306, 316–317
of mitochondrial metabolism 61
of mitosis 75–78
of NAA binding 462, 465, 469
of polysaccharide synthesis 525–526
of protein synthesis 521
of RNA synthesis 521, 546–547, 558, 560
of root growth 65, 79, 81
of sprouting 74
phenols 59, 62–63
PMB and PMBS 466, 469
polyacetylenic esters 79
rhibotoxine 319, 322–323
rifampicin 454
sarcomycin 549
tannins 58
xanthoxin 300–301
inosine,
catabolite of cytokinins **349,** 350
internal standards
abscisic acid analysis 270
cytokinin analysis 269
deuterated cytokinins 4
GC-SICM and 4
IAA analysis 264, 266
limitations 208, 262–263
successive approximations method 217
inumakilactones A and B, **67,** 78
invertase
abscisic acid and 561
auxins and 561–562
gibberellic acid and 561–562
transcription and translation 562
ion exchange chromatography
auxins 123–124
cytokinins 127–128, 151
gibberellins 125, 149
gibberellin glucosyl esters 125
gibberellin metabolites 125
indoles 123–124
zeatin metabolites 128
ionization devices, performance parameters 249
β-ionone 300
ionylidene acetic acid derivatives,
abscisic acid biosynthesis and 298–299

IPA (2iPA) – see N[6]-(Δ^2-isopentenyl)adenine or -adenosine
isatin,
  paper chromatography 159
isocoumarins, gibberellin synergists 69
N[6]-(Δ^2-isopentenyl)adenine **44, 290,** 471
  biosynthesis in tRNA 290
  catabolic degradation 350
  2,8-dihydroxy- 348
  discadenine precursor 370, 373
  ethylene production and 334
  fractionation of 127
  gas liquid chromatography 174
  gel permeation chromatography 139
  N(7)-β-D-glucopyranoside 369
  growth inhibiton 531
  high pressure liquid chromatography 179–180
  8-hydroxy- 348
  occurrence 40, 44, 292
  oxidation in tRNA 292
  paper chromatography 165
  partition coefficients 127
  phosphorylation 369
  pK_a values 125
  PVP chromatography 141
  ribosylation 369
  solvent partitioning 127
  thin layer chromatography 167
  xanthine oxidase oxidation 348
N[6]-(Δ^2-isopentenyl)adenosine **44, 290**
  adenosine amino hydrolyase and 350
  biosynthetic site 385
  catabolism, degradation 348–350
  gas liquid chromatography 173–174
  gel permeation chromatography 139–140
  high pressure liquid chromatography 174–178
  hydroxylation in vivo 348
  occurrence 43–45, 292, 385, 472
  paper chromatography 165–167
  partition chromatography 137
  5′-phosphate 292
  pK_a value 127
  quantitative analysis 187
  thin layer chromatography 165–167
isopentenylpyrophosphate 293, **294, 295, 303**
  isomerase 293
ent-isopimeradiene (sandaracopimaradiene) 305
2′-isopropyl-4′-(trimethylammonium chloride)-5′-methylphenylpiperidine-1-carboxylate – see AMO 1618
isothiocyanates
  allyl- **62,** 74
  β-phenethyl **62,** 74

juglone
  glycoside and allelopathy 74

kaempferol **58,** 64
Kartoffelfaktor 468
ent-kaura-2,16-diene 311
ent-kaura-2,16-dien-19-oic acid,
  precursor of kaurenolides 4
ent-kaurenal **303**
  biosynthesis 308, 316
  enzymatic oxidation 308
(+)-kaurene 302
ent-kaurene 35, **38, 303, 484**
  biosynthesis 303, 304–307 – see also kaurene synthetase
  carbon numbering **38**
  enzymatic oxidation 314, 317–319
  inhibitors of enzymatic oxidation 316
  high pressure liquid chromatography 178
  partition chromatography 136–137
*iso*kaurene 302
kaurene synthetase
  activity A and B 304, 317
  cyclisation mechanism 307
  GA biosynthesis and 316–317
  inhibitors 316–317
  in d-5 maize 5
  plant growth retardants and 316
  properties 304, 306
  regulation 317
  sources 305
  sub-cellular localisation 317
ent-kaurenoic acid 35, **38**
  biosynthesis **303,** 308
  6β,7β-dihydroxy- **310**
  enzymatic oxidation 308
  7β-hydroxy- 308, **309,** 310, 312
  metabolism in chloroplasts 389
  monohydroxy-derivatives 309
  partition chromatography 136–137
ent-kaurenol 35, **38**
  biosynthesis **303,** 308
  enzymatic oxidation 308, 310, 314
  7β-hydroxy- 309
  oxidation in plastids 389
  partition chromatography 136–137
ent-kaurenolides 308, 311
  7β,13,dihydroxy- 310
  7β-hydroxy- **309, 310**
2-keto-γ-methylthiobutyric acid (KMB) 324
kinetin (N[6]-furfuryladenine) **39, 471**
  α-amylase synthesis and 548–549
  biological activity 471, 475
  cell division and 556
  cell wall synthesis and 563
  deoxyisoflavone synthesis and 566
  DNA synthesis and 555

kinetin (N^6-furfuryladenine)
  ethylene production and 334
  gas liquid chromatography 174
  gel permeation chromatography 139
  IAA-induced growth and 562
  IAA peroxidase and 346
  indole-3-acetyl-L-aspartic acid and 365
  inhibition of xyloglucan turnover 531
  liposomes and 493
  partition coefficient 127
  polyphenol synthesis and 564–565
  protein binding 476
  protein synthesis and 561
  PVP chromatography 141
  ribonuclease and 559
  ribonuclease release 531
  ribosome binding and 474–475
  RNA synthesis and 560
  solvent partitioning 127
  stomatal aperture and 542–543
  thin layer chromatography 167
  tracheid formation 555
  transpiration rate and 541–542
kinetin riboside (6-furfurylamino-9-β-D-ribofuranosylpurine)
  gas liquid chromatography 174
  gel permeation chromatography 139
  high pressure liquid chromatography 179
  PVP chromatography 141
  thin layer chromatography 167

labda-8,13-dien-15-yl pyrophosphate (copalylpyrophosphate) 302, **303**, 304–305, 307, 317
labdanediol **69**, 78
lactones 67
  coumarins **58**, 67
  α-methylene 77
  terpenoid 77–78
  unsaturated 70–73
latent period of growth
  abscisic acid and 532
  auxins and 517–523
  ethylene and 532
  fusicoccin and 533
  gibberellins and 530
  metabolic processes and 517–518
  proton pump and 522, 534
  leghaemoglobin,
  IAA oxidase and 342
lecithin
  auxin complexes 496
  biosynthesis and α-amylase 550
  biosynthesis and GA_3 491
  GA_3 complex 495–496
  liposomes and 492–493
  membranes and ethylene 499
lentil factor γ 454
lignans,
  germination inhibitors 74
lignin
  hormones and biosynthesis of 564–565
linoleic acid **71**, 79
linolenic acid, **71**, 79, 318, 322
  ethylene precursor 318, 322
  lipoxidase and 322
liposomes,
  properties and hormone effects 492–496
lipoxidase 322
liquid chromatography,
  performance parameters 259
loliolide **53**, 55, 300
  occurrence in plants 53, 56
louisfieserone **62**, 74
lunularic acid 52, **53**
lupinic acid (β-zeatin-9-yl)alanine 370
  enzymatic pathway **373**
lycoricidine **63**, 75
lycoricidinol (narciclasine) **63**, 75
lycorine **63**, 75
lysine,
  NAA-binding site and 466

mandelonitrile **62**, 74
mass fragmentometry – see selected ion monitoring
mass spectrometry
  analysis by GC-MS 264
  auxins 184, 266
  chemical ionisation and zeatin metabolites 4
  GA_{20} benzyl ester 251–252
  information content 249–251
  limits of detection 249
  linear response range 249
  linked scan 4
  mass analysed kinetic energy (MIKES) 4
  selectivity 249
maytansine **73**, 80
mediator proteins
  auxins and 453–454
  cytokinins and RNA synthesis 474
melatonin
  ion exchange chromatography 123
  thin layer chromatography 161
membrane, mammalian,
  steroids and organelle structure of 498
membrane, plant cell
  auxin and ion relations 539–540
  auxin binding sites 452, 458, 461
  ethylene and 540–541
  GA_3 and 491–492, 495, 549–555
  markers 452, 458
  permeability 540
  proton pump and 451
  steroid hormones and 494

synthesis and GA_3 550
turgor and 539
membrane potential
auxin and 518
fusicoccin and 538–539
proton pump and 538–539
p-mercuribenzenesulphonate (PMBS),
auxin induced growth and 466
p-mercuribenzoate **465**
affinity labelling 466
auxin binding and 466
cysteine and 466
metabolite C – see 6′-hydroxyabscisic acid
methanethiol,
sulphur cycle and 320
methional 318, 324
methionine
adenosyl transferase 326
cytokinin biosynthesis 292
ethylene precursor 318–322, 325–326
metabolism 320–321, 329
seleno-analogue 320, 326
sulphur cycle 320
6-methoxy-2-benzoxazolinone (MBOA) **462**
IAA-induced growth and 462–463
NAA binding 462–463
supernatant factor and 462
5-methoxytryptamine,
ion exchange chromatography 123
methyl abscisate
artefact 117
isomer 117
N^6-methyladenine 45
biosynthesis in tRNA 291
gas liquid chromatography 174
9-methyl-6-benzyladenine,
biological activity 472
6-(3-methylbutylamino)-9-β-D-ribofuranosyl-purine,
gas liquid chromatography 174
6-(3-methylbut-2-enylamino)purine – see N^6-(Δ^2-isopentenyl)adenine
6-(3-methylbut-2-enylamino)-9-β-D-ribofuranosylpurine – see N^6-(Δ^2-isopentenyl)adenosine
6-(3-methylbut-2-enylamino)-9-β-D-ribofuranosylpurine 2′(3′)-phosphate,
paper chromatography 166
methyl 4-chloroindole-3-acetate
adsorption chromatography 130
solvent partitioning 120
2-methyl-4-chlorophenoxyacetic acid (MCPA **449**
IAA-binding 470
S-methylcysteine,
sulphur cycle and 320
L-2-(4-methyl-2,4-dinitroanilino)-N-methylpropionamide (MDMP),
protein synthesis and 497–498
α-(methylenecyclopropyl)glycine **63,** 76
3-methyleneoxindole(3-MeneOX) **338, 449, 533**
biological activity 448
ethylene production and 333
IAA-catabolite 337, 533
mode of action 337
reductase 337
6-(3-methyl-3-hydroxybutylamino)purine **46**
gas liquid chromatography 174
metabolite of N^6-(Δ^2-isopentenyl)adenine 3
occurrence 46
2-methylindole-3-acetic acid,
paper chromatography 159
methyl indole-3-butyrate,
paper chromatography 158
S-methylmethionine (SMM) 320–321, **329**
catabolism 329
3-methyloxindole (3-MeOX) **338**
IAA-catabolite 337–338
methyl phaeophorbide a and b **64,** 76
methyl phaseate
artefact 117
conjugate 117
6-methylpurine,
phospholipid synthesis and 550
5′-methylthioadenosine (MTA) 320, **321**
2-methylthio-6-(4-hydroxy-3-methyl-*cis*-but-2-enylamino)-9-β-D-ribofuranosylpurine – see 2-methylthio-*cis*-zeatin riboside
2-methylthio-6-(4-hydroxy-3-methyl-*trans*-but-2-enylamino)purine – see 2-methylthio-*trans*-zeatin
2-methylthio-6-(4-hydroxy-3-methyl-*trans*-but-2-enylamino)-9-β-D-ribofuranosylpurine – see 2-methylthio-*trans*-zeatin riboside
2-methylthio-N^6-(Δ^2-isopentenyl)adenine **471**
gas liquid chromatography 173–174
gel permeation chromatography 139–140
2-methylthio-N^6-(Δ^2-isopentenyl)adenosine **45, 291, 471**
gas liquid chromatography 173–174
gel permeation chromatography 139–140
high pressure liquid chromatography 180
occurrence 43, 45, 348, 472
partition chromatography 137
thin layer chromatography 168
2-methylthio-6-(3-methylbut-2-enylamino)purine – see 2-methylthio-N^6-(Δ^2-isopentenyl) adenine
2-methylthio-6-(3-methylbut-2-enylamino)-9-β-D-ribofuranosylpurine – see 2-methylthio-N^6-(Δ^2-isopentenyl)adenosine
5′-methylthioribose (MTR) 320, 321
2-methylthio-*cis*-zeatin riboside **43, 290,** 292, 348

2-methylthio-*trans*-zeatin **291**
ion exchange chromatography 128
occurrence 292
2-methylthio-*trans*-zeatin riboside **43, 291**
high pressure liquid chromatography 180
occurrence 40, 43, 292, 348
origin of methylthio-group 292
Nω-methyltryptamine,
ion exchange chromatography 123
mevalonate kinase 293
mevalonic acid **293–294, 296–298, 303**
abscisic acid precursor 295–298
cytokinin precursor 289
gibberellin precursor 301–303, 305, 309–311, 388
labelled 137, 297–298, 309–311
lactone, chromatography 136–137
terpene precursor 293–294, 302–305, 317
mimosine **64,** 76
mitomycin
α-amylase synthesis and 549
DNA synthesis and 549
momilactones A, B and C **68,** 78
monomethyl-4-chloroindole-3-acetyl-L-aspartic acid 365
morphactins,
NPA-membrane receptors and 458, 464
6-morpholinopurine
biological activity 475
ribosome binding 475
muzigadial **66,** 78
myo-inositol **73,** 80
myrosinase,
artefact formation and 115

nagilactones, A–F **67–68,** 78
naphthalene
1,4-dimethyl- **60,** 74
1,6-dimethyl- **60,** 74
2-naphthoxyacetic acid
stomatal closure and 541
transpiration rate and 541
naphthyl-1-acetic acid (NAA) 333, 448, **449**
binding 458–462
binding sites 459–461
cell wall regeneration and 562
deoxy isoflavone synthesis and 566
DNA synthesis and 556
invertase synthesis and 562
ribonuclease and 559
supernatant factor and 461–462
transpiration rate and 541
naphthyl-2-acetic acid
binding site 461
lecithin complex 496
1-N-naphthylphthalamic acid (NPA) **449**
auxin transport and 458
binding sites 458–461, 464
binding sites and transport 468
kartoffelfactor and 468
narciclasine (≡lycoridinol) **63,** 75
naringenin **58,** 65, 325
nicotinamide **63,** 75
nuclear magnetic resonance
lecithin 495
performance parameters 248
nucleoside kinase 371
null hypothesis 231
information content and 231–233, 238
verification 235

octadec-12-enoic acid
9,10-dihydroxy- **72,** 79
9,10-dihydroxy-, methyl ester **72,** 79
oestradiol **484**
mechanism of action 485–486
oestrone **69,** 78
GA biosynthesis and 361
oleamins,
pea stem elongation 79
oleic acid **71,** 79
olfactory responses
epithelial membrane and 495
gibberellin analogy 495
oosponol **59,** 69
optical rotatory dispersion (ORD)
abscisic acid 190, 270
performance parameters 249
quantitative analysis 270
β-oxidases,
GA_3 and 551
oxindole
paper chromatography 158
thin layer chromatography 160
oxindole-3-acetic acid,
IAA-catabolite **339**

palmitic acid **71,** 79
paper chromatography
auxins 144, 157–160
cytokinins 151–152, 165–166
gibberellins 160–161
identification and Rf values 154–166
indoles 157–159
performance parameters 259
parascorbic acid **59,** 70
parthenin **66,** 77
partition chromatography
abscisic acid 138, 154
auxins 132–134, 144
cytokinins 137
gibberellins 133–137, 146–149, 177
indoles 132
partition coefficients
cytokinins 125

gibberellins 124–125
ent-kaurenoic acid 125
pectin biosynthesis,
2,4-D, benzylaminopurine and 564
pectin methyl esterase
auxin effect of 516, 520
cell wall growth and 516
peroxidase 324, 337, 339
abscisic acid and 565
2,4-D and 565
ethylene and 540, 565
gibberellin and 562, 565
IAA and 565
polyphenol synthesis and 564–565
pharbitic acid (gibberthione) 36, 376, 379
phaseic acid **51**, **301**, **351**, 352–353
adsorption chromatography 130–132
biosynthesis 297–298
4′-dihydro – see dihydrophaseic acid
enzymic reduction 353
4′-*epi*dihydro- – see *epi*dihydrophaseic acid
from labelled mevalonic acid 297–298
identity with metabolite M-1 352
methyl ester as artefact 117
occurrence 51
partition chromatography 138
photosynthetic regulation 52
solvent partitioning 128
translocation in phloem 387
phaseolic acid **72**, 79
phenols
ethylene formation and 324–325
growth effects 59, 62–63
IAA peroxidase and 344–345
phenylacetic acid 21, **22**
auxin receptor site and 461
phenylacetonitrile 21, **22**
phenylalanineammonia lyase (PAL) 565
phenylaminopurine, gel permeation chromatography 139
β-phenylethyl-isothiocyanate **62**, 74
2-phenylindole, paper chromatography 158–159
phenylisobutyric acid, chloro-derivatives 316
phenylmethylsulphonylfluoride (PMSF), auxin binding techniques 468
phlorizin **58**, 65
phosphatase
effect of GA_3 551
ribotide cytokinins and 117
phospholipid synthesis, hormone effects 550–551
Phosphon D **315**, 316, 334
drought tolerance and 542
gibberellin action and 542
Phosphon S **315**, 316
phosphorylcholine 491
phosphorylcholine-cytidyl transferase (PC-C)
α-amylase synthesis 550
GA_3, abscisic acid and 491, 550
phosphorylcholine-glyceride transferase (PC-G)
α-amylase synthesis 550
GA_3, abscisic acid and 491, 550
photomorphogenesis and flavonoids 64
phyllocladene **301**, 302
phytochrome, gibberellins and 361–362
picrate complexes of hormones 143, 154
pimara-8,14-diene 304–305, **306**
pipecolic acid **63**, 76
pisatin **59**, 79
plasma membrane – see membrane, plant cell
podolactones A-E **67–68**, 78
polyphenol oxidase, GA_3 effect 551
polyphenol synthesis
2,4-D, kinetin and 564–565
peroxidase and 564–565
polyribosomes
cellulose and protein synthesis 563
IAA and 2,4-D effect 563
polysaccharides
auxins and 524–525, 527–528
in cell wall 525–529, 535–536
model cell wall 526–527, 529
nojirimycin inhibitor and 526
polymerisation 524–525
protein links 529
turnover 526–528
polysaccharide synthetase 524–525
auxin and 536–537
proton pump and 536–537
polysomes
abscisic acid and 491
GA_3 and 491
steroids and 491
polyteny, 2,4-D treatment 556
polyvinylpyrrolidone (PVP) chromatography
abscisic acid 142
auxins 140
cytokinins 141–142, 153
gibberellins 140–141
ponalactone A **68**
porphorins, methyl phaephorbides **64**, 76
portulal **69**, 78
root induction 114
prenyl transferase 293–294
progesterone **484**, 485
gas liquid chromatography 172
proline and cell wall proteins 529, 564
propylene 328, 363
Protease
in aleurone 546
GA_3 and 551
protein synthesis
actinomycin D and 561

protein synthesis
auxin and 450–451, 517–521
cytokinins and 474, 477–478
gibberellins and 489–490, 497, 561
inhibitors of 520–521
kinetin and 561
protoanemonin **60**, 70
protocatechuic acid **57**
proton pump (hypothesis) 451, 521 ff. 537 ff.
abscisic acid and 532, 539
auxins (IAA) and 451–452, 521, 534–535, 537–539
cell extension and 522–523, 531, 534–537
CO_2 fixation and 539
cytokinin and 531, 539
fusicoccin and 533–534, 538–539
gibberellin and 530, 539
ion transport and 538
latent period and 522, 534
membrane potential and 538–539
polysaccharide synthetase and 536–537
respiration and 537–539
site 539
prunasin **62**, 74
psilotin **60**, 71
psoralen **59**, 67
purine nucleosidase 371–372
purine-5′-nucleoside phosphorylase 371
purine phosphoribosyltransferase 371
β-pyrazol-1-yl-alanine **64**, 76
pyrethrosin **65**, 77
pyridoxal phosphate 319–321, 325, 335
pyrophosphomevalonate decarboxylase 317
PVP chromatography – see polyvinylpyrrolidone chromatography

quantitative analysis – see also internal standards, information content, successive approximation method and individual hormones
ABA 168, 190, 270–272, 273
bio-assay methods 241–244
cytokinins 4, 186–189, 269–270, 273
detector response 206–207, 210–216, 223, 237
detector sensitivity 210–213, 225
error 203–204, 206
estimate 203, 215–216
gibberellins 186, 241–245, 267–269, 273
IAA 186, 266, 251–252, 264–267, 273
isotope dilution 4, 186, 189–190, 264–265, 269–270
physicochemical methods 248–249 – see also individual physicochemical methods
precision 204–205
probability 204
quercetin **58**, 64

radio-immunoassay,
accuracy 246
indole-3-acetic acid 246–247
selectivity and specificity 246
ramulosin **59**, 69
raphanetan – see zeatin-N(7)-β-D-glucopyranoside
receptor protein (n-IRP) 455–457
receptors – see also under individual hormones
binding constants 446–447
criteria 445
double reciprocal plots 446–447
methodology 446–447
plasmalemma markers 452
Scatchard plots 446–447
refractometry, performance parameters 249
rhizobitoxine (L-amino-2(2′-amino-3′-hydroxypropoxy)-*trans*-3-butenoic acid) **319**
ethylene biosynthesis inhibitor 319, 322, 325–326, 333–335, 390
Rhizopus suinus, IAA from tryptophan 281
9-β-D-ribofuranosyl-6-(4-O-β-D-glucopyranosyl-3-methyl-*trans*-but-2-enylamino)purine – see glucosylribosylzeatin
N-[9-(β-D-ribofuranosyl)purine-6-carbamoyl] threonine (Ad-Co-thr) **39**, 44
9-β-D-ribofuranosylzeatin – see zeatin riboside
ribonuclease, hormones and 559
ribonucleic acid – see RNA
ribose 321
ribosomes
cytokinin binding 475–480
GA_3 and 490–492, 547
steroid hormones and 491–492
Ricinus communis
biosynthesis of diterpenes 305
enzyme systems 305
tRNA
clover leaf structure 472–473
cytokinins and 531
RNA polymerase
abscisic acid and 498
auxins and 450–455, 470, 517–518, 520
2,4-D and 558, 560–561
GA_3 and 558, 560
protein factor γ and 454
receptor protein (n-IRP) and 455
rifampicin and 454
soyabean stimulating factor 470
RNA polymerase I and II
2,4-D and 453, 455
protein mediators 453, 455
RNA synthesis
abscisic acid and 498, 560
α-amylase synthesis and 545–548
auxin and 517–521, 557–559
chain initiation 454–455

cytokinins and 478, 560
2,4-D and 453, 556–559
ethylene and 557
5-fluorouracil and 558
GA_3 and 448–449, 491, 545–547, 549, 558–560
growth and 520
IAA and 456
inhibitors 521
oestradiol and 456
receptor protein (n-IRP) and 455, 559
rifampicin and 454
root growth inhibitors 81
root growth promoters 61, 69
dihydroasparagusic acid 80
portulal 144
sesquiterpene ketones 77–78

salicylaldehyde **57**, 73
salicylic acid **57**, 61
flower induction 61
(+)-sandacaropimaradiene 305, **306**
santonin **66**, 78
sarcomycin
α-amylase synthesis and 549
DNA synthesis and 549
sayadenine **59**, 65
cytokinin-like activity 65
Scatchard plot 446–447
sclareol (and epimer) **69**, 78
scopoletin **58**, 67
selected ion monitoring (SIM) (SID) (MID) (MF)
abscisic acid 272
auxins 186, 266
cytokinins 4, 186–189, 269, 273
gibberellins 186–187, 189, 268, 273
information content 249
quantitative analysis 186, 266–272
selectivity 249
sensitivity 186, 249
selenomethionine 320, 326
sellowin A **67**
senescence
bioassay 475
ethylene and 328–329, 540
membrane permeability and 540
methionine metabolism and 329
serine, IAA precursor 282–283
serotonin, ion exchange chromatography 123
seselin **59**, 67
sinapic acid **58**
β-sitosterol **70**, 78–79, **484**, 492–493
skatole
gas liquid chromatography 170
paper chromatography 157–158
partition chromatography 132
thin layer chromatography 159–160
SKF-7997 316
SKF-525A **315**, 316
solvent partitioning
abscisic acid 128–129, 145
auxins 120–123
cytokinins 125–127, 151
gibberellins 124–125, 146–147
indoles 120–123
partition coefficients 125, 127
spectrophosphorimetry
IAA analysis 266–267, 273
performance parameters 248
starch phosphorylase,
GA_3 and 562
stearic acid glyceryl ester **72**, 79
steroid hormones,
membrane structure and 494
steroid hormone receptors 484–486
membrane sites and ribosomes 491–492
steviol **38**, 308, **309**, 310–311
acetate 311
partition chromatography 136–137
stomatal aperture
abscisic acid and 542–544
auxins and 541
Ethephon and 542
ethylene and 542
fusicoccin and 543–544
gibberellin and 542
kinetin and 541–542
strigol **66**, 77
successive approximation method
accuracy 215, 217, 221
detector sensitivity 223
hypothetical examples 221
internal standards 217
purification factor 216–217, 223, 257
successive estimates difference 218
test for variance 224
succinic acid
monoethyl ester **71**, 79
monomethyl ester **71**, 79
succinic acid-2,2-dimethylhydrazide – see B-995
N^{12}-succinyltryptamine, paper chromatography 158
sulphur cycle 320
supernatant factor (SF)
auxin binding and 461–462
chromatin and 453
composition of 462
membrane binding site and 462
syringic acid **57**

terpenoid(s)
biosynthesis 293–295
double bond stereochemistry 295

terpenoid(s)
partition chromatography 137
tetrahydrophthalamide **62**, 75
theaspirone **54**
thin layer chromatography
abscisic acid 142, 168
auxins 144–145, 157–160
cytokinins 151–152, 165–168
gibberellins 160–165
indoles 159–161, 165–168
performance parameters 259
Rf values 160–165, 167
thymidine,
thin layer chromatography 167
α-tocopherol **61**, 74
α-tomatine **70**, 78
(−)-trachylobane(*ent*-13,16-cycloatisane) 305, **306**
transcription
auxins and 451
gibberellins and 488
receptor protein and 456
transpiration rate, hormone effects 541–542
triacontanol **71**, 79
triarimol **315**, 316
2,3,6-trichlorobenzoic acid (2,3,6-TBA) **449–450**
trigonelline **64**, 76
6-(2,3,4-trihydroxy-3-methylbutylamino)purine, isolation from *Zea mays* 151
tri-0-(indole-3-acetyl)-myo-inositol **15**, 21
see also IAA myo-inositol esters
2,3,6-triiodo-benzoic acid,
membrane binding 458, 461
N,N,N-trimethyl-1-methyl-(2′,2′,6′-methylcyclohex-2′-en-1-yl)-prop-2-enylammonium iodide – see AMO 1618
trimethylsilylation 169ff.
abscisic acid 174
auxins 170
cytokinins 171–173
gibberellins 171–173, 182–184
tryptamine **18**
formation from indole 287–288
formation from tryptophan 287
gas liquid chromatography 170
IAA precursor 287–288
ion exchange chromatography 123
occurrence 287
oxidase 287–288
paper chromatography 158–159
thin layer chromatography 160
tryptamine derivatives
ion exchange chromatography 123
paper chromatography 158
thin layer chromatography 160
tryptophan **284**
aminotransferase 283–284, 286
α-N-carboethoxyacetyl **16**
α-N-carbomethoxyacetyl **16**
decarboxylase 287
glucobrassicin precursor 288–289
IAA precursor 281–288
ion exchange chromatography 123
α-N-malonyl **18**
paper chromatography 158–159
thin layer chromatography 160
tryptophol – see indole-3-ethanol
turgor pressure
growth and 515–516
ionic relations and 539
tyrosine,
NAA binding sites and 465

ultra violet fluorimetry
abscisic acid analysis 168
performance parameters 248
ultraviolet spectrometry
analysis of abscisic acid 270
performance parameters 248
ultraviolet spectrophotometry, performance parameters 248
umbelliferone **58**, 67
uracil, thin-layer chromatography 167

vanillic acid **57**
vanillin **57**, 73
vernolepin **66**, 77
violaxanthin **55**
photochemical conversion 30, 55
vitamin E (α-tocopherol) **61**, 74
vomifoliol **53**

wartburganal **66**, 78

xanthine oxidase 348
xanthium **65**, 77
xanthotoxin **59**, 67
xanthoxidase 325
xanthoxin
2-*cis*-4-*trans* **55**, **300**
2-*trans*-4-*trans*- **55**, **300**
acids **299**
adsorption chromatography 131–132
from violoxanthin 300
gas liquid chromatography 131, 174
occurrence 55, 131, 300–301
solvent partitioning 128–129
$\beta 1 \rightarrow 4$-xylanase,
GA_3 effect 551
xylem differentiation
inhibition by FUDR 554
with ABA and cytokinins 560
with IAA 554
with kinetin 555

xyloglucan **527**
auxin-reduced turnover 527–528
in model cell wall 526
release with auxin 517
xylopic acid 308, **309**
desacetyl- 311
β-xylopyranosidase,
GA_3 and 551

zeanic acid (2,8-dihydroxyquinoline-4-carboxylic acid) **63**, 75, **339**
IAA catabolite 339
in corn steep liquor 340
isomer **63**, 75
cis-zeatin **41**, **290**, **471**
biological activity 471
biosynthesis in tRNA 292
occurrence in tRNA 40–42, 292, 296
partition chromatography 137
SIM detection 186
cis-zeatin riboside **41**, 42, 290, 292
2-methylthio- **43**, 290, 292, 348
trans-zeatin **41**, **471**
adsorption chromatography 131
9-β-alanyl- 370
α-amylase and 562
biosynthesis in tRNA 290
catabolic degradation 350
cell differentiation and 560
dihydro- – see dihydrozeatin
2,8-dihydroxy- **348**
ion exchange chromatography 128
gas liquid chromatography 174
gel permeation chromatography 139, 240
glucosylation in vivo 369–370
2-halogeno- and biological activity 272
high pressure liquid chromatography 4, 180
2-hydroxy- **46**, 348
8-hydroxy- 348
ion exchange chromatography 128
isolation from *Zea mays* 154
mass spectrometry 4, 186
metabolic degradation 369
metabolic sites 385
metabolism in *Populus alba* leaves 5
occurrence 40–42, 186, 385
paper chromatography 165–166
partition chromatography 137
partition coefficients 127
picrate complex 143
precipitation methods 143
PVP chromatography 140–141
reduction in vivo 348
ribosylation in vivo 368–371
RNA synthesis and 560
silver complex 143
solvent partitioning 127
thin layer chromatography 165–167
xanthine oxidase oxidation 348
zeatin-N(7)-β-glucopyranoside (rasphanatin) **47**
GC-MS 269
ion exchange chromatography 128
isolation from *Zea mays* 150–151
isotope dilution analysis 186
occurrence 45, 47, 369
quantitative analysis 186, 269–270
zeatin-O-β-D-glucopyranoside (glucosyl zeatin) 45, **47**
gas liquid chromatography 184
isolation from *Vinca rosea* 153–154
PVP chromatography 142
zeatin metabolite 369–370
zeatin-9-glucoside
isolation from *Zea mays* 150–151
trans-zeatin riboside **41**
artefact 128
biosynthesis 348
cytokinesin I, identity as 371
gas liquid chromatography 173–174
GC-MS 186
gel permeation chromatography 139–140, 240
4-glucoside **47**, 154
high pressure liquid chromatography 179–180
ion exchange chromatography 127–128
isolation from *Zea mays* 150–151
mercury complex 143
2-methylthio – see 2-methylthio-6(4-hydroxy-3-methyl-*trans*-but-2-enylamino)-9-β-D-ribofuranosylpurine
occurrence 40–42, 292
paper chromatography 165–166
partition chromatography 137
PVP chromatography 141–142
reduction, enzymatic 368
solvent partitioning 127–128
thin layer chromatography 165–167
zeatin riboside diphosphate,
zeatin metabolite 368
zeatin riboside triphosphate,
zeatin metabolite 368
zeatin ribotide **46**
barium salt 143, 150
GC-MS 186
isolation from *Zea mays* 150–151
occurrence 40, 46, 292
thin layer chromatography 166
zeatin metabolite 368–369
β-(zeatin-9-yl)alanine (lupinic acid) 370
enzymatic pathway 373
zerumbone **66**, 77

MIX
Papier aus verantwortungsvollen Quellen
Paper from responsible sources
FSC® C105338

If you have any concerns about our products,
you can contact us on
ProductSafety@springernature.com

In case Publisher is established outside the EU,
the EU authorized representative is:
Springer Nature Customer Service Center GmbH
Europaplatz 3, 69115 Heidelberg, Germany

Printed by Libri Plureos GmbH
in Hamburg, Germany